Biomathematical and Biomechanical Modeling of the Circulatory and Ventilatory Systems

Volume 4

For further volumes:
http://www.springer.com/series/10155

Marc Thiriet

Intracellular Signaling Mediators in the Circulatory and Ventilatory Systems

 Springer

Marc Thiriet
Project-team INRIA-UPMC-CNRS REO
Laboratoire Jacques-Louis Lions, CNRS UMR 7598
Université Pierre et Marie Curie
Place Jussieu 4
Paris Cedex 05
France

ISSN 2193-1682   ISSN 2193-1690 (electronic)
ISBN 978-1-4614-4369-8   ISBN 978-1-4614-4370-4 (eBook)
DOI 10.1007/978-1-4614-4370-4
Springer New York Heidelberg Dordrecht London

Library of Congress Control Number: 2012941609

© Springer Science+Business Media New York 2013
This work is subject to copyright. All rights are reserved by the Publisher, whether the whole or part of the material is concerned, specifically the rights of translation, reprinting, reuse of illustrations, recitation, broadcasting, reproduction on microfilms or in any other physical way, and transmission or information storage and retrieval, electronic adaptation, computer software, or by similar or dissimilar methodology now known or hereafter developed. Exempted from this legal reservation are brief excerpts in connection with reviews or scholarly analysis or material supplied specifically for the purpose of being entered and executed on a computer system, for exclusive use by the purchaser of the work. Duplication of this publication or parts thereof is permitted only under the provisions of the Copyright Law of the Publisher's location, in its current version, and permission for use must always be obtained from Springer. Permissions for use may be obtained through RightsLink at the Copyright Clearance Center. Violations are liable to prosecution under the respective Copyright Law.

The use of general descriptive names, registered names, trademarks, service marks, etc. in this publication does not imply, even in the absence of a specific statement, that such names are exempt from the relevant protective laws and regulations and therefore free for general use.

While the advice and information in this book are believed to be true and accurate at the date of publication, neither the authors nor the editors nor the publisher can accept any legal responsibility for any errors or omissions that may be made. The publisher makes no warranty, express or implied, with respect to the material contained herein.

Printed on acid-free paper

Springer is part of Springer Science+Business Media (www.springer.com)

Contents

1	**Introduction**		1
2	**Signaling Lipids**		7
	2.1	Lipids in Signal Transduction	9
	2.2	Phosphoinositides	12
		2.2.1 $PI(4,5)P_2$	13
		2.2.2 $PI(3,4,5)P_3$	15
		2.2.3 $PI(5)P$	15
		2.2.4 Nuclear Phosphoinositides	16
		2.2.5 Phosphoinositide-Dependent Enzymatic Activity	18
		2.2.6 Phosphoinositide Influence on Ion Carriers	18
	2.3	(Non-Lipid) Inositol Phosphates	21
		2.3.1 $I(1,4,5)P_3$	24
		2.3.2 Main Enzymes of the Metabolism of Inositol Polyphosphates	25
	2.4	Choline-Containing Phospholipids	28
	2.5	Sphingolipids	28
		2.5.1 Ceramide	29
		2.5.2 Sphingomyelinases	29
		2.5.3 Sphingosine Kinases	30
		2.5.4 Sphingosine 1-Phosphate	31
		2.5.5 Cardiac Effects of Sphingolipids	34
		2.5.6 Sphingolipid Effects on Smooth Muscle Cells	35
		2.5.7 Sphingolipid Effects on Endothelial Cells	36
	2.6	Arachidonic Acid and Eicosanoids	38
		2.6.1 Arachidonic Acid	38
		2.6.2 Cyclooxygenases	39
		2.6.3 Lipoxygenases	40
		2.6.4 Cytochrome-P450 Monooxygenases	41
		2.6.5 Arachidonic Acid Metabolites in Lungs	42

		2.6.6	Prostaglandins	42
		2.6.7	Hydroxy- and Epoxyeicosatrienoic Acids	43
	2.7	Lipases		44
		2.7.1	Hormone-Sensitive Lipase	44
		2.7.2	Adipose Triglyceride Lipase	45
		2.7.3	Hepatic Lipase	45
		2.7.4	Endothelial Lipase	46
		2.7.5	Lipoprotein Lipase	47
		2.7.6	Lipase-H	48
	2.8	Phospholipases		48
		2.8.1	Phospholipases-A	48
		2.8.2	Phospholipases-A1	48
		2.8.3	Superfamily of Phospholipases-A2	50
		2.8.4	Phospholipase-B	65
		2.8.5	Phospholipase-C	65
		2.8.6	Phospholipase-D	69
	2.9	Phosphoinositide Kinases		70
		2.9.1	Phosphatidylinositol 3-Kinases	71
		2.9.2	Other Phosphoinositide Kinases	86
	2.10	Phosphoinositide Phosphatases		92
		2.10.1	Phosphatase and Tensin Homolog	92
		2.10.2	Myotubularins and Myotubularin-Related Phosphoinositide 3-Phosphatases	94
		2.10.3	Phosphoinositide 4-Phosphatases	94
		2.10.4	Phosphoinositide 5-Phosphatases	95
	2.11	Other Lipid Phosphatases – Phosphotransferases		101
	2.12	(Ecto)Nucleotide Pyrophosphatase–Phosphodiesterases		102
		2.12.1	Substrates and Functions of ENPPs	103
		2.12.2	Production Regulation	105
		2.12.3	ENPP1	106
		2.12.4	ENPP2 (Autotaxin)	106
		2.12.5	ENPP3	107
3	**Preamble to Cytoplasmic Protein Kinases**			109
	3.1	Protein Kinase Classification		110
		3.1.1	Superfamily 0 (Miscellaneous)	111
		3.1.2	AGC Superfamily of Protein Kinases	115
		3.1.3	CAMK Superfamily of Protein Kinases	116
		3.1.4	CK1 Superfamily of Protein Kinases	118
		3.1.5	CMGC Superfamily of Protein Kinases	119
		3.1.6	STE Superfamily of Protein Kinases	120
		3.1.7	TK Superfamily of Protein Kinases	121
		3.1.8	TKL Superfamily of Protein Kinases	123
		3.1.9	RGC Superfamily of Protein Kinases	124

	3.1.10	Superfamily of Atypical Protein Kinases	125
	3.1.11	TATA-Binding Factor-Associated Factor-1	129
3.2	Agents of Kinase Activity		130
3.3	Alternatively Spliced Variants		130
3.4	Intracellular Pseudokinases		131
	3.4.1	CASK Pseudokinase	132
	3.4.2	GCN2 Pseudokinase	132
	3.4.3	ILK Pseudokinase	132
	3.4.4	Janus Pseudokinases	132
	3.4.5	Pseudokinase Kinase Suppressors of Ras	133
	3.4.6	Sterile 20-Related Adaptors	133
	3.4.7	Tribbles Pseudokinases	134
	3.4.8	TrrAP Pseudokinase	135

4 Cytoplasmic Protein Tyrosine Kinases ... 137

4.1	Receptor and Cytosolic Protein Tyrosine Kinases		137
4.2	Family of Abl Kinases		140
	4.2.1	Abl1 Kinase (Abl)	141
	4.2.2	Abl2 Kinase (Arg)	142
	4.2.3	Abl-Binding Proteins	143
4.3	ACK Kinases		143
4.4	BrK Kinase (Protein Tyr Kinase-6; FRK Subfamily)		144
4.5	CSK Kinase		144
4.6	Focal Adhesion Kinases		145
4.7	FeS and FeR Kinases (FPS/FES Subfamily)		148
4.8	Fyn-Related Kinase		150
4.9	Janus Kinases		151
	4.9.1	Canonical and Non-Canonical Pathways of JaK Activation	151
	4.9.2	Signal Transducers and Activators of Transcription	153
	4.9.3	Other Regulation of JaK Activity	153
	4.9.4	Janus Kinases in Heart	153
	4.9.5	Inhibitors of the JaK–STAT Pathway	154
4.10	MATK Kinase or CSK Homologous Kinase (CSK Subfamily)		155
4.11	SRMS Kinase (FRK Subfamily)		156
4.12	Kinases of the SRC Family		156
	4.12.1	SRC Family Kinases during Hypoxia	157
	4.12.2	BLK Kinase (SRCB Subfamily)	158
	4.12.3	BMX Kinase (TEC Family)	159
	4.12.4	BTK Kinases (TEC Family)	160
	4.12.5	FGR Kinase (SRCA Subfamily)	160
	4.12.6	Fyn Kinase (SRCA Subfamily)	162
	4.12.7	Hemopoietic Cell Kinase (SRCB Subfamily)	163

	4.12.8	ITK Kinase (TEC Family)	163
	4.12.9	Leukocyte-Specific Cytosolic Kinase (SRCB Subfamily)	164
	4.12.10	Lyn Kinase (SRCB Subfamily)	165
	4.12.11	Src Kinase	166
	4.12.12	TEC Kinase	167
	4.12.13	TXK Kinase (TEC Family)	169
	4.12.14	Yes Kinase (SRCA Subfamily)	170
4.13	From Src to SYK		170
4.14	Spleen Tyrosine Kinase		171

5 Cytoplasmic Protein Serine/Threonine Kinases — 175

5.1	Superfamily of Ser/Thr Sterile-20-Related Kinases		175
	5.1.1	Ste20-like Kinase	178
	5.1.2	Germinal Center Kinase Family	178
	5.1.3	MAP4Ks of GCK Subfamilies 1 and 4 of the Ste20-like Kinase Superfamily	180
	5.1.4	Kinases of GCK Subfamilies 2 and 3 – MST Kinases	184
	5.1.5	STK25 of the GCK Subfamily-3	186
	5.1.6	Other Kinases of the GCK Subfamily 4	186
	5.1.7	Kinases of the GCK Subfamily 5	187
	5.1.8	Kinases of the GCK Subfamily 6	187
	5.1.9	Kinases of the GCK Subfamily 7	192
	5.1.10	Kinases of the GCK Subfamily 8	192
5.2	AGC Superfamily		194
	5.2.1	Nuclear DBF2-Related Protein Kinases and Large Tumor Suppressors	196
	5.2.2	Phosphoinositide-Dependent Kinase-1	198
	5.2.3	Protein Kinase-A Family	199
	5.2.4	Protein Kinase-X1	202
	5.2.5	Protein Kinase-Y1	203
	5.2.6	Protein Kinase-B Family	203
	5.2.7	Protein Kinase-C Family	208
	5.2.8	Protein Kinase-D Family	218
	5.2.9	Protein Kinase-G Family	218
	5.2.10	CDC42-Binding Protein Kinases	219
	5.2.11	G-Protein-Coupled Receptor Kinase Family	220
	5.2.12	Microtubule-Associated Ser/Thr Kinase Family	222
	5.2.13	P21-Activated Kinase Family	223
	5.2.14	RoCK Kinase Family	226
	5.2.15	Other Members of the ROCK Family	232
	5.2.16	P90 Ribosomal S6 Kinase Family	232
	5.2.17	P70 Ribosomal S6 Kinase Family	239

	5.2.18	Serum- and Glucocorticoid-Regulated Kinase Family	241
5.3	Diacylglycerol Receptors and Kinases		241
5.4	CAMK Superfamily		242
	5.4.1	Calmodulin-Dependent Protein Kinase (CamK) Family	242
	5.4.2	Myosin Light-Chain Kinase Family	245
	5.4.3	AMP-Activated Protein Kinases	248
	5.4.4	AMPK-Related Kinases	251
	5.4.5	Checkpoint Kinase Family	256
	5.4.6	Death-Associated Protein Kinase Family	256
	5.4.7	Death Receptor-Associated Protein-Related Apoptotic Kinase Family	257
	5.4.8	PIM Kinase Family	258
	5.4.9	Protein kinase-D Family	259
5.5	PIKK Superfamily		260
	5.5.1	Ataxia Telangiectasia Mutated and ATM and Rad3-Related Kinases	260
	5.5.2	DNA-Dependent Protein Kinase	262
	5.5.3	Target of Rapamycin	262
5.6	CK1 Superfamily		267
	5.6.1	Casein Kinase-1 Family	267
	5.6.2	Vaccinia-Related Kinase Family	268
	5.6.3	Tau–Tubulin Kinase Family	270
5.7	Family-2 Casein Kinases		271
	5.7.1	Catalytic CK2α1 and CK2α2 Subunits)	272
	5.7.2	Regulatory CK2β Subunit	273
5.8	CMGC Superfamily		273
	5.8.1	Cyclin-Dependent Kinase Family	273
	5.8.2	Homeodomain-Interacting Protein Kinase Family	274
5.9	TKL Superfamily		276
	5.9.1	Integrin-Linked Kinase	276
	5.9.2	Interleukin-1 Receptor-Associated Kinase Family	279
	5.9.3	LIM Domain-Containing Kinase Subfamily (LISK [LIMK–TESK] Family)	285
	5.9.4	Testis-Specific Kinase Subfamily (LISK [LIMK–TESK] Family)	286
	5.9.5	Receptor-Interacting Protein Kinase Family	287
5.10	Aurora Kinase Family		289
	5.10.1	Aurora Kinase-A	289
5.11	Polo-like Kinase Family		291
	5.11.1	Polo-like Kinase PLK1	292
	5.11.2	Polo-like Kinase PLK2	293
	5.11.3	Polo-like Kinase PLK3	294
	5.11.4	Polo-like Kinase PLK4	294

	5.12	With-No-K Kinase Family	294
		5.12.1 WNK1	295
		5.12.2 WNK2	296
		5.12.3 WNK3	296
		5.12.4 WNK4	297
		5.12.5 Regulation of Ion Transport in Nephron	297
	5.13	Other Types of Ser/Thr Kinases	304
		5.13.1 Doublecortin-like Kinase	304
		5.13.2 Haspin	304
		5.13.3 Liver Kinase-B1 (STK11)	305
		5.13.4 Never in Mitosis Gene-A-Related Expressed Kinases	305
		5.13.5 NEMo-like Kinase	306
		5.13.6 PTen-Induced Kinase	307
		5.13.7 Protein Kinase-R	308
		5.13.8 Slicing Factor Protein Kinase	309
		5.13.9 TANK-Binding Kinase 1	309
6	**Mitogen-Activated Protein Kinase Module**		**311**
	6.1	Main Features of MAPK Signaling	311
		6.1.1 Basic Three-Tiered MAPK Module	312
		6.1.2 Double Phosphorylation Cycle	312
		6.1.3 MAPK Pathway	312
		6.1.4 MAPK Stimuli	313
		6.1.5 MAPK Cascade Regulation	314
		6.1.6 Auto-Inhibition	316
		6.1.7 Scaffold Proteins	316
		6.1.8 Transcription Factor Substrates of MAPK Modules	317
		6.1.9 Subcellular Compartmentation	318
	6.2	MAPK Modules in Cell Metabolism	318
		6.2.1 Insulin and Glucagon Signaling	320
		6.2.2 MAPKs in Adipocytes	321
		6.2.3 MAPKs in Metabolic Syndrome	321
	6.3	Mitogen-Activated Protein Kinase Kinase Kinase Kinases	322
	6.4	Other Regulators of the MAPK Module	323
		6.4.1 Small GTPases of the Ras Hyperfamily	323
		6.4.2 Heterotrimeric G Proteins	323
	6.5	Mitogen-Activated Protein Kinase Kinase Kinases	324
		6.5.1 Family of Raf Kinases	329
		6.5.2 Moloney Sarcoma Viral Mos Proto-Oncogene Kinase (Mos)	333
		6.5.3 Family of Mixed Lineage Kinases	333
		6.5.4 Family of MAP/ERK Kinase Kinases (MEKK)	337

	6.5.5	Family of Mitogen-Activated Protein Kinase Kinase Kinases-5/6	342
	6.5.6	Mitogen-Activated Protein Kinase Kinase Kinase-7 (TAK1).......................................	346
	6.5.7	Family of Thousand and One Kinases (TAOK)	348
	6.5.8	Mitogen-Activated Protein Kinase Kinase Kinase-8 (TPL2)	348
	6.5.9	Mitogen-Activated Protein Kinase Kinase Kinase MAP3K14	349
6.6	Mitogen-Activated Protein Kinase Kinases		350
	6.6.1	MAP2K1	350
	6.6.2	MAP2K2	351
	6.6.3	MAP2K3	351
	6.6.4	MAP2K4	352
	6.6.5	MAP2K5	353
	6.6.6	MAP2K6	354
	6.6.7	MAP2K7	355
6.7	Mitogen-Activated Protein Kinases		355
	6.7.1	JNKs and P38MAPKs in Cardiovascular Diseases	358
	6.7.2	Family of Extracellular Signal-Regulated Kinases...	359
	6.7.3	Family of Jun N-Terminal Kinases.....................	367
	6.7.4	Family of P38 Mitogen-Activated Protein Kinases	371
	6.7.5	Stress-Activated Protein Kinases.....................	375
6.8	Kinase Substrates of MAPKs		376
	6.8.1	MAPK-Activated Protein Kinases	376
	6.8.2	MAPK-Interacting Ser/Thr Kinases	377
	6.8.3	Mitogen- and Stress-Activated Protein Kinases	378

7 Dual-Specificity Protein Kinases 379
   7.1 Mitogen-Activated Protein Kinase Kinase...................... 379
   7.2 Glycogen Synthase Kinase 379
   7.3 Dual-Specificity Tyrosine Phosphorylation-Regulated Kinases 380

	7.3.1	DYRK1a...	381
	7.3.2	DYRK1b...	381
	7.3.3	DYRK2	382
	7.3.4	DYRK3	382
	7.3.5	DYRK4	382
7.4	Cell Division Cycle-like Kinases.................................		382
	7.4.1	CLK1	383
	7.4.2	CLK2	383
	7.4.3	CLK3	383
	7.4.4	CLK4	383

	7.5	Wee1 Kinase	384
	7.6	Membrane-Associated Tyr/Thr Protein Kinase	384
	7.7	Raf Kinase Inhibitory Protein	385
8	**Cytosolic Protein Phosphatases**		387
	8.1	Pseudophosphatases	388
	8.2	Catalytically Active Protein Phosphatases	390
	8.3	Cytosolic Protein Serine/Threonine Phosphatases	390
		8.3.1 Protein Phosphatase-1	393
		8.3.2 Protein Phosphatase-2	401
		8.3.3 Protein Phosphatase-3	405
		8.3.4 Protein Phosphatase-4	407
		8.3.5 Protein Phosphatase-5	408
		8.3.6 Protein Phosphatase-6	408
		8.3.7 Protein Phosphatase-7	408
		8.3.8 Magnesium-Dependent Protein Phosphatase-1	409
		8.3.9 Cytosolic Protein Tyrosine Phosphatases	415
		8.3.10 Protein Tyrosine Phosphatase Regulation by Reversible Oxidation	419
		8.3.11 Classical Cytosolic Protein Tyrosine Phosphatases	420
		8.3.12 PEST Family Phosphatases in Inflammation and Immunity	435
		8.3.13 Dual-Specificity Protein Phosphatases	435
		8.3.14 Infraclass-2 Protein Tyrosine Phosphatase	458
		8.3.15 Infraclass-3 PTP Cell-Division Cycle-25 Phosphatases	459
		8.3.16 Infraclass Aspartate-Based Protein Tyrosine Phosphatases	461
9	**Guanosine Triphosphatases and Their Regulators**		465
	9.1	Heterotrimeric G Proteins (G$\alpha\beta\gamma$ GTPases)	467
		9.1.1 Set of Gα Subunits	470
		9.1.2 Set of Gβ Subunits	484
		9.1.3 Set of Gγ Subunits	485
		9.1.4 G Protein and Myogenic Response	485
	9.2	Regulators of Heterotrimeric G Proteins	486
		9.2.1 Nucleoside Diphosphate Kinases	486
		9.2.2 Regulators of G-Protein Signaling	487
		9.2.3 Gα Guanine Nucleotide-Exchange Factors	492
		9.2.4 Activators of G-Protein Signaling	492
		9.2.5 GPCR-Interacting Proteins	494
		9.2.6 G-Protein-Coupled Receptor Kinases	495
	9.3	Monomeric (Small) GTPases	496
		9.3.1 Superfamily of ARF GTPases	502
		9.3.2 CDC42 GTPase (RHO Superfamily)	515

	9.3.3	Superfamily of Rab GTPases	516
	9.3.4	Rac GTPases	531
	9.3.5	Rad GTPase	534
	9.3.6	Rag GTPases	534
	9.3.7	Ral GTPases	535
	9.3.8	Ran GTPases	536
	9.3.9	Rap GTPases	536
	9.3.10	Ras GTPases	540
	9.3.11	RHEB GTPase	553
	9.3.12	Superfamily of RHO GTPases (CDC42, Rac, and Rho)	553
	9.3.13	Family of RHO GTPases (RhoA, RhoB, and RhoC)	558
	9.3.14	Atypical RhoBTB GTPases	562
	9.3.15	Classical RhoD GTPase	563
	9.3.16	Classical RhoF GTPase	563
	9.3.17	Atypical RhoG GTPase	563
	9.3.18	Atypical RhoH GTPase	564
	9.3.19	Atypical RhoJ GTPase	564
	9.3.20	Atypical RhoQ GTPase	565
	9.3.21	RhoT GTPase	566
	9.3.22	Atypical RhoT1/2 GTPases (Miro-1/2)	566
	9.3.23	Atypical RhoU GTPase	567
	9.3.24	Atypical RhoV GTPase	567
	9.3.25	RGK (Rad–Gem/Kir–Rem) Family of GTPases	567
	9.3.26	RND Family of Atypical GTPases	568
	9.3.27	Family of RIN and RIT GTPases	570
	9.3.28	Phosphoinositide 3-Kinase Enhancer: GTPase and GAP	572
9.4	Regulators of Monomeric Guanosine Triphosphatases		572
	9.4.1	Guanine Nucleotide-Exchange Factors	573
	9.4.2	GTPase-Activating Proteins	612
	9.4.3	Guanine Nucleotide-Dissociation Inhibitors	640
10	**Other Major Types of Signaling Mediators**		**647**
10.1	Gaseous Neurotransmitters		648
10.2	Carbon Monoxide and Heme Oxygenases		650
	10.2.1	Carbon Monoxide Synthesis and its Regulation	650
	10.2.2	Vasodilation and Other Effects	653
	10.2.3	Carbon Monoxide and Nitric Oxide	655
10.3	Nitric Oxide, Nitric Oxide Synthases, Nitrite, and Nitrate		656
	10.3.1	Nitric Oxide Synthases	657
	10.3.2	Nitric Oxide Production from Nitrite and Nitrate	663
	10.3.3	Nitric Oxide and Hypoxic Vasodilation	668

		10.3.4	Reduction–Oxidation Reaction Products of Nitric Oxide	670
		10.3.5	NO Transfer	672
		10.3.6	NO Effects	673
		10.3.7	Interaction between Nitric Oxide and Hydrogen Sulfide	686
	10.4	Hydrogen Sulfide		687
		10.4.1	Hydrogen Sulfide Production	687
		10.4.2	Hydrogen Sulfide Effects	688
		10.4.3	Hydrogen Sulfide in Mitochondria	689
	10.5	NADPH and Dual Oxidases		690
		10.5.1	Structure	691
		10.5.2	Subtypes	693
		10.5.3	Activity	697
		10.5.4	NOx–ROS Axis	699
	10.6	Redox Signaling and Reactive Oxygen and Nitrogen Species		700
		10.6.1	Signaling by Reactive Oxygen Species	702
		10.6.2	Signaling by Reactive Nitrogen Species	712
	10.7	Hippo Signaling (STK3–STK4 Pathway)		720
		10.7.1	Hippo Signaling Components	721
	10.8	Coregulators		730
		10.8.1	Receptors for Activated C-Kinase	732
		10.8.2	G-Kinase-Anchoring Protein	733
		10.8.3	Scaffolds of Synapses of the Central Nervous System	734
		10.8.4	A-Kinase-Anchoring Proteins	737
		10.8.5	Annexins	747
	10.9	Transcription Factors Involved in Stress Responses		757
		10.9.1	Nuclear Factor-κB Signaling Module	758
		10.9.2	Hypoxia-Inducible Factor	788
		10.9.3	Forkhead Box Transcription Factors	792
		10.9.4	P53 Family of Transcription Factors	803
		10.9.5	Transcription Factor p27	819
11	**Signaling Pathways**			**821**
	11.1	cAMP Signaling		822
		11.1.1	Crosstalk between Calcium and cAMP Signalings	825
		11.1.2	Crosstalk between the ERK and cAMP Signaling Pathways	825
		11.1.3	cAMP Response Element-Binding Protein and Cofactors	826
		11.1.4	Adenylate Cyclases	827
		11.1.5	Phosphodiesterases	830

11.2	cGMP Signaling		846
11.3	Signaling via Cell Junctions		848
	11.3.1	Elastin–Laminin Receptor	848
	11.3.2	Adhesion Molecules	851
	11.3.3	Discoidin Domain Receptors	854
	11.3.4	Signaling via Focal Adhesions	854
	11.3.5	Signaling via Gap Junctions	854
11.4	Interactions between Ion Channels and Small GTPases		855
11.5	Calcium Signaling		858
	11.5.1	Calcium-Induced Calcium Release	859
	11.5.2	Calcium Signaling and Myocyte Contraction	860
	11.5.3	Instantaneous and Long-Lasting Responses to Calcium	861
	11.5.4	Calcium Signaling Components	862
	11.5.5	Types of Calcium Signalings	874
	11.5.6	Calcium-Mobilizing Mediators cADPR and NAADP	875
	11.5.7	Calcium and Nervous Control of Blood Circulation	876
	11.5.8	Calcium Signaling and Immunity	879
	11.5.9	Calcium Signaling and Intracellular Transport	881
	11.5.10	Calcium Signaling and Cell Fate	882
11.6	Oxygen Delivery and Hypoxia Transduction		883
	11.6.1	Hypoxia-Inducible Factor Axis	884
	11.6.2	Hypoxia-Inducible Factor-Independent Axis	890
11.7	VEGF Signaling		892
11.8	Insulin Pathway		892
11.9	Inflammasome		896
11.10	Mechanotransduction		898
	11.10.1	Basic Components of Vascular Mechanotransduction	902
	11.10.2	Mechanotransduction Signaling	902
	11.10.3	Mechanotransduction in Arterial Compartments	908

12 Conclusion 911

References 919

Notation Rules: Aliases and Symbols 1005

List of Currently Used Prefixes and Suffixes 1011

List of Aliases 1015

Complementary Lists of Notations 1045

Index 1051

Chapter 1
Introduction

> "Change begets change. Nothing propagates so fast." (C.J.H. Dickens, The Life and Adventures of Martin Chuzzlewit)

Cells have the capacity to emit as well as to receive, decode, and transmit information efficiently, and to integrate new signals. Any cell experiences numerous, simultaneous or successive, events that result from permanent, regulated communications between it and adjoining cells and the extracellular matrix as well as remote controller cells of the nervous and endocrine systems, in addition to immunocytes. Cell signaling not only governs basic cellular activities, but also coordinates actions of cell populations.

Plasmalemmal receptors that initiate signaling can be: (1) enzymes, such as receptor kinases and phosphatases, membrane-tethered guanylate cyclases, NADPH oxidases, and tranfer ATPases; (2) proteins coupled to enzymes, such as G-protein-coupled receptors and pseudo-receptor kinases; (3) some types of cytokine receptors; (4) ions channels; (5) adhesion molecules; and (6) mass-transfer receptors, which transport a first chemical messenger; in addition to (7) specialized plasmalemmal nanodomain components that participate in endocytosis, during which signaling can be launched.

Members of these various groups of plasmalemmal receptors can cluster and form the so-called *transducisome*. Any plasmalemmal molecule can undergo a strain with a more or less deep conformational change when stretched, but most of them do not send any signal. Among plasmalemmal receptors, some are mechanosensitive. Mechanical stress deforms and activates mechanosensitive molecules that can then initiate signaling by stimulating, directly or not, a second messenger.

Once the plasmalemmal receptor is activated, a cascade of chemical reactions triggers either the release of stored signaling mediators for a rapid response or gene transcription for a delayed reaction. Both event types lead to specific cell responses. In general, initial steps of signaling pathways occurs at the cell cortex upon recruitment of appropriate effectors for optimal efficiency.

Scaffold proteins serve as assembly platforms for enzymes, their regulators, and their substrates. Therefore, they have an active role in effector enzyme activation. Scaffold proteins bind to at least 2 other signaling proteins. They help to localize signaling molecules to specific cell regions. They restrain the non-specific access of enzymes to unwanted substrates. Some scaffold proteins can have other functions, such as coordination of positive and negative feedback signals or protection of activated proteins from inactivation. Scaffold proteins can have distinct functions under different conditions. Multiple scaffold–receptor complexes may exist simultaneously, thereby launching both overlapping and distinct cellular events. On the other hand, *adaptor proteins* (accessory proteins) connect proteins in order to form multimolecular complexes that do not have enzymatic activity.

Alternative splicing of messenger RNA defines different reconnections of transcribed exons, thereby generating various types of mature RNA transcripts from a given primary transcripts (splice variants). This process then leads to the synthesis of multiple stable protein isoforms, more or less functional, encoded by a single gene, under selective constraints (Vol. 1 – Chap. 5. Protein Synthesis). Although many alternative transcript variants are produced in the cell, only a small amount leads to the synthesis of stable proteins. Heterogeneous nuclear ribonucleoproteins are involved in the regulation of alternative splicing. Alternative isoforms of 10 among 26 possible primary transcript types that encode heterogeneous nuclear ribonucleoproteins can be detected [1]. Among all types of splice variants, a large number of alternative isoforms is created from homologous exons. An other large number of alternative isoforms can simply arise from the insertion or deletion of a single amino acid or a small number of residues. Small resulting differences in protein sequence and structure may cause minor changes in function. Nevertheless, alternative splicing influences the binding affinity of proteins for their partners.

Intracellular cascades of chemical reactions requires many mediators. Each signaling node can include not only a given mediator (i.e., a signaling effector) and its affector(s), but also one or several docking or scaffolding proteins. Signaling mediators — effectors, scaffolds, and adaptors — can have various names according to their discovery framework and investigation history. This book aims at presenting an exhaustive list of intracellular mediators with their effects.

In the framework of mathematical modeling, once collected the participants of a target signaling pathway and their role known, the main components of this pathway are selected without neglecting a major element, but eliminating all minor contributors. This selection enables to develop a simplified, tractable, and representative mathematical model of the explored biological process.

Signaling pathways determine cell functioning for given missions that generate specific molecular effects. The knowledge of expression patterns of various involved signaling components — from extracellular messengers that bind to their cognate receptors to trigger a reaction cascade, during which a set of effectors is activated — and the bulk activity of signaling pathways can be discovered from loss- and gain-of-function analysis and from disorders that originate from gene mutations and disturb signaling.

According to their structural and functional features, signaling mediators are classified in multiple categories. Major mediator categories are exhibited in many chapters. A first chapter presents lipid mediators and their modifying enzymes. The following chapters focus on sets of cytoplasmic protein tyrosine, serine/threonine, and dual-specificity kinases, with an emphasis on mitogen-activated protein kinase modules that include protein Ser/Thr and dual-specificity kinases, as well as protein phosphatases, in addition to others major mediators involved in cardiovascular and respiratory physiology and pathophysiology. The final chapter gives examples of several selected signaling pathways, such as those involving typical second messengers (cyclic adenosine and guanosine monophosphates and calcium ions), as well as those primed by adhesion and matrix molecules and those involved in oxygen sensing, insulin stimulation, angiogenesis, and mechanotransduction.

Among all mechanisms that contribute to cell events by modifying the structure and function of signaling mediators, phosphorylation and dephosphorylation correspond to a major process that allows cells to modulate protein activity either at their surfaces or inside their cytoplasms in response to environmental stimuli, such as conformational, positional, and nutritional signals, as well as other regulatory cues.

These signals are generated by cells themselves (intra- and autocrine regulation) or neighboring (juxta- and paracrine regulation) or remote cells (endocrine regulation). Local and remote regulation is ensured by signal transmission via the extracellular medium, whereas intracrine regulation deals with addition and removal of phosphate groups, among other post-translational modifications, that occur without secretion of signaling molecules.

Mathematical and computational biology encompasses 4 major subfields. (**Subfield 1**) Biomathematics focus on modeling and simulations of biological processes at various scales, from any cascade of chemical reactions within the cell, to the fate of a single cell in a given context as well as the collective behavior of cell populations, to the relation between different compartments of a physiological apparatus and interactions between distinct physiological systems, in addition to pharmodynamics and -kinetics. The description of the behavior of a biological process leads to its modeling and subsequent simulations. An objective can be the prediction of biological quantities, the values of which may not be adequately measured.

Last, but not least, direct observation of the dynamics of biological processes being often difficult, theoretical and numerical methods are developed to recover values of control variables and parameters by solving the *inverse problem*. In the case of an initial-value problem, a new set of equations incorporate observations at given times to predict the evolution of the explored process.

The personalized, or subject-specific, models (adapted to the patient) requires the three-dimensional reconstruction of organs of interest from a set of medical images, from which a proper mesh can be computed for mechanical analysis. In addition, the model parameters must fit available clinical measurements. In biomechanical problems, a data assimilation strategy can be adopted to reduce uncertainties of the model using partial observations of the state variables. Data assimilation of distributed mechanical systems can be based on a variational approach by minimizing a least square criterion that includes a regularization term and the difference between the observations and the model prediction [2]. The major drawback of this strategy is the iterative evaluation of the criterion, which requires many solutions of the forward problem. An alternative approach is provided by sequential algorithms, in which the model prediction is improved at every time step by analyzing the discrepancy between the actual measurements and model outputs [3].

This type of approaches can be illustrated by the estimation of the artery wall stiffness from wall displacement measurements using medical imaging and a Windkessel model to appropriately assess the boundary conditions. The parameter estimation for fluid-structure interaction problems is based on a reduced order unscented Kalman filter [4].

Another example deals with tumor growth, a multiscale phenomenon that involves the nanoscopic scale (genetic mutations and molecular pathways inside the cell), the microscopic scale (cell interactions with its environment and with adjacent normal epithelial cells, immunocytes, and vascular cells, as well as cell migration), and the macroscopic scale (metastasis). The direct problem is aimed at understanding the tumoral development, whereas the inverse problem deals with the prediction of the evolution of the tumoral cell population, i.e., targets prognosis and treatment, knowing the cell density at given times. This type of models relies on transport–reaction equations, such as:

$$\partial_t n_i + \nabla \cdot J = \mathsf{P}(n_j, c) - \mathsf{D}(n_j, c), \tag{1.1}$$

where n_i is the density of type-i cells (any tumor at a sufficiently developed stage is composed of 2 populations, proliferative and necrotic cells), J the cell flux, P and D the production and degradation rates, which depend on cell density of other species j (n_j) and nutrient concentration (c):

$$\partial_t c - \nabla \cdot (\mathcal{D}\nabla c) = \sum \bigl(\mathsf{U}(n_j) - \mathsf{C}(n_j)\bigr) c, \tag{1.2}$$

where $\mathcal{D}$ is the diffusivity and U and C the uptake and consumption rates.

Observed densities of proliferative ($n_\mathrm{p}^\mathrm{obs}$) and necrotic ($n_\mathrm{n}^\mathrm{obs}$) cells at a set of selected time ($t^\mathrm{obs} \in T^\mathrm{obs}$) are incorporated in a cost function $\mathcal{C}$ [5,6]:

$$\mathcal{C} = \frac{1}{2} \int_\Omega \sum \Bigl([n_\mathrm{p} - n_\mathrm{p}^\mathrm{obs}]^2 + [n_\mathrm{n} - n_\mathrm{n}^\mathrm{obs}]^2\Bigr) d\mathbf{x}. \tag{1.3}$$

The inverse problem consists in finding model parameters and initial data such as the solution ($n_\mathrm{p}, n_\mathrm{n}$) minimizes $\mathcal{C}$. Solving the direct problem for different sets of parameter values enable to build a data basis that can assist the solving procedure.

(**Subfield 2**) Applied mathematics in medicine and biology constitute the basis of many aspects of engineering sciences, i.e., in: (1) medical image and signal processing; (2) three-dimensional reconstruction of body's organs and virtual endoscopy;

(3) navigation tools to assist peroperating gesture as well as manipulation of materials and deployment of devices in regions of interest; (4) medical simulators for training of handling of exploration techniques as well as mini-invasive methods of medical and surgical treatments; (5) design and shape optimization of implantable medical devices; (6) medical robotics; and (7) telemedicine and -surgery.

(**Subfield 3**) Medical statistics, or biostatistics, are currently used in epidemiology, public health, forensic medicine, and clinical research. A major goal is the definition of appropriate indices for early diagnosis of chronic diseases and evaluation of prognosis, the determination of risk factors in the development of chronic pathologies, as well as optimization of treatments (administration time, duration, and dose).

(**Subfield 4**) Bioinformatics is based on the knowledge accumulated in computer science, mathematics, statistics, and statistical physics. Due to the large amount of DNA and amino acid sequences and related information to be handled, bioinformatics relies on intensive computations. This subfield thus involves algorithm and computation theory, artificial intelligence, discrete mathematics, control and system theory, databases and information systems, and statistics. Bioinformatics target sequence analysis, genome annotation, computational evolutionary biology, analysis of gene and protein expression, analysis of the structure and molecular dynamics of biological compounds, as well as between-molecule interactions.

Many processes that determine cell fate indeed rely on chemical reaction cascades. The evolution of interacting molecular species in chemical reactions can be described by deterministic and stochastic models. The present book is simply aimed at presenting many sets of molecules that serve as signaling mediators within the cell, i.e., as nodes of computational models of signaling cascades.

The present book is organized in 12 chapters, with 10 devoted to sets of cytoplasmic signaling effectors. Lipid mediators are described in chapter 2. Chapter 3 presents various categories of protein kinases that can be considered to constitute an infraphylum in the phylum of enzymes, rather than a superclass, as different types of catalytic subunits of phosphatidylinositol 3-kinase encoded by different genes (PIK3CA–PIK3CB, PIK3CD, and PIK3CG, PIK3C2A1–PIK3C2B and PIK3C2G, and PIK3C3) are currently defined by several classes, although the word "family" is more appropriate. Moreover, a class-4 is defined that contains a subset of PI3K-related kinases (target of rapamycin kinase [TOR], ataxia telangiectasia mutated kinase [ATMK], ataxia telangiectasia mutated-related kinase [ATRK], and DNA-Dependent protein kinase [DNAPK]).

The following chapters focus on subclasses of cytoplasmic non-receptor protein Tyr kinases (Chap. 4) and protein Ser/Thr kinases that incorporates many superfamilies (STE [Ste20-related], AGC, CAMK, CK1, CMGC, TK, TKL, RGC, atypical kinases, and others; Chap. 5), on components of the MAPK signaling modules (Chap. 6), and on dual-specificity protein kinases (Chap. 7).

Chapter 8 deals with the class of soluble, cytosolic phosphatases, with its 2 major subclasses of protein Tyr and Ser/Thr phosphatases. Dual-specific phosphatases form a hyperfamily. Another category of enzymes — guanosine triphosphatases — constitute another infraphylum. Two main classes are composed of small (monomeric) and trimeric GTPases with numerous subclasses and superfamilies (Chap. 9).

Chapter 10 is devoted to other categories of major signaling molecules, such as gasotransmitters, components of redox-based signaling, coregulators, and some selected transcription factors. Chapter 11 gives examples of signaling pathways. Major mediators include calcium ions and cyclic adenosine and guanosine monophosphates; their interactors and the pathway regulation are described. Among other types of pathways, this chapter highlights mechanotransduction.

Common abbreviations such as "a.k.a." and "w.r.t." that stands for "also known as" and "with respect to", respectively, are used throughout the text to lighten sentences. Latin-derived shortened expressions are also widely utilized (but not italicized despite their latin origin): "e.g." (exempli gratia) and "i.e." (id est) mean "for example" and "in other words", respectively. Rules adopted for substance aliases as well as alias meaning and other notations are given at the end of this book.

Acknowledgments The author acknowledges the patience of his family (Anne, Maud, Julien, Jean, Louis, Raphaëlle, Alrik, Matthieu, Alexandre, Joanna, Damien, and Frédéric).

Chapter 2
Signaling Lipids

Lipids include sterols, mono- and diglycerides, phospholipids, among others. They operate as structural components of cell membranes (Vol. 1 – Chap. 7. Plasma Membrane; Tables 2.1 and 2.2), contributors of energy metabolism, participants of intracellular transport (Vol. 1 – Chap. 9. Intracellular Transport), and signaling molecules.

Phospholipids are constituents of cellular membranes that are also used for emulsification in pharmaceuticals and food and preparation of liposomes for drug delivery. Phospholipids are characterized by a glycerol backbone to which a polar phosphodiester group is linked at the sn3 carbon.[1] Glycerol-based phospholipids are also named *glycerophospholipids*, or *phosphoglycerides*. Two fatty acid-derived acyl residues are connected at the sn1 and sn2 carbons. The type of polar head groups defines the class of phospholipids. Three distinct structural regions can be defined [8]: (1) a hydrophilic head at the lipid–water interface; (2) an interfacial region of intermediate polarity; and (3) a hydrophobic tail.

Lysophospholipids (LPL) are glycerophospholipids in which one acyl chain lacks.[2] Like phospholipids, these emulsifying and solubilizing agents lodge in small amounts in cellular membranes. These membrane-derived signaling molecules are generated by compartmentalized phospholipases, such as phospholipases PLA1, PLA2, and PLD, in addition to different lipase types [8]. Lysophospholipids and their receptors intervene in the reproduction, vascular development, and function of the nervous system. In particular, lysophosphatidylcholine, the most abundant

---

[1] The notation of glycerophospholipids relies on a stereochemical numbering. The numbering follows the rule of the Fischer's projection, the hydroxyl group of the second carbon of glycerol (sn2) being, by convention, on the left on a Fischer projection, the carbon at the top and bottom being at sn1 and sn3 position, respectively.

[2] Alias LPL is also used in the literature to designate a lipoprotein lipase. Hence, in the present text, the alias LPase is used for lipoprotein lipase, thereby also avoiding the confusion with lysophospholipase (LPLase). The prefix "lyso" was coined because lysophospholipids were originally found to cause hemolysis.

Table 2.1. Cellular sites — plasma and organelle membranes — of main lipids involved in organelle recognition and signaling (**Part 1**; Source: [7]). The plasma membrane contains high levels of packed sphingolipids and sterols to resist mechanical stress, as well as signaling lipids. The endoplasmic reticulum produces cholesterol, main glycerophospholipids, steryl ester, triacylglycerol (or triglyceride), ceramide, and galactosylceramide. Among glycerophospholipids (or phosphoglycerides, i.e., glycerol-based phospholipids), phosphatidylcholine, phosphatidylethanolamine, and bisphosphatidylglycerol are also called lecithin, cephalin, and cardiolipin, respectively. The first 2 species are more common than other glycerophospholipids in most cellular membranes, except cardiolipin in the inner membrane of mitochondria. The endoplasmic reticulum has low concentrations of sterols and complex sphingolipids. The Golgi body synthesizes phosphatidylcholine, sphingolipids, and glycosphingolipids. Mitochondrion manufactures about half of its phospholipids. Early endosomes have a content similar to the plasma membrane, whereas in late endosomes concentrations of sterol and bis(monoacylglycero)phosphate decrease and increase, respectively.

Lipids	Cellular location
Cholesterol	All membranes
Glycerophospholipids (or phosphoglycerides)	
Bis(monoacylglycero)phosphate	Late endosome
Diacylglycerol	Plasma membrane
Phosphatidic acid	Endoplasmic reticulum, mitochondrion
Phosphatidylcholine	Endoplasmic reticulum, Golgi body, mitochondrion, endosome, plasma membrane
Phosphatidylethanolamine	Endoplasmic reticulum, Golgi body, mitochondrion, endosome, plasma membrane
Phosphatidylglycerol	Mitochondrion
Phosphatidylinositol	Endoplasmic reticulum, Golgi body, mitochondrion, endosome, plasma membrane
Phosphatidylserine	Endoplasmic reticulum, Golgi body, mitochondrion, endosome, plasma membrane
Triacylglycerol	Endoplasmic reticulum, lipid droplet

lysophospholipids, participates in the regulation of gene transcription, cell division, monocyte chemotaxis, smooth muscle cell relaxation, and platelet activation. Lysophosphatidic acid is an intermediate mediator of transmembrane signal transduction.

Inositol-containing lipids are also strongly involved in signal transduction. Inositol lipids include *phosphoinositides*, i.e., phosphatidylinositol and its phosphorylated derivatives (in addition to inositol-containing ceramides in fungi and inositol glycolipids in pathogens). The metabolism of inositol-containing lipids is regulated by protein kinases and phosphatases.

Inositol phosphates are incorporated in this chapter (Sect. 2.3) because: (1) *myoinositol*[3] serves as a structural basis for inositol phosphates as well as lipids phosphatidylinositol (PI) and phosphatidylinositol phosphates (PIP), each category

---

[3] Among 9 stereoisomers of inositol, or cyclohexane (1,2,3,4,5,6)-hexol, myoinositol, or cis-(1,2,3,5) trans-(4,6) cyclohexanehexol, is the most prominent form among natural types. Inositol is a non-conventional carbohydrate.

Table 2.2. Cellular sites — plasma and organelle membranes — of main lipids that are involved in organelle recognition and signaling (**Part 2**; Source: [7]; PI: phosphatidylinositol). Kinases and phosphatases produce and hydrolyze specific phosphoinositides (PI(3)P on early endosomes and PI(3,5)P_2 on late endosomes). More complex sphingolipids require vesicular transport from the endoplasmic reticulum.

Lipids	Cellular location
Sphingolipids	
Ceramide	Endoplasmic reticulum, plasma membrane
Galactosylceramide	Endoplasmic reticulum
Glycosphingolipids	Endoplasmic reticulum, Golgi body
Sphingomyelin	Golgi body, endosome, plasma membrane
Sphingosine	Plasma membrane
Sphingosine 1-phosphate	Plasma membrane
Phospholipids	
Cardiolipin	Mitochondrion
PI(3)P	Early endosome
PI(4)P	Golgi body, plasma membrane
PI(3,4)P_2	Late endosome
PI(4,5)P_2	Plasma membrane
PI(3,4,5)P_3	Plasma membrane

of molecules yielding numerous secondary messengers and (2) inositol phosphates, which constitute a group of more than 30 mono- and polyphosphorylated inositols, share some kinases and phosphatases with the 7 types of phosphoinositides.

Phosphoinositides and inositol phosphates modulate the recruitment and activation of regulators, thereby participating in the control of cell growth and proliferation, apoptosis, cytoskeletal dynamics, insulin signaling, vesicular transport, and nuclear function.

2.1 Lipids in Signal Transduction

Lipids serve as signaling effectors. Many signaling lipids, their enzymes, and effectors are common to multiple lipid signaling pathways. Numerous lipids act as intra- and extracellular messengers to control cell fate. As structural components of cell membranes, they are substrates for enzymes that generate second messengers involved in cell response to stimuli. Lipids can also modulate the activity of signaling proteic mediators. Moreover, they can contribute to the recruitment at effective sites of signaling effectors.

Signaling lipids include: (1) *fatty acids*;[4] (2) *eicosanoids* and other products of *arachidonic acid* metabolism;[5] (3) *glycerolipid-derived regulators*, such as *phosphatidic acid* (PA),[6] *lysophosphatidic acid* (LPA),[7] *lysophosphatidylcholines* (LPC), or lysolecithins,[8] *monoacylglycerol* and *diacylglycerol* (DAG),[9] *anandamide*, or ^{N}arachidonoylethanolamide[10] and platelet-activating factor; as well as (4) *phosphoinositides*, or inositol phospholipids, such as the set of phosphatidylinositol mono-, bis-, and trisphosphates; (5) *sphingolipids*, such as sphingosine, sphingosylphosphorylcholine, sphingosine 1-phosphate, ceramide, and ceramide 1-phosphate.

Signaling lipids regulate cell metabolism, growth, proliferation, migration, senescence, and apoptosis. Growth factors and constituents of nutrients control the activity of enzyme-targeting lipids, such as phospholipases, prostaglandin synthases, 5-lipoxygenase (LOx), phosphoinositide 3-kinase (PI3K), sphingosine kinase (SphK), and sphingomyelinase (SMase).

Sphingomyelinase and sphingosine kinase are regulated enzymes that produce sphingosine 1-phosphate (S1P), a ligand for cognate G-protein-coupled receptors

---

[4] A fatty acid is a carboxylic acid with a long, unbranched, aliphatic (Greek αλειφαρ: oil, fat, unguent), saturated (constituents joined by single bonds [alkanes]) or unsaturated (constituents joined by double [alkenes] or triple [alkynes] bonds) chain.

[5] Arachidonic acid is a polyunsaturated ω6 fatty acid. It is a component of phospholipids, especially phosphatidylethanolamine, phosphatidylcholine, and phosphatidylinositides. It is generated from: (1) phospholipids by phospholipase-A2 and (2) diacylglycerol by diacylglycerol lipase. Arachidonic acid is synthesized as a signaling mediator by cytosolic phospholipase-A2 and as an inflammatory messenger by secretory phospholipase-A2. Arachidonic acid serves as a precursor of prostaglandin-H2 synthesized by cyclooxygenase and peroxidase, 5-hydroperoxyeicosatetraenoic acid (5HPETE) by 5-lipoxygenase, and anandamide. It can be converted into hydroxyeicosatetraenoic (HETE) and epoxyeicosatrienoic (EET) acids by epoxygenase. This second lipid messenger participates in the regulation of signaling enzymes (phospholipases PLC-γ and PLCδ and protein kinases PKCα, PKCβ, and PKCγ).

[6] Phosphatidic acid is a simple form of diacylglycerophospholipids that serves as a signaling lipid. It is generated by phospholipase-D from phosphatidylcholine, by diacylglycerol kinase (DAGK) from diacylglycerol, and by lysophosphatidic acid acyltransferase (LPAAT) from lysophosphatidic acid.

[7] Lysophosphatidic acid is a phospholipid derivative and a signaling mediator (mitogen). Among many processes, it is synthesized by lysophospholipase-D, or autotaxin, from lysophosphatidylcholine.

[8] Lysophosphatidylcholines result from the removal of one fatty acid of phosphatidylcholines, generally by phospholipase-A2. They are quickly metabolized by lysophospholipase and lysophosphatidylcholine acyltransferase.

[9] Mono- and diacylglycerol are glycerides that consist of 1 and 2 fatty acids linked to a glycerol, respectively.

[10] Anandamide (Sanskrit ananda: joy, delight, happiness, bliss, blessedness, beatitude) is a cannabinoid neurotransmitter. It is synthesized from ^{N}arachidonoylphosphatidylethanolamine (NAPE) by phospholipase-A2, phospholipase-C, and NAPE-specific phospholipase-D. It is quickly converted into ethanolamine and arachidonic acid by fatty acid amide hydrolase (FAAH).

2.1 Lipids in Signal Transduction

(Vol. 3 – Chap. 7. G-Protein-Coupled Receptors). The S1P pathway promotes, in particular, the survival of cells of the cardiovascular system.[11]

Like sphingosine 1-phosphate, extracellular *lysophospholipids*[12] interact with specific G-protein-coupled receptors[13] to influence cell proliferation, differentiation, and motility. Lysophospholipids produced by phospholipase-A2 not only exert signaling functions, but also are processed into signaling mediators (e.g., lysophosphatidylcholine is converted to platelet-activating factor). *Lysophosphatidic acid* functions as an auto- and paracrine factor that binds to a set of receptors (LPARs; Vol. 3 – Chap. 7. G-Protein-Coupled Receptors). In addition, lysophosphatidic acid modulates gene expression by binding to lipid-sensing transcription activators such as peroxisome proliferator-activated receptors.

Phosphoinositides are synthesized or degraded using regulated enzymes excited by activated plasmalemmal receptors, such as phosphatidylinositol 3-kinases that generate the second messenger phosphatidylinositol trisphosphate and phospholipase-C that hydrolyzes the messenger phosphatidylinositol (4,5)-bisphosphate into inositol trisphosphate (IP_3), a Ca^{++} mobilizer, and diacylglycerol, an activator of protein kinase-C, on the one hand, and activates protein kinase-B on the other (Sect. 5.2). Cardiac ion channel activity depends on PIP_2 density.

Membrane phospholipids are also substrates of phospholipase-A2 that generate arachidonic acid, the stem molecule for eicosanoids. *Eicosanoids* have a short half-life that ranges from seconds to minutes. The production of eicosanoids is initiated by the release of C20-polyunsaturated fatty acids such as arachidonic acid (rate-limiting stage) from phospholipids or diacylglycerol due to phospholipase-A2 in the presence of Ca^{++} ions.

Among the prostaglandin synthases, prostaglandin-H2 synthase (PGH2S) and cyclooxygenases COx1 and COx2 lead to one of the 3 families of eicosanoids: prostaglandins, prostacyclins, thromboxanes. On the other hand, 5-lipoxygenase produces leukotrienes. Certain eicosanoids can bind to intracellular receptors (i.e., transcription factors), such as farnesoid X, liver X, and peroxisome proliferator-activated receptors (Vol. 3 – Chap. 6. Receptors).

Phosphatidylcholines are phospholipids that contain choline as a head group. They operate in membrane-mediated signaling. Phospholipase-D hydrolyzes phosphatidylcholine into phosphatidic acid, hence releasing the soluble choline head group into the cytosol. Lysophosphatidylcholines derive from partial hydrolysis of phosphatidylcholines, generally by phospholipase-A2, i.e., removal of one of the fatty acid groups.

---

[11] Sphingosine 1-phosphate serves as a protective element carried by high-density lipoproteins.

[12] Lysophospholipids are phospholipids that have lost an ^{O}acyl chain. An acyl group (or alkanoyl; RCO; R: monovalent functional alkyl group) is usually derived from a carboxylic acid (R-COOH).

[13] Main lysophospholipids include lysophosphatidic acid (LPA), lysophosphatidylcholine (LPC), sphingosylphosphorylcholine (SPC), and sphingosine 1-phosphate (S1P). Lysophosphatidic acid binds to LPARs; LPC associates with GPR132 (or G2 accumulation protein [G2A]); SPC links to GPR4, GPR12, and GPR68 (a.k.a. GPR12a and ovarian cancer G-protein-coupled receptor-1 [OGR1]); and S1P targets S1PRs.

2.2 Phosphoinositides

Although phosphoinositides account for a tiny fraction of cellular phospholipids (Table 2.3), they are required for signal transduction. They indeed recruit many signaling proteins in cellular membranes.[14] The resulting signaling complexes participate not only in membrane transfer and cytoskeletal restructuring, but also cell survival, growth, and proliferation. In particular, phosphoinositides serve as mediators of calcium-mobilizing hormones and neurotransmitters.

Different phosphatidylinositol phosphates result from phosphorylation of single or multiple sites of the inositol ring of phosphatidylinositol: (1) phosphatidylinositol monophosphates PI(3)P, PI(4)P, and PI(5)P; (2) phosphatidylinositol bisphosphates PI(3,4)P_2, PI(3,5)P_2, and PI(4,5)P_2; and (3) phosphatidylinositol trisphosphate PI(3,4,5)P_3 (Tables 2.4 to 2.7).[15] Multiple inositide kinases and phosphatases that interconvert phosphoinositides intervene in the sorting of molecules to specific organelles.

Protein Tyr phosphatases usually dephosphorylate Tyr^P-containing proteins. Yet, some protein Tyr phosphatases dephosphorylate Ser^P- and Thr^P-containing proteins as well as phosphoinositides, such as phosphatase and tensin homolog deleted on chromosome 10 (PTen) and its homologs PTen2 and TPIP, as well as inositol polyphosphate 4-phosphatase-1 (IP(4)P1) and -2 (IP(4)P2) and the PTen-like phosphatase mitochondrial phosphoTyr phosphatase (PTPmt).

Table 2.3. Approximate fractions of the total amount of phosphoinositides (PI) of various members of the families of mono- (PIP), bis- (PIP$_2$), and trisphosphate (PIP$_3$) derivatives (Source: [9]).

Contributors	Fraction (%) of total PI concentration
PI, PI(4)P, PI(4,5)$_2$	∼90
PI(3)P, PI(5)P	2–5
PI(3,5)$_2$	∼2 (fibroblast)
PI(3,4)$_2$, PI(3,4,5)$_3$	∼0 at rest ↑ during stimulation (2–3)

---

[14] Numerous proteins that possess PH (pleckstrin homology), PTB (phosphotyrosine-binding), ENTH (epsin N-terminal homology), FYVE (Fab-1 [yeast ortholog of PIKFYVE], YOTB, Vac1 (vesicle transport protein), and EEA1), PX (phagocytic oxidase), FERM (4.1 protein, ezrin, radixin, and moesin), and/or CALM (clathrin-assembly lymphoid myeloid leukemia; a.k.a. ANTH [AP180 N-terminal homology]) domains recognize specific isomers of phosphoinositides.

[15] Phosphoinositides are derivatives of phosphatidylinositol in which one or more OH groups at 3-, 4-, and/or 5-position of the inositol ring is (are) phosphorylated.

2.2 Phosphoinositides

Table 2.4. Phosphatidylinositol (PI) and phosphatidylinositol phosphate (PIP) kinases (PIiK and PI(i)PjK); PI, PIP, and PIPP (phosphatidylinositol polyphosphate) phosphatases; and phospholipase-C (PLC) convert diverse types of phosphoinositides, which intervene in cell biology (MTMR: myotubularin-related phosphatase; PTen: phosphatase and tensin homolog deleted on chromosome 10; PTPmt: mitochondrial phosphoTyr phosphatase, previously named phospholipid-inositol phosphatase PLIP; SHIP: SH2-containing inositol polyphosphate 5-phosphatase; SKIP: skeletal muscle and kidney enriched inositol 5-phosphatase).

Phosphoinositide	Kinase (production)	Phosphatase (elimination)	Phospholipase and products
$PI(3)P$	PI3K	MTMR	
$PI(4)P$	PI4K	4-Pase	
$PI(5)P$	PI5K	PTPmt	
$PI(3,4)P_2$	PI3K	PTen	
	PI4K	PIP4Pase	
$PI(3,5)P_2$	PI5K	PTPmt	
		MTMR	
$PI(4,5)P_2$	PI(5)P4K	IpgD	PLC: $I(1,4,5)P_3$ and DAG
	PI(4)P5K	5-Pase	
$PI(3,4,5)P_3$	PI3K	PTen	
	PI5K	SHIP, SKIP	

Table 2.5. Phosphorylation and dephosphorylation (production and removal) of various types of phosphoinositides. **(Part 1)** Phosphoinositides (PI) and phosphatidylinositol trisphosphate ($PI(3,4,5)P_3$; PIKFYVE: FYVE finger-containing phosphoinositide kinase, or type-3 phosphatidylinositol 3-phosphate and phosphatidylinositol 5-kinase; IP(5)P: inositol polyphosphate 5-phosphatase PTen: phosphatase and tensin homolog deleted on chromosome 10 [ten], a phosphatidylinositol 3-phosphatase).

Substrate	Product	Enzyme
PI	$PI(3)P$	Class-3 PI3K
	$PI(4)P$	PI4K
	$PI(5)P$	PIP5K3 (PIKFYVE)
$PI(3,4,5)P_3$	$PI(3,4)P_2$	IP(5)P
	$PI(4,5)P_2$	PTen

2.2.1 $PI(4,5)P_2$

The inositol region of phosphoinositides is used as a specific docking site for lipid-binding effectors of signaling cascades. Phosphatidylinositol (4,5)-bisphosphate is hydrolyzed by phospholipase-C into (1,2)-diacylglycerol and inositol (1,4,5)-trisphosphate. Both products act as second messengers that activate conventional protein kinase-C and release calcium ion from internal stores through IP_3R,

Table 2.6. Phosphorylation and dephosphorylation (production and removal) of various types of phosphoinositides. (**Part 2**) Phosphatidylinositol monophosphates (PIP; PIPK: phosphatidylinositol phosphate kinase). The myotubularin superfamily of phosphatases — myotubularins (MTM) and myotubularin-related proteins (MTMR) — are inositol 3-phosphatases that dephosphorylate PI3P and PI(3,5)P_2.

Substrate	Product	Enzyme
PI(3)P	PI(3,4)P_2	PI4K
	PI(3,5)P_2	PIP5K3 (PIKFYVE)
		or PI(3)P5K (type-3 PIPK)
	PI	MTM, MTMR
PI(4)P	PI(3,4)P_2	Class-1 and -2 PI3Ks
	PI(4,5)P_2	PI(4)P5K (type-1 PIPK)
	PI	
PI(5)P	PI(3,5)P_2	PI3K
	PI(4,5)P_2	PI(5)P4K (type-2 PIPK)

Table 2.7. Phosphorylation and dephosphorylation (production and removal) of various types of phosphoinositides. (**Part 3**) Phosphatidylinositol bisphosphates PIP$_2$ (PIC: phosphoinositidase-C [or phospholipase-C (PLC)]).

Substrate	Product	Enzyme
PI(3,4)P_2	PI(3,4,5)P_3	PI5K
	PI(3)P	4PTase
	PI(4)P	PTen
PI(3,5)P_2	PI(3)P	5PTase
	PI(5)P	MTM, MTMR
PI(4,5)P_2	PI(3,4,5)P_3	Class-1/2 PI3Ks
	PI(4)P	5PTase
	PI(5)P	4PTase
	IP$_3$, DAG	PLC
	IP$_3$, IP$_6$,	(PIC)
	IP$_7$, IP$_8$	

respectively. Diacylglycerol that remains within the membrane to act as a PKC coactivator can also stimulate some of the canonical transient receptor potential channels (TRPC), independently of PKC enzyme.

Alternatively, PI(4,5)P_2 can be converted to phosphatidylinositol (3,4,5)-trisphosphate by phosphoinositide 3-kinase that is regulated by plasmalemmal receptors (GPCRs and RTKs; Vol. 3 – Chaps. 7. G-Protein-Coupled Receptors and 8. Receptor Kinases).

In addition, PIP$_2$ can also be hydrolyzed by phospholipases-D to generate phosphatidic acid, an activator of signaling mediators. Phosphatidic acid controls cell proliferation and survival via Son of sevenless and target of rapamycin, respectively. Last, but not least, PI(4,5)P_2 controls many ion channels and exchangers (Table 2.8).

2.2 Phosphoinositides

Table 2.8. Activity of PI(4,5)P$_2$ (Sources: [10, 11]). Phospholipase-D (PLD) has 2 isoforms (PLD1–PLD2). The PLD1 subtype localizes primarily to vesicles; PLD2 mainly to the plasma membrane. Phospholipase-D interacts with PI(4)P 5-kinase-α(PI(4)P5Kα). Consequently, a local generation of PI(4,5)P$_2$ contributes to the regulation of PLD activity. Phospholipase-D hydrolyzes phosphatidylcholine to produce phosphatidic acid (PA), a lipidic second messenger. An intracellular effector of phosphatidic acid, the cytosolic enzyme sphingosine kinase (SphK), phosphorylates sphingosine to generate sphingosine 1-phosphate (S1P), another lipidic second messengers. This messenger stimulates cell proliferation. Transient receptor potential (TRP) channels open or close when they are bound to PI(4,5)P$_2$. Once receptors coupled to phospholipase-C are stimulated by agonist binding, PI(4,5)P$_2$ is degraded, and then TRPV1 opens and TRPM4, TRPM7, and TRPM8 close.

Effect	Mechanism of action
Generation of second messengers	Calcium release via IP$_3$ Synthesis of PIP$_3$ PLD–PA–SphK–S1P axis
Ion carrier activity	Inhibition of TRPV1 Activation of TRPM4/7/8 Inhibition of K$^+$ channels
Intracellular transport	Endo- and exocytosis Phagocytosis

2.2.2 PI(3,4,5)P$_3$

The PI(3,4,5)P$_3$ concentration is negligible in resting cells, but rises transiently in response to activated GPCRs, growth factor RTKs, or cytokine receptors. Agent PI(3,4,5)P$_3$ modulates cell transport, growth, proliferation, and motility (Table 2.9). It indeed recruits guanine nucleotide-exchange factors and GTPase-activating proteins (Sect. 9.4) that regulate the actin cytoskeleton, as well as phosphoinositide-dependent protein kinase PDK1 (Sect. 5.2.2) and protein kinase-B (Sect. 5.2.6).

Mediator PI(3,4,5)P$_3$ is dephosphorylated by 2 types of phosphatases. Phosphatase and tensin homolog deleted on chromosome 10 (PTen) regenerates PI(4,5)P$_2$ (Sects. 2.10.1 and 8.3.13.1). 5-Phosphatases, such as SH2 domain-containing inositol 5-phosphatases SHIP1 and SHIP2, or IP(5)Pd and IP(5)PL1, respectively, produce PI(3,4)P$_2$.

2.2.3 PI(5)P

Phosphoinositide PI(5)P exists at low concentrations. It is converted by PI(5)P4K2 into PI(4,5)P$_2$. However, the latter is mainly produced from PI(4)P. Conversion of PI(5)P into PI(4,5)P$_2$ by PIP4K2β impedes insulin signaling, as PIP4K2β activates PI(3,4,5)P$_3$ 5-phosphatases [12].

Table 2.9. Activity of PI(3,4,5)P$_3$ (Sources: [10]). This signaling mediator is, in particuler, an effector of insulin signaling to regulate energy uptake and storage. Agent PI(3,4,5)P$_3$ binds to various signaling mediators such as members of the TEC Tyr kinase family, Bruton Tyr kinase (BTK), and interleukin-2-inducible T-cell kinase (ITK). These kinases support the activity of phospholipase-Cγ (PLCγ), thereby fostering Ca^{++} signaling. Agent PI(3,4,5)P$_3$ also stimulates phosphoinositide-dependent kinases PDK1 and PDK2 and protein kinase-B (PKB). The latter activates numerous effectors (B-cell lymphoma and leukemia [BCL2] antagonist of cell death [BAD], forkhead box transcription factor FoxO, glycogen synthase kinase GSK3, P70 ribosomal S6 kinase [S6K], and target of rapamycin [TOR]). It also activates monomeric CDC42, Rac, and Rho GTPases to remodel the cytoskeleton as well as to produce (via Rac) the free radical, superoxide anion (O2$^-$).

Event	Mechanism of action
Cytoskeleton restructuring	Activation of CDC42, Rac, and Rho GTPases
Calcium signaling	Triggering of BTK (ITK)–PLCγ–IP$_3$ axis
Cell metabolism	Glycogen metabolism via GSK3 Lipid synthesis Protein synthesis via S6K and TOR
Cell fate	Cell survival via BAD and TOR, hormonal modulation of apoptosis Cell growth via TOR Inflammation

Bacterial IpgD 4-phosphatase can translocate into mammalian cells during bacterial invasion. It converts PI(4,5)P$_2$ into PI(5)P and provokes PKB activation.[16]

Agent PI(5)P enhances the activity of various myotubularins (MTM1, MTMR3, and MTMR6) toward their preferred substrate PI(3,5)P$_2$. The mitochondrial phosphoTyr phosphatase (PTPmt), a PTen-like phosphatase, produces PI(5)P by dephosphorylating PI(3,5)P$_2$ agent.

2.2.4 Nuclear Phosphoinositides

The nuclear envelope is a double membrane that contains phospholipids and proteins. Lipids also reside in the intranuclear space; phospholipids exist in the nucleus as proteolipid complexes. Polyphosphoinositol lipids, together with their synthesizing enzymes, form an intranuclear phospholipase-C signaling complex that generates diacylglycerol and inositol (1,4,5)-trisphosphate from phosphatidylinositol (4,5)-bisphosphate [28]. Diacylglycerol can recruit protein kinase-C to the nucleus to phosphorylate intranuclear proteins, whereas IP$_3$ can mobilize

---

[16]Expression of IpgD in mammalian cells causes significant loss of PI(4,5)P$_2$ and important changes in actin cytoskeleton organization, cell morphology, and adhesion.

Ca^{++} from the space between the 2 nuclear membranes, thereby increasing the nucleoplasmic Ca^{++} level. Furthermore, PI(4,5)P$_2$ can participate in RNA splicing. The PI–PLCβ1[17] and PI–PLCδ4 complexes, diacylglycerol kinase, and PI and PIP kinases, as well as diacylglycerol, multiple PKC isoforms, phospholipase-A2, phosphoinositide 3-kinase, and sphingolipids, in addition to nuclear Ca^{++}, contribute to nuclear lipid signaling. Enzymes of lipid synthesis localize to both the nuclear envelope and nucleoplasm [28]. Both type-1 PI(4)P5K and type-2 PI(5)P4K kinases reside in the nucleus. Phosphatidylcholine, phosphatidylethanolamine, and phosphatidylserine lodge within the nucleus. Nuclear phosphatidylglygerol contributes to PKC translocation and activation [28].

Phosphatidylcholine provides an additional source of diacylglycerol. The 2 distinct nuclear DAG pools generated from different phospholipids are assigned to subnuclear locations. Diacylglycerol kinase targets preferentially PI–PLC-derived DAG because of DAG accessibility rather than actual DAG specificity of DAG kinase [28]. The relative contributions of phosphatidylcholine and phosphoinositides to overall nuclear DAG concentration differ between distinct cell types.

Diacylglycerol kinase-ζ (DAGKζ) is a nuclear isoform following PKC-mediated phosphorylation (PKCα and/or PKCγ) [28]. Other isoforms of diacylglycerol kinase also translocate to the nucleus. Diacylglycerol kinase regulates nuclear PKC activity, as it removes the PKC activator diacylglycerol.

Phosphoinositide transfer protein-α (PITPα)[18] that also localizes in the nucleus can have several functions, as it can bring phosphoinositides into the nucleus from the envelope, transport phosphoinositides to different sites within the nucleus, and/or serve to present more substrate to phosphoinositide kinases [28].

Insulin-like growth factor-1 (IGF1) stimulates nuclear inositide metabolism [28]. It transiently activates the nuclear PI–PLCβ1 complex and causes PKC translocation to the nucleus. The PI–PLC complex can also be phosphorylated (inhibited) by PKC (negative feedback loop). Extracellular signal-regulated kinases ERK1 or ERK2 can translocate into the nucleus and phosphorylate the PI–PLCβ1b complex.

In cardiomyocytes, whereas PKCα, PKCβ2, and PKCζ reside predominantly in the perinuclear region, PKCδ and PKCε shuttle between the cytosol and the nucleus. However, all PKC isoforms can translocate between these 2 main cellular compartments [28]. Isozyme PKCβ2 can phosphorylate lamin-B.

---

[17] In the cytoplasm, the PI–PLCβ1 complex serves as a G-protein-controlled enzyme. Splice variants PLCβ1a and PLCβ1b are predominantly cytosolic and nuclear in some cell types, respectively [28].

[18] Phosphatidylinositol transfer protein-α belongs to a family of phospholipid-binding proteins. This protein regulates vesicular transfer and lipid-mediated signaling. It can transport phosphatidylinositol and phosphatidylcholine between membranes through aqueous media, thereby controlling the phospholipid composition of membranes. It is involved in the presentation of phosphatidylinositol to phosphatidylinositol-specific kinases and phospholipases. Once phosphorylated by protein kinase-C, it enables relocalization from the nucleus and the cytosol to the Golgi body membrane, where it colocalizes with its isoform PITPβ [29].

Nuclear PI3KR1 (P85 regulatory subunit) can be activated by nuclear guanosine triphosphatase PIKE (Sect. 9.3.28). The PI–PLCγ1 complex can act as a guanine nucleotide-exchange factor for PIKE agent. Besides, PI(3,4,5)P$_3$-binding protein (PIP$_3$BP) can translocate to the nucleus. In addition, type-2 PI3Ks — PI3KC2α and PI3KC2β — contain a nuclear localization sequence [28].

Phospholipase-A2 can translocate to the nucleus or nuclear envelope. Moreover, prostanoid receptors have a perinuclear distribution. Arachidonic acid (Sect. 2.6) can then serve near the nucleus. Phosphatidylcholine is a potential PLA2 substrate.

2.2.4.1 Cell Cycle and Nuclear Lipids

The concentration of nuclear DAG often increases close to the beginning of S phase. The PI–PLCβ1 complex is able to stimulate the cell cycle progression. A pulse of PI–PLC activation appears approximately at the G1–S transition. The PI–PLCβ1 complex yields nuclear DAG generation and PKC activation. It induces an increase in constitutive level of the CcnD3–CDK4 complex and retinoblastoma protein phosphorylation and its dissociation from E2F factor [28]. However, the absence of PI–PLCβ1 does not disturb the cell cycle progression, as most DAG derives from phosphatidylcholine.

The nuclear DAG level also increases at the G2–M transition. Kinase PKCβ2 indeed intervenes in nuclear disassembly, as it phosphorylates lamin-B. Conversely, DAGKζ suppresses nuclear DAG in non-proliferating cells.

2.2.5 Phosphoinositide-Dependent Enzymatic Activity

Phosphatase PTen dephosphorylates only PIP$_3$ into PI(4,5)P2. Activation of PTen that depends on PI(4,5)P$_2$ thus yields a positive feedback.

Myotubularin that produces PI(5)P by dephosphorylating PI(3,5)P$_2$ is upregulated by PI(5)P agent. Phosphoinositides that act as signaling molecules are involved in the coupling of voltage sensing and enzymatic activity of voltage-sensing phosphoinositide phosphatase (VSP).

2.2.6 Phosphoinositide Influence on Ion Carriers

Localized PIP$_2$ depletion can occur close to receptor-activated phospholipase-C, thereby inactivating adjacent K$_{IR}$ or K$_V$ channels (Table 2.10). However, activated phospholipase-C can be associated with activated phosphoinositide (PIK) and phosphoinositide phosphate (PIPK) kinases that result from stimulation of PLC-coupled receptors. Consequently, the PIP$_2$ concentration can locally rise.

Table 2.10. Regulation mediated by PIP_2 of cardiac ion carriers (Source: [30]; GIRK: $G\beta\gamma$-regulated inwardly rectifying K^+ channel; HCN: hyperpolarization-activated, cyclic nucleotide-gated K^+ channel; hERG: human ether-à-go-go related channel; K_{IR}: inward rectifier K^+ channel; IKr: rapidly activating delayed rectifier K^+ current; IKs: slowly activating delayed rectifier K^+ current; If: hyperpolarization-activated cation current; minK: auxiliary, kinetics-slowing, inactivation-inhibiting, conductance-raising β subunit MiRP: MinK-related peptide; NCX: Na^+–Ca^{++} exchanger; NSCC: Ca^{++}-activated non-selective cation current; ROC: receptor-operated Ca^{++} current; SOC: store-operated Ca^{++} current; SuR: sulfonylurea receptor; TRP: transient receptor potential current activated by several types of stimuli; TRPC: TRP canonical; TRPM: TRP melastatin).

Protein	PIP_2 effect	Function
Potassium channels		
$K_{IR}2.1$	Channel opening	Resting membrane potential maintenance (i_{K1})
$K_{IR}3.1/4$ (GIRK1/4)	$G\beta\gamma$-mediated regulation change	Parasympathetic stimulation ($i_{K_{ACh}}$)
$K_{IR}6.2$ and SUR2a	ATP-binding exclusion	ATP-induced closing ($i_{K_{ATP}}$) Repolarization reserve
$K_V11.1$ (hERG) minK, and MiRP1	Activation shift (hyperpolarization) Slow deactivation	Promote repolarization (i_{Kr}) Maintenance of pacemaker automaticity
$K_V7.1$ and minK	Prevention of inhibition	Repolarization (i_{Ks})
Potassium and sodium channels		
HCN2/4	Activation shift (toward depolarization)	Spontaneous diastolic depolarization (If)
Calcium and sodium channels		
NCX1	Prevention of auto-inhibition (binding to NCX inhibitor)	Ca^{++} extrusion or uptake ($i_{Na/Ca}$) Spontaneous diastolic depolarization
Non-selective polymodal channels		
TRPC1–C7	Inhibition?	Pacemaker current? Stretch activation?
TRPM4	Depletion-induced desensitization	Depolarization

2.2.6.1 Inwardly Rectifying Potassium Channels

In heart, 3 major families of inwardly rectifying K^+ channel — $K_{IR}2$, $K_{IR}3$, and $K_{IR}6$ — require PIP_2 for activity.[19] Channels of the K_{IR} class open after PIP_2 binding, whereas they close under PIP_2 depletion [30]. Most K_{IR} channels respond relatively specifically to $PI(4,5)P_2$ lipid. However, K_{ATP} channels are sensitive to $PI(3,4)P_2$, $PI(3,5)P_2$, $PI(4,5)P_2$, and PIP_3 molecules.

2.2.6.2 Repolarizing Potassium Channels

Voltage-gated K^+ channels (K_V) are responsible for cardiomyocyte repolarization. Both $K_V11.1$[20] and $K_V7.1$[21] that are involved in rapid and slow phase of repolarization are activated by PIP_2 [30].

2.2.6.3 Pacemaker Channels

Hyperpolarization-activated, cyclic nucleotide-gated channel (HCN) is regulated by PIP_2 lipid [30]. The latter shifts HCN voltage dependence toward depolarized potentials, hence increasing the spontaneous firing rate.

2.2.6.4 Sodium–Calcium Exchanger

Sodium–calcium exchanger (NCX) contains an inhibitory domain that auto-inhibits the channel. Lipid PIP_2 binds to the inhibitory sequence of NCX1 exchanger, thereby preventing its auto-inhibition and activating the carrier [30].

---

[19] Members of the $K_{IR}2$ family maintain resting transmembrane potential in atrio- and ventriculomyocytes. Channels of the $K_{IR}3$ family in atrial and pacemaker (sinoatrial node) myocytes are muscarinic K^+ channels (K_{ACh}) targeted by the parasympathetic control. Suppression of PIP_2 binding to K_{ACh} channels retards the stimulation of $G\beta\gamma$ or Na^+ on channel activity. The $K_{IR}6$ family encompasses K_{ATP} channel that is inhibited by intracellular ATP. Under normal conditions, K_{ATP} channel does not contribute to repolarization. The apparent affinity of K_{ATP} for ATP is indeed strongly reduced by PIP_2 agent. When cardiomyocytes are exposed to metabolic stress, K_{ATP} channel opens, thereby reducing action potential duration [30].

[20] The $K_V11.1$ channel is regulated by PKA phosphorylation and cAMP binding. Phosphorylated $K_V11.1$ associates with 14-3-3 proteins to raise its activity. Lipid PIP_2 can influence the response to cAMP and PKA agents. Lipid PIP_2 causes a hyperpolarizing shift in the voltage dependence of activation and slows deactivation [30].

[21] A-kinase anchoring protein AKAP9 modulates $K_V7.1$ activation by protein kinase-A. Binding of PIP_2 relieves $K_V7.1$ auto-inhibition.

2.2.6.5 Transient Receptor Potential Channels

Canonical transient receptor potential, low conductance, relatively non-selective cation channels, are activated by receptors coupled to phospholipase-C. They are also regulated by stretch,[22] diacylglycerol, and IP_3R-induced Ca^{++} store depletion. The TRPC3 channel can yield a pacemaker current modulated by IP_3 or PIP_2.[23] Channel TRPC4 can be inhibited by PIP_2 binding, whereas TRPC6[24] and TRPC7 are activated by PIP_2. In the sinoatrial node, TRPM4 channel is a calcium-activated, non-selective cation channel activated by IP_3 [30].

2.3 (Non-Lipid) Inositol Phosphates

Inositol (1,4,5)-trisphosphate (IP_3), a Ca^{++}-mobilizing second messenger, undergoes a catabolism that terminates its action. Its metabolism is governed by a complicated inositol phosphate pathway (Tables 2.11 to 2.14). The inositol phos-

Table 2.11. Phosphorylation and dephosphorylation (production and removal) of various types of inositol phosphates. (**Part 1**) Inositol monophosphates (Source: [10]). The free inositol synthesized by inositol monophosphatase (IMPase), which is inhibited by the lithium ion (Li^+), is reused to resynthesize phosphorylated phosphatidylinositol, which can relocalize to the plasma membrane for phosphoinositide signaling.

Substrate	Product	Enzyme
I(1)P	Inositol	IMPase
I(3)P	Inositol	IMPase
I(4)P	Inositol	IMPase

Table 2.12. Phosphorylation and dephosphorylation of various types of inositol phosphates. (**Part 2**) Inositol bisphosphates (Source: [10]; IPMK: inositol phosphate multikinase [a.k.a. inositol (1,3,4,6)-tetrakisphosphate 5-kinase]).

Substrate	Product	Enzyme
$I(1,3)P_2$	I(1)P	IP(3)Pase
$I(1,4)P_2$	I(4)P	IP(1)Pase
$I(3,4)P_2$	I(3)P	IP(4)Pase
$I(4,5)P_2$	$I(1,4,5)P_3$	IPMK

---

[22] Protein TRPC1 is a stretch-activated channel in heart.
[23] The TRPC3 channel associates with the sodium–calcium exchanger NCX1 in the sarcolemma. Therefore, Ca^{++} entry through TRPC3 partly depends on activity mode of NCX1 exchanger.
[24] The TRPC6 channel is activated directly and selectively by α1a-adrenoceptors.

Table 2.13. Phosphorylation and dephosphorylation of various types of inositol phosphates. (**Part 3**) Inositol trisphosphates (Source: [10]; IP_36K: $I(1,3,4)P_3$ 6-kinase; IPMK: inositol phosphate multikinase; OCRL: oculocerebrorenal syndrome of Lowe [or inositol polyphosphate 5-phosphatase IP(5)Pf encoded by the INPP5F gene]).

Substrate	Product	Enzyme
$I(1,3,4)P_3$	$I(1,3,4,6)P_4$	$IP_3(5,6)K$ ($I(1,3,4)P_35K/6K$)
	$I(1,3,4,5)P_4$	$IP_3(5,6)K$ ($I(1,3,4)P_35K/6K$)
	$I(1,3)P_2$	IP(4)Pase
	$I(3,4)P_2$	IP(1)Pase
$I(1,4,5)P_3$	$I(1,3,4,5)P_4$	IMPK, IP_33K
(IP_3)	$I(1,4)P_2$	5PTase1 (IP(5)Pa), OCRL (IP(5)Pf)

phate metabolism produces free inositol that enables synthesis of phosphoinositides, which can serve in further signaling. In addition, it generates a collection of inositol polyphosphates that also contribute to intracellular signaling.

Inositol (1,4,5)-trisphosphate is converted to numerous inositol phosphates, such as inositol tetrakisphosphate (IP_4), pentakisphosphate (IP_5), hexakisphosphate (IP_6), and pyrophosphate (diphosphoinositol phosphate [PPIP]) owing to inositol phosphate kinases (IPK; Fig. 2.1).

In addition to PI3K kinases, inositol polyphosphate multikinase (IPMK)[25] generates the second messenger phosphatidylinositol (3,4,5)-trisphosphate as well as water soluble inositol phosphates. Enzyme $P110_{PI3K}$ phosphorylates (activates) IPMK kinase [13]. Therefore, growth factors stimulate the PI3K–PKB axis and generate PIP_3 via the sequential activations of $P110_{PI3K}$ and IPMK kinases. As inositol phosphates prevent PKB signaling, IPMK either inhibits or stimulates PKB via its inositol phosphate kinase or PI3K activity, respectively.

Inositol hexakisphosphate is a coactivator of nucleoporin Gle1 that is involved in mRNA export from the nucleus to the cytoplasm. Protein Gle1 interacts with the nuclear-pore complex NuP155 and activates Dead-box ATPase DBP5 in nuclear-pore complex to export mRNA into the cytoplasm [14, 15]. Agent IP_6 binds Gle1 to potentiate Gle1 stimulation of DBP5. It contributes to the conformation of RNA-editing adenosine deaminase ADAR2 [16]. Inositol hexakisphosphate also acts as a cofactor for adenosine deaminase that acts on transfer RNA (i.e., that edits tRNA) ADAT1.

Inositol phosphates and pyrophosphates participate in various cellular activities. Inositol trisphosphate (IP_3) is phosphorylated into inositol tetrakisphosphate that is required for the activation of ITK kinase after T-cell receptor stimulation in CD4+, CD8+ thymocytes [17]. Inositol tetrakisphosphate (IP_4), pentakisphosphate (IP_5),

---

[25] A.k.a. inositol (1,3,4,6)-tetrakisphosphate 5-kinase.

Table 2.14. Phosphorylation and dephosphorylation of various types of inositol phosphates. (**Part 4**) Inositol tetrakisphosphates and higher phosphorylated derivatives, i.e., inositol penta- (IP_5), hexa- (IP_6; phytic acid or phytate; concentration 10–100 μmol in mammalian cells), hepta- (IP_7), and octakisphosphatekisphosphate (IP_8; a.k.a. bispyrophosphorylated inositol phosphate and bis-diphosphoinositol tetrakisphosphate [$(PP)_2IP_4$]; Source: [10]; DIPP: diphosphoinositol phosphate phosphohydrolase; IP_45K: $I(1,3,4,6)P_4$ 5-kinase; IPMK: inositol phosphate multikinase; MIPP: multiple inositol phosphate phosphatase [$I(1,3,4,5,6)P_5/IP_6$ 3-phosphatase]; OCRL: oculocere-brorenal syndrome of Lowe). Enzyme IP_6K, an $IP_5/IP_6/IP_7$ 5-kinase, generates diphosphoinositol phosphates (PPIP). Diphosphate is either at C4, C5, or C6, hence the designation (4)$PPIP_5$ (or $IP_{7_{(4)}}$], (5)$PPIP_5$ (or $IP_{7_{(5)}}$), and (6)$PPIP_5$ (or $IP_{7_{(6)}}$), respectively. Inositol hexakisphosphate (IP_6) and diphosphoinositol pentakisphosphate ($PPIP_5$) kinase-1, or diphosphoinositol pentakisphosphate kinase-1 ($PPIP_5K1$; a.k.a. histidine acid phosphatase domain-containing protein HisPPD2a, VIP1 homolog), is a bifunctional IP_6/IP_7 kinase that yields (i)$PPIP_5$ ($IP_{7_{(i)}}$) and $(PP)_2IP_4$ (IP_8).

Substrate	Product	Enzyme
$I(1,3,4,5)P_4$	$I(1,3,4,5,6)P_5$	IMPK
	$I(1,3,4)P_3$	5PTase1 (IP(5)Pa), OCRL (IP(5)Pf)
		IP(5)Pd (SHIP1)
$I(1,3,4,6)P_4$	$I(1,3,4,5,6)P_5$	IMPK ($I(1,3,4,6)P_45K$ $I(1,3,4,5,6)P_5$1Pase)
$I(1,4,5,6)P_4$	$I(1,3,4,5,6)P_5$	$I(1,4,5,6)P_43K$
$I(3,4,5,6)P_4$	$I(1,3,4,5,6)P_5$	$I(3,4,5,6)P_41K$ (or $I(1,3,4,5,6)P_5$1Pase)
IP5	IP_6	IMPK, $IP(1,3,4,5,6)P_52K$ (IPK1)
	$PPIP_4$	IP_6K
	$I(1,4,5,6)P_4$	PTen
	$I(3,4,5,6)P_4$	$I(1,3,4,5,6)P_5$1Pase
IP6	IP7	IP_6K
	IP5	MIPP
IP7 ($PPIP_5$)	IP8	PPIP5K
	IP6	DIPP
IP8 ($[PP]_2IP_4$)	IP7	DIPP

and hexakisphosphate (IP_6) modulate the activities of several chromatin-remodeling complexes, either preventing or favoring nucleosome mobilization [18].

Inositol pyrophosphate synthases target IP_6 and IP_7 kinases to regulate the actin-related protein ARP2–ARP3 complexes [19].[26] Inositol pyrophosphates diphospho-inositol pentakisphosphate (IP_7) and bisdiphosphoinositol tetrakisphosphate (IP_8) contribute to the regulation of the phosphorylation of many proteins [21]. Heat

---

[26]In Saccharomyces cerevisiae, inositol heptakisphosphate (IP_7) stimulates the cyclin-dependent kinase inhibitor Pho81 of the phosphate-responsive (Pho) signaling pathway that represses Pho80–Pho85 cyclin–CDK complex [20].

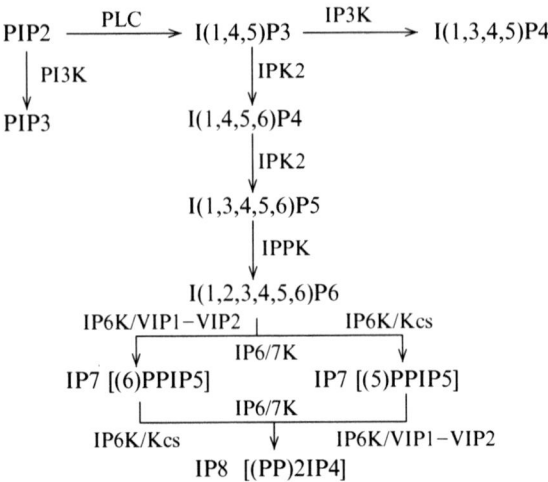

Fig. 2.1 Inositol phosphates and inositol phosphate kinases (IP_3: inositol trisphosphate; IP_4: inositol tetrakisphosphate; IP_5: inositol pentakisphosphate; IP_6: inositol hexakisphosphate; IP_6/IP_7K: inositol hexakisphosphate and heptakisphosphate kinase of the $IP_6K/Kcs1$ and $IP_6K/VIP1$–VIP2 categories; IP_7: inositol heptakisphosphate [monopyrophosphorylated inositol phosphate or diphosphoinositol pentakisphosphate ($PPIP_5$)]; IP_8: inositol octakisphosphate [bispyrophosphorylated inositol phosphate or bisdiphosphoinositol tetrakisphosphate ($[PP]_2IP_4$)]; IPK2: inositol polyphosphate 6-/3-/5-kinase; IPPK: inositol (1,3,4,5)-pentakisphosphate 2-kinase [a.k.a. IP_52K and IPK1]; PI3K: phosphatidylinositol 3-kinase; PIP_2: phosphatidylinositol bisphosphate; PIP_3: phosphatidylinositol trisphosphate).

shock protein HSP90 binds to inositol hexakisphosphate kinase IP_6K2 that produces IP_7 and mediates apoptosis to inhibit IP_6K2 and promote cell survival [22]. In addition, inositol pyrophosphates are involved in vesicular transport.[27]

Protein kinase-B and calcium released through inositol trisphosphate receptors are involved in pathways that lead to either cell survival or death. Once it is activated by survival signals, PKB phosphorylates IP_3Rs to reduce Ca^{++} flux from the endoplasmic reticulum to mitochondria [23]. Large IP_3-induced Ca^{++} release can indeed promote apoptosis. Kinases PKA and CamK2 also target IP_3Rs, but cause low levels of IP_3R phosphorylation.

2.3.1 $I(1,4,5)P_3$

Inositol (1,4,5)-trisphosphate (IP_3) is dephosphorylated by type-1 inositol polyphosphate 5-phosphatase to form $I(1,4)P_2$ or phosphorylated by IP_3 3-kinase to generate $I(1,3,4,5)P_4$. Inositol trisphosphate triggers Ca^{++} influx, especially from its

---

[27]On the other hand, HSP90 links to other protein kinases, such as PKB, PDK1, CDK4, CDK6, and CDK9, that are then stabilized and activated.

Table 2.15. Role of the IP_3–Ca^{++} pathway (Source: [10]).

Opening of Ca^{++}-sensitive ion channels
Cell contraction (stress fibers, sarcomeres)
Exocytosis (fluid secretion, aldosterone release)
Synapse remodeling
Metabolism (liver)
Cell differentiation
Chemotaxis
Platelet aggregation
Fertilization

intracellular stores, to launch a set of cellular processes (Table 2.15). Three isoforms of IP_3 3-kinase exist (IP_33Ka–IP_33Kc). Both IP_33Ka and IP_33Kb are activated by Ca^{++}–calmodulin, IP_33Kb being the most sensitive. Protein kinase-A activates IP_33Ka, but inhibits IP_33Kb [10].

2.3.2 Main Enzymes of the Metabolism of Inositol Polyphosphates

2.3.2.1 Kinases of Inositol Polyphosphates

Inositol phosphate kinases include: (1) IPK1 (a.k.a. IP52K and IPPK), an IP_5 2-kinase that synthesizes IP_6; (2) inositol polyphosphate 3-/5-/6-kinase IPK2 that produces $I(1,3,4,5,6)P_5$; (3) IP_3 3-kinase (IP_33K) that manufactures $I(1,3,4,5)P_4$ from $I(1,4,5)P_3$; (4) $IP_3(5/6)K$, an $I(1,3,4)P_3$ 5- and 6-kinase that forms $I(1,3,4,5)P_4$ and $I(1,3,4,6)P_4$; (5) IP_6K, an IP_5, IP_6, and IP_7 5-kinase that generates diphosphoinositol phosphates (i.e., PPIPs); and (6) inositol hexakisphosphate (IP_6) and diphosphoinositol pentakisphosphate ($PPIP_5$) kinase-1 (IP_6K or $PPIP_5K1$),[28] which is a bifunctional IP_6 and IP_7 kinase that yields (i)$PPIP_5$ ($IP_{7_{(i)}}$)[29] and $(PP)_2IP_4$ (IP_8). In addition to $PPIP_5K1$, another $PPIP_5K2$ subtype exist.[30]

---

[28] A.k.a. diphosphoinositol pentakisphosphate kinase-1 ($PPIP_5K1$), histidine acid phosphatase domain-containing protein HisPPD2a, and VIP1 homolog.

[29] Diphosphate is either at C4, C5, or C6, hence the designation (4)$PPIP_5$ (or $IP_{7(4)}$], (5)$PPIP_5$ (or $IP_{7_{(5)}}$), and (6)$PPIP_5$ (or $IP_{7_{(6)}}$), respectively.

[30] Diphosphoinositol pentakisphosphate kinase-2 is also called histidine acid phosphatase domain-containing protein-1, inositol hexakisphosphate (IP_6) and diphosphoinositol pentakisphosphate ($PPIP_5$) kinase-2, IP_7K2, and VIP1 homolog-2 (VIP2).

Kinase IP$_6$K possesses 3 isoforms (IP$_6$K1–IP$_6$K3). The I(3,4,5,6)P$_4$ 1-kinase, or I(1,3,4,5,6)P$_5$ 1-phosphatase, convert I(3,4,5,6)P$_4$, a regulator of chloride channels. Inositol phosphate multikinase (IPMK) transforms I(4,5)P$_2$ into I(1,4,5)P$_3$, I(1,4,5)P$_3$ into I(1,3,4,5)P$_4$, I(1,3,4,5)P4 into IP$_5$, and IP$_5$ into PPIP$_4$.

The PPIP$_5$ kinase targets PPIP$_5$ (IP$_7$) to form (PP)$_2$IP$_4$ (IP$_8$). On the other hand, diphosphoinositol phosphate phosphohydrolase (DIPP) dephosphorylates both PPIP$_5$ (IP$_7$) and (PP)$_2$IP$_4$ (IP$_8$).

Inositol phosphate kinases can transduce cues received by GPCRs [24]. They are required for tissue development and cell adaptation. Activated Gα_q stimulates the formation of metabolites downstream from IP$_3$ (down to IP$_8$). Messengers IP and PPIP, which depend on IPK, can be involved in RNA production, mRNA export, chromatin remodeling, protein phosphorylation, and immunity.

Inositol (1,3,4)-trisphosphate [I(1,3,4)P$_3$] 5/6-kinase (ITPK1) synthesizes inositol tetrakisphosphates I(1,3,4,5)P$_4$ and I(1,3,4,6)P$_4$. It is not only phosphorylated, but also acetylated (Lys340, Lys383, and Lys410) by acetyltransferases CBP or P300 that reduce the ITPK1 half-life and enzyme activity [25]. On the other hand, ITPK1 is deacetylated by silent information regulator-2 (SIRT1).

2.3.2.2 Inositol Phosphate Phosphatases

Inositol phosphate phosphatases encompass, in particular: (1) IP(1)Pase, encoded by the INPP1 gene, which is an I(1,4)P$_2$ and I(1,3,4)P$_3$ 1-phosphatase; (2) 5PTase, an I(1,4,5)P$_3$ and I(1,3,4,5)P$_4$ 5-phosphatase; (3) PTen, a 3-phosphatase that also dephosphorylates IP$_5$; and (4) DIPP that degrades PPIPs.

5-Phosphatase-1 hydrolyzes I(1,4,5)P$_3$ and I(1,3,4,5)P$_4$ inositol phosphates. This enzyme is phosphorylated (inhibited) by Ca^{++}–calmodulin-dependent protein kinase CamK2 [10]. 5-Phosphatase-2 hydrolyzes inositol phospholipids.

Inositol Monophosphatases

Inositol monophosphatase (IMPase) dephosphorylates inositol monophosphate to generate free inositol, a precursor of membrane phosphatidylinositols and its phosphorylated derivatives, the phosphatidylinositol phosphates. Cells also produce inositol using glucose 6-phosphate that is isomerized by inositol 1-phosphate synthase-A1 (ISynA1)[31] to produce inositol 1-phosphate, which is dephosphorylated by inositol monophosphatase to generate free inositol.

---

[31] A.k.a. myoinositol 1-phosphate synthase-A1 (MIP) and inositol 3-phosphate synthase-1 (InoS1, Ino1, or IPS).

2.3 (Non-Lipid) Inositol Phosphates

Table 2.16. Substrate specificity of Mg^{++}-dependent inositol monophosphatase IMPase1 and IMPase2. Relative catalytic activities are scaled between 1 and 10 (Source: [26]).

Type	IMPase1	IMPase2
I(1)P	10	2
I(2)P	2	1
I(3)P	10	2
I(4)P	10	2
I(5)P	10	2
I(6)P	10	2

Three subtypes of inositol 1 (or 4)-monophosphatases are encoded by 3 distinct genes: lithium-sensitive IMPase1 (IMP1 or isoform-A1 [IMPa1]); IMPase2 (IMP2 or isoform-A2 [IMPa2]); and IMPase3 (IMP3 or isoform-A3 [IMPa3]).[32] Subtype IMPa2 homodimerizes, but does not heterodimerize with IMPa1 isoform [26]. In addition, IMPa2 has a significantly lower activity on inositol monophosphate than IMPa1 (Table 2.16). Calbindin-1, a Ca^{++}-binding protein, interacts with IMPase (apparent equilibrium dissociation constant 0.9 μmol) and activates IMPase up to 250-fold, according to H^+ and substrate concentrations [27].

Inositol Polyphosphate Phosphatases

Inositol polyphosphate 1-phosphatase (IP(1)Pase) dephosphorylates both $I(1,4)P_2$ and $I(1,3,4)P_3$. Inositol polyphosphate 3-phosphatase (IP(3)Pase) dephosphorylates $I(1,3)P_2$ and PI3P. Inositol polyphosphate 4-phosphatase (IP(4)Pase; Sect. 2.10.3) dephosphorylates both $I(3,4)P_2$ and $I(1,3,4)P_3$, as well as $PI(3,4)P_2$ lipid.

Inositol polyphosphate 5-phosphatases (IP(5)Pase; Sect. 2.10.4) constitute a family of 10 mammalian members. These enzymes regulate many important cellular functions, such as hematopoietic cell proliferation and activation, insulin signaling, endocytosis, and actin polymerization.

Among members of the IP(5)Pase family, inositol polyphosphate 5-phosphatase OCRL (or IP(5)Pf) hydrolyzes both $I(1,4,5)P_3$ and $I(1,3,4,5)P_4$ inositol phosphates and $PI(4,5)P_2$ and $PI(3,4,5)P_3$ phosphoinositides. Src homology-2 (SH2) domain-containing inositol phosphatases (SHIP1 and SHIP2), i.e., IP(5)Pd and IP(5)PL1, target $I(1,3,4,5)P_4$ and $PI(3,4,5)P_3$.

---

[32] A.k.a. inositol monophosphatase domain-containing protein IMPaD1.

2.4 Choline-Containing Phospholipids

Choline-containing phospholipids (phosphatidylcholine, sphingomyelin, and ether phospholipids[33] with a choline head group) are the most abundant phospholipids (~50% of the phospholipid pool) in cellular membranes [31]. They serve as structural components of membranes as well as signaling molecules and precursors of secondary messengers. Choline enters cells through specific membrane transporters and can be phosphorylated by choline kinase.

2.5 Sphingolipids

Sphingolipids constitute a class of lipids that derive from sphingosine, an amino alcohol with an unsaturated hydrocarbon chain that can be phosphorylated. Ceramide is the structural unit common to all sphingolipids. Sphingolipids participate in signal transmission and cell recognition, as they can serve as surface markers.

Two types of sphingolipids exist according to head group type: (1) sphingomyelins (or ceramide phosphorylcholines) that reside in cell membranes and have a phosphorylcholine or phosphoethanolamine linked to sphingosine bonded to a fatty acid (ceramide) and (2) glycosphingolipids that are ceramides with at least one oligosaccharide. Glycosphingolipids comprise: (1) cerebrosides such as myelin that possess a single glucose or galactose, sulfatides being sulfated cerebrosides, and (2) gangliosides with at least 3 oligosaccharide glucids, including one (monosialogangliosides) or more (polysialogangliosides) sialic acids. Glycosphingolipid sulfates[34] are glycosphingolipids that contain a sulfate ester group attached to the carbohydrate moiety.

The ganglioside family encompasses mono- (GM1–GM3), di- (GD1a/b and GD2–GD3), tri- (GT1b), and quadrisialoganglioside (GQ1b). Major signaling sphingolipids include ceramide, ceramide 1-phosphate, sphingosine, sphingosine 1-phosphate, sphingosine phosphorylcholine (SPC), and lysosulfatides (sulfogalactosylsphingosine) that have lost their fatty acid constituent.

---

[33] Ether phospholipids, major components of membranes, are lipids containing an alkyl group, rather than an acyl group, linked to one of the oxygen atoms of glycerol. At least one carbon atoms on glycerol is bonded to an alkyl chain via an ether linkage, not the usual ester linkage. Ether phospholipids have either a 1-^{O}alkyl (plasmanyl) or a 1-^{O}alk-1′-enyl (plasmenyl) linkage at the sn1 position of glycerol. In cells, these lipids belong almost exclusively to the choline and ethanolamine glycerophospholipid categories. They abound in the heart, kidney, and central nervous system. The body's organs can contain anabolic acyltransferases, phosphotransferases, and catabolic lipases that are selective for plasmanyl and plasmenyl phospholipids. Ether phospholipids are synthesized in peroxisomes and microsomes.

[34] A.k.a. sulfatides or sulfoglycosphingolipids.

Members of the orosomucoid (OrM)[35] family are encoded by the ORM and ORMDL genes. Proteins of the OrM1-like subfamily include 3 identified members (OrMdL1, OrMdL2 [adoplin-2], and OrMdL3).[36] Members of the ORM gene family encode transmembrane proteins located in the endoplasmic reticulum, where the synthesis of sphingolipids begins. Proteins of OrM1-like family are inhibitors of sphingolipid synthesis, as they form a complex with serine palmitoyltransferase, the first and rate-limiting enzyme in sphingolipid production in the endoplasmic reticulum.[37] Phosphorylation of OrM1-like proteins relieves their inhibitory activity [32].

2.5.1 Ceramide

Ceramide not only exerts biological effects, but also can be processed into bioactive sphingolipids. Ceramide synthesized by *sphingomyelinases* is indeed deacylated by neutral *ceramidase* into sphingosine. The latter is then phosphorylated into sphingosine 1-phosphate by *sphingosine kinases*. Ceramide and sphingosine are primarily antiproliferative and pro-apoptotic, whereas S1P promotes cell proliferation and impedes apoptosis.

2.5.2 Sphingomyelinases

Sphingomyelinases hydrolyze ubiquitous membrane-associated sphingomyelin into ceramide and phosphocholine. In the cardiovascular system, sphingomyelinases act in cardiomyocytes and vascular endothelial and smooth muscle cells not only to contribute to the regulation of cell proliferation and death, but also contraction of cardiac and vasculomyocytes [33].

Three main types of sphingomyelinases exist according to optimal pH for their activity [33]: (1) *alkaline sphingomyelinases* that are synthesized in intestinal mucosa and liver [33]; (2) *acidic sphingomyelinases*; and (3) Mg^{++}-dependent, membrane-associated, *neutral sphingomyelinases* (nSMase; optimal pH 7.4). Acidic sphingomyelinases that are encoded by the single sphingomyelin phosphodiesterase SMPD1 gene and generated from a single precursor are subdivided into 2 functionally distinct forms: (1) Zn^{++}-bound, *lysosomal*, acidic ($_{l,ac}$SMase; $4.5 < pH$ optimum < 5) and (2) Zn^{++}-dependent, *secreted*, acidic ($_{s,ac}$SMase).

---

[35] A.k.a. α1-acid glycoprotein (AGP).

[36] The ORMDL3 gene is implicated in the development of childhood asthma.

[37] In Saccharomyces cerevisiae, the SPOTS complex comprises Ser palmitoyltransferase, OrM1 and OrM2, activator of Ser palmitoyltransferase Tsc3, and phosphoinositide phosphatase SAc1 [32].

These types of enzymes have common activators, but differ in their enzymatic properties and subcellular locations [33]. Neutral sphingomyelinases reside in the endoplasmic reticulum and Golgi body. They also hydrolyze sphingomyelin in the inner leaflet of the plasma membrane. Lysosomal acidic sphingomyelinase can relocate to the outer leaflet of the plasma membrane. In human lymphocytes, members of the TNFR superfamily (Vol. 3 – Chap. 11. Receptors of the Immune System) actually trigger $_{l,ac}$SMase translocation from lysosomes to the extracellular surface of the plasma membrane. Consequently, several types of sphingomyelinases act on both sides of the plasma membrane, i.e., in the cytosol and extracellular matrix.

In humans, once stimulated by inflammatory cytokines, such as interferon-γ and interleukin-1β, endothelial cells secrete large amounts of $_{s,ac}$SMase with reduced $_{l,ac}$SMase amount. Cytokine-induced s,acSMase release actually antagonizes lysosomal axis with a common precursor [33].

Three genes (SMPD2–SMPD4) encode neutral sphingomyelinases (nSMase1–nSMase3). Paralog nSMase2 (encoded by the SMPD3 gene) is ubiquitous. Isoform nSMase3, highly expressed in the heart, is targeted by the complex formed by TNFR1 receptor and adaptor factor associated with neutral sphingomyelinase activation (FAN) [33]. Neutral sphingomyelinase-derived ceramide and activated PKCε, can trigger hypoxic pulmonary vasoconstriction.

Translocated $_{l,ac}$SMase localizes to sphingolipid-rich membrane rafts and releases extracellularly oriented ceramide. Ceramide-enriched plasmalemmal platforms cluster receptors for apoptosis initiation as well as, in vascular endothelial and smooth muscle cells, contribution of TNFSF6-induced impairment of the vasodilator response and muscarinic M_1 receptor-mediated constriction [33].

Neutral sphingomyelinase, but not acid sphingomyelinase, that produces ceramide and activation of PKCζ is an early signaling event in acute hypoxia-induced pulmonary vasoconstriction [34].

2.5.3 Sphingosine Kinases

Sphingosine kinases SphK1 and SphK2 control cell fate, especially in the cardiovascular system (Table 2.17). They are phosphorylated (activated) by extracellular signal-regulated protein kinases ERK1 and -2 (Sect. 6.7.2.2). Sphingosine kinase SphK2 resides in the nucleus. Its nuclear export happens after phosphorylation by protein kinase-D [35]. Enzyme S1P phosphohydrolase-1 intervenes in the recycling of sphingosine into ceramide.

Paralogs SphK1 and SphK2 contain several splice variants. Isoforms SphK1 and SphK2 might have distinct subcellular locations, hence directing spatially restricted S1P production. In addition, SphK2 partly antagonizes SphK1, as SphK2 also contributes to conversion of sphingosine into ceramide [35].

In addition to PKCε, sphingosine kinases are stimulated by G-protein-coupled receptor ligands (e.g., acetylcholine, histamine, and S1P), receptor Tyr and Ser/Thr

2.5 Sphingolipids

Table 2.17. Effect of sphingosine kinases SphK1 and SphK2 via S1P in various cell types of the cardiovascular system (Sources: [35, 36]; −: negative effect; SMC: smooth muscle cell). Sphingosine 1-phosphate also acts in an auto- and paracrine manner via its cognate receptors.

Type	Endothelial cell	Cardiomyocyte	Fibroblast	SMC
SphK1 (S1P)	Survival Proliferation Migration Angiogenesis Permeability	Survival Hypertrophy Inotropy − Chronotropy −	Survival Proliferation Migration	Survival Contraction Migration inhibition
SphK2	Survival Apoptosis	Survival Apoptosis	Survival Apoptosis	Survival Apoptosis

kinase agonists (e.g., EGF, PDGF, VEGF, TGFα, and TGF-β), immunoglobulin receptors, interleukins, and estrogens [35]. Moreover, sphingosine kinase-1 is stimulated by HIF1α and HIF2α, but reactive oxygen species lead to SphK1 degradation [35].

Sphingosine kinases are also stimulated by δ1-catenin–δ2-catenin, aminocyclase-1, and eukaryotic elongation factor-1A [35]. On the other hand, they are inhibited by SphK1-inteacting protein (SKIP), platelet endothelial adhesion molecule PECAM1, and LIM-only factor FHL2 (or SLIM3). Both SphK1 and SphK2 bind calmodulin that allows their translocation from the cytosol to the plasma membrane.

After injury, fibroblasts transform into myofibroblasts and produce extracellular matrix. The TGFβ-stimulated production of collagen by cardiac fibroblasts is mediated by S1P produced by SphK1 and released to act as an auto- and paracrine agent via $S1P_2$ receptors [37].

2.5.4 Sphingosine 1-Phosphate

Sphingosine 1-phosphate is a bioactive lipid metabolite that causes intracellular calcium mobilization and intervenes in cell survival, proliferation, migration, and contraction (Table 2.17). In general, it operates via ubiquitous S1P G-protein-coupled receptors $S1P_1$ to $S1P_3$ (Vol. 3 – Chap. 7. G-Protein-Coupled Receptors); Table 2.18). Sphingosine 1-phosphate can be dephosphorylated by S1P phosphatase into sphingosine or degraded by S1P lyase into hexadecanal and phosphoethanolamine.

2.5.4.1 HDL-Associated Sphingosine 1-Phosphate

High-density lipoproteins carry not only cholesterol in blood, but also sphingosine 1-phosphate. Both HDL-associated and free S1Ps differ in their signaling

Table 2.18. Signaling pathways from S1P receptors in heart and blood vessels (Sources: [36, 38]; ACase: adenylate cyclase; CMC: cardiomyocyte; GSK: glycogen synthase kinase; K$_{ACh}$: acetylcholine-activated inwardly rectifying potassium channel; NOS: nitric oxide synthase; PI3K: phosphatidylinositol 3-kinase; PKA, PKB, PKC: protein kinases-A, -B, and C; PTen: phosphatase and tensin homolog deleted on chromosome 10; VSMC: vascular smooth muscle cell). Receptors of S1P include S1P$_1$ to S1P$_3$. Receptor S1P$_1$ couples to Gi protein and S1P$_2$ and S1P$_3$ to Gi, Gq, and G12/13 proteins. Prolonged stimulation of Gq-coupled receptors (e.g. α1-adrenergic and endothelin-1 receptors) provokes hypertrophy. Subunits G12 and G13 also inhibit JNK and cause complementary hypertrophy.

Type	Pathway	Effect
S1P$_1$	Gi–ACAse	Negative inotropy via cAMP and PKA inhibition of Ca$_V$1 channel
	Gβγ–K$_{ACh}$	Bradycardia
	PI3K–PKB–GSK3β	Cardioprotection by GSK3 inhibition
	Gi–Rac1–NOS3–NO	Endothelium-dependent vasodilation
	Gi–PI3K	Inhibition of endothelium permeability
S1P$_2$	G12/13–RhoA	Fibroblast proliferation
	Gβγ–PKB–NOS	Cardioprotection
	RhoA–RoCK–PKC	Basal vasomotor tone (VSMC)
	Rho–PTen	PI3K inhibition; endothelium permeability
S1P$_3$	Gq–PLC	CMC hypertrophy
	G12/13–JNK	CMC concomitant hypertrophy
	Gβγ–PKB–NOS	Cardioprotection
	Rac1–NOS3–NO	Endothelium-dependent vasodilation
	RhoA–RoCK–PKC	VSMC contraction

properties. Circulating S1P (100–1000 nmol) is detected in leukocytes, platelets, and erythrocytes as well as bound to albumin, LDLs, VLDLs, and mainly HDLs [39]. High-density lipoproteins actually have a high affinity for S1P. Sources of plasmatic S1P are hematopoietic (erythrocytes more than platelets) [40]) and endothelial cells. Sphingosine 1-phosphate attached to HDLs mediates, partly or entirely, indirectly or directly, some functions of HDL, such as endothelial nitric oxide production and vasodilation.[38]

2.5.4.2 HDL–S1P Association in Endothelium

Sphingosine 1-phosphate particularly targets endothelial cells. Whereas free S1P can have vasoconstrictive and pro-inflammatory effects, HDL–S1P promotes vasorelaxation via nitric oxide synthesized by NOS3 as well as prostacyclin (PGI2) in

---

[38] High-density lipoproteins operate in the reverse cholesterol transport, as it carries it from cells to the liver. They also exert potent anti-inflammatory, anti-oxidative, anti-apoptotic, and vasodilatory effects.

2.5 Sphingolipids

both vascular endothelial and smooth muscle cells, as it stimulates cyclooxygenase-2 and P38MAPK.[39] The HDL–S1P complex also influences endothelial barrier function, angiogenesis, and endothelial precursor cell responsiveness.

High-density lipoproteins are able to induce capillary formation via the Ras–Raf–ERK and PKB–ERK–NOS3 pathways. Receptors of S1P targeted by HDL-associated S1P augment both endothelial cell motility and endothelial barrier integrity [39]. Between-endothelial cell junctions are strengthened by $S1P_1$ and $S1P_3$, but weakened by $S1P_2$ receptor. Moreover, S1P-mediated Ca^{++} influx stabilizes the endothelial barrier, as it activates NOS3 and produced NO favors the sealing of endothelial cells. Protection against apoptosis also promotes endothelial integrity. Furthermore, HDLs stimulate endothelial progenitor cells for endothelial repair and differentiation of human peripheral mononuclear cells into endothelial progenitor cells to enhance ischemia-induced angiogenesis.

High-density lipoproteins reduce oxidative stress, as they carry anti-oxidative enzymes, such as paraoxonase-1 and -3 and platelet-activating factor acetylhydrolase (PAFAH), that counteract protein oxidation, especially that of LDLs [39]. In addition, HDL-associated S1P, particularly small dense HDL3 that has the highest S1P/sphingomyelin ratio, inhibits ROS production via $S1P_3$-dependent NADPH oxidase inhibition and antagonizes oxidized LDL-induced apoptosis.

High-density lipoproteins restrain leukocyte adhesion to the vascular endothelium, as they reduce the expression of endothelial adhesion molecules, such as VCAM1, ICAM1, and E-selectin [39]. They also lower the density of α_M-integrin on monocytes and chemokine CCL2 in vascular smooth muscle cells. Consequently, HDLs repress vascular inflammation and promote atheroma stabilization. They hamper cell transmigration, as they cause NO production mediated by SRb1 as well as PGI2 in addition to anti-adhesive, HDL-associated S1P that targets $S1P_3$ receptor. On the other hand, free S1P generated by sphingosine kinase-1 activated by TNFα stimulates the expression of VCAM1 and ICAM1 adhesion molecules.

2.5.4.3 HDL–S1P in Cardiomyocytes

In cardiomyocytes, HDL-associated S1P activates STAT3 via $S1P_2$ receptors. Activation of STAT3 follows stimulation of $S1P_2$ receptor, extracellular signal-regulated kinases ERK1 and ERK2, and Src kinase [41]. Prostacyclin, the production of which by COx2 is heightened by HDL-associated S1P, targets IP receptor to activate cAMP and preclude cardiomyocyte hypertrophy as well as EP receptor to protect

---

[39]Prostacyclin tethers its G-protein-coupled receptors IP and EP (Vol. 3 – Chap. 7. G-Protein-Coupled Receptors) to increase the concentration of intracellular cAMP and activate potassium channels. Prostacyclin not only provokes vasodilation, but also impedes the migration of vascular smooth muscle cells as well as activation of platelets. Moreover, it suppresses the production of pro-inflammatory cytokines.

cardiomyocytes from damage caused by oxidative stress via mitochondrial ATP-dependent potassium channels [39]. On the other hand, free S1P favors production by COx2 of inflammatory prostaglandins such as PGE2.

Some HDL effects in the cardiovascular system depend completely or partially on S1P, whereas others are independent of their S1P content [39]. Cholesterol efflux in macrophages, a major element of reverse cholesterol transport in arterial walls, does not depend on S1P. Vasodilation dependent on NOS3 mediated by HDL partly relies on S1P agent. Last, but not least, HDL can counteract S1P. High-density lipoproteins can indeed prevent S1P-induced adhesion molecule expression.

2.5.5 Cardiac Effects of Sphingolipids

2.5.5.1 Cardiomyocyte Fate: Apoptosis or Survival

Elevated ceramide content is correlated with cardiomyocyte death after ischemia–reperfusion injury. Cardioprotective pre- and postconditionings that consist of applied transient episodes of ischemia–reperfusion before and after sustained ischemia limit apoptosis. Limited accumulation of ceramide with increased S1P content can mediate protection during ischemic preconditioning, but a large amount that renders NOS3 unavailable favors apoptosis after relatively prolonged hypoxia–reoxygenation [33].

Early nSMase activation occurs in cardiac glutathione deficiency, as glutathione inhibits sphingomyelinase. Reduced glutathione is a cofactor of glutathione peroxidase to reduce intracellular reactive oxygen species, as it is oxidized to a disulfide-linked dimer [33]. In sustained oxidative stress, this dimer cannot be recycled and exit out of the cell, hence causing a glutathione deficiency.

In chronic heart failure, pro-inflammatory cytokines (e.g., TNFα and IL1β) trigger $_{S,Ac}$SMase secretion from endothelial cells that combines with stimulation of reactive oxygen species on enzyme activity [33].

Activation of sphingosine kinase produces cardioprotective sphingosine 1-phosphate, especially in response to acute ischemia–reperfusion injury. Auto- and paracrine regulator S1P activates prosurvival PI3K pathway that activates PKB and inactivates GSK3β. In addition, nitric oxide synthase contributes to S1P-mediated cardioprotection. Ligand-bound S1P receptors trigger Gi- and RhoA-mediated hypertrophy, but more slowly and less efficiently than Gq–PLC signaling primed by norepinephrine and endothelin [36]. Sphingosine 1-phosphate also influences the electrophysiological and contractile behavior of cardiomyocytes. Transforming growth factor-β stimulates sphingosine kinase SphK1. Released S1P elicits autocrine and paracrine activation of S1P$_2$ receptors and subsequent collagen production.

2.5.5.2 Regulation of Cardiac Pacemaker and Myocyte Activity

Acetylcholine released during parasympathetic stimulation slows heart rate via activation of Gi-coupled muscarinic receptors M2 of sinoatrial node cells. Subunits G$\beta\gamma$ directly activate atrial muscarinic K$^+$ channel (K$_{IR}$3; K$_{ACh}$ current; Vol. 3 – Chap. 3. Main Classes of Ion Channels and Pumps). Members of the G-protein-gated inwardly rectifying K$^+$ channel (GIRK1–GIRK5 or K$_{IR}$3.1–K$_{IR}$3.5) function as highly active heteromultimers or low to moderately active homomultimers responsible for acetylcholine-induced bradycardia during vagal activity.

β-Adrenergic signaling increases the heart frequency. β-Adrenoceptors are coupled with the stimulatory Gs subunit that stimulates the activity of adenylate cyclase to produce cAMP messenger. This mediator fosters the activity of protein kinase-A that phosphorylates different cardiac regulators with positive inotropic and/or lusitropic effects. Kinase PKA phosphorylates Ca$_V$1 channels and ryanodine receptors to prime Ca^{++} influx from the extracellular space and intracellular stores. Among other substrates, phospholamban stimulates Ca^{++} reuptake into the sarcoplasmic reticulum via SERCA2a pump. Troponin-I facilitates actin–myosin detachment. In the sinoatrial node, PKA enhances the pacemaker activity, as it phosphorylates Ca$_V$1 and delayed rectifier K$^+$ channels (K$_V$).

On the other hand, cardiac protein phosphatases PP1 and PP2 (Sect. 8.3) contribute to the regulation of Ca^{++} handling and myofilament activity. Both kinases and phosphatases control activities of Ca$_V$1 and delayed rectifier K$^+$ channels in nodal cells. Phosphatase PP1 is regulated by subunits of the G12/13 subclass as well as Rho and its effector RoCK (Sects. 5.2.14 and 9.3). In sinoatrial cells as well as atrial and ventricular myocytes, PP2 phosphatase, which dephosphorylates PKA, myosin-binding protein-C, and troponin-I, connects to P21-activated kinase PAK1 to form a regulatory complex [42] (Sect. 5.2.13.1). Activation of PP2 by PAK1 increases myofilament sensitivity to Ca^{++} in cardiomyocytes. Moreover, PAK1 has a negative chronotropic effect. In addition to CDC42 and Rac1, sphingosine directly causes autophosphorylation (activation) of both PAK1 and PAK2 kinases.

2.5.6 Sphingolipid Effects on Smooth Muscle Cells

2.5.6.1 Smooth Muscle Cell Fate

Receptors S1P$_1$ and S1P$_3$ promote proliferation of smooth muscle cells in balloon injury models, whereas S1P$_2$ antagonizes this effect. In cultured vascular smooth muscle cells, apolipoprotein-C1-enriched high-density lipoproteins stimulate nSMase and prime apoptosis via the release of cytochrome-C from mitochondria and caspase-3 activation [33]. On the other hand, oxidized low-density lipoproteins and tumor-necrosis factor-α that activate nSMase2 in these cells provokes cell proliferation owing to the conversion of ceramide into S1P and activation of ERK1 and ERK2 kinases.

2.5.6.2 Contraction–Relaxation

Sphingosine 1-phosphate generates contraction of vascular smooth muscle cells mainly via $S1P_3$ receptors. However, $S1P_2$ is responsible for the maintenance of the basal vasomotor tone in resistance arteries of the mesenteric and renal trees. In human airway smooth muscle cells, S1P also elicits constriction via $S1P_2$ receptors.

Sphingosine 1-phosphate receptors are involved in cytoskeleton rearrangement. Receptor $S1P_1$ couples to Rac GTPase via Gi and $S1P_3$ to RhoA via G12/13 subunit. Both RhoA and Rac1 intervene in cardiac remodeling. Effectors of RhoA comprise RoCK kinase and diaphanous (Dia; Vol. 1 – Chap. 6. Cell Cytoskeleton). Kinase RoCK phosphorylates myosin light-chain phosphatase to promote contraction of vascular smooth muscle cells. Activated Rac1 causes cytoskeleton reorganization via WASP, WaVe, IqGAP, and PAK1, as well as ROS production.

Smooth muscle cell contraction is mainly triggered by a rise in cytosolic Ca^{++} concentration that promotes binding of Ca^{++} to calmodulin. The Ca^{++}–calmodulin complex activates myosin light-chain kinase. Phosphorylation of myosin light chain provokes interaction of myosin-2 with actin, hence smooth muscle contraction. In smooth muscle cells, ligand-bound G-protein-coupled receptors activate the phosphoinositide cascade that causes Ca^{++} release from the sarcoplasmic reticulum. In addition, activation of various Ca^{++} channels prime Ca^{++} influx from the extracellular medium. In vascular smooth muscle cells, once GTPase RhoA is activated by S1P receptors, RoCK phosphorylates (inactivates) myosin light-chain phosphatase and sensitizes smooth muscle cell to Ca^{++}.

2.5.7 Sphingolipid Effects on Endothelial Cells

2.5.7.1 Endothelium Integrity and Permeability

Whereas $S1P_1$ activates the Gi–PI3K pathway to prevent permeability of vascular endothelium, $S1P_2$ precludes the PI3K pathway via the Rho–RoCK–PTen–Rac axis, hence raising endothelial permeability by disrupting adherens junction [38]. The balance between $S1P_1$ and $S1P_2$ in endothelial cells contributes to the regulation of blood vessel wall permeability.

2.5.7.2 Vasomotor Tone Regulation

When smooth muscle cells are exposed to NO, cGMP-dependent protein kinase PKG is activated. Subsequently, PKG activates myosin light-chain phosphatase, thereby relaxing smooth muscle cells.

Ceramide

Mechanical stimulation can generate ceramide in caveolae and NOS3 phosphorylation (activation), hence vasorelaxation. On the other hand, in aging rat arteries, activated nSMase produces ceramide that excites protein phosphatase-2A, thereby diminishing NOS3 phosphorylation [33].

Sphingosine 1-Phosphate

A given blood vessel responds to S1P with vasodilation or vasoconstriction according to the context, i.e., S1P concentration, targeted receptor subtypes, vascular bed type, etc. Moreover, like numerous vasoactive substances, S1P regulates vascular tone by modulating stress fiber state in smooth muscle cells or by stimulating endothelial cell release of agents that regulate SMC contraction level. Factor S1P can cause both vasorelaxation from adjoining endothelial cells via the Rac1–NOS3 pathway and nitric oxide release and vasoconstriction in smooth muscle cells via the RhoA–RoCK pathway. Resulting effect depends on the context (vascular bed, environmental conditions, S1P concentration, S1P receptor subtype expression pattern, etc.). Agent S1P activates the NOS3–NO axis mainly via $S1P_1$ receptors [40].

Vascular endothelial growth factor can regulate $S1P_1$ density and magnitude of NOS3 response to S1P via protein kinase-C.[40] Reactive oxygen species also raise $S1P_1$ expression, and hence S1P-dependent NOS3 activation.

Both endothelial nitric oxide synthase that is dually acylated by saturated fatty acids myristate and palmitate[41] and $S1P_1$ reside mainly in caveolae, invaginated domains relatively enriched in cholesterol and sphingolipids of the plasma membrane. Enzyme NOS3 is phosphorylated by phosphoinositide 3-kinase-β and protein kinase-B (Ser117) downstream from a pathway that incorporates Gβγ, small GTPase Rac1, and AMP-activated protein kinase [40].[42]

---

[40] Similarly, GPCR stimulation that leads to activation of small GTPase RhoA can not only excite Rho effector kinase RoCK, but also specific PKC isozymes. Kinases RoCK and PKC inhibit MLCP either independently or synergistically to promote smooth muscle cell contraction.

[41] Irreversible myristoylation of NOS3 (Gly2) occurs cotranslationally, whereas reversible palmitoylation of NOS3 happens post-translationally (Cys15 and Cys26). Ubiquitous caveolin-1 and muscle-restricted caveolin-3 interact with and regulate NOS3 in vascular endothelial cells and cardiomyocytes, respectively, hence precluding NOS3 activity. Once endothelial cells stimulated, calmodulin–NOS3 interaction replaces linkage with inhibitory caveolin to increase NO production. Prolonged stimulation can lead to NOS3 translocation from caveolae to intracellular membranes.

[42] Both AMPK and Rac1 are inhibited by caveolin.

2.6 Arachidonic Acid and Eicosanoids

Phospholipids are hydrolyzed by phospholipase-A2 (Sect. 2.8.1) into arachidonic acid, a precursor of eicosanoids (Tables 2.19 and 2.20).

Eicosanoids are generated by cyclo- and lipoxygenases as well as cytochrome-P450-based monooxygenases. Eicosanoids interact with receptors and ion channels and pumps [43] (Table 2.21).

2.6.1 Arachidonic Acid

In heart, arachidonic acid metabolism is mainly associated with cardiomyocytes, and vascular endothelial and smooth muscle cells that are interconnected by intercellular communications. In inflammation, neutrophils and macrophages can infiltrate myocardium and contribute to the production of eicosanoids and other lipidic messengers, as well as platelets and fibroblasts. Arachidonic acid increases both voltage-dependent activation and inactivation kinetics of delayed rectifier K^+ channel $K_V 1.1$ (Table 2.21).

Table 2.19. Production of arachidonic acid (AA) and its derivatives from membrane phospholipids by phospholipase-A2 (PLA2) and eicosanoid-producing enzymes (Source: [43]). Arachidonate-containing plasmalogen phospholipids are hydrolyzed by PLA2 to generate free arachidonic acid. The latter is metabolized by cyclooxygenases (COx), lipoxygenases (LOx), and cytochrome-P450 monooxygenases (CyP450) to produce prostaglandins (PG), leukotrienes (Lkt), and epoxyeicosatrienoic (EET) and hydroxyeicosatrienoic (HETE) acids. Arachidonate-containing diacyl phospholipids are processed by PLA2α or PLA2β that release arachidonic acid. On the other hand, PLA2γ mainly catalyzes the production of 2-arachidonoyl lysolipids (2ALL). Subsequently, lysophospholipase-D (lysoPLD) and lysophospholipase (LPLase) synthesize 2-arachidonoyl glycerol (2AG) and arachidonic acid from 2-arachidonoyl phospholipids, respectively. 2-Arachidonoylglycerol is targeted by cyclooxygenases, lipoxygenases, and monoacylglycerol lipase (MAGL) to produce prostaglandin glycerol esters (PGE), leukotriene glycerol esters (LGE), and arachidonic acid, respectively.

Substrate	Reaction cascade
Plasmalogen phospholipids	PLA2β/PLA2γ–AA
Diacyl phospholipids	PLA2α/PLAβ–AA PLA2γ–2ALL–lysoPLD–2AG–COx–PGE PLA2γ–2ALL–lysoPLD–2AG–LOx–LGE PLA2γ–2ALL–lysoPLD–2AG–MAGL–AA PLA2γ–2ALL–LPL–AA
Arachidonic acid	AA–COx–prostaglandins AA–LOx–leukotrienes AA–CyP450–EETs/HETEs

2.6 Arachidonic Acid and Eicosanoids

Table 2.20. Eicosanoid synthesis from arachidonic acid (AA; Source: [43]). Prostaglandins PGG2 and PGH2 are synthesized by COx1 and COx2. Prostaglandin PGH2 is further metabolized to PGE2, PGF2α, and PGI2 by corresponding synthases. Arachidonic acid is oxidized into 12-hydroperoxyeicosatrienoic acid (12HpETE) by 12-lipoxygenase (12LOx) that is then converted into 12-hydroxyeicosatrienoic acid (12HETE). Cytochrome-P450 epoxigenase and hydroxylase form (14,15)-epoxyeicosatrienoic acid ((14,15)EET), and 16- and 20-HETEs, respectively. Epoxide hydrolase (EH) catalyzes the conversion of (14,15)EET to (14,15)-dihydroxyeicosatrienoic acid ((14,15)DHET).

Upstream enzymes	Pathways
Cyclooxygenases	AA–COx–PGG2–PGH2–PGE2 AA–COx–PGG2–PGH2–PGF2α AA–COx–PGG2–PGH2–PGI2
Lipoxygenases	AA–12LOx–12HpETE–12HETE
Cytochrome-P450 epoxygenase	AA–(14,15)EET–(14,15)DHET
Cytochrome-P450 hydroxylase	AA–16HETE AA–20HETE

Table 2.21. Effect of arachidonic acid and epoxyeicosatrienoic and hydroxyeicosatrienoic acids on ion channels (Source: [43]).

Molecule	Effect
Arachidonic acid	$K_V1.1$ kinetics increase
Epoxyeicosatrienoic acids	Ca_V1 activity enhancement K_{ATP} sensitivity reduction [(11,12)EET] Na^+ channel-gating attenuation
Hydroxyeicosatrienoic acids	BK channel inhibition (20HETE)

2.6.2 Cyclooxygenases

Cells such as cardiomyocytes produce 2 types of cyclooxygenases, constitutively expressed COx1 and inducible COx2, that have distinct functions. Cyclooxygenases form prostaglandin-G2 and then prostaglandin-H2 (Table 2.20).

Enzyme COx2 is a mediator of inflammation that can ensure cardioprotection. It is upregulated during ischemic preconditioning via kinases PKC and Src, transcription factor NFκB, as well as NOS2 [43]. PGF2α activates its Gq-coupled prostanoid FP receptor that stimulates ERK2, JNK1, and cytosolic Tyr kinases and provokes cardiac hypertrophy. Kinase JNK1 can suppress hypertrophic signaling via phosphorylation of nuclear factor of activated T-cell that hinders its nuclear translocation [43]. Agent PGF2α also raises glucose transporter GluT1 expression and thus glucose uptake. PGI2 binds to its Gs-coupled prostanoid IP receptor and represses maladaptive cardiac hypertrophy in response to pressure overload.

2.6.3 Lipoxygenases

Lipoxygenases are iron-containing enzymes that catalyze dioxygenation of polyunsaturated fatty acids in lipids. Numerous lipoxygenase isozymes are involved in the metabolism of prostaglandins and leukotrienes. Well known types includes arachidonate 5-[43] and 12-lipoxygenase, in addition to erythroid cell-specific arachidonate 15-lipoxygenase.[44]

Lipoxygenases catalyze oxidation of arachidonic acid to produce hydroperoxyeicosatetraenoic acids that are then reduced to form hydroxyeicosatrienoic acids (HETE). In addition, lipoxygenases promote insulin-stimulated glucose uptake via GluT4 transporter, as it favors suitable actin rearrangement [43].

Lipoxygenase-mediated production of HETEs participates in the maintenance of the myocardial function. However, overexpression of 12-lipoxygenase, the most abundant type in myocardium, leads to cardiac hypertrophy and fibrosis [43].

Enzyme 15-lipoxygenase-1 oxidizes unsaturated fatty acids such as arachidonic acid at the cell membrane, to generate active hydroperoxy and epoxy metabolites. It is produced by macrophages, eosinophils, and mastocytes, among others. Epithelial 15-lipoxygenase-1 generates intracellular 15-hydroxyeicosatetraenoic acid. In particular, 15HETE acts in interleukin-13-induced mucin Muc5ac production by human airway epithelial cells. Product 15HETE can bind to phosphatidylethanolamine to form the 15HETE–PE complex. This complex as well as 15LOx1 interact with phosphatidylethanolamine-binding protein PEBP1,[45] thereby dissociating PEBP1, a Raf inhibitor,[46] from cRaf kinase.[47] Both the 15LOx1–PEBP1 and the 15HETE–PE–PEBP1 complexes then activate the cRaf–ERK pathway, especially in asthma [44].[48]

---

[43] A.k.a. arachidonic acid 5-lipoxygenase, 5δ-lipoxygenase, and leukotriene-A4 synthase.

[44] A.k.a. arachidonate ω6-lipoxygenase.

[45] Enzyme 15LOx1 binds to phospholipids such as phosphatidylinositols and some phosphatidylethanolamines to generate 15HETE and 13-hydroxyoctadecadienoic acids (13HODE).

[46] A.k.a. Raf kinase inhibitory protein (RKIP).

[47] Molecules PEBP1, cRaf, and ERK are membrane-associated proteins.

[48] Concentrations of both 15LOx1 and its products 15HETE increase in asthma and eosinophilic inflammation. In addition, 15LOx1 synthesis is stimulated by IL13.

2.6.4 Cytochrome-P450 Monooxygenases

Membrane-bound, heme-containing, cytochrome-P450-based monooxygenases catalyze the insertion of a single atom of oxygen as a part of multicomponent electron transfer complexes. Cytochrome-P450-containing monooxygenases primarily reside in the inner membrane of mitochondria and endoplasmic reticulum.

They oxidize several endogenous lipids, such as arachidonic, retinoic, and linoleic acids. From arachidonic acid, CyP450 forms a series of fatty acid epoxides, such as (5,6)-, (8,9)-, (11,12)-, and (14,15)-EETs, as well as alcohols such as midchain and ω-terminal HETEs. They also catalyze oxygen- and NADPH-dependent metabolism (oxidation, peroxidation, and/or reduction) of xenobiotics.

Multiple cytochrome-P450 types exist.[49] Cytochrome-P450 monooxygenases require O_2, CyP450 reductase, and NADPH cofactor. Some CyP450 types, such as CyP450-2b, CyP450-2c, and CyP450-2j, are primarily arachidonic acid epoxygenases. Others are chiefly arachidonic acid hydroxylases, such as CyP450-4a and CyP450-4f family members.

Most cytochrome-P450-based enzymes are predominantly synthesized in the liver. However, some cytochrome-P450-containing enzymes, such as members of the CyP450-2j and -4a subfamilies are predominantly detected in the heart, vasculature, kidney, lung, and gastrointestinal tract. Some cytochrome-P450 monooxygenases are expressed constitutively, whereas others are induced.

CyP450 ω-hydroxylases (CyP450-4a1, -4a2, and -4f) produce 20HETE, a potent vasoconstrictor that inhibits Ca^{++}-sensitive K^+ channels [43]. Product 20HETE thus aggravates ischemia damage.

On the other hand, epoxyeicosatrienoic acids ensure cardioprotection. Epoxygenase CyP450-2j2 attenuates ischemia–reperfusion injury, as produced EETs modulate K_{ATP} channel activity and MAPK signaling [43] (Table 2.22).

Cytochrome-P450 epoxygenases produce (5,6)EET and (11,12)EET that raise cAMP and intracellular Ca^{++} concentration, as Ca^{++} influx from the extracellular medium through Ca_V1 channels that are stimulated by EETs activates ryanodine receptor [43]. Agent (11,12)EET lowers K_{ATP} channel sensitivity to ATP (Table 2.21). Several EETs attenuate Na^+ channel gating. Agent 20HETE inhibits high-conductance, Ca^{++}-activated, voltage-gated K^+ channel (BK).

---

[49]CyP450-1 to CyP450-5, CyP450-7 and CyP450-8, CyP450-11, CyP450-17, CyP450-19 to CyP450-21, CyP450-24, CyP450-26 and CyP450-27, CyP450-39, CyP450-46, CyP450-51. Cardiomyocytes possess CyP450-1a and CyP450-1b, CyP450-2a to CyP450-1c, CyP450-2e and CyP450-2j, CyP450-4a, and CyP450-11 [43].

Table 2.22. Effects of epoxyeicosatrienoic acids (Source: Wikipedia). Epoxyeicosatrienoic acids activate large-conductance, calcium-activated potassium channels in vascular smooth muscle cells, thereby causing hyperpolarization of the membrane potential and vasodilation. They also influence Ca_V1 channel activity in cardiomyocytes. On the other hand, 20HETE, the activity of which is impeded by nitric oxide, inhibits calcium-activated potassium channels and activates Ca_V1 channels in vascular smooth muscle cells.

Calcium release from intracellular stores
Increased sodium–hydrogen antiporter activity
Increased cell proliferation
Decreased cyclooxygenase activity
Decreased release of insulin and glucagon
Vasodilation
Increased risk of tumor adhesion on endothelial cells
Decreased platelet aggregation
Cardioprotection after ischemia–reperfusion

2.6.5 Arachidonic Acid Metabolites in Lungs

Arachidonic acid metabolites derived from action of cyclo-, lipo-, and CyP450 monooxygenases achieve various functions in lungs, particularly pulmonary vascular and bronchial smooth muscle tone as well as airway epithelial ion transport. Cytochrome P450-derived arachidonate metabolites contribute to hypoxic pulmonary vasoconstriction, regulation of bronchomotor tone, control of the composition of airway lining fluid, and limitation of pulmonary inflammation [45].

Lung cells produce arachidonic acid from membrane phospholipid stores via phospholipases (e.g., cytosolic PLA2), both constitutively and on chemical or mechanical stimuli. Free arachidonic acid can then enter into one of the 3 following metabolic pathways [45]: (1) the prostaglandin-H synthase pathway that produces thromboxane and prostacyclin; (2) the lipoxygenase pathway that synthesizes leukotrienes, midchain hydroxyeicosatetraenoic acids, and lipoxins; and (3) the cytochrome P450 monooxygenase pathway that manufactures midchain and ω-terminal HETEs and cis-epoxyeicosatrienoic acids.

2.6.6 Prostaglandins

Prostaglandin-E2 is synthesized in the airway epithelium. It is a potent bronchodilator. It also has anti-inflammatory effects. Prostacyclin (PGI2) is produced in the pulmonary artery endothelium. It has vasodilatory, bronchodilatory, and platelet anti-aggregatory effects. On the other hand, prostaglandin-F2α and thromboxane (TXA2) are potent bronchoconstrictors. These vasoconstrictors also enhance platelet aggregation.

Widespread prostaglandin-F2α that causes renin secretion and vasoconstriction, among other functions, is synthesized by prostaglandin-F synthase of the aldoketo reductase (AKR) family.[50]

Products of prostaglandin-H synthases PGHS1 and PGHS2 limit lung inflammation in response to inhaled allergens and lipopolysaccharides [45]. On the other hand, lipoxygenase products of arachidonic acid contribute to airway inflammation, bronchoconstriction, increased mucous secretion, and vascular permeability in asthma.

2.6.7 Hydroxy- and Epoxyeicosatrienoic Acids

Epoxyeicosatrienoic acids participate in the relaxation of vascular smooth muscle cells by the so-called endothelium-derived hyperpolarizing factor, especially in the coronary bed. On the other hand, 20HETE, the dominant arachidonic acid metabolite in the renal cortex that promotes natriuresis, is one of the most potent vasoconstrictors.

Human peripheral lung microsomes that convert arachidonic acid to 20HETE are enriched in CyP450-4a enzymes [45]. In human lungs, CyP450-2j enzymes are detected in both ciliated and non-ciliated airway epithelial cells, bronchial and vascular smooth muscle cells, endothelial cells, and alveolar macrophages. Members of the CyP450-1a, -2b, -2e, -2f, -3a, and -4b subfamilies have also been observed in human lungs [45].

Arachidonic acid precludes Cl^- secretion in human airways [45]. Epoxyeicosatrienoic acids hamper tracheal Ca^{++}-sensitive Cl^- flux. They thus modulate the composition of the airway lining fluid and gel layer (mucus).

Both (5,6)EET and (11,12)EET cause hyperpolarization of the plasmalemmal potential and relaxation of rabbit tracheal smooth muscle cells [45]. Epoxyeicosatrienoic acids activate large-conductance, Ca^{++}-activated K^+ channels.

Lipid 20HETE decreases the vasomotor tone of human pulmonary arteries [45]. The CyP450-4a–20HETE axis reduces subacute hypoxic pulmonary vasoconstriction. Conversely, a subacute exposure to hypoxia hinders 20HETE formation. On the other hand, the production of prostacyclin and nitric oxide rises during subacute or chronic hypoxia.

---

[50]The PGF2 synthesis can be produced from [46]: (1) the 9-keto group of PGE2 by PGE 9-ketoreductase, which also has 20α-hydroxysteroid dehydrogenase activity; (2) the 11-keto group of PGD2 by PGD2 11-ketoreductase; and (3) the (9,11)-endoperoxide group of PGH2. The enzyme PGD2 11-ketoreductase catalyzes the reduction of PGD2 into 9α, 11βPGF2, a stereoisomer of PGF2α, which has biological function like PGF2α, as well as that of PGH2 to PGF2α. Various prostaglandin-F synthases working with cofactor NADPH pertain to the AKR5A, AKR1B, and AKR1C subfamilies among 15 different families (AKR1–AKR15). In addition to members of the AKR family, some enzymes of the thioredoxin family can also contribute to PGF2 synthesis.

2.7 Lipases

Lipase activity can be achieved by numerous proteins, such as hormone-sensitive, lipoprotein, endothelial, gastric, hepatic, pancreatic, carboxyl ester, lysosomal acid, and monoacylglycerol lipase, as well as triacylglycerol hydrolases TGH1 and TGH2. Several other types of lipases exist, such as phospholipases and sphingomyelinases, but they are usually not considered as conventional lipases.

Triglyceride lipases hydrolyze linkages of triglycerides. Lipases produce free fatty acids used as an energy source. In particular, fatty acids are mobilized from triglyceride stores during exercise to supply working muscle with energy, especially during prolonged exercise, when carbohydrate reserves get depleted.

Nutritional glucids and lipids in excess are efficiently converted into triglycerides. Most of the body's energy reserves are stored in white adipose tissue (Vol. 2 – Chap. 1. Remote Control Cells). In normal conditions, an equilibrium exists between lipid synthesis and degradation. Catabolism of triglyceride storage depots and mobilization of free fatty acids in adipocytes and other cell types depend on lipases. Two lipases operate successively: hormone-sensitive (HSL) and adipose triglyceride lipase (ATGL).

Mobilization of fatty acids from adipose tissue is controlled by hormones, cytokines, and adipokines. Catecholamines (β-adrenergic stimuli), atrial natriuretic peptide, growth hormone, and leptin activate lipases to raise fatty acid release into the blood circulation. On the other hand, insulin inhibits hormone-sensitive lipase.

2.7.1 Hormone-Sensitive Lipase

Hormone-sensitive lipase (HSL) is synthesized in multiple tissue and cell types, such as white and brown adipose tissue, skeletal muscle, heart, steroidogenic tissues, and intestine, as well as pancreatic β cells and macrophages (Table 2.23).

Hormone-sensitive lipase is able to hydrolyze tri-, di- and monoacylglycerols as well as cholesterol and retinyl esters. It most efficiently hydrolyzes diglycerides. It is

Table 2.23. Hormone-sensitive lipase and its products in different cell types (Source: [47]).

Tissue Cell	Major product	Function
Adipose tissue	Fatty acids	Export for oxidation
Muscle, heart	Fatty acids	Oxidation
β Cells	Fatty acids	
Adrenals	Cholesterol	Substrate for steroidogenesis
Ovaries, testes	Cholesterol	Substrate for steroidogenesis
Ovaries, placenta	Steroids	Transcriptional control
Mammary gland	Cholesterol	Milk component, membranogenesis
Macrophage	Cholesterol	Export (via high-density lipoprotein)

phosphorylated (activated) by protein kinase-A and AMP-activated protein kinase. Conversely, it is inactivated by PP2 and PPM1 phosphatases. Phosphorylation by protein kinase-A of HSL and lipid droplet-associated perilipin trigger the translocation of HSL from the cytoplasm to the lipid droplet.[51]

2.7.1.1 HSL in Macrophages

Cholesterol of lipoproteins is stored in lipid droplets as cholesterol esters after re-esterification by acyl-CoA:cholesterol acyltransferase ACAT1. Hydrolysis of cholesterol esters is the initial step toward elimination of cholesterol. Hormone-sensitive lipase and cholesteryl ester hydrolase (CEH) hydrolyze cholesterol esters in macrophages.

Cholesterol homeostasis, i.e., balance between synthesis and hydrolysis, is actually achieved by an equilibrium between the activities of ACAT1 that catalyzes the formation of cholesteryl esters and cytosolic neutral cholesterol ester hydrolase (nCEH) that supports the remobilization of free cholesterol.

Neutral cholesterol ester hydrolase removes cholesterol from macrophages in cooperation with hormone-sensitive lipase that also has a neutral cholesteryl ester hydrolase activity [48].

2.7.2 Adipose Triglyceride Lipase

Adipose triglyceride lipase (ATGL) promotes the catabolism of stored lipids in adipose and non-adipose tissues. This triacylglycerol hydrolase specifically performs the first step of lipolysis. It indeed generates diglycerides and fatty acids. Therefore, efficient lipolysis depends on the coordinated action of lipases that operate sequentially: ATGL forms diglycerides that are subsequently hydrolyzed by hormone-sensitive lipase.

2.7.3 Hepatic Lipase

Hepatic lipase (LipC)[52] is expressed not only in the liver, but also adrenal glands. It aims at converting intermediate-density lipoproteins to LDL particles. It possesses the dual function of triglyceride hydrolase and ligand-bridging factor for receptor-mediated lipoprotein uptake.

---

[51] Perilipin acts as a protective coating of lipid droplets from hormone-sensitive lipase. Perilipin is hyperphosphorylated by PKA, for example after β-adrenergic receptor activation.
[52] A.k.a. hepatic triacylglycerol lipase (HTGL) and possibly LipH. The LipH alias should not be used to avoid confusion with lipase-H (Sect. 2.7.6).

Hepatic lipase, like LPase, facilitates the interaction of lipoproteins with cell-surface proteoglycans and receptors, thereby enhancing binding and uptake of different lipoproteins as well as cholesteryl esters.

2.7.4 Endothelial Lipase

Endothelial lipase (EL), or endothelial cell-derived lipase (EDL), encoded by the LIPG gene, is synthesized, secreted, then either anchored to the plasma membrane or released by the endothelial cells [49]. It is also detected in hepatocytes and macrophages. It has high-sequence homology with lipoprotein lipase (LPase). In fact, endothelial lipase is an additional member of the triacylglycerol lipase family with lipoprotein lipase (LPase) and hepatic lipase (HL). They intervene in metabolism of high-density lipoproteins.

Endothelial lipase is produced in various organs, such as the liver, lung, kidney, and placenta, but not skeletal muscle [49]. Expression of endothelial lipase rises during inflammation.

Unlike LPase and HL, endothelial lipase is mainly a phospholipase with a slight triacylglycerol lipase activity. It is aimed at binding HDLs and taking up HDL-carried cholesterol esters. It reduces the plasma concentration of HDL–cholesterol (HDL–C) and its major protein apolipoprotein-A1 [49]. A low concentration of plasma HDL–C is major risk factor for the development of arterial diseases (Vol. 6 – Chap. 7. Vascular Diseases). Endothelial lipase also contributes to the cellular uptake of lipoproteins, independent of its catalytic activity.

Administration of endothelial lipase-coding adenovirus to cells raises HDL binding and uptake 1.5 fold as well as HDL–CE uptake 1.8 fold [50]. On the other hand, lipoprotein lipase has a less pronounced effect (1.1-fold and 1.3-fold increase in HDL and HDL–CE uptake, respectively).[53] Inhibition of EL enzymatic activity by tetrahydrolipstatin (THL) enhances its effect on lipid uptake (5.2-fold, 2.6-fold, 1.1-fold increase in HDL binding and HDL and CE uptake, respectively). In the presence of THL, plasmalemmal EL concentration is augmented. Conversely, HDL and free fatty acids reduce the amount of cell surface-bound endothelial lipase [50].

Like in macrophages, on which lipoprotein lipase mediates HDL–CE uptake, in addition to ScaRs (Vol. 3 – Chap. 4. Membrane Compound Carriers), endothelial lipase alone is able to stimulate HDL–CE uptake, independently of ScaRb1 scavenger receptor.

Statins that lower LDL–C level, hence used to treat hypercholesterolemia, also increase plasma HDL–C concentration. They selectively inhibit 3-hydroxy 3-methyl glutaryl coenzyme-A reductase (HMGCoAR), a rate-limiting enzyme of

---

[53] High-density lipoproteins receive cholesterol from cell membranes. Cholesterol is then esterified by lecithin–cholesterol acyltransferase.

the cholesterol synthesis. They also suppress the synthesis of isoprenoid intermediates, such as geranylgeranyl and farnesyl pyrophosphate, which serve to attach lipids to various signaling molecules [51]. In macrophages, statins increased apoA1 level, as they prevent membrane translocations of small GTPases CDC42, Rac, Ras, and RhoA and activates extracellular signal-regulated kinases ERK1 and ERK2 as well as P38MAPK and PPARγ (nuclear receptor NR1c3) [52]. They stimulate NR1c3 via COx2-dependent increase in prostaglandin PGJ2 level via the RhoA–P38MAPK and CDC42–P38MAPK pathways and RhoA- and CDC42-independent ERK1/2 pathway. Moreover, statins also provoke the synthesis of fatty acid translocase CD36,[54] ATP-binding cassette transporter ABCa1, as well as scavenger receptor ScaRb1 (or SRB1; a.k.a. CD36L1). The latter has high affinity for HDL–C and mediates cholesterol efflux and uptake. Transporter ABCa1 causes lipid efflux from cells to lipid-poor apolipoproteins. Translocation of ABCa1 to the plasma membrane thus participates in HDL plasma level, independently of ScaRb1 [53]. Furthermore, statins suppress the basal and TNFα-induced expression of endothelial lipase in human umbilical vein endothelial cells [51].

2.7.5 Lipoprotein Lipase

Lipoprotein lipase hydrolyzes lipids in lipoproteins, such as those in chylomicrons and very-low-density lipoproteins, into free fatty acids and monoacylglycerol. It requires ApoC2 cofactor. Lipoprotein lipase specifically lodges in endothelial cells of capillaries.

Insulin causes LPase synthesis in adipocytes. It also favors its insertion in the capillary endothelium. Different LPase isozymes exist in separate tissues, with distinct regulation modes. Lipoprotein lipase in adipocytes is activated by insulin, but not that in striated myocytes.

Lipoprotein lipase is a homodimer that has the dual function of triglyceride hydrolase and ligand-bridging factor for receptor-mediated lipoprotein uptake. Lipoprotein lipase actually mediates HDL–CE uptake by macrophages and hepatocytes. Lipoprotein lipase links to lipoproteins and anchors them to the plasma membrane. It indeed has a high affinity for cell-surface proteoglycans, especially heparan sulfate proteoglycans that tether molecules. Therefore, LPase recruits lipoproteins to the cell membrane and promotes cholesterol ester uptake from lipoproteins. In addition, complexes composed of HSPG, LPase, and lipoproteins can be internalized.

---

[54] A.k.a. thrombospondin receptor, platelet collagen receptor, and glycoprotein GP3b or GP4.

2.7.6 Lipase-H

The secreted enzyme lipase-H (LipH)[55] hydrolyzes phosphatidic acid into lysophosphatidic acid. This lipid mediator operates in platelet aggregation, smooth muscle contraction, and cell proliferation and motility.

2.8 Phospholipases

Inositol phosphates are intracellular second messengers that regulate various cell processes, from calcium signaling to chromatin remodeling (e.g., digestion, inflammation, membrane remodeling, and intercellular signaling via the degradation of phospholipids). Phospholipases target phospholipids of cellular membranes. Phospholipases are involved with phospholipids in the transmission of ligand-bound receptor signaling.

Four main types of phospholipases exist (PLA–PLD). Activated phospholipases particularly stimulate protein kinase-C that is maximally active in the presence of diacylglycerol and calcium ions released from its intracellular stores by inositol trisphosphate.

2.8.1 Phospholipases-A

Phospholipase-A hydrolyzes one of the acyl groups of phosphoglycerides or glycerophosphatidates. Phospholipases-A1 and -A2 target the acyl group at the 1- and 2-position, respectively.

2.8.2 Phospholipases-A1

Phospholipase-A1 (PLA1) is also called phosphatidylcholine 1-acylhydrolase. It produces 2-acyl lysophospholipids and fatty acids. Extracellular PLA1s that belong to the pancreatic lipase gene family include: (1) phosphatidylserine-specific PLA1 (psPLA1); (2) membrane-associated phosphatidic acid-selective PLA1s ($_m$paPLA1α and $_m$paPLA1β); (3) hepatic lipase (Sect. 2.7.1); (4) endothelial lipase (Sect. 2.7.4); and (5) pancreatic lipase-related protein PLRP2 [54].

---

[55] A.k.a. phosphatidic acid-selective phospholipase-A1α (PAPLA1α), membrane-associated phosphatidic acid-selective phospholipase-A1α (MPAPLA1 and mPA-PLA1α), phospholipase-A1b member-B (PLA1b), and lipid defect lipase (LpDL)-related protein (LpDLR); not hepatic lipase with possible alias LipH rather than LipC or HTGL.

The 3 first-mentioned PLA1s differ from other members by their structures and substrate specificities. They indeed exhibit only a PLA1 activity, whereas hepatic and endothelial lipases and PLRP2 are triacylglycerol-hydrolases with a PLA1 activity [54]. Both psPLA1 and $_m$paPLA1s specifically hydrolyze phosphatidylserine and phosphatidic acid, respectively, producing lysophosphatidylserine and lysophosphatidic acid that serve as lipid mediators.

Activity of phospholipase-A1 is detected in many types of tissues and cells. Phosphatidylserine-specific PLA1 is a secreted enzyme that targets phosphatidylserine, which is normally located in the inner leaflet of the lipid bilayer, hence of reduced accessibility. Nevertheless, psPLA1 processes phosphatidylserine exposed on the surface of cells, such as apoptotic cells and activated platelets [55]. Generated 2-acyl lysophosphatidylserine is a lipid mediator for mastocytes, T lymphocytes, and neurons.

2.8.2.1 Pancreatic Lipase-Related Protein PLRP2

Dietary triglycerides, the predominant nutritional lipids, yield a major nutritional energy source. The intestine produces very-low-density lipoproteins and chylomicrons to transport lipids and lipid-soluble vitamins into blood.[56] Vitamin A-rich chylomicron remnants result from the selective removal of triglyceride catalyzed by lipoprotein lipase. Dietary triglycerides are precursors of cellular membrane components, prostaglandins, thromboxanes, and leukotrienes. They improve the palatability of food.[57] Dietary fats are particularly important for newborns and infants during the first 6 months of life. Efficient digestion of dietary triglycerides relies on their cleavage into fatty acids and monoacylglycerols.

Adult humans require pancreatic triglyceride lipase and colipase for efficient dietary lipid digestion. On the other hand, in newborns and infants, pancreatic triglyceride lipase does not contribute to nutritional lipid digestion. Newborns and suckling infants express colipase, but not pancreatic triglyceride lipase until the suckling–weanling transition. Therefore, colipase interacts with another lipase

---

[56]The assembling of very-low-density lipoproteins and chylomicrons is constitutive and occurs during the postprandial state, respectively. Chylomicrons consist of a core of triglycerides and cholesterol esters and a monolayer of phospholipids, cholesterol, and proteins. The fatty acid composition of triglycerides, but not phospholipids, in chylomicrons reflects the fatty acid composition of dietary lipids.

[57]Palatability is related to palatable food and fluid, i.e., food and fluid sufficiently agreeable in flavor to the taste (to the palate) to be eaten. The palatability of food and fluid, unlike flavor, varies with the feeding–fasting state: it decays after consumption and rises during fasting. Palatability can create a hedonic hunger independently of homeostatic needs. The taste-reactivity response associated with the palatability of substances and fluids is determined by opioid receptor-related processes in the nucleus accumbens and ventral pallidum (opioid eating site) [56]. On the oher hand, the assignment of opioid-mediated reward or inventive value of nutrition (i.e., the decision to engage food-related action that is based largely on the degree to which the goal, or reward, is desired) involves the basolateral amygdala.

in newborns. Efficient lipid digestion in newborns actually requires colipase that interacts with pancreatic lipase-related protein-2 (PLRP2), a homolog of pancreatic triglyceride lipase [57].

2.8.3 Superfamily of Phospholipases-A2

In mammals, the genome encodes more than 30 phospholipases-A2 (PLA2) or related enzymes. Phospholipase-A2 hydrolyzes glycerophospholipids to yield fatty acids and lysophospholipids. In particular, processing of phosphatidylcholine produces free fatty acids and lysophosphatidylcholine, which potentiate diacylglycerol activity.

In most mammalian cells, PLA2 catalyzes the breakdown of membrane phospholipids into arachidonic acid, which is involved in signaling. Arachidonic acid is the precursor of eicosanoids (Vol. 5 – Chaps. 8. Smooth Muscle Cells and 9. Endothelium), such as leukotrienes and prostaglandins. Eicosanoids can also be synthesized from diacylglycerol formed by phospholipase-C.

About one third of enzymes that possess PLA2 or related activity are secreted subtypes with distinct localizations and enzymatic properties. They act on extracellular phospholipids. Intracellular PLA2s intervene in signal transduction and membrane homoeostasis.

Numerous categories (g_iPLA2) and subcategories of PLA2s have been defined within the PLA2 superfamily.[58] In humans, the phospholipases-A2 superfamily can be split into the following groups (Table 2.24): group 1 encoded by the gene PLA2G1B; group 2 by the genes PLA2G2A and PLA2G2C (possible pseudogene) to PLA2G2F; group 3 by the PLA2G3 gene; group 4 by the genes PLA2G4A to PLA2G4F; group 5 by the PLA2G5 gene; group 6 by the PLA2G6 gene; group 7 by the PLA2G7 gene; group 10 by the PLA2G10 gene; and group 12 by the PLA2G12A and PLA2G12B genes. Enzymes g4PLA2s and g6PLA2 operate in the perinuclear membranes, where arachidonic acid-metabolizing enzymes (cyclooxygenases and prostaglandin synthases) reside.

Four main families of PLA2s include (Table 2.25): (1) low-molecular-weight, secretory PLA2s ($_s$PLA2); (2) cytosolic, calcium-dependent PLA2s (cPLA2) such as g4PLA2s; (3) intracellular, calcium-independent PLA2s (iPLA2) such as g6PLA2; and (4) lipoprotein-associated phospholipase-A2 (lpPLA2), or platelet-activating factor acid hydrolases (PAFAH), such as g7PLA2. Enzyme iPLA2 is observed in endothelial cells and contributes to endothelium-dependent vasoconstriction induced by acetylcholine.

---

[58] Enzymes PLA2 are components of old world (group 1) and new world (group 2), and bee venom (group-3) [59]. Phospholipases-A2 secreted by macrophages are members of the groups 4A and 5. Fifteen groups of PLA2 have been defined.

2.8 Phospholipases

Table 2.24. Groups of phospholipases-A2 (Source: [60]; cPLA2: intracellular, calcium-dependent PLA2; iPLA2: intracellular, calcium-independent PLA2; sPLA2: secreted PLA2).

Group	Sources
	Small secreted phospholipases-A2
1A	Old world snake
1B	Human, pig (pancreas)
2A	Human (synovium), new world snake
2B	New world snake
2C	Rodent
2D	Human, Murinae (pancreas, spleen)
2E	Human, Murinae (brain, heart, uterus)
2F	Human, Murinae (testis, embryo)
3	Human, Murinae, lizard, bee
5	Human, Murinae (heart, lung, macrophage)
9	Marine snail Conus venom (conodipine-M)
10	Human (spleen, thymus, leukocyte)
11A	Green rice shoots (PLA2-1)
11B	Green rice shoots (PLA2-2)
12	Human, Murinae
13	Parvovirus
14	Symbiotic fungus, bacteria
G7a	PAFAH1 (lpPLA2)
	Intracellular phospholipases-A2
4	Human (cPLA2)
	(4A–4F)
6	Human (iPLA2)
	(6A1/2, 6B–6H)
G7b	Human (PAFAH2)
G8a/b	Human (PAFAH1b [PAF1b] subunits)
15	Human (lPLA2)

These families can be subdivided into subfamilies. For example, the family of secreted PLA2 can contain 5 groups: group-1 (g1$_S$PLA2), 2 (g2$_S$PLA2), 3 (g3$_S$PLA2), 5 (g5$_S$PLA2), and 10 (g10$_S$PLA2).

Several members of the cPLA2 and iPLA2 families also possess PLA1, lysophospholipase, or triglyceride lipase activity, whereas $_S$PLA2s have only a PLA2 activity [62]. The conserved catalytic center of mammalian extracellular $_S$PLA2s contains a His–Asp dyad and a Ca^{++}-binding loop, whereas that of intracellular cPLA2s and iPLA2s is characterized by a Ser–Asp dyad.[59]

---

[59]Dyad, which means a couple or pair, is used in biology to define not only a pair of sister chromatids joined by the centromere before the separation during anaphase of mitosis and

Table 2.25. Phospholipases-A2, their subcellular localization, and Ca^{++}-dependent activity (Source: [61]; cPLA2: intracellular, calcium-dependent PLA2; iPLA2: intracellular, calcium-independent PLA2; lPLA2: lysosomal PLA2; lpPLA2: lipoprotein-associated PLA2; PAFAH: platelet-activating factor acid hydrolase; $_s$PLA2: secreted PLA2; ER: endoplasmic reticulum). Both arachidonic acid (AA) and docosahexaenoic acid (DHA) increase entry of extracellular Ca^{++} into neurons via NMDA-type glutamate receptors. Arachidonic acid and docosahexaenoic acid increases and inhibits glutamate-induced prostaglandin release from astrocytes, respectively. Unesterified DHA can promote Ca^{++} release from the endoplasmic reticulum. Neuroprotectin-D1 is a metabolite of docosahexaenoic acid. Cell-specific and agonist-dependent events coordinate translocation of cPLA2 to the nuclear envelope and membranes of the endoplasmic reticulum and Golgi body. Calcium ion attenuates iPLA2β activity that promotes iPLA2β–calmodulin interaction, which impedes iPLA2 activity.

Family	Calcium effect	Distribution	Fatty acid preference
$_s$PLA2	Stimulation	Ubiquitous	None
cPLA2	Translocation (membrane binding)	Ubiquitous	AA
iPLA2	Activation by Ca^{++} release from ER (but not Ca^{++} binding)	Brain, heart, skeletal muscle, pancreas, testis, placenta	DHA
lPLA2		Brain, Heart, lung, liver, kidney, spleen, thymus, macrophages	Oleate, linoleate
PAFAH (lpPLA2)		Ubiquitous	AA, DHA

Whereas cPLA2s preferentially release arachidonic acid, iPLA2s liberate docosahexaenoic acid (DHA), a polyunsaturated fatty acid [63]. Arachidonic acid is a precursor of eicosanoids (Table 2.26). Docosahexaenoic acid is a precursor of neuroprotectin-D, a docosanoid.[60] It also contributes to prostanoid production.

2.8.3.1 Secreted Phospholipases-A2

Many secreted forms of phospholipases-A2 (Table 2.27) hydrolyze phosphatidylcholine and lysophosphatidylcholine. Extracellular forms of phospholipases-A2

---

anaphase 2 of meiosis, but also to describe protein morphology. In chemistry, it designates an element with a valence of 2 and in mathematics, it refers to a second rank tensor formed from a pair of vectors.

[60] On the other hand, neuroprotectin-A and -B are bicyclohexapeptides.

2.8 Phospholipases

Table 2.26. Signaling axes activators of PLA2 enzymes and effects (DAG: diacylglycerol; ERK: extracellular signal-regulated protein kinase; PDK: phosphoinositide-dependent kinase; PG: prostaglandin; PI3K: phosphatidylinositol 3-kinase; PIP_3: phosphatidylinositol triphosphate; PKB/C: protein kinase B/C; PLA2/C: phospholipase A2/C; Tx: thromboxane).

PLC–DAG–PKCα–cRaf–ERK1/2–cPLA2–AA–PGs/Tx

PLC–DAG–PKCα–iPLA2–AA
$PI3K–PIP_3$–PDK1–PKB–PKCα–iPLA2–AA

Table 2.27. Secreted forms of phospholipase-A2 (Sources: [58, 251]; npsPLA2: non-pancreatic secretory PLA2; PLA2S: synovial PLA2; PPLA2: pancreatic PLA2; SPLASH: secretory-type PLA2, stroma-associated homolog). Pancreatic, group-1b phospholipase-A2 (PLA2-1b) interacts with PLA2-2a. Membrane-associated, group-2a phospholipase-A2 is detected in platelets and synovial fluid. Element PLA2G2C is a possible pseudogene.

Type	Gene	Other names
sPLA2-1b	PLA2G1B	Conventional sPLA2 G1b-PLA2, PLA2a, PPLA2
sPLA2-2a	PLA2G2A	Conventional sPLA2 G2a-PLA2, G2c-sPLA2, npsPLA2, sPLA2, PLA2b, PLA2L, PLA2S, PLAS1, MOM1 (inflammatory sPLA2)
sPLA2-2d	PLA2G2D	G2d-PLA2, G2d-sPLA2, sPLA2S, SPLASH
sPLA2-2e	PLA2G2E	G2e-PLA2, G2e-sPLA2
sPLA2-2f	PLA2G2F	G2f-PLA2, G2f-sPLA2
sPLA2-3	PLA2G3	Atypical sPLA2 G3-PLA2, G3-sPLA2
sPLA2-5	PLA2G5	Conventional sPLA2 without group-1 and -2 properties G5-PLA2, G5-sPLA2
sPLA2-10	PLA2G10	Conventional sPLA2 with group-1 and -2 properties G10-PLA2, G10-sPLA2
sPLA2-12a	PLA2G12A	Atypical sPLA2 G12a-PLA2, G12-sPLA2, ROSSY
sPLA2-12b	PLA2G12B	G12b-sPLA2, G13-PLA2, G13-sPLA2, FKSG71

require Ca^{++} ion. A secretory PLA2-binding protein that has a modest affinity for sPLA2-1b with respect to sPLA2-2a may possess a clearance function for circulating sPLA2 as well as signaling functions [58]. These enzymes operate in digestion, reproduction, skin homoeostasis, and host defense, as well as in inflammation, tissue injury, and atherosclerosis.

Family of Secreted Phospholipases-A2

Eleven $_s$PLA2s (secreted phospholipase-A2-1B, -2A, -2C–2F, -3, -5, -10, and -12A–12B) have been identified in mammals. They are subdivided into conventional $_s$PLA2 groups 1, 2, 5, and 10 and atypical groups 3 and 12. Subtype $_s$PLA2-2c is absent in humans (pseudogene) [62].

Effects of Secreted Phospholipases-A2

Secreted phospholipases-A2 can act as auto- or paracrine enzymes on plasma membranes to release not only arachidonic acid,[61] but also various types of saturated and mono- and polyunsaturated fatty acids, such as ω3-eicosapentaenoic acid (EPA) and docosahexaenoic acid (DHA), which are precursors of anti-inflammatory lipid mediators [62]. However, the action of $_s$PLA2s on lipid mediator synthesis depends on the coordinated activation of cPLA2α. The other products, i.e., lysophospholipids, such as lysophosphatidylcholine and lysophosphatidic acid, have their own activities.

In fact, in humans, group-2A and -10 PLA2 release arachidonic acid mainly during their secretion on the inside rather than on the outside of the cell membrane [60]. Moreover, they may promote the exocytosis of substances during their own release, in particular the degranulation of mastocytes.

Moreover, $_s$PLA2s can also act on phospholipids of extracellular vesicles, pulmonary surfactant,[62] and lipoproteins, as well as foreign phospholipids, such as dietary phospholipids and microbial membranes.

In mammals, G2A, G2D, and G5 PLA2s inhibit prothrombinase activity, as they bind to clotting factor Xa (potent anticoagulant activity) [60].

Receptors of Secreted Phospholipases-A2

In addition to their enzymatic action, $_s$PLA2s have a catalysis-independent function that relies on their ligand-like action, as they bind to numerous plasmalemmal

---

[61] Several $_s$PLA2s can liberate arachidonic acid from the plasma membrane, thereby leading to an augmented production of eicosanoids, with the following potency order [62]:

$$_sPLA2\text{-}10 > {_sPLA2\text{-}5} > {_sPLA2\text{-}3} > {_sPLA2\text{-}2f} > {_sPLA2\text{-}2a} > {_sPLA2\text{-}1b} > ...$$
$$..._sPLA2\text{-}2d > {_sPLA2\text{-}2e} > {_sPLA2\text{-}12a}.$$

Among $_s$PLA2s that release arachidonic acid, $_s$PLA2-3, -5, and -10 bind phosphatidylcholine, a major phospholipid of the outer leaflet of the plasma membrane. Group-2, AA-releasing $_s$PLA2s require cofactors such as heparan sulfate proteoglycans.

[62] Pulmonary surfactant synthesized by alveolar type-2 pneumocytes is aimed at lowering surface tension along the alveolar epithelium and maintaining alveolar stability during alveolus deformation responsible for breathing.

receptors and soluble interactors, especially the catalytically weak or inactive group-12 $_S$PLA2s, which are widespread.

Different types of plasmalemmal receptors exist for $_S$PLA2s, such as N- and M-type receptors [64]. N-Type receptors of $_S$PLA2s exist in the nervous system. M-Type receptors of $_S$PLA2s that belong to the C-type lectin subclass[63] are expressed in various tissues.

Receptor PLA2R1 (CLec13c) interacts with $_S$PLA2 with high to moderate affinity, thereby hindering enzymatic activity of $_S$PLA2s [62]. Furthermore, the clearance receptor PLA2R1 internalizes bound $_S$PLA2 into phagolysosomes, in which the enzyme is rapidly degraded. However, in humans, the PLA2R1–$_S$PLA2 interaction specificity is not conserved.[64]

Secreted PLA2s can tether not only to the PLA2R1 member of the mannose receptor family of C-type lectins, but also to glycosaminoglycans to possibly release cytokines, leukotrienes, and platelet-activating factor. In humans, $_S$PLA2 receptor transcripts can generate membrane-bound $_S$PLA2 receptors as well as secreted, soluble $_S$PLA2 receptors [65].

Secreted Phospholipases-A2 in the Cardiovascular Apparatus

Secreted phospholipase-A2-2A that requires calcium ion (millimolar concentration) for its enzymatic activity markedly increases in acute phase of inflammation, especially in cardiovascular diseases. Inflammatory cytokines, such as interleukin-1β and -6, interferon-γ, and tumor-necrosis factor-α, stimulate its synthesis in vascular smooth muscle cells and hepatocytes. Enzyme $_S$PLA2-2a hydrolyzes ester bonds in glyceroacyl phospholipids in cell membranes as well as both low- and high-density lipoproteins to release fatty acids and lysophospholipids. In addition, it contributes to an augmented production of interleukin-6 and cyclooxygenase-2 by lung macrophages [66] and mastocytes [67], respectively. Moreover, it stimulates differentiation of monocytes into dendritic cells [68].

Among the mammalian secretory $_S$PLA2s that hydrolyze lipoprotein phosphatidylcholine to lysophosphatidylcholine and free fatty acids in the arterial wall from the phosphatidylcholine-rich outer leaflet of the plasma membrane, $_S$PLA2-2f, -3,

---

[63]C-Type lectin CLec13c is the 180-kDa phospholipase-A2 receptor PLA2R1.
[64]Unlike neurotransmitters, hormones, growth factors, and cytokines, for which the ligand–receptor tethering is generally highly specific and well conserved among animal species, several classes of pattern recognition receptors also broadly recognize microbial components. In addition, PLA2R1 possesses only a short cytoplasmic region without known signaling module for connection with enzymes or adaptors, except a motif for endocytosis. Nevertheless, PLA2R1 may link to a signal-transmitting subunit to form a functional receptor complex. Upon binding with their ligands, some C-type lectins can cooperate with Toll-like receptors. In addition, PLA2R1 may be associated with the activation of MAPK and NFκB mediators. In addition, PLA2R1 can support senescence of human fibroblasts [62]. Production of $_S$PLA2-2a rises during cell senescence.

Table 2.28. Cytosolic, calcium-dependent types of phospholipase-A2 (Source: [58]). Group-4, cytosolic cPLA2s comprise calcium-dependent and -independent isoforms.

Type	Gene	Other names
cPLA2-4a	PLA2G4A	G4a-PLA2, cPLA2α
cPLA2-4b	PLA2G4B	G4b-PLA2, cPLA2β
cPLA2-4c	PLA2G4C	G4c-PLA2, cPLA2γ
cPLA2-4d	PLA2G4D	G4d-PLA2, cPLA2δ
cPLA2-4e	PLA2G4E	G4e-PLA2, cPLA2ε
cPLA2-4f	PLA2G4F	G4f-PLA2, cPLA2ζ

-5, and -10 have much more potent phosphatidylcholine-hydrolyzing activity than that of the others, including pro-inflammatory $_s$PLA2-2a, thereby contributing to the development of atherosclerosis [69].[65]

2.8.3.2 Cytosolic Phospholipase-A2

Cytosolic PLA2 types participate in cell signaling (Tables 2.28 and 2.29). Two categories of cytosolic PLA2s exist whether they depend on Ca^{++} ions (cPLA2), like secreted PLA2s (but are larger with different 3D structures than $_s$PLA2s), or not (Ca^{++}-independent PLA2s [iPLA2]).

Group-4, cytosolic phospholipase-A2 types (g4aPLA2–g4fPLA2 or cPLA2-4a–cPLA2-4f) hydrolyze membrane phospholipids to release arachidonic acid that is subsequently metabolized into eicosanoids. Lysophospholipids that are also produced can be converted into platelet-activating factor. They are activated by an increased intracellular Ca^{++} concentration and on phosphorylation.

Group-4 cPLA2 is constituted by: (1) calcium-independent PLA2s such as cPLA2-4c that is constitutively associated with membranes and functions for PGE2 production and (2) calcium-dependent cPLA2 such as cPLA2-4a enzyme. The first discovered cytosolic PLA2 was 85 kDa cPLA2-4a. It bears Ca^{++}-directed translocation to perinuclear membranes.

Calcium-dependent phospholipases-A2 intervene in excitotoxic functions in neurons, whereas $_s$PLA2-2a operate in response to inflammation in glial cells [63]. In addition, cPLA2 and $_s$PLA2 influence cell membrane properties and dynamics.

---

[65]Group-2A $_s$sPLA2 hydrolyzes phosphatidylcholine without discrimination of liberated fatty acids in disordered lipid phases with free cholesterol, sphingomyelin, and phosphatidylcholine, but at a lower rate than group-5 and -10 $_s$PLA2s. Group-5 and group-10 $_s$PLA2 isozymes release preferentially oligounsaturated (e.g., linoleate) and polyunsaturated (e.g., arachidonate) fatty acids from hydrolysis of phosphatidylcholine, respectively [70]. Accumulation of endogenous free cholesterol and sphingomyelin relative to phosphatidylcholine and the lipid phase state (ordered vs. disordered) affects the reaction rate of $_s$PLA2 hydrolysis of lipoproteins.

2.8 Phospholipases

Table 2.29. Cytosolic, calcium-independent types of phospholipase-A2 (Sources: [60, 62, 71]; ATGL: adipose triglyceride lipase; GS2: gene sequence-2; GS2L: GS2-like phospholipase; INAD: infantile neuroaxonal dystrophy; kREH: keratinocyte retinyl ester hydrolase; NBIA: neurodegeneration with brain iron accumulation; NRE: neuropathy target esterase-related esterase; NTE: neuropathy target esterase; NTEL1: NTE-like esterase-1; NTER1: liver NTE-related protein-1; Park: parkinsonism; PEDFR: pigment epithelium-derived factor [serpin-F1] receptor; PnPLA: patatin-like phospholipase-A2 domain-containing protein; TTS: transport-secretion protein). Membrane-bound phospholipase iPLA2-6b, a.k.a. calcium-independent phospholipase-A2γ and intracellular, membrane-associated, Ca^{++}-independent phospholipase-A2γ, possesses mitochondrial and peroxisomal localization signals. Phospholipases PnPLA6 and PnPLA7 contain a transmembrane domain.

Type	Gene	Group	Other names
PnPLA1	PNPLA1	G6a1	iPLA2 (iPLA2α)
PnPLA2	PNPLA2	G6e	iPLA2ζ, ATGL, desnutrin, PEDFR, TTS2, TTS2.2
PnPLA3	PNPLA3	G6d	iPLA2ϵ, adiponutrin (Adpn), acylglycerol ^{O}acyltransferase
PnPLA4	PNPLA4	G6f	iPLA2η, GS2, kREH
PnPLA5	PNPLA5	G6g	GS2L
PnPLA6	PNPLA6	G6c	iPLA2δ, NTE
PnPLA7	PNPLA7	G6h	NRE, NTEL1, NTER1
PnPLA8, iPLA2-6b	PNPLA8	G6b	iPLA2γ
PnPLA9, iPLA2-6a	PLA2G6	G6a2	iPLA2β, INAD1, NBIA2a, NBIA2b, Park14

Intracellular Calcium-Independent Phospholipases-A2 – Patatin-like Phospholipases

Patatins constitute a family of soluble glycoproteins of mature potato tubers. These storage proteins operate as non-specific lipid acyl hydrolases that have the serine lipase Gly-X-Ser-X-Gly motif. Members of the family of patatins cleave fatty acids from membrane lipids and act as defense against plant parasites.

Nine patatin-like phospholipases (PnPLA1–PnPLA9) characterized by a catalytic Ser–Asp dyad are encoded by the human genome [72, 73]. Mammalian PnPLAs target diverse substrates, such as triacylglycerols, phospholipids, and retinol esters.[66]

In humans, patatin-like hydrolases that do not have phospholipase activity participate in the regulation of adipocyte differentiation. In human preadipocytes, several members are indeed differentially regulated during cell differentiation [72]. The expression of PnPLA2, PnPLA3, and PnPLA8 is upregulated during adipocyte differentiation.

---

[66]The set of lipid hydrolases that cut ester and amide bonds of fatty acids of lipids includes triacylglycerol hydrolases (lipases), phospholipases, ceraminidases, and cholesterol ester and retinol ester hydrolases.

Three patatin-like proteins (PnPLA2–PnPLA4) link to the plasma membrane. They have catabolic (triacylglycerol) and anabolic (transacylation) activities [72]. Isoforms PnPLA4, PnPLA7, and PnPLA8 are widespread. Structure, enzymatic activity, or function of PnPLA1 have not been deeply investigated.

PnPLA2

Subtype PnPLA2 acts coordinately with hormone-sensitive lipase in the catabolism of triglycerides. The highest expression levels of this widely distributed protein are found in white and brown adipose tissue [73]. In the cell, it localizes to lipid droplets, where it catalyzes the initial step in triacylglycerol hydrolysis, thereby yielding diacylglycerol and fatty acids.

The activator of PnPLA2, Comparative gene identification CGI58, or α/β-hydrolase domain-containing protein α/βHD5, enhances PnPLA2-mediated triacylglycerol hydrolysis without affecting hormone-sensitive lipase activity.[67] This activator binds to perilipin, a lipid droplet protein. When protein kinase-A phosphorylates perilipin, CGI58 dissociates from perilipin and colocalizes with PnPLA2 [73].

Subtypes PnPLA2, PnPLA3, and PnPLA4 possess high triacylglycerol lipase and acylglycerol transacylase activities, in addition to PLA2 activity [60].

PnPLA3

Patatin-like phospholipase-A domain-containing protein PnPLA3, or adiponutrin, localizes to cellular membranes of adipocytes. It serves as an indicator of the nutritional state. Its concentration rises during adipocyte differentiation as well as upon insulin stimulation, but lowers during fasting [73]. In vitro, human PnPLA3 has triacylglycerol hydrolase, acylglycerol transacylase, and modest PLA2 activity. In vivo, these activities are minor.

PnPLA4

Despite its wide distribution, PnPLA4 is predominantly expressed in both muscle, heart, and adipose tissue. It may function as acylglycerol and retinol transacylase, triacylglycerol hydrolase, and PLA2 [73].

---

[67] Mutations in both PNPLA2 and ABHD5 genes cause neutral lipid storage disease characterized by the accumulation of triacylglycerol in tissues. Mutations in the ATGL gene provoke neutral lipid storage disease with myopathy.

PnPLA5

Isoform PnPLA5 is produced at low levels in the brain, lung, brown and white adipose tissues, and hypophysis (pituitary gland). In vitro, it may operate as a triacylglycerol hydrolase [73]. Like PnPLA3, it is regulated by the nutritional status.

PnPLA6

Patatin-like phospholipase-A domain-containing protein PnPLA6 is an essential lysophospholipase in the brain [73].[68]

It localizes to the endoplasmic reticulum and Golgi body. In vitro, it acts as an esterase and hydrolase against membrane lipids, preferentially lysophospholipids.

In lymphocytes, PnPLA6 hydrolyzes numerous lysophospholipids and is involved in membrane trafficking [72]. Both PnPLA6 and PnPLA7 are integral membrane proteins that can be regulated by cAMP or cGMP nucleotides [72].

PnPLA7

Subtype PnPLA7 is an insulin-regulated lysophospholipase in muscle and adipose tissue [73]. In humans, PnPLA7 is primarily synthesized in white adipose tissue, pancreas, and prostate. Its expression is strongly induced by fasting.

PnPLA8

Subtype PnPLA8 is a myocardial phospholipase maintaining mitochondrial integrity [73]. Its expression is prominent in the human heart, but also, to a lower extent, in other tissues. It localizes to mitochondria and peroxisomes. It exhibits both PLA1 and PLA2 activity for saturated or monounsaturated fatty acids of phospholipids.

PnPLA9

Subtype PnPLA9 is activated during apoptosis [72]. It hydrolyzes (as an acetyl hydrolase) platelet-activating factor. It mainly localizes to the cytoplasm; upon stimulation, it translocates to membranes of the perinuclear area [73].

---

[68] This enzyme is targeted by toxic organophosphates. These compounds cause organophosphate-induced delayed neuropathy characterized by degeneration of long axons in the spinal cord and peripheral nerves. In humans, mutations in the PNPLA6 gene cause a motor neuron disease characterized by progressive spastic paraplegia and distal muscle wasting.

It may intervene in the differentiation of bone marrow stromal cells, as, in the absence of PnPLA9, undifferentiated bone marrow stromal cells evolve toward adipogenesis rather than osteogenesis [73].

Subtype PnPLA9 operates in phospholipid processing, eicosanoid formation upon arachidonic acid release, protein expression, acetylcholine-mediated endothelium-dependent relaxation of the vasculature, secretion, and lymphocyte proliferation, in addition to apoptosis [60].

Intracellular Calcium-Dependent Phospholipases-A2

Enzyme cPLA2 was first detected in platelets and macrophages. The cPLA2 family, or group-4 PLA2, includes at least 6 isoforms. Subtype cPLA2α is the most ubiquitous enzyme in this group. Its regulation involves subcellular localization, intracellular Ca^{++} content (Ca^{++}-dependent lipid binding), phosphorylation, proteic interaction, and cleavage. Activated cPLA2 specifically cleaves the acyl ester bond of phosphatidylcholine, thereby generating lysophosphatidylcholine and release free fatty acid such as arachidonic acid.

In non-neural cells, cPLA2 targets preferentially phospholipids in endoplasmic reticulum and Golgi membranes [63]. It can also process other subcellular membranes, such as mitochondrial, nuclear, and plasma membranes. Processing of membrane phospholipids changes physical properties of cellular membranes. Enzyme PLA2 decreases the liquid-disordered and increases the liquid-ordered phase [63].

In neurons and glial cells, intracellular Ca^{++}-dependent phospholipases-A2 (cPLA2) are coupled to Ca^{++}-mobilizing receptors and oxidative signaling [63]. In astrocytes, cPLA2 can be activated via G-protein-coupled $P2Y_2$ receptor that stimulates MAPK and PKC kinases, as well as angiotensin-2 AT_1 receptor. In neurons, activated cPLA2 stimulates the NMDA-type glutamate receptor and its signaling via NADPH oxidase that resides in particular in synapses, hence the production of reactive oxygen species. The latter activate the signaling axis that causes phosphorylation of extracellular signal-regulated kinases. Reactive oxygen species may modulate synaptic function via cPLA2 and the release of arachidonic acid preferentially from phosphatidylcholine. Arachidonic acid can inhibit both pre- and postsynaptic potassium channels [63].

Activation of neuronal receptors can stimulate cPLA2 though Ca^{++} influx from the extracellular space or via the coupling to Gi-associated Gβγ subunit [61]. On the other hand, Ca^{++} ion attenuates iPLA2β activity that promotes iPLA2β–calmodulin interaction, which impedes iPLA2 activity. In unstimulated cells, SERCA pump works continuously to fill the endoplasmic reticulum with Ca^{++} ions and calmodulin inhibits iPLA2 protein. When receptors of the neuronal plasma membrane (purinergic, muscarinic, or metabotropic glutamatergic receptors) are stimulated, Ca^{++} can be released from the endoplasmic reticulum through inositol trisphosphate and ryanodine receptors (Ca^{++}-induced Ca^{++} release).

cPLA2-4a – Phospholipase-A2α

Cytosolic, Ca^{++}-binding, phospholipase-A2α contributes to limit maladaptive hypertrophy of cardiomyocytes that results from pressure overload [43].[69] Elevated intracellular Ca^{++} concentration causes translocation of PLA2α to cellular membranes. Calmodulin-dependent kinase CamK2 and mitogen-activated protein kinase phosphorylate PLA2α to modulate the release of arachidonic acid [43]. In addition, PLA2α precludes activation of adenylate cyclase and protein kinase-A induced by β2-adrenoceptors.

The cPLA2α activity is regulated by phosphorylation and nitrosylation. Enzyme cPLA2α is phosphorylated by extracellular signal-regulated kinases ERK1 and ERK2, P38MAPK, Ca^{++}–calmodulin-dependent protein kinase CamK2, and MAPK-interacting kinase MNK1 [74]. Messengers ATP and UTP stimulate cPLA2α phosphorylation via PKC-dependent and -independent ERK pathways. In noradrenaline-stimulated vascular smooth muscles cells, phosphorylation of cPLA2α (Ser515) by CamK2 is required for the phosphorylation by ERK1 and ERK2 (Ser505). Enzyme cPLA2α can colocalize with cyclooxygenase-2 in the plasma membrane [74]. In macrophages, production of prostaglandins via the cPLA2–COx2 axis regulates expression of the Nos2 gene. In human epithelial cells, nitric oxide produced by inducible nitric oxide synthase (NOS2) enhances cPLA2α activity via S-nitrosylation.

cPLA2-4b – Phospholipase-A2β

In the myocardium, the predominant PLA2 isoform is Ca^{++}-independent PLA2β that possesses calmodulin- and ATP-binding domains, hence is sensitive to local Ca^{++} concentration and energetic status [43]. Ca^{++}–calmodulin inhibits PLA2β. Enzyme PLA2β also hydrolyzes fatty acylCoA.[70] Hypoxia primes PLA2β activity in cardiomyocytes.

cPLA2-4c – Phospholipase-A2γ

A second major PLA2 isoform in myocardium corresponds to Ca^{++}-independent PLA2γ. Multiple splice variants of PLA2γ have been detected. Enzyme PLA2γ sequentially hydrolyzes diacyl, arachidonate-containing phospholipids [43]. Afterward, PLA2α or lysophospholipase hydrolyzes 2-arachidonoyl lysophosphatidylcholine to generate a large amount of arachidonic acid for eicosanoid production

---

[69] Phospholipase-A2α antagonizes insulin growth factor 1-mediated phosphorylation of PKB, GSK3β, and P38MAPK. It indeed allows PDK1 recruitment and PKCζ activation that prevent IGF1-mediated hypertrophic growth.

[70] AcylCoA accumulates quickly and strongly during myocardial ischemia. AcylCoA acts as both a substrate for and regulator of PLA2β, because acylCoA relieves calmodulin inhibition of PLA2β activity.

in the myocardium. On the other hand, 2-arachidonoyl lysophosphatidylcholine is targeted by the nucleotide pyrophosphatase–phosphodiesterase to generate the endocannabinoid 2-arachidonoyl glycerol, a substrate for cyclooxygenase-2 (but not COx1), and 12- and 15-lipoxygenases.

cPLA2-4d – Phospholipase-A2δ

The PLA2G4D mRNA can be identified in the placenta [75]. Upon stimulation, cPLA2δ that localizes in the cytoplasm in the resting state translocates to the perinuclear region, like cPLA2α, but the motion takes a longer time (5 mn instead of 1 mn for cPLA2α) [75].[71]

cPLA2-4e – Phospholipase-A2ε

A transcript of the PLA2G4E gene is observed predominantly in the heart, skeletal muscle, thyroid, and testis, and, at low expression levels, in the brain and stomach [75]. Enzyme cPLA2ε is partly associated with lysosomes, but neither with the endoplasmic reticulum, Golgi body, nor mitochondria. Stimulation does not cause redistribution of cPLA2ε and cPLA2ζ until at least 10 mn [75].[72]

cPLA2-4f – Phospholipase-A2ζ

The PLA2G4F mRNA is detected strongly in thyroid, moderately in stomach, and very weakly in large intestine and prostate [75].[73]

2.8.3.3 Platelet-Activating Factor Acid Hydrolases

Platelet-activating factor acid hydrolases (PAFAH), or lipoprotein-associated phospholipases-A2 (lpPLA2), are classified as [61]: g7aPLA2, or lipoprotein-associated PLA2;[74] g7bPLA2, or PAFAH2; g8aPLA2, or PAF1b; and g8bPLA2, which heterodimerizes with g8aPLA2 (Table 2.30).

---

[71] Enzyme cPLA2δ does not have a notable preference among 4 tested substrates (palmitoyl arachidonoyl-phosphatidylcholine [PAPC] and -phosphatidylethanolamine [PAPE] and palmitoyl linoleoyl-phosphatidylcholine [PLPC] and -phosphatidylethanolamine [PLPE]).

[72] Like cPLA2δ, cPLA2ε does not have a marked preference among the 4 tested substrates (footnote 71).

[73] Activity of cPLA2ζ for 4 tested substrates is much higher than that of cPLA2δ and cPLA2ε, but is weak for lysophosphatidylcholine [75] (footnote 71).

[74] A.k.a. plasma platelet-activating factor acetylhydrolase, 1-alkyl 2-acetylglycerophosphocholine esterase, and LDL-associated phospholipase-A2 (ldlPLA2 or lpPLA2).

2.8 Phospholipases

Table 2.30. Platelet-activating factor acid hydrolases and lysosomal group-15 phospholipase-A2 (Source: [60] LDL: low-density lipoprotein; lPLA2: lysosomal PLA2; lpPLA2: lipoprotein-associated PLA2; PL: phospholipid).

Group	Aliases	Features
G7a	lpPLA2	Secreted, plasma PAFAH, catalytic Ser–His–Asp triad, hydrolysis of short- and medium fatty acids from diacylglycerols and triacylglycerols, hydrolysis of oxidized PL in LDLs, PLA1 activity
G7b	PAFAH2	Intracellular, myristoylated
G8a	PAFAH1bα1	Intracellular, Ser–His–Asp triad, homodimer or g8aPLA2–g8bPLA2 heterodimer
G8b	PAFAH1bα2	Intracellular, Ser–His–Asp triad, homodimer or g8aPLA2–g8bPLA2 heterodimer
G15	lPLA2, LLPL	Ser–His–Asp triad, glycosylated

Enzyme PAFAH cleaves the acetyl group from the sn2 position of platelet-activating factor. Furthermore, this enzyme can cleave oxidized lipids in the sn2 position of 9-carbon-long chains [61]. Unlike other PLA2s, PAFAH access substrate in the aqueous phase in a Ca^{++}-independent fashion.

Type-1 PAFAH (PAFAH1b) is a complex of 2 catalytic subunits (PAFAH1bα1–PAFAH1bα2), a regulatory β-subunit, and a γ-subunit. In humans, these subunits are encoded by the PAFAH1B1 to PAFAH1B3 genes that produce platelet-activating factor acetylhydrolase isoform-1B subunit-1 to -3, or PAFAH α-, β-, and γ-subunit, respectively [76]. The catalytic subunits G8aPLA2 and G8bPLA2 form catalytically active homo- and heterodimers. The g8bPLA2–g8bPLA2 homodimer preferentially targets platelet-activating factor (1-^{O}alkyl 2-acetyl glycero 3-phosphocholine) and 1-^{O}alkyl 2-acetyl glycero 3-phosphoethanolamine; the g8aPLA2–g8aPLA2 homodimer and the g8aPLA2–g8bPLA2 heterodimer have a higher activity toward 1-^{O}alkyl 2-acetyl glycero 3-phosphatidic acid [60].

Type-2 PAFAH, or serine-dependent phospholipase-A2 (sdPLA2), is a single polypeptide encoded by the Pafah2 gene. It is myristoylated at its N-terminus. It resides both in the cytosol and membranes. It indeed translocates from the cytosol to membranes in the presence of oxidants and from membranes to the cytosol in the presence of anti-oxidants [60].

Plasma PAFAH, or lpPLA2 (g7aPLA2), has a structure close to that of PAFAH2. It associates with plasma lipoproteins. In the blood, it travels mainly with low-density lipoproteins; less than 20% is connected to high-density lipoproteins. It is produced by inflammatory cells. This secreted enzyme catalyzes the degradation of PAF to inactive products lysoPAF and acetate. It also hydrolyzes oxidized phospholipids in LDL particles, hence its name oxidized lipid lipoprotein-associated PLA2 subtype.

Table 2.31. Intracellular PLA2 isoforms in endothelial cells in humans (EC; Source: [77]). Group-2 sPLA2s are not detected in quiescent endothelial cells; thay are produced upon cytokine stimulation. Group-6A iPLA2s contribute to regulation of endothelial cell proliferation.

Endothelial cell type	Group(s)	PLA2 isoforms
Aortic ECs	4, 5	cPLA2, $_s$PLA2
Bone marrow ECs	2	$_s$PLA2
Brain microvascular ECs	4	cPLA2
Coronary artery ECs	2, 4–6	cPLA2, iPLA2, $_s$PLA2
Umbilical vein ECs	4–6	cPLA2, iPLA2, $_s$PLA2

2.8.3.4 Lysosomal Phospholipase-A2

Lysosomal PLA2, or group-15 PLA2, is a calcium-independent enzyme that has its maximal activity in an acidic medium (optimal enzymatic activity at pH 4.5), hence its other label acidic iPLA2 enzyme (Table 2.30). This enzyme was first named 1-^{O}acylceramide synthase (ACS). It possesses Ca^{++}-independent PLA2 and transacylase activities [60]. It preferentially hydrolyzes phosphatidylcholine and phosphatidylethanolamine.

2.8.3.5 Phospholipases-A2 in Endothelial Cells

In endothelial cells, phospholipases-A2 can be involved in blood–brain barrier function, adhesion and transmigration of inflammatory cells, and angiogenesis (Table 2.31). In the vasculature and blood, PLA2s are active in leukocytes, macrophages, platelets, smooth muscle cells, preadipocytes, and fibroblasts.

Kinases ERK1 and ERK2 phosphorylate (activate) cPLA2 concomitantly with Ca^{++} influx. The latter fosters cPLA2 translocation to cellular membranes (endoplasmic reticulum, Golgi body, and nuclear envelope) [77]. Ceramide 1-phosphate is an allosteric activator of cPLA2α. Protein kinase Cα phosphorylates (activates) either iPLA2 or its regulator. This kinase can be activated by diacylglycerol and the PI3K–PDK1–PKB axis.

Free arachidonic acid is oxidized to prostaglandins by cyclooxygenases COx1 and COx2,[75] hydroxyeicosanoic acid by lipoxygenases, or epoxyeicosatrienoic acids by P450-dependent epoxygenase. Lysophospholipid, the other product of PLA2, can generate the precursor of platelet-activating factor, another potent inflammatory mediator.

---

[75]Prostaglandins PGE2 and PGF2α stimulates in an auto- and paracrine manner endothelial cell proliferation via G-protein-coupled E prostanoid receptors (EP1–EP4).

2.8.4 Phospholipase-B

Phospholipase-B (PLB) attacks lysolecithin, releasing glycerylphosphorylcholine and fatty acid.[76] Certain phospholipase B types hydrolyze phosphatidylcholine, phosphatidylinositol, and phosphatidylethanolamine.

2.8.5 Phospholipase-C

Phospholipase-C (PLC) cleaves phosphatidylinositol (4,5)-bisphosphate at the plasma membrane to produce 2 second messengers: inositol (1,4,5)-trisphosphate and diacylglycerol. The latter remains bound to the membrane, where it can activate protein kinase-C to phosphorylate its substrates. The former is released as a soluble molecule into the cytosol, where it targets IP_3 receptors, particular those in the endoplasmic reticulum to raise the cytosolic concentration of calcium.

Messenger IP_3 can be phosphorylated by kinases to produce inositol hexakisphosphate IP_6, mono- and bispyrophosphorylated inositol phosphates $PPIP_5$ and $(PP)_2IP_4$. Agent $PPIP_5$ is implicated in gene expression and protein phosphorylation, among other functions. Inositol (1,3,4,5)-tetrakisphosphate (IP_4) is the polar head group of phosphatidylinositol (3,4,5)-trisphosphate, which allows the PIP_3 effector to distinguish PIP_3 from its PIP_2 precursor. Phosphorylation of IP_3 into IP_4 promotes PLCγ phosphorylation [17].

Thirteen types of mammalian phospholipases-C are classified into 6 groups (PLCβ–PLCη) according to the structure and activation mechanisms. Subtype-specific domains indeed contribute to specific regulations. Each isoform type (PLCβ1–PLCβ4, PLCγ1–PLCγ2, PLCδ1–PLCδ3, PLCε1, PLCζ1, or PLCη1–PLCη2) require calcium for catalysis.

Isoform PLCβ is activated by G proteins of the Gq subclass (Gq, G11, and G14–G16) as well as Gβγ liberated from Gi subunit; PLCγ by receptor Tyr kinases,[77] PLCδ by high calcium levels; and PLCε by small GTPase Ras;[78] PLCη has a high sensitivity to the calcium concentration (PLCη2 being activated at calcium level $O[10\,nmol]$). Sperm PLCζ can cause a large, transient increase in free Ca^{++} concentration in an ovum (or unfertilized oocyte) upon encountering a spermatozoon (i.e., a motile sperm cell). Activity of isozymes PLCβ and PLCγ are stimulated by bradykinin and platelet-derived growth factor, respectively.

Isoform PLC-β1 is highly expressed in the central nervous system; PLC-β2 slightly in hematopoietic cells (but at greater value than in other tissues); PLCβ3

---

[76]Phospholipase-B is also known as lysolecithin acyl hydrolase or lysolecithinase.

[77]Various growth factors, such as platelet-derived, epidermal, and nerve growth factor can activate PLCγ isozyme.

[78]Isozyme PLCε can be activated by both activated GPCRs and RTKs with distinct activation mechanism.

Table 2.32. Tissue distribution of PLC isoforms in the cardiovascular and ventilatory systems as well as related organs (Source: [78]).

Type	Localization
PLCβ1	Brain, nerve, kidney, lung, adipose tissue
PLCβ2	Brain, ganglia, kidney, lung, blood, bone marrow, lymph node
PLCβ3	Brain, nerve, kidney, lymph node, adipose tissue
PLCβ4	Brain, nerve, ganglia, heart, blood, kidney, lung
PLCγ1	Brain, ganglia, heart, blood, bone marrow, lymph node, kidney, lung, adipose tissue
PLCγ2	Brain, ganglia, nerve, heart, blood, bone marrow, lymph node, kidney, lung, adipose tissue
PLCδ1	Brain, ganglia, heart, blood, bone marrow, kidney, lung
PLCδ3	Brain, heart, blood, lymph node, kidney, lung
PLCδ4	Brain, heart, kidney, lung
PLCε	Brain, heart, kidney, lung, blood
PLCζ	Brain
PLCη1	Brain, lung
PLCη2	Brain, kidney, lung, lymph node

in the brain, liver, and parotid gland; PLCβ4 particularly in the cerebellum and retina [78] (Table 2.32). Subtype PLCγ1 abounds in the embryon cerebral cortex; PLCγ2 localizes primarily to hematopoietic cells. Isozyme PLCδ1 abounds in the brain, heart, lung, skeletal muscle, and testis and PLCδ3 in the brain, heart, and skeletal muscle; PLCδ4 resides especially in the brain, kidney, skeletal muscle, and testis. PLCε is detected in the brain, lung, and colon, with the highest expression in heart. PLCζ1 is only found in the testis. The highest level of PLCη1 is observed in the brain and kidney and, at smaller levels in the lung, spleen, intestine, thymus, and pancreas; PLCη2 is identified in the brain and intestine.

2.8.5.1 Phospholipase-Cβ

Phospholipases-Cβ constitute a family that includes PLCβ1 to PLCβ4. They possess an elongated C-terminus. These isoforms are regulated by heterotrimeric guanine nucleotide-binding (G) proteins. They have a high GTPase-stimulating activity. Dimer Gβγ interacts with PLCβ1 to PLCβ3, but exhibits a high affinity only for PLCβ2 [78]. In platelets, various signals activate PLCβ isozymes via Gq (but not Gβγ).

Many signaling cascades relies on inositol lipid hydrolysis catalyzed by phospholipase-Cβ[79] that is activated by Gq subunit of heterotrimeric G proteins, and, in turn,

---

[79] Phospholipase-C hydrolyzes phosphatidylinositol (4,5)-bisphosphate to the second messengers inositol (1,4,5)-trisphosphate and diacylglycerol under the action of many neurotransmitters, hormones, growth factors, among others.

inactivates Gq, thus limiting the signaling duration, a procedure that is required when signal transduction necessitates high amplification and rapid activation–deactivation cycles [79].[80]

Isoform PLCβ2 can be activated by $G\alpha_q$ and Gβγ subunits of heterotrimeric G proteins as well as small, monomeric guanosine triphosphatase Rac. These 3 activators use overlapping and distinct mechanisms to recruit PLCβ2 to the plasma membrane and activate signaling [80]. Activation of PLCβ2 by Gq protomer causes PLCβ2 association with fast-diffusing plasmalemmal constituents (PLCβ2 movement within the membrane similar to that of lipids), thereby searching for relatively localized substrates. PLCβ2 substrate sampling is carried out over midsize zone prior to dissociation. Stimulation of PLCβ2 by Gβγ dimer generates pure lateral diffusion 3- to 5-fold faster than lipids, hence transient linkage to the membrane that allow PLCβ2 to scan the plasma membrane over large regions to interact with dispersed substrates. Excitation of PLCβ2 by Rac2 induces slow lateral diffusion and much faster PLCβ2 interaction with slowly diffusing molecules in the membrane, hence creating spatially restricted signaling.

2.8.5.2 Phospholipase-Cγ

Phospholipase-Cγ hydrolyzes PI(4,5)P$_2$ into IP$_3$ and DAG messengers. Two mammalian PLCγ isoforms (PLCγ1 and PLCγ2) have been identified. They are activated by receptor Tyr kinases as well as other types of receptors. Isoform PLCγ1 can be stimulated by GPCRs, such as angiotensin-2 and bradykinin receptors, cytokine receptors, and immunoreceptors such as T-cell receptors [78]. Isozyme PLCγ2 is activated by immunoglobulin and adhesion receptors on B lymphocytes, platelets, and mastocytes.

During cell proliferation caused by growth factors, i.e., during phenotypic change from a differentiated contractile phenotype to a proliferative phenotype, marker genes of vascular smooth muscle cells are repressed, partly following translocation of extracellular signal-regulated kinases ERK1 and ERK2 to the nucleus that phosphorylate (activate) ELk1 transcription factor. In particular, upon stimulation by platelet-derived growth factor and subsequent activation of PLCγ1, ERK-binding, 15-kDa phosphoprotein enriched in astrocytes PEA15 is phosphorylated and, then, supports the nuclear translocation of ERK1 and ERK2 [81].[81]

---

[80] A first domain of PLCβ binds to activated Gq and a second domain accelerates guanosine triphosphate hydrolysis by Gq, thereby causing the dissociation of the signaling complex and termination of signal propagation.

[81] Protein PEA15 is an ERK-binding partner that acts as a cytoplasmic anchor to prevent ERK1/2 nuclear translocation. Its phosphorylation releases ERK1 and ERK2, thereby enabling ERK translocation to the nucleus and subsequent activation of ELk1 factor.

In humans, naive T lymphocytes[82] do not express vitamin-D receptor. Moreover, they produce a very low amount of phospholipase-Cγ1, thereby displaying a low T-cell antigen receptor responsiveness. They are thus not able to deliver a specific antigenic response. Vitamin-D controls T-cell receptor signaling that primes the synthesis of vitamin-D receptor via mitogen-activated protein kinase P38MAPK, upregulates PLCγ1 expression (~75 fold) and causes a greater TCR responsiveness required for subsequent classical T-lymphocyte activation [82].

2.8.5.3 Phospholipase-Cδ

Three isozymes have been identified in the PLCδ family (PLCδ1, PLCδ3, and PLCδ4). Phosphatidylserine activates PLCδ1, but not PLCδ4 [83]. PLCδ1 and PLCδ4 have similar activities and dependence on Ca^{++} cofactor. Although all PLC isozymes require calcium for activity, PLCδ isozymes are among the most sensitive to calcium [78].

Rho GTPase-activating protein RhoGAP7 (or P122RhoGAP) binds and activates PLCδ1 to suppress the formation of stress fibers and focal adhesions [78]. In addition, PLCδ1 shuttles between the cytoplasm and nucleus. Isozyme PLCδ1 is involved in osmotic response.

Differences exist between the C2 domain of PLCδ4 and those of PLCδ1 and PLCδ3. The C2 domain of the former that has Ca^{++}-independent membrane affinity allows localization to various cellular membranes, including the nuclear envelope and plasma membrane and engages in between-protein interaction [83]. It interacts with glutamate receptor-interacting protein GRIP1. The C2 domain of PLCδ1 and PLCδ3 support linkage specifically to the plasma membrane with a strong Ca^{++}-dependent membrane affinity.

In the regenerating liver, PLCδ4 is detected in the nucleus during the S phase of the cell cycle, whereas PLCδ1 mostly resides in the cytosol [83].

Phospholipase-Cδ4 is expressed in many tissues. Its transcription is primed by growth factors as well as bradykinin and lysophosphatidic acid. Unlike other PLCδ isozymes, PLCδ4 has 3 alternative splice variants ($PLCδ4_{alt1}$–$PLCδ4_{alt3}$) [83]. Variant $PLCδ4_{alt3}$ is catalytically inactive. It may act as an inhibitor of PLC activity. The alternative splice variants $PLCδ4_{alt1}$ and $PLCδ4_{alt2}$ are expressed mainly in the testis and $PLCδ4_{alt3}$ primarily in some neuron types.

---

[82]T lymphocyte is an abbreviation for thymic lymphocytes, as thymus is the organ where these cells mature. T lymphocytes can be distinguished from other lymphocyte types, such as antigen-presenting B lymphocytes (B: bone marrow) and natural killer cells by the presence of special plasmalemmal receptors, the T-cell receptors. Following T-cell development, matured, naive T lymphocytes (that have never been exposed to any antigen) leave the thymus and begin to spread throughout the body, especially the lymph nodes. Activated T lymphocytes can then mature into natural killer T cells, CD4+ or CD8+ memory T cells, CD4+ regulatory T cells, CD4+ T helper cells, and CD8+ cytotoxic T cells.

2.8.5.4 Phospholipase-Cε

Phospholipase-Cε (or 1-phosphatidylinositol (4,5)-bisphosphate phosphodiesterase-ε1) is the largest identified PLC isozyme. Small GTPases hRas, Rap (Rap1a and Rap2b upon stimulation by growth factor RTKs and GPCRs, respectively [78]), and Rho activates PLCε isozyme. Subtype PLCε is activated by Gβγ subunit released upon activation by $G\alpha_{12/13}$-, but not $G\alpha_q$-coupled receptors [84].

Dermal fibroblasts and epidermal keratinocytes expressed pro-inflammatory cytokines via PLCε once stimulated with cytokines (IL4, IL17, Ifnγ, and TNFα) released by infiltrated, CD4+ T lymphocytes [85].

Two spliced variants have been identified (PLCε1a–PLCε1b). The PLCε1a transcript is expressed in various tissues, but not in blood leukocytes; PLCε1b has a limited expression in the lung, spleen, and placenta.

2.8.5.5 Phospholipase-Cη

Phospholipase-Cη has been identified as a sperm-specific isozyme. It contains a nuclear localization signal that promotes its accumulation in the nucleus [78]. On the other hand, phospholipase-Cη and adaptor B-cell linker cause preB-cell differentiation.

The family of PLCη consists of 2 enzymes (PLCη1 and PLCη2). They are more sensitive to Ca^{++} than other PLC isozymes. They mediate G-protein-coupled receptor signaling. Both enzymes are expressed in neurons and neuroendocrine cell lines [86].

2.8.5.6 Phospholipase-Cζ

Two PLCζ isozymes (PLCζ1 and PLCζ2) exist in humans. Both PLCζ isoforms are highly expressed in cerebral neurons. They may then be involved in the regulation of neural or neuroendocrine systems [78].

Isozyme PLCζ1, the smallest known mammalian PLC isotype, elicits Ca^{++} oscillations following phosphatidylinositol (4,5)-bisphosphate hydrolysis and activation of embryogenesis. It is approximately 100 times more sensitive to Ca^{++} than the closely related PLCδ1 isoform [87].

2.8.6 Phospholipase-D

Phospholipase-D hydrolyzes phosphatidylcholine and releases phosphatidic acid. Phosphatidic acid controls various cellular functions, such as cell proliferation and survival via signaling mediated by Son of sevenless and target of rapamycin, respectively, as well as cell migration. Phosphatidic acid is hydrolyzed into diacylglycerol by phosphatidic acid phosphomonoesterase.[83]

Phosphatidic acid activates some guanine nucleotide-exchange factors such as Dedicator of cytokinesis-2 (DOCK2). Activator DOCK2 is a RacGEF that mediates neutrophil chemotaxis. Upon PIP_3 formation, DOCK2 is recruited to the plasma membrane. Subsequently, PLD is activated and produces phosphatidic acid. Moreover, phosphatidic acid binds directly to SOS protein.

Phospholipase-D also concurs to Rac recruitment to the plasma membrane. It binds to GRB2 that recruits Son of sevenless. In addition, phosphatidylinositol triphosphate helps to unmask the GEF activity of Son of sevenless for Rac.

Platelets contain protein kinase-C and phospholipase-C and -D. Phospholipase-D is rapidly activated once platelets are stimulated to aggregate and secrete their materials. Phospholipase-D activation requires protein kinase-C, but not $\alpha_{2B}\beta_3$-integrin and fibrinogen binding [88]. Yet, both $\alpha_{2B}\beta_3$-integrin and fibrinogen are necessary for full platelet aggregation. The main adhesion receptor, $\alpha_{2B}\beta_3$-integrin, can shift from a low- to high-affinity state for its ligands to enable platelet adhesion and aggregation. In response to stimulation of G-protein-coupled- and immunoreceptors, phospholipases cleave membrane phospholipids to generate soluble second messengers. In adherent platelets, whereas phosphatidylinositol 3-kinase activation is controlled by both $\alpha_{2B}\beta_3$-integrin engagement and released ADP, phospholipase-C is stimulated only by $\alpha_{2B}\beta_3$-integrin–fibrinogen interaction [89].

Two PLD isoforms exist (PLD1–PLD2). In addition to phospholipase-C, phospholipase-D1 and its product phosphatidic acid intervene in $\alpha_{2B}\beta_3$-integrin activation in response to agonists and platelet aggregation under high shear mediated by glycoprotein-1B [90]. Glycoprotein-1B serves as a receptor for von Willebrand factor and a component of the GP1b–GP5–GP9 complex on platelets.[84]

Phospholipase-D2 binds directly to Rac2 GTPase, a regulator of actin cytoskeletal remodeling, and functions as a guanine nucleotide-exchange factor [91]. In leukocytes, PLD2 thus intervene in cell polarization and adhesion, chemotaxis, and phagocytosis.

2.9 Phosphoinositide Kinases

Phosphatidylinositol phosphate kinases (PIPK) phosphorylate 3 hydroxyl groups (3–5) of phosphatidylinositol among 5 because of hindrance. These kinases phosphorylate PIP and PIP_2 phosphoinositides. The set of PIPKs is subdivided into 3 groups with significant sequence homology, but distinct substrate specificities, subcellular locations, and functions.

---

[83]Phosphatidic acid and diacylglycerol are interconvertible owing to phosphatidic acid phosphohydrolase and diacylglycerol kinase. Fixation of serine, choline, ethanolamine, or inositol on phosphatidic acid leads to phosphatidylserine, phosphatidylcholine, phosphatidylthanolamine, or phosphatidylinositol.

[84]Glycoproteins GP1b, GP5, and GP9 are also called CD42, CD42d, and CD42a, respectively.

Kinases and phosphatases that regulate phosphoinositide pathways participate in cell fate. Both PIP_2 and PIP_3 control actin polymerization. In particular, phosphoinositides and phosphoinositide kinases and phosphatases are involved in cell transport (Vol. 1 – Chap. 9. Intracellular Transport). Moreover, plasmalemmal gradients of PIP_3 organized by phosphoinositide kinases and phosphatases are achieved during cell migration (Vol. 2 – Chap. 6. Cell Motility).

2.9.1 Phosphatidylinositol 3-Kinases

Phosphoinositide 3-kinases (PI3K) phosphorylate different phosphoinositides to produce $PI(3)P$, $PI(3,4)P_2$, $PI(3,5)P_2$, and $PI(3,4,5)P_3$ once stimulated by their activators.[85] They operate early in intracellular signal transduction. They contribute to the recruitment and activation of several protein kinases, such as phosphoinositide-dependent protein kinase PDK1 that phosphorylates (activates) AGC protein kinases (Sect. 5.2), especially protein kinase-B (Table 2.33). Second messenger $PI(3,4,5)P_3$ interacts with kinases of the TEC family (Sect. 4.12.12), GTPase-activating proteins, guanine nucleotide-exchange factors, and scaffold proteins. In addition to their lipid kinase activity, several PI3Ks also have protein kinase activity.

Heterodimeric PI3K contains a subunit that has enzymatic activity. The superclass of phosphatidylinositol 3-kinases includes 8 PI3K isoforms divided into 5 categories (1A–1B and 2–4) and regrouped into 3 classes (C1–C3). PI3K-related kinases are actually grouped in a class 4 that includes target of rapamycin (TOR), ataxia telangiectasia mutated (ATMK), ataxia telangiectasia mutated-related (ATRK), and DNA-dependent (DNAPK) protein kinase.

Signaling via PI3K is implicated in metabolic control, actin remodeling, intracellular vesicular transport, and cell growth, survival, proliferation, and migration, hence, immunity, angiogenesis, and cardiovascular homeostasis. Furthermore, it is one of the most frequently deregulated pathways in cancer.

---

[85] Phosphoinositide 3-kinases phosphorylate the 3-hydroxyl group of the inositol ring of 3 species of phosphatidylinositol substrates; PI, PI(4)P, and $PI(4,5)P_2$ lipids. A fourth 3-phosphoinositide species $PI(3,5)P_2$ is generated by 5-phosphorylation of PI(3)P by phosphatidylinositol 3-phosphate 5-kinase (PIP5K and PIP5K3; a.k.a. FYVE finger-containing phosphoinositide kinase [PIKfyve] and yeast formation of aploid and binucleate cells FAB1) [94]. Lipid $PI(3,5)P_2$ then interacts with 4 mammalian WD40 repeat protein interacting with phosphoinositides (WIPI) proteins. Kinase PIP5K3 is involved in endosomal and/or lysosomal transfer. Resulting 3-phosphoinositides coordinate the localization and function of multiple effector proteins (kinases, adaptors, guanine nucleotide-exchange factors, and GTPase-activating proteins) that bind these lipids through specific lipid-binding domains (pleckstrin homology [PH], phox homology [PX], and FYVE domain).

Table 2.33. Targets of antagonists phosphatidylinositol 3-kinase (PI3K) and phosphatase and tensin homolog deleted on chromosome 10 (phosphatidylinositol 3-phosphatase [PTen]). Phosphoinositide 3-kinases phosphorylate PI, PI(4)P, and PI(4,5)P_2 to produce PI(3)P, PI(3,4)P_2, and PI(3,4,5)P_3. Lipid PI(3,5)P_2 is synthesized by PIP5K3 (PIKfyve). These lipidic mediators interact with multiple proteic effectors (kinases, adaptors, guanine nucleotide-exchange factors, and GTPase-activating proteins). Lipid phosphatases (PTen, inositol polyphosphate 4-phosphatase INPP4, phosphatidylinositol polyphosphate 5-phosphatase INPP5e, SH2 domain-containing inositol 5-phosphatase SHIP2, myotubularin phosphatases, a dual phosphatidylinositol 3-phosphatase and (3,5)-bisphosphatase) antagonize PI3K kinases (BAD: BCL2 antagonist of cell death; BTK: Bruton Tyr kinase; eIF4eBP1: eukaryotic translation initiation factor-4e-binding protein; FoxO: type-O forkhead box transcription factor; IKK: IκB kinase; GSK: glycogen synthase kinase; GyS: glycogen synthase; ITK: interleukin-2-inducible T-cell kinase; NFAT: nuclear factor of activated T cells; NOS: nitric oxide synthase; PDK: phosphoinositide-dependent kinase; PFK: phosphofructokinase; PKB(C): protein kinase-B(C); PKB1S1: PKB1 substrate-1; RLK: resting lymphocyte kinase; SGK: serum and glucocorticoid-regulated kinase S6K: p70 ribosomal S6 kinase; TOR: target of rapamycin; TEC: Tyr kinase expressed in hepatocellular carcinoma; (+): activation; (−): inactivation).

Target	Activity
PDK1	Phosphorylation (+) of PKB, PKC, SGK1, S6K
GSK3	Phosphorylation (−) of β-catenin, cyclin-D, GyS, Jun, NFAT
PKB	Phosphorylation of BAD, FoxO3α, DM2, Rho, PKB1S1, cRaf, NOS3, IKK, GSK3β (−), PFK2, PKC, TOR, TEC (TEC, BTK, ITK, RLK)
TOR	Phosphorylation of eIF4eBP1 (−), S6K (+)

2.9.1.1 PI3K Classes

Phosphatidylinositol 3-kinases correspond to main nodes in growth factor signaling from receptor Tyr kinases to effectors. They activate distinct signaling pathways that govern various cellular processes, such as glucose homeostasis and protein synthesis, in addition to cell fate.

The superclass of 8 PI3K isoforms is subdivided into 3 functional classes based on protein domain structure, lipid substrate specificity, and associated regulatory subunits (Table 2.34): class-1 PI3K enzymes with catalytic subunits $PI3K_{c1\alpha}$ to $PI3K_{c1\delta}$; class-2 PI3K enzymes ($PI3K_{c2\alpha}$–$PI3K_{c2\gamma}$); and class-3 PI3K enzyme that corresponds to the single, ubiquitous, Ca^{++}–calmodulin-activated vacuolar protein sorting VPS34 that connects to regulatory VPS15 (or $PI3K_{r4}$) subunit. It can be better named type-3 $PI3K_{c3}$ catalytic subunit.

Some PI3Ks are ubiquitous or widespread ($PI3K_{c1\alpha}$, $PI3K_{c1\beta}$, $PI3K_{c2\alpha}$, $PI3K_{c2\beta}$, and $PI3K_{c3}$), whereas others have a restricted distribution ($PI3K_{c1\gamma}$, $PI3K_{c1\delta}$, and $PI3K_{c2\gamma}$; Table 2.35).

Three genes encode regulatory subunits $PI3K_{r1}$ to $PI3K_{r3}$ [93]. The PIK3R1 gene encodes via distinct promoter usage not only $^{P85\alpha}PI3K_{r1}$, but also $^{P55\alpha}PI3K_{r1}$ and $^{P50\alpha}PI3K_{r1}$ [94].

2.9 Phosphoinositide Kinases

Table 2.34. Classes and types of phosphoinositide 3-kinases and their genes. The 4 class-1 PI3K isoforms (PI3K$_{c1\alpha}$–PI3K$_{c1\gamma}$) are synthesized in all mammalian cell types at various concentrations. Class-2 PI3Ks do not have regulatory subunits.

Gene	Subunit type	Name	Other alias
	Class-1A PI3K subunits		
PIK3CA	Catalytic	PI3K$_{c1\alpha}$	P110α, PI3Kα
PIK3CB	Catalytic	PI3K$_{c1\beta}$,	P110β, PI3Kβ
PIK3CD	Catalytic	PI3K$_{c1\delta}$	P110δ, PI3Kδ
PIK3R1	Regulatory	PI3K$_{r1}$	P85α, P55α, P50α
PIK3R2	Regulatory	PI3K$_{r2}$	P85β
PIK3R3	Regulatory	PI3K$_{r3}$	P55γ
	Class-1B PI3K subunits		
PIK3CG	Catalytic	PI3K$_{c1\gamma}$	P110γ, PI3Kγ
PIK3R5	Regulatory	PI3K$_{r5}$	P101
PIK3R6	Regulatory	PI3K$_{r6}$	P87 (P84)
	Class-2 PI3K subunits		
PIK3C2A	Catalytic	PI3K$_{c2\alpha}$	
PIK3C2B	Catalytic	PI3K$_{c2\beta}$	
PIK3C2G	Catalytic	PI3K$_{c2\gamma}$	
	Class-3 PI3K subunits		
PIK3C3	Catalytic	PI3K$_{c3}$	VPS34
PIK3R4	Regulatory	PI3K$_{r4}$	VPS15 (P150)
	Class-4 PI3K-related kinases		
	Target of rapamycin kinase (TOR		
	Ataxia telangiectasia mutated kinase (ATMK)		
	Ataxia telangiectasia mutated-related kinase (ATRK)		
	DNA-Dependent protein kinase (DNAPK)		

Class-1 PI3K Enzymes

Class-1 PI3Ks are involved in cell size. All class-1 PI3Ks are able to phosphorylate PI to PI(3)P, PI(4)P to PI(3,4)P$_2$, and PI(4,5)P$_2$ to PI(3,4,5)P$_3$, but PI(4,5)P$_2$ is the preferred lipid substrate in vivo (Table 2.36).

In resting cells, class-1 PI3Ks reside mainly in the cytosol. Upon stimulation, they are recruited to cellular membranes via interactions with receptors or adaptors. Once receptors bind to cognate ligands, class-1 PI3Ks mainly generate PI(3,4,5)P$_3$ that recruits and activates various effectors, such as protein kinase-B, 3-phosphoinositide-dependent kinase PDK1, monomeric GTPases. Therefore, numerous targets are phosphorylated, such as glycogen synthase kinase GSK3β, target of rapamycin, S6 ribosomal kinase (^{P70}RSK or S6K), nitric oxide synthase NOS3, and several anti-apoptotic effectors.

Table 2.35. Distribution and effects of PI3Ks (Source: [92]). Both PI3K$_{c1\delta}$ (class 1A) and PI3K$_{c1\gamma}$ (class 1B) are synthesized predominantly (but not exclusively) in leukocytes.

Type	Distribution	Effects
PI3K$_{c1\alpha}$	Ubiquitous	Development, myocardial contractility
PI3K$_{c1\beta}$	Ubiquitous	Development, insulin signaling, cell motility, phagocytosis, blood coagulation
PI3K$_{c1\gamma}$	Leukocytes	T-cell development and migration, neutrophil, macrophage, dendritic cell migration, mastocyte degranulation, neutrophil burst, insulin secretion, myocardial contractility
PI3K$_{c1\delta}$	Leukocytes	Immunity, cytokine receptor signaling, B-cell receptor-mediated antigen presentation
PI3K$_{c2\alpha}$	Widespread	Insulin signaling, vesicular transport, cell survival, vascular smooth muscle cell contraction
PI3K$_{c2\beta}$	Widespread	Cell migration, liver growth, vesicular transport
PI3K$_{c2\gamma}$	Hepatocyte	Liver regeneration
PI3K$_{c3}$	Ubiquitous	Toll-like receptor signaling, receptor-independent membrane trafficking

Table 2.36. Products of members of the PI3K superclass (Source: [92]).

Type	Products (in vitro)
PI3K$_{c1\alpha}$	PI(3)P, PI(3,4)P$_2$, PI(3,4,5)P$_3$
PI3K$_{c1\beta}$	PI(3)P, PI(3,4)P$_2$, PI(3,4,5)P$_3$
PI3K$_{c1\gamma}$	PI(3)P, PI(3,4)P$_2$, PI(3,4,5)P$_3$
PI3K$_{c1\delta}$	PI(3)P, PI(3,4)P$_2$, PI(3,4,5)P$_3$
PI3K$_{c2\alpha}$	PI(3)P, PI(3,4)P$_2$
PI3K$_{c2\beta}$	PI(3)P, PI(3,4)P$_2$
PI3K$_{c2\gamma}$	PI(3)P, PI(3,4)P$_2$
PI3K$_{c3}$	PI(3)P

Three class-1A catalytic subunits exist with isoforms: PI3K$_{c1\alpha}$, PI3K$_{c1\beta}$, and PI3K$_{c1\delta}$, which are encoded by the PIK3CA, PIK3CB, and PIK3CD genes, respectively.[86]

Phosphoinositide 3-kinases of the class-1A are composed of: (1) a regulatory subunit among 5 isoforms ($^{P50\alpha}$PI3K$_{r1}$, $^{P55\alpha}$PI3K$_{r1}$, $^{P85\alpha}$PI3K$_{r1}$, PI3K$_{r2}$, and

---

[86]All catalytic subunits of class-1 PI3Ks contain 4 major domains from N- to C-terminus: Ras-binding (RBD); membrane phospholipid-binding C2; helical, protein-binding PIK; and kinase domains. Small GTPase Ras activated by GPCRs may be an important axis for class-1 PI3K activation. Catalytic PI3K$_{c1\gamma}$ subunit contains 2 G$\beta\gamma$ interaction sites.

PI3K$_{r3}$) that potentially generate to 15 distinct subunit combinations) that contains SH2 and SH3 domains, which interact with other signaling proteins; and (2) a catalytic subunit among 3 isoforms (PI3K$_{c1\alpha}$, PI3K$_{c1\beta}$, and PI3K$_{c1\delta}$) that phosphorylates inositol.

Isoform PI3K$_{c1\alpha}$, but not PI3K$_{c1\beta}$, acts via insulin receptor substrate. Catalytic subunits PI3K$_{c1\alpha}$ and PI3K$_{c1\beta}$ are ubiquitous, whereas PI3K$_{c1\gamma}$ is synthesized primarily in hematopoietic cells, myocytes, and pancreatic cells. Both PI3K$_{c1\delta}$ and PI3K$_{c1\gamma}$ are highly enriched in leukocytes.

The regulatory subunit yields at least 3 functions to PI3Ks: (1) stabilization; (2) inactivation of kinase activity in the basal state, and (3) recruitment to Tyr^P residues of receptors and adaptors that relieve inhibition mediated by the regulatory subunit. Class-1A PI3K enzymes, in fact, consist of an inhibitory (among 5 distinct P85 isoforms) and a catalytic subunit. Binding of the regulatory subunit to activated receptor Tyr kinases or adaptor proteins, such as insulin receptor substrates IRS1 and IRS2, relieves the basal repression of the catalytic subunit.

Class-1A PI3Ks are stimulated by activated receptor Tyr kinases and cytokine receptors. Class-1A PI3Ks are involved in cell growth, division, and survival in response to activated growth factor receptors.[87]

A small fraction (10%) of total class-1 PI3K activity suffices to maintain cell survival and proliferation at least in hematopoietic progenitors and mouse embryonic fibroblasts [96]. In the case of persistent selective or combined inhibition of PI3K isoforms, the remaining isoforms, even at very low concentrations, ensure signaling downstream of Tyr kinases and small GTPase Ras due to functional redundancy of class-1A PI3K isoforms.

Whereas class-1A enzymes are preferentially activated by protein Tyr kinases, the class-1B enzyme is linked to G-protein-coupled receptors. However, most class-1 PI3K subunits may be activated by GPCRs, either directly via G$\beta\gamma$ subunits, such as PI3K$_{c1\beta}$ and PI3K$_{c1\gamma}$, or indirectly via Ras GTPase, in particular.[88] Enzyme PI3Kβ can actually effectively couple to G-protein-coupled receptors.

The catalytic PI3K$_{c1\gamma}$ subunit, which is encoded by the PIK3CG gene is the sole class-1B catalytic member. Kinase PI3Kγ consists of a catalytic and a regulatory subunit: (1) PI3K regulatory subunit-5 (PI3K$_{r5}$)[89] or (2) PI3K regulatory subunit-6 (PI3K$_{r6}$).[90] Therefore, 2 PI3Kγ heterodimers exist: PI3K$_{c1\gamma}$–PI3K$_{r5}$ and PI3K$_{c1\gamma}$–PI3K$_{r6}$. Kinase PI3Kγ elicits cardiovascular and immunological responses.[91]

---

[87] Class 1A PI3K also operates in tumor environment as it regulates tumor vasculature [95]. Tumor angiogenesis is regulated by vascular growth factors secreted by tumor cells and tumor environment. These growth factors target endothelial growth factor receptors, such as VEGFR1 to -3, Tie1 and -2, FGFR1 and -2, PDGFRβ, and HER1 to -4 (Vol. 3 – Chap. 8. Receptor Kinases).

[88] Small GTPase Ras can be activated by RTKs and GPCRs.

[89] Encoded by the PIK3R5 gene.

[90] A.k.a. P87$_{PI3K}^{PIKAP}$ and P84$_{PI3K}$, which is encoded by the PIK3R6 gene.

[91] Defense mechanisms of the immune system, such as chemotaxis, release of antimicrobial substances from granules, and production of reactive oxygen species are controlled by chemokines,

Table 2.37. Activators of class-1 PI3Ks (Sources: [92,94]; Gq: type-q Gα subunit of heterotrimer Gαβγ coupled to GPCR [G-protein-coupled receptor]; RTK: receptor Tyr kinase; −: inhibition). Kinase PI3K is activated by different Toll-like receptors (TLR) in eosinophils, macrophages, neutrophils, and dendritic cells. Ligands bind to immune, cytokine, and G-protein-coupled receptors to prime signal transduction via adaptors, subunits of heterotrimeric G proteins, monomeric Ras GTPases, and protein Tyr kinases that recruit PI3Ks to the plasma membrane and activate it. Once activated, PI3Ks phospharylate their lipidic substrates, hence transforming them into potent signaling mediators.

Catalyzer	Regulators
$PI3K_{c1\alpha}$	RTK, Ras; Gq (−)
$PI3K_{c1\beta}$	RTK, Gβγ, Rab5 (endosomes)
$PI3K_{c1\delta}$	RTK, TLR, rRas2
$PI3K_{c1\gamma}$	Gβγ, TLR, Ras

Table 2.38. Effectors of PI3Ks, signal transduction pathways, and resulting effects (Source: [92]; BCLxL: B-cell lymphoma-extra large protein; BTK: Bruton Tyr kinase; GAP: GTPase-activating protein; GEF: guanine nucleotide-exchange factor; ITK: interleukin-2-inducible T-cell kinase; NFκB: nuclear factor κ light chain-enhancer of activated B cells; PDK: phosphoinositide-dependent kinase; PKB: protein kinase-B; PKC: protein kinase-C; PLC: phospholipase-C; PREx: PIP_3-dependent Rac exchanger; PTK: protein Tyr kinase; SwAP70: switch-associated protein-70 (SWAP switching B-cell complex 70-kDa subunit); S6K: P70 ribosomal S6 kinase).

PI^P effector	Mediators	Effects
Centaurins (ArfGAPs)	CDC42 Rac	Cytoskeleton organization
PREx, SWAP70 (RacGEFs)	CDC42 Rac	Cytoskeleton organization
TEC family (PTK) (BTK, ITK)	CDC42, Rac Ca^{++}	Cytoskeleton organization Degranulation
PDK	PLCγ–Ca^{++} PLCγ–PKC PKB–S6K PKB–BCLxL PKB–NFκB	Degranulation Microfilament and -tubule assembly Cell adhesion Cell proliferation Cell survival Inflammation

The regulatory subunits $PI3K_{r5}$ and $PI3K_{r6}$ have a distinct tissue distribution. They relay signals from Gβγ dimer and Ras GTPase to $PI3K_{c1\gamma}$ (Table 2.37). Lipid $PI(3,4,5)P_3$ generated by $PI3K_{c1\gamma}$–$PI3K_{r5}$, unlike that generated by $PI3K_{c1\gamma}$–$PI3K_{r6}$, is rapidly endocytosed with vesicles [94].

Class-1B PI3Ks link G-protein-coupled receptors to signaling phospholipids for cell growth, division, survival, and migration (Table 2.38). Activation of $PI3K_{c1\gamma}$

---

chemotactic peptides, and nucleosides. These agents signal via G-protein-coupled receptors and phosphoinositide 3-kinase-γ.

2.9 Phosphoinositide Kinases

by stimulated GPCRs is amplified by simultaneous binding of G$\beta\gamma$ subunit of G protein to PI3K$_{r5}$ subunit. Interaction of PI3K$_{r6}$ with G$\beta\gamma$ subunit differs from that of PI3K$_{r5}$ subunit. Subunit PI3K$_{r5}$ strongly interacts with G$\beta\gamma$, whereas PI3K$_{r6}$ only weakly interacts, hence failing to sensitize PI3K$_{c1\gamma}$ for its activation by G$\beta\gamma$ and limiting its membrane recruitment. On the other hand, Ras^{GTP} connects to PI3K$_{c1\gamma}$, thereby causing translocation of the PI3K$_{c1\gamma}$–PI3K$_{r6}$ dimer and activation [97]. Small GTPase Ras, in cooperation with G$\beta\gamma$, hence yields PI3Kγ signaling specificity, because Ras: (1) can recruit PI3K$_{c1\gamma}$ (as well as PI3K$_{c1\gamma}$–PI3K$_{r6}$ and PI3K$_{c1\gamma}$–PI3K$_{r5}$) to membranes and (2) can act as a costimulator of PI3K$_{c1\gamma}$–PI3K$_{r6}$, whereas G$\beta\gamma$ suffices to activate PI3K$_{c1\gamma}$–PI3K$_{r5}$.

Class-1 phosphoinositide 3-kinases target not only lipids, but also proteins. They indeed allow autophosphorylation of both catalytic and regulatory subunits as well as activation of mitogen-activated protein kinase (Chap. 6).

Small GTPase Ras can activate PI3Kα and PI3Kγ. Moreover, rRas2 may contribute to the activation of PI3K$_{c1\delta}$ [94]. In addition to Ras, other small GTPases can be recruited to PI3K. Protein PI3Kβ can be activated by active Rab5 in clathrin-coated vesicles. Class-1 PI3Ks can interact with small CDC42, Rac, and Rho GTPases. The PI3K kinases can act both downstream from monomeric GTPases as well as upstream from them via guanine nucleotide-exchange factors and GTPase-activating proteins. These links create possible feedback loops [94].

Phosphoinositides PI(3,4,5)P$_3$ and PI(3,4)P$_2$ coordinate the recruitment and function of many effectors, such as protein kinases (e.g., PKB and BTK), adaptors (e.g., GAB2), and regulators of small GTPases (GAPs and GEFs). The PI3K kinases can regulate ARF, Rac, and Ras GTPases, as they can control their guanine nucleotide-exchange factors and GTPase-activating proteins. All PI3K isoforms promote Rac activation. On the other hand, RhoA is inhibited by PI3Kδ, but stimulated by PI3Kα subtype.

Membrane ruffling caused by actin remodeling upon growth factor stimulation (e.g., epidermal and platelet-derived growth factors) is mediated by various signaling molecules, such as phosphatidylinositol (3,4,5)-trisphosphate, ^{F}actin- and PI(3,4,5)P$_3$-binding guanine nucleotide-exchange factor SWAP70 (70-kDa switch-associated protein), and Rac1 GTPase [98, 99]. Activator SWAP70 is a RacGEF that cooperates with activated Rac1 GTPase.

Members of the DBL epifamily of Rho guanine nucleotide-exchange factors can serve as direct effectors of heterotrimeric G proteins. Subunits G12/13, Gq, and G$\beta\gamma$ directly bind and regulate RhoGEF regulators. Guanine nucleotide-exchange factors of the PIP$_3$-dependent Rac exchanger (PREX) family activate Rac GTPase. Protein PREx1 is widely expressed in the central nervous system, whereas PREx2 is specifically produced in Purkinje neurons of the cerebellum [100].

Class-1 PI3Ks can serve as scaffolds. Subunit PI3K$_{c1}$ binds to phosphodiesterase PDE3b and protein kinase-C [94]. Enzymes PI3Kα, PI3Kβ, and PI3Kγ, as well as PTen phosphatase, are synthesized in cells of the cardiovascular system (cardiomyocytes, fibroblasts, and vascular endothelial and smooth muscle cells), in which they modulate metabolism, survival, hypertrophy, contractility, and mechanotransduction [101].

Active calpain small subunit-1 (CapnS1) of calpain heterodimers interacts with class-1A PI3Ks in stimulated cells [109]. Calpain-1 and -2 cleave $PI3K_{c1\alpha}$ isoform, thereby terminating PI3K signaling.

Class-2 PI3K Enzymes

Class-2 phosphoinositide 3-kinase is made of a catalytic subunit among 3 isoforms ($PI3K_{c2\alpha}$–$PI3K_{c2\gamma}$) without regulatory subunit. Class-2 PI3Ks are encoded by the PIK3C2A, PIK3C2B, and PIK3C2G genes. They are mainly associated with cellular membranes and nucleus. Isozymes $PI3K_{c2\alpha}$ and $PI3K_{c2\beta}$ have a broad, but not ubiquitous, tissue distribution, whereas the expression pattern of $PI3K_{c2\gamma}$ is more restricted [94]. Class-2 PI3Ks do not have regulatory subunits.

Class-3 PI3K Enzyme

Class-3 PI3K can phosphorylate only PI substrate. It complexes to a single regulatory subunit encoded by the PIK3R4 gene. Class-3 PI3K is actually represented by a single known heterodimer with a catalytic subunit ($PI3K_{c3}$) encoded by the PIK3C3 gene and a regulatory ($PI3K_{r4}$) subunit.

Class-3 PI3K heterodimer assembly at membranes is promoted by small Rab5 and Rab7 GTPases that connect to $PI3K_{r4}$ protomer. It is antagonized by phosphoinositide 3-phosphatase myotubularin-6 [94].

2.9.1.2 PI3Kα Enzyme (Class 1A)

Subunit $PI3K_{c1\alpha}$ with Tyr kinase activity that targets Ras signaling has specific functions in growth factor, insulin, and leptin signaling [102]. This function relies on the selective recruitment of adaptors, insulin receptor substrates, transmitting signals from receptors of insulin and insulin-like growth factor-1 mainly to IRS-bound PI3K to activate the PKB and ERK pathways.[92] Isoform $PI3K_{c1\alpha}$ is the primary insulin-responsive PI3K, at least in cultured cells, whereas $PI3K_{c1\beta}$ is dispensable, but sets a threshold for $PI3K_{c1\alpha}$ activity [103].

2.9.1.3 PI3Kβ Enzyme (Class 1A)

Ubiquitous class-1A PI3Kβ isoform has a $PI3K_{c1\beta}$ subunit with both kinase-dependent and -independent functions. Synthesis of $PI3K_{c1\beta}$ in the liver is required for insulin sensitivity and glucose homeostasis, but does not signicantly affect protein kinase-B phosphorylation in response to receptor activation by insulin and epidermal growth factor [104].

---

[92]Other broadly expressed $PI3K_{c1\beta}$ does not contribute to IRS-associated PI3K activity.

Subunit $PI3K_{c1\beta}$ also regulates cell proliferation independently of its kinase activity. However, its kinase activity is required for the signaling triggered by lysophosphatidic acid attached to its G-protein-coupled receptor. In addition, $PI3K_{c1\beta}$, but not $PI3K_{c1\alpha}$, intervenes in tumorigenesis with a concomitant increase in PKB phosphorylation.

Lipid PTen phosphatase opposes kinase activity of class-1A PI3Ks. The basal pool of $PI(3,4,5)P_3$ catalyzed by $PI3K_{c1\beta}$ that is not disturbed by stimulation by insulin and growth factors supports oncogenic transformation in the absence of PTen, at least in some tumor models.

Among its kinase-independent functions, PI3Kβ participates in sensing of double-strand breaks. Enzyme PI3Kβ regulates binding of nibrin, a sensor of damaged DNA and a member of the double-strand DNA break-repair complex [105].

2.9.1.4 PI3Kγ (Class 1B)

Phosphoinositide 3-kinase-γ is a heterodimer composed of $PI3K_{r5}$ or $PI3K_{r6}$ regulatory and $PI3K_{c1\gamma}$ catalytic subunit. It converts $PI(4,5)P_2$ into $PI(3,4,5)P_3$ at the inner leaflet of the plasma membrane. The lipid kinase activity is stimulated by the combined interaction of regulatory and catalytic subunit with activated Gβγ dimer and Ras GTPase.

Kinase PI3Kγ is activated by Gi-coupled receptors stimulated by chemokines (e.g., chemokines CCL2,[93] CCL3,[94] CCL5,[95] CXCL2,[96] and CXCL8),[97] proinflammatory lipids (PAF and LTb4), bacterial products, and vasoactive molecules (C5a, ADP, and angiotensin-2). Activated PI3Kγ acts cooperatively with other signaling pathways. These signaling pathways involve Src kinases, class-1A PI3Ks, MAPKs, etc.

Agent PIP_3 and its dephosphorylation product $PI(3,4)P_2$ are signaling lipids that support the plasmalemmal localization and regulation of several effectors (Table 2.39). The transduction magnitude and duration via these effectors is regulated by inositol polyphosphate 5-phosphatases such as SHIP, 3-phosphatases such as PTen, and inositol polyphosphate 4-phosphatases. The PI3Kγ pathway regulates the function of: (1) immunocytes (macrophages, monocytes, mastocytes, neutrophils, dendritic cells, and T lymphocytes); (2) vascular endothelial cells and smooth muscle cells that determine the vasomotor tone; and (3) platelets involved in blood coagulation.

---

[93] A.k.a. monocyte chemoattractant protein-1 (MCP1).

[94] A.k.a. macrophage inflammatory proteins MIP1α.

[95] A.k.a. T-cell-specific protein Regulated upon activation, normal T-cell expressed, and secreted (RANTES). It attracts T lymphocytes, eosinophils, and basophils.

[96] A.k.a. MIP2.

[97] A.k.a. interleukin-8

Table 2.39. Phosphoinositide 3-kinase-γ effectors downstream from PI(3,4,5)P$_3$ and its dephosphorylation product PI(3,4)P$_2$ (GEF: guanine nucleotide-exchange factors; GAP: GTPase-activating protein; Source: [106]).

Direct	Indirect
Ser/Thr kinases	Protein kinases
Tyr kinases	Lipid messengers
RacGEFs	Transcription factors
ArfGEFs	Integrins
ArfGAPs	Small GTPases
Adaptors	NADPH oxidase
	cAMP

2.9.1.5 PI3Kδ (Class 1A)

Isozymes PI3Kγ and PI3Kδ that are relatively specific to leukocytes orchestrate innate and adaptive immunity, especially in respiratory diseases (allergic asthma and chronic obstructive pulmonary diseases [COPD]) and cardiovascular pathologies (atherosclerosis and myocardial infarction).

Phosphoinositide 3-kinases contribute to the genesis of asthma, as they regulate inflammatory mediator activation, inflammatory cell recruitment, and immunocyte function. In particular, they control the expression of interleukin-17 that is an important cytokine in airway inflammation. Signaling mediated by PI3Kδ promotes IL17 expression via nuclear factor-κB [107].

Glucocorticoid function is markedly impaired in lungs of patients with chronic obstructive pulmonary disease. Oxidative stress causes phosphorylation of PKB in monocytes and macrophages that is associated with a selective upregulation of PI3Kδ [108].

2.9.1.6 Class-2 Phosphatidylinositol 3-Kinases

Class-2 PI3Ks are stimulated by extracellular signals, such as integrin engagement, growth factors, and chemokines [93]. They can phosphorylate PI and PI(4)P, but not PI(4,5)P$_2$. Class-2 PI3Ks can also be stimulated by protein kinases and GPCRs, after recruitment from the small cytosolic pool to the plasma membrane via membrane-associated adaptor proteins [94]. Class-2 PI3Ks may also be recruited and activated by membrane-bound GTPases such as monomeric RhoJ GTPase.

2.9.1.7 Class-3 Phosphatidylinositol 3-Kinase

Class-3 phosphatidylinositol 3-kinase (PI3KC3) produces phosphatidylinositol 3-phosphate, thereby intervening in membrane trafficking, endocytosis, phagocytosis, autophagy, nutrient sensing by the target of rapamycin pathway, and cell signaling,

2.9 Phosphoinositide Kinases

Table 2.40. The PI3KC3 complexes and their subcellular localization (Source: [94]). Enzyme PI3KC3 participates in various vesicular transfer types that always deliver cargos to lysosomes. Cup-shaped omegasome is a double-membrane structure that forms upon amino acid starvation, expands, sequesters cytoplasmic material, and eventually forms an autophagosome (AMBRA: activating molecule in beclin-1-regulated autophagy protein; Atg14L: autophagy-specific PI3K complex Atg14-like protein; beclin: BCL2-interacting protein; BIF: BAX-interacting factor; EEA: early endosomal antigen; Rubicon: RUN domain and Cys-rich domain-containing, beclin-1-interacting protein; UVRAG: ultraviolet radiation resistance-associated gene product).

Role	Complex components	Locus
Autophagy	Beclin-1 UVRAG (or Atg14L) BIF1, AMBRA1, Rubicon	Endoplasmic reticulum (omegasome)
Endocytosis	Rab5, EEA1	Cell cortex (early endosome)
Phagocytosis	Dynamin, Rab5	Cell cortex (early phagosome)

especially downstream from G-protein-coupled receptors. It acts in the signaling primed by activated Gq-coupled receptors such as muscarinic M_1 receptor [94]. It can control amino acid-dependent activation of S6K1 kinase.

It is implicated in endosome fusion during intracellular transport. Kinase PI3KC3 participates in transport to lysosomes via multivesicular bodies, from endosomes to the trans-Golgi network via retromers (proteic complexes that contribute to the recycling of plasmalemmal receptors), and phagosome and autophagosome maturation. In addition, PI3KC3 is involved in nutrient sensing in the target of rapamycin pathway.

Enzyme PI3KC3 alternates between a closed cytosolic form and an open form on membranes. It is, in fact, located mainly on intracellular membranes [93, 110]. It possesses a constricted adenine-binding pocket. Both the phosphoinositide-binding loop and C-terminus of PI3KC3 mediate catalysis on membranes [110].

Kinase PI3KC3 synthesizes PI(3)P that is specifically recognized by proteins that contain FYVE or PX binding domains and initiate the assembly of complexes at membranes of endosomes, phagosomes, or autophagosomes. Regulators, such as Rab5 and Rab7, bind to $PI3K_{r4}$ and enable activation of the $PI3K_{r4}$–$PI3K_{c3}$ complex at membranes [110].

Proteins associated with $PI3K_{c3}$ regulate, directly or not, $PI3K_{c3}$ activity (Table 2.40). Agent BAX-interacting factor-1 (BIF1) stimulates $PI3K_{c3}$ in the beclin-1–UVRAG complex. Subunit $PI3K_{c3}$ actually forms different functional complexes [94]. A complex devoted to autophagy contains beclin-1 (coiled-coil, moesin-like, B-cell lymphoma protein-2 [BCL2]-interacting protein), a protein essential for autophagy, either subunit of autophagy-specific PI3K complex, autophagy-related protein-like agent Atg14L (or beclin-1-associated autophagy-related key regulator [barkor]) or ultraviolet radiation resistance-associated gene product (UVRAG), BAX-interacting factor BIF1, activating molecule in beclin-1-regulated autophagy protein AMBRA1, and RUN domain and Cys-rich domain-containing, beclin-1-

interacting protein Rubicon. The composition of this complex can vary, as the presence of some components excludes the presence of others. Components Atg14L and UVRAG bind to the beclin-1–PI3K$_{c3}$ complex in a mutually exclusive manner, whereas Rubicon binds only to a subpopulation of UVRAG complexes [111]. Effectors include WD repeat domain-containing phosphoinositide-interacting protein WIPI1, autophagy-linked FYVE protein (ALFY), and double FYVE-containing protein-1 (DFCP1).

The PI3K$_{r4}$–PI3K$_{c3}$ molecule linked to endosomes, which acts as a regulator of vesicular transfer, can complex with MTM1 myotubularin, a phosphatidylinositol 3-phosphatase, and (or) UVRAG [94]. The latter increases PI3K$_{c3}$–VPS15 binding [112]. Effectors comprise early endosome antigen EEA1, endosomal sorting complexes required for transport ESCRT2 and ESCRT3, hepatocyte growth factor-regulated Tyr kinase substrate (HRS), sorting nexins, and Rab5 effector rabenosyn-5.[98] The degradative pathway comprises early Rab5+ endosomes, multivesicular bodies, and late Rab7+ endosomes before fusion with lysosomes. In early endosomes, Rab5 and PI(3)P bind and assemble effectors, such as EEA1, HRS, and ESCRT proteins. In late endosomes, Rab5 is replaced by Rab7 that recruits its effector Rab7-interacting lysosomal protein (RILP) for acquisition of the lysosomal marker lysosome-associated membrane protein LAMP1 and fusion with lysosomes.

The PI3K$_{r4}$–PI3K$_{c3}$ complex intervenes in phagocytosis. Several waftures of PI(3)P accumulation occur during phagocytosis with an early transient peak immediately after particle internalization and a following wave during phagosome maturation [94]. Production of PI3P depends on class-1A PI3Ks; phagosome maturation on PI3KC3 kinase. In nascent phagosomes, PI3KC3 interacts with large GTPase dynamin that recruits small GTPase Rab5 during phagosome maturation. Synthesized PI3P is also required for activation of NADPH oxidase assembled on phagosomes.

2.9.1.8 Nuclear PI3K

Although PI3K and its effector protein kinase-B are predominantly located in the cytoplasm, they also operate in the nucleus. Nuclear PI3K and its effectors intervene in various activities, such as cell survival, differentiation, and proliferation, as well as pre-mRNA export from the nucleus and splicing. Splicing of pre-mRNA and export of mRNA are coupled.

Most enzymes of the phosphoinositol lipid metabolism (phospholipase-C, phosphatidylinositol phosphate kinases, phosphoinositide-dependent kinase-1, PTen, and protein kinase-B) reside in the nucleus. Phosphoinositides associate with nuclear speckles. These nuclear subcompartments contain small ribonucleoproteins, splicing factors, and elements of nuclear phosphoinositide metabolism.

---

[98] A.k.a. zinc finger, FYVE domain-containing protein ZFYVE20.

Nuclear protein THO complex subunit THOC4[99] is required for cell cycle progression owing to phosphorylation and phosphoinositide binding. The mRNA export factor and transcriptional regulator THOC4 is a protein of nuclear speckles that interacts with nuclear PI(4,5)P_2 and PI(3,4,5)P_3 [113]. During the cell division cycle, PI3K is involved in both G1–S transition and S-phase progression. Molecular chaperone THOC4 regulates PI3K subnuclear residency. Nuclear PKB phosphorylates THOC4, and hence controls THOC4 binding to phosphoinositides.

2.9.1.9 PI3K Signaling

Phosphatidylinositol 3-kinase is phosphorylated (activated) by various receptor and receptor-associated protein Tyr kinases. Enzyme PI3K is stimulated by receptors of hormones (e.g., insulin) and growth factors (EGF, HGF, IGF1, NGF, PDGF, and SCF; Vols. 2 – Chap. 3. Growth Factors and 3 – Chap. 8. Receptor Kinases). Kinase PI3K phosphorylates phosphatidylinositols at position 3 of the inositol ring. In response to various stimuli, phosphatidylinositol 3-kinase phosphorylates phosphatidylinositol (4,5)-bisphosphate to generate phosphatidylinositol (3,4,5)-trisphosphate that binds to pleckstrin homology (PH) domains of signaling proteins as well as phox homology (PX) domains of certain proteins.

Phosphatidylinositol 3-kinase generates substrates for phospholipase-Cγ that produces second messengers diacylglycerol and inositol trisphosphate. Effectors also include Ser/Thr kinases, protein kinase-B, and phosphatidylinositol-dependent protein kinase-1 (PDK1), various isoforms of protein kinase-C (PKC), and members of the TEC family of Tyr kinases, such as TEC (Tyr kinase expressed in hepatocellular carcinoma), BTK (Bruton Tyr kinase), ITK (interleukin-2-inducible T-cell kinase), and RLK (resting lymphocyte kinase, or TXK) [114].

The PI3K–PKB pathway is involved in cell growth, survival, metabolism, and migration, as well as tumorigenesis. Kinase PKB is recruited to the plasma membrane, as it binds to PI3K products PI(3,4)P_2 and PI(3,4,5)P_3 owing to its PH domain. After PKB translocation to the plasma membrane, the formation of a complex with its upstream kinases at the plasma membrane is facilitated by scaffold proteins. Activation of PKB can require a sequential phosphorylation by 2 phosphoinositide-dependent kinases, PDK1 and PDK2 (or TOR complex-2). For full activation, PKB indeed needs to be phosphorylated by PDK1 and PDK2 (Thr308 and Ser473, respectively).

Phosphatidylinositol 3-kinase stimulated by growth factors and chemoattractants intervenes in cell migration, as it activates Rac and CDC42 (Sect. 9.3) at the leading edge of migrating cells (Vol. 2 – Chap. 6. Cell Motility).[100] Small Rac and CDC42

---

[99] A.k.a. basic region-leucine zipper (bZIP)-enhancing factor BEF, Ally of AML1 and LEF1 (Aly), and REF.
[100] Activated GTPases Rac and CDC42 cause actin reorganization to form lamellipodia and filopodia, respectively.

GTPases act via their effectors of group-1 P21-activated protein kinases. Kinase PI3K activates a signaling pathway that includes Rac1, PDK1, and PKB effectors. Small Rac1 GTPase not only provokes phosphorylation by PAK1s, but also causes an additional kinase-independent PAK1 activity: PAK1 operates as a scaffold for PKB recruitment to the plasma membrane and its activation by PDK1 [115].

In T-lymphocytes, the PI3K pathway is implicated in growth, proliferation, cytokine secretion, and survival. T lymphocytes produce all 3 class-1A PI3K isoforms, which are regulated by receptor Tyr kinases, and class-1B PI3K isoform, which is activated by G-protein-coupled receptors. Chemokines binds to cognate GPCRs. T-cell receptor, costimulatory receptors of the CD28 family, and cytokine receptors activate class-1A PI3Ks.

Among class-1 PI3K heterodimeric lipid kinases (with regulatory and catalytic subunits), class-1A PI3Ks with a P85 regulatory subunit ($^{P85\alpha}PI3K_{r1}$ or $^{P85\beta}PI3K_{r2}$ adaptor) is activated by receptor Tyr kinases, whereas class-1B enzyme (PI3Kγ) is stimulated by Gβγ subunits of heterotrimeric G proteins.

In addition to serve as a regulatory subunit for $PI3K_{c1}$ protomer of PI3Ks, $P85_{PI3K}$ subunit also interacts with other proteins, such as the small CDC42 GTPase, nuclear receptor corepressor, receptor Tyr protein phosphatase PTPRj,[101] as well as the short splice variant of the transcription factor X-box-binding protein $XBP1_S$ of the ATF–CRE (activating transcription factor–cAMP response element)-binding protein family [116].[102]

PI3K Signaling in the Cardiovascular System

Signaling via PI3K plays a prominent role in the cardiovascular system. It regulates the activity of vascular endothelial and smooth muscle cells as well as platelets and leukocytes. Its effectors include mainly protein kinase-B (Sect. 5.2.6), nitric oxide synthase NOS3 (Sect. 10.3.1), target of rapamycin (Vol. 2 – Chap. 2. Cell Growth and Proliferation and Sect. 5.5.3), and forkhead transcription factor FoxO (Sect. 10.9.3.3). It thus controls various processes, such as vascular tone, angiogenesis, endothelial cell differentiation, and endothelial cell–leukocyte interactions. Protein kinase-B operates in the nucleus to increase the gene transcription.

Cardiac Effects of PI3K

Cardiac effects of PI3K are summarized in Table 2.41. Anti-apoptotic action of growth factors relies almost exclusively on the PI3K–PKB pathway. Cardioprotection use also this pathway downstream from GPCRs of adrenomedullin, ghrelin, and urocortin, as well as β2-adrenoceptors.

---

[101] A.k.a. RPTPη and CD148.

[102] Together with ATF6, $XBP1_S$ orchestrates the unfolded protein response during endoplasmic reticulum stress (Vol. 1 – Chap. 5. Protein Synthesis).

Table 2.41. Cardiac effects of PI3K (Source: [101]).

Effect	Mediators
Cell survival	PKB, BAD, caspase-3, FoxO, IκB kinase
Glucose uptake	GluT
CMC hypertrophy	PKB1, S6K
CMC contractility	PDE4
Mechanotransduction	PKB, GSK3β
Angiogenesis	PKB

Cardiomyocyte adaptive and maladaptive hypertrophy requires PI3K. Kinase PI3Kα controls adaptive growth, whereas PI3Kγ is involved in maladaptive hypertrophy.

The cAMP–PKA pathway has positive chronotropic, inotropic, and lusitropic effects. Kinase PI3Kγ targets phosphodiesterase that restricts cAMP activity, whereas PI3Kα favors cardiomyocyte contractility.

Kinase PI3K regulates several ion channels and exchangers, such as voltage-gated Ca_V1 and inward (K_{IR}) and delayed rectifier (K_V) K^+ channels. Kinase PKB2, but not PKB1, is implicated in insulin-stimulated glucose uptake and metabolism. In addition, stretch activates PKB and GSK3β. Lastly, PI3Kγ contributes to the recruitment of endothelial progenitor cells.

Vascular Effects of PI3K

Protein and lipid PI3K kinases operate also in vascular compartments, upon receptor Tyr kinase, G-protein-coupled receptor, or Ras activation. Ubiquitous PI3Kα and PI3Kβ abound in the vasculature, whereas PI3Kδ and PI3Kγ are mainly restricted to leukocytes.

Endothelial Cells

The PI3K–PKB pathway intervenes in several endothelial functions, such as vascular tone regulation, angiogenesis, and extravasation (Table 2.42). Kinase PI3K indeed promotes nitric oxide release, endothelial progenitor cell recruitment, and cell migration [117]. Endothelial PI3Kγ hinders and PI3Kδ favors neutrophil adhesion to endothelium. Subtype PI3Kα operates in VEGFa-dependent migration of endothelial cells. Sphingosine 1-phosphate-dependent endothelial migration requires both PI3Kβ and PI3Kγ.

Table 2.42. Effects of PI3K on vascular endothelium (Source: [117]; EPC: endothelial progenitor cell).

Process	Targets
Angiogenesis	RhoA (cell migration)
Endothelial permeability	hRas
NO release	PKB (NOS3 phosphorylation)
Leukocyte extravasation	Adhesion molecules
Cell survival	BAD, FoxO
EPC differentiation	HDAC3

Table 2.43. PI3K Effects on vascular smooth muscle cells (Source: [117]).

Process	Effects
Vascular tone	Ca^{++} influx
	$Ca_V\beta 2a$ subunit phosphorylation
	Ca^{++}-permeable, non-selective cation channel NSCC2 gating
	Store-operated Ca^{++} channel opening
Remodeling	Activation of target of rapamycin

Vascular Smooth Muscle Cells

Kinase PI3Kγ regulates contractility and proliferation of vascular smooth muscle cells, thereby causing vasoconstriction and neointimal hyperplasia [117] (Table 2.43).

PI3K Effects on Platelets

Different class-1 PI3Ks (PI3Kα–PI3Kγ) are synthesized in thrombocytes (Vol. 5 – Chap. 4. Blood Cells). Phosphoinositide 3-kinases promote thrombosis, as it enhances calcium release and activation of $\alpha_{2B}\beta_3$-integrins that promote platelet anchoring to the vessel wall despite blood flow (Vol. 5 – Chap. 10. Endothelium). Isoform PI3Kβ is the main contributor to formation and stability of integrin bonds. Class-2 PI3Ks intervene in signaling from platelet environment. The ADP–P2Y$_{12}$–PI3Kβ axis, in cooperation with PI3Kγ, enhances long-term Ca^{++} mobilization induced by Gs-coupled thrombin receptor. In addition, pro-aggregant IGF1 signals via PI3Kα.

2.9.2 Other Phosphoinositide Kinases

Type-1 phosphatidylinositol phosphate kinases correspond to PI(4)P5Ks that phosphorylate PI(4)P to form PI(4,5)P$_2$. Type-2 PIPKs are PI(5)P4Ks that phosphorylate PI(5)P. Type-3 PIPKs are PI(3)P5Ks that phosphorylate PI(3)P to PI(3,5)P$_2$.

Lipid kinases, such as phosphatidylinositol 3-kinase, phosphatidylinositol 4-kinase, and phosphatidylinositol 4-phosphate 5-kinase, are implicated in membrane trafficking. Phosphatidylinositol 4- and 5-kinases convert PI(5)P and PI(4)P to PI(4,5)P$_2$. The latter is a major substrate for activated phospholipase-C that hydrolyzes PI(4,5)P$_2$ into inositol (1,4,5)-trisphosphate and diacylglycerol. Lipid PI(4,5)P$_2$ is also a substrate for several other phosphoinositide kinases, such as PI(4)P5K and PI3K that produce PI(4,5)P$_2$ and PI(3,4)P$_2$, respectively. Signaling molecule PI(4,5)P$_2$ controls cytoskeletal organization as well as activity of enzymes (e.g., phospholipase-D) and ion transporters and channels (e.g., inwardly rectifying potassium channels K$_{IR}$1.1,[103] K$_{IR}$2.1,[104] and K$_{IR}$3.2,[105] voltage-gated potassium channels K$_V$7.2[106] and K$_V$7.3[107], cyclic nucleotide-gated [CNG] ion channel, and transient receptor potential channels) [118]. Channel TRPM4 is implicated in PI(4,5)P$_2$ sensing.

Signaling lipid PI(3,5)P$_2$ in endosomes regulates retrograde transport to the trans-Golgi network. The PI3P 5-kinase (PI(3)P5K)[108] produce PI(3,5)P$_2$. Scaffold Vac14[109] recruits PI(3,5)P$_2$ regulators: the lipid kinase PI(3)P5K, lipid 5-phosphatase suppressor of actin mutations-like phosphatase SAc3,[110] and PI(3)P5K activator Vac7 and inhibitor Atg18 for fast and transient conversion of PI(3)P into PI(3,5)P$_2$ [119].

2.9.2.1 Phosphatidylinositol 4-Kinase

Ubiquitous phosphatidylinositol 4-kinase (PI4K) catalyzes the phosphorylation of membrane phosphatidylinositol to generate phosphatidylinositol 4-monophosphate (PI4P). The latter operates in various signaling events as well as vesicular trafficking and lipid transport. In particular, it serves as a precursor in the phosphoinositide 3-kinase–phospholipase-C pathway. In addition, PI4P recruits PH domain-containing proteins involved in intracellular transport and clathrin AP1 and AP3 adaptor complexes.

---

[103] A.k.a. ROMK1.
[104] A.k.a. IRK1, LQT7, and SQT3.
[105] A.k.a. GIRK2 and K$_{ATP}$2.
[106] A.k.a. K$_V$LQT2.
[107] A.k.a. K$_V$LQT3.
[108] A.k.a. Fab1 or PIKFYVE.
[109] A.k.a. ArPIKFYVE.
[110] A.k.a. Fig4. The SAc domain is homologous to yeast lipid phosphatase SAc1. Several SAc domain-containing proteins have been identified in yeast. Four inositol polyphosphate 5-phosphatases have been detected in Saccharomyces cerevisiae: Inp51p, Inp52p, Inp53p, and Inp54p. Each enzyme possesses a 5-phosphatase domain that hydrolyzes PI(4,5)P$_2$ to form PI(4)P. However, Inp52p and Inp53p also express a polyphosphoinositide phosphatase domain associated with the SAc1-like domain. Protein Fig4p is composed only of the SAc domain. In addition to synaptojanins, 3 SAc domain-containing proteins (SAc1–SAc3) exist in humans.

Table 2.44. The PI4K isozymes in mammalian cells (Source: [120]; ER: endoplasmic reticulum; MVB: multivesicular body; TGN: trans-Golgi network). They localize to distinct membrane compartments and have specific roles in vesicular trafficking and Golgi body function, where PI(4)P rather than PI(4,5)P$_2$ is the lipid regulator. The PI4K enzymes are divided into structurally and biochemically distinct PI4K2 and PI4K3 subfamilies. Type-2 kinases are relatively sensitive to Ca^{++} and adenosine. Type-3 isoforms share significant amino acid sequence similarity with the catalytic domains of members of the protein- and phosphoinositide 3-kinase superclass.

Type	Location	Ki (adenosine; mmol)	Km (ATP; mmol)	Km (PI; mmol)
PI4K2α	Plasma membrane TGN, endosomes	10–70	10–50	20–60
PI4K2β	Plasma membrane TGN, endosomes	10–70	10–50	20–60
PI4K3α	Plasma membrane, ER, Golgi, MVB, mitochondrion, nucleolus nervous system	O[1]	∼700	∼100
PI4K3β	Golgi, nucleus, plasma membrane	O[1]	∼400	∼100

Multiple PI4K isoforms exist in eukaryotic cells (Table 2.44). Types PI4K can be distinguished by their distinct phosphatidylinositol kinase activities: (1) type-1 corresponds to phosphoinositide 3-kinase, whereas (2) types-2 and -3 exclusively use phosphatidylinositol and phosphorylate only at the 4-OH position of the inositol ring to produce phosphatidylinositol 4-monophosphate. Type-2 PI4Ks are membrane-bound kinases owing to palmitoylation. However, a significantly larger fraction of PI4K2β than that of PI4K2α is cytosolic.

Isoforms PI4K2α and PI4K2β that constitute the type-2 PI4K family are structurally and chemically distinct from type-3 PI4Ks that are members of the PI3K superclass. Type-2 PI4Ks localize to organelle and plasma membranes.

Type-3 phosphatidylinositol 4-kinases favor synthesis of plasmalemmal phosphoinositides during phospholipase-C activation and Ca^{++} signaling. Phosphatidylinositol 4-kinase-3α resides at the endoplasmic reticulum and Golgi body, whereas phosphatidylinositol 4-kinase-3β localizes primarily to the Golgi body and in the pericentriolar zone.

Phosphatidylinositol 4-Kinase-2α

Epidermal growth factor and Gi/o protein enhances PI4K2α activity. In response to EGF, PI4K2α indeed connects to epidermal growth factor receptor. In addition, it also links to receptor HER2 as well as activated T-cell coreceptor CD4 [121]. It can form a complex with: (1) scaffold tetraspanins of the transmembrane 4 superfamily (TM4SF) — Tspan24 (CD151), Tspan28 (CD81), and Tspan30 (CD63) — that anchor and regulate other receptors;[111] (2) protein kinase-Cμ and type-1 phosphatidylinositol 4-phosphate 5-kinase, as PKCμ autophosphorylates and serves as a scaffold for recruitment of phosphoinositide kinases to membrane;[112] and (3) epidermal growth factor receptor, phosphatidylinositol transfer protein PITPα, and phospholipase-Cγ1.[113] Kinase PI4K2α is required in recruitment of heterotetrameric adaptor AP1 to the Golgi body and AP3 to endosomes [120].

Phosphatidylinositol 4-Kinase-2β

Unlike PI4K2α, a significant proportion of PI4K2β is cytosolic and substantially less active than the membrane-associated fraction. Cytosolic PI4K2β is recruited to the plasma membrane in response to platelet-derived growth factor and active small GTPase Rac1 [125]. Calcium inhibits both type-2 PI4Ks.

Phosphatidylinositol 4-Kinase-3α

Phosphatidylinositol 4-kinase-3α is mainly located in the endoplasmic reticulum of mammalian cells, but also resides in the plasma membrane as well as pericentriolar area over the Golgi body, mitochondria, multivesicular bodies, and nucleolus. Phosphatidylinositol 4-kinase-3α actually possesses nuclear localization signals. In the nucleolus, it can participate in DNase- and RNase-sensitive complexes [126].

Phosphatidylinositol 4-kinase-3α is responsible for the generation of the plasmalemmal pool of PI(4)P [127]. In particular, PI4K3α is required for the production of PI(4)P, PI(4,5)P$_2$, and Ca^{++} signaling during angiotensin-2 stimulation [128]. Phosphatidylinositol 4-kinase-3α forms a signaling complex with P2X$_7$ ion channels [120].

Phosphatidylinositol 4-kinase-3α can function with membrane-associated *phosphatidylinositol transfer protein* PITPm that lodges in the endoplasmic reticulum and Golgi body.[114]

---

[111] Tetraspanins link PI4K to $α_3β_1$-integrin in focal complexes at cell periphery rather than in focal adhesions [122]. Tetraspanins contribute to modulation of intracellular calcium concentration, and Tyr phosphorylation, as well as cell proliferation.

[112] Protein kinase-Cμ, also named protein kinase-D, that is located in the Golgi body forms in vivo a complex with a phosphatidylinositol 4-kinase and phosphatidylinositol-4-phosphate 5-kinase and acts as a scaffold for assembly of enzymes involved in phosphoinositide synthesis [123].

[113] Stimulation by EGF of PLCγ requires PITP via association of PITP with PI4K, EGFR, and PLCγ [124].

[114] The family of phosphatidylinositol transfer proteins (PITP) is one of the 2 families of phosphoinositide transfer proteins that can bind to and exchange one molecule of either phosphatidylinositol or phosphatidylcholine and facilitate the transfer of these lipids between different membrane compartments. (The other family of proteins involved in the transfer of

Phosphatidylinositol 4-Kinase-3β

Phosphatidylinositol 4-kinase-3β is primarily located in the Golgi body, but can be detected in the nucleus owing to its nuclear localization and export signals [120]. Recruitment of PI4K3β to the Golgi body is regulated by small GTPase ARF1 that is regulated by calcium-binding protein frequenin (Freq or neuronal calcium sensor NCS1). Active Rab11^{GTP} also binds PI4K3β. Kinase PI4K3β is phosphorylated (Ser268) by protein kinase-D to ensure its recruitment to Golgi body and activity.

Phosphatidylinositol 4-kinase-3β intervenes in the endoplasmic reticulum-to-Golgi body transport of ceramide [128] as well as Golgi-to-plasma membrane trafficking. Its lipid products, together with small GTPases Rab11 and ARFs, recruit clathrin adaptors and other effectors to promote budding and cleavage of Golgi-derived transport vesicles. Among proteins that possess PH domains, which specifically recognize PI(4)P, there are lipid transport proteins, such as oxysterol-binding proteins, phosphoinositol 4-phosphate adaptor protein FAPP2, and ceramide transport protein. Kinase PI4K3β also regulates exocytosis at the plasma membrane. In addition, both PI4K3α and PI4K3β isoforms operate during hepatitis C virus infection [129].

2.9.2.2 Phosphatidylinositol 5-Kinase

In polarized cells, phosphatidylinositol 5-kinase (PI5K) stimulates delivery from the trans-Golgi network to apical membrane of raft-associated proteins, but not that of non-raft-associated apical or basolateral proteins [130]. This transport is carried out via Arp2/3 complex, i.e., using both actin filaments and microtubules. The Golgi body indeed contains a pool of phosphatidylinositol 4-phosphate that limits delivery to apical membrane of raft-associated proteins. Increased phosphatidylinositol (4,5)-bisphosphate level caused by PI5Ks results in generation of nucleating branches of actin filaments, the so-called *actin comets*, that transport vesicles in the cytoplasm.

---

phosphatidylinositol and phosphatidylcholine is the family of Sec14p proteins.) Members of the PITP family participate in signal transduction and membrane trafficking. In humans, the PITP family contains several members: PITPα, PITPβ, cytoplasmic type-1 PITP (PITPc), membrane-associated type-1 and -2 PITPs (PITPm1 and PITPm2), PITPNM family member 3, or membrane-associated type-3 PITP (PITPm3). The N-terminus targets lipids. In addition, phosphatidylinositol transfer protein PITPα maintains the PI(4,5)P$_2$ pool during PLC activation. Protein PITPα operates with phosphatidylinositol phosphate 5-kinase. It also associates with class-1 and -3 PI3K and type-2 PI4K enzymes.

2.9.2.3 Phosphatidylinositol Phosphate 4-Kinase

Phosphatidylinositol 5-phosphate 4-kinases regulate the cellular concentration of phosphatidylinositol 5-phosphate (PI(5)P). Three PIP4K (or PI(5)P4K) isoform exist (PIP4Kα–PIP4Kγ). Isoform PIP4Kα resides predominantly in the cytosol and can be recruited to the plasma membrane. Subtype PIP4Kβ is located in the nucleus. Isozyme PIP4Kγ is the predominant PIP4K in kidney [131].

In the brain, all 3 isoforms are produced with different spatial distribution. In spleen, PIP4Kα is the most abundant isoform. In heart and skeletal muscles, PIP4Kβ is synthesized at high levels. Isoform PIP4Kγ is highly expressed in kidneys, brain, heart, ovary, and testis. It contributes to actin remodeling during endocytosis. In kidneys, PIP4Kγ is primarily distributed in epithelial cells in the thick ascending limb and intercalated cells of the collecting duct, where it regulates vesicular transport [131].

2.9.2.4 Phosphatidylinositol Phosphate 5-Kinase

Phosphatidylinositol 4-phosphate 5-kinase (PI(4)P5K) synthesizes phosphatidylinositol (4,5)-bisphosphate, a precursor in phosphoinositide signaling.[115] In addition, it controls actin polymerization and operates in cell adhesion and migration independently of its catalytic activity.

Phosphatidylinositol 4-phosphate 5-kinases constitute a superfamily with several kinase types: type-1 (PIP5K1), -2 (PIP5K2), and -3 (PIP5K3) PI(4)P 5-kinases. In addition, PIP5K1 possesses isoforms, such as PIP5K1α and PIP5K1β [132]. In erythrocytes, both PIP5K1α and PIP5K2 exist. Lipid PIP5K1, but not PIP5K2, can be stimulated by phosphatidic acid and can use PI4P in membranes as a substrate.

Different phosphatidylinositol phosphate 5-kinase generate functionally distinct PIP_2 pools. Isokinase PIP5K1, but not PIP5K2, is implicated in the secretion of neurotransmitters and regulation of monomeric GTPases [132]. Isoform PIP5K1α tethers to Rac, PIP5K1β serves in endocytosis, and PIP5K1γ targets focal adhesions.

Isoform PI(4)P5K1α, but not PI(4)P5K1β, serves as a scaffold protein that localizes Rac1 to the plasma membrane at sites of integrin activation [133]. Small Rac1 GTPase links activated integrins to the regulation of cell migration.

In the brain, the major $PI(4,5)P_2$-producing enzyme corresponds to phosphatidylinositol 4-phosphate 5-kinase-γ_{661} [134]. Lipid $PI(4,5)P_2$ recruits the components of clathrin-mediated endocytosis that retrieve synaptic vesicles at nerve terminals. The AP2 complex binds to and activate $PIP5K\gamma_{661}$.

---

[115] Agent $PI(4,5)P_2$ is hydrolyzed by phospholipase-C into second messengers diacylglycerol and inositol trisphosphate as well as phosphorylated by PI3K into $PI(3,4,5)P_3$. Lipid $PI(4,5)P_2$ also modulates the activity of phospholipase-D and many actin-binding proteins. It is also required for Ca^{++}-regulated secretion of neurotransmitters.

2.10 Phosphoinositide Phosphatases

Phosphoinositide phosphatases (Table 2.45) dephosphorylate diverse positions (1, 3, 4, and 5) of the inositol ring (1-, 3-, 4-, and 5-phosphatases). They belong to 2 families of enzymes: protein Tyr phosphatases and inositol 5-phosphatase isoenzymes. There are multiple phosphoinositide lipid phosphatases in humans.

Lipid phosphatases — 3′-phosphatase PTen and 5′-phosphatase SHIP — control PI3K-dependent T lymphocyte signaling, i.e., immune system development and responsiveness as well as prevention of lymphocyte proliferation and autoimmunity. Phosphatases PTen and SHIP convert $PI(3,4,5)P_3$ into $PI(4,5)P_2$ and $PI(3,4)P_2$, respectively.

2.10.1 Phosphatase and Tensin Homolog

Phosphatase and tensin homolog deleted in chromosome 10 (PTen) is a dual-specificity protein–lipid phosphatase that has protein Tyr phosphatase activity and hydrolyzes phosphorylated position 3 of $PI(3)P$, $PI(3,4)P_2$, and $PI(3,4,5)P_3$, as well as inositol (1,3,4,5)-tetrakisphosphate.[116] Substrate $PI(3,4,5)P_3$ counteracts phosphoinositide 3-kinase activity. Phosphatase PTen can indeed be considered

Table 2.45. Phosphoinositide phosphatases (Source: [118]; MTM: myotubularin; MTMR: myotubularin-related phosphatase; OCRL: oculocerebrorenal syndrome of Lowe phosphatase; PTen: phosphatase and tensin homolog deleted on chromosome 10; SHIP: SH-containing inositol phosphatase; SKIP skeletal muscle- and kidney-enriched inositol phosphatase; Synj: synaptojanin).

Type	Function	Substrate(s)
PTen	3-Phosphatase	$PI(3,4)P_2$, $PI(3,4,5)P_3$
MTM1	3-Phosphatase	$PI(3)P$, $PI(3,5)P_2$
MTMR1–8	3-Phosphatase	$PI(3)P$, $PI(3,5)P_2$
4-Phosphatase type 1	4-Phosphatase	$I(3,4)P_2$, $I(1,3,4)P_3$, $PI(3,4)P_2$
4-Phosphatase type 2	4-Phosphatase	
OCRL1	5-Phosphatase	$PI(4,5)P_2$
SHIP1	5-Phosphatase	$I(1,3,4,5)P_4$, $PI(3,4,5)P_3$
SHIP2	5-Phosphatase	$I(1,3,4,5)P_4$, $PI(3,4,5)P_3$
Synj1	5-Phosphatase	$PI(4,5)P_2$, $PI(3,4,5)P_3$
SKIP	5-Phosphatase	$PI(4,5)P_2$, $PI(3,4,5)P_3$

---

[116] PTen Phosphatase possesses a PIP_2-binding polybasic sequence, a phosphatase region, a C2 motif, a PDZ (PSD95–DLg–ZO1) domain, as well as multiple phosphorylation sites.

essentially as a PI(3,4,5)P$_3$ phosphatase that produces PI(4,5)P$_2$. The balance between PTen and PI3K controls PI(3,4,5)P$_3$ concentration in the plasma membrane that regulates cell survival and proliferation.

Although PTen functions predominantly as a lipid phosphatase, it dephosphorylates focal adhesion kinase and adaptor SHC to inhibit cell migration [101].

Binding of PTen to the plasma membrane is needed to maintain an appropriate level of phosphatidylinositol (3,4,5)-trisphosphate. Phosphatase PTen binds to negatively charged phosphatidylserine in the inner plasmalemmal leaflet and phosphatidylinositol (4,5)-bisphosphate. Phosphatase PTen actually switches between 2 states that regulate its location (membrane attachment and detachment), as well as its function and degradation. Phosphorylation of PTen regulates its membrane binding [135].[117]

Ubiquitous phosphatase PTen is rapidly degraded (half-life 2–4 h according to the cell type). Phosphatase PTen is involved in many pathways that can stimulate cell growth or apoptosis. It particularly intervenes in the development of the nervous system. Phosphatase PTen is an important regulator of class-1 PI3Ks. Leptin effect in pancreatic β cells requires coincident PI(3,4,5)P$_3$ generation and actin depolymerization owing to both the lipid and protein phosphatase PTen activities. Phosphatase PTen suppresses many tumor types. It functions as a tumor suppressor by precluding the protein kinase-B pathway. It indeed stops cell division and causes apoptosis when necessary.

Enzyme PTen is activated by transcription factors early growth response EGR1 and P53.[118] Upon activation by IGF1, PKB phosphorylates EGR1. Afterward, EGR1 is sumoylated by alternate reading frame protein (ARF) in the nucleolus. The PKB–EGR1–ARF pathway of PTen transcription is characterized by inhibitory feedback, as PTen inhibits its own production [136].

Nucleus–cytosol partitioning of PTen results from PTen ubiquitinination and deubiquitinination by ubiquitin-specific peptidase (deubiquitinase) USP7 [137].[119] Promyelocytic leukemia protein (PML) opposes HAUSP activity via adaptor death domain-associated protein and excludes PTen from the cell nucleus. Conversely, PML function is disrupted by the PML–RARα fusion oncoprotein that restores PTen nuclear location that is required for repression of tumor initiation and development.

---

[117] Phosphorylation of PTen induces a conformation with an inactive phosphatase domain. Phosphorylation-dependent interaction between some PTen residues prevents membrane binding. Binding to PI(4,5)P$_2$ activates the phosphatase domain. Dephosphorylation allows PTen to associate with the plasma membrane. However, the accurate PTen phosphorylation–dephosphorylation process remains to be discovered.

[118] Early growth response-1 also activates P53, P73, P300–CBP, and other pro-apoptotic molecule genes. Transcription factor EGR1 is upregulated by growth factors to stimulate cell proliferation. Transcription factors EGR1 and P53 are able to induce mutual transcription.

[119] A.k.a. herpes virus-associated ubiquitin-specific peptidase (HAUSP)

2.10.2 Myotubularins and Myotubularin-Related Phosphoinositide 3-Phosphatases

Myotubularin MTM1[120] (Sect. 8.3.13.7) is either considered as a potent phosphoinositide 3-phosphatase [118] or an enzymatically inactive homolog of phosphoinositide phosphatase (pseudophosphatases; Sect. 8.1) that can still bind 3-phosphorylated inositol lipids and acts as an adaptor [138].

The myotubularin-related (MTMR) family of phosphoinositide 3-phosphatases includes 14 members in humans (MTM1, MTMR1–MTMR13) [118]. These phosphatases use inositol phospholipids rather than phosphoproteins (Tyr^P) as substrates.

Myotubularin MTM1 and MTMR phosphatases dephosphorylate specifically PI(3)P and PI(3,5)P_2 that are involved in endocytosis to generate PI and PI(5)P, respectively. However, the myotubularin superfamily contains a subset of 6 proteins (MTMR5, MTMR9–MTMR13) that are catalytically inactive and function as adaptors for the active members as well as regulators.

Phosphatase MTMR6 interacts with Ca^{++}-activated K^+ channel $K_{Ca}3.1$ that can be indirectly activated by PI(3)P lipid. Calcium-activated potassium channels maintain the membrane potential that generates Ca^{++} influx in activated naive and memory CD4+ T lymphocytes. Phosphatase MTMR6 regulates T-cell activation by dephosphorylating PI(3)P agent.

Regulation via Interactions between Myotubularins

Myotubularins are generally ubiquitous. They are not functionally redundant. Active and inactive myotubularins can operate in the same signaling pathways. Active and inactive myotubularins associate to form regulators, such as MTM1 and pseudophosphatase MTMR12, MTMR2 and pseudophosphatase MTMR5 or MTMR13, and MTMR7 and pseudophosphatase MTMR9 [139].

In MTMR2–MTMR5 and MTM1–MTMR12 couples, the pseudophosphatase regulates the subcellular localization of its active partner. Pseudophosphatases can also enhance the catalytic activity of active myotubularins (e.g., MTMR2, MTMR5, MTMR9, and MTMR7). Lastly, pseudophosphatases can act as adaptors that bind to corresponding ligands to coordinate components of a signaling cascade.

2.10.3 Phosphoinositide 4-Phosphatases

The family of phosphoinositide 4-phosphatases includes 2 ubiquitous isoforms. Type-1 and -2 PI(4,5)P_2 4-phosphatases convert PI(4,5)P_2, and, to a much lesser extent, I(3,4)P_2 and I(1,3,4)P_3 into PI(5)P. Both enzymes are located in late

---

[120]The myotubularin gene Mtm1 has been identified as a gene mutated in myotubular myopathy.

endosomal and lysosomal membranes. Alternatively spliced variants of type-1 and -2 4-phosphatases exist (α and β). Type-1α 4-phosphatase is expressed in megakaryocytes and fibroblasts using transcription factor GATA1 to regulate their proliferation [118]. Type-1α 4-phosphatase modulates PI3K signals in endosomes and at the plasma membrane. It decreases PI(3,4)P_2 level at the plasma membrane upon growth factor stimulation.

2.10.3.1 Inositol Polyphosphate 4-Phosphatases

Type-1 and -2 inositol polyphosphate 4-phosphatases are widespread. They remove the 4-position phosphate from phosphatidylinositol (3,4)-bisphosphate.[121]

In the central nervous system, inositol polyphosphate 4-phosphatase-A (IP(4)Pa) is a suppressor of glutamate excitotoxicity [140]. Enzyme IP(4)Pa localizes to postsynaptic densities. It regulates the synaptic localization of NMDA-type glutamate receptors and excitatory postsynaptic current through these receptors.

Inositol polyphosphate 4-phosphatase-2 (or IP(4)Pb) is a regulator of the PI3K signaling. Stimulation of platelets by thrombin causes a rapid, transient formation by phosphatidylinositol 3-kinase of PI(3,4)P_2 and PI(3,4,5)P_3. Once platelets are aggregated, a sustained, delayed accumulation of PI(3,4)P_2 results from the production of PI(3)P, and its phosphorylation by PI(3)P4K kinase. On the other hand, inositol polyphosphate 4-phosphatase IP(4)Pb hydrolyzes I(3,4)P_2, I(1,3,4)P_3, and PI(3,4)P_2. Like inositol polyphosphate 5-phosphatase that complexes to the P85 subunit of PI3K and produces PI(3,4)P_2, IP(4)Pb associates with the P85 subunit of PI3K in the cytosol of human platelets [141]. In addition, in activated thrombocytes, IP(4)Pa is degraded by calpain.

2.10.4 Phosphoinositide 5-Phosphatases

Phosphoinositide 5-phosphatases are classified into 4 types. Type-1 phosphoinositide 5-phosphatases hydrolyze only water-soluble I(1,4,5)P_3 and I(1,3,4,5)P_4 agents. They are prenylated and anchored to the membrane.

Type-2 phosphoinositide 5-phosphatases process not only water-soluble inositol phosphates, but also PI(4,5)P_2 and PI(3,4,5)P_3 substrates. These phosphatases are subdivided into 4 subgroups. The subgroup 1 is constituted by 5-phosphatase-2, or GAP domain-containing inositol 5-phosphatase (GIP). The GIP category of type-2 phosphoinositide 5-phosphatases also includes IP(5)Pf enzyme that contains a

---

[121] Phosphatidylinositol (4,5)-bisphosphate is hydrolyzed by phospholipase-C to produce soluble inositol trisphosphate and diacylglycerol. Phosphatidylinositol (3,4,5)-trisphosphate is dephosphorylated either by PTen phosphatase to form PI(4,5)P_2 or by 5-phosphatase to generate PI(3,4)P_2.

Table 2.46. Inositol polyphosphate 5-phosphatases (IP(5)P; **Part 1**; IP$_3$: inositol (1,4,5)-trisphosphate; IP$_4$: inositol (1,3,4,5)-tetrakisphosphate; GIP: GAP domain-containing inositol 5-phosphatase; OCRL: oculocerebrorenal syndrome of Lowe protein; pharbin: 5-phosphatase that induces arbolization; PI: phosphatidylinositol; PI(i)P: phosphatidylinositol (i)-monophosphate; PI(i, j)P$_2$: phosphatidylinositol (i, j)-bisphosphate; PI(3,4,5)P$_3$: phosphatidylinositol (3,4,5)-trisphosphate; PIB5Pa: phosphatidylinositol (4,5)-bisphosphate 5-phosphatase-A; PIPP: proline-rich inositol polyphosphate 5-phosphatase; SHIP: SH2 domain-containing inositol 5-phosphatase; SKIP: skeletal muscle- and kidney-enriched inositol 5-phosphatase; Synj: synaptojanin). Various phosphatases hydrolyze phosphoinositides at different position of the phosphate group (1-, 3-, 4-, or 5-phosphatases). Among many phosphoinositide phosphatases, 5-phosphatases constitute a relatively large family. They are characterized by their substrate specificity (type-1 inositol 5-phosphatases hydrolyze only water-soluble substrates such as I(1,4,5)P$_3$ and I(1,3,4,5)P$_4$). Type-2 inositol 5-phosphatases hydrolyze water-soluble inositol phosphates as well as PI(4,5)P$_2$ and PI(3,4,5)P$_3$. They are classified into 4 subgroups as well as 3 categories (GIP, SKIP, and SCIP).

Type	Members	Alias	Substrates
		Type-1 to type-4 IP(5)Ps	
1	IP(5)Pb	GIP	IP$_3$, IP$_4$,
	IP(5)Pf	OCRL	PI(4,5)P$_2$, PI(3,4,5)P$_3$
2	Synj1		PI(4,5)P$_2$ (type-2 5-phosphatase domain),
	Synj2		PI(3)P, PI(4)P, PI(3,5)P$_2$ (SAc1 domain)
	IP(5)Pj	PIB5Pa, PIPP	IP$_3$, IP$_4$, PI(4,5)P$_2$
	IP(5)Pk	SKIP	IP$_3$, IP$_4$, PI(4,5)P$_2$
	Pharbin		PI(4,5)P$_2$, PI(3,4,5)P$_3$
3	IP(5)Pd	SHIP1	PI(3,5)P$_2$, PI(3,4,5)P$_3$
	IP(5)PL1	SHIP2	
4			PI(3,5)P$_2$, PI(4,5)P$_2$, PI(3,4,5)P$_3$

GAP domain (Tables 2.46 and 2.47). Subgroups comprise IP(5)Pf,[122] synaptojanin-1 and -2, IP(5)Pj,[123] and IP(5)Pk.[124] Neuronal synaptojanin-1 and ubiquitous synaptojanin-2 form the SCIP category (SCIP: SAc1 domain-containing inositol phosphatases) of type-2 phosphoinositide 5-phosphatases.

Type-3 phosphoinositide 5-phosphatases target phosphate at the 5-position of phosphoinositides and inositol polyphosphates that have a 3-position phosphate group, such as PI(3,5)P$_2$ and PI(3,4,5)P$_3$. They include SH2 domain-containing inositol phosphatases SHIP1 and SHIP2, or IP(5)Pd and IP(5)PL1 enzymes. Hence, type-3 phosphoinositide 5-phosphatases correspond to the SHIP category of type-2

---

[122] A.k.a. oculocerebrorenal syndrome of Lowe (OCRL) phosphatase.

[123] A.k.a. phosphatidylinositol (4,5)-bisphosphate 5-phosphatase PIB5Pa and proline-rich inositol polyphosphate 5-phosphatase (PIPP).

[124] A.k.a. skeletal muscle and kidney enriched inositol phosphatase (SKIP).

Table 2.47. Inositol polyphosphate 5-phosphatases (IP(5)P; **Part 2**; IP$_3$: inositol (1,4,5)-trisphosphate; IP$_4$: inositol (1,3,4,5)-tetrakisphosphate; GIP: GAP domain-containing inositol 5-phosphatase; OCRL: oculocerebrorenal syndrome of Lowe protein; PI: phosphatidylinositol; PI(*i*)P: phosphatidylinositol (*i*)-monophosphate; PI(*i, j*)P$_2$: phosphatidylinositol (*i, j*)-bisphosphate; PI(3,4,5)P$_3$: phosphatidylinositol (3,4,5)-trisphosphate; SHIP: SH2 domain-containing inositol 5-phosphatase; SCIP: SAc1 (suppressor of actin mutation protein homolog) domain-containing inositol 5-phosphatase [SAc N-terminus is homologuous to that of the yeast phosphatidylinositol phosphatase SAc]; Synj: synaptojanin).

Type	Members	Alias	Substrates
Categories GIP, SHIP, and SKIP			
GIP	IP(5)Pb	GIP, 5PTase	IP$_3$, IP$_4$,
	(IP(5)Pf)	OCRL	PI(4,5)P$_2$, PI(3, 4,5)P$_3$
SHIP	IP(5)Pd	SHIP1	PI(3, 4,5)P$_3$
	IP(5)PL1	SHIP2	
SCIP	Synj1		PI(3)P, PI(4)P,
	Synj2		PI(3,5)P$_2$, PI(4,5)P$_2$
	IP(5)Pf	SAc2	PI(4,5)P$_2$, PI(3,4,5)P$_3$

phosphoinositide 5-phosphatases. The SHIP phosphatases that possess an SH2 domain at the N-terminus complex with intracellular signaling adaptors, such as GRB2 and SHC proteins. They target PI(3,4,5)P$_3$. Whereas IP(5)Pd localizes to hematopoietic cells, IP(5)PL1 resides in non-hematopoietic cells, where it antagonizes signaling primed by EGF, PDGF, and insulin.

Type-4 5-phosphatases hydrolyze PI(3,5)P$_2$, PI(4,5)P$_2$, and PI(3,4,5)P$_3$.

Mitochondrial phosphoTyr phosphatase PTPmt1,[125] is a PTen-like phosphatase with a transmembrane domain. It has a highly specific 5-phosphatase activity against PI(5)P, but it is unrelated to the inositol polyphosphate 5-phosphatase family. The PTPmt1 phosphatase is involved in the regulation of adenosine triphosphate production and insulin secretion in pancreatic β cells.

2.10.4.1 Inositol Polyphosphate 5-Phosphatases

Inositol and phosphatidylinositol polyphosphate 5-phosphatases selectively remove the phosphate at position 5 of the inositol ring from its substrates.

Type-1 inositol 5-phosphatase dephosphorylates I(1,4,5)P$_3$ and I(1,3,4,5)P$_4$, but does not target any phosphoinositide [118].

Type-2 inositol 5-phosphatases comprise 9 different isoforms in humans. They contain many binding domains (SH2, proline-rich, GAP, CAAX, or sterile α motif [SAM] domains). Inositol 5-phosphatase activity depends on Mg^{++} ion.

---

[125] Previously named phospholipid-inositol phosphatase (PLIP).

Type-2 5-phosphatase, proline-rich inositol polyphosphate 5-phosphatase (PIPP), or IP(5)Pj, localizes to membrane ruffling zones. This phosphatase hydrolyzes phosphate at the 5-position of I(1,4,5)P$_3$, I(1,3,4,5)P$_4$, and PI(4,5)P$_2$ [142].

Type-4 inositol polyphosphate 5-phosphatase[126] has a high affinity for PIP$_3$, but can also target PI(3,5)P$_2$. It prevents protein kinase-B phosphorylation [118]. It also hydrolyzes I(1,4,5)P$_3$ and I(1,3,4,5)P$_4$ more effectively than PI(4,5)P$_2$ [142].

2.10.4.2 Synaptojanins

Synaptojanins complex with dynamin, amphiphysin, and GRB2 adaptor to promote synaptic vesicle recycling. Both synaptojanins have 3 domains: an N-terminal SAc1 homology region, a central type-2 inositol 5-phosphatase domain, and a C-terminal proline-rich region. Hence, synaptojanins contain 2 phosphatase domains: (1) one common to all type-2 inositol 5-phosphatases that can hydrolyze PI(4,5)P$_2$ into PI(4)P, but not PI(3,5)P$_2$, as well as PI(3,4,5)P$_3$ and water-soluble I(1,4,5)P$_3$ and I(1,3,4,5)P$_4$ substrates, and (2) a SAc domain that dephosphorylates PI(3)P, PI(4)P, and PI(3,5)P$_2$, but not PI(4,5)P$_2$ [142, 143]. The proline-rich domain enables interaction with SH3 domain-containing proteins.

Synaptojanin-1 and -2 participate in the regulation of clathrin-mediated endocytosis. Synaptojanin-1 promotes uncoating of clathrin-coated vesicles, whereas synaptojanin-2 supports early steps of clathrin-mediated endocytosis. Cofactor PI(4,5)P$_2$ is used by several proteins required for the formation of clathrin-coated pits, such as dynamin, Adaptor proteic complex AP2, synaptotagmin, and epsin. Localized PI(4,5)P$_2$ turnover due to synaptojanin-2 may assist the exchange of clathrin subunits that occurs during coated pit invagination and constriction [143].

Synaptojanin-1

Synaptojanin-1 is a nerve terminal protein that participates with dynamin in synaptic vesicle recycling. In vitro, synaptojanin-1 targets PI(3)P, PI(4)P, PI(3,5)P$_2$, and PI(4,5)P$_2$ [118]. Polyphosphoinositide phosphatase activity is independent of Mg^{++} ions.

Synaptojanin-2

Synaptojanin-2 has several isoforms (Synj2a and Synj2b1–Synj2b2)[127] that are linked to different protein interactors to direct proteins to different subcellular com-

---

[126] A.k.a. pharbin (5-phosphatase that induces arbolization) and 72-kDa inositol 5-phosphatase.

[127] Synaptojanin-2A, -2B1, and -2B2 correspond to synaptojanin-2α, -2δ, and -2β, respectively. In humans, mRNAs have been identified for only 2 isoforms: synaptojanin-2A and -2B2. However, a human mRNA similar to synaptojanin-2ε has also been detected.

partments. In particular, Synj2a, but not Synj2b, binds to synaptojanin-2-binding protein Synj2BP[128] to control the intracellular distribution of mitochondria [144].[129] Synaptojanin-2 is required for formation of clathrin-coated pits and lamellipodia during cell migration [118].

In addition to the proline-rich domain, the C-terminus contains a Rac1-binding domain [143]. Small Rac1 GTPase regulates synaptojanin-2 translocation from the cytoplasm to the plasma membrane.

Ubiquitous synaptojanin-2 binds to amphiphysin, endophilin, and GRB2 adaptor. The interaction with endophilin and amphiphysin depends on alternatively spliced isoform type [143]. Amphiphysin links to both synaptojanin-2B subtypes, endophilin only binds to synaptojanin-2B2. Adaptor GRB2 can connect synaptojanin-2 to many binding partners. Synaptojanin-2 also tethers to SH3 domain-containing kinase-binding protein SH3KBP1,[130] a scaffold protein that coordinates events in clathrin-mediated endocytosis.

2.10.4.3 SAc Domain-Containing 5-Phosphatases

In addition to synaptojanins, 3 suppressor of actin (SAc) domain-containing 5-phosphatases (SAc1–SAc3), which exist in humans, contribute to the organization of intracellular actin and organelle membranes such as that of the Golgi body.

SAc1 5-phosphatase executes some cell chores, such as organization of Golgi membranes and mitotic spindles [145]. In addition to its enzymatic activity, endoplasmic reticulum localization is needed for its functioning. Mammalian SAc1 is an integral membrane protein with an anchor that primarily resides in endoplasmic reticulum membranes. The CoP1–SAc1 complex captures SAc1 in the Golgi body to recycle it back to the endoplasmic reticulum in CoP1-associated vesicles. Sac1 5-phosphatase dephosphorylates $PI(3)P$, $PI(4)P$, and $PI(3,5)P_2$.

SAc2 5-phosphatase has a different substrate specificity than that of SAc1, as it can dephosphorylate the 5-position of $PI(4,5)P_2$ and $PI(3,4,5)P_3$ [142]. It hydrolyzes most effectively $PI(4,5)P_2$, with an affinity for $PI(4,5)P_2$ that is comparable with that of type-2 5-phosphatases.

Mammalian magnesium-activated *SAc3 5-Phosphatase* (or Fig4), a $PI(3,5)P_2$ 5-phosphatase, localizes to the vesicular membrane of the late endosome–lysosome compartment. Signaling $PI(3,5)P_2$ is produced on endosomes to regulate the retrograde traffic to the trans-Golgi network and control the vesicle size. Vacuole-associated protein Vac14 serves as a scaffold for regulators of $PI(3,5)P_2$, such as 5-phosphatase SAc3, phosphatidylinositol 3-phosphate 5-kinase (PI(3)P5K), Vac7, and WD repeat domain-containing phosphoinositide-interacting protein WIPI1 (or Atg18) that allows a rapid interconversion of PI3P and $PI(3,5)P_2$ [146]. Both Vac7

---

[128] A.k.a. outer mitochondrial protein OMP25.

[129] Overexpression of OMP25 causes a perinuclear clustering of mitochondria.

[130] A.k.a. 85-kDa CBL-interacting protein CIN85.

and Vac14 favor PI(3)P5K activity for PI(3,5)P$_2$ synthesis. In addition, Vac14 (but not Vac7) recruits 5-phosphatase SAc3 responsible for PI(3,5)P$_2$ dephosphorylation to vesicle [147].

2.10.4.4 Skeletal Muscle and Kidney Enriched 5-Phosphatase

Skeletal muscle and kidney enriched 5-phosphatase (SKIP) abounds particularly in skeletal muscles, heart, and kidney. Phosphatase SKIP hydrolyzes PI(4,5)P$_2$ and PI(3,4,5)P$_3$, but substrate specificity can be cell-type specific [118]. It acts as an inhibitor in the regulation of the actin cytoskeleton. Phosphatase SKIP precludes translocation of glucose transporter GluT4 in response to insulin.

2.10.4.5 Proline-Rich Inositol Polyphosphate 5-Phosphatase

Proline-rich inositol polyphosphate 5-phosphatase (PIPP) is observed in membrane ruffles. Its substrates are I(1,4,5)P$_3$, I(1,3,4,5)P$_4$4, PI(4,5)P$_2$, and PI(3,4,5)P$_3$ [118]. Dephosphorylation of PI(3,4,5)P$_3$ modulates activity of protein kinase-B and glycogen synthase kinase-3β, as well as microtubule polymerization.

2.10.4.6 SH2-Containing Inositol Polyphosphate 5-Phosphatases

SH2-containing inositol polyphosphate 5-phosphatases comprise 2 isoforms (IP(5) Pd and IP(5)PL1 or SHIP1–SHIP2) that operate in the immune response. Their substrates are I(1,3,4,5)P$_4$, PI(4,5)P$_2$, and PI(3,4,5)P$_3$. In vivo, they essentially target PI(3,4,5)P$_3$ to produce PI(3,4)P$_2$ that interacts with specific adaptors: 32 kDa B-lymphocyte-associated adaptor molecule (BAM32) and tandem PH domain-containing proteins TaPP1 and TaPP2 [118]. Many SHIP interactors are cytoskeletal proteins. The SHIP phosphatases inhibits membrane ruffling. Both types impedes the MAPK and PKB pathways after receptor stimulation.

Inositol polyphosphate 5-phosphatase IP(5)Pd (or SHIP1) expressed in hematopoietic cells interacts with GRB2 and SHC [118]. It is implicated in myeloid cell survival. Phosphatase IP(5)Pd regulates levels of PI(3,4,5)P$_3$ and PI(3,4)P$_2$ in stimulated platelets to favor platelet contractility, thrombus organization and fibrin clot retraction. The IP(5)Pd phosphatase is phosphorylated in response to TCR–CD28 binding [148]. It inhibits TEC kinase. It could act in CD4+, CD25+ T$_{Reg}$-cell development. It also influences PI(3,4,5)P$_3$ metabolism in B lymphocytes and mastocytes. In mastocytes, the high-molecular-weight form of IP(5)Pd (IP(5)Pd$_{hMW}$) binds to filamin in ^{F}actin to accumulate in the vicinity of FcϵR1 receptor. Phosphatase IP(5)Pd binds to diverse receptors (erythropoietin receptor, FcγR2b1, and FcϵR1) directly or via adaptors. It is Tyr phosphorylated in response to various stimuli by growth factors, cytokines, and B-cell receptors.

Widespread inositol polyphosphate 5-phosphatase IP(5)PL1 (or SHIP2) is involved in insulin signaling [118]. Upon insulin stimulation, IP(5)PL1 translocates to the plasma membrane, where it inhibits insulin-specific subcellular redistribution of protein kinase-B2. This phosphatase also influences the metabolism of PI(3,4,5)P$_3$ and PI(4,5)P$_2$ in fibroblasts. It associates directly or via adaptors with the receptors of EGFR, HGFR, and CSF1, as well as FcγR2B1. It is phosphorylated by EGF, PDGF, or insulin. In addition, IP(5)PL1 forms a complex with actin-binding protein filamin-C to regulate PI3K signaling directed to the actin cytoskeleton.

2.10.4.7 Voltage-Sensing Phosphoinositide Phosphatase

Sensing of transmembrane electrical potential is not only carried out by voltage-gated ion channels, but also by voltage-sensing phosphoinositide phosphatase (VSP), or phosphoinositide-converting protein that contains a voltage sensor domain (Ci-VSP). Phosphatase VSP targets negatively charged phosphoinositides. Transmembrane voltage regulates its enzymatic activity. Upon activation by membrane depolarization, VSP acts as a 5-phosphatase that converts PI(4,5)P$_2$ and PI(3,4,5)P$_3$ into PI(4)P and PI(3,4)P$_2$ [149].

2.11 Other Lipid Phosphatases – Phosphotransferases

Phospholipids and sphingolipids participate in signal transduction and intracellular membrane trafficking as well as cell growth and survival. Many enzymes that are often tightly associated with cell membranes, are responsible for bioactive lipid metabolism. Lipid phosphates can initiate signaling cascades.[131] Therefore, phosphatases that degrade lipid phosphates on the cell surface as well as within the cell regulate cell signaling. They either attenuate signaling by the lipid phosphates or produce new active compounds, such as ceramide and sphingosine [150].

Five subclasses of homologous proteins termed lipid phosphatases or phosphotransferases (LPT) are characterized by a core domain with 6 transmembrane-spanning α-helices connected by extramembrane loops, 2 of which interact to form the catalytic site. Members of the LPT class are located in membranes. Their active site faces the lumen of endomembrane compartments or extracellular face of the plasma membrane. The LPT class includes lipid phosphate phosphatases (LPP),

---

[131] Lysophosphatidate and sphingosine 1-phosphate that can be produced by activated platelets stimulate their cognate G-protein-coupled receptors, in particular for tissue repair. Inside the cell, phosphatidate, ceramide 1-phosphate, lysophosphatidic acid, and sphingosine 1-phosphate act as messengers to stimulate cytoskeletal rearrangement, Ca^{++} transients, up to cell division.

sphingosine phosphate phosphatases (SPP), sphingomyelin synthases (SMS), lipid phosphatase-related proteins (LPR),[132] and type-2 candidate sphingomyelin synthases (CSS2) [138].[133]

Lipid phosphate phosphatases hydrolyze lysophosphatidic acid (LPA), sphingosine 1-phosphate (S1P), and structurally related substrates. Lipid phosphate phosphatases can dephosphorylate intracellular substrates to control intracellular lipid metabolism and signaling. However, they mainly regulate cell surface receptor-mediated signaling by LPA and S1P [138]. Isoenzymes LPP1 and LPP3 are widely expressed in human tissues, whereas LPP2 has lower concentration and more restricted localization.

Sphingosine phosphate phosphatases are intracellular, S1P-selective phosphatases that operate in sphingolipid metabolism to control cell growth and survival. Isoenzyme SPP2 has a more widespread expression than that of SPP1. Enzymes SPPs regulate cell survival, apoptosis, and migration [138].

Related sphingomyelin synthases catalyze interconversion of phosphatidylcholine to ceramide by transferring a phosphocholine group, thereby generating sphingomyelin and diacylglycerol to regulate bioactive lipid mediators. Two isoenzymes exist, SMS1 and SMS2, that are mainly located in the Golgi body and plasma membrane, respectively.

The LPR–PRG subgroup that may interact with LPA, S1P, or related lipidic signaling molecules includes 4 members with splice variants. They have restricted expression patterns. Nonetheless, LPR4 (PRG2) is more widely expressed than LPR3 (PRG1) and LPR2 (PRG4) than LPR1 (PRG3) [138]. Enzyme LPR3 is most strongly expressed in neurons, where it raises extracellular LPA breakdown.

Type-2 candidate sphingomyelin synthase CSS2β is widely expressed in mammalian tissues, whereas CSS2α is more restrictedly expressed. The CSS2 enzymes operate in nuclear phospholipid metabolism.

2.12 (Ecto)Nucleotide Pyrophosphatase–Phosphodiesterases

Ectonucleotide pyrophosphatase–phosphodiesterases target nucleotides (e.g., ATP). In addition, certain ENPP family members process lipids [151] (Table 2.48). Extracellular nucleotides are processed by ectoATPase (extracellular ATP and ADP), ecto-ATP diphosphohydrolase (ectoATPDase; extracellular ADP), and ecto-5′-nucleotidase (extracellular AMP), among which are ENPPs (Vol. 3 – Chap. 1. Signal Transduction).

---

[132] A.k.a. plasticity-related gene products (PRG).

[133] Three lipid phosphate phosphatases and a splice variant dephosphorylate phosphatidic acid, lysophosphatidic acid, ceramide 1-phosphate, and sphingosine 1-phosphate [150]. In addition, sphingosine 1-phosphate has 2 specific S1P phosphatases (SPP). Lipid phosphates are also substrates of 4 lipid phosphate phosphatase-related proteins.

2.12 (Ecto)Nucleotide Pyrophosphatase–Phosphodiesterases

Table 2.48. Substrates and functions of ENPP family enzymes (Source: [151]). Enzymes ENPP1 to ENPP3 have a wide substrate specificity. They intervene in many cell activities, such as bone mineralization, calcification of ligaments and joint capsules, modulation of purinergic receptor signaling, nucleotide recycling, and cell motility.

Type	Substrate	Functions
ENPP1	Nucleotides	Prevention of excessive bone calcification
ENPP2	Lysophospholipids	Vasculature and neural tube formation
		Lymphocyte migration
ENPP3		Basophil marker
ENPP4		
ENPP5		
ENPP6	Glycerophosphorylcholine	
	Lysophosphatidylcholine	
	Sphingosylphosphorylcholine	
ENPP7	Sphingomyelin	
	Lysophosphatidylcholine	
ENPP8		

The family of nucleotide pyrophosphatase–phosphodiesterases or ectonucleotide pyrophosphatase–phosphodiesterases is constituted of 8 members (NPP1–NPP8 or ENPP1–ENPP8). Similar to most of the molecules, a large number of aliases results from the fact that ENPPs have been discovered independently several times with possibly updated functions (Table 2.49). Isoenzymes ENPP1 to ENPP3 are observed in almost all tissues, although isoforms are usually confined to specific cell types [152].

Each member possesses a catalytic domain. Members ENPP4 to ENPP7 consist of only this single catalytic domain. Isoforms ENPP4 and ENPP5 are type-1 transmembrane proteins (Vol. 1 – Chap. 7. Plasma Membrane) with a short intracellular C-terminus and a smaller extracellular region that contains only a phosphodiesterase motif. Other members — ENPP1 to ENPP3 — are type-2 transmembrane metalloenzymes. They have a short intracellular N-terminus, a single transmembrane region, and an extracellular domain that contains the catalytic site. These members incorporate 2 additional domains with respect to ENPP4 to ENPP7: a somatomedin-B-like and nuclease-like domain in the N- and C-termini, respectively [151]. Enzymes ENPP1 and ENPP2 are disulfide-bound homodimers, whereas ENPP3 is monomeric [152].

2.12.1 Substrates and Functions of ENPPs

Each ENPP has distinct substrate ranges and functions, particularly related to nucleotide recycling, modulation of purinergic receptor signaling, control of extracellular pyrophosphate level, stimulation of cell motility, especially during

Table 2.49. Aliases of ectonucleotide pyrophosphatase–phosphodiesterases (Atx: autotaxin [autocrine motility factor], CD39L: CD39 antigen-like; ectoATPase: ecto-adenosine triphosphatase; ectoATPDase: ecto-ATP diphosphohydrolase; GP130^{RB13-6}: 130-kD glycoprotein recognized by the monoclonal antibody RB13-6; LALP1: lysosomal apyrase-like protein-1; LALP70, LAP70, LysAL1: lysosomal apyrase-like protein of 70 kDa; LCAA: lymphoid cell activation antigen; MAFP: major aFGF-stimulated protein; M6S1: membrane component chromosome 6 surface marker 1; NPP [NPPS]: nucleotide pyrophosphatase; NTPDase: nucleoside triphosphate diphosphohydrolase; PC1: plasma cell membrane glycoprotein-1; PCA1: plasma cell differentiation antigen-1; PCPH: proto-oncogene carcinogenesis promotion hamster; PDi: phosphodiesterase-I; PDNP: phosphodiesterase nucleotide pyrophosphatase; UDPase: uridine-diphosphatase).

Type	Other aliases
ENPP1	NPP1, NPPS, CD39, ecto-apyrase, ectoATPase1, ectoATPDase1, NTPDase1, NPPγ, PDNP1, MAFP, M6S1, LCAA, PCA1, PC1
ENPP2	NPP2, Atx, CD39L1, ectoATPase2, ectoATPDase2, NTPDase2, PDiα, NPPα, PDNP2
ENPP3	NPP3, CD39L3, ecto-apyrase-3, ectoATPase3, ectoATPDase3, NTPDase3, HB6, B10, PDiβ, NPPβ, PDNP3, GP130^{RB13-6}
ENPP4	NPP4, UDPase, NTPDase4, LALP70, LAP70, LysAL1
ENPP5	NPP5, CD39L4, NTPDase5, ER-UDPase, PCPH
ENPP6	NPP6, CD39L2, NTPDase6
ENPP7	NPP7, LALP1, NTPDase7
ENPP8	NPP8, NTPDase8

angiogenesis, and possibly regulation of insulin receptor signaling and activity of ectokinases (Table 2.48). Isozyme ENPP1 catalyzes nucleotides that generate pyrophosphate [151]. Subtype ENPP2 (or autotaxin) hydrolyzes lysophosphatidylcholine into lysophosphatidic acid. Isoform ENPP6 has lysophospholipase-C activity for choline-containing glycerophosphodiesters. Isoform NPP7 has alkaline sphingomyelinase activity and catalyzes lysophosphatidylcholine.

Isozymes ENPP1 to ENNP3 are classified as alkaline ectonucleotide pyrophosphatase and phosphodiesterase-I. They catalyze the hydrolysis pyrophosphate and phosphodiester bonds in a 2-step mechanism. Two essential divalent metal ions are required for the formation of a nucleotidylated active-site threonine intermediate and subsequent release of nucleoside 5′-monophosphate [152].

The catalytic activity of ENPP1 to ENNP3 is inhibited, at least in vitro, by heparin and heparan sulfate [152]. At low ATP concentration, autophosphorylation of these enzymes may occur and inhibit their enzymatic activity.[134]

---

[134] This autoregulatory mechanism reverses upon local accumulation of nucleotides (>5 μmol) [152].

Table 2.50. Regulators of ENPP expression (Source: [152]; +, −: increased, decreased ENPP synthesis; BMP: bone morphogenic protein; FGF: fibroblast growth factor; Ifn: interferon; IGF: insulin-like growth factor; IL: interleukin; PDGF: platelet-derived growth factor; TGF: transforming growth factor).

Regulators	Cell type (effect on ENPP isoform synthesis)
Angiotensin-2	Vascular smooth muscle cell (ENPP3 −)
Retinoic acid	Epithelial cell (ENPP2 +)
FGF2	Vascular endothelial and smooth muscle cells (ENPP2 +)
IGF1	Chondrocyte (ENPP1 −)
PDGF	Vascular smooth muscle cell (ENPP3 −)
TGFβ	Chondrocyte, synoviocyte, osteobalst, hepatocyte (ENPP1 +)
BMP2	Mesenchymal progenitor (ENPP2 +)
IL1β	Chondrocyte (ENPP1 −), synoviocyte (ENPP1/2 −)
IL4	Synoviocyte (ENPP1/2 −)
Ifnγ	Synoviocyte (ENPP2 −)

2.12.2 Production Regulation

Synthesis of ENPPs appears to be constitutive in some cell types, such as plasmocytes and chondrocytes, but inducible in others [152]. Production of ENPPs is regulated by many agents such as growth factors, among which are cytokines (Table 2.50).

Transforming growth factor-β upregulates the expression of NPP1 in chondrocytes, synoviocytes, osteoblasts, and hepatocytes [152]. In chondrocytes, this effect is associated with an enhanced transport to the plasma membrane. However, this effect is antagonized by interleukin-1β. The TGFβ receptors activate Jun N-terminal kinase via activated mitogen-activated protein kinase kinases MAP2K4 and MAP2K7. On the other hand, inhibitory iSMAD connects to the TβR1 receptor, thereby preventing TβR1 association with rSMADs.

Kinases MAP2K4 and MAP2K7 are also activated by IL1 receptor-associated kinases (IRAK) stimulated by interleukin-1β. Activated glucocorticoid receptor sequesters the transcription factor NFκB [152]. Receptors of TGFβ and angiotensin-2 transactivate epidermal growth factor receptor using a metallopeptidase, hence priming the extracellular signal-regulated protein kinase pathway. The MAPK enzymes phosphorylate SMAD1 to SMAD3, thus inhibiting their nuclear translocation.

Active soluble forms of ENPP1 to ENPP3 are detected in blood after cleavage close to the transmembrane domain (ENPP1) or more distantly (ENPP2 and ENPP3). This cleavage generates catalytic active soluble fragments.

2.12.3 ENPP1

Enzyme ENPP1 is a type-2 transmembrane glycoprotein that cleaves nucleotides and nucleotide sugars. The short intracellular domain of ENPP1 has a basolateral membrane-targeting signal. Expression of NPP1 is constitutive or induced by transforming growth factor-β and glucocorticoids, at least in some cell types [152].

Production of ENPP1 is low or absent on most human T or B lymphocytes, although it can be upregulated upon activation of T lymphocytes [152]. It is detected in antibody-secreting plasmocytes. Human ENPP1 is highly expressed in bone and cartilage cells. Moderate levels are found in heart, liver, placenta, and testis. It also lodges in the distal convoluted tubules of the kidney, brain capillary endothelium, among other tissues.

2.12.4 ENPP2 (Autotaxin)

Autotaxin (Atx) is also called ectonucleotide pyrophosphatase–phosphodiesterase-2 (ENPP2) as well as extracellular lysophospholipase-D (lysoPLD). In humans, it is mainly expressed in the brain, small intestine, ovary, and placenta [152].

Enzyme ENPP2 is an ectophosphodiesterase that synthesizes lysophosphatidic acid, in particular in various biological fluids, such as blood and cerebrospinal fluid. It was originally isolated as an autocrine motility stimulating factor, hence its name. Agent ENPP2 may be either a type-2 membrane protein that is cleaved to create a soluble or secreted protein. In humans, 3 alternatively spliced variants (ENPP2β–ENPP2δ) have been identified.

Lysophosphatidic acid, the simplest phospholipid, is a second-generation lipid mediator that binds to its cognate G-protein-coupled receptors (Vol. 3 – Chap. 7. G-Protein-Coupled Receptors). It can provoke cell proliferation and migration, cytokine and chemokine secretion, platelet aggregation, smooth muscle contraction, transformation of smooth muscle cells, and neurite retraction, and prevents apoptosis.

Lysophosphatidic acid is synthesized both inside and outside the cell. It is produced by a degradative reaction from phospholipids via at least 2 routes. In cells, LPA is converted from phosphatidic acid. In biological fluids, LPA is obtained from lysophospholipids such as lysophosphatidylcholine.

Ectoenzyme ENPP2 hydrolyzes various lysophospholipids, such as lysophosphatidylcholine, lysophosphatidylethanolamine, and lysophosphatidylserine [151]. It also processes sphingosylphosphorylcholine to produce sphingosine 1-phosphate. Its substrate specificity is influenced by some divalent cations, such as Co^{++} or Mn^{++} that can replace Zn^{++} ions, which bind in the vicinity of the catalytic site.

Protein ENPP2 exerts its activity mainly via lysophosphatidic acid. Yet, it has some catalytic-independent functions. It actually possesses a modulator of oligodendrocyte remodeling and focal adhesion organization (MORFO) for reorganization of focal adhesions [151].

2.12.5 ENPP3

Enzyme ENPP3 localizes to the apical surface of polarized cells. It is found in glial precursors (at least in rats, but not mature astrocytes). It can serve as a marker of activated basophils, in addition to mastocytes. It also lodges on hepatocyte apical membrane. It can be detected in human prostate as well as uterus and colon, but not liver, pancreas, and small intestine.

Chapter 3
Preamble to Cytoplasmic Protein Kinases

Among kinases (or phosphotransferases) that transfer phosphate groups from donor molecules such as ATP or GTP to specific substrates include protein kinases that modify the activity of their proteic substrates. Kinases are used to transmit signals in cells. Other types of kinases act on amino acids, lipids, carbohydrates, and nucleotides, that are then used in signaling pathways or cell metabolism.

Protein kinases are encoded by one of the largest gene families. About 2% of the human genome encodes 538 identified kinases. They relay extracellular signals by transferring phosphate groups from ATP or GTP to free hydroxyl groups of specific amino acids. When they reversibly catalyze the transfer of phosphate groups to target proteins, protein kinases change the substrate activity, location, or binding capacity with other substances, thereby regulating metabolic and signaling pathways.

Protein phosphorylation is the most widespread post-translational modification in signal transduction. The number of substrates and phosphorylation sites (from a few sites to hundreds of sites) varies according to the kinase type.[1] Many substrates must be phosphorylated at multiple sites to change their function. Phosphoryl transfer by kinases on specific substrates regulates cell metabolism, transport, growth, division, differentiation, motility, as well as muscle contraction, immunity, learning, and memory (Table 3.1). Adaptor and scaffold proteins enable signal transduction in kinase pathways, as they serve as platforms for kinases to achieve access to their substrates, which also are frequently kinases.

---

[1]The amino acid composition in the neighborhood of the target residue plays a role. Adjoining proline-directed kinases may regulate a larger number of proteins than other non-Pro-directed kinases [153].

Table 3.1. Examples of cell activities and involved protein kinase families (Source: [154]; AATK: apoptosis-associated Tyr kinase; Axl: adhesion-related receptor Tyr kinase; CamK: calmodulin-dependent kinase; DAPK: death-associated protein kinase; EGFR: epidermal growth factor receptor Tyr kinase; EPH: erythropoietin-producing hepatocyte receptor Tyr kinase; HUNK: hormonally up-regulated Neu-associated kinase; IKK: IκB kinase; JaK: Janus (pseudo) kinase; IRAK: interleukin-1 receptor-associated kinase; MAPK: mitogen-activated protein kinase; NTRK: neurotrophic Tyr receptor kinase; PDGFR: platelet-derived growth factor receptor Tyr kinase; RIPK: receptor-interacting protein kinase; RSK: P90 ribosomal S6 kinase; Src: sarcoma-associated (Schmidt-Ruppin A2 viral oncogene homolog) kinase; TAOK: thousand and one kinases (MAP3K16–MAP3K18); TEC: Tyr kinase expressed in hepatocellular carcinoma; TIE: Tyr kinase with Ig and EGF homology domains (angiopoietin receptor); VEGFR: vascular endothelial growth factor receptor).

Function	Types
Angiogenesis	PDGFR, VEGFR, TIE, Axl, IKK, IRAK, RIPK, Src, TEC
Apoptosis	AATK, DAPK, RIPK
Calcium signaling	CamK1/2
EGF signaling	EGFR, HUNK, RSK, Src, TAOK
Hemopoiesis,	JaK
Immunity	JaK
Nervous control	EPH, NTRK
Signaling	MAPK modules

3.1 Protein Kinase Classification

The set of protein kinases is constituted of diverse enzymes according to their regulation, activation, and substrate recognition. Phosphorylation rate depends on the kinase location. The magnitude order of the different stages of protein phosphorylation (ATP binding, substrate binding, ADP release, and substrate release) depends on kinase type.

According to the targeted residues, protein kinases are classified into: (1) tyrosine (Tyr) kinases; (2) serine/threonine (Ser/Thr) kinases that phosphorylate serine or threonine; and (3) dual-specificity kinases that are able to phosphorylate the 3 types of residues. Among specific protein kinases, Ser/Thr kinases are the most abundant; in humans, more than 80% of protein kinase genes encode Ser/Thr kinases.

On the other hand, protein phosphatases (Sect. 8.2) remove phosphate groups from amino acid residues of proteins phosphorylated by protein kinases.[2]

---

[2]The numbers of Tyr kinases and phosphatases are similar; but Ser/Thr phosphatases are less numerous than Ser/Thr kinases.

3.1 Protein Kinase Classification

Table 3.2. Identified protein kinase sets and subsets in 2002 (Sources: [154, 155]).

Superfamily	Families number	Subfamilies number	Kinases number
AGC	14	21	63
CAMK	17	33	74
CK1	3	5	12
CMGC	8	24	61
Other	37	39	83
STE	3	13	47
Tyr kinase	30	30	90
Tyr kinase-like	7	13	43
RGC	1	1	5
Atypical-PDHK	1	1	5
Atypical-Alpha	1	2	6
Atypical-RIO	1	3	3
Atypical-A6	1	1	2
Atypical-Other	7	7	9
Atypical-ABC1	1	1	5
Atypical-BRD	1	1	4
Atypical-PIKK	1	6	6

The class of protein Ser/Thr kinases is subdivided into several superfamilies (Table 3.2): (0) superfamily of *miscellaneous* protein Ser/Thr kinases that encompasses numerous families; (1) **AGC** superfamily based on protein kinases-A, -C, and -G; (2) **CAMK** superfamily of Ca^{++}–calmodulin-dependent kinases; (3) **CK1** superfamily of casein kinases-1;[3] (4) **CMGC** superfamily of Ser/Thr protein kinases includes kinases that intervene in cell proliferation in response to mitogenic signals, i.e., cyclin-dependent kinases (CDK), mitogen-activated protein kinases (MAPK), glycogen synthase kinases (GSK), and cyclin-dependent kinase-like kinases (CDKLK); (5) **STE** superfamily of Sterile-20 homolog kinases; (6) **TK** superfamily of Tyr kinases; (7) **TKL** superfamily of Tyr kinase-like kinases that are Ser/Thr protein kinases; (8) **RGC** superfamily of receptor guanylate cyclases that constitute a small set of kinases similar in sequence to the TK superfamily; and (9) *atypical kinase* superfamily.

3.1.1 Superfamily 0 (Miscellaneous)

Kinases that do not belong to defined kinase superfamilies are grouped into superfamily 0 (miscellaneous). In addition to numerous Sugen kinases, this superfamily encompasses multiple families (Table 3.3) [154]:

---

[3]Casein is a convenient substrate for experimental examination of kinase activity.

Table 3.3. Superfamily 0 of other protein kinases (Sugen nomenclature; Source: [156]).

Family	Members
Aurora	AurA–AurC (or Aur1–Aur3)
BUB	BUB1, BUBR1 (or BUB1β)
CAMKK	CamKK1–CamKK2
CDC7	CDC7
CK2	CK2α1–CK2α2
IKK	IKKα, IKKβ, IKKε, TBK1
IRE	IRE1–IRE2
MOS	MOS
NAK	AAK1, BIKE, GAK, MPSK1
NEK	NEK1–NEK11
NKF1	SBK, SgK069, SgK110
NKF2	PINK1
NKF4	CLIK1, CLIK1L
PEK	GCN2, HRI, PERK, PKR
PLK	PLK1–PLK4
TLK	TLK1–TLK2
TOPK	PBK
TP53RK	PRPK
TTK	TTK
ULK	Fused, ULK1–ULK3
VPS15	PIK3R4
WEE	Myt1, Wee1, Wee1b
WNK	WNK1–WNK4
Other	KIS, SgK496

Family **1** contains Aurora kinases (AurAAurC). (Family **2**) Kinase NAK intervenes during cell division that leads to distinct cell lineages via its interaction with membrane-bound, Notch antagonist Numb that determines cell fate during development and contributes to the establishment of cell identity. The NAK family includes SNF1-subfamily member *adaptor-related proteic complex-2 (AP2)-associated protein kinase-1* (AAK1), *bone morphogenic protein-2 (BMP2)-inducible kinase* (BMP2K or BIKE), *cyclin-G-associated kinase* (GAK), and *myristoylated and palmitoylated Ser/Thr kinase-1* (MPSK1).[4] Cyclin-G-associated kinase interacts with cyclin-G1 and cyclin-dependent kinase CDK5 [251].

Family **3** comprises *guanylate cyclases*, i.e., atrionatriuretic peptide receptors (NPR1–NPR3, a.k.a. ANPRa–ANPRc and GuCy2a–GuCy2b), guanylate cyclase-2C (GuCy2c), -2D (GuCy2d, or CyGd), and -2F (GuCy2f or CyGf).[5]

---

[4] Kinase MPSK1 is also named Ser/Thr kinase STK16, protein kinase expressed in day 12 fetal liver (PKL12), and kinase related to (Saccharomyces) cerevisiae and (Arabidopsis) thaliana (KRCT).

[5] A.k.a. heat stable enterotoxin receptor (HSER).

3.1 Protein Kinase Classification

Family **4** encompasses *budding uninhibited by benzimidazole kinases* with BUB1 and BUB1β, or BUBR1 (Vol. 2 – Chap. 4. Cell Survival and Death); family **5** *calcium–calmodulin-dependent protein kinase kinases*, i.e., upstream kinases of CamK enzymes (CamKK1–CamKK2); family **6** *cell division cycle kinase*-7 (CDC7); and family **7** *casein kinases*-2 (CK2α1–CK2α2; Sect. 5.7).

Family **8** consists of *CLP36-interacting kinase*.[6] It interacts with actinin-1, -2, and -4. It localizes to actin stress fibers. Hence, one type of CLP36-interacting kinase is also called PDLIM1-interacting kinase-1-like kinase (PDIK1L). Other members are CLIK1 (or STK35) and CLIK1L (or PDIK1L).

Family **9** is composed of kinases that phosphorylate eukaryotic initiation factor eIF2α to inhibit the initiation of translation, i.e., the group-F pseudokinase *general control non-derepressible*-2 (GCN2), *heme-regulated inhibitor kinase* (HRI), *eukaryotic translation initiation factor-2α kinase*-3 (eIF2aK3),[7] and *double-stranded RNA-dependent eIF2α kinase* (PKR), as well as, on the same dendogram branch, *TBC domain-containing kinase* (TBCK).

Eukaryotic initiation factor eIF2 is required in the initiation of translation. It mediates the binding of initiator $tRNA^{Met}$ to ribosome, as it forms a ternary complex with GTP and $tRNA^{Met}$ ($eIF2^{GTP}$–$tRNA^{Met}$; Vol. 1 – Chap. 5. Protein Synthesis). This complex binds to the 40S ribosomal subunit to form the 43S pre-initiation complex (PIC) with initiation factors eIF1, eIF1A, and eIF3. The PIC complex then binds mRNA that has previously been unwound by eIF4 to build the 48S complex on the mRNA. Upon base pairing of the start codon with $tRNA^{Met}$, GTPase-activating protein eIF5 is recruited to the complex. Following $eIF2^{GTP}$ hydrolysis into $eIF2^{GDP}$, eIF2 is released from the 48S complex and translation begins after recruitment of the 60S ribosomal subunit and formation of the 80S initiation complex. Activity of eIF2 is regulated by both guanine nucleotide-exchange and phosphorylation. Once phosphorylated, eIF2 has elevated affinity for its guanine nucleotide-exchange factor eIF2b. The latter exchanges GDP for GTP only in unphosphorylated eIF2. Therefore, phosphorylated eIF2 inhibits the GEF function of its subunit eIF2b and becomes inactive. Activated kinases thus phosphorylates eIF2 to stop translation. Kinases that target eIF2 comprise GCN2 in response to amino acid deprivation, PERK during endoplasmic reticulum stress, PKR in the presence of double-stranded RNA, and HRI upon hemoglobin deficiency. whereas, under normal conditions, eIF2α interacts with eIF2b, hence allowing guanine nucleotide-exchange for protein translation (productive interaction), these activated kinases phosphorylate and sequester eIF2α (non-productive interaction).

Family **10** is constituted by *inhibitor of NFκB kinases* (IKKα, IKKβ, and IKKε) as well as *tumor-necrosis factor receptor* (TNFR)-*associated factor* (TRAF) *family member-associated NFκB activator* (TANK)-*binding kinase* TBK1 (Sect. 5.13.9)

---

[6] Protein CLP36 is also termed PDZ and LIM domain protein PDLIM1, CLIM1, and enigma PDZ and LIM domain-1 protein Elfin.

[7] A.k.a. PKR endoplasmic reticulum-related kinase [PERK] and pancreatic eIF2α kinase (PEK).

that is similar to IKKs and can mediate activation of nuclear factor κ light chain enhancer of activated B cells (NFκB; Sect. 10.9.1) in response to some growth factors.

Family **11** embodies *inositol-requiring kinases* (IRE1–IRE2) and *ribonuclease-L* (RNaseL). Protein RNaseL is a component of the interferon-regulated endogenous, antiviral, RNA-degradation 2-5A substance. The 2-5A substance mediates the antiviral and antiproliferative effects of interferons. Interferons activate genes that encode several double-stranded RNA-dependent synthases. These enzymes generate 5-triphosphorylated (2,5)-phosphodiester-linked oligoadenylates (2-5A) from ATP. Double-stranded RNA that is frequently produced during viral infection stimulates (2,5)-oligoadenylate synthases, hence the synthesis of 2-5A substance. The effects of 2-5A in cells are transient, as it is targeted by phosphodiesterases and phosphatases. Nevertheless, 2-5A substance connects to and activates RNaseL that cleaves single-stranded RNA and can initiate apoptosis.

(Family **12**) *Haspin*,[8] a histone-3 Thr kinase, maintains centromeric cohesion during mitosis (Sect. 5.13.2).

(Family **13**) Protein Ser/Thr kinase *Mos* (Moloney murine sarcoma viral oncogene homolog), a proto-oncogene product, activates and stabilizes the cell cycle CcnB–CDK1 complex.

Family **14** regroups *kinase-interacting stathmin* (KIS)[9] and *dual-specificity Thr/Tyr kinase* (TTK, or TTK1).[10] The latter is orthologous to the yeast monopolar spindle kinase (Mps1) that is required for spindle pole duplication and the spindle checkpoint. Kinase TTK localizes to the kinetochore in prometaphase and moves to the centrosome upon chromosomal alignment at the equator of the dividing cell. Centrosomes and kinetochores are primary regulators of microtubule dynamics, the spatial and temporal control of which ensures successful cell division (Vol. 2 – Chap. 2. Cell Growth and Proliferation).

Family **15** *Never in mitosis gene-A (NIMA)-related kinases* (NeK1–NeK11) are involved in the response to DNA damage. Family **16** is the group G of pseudokinases *nuclear receptor-binding proteins* (NRBP1–NRBP12).

Family **17** incorporates MAP2K-like *PDZ-binding protein kinase* (PBK);[11] family **18** *phosphoinositide 3-kinase regulatory subunit* $PI3K_{r4}$; family **19** *PTen-induced putative kinase* PINK1;[12] family **20** *Polo-like kinases* (PLK1–PLK4; Vol. 2 – Chap. 2. Cell Growth and Proliferation); family **21** pseudokinase *P53-related protein kinase* (PRPK);[13] family **22** *SH3 binding domain-containing kinase* SBK1;

---

[8] A.k.a. germ cell-specific nuclear protein kinase GSG2.

[9] A.k.a. U2AF homology motif (UHM) kinase UHMK1.

[10] A.k.a. phosphotyrosine picked Thr kinase (PYT).

[11] A.k.a. T-lymphokine-activated killer T-cell-originated protein kinase (TOPK) and spermatogenesis-related protein kinase (SPK).

[12] A.k.a. BRPK and PARK6.

[13] A.k.a. TP53-regulating kinase (TP53RK) and Bud32 in yeast.

Family **23** is made up of regulators of the *SCY1-like family of kinase-like* proteins (group-G pseudokinases SCYL1–SCYL3). Protein SCY1-like-1 (SCYL1; SCY: Suppressor of GTPase mutant in Saccharomyces cerevisiae))[14] is a transcriptional regulator that activates transcription of the telomerase reverse transcriptase and DNA polymerase β genes. Protein SCYL2[15] is a regulator of clathrin-mediated SNARE sorting. Protein SCYL3[16] is a regulator of interactions between the cytoplasmic domains of transmembrane proteins and actin via ezrin.

Family **24** includes group-G *pseudokinase Slob*; family **25** *Tousled-like kinases* (TLK1–TLK2, or PKUβ–PKUα); family **26** *uncoordinated-51-like kinases* (ULK1–ULK4)[17] and *Fused* of the Hedgehog pathway; family **27** *CDK inhibitory kinases* Wee1a, Wee1b, and membrane-associated Tyr (Y)- and Thr (T)-specific CDK1-inhibitory kinase PKMYT1 (Vol. 2 – Chap. 4. Cell Survival and Death and Sect. 7.5); and family **28** *With-no-K (Lys) kinases* (WNK1–WNK4; Sect. 5.12).

3.1.2 AGC Superfamily of Protein Kinases

The AGC superfamily of protein kinases contains cyclic nucleotide-dependent kinases of the **PKA** (PKAα–PKAγ; Sect. 5.2.3)) and **PKG** (PKG1–PKG2; Sect. 5.2.9) families, those of the **PKC** (PKCα–PKCι; (Sect. 5.2.7) and **PKB** (PKBα–PKBγ; Sect. 5.2.6) families, as well as other families (Table 3.4) [154].

The other families are constituted by:

1. citron Rho-interacting Ser/Thr kinase-21 (CRIK);
2. dystrophia myotonica protein kinases (DMPK1–DMPK2);
3. GPCR kinases (GRK1–GRK7);
4. large tumor suppressors (LaTS1–LaTS2) and microtubule-associated Ser/Thr-protein kinases (MAST1–MAST4 and MASTL);
5. myotonic dystrophy kinase-related CDC42-binding protein kinases (MRCKα–MRCKβ);[18]
6. mitogen- and stress-activated protein kinases (MSK1 and MSK2);[19]
7. nuclear DBF2 (cell cycle protein kinase in Saccharomyces cerevisiae)-related protein kinases (NDR1–NDR2; Sect. 5.2.1);[20]
8. phosphoinositide-dependent kinase PDK1 (Sect. 5.2.2);

---

[14] A.k.a. N-terminal kinase-like protein (NTKL), telomerase regulation-associated protein kinase (TAPK), telomerase transcriptional elements-interacting factor (TEIF), and teratoma-associated tyrosine kinase (TRAP).
[15] A.k.a. 104-kDa coated vesicle-associated kinase (CVAK104).
[16] A.k.a. protein associating with the C-terminus of ezrin PACE1.
[17] Protein ULK4 belongs to the group-E pseudokinase.
[18] A.k.a. CDC42-binding protein kinases CDC42BPα–CDC42BPβ.
[19] A.k.a. ribosomal protein S6 kinases (RSKα5–RSKα4).
[20] A.k.a. protein Ser/Thr kinases STK38 and STK38L.

Table 3.4. The AGC superfamily of protein kinases (Sugen nomenclature; Source: [156]). A given family can generate distinct subfamilies, each possessing diverse members (e.g., DMPK family to CRIK, GEK, and ROCK subfamilies [GEK to DMPK and MRCK] and RSK family to MSK, RSK, RSKL, SgK494, and S6K subfamilies).

Family	Members
DMPK	CRIK, DMPK1–DMPK2, MRCKα–MRCKβ, ROCK1–ROCK2
GRK	GRK1–GRK7
MAST	MAST1–MAST4, MASTL
NDR	LATS1–LATS2, NDR1–NDR2
PDK1	PDK1
PKA	PKAα–PKA–γ, PRKX–PRKY
PKB	PKBα–PKBγ
PKC	PKCα–PKCι
PKG	PKG1–PKG2
PKN	PKN1–PKN3
RSK	MSK1–MSK2, RSK1–RSK4, RSKL1–RSKL2, SgK494, S6Kα–S6Kβ
SGK	SGK1–SGK3
YANK	YANK1–YANK3

9. protein kinase-N (PKN1–PKN3), a.k.a. protein kinase-C-related (-like) kinases (PRK1–PRK3 or PRKCL1–PRKCL3) or protease-activated kinases (PAK1–PAK3; not P21-activated kinases);
10. protein kinases X- (PRKX) and Y-linked (PRKY);
11. Rho-associated, coiled-coil-containing protein kinases, effectors of small Rho GTPase (RoCK1–RoCK2; Sect. 5.2.14);
12. P90 (RSK1–RSK4 and RSKL1–RSKL2) and P70 ribosomal S6 kinases (S6Kα–S6Kβ or S6K1–S6K2; Sect. 5.2.17);
13. serum- and glucocorticoid-regulated kinases (SGK1–SGK3); and
14. kinases YANK1 to YANK3 (STK32a–STK32c).

3.1.3 CAMK Superfamily of Protein Kinases

The CAMK superfamily of protein kinases encompasses:

1. calcium–calmodulin-dependent protein kinases CamK (Table 3.5) [154]:
2. AMP-activated protein kinases (AMPKα1–AMPKα2);
3. AMPK-related protein kinases (ARK5), or NUAK family SNF1-like kinase NUAK1, and SNF1 and AMPK-related kinase SNARK, or NUAK2;
4. brain-selective kinases (BrSK1–BrSK2);
5. calcium–calmodulin-dependent serine protein kinase (CASK);
6. checkpoint kinases (ChK1–ChK2);
7. death-associated protein kinases (DAPK1–DAPK3);

3.1 Protein Kinase Classification

Table 3.5. CAMK superfamily of protein kinases (Sugen nomenclature). Trio possesses 2 functional GEF domains, one specific for Rho and the other for Rac GTPase, and a protein Ser/Thr kinase region, hence its name — triple functional domain. This guanine nucleotide-exchange factor thus belongs to the CAMK superfamily. It interacts with small GTPase Rac, P21-activated kinase, Abl protein Tyr kinase, PTPRf transmembrane protein Tyr phosphatase, and adaptors Non-catalytic region of Tyr kinase adaptor NCK1 and Enabled homolog (EnaH).

Family	Members
CAMK1	CamK1α–CaMK1δ, CamK4
CAMK2	CamK2α–CaMK2δ
CAMKL	AMPKα1–AMPKα2, BrSK1–BrSK2, CHK1, HUNK, LKB1, MARK1–MARK4, MELK, NIM1, NuaK1–NuaK2, PASK, QIK, QSK, SIK, SNRK
CASK	CASK
DAPK	DAPK1–DAPK3, DRAK1–DRAK2
DCAMKL	DCAMKL1–DCAMKL3
MAPKAPK	MAPKAPK2–MAPKAPK3, MAPKAPK5, MNK1–MNK2
MLCK	caMLCK, skMLCK, smMLCK, SgK085, Ttn
PHK	PhKγ1–PhKγ2
PIM	PIM1–PIM3
PKD	PKD1–PKD3
PSK	PSKH1–PSKH2
RAD53	CHK2
RSK-B	MSKβ1–MSKβ2, RSKβ1–RSKβ4
STK33	STK33
Trb	Trb1–Trb3
Trio	Obscn, Obscnβ, SPEG, SPEGβ, Trad, Trio
TSSK	TSSK1–TSSK4, TSSK1b, TSSK6/SSTK
VACamKL	VACamKL

8. DAPK-related apoptosis-inducing kinases (DRAK1–DRAK2);
9. doublecortin- and CamK-like kinases (DCamKL1–DCamKL3);
10. Duo;[21]
11. hormonally upregulated Neu-associated kinase (HUNK);
12. mitogen-activated protein kinase (MAPK)-activated protein kinases, which form the MAPKAPK2 and MAPKAPK5 sets;
13. microtubule-associated protein (MARK);
14. maternal embryonic leucine zipper kinase (MELK);
15. myosin light chain kinases (cMLCK, skMLCK, and smMLCK: isoforms in cardiomyocytes, skeletal myocytes, and smooth muscle cells);
16. MAPK-interacting kinases (MNK1–MNK2);
17. some mitogen- and stress-activated protein kinases (MSKβ1–MSKβ2);

---

[21] A.k.a. Duet and Trad.

18. Wee1 inhibitor kinase mitotic inducer (Nim1, or Cdr1);
19. obscurin (Obscn), a cytoskeletal calmodulin and titin-interacting RhoGEF as well as a Ser/Thr kinase that belongs to the family of giant sarcomeric signaling proteins;
20. certain P90 ribosomal protein S6 kinases (RSK1β–RSK4β);
21. PAS domain-containing protein Ser/Thr kinase (PASK);
22. phosphorylase kinase-γ (PhKγ1–PhKγ2);
23. proto-oncogene protein Ser/Thr kinases (PIM1–PIM3);
24. protein kinase-D family members (PKD1–PKD3);
25. protein serine kinase-H family members (PSKH1–PSKH2);
26. salt-inducible kinase SIK1;[22]
27. salt-inducible kinase SIK2;[23]
28. salt-inducible kinase SIK3;[24]
29. SNF1 (plant AMPK)-related protein Ser/Thr kinase (SNRK);
30. smooth muscle cell preferentially expressed gene product (SPEG);[25]
31. small protein Ser/Thr kinase (SSTK, or TSSK6);
32. protein Ser/Thr kinases STK11 (or LKB1) and STK33;
33. Tribbles homologs (Trb1–Trb3); Trio;
34. testis-specific serine kinases (TSSK1, TSSK1b, TSSK2–TSSK4, and TSSK6);
35. titin (Ttn, or connectin); and
36. vesicle-associated calmodulin-binding protein (VACamKL or CamKV).

3.1.4 CK1 Superfamily of Protein Kinases

Kinases of the CK1 superfamily phosphorylate many substrates such as regulators of cell differentiation, proliferation, chromosome segregation, and circadian rhythms. The CK1 superfamily of protein kinases includes the families of (Table 3.6) [154]:

1. CK1 kinases (CK1α1–CK1α2, CK1γ1–CK1γ3, and CK1δ–CK1ε; Sect. 5.6.1);
2. Tau tubulin kinases (TTbK1–TTbK2); and
3. vaccinia-related kinases (VRK1–VRK3; Sect. 5.6.2).

---

[22] Salt-inducible kinase SIK1 is specifically produced in adrenal glands, SIK2 is specific to adipose tissue, SIK3 is ubiquitous. These 3 enzymes constitute a protein Ser/Thr kinase subfamily of the AMP-activated protein kinase (PKA) family. Kinase SIK1 travels between the cytoplasm and nucleus under ACTH–PKA stimulation to regulate steroidogenic gene expression. Kinase SIK2 can phosphorylate insulin receptor substrate IRS1 [157].

[23] A.k.a. SNF1-like kinase SNF1LK2 and Qin-induced kinase (QIK).

[24] A.k.a. QSK kinase.

[25] A.k.a. aortic preferentially expressed protein APEG1.

3.1 Protein Kinase Classification

Table 3.6. CK1 superfamily of protein kinases (Sugen nomenclature; Source: [156]).

Family	Members
CK1	CK1α1–CK1α2, CK1γ1–CK1γ3, CK1δ–CK1ε
TTBK	TTbK1–TTbK2
VRK	VRK1–VRK3

Table 3.7. CMGC superfamily of protein kinases (Sugen nomenclature; Source: [156]).

Family	Members
CDK	CCRK, CDK1–CDK11, ChED, CRK7, PCTAIRE1–PCTAIRE3, PFTAIRE1–PFTAIRE2, PITSLRE
CDKL	CDKL1–CDKL5
CLK	CLK1–CLK4
GSK	GSK3α–GSK3β
MAPK	ERK1–ERK5, ERK7, JNK1–JNK3, NLK, P38α–P38δ
DYRK	DYRK1a–DYRK1b, DYRK2–DYRK4, HIPK1–HIPK4, PRP4K
RCK	ICK, MAK, MOK
SRPK	MSSK1, SRPK1–SRPK2

3.1.5 CMGC Superfamily of Protein Kinases

The CMGC superfamily of protein kinases comprises (Table 3.7) [154]:

1. cyclin-dependent kinases (CDK1–CDK11);
2. CDK-like kinases (CDKL1–CDKL5);
3. cholinesterase-related cell division controller (ChED);[26]
4. CDK1 (CDC2)-like kinases (CLK1–CLK4);
5. dual-specificity Tyr phosphorylation-regulated kinases (DYRK1a–DYRK1b and DYRK2–DYRK4);
6. extracellular signal-regulated kinases (ERK1–ERK5 and ERK7);
7. glycogen synthase kinases (GSK3α–GSK3β);
8. homeodomain-interacting protein kinases (HIPK1–HIPK4);
9. intestinal cell MAK-like kinase (ICK);[27]
10. Jun N-terminal kinases (JNK1–JNK3);
11. protein Ser/Thr male germ cell-associated kinase (MAK);
12. MAPK–MAK–MRK overlapping kinase (MOK);[28]
13. protein Ser/Thr kinase MSSK1;[29]

---

[26] A.k.a. cell division cycle 2-like CDC2L5 and CDC2-related protein kinase-5.

[27] A.k.a. male germ cell-associated kinase (MAK)-related kinase (MRK) and LCK2.

[28] A.k.a. renal cell carcinoma antigen RAGE or RAGE1.

[29] A.k.a. SFRS protein kinase, serine (S)/arginine (R)-rich protein-specific kinase SRPK3, and STK23;

14. Nemo-like kinase (NLK);
15. proteins characterized by a substitution of serine (S) by cysteine (C) in the PSTAIRE motif, a hallmark of members of the CDK1–CDC28-related protein kinase family (PCTAIRE1–PCTAIRE3 or PCTK1–PCTK3);
16. protein Ser/Thr kinases (PFTAIRE1–PFTAIRE2 or PFTK1–PFTK2);
17. protein Ser/Thr kinase PITSLRE;[30]
18. protein Ser/Thr kinase PRP4 or PRP4K;[31]
19. P38MAPKs (P38MAPKα–P38MAPKδ); and
20. serine/arginine (S/R) splicing factor protein kinases (SRPK1–SRPK2).

3.1.6 STE Superfamily of Protein Kinases

The STE superfamily includes many protein kinases involved in mitogen-activated protein kinase modules. These signaling modules correspond to a phosphorylation cascade, in which MAPK is phosphorylated (activated) by a MAPK kinase (MAP2K) that itself is activated by a MAP2K kinase (MAP3K). The latter can be stimulated by a MAP3K kinase (MAP4K; Chap. 6).

The STE superfamily of protein kinases includes (Table 3.8) [154]:

1. germinal center kinase (GCK; Sect. 5.1.2);
2. general control non-derepressible pseudokinase-2 (GCN2b);
3. HPK1/GCK-related kinase (HGK);
4. hematopoietic progenitor kinase-1 (HPK1; Sect. 5.1.3.1);
5. kinases homologous to sterile 20 (KHS1–KHS2; Sect. 5.1.3.5);
6. lymphocyte-oriented kinase (LOK; Sect. 5.1.7);
7. mitogen-activated protein kinase kinase (MAP2K1–MAP2K7; Sect. 6.6);
8. some mitogen-activated protein kinase kinase kinases that are split into several subsets: MEKK (MAP/ERK kinase kinases; MAP3K1–MAP3K4; Sect. 6.5.4); ASK (apoptosis signal-regulating kinases; MAP3K5–MAP3K6; Sect. 6.5.5); TAK1 (transforming growth factor-β-activated kinase; MAP3K7; Sect. 6.5.6); and TPL2 (tumor progression locus-2; MAP3K8; Sect. 6.5.8);[32]
9. Thousand and one kinases (TAOK1–TAOK3), also related to the previous families (MAP3K16–MAP3K18; Sects. 5.1.10 and 6.5.7);
10. mammalian sterile 20 (twenty)-like kinases (MST1–MST4);
11. subclass-3 myosins (Myo3a–Myo3b) that constitute GCK subfamily 7;
12. NCK-interacting kinase (NIK);
13. oxidative stress-responsive kinase (OSR1);

---

[30] A.k.a. cell division cycle-2-like protein CDC2L2, or CDC2L3, CDK11^{P46}, CDK11^{P58}, CDK11^{P110}, and galactosyltransferase-associated ^{P58}GTA protein kinase.

[31] A.k.a. PRP4 protein kinase homolog (PR4H) and PRP4 pre-mRNA processing factor-4 homolog-B (PRPF4B).

[32] A.k.a. cancer Osaka thyroid proto-oncogene product (COT).

3.1 Protein Kinase Classification

Table 3.8. STE superfamily of protein kinases (Sugen nomenclature).

Family	Members
	STE7
MAP2K	MAP2K1–MAP2K7
	STE11
MAP3K	MAP3K1–MAP3K8
	STE20
STE20	GCK, HPK1, KHS1–KHS2, LOK, MST1–MST4, Myo3a–Myo3b, OSR1, PAK1–PAK6, SLK, STLK3, STLK5–STLK6, TAOK1–TAOK3, YSK1, ZC1/HGK, ZC2/TNIK, ZC3/MINK, ZC4/NRK
	STE-Unique
COT	COT
GCN2	GCN2
NIK	NIK

14. P21-activated kinases (PAK1–PAK6; Sect. 5.2.13);
15. sterile 20-like kinase (SLK);
16. putative protein Ser/Thr kinases Ste20-like (STLK);
17. sterile 20-like oxidant stress response kinase (SOK1);[33] and
18. the subfamily of ZC (zebrafish homologous) kinases of the germinal center kinase family (ZC1–ZC4 or HGK/MAP4K4, TNIK, MinK/MAP4K6, and NRK, respectively; Sect. 5.1.6).

3.1.7 TK Superfamily of Protein Kinases

The TK superfamily is constituted by kinases involved in inter- and intracellular signal transduction for tissue development and homeostasis, transcriptional control, proliferation or differentiation decisions, and cell shape and motility. Members of the TK superfamily contribute to signaling cascades as receptors or intracellular effectors in immune response (Janus kinases activated from cytokine receptors), glucose metabolism (insulin receptor), embryogenesis and maintenance of stem cells, as well as angiogenesis (vascular growth factor receptors).

The TK superfamily encompasses (Table 3.9 and Vol. 3 – Chap. 8. Receptor Kinases) [154]:

1. apoptosis-associated tyrosine kinases (AATyK1–AATyK3);[34]
2. Abelson kinases (Abl1–Abl2; Sect. 4.2);

---

[33] A.k.a. yeast Sps1/Ste20-related kinase YSK1, and Ser/Thr kinase STK25.

[34] A.k.a. lemur Tyr kinases (Lmr1–Lmr3).

Table 3.9. TK superfamily of protein kinases (Sugen nomenclature).

Family	Members
\multicolumn{2}{c}{Cytosolic kinases}	
ACK	ACK, TNK1
ABL	Abl, ARG
CSK	CSK, CTK
FAK	FAK1, FAK2 (PYK2)
FER	Fer, Fes
JAK	JAK1–JAK3, TYK2
SRC	BLK, BrK, FGR, FRK, Fyn, HCK, LcK, Lyn, Src, SRM, Yes
SYK	SYK, ZAP70
TEC	BMX, BTK, ITK, TEC, TXK
\multicolumn{2}{c}{Receptor kinases}	
AATYK	AATyK1–AATyK3
ALK	ALK, LTK
AXL	Axl, Mer, Tyro3
DDR	DDR1–DDR2
EGFR	HER1–HER2, HER4
EPH	EPHa1–EPHa8, EPHb1–EPHb4, EPHb6, EPHb10
FGFR	FGFR1–FGFR4
IR	IGF1R, InsR, IRR
MET	Met, RON
MUSK	MuSK
NTRK	NTRK1–NTRK3
PDGFR	CSFR (Kit), CSF1R (M-CFSR), FLT3, PDGFRa–PDGFRb
PTK7	PTK7
RET	RET
ROR	ROR1–ROR2
SCFR	SCFR
SEV	ROS
RYK	RYK
TIE	TIE1–TIE2
VEGFR	VEGFR1–VEGFR3

3. ALK–LTK family kinases (anaplastic lymphoma kinase and leukocyte Tyr kinase);
4. Axl–Mer–Sky family kinases (Axl, Mer, Tyro3/Sky);
5. Fyn-related kinases (BRK – Sect. 4.4, BSK, FrK – Sect. 4.8, GTK, IYK, PTK5, RAK, and SRMS – Sect. 4.11);
6. colony stimulating factor-1 receptor (CSF1R or Fms);
7. discoidin domain receptors (DDR1–DDR2);
8. ephrin receptors (EPHa1–EPH8a, EPHa10, EPHb1–EPHb4, and EPHb6);
9. focal adhesion kinases (FAK1–FAK2; Sect. 4.6);
10. feline sarcoma oncogene kinases (Fer and Fes; Sect. 4.7);

3.1 Protein Kinase Classification

11. fibroblast growth factor receptors (FGFR1–FGFR4);
12. Fms-like tyrosine kinase receptor (FLT3);
13. human epidermal growth factor receptors (HER1–HER4);
14. hepatocyte growth factor receptor (HGFR or Met);
15. insulin and insulin-like growth factor receptors (InsR, IRR, and IGF1R);
16. Janus kinases (JaK1–JaK3 and Tyk2 as well as JaK1b–JaK3b and Tyk2b);
17. macrophage-stimulating-1 receptor (MSt1R or Ron);
18. muscle-specific kinase (MuSK);
19. neurotrophic Tyr receptor kinases (NTRK1–NTRK3);
20. platelet-derived growth factor receptors (PDGFRα–PDGFRβ);
21. protein tyrosine kinase-7 (PTK7);
22. rearranged during transfection (ReT);
23. receptor tyrosine kinase-like orphan receptors (Ror1–Ror2);
24. Ros receptor-like tyrosine kinase (Ros);
25. receptor-like tyrosine kinase (RYK);
26. stem cell factor receptor (SCFR or Kit);
27. SRC family kinases (Src, CSK, Chk, BLK, FGR, Fyn, HCK, LcK, Lyn, and Yes; Sect. 4.12);
28. spleen tyrosine family kinases (SYK and ZAP70; Sect. 4.14);
29. TEC (Tyr kinase expressed in hepatocellular carcinoma) family kinases (TEC, TXK, BTK, BMX/ETK, ITK; Sect. 4.12);
30. angiopoietin receptors Tyr kinase with immunoglobulin and EGF repeats (TIE1–TIE2); and
31. vascular endothelial growth factor receptors (VEGFR1–VEGFR3 or FLT1–FLT2 and FLT4).

3.1.8 TKL Superfamily of Protein Kinases

The TKL superfamily contains many families (Table 3.10) [154]:

1. protein Ser/Thr ankyrin repeat domain-containing kinase AnkRD3,[35]
2. integrin-linked kinase (ILK; Sect. 5.9.1);
3. IRAK (interleukin-1 receptor-associated kinases; IRAK1–IRAK4; Sect. 5.9.2);
4. kinase suppressor of Ras (KSR1–KSR2);
5. LISK (for LIMK/TESK) that regroups subfamilies of LIM domain-containing kinases (LIMK1–LIMK2; Sect. 5.9.3) and of testis-specific kinases that contain an N-terminal protein kinase domain closely related to those of LIMK kinases (TeSK1–TeSK2; Sect. 5.9.4);

---

[35] A.k.a. AnkK2, PKCδ-interacting protein kinase (DIK or PKK) and receptor-interacting protein Ser/Thr kinase RIP4 or RIPK4.

Table 3.10. TKL superfamily of protein kinases (Sugen nomenclature).

Family	Members
IRAK	IRAK1, IRAK4
LISK	LIMK1–LIMK2, TESK1–TESK2
LRRK	LRRK1–LRRK2
MLK	DLK, HH498, ILK, LZK, MLK1–MLK4, MLK7, TAK1
MLKL	MLKL
RAF	A-RAF–C-RAF, KSR1–KSR2
RIPK	AnkRD3, RIPK1–RIPK3, SgK288
STKR	ActR2a–ActR2b, ALK1–ALK2, ALK4, ALK7, AMHR2, BMPR1a–BMPR1b, BMPR2, TGFβR1–TGFβR2

6. leucine rich-repeat kinases (LRRK1-LRRK2);[36]
7. MAPK module components, such as dual leucine zipper-bearing kinase (DLK or MAP3K12), leucine zipper-bearing kinase (LZK or MAP3K13), mixed lineage kinases (MLK1–MLK4), Raf (aRaf–cRaf), TGFβ-activated kinase (TAK1 or MAP3K7), and sterile-α motif- and leucine zipper-containing kinase (ZAK or MLK7), as well as kinase associated with clone HH498; Chap. 6);
8. receptor-interacting protein kinases (RIPK1–RIPK3; Sect. 5.9.5); and
9. protein Ser/Thr receptor kinases (STRKs; Vol. 3 – Chap. 8. Receptor Kinases), i.e., activin and TGFβ receptors (TβR1 (ALK5)–TβR2, anti-Müllerian hormone receptor AMHR2 [or MISR2], bone morphogenetic protein receptors [BMPR1a–BMPR1b (ALK3 and ALK6) and BMPR2], activin receptors [AcvR2a–AcvR2b]), and related activin receptor-like kinases (ALK1–ALK2 [AcvRL1 and AcvR1a], ALK4 [AcvR1b], and ALK7 [AcvR1c]).

3.1.9 RGC Superfamily of Protein Kinases

The small RGC superfamily of receptor guanylate cyclases contains enzymes similar in sequence to the TK superfamily. Photoreceptor membrane guanylate cyclase actually has Mg^{++}-dependent Ser autophosphorylating kinase activity.

---

[36] A.k.a. RIPK6 and RIPK7 as well as proteins of the ROCO family —ROCO1 and ROCO2 — characterized by the presence of a Ras GTPase-like domain ROC (Ras of complex protein) immediately followed by a COR (C-terminal of ROC) domain.

3.1.10 Superfamily of Atypical Protein Kinases

Atypical protein kinases (aPK) are small. They lack sequence similarity to that of usual protein kinases, but possess a kinase activity.

3.1.10.1 Family of α Protein Kinases

The small family of atypical α kinases contains eukaryotic elongation factor-2 kinases eEF2K, 2 multipass transmembrane ion channel kinases of the transient receptor potential cation channel superfamily, melastatin-related TRP cation channel TRPM6[37] and TRPM7,[38] with an α-kinase fused to an ion channel, and 3 α-protein kinases (αK1–αK3 or AlpK1–AlpK3).

3.1.10.2 Family A6 of Protein Pseudokinases

Founding members of the A6 family are actin-binding protein Tyr kinase PTK9, or A6,[39] and PTK9-like protein (PTK9L).[40] These proteins interact with PKCζ and bind ATP, but do not have kinase activity [158].

3.1.10.3 ABC1 Family of Protein Kinases

The ABC1/ADCK (ABC1: activity of bc1 complex [complex-3 or ubiquinol–cytochrome-C reductase] of the respiratory chain; ADCK: aarF domain-containing kinase [ADCK1–ADCK5]) family of ABC1 domain-containing atypical kinases is constituted by ABC1 kinases that are not related to the ABC family of ATPase transporters. They can be involved in coenzyme-Q synthesis.[41] ABC1 domain-containing kinase ADCK3 serves as a chaperone involved in ubiquinone synthesis.[42] The CABC1/ADCK3 gene is the human homolog of the yeast ABC1/COQ8

---

[37] A.k.a. channel kinase ChaK2.
[38] A.k.a. ChaK1.
[39] A.k.a. twinfilin homolog-1 (Twf1).
[40] A.k.a. actin-binding twinfilin homolog-2 (Twf2) and A6-related protein (A6R or A6RP).
[41] Coenzyme-Q serves in oxidative phosphorylation. Coenzyme-Q10 (CoQ10, ubiquinone), a lipophilic component of the inner mitochondrial membrane, distributes electrons between various dehydrogenases and cytochromes of the respiratory chain. It also operates in anti-oxidant defenses.
[42] Another alias CABC1 stands for chaperone, ABC1 activity of bc1 complex homolog. Mutations of the Cabc1 gene in ubiquinone-deficient patients is associated with progressive neurological disorder characterized by cerebellar atrophy and seizures [159].

gene, itself the homolog of the UBIB[43] and aarF[44] genes required for the first mono-oxygenation step of ubiquinone synthesis in the prokaryotes Escherichia coli and Providencia stuartii, respectively.

3.1.10.4 BCR Protein Kinases

Breakpoint cluster region protein (BCR) is the fusion partner of Abl kinase in chronic myelogenous leukemia, hence its alias CML. Protein BCR has protein kinase activity used for auto- and transphosphorylation. In humans, active BCR-related gene product (ABR), which is about 70% identical in amino acid sequence, lacks the N-terminal kinase domain, but possesses the conserved GTPase activator domain.

3.1.10.5 Family of Bromodomain (BrD)-Containing Protein Kinases

The BRD family of atypical bromodomain-containing kinases is composed of BrD2[45] to BrD4, testis-specific BrDT, or BrD6 kinase. They are transcription factor kinases. Kinase BrD2 targets cell cycle-regulating transcription factor E2F in the presence of acetylated histones to transactivate the promoters of E2F-dependent cell cycle genes.

3.1.10.6 Fas (TNFRSF6a)-Activated Ser/Thr Kinases

The Fas-activated Ser/Thr kinases (FASTK) bind to cytotoxic granule-associated RNA-binding protein T-cell-restricted intracellular antigen-1 (TIA1), or nucleolysin TIA1 isoform P40, an effector of apoptosis that regulates alternative splicing of the mRNA encoding apoptosis-promoting Fas receptor, i.e., TNFRSF6a receptor. It is dephosphorylated (activated) during Fas (TNFSF6)-mediated apoptosis. It is able to autophosphorylate and phosphorylate nuclear TIA1 protein.

3.1.10.7 G11 Kinase (STK19)

This family contains a single protein that is a Ser/Thr kinase (STK19) for α casein and histone. It is expressed in numerous cell types, particularly epithelial cells, hepatocytes, monocytes, and lymphocytes [161]. Within the cell, it localizes

---

[43] I.e., ubiquinone biosynthesis.

[44] I.e., chromosomal $2'$-^{N}acetyltransferase [AAC($2'$)1a] involved in the O-acetylation of peptidoglycan. The aarF locus is required ubiquinone production [160].

[45] A.k.a. really interesting new gene protein RING3. It is an autophosphorylating nuclear protein.

predominantly in the nucleus. The G11 gene is located in the DNA segment that contains the genes of the human major histocompatibility complex, hence its name in humans HLA-RP1, or simply RP1 protein.

3.1.10.8 H11 Kinase

Protein HSPb8, or H11, of the category of small heat shock proteins, belongs to the HSP20 (HSPβ6)/α-crystallin family. Kinase H11 can bear Mn^{++}-dependent Ser/Thr autophosphorylation.

3.1.10.9 PDHK Family of Protein Kinases

The PDHK family of ubiquitous, mitochondrial, atypical kinases is composed of pyruvate dehydrogenase kinases. These kinases phosphorylate (inactivate) E1 subunit of the pyruvate dehydrogenase complex that catalyzes the oxidative decarboxylation of pyruvate, a regulatory step in oxidative metabolism (cellular respiration and oxidative phosphorylation; Vol. 1 – Chap. 4. Cell Structure and Function), hence controlling the balance between glucose and lipid oxidation according to substrate supply.

The PDHK family consists of 5 members, PDHK1 to PDHK4 and mitochondrial 3-methyl 2-oxobutanoate dehydrogenase kinase, the branched chain ketoacid dehydrogenase kinase (BCKDK).

3.1.10.10 PIKK Family

The PIKK family of atypical phosphatidylinositol 3-kinase-related kinases is involved in the coordination of cell responses to DNA damage and participates in the control of cell growth, gene expression, and V(D)J recombination. Kinases of the PIKK family have catalytic domains related to phosphatidylinositol (3 or 4)-kinases, but phosphorylate proteins rather than lipids.

Members of the PIKK family contain: (1) a N-terminal FRAP (TOR)–ATMK–TrrAP (FAT) region (FATN); (2) a phosphatidylinositol kinase domain (similar to that of PI3K and PI4K); (3) a PIKK-regulatory sequence (PRD); and (4) a C-terminal FAT (FATC) motif.

The PIKK family is constituted of 6 members, among which 5 (ATMK, ATRK, TOR, DNAPK, and SMG1) are protein Ser/Thr kinases. Target of rapamycin (TOR) is a modulator of cell growth. It is also called FK506-binding protein-12–rapamycin associated protein (FRAP).[46] Other members include ataxia telangiectasia mutated kinase (ATMK) and ATM and RAD3-related kinase (ATRK) as well as DNA-dependent protein kinase (DNAPK), SMG1, and TRRAP (Table 3.11).

---

[46] A.k.a. FRAP1 and FRAP2 as well as rapamycin target protein RapT1.

Table 3.11. Members of the PIKK family of atypical protein kinases (ATMK: ataxia telangiectasia mutated kinase; ATRK: ataxia telangiectasia and Rad3-related kinase; DNAPK: DNA-dependent protein kinase; SMG1: suppressor with morphological effect on Genitalia; TOR: target of rapamycin; TrrAP: Transformation/transcription domain-associated protein).

Member	Function
ATMK	DNA damage response
ATR	DNA damage response
DNAPK	DNA damage response
SMG1	Non-sense-mediated mRNA decay, embryogenesis
TOR	Cell growth
TRRAP	Transcription activation

Kinase Suppressor with morphological effect on Genitalia SMG1[47] is involved in embryogenesis, DNA damage responses, telomere maintenance, and non-sense-mediated mRNA decay, as it phosphorylates the regulator of non-sense transcripts, or ATP-dependent RNA helicase Upstream frameshift protein UpF1.

The other member Transformation/transcription domain-associated protein (TrrAP, TrAP, or Tra1)[48] is an adaptor and transcription factor coactivator connected to chromatin complexes with histone acetyltransferase activity. It participates in E2F1-, E2F4-, P53-, and MyC-mediated transcription activation. It possesses significant sequence similarities to TOR, ATM, and ATR kinases, and the catalytic subunit of DNAPK enzyme. However, unlike the other members of the PIKK family, PAF400 is not a protein kinase due to the lack of kinase motif and autophosphorylation activity.

3.1.10.11 RIO Family of Protein Kinases

The RIO family (RiO kinase: right open reading frame) encompasses 3 families of atypical Ser/Thr kinases with a single member in each. The RIO family actually embodies 3 members (RiOK1–RiOK3) with distinct functions that can operate in cell cycle progression and chromosome maintenance [158].

---

[47] A.k.a. λ/ιprotein kinase-C-interacting protein LIP.
[48] A.k.a. 350- or 400-kDa P300–CBP-associated factor (PCAF)-associated factor (PAF350/400) and 400-kDa subunit of the PCAF histone acetylase complex (PAF400). The PCAF histone acetylase complex contains more than 20 associated proteic components.

3.1.11 TATA-Binding Factor-Associated Factor-1

TATA-binding factor-associated factor TAF1[49] is a protein kinase and a constituent of the transcription initiation factor TF2d that contributes to basal transcription initiation. It specifically phosphorylates general transcription factor-2F subunit GTF2f11[50] that intervenes in transcription by RNA polymerase-2 and TF2a.

3.1.11.1 TIF Family of Protein Kinases

The TIF family of atypical kinases includes 3 transcription intermediary factors (TIF1α–TIF1γ)[51] that are encoded by the TRIM24 (Tif1A), TRIM28 (Tif1B), and TRIM33 (Tif1G) genes. They are involved in transcription regulation.

Like TAF and BRD kinases, TIF proteins contain bromodomains. They are able to autophosphorylate as well as phosphorylate the transcription initiation factor TATA box-binding protein (TBP)-associated factor (TAF) subunits TAF6,[52] TAF7,[53] and TAF11[54] of the transcription initiation TF2d factor, a component of the set of general transcription factors, which with regulatory proteins and RNA polymerase-2 constitute the RNA polymerase-2 holoenzyme.[55]

Protein TIF1α (TriM24) participates in transcriptional control, as it interacts with several nuclear receptors, such as androgen (AR, or NR3c4), estrogen (ERα; a.k.a. NR3a1 and EsR1), glucocorticoid (GR, or NR3c1), mineralocorticoid (MR, or NR3c2), retinoic acid (RARα, or NR1b1), and vitamin-D3 (NR1i1) receptors [251]. Protein TIF1β (TriM28) connects to glucocorticoid receptor NR3c1, nuclear receptor corepressor NCoR1, heterochromatin protein P25β,[56] and heterochromatin-like protein-1,[57] among others [251]. Protein TIF1γ (TriM33) links to thyroid hormone and retinoic acid receptor-associated corepressor NCoR1.

Protein TIF1α that, in addition to autophosphorylates, selectively phosphorylates the TAF6, TAF7, and TAF11 components of the RNA polymerase-2–TF2d

---

[49] A.k.a. TAF2-250.

[50] A.k.a. RAP74.

[51] A.k.a. tripartite motif-containing protein TriM24, TriM28, and TriM33, respectively.

[52] A.k.a. TF2e, TAF2-70, TAF2-80, and TAF2-85 subunit of the RNA polymerase-2–TF2d complex.

[53] A.k.a. TAF2-55 subunit of the RNA polymerase-2–TF2d complex.

[54] A.k.a. 28-kDa subunit TAF2-28 of the RNA polymerase-2–TF2d complex.

[55] Subunits in the TF2d complex include: TATA box-binding protein (TBP), TAF1 (TAF2-250), TAF2 (150-kDa cofactor of initiator function CIF150), TAF3 (TAF2-140), TAF4 (TAF2-130/135), TAF4b (TAF2-105), TAF5 (TAF2-100), TAF6 (TAF2-70/80/85), TAF7 (TAF2-55), TAF8 (TAF2-43), TAF9 (TAF2-31/32), TAF9b (TAF2-31L), TAF10 (TAF2-30), TAF11 (TAF2-28), TAF12 (TAF2-20/15), TAF13 (TAF2-18), and TAF15 (TAF2-68).

[56] A.k.a. chromobox homolog CBx1.

[57] A.k.a. chromobox homolog CBx3.

complex [162], serves as ubiquitin–protein ligase. Protein TIF1α indeed promotes P53 ubiquitination and degradation [163]. This protein kinase is phosphorylated upon interaction with liganded nuclear receptors.

Like TIF1α, TIF1γ (or ectodermin) operates as a ubiquitin–protein ligase that targets SMAD4 for ubiquitination, nuclear exclusion, and proteasomal degradation [164].[58] In addition, SMAD2/3–TIF1γ and SMAD2/3–SMAD4 complexes act as complementary effectors in the control of hematopoietic cell fate by the TGFβ–SMAD pathway [166].[59]

In humans, a fourth member of the TIF family exists: TriM66 protein. However, it does not contain the conserved domains of the structure of other TIF1 family members [154].

3.2 Agents of Kinase Activity

Any protein kinase must recognize its phosphorylation sites among multiple amino acids that compose a target protein. Multiple features allow specific functioning: the structure of the catalytic site, the interactions between the kinase and its substrate due to complementary sequences, hydrogen bonding, surface charge and hydrophobicity, the formation of complexes with scaffolds and adaptors,[60] competition between substrates,[61] and error correction.[62]

3.3 Alternatively Spliced Variants

Alternatively spliced transcript variants encode protein kinase and phosphatase isoforms with modified substrate affinities, subcellular localizations, and activities,

---

[58] Signaling primed by TGFβ that relies on the assembly of the SMAD complex enables the formation of the mesoderm and endoderm, but is precluded for ectoderm specification. Protein TIF1γ restricts the mesoderm-inducing activity of TGFβ signals. When SMAD4 is monoubiquitinated (Lys519), its association with SMAD2^P is prevented [165]. Deubiquitinase USP9x counteracts the activity of SMAD4 monoubiquitin ligase TIF1γ.

[59] Ubiquitin–protein ligase TIF1γ selectively binds receptor-phosphorylated SMAD2 and SMAD3 in competition with SMAD4 factor. Rapid and strong binding of TIF1γ to SMAD2 and SMAD3 occurs in hematopoietic, mesenchymal, and epithelial cell types in response to TGFβ factor. In humans, TGFβ inhibits the proliferation of hematopoietic stem and progenitor cells and stimulates erythroid differentiation. Protein TIF1γ primes the differentiation, whereas SMAD4 triggers the antiproliferative response. Both SMAD2 and SMAD3 participate in both responses.

[60] Scaffold proteins coordinate interactions of kinases with other kinases and phosphatases. Scaffolds also can recruit regulators of kinases.

[61] Any substrate can act as a competitive inhibitor for other substrates.

[62] Iterative control of multiple phosphorylations on a given targeted protein minimizes erratic phosphorylation. Any wrong phosphorylation (on an inappropriate site) can be corrected by active phosphatase.

hence adding a tier in diversity. In mice at least 75% of the multi-exon phosphoregulator (kinase and phosphatase) loci are able to undergo alternative splicing, in particular alternative transcription initiation and/or termination [167]. However, alternatively generated catalytic and non-catalytic domain structures also give rise to isopeptides. Moreover, 69% of loci can produce more than one isoform.

Many of the receptor kinases and phosphatases, either secreted[63] or transmembrane[64] forms, likely possess dominant negative subtypes. Dominant negative isoforms are able to compete for ligand binding and signaling repression (e.g., VEGFR1, HER2, EPHa7, and NTRK2). Tethered isoforms have the potential to prevent the cell response to ligand, whereas secreted isoforms are potentially able to dampen the response.

Enzyme variants can have an altered or removed catalytic domain, hence an modified activity and substrate specificity. However, regulation of alternative promoters, terminators, and splice junctions allows a cell to produce either alternative peptides with a slightly different activity or the same peptide that can serve in a different context.

3.4 Intracellular Pseudokinases

Pseudokinases possess a structurally related protein kinase domain that is unable to phosphorylate, but can regulate other protein kinases. Forty-eight detected human proteins have a kinase-like domain that lacks at least one of the conserved catalytic residues. Therefore, these enzymes are inactive and termed *pseudokinases*. However, although they lack the ability to phosphorylate substrates, pseudokinases still can regulate cellular processes.

Identified receptor and cytosolic pseudokinases include receptor tyrosine kinases, such as EGFR family member HER3, ephrin receptors EPHa10 and EPHb6, orphan receptor PTK7[65] (Vol. 3 – Chap. 8. Receptor Kinases), as well as kinase suppressor of Ras, Tribbles homolog TRIB3, a negative regulator of NFκB (Sect. 10.9.1), eukaryotic translation initiation factor kinase eIF2αK4, transformation/transcription domain-associated protein (TrrAP), integrin-linked kinase (ILK), and calcium–calmodulin-dependent Ser protein kinase (CASK) [168].

---

[63]Secreted isoforms include those of protein tyrosine kinases ALK, CSF1R, EGFR, EPHa1, EPHa3, EPHa5, EPHa7, EPHa10, EPHb1, VEGFR1, VEGFR2, FLT3 (a.k.a. STK1 and FLK2), InsR, InsRR, HGFR, and PTK7, as well as those of protein Tyr phosphatases PTPRc, PTPRd, PTPRg, PTPRn, PTPRn2, PTPRo, PTPRr, PTPRs, PTPRu, and PTPRz1 [167].

[64]Transmembrane isoforms comprise those of kinases Axl, BMPR1a, CSF1R, EPHa4, EPHa5, EPHa6, EPHa7, NTRK2, NTRK3, and PDGFRα, as well as those of phosphatases PTPRk, PTPRm, PTPRu [167].

[65]A.k.a. colon carcinoma kinase CCK4.

3.4.1 CASK Pseudokinase

Calcium–calmodulin-dependent Ser protein kinase (CASK) is a pseudokinase that contains an N-terminal pseudokinase domain similar to calcium–calmodulin-dependent protein kinase domain, an inactive C-terminal guanylate kinase sequence, several PDZ motifs, and 2 SH3 regions. It forms complexes with regulators of pre- and postsynaptic processes [168].

3.4.2 GCN2 Pseudokinase

General control non-derepressible kinase GCN2 that is also labeled eukaryotic translation initiation factor-2α kinase eIF2αK4 possesses an N-terminal pseudokinase domain prior to its catalytic domain. It is one of the translation initiation factor eIF2α kinases that are activated in response to nutrient deprivation to lower protein synthesis [168].

3.4.3 ILK Pseudokinase

Integrins of the cell surface bind to extracellular matrix and cause assembly of integrin interactome that transduces mechanical and chemical signals. Integrin-linked (pseudo)kinase (ILK) binds to cytoplasmic domains of β-integrins. It also regulates actin dynamics, as it recruits actin-binding regulators such as α- and β-parvins. It actually possesses a N-terminal domain with ankyrin repeats and a C-terminus that interact with LIM domain-containing adaptors of the PINCH- and parvin family, respectively, to form the ILK–PINCH–parvin (IPP) scaffold complex (Sect. 5.9.1) [168]. The IPP complex contributes to the control of actin dynamics. The body's development requires integrin-mediated interactions between cells and their extracellular environment. Whereas ILK is dispensable for mammalian development, its interaction with α-parvin is needed for kidney development [169].

3.4.4 Janus Pseudokinases

Janus tyrosine kinases (JaK1–JaK3 and Tyk2) have a C-terminal functional catalytic domain (JH1) that is preceded by a pseudokinase (JH2) domain [168]. They are associated with cytokine receptors. Upon cytokine binding, they are activated and phosphorylate the receptor cytoplasmic domain, hence creating recruitment sites for signaling effectors such as signal transducers and activators of transcription.

3.4 Intracellular Pseudokinases 133

The latter are also phosphorylated by JaKs to form dimers that translocate to the nucleus.

3.4.5 Pseudokinase Kinase Suppressors of Ras

Many pseudokinases (KSR, TrrAP, Trb, ILK, and CASK) act as organizers, i.e., scaffolds. Kinase suppressors of Ras KSR1 and KSR2 are scaffolds that, in particular, bridge MAP2K kinases to Raf (MAP3K) kinase. Scaffolds KSR1 and KSR2 form a plasmalemmal regulatory complex with components of the MAPK module (Chap. 6) to coordinate signal propagation. Phosphatase PP3 selectively interacts with KSR2 agent. Adaptor KSR2 contributes to Ca^{++}-mediated extracellular signal-regulated kinase (ERK) signaling [170].

3.4.6 Sterile 20-Related Adaptors

Pseudokinase Ste20-related adaptor (StRAd) allosterically regulates the catalytic function of the protein kinase serine/threonine kinase STK11 (or LKB1), an upstream kinase of adenine monophosphate-activated protein kinase (AMPK) [171]. Kinase STK11,[66] known as a tumor suppressor, prevents cell growth and proliferation by activating numerous kinases, such as AMPK and AMPK-related kinases (Sect. 5.4.3 and 5.4.4).[67] Hence, unlike many kinases that are activated by phosphorylation, STK11 is activated by StRAd pseudokinase to hold affector STK11 kinase in an active conformation (allosteric mechanism of activation).

Ste20-related adaptors StRAdα[68] and StRAdβ[69] heterotrimerize with: (1) isoforms of scaffold mouse protein Mo25 (Mo25α and -β) for AMP-activated protein kinase kinases (AMP2K) and (2) Ser/Thr kinase STK11 to control STK11 activity [168]. They hence serve as STK11-specific adaptors, substrates, and activators. Kinase STK11 is actually not activated by phosphorylation by a kinase, but by StRAd binding. Removal of Stradβ by small interfering RNA abrogates STK11-induced G1-phase arrest [172]. Catalytically active complexes formed by STK11, StRAdα or StRAdβ, and Mo25α or Mo25β are required for full kinase activity [173]. Both AMP2K1 and AMP2K2 can contain STK11, Stradα, and

---

[66] Alternative splice variants of Stk11 gene transcript exist: STK11 long (STK11$_L$) and short (STK11$_S$).
[67] Kinase AMPK precludes cell growth and proliferation when nutrient and energy availability is scarce.
[68] A.k.a. protein kinase StLK5 and Lyk5.
[69] A.k.a. StLK6.

Mo25α to phosphorylate (activate) AMPK enzyme. Heterotrimer ability to activate STK11 has the following efficiency order:

$$\text{STK11–StRAd}\beta\text{–Mo25}\beta < \text{STK11–StRAd}\alpha\text{–Mo25}\beta \sim$$
$$\sim \text{STK11–StRAd}\beta\text{–Mo25}\alpha < \text{STK11–StRAd}\alpha\text{–Mo25}\alpha.$$

The STK11–StRAd–Mo25 complexes not only phosphorylate both AMPKα1 and AMPKα2, but also all the 12 other members of the AMPK-related protein kinase family (i.e., BrSK1–BrSK2, MARK1–MARK4, MELK, NuaK1–NuaK2, QIK, QSK, and SIK) [174].

3.4.7 Tribbles Pseudokinases

Isoforms of Trb proteins[70] bind to the CDC25 homolog String that regulates cell division and DNA-damage repair to promote its ubiquitination and proteasome-mediated degradation [168]. Under fasting conditions in adipose tissue, Trb3 impedes lipid synthesis, as it provokes the degradation of the rate-limiting enzyme acetyl-coenzyme-A of fatty acid synthesis by interacting with COP1 E3-ubiquitin ligase.

3.4.7.1 Trb1 Protein

Protein Trb1[71] controls mitogen-activated protein kinase (MAPK) modules (Chap. 6). The Trb proteins indeed bind to MAP2Ks, specifically MAP2K1 and MAP2K4 [175]. These interactions not only regulate MAP2K activity, but also modulate Trb1 expression level by influencing Trb1 turnover and/or stability.

Protein Trb1 has a Ser/Thr kinase-like domain with mutations at critical positions for catalytic activity, hence it is a pseudokinase. It resides in the nucleus. Its highest expression levels are detected in peripheral blood lymphocytes as well as muscle, thyroid gland, pancreas, and bone marrow.

Protein Trb1 regulates AP1 activation [176]. Expression of Trb1 is upregulated by inflammatory stimuli in various cell types [177]. In addition, Trb1 interacts with 12-lipoxygenase [176]. It is also a major regulator of vascular smooth muscle cell proliferation and chemotaxis via the MAP2K4–JNK pathway [178].[72]

---

[70]The Trb proteins are mammalian orthologs of Drosophila protein Tribbles.

[71]A.k.a. stress kinase inhibitory protein SKIP1.

[72]In vascular smooth muscle cells, JNK and P38MAPKs operate in responses primarily primed by cell stresses (heat, hypoxia, chemical, oxidative stress, etc.) and pro-inflammatory cytokines, whereas ERK kinases predominantly respond to mitogenic stimuli such as growth factors (PDGF), oxidized low-density lipoproteins, and angiotensin-2. Inflammatory cytokines and chemokines can lead to migration and proliferation of vascular smooth muscle cells to build a neointima.

3.4.7.2 Trb2 Protein

Protein Trb2 lodges mainly in the cytoplasm [177]. It is abundantly expressed in circulating leukocytes and kidney mesenchymal cells.[73] The regulation of its expression is cell-type specific. It inhibits C/EBP family proteins via proteolysis [179]. It can impede adipocyte differentiation. It also influences monocyte function, as it hampers mitogen-activated protein kinases MAP2K1 and MAP2K7 (or JNKK2). It is expressed at high levels in vulnerable regions of atherosclerotic plaques. Its production is indeed enhanced by oxidized low-density lipoproteins in macrophages.

3.4.7.3 Trb3 Protein

Protein Trb3 serves as an inhibitor for transcription factors ATF4 and NFκB component RelA (or P65) as well as insulin-dependent activation of protein kinase-B [177].

3.4.8 TrrAP Pseudokinase

Transformation/transcription domain-associated protein TrrAP is a pseudokinase that serves as a constituent of most histone acetyltransferase complexes. It also connects to several transcription factors to control the repair of double-strand DNA breaks [168].

---

[73]Protein Trb2 during gestation is detected in many organs, such as kidney, mesonephros, heart, blood vessels, spinal cord and ganglia, eye, thymus, muscle, bone, tongue, and testis [179].

Chapter 4
Cytoplasmic Protein Tyrosine Kinases

Protein tyrosine kinases (PTK), i.e., enzymes that catalyze the phosphorylation of Tyr residues of proteins. are mainly associated with growth factor signaling. They actually modulate multiple cellular events, such as differentiation, growth, metabolism, and apoptosis. On the other hand, protein serine/threonine kinases are principally related to second messengers, such as cyclic nucleotides cAMP (Sect. 11.1) and cGMP (Sect. 11.2), lipidic and related mediators diacylglycerol and inositol trisphosphate (Chap. 2), and calmodulin (Sect. 11.5.3).

Protein kinase activity is strongly regulated by phosphorylation, either by the kinase itself (autophosphorylation), or kinase-bound activators or inhibitors. Because enzymes can have different phosphorylation sites, phosphorylation and dephosphorylation either regulate the enzyme activity or directly launch enzyme activation or inhibition.

4.1 Receptor and Cytosolic Protein Tyrosine Kinases

The class of about 100 detected protein Tyr kinases includes 2 main subclasses (Tables 4.1 and 4.2): (1) transmembrane (plasmalemmal) *receptor tyrosine kinases* (RTK; Vol. 3 – Chap. 8. Receptor Kinases) that can be grouped into 19 families and (2) intracellular *non-receptor tyrosine kinases* (NRTK) that can be subdivided into 10 families.

Receptor protein Tyr kinases contain an extracellular ligand-binding domain, transmembrane region, and intracellular cytoplasmic kinase domain. The RTK subclass with 58 members is constituted by growth factor receptors (e.g., epidermal [EGFR], fibroblast [FGFR], insulin-like [IGFR], platelet-derived [PDGF] growth factor receptor). Non-receptor protein Tyr kinases recruited in signaling pathways following ligand binding to receptor comprise 32 known kinases that include Janus, SRC, and TEC kinase families (Tables 4.3 and 4.4).

Table 4.1. Human protein tyrosine kinases. (**Part 1**) Receptor and cytoplasmic (non-receptor) families (BMX: bone marrow Tyr kinase gene in chromosome-X product; BRK: breast tumor kinase; BTK: Bruton Tyr kinase; CSK: C-terminal Src Tyr kinase; InsRR: insulin receptor-related receptor; ITK: interleukin-2-inducible T-cell kinase; LTK: leukocyte Tyr kinase; MATK: megakaryocyte-associated Tyr kinase; HGFR: hepatocyte growth factor receptor [a.k.a. mesenchymal–epithelial transition factor (MET)]; MST1R: macrophage stimulating-1 factor receptor; Smrs: src-related kinase lacking C-terminal regulatory Tyr and N-terminal myristylation; TEC: family of NRTKs in T and B lymphocytes and hepatocytes; TNK: Tyr kinase inhitor of nuclear factor-κB).

Receptor families	Non-receptor families
Apoptosis-associated Tyr kinases (Aatyk1–Aatyk3)	Abelson kinases (Abl1–Abl2)
Anaplasic lymphoma kinase (ALK, LTK)	Activated CDC42-associated kinases (ACK1–ACK2 or TNK2–TNK1)
Adhesion-related kinases (Axl, Eyk, Mer, Tyro3, Rek)	SRC Tyr kinases (CSK, MATK)
Discoidin domain kinases (DDR1–DDR2)	Focal adhesion kinases (FAK1, FAK2)
EGF receptors (HER1–HER4)	Feline sarcoma oncogene kinases (FeR, FeS)
EPH receptors (EPHa1–a10; EPHb1–b6; EPHx)	Fyn-related kinases (BRK, FRK, Smrs)
FGF receptors (FGFR1–FGFR4)	Janus kinases (JaK1–JaK3, TyK2)
Insulin and IGF receptors (InsR, InsRR, IGF1R)	SrcA (FGR, Fyn, Src, Yes1) SrcB (BLK, HCK, LCK, Lyn)
MET (HGFR, MST1R)	Spleen Tyr kinases (SYK, ZAP70)
Muscle-specific kinase (MUSK)	TEC kinases (BMX, BTK, ITK, TEC)

Src homology SH2 domains of intracellular protein Tyr kinases can keep these enzymes in an auto-inhibited conformation. However, SH2 domain of intracellular protein Tyr kinases can enhance substrate recognition and catalytic activity of FeS and Abl kinases that lack this between-domain auto-inhibitory interaction [180]. SH2 domain is involved in substrate recruitment and couples substrate recognition to kinase activation. Moreover, SH2 domain interaction with the kinase domain stabilizes the active kinase conformation.

4.1 Receptor and Cytosolic Protein Tyrosine Kinases

Table 4.2. Human protein tyrosine kinases. (**Part 2**) Receptors (TEK: Tyr endothelial kinase; TIE: Tyr kinase with Ig and EGF homology receptor (angiopoietin receptor); TRK: neurotrophin receptor).

Receptor PTK families (Cont.)	Members
PDGF receptors	PDGFRα–PDGFRβ CSF1R, FLT3, SCFR
Protein Tyr kinase-7	PTK7
Rearranged during transfection	ReT
Receptor Tyr kinase-like orphan receptor	ROR1–ROR2
Ros UR2 sarcoma virus oncogene homolog-1	Ros1
Related to receptor Tyr kinase	RYK
Angiopoietin receptors	TEK, TIE
Tropomyosin receptor kinase (TRK)	NTRK1–NTRK3
VEGF Receptors	VEGFR1–VEGFR3

Table 4.3. Partners and substrates of cytoplasmic protein Tyr kinases (**Part 1**; Sources: multiple).

Type	Expression pattern	Partners and substrates
Abl1/2	Ubiquitous	NTRK1, EPHb2, BCR, TOR, ATMK, PAK2, GPX1, Rad9a, Rad51, RB, BrCa1, TeRF1, NCK1, DOK1, PAG1, BCAR1, CASL, CBL, CRKL, SHC, GRB2/10, PSTPIP1, ArgBP2, Vav1, RFX1, HDM2, P73, spectrin-α1
ACK1/2	Ubiquitous	EGFR, Src, Fyn, CDC42 ALX, GRB2, NCK, DblGEF, RasGRF1
BrK	Epithelial cell	EGFR, PI3K, PTen, PKB
FRK		CRK2, KhdRBS1, STAP2, SHB, paxillin
BTK	B and mast cells	TNFSF6, PI(4)P5K, PLCγ2 PI(3,4,5)P$_3$
CSK	Hematopoietic cell	InsR, IGF1R, IRS1 FAK, PKA, G protein, PTP, paxillin CBP, DOK, G3BP, SIT, caveolin-1
FAK1/2	Ubiquitous	Src, PI3K GRB2, SHC, BCAR1, PtdIns, paxillin
FeR, FeS	Granulocyte Macrophage	EGFR Dynamin-1–3, catenin, cortactin, E-cadherin, TMF1

Table 4.4. Partners and substrates of cytoplasmic protein Tyr kinases (**Part 2**; Sources: multiple; HSP: heparan sulfate proteoglycan [a.k.a. CD44 and phagocytic glycoprotein PGP1]).

Type	Expression pattern	Partners and substrates
JaK1–3, TyK2	Ubiquitous	Cytokine receptors STATs CIS, SOCS, SSI
Src	Ubiquitous	GluN2a, HSP, EPHb2, β3AR, EGFR, PDGFR, CSF1R, AR ERα and -β, NR1b1, HNF1a, cRaf, PKCζ, PLD2, FAK2, ARNT, AHR, dystroglycan, KhdRBS1, PDE6γ, STAT1/3, SRE, DDEF1, NCoA6 GRB2, Dab2, EPS8, BCAR1, SHB, WASP, RICS, P120RasGAP, GNB2L1, ND2, Mucin-1
SYK, ZAP70	Hematopoietic cells	EpoR, CSF3R, Fc BTK, FGR, Fyn, LCK, Lyn, Src, FAK1/2, PKCα, PRKD1, PTP6, PLCγ1/2 Cortactin, paxillin, and tubulin Vav, GRB2, SHC, CRKL, SLAP, TRAF6, BCAP, BAnk, BLnk, LCP2, LAB, LAT, LAX
TEC	T lymphocyte B lymphocyte (BTK)	LCK, PI3K, PLCγ1 EEA1

4.2 Family of Abl Kinases

Viral Abelson murine leukemia oncogene homolog is expressed with 2 paralogs — Abl1 and Abl2 —, which represent the mammalian members of the Abelson family of non-receptor protein Tyr kinases. Kinases Abl1 and Abl2 (or Arg) have redundant functions. Ubiquitous Abl1 and Abl2 can heterodimerize.

Several types of post-translational modifications enable the control of Abl catalysis, subcellular localization, and stability. Binding partners yield additional regulation of Abl activity, substrate specificity, and signaling.

Each Abl subtype contains an SH3–SH2–TK (SH: Src homology; TK: Tyr kinase) domain cassette that confers autoregulated kinase activity. This cassette is coupled to an actin-binding and -bundling domain that couples phosphoregulation to actin filament reorganization.

Kinase Abl1[1] is a cytosolic and nuclear protein Tyr kinase that contributes to cell differentiation, division, and adhesion, as well as stress response. Protein Abl2 is detected exclusively in the cytoplasm.

Kinase Abl1 possesses nuclear localization signals and a DNA-binding domain used to mediate DNA-damage repair. On the other hand, Abl2 has additional binding motifs for actin and microtubules that enable cytoskeletal remodeling.

4.2.1 Abl1 Kinase (Abl)

Kinase Abl1 (a.k.a. JTK7 and P150) possesses multiple domains, such as SH1 (Tyr kinase), SH2, and SH3 domains, as well as a large C-terminus that contains proline-rich sequences, DNA- and actin-binding motifs, and nuclear localization and export signals.

Messenger RNA ABL1 can be alternatively spliced at the level of the N-terminus to generate Abl1a and Abl1b isoforms [181]. Kinase Abl1 operates in the regulation of cytoskeletal dynamics and cell proliferation and survival. Altered Abl1 exhibits constitutive Tyr kinase activity and provokes leukemias in humans.

In adult humans, the highest Abl1 expression levels are observed in cartilage, adipocytes, and ciliated epithelium. In human fetuses, its highest concentrations are detected in muscle, osteoblasts, endothelial cells, and neovasculature at sites of endochondral ossification in the umbilical cord stroma [181].

Kinase Abl1 regulates pro- or anti-apototic response according to cell type. Activated by DNA damage or oxidative stress, it accumulates in the nucleus to prime a pro-apoptotic response in fibroblasts [181]. It can also be activated in DNA repair that depends on DNA-dependent protein kinase (DNA-PK; Sect. 5.5.2) and ataxia telangiectasia mutated kinase (ATMK; Sect. 5.5.1) as well as G1–S checkpoint.

Enzyme Abl1 undergoes phosphorylation by CDK1. It associates with retinoblastoma protein. It phosphorylates RNA polymerase-2. It can cause cell cycle arrest in response to cell stress. It then activates the stress-activated protein kinase pathway.

In T lymphocytes, upon TCR activation, Abl1 phosphorylates LAT[2] adaptor and ZAP70[3] kinase [181]. Kinase Abl1 intervenes in proliferation of stimulated splenic B lymphocytes as well as that of fibroblasts in response to platelet-derived growth factor. It interacts with B-cell coreceptor CD19 on follicular dendritic cells and B lymphocytes.

Kinase Abl1 is activated by stimulated receptors of growth factors (PDGFR and EGFR), ephrins (EPHb2), and semaphorin-6D, as well as RPTK MuSK (Vol. 3 – Chap. 8. Receptor Kinases) and integrin engagement to reorganize the

---

[1] Oncoproteins Gag–Abl and Bcr–Abl that are synthesized after mutations in the ABL1 gene are detected in preB lymphomas associated with Abelson murine leukemia virus in mice, and chronic myelogenous leukemia and some forms of acute lymphocytic leukemia in humans, respectively.

[2] I.e., linker of activated T lymphocytes.

[3] I.e., ζ-associated protein.

cytoskeleton [181]. It is also activated by DNA damage and oxidative stress. On the other hand, in addition to auto-inhibition due to intramolecular interactions and folding, Abl1 is inhibited by phosphatidylinositol (4,5)-bisphosphate as well as dephosphorylation by PTPn12 (Sect. 8.3.11.11). Moreover, it is degraded by ubiquitin-dependent proteasome.

Protein Abl1 can heterodimerize with Abl2. It also interacts with adaptor CRK, docking proteins DOK1 and DOK2, Abl interactors AbI1 and AbI2, CDK5 and Abl enzyme substrate CABLES1, Ras and Rab interactor RIn1, proline/serine/threonine phosphatase-interacting protein PSTPIP1, phospholipase-Cγ, anti-oxidant enzyme peroxiredoxin-1, ^{G}actin, and WAVe1 [181].

Kinase Abl1 stimulates the activity of the RNA-binding protein heterogeneous nuclear ribonucleoprotein hnRNPe2 that prevents the translation of the mRNA of the transcription factor C/EBPα, which is involved in leukocyte differentiation. MicroRNA-328 is a double-duty molecule that acts as both a RNA silencer to prevent protein synthesis and competitive inhibitor of translation-inhibiting hnRNPe2 to promote protein synthesis [182].[4]

4.2.2 Abl2 Kinase (Arg)

Abelson-related kinase Abl2 is another cytoplasmic kinase that shares structural and sequence homology with Abl1. Kinase Abl2 contains Tyr kinase domains (SH1–SH3) and a C-terminus with proline-rich sequences and binding domains for ^{F}actin and tubulin [181].

Kinase Abl2 contributes to linkage between plasmalemmal receptors (T-cell receptors, MuSK, EPHb2, and platelet-derived growth factor receptor, as well as integrins) and cytoskeleton and/or cell proliferation signaling. It can be identified in actin cytoskeletal structures and focal adhesions, at least, under certain conditions. In fact, Abl2 is not detected in stress fibers. It is specifically required for integrin-mediated adhesion to laminin-1 and semaphorin7A. Abl Kinases target adaptor P130CAS (or BCAR1) and structurally related docking protein CASL[5] that both interact with FAK1 in focal adhesions. Adaptor CRK associates with both Abl1 and Abl2 kinases. Other potential Abl partners that are components of the actin cytoskeleton and focal adhesions include FAK, paxillin, vinculin, talin, and tensin. In addition, target SH3 domain-binding protein SH3BP1 has GTPase-activating protein activity for Rac GTPases involved in the regulation of the actin cytoskeleton [183].

---

[4]MicroRNA-328, like miRs, tethers to Argonaute proteins to form a miR–ribonucleoprotein complex for post-transcriptional gene silencing. It also simultaneously binds to regulatory RNA-binding protein hnRNPe2 and mRNA of the survival factor PIM1 to prevent hnRNPe2 from precluding translation of mRNAs.

[5]A.k.a. CAS2, human enhancer of filamentation HEF1 or neural precursor cell-expressed developmentally downregulated protein NEDD9.

Kinase Abl2 is activated by autophosphorylation as well as phosphorylation by Abl1 and Src kinases [181]. Association of Abl2 with Ras effector Rin1 can also stimulate Abl2. Kinase Abl2 interacts with adaptors CRK and Arg (Abl2)-interacting protein ArgBP2,[6] pro-apoptotic protein Siva-1, and ubiquitin for degradation.

4.2.3 Abl-Binding Proteins

Binding partners of Abl1 and Abl2 kinases such as ArgBP2[7] are substrates of both enzymes. Protein ArgBP2 is widespread in human organs and extremely abundant in the heart. In epithelial cells, ArgBP2 and Abl are located near stress fibers and in the nucleus. Protein ArgBP2 then links Abl kinases to the actin cytoskeleton. This adaptor assembles signaling complexes in stress fibers. In cardiomyocytes, ArgBP2 is located in Z discs of sarcomeres, where it can be targeted by signaling mediators [183].

4.3 ACK Kinases

Activated CDC42-associated kinases ACK1 and ACK2[8] are intracellular protein Tyr kinases. Small GTPase CDC42^{GTP} binds to the CDC42–Rac interactive-binding domain (CRIB) of ACK in a GTP-dependent manner. Kinase ACK indeed possesses a Tyr kinase core, an SH3 domain, a CDC42-binding region, a RALT homology region,[9] and a proline-rich region.

Ubiquitous ACK is highly synthesized in the brain, thymus, and spleen [184]. Activation of integrins as well as EGFR and PDGFR receptors leads to ACK recruitment and its Tyr phosphorylation (activation).

The ACK1 splice variant is activated from chondroitin sulfate proteoglycan. It can regulate cell motility. Isoform ACK2 can regulate cell spreading, motility, and integrin-mediated adhesion [185]. Kinase ACK2 represses focal adhesion complex organization.

Both ACK1 and ACK2 can associate with clathrin to regulate receptor-mediated endocytosis [185]. Epidermal growth factor activates ACK1 that, once phosphorylated, activates RhoGEF21 agent. Activated RhoGEF21 subsequently stimulates

---

[6] A.k.a. sorbin and SH3 domain-containing protein SorbS2.
[7] A.k.a. sorbin and SH3 domain-containing protein SorbS2.
[8] A.k.a. Tyr non-receptor kinase TNK2 and TNK1, respectively.
[9] Protein receptor-associated late transducer (RALT), the expression of which results from the activation of the Ras–Raf–ERK pathway, is a labile inhibitor of HER2 mitogenic signals. It is also called ErbB (HER) receptor feedback inhibitor ErRFI1. This mitogen-induced signal transducer binds to the HER2 kinase domain.

Rho GTPases. It contributes to cytoskeletal rearrangements. Kinase ACK2 supports phosphorylation of sorting nexin SNx9 (or SH3PX1) by epidermal growth factor receptor, thereby promoting EGFR degradation.

Phosphorylated ACK1 binds to GRB2 and NCK adaptors. In addition, ACK1 phosphorylates RasGRF1 (a GEF) that targets hRash, kRas, and nRas. Isoform ACK2 binds to Src kinase and ALX (adaptor in lymphocytes of unknown function [X]) adaptor.[10]

4.4 BrK Kinase (Protein Tyr Kinase-6; FRK Subfamily)

Protein Tyr kinase-6 (PTK6), or breast tumor kinase (BrK), contains an SH2, an SH3, and a catalytic domain. A splice variant comprises a SH3 domain and a short C-terminus.

In mammary epithelial cells, PTK6 enhances mitogenic responses to epidermal growth factor [186]. Expression of PTK6 in breast carcinoma cells causes cell proliferation owing to an elevated EGF-dependent phosphorylation of EGFR-related receptor HER3 and potentiated recruitment of phosphoinositide 3-kinase to HER3 receptor. Protein PTK6 phosphorylates KH domain-containing, RNA-binding signal transduction-associated protein KhdRBS1 (or Sam68) and related proteins to prevent its binding to RNA, thereby impeding the ability of KhdRBS1 to increase expression from Rev-responsive element-containing genes.

Kinase PTK6 regulates protein kinase-B. It also phosphorylates paxillin and promotes Rac1 activation via CRK2, hence assisting in chemotaxis. Protein PTK6 can form complexes with EGFR (HER1), HER3, signal-transducing adaptor protein STAP2, and paxillin, in addition to KhdRBS1 and PKB [186]. Its binding partners also include insulin receptor substrate IRS4 in resting and insulin-like growth factor-1-stimulated cells [187].

4.5 CSK Kinase

Carboxy-terminal Src Tyr kinase (CSK) phosphorylates a regulatory Tyr residue at the C-terminus of kinases of the SRC family, thereby inactivating Src kinases. Kinase CSK is especially produced in hematopoietic cells [188]. It blocks signaling downstream from T-cell receptors mediated by SRC family Tyr kinases.

Protein kinase-A phosphorylates (Ser364) CSK, thereby heightening its kinase activity (2–4-fold) [188]. Activity of CSK is also moderately enhanced (2-fold) by $G\beta\gamma$ subunit. Kinase CSK also forms a complex with the RasGAP-binding

---

[10] Adaptor ALX, also called hematopoietic SH2 domain-containing protein HSH2, has a hematopoietic-specific expression pattern.

protein GTPase-activating protein (SH3 domain)-binding protein G3BP. In addition, CSK interacts with receptors of insulin and insulin-like growth factor-1, insulin receptor substrate-1, focal adhesion kinase, protein Tyr phosphatase, paxillin, palmitoylated transmembrane adaptor phosphoprotein associated with glycosphingolipid microdomain PAG1, or CSK-binding protein (CBP) in membrane rafts, caveolin-1, leukocyte-specific protein Tyr kinase-interacting molecule (LIME), SHP2 (PTPn11)-interacting transmembrane adaptor (SIT), and members of the Downstream of Tyr kinase docking protein (DOK) family.

The DOK proteins participate in the regulation of signaling mediated by receptor and non-receptor kinases. Protein DOK1 precludes cell proliferation initiated by growth factors, as it tethers different signaling components to the cell membrane. It attenuates platelet-derived growth factor action by recruiting CSK to active Src kinases. Moreover, DOK1 impedes PDGF-induced activation of the Ras–MAPK cascade by acting on RasGAP and other DOK1-interacting proteins [189]. Adaptor DOK3 rapidly bears Tyr phosphorylation in response to immunoreceptor stimulation and subsequently recruits inhibitors SHIP phosphatase and CSK kinase [190].

4.6 Focal Adhesion Kinases

Focal adhesion kinases — FAK1,[11] and FAK2 —[12] are encoded by the Fak1 and PTK2B genes. Four transcript variants encode 4 isoforms. These cytosolic protein Tyr kinases localize to focal adhesions that create attachments with the extracellular matrix or with other cells. Cells synthesize *FAK-related non-kinase* (FRNK) from an independent transcript that corresponds to the FAK C-terminus (and hence lacks the kinase activity and autophosphorylation site) to prevent FAK activation.

The primary FAK function is the transmission of signals emitted by integrins. Kinase FAK also participates in growth factor receptor-mediated signaling. It transduces signals to regulate cell survival, proliferation, adhesion, migration, and death. Its activity is modulated by phosphorylation and dephosphorylation. Kinase FAK is phosphorylated in response to integrin engagement, growth factor stimulation, and mitogenic neuropeptides.

Focal adhesion kinase is a scaffold and Tyr kinase that binds to itself and cellular partners through its four-point-one, ezrin, radixin, moesin (FERM) domain.[13]

---

[11] A.k.a. protein Tyr kinase-2 (PTK2), focal adhesion kinase with aliases FAdK and FADK1, as well as PP125FAK and FRNK. In fact, FAK-related non-kinase (FRNK) corresponds to the C-terminus of FAK, hence lacks its N-terminal and central catalytic domain, but retains the focal adhesion targeting sequence.

[12] A.k.a. protein tyrosine kinase-2β (PTK2β), calcium-dependent tyrosine kinase (CaDTK), cell adhesion kinase-β (CAKβ), related adhesion focal tyrosine kinase (RAFTK), proline-rich tyrosine kinase PYK2, and focal adhesion kinase FAdK2.

[13] In humans, the larger set of FERM domain-containing proteins include certain kinases (FAKs and JAKs), myosins (Myo7, Myo10, and Myo15), phosphatases (PTPn13), guanine

Regulators convert auto-inhibited autocatalytic Tyr kinase FAK into its active state by binding to its FERM domain.[14] Interactors of FAK support this protein in its coordination of diverse cellular functions in which it is involved.

Partners of FAK that bind to the FERM domain comprise PI(4,5)P$_2$ generated by phosphatidylinositol 4-phosphate 5-kinase PI(4)P5K1γ,[15] epidermal (EGFR), hepatocyte (HGFR), platelet-derived (PDGF), and vascular endothelial (VEGF) growth factor receptors, protein Tyr phosphatase PTPn21,[16] ezrin, BMX kinase,[17] actin-related protein ARP3, receptor for activated kinase RACK1, insulin receptor substrate IRS1, and MAPK8-interacting protein MAPK8IP3,[18] in addition to β subunits of integrins [191].

Neural Wiskott-Aldrich syndrome protein, an effector of CDC42 GTPase, is a substrate of FAK enzyme. Together with ARP3, RACK1, and phosphodiesterase PDE4d5, FAK contributes to the assembly of nascent integrin-based adhesions.

Merlin, related to the ERM (ezrin, radixin, and moesin) proteins, is phosphorylated (inactivated) by P21-activated kinase activated upon integrin engagement. Conversely, cadherin engagement inactivates PAK kinase. Merlin can accumulate in the nucleus, where it binds to and inhibits the ubiquitin ligase CRL4–DCAF1 complex (CRL4^{DCAF1}),[19] hence the expression of numerous genes [192].

Phosphorylation of specific residues of FAK regulate distinct cellular processes. Phosphorylation (Tyr861 and Tyr925) regulates cell migration of endothelial cells and proliferation, respectively [193] (Table 4.5). Focal adhesion kinase is activated by multiple pro-angiogenic growth factors, such as VEGF, EGF, and FGF2. Enzyme FAK integrates integrin and growth factor signaling to promote cell motility. Kinase FAK not only operates as a kinase, but also as a scaffold, as it yields Tyr^P-binding sites for signaling mediators.

---

nucleotide-exchange factors, ezrin, radixin, moesin, kindlins, talins, and Krev interaction trapped (KRIT) proteins. FERM domain-containing proteins that can shuttle between the cell cortex and the nucleus comprise ezrin, radixin, moesin, merlin, kindlin-2, myosin-10, and protein Tyr phosphatase PTPn3, PTPn13, and PTPn14, in addition to FAK kinase [191].

[14]Displacement of the auto-inhibitory interactions primes activation of the catalytic activity. The auto-phosphorylation site (Tyr397) in FAK lodges in a segment between the FERM and kinase domains. Phosphorylation by Src kinase (Tyr576 and Tyr577) that binds to Tyr397^P leads to full activation.

[15]Phosphatidylinositol (4,5)-bisphosphate triggers a FAK conformational changes and can then activate FAK enzyme. Kinase PI(4)P5K1γ is involved in FAK activation downstream from both stimulated integrins and growth factor receptors.

[16]Phosphatase PTPn21 is required for FAK activation upon EGF stimulation.

[17]A.k.a. epithelial and endothelial tyrosine kinase (ETK).

[18]A.k.a. JNK-interacting protein JIP3 and JNK–SAPK-associated protein JSAP1.

[19]The cullin-4A–RING (really interesting new gene) domain-containing ubiquitin ligase CRL4 is a proteic complex that comprises cullin-4A (Cul4), RING-box protein RBx1 (a.k.a. Regulator of cullins ROC1 and RING finger protein RNF75), and adaptor DNA-damage-binding protein DDB1. Proteins bind DDB1 adaptor to recruit specific targets to CRL4 for ubiquitination. The CRL4 substrate recognition module relies on DDB1- and CUL4-associated factors (DCAF).

Table 4.5. Structural features and binding sites of focal adhesion kinases and partners (Sources: [191,194]; FAT: focal adhesion-targeting domain; FERM: band 4.1 [four], ezrin, radixin, and moesin homology N-terminal domain; Pro: proline-rich Src-homology 3 (SH3) binding motif; TK: tyrosine kinase domain). Focal adhesion kinase associates with adaptors, structural proteins, kinases and phosphatases, as well as small GTPase regulators. Src-mediated phosphorylation of FAK (Tyr925) creates a binding site for GRB2 adaptor to prime the extracellular signal-regulated (ERK) pathway. The FERM domain of FAK, in particular, binds to cytoskeleton- and membrane-linked proteins and phospholipids, mainly at adhesion sites, where it modulates cortical actin and nascent adhesions by binding to the ARP2–ARP3 complex and the molecular scaffold RACK1, in addition to optimal signaling from growth factor receptors (EGFR, HGFR, and PDGFR), but also in the nucleus, where it connects to P53 and its regulator DM2.

Domain	Partners (pathways)
	N-terminus
FERM	FAK, EGFR, HGFR, PDGFR, VEGFR, BMX, PI3K, ezrin, β-integrin, ARP3, RACK1, IRS1, MAPK8IP3, PI(4,5)P$_2$, P53, DM2
Pro1	GRB7, PLCγ
	Central domain
TK	Src, PTen, PTPn12, SHC (Tyr397 autophosphorylation site) (FAK–Src–CAS complex–JNK)
	C-terminus
Pro2	CAS
Pro3	ASAP1 (ArfGAP), RhoGAP26, GRB2
FAT	paxillin, talin, (indirectly integrin, APAP1, Rho(Arh)GEF6/7)

Focal adhesion kinases mediate several integrin signaling pathways. They can perform autophosphorylation on a Tyr residue. Autophosphorylation of FAK creates a binding site for kinases of the SRC family. This complex allows Tyr phosphorylation of focal adhesion-associated proteins. The FAK kinases interact with Src and PI3K kinases, GRB2 adaptor, integrin signaling mediator CRK-associated substrate CAS (or BCAR1), and paxillin.

Hence, activated FAK controls cell adhesion and motility [195]. Tyrosine-phosphorylated FAK promotes interactions with certain proteins, which allows connections to the Rac and Rho GTPases and extracellular signal-regulated kinase isoform ERK2 cascade. It acts on the assembly and maturation of focal contacts. Activity of the FAK–Src axis promotes phosphorylation of phosphatidylinositols. Kinase Src causes disassembly of focal contacts, as it activates calpain and extracellular matrix metalloproteinases; it also regulates cadherin-mediated cellular junctions. Signaaling vi FAK to Rho GTPases regulates changes in actin and microtubules in cell protrusions of migrating cells.

Mechanical signals from the extracellular matrix sent to fibroblasts are sensed by integrin-based cell–matrix adhesions and transmitted by several types of molecules, among which focal adhesion kinases. The latter actually contributes to the activation of cardiac fibroblasts subjected to mechanical stress [196]. Mechanical stimuli provoke proliferation of cardiac fibroblasts and their differentiation into myofibroblasts and raise the production of matrix components, such as MMP2 and collagen. Enzyme FAK works via target of rapamycin.

Focal adhesion kinase FAK2 is involved in Ca^{++}-induced regulation of ion channels and activation of the MAPK signaling. Four transcript variants generate 2 different isoforms. It serves as a signaling mediator between neurotransmitters and neuropeptide-activated receptors that cause calcium influx on the one hand and signals that regulate neuronal activity on the other. In response to increased intracellular calcium concentration, nicotinic acetylcholine receptor activation, membrane depolarization, or protein kinase-C activation, FAK2 undergoes rapid Tyr phosphorylation (activation).

In vascular smooth muscle cells, focal adhesion kinase and its inhibitor FAK-related non-kinase (FRNK) regulate cell spreading and migration during vasculo- and angiogenesis as well as vascular remodeling [193].[20] Phosphorylation of FRNK on Tyr residues leads to inhibition of VSMC spreading and migration.

In cardiomyocytes, hypoxia not only activates SRC family kinases Src and Fyn and the Ras–MAPK pathway to target stress-activated protein kinase and P38MAPK, but also focal adhesion kinase and paxillin [197]. Hypoxia-induced FAK activation leads to association with adaptors SHC and GRB2 and cytoplasmic Tyr kinase Src. Furthermore, hypoxia causes subcellular translocation of FAK from perinuclear sites to focal adhesions.

4.7 FeS and FeR Kinases (FPS/FES Subfamily)

The mammalian FPS/FES (fujinami poultry sarcoma/feline sarcoma)[21] gene encodes protein Tyr kinase FeS that is specifically expressed during hematopoiesis, principally in cells of the granulocyte-macrophage lineage. Another Tyr kinase, antigenically related to FeS, is encoded by FER gene (FeS-related gene). Hence, FeR is considered as a member of the FPS/FES family of non-receptor protein Tyr kinases [198].[22]

---

[20] FAK-related non-kinase is restricted to arterial smooth muscle cells. In response to vascular injury, FRNK production raises.

[21] Oncogenes fps and fes have been isolated as from avian and feline retroviruses that cause tumors. Retroviral fps and fes oncogenes and cellular homologs correspond to the same FES gene. Viral fes oncogenes encode chimeric proteins with unregulated tyrosine kinase activity that abrogates the need for stimuli to influence cell fate and cause cell transformation.

[22] Kinases FeS and FeR are the only known members of this distinct family of intracellular protein Tyr kinases.

4.7 FeS and FeR Kinases (FPS/FES Subfamily)

Cytoplasmic FeS and FeR kinases have an N-terminal FCH domain[23] for possible microfilament association, 3 regions that potentially regulate oligomerization, a central Src homology-2 domain for binding to phosphorylated (Tyr^P-containing) proteins, and a C-terminal catalytic domain [199]. The FCH domain is implicated in the regulation of cytoskeletal rearrangement and vesicular transport. It can bind to microtubules and its SH3 domain can link to actin-reorganizing Wiskott-Aldrich syndrome proteins.[24]

Kinases FeS and FeR are dispensable for hematopoiesis, as they have redundant activities with other kinases. They are mainly involved in the regulation of inflammation, particularly mastocyte migration and leukocyte diapedesis, and innate immunity. They are actually involved in between-cell and cell–matrix interactions, possibly via rearrangement of the cytoskeleton and crosstalks between integrins of focal adhesions, cell adhesion molecules of adherens junctions, and receptors for growth factors and cytokines [199].

Kinases FeS and FeR participate in signaling downstream from receptors for cytokines (e.g., gmCSF [CSF2], EpoR, IL3R–IL6R, and IL11R), growth factors (e.g., PDGF), and immunoglobin-G. Kinases FeS and FeR can also be activated downstream from IgE receptors on mastocytes and glycoprotein GP6 collagen receptor on platelets [199]. They control cytoskeletal rearrangements and signaling that accompany receptor–ligand, cell–matrix, and cell–cell interactions. They participate in the regulation of inflammation and innate immunity. Kinase FeS indeed abounds in macrophages and neutrophils.

Cell migration requires cycles of actin polymerization and depolymerization. These cycles rely on coordinated interactions between the plasma membrane and actin regulators. Cortactin is an activator of actin polymerization mediated by the actin-related protein ARP2–ARP3 complex. Cortactin thus contributes to cytoskeletal reorganization during cell migration and vesicular transport. Cortactin localizes in lamellipodia and endosomes at the leading edge of moving cells. Tyrosine phosphorylation of cortactin inhibits its actin crosslinking activity and promotes its proteolytic degradation by calpain. Tyrosine phosphorylation of cortactin after engagement of PDGF and FcεR1 receptors is caused, at least partly, by FeR kinase.

Phosphatidic acid can be produced by phospholipase-D for cell proliferation, survival, and migration. Phosphatidic acid binds to the Tyr kinase FeR and enhances its ability to phosphorylate cortactin [200]. Kinase FeR favors lamellipodium formation and cell migration. This effect depends on PLD activity and phosphatidic acid–FeR interaction.

Enzyme FeR regulates intercellular adhesion and transmits signals initiated by growth factor receptors from the plasma membrane to the cytoskeleton. Kinase FeR is a cell volume-sensitive kinase that acts downstream from Fyn that targets

---

[23] Alias FCH stands for FPS/FeS/FeR/CIP4 (CDC42-interacting protein-4) homology.

[24] The FCH domain-containing proteins include, in addition to FeS–FeR family of PTKs, RhoGAP proteins, and proteic adaptors.

α-, β-, and δ1 catenins [201]. It also phosphorylates TATA element modulatory factor TMF1 that can bind TATA element in RNA polymerase-2 promoters. Nuclear kinase FeR can modulate the suppressive activity of TMF during cell growth and differentiation [202].

In fibroblasts, FeR reduces cell adhesion to matrix, as it phosphorylates docking protein FAK-associated CRK-associated substrate CAS1[25] by activated protein Tyr phosphatase and increased Tyr phosphorylation of adherens junction proteins catenin-δ1[26] and β-catenin [199]. Conversely, protein Tyr phosphatases that dephosphorylate PTK substrates in adherens junctions and focal adhesions include at least plasmalemmal Tyr phosphatases PTPRa and PTPRs as well as cytoplamic Tyr phosphatases PTPn1, PTPn11, and PTPn12.

4.8 Fyn-Related Kinase

Like Tyr kinases Wee1 and Abl1, Fyn-related kinase[27] is a nuclear protein. It is composed of SH2 and SH3 domains at the N-terminus and Tyr residues within the kinase domain and near the C-terminus. It is expressed primarily in epithelial cells.

It operates during G1 and S phases of the cell cycle to suppress cell proliferation. Kinase FRK interacts with retinoblastoma protein [203]. It also forms a signaling pathway with SH2-domain-containing adaptor SHB.[28] This signaling axis involves focal adhesion kinase and insulin receptor substrates IRS1 and IRS2. In endothelial cells, SHB both promotes apoptosis under anti-angiogenic condition, but also participates in mitogenicity, spreading, and tubular morphogenesis [204]. The FRK–SHB pathway regulates cell apoptosis, proliferation, and differentiation.

The FRK family of Tyr kinases includes BrK, FRK, and SRMS [205]. Several FRK family kinases preclude the Ras pathway primed by activated receptor Tyr kinases. They phosphorylate KH-domain-containing, RNA-binding, signal transduction associated protein KhdRBS1 (or Sam68) and adaptor signal-transducing adaptor protein STAP2 (or BKS). Kinase FRK phosphorylates PTen, thereby avoiding PTen anchorage to Ub ligase NEDD4-1 and degradation [206]. As a positive regulator of PTen, FRK suppresses PKB signaling.

---

[25] A.k.a. P130CAS and breast cancer anti-estrogen resistance BCAR1.

[26] A.k.a. P120-catenin or cadherin-associated Src substrate P120CAS.

[27] A.k.a. B-cell Src-homology Tyr kinase BSK, gut Tyr kinase GTK, intestinal Tyr kinase IYK, PTK5, and RAK.

[28] Ubiquitous SHB adaptor generates signaling complexes in response to tyrosine kinase activation. It indeed mediates some responses from receptors of platelet-derived growth factor, fibroblast growth factor, neural growth factor (NTRK1), and interleukin-2, as well as T-cell receptor and focal adhesion kinase.

4.9 Janus Kinases

Janus kinases[29] (JaK) are activated by cytokine receptors (Table 4.6). Intracellular Janus kinases are involved in the JaK–STAT pathway (STAT: signal transducer and activator of transcription).

The Janus kinase family includes 4 members (JaK1–JaK3 and Tyr kinase TyK2). Kinase JaK3 is strictly expressed in hematopoietic cells, whereas the other members of the JaK family (JaK1, JaK2, and TyK2) are ubiquitous.

Kinases JaK1 and JaK2 are involved in interferon-2 (interferon-γ) signaling (Vol. 2 – Chap. 3. Growth Factors). Kinases JaK1 and TyK2 associate with interferon-1. Kinase JaK2 is also an effector of prolactin receptor among others. Kinase Jak3 connects exclusively to a single cytokine receptor subunit (γ_C) and predominantly promotes lymphopoiesis, whereas the other members of the Janus kinase family interact with many different cytokine receptor subunits.[30]

4.9.1 Canonical and Non-Canonical Pathways of JaK Activation

Kinase JaK2 possesses a basal activity in the absence of dimerization and transautophosphorylation. Upon cytokine-induced receptor dimerization (*canonical pathway*), activated JaK that is pre-associated via receptor subunits dimerizes, transautophosphorylates (Tyr1007 and Tyr1008), and acquires high activity, as its enzymatic efficiency with respect to ATP increases by at least 4 orders of magnitude and its substrate recognition spectrum broadens. Besides autophosphorylation, JaK phosphorylates cytoplasmic domain of receptors that can then serve as docking sites for scaffold and signaling proteins such as signal transducers and activators of transcription.

In addition, JaKs can be activated in *non-canonical pathways*. These signaling cascades entail [208]: (1) G-protein-coupled receptors, such as AT_1 angiotensin-2 receptor that operates via PKCδ, NADPH oxidase, and possibly FAK2 and Src kinases to activate JaK2; Gq/11-protein-coupled angiotensin receptor Mas1 (a protooncogene product) that responds to angiotensin-3 and angiotensin$_{(1--7)}$ (but not angiotensin-2);[31] B_2 bradykinin receptor, CCK_2 cholecystokinin receptor; opioid receptors; platelet-activating factor receptor; and CCR2 and CXCR4 chemokine

---

[29] Janus kinases were initially named "just another kinase" (JAK).

[30] Kinases JaK are characterized by 7 JaK homology (JH) domains: JH1 kinase and JH2 pseudokinase regulatory domains as well as the JH3–JH4 and JH4–JH7 regions, the latter being involved in interactions of JaKs with receptors as well as auto-inhibition. In the inactive state, the JH2 domain may interact with the JH1 domain to impede the catalytic activity.

[31] Human angiotensinogen possesses 452 amino acids. Angiotensin activity depends mainly on the first 12 amino acids. Heptapeptide angiotensin$_{(1--7)}$ has a baroreflex effect that counteracts that of angiotensin-2.

Table 4.6. Activation of JaKs and STATs by cytokines (Source: [207]). Type-1 and -2 cytokine classification is related to 3D structure. Type-1 cytokine possesses either a short (~15 amino acids) or long (~25 amino acids) chain. Cytokines IL5 and M-CSF (or CSF1) bind as dimers, whereas other type-1 cytokines bind as monomers. Type-2 cytokines have different structures (βc, γc: common cytokine receptor β and γ chain; CNTF: ciliary neurotrophic factor; CT1: cardiotrophin-1; gCSF: granulocyte colony-stimulating factor; gmCSF: granulocyte–monocyte colony-stimulating factor; IL: interleukin; mCSF: macrophage colony-stimulating factor; OsM: oncostatin-M; LIF: leukemia-inhibitory factor; SCF: stem cell factor; TSLP: thymic stromal lymphopoietin; TyK2: tyrosine kinase-2).

Cytokines	JaKs	STATs
Type-1 cytokines		
Short-chain cytokines that share γc		
IL2/7/9/15/21	JaK1/3	STAT3/5a/5b
IL4	JaK1/3	STAT6
Short-chain cytokines that share βc		
IL3/5, gmCSF (CSF2)	JaK2	STAT5a/5b
Short-chain cytokines; receptors with Tyr kinase domain		
mCSF (CSF1)	JaK1, TyK2	STAT1/3/5a/5b
SCF		STAT5a/5b
Other short-chain cytokines		
IL13	JaK1/2, TyK2	STAT6
TSLP		STAT5a/5b
Long-chain cytokines that share GP130		
IL6/11, OsM, CNtF, LIF, CT1	JaK1/2, TyK2	STAT3
Other long-chain cytokines		
IL12	JaK2, TyK2	STAT4
Growth hormone	JaK2	STAT3/5a/5b
Prolactin, Epo, Tpo	JaK2	STAT5a/5b
Leptin	JaK2	STAT3
gCSF (CSF3)	JaK1/2, TyK2	STAT1/3/5a/5b
Type-2 cytokines		
Ifnα/β	JaK1, TyK2	STAT1/2/4
Ifnγ	JaK1/2	STAT1
IL10	JaK1, TyK2	STAT3
IL20, IL22		STAT3

receptors; (2) oxidative stress, as hydrogen peroxide can indirectly stimulate JaK2, but oxidative stress attenuates or inhibits JaK activity downstream from cytokine stimulation; (3) hyperglycemia that involves PKC and subsequent activation of NADPH oxidase and generation of ROS that inhibit Tyr phosphatase PTPn6 (or SHP1); and (4) possibly hyperosmolarity.

4.9.2 Signal Transducers and Activators of Transcription

The family of signal transducers and activators of transcription (STAT) includes 7 members (STAT1–STAT4, STAT5a–STAT5b, and STAT6). These transcription factors are phosphorylated (activated) by Tyr kinases (RTKs and NRTKs, such as JaKs, Src, and Abl) in response to many cytokines, growth factors, and hormones. Their phosphorylation leads to STAT homo- or heterodimerization. These DNA-binding proteins particularly mediate interferon-dependent gene expression. However, phosphorylated STAT dimers as well as unphosphorylated STATs can stimulate gene expression, but using distinct mechanisms [209]. In the absence of Tyr phosphorylation, STAT1 is able to activate the expression of a gene that encodes 20S proteasome subunit-β9 (PsmB9).[32] Unphosphorylated STAT3 binds to unphosphorylated NFκB, hence competing with IκB. The unphosphorylated STAT3–NFκB complex then accumulates in the nucleus to activate target genes.

4.9.3 Other Regulation of JaK Activity

Oxidants such as nitric oxide can reversibly inhibit JaK kinases. PKCδ phosphorylates (inactivates) JaK2 kinase. Cytokine-induced JaK2 phosphorylation (Tyr570) and autophosphorylation at Tyr913 result from a negative feedback [208]. On the other hand, autophosphorylation at Tyr813 enhances JaK2 catalytic activity by creating a docking site for adaptor SH2B1 that selectively improves JaK2 activity. In addition, JaK2 nitration (Tyr1007 and Tyr1008) prevents its activity.

Therefore, phosphorylation of some target amino acid residues and nitration are post-translational inhibitory modification. On the other hand, phosphorylation of other target amino acid residues and interaction with adaptor proteins represent stimulatory controls. In leptin and insulin signaling, insulin receptor substrates IRS1 and IRS2 are recruited to the SH2B1–JaK2 complex. Their subsequent phosphorylation by JaK2 promotes their association with PI3K for PKB activation [208].

4.9.4 Janus Kinases in Heart

In cardiomyocytes, JaK1 and JaK2 are predominant members of the JaK family. Numerous members of the type-1 cytokine receptors[33] that activate JaK1 and/or JaK2, hence STAT3, protect heart from acute and chronic oxidative stress owing

---

[32] A.k.a. large multifunctional peptidase LMP2.
[33] Ligands of these receptors encompass interleukins of the IL6 family (IL6, IL11, LIF, cardiotrophin-1, and oncostatin-M), growth hormone, erythropoietin, and granulocyte colony-stimulating factor (CSF3), in addition to insulin.

to 2 intracellular signaling cascades. These pathways that are activated by most of these cytokines and rely on JaK1 and JaK2 kinases include: (1) reperfusion injury salvage kinase (RISK) pathway and (2) JaK–STAT signaling [208]. The RISK pathway involves activation of phosphatidylinositol 3-kinase and extracellular signal-regulated protein kinases ERK1 and ERK2, hence restraining mitochondrial permeability transition pore. Signaling from the JaK–STAT axis entails activation of transcription factor STAT3, upregulation of cyclooxygenase-2 and nitric oxide synthase-2 that also inhibits mitochondrial permeability transition pore, vascular endothelial growth factor, anti-oxidants manganese superoxide dismutase and metallothioneins MT1[34] and MT2 (or MT2a), as well as matrix metalloproteases for repair and scar formation [208]. Whereas auto- and paracrine action of IL6 initiates heart preconditioning, the JaK–STAT pathway ensures delayed preconditioning, the so-called second window of protection (SWOP).

Although Janus kinases such as JaK2 may be activated by acute oxidative stress to prime an anti-oxidative stress program, they can also be inhibited by oxidative stress in cardiomyocytes, especially JaK1 kinase.

4.9.5 Inhibitors of the JaK–STAT Pathway

Cytokines bind to their cognate plasmalemmal receptor to trigger the JaK–STAT pathway. However, various molecules restrain the signaling cascade [210]. Inhibitors of JaK–STAT signaling include: (1) cytokine-inducible SH2-containing proteins (CIS), (2) suppressors of cytokine signaling (SOCS), and (3) STAT-induced STAT inhibitors (SSI). Many JaK–STAT inhibitors (CIS, SOCS1–SOCS7) are expressed in mammalian cells stimulated by cytokines, such as ciliary neurotrophic factor, erythropoietin, growth hormone, granulocyte (CSF3)and granulocyte–macrophage (CSF2) colony-stimulating factors, interferon-α and -γ, interleukins, leptin leukemia inhibitory factor (LIF or cholinergic differentiation factor [CDF]), prolactin, and tumor-necrosis factor-α.

Suppressors of cytokine signaling can bind to receptors, especially insulin and IGF1 receptors. Cytokine-inducible SH2-containing protein impedes signaling from IL2, IL3, and erythropoietin. Agents SOCS1, SSI1, and JaK-binding protein (JAB) bind to and hamper JaK kinases. Protein SOCS1 inhibits IL1, IL4, IL6, and Ifnγ signaling [211]. It also suppresses Toll-like receptor TLR4 signaling [212]. Interactions between SOCS1 and TLR are mediated via Ifnα and -β [213]. Protein SOCS3 is expressed in response to IL2 in T cells and blood lymphocytes [214]. It interacts with IL2 receptor complex and phosphorylates JaK1 and IL2Rβ. It also inhibits growth hormone signaling by binding to GH receptor, then interacting with and hindering receptor-bound JaK2. Both SOCS1 and SOCS3 impede effects of the

---

[34]Many metallothionein MT1 isoforms exist (MT1a–MT1b, MT1e–MT1h, MT1l, MT1m, and MT1x), besides pseudogene products (MT1iP–MT1jP) [251].

IL6 family of cytokines, as they suppress STAT3 phosphorylation, whereas CIS or SOCS2 do not hamper this signaling pathway. Factors SOCS1 and SOCS3 regulate IL6 and Ifnγ signaling. Protein SOCS3 prevents Ifnγ response in cells stimulated by IL6 [215].

Obesity is associated with an increase in SOCS1 and SOCS3 in liver, muscle, and, to a lesser extent, adipose tissue [216]. Phosphorylation of insulin receptor is partly impaired and that of insulin receptor substrates IRS1 and IRS2 is almost completely suppressed.[35] Although both SOCS1 and SOCS3 bind to insulin receptor in an insulin-dependent manner, they do not affect insulin-dependent insulin receptor autophosphorylation. Therefore, SOCS1 and SOCS3 link insulin resistance to cytokine signaling, as they inhibit insulin signaling. Moreover, SOCS1 and SOCS6 inhibit insulin-dependent activation of ERK1, ERK2, and PKB in vivo [217]. The SOCS proteins contribute to the pathogenesis of type-2 diabetes.

4.10 MATK Kinase or CSK Homologous Kinase (CSK Subfamily)

Megakaryocyte-associated tyrosine kinase (MATK) of the CSK subfamily is also designated as CSK homologous kinase (CHK).[36] Kinases CSK and MATK not only share significant sequence identity, but also can inactivate SRC family kinases by phosphorylating their regulatory Tyr in the C-terminus [218]. In addition, MATK can also inhibit SRC family kinases via a non-catalytic mechanism by binding to active SRC family kinases to form stable MATK–SFK complexes.

Kinase MATK contains Src homology binding domains (SH2 and SH3), a kinase sequence, and a unique N-terminus, but lacks myristylation signals, negative regulatory phosphorylation site, and autophosphorylation motif. Therefore, it is not regulated by Tyr phosphorylation. It interacts with many partners via its SH2 domain, whereas the SH3 domain governs its subcellular localization. Its recruitment to the plasma membrane is associated with inhibition of Lyn kinase.

Enzyme MATK can be detected in the cytosol, at the plasma membrane, as well as in the nucleus. However, it lacks the N-terminal fatty acid acylation domain for membrane anchoring. However, it needs to be positioned in proximity to SRC family kinases to phosphorylate these kinases that reside at the plasma membrane, endosomes, and perinuclear regions, using its SH2 domain to bind to specific phosphorylated Tyr residues as well as SH3 motif.

---

[35]Overexpression of SOCS3 in liver markedly decreases phosphorylation of both IRS1 and IRS2, whereas SOCS1 overexpression preferentially inhibits IRS2 phosphorylation.

[36]A.k.a. CSK-type kinase (CTK), hematopoietic consensus Tyr-lacking kinase (HYL), leukocyte C-terminal Src kinase-related protein (LSK), and brain-associated Tyr kinase (BATK).

Kinase MATK abounds in megakaryocytes and brain cells. In human resting monocytes, its expression is induced upon stimulation by interleukins-3 and -4. In megakaryoblasts, its production rises upon stem cell factor excitation.

Three alternatively spliced transcript variants generate different isoforms. The conserved lysine that binds ATP corresponds to Lys221 in ^{P52}MATK and Lys262 in ^{P56}MATK [218]. Isoform ^{P52}MATK abounds in neurons, astrocytes, and oligodendrocytes, especially in the hippocampus, substantia nigra, and cortex. Expression of ^{P56}MATK is specific to hematopoietic cells.

Kinase MATK contributes to the regulation of megakaryocytopoiesis. In hematopoietic cells, it phosphorylates (inactivates) SRC family kinases. It hence inhibits T-cell proliferation. In addition to suppression of cell growth and proliferation, MATK is also implicated in the regulation of chromosome movement during mitosis [218].

Protein MATK interacts with stem cell factor receptor (SCFR) [219] as well as, in the nervous system, neurotrophic Tyr kinase receptor NTRK1 [220]. Kinase MATK also connects to receptor Tyr kinases HER2, focal adhesion kinase-related kinase FAK2, and scaffold paxillin [218].

4.11 SRMS Kinase (FRK Subfamily)

Non-receptor Tyr kinase Src-related kinase lacking regulatory and myristylation sites SRMS (or SRM) has SH2, SH2′, and SH3 domains and a Tyr residue for autophosphorylation in the kinase domain, but lacks N-terminal glycine for myristylation and a C-terminal Tyr that, once phosphorylated, suppresses kinase activity [221]. The expression of 2 transcripts is ubiquitous, but depends on cell type and developmental stage.

4.12 Kinases of the SRC Family

The SRC family (Src stands for sarcoma) of intracellular protein Tyr kinases are signaling proteins associated with cell adhesion, growth, proliferation, differentiation, survival, and migration, especially during angiogenesis (Vol. 5 – Chap. 10. Vasculature Growth) as well as regulation of ion channels in neuronal cells. Constitutive activation and overexpression of SRC family kinases (SFK) cause many types of cancers, such as breast and colon carcinomas.

The activity of SRC family kinases is controlled mainly by phosphorylation of 2 conserved tyrosines: (1) autophosphorylation site (Y_A) in the activation loop and (2) regulatory tyrosine located near the C-terminal tail (Y_T). The latter site is targeted by CSK and MATK kinases. Autophosphorylation at (Y_A) causes full activation, whereas phosphorylation at Y_T inactivates tha enzyme. The SRC family members have regions with similar sequences, the so-called Src homology (SH) domains: the SH1 domain is a catalytic sequence, whereas SH2 and SH3 domains are

4.12 Kinases of the SRC Family

protein-binding motifs. The SH2 domain usually binds phosphotyrosine-containing proteins and SH3 interacts with cytoskeletal proteins. Upon Y_T phosphorylation, SFKs adopt an inactive conformation stabilized by 2 major intramolecular interactions: (1) binding of phosphorylated Y_T to the SH2 domain and (2) binding of a segment linking SH2 and kinase domains (SH2–kinase linker) to the SH3 domain.

The SRC family of Tyr kinases includes include many members, which are reversibly coupled to the inner leaflet of the plasma membrane: Src, Yes, FGR, Fyn, LCK, Lyn, BLK, and HCK enzymes. Cytoplasmic kinases of the SRC family are either ubiquitous (e.g., Fyn, Src, and Yes) or have a restricted expression pattern (BLK, FGR, HCK, LCK, and Lyn).

The SRC family can be decomposed into subfamilies: (1) the SRCA subfamily that comprises FGR, Fyn, Src, and Yes; (2) the SRCB subfamily that consists of BLK, HCK, LCK, and Lyn; and (3) the FRK subfamily. The TEC family constitutes a subset of Src kinases that is composed of BMX, BTK, ITK, TEC, and TXK kinases. Kinases of the TEC family possess a pleckstrin- and TEC-homology domain.[37]

Cytosolic Src kinase is often self-restrained to an inactive conformation. The balance between its constitutively active Tyr kinases (CSK and MATK) and phosphatases (e.g., PTPRa) in both basal and stimulated states is partially responsible for the state of Src activation.

Src kinase becomes activated during the transit from the perinuclear region to the plasmalemma, which requires the actin cytoskeleton and RhoB-associated endosomes.[38] Disruptions in actin filaments inhibit both the membrane translocation and activity of Src [223]. The active Src colocalizes with RhoB in the perinuclear region. Fibronectin is a main Src-activating extracellular stimulus. Src- and RhoB-containing structures are associated with Src-promoted, polymerized actin.

4.12.1 SRC Family Kinases during Hypoxia

Kinases of the SRC family are highly expressed in pulmonary arteries, where they are implicated in vascular smooth muscle cell contraction. Hypoxic pulmonary vasoconstriction corresponds to an acute, regional, adaptive process that directs blood flow away from poorly ventilated regions of the lung to support the ventilation–perfusion matching. In isolated pulmonary arteries, hypoxic pulmonary vasoconstriction is a biphasic process characterized by a first stage with a transient

---

[37]Protein Tyr kinase Lyn is a member of SRC-related family found in hematopoietic cells (macrophages, platelets, and B lymphocytes). The TEC kinases are also expressed in many hematopoietic cells. In fibroblasts in particular, using the TEC-homology domain, TEC binds to Lyn kinase. The latter then phosphorylates TEC protein, which thus acts downstream from Lyn in intracellular signaling pathways [222]. Kinases TEC and BTK can be activated in response to cytokines, such as IL3 and IL6, probably via Janus kinases.

[38]Protein RhoB is an endosome-associated Rho GTPase (Sect. 9.3.13).

elevation in intracellular calcium level and a second sustained phase with a gradually rising then stabilized intracellular calcium level. Hypoxia hastens Src kinases that triggers RoCK kinase activation (Sect. 5.2.14) that mediates Ca^{++} sensitization during the second phase of vasoconstriction [224]. Subsequent inhibition of myosin light-chain phosphatase via phosphorylation of $PP1_{r12a}$ by RoCK kinase leads to increased phosphorylation of myosin light chain MLC20.[39] In cultured pulmonary artery smooth muscle cells, hypoxia not only enhances phosphorylation of both $PP1_{r12a}$ and MLC20, but also causes translocation of RoCK kinase from the nucleus to the cytoplasm.

Both Src and Fyn kinases are activated during hypoxia in cardiomyocytes [225, 226]. Hypoxia also increases production of reactive oxygen species and subsequent Src phosphorylation to promote the synthesis of hypoxia-inducible factor-1α (Sect. 10.9.2) and plasminogen activator inhibitor-1 in rodent aortic smooth muscle cells [227].

Engagement of B-cell receptor immediately activates receptor-associated kinases of the SRC family (BLK, Fyn, Lyn, and LCK). This response primes phosphorylation of cellular substrates. Triggered signaling cascades are composed of effectors Ras GTPase-activating protein, phosphatidylinositol 3-kinase, phospholipases PLCγ1 and PLCγ2, and extracellular signal-regulated protein kinase. Three distinct sites allow the interaction of these kinases with effectors. The unique N-terminal domain of Lyn mediates association with PLCγ2, MAPK, and RasGAP; Src homology SH3 domain with PI3K; and SH2 domain with a relatively small proportion of PI3K, PLCγ2, MAPK, and RasGAP [228]. Kinases BLK, Lyn, and Fyn differ in their ability to bind MAPK and PI3K kinases. Therefore, they preferentially bind and subsequently phosphorylate their effectors.

4.12.2 BLK Kinase (SRCB Subfamily)

B-lymphoid Tyr kinase, also designated as B-lymphocyte specific Tyr kinase,[40] is specifically expressed in the B-cell lineage [230]. In humans, 2 BLK transcripts arise from the transcription of gene Blk by 2 distinct promoters that can be regulated by different transacting factors [231]. Kinase BLK interacts with ubiquitin ligase to be degraded.

---

[39]The phosphorylation status of MLC20 (at Ser19) involves a balance between Ca^{++}–calmodulin-dependent phosphorylation by MLCK kinase and dephosphorylation by MLCP phosphatase. Concomitant phosphorylation of $PP1_{r12a}$ by RoCK kinase together with elevation of intracellular Ca^{++} concentration greatly enhance MLC20 phosphorylation.

[40]Alias Blk also designates a pro-apoptotic BH3-only protein that is also called Bik (BCL2-interacting killer) or Nbk. It is expressed in hemopoietic and venous (but not arterial) endothelial cells. In humans, Blk localizes to the mitochondrial membrane. Protein Blk contains a BH3 domain and can then interact with anti-apoptotic proteins BCL2 and BCLxL [229].

4.12.3 BMX Kinase (TEC Family)

Bone marrow Tyr kinase gene in chromosome-X product (BMX)[41] is characterized by an N-terminal pleckstrin homology domain, Src homology SH3 and SH2 segments, and a catalytic kinase motif. It belongs to the TEC family that also includes TEC, TXK, ITK, and BTK, which are marked by an N-terminal TEC homology domain located downstream from the PH domain.

Kinase BMX mediates activation of Rho GTPase by Gα12 and -13 upon stimulation by hormones and neurotransmitters. It interacts with P21-activating Ser/Thr kinase PAK1 [232]. The latter maps small GTPases Rho, CDC42, and Rac1 onto cytoskeleton organization and nuclear signaling on the one hand, and associates with FAK and non-receptor protein Tyr phosphatase PTPN21 to increase STAT3 activation, and RUN and FYVE domain-containing protein RUFY1 that is involved in vesicular transport (early endosomes) on the other.

Kinase BMX operates in endothelial cells and lymphocytes, as well as cardiomyocytes [233]. It indeed participates in pressure overload-induced hypertrophic growth. Tumor-necrosis factor can provoke angiogenesis. This factor activates BMX specifically via TNFR2 [234]. Kinase BMX forms pre-existing complex with TNFR2 (independently from TNF). Activated BMX then mediates TNF-induced migration of endothelial cells and tube formation.

In endothelial cells, the endocannabinoid anandamide connects to cannabinoid receptor-1 (CB1R) and G-protein-coupled receptor GPR55 to initiate distinct signaling pathways, whether integrins are inactive or not, respectively [235]. Signaling primed by CB1R includes Gi-protein-mediated activation of spleen Tyr kinase to translocate NFκB and inhibit phosphoinositide 3-kinase. Yet, PI3K operates in GPR55-triggered signaling. On the other hand, integrin aggregation provokes CB1R splitting from integrins and subsequent SYK inactivation. Therefore, SYK cannot further inhibit GPR55-triggered signaling. Consequently, anandamide generates Ca^{++} influx from the endoplasmic reticulum (ER) via the GPR55–PI3K–BMX–PLC pathway and activates nuclear factor of activated T-cells.

In lymphocytes, TLR2, TLR4, and $\alpha_5\beta_1$ integrin signal via MyD88 and focal adhesion kinase [236]. Toll-like receptors recruit adaptors, such as Mal, MyD88, TRIF, and TRAM to activate 2 distinct pathways, the TIRAP–Mal–MyD88–NFκB–AP1[42] or TRIF–TRAM–IRF axes.[43] Kinase BMX is involved in crosstalk between the integrin–FAK and MyD88 pathways. In macrophages, BMX regulates

---

[41] A.k.a. epithelial and endothelial Tyr kinase ETK, NKT38, PSCTK2, and PSCTK3.

[42] Toll–IL1R domain-containing adaptor (TIRAP) together with adaptor Mal recruits MyD88 to plasmalemmal nanodomains rich in phosphatidylinositol (4,5)-bisphosphate to launch TLR2 and TLR4 signaling. Turnover of PIP$_2$ is controlled by integrins via activation of ARF6 GTPase.

[43] Toll–IL1R domain-containing adaptor inducing Ifnβ (TRIF) and TRIF-related adaptor leads to a signaling that targets the transcription factors interferon regulatory factors (IRF), which regulate transcription of interferons.

IL6 production caused by Toll-like receptors independently of P38MAPK and NFκB [237]. On the other hand, BTK intervenes in production of inflammatory cytokine tumor-necrosis factor-α, but not IL6.

4.12.4 BTK Kinases (TEC Family)

Bruton Tyr kinase (BTK),[44] a member of the TEC family of cytoplasmic protein Tyr kinases, contributes to B-cell maturation as well as mastocyte activation via IgE receptors. The Btk gene is located on the X chromosome. Kinase BTK is constitutively associated with phosphatidylinositol 4-phosphate 5-kinase that synthesizes PI(4,5)P$_2$ [238]. The BTK–PI(4)P5K complex localizes to membrane rafts, where BTK binds phosphatidylinositol (3,4,5)-trisphosphate, then phosphorylates phospholipase-Cγ2.

Kinase BTK is a dual-function regulator of apoptosis, as it promotes stress-induced apoptosis but prevents TNFSF6-activated apoptosis in B cells [239]. When B lymphocytes are exposed to reactive oxygen species, BTK lowers the antiapoptotic activity of transcription factor STAT3. On the other hand, BTK associates with TNFRSF6a death receptor and impairs its interaction with Fas (TNFRSF6a)-associated protein with death domain FADD that helps TNFRSF6a to recruit and activate caspase-8, thereby precluding the assembly of the death-inducing signaling complex.

4.12.5 FGR Kinase (SRCA Subfamily)

FGR (viral feline Gardner-Rasheed sarcoma oncogene homolog) kinase is a member of the SRCA subfamily of protein Tyr kinases. Protein FGR contains in its N-terminus sites for myristylation and palmitylation. Its catalytic domain as well as SH2 and SH3 motifs are involved in protein–protein interactions.

Leukocytes express 3 different subfamilies of Tyr kinases: (1) SRC, (2) SYK, and (3) FAK family kinases. In leukocytes of the myeloid lineage, detected SRC family kinases include FGR, HCK, and Lyn. Enzymes SYK and FAK2 are predominant members of their respective subfamilies. On the other hand, in leukocytes of the lymphoid lineage, ZAP70 and FAK are the main members. Kinases of the SRC family FGR, HCK, and Lyn, as well as SYK, are activated in stimulated

---

[44] Kinase BTK (Bruton agammaglobulinemia Tyr kinase) is also named agammaglobulinemia Tyr kinase (AT or ATK), X-linked agammaglobulinemia Tyr kinase (AgmX1 or XlA), and B-cell progenitor kinase (BPK), as well as IMD1 and PSCTK.

4.12 Kinases of the SRC Family 161

Table 4.7. Integrins produced by phagocytes and their ligands (Source: [240]; CR3: complement receptor-3; C3bi: fragment of complement component C3; GP: glycoprotein; HW: high-molecular-weight kininogen; ICAM: intercellular adhesion molecule; LFA: lymphocyte function-associated antigen; Mac1: macrophage-1 antigen; NIF: neutrophil inhibitory factor; VCAM: vascular cell adhesion molecule; VLA: very late antigen). Phagocytes are the leukocytes that ingest harmful foreign particles, from pathogens to dying cells. They are called professional or non-professional whether they phagocytize effectively or not. Professional phagocytes include neutrophils, monocytes, macrophages, and dendritic and mast cells that have surface receptors to detect foreign bodies. Non-professional phagocytes comprise lymphocytes, epithelial and endothelial cells, fibroblasts, and mesenchymal cells. Phagocytes are recruited by chemoattractants (cytokines, chemokines, clotting peptides, complement components, and invading microbe products) released by invaders or phagocytes already at work. Chemotaxis of phagocytes implies crossing of the endothelial barrier for circulating cells and displacement through a matrix of transmigrated leukocytes and tissue-resident phagocytes, hence transient cell adhesion. Contact with foreign bodies trigger a chemical attack with reactive oxygen species and possibly delayed antigen presentation to lymphocytes close to working site (macrophages) or remotely in lymph nodes (dendritic cells). Extracellular killing relies on interferon-γ released by CD4+ and CD8+ T lymphocytes, natural killer (NK) and NKT cells, B lymphocytes, monocytes, macrophages, and dendritic cells to stimulate macrophages to produce and deliver toxic level of nitric oxide.

Integrins	Other Alias	Ligands
	β_1 Integrins	
$\alpha_4\beta_1$	CD49d–CD29, VLA4	Fibronectin, VCAM1
$\alpha_5\beta_1$	CD49e–CD29, VLA5	Fibronectin
$\alpha_6\beta_1$	CD49f–CD29, VLA6	Laminin
	β_2 Integrins	
$\alpha_L\beta_2$	CD11a–CD18, LFA1	ICAM1, ICAM2, ICAM3
$\alpha_M\beta_2$	CD11b–CD18, CR3, Mac1	ICAM1, ICAM2, fibrinogen, factor X, HW, β-glucan, heparin, NIF, C3bi, elastase, oligodeoxynucleotides
$\alpha_X\beta_2$	CD11c–CD18, GP150–95	Fibrinogen, C3bi, lipopolysaccharides
$\alpha_D\beta_2$	CD11d–CD18	ICAM3
	β_3 Integrins	
$\alpha_v\beta_3$	CD51–CD61	Vitronectin, entactin

granulocytes (mainly FGR and Lyn that relocalize to the actin cytoskeleton) and macrophages.[45] Resulting signaling events increase the number and affinity of cell adhesion molecules, especially integrins (Table 4.7).

---

[45] Pro-inflammatory factors that activate granulocytes include tumor-necrosis factor-α, granulocyte–macrophage and granulocyte colony-stimulating factor, chemokines (interleukin-8, CCL3, CXCL2, and CXCL6), platelet-activating factor, and leukotriene-B4, as well as pathogen-derived factors such as lipopolysaccharide (LPS) and formylated-methionineleucine-phenylalanine (fMLP).

Kinase FGR, as well as HCK and Lyn, are dispensable for development of myeloid cells as well as most of their functions, but they are indispensable for integrin-mediated signaling [240]. Integrin-mediated adhesion is a potent costimulus for neutrophil activation. Integrin clustering consecutive to bound ligands primes signaling that culminates in actin cytoskeletal rearrangement and neutrophil migration, degranulation (i.e., release of granule constituents, such as hydrolytic enzymes, metal-binding proteins, and peroxidases), and oxidative burst (i.e., rapid release of reactive oxygen species after assembly of NADPH oxidase subunits at the plasma membrane).

4.12.6 Fyn Kinase (SRCA Subfamily)

Kinase Fyn, a member of the SRCA subfamily of protein Tyr kinases, is primarily located on the cytoplasmic leaflet of the plasma membrane. It phosphorylates various substrates to regulate their activity and/or to generate a binding site for signaling effectors. Protein Fyn participates in signaling initiated by T- and B-cell receptors, excited integrins, growth factors, activated platelet, and gated ion channels [241]. Alternatively spliced variants exist. Most tissues express FynB, whereas T lymphocytes synthesize FynT.

Activated Fyn autophosphorylates (Tyr417). Kinase Fyn binds to P85 subunit of phosphatidylinositol 3-kinase and Fyn-binding protein (FyB). Kinase CSK phosphorylates (inactivates) Fyn (Tyr528). Transmembrane protein Tyr phosphatase receptors PTPRa, PTPRc, and PTPRf as well as protein Tyr phosphatase PTPn6 dephosphorylate Fyn (Tyr528) [241]. Phosphatase PTPn5 also dephosphorylates Fyn (Tyr417). Depalmitoylated Fyn is released from membrane, hence hindering Fyn-mediated phosphorylation of membrane-bound substrates.

Protein Fyn interacts with [241]: (1) receptor protein Tyr kinases;[46] (2) immunocyte receptors;[47] (3) interleukin receptors (IL2R–IL5R and IL7R); (4) Tyr kinases (BTK, FAK, JAK2, and PTK2b); (5) Ser/Thr kinases (PKCδ and PKCη); (6) Tyr phosphatases (endoplasmic reticulum phosphatase StEP61 as well as PTPn11 and PTPRa); (7) adaptors and scaffolds;[48] (8) cell surface and/or GPI-linked proteins;[49]

---

[46]These receptors include PDGFR, EGFR, CSF1R (or M-CSFR), VEGFR1, and ephrin receptors (EPHa4, EPHa8, and EPHb3), as well as receptor Tyr kinase MuSK (muscle-specific kinase) and neurotrophic Tyr kinase receptor NTRK2 (or TrkB).

[47]These receptors comprise TCRζ subunit, B-cell receptor (Igα and -β), FcRγ, FcεR2 (CD23), Fcµ, and cytotoxic T-lymphocyte antigen CTLA4 (or CD152).

[48]Kinase Fyn is able to bind Fyn-binding protein (or SLAP130), SH2 domain protein 1A (SAP), insulin receptor substrate IRS1, docking protein DOK1 and DOK4, Src kinase-associated phosphoprotein SKAP1 [or SKAP55], receptor for activated C-kinase (RACK), target of Myb1-like TOM1L1 (or Srcasm), CAS proteins (P130CAS, CASL, and embryonal Fyn substrate [Efs or Sin]), Disc large homolog DLG4, caveolin-1, and SHC.

[49]Kinase Fyn targets contactin, CD2, CD19, CD20, CD28, CD36, CD43, CD55, and CD90.

(9) ion channels;[50] (10) cytoskeletal and adhesion proteins;[51] (11) GTPase-activating proteins (RhoGAP32, RhoGAP35 and RasA1 [a RasGAP]); and (12) other proteins (CBL and Itch ubiquitin ligases, PI3K, Rapsyn, KH domain-containing, RNA-binding, signal transduction-associated proteins KhdRBS1 and KhdRBS2, BCL3, and TNFSF6).

Among these partners, Fyn particularly targets regulators of the cytoskeletal structure in both hematopoietic and non-hematopoietic cells [242]. Fyn Kinase binds to Wiskott-Aldrich syndrome protein. Members of the WASP family associate with numerous signaling molecules to depolymerize actin directly and/or serve as adaptors or scaffolds for these mediators to regulate the cytoskeleton dynamics.

4.12.7 Hemopoietic Cell Kinase (SRCB Subfamily)

Hemopoietic cell kinase (HCK) is predominantly expressed in hemopoietic cells. It can help to couple Fc receptor to activation of the respiratory burst (or oxidative burst), i.e., the rapid release of reactive oxygen species. In addition, it can intervene in neutrophil migration and degranulation. Two alternatively spliced variants exist with different subcellular locations.

Kinase HCK interacts with Breakpoint cluster region protein (BCR) [243], actin-binding proteins WASP, WASP-interacting protein (WIP), phagocytosis promotor Engulfment and cell motility ElMo1 [244] adaptor/Ub ligase CBL, GTPase-activating proteins RasGAPs RasA1 and RasA3 [245], RapGEF1 [246], granulocyte colony-stimulating factor receptor [247], and adamlysin ADAM15 [248].[52]

4.12.8 ITK Kinase (TEC Family)

Intracellular Tyr kinase or IL2-inducible T-cell kinase (ITK) is expressed in T lymphocytes.[53] It is produced in T, NK, and mast cells. Kinase ITK pertains

---

[50]Targeted ion carriers include ionotropic glutamate receptor NMDAR, ionotropic nicotinic acetylcholine receptors nAChR, Cl^- channel, Ca^{++} channels RyR and IP_3R, K^+ channel $K_V1.5$, and anion exchanger.

[51]This set of molecules consists of microtubule-associated protein MAP2, dynein light chain, WASP, β-catenin, Tau, α-tubulin, $α_6β_4$-integrin, β-adducin, annexin-6, neural cell adhesion molecule NCAM120 and NCAM140, cadherin, axonin-1, myelin-associated glycoprotein (also called sialic acid-binding Ig-like lectin SIgLec4A or MAG).

[52]Kinases LCK, Fyn, Abl, and Src are also able to associate with ADAM15, or metargidin, in vitro. Kinase LCK phosphorylates ADAM15, but at a lower extent than HCK. Adamalysin ADAM15 can interact specifically with HCK and adaptor GRB2 in hematopoietic cells.

[53]A.k.a. PSCTK2, Tyr kinase expressed mainly in T lymphocytes (EMT), T-cell-specific kinase (TCSK, TLK, and TSK), and LYK.

to the TEC family, the member of which are involved in signaling emitted by cytokine receptors, immunoreceptors, and other lymphoid cell surface receptors. Adhesion molecule CD2 on the surface of T lymphocytes and natural killer cells and costimulator CD28, a constitutively expressed B7 receptor on naive T lymphocytes provokes phosphorylation (activation) of ITK by LCK [249].

Kinase ITK interacts with CD2, kinases Fyn and Src, phospholipase-Cγ1, suppressor of cytokine signaling protein SOCS1, as well as adaptors GRB2, CBL, and lymphocyte cytosolic protein LCP2,[54] and linker of activated T cells (LAT),[55] in addition to DNA-binding protein KhdRBS1 (or Sam68) and heterogeneous nuclear ribonucleoprotein hnRNPk, Wiskott-Aldrich syndrome protein that activates the ARP2–ARP3 complex for actin nucleation, karyopherin-α2 (KpnA2) involved in nuclear transport of proteins, and peptidylprolyl isomerase-A (PPIa; a.k.a. cyclophilin-A) that accelerates protein folding [250, 251].

4.12.9 Leukocyte-Specific Cytosolic Kinase (SRCB Subfamily)

Leukocyte-specific cytoplasmic protein Tyr kinase (LCK), a member of the SRCB subfamily, is observed in lymphocytes, where it phosphorylates signaling mediators. It associates with the cytoplasmic tails of CD4 and CD8 coreceptors on T helper and cytotoxic T cells, respectively, to assist signaling from T-cell receptors.

Kinase LCK phosphorylates (activates) kinase ZAP70 that, in turn, phosphorylates transmembrane adaptor linker of activated T cells (LAT) that is a docking site for various proteins, such as SHC, GRB2, SOS, PI3K, and PLC. The corresponding cascade provokes Ca^{++} influx and activation of MAPK module to activate NFAT, NFκB, and AP1 transcription factors.

Protein LCK phosphorylates numerous substrates, such as SCFR, CD3, and IL2R receptors, CD44 that participates in lymphocyte activation, CD48, cell-surface antigens, such as B-cell membrane-spanning 4-domain MS4A1 and thymocyte protein THY1, TNFRSF6a receptor and its TNFSF6 ligand, cytotoxic T-lymphocyte-associated antigen-4, TCR-interacting molecule TRAT1, Notch1, membrane enzymes, such as Axl and PTPRC, kinases ZAP70, ITK, FAK2, PI3K, PI4K-α, and PKCθ, protein Tyr phosphatase PTPn6, PTPn11 (SHP2)-interacting transmembrane adaptor SIT1, cRaf kinase, RasGAP RasA1, RhoGAP17, Vav1

---

[54] Adaptor LCP2 is also termed SH2 domain-containing leukocyte protein of 76kDa (SLP76). It binds to GRB2 and transduces TCR signals to activate T lymphocytes. Once phosphorylated by SYK, LCP2 connects to PI3K as well as guanine nucleotide-exchange factors Vav or NCK adaptor.

[55] The CD4 coreceptor on the surface of T helper and regulatory T cells, monocytes, macrophages, and dendritic cells as well as CD8 coreceptor on cytotoxic T cells are aimed at assisting the T-cell receptor complex. When T-cell receptors are engaged by specific antigens presented by MHC, LCK phosphorylates CD3 and ζ-chain of the TCR complex, hence allowing binding of ZAP70 to TCR. Kinase LCK then phosphorylates (activates) ZAP70 that, in turn, phosphorylates LAT. This adaptor docks the SHC–GRB2–SOS complex, PI3K, and PLC.

GEF, adpribosylation factor-related protein ArfRP1, CSK-binding protein PAG1, Src-associated phosphoprotein ScAP1, phospholipase-C, KhdRBS1, CRK- associated substrate-related protein NEDD9 (or CASL), adamlysins ADAM10 and ADAM15, paxillin, PECAM1, catenin-δ1, and Lnk, LCP2, and SHC adaptors, CBL adaptor and Ub ligase, UbE3a ubiquitin ligase, and sequestosome Sqstm1 [251].

4.12.10 Lyn Kinase (SRCB Subfamily)

Kinase Lyn (viral yes-1 Yamaguchi sarcoma-related oncogene homolog) belongs to the SRCB subfamily of protein Tyr kinases. It is mainly expressed in hematopoietic and nervous cells.

Kinase Lyn can be recruited with G-protein subunit Gi to plasmalemmal nanoclusters supported by CD59 (a.k.a. protectin and membrane inhibitor of reactive lysis [MIRL]), a complement regulatory protein. Once in this nanocluster, Lyn can transiently activate phospholipase-Cγ that produces inositol trisphosphate from phosphatidylinositol bisphosphate.

Upon B-cell receptor activation, Lyn is rapidly phosphorylated. Activated Lyn triggers a cascade of signaling events that starts with phosphorylation of immunoreceptor Tyr-based activation motifs (ITAM) of receptors to recruit and stimulate effectors, such as SYK, PLCγ2, and PI3K. Other interacting kinases include TyK2, TEC, BTK, FAK1 and FAK2, JaK2, ERK1, CK2, S6K, and CDK1, -2, and -4 [251].

Kinase Lyn also transmits inhibitory signals via phosphorylation of immunoreceptor Tyr-based inhibitory motifs (ITIM) in regulators, such as inhibitor for B-cell receptor CD22, pair of Ig-like receptors PIR-B, FcαR, and FCγR2b1 to recruit and stimulate phosphatases, such as SHIP1 and PTPn6 (SHP1; immunotolerance) [252].

Lyn Kinase also interacts with erythropoietin receptor, cytokine receptors SCFR (Kit or CD117), CSF1R and CSF3R, glycoprotein receptor for collagen GP6 on platelet, cell adhesion molecule CD24 that resides on most B lymphocytes, integral membrane protein CD36 that binds many ligands, mucin-1, PECAM1, membrane neutral metallo-endopeptidase neprilysin, PDE4a, protein Tyr phosphatase receptor PTPRc, Tyr phosphatase PTP6, inhibitory $PP1_{r8}$ and $PP1_{r15a}$ subunits of protein phosphatase-1, inositol polyphosphate-5-phosphatase Inpp5d (SHIP1), Src kinase-associated phosphoproteins ScAP1 and ScAP2, adaptor lymphocyte cytosolic protein LCP2, cytotoxic T-lymphocyte antigen CTLA4, PML-RARA regulated adaptor molecule PRAM1, CAS-Br-M ecotropic retroviral transforming sequence (CBLC), docking protein DOK1, adaptors Gab2 and Gab3, SHC, breast cancer anti-estrogen resistance BCAR1 (or p130CAS), neural precursor cell expressed, developmentally downregulated molecule NEDD9, B-cell scaffold protein with ankyrin repeats BAnk1, and ion channel TRPV4 [251].

4.12.11 Src Kinase

Src Kinase (or cSrc: cellular Src) is encoded by the SRC proto-oncogene. Mutations of the SRC gene yield cancers. Two splice variants exist. The protein consists of 3 domains: N-terminal SH3 and central SH2 protein-binding sites and kinase domain. The SH2 and SH3 domains cooperate in Src auto-inhibition. Various mechanisms activate Src: (1) C-terminus dephosphorylation by a protein Tyr phosphatase; (2) binding of the SH2 domain by a phosphoprotein such as focal adhesion kinase; and (3) binding of the SH3 domain.

Src corresponds to the prototype of the cytoplasmic, membrane-associated, Tyr kinases that acts as a cotransducer of mitogenic signals emanating from numerous Tyr kinase growth factor receptors, such as those for platelet-derived, epidermal, and fibroblast FGF2 growth factors, and colony-stimulating factor CSF1. Interactions between Src and these receptors are bidirectional, i.e., Src phosphorylates (activates) these receptors and vice versa.

Preferred substrates of Src include almost exclusively molecules that associate with the actin cytoskeleton or focal adhesions, such as cortactin, RhoGAP35, and CAS (or BCAR1) [253]. On the other hand, preferential substrates of growth factor receptors such as EGFR comprise the receptor itself, phospholipase-Cγ, and SHC and DOK1 adaptors. Major mitogenic signaling proceeds directly from the receptor and uses the SHC–GRB2–SOS–Ras–Raf–MAPK–ELK1 axis. Cell proliferation needs actin cytoskeleton, hence Src kinase and its substrates. Kinase Src then serves as transducers of growth signals and/or monitors of anti-apoptotic conditions.

Signaling from PDGF involves 43 potential Src kinase substrates [254]. In particular, PDGF provokes the phosphorylation by Src of calcium-dependent, non-lysosomal cysteine protease calpain-2, epidermal growth factor receptor pathway substrate EPS15, ubiquitin-associated protein UbAP2l, RNA-binding motif protein RBM10, Far upstream element (FUSE)-binding protein FUBP1, TRK-fused gene product TFG, and transcriptional cocontroller Tripartite motif-containing protein TriM28.

Kinase Src contributes to angiotensin-2-induced signal transduction and cardiac hypertrophy [255]. Moreover, Src can be activated during myocardial ischemia. Although, CSK that phosphorylates (inactivates; Tyr527) Src[56] is upregulated in failing left ventricles, Src is also markedly upregulated with respect to normal subjects. Increased CSK expression in response to hypertrophic stimuli seems inconsistent with its antihypertrophic effect. However, it can imply that these kinases are required for a fine-tuning of the response.

---

[56]Autophosphorylation (Tyr416) increases and phosphorylation (Tyr527) suppresses Src activity, respectively.

4.12 Kinases of the SRC Family

Table 4.8. Extracellular stimuli that regulate cell fate exert their action by interacting with plasmalemmal receptors to generate intracellular signaling. Various extracellular signals activate TEC family kinases. There are 3 consecutive activation steps: (1) recruitment to the membrane; (2) phosphorylation by SRC family kinases; and (3) autophosphorylation. Negative feedbacks exist. According to cell and signal types, several pathways can be triggered to regulate cell proliferation, differentiation, apoptosis, migration, and adhesion. One set of receptors, the receptor protein Tyr kinases (RTK), includes receptors for growth factors (e.g., PDGFR, EGFR, and CSF1R). A second set is constituted by cytokine receptors (e.g., IL2R–IL7R, IL9R, IL11R–IL13R, IL15R, IL21R, IL23R, IL27R, CSF2R, CSF3R, EpoR) that correspond to aggregation of receptor subunits to form functional receptors, without enzymatic activity, but by recruiting cytoplasmic Tyr kinases. In both receptor sets, signaling primed by extracellular interaction between ligands and receptors triggers protein phosphorylation.

Receptor type	Mediators
Cytokine receptor	JaK, STAT
Fc receptor	WASP, SLP76, SYK, PLCγ1/2, PKC, SRC family kinase
Toll-like receptor	IRAK1, MAL, MYD88, TRIF, TRAM
Death receptor	FADD
GPCR	G protein
RTK	PI3K, SRC family kinase
Integrin	FAK, cytoskeletal proteins
	NRPTK, CBL, PI3K, PtdIns, PLCγ,
	DAG, PKC, IP$_3$, Ca^{++}, PLD, Arf
	Vav, Ras, ERK1/2, JNK
	CDC42, Rac, Rho, JNK, cytoskeletal proteins

4.12.12 TEC Kinase

Non-receptor protein Tyr kinase expressed in hepatocellular carcinoma (TEC), or PSCTK4, is encoded by the Tec gene. It is involved in intracellular signaling from cytokine receptors, lymphocyte immunoreceptors, G-protein-coupled receptors, and integrins (Table 4.8). Kinase TEC is indeed able to interact with Gα12 subunit of heterotrimeric G protein, SCFR cytokine receptor, docking protein DOK1, Janus kinase JaK2, suppressor of cytokine signaling SOCS1, and RhoGEF12. Kinase TEC thus contributes to the regulation of the immune response.

Kinase TEC is a potent activator of the transcription of cytokine genes, such as IL2 and IL4. It favors the activity of nuclear factor of activated T cells by, at least, enhancing NFAT nuclear import [256]. It is characterized by a unique subcellular localization in small vesicles at the plasma membrane upon signaling from T-cell receptor (but not other family members) [257]. Kinase TEC colalizes with kinase LCK, PLC1, and early endosomal antigen-1 marker (EEA1).

In addition to TEC kinase (PSCTK4), the *TEC family* includes protein Tyr kinases BMX (PSCTK2/3), BTK (PSCTK1), ITK (PSCTK2), and TXK (PSCTK5) that are encoded by the Bmx, Btk, Itk, and TXK genes, respectively. Kinases of the TEC family possess several between-protein interaction and localization domains, such as a SH2 domain for interactions with phosphotyrosine moieties, a SH3 domain

Table 4.9. TEC family kinase binding partners (Source: [259]).

Kinase	Partners
BMX	STAT3
BTK	Factin, TNFRSF6a
	PKCβ1/2, Fyn, HCK, Lyn
	Gαq/12, Gβγ, Vav
	BAP135, EWSR1, CBL, SH3BP5, KhdRBS1, WASP, SLP65
ITK	CD28, Gαq, Gβγ, PLCγ
	GRB2, hnRNPk
TEC	SCFR, CD28, PI3K(p55γ)
	BRDG1/STAP1, SLP76, Vav

Table 4.10. Signaling axes that involve TEC family kinases (Source: [259]).

Receptor type	Signaling pathways
GPCR	BTK–JNK–Jun
	TEC–CDC42/Rac/Rho–cytoskeleton
RTK	GRB2–SOS–Ras–Raf–ERK–Fos
	PI3K–BTK–CDC42/Rac/Rho–cytoskeleton
Cytokine R	BMX
	JaK–STAT
BCR	Lyn–BTK
	SYK–BLNK–PLCγ–DAG/IP$_3$–PKC–Ca^{++}–NFAT
	SYK–Vav–CDC42/Rac/Rho
TCR	ZAP–LAT–SLP76–PLCγ–DAG/IP$_3$–PKC–Ca^{++}–NFAT
	LCK–ITK–PLCγ–DAG/IP$_3$–PKC–Ca^{++}–NFAT

for linkage with proline-rich sequences, and a PH domain (cysteine-rich string in TXK) for membrane recruitment, as it binds to PI3K products PI(3,4)P$_2$ and PI(3,4,5)P$_3$.

The main goal of TEC family kinases is the activation of phospholipase-Cγ1 to produce inositol trisphosphate and diacylglycerol, hence causing Ca^{++} influx and activating calcium-dependent enzymes such as PP3, in addition to protein kinase-C and Ras GTPase. Kinases of the TEC family participate in the control of cell survival, proliferation, differentiation, migration, and apoptosis (Tables 4.9, 4.10, and 4.11).

Kinases of the TEC family are regulators of lymphocyte activation and effector function. In T lymphocytes, they comprise TEC, ITK, and TXK, whereas in B lymphocytes the main member is BTK. Unlike in B lymphocytes, in which BTK is indispensable for development and activation, TEC kinases modulate function of activated T lymphocytes. Kinases of the TEC family are indeed less effective than

Table 4.11. Expression sites and associated receptors of members of the TEC family (Source: [260]).

Kinase	Expression	Stimulus
BMX	Bone marrow, heart, lung, testis, colon macrophage, neutrophil	IL-6R G12/13
BTK	Bone marrow, spleen, lymph node B, myeloid, erythroid, and mast cells, megakaryocyte	BCR, FcεR1, GP130, IL5R, IL6R, IL10R, CD19, CD28, CD38, CD40
ITK	Thymus, spleen, lymph node T, NK, and mast cells	TCR, FcεR1, CD2, CD28
TXK	Thymus, spleen, lymph node, tonsil, testis T and myeloid cells	TCR
TEC	Bone marrow, spleen, thymus T, B, and myeloid cells	BCR, TCR, GP130, CD28, CD38, IL3R, IL6R, EpoR, TpoR, SCFR, G-CSFR

SRC or SYK family kinases. Whereas B lymphocytes are able to synthesize BTK and TEC and T lymphocytes ITK, TXK, and TEC, mastocytes produce BTK, ITK, TEC, and TXK.

4.12.13 TXK Kinase (TEC Family)

Non-receptor protein Tyr kinase mutated in X-linked agammaglobulinemia (TXK)[57] is a member of the TEC subset of the SRC family. It is synthesized in T lymphocytes and some types of the myeloid cell lineage. Like other members of the TEC family, TXK is associated with membrane rafts and can be phosphorylated (activated) by kinases of the SRC family.

Naive T lymphocytes differentiate into T helper T_{H1} or T_{H2} cells. Both cytokine receptor and T-cell receptor-mediated signaling direct this differentiation. Interleukin-12 and interferon-γ determines T_{H1} differentiation, whereas IL4 leads

---

[57] A.k.a. Tyr kinase-like kinase (TKL), Bruton Tyr kinase-like kinase (BTKL), resting lymphocyte kinase (RLK), PSCTK5, and protein Tyr kinase PTK4. Defects in the TXK gene cause agammaglobulinemia, an X-linked immunodeficiency characterized by failure to produce mature B lymphocytes, hence associated with a downfall of rearrangement of Ig heavy chains.

to T$_{H2}$ development. In addition, distinct signaling effectors mediate cytokine expression in T$_{H1}$ and T$_{H2}$ cells. Protein TXK importantly contributes to T$_{H1}$ cytokine production [262].

4.12.14 Yes Kinase (SRCA Subfamily)

Kinase Yes1 (viral yes-1 Yamaguchi sarcoma oncogene homolog) belongs to the SRCA subfamily. It is detected in cerebellar Purkinje cells. It colocalizes with occludin in tight junctions in epithelial and endothelial cells [263].

Kinase Yes1 interacts with many identified partners [251]: (1) plasmalemmal receptors, such as colony-stimulating factor-1 receptor, CD36 of the class-B scavenger receptor family,[58] and CD46, a complement regulatory protein; (2) kinases, either plasmalemmal such as EPHb2 or cytoplasmic, such as Fyn, JaK2, and FAK1, as well as P13K; (3) phosphatases such as PP2; (4) adhesion molecules, such as PECAM1, a label to destroy aged neutrophils by macrophages, catenin-δ1 (encoded by the CTNND1 gene) that operates in cell adhesion and signal transduction, cadherin-1, and occludin at tight junctions; (5) partners of small GTPases such as RasGAP RasA1; (6) adaptor, docking, and scaffold proteins such as DOK1; and (7) other proteins that have diverse functions, such ribosomal protein RPL10, a constituent of large ribosomal 60S subunit that binds to Jun transcription factor, ion channel TRPV4, sodium–hydrogen antiporter-3 regulator SLC9a3R1 (or NHERF), and nephrin, a transmembrane protein involved in the renal filtration barrier.

4.13 From Src to SYK

Intracellular signals delivered by kinases of the SRC family, SYK, and ZAP70, are coordinated by adaptors or linkers that mediate protein–protein and, in some cases, protein–lipid interactions. Adaptors allow immunoreceptors and assigned kinases to come into contact with their targets.[59]

Two families of cytosolic Tyr kinases, the Src and SYK families, are required for T-cell receptor signaling. T-cell receptor engaged with processed antigen fragments presented by antigen-presenting cells primes the sequential activation of Src (LCK and Fyn) and SYK (SYK and ZAP70) kinases. Upon TCR activation, LCK is activated and phosphorylates the intracellular ITAM segment of CD3 of the TCR–CD3

---

[58] Receptor CD36 binds many ligands, such as collagen, thrombospondin, lipoproteins, long-chain fatty acids, and oxidized phospholipids and low-density lipoproteins.

[59] During T-cell activation, adaptors lymphocyte cytosolic protein LCP2 (or adaptor SLP76) and linker of activated T lymphocytes (LAT) are required for proper tyrosine phosphorylation and activation of PLCγ, calcium influx, and Ras stimulation. In activated B lymphocytes, specific adaptor B-cell linker protein (BLnk) is needed for tyrosine phosphorylation of PLCγ and activation of c-Jun N-terminal kinase.

complex. Phosphorylation of ITAM allows binding of ζ-chain-associated protein kinase ZAP70 to CD3. This kinase then phosphorylates adaptor LCP2 (also known as SLP76) that interacts with both adaptor GRB2 and phospholipase-Cγ1.

4.14 Spleen Tyrosine Kinase

The SYK family of non-receptor protein Tyr kinases is composed of 2 members: spleen Tyr kinase (SYK) and ζ-associated protein-70 (ZAP70). Kinase SYK is produced in hematopoietic cells (B lymphocytes, mastocytes, neutrophils, macrophages, platelets, etc.) as well as other cell types, whereas expression of ZAP70 is restricted to T and natural killer cells. Alternatively spliced variant SYKb is less active than SYK protein.

Kinase SYK intervenes in both innate and adaptive immunity. It also participates in diverse processes, such as cell adhesion, osteoclast maturation,[60] platelet activation,[61] and vascular development. In particular, SYK is required for the separation of newly formed lymphatic vessels from blood vessels [264].

It is activated by immunoreceptors of the adaptive immune response (B- [BCR] and T-cell [TCR] receptors as well as various Ig receptors [FcR]) as well as C-type lectins[62] and integrins.[63]

Enzyme SYK is recruited to aggregated immune recognition receptors with dually phosphorylated immunoreceptor Tyr-based activation motif (ITAM), such as B-cell antigen receptor, IgE receptor (FcεR1), multiple IgG receptors (FcγR1–R3), T-cell antigen receptor, DAP12 coreceptor associated with non-inhibitory Ly49D and H receptors on NK cells, and FcR γ-chain component of platelet collagen receptor [265].[64]

The resulting conformational change stimulates its autophosphorylation. Activation of SYK is enhanced via phosphorylation by members of the SRC family. Phosphorylation of SYK (Tyr of linker region between SH2 and catalytic domain) allows its interactions with effectors, such as guanine nucleotide-exchange factor Vav, phospholipase-Cγ2, and CBL ubiquitin ligase.

---

[60] Kinase SYK contributes to bone resorption by osteoclasts.

[61] In thrombocytes, the collagen receptor glycoprotein GP6 signals via SYK enzyme.

[62] Kinase SYK participates in innate recognition of fungi and other pathogens by C-type lectins. Activation of SYK by C-type lectins activates the CARD9–BCL10–MALT1 pathway (CARD: caspase-recruitment domain-containing protein; BCL: B-cell lymphoma protein; MALT: mucosa-associated lymphoid tissue lymphoma translocation protein) Kinase SYK also activates the NLRP3 inflammasome (NLRP3: NLR family, pyrin domain-containing protein-3) following fungal infection [264].

[63] Kinase SYK mediates integrin signaling in neutrophils, macrophages, and platelets as well as signaling by P-selectin glycoprotein ligand PSGL1 (the major selectin receptor on leukocytes) via ITAM-containing adaptors.

[64] Binding of SYK to phosphorylated ITAMs or ITAM-like sequences in C-type lectins relieves SYK from auto-inhibition and triggers its association with signaling molecules.

Kinase SYK also associates with IL2 receptor β-chain as well as receptors for G-CSF, GM-CSF, IL3, IL5, IL15, and erythropoietin, in addition to Toll-like receptor TLR4 [265]. Numerous signaling mediators and adaptors also link to SYK [265]. Partners of SYK include [251] (Table 4.12): (1) receptors, such as EpoR, CSF3R, Fc fragment of IgG FCGR2A, Fc fragment of IgE FCER1, GFCRL3, MS4A2, CD3E, CD4, CD19, CD22, CD72, CD79A/B, and CR3; (2) cytoskeleton components and cytoskeleton- and cell adhesion-associated molecules, such as cortactin, paxillin, and tubulin-α1a; (3) kinases, such as BTK, FGR, Fyn, LCK, Lyn, Src, FAK1/2, PKCα, PRKD1 (PKCμ), and ribosomal protein S6 kinases S6K1 and S6K2; (4) phosphatases such as PTP6; (5) phospholipases, such as PLCγ1 and PLCγ2; (6) chaperone α-synuclein; (7) partners of small GTPases such as Vav GEF; and (8) adaptors and scaffolds, such as GRB2, SHC, CRKL, SLAP, TRAF6, BCAP, BAnk, BLnk, SiRPB1, PAG1, SLA, SH3BP2, hematopoietic cell-specific Lyn substrate HCLS1 (or HS1), LCP2 (or SLP76), and SH3P7, as well as adaptor/Ub ligase CBL; (5) transmembrane adaptors in membrane rafts, such as linker for activation of T cells (LAT) in T and NK cells, mastocytes, and platelets), linker for activation of B cells (LAB)[65] in B and NK cells, monocytes, and mastocytes, and linker for activation of X cells (LAX) in both B and T cells as well as monocytes.

In phagocytosis of pathogens and apoptotic cells, SYK operates downstream from receptor FcγR and integrin-$\alpha_M\beta_2$ (also called complement receptor CR3) in macrophages,[66] NK-cell receptor-like C-type lectin Dectin-1 in dendritic cells, and apoptotic cell-recognizing receptor [266].

Kinase SYK operates in signal transduction primed by activated immunoreceptors, such as B-cell receptor, Fc receptors, and activating receptors of NK cells [266]. Ligation of immunoreceptor actually leads to activation of different members of the SRC family kinase according to the cell type that phosphorylate (activate) the immunoreceptor. The latter then attracts and activates SYK, as SYK undergoes conformational changes and autophosphorylation. Activated SYK phosphorylates various substrates, such as adaptor SH2-domain-containing leukocyte protein SLP76 and the Vav family of guanine nucleotide-exchange factors or B-cell linker protein (BLnk).

Kinase SYK, indeed, acts not only in immunoreceptor signaling, but also cue transduction from integrins in certain hematopoietic cells, such as platelets,

---

[65] A.k.a. linker for activation of T cells LAT2 and non-T-cell activation linker NTAL.

[66] Complement-mediated phagocytosis is initiated by binding of complement components to their corresponding receptors CR1, CR3 ($\alpha_M\beta_2$-integrin), and CR4 ($\alpha_X\beta_2$-integrin). Complement component C3bi binding to CR3 induces the most effective phagocytosis. The complement system is composed of 3 distinct pathways – classical, mannose-binding lectin, and alternative pathway – that generate protease C3-convertase. Each pathway comprises 3 types of processes: (1) opsonization of pathogens toward engulfment by phagocytes that have receptors for complements (C3b, C3bi); (2) recruitment of inflammatory cells (C3a, C5a); and (3) killing of pathogens (C5b, C6, C7, C8, C9).

4.14 Spleen Tyrosine Kinase

Table 4.12. Direct and indirect partners of spleen Tyr kinase and effects (Source: [264]; BCR: B-cell receptor; BCL: B-cell lymphoma; BLnk: B-cell linker protein; CARD: caspase-recruitment domain; CBL: ubiquitin ligase Casitas B-lineage lymphoma; DAG: diacylglycerol; ERK: extracellular signal-regulated kinase; FAK: focal adhesion kinase; FcR: (Ig) Fc receptor; JNK: Janus kinase; LAT: linker for activation of T cells; LCP: lymphocyte cytosolic protein; MALT: mucosa-associated lymphoid tissue lymphoma translocation protein; NFAT: nuclear factor of activated T cells; NFκB: nuclear factor-κB; NLRP: NLR family, pyrin domain-containing protein; PI3K: phosphatidylinositol 3-kinase; PKB(C): protein kinase-B(C); PLC: phospholipase-C; PTPn: protein Tyr phosphatase non-receptor; RasGRP: RAS guanyl-releasing protein; ROS, reactive oxygen species; TCR: T-cell receptor; TEC: tyrosine kinase expressed in hepatocellular carcinoma).

Receptors	BCR, TCR, FcR, C-type lectin, integrin
Binding partners	Vav, PLCγ, PI3K, LCP2, BLnk
Signaling effectors	Ca^{++}, DAG, LAT, CARD9–BCL10–MALT1, NLRP3, Rho, RasGRP, PKB, PKC, FAK2, TEC, ERK, P38MAPK, JNK, NFAT, NFκB
Inhibitors	CBL, PTPn6
Effects	Cell adhesion, differentiation, proliferation, survival, Cytoskeletal reorganization, Production of reactive oxygen species, Immunity, cytokine release, inflammasome activation, Transition from proB cell to preB cell, Early thymocyte development (DN3–DN4 transition), Activity of macrophage, mastocyte, neutrophils, Lymphangiogenesis, platelet activation, Bone resorption

neutrophils, and monocytes, in addition to osteoclasts [266].[67] Molecules DAP12 and FcγR are able to associate with SYK and mediate $β_2$-integrin signaling in neutrophils and macrophages. In addition, $α_{2B}β_3$-integrin that is expressed in cells of the megakaryocyte lineage is involved in platelet aggregation and granule secretion when platelets bind to fibrinogen. Integrin-$α_{2B}β_3$–fibrinogen binding leads to activation not only of HCK, FGR, Lyn, and Src kinases, but also of SYK [266]. Neutrophils also need SYK for degranulation upon integrin–ligand binding as well as main SRC family kinases that exist in the myeloid lineage to produce reactive oxygen species.

Kinases Src and SYK cooperatively transmit signals downstream of excited integrins. A kinase of the SRC family that can associate with integrins is activated, adaptors undergo phosphorylation, and then SYK is recruited and stimulated.

---

[67]Osteoclasts are multinucleated cells generated from macrophage fusion that degrade bone by releasing acidic granule content with degradative enzymes.

Chapter 5
Cytoplasmic Protein Serine/Threonine Kinases

Numerous protein serine/threonine kinases operate in signal transduction, either as initiating nodes, such as *receptor* (RSTK; Vol. 3 – Chap. 8. Receptor Kinases), or as intermediate nodes of signaling cascades, *non-receptor* (intracellular; NRSTK) protein Ser/Thr kinases. Typically, second messengers activate protein Ser/Thr kinases, whereas extracellular signals stimulate protein Tyr kinases (Chap. 4). For example, IκB kinases (IKK) of IκB inhibitors of nuclear factor-κB are serine kinases (Sect. 10.9.1).

Non-receptor protein Ser/Thr kinases include numerous kinases, such as cAMP-dependent protein kinase-A (PKA), protein kinase-B (PKB or Akt), protein kinase-C (PKC), cGMP-dependent protein kinase-G (PKG), calmodulin-dependent kinases, protein kinase-D (PKD) RoCK kinases, P90 (RSK) and P70 (S6K) ribosomal S6-kinases, P21-activating kinase, Bcr proteins, etc. Certain enzymes have been considered to be Ser/Thr kinases despite their dual specificity, such as mitogen-activated protein kinases and glycogen synthase kinase-3.[1]

5.1 Superfamily of Ser/Thr Sterile-20-Related Kinases

The superfamily of Sterile-20-like kinases includes many kinases that are related to Ser/Thr kinase Ste20 in Saccharomyces cerevisiae. Among these kinases, several are mitogen-activated protein kinase kinase kinase kinases (MAP4Ks) and some mitogen-activated protein kinase kinase kinases (MAP3Ks; Chap. 6). Many Ste20-like family kinases actually function as regulators of MAPK cascades. Therefore,

---

[1] Phosphorylation at different residues controls the enzyme activity. In some cases, phosphorylation of Ser or/and Thr residues can cause inactivation, whereas phosphorylation of Tyr residues can increase activity.

Table 5.1. The Ste20-like kinase superfamily and its P21-activated kinase (PAK) and germinal center kinase (CGK) families (Source: [251, 267]; CRK: avian sarcoma virus CT10 oncogene homolog (chicken tumor virus regulator of kinase); CRKL: CRK-like adaptor; GLK: GCK-like kinase; GRB: growth factor receptor-bound protein; GCKR: GCK-related kinase; HGK: HPK1/GCK-related kinase; HPK: hematopoietic progenitor kinase; MinK: Misshapen-like kinase; MAP3K: mitogen-activated protein kinase kinase kinase; NCK: non-catalytic region of tyrosine kinase adaptor; NIK: NCK-interacting kinase; cRaf: v-Raf-1 murine leukemia viral oncogene homolog; TRAF: tumor-necrosis factor-receptor associated factor).

Name	Other Alias(es)	Substrates Effectors Affectors	Adaptors Coactivators
GCK	MAP4K2	MAP3K1, MAP2K4 Rab8A	TRAF2
GCKR	MAP4K5, KHS	MAP3K1	CRK, CRKL, TRAF2
GLK	MAP4K3		
HPK1	MAP4K1	MAP3K7, MAP3K11 EGFR, PP4	CRK, CRKL, GRB2, GrAP2, NCK1, Dbnl, BLnk, LAT, HCls1
NIK	MAP4K4, HGK	MAP3K1, MAP3K7 Integrin-β1	NCK1, GBP3, RasA1, BrCa1
MinK1			NCK1
PAK1	PAKα	cRaf HER2, CDC2L2, CDK5 BMX, LIMK1, MLCK Abl3, PKB CDC42, Rac1, RhoJ Dncl1, ArpC1B	NCK1/2 PAK1IP1 IRSp53 RhoGEF2/7
PAK2	PAKγ	Abl1/3, CDC42, Rac1/2, MLC2, MHC10 MCM3	SH3KBP1, APAP1/2 DOCK2
PAK3	PAKβ	cRaf Rac1	RhoGEF7

Ste20-like kinases regulate multiple cellular processes, such as cell cycle, growth, differentiation, apoptosis, and stress responses. In addition, certain Ste20-like kinases contribute to cellular volume sensing.

The Ste20-like kinase superfamily is subdivided into 2 structurally and functionally distinct families (Table 5.1): (1) family of P21-activated kinases (PAK; Sect. 5.2.13) that are regulated by small GTPases Rac1 and CDC42 (Sect. 9.3) and (2) family of germinal center kinases (GCK).[2]

---

[2]Not glucokinase (GK) that is also labeled GcK in the literature.

Members of the PAK family contains a C-terminal catalytic domain and an N-terminal binding site CRIB (CDC42–Rac1 interactive binding domain) for small GTPases Rac1 and CDC42. The PAK kinases (Sect. 5.2.13) indeed activate small GTPases Rac and CDC42 to regulate cytoskeletal organization and cell motility. They are also implicated in the activation of the JNK pathways.

On the other hand, GCK family members have an N-terminal kinase domain and a C-terminal regulatory region (without CRIB consensus sequence). Many GCK kinases activate Jun N-terminal kinases or P38MAPKs. Others are activated under some stress conditions or by apoptotic agents.

Certain kinases phosphorylate ion transporters. Furthermore, many of these kinases contribute to numerous pathways excited by the transforming growth factor-β and integrins activated by EPH receptors. They are also involved in lymphocyte aggregation, as they regulate cell clustering via $\alpha_L\beta_2$-integrin.[3] They participate in T-cell receptor signaling.

The set of dual-specificity mitogen-activated protein kinases includes 3 families associated with ERK, JNK, and P38MAPK kinases. Whereas extracellular signal-regulated kinases are activated in response to ligand binding to cognate receptor Tyr kinases or G-protein-coupled receptors via Ras GTPases, JNKs are stimulated in response to cell stresses, such as osmotic and heat shock, ionizing electromagnetic waves, oxidative stress, protein synthesis inhibitors, and pro-inflammatory cytokines. On the other hand, P38MAPKs also respond to pro-inflammatory agents (e.g., interleukin-1, tumor-necrosis factor-α, lipopolysaccharides) and environmental stress (osmotic shock). Both JNKs and P38MAPKs that are targeted by many GCKs are excited by ischemia–reperfusion events and hypertension as well as humoral factors, such as angiotensin-2 and endothelin. Therefore, JNKs and P38MAPKs are poorly activated by mitogens, but potently by environmental stresses. Members of the JNK and P38MAPK subfamilies stimulate several transcription factors, especially Activator protein-1.

Activatory kinases of JNKs are MAP2K4 and MAP2K7, whereas P38MAPK activators are MAP2K3 and MAP2K6. One signaling cascade tier upstream from MAP2Ks, MAP3K1, MAP3K10, and MAP3K11 are specific stimulators of the JNK pathway. In addition, members of the RHO superfamily of the RAS hyperfamily (CDC42 and Rac1) can activate the JNK and P38MAPK cascades, the mediators of which contain a CDC42–Rac-interacting and -binding (CRIB) domain, such as MAP3K1, MAP3K4, MAP3K10, and MAP3K11.

---

[3]Integrin-$\alpha_L\beta_2$-integrin, or lymphocyte function-associated antigen LFA1 is located on all T-cell types as well as B cells, macrophages, and neutrophils. This integrin is involved in cell recruitment to the site of infection. It binds to ICAM1 on antigen-presenting cells.

5.1.1 Ste20-like Kinase

Ste20-like kinase (SLK), or STK2, is a microtubule-associated protein that provokes actin stress fiber disassembly. Kinase SLK redistributes with vinculin and colocalizes with CK2 in structures reminiscent of membrane lamellipodia and ruffles during cell spreading. Kinase Src activates casein kinase CK2 that phosphorylates SLK (Ser347 and Ser348) to reduce SLK activity [268].

Kinase SLK mediates apoptosis and actin stress fiber degradation via distinct domains released after caspase-3 cleavage [269]. Activated N-terminal kinase domain promotes apoptosis and cytoskeletal rearrangement, whereas C-terminus disassembles actin stress fibers. Kinase SLK interacts with C-terminal linking and modulating protein (CLAMP) that is also designated as PDZ domain-containing protein PDZK1, or sodium–hydrogen exchanger member-3 regulator NHERF3 [251].

5.1.2 Germinal Center Kinase Family

Many GCK family members, such as MAP4K1 to MAP4K5, activate the JNK pathway. Adaptors are crucial for signal transduction in cells because they mediate rapid, selective formation of multiproteic complexes. Adaptors of the CRK family are composed of Src homology domains SH2 and SH3 to specifically tether phosphorylated tyrosyl- and proline-rich sequence-containing proteins, respectively [270]. Adaptors CRK and CRKL connect guanine nucleotide-exchange factors C3G (Rap1a/b GEF) and dedicator of cytokinesis DOCK1, as well as EGFR pathway substrate Eps15. Adaptors CRK and CRKL also interact with other small adaptors NCK and GRB2, Abl family kinases (Sect. 4.2), SOS1 and SOS2 RasGEFs. Adaptors CRK and CRKL thus intervene in integrin signaling, B-cell receptor activation, and JNK activation.

Among GCK family members, MAP4K1 and MAP4K5 are signaling partners of CRK family adaptors (Table 5.2). Whereas MAP4K5 binds exclusively to CRK family proteins, MAP4K1 associates with both GRB2 and NCK (weakly) [270]. Kinase MAP4K4 interacts with NCK adaptor to link protein Tyr kinase to JNK. Enzymes MAP4K2 and MAP4K5 mediate TNF-induced JNK activation via TRAF2 TNFR-associated factor.

The GCK family can be subdivided into 2 groups based on structural and functional properties [271]. Group-1 GCKs include hematopoietic progenitor kinase HPK1 (or MAP4K1), germinal center kinase (GCK or MAP4K2), GCK-like kinase (GLK or MAP4K3), GCK-related kinase (GCKR or MAP4K5), and NCK-interacting kinase (NIK or MAP4K4)[4] Their C-termini possess 2 PEST

---

[4]A.k.a. hepatocyte progenitor kinase-like–germinal center kinase-like kinase (HGK) and MAP4K4. Alias NIK that stands for NCK-interacting kinase is also used to designate NFκB-inducing kinase, a synonym for MAP3K14 (a.k.a. HS and HSNIK).

5.1 Superfamily of Ser/Thr Sterile-20-Related Kinases

Table 5.2. The CGK kinases and MAPK modules in the JNK pathway (Source: [272]). The MAPK tiered modules encompass MAP4Ks of the group-1 GCKs, MAP3Ks, MAP2Ks, and MAPKs. The MAPK effector set targeted by MAP4Ks are JNK kinases. Group-1 GCKs represented by PAK isoforms are also able to activate JNKs. The activation of the JNK pathway is mediated by adaptors, small GTPases, and TNF receptor-associated factors (TRAF).

Stimuli	Mitogens	Cytokines	Cellular stress
Adaptors	CRK, CRKL GRB2, NCK	TRAF	CDC42, Rab8, Rac1, Ras, Rho
Tier 1	MAP4Ks/CGKs and PAKs		
Tier 2	MAP3Ks		
Tier 3	MAP2Ks		
Tier 4	MAPKs – JNKs		

(proline–glutamic acid–serine–threonine) motifs and at least 2 polyproline consensus binding sites for SH3-containing proteins as well as a hydrophobic, leucine-rich sequence and a C-terminal (CT) region. These enzymes activate selectively SAPKs/JNKs.

Group-2 GCKs comprise Ste20-like oxidant stress-activated kinase SOK1 (or STK25);[5] mammalian sterile-20 (twenty)-like kinases MST1 (STK4 or kinase responsive to stress KRS2), MST2 (KRS1 or STK3), and MST3 (STK24); and lymphocyte-oriented kinase (LOK or STK10). These enzymes do not activate any MAPK pathways. Members of the group-2 GCKs — STK3 and STK4 — can be activated by both phosphorylation and dephosphorylation, as they are stimulated by PP2 phosphatase.

The GCK family can be also broken into a higher number of subgroups according to structure and biochemical properties [272]. The HPK1/GCK (MAP4K1–MAP4K2) and NIK (MAP4K4) subgroups are constituted by MAP4K enzymes. The MST/SOK subgroup includes mammalian Ste20-like proteins MST1 to MST3, SOK1, and proline–alanine-rich Ste20-related kinase (PASK). Other branches corresponds to LOK and SLK on the one hand and PSK on the other.

Analyses on human kinome led to the reclassification of the GCK family members into 8 phylogenetically distinct subfamilies. The GCK subfamily 1 comprises MAP4K1 to MAP4K3 and MAP4K5 enzymes. The GCK subfamily 2 encompasses STK3 (KSR1 or MST2) and STK4 (KSR2 or MST1) that are activated during apoptosis. The GCK subfamily 3 consists of 3 mammalian kinases, STK24,

---

[5] Other alias: YSK1.

STK25 (SOK1 or YSK1), and MST3- and SOK1-related kinase (MASK or MST4). Unlike other GCK family kinases, neither STK24 nor STK25 activate the JNK or P38MAPK pathways. They are stimulated by oxidant stress and hypoxia. However, they do not participate in apoptosis. The GCK subfamily 4 is composed of MAP4K4 (NIK), MAP4K6 (MinK), NRK, and TNIK enzymes. The GCK subfamily 5 is constituted by SLK and STK10 (LOC). The GCK subfamily 6 includes OxSR1 that targets both JNK and P38MAPK and STK39 (PASK, SPAK, or DCHT) that only activates P38MAPK. The GCK subfamily 7 contains unconventional subclass-3 myosins Myo3a and Myo3b that have a kinase domain N-terminal to the conserved N-terminal motor domains. They are expressed in photoreceptors. The GCK subfamily 8 is made up of TAOK1 (PSK2 or KFCb), TAOK2 (PSK1, MAP3K17, or KFCc), and TAOK3 (JIK or KFCa).

Most identified GCK family kinases activate either JNK and/or P38MAPK signaling cascades. The GCK subfamily-3 kinases STK24, STK25, and MASK, GCK subfamily-5 STK10, and GCK subfamily-6 OxSR1 do not stimulate these pathways. Four GCK family kinases, GCK subfamily-1 MAP4K1, GCK subfamily-2 STK3 and STK4, and GCK subfamily-5 SLK directly or indirectly contribute to apoptosis.

5.1.3 MAP4Ks of GCK Subfamilies 1 and 4 of the Ste20-like Kinase Superfamily

The MAP4K1/GCK (HPK1/GCK) subfamily of mammalian Ste20-like kinases, or GCK subfamily 1, includes MAP4K1 to MAP4K3 and MAP4K5, whereas MAP4K4 and MAP4K6 belong to the GCK subfamily 4 (Table 5.3).

5.1.3.1 MAP4K1

Mitogen-activated protein kinase kinase kinase kinase MAP4K1 is also termed hematopoietic progenitor kinase HPK1 as, in mammals, it is a hematopoietic cell-specific Ste20-like protein kinase. It is characterized by an N-terminal kinase domain, an intermediate region, and a C-terminal Citron-homology domain. It is a potent and highly specific activator of Jun N-terminal kinases in various cell types [273].

Enzyme MAP4K1 is involved in various signaling pathways initiated by receptors of epidermal growth factor, transforming growth factor-β, erythropoietin, and prostaglandin-E2, as well as B- (BCR) and T-cell (TCR) receptors. In lymphocytes, MAP4K1 activity rises in response to TCR or BCR crosslinking [273].

5.1 Superfamily of Ser/Thr Sterile-20-Related Kinases

Table 5.3. Mitogen-activated protein kinase kinase kinase kinases (MAP4K) of GCK subfamilies 1 and 4 of the Ste20-like kinase superfamily (GCK: germinal center kinase; GCKR: GCK-related kinase; GLK: GCK-like kinase; HGK: HPK1/GCK-related kinase; HPK: hepatocyte progenitor kinase; HPK1: hematopoietic progenitor kinase; KHS: kinase homologous to Ste20; MinK: Misshapen-like kinase; NESK: NIK-like embryo-specific kinase; NIK: non-catalytic region of tyrosine kinase adaptor (NCK)-interacting kinase [not NFκB-inducing kinase, i.e., MAP3K14]; NRK: NIK-related kinase; Rab8IP: Rab8-interacting protein; Rab8IPL: Rab8-interacting protein-like protein; TNIK: TRAF2 and NCK-interacting kinase).

Member	Aliases
GCK subfamily 1	
MAP4K1	HPK1
MAP4K2	GCK, Rab8IP
MAP4K3	GLK, Rab8IPL1
MAP4K5	GCKR, KHS1
GCK subfamily 4	
MAP4K4	HPK, HGK, NIK
MAP4K6	MinK
NRK	NESK
TNIK	

Activation of MAP4K1 depends on autophosphorylation (Thr165) and phosphorylation by protein kinase-D (Ser171) [273]. Kinase MAP4K1 may also be activated by Ser/Thr protein phosphatase-4 [277] (Sect. 8.3.4). Furthermore, PP4-induced activation is accompanied by an increase in MAP4K1 expression. Phosphatase PP4 increases the MAP4K1 half-life. Activity of MAP4K1 is also regulated by ubiquitination followed by degradation. Phosphatase PP4 precludes MAP4K1 ubiquitination. Stimulation by TCRs enhances PP4–MAP4K1 interaction.

In T lymphocytes and myeloid cells, MAP4K1 activity is stimulated after exposure to immunosuppressive PGE2 prostaglandin [273]. In T lymphocytes, MAP4K1 impedes PGE2-induced transcription of the FOS gene. Conversely, PGE2 causes HPK1 activation via the cAMP–PKA axis.

Kinase MAP4K1 can stop T-cell activation and provoke T-cell apoptosis [273]. It is also implicated in monocyte differentiation.

Protein MAP4K1 interacts with many proteins: MAP3K7,[6] MAP3K11,[7] EGFR, HCLS1,[8] BLnk,[9] CLnk,[10] DbnL,[11] NCK1, GRB2, CRK, CRKL,[12] GRAP2,[13] LCP2[14] LAT[15] and DAPP1[16] (Table 5.1) [251, 273].

Furthermore, MAP4K1 contributes to the activation of NFκB transcription factors. Full length MAP4K1 interacts with IKKβ in T lymphocytes to activate NFκB upon TCR stimulation, using its substrate and adaptor caspase activation and recruitment domain (CARD)-containing protein family CARD11, or CARMA1 [273].

5.1.3.2 MAP4K2

Mitogen-activated protein kinase kinase kinase kinase MAP4K2 is also identified as germinal center kinase GCK as well as BL44, Rab8IP, as it is an interacting protein

---

[6] A.k.a. TGFβ-activated kinase TAK1.

[7] A.k.a. mixed lineage kinase MLK3.

[8] I.e., cortactin-like protein hematopoietic cell-specific Lyn substrate.

[9] I.e., B-cell linker protein.

[10] I.e., cytokine-dependent hematopoietic cell linker. A.k.a. mastocyte immunoreceptor signal transducer (MIST).

[11] I.e., drebrin-like protein. A.k.a. 55-kDa HPK1-interacting protein (HIP55), SH3 domain-containing protein SH3P7, and filamentous actin-binding protein ABP1. It activates HPK1 and JNK during TCR signaling, particularly regulating the JNK1 cascade [274, 275].

[12] Kinase MAP4K1 interacts using its proline-rich domains with growth factor receptor-bound adaptor GRB2, CT10 regulator of kinase (CRK), and CRK-like adaptor (CRKL).

[13] I.e., GRB2-related adaptor protein-2. A.k.a. GRB2-related adaptor downstream of SHC (GADS), GRB2-like protein (GRB2L), SH2 and SH3 domain-containing (SH3–SH2–SH3) adaptor Mona, growth factor receptor-binding protein, hematopoietic cell-associated adaptor protein, GRB2-related protein with insert domain (GRID), and GRB2-related protein of the lymphoid system (GRPL). It acts in T-cell signaling via interactions with LCP2 and LAT adaptors [276]. Engagement of the TCR–CD3 complex activates SRC family kinases Fyn and LCK that rapidly phosphorylate CD3 protein. This phosphorylation yields a binding site for ZAP70 kinase. The latter is subsequently phosphorylated (activated) by LCK kinase. Activated SRC family kinases and ZAP70 then phosphorylate phospholipase-Cγ1, Vav, CBL, LCP2, and LAT, thereby initiating signaling cascades that lead to cytokine gene transcription via nuclear factor of activated T cell that targets NFAT response elements.

[14] I.e., lymphocyte cytosolic protein, or 76-kDa SH2 domain-containing leukocyte protein (SLP76).

[15] I.e., linker of activated T cells. It localizes in specific plasmalemmal compartments, glycolipid-enriched nanodomains. Its phosphorylated form acts as a docking protein for several adaptors, such as GRAP2 and GRB2. These LAT-associated adaptors recruit various signaling molecules to glycolipid-enriched nanodomains, where they are activated. Both LAT and LCP2 adaptors enable the assembly of phosphorylation-dependent and -independent signaling complexes, as they interact with SH2 and SH3 domain-containing enzymes and adaptors. Therefore, TCR signaling leads to JNK and ERK activation, intracellular Ca^{++} release, and interleukin-2 production [275].

[16] I.e., dual adaptor of phosphoTyr and 3-phosphoinositides. A.k.a. 32-kDa B-cell adaptor molecule (BAM32).

of small GTPase Rab8. It is detected in many tissues. However, its expression in lymphoid follicles is restricted to the cells of the germinal center, where it can participate in B-lymphocyte differentiation. It can be activated by TNFα. It interacts with TNF receptor-associated factor TRAF2 to activate MAP3K1.

5.1.3.3 MAP4K3

Mitogen-activated protein kinase kinase kinase kinase MAP4K3 is also tagged GLK and Rab8IPL1. Activity of MAP4K3 is regulated by amino acids, but not insulin. It is required to activate P90 ribosomal S6 kinase and for maximal phosphorylation of eukaryotic initiation factor 4E-binding protein 4E-BP1 [278].

5.1.3.4 MAP4K4

Mitogen-activated protein kinase kinase kinase kinase MAP4K4[17] specifically activate MAPK8 (JNK1) via the MAP3K1–MAP2K4 and MAP3K7–MAP2K7 cascades. It interacts also with GTPase activating protein RasA1 guanylate-binding protein GBP3, adaptor NCKα (or NCK1), $β_1$-integrin, and breast cancer early onset protein BrCa1 that operates in DNA-damage repair, ubiquitination, and transcriptional regulation [251].[18]

5.1.3.5 MAP4K5

Mitogen-activated protein kinase kinase kinase kinase MAP4K5 is similar to Ste20 kinase, hence its aliases GCKR (GCK-related kinase) and KHS (or KHS1: kinase homologous to Ste20). It activates Jun N-terminal kinases in the framework of stress response. Two alternatively spliced transcript variants encode the same protein. It interacts with CRK, CRKL, and TRAF2 (Table 5.1).

---

[17] A.k.a. hepatocyte progenitor kinase-like/germinal center kinase-like kinase (HPK/GCK-like kinase [HGK]) and NCK-interacting kinase (NIK; not NFκB-inducing kinase [NIK]).

[18] Kinase MAP3K14, or NFκB-inducing kinase (NIK), phosphorylates (activates) Na^+/H^+ exchanger NHE1 (or SLC9a1), hence participating in the regulation of solute and water transport. Kinase MAP3K14 interacts with MAP3K7 and MAP3K8, receptor-interacting Ser/Thr kinase RIPK1, IκB kinase-α and -β, cRaf kinase inhibitor protein (RKIP or phosphatidylethanolamine-binding protein PEBP1), TNF receptor-associated factors TRAF1 to TRAF3 and TRF6, NFκB1a and NFκB2, CASP8 and FADD-like apoptosis regulator (CFLAR), and caspase-8 and -10 [251].

Table 5.4. Members of GCK subfamilies 2, 3, and 5 of the Ste20-like kinase superfamily (KSR: kinase suppressor of Ras; LOK: lymphocyte-oriented kinase; MASK: MST3- and SOK1-related kinase; MST: mammalian Ste20-like protein kinase; SLK: Ste20-like kinase; SOK: Ste20-like oxidant stress-activated kinase; STK: Ser/Thr kinase).

Member	Aliases
GCK subfamily 2	
STK3	KSR1, MST2
STK4	KSR2, MST1
GCK subfamily 3	
STK24	MST3
STK25	SOK1
MASK	MST4
GCK subfamily 5	
STK2	SLK
STK10	LOK

5.1.3.6 MAP4K6 – Misshapen-like kinase

Mitogen-activated protein kinase kinase kinase kinase MAP4K6, or protein Ser/Thr Misshapen-like kinase MinK1,[19] activates JNK and P38MAPK kinases. It interacts with the adaptor Non-catalytic region of Tyr kinase NCK1 to transmit signals from receptor Tyr kinases, such as EGFR, PDGFR-B, and EPHb1.

5.1.4 Kinases of GCK Subfamilies 2 and 3 – MST Kinases

Subfamily-2 GCKs are MST1, or STK4, and MST2, or STK3, whereas subfamily-3 GCKs correspond to MST3, or STK24, and MST4 (Table 5.4).

5.1.4.1 STK3 (MST2)

Protein Ser/Thr kinase STK3[20] possesses a regulatory C-terminus that modulates its kinase activity. It contains at least one caspase-3 cleavage site. In response to apoptotic stimuli, STK3 is indeed cleaved by caspase-3 to release the activated N-terminal kinase domain that is much more active and stimulates P38MAPK and JNK kinases [279]. Moreover, full-length STK3 is highly susceptible to dephosphorylation with respect to caspase-cleaved form. Activity of STK3 is enhanced by autophosphorylation following homodimerization.

---

[19] A.k.a. B55, ZC3, and YSK2.

[20] A.k.a. mammalian Sterile-20-like protein kinase MST2 and kinase response to stress KRS1.

Targets of STK3 comprise Large tumor suppressors LaTS1 and LaTS2 kinases (Hippo pathway; Sect. 10.7), connector enhancer of kinase suppressor of Ras CnKSR1, Ras association (RalGDS/AF6) domain-containing protein family members RASSF1 and RASSF2, Salvador homolog Sav1 (or WW45), thyroid transcription factor TTF1, and forkhead box FoxA (or hepatocyte nuclear factor HNF3) [251, 279]. Transcription factor TTF1 is required for thyroid-specific production of thyroglobulin and thyroperoxidase.

Respiratory epithelium-specific transcription is also mediated by thyroid transcription factor TTF1 and transcription factors of the FoxA subfamily. These transcription factors are needed for lung-specific synthesis of surfactant protein-A, -B, and -C, and small, secreted, dimeric secretoglobin-1A (Scgb1A1).[21] The SCGB1A1 promoter is regulated by FoxA1 and FoxA2, but neither FoxJ1 nor TTF1 [280].

Activity of STK3 is regulated by phosphorylation, proteolysis, and other types of between-protein interactions. Kinase STK3 is activated by stressors, such as heat shock, toxin and phosphatase okadaic acid and anti-fungal, anti-hypertensive staurosporine. Kinase cRaf binds to STK3 and prevents dimerization and autophosphorylation.

5.1.4.2 STK4 (MST1)

Protein Ser/Thr kinase STK4[22] is widely expressed. This member of the Sterile-20 superfamily of cytoskeletal, stress-activated, and apoptotic kinases is stimulated by STK4K kinase that phosphorylates Thr residues (Thr183 and Thr187). Homodimerization enhances its activation via autophosphorylation (Thr177 and Thr387) [281]. Full-length STK4 ($STK4_{FL}$) shuttles between the nucleus and cytosol.

It is activated by cellular stressors to activate the stress-induced mitogen-activated protein kinase cascade. It provokes chromatin condensation, as it phosphorylates histone-2B (Ser14). It also causes apoptosis, as it is activated upon engagement of the pro-apoptotic receptor TNFRSF6 of the TNFR superfamily. In response to apoptotic stimuli, STK4 is cleaved by caspase-3 (AsP326) to release the N-terminal kinase domain that is much more active. Cleaved $STK4_{36}$ localizes mostly to the nucleus.

Owing to the SARAH domain, it interacts with itself and other proteins, such as Ras association (RalGDS/AF6) domain family members RASSF1A and RASSF5 that associates with DNA-repair protein XPA and activated Ras, Rap1, and Ras-like small GTPases, respectively, and Salvador homolog Sav1 that interacts with 14-3-3-σ (stratifin) [279]. Other STK4 Substrates include Large tumor suppressor LaTS1 kinase that forms a complex with CDK1 in early mitosis, connector enhancer of kinase suppressor of Ras CnKSR1, and protein kinase interferon-inducible double

---

[21] A.k.a. Clara cell-specific 10-kDa protein (CC10).
[22] A.k.a. mammalian Ste20-like protein kinase MST1, kinase response to stress KRS2, and Kas2.

stranded RNA-dependent inhibitor (PRKRIR or death-associated protein DAP4). Moreover, STK4 phosphorylates transcription factor FoxO3 (Ser207) to cause its dissociation from 14-3-3 protein.

5.1.4.3 STK24 (MST3)

Protein Ser/Thr kinase STK24[23] belongs to the GCK family (GCK subfamily 3) of Sterile-20-like kinases or SPS1 group. It participates in the response to cellular stress. The proteolytic cleavage of its C-terminus that possesses nuclear localization and export signals provokes nuclear translocation of its kinase domain.

5.1.4.4 MASK (MST4)

Ubiquitous protein Ser/Thr kinase MST4[24] belongs to the GCK subfamily 3. It contains an N-terminal kinase domain and a C-terminal regulatory domain that mediates homodimerization and prevents its kinase activity. It is located in the Golgi body, where it is specifically activated by binding to the Golgi matrix protein GM130. It is cleaved by caspase-3 to generate a C-terminally truncated form. Several alternatively spliced transcript variants exist. Both full-length MASK and its C-terminally truncated form can induce apoptosis.

5.1.5 STK25 of the GCK Subfamily-3

In addition to STK24 (MST3) and MASK1 (MST4), the GCK subfamily 3 comprises STK25 kinase.[25] It interacts with STK24, PP2, apoptosis-related protein-15, or programmed cell death protein PdCD10, striatins Strn, Strn3, and Strn4, suppressor of IKKε, T-complex protein-1, and sarcolemma-associated protein [251].

5.1.6 Other Kinases of the GCK Subfamily 4

The GCK subfamily 4 is composed of MAP4K4 (Sect. 5.1.3.4), or non-catalytic region of Tyr kinase (NCK)-interacting kinase (NIK), and MAP4K6 (Misshapen-like kinase [MinK]; Sect. 5.1.3.6)) as well as NIK-related kinase (NRK) and TRAF2 and NCK-interacting kinase (TNIK).

---

[23] A.k.a. mammalian Ste20-like protein kinase MST3 and MST3b.
[24] A.k.a. MST3 and SOK1-related kinase (MASK).
[25] A.k.a. Ste20-like oxidant stress-response kinase SOK1 and yeast Ste20-related kinase YSK1.

5.1.6.1 NIK-Related Kinase (NRK)

NCK-interacting kinase (NIK)-related kinase (NRK)[26] of the germinal center kinase family activates Jun N-terminal kinase. It also phosphorylates (inactivates) actin-depolymerizing factor cofilin (Ser3) [282].

5.1.6.2 TRAF2 and NCK-Interacting Kinase (TNIK)

TRAF2 (TNF receptor-associated factor-2) and NCK (non-catalytic region of Tyr kinase adaptor)-interacting kinase (TNIK) is characterized by an N-terminal kinase domain and a C-terminal GCK domain that serves as a regulator [283]. In addition to activating the JNK pathway, TNIK causes disruption of ^{F}actin structures. Kinase TNIK is able to phosphorylate gelsolin. It also interacts with Na^+-H^+ exchanger SLC9a1 (or NHE1) and TRAF2 [251].

5.1.7 Kinases of the GCK Subfamily 5

The GCK subfamily 5 contains STK10, or lymphocyte-oriented kinase (LOK), and SLK (Sect. 5.1.1). Kinase STK10 is expressed predominantly in lymphoid organs, such as the spleen, thymus, and bone marrow, i.e., lymphocytes. Its non-catalytic domain does not exhibit any similarity to that of other known members of the subfamily. It possesses a proline-rich motif with SH3 domain-binding potential followed by a long coiled-coil structure at the C-terminus [284]. Nevertheless, a close sequence homology exists with SLK of the same subfamily that serves as a Polo-like kinase kinase (PL2K).

Kinase STK10 actually phosphorylates Polo-like kinase PLK1. It then cooperates with SLK to regulate PLK1 activity in cells [285]. Kinase STK10 phosphorylates itself as well as substrates, such as myelin basic protein and histone-2A, but does not target any MAPK (ERK, JNK, or P38MAPK). In T lymphocytes, STK10 prevents interleukin-2 expression via the MAP2K1 pathway.

5.1.8 Kinases of the GCK Subfamily 6

Members of the germinal center kinase subfamily-6 comprise oxidative stress-responsive kinase OxSR1 as well as protein Ser/Thr kinases STK37 and STK39 that

---

[26] A.k.a. NIK-like embryo-specific kinase (NESK).

Table 5.5. Members of subfamilies 6, 7, and 8 of the Ste20-like kinase family (DPK: dendritic cell-derived protein kinase; JIK: Jun N-terminal kinase (JNK)-inhibitory kinase; KFC: kinase from chicken; Myo: myosin; OxSR1, OSR1: oxidative stress-responsive kinase; PASK: PAS domain-containing Ser/Thr kinase; PSK: prostate-derived Sterile-20-like kinase; SPAK: Ste20-related proline/alanine-rich kinase; STK: Ser/Thr kinase; TAOK: thousand and one amino acid protein kinase).

Member	Aliases
	GCK subfamily 6
OxSR1	OSR1
STK37	PASK, paskin
STK39	SPAK
	GCK subfamily 7
Myo3a	
Myo3b	
	GCK subfamily 8
MAP3K16	TAOK1, KFCb, PSK2, MAR2K
MAP3K17	TAOK2, KFCb, PSK1,
MAP3K18	TAOK3, KFCa, JIK, DPK

are involved in the regulation of ion homeostasis and volume control in mammalian cells as well as extracellular ion and water balance, hence the maintenance of blood pressure (Table 5.5). They indeed activate ubiquitous Na^+–K^+–$2Cl^-$ cotransporter NKCC1 in addition to P38MAPK enzymes.

In fact, OxSR1 and STK39 interact with different members of the SLC12A family: KCC3 (or SLC12a6), NKCC1 (or SLC12a2), NKCC2 (or SLC12a1), and NCC (or SLC12a3) [286].[27]

5.1.8.1 Regulation of Activity

Both OxSR1 and STK39 are phosphorylated (activated) by with-no-lysine protein Ser/Thr kinase WNK1 and WNK4 [287]. In addition to their similar kinase domains, these kinases share 2 conserved C-terminal PF1 and PF2 domains. The PF1 domain controls the catalytic activity, whereas the PF2 domain is responsible for between-protein interactions.

---

[27] Sodium–chloride cotransporter (NCCT) is also called Na^+–Cl^- (NCC) symporter and thiazide-sensitive sodium–chloride cotransporter (TSC).

5.1 Superfamily of Ser/Thr Sterile-20-Related Kinases

5.1.8.2 OxSR1

Oxidative stress-responsive kinase OxSR1, or OSR1,[28] is widely expressed with high levels in the heart and skeletal muscle. Kinase OSR1 was named for its similarity to Ste20-like oxidant stress response kinase STK25 (or SOK1) of the GCK subfamily 3 (without a functional link to oxidative stress). It is sensitive to osmotic stress.

Oxidative stress-responsive kinase-1 phosphorylates PAK1 and impedes its responsiveness to CDC42. It also phosphorylates TNFRSF19L, a tumor-necrosis factor receptor (TNFR) superfamily member,[29] which operates in T-cell proliferation [286].

Both OxSR1 and STK39 phosphorylate (activate) Na^+–K^+–$2Cl^-$ cotransporter-1 (or SLC12a2) and Na^+–Cl^- cotransporter, or SLC12a3, in response to changes in cell volume or intracellular ion concentrations [286].

5.1.8.3 STK37 (PASK)

Kinase STK37[30] interacts with STK16, RNA U3 small nucleolar-interacting protein-1 (fibrillarin), proteasome subunit HC5,[31] and WD repeat domain-containing protein WDR5 that can facilitate formation of multiprotein complexes [251].

5.1.8.4 STK39 (SPAK)

Kinase STK39[32] has a tissue distribution that overlaps that of OxSR1. It indeed abounds in neurons and epithelia specialized in transport.

Kinase STK39 interacts with kinases P38MAPKα (MAPK14), STK39, WNK4, and apoptosis-associated Tyr kinase (AATK), as well as gelsolin, myelin basic protein, otoferlin, Na^+–K^+–$2Cl^-$ cotransporter SLC12a2, and heat shock protein HSPH1 (HSP105 or HSP110) [251,286]. The redundant STK39 and OxSR1 kinases regulates shrinkage-induced activation of ubiquitous Na^+–K^+–$2Cl^-$ cotransporter SLC12a2 [286].

---

[28] The OSR1 alias also stands for odd skipped-related protein-1. This Ste20 family protein kinase is also a transcription factor that directs mesenchymal precursors toward kidney precursor cells and differentiation into nephrons. The Osr1 gene encodes a zinc finger transcription factor that limits endoderm differentiation from mesendoderm, balancing differentiation into nephric and blood vascular progenitor cells [288]. Kidney and vasculature are thus not only functionally linked, but also developmentally.

[29] A.k.a. receptor expressed in lymphoid tissues (RELT).

[30] A.k.a. PAS domain-containing Ser/Thr kinase (PASK) and paskin.

[31] A.k.a. prosome macropain subunit-β1.

[32] A.k.a. Ste20-related proline/alanine-rich kinase (SPAK).

Table 5.6. Some properties of cation–Cl^- cotransporters: N^+–K^+–$2Cl^-$ (NKCC), K^+–Cl^- (KCC), and Na^+–Cl^- (NCC) cotransporters (Source: [290]; DCT: distal convoluted tubule; RBC: red blood cell; TAL: thick ascending limb of Henle's loop).

Property	NKCC	KCC	NCC
Cl^- flux direction	Influx	Efflux	Influx
ATP	Stimulation	Inhibition	
Tissue distribution	NKCC1: Kidney, heart, lung, brain, etc. NKCC2: Kidney (TAL)	KCC1: Brain, heart, kidney, liver, lung, etc. KCC2: Brain	NCC1: Kidney (DCT), bladder
Cell distribution	RBC, smooth, skeletal, cardiac myocytes, fibroblasts, epithelial cell (respiratory)	RBC, neuron, epithelial cell	Renal cell
Intracellular Cl^-	Inhibition	Stimulation	
Activation	WNK1/4 OxSR1, STK37/39	WNK1/4 STK37	WNK1/4 OxSR1, STK39

Kinases of the GCK Subfamily 6 and Salt and Water Balance

Ste20-like protein kinases of the GCK subfamily 6, oxidative stress-responsive kinase OxSR1 and STK39 bind to volume-sensitive Na^+–K^+–$2\,Cl^-$ cotransporters NKCC1 and NKCC2 of the solute carrier SLC12 subclass (Vol. 3 – Chaps. 2. Membrane Ion Carriers and 4. Membrane Compound Carriers). Kinase STK37 regulates the activity not only of Na^+–Ka^+–$2Cl^-$, but also K^+–Cl^- cotransporters (KCC). These 2 types of cotransporters intervene in epithelial transport, intra- and extracellular ion homeostasis in the central nervous system, and systemic salt and water balance, as well as cell-volume regulation [289] (Table 5.6).

The regulatory volume decrease that results from swelling provokes activation of KCC cotransporters to cause KCl efflux and osmotically associated water motion. Conversely, regulatory volume increase that follows cell shrinkage is mainly elicited by activated NKCC cotransporters that generate uptake of NaCl and KCl. Activation of KCCs and NKCCs is conferred by dephosphorylation and phosphorylation by volume-sensitive kinases and phosphatases, respectively. Phosphorylation of NKCCs and KCCs provokes their activation and inactivation, respectively, whereas dephosphorylation of these cotransporters precludes NKCC and activates KCC.

Kinase STK37 cooperates with OSR1 to control the activity of NKCCs and KCCs. They are themselves regulated by P38MAPK, apoptosis-associated Tyr kinase (AaTyK; Vol. 3 – Chap. 8. Receptor Kinases) and with no lysine (K) family kinase WNK4. The latter works with STK37 to regulate NKCC1 and KCC2 [289]. Isozymes WNK1 and WNK4 actually phosphorylate (activate) STK37 and OSR1 that can then phosphorylate NKCC1.

Kinase WNK1 also regulates the functioning of OSR1, STK39, and NKCC [291]. Enzyme WNK1 phosphorylates (activates) both OSR1 and STK39 kinases. Activated OSR1 and STK39 phosphorylate (activate) NKCC1 and NKCC2 cotransporters.[33]

Cell-volume sensor modulates concentration of osmolyte glycerol. Glycerol accumulates by increased glycerol synthesis and reduction in glycerol efflux. Glycerol synthesis is partly regulated by Ste20-related kinases such as STK37. Cell shrinkage and swelling induce conformational changes in cytoskeletal and/or proteic sensors that can then activate and inactivate STK37, respectively.

Both OxSR1 and STK39 are activated by with-no-lysine kinases WNK1 and, in turn, phosphorylate (activate) thiazide diuretic-sensitive sodium–chloride cotransporter (NCC) [292]. This sodium-chloride symporter of the SLC12 subclass removes Na^+ and Cl^- from the distal convoluted tubule of the kidney. Mutations of Wnk1 and Wnk4 genes give rise to NCC hyperactivity, increased renal salt reabsorption, and familial hyperkalemic hypertension (pseudohypoaldosteronism type 2). Kinase WNK3 also activates NCC, but antagonizes WNK4 that directly suppresses NCC activity. Calcium- and DAG-regulated Ras guanine nucleotide-releasing protein RasGRP1 primes the MAPK cascade to activate extracellular signal-related kinases ERK1 and ERK2 and regulate NCC carrier. Angiotensin-2 and aldosterone also contribute to the regulation of the WNK4–STK39–NCC pathway.

Epithelial Transport

Certain epithelia, such as the respiratory epithelium and nephron coating that produce periciliary fluid and mucus layers for airway clearance and control blood volume and ion concentrations, respectively, transport fluid and solutes. Epithelial transport requires balance between fluxes across apical and basolateral cell membranes. In respiratory conduits, this material transport is mediated by activated basolateral NKCC1s and apical CFTRs.

---

[33]Certain sites phosphorylated in NKCC1 are conserved in NKCC2 and sodium–chloride cotransporter, another SLC12 subclass member. These sites in NKCC2 are phosphorylated in response to vasopressin.

Cell Volume and Division Cycle

Cell cycle progression can be linked to changes in cell volume. Therefore, volume-sensitive anion channels and cation–chloride cotransporters can contribute to cell cycle regulation. Inhibition of activity of volume-regulated anion channel (VRAC) causes G0–G1 arrest [289]. Conversely, activated VRAC causes cell shrinkage, an element of cell apoptosis. Ste20-like kinases could operate as a scaffold to recruit signaling mediator and coordinate the activity of volume-sensitive anion channels and cation–Cl$^-$ cotransporters with cell cycle events.

5.1.9 Kinases of the GCK Subfamily 7

Kinases of the GCK subfamily-7 include actin-dependent, non-processive nanomotors, the unconventional myosin-3A and -3B. These myosins have a kinase region N-terminal to the conserved N-terminal motor domain. They can undergo autophosphorylation in the C-terminal end of the motor domain that activates the kinase activity, without influencing the ATPase activity [293]. Moreover, human myosin-3A and -3B display an actin-translocating activity.

The MYO3A gene expression is highly restricted, with the strongest level in the retina and cochlea. Myosin-3A interacts with the glutamate receptor ionotropic ^{N}methyl^{D}aspartate-like protein-1A, or GRINL1A complex gene locus product [251]. Isoform Myo3b is a shorter subclass-3 myosin that is expressed in the kidney, retina, and testis. Myosin-3B has several splice variants containing either one or two calmodulin binding (IQ) motifs in the neck domain and one of 3 predominant tail variations: a short tail ending just past the second IQ motif or 2 alternatively spliced longer tails [294].

5.1.10 Kinases of the GCK Subfamily 8

Members of this kinase set include rat Thousand and one amino acid protein kinase (TAO1) and kinase from chicken (KFCb). In humans, these kinases are preferentially referred to as TAOK1, TAOK2, and TAOK3 [251]. However, the straightforward names are mitogen-activated protein kinase kinase kinases MAP3K16, MAP3K17, and MAP3K18 (Sect. 6.5). Kinases of the GCK subfamily 8 have many similarities, such as a N-terminal kinase domain, a serine-rich region, and a coiled-coil configuration within the C-terminus. All subfamily-8 kinases are able to activate the P38MAPK pathway via the specific activation of MAP2K3, in addition to MAP2K6.

Whereas TAOK3 is ubiquitous with high levels in hematopoietic cells such as peripheral blood leukocytes and related tissues (thymus and spleen) and significantly in some organs (prostate, testes, and pancreas; at moderate to low level elsewhere),

5.1 Superfamily of Ser/Thr Sterile-20-Related Kinases

concentrations of TAOK1 and TAOK2 are small in peripheral blood leukocytes. Concentration of TAOK1 is high in testes and significant in skeletal muscle and placenta, but undetectable in the liver, kidney, and ovary. Enzyme TAOK2 is ubiquitous with high levels in the brain and testes.

Prostate-Derived Sterile-20-like Kinases

Human prostate-derived sterile 20-like kinase PSK1 influences actin cytoskeletal organization and binds to microtubules to regulate their arrangement and stability. On the other hand, PSK2 lacks a microtubule-binding site. It activates Jun N-terminal kinase and induces apoptotic morphological changes (cell contraction, membrane blebbing with caspases, and apoptotic body formation). The PSK isoforms are members of the GCK subfamily-8.[34]

5.1.10.1 Thousand and One Amino Acid Kinase TAOK1 (MAP3K16)

The Ste20-related, GCK-like, protein Ser/Thr kinase Thousand and one kinase TAOK1 (or MAP3K16)[35] binds and regulates microtubules. It is highly expressed in the brain as is the other TAOK family member MAP3K17. It activates MAP2K3, MAP2K4, and MAP2K6 in vitro, but only MAP2K3 in cells [295].

Apoptotic stimuli increase the catalytic activity of MAP3K16 and JNK kinases. The latter is activated by MAP3K16 enzyme. Kinase MAP3K16 regulates apoptotic morphology that involves dynamic reorganization of the actin cytoskeleton via its downstream targets JNK and RoCK1 [296]. Protein MAP3K16 indeed stimulates the cleavage of RoCK1 kinase. Activity of RoCK1 is required for MAP3K16 to cause cell contraction and membrane blebbing. Enzyme MAP3K16 is itself a substrate for caspase-3.

5.1.10.2 Thousand and One Amino Acid Kinase TAOK2 (MAP3K17)

The Ste20-related, GCK-like, protein Ser/Thr kinase Thousand and one kinase TAOK2 (or MAP3K17)[36] lacks a microtubule-binding domain. It specifically interacts with MAP2K3 and MAP2K6 [251]. It can activate JNK and P38MAPK, but not ERK1 and ERK2 [297].

---

[34] Rat homologs of PSK1 and PSK2 are thousand and one amino acid protein kinases TAO2 and TAO1, respectively.

[35] A.k.a. microtubule affinity-regulating kinase kinase (MARKK or MAR2K), prostate-derived Sterile-20-like kinase PSK2, and kinase from chicken homolog KFCb.

[36] A.k.a. prostate-derived Sterile-20-like kinase PSK1 (or PSK), kinase from chicken homolog KFCc, as well as TAO1, and TAO2.

5.1.10.3 Thousand and One Amino Acid Kinase TAOK3 (MAP3K18)

The Ste20-related, GCK-like, protein Ser/Thr kinase Thousand and one kinase TAOK3 (or MAP3K18)[37] Instead of stimulating the JNK pathway, MAP3K18 hampers the basal activity of JNK. Its inhibition of JNK signaling appears between EGFR and small GTPases Rac1 and CDC42 [298]. Kinase MAP3K18 thus demonstrates that a Ser/Thr kinase can be inhibited by activated RTKs such as EGFR, as it impedes the JNK pathway.[38] Kinase MAP3K18 also interacts with endoplasmic reticulum to nucleus signaling protein ERN1 and TNF receptor-associated factor TRAF2 [251].

5.2 AGC Superfamily

The AGC superfamily of protein Ser/Thr kinases stands for protein kinases-A, -G, and -C. Members of the AGC superfamily display strong homology to PKA, PKC, and PKG kinases. About 70 among the 518 protein kinases encoded by the human genome are members of the AGC superfamily. All share a conserved catalytic kinase domain.

Members of the AGC superfamily are regulated by various mechanisms, but most require phosphorylation and conformational changes to be activated. In addition to the kinase domain, other functional regions allow regulation of the activity and localization of AGC kinases. Activation of many AGC kinases involves phosphorylation of 2 regulatory motifs: the activation segment in the catalytic domain and hydrophobic motif in a non-catalytic region. Several AGC kinases also contain another phosphorylation site that promotes their integrity, the turn motif. Some substrates are phosphorylated by a single AGC kinase, others by multiple AGC kinases.

The superfamily of AGC kinases includes cAMP-dependent protein kinase-A, protein kinases-B, several types of protein kinase-C, cGMP-dependent protein kinase-G, ribosomal S6-kinases S6K (^{P70}RSK) and RSK (^{P90}RSK), mitogen- and stress-activated protein kinases (MSK), serum- and glucocorticoid-induced protein kinase (SGK), 3-phosphoinositide-dependent kinase PDK1, and G-protein-coupled receptor kinases (GRK) [299], as well as nuclear Dbf2-related kinase (NDR or STK38), large tumor suppressors LaTS1 and LaTS2 of the Hippo pathway, members of Rho-activated kinases, such as Rho-associated coiled-coil-containing protein ki-

---

[37] A.k.a. dendritic cell-derived protein kinase (DPK), JNK-inhibitory kinase (JIK), and kinase from chicken homolog KFCα.

[38] Numerous molecules preclude the JNK signaling pathway, such as glutathione S-transferase Pi, scaffold JNK-inhibitory protein JIP1, thioredoxin that inhibits MAP3K5, and G-protein pathway suppressor GSP2 that blocks the JNK activation by TNFα.

5.2 AGC Superfamily

Table 5.7. Kinases of the AGC superfamily and selected substrates (**Part 1**; Source: [300]; ATF: activating transcription factor; CREB: cAMP responsive element-binding protein; DMPK: myotonic dystrophy-associated protein kinase; GRK: G-protein-coupled receptor kinase; LaTS, large tumor suppressor; LIMK: LIM domain kinase; MAST: microtubule-associated Ser/Thr kinase; MRCK: myotonic dystrophy kinase-related CDC42-binding kinase; MSK: mitogen- and stress-activated kinase; PPp1R12a: protein phosphatase-1 regulatory subunit-12A [or MYPT1: myosin phosphatase-1]; NDR: nuclear Dbf2-related kinase [STK38]; NFκB: nuclear factor-κB; NHE3: Na^+–H^+ exchanger; NuR: orphan nuclear receptor; PDGFR, platelet-derived growth factor receptor; PDK1: 3-phosphoinositide-dependent kinase; PP1: phosphatase-1; PTen, phosphatase and tensin homolog deleted on chromosome 10; RSK: P90 ribosomal S6 kinase; S6K: P70 ribosomal S6 kinase; SGK: serum- and glucocorticoid-induced protein kinase; SRF: serum response factor; STK: Ser/Thr protein kinase; YAP: yes-associated protein).

Kinase	Substrates
DMPK	Phospholamban, phospholemman, PPp1R12a, SRF
GRK	Ezrin, GPCRs, PDGFR, P38MAPKα, synucleins
LaTS	YAP1
MAST	NHE3, PTen
MRCK	LIMK, PP1δ
MSK	ATF1, CREB, histone-3, STK11, NFκB, NuR77
NDR (STK38)	
PDK1	PKB, PKC, PKN1, RSK, S6K, SGK

nase (RoCK), citron Rho-interacting kinase (CRIK), myotonic dystrophy-associated protein kinase (DMPK), and CDC42-binding protein kinase-α(CDC42BPα)[39] (Tables 5.7 to 5.9) [300].

Members of the AGC superfamily have important functions in normal and pathological cellular processes, as they regulate cell survival, metabolism, motility, growth, division, and differentiation. These kinases indeed form a set of enzymes stimulated by growth factors and insulin. Multiphosphorylation primes kinase activation. However, occupancy of the nucleotide-binding site by phosphodonor ATP or, paradoxically, ATP-competitive inhibitor, induces hyperphosphorylation of PKB and PKC [301, 302]. Nucleotide binding (intrinsic mechanism), rather than ATP-dependent phosphotransfer launched by autophosphorylation (extrinsic mechanism that arises as a functional consequence of intrinsic kinase activity), stimulates enzymatic activity. Once phosphorylated, PKB and S6K intervene in PI3K pathway, PKC in calcium signaling, RSK and MSK in MAPK pathway, and PRK in Rho signaling.

---

[39] A.k.a. myotonic dystrophy kinase-related CDC42-binding kinase MRCKα, Sugen kinase SGK494, protein kinase-X (PKX1 or PRKX) and -Y (PRKY), microtubule-associated Ser/Thr kinases (MAST), and group-E pseudokinases ribosomal S6 kinase-like RSKL1 and RSKL2.

Table 5.8. Kinases of the AGC superfamily and selected substrates (**Part 2**; Source: [300]; AID: activation-induced cytidine deaminase; AS160: Akt [PKB] substrate of 160 kDa; BAD: BCL2 antagonist of cell death; CKI: cyclin-dependent kinase inhibitor; CFTR: cystic fibrosis transmembrane conductance regulator; CREB: cAMP responsive element-binding protein; FHoD: formin homology domain protein; FOxO: forkhead box protein O; GP: glycoprotein; GRK: G-protein-coupled receptor kinase; GSK: glycogen synthase kinase; HDM: human double minute; HSL: hormone-sensitive lipase; IRS: insulin receptor substrate; $K_{Ca}2.2$: K^+ intermediate–small conductance Ca^{++}-activated channel [SK2]; MARCK: myristoylated Ala-rich C-kinase substrate; NFAT: nuclear factor of activated T cells; PDCD: programmed cell death; PDE: phosphodiesterase; PFK2: 6-phosphofructo-2-kinase; PGC: PPARγ coactivator; PKA: cAMP-dependent protein kinase-A; PKC: protein kinase-C; PKG: cGMP-dependent protein kinase-G; PKN: protein kinase-C-related kinase; PLC: phospholipase; PRAS40: Pro-rich Akt [PKB] substrate of 40 kDa [PKB1S1 or Akt1S1]; PRK: protein kinase; PTPn6: protein Tyr phosphatase non-receptor-6 [a.k.a. SH2 phosphatase SHP1]; SKP: S-phase kinase-associated protein; TRPC: transient receptor potential cation channel; TSC: tuberous sclerosis complex; VASP: vasodilator-stimulated phosphoprotein; WNK: with no K [Lys] kinase).

Kinase	Substrates
PKA	AID, BAD, calpain-2, CFTR, CREB, GRK2, GSK3, HSL, $K_{Ca}2.2$, NFAT2, VASP
PKB	AS160, BAD, FOxO3, GSK3, HDM2, CKI1b, PFK2, PGC1α, PDCD4, PRAS40, bRaf, SKP2, TSC2, WNK1
PKC	Adducin-1, GP130, GRK2, GSK3, IRS1, MARCK, PKD, PDE3a, cRaf, PTPn6
PKG	FHoD1, PDE5, PLCβ3, TRPC3, VASP
PRK	Polycystin-1
PKN1	Histone-3, MARCK, Tau

5.2.1 Nuclear DBF2-Related Protein Kinases and Large Tumor Suppressors

Nuclear DBF2-related (NDR) protein kinases and large tumor suppressors (LaTS) constitute a set (LATS and NDR families) of the AGC superfamily of protein kinases. They control morphological changes, mitosis, cytokinesis, and apoptosis. Kinases of the NDR family are regulated by phosphorylation by kinases and dephosphorylation by phosphatase.

There are 4 related kinases: NDR1, NDR2, LaTS1, and LaTS2.[40] Kinases NDR and LaTS possess an auto-inhibitory sequence in the kinase domain and an N-terminal regulatory motif. They are activated upon phosphorylation of their activation segment and hydrophobic motif.

Kinases NDR1 and NDR2 contribute to centrosome duplication. Kinases LaTS1 and LaTS2 participate in the Hippo pathway (Sect. 10.7) that controls cell size in organs [300].

---

[40] NDR1 is also labeled as protein Ser/Thr kinase-38 (STK38), and NDR2 as STK38L.

5.2 AGC Superfamily

Table 5.9. Kinases of the AGC superfamily and selected substrates (**Part 3**; Source: [300]; ATF: activating transcription factor; BAD: BCL2 antagonist of cell death; CKI: cyclin-dependent kinase inhibitor; CCT: chaperonin-containing T-complex polypeptide; CRMP: collapsin response mediator protein; DAPK: death-associated protein kinase; eEF2K: eukaryotic elongation factor-2 kinase; eIF: eukaryotic initiation factor; ENaC: epithelial Na^+ channel; FOxO: forkhead box protein O; GSK: glycogen synthase kinase; IRS: insulin receptor substrate; LIMK: LIM domain kinase; MAD: MAX dimerization protein; MARCK: myristoylated Ala-rich C-kinase substrate; MAP3K: mitogen-activated protein kinase kinase kinase; MBS: myosin-binding subunit; MLC: myosin light chain; PPp1R12a: protein phosphatase-1 regulatory subunit-12A [or MyPT1: myosin phosphatase-1]; NDRG: N-Myc downregulated protein; NEDD4L: neural precursor cell expressed, developmentally downregulated 4-like isoform-2; NFAT: nuclear factor of activated T cells; PDCD: programmed cell death; Raptor: regulatory associated protein of TOR; RoCK: Rho-associated coiled-coil-containing protein kinase; RPS6, ribosomal protein-S6; RSK: P90 ribosomal S6 kinase; S6K: P70 ribosomal S6 kinase; SGK: serum- and glucocorticoid-induced protein kinase; SKAR: S6K1 Aly/REF-like target; STK: Ser/Thr protein kinase; TOR: target of rapamycin; TSC: tuberous sclerosis complex).

Kinase	Substrates
RoCK	Adducin-1, CRMP2, ezrin, radixin, moesin, LIMK, MARCK, MBS, MLC, PPp1R12a, RhoE, vimentin
RSK	ATF4, BAD, CCT, DAPK, eEF2K, eIF4b, GSK3β, STK11, NFAT3, MAD1, CKI1b, Raptor, RPS6, TSC2
S6K	CCT, eEF2K, eIF4b, GSK3β, IRS1, MAD1, PDCD4, RPS6, SKAR, TOR
SGK	ENaC, FOxO3, MAP3K2, NDRG1, NEDD4L

Kinases of the NDR and LATS families interact with Mps-one binder (MOB) proteins to promote autophosphorylation of the activation segment and partial activation [300]. Six MOB proteins (MOB1a–MOB1b, MOB2, and MOB3a–MOB3c) are components of signaling pathways that control mitotic exit and centrosome duplication, hence cell proliferation, as well as apoptosis. Subtype MOB1a activates NDR1 and NDR2 kinases as well as LaTS1 and LaTS2 kinases [300]. Protein MOB2 binds to NDR1 and NDR2, but not to LaTS1 enzyme. In fact, 3 among the 6 MOB family members neither bind to nor activate NDR1, NDR2, LaTS1, or LaTS2 kinases [303]. Molecule MOB2 competes with MOB1a and MOB1b for NDR binding. Moreover, MOB2 inhibits NDR1 and NDR2 kinases, unlike MOB1a and MOB1b isoforms.

Kinases NDR1 and NDR2 are also regulated by calcium-binding proteins of the S100 family following Ca^{++} import. Maximal activation occurs by phosphorylation of the hydrophobic motif of NDR and LATS by sterile-20-like kinases. Kinases MST1 (or STK4) and MST2 (or STK3) phosphorylate LaTS1 and LaTS2, whereas MST3 (or STK24) phosphorylates NDR1 and NDR2 [300]. The Ras association domain-containing protein RASSF1 activates MST1 and MST2.

5.2.2 Phosphoinositide-Dependent Kinase-1

Phosphoinositide-dependent Ser/Thr kinase-1 (PDK1) pertains to the AGC superfamily. It has multiple substrates that enable the transduction of extracellular signals. It participates in the regulation of the activity of protein kinases during cell signaling. Homodimerization of PDK1 is spatially and temporally regulated to control its activity. The spatial distribution of PDK1 homodimers differs from that of PDK1–PKB heterodimers.

Constitutively active PDK1 phosphorylates the activation segment of at least 23 AGC kinases: protein kinase-B, ribosomal S6K and RKS kinases, various PKC isoforms, PKC-related protein kinases PKN1 and PKN2, P21-activated kinase, and serum and glucocorticoid-regulated kinase (SGK) [304].

Kinase PDK1 possesses a pleckstrin homology (PH) domain at its C-terminus that binds phospholipids with high affinity. It preferentially binds phosphatidylinositol (3,4,5)-trisphosphate.[41] It also has a high affinity for phosphatidylinositol (3,4)-bisphosphate[42] and (4,5)-bisphosphate.[43] It interacts with I(1,3,4,5)P$_4$ and I(1,3,4,5,6)P$_5$ with submicromolar dissociation constant.

Activity of PDK1 is governed by conformational changes. This activity is focused on protein–PDK1 and lipid–PDK1 interactions that occur with a given dynamics. Post-translational modifications such as phosphorylation of Ser, Thr, and Tyr residues, cause conformational changes, can stabilize the protein, and enable tethering to other proteins or specific domains of the plasma membrane and nucleus.

As PDK1 has a high basal level of activity, it is considered constitutively active, ready to phosphorylate substrates that have been targeted by various extracellular signals, owing to its intrinsic ability to transautophosphorylate (Ser241) its own activation segment.

Phosphorylation causes a mild (2- to 3-fold) increase in activity, but provokes membrane recruitment of PDK1 enzyme. Five phosphorylated Ser residues have been identified in PDK1, with autophosphorylation site (Ser241), as well as Tyr residues (Tyr9, Tyr373, and Tyr376) by ReT receptor Tyr kinase as well as cytosolic Abl, FAK2, and Src kinases in response to insulin, angiotensin-2, and oxidative stress [304]. Kinase PDK1 is sequestered by 14-3-3 proteins. It associates with HSP90 that increases its stability.

Homodimerization influences the behavior of intracellular kinases, as it can provoke the nuclear translocation and modify (elevate or lower according to the enzyme type) the kinase activity. Homodimerization of PDK1 depends on its PH domain [305]. A basal pool of PDK1 homodimers exist in the cytosol and at the plasma membrane. Upon stimulation such as PI3K-based signaling, the plasmalemmal concentration of PDK1 rises. The dimer is not required for autophosphorylation.

---

[41] Dissociation constant $K_d = 1.6$ nmol.
[42] $K_d = 5.2$ nmol.
[43] $K_d = 24$ nmol.

Kinase PDK1 may exist as a conformational isomer, or conformer, loaded with ATP, that enable the autophosphorylation (Ser241) to destabilize the homodimer and become potentially active upon dissociation of the dimer. Increased amount of active monomer can support an augmented interaction with its PKB substrate. Monomeric PDK1 is the active conformation of the kinase; PDK1 dimer correponds to an autoinhibitory conformation [305]. In stimulated cells, the translocation of PDK1 to the plasma membrane that enables PKB activation is also associated with an elevated population of PDK1 homodimers. The dimerizarion may result from a negative feedback loop.

Conformational change caused by binding to 3-phosphoinositides, especially $PI(3,4,5)P_3$, promotes PKB phosphorylation by PDK1 [300]. Colocalization of PKB and PDK1 is facilitated not only by phosphoinositides, but also scaffold proteins, such as growth factor receptor-bound protein GRB14 and coiled-coil and C2 domain-containing protein CC2D1a,[44] and TNF receptor-associated factor TRAF6 Ub ligase. In addition, it binds to A-kinase-anchoring protein for RSK activation [300].

In cardiomyocytes, PDK1 coordinates cell survival and β-adrenergic response via the PI3Kγ–PDK1 axis [306]. Kinase PDK1 prevents cardiomyocyte apoptosis and preserves responsiveness to β-adrenergic stimulation that controls cardiac chronotropy and inotropy, as it avoids β1-adrenoceptor endocytosis primed by β-adrenoceptor kinase-1 (βARK1 or GRK2). It actually accommodates β-adrenergic response to prevent cardiac decompensation. Striated muscle-specific PDK1, especially myocardium-specific PDK1, is also necessary for cardimyocyte growth and its response to hypoxia.

5.2.3 Protein Kinase-A Family

Cyclic adenosine monophosphate (cAMP)-dependent protein kinases, or protein kinases-A, are tetrameric, inactive holoenzyme composed of 2 regulatory (PKA_r) and 2 catalytic (PKA_c) subunits. Binding of cAMP to PKA_r releases PKA_c from PKA_r. Once the holoenzyme is dissociated and PKA_c liberated, PKA is able to phosphorylate its substrates.

Ubiquitin ligases *Praja* (Pja1–Pja2) are binding partners of PKA enzyme.[45] An elevated cAMP concentration promotes ubiquitination of the regulatory subunit by Praja-2 ligase and its subsequent degradation, thereby improving PKA signaling [307].

The catalytic subunit of protein kinase-A has 3 main conformations (apo, intermediate, and closed). The transition from apo (major state in the absence of ligand) to intermediate and from intermediate to closed conformation is triggered

---

[44] A.k.a. five repressor element under dual repression-binding protein FREUD1.
[45] A.k.a. RING finger domain-containing proteins RNF70 and RNF131.

by ligand binding [308]. The catalytic subunit of protein kinase-A is inhibited by 2 types of regulatory subunits whether they require adenosine triphosphate or not. Protein kinase-A can thus have isoform-specific activators and antagonists for PKA [309]. Kinase PKA also requires phosphorylation of its activation segment (Thr197) by PDK1 kinase [300].

5.2.3.1 PKA Subtypes

Two major isoforms exist with different regulatory subunits: type-1 and type-2 cAMP-dependent protein kinases.

PKA1

Protein kinase-A1 possesses type-1 regulatory subunits that dimerize. These subunits have a very low affinity for A-kinase-anchoring proteins. It has a higher cAMP-binding affinity, thereby responding to a lower cAMP concentration in the cytoplasm. Protein kinase-A1 is mainly free in the cytoplasm.

PKA2

The regulatory dimer of protein kinase-A2 is made up of type-2 regulatory subunits. These subunits have a much higher affinity for AKAP proteins, thereby localizing to more precise sites. The AKAP scaffolds are involved in the spatial organization of signaling pathways, as they bring PKA into contact with its substrates. Specificity of PKA phosphorylation is, at least partly, achieved by the coupling of PKA_r to AKAPs, which tether to different types of substrates.

5.2.3.2 PKA Effects

Activation of Gs-coupled receptors mobilizes localized pulses of cAMP messenger upon stimulation of adenylate cyclases. Protein kinase-A regulate specific and non-specific cellular processes. Specific substrates include AMPA-type glutamate receptor, fructose (2,6)-bisphosphatase, salt-inducible kinase-2 (inhibition), hormone-sensitive lipase, phosphorylase kinase, transcription factor cAMP response element-binding protein (CREB), protein phosphatase PP1 inhibitor I1, cystic fibrosis transmembrane conductance regulator (CFTR), aquaporin-2, and vasodilator-stimulated phosphoprotein (VASP).

PKA and Gene Transcription

Kinase PKA contributes to the regulation of gene transcription primed from G-protein-coupled receptors. It phosphorylates (activates) transcription factors CREB and nuclear factor-κB [300]. Its subcellular localization is directed by AKAP scaffolds that also determine the magnitude and duration of PKA signaling.

PKA and Transfer of Signaling Mediators

The antidiuretic hormone vasopressin stimulates V_2 receptors (or AVPR2) in principal cells of collecting ducts of the nephron and causes the translocation of TRPC3 channels to the apical membrane via the activated AC–cAMP–PKA cascade [310].

Acute adaptation of epithelial sodium channel at the apical surface of epithelial cells of the nephron cortical collecting duct results mainly from exocytosis and ENaC insertion into the apical membrane owing to soluble ^{N}ethylmaleimide-sensitive factor attachment protein receptors and SNARE-binding protein complexin. A rapid stimulation of sodium transport is elicited by vasopressin that increases the intracellular cAMP concentration to activate protein kinase-A [311].

Phosphorylation by PKA of aquaporin-2 (Ser256) is necessary, but not sufficient, for Aqp2 translocation and insertion into the plasma membrane [312]. On the other hand, prostaglandin-E2 and dopamine induce Aqp2 internalization independently of its dephosphorylation, but upon activation of cAMP production.

In cardiomyocyte, myopodin is an actin-bundling protein that shuttles between the nucleus and Z-disc according to the cell differentiation status and stress context. Myopodin forms a Z-disc signaling complex with α-actinin, PP3, Ca^{++}–calmodulin-dependent kinase CamK2, muscle-specific A-kinase anchoring protein AKAP6, and myomegalin (PDE4dIP) [313]. In myoblasts, phosphorylation of myopodin by protein kinase-A or CamK2 causes binding of myopodin to 14-3-3 protein, importin-α binding, and nuclear import. Conversely, dephosphorylation of myopodin by PP3 abrogates 14-3-3β binding. In adult cardiomyocytes, PKA activation or PP3 inhibition releases myopodin from the Z-disc and induces its nuclear import [313].

Density of cardiac $Na_V1.5$ channel on the sarcolemma of cardiomyocytes rises upon activation of protein kinase-A [314]. In the absence of PKA stimulation, $Na_V1.5$ channels reside in the perinuclear region as well as beneath the cell surface. Activated PKA promotes $Na_V1.5$ transfer from both regions to the plasma membrane.

Voltage-gated $K_V4.2$ channel is the α subunit of the channel that generates the A-type K^+ current in CA1 pyramidal neurons of the hippocampus to regulate dendritic excitability and adaptivity (i.e., synaptic change in strength — the so-called synaptic plasticity).[46] Phosphorylation of $K_V4.2$ by PKA enables $K_V4.2$ transfer via its interaction with K^+ channel-interacting protein KChIP4a [315].

The regulated incorporation of AMPA-type glutamate receptors (Vol. 3 – Chap. 2. Membrane Ion Carriers) into synapses for synaptic adaptivity implicates PKA that phosphorylates ^{AMPA}Glu subunits GluR4 and GluR1 and calcium–calmodulin-dependent kinase [316].

PKA and Cell Fate

Protein kinase-A influences migration and apoptosis of neutrophils and other cell types. Various inflammatory mediators accelerate or delay neutrophil apoptosis, thereby controlling the neutrophil number in tissues. Activation of type-1 cAMP-dependent protein kinases, but not type-2, reduces caspase-3-mediated apoptosis in neutrophils [317].

5.2.4 Protein Kinase-X1

Protein kinase-X1 (PKX1 or PrKX) is a cAMP-dependent protein kinase of the AGC superfamily of Ser/Thr protein kinases. It is related to PKA kinase. It is encoded by the Pkx1 gene on X chromosome. On chromosome Y, a gene encodes for homologous protein PKY1. Kinase PKX1 is synthesized in most adult and embryonic tissues [318]. However, it is mainly expressed during embryogenesis with high levels in the brain, heart, kidney, and lung. Concentration of PKX1 in adult tissue compared with 20-wk-old human fetal tissue is much lower (brain and kidney) or non-detectable (liver and heart). Kinase PKX1 is also expressed in macrophages and granulocytes and contributes to their maturation.

Its substrates include PKA_{r2} regulatory subunit, SMAD6, and TRPP1, or G-protein-coupled receptor-like polycystin-1. It is inhibited by the heat stable protein kinase inhibitor (PKI). Under unstressed conditions, it forms a heterotetrameric holoenzyme with PKA_{r1} regulatory dimer. It can be released from regulatory subunits upon binding of cAMP to these regulatory subunits.

Kinase PKX1 autophosphorylates. Autophosphorylation of PKX1 and phosphorylation is inhibited by PKA regulatory subunit isoforms $PKA_{r1\alpha}$ and $PKA_{r1\beta}$. The Prkx gene is expressed during maturation of monocytes, macrophages, and granulocytes under control of $PKC\beta$ [319].

---

[46] In rheology, plasticity defines a material that undergoes a non-reversible deformation in response to applied loads. In this context, plasticity is related to the lack of adjustability. This term is thus avoided.

5.2.5 Protein Kinase-Y1

Protein kinase Y-linked (PKY1 or PRKY) is an Y-homolog of protein kinase PRKX. This cAMP-dependent Ser/Thr protein kinase is encoded by a gene that is located on chromosome Y.

5.2.6 Protein Kinase-B Family

Protein kinase-B (or Akt) has 3 isoforms (PKB1–PKB3 or PKBα–PKBγ). It serves as a mediator of growth factor signaling for cell survival and proliferation. Constitutively active PKB prevents apoptosis upon growth factor withdrawal, as it: (1) maintains transporters of and receptors for extracellular substances (glucose, amino acids, low-density lipoprotein, etc.), even in the transient absence of growth factors and (2) phosphorylates (inactivates) several pro-apoptotic mediators. Tumor cells do not initiate apoptosis upon growth factor withdrawal, as they augment PI3K and reduce PTen activity, so that PKB remains hyperactive.

Protein kinase-B phosphorylates numerous regulators that control: (1) glucose transport, glycolysis, glycogen synthesis, and suppression of gluconeogenesis to protein synthesis; (2) cell growth (size enlargement); (3) cell cycle progression; (4) cell survival, repression of apoptosis, and preservation of mitochondrial integrity.

In the absence of growth factors, inactive PKB is located in the cytoplasm. Kinase PKB is activated by liganded growth factor receptors via phosphatidylinositol 3-kinase. Because it phosphorylates phosphatidylinositol (4,5)-bisphosphate, activated PI3K generates a pool of plasmalemmal phosphatidylinositol (3,4,5)-trisphosphate that triggers PKB recruitment from the cytosol to plasma membrane and binds PKB. However, PKB linkage to plasma membrane is necessary but not sufficient to induce its hyperphosphorylation. Recruitment of PKB to the plasma membrane and its activation depend on cholesterol- and sphingolipid-rich PIP_3-containing membrane nanodomains (rafts) [320]. Sphingolipid and cholesterol repress lateral diffusion of raft-associated proteins. Upon plasmalemmal contact, PKB is phosphorylated (activated) by phosphoinositide-dependent kinases PDK1 and PDK2 also located in membrane rafts. Kinase PKB is then released from the plasma membrane and phosphorylates both cytosolic and nuclear targets.

Protein kinase-B isoforms have signaling specificity in the regulation of metabolism, cell growth, proliferation, and apoptosis. Insulin activates both PKB1 and PKB2 in adipocytes. At the plasma membrane, PKB2 regulates incorporation of

GluT4 glucose transporter, as it phosphorylates 160-kD GTPase-activating protein RabGAP Akt (PKB) substrate AS160 for GluT4 membrane recruitment [321].[47]

5.2.6.1 PKB Activation

Full PKB activation requires successive events: (1) membrane translocation that depends on phosphatidylinositol (1,4,5)-trisphosphate and (2) phosphorylation by phosphoinositide-dependent kinase-1 (Thr308) and target of rapamycin complex-2 (Ser473). These events can be primed by ATP-competitive PKB inhibitors upon occupancy of the ATP-binding domain that produces a PKB conformational change, thereby promoting membrane association and access of cognate residues (Thr308 and Ser473) for phosphorylation [301].[48]

Activation of PKB requires its phosphorylation on the activation loop by PDK1 kinase and on the hydrophobic motif by the TORC2 complex. In addition, IKKε and tumor-necrosis factor receptor-associated factor family member (TRAF)-associated NFκB (TANK)-binding kinase TBK1 phosphorylate PKB on both the hydrophobic motif and the activation loop in reponse to PI3K signaling [323].

5.2.6.2 Nuclear PKB

In addition to the plasma membrane and cytosol, the nucleus and mitochondria represent 2 other subcellular target sites for PKB action. Protein kinase-B can be activated at the plasma membrane or in the nucleus, eventually after PI3K and PDK1 translocation. PKB Targets that cause cell proliferation or prevents apoptosis reside within the nucleus (nuclear pool of PI(3,4,5)P$_3$ generated by type-1 PI3K, regulatory GTPase PIKE, PI(3,4,5)P$_3$-specific phosphatases SHIP2 and PTen, nucleophosmin,[49] S6K, forkhead transcription factors, adaptor BRCA1, cyclin-dependent kinase inhibitors CKI1a and CKI1b, ubiquitin ligase double minute DM2, kinases PIM1, Wee, and ERKs, and tuberin) [324]. Nuclear PKB accumulation

---

[47] GTPase-activating protein AS160 (a.k.a. TBC1 (TRE2/USP6–BUB2–CDC16) domain family member TBC1d4) is phosphorylated (inhibited) by PKB to activate Rab2a, Rab8a, Rab10, and Rab14 GTPases to translocate glucose transporter GLUT4 to the plasma membrane [322].

[48] Rapamycin that precludes the PI3K–PKB-TORC1 pathway by phosphorylating TSC2, can also, in tumors, activate PKB by feedback regulation of insulin-like growth factor-1 signaling. ATP-competitive inhibitors do not operate via feedback such as the PKB–TORC1–S6K axis (extrinsic feedback). Inhibitor-induced (intrinsic feedback) hyperphosphorylated PKB can become again active and phosphorylates its effectors after dissociation of ATP-competitive PKB inhibitor. However, phosphatases (that target Thr308P and Ser473P) can be active in cells.

[49] A.k.a. Nucleolar phosphoprotein B23, and numatrin.

can serve to modulate localization of these effectors. Moreover, PKB regulator can translocate to the nucleus, such as PI3K, PDK1, and PTen. Kinase PI3K resides in the nucleus upon stimulation by insulin and growth factors (IGF1, PDGF, and NGF).

In cardiomyocytes, in response to cardioprotective stimuli such as atrial natriuretic peptide, both PI3K and PDK1 concomitantly accumulate in the nucleus. Cardioprotective and proliferative PKB effects rely on PIM1 kinase[50] (Sect. 5.4.8) of the calmodulin-dependent protein Ser/Thr kinase-related (CAMK) superfamily [324]. In general, PIM1 production (upon stimulation by hormones, cytokines, and mitogens) and loss promotes cell survival and death, respectively.

Related isoforms, PIM2 and PIM3, have overlapping as well as distinct functions in cell growth and survival. Furthermore, mitochondrial PIM1 preserves mitochondrial integrity.

5.2.6.3 Mitochondrial PKB

Activated PKB translocates not only to the nucleus, but also to mitochondria. In mitochondria, PKB inhibits opening of the permeability transition pore to maintain mitochondrial integrity [324].[51] Protein kinase-B phosphorylates BAD (Ser136; Vol. 2 – Chap. 4. Cell Survival and Death) that then dissociates from BCLxL to inhibit BAX–BAK mediated pore formation. In addition, PKB phosphorylates (inhibits) BAX as well as GSK3β to prevent opening of permeability transition pore. Protein kinase-B also phosphorylates (inhibits) mitochondrial serine peptidase HTRa2[52] that is released into the cytosol in response to stress, where it binds to and inhibits inhibitor of apoptotic proteins to preclude their caspase inhibition [324] (Table 5.10). Furthermore, PKB phosphorylates (inactivates) procaspase-9.

Mitochondrial hexokinases — ubiquitous HK1 and insulin-sensitive cell-specific HK2 — preserve mitochondrial integrity and protect against oxidative stress. Moreover, HK2 impedes BAX to bind to mitochondria, thereby releasing cytochrome-C and apoptosis-inducing factor with subsequent activation of caspase-9 and -3 and apoptosis [324].[53]

---

[50]In cardiomyocytes, PIM1 synthesis is reduced within a few weeks after birth, but reappears in some pathological circumstances to elicit cardioprotection [324]. Many PIM1 target molecules regulate cell cycle progression and apoptosis. Furthermore, PIM1 has positive inotropic effects and impedes maladaptive hypertrophic remodeling.

[51]The permeability transition pore is composed of voltage-dependent anion channel, adenine nucleotide translocase, cyclophilin-D, and hexokinase-2.

[52]High temperature requirement-A endopeptidase in the periplasmic space of bacteria is required for bacterial thermotolerance. This protein removes damaged and denatured proteins in bacteria exposed to elevated temperature or oxidative stress. Mammalian homolog of HTRa is also called Omi.

[53]Once cytochrome-C is released from mitochondria, it forms apoptosome, an APAF1–dATP–procaspase-9 complex (Vol. 2 – Chap. 4. Cell Survival and Death). Deoxyadenosine triphosphate (dATP) is a nucleotide precursor. Protein APAF1 contains a dATP as a cofactor. Apoptosome forms when cytochrome-C binds to APAF1 and induces dATP hydrolysis into dADP with subsequent

Table 5.10. Protein kinase-B and mitochondrial integrity (Source: [324]). Protein kinase-B preserves mitochondrial integrity against stress, as it stimulates or inhibits various agents at different subcellular sites (AIF: apoptosis-inducing factor; BAD: BCL2 antagonist of cell death; BAK: BCL2-antagonist/killer-1; BAX: BCL2-associated X protein; GSK: glycogen synthase kinase; HK: hexokinase; IP_3R: IP_3 receptor; NCX: Na^+–Ca^{++} exchanger; NHE: Na^+–H^+ exchanger). Protein kinase-B favors Ca^{++} outflux and hampers Ca^{++} influx, as Ca^{++} overload causes opening of permeability transition pores. Protein kinase-B phosphorylates (inactivates) NHE that extrudes H^+ and imports Na^+. Na^+ influx precludes Ca^{++} outflux through NCX and can even stimulate Ca^{++} influx (reverse-mode). Protein kinase-B also phosphorylates (inhibits) type-1 IP_3R.

Stimulation	Inhibition
	Mitochondrion
HK	GSK3β
	AIF, BAD, BAX, caspase-9, HTRa
	Plasma membrane
NCX	NHE
	Endoplasmic reticulum
	IP_3R

In cardiomyocytes, BCL2-related BH3-only protein BNIP3L[54] localizes at both the sarcoplasmic reticulum and mitochondria. Pro-apoptotic protein BNIP3L increases Ca^{++} content of the sarcoplasmic reticulum up to Ca^{++} overload. In mitochondria, BNIP3L induces mitochondrial outer membrane permeabilization in coordination with BAK and BAX [324].

5.2.6.4 PKB Effectors

Protein kinase-B mediates survival and proliferation signaling via various effectors. Isozymes PKB1 to PKB3 prime non-redundant signalings. Protein kinase-B1 can block cancer cell migration by inhibition of transcription factor nuclear factor of activated T cells and extracellular signal-regulated kinase [326]. Hyperproliferation and anti-apoptotic activities caused by stimulation of insulin-like growth factor-1 receptor are reversed by PKB2 downregulation [327].

---

nucleotide exchange on APAF1 [325]. After apoptosome formation, procaspase-9 is cleaved and activated to stimulate caspase-3.

[54] BCL2/adenovirus-E1b 19-kDa protein-interacting protein (BNIP3)-like protein is also called NIP3-like protein (NIP3L) and NIP3-like protein-X (NIX).

Cell survival mediated by PKB relies on glucose metabolism. This kinase type increases surface concentration of glucose transporters, stimulates the mitochondrial association of hexokinase [328], and phosphorylates phosphofructokinase PFK2 to increase the production of fructose (2,6)-bisphosphate that regulates glycolysis.

Furthermore, PKB promotes the plasmalemmal residence of multiple transporters and receptors relevant to extracellular molecules required for metabolic homeostasis, such as amino acids, cholesterol, and iron, in addition to glucose [329].

5.2.6.5 PKB1 Isoform

Isoform PKB1[55] is implicated in cell metabolism, survival, and migration. It is activated by platelet-derived growth factor via phosphatidylinositol 3-kinase. In the developing nervous system, PKB1 intervenes in growth factor-induced neuronal survival. In the vascular system, it is the main isoform expressed in endothelial cells, in which it promotes endothelial nitric oxide synthase (NOS3). It contributes to placental angiogenesis.

In acute inflammation, PKB1 regulates leukocyte migration into inflamed tissues, especially neutrophil and monocyte infiltration [330]. It heightens vascular permeability, hence causing edema and leukocyte extravasation, via NOS3 and VE-cadherin. The latter partly dissociates from the cortical cytoskeleton. Endothelial-derived NO mediates microvascular permeability in the early phase. Kinase PKB1 is then required to ensure propermeability effect on postcapillary venules of pro-inflammatory mediators, such as bradykinin, histamine, tachykinins, complement, and nitrogen oxide species.

Activated PKB1 isozyme controls antiviral immunity via interferon-stimulated gene expression, as PKB1 phosphorylates (inactivates; Ser209) EMSY transcriptional repressor [331].[56] The latter binds to BrCa2, but not BrCa1, with which it cooperates in transcriptional repression by binding to gene promoters with interferon-stimulated response element (ISRE).

5.2.6.6 PKB2 Isoform

Isoform PKB2[57] serves in signal transduction from insulin receptor. Abnormal PKB2 functioning causes impaired glucose tolerance and insulin resistance.

---

[55] A.k.a. Akt1, PKBα, and RACα Ser/Thr kinase.

[56] Interferons constitute a family of secreted proteins with 3 interferon categories: Ifn1 (13 Ifnα, 3 Ifnβ, Ifnβ2 being IL6, 1 Ifnκ, and 1 Ifnω1), Ifn2 (1 Ifnγ), and Ifn3 (3 Ifnλ[IL28 and IL 29]). They function in an auto- and paracrine manner to regulate innate and adaptive immunity.

[57] A.k.a. Akt2, RACβ Ser/Thr kinase, and PKBβ.

5.2.6.7 PKB3 Isoform

Isoform PKB3[58] is stimulated by platelet-derived growth factor, insulin, and insulin-like growth factor IGF1. It is highly expressed in the brain. Alternatively spliced transcript variants generate distinct subtypes.

5.2.6.8 PKB in Cell Migration

Two PKB isoforms, PKB1 and PKB2, intervene in cell motility with different roles and in a cell type-dependent manner. PDGF-induced cell motility mainly involves PKB1 that forms proteic complex with its upstream kinases PDK1 and PDK2 at the plasma membrane after translocation. The proteic complex formation is favored by group-1 P21-activated kinases PAK1 and PAK2 [115] (Sect. 5.2.13). Recruitment of PKB to the plasma membrane by binding to 3-phosphoinositides followed by PDK-mediated activation is indeed not sufficient for efficient PKB translocation and activation that require the PAK–PDK1–PKB complex. This complex restricts both spatially and temporally growth factor-initiated signaling and can contribute to PKB signaling specificity, as PKB has many substrates. On the other hand, PKB phosphorylates (moderately activates) PAK1 that then dissociates from NCK and focal adhesions. In addition, PDK1 phosphorylates (activates) PAK1 kinase.

5.2.7 Protein Kinase-C Family

The family of typically membrane-localized, lipid-activated protein kinase-C represents about 2% of the human kinome with 12 known genes. Members of the PKC family transduce signals that are transmitted via the hydrolysis of membrane phospholipids.

This family includes multiple enzymes that have specific tissue expression and contribute to the spatial organization of signal propagation owing to their compartmentalized action (localized second messenger production and interaction with membrane-anchored small GTPasess, spatial organizers such as scaffold proteins, and accessory proteins).

These kinases can be divided into 3 main groups: (1) conventional, or classical, PKCs (cPKC) diacylglycerol- and calcium-sensitive isoforms (PKCα, PKCβ1, PKCβ2, and PKCγ); (2) novel (nPKC) calcium-independent isoforms (PKCδ, PKCε, PKCη, and PKCθ) that are activated by phosphatidylserine and diacylglycerol; and (3) atypical (aPKC) PKCs (PKCζ and PKCλ/ι) that are activated neither by calcium nor diacylglycerol, but by phospholipid cofactors as well as stimulus-induced phosphorylation. In addition, atypical PKCs can be allosterically activated

---

[58] A.k.a. Akt3, RACγ Ser/Thr kinase, PKBγ, and STK2.

by the Par6–CDC42 complex[59] that specifies cell polarity [332]. A fourth family encompasses protein kinase-N (PKN1–PKN3). Small GTPases Rho or Rac tether to PKN kinases and disengages their pseudosubstrate to activate the kinase function. The PKC isozymes are expressed in all tissues.[60]

Conventional PKCs possess 4 common, homologous domains (C1–C4) that are interspaced with isozyme-unique, variable (V) domains. The PKC isozymes have a protein kinase region linked to C1 and C2 domains. The C1 domain binds to diacylglycerol. The C2 domain connects to negatively charged lipids such as phosphatidylinositol (4,5)-bisphosphate and Ca^{++} cations. Domains C1 and C2 are regulatory membrane-targeting motifs. Domains C3 and C4 contain ATP- and substrate-binding sites.[61] *Novel PKCs* have C2 domains that lack calcium-coordinating acidic residue side chains that confer calcium sensitivity, but serve in between-protein interactions. *Atypical PKCs* lack a calcium-sensitive C2 domain and contain only a single cysteine-rich zinc finger structure that does not bind DAG.

Activity of PKCs requires priming phosphorylations at 3 sites within the kinase domain. Priming phosphorylations depend on the conformation of the nucleotide-binding pocket, rather than autophosphorylation, i.e., intrinsic kinase activity [302]. Removal of ATP from the nucleotide-binding site of PKC induces rapid dephosphorylation of all PKC phosphorylated residues. Occupancy of ATP-binding pocket by PKC inhibitors can also trigger phosphorylation of inactive PKC.

Activation of PKC is primed by phosphorylation that is followed by the recruitment of PKC to the membrane by Ca^{++} as well as both PIP_2 and DAG for effective activation.[62] The PKC isoforms are activated by at least 2 phosphorylations: one of their activation segment and another of their hydrophobic motif (when present). Phosphoinositide-dependent protein kinase PDK1 phosphorylates the activation segment, whereas target of rapamycin complex-2 phosphorylates the hydrophobic motif of at least some isoforms [300]. However, these modifications alone are not sufficient for activation, as phosphorylated PKCs can be maintained in an inactive state, until the release of the pseudosubstrate.

Lipid mediators are not selective for a given PKC isoform. Products of phosphatidylinositol 3-kinase indeed activate novel (PKCε and PKCη) as well as atypical PKCs (PKCζ and PKCι).

---

[59] Only active $CDC42^{GTP}$ interacts with partitioning defective protein Par6 to regulate aPKCs.

[60] In heart, the distribution of PKC isozymes differs between atria and ventricles [333]. Calcium-dependent isoforms PKCα, -β1, and -β2 reside predominantly in ventricles, whereas PKCδ and PKCζ are mainly located in atria. Novel PKCε and atypical PKCι are evenly distributed in atria and ventricles. In cardiac fibroblasts, PKC isozymes also include PKCα, PKCβ1, PKCβ2, PKCδ, PKCε, and PKCζ. Enzymes PKCδ and PKCζ that act downstream from TGFβ1 impede and favor fibroblast proliferation, respectively [333].

[61] The C1 domain consists of tandem cysteine-rich zinc finger structures that bind diacylglycerol and phospholipid. The C2 domain binds anionic phospholipids in a calcium-dependent manner.

[62] Inactive PKC is associated with heat-shock protein HSP90 at the membrane. Activation of plasmalemmal receptors causes the production of diacylglycerol in the membrane and an increase in the cytoplasmic Ca^{++} concentration.

Activated by diacylglycerol and calcium ions, PKCs are involved in signaling initiated by certain hormones, growth factors, and neurotransmitters. Diacylglycerol causes PKC translocation from the cytosol to the plasma membrane, where it binds calcium ions and plasmalemmal phosphatidylserine. These events relieve PKC autoinhibition. Kinase PKC moves to the cytoskeleton, perinuclear sites, and nucleus.

Phosphorylation of various PKC substrates leads to either increased or decreased activity.[63]

Activity of all of the members of the PKC family is influenced by guanine nucleotide-binding proteins, either indirectly via G-protein-coupled receptors that signal via heterotrimeric G proteins to control allosterically phospholipase-C and subsequent generation of diacylglycerol for PKC activation or directly via small GTPase activators. Atypical PKCs are regulated by small GTPases CDC42 and Rac that act via partitioning defective Par6 [332].

Isoforms PKCβ1 and PKCε control tight junction integrity. Kinase PKCη is expressed predominantly in epithelia, where it regulates tight junctions that constitute a barrier between apical and basolateral membranes [334]. Tight junction is made of at least 3 types of transmembrane proteins, occludin, claudins,[64] and junctional adhesion molecules. Isozyme PKCη phosphorylates occludin for tight junction integrity.

5.2.7.1 PKC-Anchoring Proteins

Protein kinase-anchoring and scaffold proteins are essential components of cell signaling. Isozyme specificity results, at least partly, by association with specific anchoring proteins. A heterogeneous set of PKC-binding proteins actually influence PKC subcellular localization, substrate availability, exposure to allosteric activator, and activation-dependent relocation within cells. These proteins include: (1) PKC-binding proteins and PKC substrates that interact with PKC prior to phosphorylation [335]; (2) cytoskeleton–vesicle interacting proteins; (3) A-kinase-anchoring proteins; and (4) receptors for activated C-kinase that are non-substrate proteins that bind to active PKC.

The PKC isoforms not only can be recruited to scaffolds with other transducers, but also can control the behavior of signaling mediator complexes, as they influence their assembly or disassembly and subcellular localization, without incorporating these complexes [332].

Substrates that interact with C-kinase (STICK), such as myristoylated alanine-rich C-kinase substrate (MARCKS) and its related protein MacMARCKS, as well

---

[63] Phosphorylation of the epidermal growth factor receptor by PKC prevents EGFR activity.

[64] Intracellular domains of occludin and claudins interact with proteins ZO1, ZO2, and ZO3 that form a platform for recruitment of scaffold proteins cingulin, Par3, Par6, etc. Tight junction is thus anchored to the perijunctional actomyosin ring.

as α- and γ-adducins, among others, are not only PKC-interacting proteins, but also PKC substrates [335]. Binding parters of PKC that are PKC substrates and directly bind phosphatidylserine include talin, vinculin, MARCKS, β-adducin homolog, and AKAP5 and AKAP12 [336]. Sdr-related gene product that binds to C-kinase (SBRC) is a substrate of PKC (PKCα and PKCδbut not to PKCζ) that binds to PKC regulatory domain in a phosphatidylserine-dependent manner and is phosphorylated after PKC activation.

Localization of PKC to different intracellular sites is directed by specific PKC-binding proteins that can discriminate between different PKC isoforms. PDZ (PSD95 [DLg4], Disheveled, and ZO1) domain-containing proteins mediate between-protein interactions during receptor and ion channel clustering and recruit kinases and phosphatases to their membrane-associated substrates. In addition, certain PDZ domains can homo-oligomerize to form complexes at specific subcellular sites.

Protein interacting with C-kinases (PICK), a protein kinase-Cα-binding protein, is a PDZ domain-containing PKC substrate that interacts with a PDZ-binding motif (QSAV) at C-terminus of PKCα to selectively recruit PKCα to suitable subcellular sites [337].

Several different *Receptor for inactive C-kinases* (RICK) can anchor a given PKC isozyme to different subcellular sites in the inactive state. Bruton Tyr kinase (Sect. 4.12.4) binds to all the inactive PKC isozymes in a lipid-dependent fashion [336]. Kinase BTK thus acts as a RICK.

Receptor for activated C-kinase (RACK) is a membrane-associated anchoring protein responsible for the binding of active forms of the PKC family, hence directing PKC isoform subcellular localization. Different RACKs exist for different isoforms of PKC, thereby preserving the specificity of signaling cascades. Protein RACK1 is not a substrate for PKC, but forms an active PKC–RACK1 complex that raises 7-fold PKC-substrate phosphorylation [336]. Protein RACK1 colocalizes with activated PKCβ2 to the perinucleus in cardiomyocytes.

5.2.7.2 Atypical PKCs as Kinases and Dockers

Atypical PKCs intervene in signaling responses as kinases or binding proteins. The V1 sequence of aPKCs serves as binding site for proteins that regulate apoptosis and cell polarity such as scaffold sequestosome-1; PKC, apoptosis, WT1 regulator (PAWR);[65] cell-polarity partitioning-defective protein Par6; and Partitioning defective-3 homolog (Par3).[66]

---

[65] A.k.a. prostate apoptosis response protein PAR4. Protein PAWR causes apoptosis of cancer cells, but not normal cells.
[66] Mammalian homolog Par3 is also named atypical PKC isotype-specific interacting protein (ASIP) and Bazooka.

Atypical PKCs — PKCζ and PKCι — bind PDK1 (Sect. 5.2.2). They have a single zinc-finger domain in their regulatory region that binds PKC, apoptosis, WT1, regulator (PAWR). The latter selectively inhibits both aPKCs [347]. The PB1 (V1) region interacts with scaffolds sequestosome-1 and Par6 that serve to connect PKCζ to nuclear factor-κB and to control cell polarity, respectively. Protein Par6 forms a complex with active CDC42^{GTP}, Par3, and aPKC to contribute to the formation of tight junctions that establish cell polarity in epithelial cells [338, 339]. Protein Par6 hence serves as an adaptor that links CDC42 and aPKCs to Par3 agent.

Atypical protein kinases PKCζ and PKCι are located in the cytosol, where they can associate with endosomes via sequestosome-1 and tight junctions via Par6 protein. Upon nerve growth factor stimulation, sequestosome-1 facilitates polyubiquitination of TRAF6 and formation of the Sqstm1-TRAF6-IKKβ-PKCι signaling complex that activates the NFκB pathway [340].

5.2.7.3 PKC in Cell Interactions with Environment

The PKC isoforms are involved in homo- and heterotypic intercellular recognition as well as cell–extracellular matrix interactions, particularly during cell migration. Isozyme PKCθ is highly synthesized by T lymphocytes. It regulates recognition by effector T lymphocytes of antigen-presenting cells via T-cell receptor that recognizes a processed antigen bound to a major histocompatibility complex class-1 or -2 molecule on antigen-presenting cells [332]. This recognition triggers assembly of an immunological synapse with plasmalemmal subdomains[67] asymmetric T-cell division and differentiation into CD4+ T helper cell (T_{H1}, T_{H2}, and T_{H17}). Subtype PKCθ contributes to activation of transcription factors, such as nuclear factor-κB, nuclear factor of activated T cells, and activator protein AP1, as well as other transcription factors involved in the differentiation of CD4+ helper T cells.

In polarized cells characterized by subapical tight junctions, apically located atypical protein kinase-C isoforms can regulate the formation, composition, and localization of polarity complexes via their scaffold and catalytic activities. In cell apex, aPKCs complex with polarity proteins, the partitioning defective proteins Par3 and Par6 (Par3–Par6–aPKC complex). In addition, the PAR6–aPKC heterodimer is targeted by CDC42 and Rac1 GTPases to relieve PAR6-induced inhibition of aPKCs [332]. Atypical PKCs phosphorylate various proteins, such as protein associated with LIN-7 PALS1, PALS1-associated tight junction protein (PATJ), crumbs protein homolog CRB3, Par1, Par3, and Par6, basolateral domain proteins in epithelial polarized cells, such as mammalian lethal giant larva proteins

---

[67]These plasmalemmal subdomains contain a supramolecular activation complex (SMAC) that is constituted by concentric domains, i.e., a central (cSMAC) with TCR, scaffold CARD-containing MAGUK protein-3 (CarMa1), and LcK and PKCθ kinases, peripheral (pSMAC) with $\alpha_4\beta_7$-integrin, and distal (dSMAC) SMAC with PTPRc receptor phosphoTyr phosphatase.

LGL1 and LGL2, Disc large homolog DLg1 and Scribble homolog; and 14-3-3 scaffolds [332].[68]

Isoform PKCα promotes transfer of activated hepatocyte growth factor receptor along microtubules from peripheral early endosomes, where HGFR remains active, to perinuclear endocytic recycling compartment, where it can efficiently activate signal transducer and activator of transcription STAT3, overcome the action of phosphoTyr phosphatases more strongly than in early endosomes, and cause nuclear accumulation of phosphorylated STAT3 [332].

The PKC subtypes contribute to cytoskeletal dynamics during cell migration. They interact with numerous plasmalemmal proteins, e.g., PKCα with syndecan-4 and $β_1$-integrins. Atypical PKCs that pertain to activated Par complex or exocyst complex are required for cell migration. In migrating kidney cells, PKCζ and PKCι complex with kidney and brain protein (KIBRA) and the exocyst complex and localize to the leading edge of moving cells, where they phosphorylate (activate) components of the MAPK module, such as Jun N-terminal kinase JNK1 and extracellular signal-regulated kinases ERK1 and ERK2 that all phosphorylate cytoskeletal paxillin [332]. In migrating astrocytes α- and β-integrins activate small GTPase CDC42 at the leading edge that then recruits and activates the Par6–PKC complex. Atypical PKCs then phosphorylate (inactivate) glycogen synthase kinase GSK3β to reorientate the centrosome, i.e., the main microtubule-organizing center, for directed cell migration, as it causes GSK3β substrate adenomatous polyposis coli to connect to microtubule plus-ends at the leading edge [332]. Moreover, atypical PKC acts via the microtubule nanomotor dynein to determine the polarity of the centrosome and Golgi body.

5.2.7.4 Protein Kinase-Cα (Conventional PKC)

Protein kinase-Cα, encoded by the PRKCA gene, is activated in response to hormones, mitogens, and neurotransmitters. In unstimulated cells, PKCα resides primarily in the cytoplasm. Upon stimulation by signals that promote the hydrolysis of membrane phospholipids, PKCα translocates to the plasma membrane as well as the nuclear envelope. This kinase operates in many cellular processes, such as cell cycle checkpoint, cell adhesion and volume control, as well as transformation.

Protein kinase-Cα is located in focal contacts in non-stimulated fibroblasts and, once activated, translocates to the perinuclear region. In the myocardium, PKCα regulates contractility and hypertrophy.

Among classical (PKCα and PKCβ) and novel (PKCδ and PKCθ) isoforms synthesized by platelets, PKCα regulates thrombus formation [341]. Dense-granule genesis and secretion depend on PKCα.

---

[68] Once phosphorylated by aPKCs at tight junctions, LGL1 and LGL2 are excluded from cell apex, thus restricted to the basolateral region, where they can form a complex with membrane-bound proteins Scribble homolog and Disc large homolog DLg1 [332].

5.2.7.5 Protein Kinase-Cβ (Conventional PKC)

Protein kinase-Cβ, encoded by the PRKCB1 (or PRKCB2) gene, is involved in various cellular functions, such as B-cell activation, apoptosis, endothelial cell proliferation, and intestinal sugar absorption. Both isoforms – PKCβ1 and PKCβ2 – are produced from the same gene. Activated PKCβ triggers expression of early growth response EGR1 and pro-inflammatory and procoagulant molecules that lead to acute vascular stresses.

Protein Kinase-Cβ1

Protein kinase-Cβ1 interacts with many substances, such as kinase receptor-interacting protein kinase RIPK4, β-adrenergic receptor kinase AdRβK1 (or GRK2), PDPK1, CDK1, and BTK, as well as colony-stimulating factor-2 receptor-β, G-protein subunits $G\alpha_{12}$ and $G\alpha_{13}$, receptor for activated C kinase RACK1,[69] regulator of G-protein signaling RGS2, RasGRP3, Rho activator TIAM1, NADPH oxidase organizer-2,[70] actin filament-associated protein (AFAP), lamin-B1, neural cell adhesion molecule NCAM1, and 14-3-3γ [251].

Protein Kinase-Cβ2

Protein kinase-Cβ2 associates with fibrillar structures in resting cardiomyocytes and, once activated, translocates to the perinucleus and cell cortex. Filamentous actin binds PKCβ2, but not PKCβ1.

Protein kinase-Cβ2 is a major mediator of TNFα-induced vascular endothelial cell apoptosis. In addition, PKCβ2, as well as PKCε, but neither PKCα nor PKCδ, yields full activation of voltage-gated calcium channels $Ca_V2.2$ via the stimulatory site (Thr422). Isozyme PKCε is better at attenuating channel activity via the inhibitory site (Ser425). In any case, the stimulating effect of PKCβ2 or PKCε is dominant over their inhibitory action [342]. Whereas PKCα intervenes in the potentiation of $Ca_V2.3$ flux by muscarinic M1 receptor agonist, PKCβ2 and PKCε contribute to the potentiation of $Ca_V2.3$ current by phorbol-12-myristate, 13-acetate [343]. The differential response of Ca_V fluxes to chemical stimuli is attributed to: (1) type of activated PKC isozyme and (2) PKC-selective phosphorylation sites in drug-binding, channel pore α1 subunit of Ca_V channels.

---

[69] A.k.a. Gβ polypeptide-2-like-1.

[70] A.k.a. neutrophil cytosolic factor NCF1.

5.2.7.6 Protein Kinase-Cγ (Conventional PKC)

Protein kinase-Cγ, encoded by the PRKCG gene, has a localization restricted to neurons, where it operates in long term potentiation (LTP) and depression (LTD).

This kinase interacts with ionotrophic, AMPA-type glutamate receptor GluR4, $GABA_A$ receptorα1 and -α4, actin filament-associated protein (AFAP), Rho activator TIAM1, regulator of G-protein signaling RGS2, and 14-3-3γ [251].

5.2.7.7 Protein Kinase-Cδ (Novel PKC)

Protein kinase-Cδ contributes to cell growth inhibition and apoptosis, as well as mitogenesis, differentiation, tumor progression, and tissue remodeling during inflammation [344]. Its lipid cofactors that interact with the C1 domain, promote PKCδ translocation to membrane and induce a conformational change that removes auto-inhibition by expelling the N-terminal auto-inhibitory domain from the substrate-binding pocket. It can also be activated from phosphorylation by SRC family kinases and Abl1 in cells subjected to oxidative stress as well as caspase-mediated proteolysis during apoptosis [344].

Protein kinase-Cδ bears a series of sequential priming phosphorylations (Thr505, Ser643, and Ser662). Phosphorylated PKCδ (at a Tyr residue) does not translocate to membranes. Phosphorylation mediated by Src (Tyr64 and Tyr565) causes inactivation [344]. The single known RACK for PKCδ is the 32-kDa PKC-interacting glycoprotein complement component-1, q subcomponent-binding protein C1qBP[71]

Protein kinase-Cδ can exert kinase-independent effects, as it can act as a signal-regulated scaffold to translocate signaling mediators that associate with PKCδ, such as SRC family kinases, lipid or protein phosphatases, and adaptor proteins such as Shc [344]. In vascular smooth muscle cells, inactive PKCδ mimicks the effect of wild-type PKCδ and induces apoptosis.

Protein kinase-Cδ interacts with plasmalemmal actin filament crosslinker myristoylated alanine-rich PKC substrate MARCKS, axon constituent growth-associated protein GAP43, kinases, such as SRC family kinases (Src, Fyn, Lyn), Abl1, PI3K, DNAPK, TOR, and receptor-interacting kinase RIPK4,[72] insulin receptor and IRS1, protein phosphatases, such as PP2 and PTPn6, adaptor SHC, actin, STAT3, and NADPH oxidase [344].

---

[71] A.k.a. receptor for globular heads of complement component C1q (gC1qR) and mitochondrial matrix protein P32RACK. Protein kinase-Cδ only binds to P32RACK in the presence of PKC activators (lipid cofactors). Isoforms PKCα and PKCζ need to be activated. Other PKC isoforms, such as PKCβ, PKCε, and PKCθ, bind to P32RACK regardless of the presence of PKC activators. Furthermore, PKCμ binds even better in their absence [345].

[72] A.k.a. PKCδ-interacting kinase DIK.

5.2.7.8 Protein Kinase-Cε (Novel PKC)

In cardiomyocytes, PKCε, encoded by the PRKCE gene, is located in the nucleus and perinuclear region before stimulation to translocate upon activation to sarcomeres and intercellular contacts. Isozyme PKCε, but not other PKC isoforms, binds to filamentous actin and synaptosomes.

This kinase interacts with voltage-dependent anion channel VDAC1, adenine nucleotide translocator SLC25a4, CFTR channel, receptor for activated C kinase RACK1, cRaf kinase, Rho activator TIAM1, RasGRP3, $G\alpha_{12}$ and $G\alpha_{13}$, glutamate decarboxylase-1 and -2, hexokinase-2, actin-α1, actin filament-associated protein (AFAP), myosin heavy chain-9, keratin-1, -8, and -18, connexin-43, PI3K-binding protein centaurin-α coatomer protein complex subunit-β2, and 14-3-3ζ [251].

5.2.7.9 Protein Kinase-Cη (Novel PKC)

Protein kinase-Cη, encoded by the PRKCH gene, is involved in neuron channel activation, apoptosis, cardioprotection from ischemia, heat shock response, insulin exocytosis, as well as lipopolysaccharide-mediated signaling in activated macrophages. It interacts with cyclin-dependent kinase CDK2.

5.2.7.10 Protein Kinase-Cθ (Novel PKC)

Protein kinase-Cθ, encoded by the PRKCQ gene, is required for the activation of the transcription factors NFκB and AP1. It interacts with kinases PKB, Fyn, LCK, and IKKβ, basic helix–loop–helix transcription factor TAL1, thioredoxin-like-2, guanine nucleotide-exchange factor Vav1, and 14-3-3γ [251].

Protein kinase-Cθ is expressed in T lymphocytes, in which it fosters activation of mature effector T lymphocytes. During T-cell activation, PKCθ is recruited to the interface between the T lymphocyte and activating antigen-presenting cells [346]. However, in regulatory T lymphocytes, PKCθ, an inhibitor, is sequestered from the activating cellular interface.

5.2.7.11 Protein Kinase-Cζ (Atypical PKC)

Atypical protein kinase-Cζ, encoded by the PRKCZ gene, phosphorylates transcription factor $RelA_{NF\kappa B}$ (Ser311), cytoskeletal lethal giant larva homologous protein LGL1 that associates with non-muscle myosin-2 heavy chain, and Crumbs homolog CrB1 expressed specifically in human retina and brain, which organizes an intracellular protein scaffold for the assembly of zonula adherens [347].

Atypical PKCs can be regulated by certain lipid messengers, such as ceramide and phosphatidylinositol (3,4,5)-trisphosphate. Second messengers stimulate PKC by binding to the regulatory domain, hence translocating the enzyme from cytosol to membrane and producing a conformational change that relieves PKC autoinhibition.

Protein kinase-Cζ is ubiquitous with its highest level in the lung. The brain-specific isoform PKMζ is generated from an alternatively spliced transcript. It lacks the regulatory region of full-length PKCζ, hence being constitutively active. In fact, PKMζ corresponds to the catalytic domain of PKCζ.

GluT4+ vesicle contains VAMP2 that binds to the Stx4–SnAP23 complex formed by syntaxin-4 and synaptosome-associated protein SnAP23 via syntaxin-binding protein StxBP3.[73] Protein StxBP3 prevents GluT4 vesicle binding to the plasma membrane in the basal state. Once stimulated by insulin, PKCζ removes the StxBP3 clamp and triggers GluT4 vesicle translocation to and docking in the plasma membrane [348].

PKCζ and Its Splice Variant PKMζ

Learning and memory relies on long-term potentiation mediated by the excitatory neurotransmitter glutamate and maintenance of persistent synaptic connections. Long-term memory and glutamate-induced dendritic protein synthesis relies on full-length protein kinase-Cζ and its N-terminal truncated alternative transcript, brain-specific, constitutively active, protein kinase–Mζ. In dendritic spines and shafts, peptidyl-prolyl isomerase Protein interacting with NIMA1 (PIN1) prevents synthesis of PKMζ as well as dendritic translation [349]. Conversely, PKMζ phosphorylates (Ser16; inhibits) PIN1. Glutamate signals preclude PIN1 activity, thereby enabling PKMζ synthesis. A positive feedback is hence created, as PKMζ impedes PIN1 function, which enables long-term potentiation to perpetuate mental representations of history.

PKCζ-Interacting Proteins

Interactors of PKCζ comprise fasciculation and elongation protein-ζ(FEZ1), or zygin-1, and scaffold sequestosome-1. The former mainly resides in the plasma membrane, where it can associate with PKCζ to be phosphorylated; it is involved in axon guidance. The latter generates complexes between PKCζ and other proteins. Sequestosome-1 is a ubiquitin-binding protein in neuronal and glial cells that regulates NFκB activation in response to nerve growth factor, interleukin-1, and tumor-necrosis factor, i.e., in both immune and nervous systems, via the formation of PKCζ–Sqstm1–TRAF6–zyxin or ajuba complex [350]. Counteracting sequestosome-1 shuttles substrates for proteasomal degradation [351].

---

[73] A.k.a. Munc18c or Munc18-3.

5.2.7.12 Protein Kinase-Cι (PKCλ; Atypical PKC)

Protein kinase-Cι is the human homolog of mouse PKCλ. Like PKCζ, PKCι is implicated in cell proliferation and survival via transcription factors AP1 and NFκB, respectively. Isoform PKCι is activated upon PDGFR stimulation by the PI3K pathway. Atypical PKC isotypes contribute to activation of mitogen-activated protein kinase.

PKCι-Interacting Proteins

A selective activator of PKCι — (PKC)λ-interacting protein (LIP) — does not even bind to the highly related PKCζ. On the other hand, both PKCζ and PKCι participate in the regulation of NGF-induced cell differentiation via PKCζ-binding protein Sqstm1 [352]. Shuttle Sqstm1 and PKCι colocalize with Rab7 GTPase in late endosomes regardless of PKCι phosphorylation state.

5.2.8 Protein Kinase-D Family

The family of Ser/Thr protein kinase-D includes PKD1 (or atypical PKCμ), PKD2, and PKD3 (or PKCν). As PKD1 was originally reported to be an atypical isoform of PKC, PKD was incorporated in the PKC family. However, PKD differs from PKC by amino acid composition, domain structure, regulation, and substrate specificity. In particular, the catalytic domain of PKDs has sequence homology with myosin light-chain kinase and calmodulin-dependent kinase. The 3 isoforms of PKD are now classified in a proper family within the CAMK superfamily.

The PKD kinases operate downstream from activated G-protein-coupled receptors and receptor Tyr kinases that transmit signal via phospholipase-C. They regulate cellular processes, such as proliferation, vesicular transport, apoptosis, and stress response, as well as immunity.

5.2.9 Protein Kinase-G Family

Isozymes of protein kinase-G — PKG1 and PKG2 —, or cGMP-dependent protein kinase are encoded by 2 genes. They are activated by cyclic guanosine monophosphate. This second messenger is produced by soluble and transmembrane guanylate cyclases activated by nitric oxide and natriuretic peptides, respectively. Type-1 homodimeric PKG is widespread. On the other hand, type-2 dimeric PKG has a limited tissue distribution.

Two PKG1 isoforms exist: PKG1α and PKG1β. Subtype PKG1α is activated at about 10-fold higher cGMP concentrations than PKG1β isoform. Both PKG1 subtypes are homodimers. Each subunit is composed of 3 functional domains: (1) N-terminal leucine zipper autoinhibitory domain that mediates homodimerization, suppression of kinase activity in the absence of cGMP, and interactions with proteins, as well as autophosphorylation sites that control PKG basal activity; (2) cGMP-binding regulatory motif; and (3) C-terminal kinase sequence.

Enzyme PKG1 is observed at high concentration in all types of smooth muscle cells (both isoforms) and platelets (predominantly PKG1β) and at lower level in vascular endothelial cells and cardiomyocytes.

Enzyme PKG2 resides in renal cells, zona glomerulosa cells of the adrenal cortex, Clara cells in distal airways, among other cell types (but not in cardiac and vascular myocytes).

Kinase PKG1 participates also in vascular tone control, platelet activation, and synaptic remodeling. In vascular smooth muscle cells, PKG1 has an antiproliferative function, in addition to relaxation. Kinase PKG phosphorylates myosin light-chain phosphatase that dephosphorylates myosin light chains to initiate smooth muscle relaxation. Calcium influx, mainly via store-operated Ca^{++} channels in non-excitable cells, controls gene regulation, cell contraction, exocytosis, and apoptosis. Vascular endothelial cells respond to numerous stimuli by secreting vasoactive factors. Most often, the initial response of endothelial cells involves Ca^{++} release from intracellular stores. Store depletion activates Ca^{++} entry. Elevated cGMP level attenuates the store-operated Ca^{++} import via PKG in vascular endothelial cells [353]. Kinase PKG also modulates calcium entry by acting on Ca_V1 and mechanosensitive Ca^{++} channels, thereby changing the intracellular calcium concentration for the activity of nitric oxide synthases NOS1 and NOS3. In addition, PKG directly stimulates NOS3 and soluble guanylate cyclases, and inhibits phosphodiesterase PDE5 that selectively degrades cGMP.

Kinase PKG1 phosphorylates transcription factors to prime Fos promoter response. In addition, PKG1 colocalize with SMAD1 at the Id1 promoter on BMP2 stimulation [354].

5.2.10 CDC42-Binding Protein Kinases

CDC42-binding protein kinase-α (CDC42BPα)[74] possesses a CDC42–Rac-interactive binding (CRIB) domain. It hence can bind CDC42 to promote cytoskeletal reorganization and nuclear displacement to cell rear during cell migration. It phosphorylates myosin light chain to activate actomyosin filament contraction.

---

[74] A.k.a. myotonic dystrophy kinase-related CDC42-binding kinase (MRCK) and myotonic dystrophy protein kinase-like kinase (DMPKL).

5.2.11 G-Protein-Coupled Receptor Kinase Family

G-protein-coupled receptor Ser/Thr kinases (GRK1–GRK7) recognize active GPCRs and phosphorylate these receptors for desensitization. Multiple GRKs (GRK2, GRK3, GRK5, and GRK6) are located in the central nervous system.

G-protein-coupled receptor kinases are classified into 3 categories: (1) rhodopsin kinase (RK or GRK1); (2) β-adrenergic receptor kinases (βARK1–βARK2, i.e., GRK2 and GRK3, respectively); and (3) GRK4 category (GRK4–GRK6). Both category-2 and -3 GRKs bind to $PI(4,5)P_2$ lipid.

Once they are phosphorylated by GRKs, GPCRs bind to arrestins that prevent reassociation of G proteins with their receptors, hence activation of G proteins by GPCRs despite continuous binding of agonist. Phosphorylated Ser and Thr residues in the cytoplasmic loops and tails of activated GPCRs indeed act as binding sites for arrestins. Phosphorylation and desensitization of GPCRs can be followed by endocytosis and resensitization.

Unlike most AGC kinases, GRKs are not activated by phosphorylation of the activation segment or hydrophobic motif. Instead, activation results from a conformational rearrangement triggered by GRK interaction with activated GPCRs at the plasma membrane [300].

Phosphorylation of G-protein-coupled receptors by GRK kinases governs interactions of GPCRs with β-arrestins that control transfer of and signaling from these receptors. β-Arrestins tether to GPCRs and end G-protein activation after phosphorylation of specific Ser residues by GRK kinases as well as redirect the signaling to the MAPK module. β-Arrestin-1 is also recruited to receptor Tyr kinases such as insulin-like growth factor IGF1 receptor, thereby regulating receptor degradation and signaling. β-Arrestin-1 serves as an adaptor to link DM2 ubiquitin-ligase to IGF1R, thereby assisting in IGF1R degradation. On the other hand, IGF1R complexed to β-arrestin-1 activates the MAPK module. Interaction between β-arrestin and IGF1R supported by GRK2 and GRK6 is transient and stable, respectively [355]. Whereas GRK5 and GRK6 phosphorylate Ser1291 of IGF1R and promote IGF1- and β-arrestin-mediated ERK and late-phase PKB activation, GRK2 that targets Ser1248 of IGF1R represses ERK activation, but mildly supports early phase PKB signaling [355]. The stable βArr–IGF1R association enables increased ERK phosphorylation. In addition, GRK2 reduces and GRK6 enhances IGF1R degradation. In summary, distinct GRKs generate different RTK phosphorylation patterns that create different functional outcomes mediated by recruited β-arrestins.

G-protein-coupled receptor kinases interact also with several other proteins, such as calmodulin, caveolin, Raf kinase inhibitor protein, and GPCR kinase interactors APAP1 and APAP2, as well as kinases PI3K, PKB, and MAP2Ks.

5.2.11.1 GRK1

Kinase GRK1 interacts with caveolin-1, lipid-binding synuclein-α, -β, and -γ that are primarily expressed in the nervous system, neuronal calcium-binding protein recoverin that is mainly observed in photoreceptors, and rhodopsin [251]. Membrane localization of GRK1 relies on its farnesylation [300].

5.2.11.2 GRK2

Kinase GRK2, or β-adrenergic receptor kinase-1, interacts not only with β2 adrenoceptors, but also other GPCRs, such as endothelin receptor ETRb (or ET$_B$), angiotensin receptor-1, chemokine receptor CCR4, and follicle-stimulating hormone receptor. It also connects to Src kinase and protein kinase-Cβ1, Gα_q and Gα_{15} subunits, phosphodiesterase-6G, APAP1 and APAP2, caveolin-1, rhodopsin, phosducin, and phosducin-like protein [251].

Binding of GRK2 to both Gq and G$\beta\gamma$ subunits prevents their interaction with effectors [300]. Kinase PKA phosphorylates (Ser685) GRK2, thereby enhancing its ability to bind to G$\beta\gamma$. On the other hand, ERK1 phosphorylates GRK2 (Ser670) and impedes GRK2–G$\beta\gamma$ binding.

5.2.11.3 GRK3

Kinase GRK3, or β-adrenergic receptor kinase-2, interacts with GPCRs, such as chemokine receptors CCR4 and CXCR4, as well as APAP1 [251].

5.2.11.4 GRK4

Kinase GRK4 links to follicle-stimulating hormone receptor and calmodulin-1 (or phosphorylase kinase-δ). The latter operates in signal transduction and synthesis and release of neurotransmitters [251]. Membrane localization of GRK4 relies on its palmitoylation [300].

5.2.11.5 GRK5

Kinase GRK5 associates with angiotensin receptor-1, arginine vasopressin receptors AVpR1a (V$_{1A}$) and AVpR2 (ADHR), neurokinin-1 receptor (substance-P or tachykinin receptor TacR1), oxytocin receptor (OxtR) on myoepithelial cells of the mammary gland, as well as in uterine myometrium and endometrium at the end of pregnancy [251]. Like GRK1, GRK5 connects to synuclein-α, -β, and -γ, recoverin, caveolin-1, and rhodopsin, in addition to calmodulin-1 and caldendrin (calcium-binding protein-1) [251]. Isoform GRK5 binds PI(4,5)P$_2$ to tether to the plasma membrane [300].

5.2.11.6 GRK6

Kinase GRK6 tethers to β2-adrenoceptor and follicle-stimulating hormone receptor, synucleins-α, -β, and -γ, recoverin, and serine peptidase PrsS23 (or Sig13) [251]. Membrane localization of GRK6 relies on its palmitoylation [300].

5.2.11.7 GRK7

Kinase GRK7 phosphorylates light-sensitive, membrane-bound G-protein-coupled receptors cone opsins in photoreceptors of the retina and initiates their deactivation. Membrane localization of GRK7 relies on its farnesylation [300].

5.2.12 Microtubule-Associated Ser/Thr Kinase Family

Microtubule-associated Ser/Thr kinases constitute a family with 5 known members (MAST1–MAST4 and related MASTL). Kinase MAST2 interacts with phosphatase and tensin homolog [300]. Microtubule-associated Ser/Thr kinase-like protein MASTL is the human ortholog of Greatwall kinase.

During the cell division cycle, mitotic regulator MASTL promotes mitotic entry, as it inhibits protein phosphatase-2 that dephosphorylates CcnB–CDK1 complex [356]. Inhibition of PP2 is mandatory for CcnB–CDK1 activation at mitotic entry. The CcnB–CDK1 activity is controlled during G2 phase by inhibitory phosphorylations (Thr14 and Tyr15) by kinases Wee1 and PKMYT1 membrane-associated Tyr-Thr protein kinase. Upon mitotic entry, activatory phosphatase CDC25 relieves this inhibition. Phosphatase PP2 also regulates CcnB–CDK1 activity, as kinases Wee1 and PKMYT1 and CDC25 phosphatase are PP2 substrates. At mitotic exit, the CcnB–CDK1 complex is again inhibited by the ubiquitin-dependent degradation of cyclin-B. This inactivation results from PP2 reactivation. G2 Arrest results from dephosphorylation of the CcnB–CDK1 complex as well as its inhibition by Wee1 and PKMYT1 in the absence of CDC25-induced dephosphorylation. The ability of cells to remain arrested in mitosis by the spindle assembly checkpoint is proportional to the MASTL amount. When MASTL is slightly reduced, cells arrest at prometaphase. More complete depletion in MASTL correlates with premature dephosphorylation of the CcnB–CDK1 complex, inactivation of the spindle assembly checkpoint, and subsequent exit from mitosis with severe cytokinesis defects. The balance between CcnB–CDK1 dimer and PP2 is tightly regulated for correct mitotic entry and exit by MASTL kinase [356].

5.2.13 P21-Activated Kinase Family

The P21 (Ras-like cyclin-dependent kinase inhibitor-1A)-activatedl kinases (PAK) serve as regulators that intervene in cytoskeletal organization as well as cell survival, apoptosis, and angiogenesis. The PAK proteins are effectors that link small GTPases to cytoskeleton reorganization and nuclear signaling.

The PAK family includes 6 known members (PAK1–PAK6). Members of the PAK family have been classified into 2 subfamilies: subfamily 1 (or A; PAK1–PAK3) and 2 (or B); PAK4–PAK6) that do not contain the auto-inhibitory domain of group-1 PAKs. Subfamily-2 PAKs are expressed in the brain, where they participate in locomotion as well as learning and memory [360].

P21-Activated kinases are effectors of the Rac and CDC42 GTPases. In fact, they are CKI1a-, CDC42-, and Rac1-activated kinases. Binding of active, monomeric GTPase Rac^{GTP} or CDC42^{GTP} as well as sphingosine-related lipids causes conformational changes. The PAK proteins are then potently activated by autophosphorylation at multiple sites. However, PAK can be phosphoryled by other kinases [357].

On the other hand, Mg^{++}–Mn^{++}-dependent protein phosphatases of the family of protein Ser/Thr phosphatase-2C, the PP2C domain-containing protein phosphatases PPM1e and PPM1f,[75] inactivates PAK [358]. Magnesium-dependent protein phosphatase PPM1a also inhibits PAK1 kinase. In addition, PP2 can form a complex with PAK1 and PKCζ to dephosphorylate troponin-I and -T in cardiomyocytes.

5.2.13.1 PAK1

Kinase PAK1[76] exists as an homodimer under basal conditions. The activity of PAK1 is governed by the phosphorylation–dephosphorylation cycle (Ser144) and binding of certain lipids, such as sphingosine, dihydro- and phytosphingosine, monosialoganglioside (GM3), phosphatidylinositol, lysophosphatidylinositol, dimyristoylphosphatidic acid, and phosphatidic acid [357]. Small GTPases CDC42, Rac1 to Rac3, RhoJ, RhoQ, and RhoU activate PAK1 by disrupting homodimerization and potentiating phosphorylation.

Isoform PAK1 participates in polarized cytoskeletal regulation, cell morphology, motility, survival, apoptosis, and transformation. It is located in regions of cytoskeletal activity (e.g., ruffles, lamellipodia, etc.) and focal adhesions [357].

---

[75] A.k.a. partners of PAK-interacting exchange factor (PIx) POPx1 and POPx2, respectively. These phosphatases are binding partners of the guanine nucleotide-exchange factors RhoGEF6 (PIxα) and RhoGEF7 (PIxβ). Among the 3 known isoforms – RhoGEF7 (a.k.a. P50Cool1, P^{85}Cool1, and β-PIx; encoded by the ARHGEF7 gene) and RhoGEF6 (a.k.a. P^{90}Cool2 and α-PIx; encoded by the ARHGEF6 gene), only RhoGEF6 has a significant GEF activity [361].

[76] A.k.a. protein kinase-N1 (PKN1), protein kinase-Nα (PKNα), and protein kinase PrK1, and protein kinase C-related kinase-1 (PrKCl1).

Various active forms of PAK1 induce cytoskeletal reorganization by kinase- dependent and -independent effects, as they can also act as scaffolds that recruit adaptors NCK and GRB2 as well as guanine nucleotide-exchange factors RhoGEF6 and RhoGEF7. Kinase PAK1 attracts via these Rac–CDC42GEF the G-protein-coupled receptor kinase-interacting protein APAP1[77] that links PAK1 to focal adhesions via paxillin.

The PAK1 activity can rise upon binding of regulators, such as RhoGEF6 and −7 and calcium- and integrin-binding protein (CIB) [357]. The PAK1 function is also enhanced by guanylate cyclase[78] to promote cell migration and lamellipodium formation, $G\beta/\gamma$ subunit of G protein for directional sensing and migration, and PI3K. Conversely, cytoskeletal protein Merlin,[79] insulin receptor substrate partner nischarin,[80] apoptosis mediator CDK11 kinase,[81] estrogen-receptor signaling mediator cysteine-rich inhibitor of PAK1 (CRIPAK), and PAK-interacting protein PAK1IP1 (or PIP1) prevent PAK1 activity.

Substrates of PAK1 comprise myosin light-chain kinase (MLCK; inactivation) and LIM domain-containing protein kinase LIMK1 and LIMK2 (activation) [357] (Table 5.1). Phosphorylated MLCK cannot phosphorylate the myosin regulatory light chain, thereby impeding cell contractility. The LIMK kinases regulate cofilin cycling.

Other targets include cytoskeletal proteins, such as myosin-6, myosin heavy chains-1E and -2B, filamin-A, caldesmon, desmin, vimentin, regulators of microtubule dynamics stathmin and tubulin cofactor-B, dynein light chain-1, and linker of synaptic vesicles synapsin-1 [357]. Protein PAK1 operates in GTPase bearing, as it interacts with GTPase regulators and effectors, such as heterotrimeric G-protein subunit Gz, RhoGDIα, and RhoGEF2 and RhoGEF8, in addition to RhoGEF6 and -7, APAPs, and Merlin. Last but not least, PAK1 interacts with components of the MAPK module, i.e., cRaf, MAP3K1, and MAP2K1, as well as Aurora-A kinase and Polo-like kinase-1.

Protein PAK1 is phosphorylated by CDK1 and CDK5 (Thr212; inactivation) to modulate the microtubule dynamics and cell spreading, as well as PDK1 (Thr423; activation), PKG (Ser21) to inhibit NCK binding, PKB1 (Ser21; activation) via scaffold Arg protein Tyr kinase-binding protein ArgBP2, PKB2 (inhibition), BMX, and JAK2 (Tyr153, Tyr200, and Tyr284; activation) kinases [357].

Kinase PAK1 interacts with NCK adaptor to regulate endothelial cell migration, GRB2 to couple PAK1 to EGFR, and inhibitor ABI gene family member ABI3 [357].[82]

---

[77] A.k.a. Cool-associated, Tyr-phosphorylated CAT1.

[78] Rac-Activated PAK1 elevates the activity of guanylate cyclases (positive feedback loop).

[79] The portmanteau Merlin stands for moesin-ezrin-radixin-like protein. It is also called neurofibromin-2 or schwannomin.

[80] A.k.a. IRAS.

[81] A.k.a. CDC2L2 protein and PITSLRE.

[82] A.k.a. New molecule including SH3 (NeSH).

Target transcription factors consist of FoxO1, estrogen receptor-α (or NR3a1), epithelial-specific E74-like factor ELF3 of the ETS transcription factor family, Snai that can contribute to epithelial–mesenchymal transition, STAT5a, histone-3, SHARP that interacts with recombination signal-binding protein for immunoglobulin κJ region (RBPJ) and represses Notch pathway, and transcriptional repressor CtBP1.

In addition to small RhoA GTPase, PAK1 interacts with neurogenic differentiation NeuroD2 factor, A-kinase-anchoring protein AKAP9, phosphoinositide-dependent protein kinase PDK1, phospholipase-D1, coiled-coil domain-containing protein CCDC85b (or δ-interacting protein-A [DIPa]), pyruvate dehydrogenase kinase-1, light and heavy polypeptidic neurofilament (NeFL and NeFH), 62-kDa cerebellar degeneration-related protein CDR2, actinin-α1, and vimentin [251].

5.2.13.2 PAK2

Cardiolipin-activated PAK2[83] is activated by proteolytic cleavage during caspase-mediated apoptosis. It interacts with CDC42, SH3-domain kinase-binding protein SH3KBP1, and Abl kinase (Table 5.1).

Others partners include cytosolic protein Tyr phosphatase PTPn13, 3-phosphoinositide-dependent protein kinase PDK1, protein kinase-B, mitogen-activated protein kinase kinase kinase MAP3K2, monomeric RhoA GTPase, and NCK1 adaptor [251].

5.2.13.3 PAK3

Kinase PAK3 [84] forms the GTP–CKI1a—CDK1–Rac1 complex (Table 5.1).

Two kinases — PAK3 and SGK2 — intervene in viability and proliferation of epithelial cells, in particular, following P53 inactivation. Loss of P53, SGK2, or PAK3 alone has little effect on cell survival, whereas loss of P53 together with that of either SGK2 or PAK3 leads to cell death [362].

5.2.13.4 PAK4

Kinase PAK4 intervenes in filopodium formation. Multiple alternatively spliced transcript variants encode distinct isoforms. Protein PAK4 is an effector of CDC42 in the regulation of cytoskeleton reorganization. It may protect cells from apoptosis, preclude cell adhesion, and promote cell migration.

Protein PAK4 interacts specifically with CDC42^{GTP}. It also targets β5-integrins and LIMK1 kinase. It weakly activates JNK kinases.

---

[83] A.k.a. protein kinase-N2 (PKN2), protein kinase PrK2, and protein kinase-C-like-2 (PrKCl2).

[84] A.k.a. protein kinase-N3 (PKN3) and protein kinase-Nβ (PKNβ).

5.2.13.5 PAK5 or PAK7

Although PAK5, or PAK7, contains, like subfamily-1 PAKs, an auto-inhibitory domain, it does not possess, unlike subfamily-1 PAKs, a PAK inhibitor domain [359]. Moreover, it is activated by CDC42, but not as strongly as subfamily-1 PAK members. The Pak5 gene possibly belongs to a cluster of genes regulated by miR15a and miR16 agents.

Isoform PAK5 is predominantly expressed in the brain. It interacts with CDC42 to promote neurite outgrowth. It stabilizes the microtubule network. Isoform PAK5 localizes to mitochondria, where it has a strong pro-survival activity [363]. Upon growth factor stimulation or cell cycle progression, PAK5 transits from mitochondria to the nucleus. Its binding to CDC42 and RhoD triggers its relocalization.

Kinase PAK5 can be activated by MAP3K6, an activator of P38MAPK. On the other hand, PAK5 weakly interacts with activated Rac. In addition, PAK5 inhibits microtubule affinity-regulating kinase MARK2. In the mitochondrion, active phosphorylated $PAK5^P$ can phosphorylate BAD, thereby promoting cell survival. PhosphoPAK5 not only phosphorylates BAD (Ser112), but also favors phosphorylation (Ser136) possibly by PKB [363].

5.2.13.6 PAK6

Isoform PAK6 is highly expressed in testis and prostate. It interacts with the androgen receptor (NR3c4), a steroid hormone-dependent transcription factor that operates in male sexual differentiation and development.

5.2.14 RoCK Kinase Family

Rho-associated, coiled-coil-containing protein kinases (RoCK1–RoCK2), also simply called Rho kinases, are effectors of Rho GTPases. The RoCK kinases consist of a N-terminal kinase domain, a coiled-coil region with a Rho-binding sequence (RBD), and pleckstrin homology (PH) motif with an internal cysteine-rich region (CRD) at the C-terminus. The auto-inhibitory region includes RBD and PH domains. Both domains can bind independently to N-terminal kinase region to inhibit the kinase activity [364]. The RoCK enzymes are dimers. Binding of Rho^{GTP} induces a conformational change that relieves RoCK auto-inhibition by the C-terminus.

The RoCK proteins are implicated in cell adhesion, motility, growth, contraction, and cytokinesis. They are activated by autophosphorylation or RhoA (Sect. 9.3). Activation of RoCK, which depends on integrins, induces interaction with Rho and RoCK translocation to the plasma membrane. The RoCK kinases enhance the formation of focal adhesions and assembly of stress fibers. Moreover, RoCKs phosphorylate (inhibit) myosin light-chain phosphatase (MLCP) which

5.2 AGC Superfamily 227

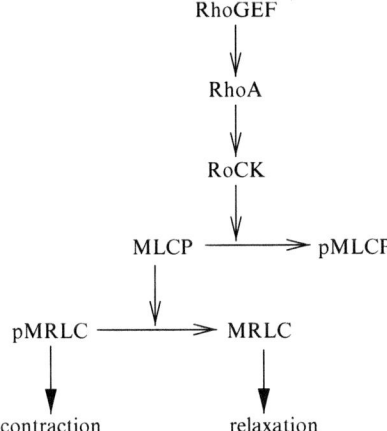

Fig. 5.1 RoCK kinase and protein phosphatase regulate the activity of myosin-2, the nanomotor in smooth muscle and non-muscle cells. Myosin-2 is regulated by phosphorylation and dephosphorylation of its myosin regulatory light chain (MRLC) by Ca^{++}–calmodulin-regulated myosin light-chain kinase (MLCK) and myosin light-chain phosphatase (MLCP). Phosphorylated MRLC (pMRLC) activates actin–myosin-2 binding. Myosin light-chain phosphatase dephosphorylates MRLC, thus inducing relaxation. Activated small GTPase RhoA stimulates Rho kinase, which phosphorylates MLCP (pMLCP), thereby inhibiting MLCP activity and increasing MRLC phosphorylation, favoring contraction. The Rho–RoCK pathway contributes to the tonic phase of contraction in smooth muscle cells.

dephosphorylates the regulatory light chain of myosin-2 (Fig. 5.1). Hence, RoCKs regulate the strength and the speed of actin–myosin cross-bridging in smooth muscle and non-muscle cells.

Some lipids, especially arachidonic acid, can activate RoCKs independently of Rho GTPase [364]. Protein oligomerization may also regulate RoCK activity, possibly via N-terminal transphosphorylation.

Kinase RoCK phosphorylates ezrin, radixin, and moesin proteins that are involved in interaction between actin filaments and the plasma membrane. It also phosphorylates LIMK kinases, thereby precluding cofilin-mediated actin filament disassembly.

Activated RoCK kinases by lysophosphatidic acid[85] reduce the peak current amplitude of $Ca_V 3.1$ and $Ca_V 3.3$ channels without affecting the voltage dependence of activation and inactivation [365]. They induce depolarizing shifts for activation and inactivation in $Ca_V 3.2$ channels.

---

[85]Lysophosphatidic acid (LPA) activates 5 different G-protein-coupled receptors (Vol. 3 – Chap. 7. G-Protein-Coupled Receptors) coupled to: (1) $G\alpha_q$ (LPA1–LPA5) that activates phospholipase-C and protein kinase-C; (2) $G\alpha_i$ (LPA1–LPA3) that inhibits protein kinase-A; (3) $G\alpha_s$ (LPA4) that stimulates PKA; or (4) $G\alpha_{12/13}$ (LPA1–LPA2 and LPA4–LPA5) that activate RhoA GTPase.

The RhoA–RoCK axis modulates ion channel activity through actin reorganization. It actually hampers the activity of several potassium channels, such as $K_V1.2$ [366], $K_V11.1$ [367], and inward rectifier $K_{IR}2.1$ [368]. Moreover, RhoA and RoCK impede actin polymerization and vasoconstriction caused by uridine triphosphate. The latter provokes depolarization resulting from $K_V1.2$ channel activity in smooth muscle cells of cerebral arterioles. Activated PKA or PKG reduce inhibition of $K_V1.2$ activity by RhoA and RoCK [369]. In addition, RhoA mediates inhibition of RhoGDIα on Ca_V1 channels in ventriculomyocytes [370]. On the other hand, RhoA and RoCK increase the activity of epithelial sodium channel, as they activate phosphatidylinositol 4-phosphate 5-kinase and increase channel insertion into the plasma membrane [371]. They also enhance the functioning of volume-regulated anion channels in pulmonary artery endothelial cells [372].

Two isoforms — RoCK1[86] and RoCK2 —[87] are encoded by the Rock1 and Rock2 genes.[88] They can be differentially regulated according to circumstances.

Kinase RoCK1 is cleaved by caspase and activated during apoptosis, whereas RoCK2 can be cleaved and activated by granzyme-B[89] to cause caspase-independent membrane blebbing.

Both paralogs participate in the regulation of cytokinesis, cell morphology, contraction, motility, and adhesion, as well as endothelial barrier function and phagocytosis, as they favor actin–myosin-mediated contractility. They also indirectly influence gene transcription as well as cell growth, proliferation, and survival. They thus contribute to neuronal morphogenesis, smooth muscle contraction, immune cell chemotaxis, and epithelial sheet movements.

The RoCK kinases phosphorylate various substrates, such as myosin light-chain phosphatase, myosin light chain, ezrin, radixin, moesin, and LIMK kinases to favor the formation of actin stress fibers and focal adhesions.

In smooth muscle cells, they assist agonist-induced muscle contraction by Ca^{++}-sensitization that results from phosphorylation of myosin-binding subunit of myosin light-chain phosphatase.

They participate in cell migration by enhancing cell contractility. They are involved in tail retraction of monocytes. Last, but not least, they contribute to the control of cell size and regulation of distance between the 2 centrioles [364].

In non-myocytes, actomyosin contractility depends on the phosphorylation state of regulatory myosin light chain of nanomotor myosin-2, i.e., the balance between Ca^{++}-dependent myosin light-chain kinase (MLCK) and phosphatase (MLCP). Kinase RoCK phosphorylates myosin light chain (Ser19) like MLCK, even in

---

[86] A.k.a. RoCKβ, p160RoCK, and Rho-associated protein kinase-1.

[87] A.k.a. RoCKα and Rho kinase-2.

[88] Paralogs RoCK1 and RoCK2 share overall 65% identity with 92% identity in their kinase domains.

[89] Granzymes are serine peptidases released from cytotoxic T and natural killer cells. The multimer granzyme-B–perforin–granulysin can enter a cell through transfer receptors such as the mannose 6-phosphate receptor.

the absence of Ca^{++} ion. Phosphorylation of myosin light chain leads to its interaction with actin that activates myosin ATPase and causes cell contraction. Phosphorylation of myosin-binding subunit (MBS or myosin phosphatase target subunit MyPT; Ser854, Thr697, and Thr855) of myosin light-chain phosphatase by RoCKs prevents the dephosphorylation of myosin light chain, thereby favoring stress-fiber formation and cell contractility [364].

In endothelial cells, RoCKs disrupt the integrity of both tight and adherens junctions, as they provoke contraction of stress fibers. They thus raise endothelial permeability. However, in epithelial cells, RoCK activation favors tight-junction integrity, hence enhancing barrier function. Nonetheless, activated RoCK may also disturb E-cadherin-mediated adherens junctions by initiating actomyosin bundle contraction.

Small GTPases of the RHO superfamily (CDC42, Rac1, and RhoA to RhoC), under the control of specific guanine nucleotide-exchange factors (RhoGEF) and GTPase-activating proteins (RhoGAP),[90] regulate angiogenesis, as they intervene in extracellular matrix remodeling, cell migration, proliferation, and survival, and morphogenesis. Vascular endothelial growth factor activates RhoA GTPase that intervenes in endothelial migration during angiogenesis. In general, morphogenesis of a duct network depends on the frequency and geometry of branchings, hence on the control of endothelial sprouting (i.e., initiation of new branches from existing vessels in response to chemical messengers).[91] The RoCK kinases limit endothelial sprouting and heightens the length of vessels formed upon VEGF stimulation [374]. In addition to elevated matrix stiffness, the activity of myosin-2 supported by RoCK enzymes prevents endothelial sprouting, as they lower the activation of VEGFR2 receptor [375].[92] Myosin-2 localizes to the cortex of endothelial cells. Local depletion of cortical myosin-2 precedes pseudopodium, hence branch initiation [375]. The RhoA–RoCK axis determines branching sites of tip cells by regulating tension in the cortex of endothelial cells. On the other hand, RhoA at the trailing edge promotes tail retraction to assist migration of stalk cells [373] (Table 5.11). In addition, RoCK kinase prevents activity of endothelial nitric oxide synthase, hence that of NO in the development and maintenance of the microvasculature [373].

---

[90] Both RhoGAP5 and RhoGAP24 are implicated in the regulation of angiogenesis. Pleckstrin homology (PH) domain-containing, family-G (with RhoGEF domain) member PlekHg5 (a.k.a. in mice synectin-binding RhoA exchange factor Syx) is a RhoGEF that acts in angiogenic sprouting [373]. In addition, RhoGEF31 also participates in the regulation of angiogenesis.

[91] Tip endothelial cells branch off after emitting directional protrusion of pseudopodes.

[92] Once VEGFR is activated, Delta-like ligand DLL4 in a given endothelial cell confers a phenotype of tip cell for the growth of blood vessels and activates Notch in adjacent endothelial cells for stalk cell specification. Activated Notch suppresses VEGFR action, thereby hampering their conversion into tip cells (Vol. 5 – Chap. 10. Vasculature Growth).

Table 5.11. Monomeric Rho GTPase and its effector RoCK kinase in angiogenesis (Source: [373]).

RhoA–RoCK	Regulation of tension in the cortex of endothelial cells, inhibition of tip cell protrusion, hence of vascular sprouting Trailing edge retraction, hence supporting migration of stalk cells
VEGFR2–RoCK	Rho-independent activation of RoCK enzymes Negative feedback exerted by RoCK1/2 on VEGFR2

5.2.14.1 RoCK1 Kinase

Ubiquitous Rho-activated protein kinase RoCK1 has a slightly higher expression in the human brain, spinal cord, retina, and lung [376]. Kinase RoCK1 is located in the cytoplasm, accumulating particularly in the perinuclear region and Golgi body. It colocalizes with actin stress fibers. It associates with centrosomes. It can contribute to centrosome positioning and centrosome-dependent exit from mitosis [376].

The major cellular function of RoCK1 is the regulation of cell morphology by controlling functioning of actin–myosin and intermediate filaments. Isokinase RoCK1, but not RoCK2, is inhibited by RhoE and Gem GTPases [376] (Sects. 9.3.25 and 9.3.26). Isoform RoCK1 could play a more important part in the regulation of the actin–myosin cytoskeleton and phagocytosis, as it differs from RoCK2 in phosphatidylinositol binding and subcellular localization.

The activity of RoCK1 is regulated by binding of $RhoA^{GTP}$, $RhoB^{GTP}$, or $RhoC^{GTP}$ via 3 sequences: Rho-binding domain (RBD), Rho-interacting domain (RID), and possibly a leucine zipper homology region HR1 [376]. Activation can also result from caspase-mediated proteolysis that generates an active fragment lacking the C-terminal auto-inhibitory domain. Cyclin-dependent kinase inhibitor CKI1a hinders RoCK1 activity, when CKI1a is located in the cytoplasm.

Substrates of RoCK1 comprise RhoE and PTen phosphatase. Kinase RoCK1 also phosphorylates LIMK1 (Thr508) and LIMK2 (Thr505), thereby enhancing the ability of LIMKs to phosphorylate actin-depolymerizing cofilin (Ser3). Phosphorylation by RoCK of MLCP, MLC, and LIMK heightens actin-filament assembly and actomyosin filament contraction. Kinase RoCK1 also interacts with RhoA and RhoE GTPases, cofilin-1, profilin-2, GRB2-associated-binding protein GAB1, and CDC25a phosphatase [251]. Isoform RoCK1 phosphorylates (activates) Na^+–H^+ exchanger NHE1 (SLC9a1). It participates in RhoA-induced reorganization of actin filaments. It also binds to ezrin, radixin, and moesin to contribute to fibroblast migration [364].

5.2.14.2 RoCK2 Kinase

Ubiquitous RoCK2 is the main type in smooth muscle cells and neurons. It is activated by association with Rho^{GTP}. Kinase RoCK2 stabilizes filamentous actin and enhances myosin ATPase activity. It promotes the formation of contractile actin–myosin bundles (stress fibers) and integrin-containing focal adhesions. It phosphorylates (inhibits) myosin light-chain phosphatase, and thus increases myosin-2 activity even at a constant intracellular calcium level (*calcium sensitization*), and causes sustained vasoconstriction and bronchospasm.

Kinase RoCK2 localizes in the cytoplasm, especially close to the plasma membrane and in the perinuclear region. Linkage to Rho^{GTP} primes its translocation to membranes. In some circumstances such as growth factor stimulation, RoCK2 translocates to the nucleus, where it phosphorylates P300 acetyltransferase and CDC25a phosphatase [377]. During late mitosis, it localizes with vimentin (i.e., intermediate filaments) at the cleavage furrow.

Like RoCK1, RoCK2 is regulated by binding of GTP-loaded RhoA, RhoB, or RhoC. Activation also happens by granzyme-B-mediated proteolysis that generates an active fragment lacking the C-terminal auto-inhibitory motif [377] (RoCK2$^{\Delta^{CT}}$). Furthermore, binding of PI(3,4,5)P$_3$ or PI(4,5)P$_2$ to the pleckstrin homology domain heightens RoCK2 activity. Cyclin-dependent kinase inhibitor CKI1a, when it resides in the cytoplasm, hinders RoCK activity. Kinase cRaf, particularly after stimulation of the pro-apoptotic TNFRSF6a pathway, inhibits RoCK2 enzyme.

Kinase RoCK2 also associates with activator arachadonic acid and multiple cytoskeletal proteins and their regulators, such as myosin regulatory light chain, myosin-binding subunit of the myosin light-chain phosphatase, smooth muscle-specific PP1$_{r14}$ inhibitory subunit of myosin phosphatase, calcium-binding myosin-ATPase inhibitor calponin-1, plasma membrane–cytoskeleton connectors ezrin, radixin, and moesin, actin-binding protein profilin-2, α-adducin (Add1), plasmalemmal actin-filament crosslinker myristoylated alanine-rich protein kinase-C substrate (MARCKS), LIMK1 and LIMK2, the sarcomeric and intermediate filament constituents desmin and vimentin, cardiac troponin-I and -T, neurofilament light polypeptide, astrocyte-specific intermediate filament protein glial fibrillary acidic protein (GFAP), dihydropyrimidinase-like protein Dpysl2[93] that modulates RhoA-induced neuronal morphology [377].[94]

Adducin is a filamentous actin-binding protein that promotes spectrin–actin assembly beneath the plasma membrane. Phosphorylation of α-adducin by RoCK2 enhances linkage of adducin to actin filaments. α-Adducin phosphorylated by RoCK2 can be dephosphorylated by MLCP. Ezrin–radixin–moesin complexes crosslink actin filaments to plasmalemmal proteins. Phosphorylation of radixin

---

[93] A.k.a. collapsin response mediator protein CRMP2.

[94] Neuronal collapsin response mediator protein CRMP2 is involved in semaphorin-3A- and lisophosphatidic acid-induced collapse of growth cones that are dynamic, actin-supported extensions of a developing axon seeking its synaptic target.

by RoCK2 (Thr564) disrupts the interaction between its N- and C-termini [364]. Like protein kinase-Cθ, RoCK2 can phosphorylate moesin (Thr558), thereby antagonizing MLCP. Kinase RoCK2 phosphorylates several ^{F}actin-binding proteins to decrease their actin-binding capacity, such as smooth muscle calponin-1 that hampers actin-activated myosin ATPase activity.

Vimentin, glial fibrillary acidic protein, and neurofilament light protein are intermediate-filament proteins that can be phosphorylated by various kinases. Vimentin phosphorylation induces filament disassembly.

Additional partners include STK16[95] and IKKε [251]. Other RoCK2 substrates encompass eukaryotic translation elongation factor eEF1α1 expressed in the brain, lung, liver, kidney, pancreas, and placenta and responsible for the enzymatic delivery of aminoacyl tRNAs to the ribosome, PTen, and Na^+–H^+ exchanger isoform-1 [377].

5.2.15 Other Members of the ROCK Family

In addition to RoCK kinases, members of the ROCK family include citron Rho-interacting kinase (CRIK), myotonin protein kinase (DMPK), and myotonic dystrophy kinase-related CDC42-binding kinase (MRCK).[96] They control the cytoskeleton, hence cell shape and motility (Sect. 9.3.12).

Kinase CRIK can bind Rho^{GTP}; DMPK interacts with Rac^{GTP}; and MRCK connects to diacylglycerol that provokes a conformational change allowing its binding to CDC42^{GTP} [300]. Kinase RoCK is also activated in apoptotic cells by cleavage of its auto-inhibitory C-terminus by caspase-3. The hydrophobic motif is mainly aimed, at least in RoCK, at promoting dimerization.

5.2.16 P90 Ribosomal S6 Kinase Family

Ribosomal protein S6 kinases constitute a family of enzymes that may be considered as belonging to distinct superfamilies of protein Ser/Thr kinases, the AGC and CAMK superfamilies (Table 5.12) [378]. Kinase isoforms RSKi ($i = 1,\ldots,4$) and MSKi ($i = 1,\ldots,2$) contain 2 kinase domains, a N-terminal (NTK) and a secondary C-terminal (CTK) kinase motif that pertain to the AGC and CAMK superfamily, respectively [379]. The CTK domain activates the NTK sequence. The latter phosphorylates substrates.

---

[95] A.k.a. TGFβ-stimulated factor TSF1, myristoylated and palmitoylated Ser/Thr kinase MPSK1, protein kinase expressed in day-12 fetal liver (PKL12), and kinase related to Saccharomyces cerevisiae and Arabidopsis thaliana KRCT.

[96] A.k.a. CDC42BPα.

Table 5.12. Ribosomal protein S6 kinases (RPS6K) of the AGC superfamily. Ubiquitous kinase isoforms RSKi ($i = 1,\ldots,4$; RSK: 90-kDa ribosomal S6 kinase) and MSKi ($i = 1,\ldots,2$; MSK: mitogen- and stress-activated protein kinase) of the AGC superfamily. The RSK proteins, or MAPKAPK1s, and MSK work in signal transduction downstream from MAPK cascades. The former is activated by members of the extracellular signal-regulated kinase (ERK) family in response to growth factors, many polypeptidic hormones, neurotransmitters, chemokines, and other stimuli. The latter is activated by ERKs in response to same stimuli as well as P38MAPKs in response to various cellular stresses and pro-inflammatory cytokines.

Name	Alias	Other alias
RPS6K, 52kDa, polypeptide-1	RPS6Kδ1 (RPS6Kc1)	RSKL1
RPS6K-like 1	RPS6KL1	RSKL2
RPS6K, 70kDa, polypeptide-1	RPS6Kβ1	S6K1
RPS6K, 70kDa, polypeptide-2	RPS6Kβ2	S6K2
RPS6K, 70kDa, polypeptide-3	RPS6Kβ3	S6K3
RPS6K, 90kDa, polypeptide-1	RPS6Kα1	RSK1
RPS6K, 90kDa, polypeptide-2	RPS6Kα2	RSK3
RPS6K, 90kDa, polypeptide-3	RPS6Kα3	RSK2
RPS6K, 90kDa, polypeptide-4	RPS6Kα4	RSKb, MSK2
RPS6K, 90kDa, polypeptide-5	RPS6Kα5	RLPK, MSK1
RPS6K, 90kDa, polypeptide-6	RPS6Kα6	RSK4

The MAPK-activated protein kinases (MAPKAPK) that respond to extracellular stimuli on MAPK-induced phosphorylation (activation) constitute a category with 5 MAPKAPK subfamilies of [380]: (1) P90 ribosomal S6 kinase (RSK1–RSK4); (2) mitogen- and stress-activated kinase (MSK1–MSK2); (3) MAPK-interacting kinases (MNK1–MNK2); (4) MAPK-activated protein kinase-2 and -3 (MK2–MK3);[97] and (5) MK5.[98] These enzymes are involved in regulation of nucleosome and gene expression, mRNA stability and translation, and cell proliferation and survival (Table 5.13).

Ribosomal protein S6 kinases possess 2 non-identical kinase domains to phosphorylate S6 ribosomal protein, hence to increase protein synthesis and favor cell proliferation. They can phosphorylate various substrates, such as eukaryotic translation initiation factor eIF4b (by RPS6Kβ2 [or S6K1]), members of mitogen-activated protein kinase modules (by members of the RPS6Kα [RSK] set), as well as transcription factors CREB1 and Fos (by RPS6Kα4 [or RSKb]).

Ribosomal protein S6 kinase-δ1[99] binds to sphingosine kinase-1, phosphatidylinositol 3-phosphate, and anti-oxidant peroxiredoxin Prdx3 (thioredoxin-dependent peroxide reductase) [381, 382]. It contains a phox homology (PX) motif and

---

[97] The gene Mk3 has been detected in Mus musculis, but not in humans.

[98] A.k.a. P38-regulated/activated protein kinase (PRAK).

[99] A.k.a. RPS6Kc1, RSKL1, 118-kDa ribosomal S6 kinase-like protein with 2 PSK domains protein (RPK118), and sphingosine kinase SphK1-binding protein.

Table 5.13. Activation pathways and effects of P90 ribosomal S6 kinases (RSK) and mitogen- and stress-activated kinases (MSK) that constitute 2 MAPK-activated protein kinase (MAPKAPK) sets (Source: [380]).

Stimuli	Pathways	Effects
	RSK1–RSK4	
Cell stresses	MAP3K2/3/8–MAP2K5–ERK4(5)	Gene transcription,
Growth factors	MAP3K2/3/8–MAP2K5–ERK4(5)	cell survival, growth,
	a/b/cRaf, Mos–MAP2K1/2–ERK1/2	and proliferation
Mitogens	a/b/cRaf, Mos–MAP2K1/2–ERK1/2	
	MSK1–MSK2	
Growth factors, mitogens	a/b/cRaf, Mos–MAP2K1/2–ERK1/2	Gene transcription, nucleosomal response
Cytokines, cell stresses	MAP3K1–5/7/8/10/12/17/18– –MAP2K3/6–P38MAPK	
	MNK1–MNK2	
Growth factors, mitogens	a/b/cRaf, Mos–MAP2K1/2–ERK1/2	mRNA translation
Cytokines, cell stresses	MAP3K1–5/7/8/10/12/17/18– –MAP2K3/6–P38MAPK	
	MK2–MK3 (MAPKAP2–MAPKAP3)	
Cytokines, cell stresses	MAP3K1–5/7/8/10/12/17/18– –MAP2K4/7–JNK1–3	Gene transcription, actin remodeling, cytokine synthesis, cell cycle control, cell migration
	MK5 (MAPKAP5) ER2K3/4–ERK3/4	Actin remodeling

2 pseudokinase domains. On early endosomes, RPS6Kδ1 colocalizes with sphingosine kinase-1, which synthesizes lipid messenger sphingosine 1-phosphate. It is implicated in sphingosine 1-phosphate-mediated signaling.

P90RSK Family

The family of 90-kDa protein Ser/Thr ribosomal S6-kinases regulates cell growth, survival, proliferation, and motility. The RSK family consists of 4 isoforms (RSK1–RSK4) and 2 structurally related homologs, RSK-like protein kinase or

Table 5.14. Members the RSK family of the AGC superfamily. Kinases MAPKAPK2, -3, and -5 pertain to the CAMK superfamily. Five MAPKAPK subfamilies include the RSK, MSK, MINK, MK2, and MK5 groups (MK [MAPKAPK]: mitogen-activated protein kinase-activated protein kinase; MNK [MINK]: MAPK-interacting protein serine/threonine kinase, or MAPK signal-integrating kinase; MSK: nuclear mitogen-and stress-activated protein kinase; RPS6K [RSK]: ribosomal protein S6 kinase; NA: not applicable).

Gene	Protein	Alias in MAPK module
RSK subfamily — RPS6K		
RPS6KA1	RSK1	MAPKAPK1a, RPS6Kα1
RPS6KA2	RSK3	MAPKAPK1c, RPS6Kα3
RPS6KA3	RSK2	MAPKAPK1b, RPS6Kα2
RPS6KA6	RSK4	NA
MSK subfamily — RPS6K		
RPS6KA4	RSKb	MSK2, RPS6Kα5
RPS6KA5	RLPK	MSK1, RPS6Kα4
Other subfamilies of MAPK-activated protein kinases (members of the CAMK superfamily)		
MINK subfamily		
MKNK1	MNK1	
MKNK2	MNK2	
Mapkapk2	MK2	MAPKAPK2
Mapkapk5	MK2	MAPKAPK2

mitogen- and stress-activated kinase-1 (RLPK or MSK1) and -2 (RSKb or MSK2; Table 5.14).[100]

The P90 ribosomal S6-kinases act downstream from the Ras–ERK pathway in response to growth factors, peptidic hormones, neurotransmitters, and chemokines. Ligand-bound receptors cause autophosphorylation and create docking sites for adaptors, such as growth factor receptor-bound protein GRB2 that links the receptor to the guanine nucleotide-exchange SOS factor. Son of sevenless stimulates Ras GTPase that recruits Raf kinases to the plasma membrane for activation. Activated Raf phosphorylates both MAP2K1 and MAP2K2 enzymes. Activated MAP2Ks then phosphorylate extracellular signal-regulated kinases ERK1

---

[100] Mitogen- and stress-activated kinases (Chap. 6) are regulated by common and distinct upstream activators with respect to RSKs. They activate different targets. Whereas RSK activation result from phosphorylation at 6 sites, RSK-like kinases possess a single phosphorylation site. Moreover, MSKs are constitutively localized to the nucleus, whereas RSKs are predominantly cytosolic in inactivated cells. MSK1 and MSK2 are phosphorylated by ERK1 and ERK2, as well as by P38MAPK activated by MAP2K3 and MAP2K6, but not by PDK1. The MSK enzymes require additional autophosphorylation to become fully active.

and ERK2. The P90 ribosomal S6-kinases are targeted not only by ERK1 and -2, but also by 3-phosphoinositide-dependent kinase PDK1.[101] In dendritic cells, RSKs can be activated also by P38MAPK. On the other hand, protein phosphatase-2Cδ binds to and dephosphorylates RSKs. Phosphorylation mediated by ERK1 and ERK2 of PPM1d dissociates the RSK–PPM1d complex.

Inactive RSKs are located in the cytoplasm and nucleus, whereas activated RSKs remain linked to the plasma membrane, in the cytosol, or translocate to the nucleus. Phosphorylation of RSK occurs at the plasma membrane and in the cytoplasm and nucleus [383]. In addition, a fraction of activated ERKs and RSKs enters the nucleus to prime immediate-early gene expression. Nevertheless, constitutively active RSK4 is predominantly cytosolic.

RSKs regulate various transcription factors, such as cAMP-responsive element-binding protein, serum response factor, ETS transcription factor ER81, estrogen receptor-α, nuclear factor-κB, nuclear factor of activated T-cell NFAT3, cytoplasmic, calcineurin-dependent NFAT4, and transcription initation factor TIF1A2. RSKs also phosphorylate eukaryotic translation initiation factor-4B that stimulates heterotrimeric eIF4f[102] and promotes interaction between eIF4b and eIF3 proteins.

In addition, RSKs favors cell survival by inactivating some pro-apoptotic proteins such as BCL2 antagonist of cell death (BAD), death-associated protein kinase (DAPK), whereas it phosphorylates CCAAT/enhancer binding protein-β that inhibits caspases [383]. Both RSK1 and RSK2 also regulate the cell division cycle, as they phosphorylate cyclin-dependent kinase inhibitor CKI1b for 14-3-3-mediated cytosolic sequestration, thereby promoting G1-phase progression.[103] Kinase RSK1 phosphorylates (inactivates) nitric oxide synthase-1. Kinase RSK2 phosphorylates Na^+–H^+ exchanger isoform NHE1, which regulates intracellular pH. Like SGK1, PKB, and AMPK, RSK phosphorylate PKB substrate AS160 that mediates translocation of glucose transporter GluT4 to the plasma membrane.

Inactive RSK1 is able to bind to the PKA regulatory subunit to attenuate the interaction between the regulatory and catalytic subunits. Activated RSK1 associates with PKA catalytic subunit and reduces the ability of cAMP to stimulate PKA [383]. Interaction between RSK1 and PKA allows RSK1 colocalization with A-kinase-anchoring proteins, whereas PKA–AKAP interaction supports nuclear accumulation of active RSK1.

RSK1, -2, and -3 are ubiquitously expressed, but with tissue-specific concentrations [383]. RSK1 is predominantly expressed in kidney, lungs, and pancreas. Both

---

[101] PDK1 mediates PI3K effect. PI3K indirectly regulates PDK1 by phosphorylating phosphatidylinositols, as PDK1 interacts with membrane phospholipids, such as phosphatidylinositol (3,4)-bisphosphate and phosphatidylinositol (3,4,5)-trisphosphate. However, PDK1 can be constitutively active and does not needs phosphatidylinositol binding for phosphorylation of most of its substrates, especially in the cytosol.

[102] The eIF4f complex is made of cap-binding protein eIF4e, scaffolding protein eIF4g, and helicase eIF4a.

[103] G1-phase progression is impeded by nuclear CKI1b that inactivates cyclin-E–CDK2 or cyclin-A–CDK2 complexes.

RSK2 and RSK3 are highly produced in skeletal muscle, heart, and pancreas. In the brain, RSK1 is mainly synthesized in the cerebellum; RSK2 in the cerebellum and occipital and frontal lobes; and RSK3 in the medulla. Defective RSK2 gene causes Coffin-Lowry syndrome. Aberrant RSK activation has been detected in some cancer types.

Mitogen- and stress-activated kinases activated by ERK1 and ERK2 as well as P38MAPK (but not PDK1) phosphorylate transcription factors cAMP-responsive element-binding protein and activating transcription factor ATF1, as well as histone-3. Mitogen- and stress-activated kinases participate in full stimulation of several immediate-early genes, such as Fos, JunB, MAPK phosphatase MKP1, and NURR1.

During evolution, a unique feature of kinases of the RSK family arises from fusion of genes for 2 distinct protein kinases generating a kinase with a C-terminal kinase domain (CTKD) that receives an activating signal from ERK and transmits this cue to the N-terminal kinase domain (NTKD).

5.2.16.1 RSK1 Kinase

Kinase RSK1[104] is expressed in many tissues, with higher levels in the brain, lung, kidney, liver, pancreas, skeletal muscle, spleen, and thymus. It resides in the cytoplasm. Upon mitogenic stimulation, it moves toward the plasma membrane, where it accumulates transiently, and then translocates into the nucleus, where it phosphorylates several substrates involved in gene transcription [384].

Enzyme RSK1 phosphorylates ribosomal protein-S6 in response to activated Ras–ERK pathway. It regulates several transcription factors, such as SRF, Fos, NFAT4, ETS variant gene product ETV1, and microphthalmia-associated transcription factor (MiTF), as well as testicular receptor TR3 (or nuclear receptor NR4a1 of the NUR family [NuR77]) and estrogen receptor-α (NR3a1) [384]. It also interacts with transcriptional coactivator CREB-binding protein (CBP), cyclin-dependent kinase inhibitor CKI1b, Na^+–H^+ exchanger NHE1, and GSK3 kinase. In response to mitogenic signaling, RSK1 phosphorylates (inhibits) nitric oxide synthase NOS1. Kinase RSK1 also regulates hamartin (tuberous sclerosis-1) TSC1 and tuberin (tuberous sclerosis-2) TSC2, ribosomal protein-S6, calcium–calmodulin-dependent eukaryotic elongation factor-2 kinase eEF2K, STK11 (or LKb1), a kinase of AMPK, and translation initiation factor eIF4b. Therefore, RSK1 participates in cell growth and proliferation. Kinase RSK1 also phosphorylates filamin-A, hence contributing to cytoskeleton reorganization. Other partners include protein kinase PKA, transducer of HER2 TOB1, and calcium-regulated heat stable protein-1 (CaRHSP1) [251].

---

[104] A.k.a. ribosomal protein S6 kinase-α1 and ribosomal protein S6 kinase, 90-kDa, polypeptide-1 (RPS6Kα1).

Protein RSK1 controls cell survival, as it phosphorylates (inactivates) proapoptotic protein BAD and death-associated protein kinase (DAPK). It also targets transcription factors CCAAT/enhancer-binding proteins C/EBPβ and nuclear factor-κB [384].

Kinase RSK1 as well as ERK1 and ERK2 form an inactive complex in quiescent cells. Upon mitogenic stimulation, ERK1 and ERK2 phosphorylate RSK1, that, in turn, autophosphorylates to create a docking site for phosphoinositide-dependent protein kinase PDK1 [384]. On the other hand, RSK1 is sequestered by 14-3-3β protein.

5.2.16.2 RSK2 Kinase

Kinase RSK2[105] is implicated in the control of cell growth and differentiation. Like RSK1, RSK2 phosphorylates ribosomal protein-S6 in response to mitogenic stimulation of the Ras–ERK pathway.

Enzyme RSK2 interacts with CREB-binding protein transcriptional coactivator and histone acetyltransferase (CBP), ERK1, ERK2, P38MAPKα, 3-phosphoinositide-dependent protein kinase PDK1, and death effector domain-containing phosphoprotein enriched in astrocytes PEA15 [251].

5.2.16.3 RSK3 Kinase

Kinase RSK3[106] is predominantly expressed in the lung, heart, skeletal muscles, and pancreas [385]. Three RSK3 splice variants exist (RSK3a–RSK3c).

Enzyme RSK3 intervenes during cell cycle progression [385]. It translocates to the nucleus during G0–G1 transition. It phosphorylates Fos and histones H1, H2A, H2B, H3, and H4.

Kinase RSK3 contains 6 phosphorylation sites. Maximal activation of growth factor-stimulated RSK3 requires the participation of both ERK1 and ERK2 and PDK1 kinases. Inactive RSK3 forms a complex with ERK1 and ERK2 and remains associated with these active ERKs following mitogen stimulation [385]. It interacts with PPM1d phosphatase. Kinase RSK3 also phosphorylates protein phosphatase PP1 and ribosomal 40S subunit. Lastly, RSK3 interacts with CREB-binding protein [251].

---

[105] A.k.a. insulin-stimulated protein kinase ISPK1, mental retardation X-linked gene product MRX19, S6Kα3, and ribosomal protein S6 kinase, 90-kDa, polypeptide-3 (RPS6Ka3).

[106] A.k.a. ribosomal protein S6 kinase-α2, protein kinase MPK9, and ribosomal protein S6 kinase, 90-kDa, polypeptide-2 (RPS6Ka2).

5.2.16.4 RSK4 Kinase

Kinase RSK4[107] resides in many tissues, with high levels in fetal tissues as well as adult brain and kidney [384].

Enzyme RSK4 participates in P53-dependent growth arrest. It also promotes protein synthesis. In addition, RSK4 acts as an inhibitor of receptor Tyr kinase signaling during embryogenesis [384].

Kinase RSK4 possesses 6 phosphorylation sites.[108] Activity of RSK4 is constitutively high even in the absence of growth factor stimulation [384]. Mitogenic stimulation further heightens RSK4 activity.

Protein RSK4 can associate with its activating extracellular signal-regulated kinases ERK1 and ERK2 via a docking (D) domain near its C-terminus [384]. It interacts with PPM1d phosphatase. In addition, it phosphorylates ribosomal protein S6 peptide and glycogen synthase kinase GSK3β. It also interacts with the nuclear receptor NR4a1 [251].

5.2.16.5 Kinases RSKL1 and RSKL2

Ribosomal S6 kinase-like proteins RSKL1 and RSKL2 are classified into group-E pseudokinases, as they lack the essential catalytic motifs in the kinase domain (Vol. 3 – Chap. 8. Receptor Kinases and Sect. 3.4). Pseudokinase domains of RSKL1 are used to bind sphingosine kinase-1 as well as peroxiredoxin-3 [300]. Isoform RSKL1, but not RSKL2, can interact with PI(3)P, and hence localize in early endosomes. In addition, RSKL1 and RSKL2 can connect to microtubules and intervene in intracellular transfer of cargos.

5.2.17 P70 Ribosomal S6 Kinase Family

The family of 70-kDa Ser/Thr ribosomal S6 kinases of the AGC superfamily includes S6K1 and S6K2 that are encoded by the RPS6KB1 and RPS6KB2 genes, respectively.

Both paralogs are ubiquitous. They participate in the regulation of cell metabolism, survival, and growth. These monomeric kinases are activated in cardiac, skeletal, and smooth muscle in response to a hypertrophic stimulus. Both paralogs have a catalytic activity that is controlled via multisite phosphorylation.

Activity of S6Ks is controlled by a set of phosphorylations of their activation segment and hydrophobic motif in response to insulin and growth factors. They are activated by target of rapamycin, a signaling node for many pathways that

---

[107] A.k.a. ribosomal protein S6 kinase, 90-kDa, polypeptide-6 (RPS6Ka6).
[108] Ser232, Thr368, Ser372, Ser389, Thr581 and Ser742.

mainly rely on PI3K, PTen, and PKB enzymes. The S6K isoforms are phosphorylated (Thr389) by TORC1 complex using proline-rich Akt (PKB) substrate of 40 kDa (PRAS40), a TORC1 inhibitor, and tuberous sclerosis complex TSC2 and afterward by PDK1. As their name suggests, they phosphorylate 40S ribosomal protein subunit-S6 (RPS6). Ribosomal S6 kinases S6K1 and S6K2 augment protein synthesis at ribosomes and favor cell proliferation. Both isoforms have also been shown to phosphorylate eukaryotic translation initiation factor eIF4b [386].

5.2.17.1 S6K1

Paralog S6K1[109] possesses 2 isoforms encoded by alternatively spliced transcript variants, the short ($S6K1_S$ or $^{P70}S6K1$) and long ($S6K1_L$ or $^{P85}S6K1$) splice variants [386].

Kinase S6K1 is activated by various growth factor pathways. In addition, it is sensitive to intracellular nutrient conditions. A sudden increase in cellular amino acid level primes S6K1 activation, whereas a lowered level precludes S6K1 activity. Similarly, when ATP levels drop, S6K1 is inhibited [386].

Glycogen synthase kinase-3 and target of rapamycin cooperate to control the activity of S6K1 [387].

Paralog S6K1 phosphorylates programmed cell death protein PDCD4, an inhibitor of eIF4a and eIF4g interaction. Moreover, it phosphorylates TOR (retrograde signaling). Kinase S6K1 can directly influence cell growth by acting on its specific substrates, such as S6K1-specific Aly/REF-like target (SKAR) that is also known as polymerase DNA-directed, δ-interacting protein PolDIP3 or PDIP46, a nuclear protein that intervenes at mRNA processing stage [386]. Under certain circumstances, S6K1 can phosphorylate glycogen synthase kinase GSK3 and BCL-2 associated death promoter (BAD; inhibition). Insulin-stimulated S6K1 phosphorylates (inactivates) insulin receptor substrate IRS1 (negative feedback loop). Additional S6K1 interactors include PDK1, small GTPases CDC42 and Rac1, the eIF3 complex, and isoforms PKCζ and PKCι.

5.2.17.2 S6K2

Messenger RNA of S6K2 generates 2 isoforms: $S6K2_S$ and $S6K2_L$. Unlike S6K1, S6K2 contains an additional C-terminal nuclear localization signal [386]. Paralog S6K2 is regulated by many S6K1 interactors (PI3K, PKB, PKCζ, Rac1, and CDC42). However, S6K2 has unique functions that can complement those of S6K1

---

[109] A.k.a. 70-kDa ribosomal protein S6 kinase-1, ribosomal protein S6 kinase, polypeptide 1, and Ser/Thr kinase STK14α.

kinase. It forms a complex with bRaf and PKCε upon signaling launched by fibroblast growth factor FGF2. The extracellular-signal-regulated kinase pathway can activate S6K2.

5.2.18 Serum- and Glucocorticoid-Regulated Kinase Family

Serum- and glucocorticoid-regulated kinases constitute a family of 3 members: SGK1 (or SGK), SGK2 (or hSGK2); and SGK3 (a.k.a. CISK and SGKL). They can activate certain potassium, sodium, and chloride channels. Kinases SGK1 and SGK2 have a single identifiable functional domain, its catalytic domain, but SGK3 possesses a phosphoinositide-binding Phox homology (PX) domain at its N-terminus that interacts with phosphatidylinositol 3-phosphate to mediate the endosomal association of SGK3, which is essential for its phosphorylation and activation by PDK1 kinase.

Kinase SGK1 can stimulate sodium transport into epithelial cells, as it enhances the density and stability of epithelial sodium channels. It may also have overlapping functions with PKB isoforms. Isoform SGK1 is phosphorylated (Ser422) by TORC2 complex, then by PDK1 kinase [300].

5.3 Diacylglycerol Receptors and Kinases

There are many diacylglycerol receptors (DAGR): protein kinase-C (PKC) and -D (PKD),[110] chimaerins,[111] and others.

Receptors for activated C-kinase (RACK) bind to and position newly activated PKCs to discrete cellular locations. Peptides associated with RACK–PKC interaction act as translocation activators or inhibitors. Both PKCδ and PKCε are implicated in the evolution of the cardiac function after myocardial infarction [390]. They are also implicated in vasculogenesis. Enzymes PKCα and PKCε control integrin signaling to extracellular signal-regulated protein kinase [391].

The family of protein Ser/Thr kinase-D includes PKD1 (or PKCμ), PKD2, and PKD3 (or PKCν). They are located in the cytosol, nucleus, Golgi body, and plasma membrane.

---

[110]Function of PKD is determined by its intracellular location and cell context [388]. The release of reactive oxygen species from the mitochondria activates PKD.

[111]Chimaerins have a DAG-binding C1 domain and a GTPase-activating protein (GAP) domain that targets Rho GTPases. They block Rac signaling, cell proliferation, and cytoskeletal reorganization [389].

Diacylglycerol kinases[112] phosphorylate diacylglycerol to stop its signaling and produce phosphatidic acid. They then modulate PKC signaling because protein kinase-C is activated by diacylglycerol, and Ras signaling, as well as protein recruitment to membrane domains.

5.4 CAMK Superfamily

5.4.1 Calmodulin-Dependent Protein Kinase (CamK) Family

Calmodulin-dependent Ser/Thr protein kinases (CamK) require activated calmodulin for their activity. Like protein kinases PKA and PKC, CamK phosphorylates nitric oxide synthase (Sect. 10.3.1) [392].

Two categories of CamKs include: (1) specialized CamKs such as myosin light-chain kinase that phosphorylates myosin for muscle contraction and (2) multifunctional CamKs such as type-1 and -4 calmodulin-dependent protein kinases. Members of the CamK family include CamK1, CamK2, and CamK4 that are encoded by the genes of the CAMK1 (CAMK1, CAMK1D, and CAMK1G), CAMK2 (CAMK2A, CAMK2B, CAMK2D, and CAMK2G), and CAMK4 subfamily, respectively (Table 5.15).

Calmodulin-dependent protein kinases contain catalytic, regulatory, and association domains. These enzymes assemble into a dodecameric holoenzyme. Several CamKs can aggregate into a homo- or hetero-oligomer. In the absence of Ca^{++}–calmodulin, the catalytic domain is auto-inhibited by the regulatory domain. Upon activation by Ca^{++}–calmodulin, activated CamK autophosphorylates. To gain maximal activity, CamK1 and -4 can be further phosphorylated by calmodulin-dependent protein kinase kinase (Cam2K).[113] Two distinct Cam2Ks exist: Cam2Kα and Cam2Kβ.[114]

5.4.1.1 Calcium–Calmodulin-Dependent Protein Kinase-1

Calcium–calmodulin-dependent protein kinase-1 is expressed by many cell types. Kinase CamK1 phosphorylates many substrates, such as synapsin-1 and -2, cystic fibrosis transmembrane conductance regulator, and transcription factors CREB and ATF1. Isozyme CamK1 is involved in angiotensin-2 and K^+ stimulation of

---

[112] Diacylglycerol kinases are classified into five subtypes according to their structure.

[113] Kinase Cam2K targets Thr177 and Thr196 in CamK1 and -4, respectively.

[114] The human CAM2KB gene that maps to chromosome 12q24.2, is organized into 18 exons and 17 introns. Alternative splicing generates several CAM2KB transcripts that have a distinct ability to undergo autophosphorylation and to phosphorylate both CamK1 and -4 [393].

Table 5.15. Calmodulin-dependent protein kinases. Alternatively spliced transcript variants encode distinct isoforms. Disc large homolog DLg1 is a member of the membrane-associated guanylate kinase family of scaffolds that control clustering of postsynaptic ion channels in neurons and in adherens junction formation in epithelial cells. Regulatory subunit-1 (P35) of cyclin-dependent kinase-5 (CDK5$_{r1}$) is a neuron-specific CDK5 activator. Synapsins regulate neurotransmitter release at synapses. Small GTPase Ras-associated with diabetes (Rad) inhibits CamK2 in cardiomyocytes (CMC). Phospholamban inhibits sarcoplasmic reticulum Ca^{++} ATPase SERCA. In cardiomyocytes, CamK2 has both positive inotropic and lusitropic effects.

Gene	Isoforms	Distribution	Interaction
CAMK1		Many types	Cam2K; synapsins; CFTR; CREB, ATF1
CAMK1D			
CAMK1G			
CAMK2A	2	Neuron	Actinin-α4, DLg1, CDK5$_{r1}$
CAMK2B	8	Neuron	Actinin-α4
CAMK2D	3	Many types	Phospholamban, RyR, Ca$_V$1
		CMC	HDAC4
CAMK2G	6	Many types	Rad, phospholamban
		CMC	RyR and Ca$_V$1
CAMK4		T lymphocyte	Cam2K, PP2
		Neuron	
		Male germ cell	

aldosterone synthase gene CYP11B2 transcription in the adrenal glomerulosa to produce aldosterone [394]. Subtype CamK1 activation cascade involves calcium–calmodulin and Cam2K. Protein kinase-A phosphorylates (inactivates) Cam2K [395]. Kinase Cam2K is also phosphorylated by CamK1 (negative feedback). Regulation by PKA of the Cam2K–CamK1 cascade can modulate the balance between cAMP and Ca^{++} signals.

5.4.1.2 Calcium–Calmodulin-Dependent Protein Kinase-2

Calcium–calmodulin-dependent protein kinase-2 phosphorylates a large number of substrates, such as ion channels and intracellular mediators of signal transduction. For example, it increases PLA2 activity. The canonical activation of CamK2 requires Ca^{++} influx and calmodulin binding. However, CamK2 can be activated by oxidation.

Kinase CamK2 is made of catalytic, regulatory, and binding domains to form oligomers and act on its targets. Four isoforms of CamK2 (CamK2α–CamK2δ) are encoded by different genes. Isoforms CamK2α and CamK2β are almost exclusively produced in the central nervous system, where they intervene in synaptic evolvability, whereas CamK2γ and CamK2δ are more ubiquitous. Numerous splice variants have been recognized: 13 for CamK2γ and 15 for CamK2δ.

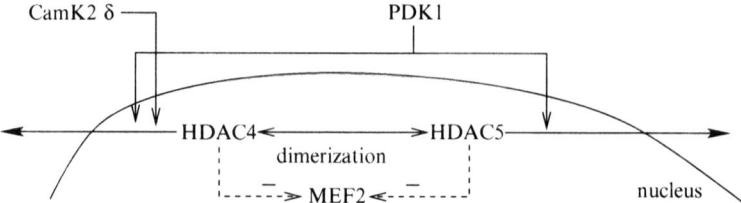

Fig. 5.2 Calcium–calmodulin-dependent protein kinase-2δ regulates gene transcription in cardiomyocyte (Source: [396]). Isoform CamK2δ phosphorylates histone deacetylase HDAC4 for nuclear export to relieve its repression on gene transcription. Nuclear HDAC4 can form dimer with HDAC5. Both HDAC4 and HDAC5 impedes activity of transcription factor myocyte enhancer factor MEF2. Phosphoinositide-dependent kinase PDK1 is another kinase that phosphorylates class-2a HDACs, i.e., both HDAC4 and HDAC5.

CamK2 Kinase in Smooth Muscle Cell

Vascular smooth muscle cells synthesize CamK2γ and CamK2δ isoforms and their variants. Isoform CamK2δ is coupled to ERK1 and ERK2 signaling. Isoform CamK2δ2 specifically regulates smooth muscle cell migration.

CamK2 Kinase in Cardiomyocyte

In the heart, CamK2β is expressed at a low level. Cardiac CamK2γ and CamK2δ are implicated in calcium homoeostasis, cell cycle progression, protein secretion, cytoskeletal organization, cell apoptosis, gene expression, and excitation–contraction coupling, as they phosphorylate phospholamban, ryanodine receptor, and Ca_V1 channel (Vols. 3 – Chap. 3. Main Classes of Ion Channels and Pumps and 5 – Chap. 7. Heart Wall).

Calcium–calmodulin-dependent protein kinase-2δ is the major isoform synthesized in the heart that is implicated in transduction of calcium signals. Isoform CamK2δ participates in maladaptive cardiac remodeling (heart hypertrophy) under pressure overload [396]. Enzyme CamK2δ targets transcriptional repressor histone deacetylase HDAC4 that is a substrate of stress-responsive protein kinases and suppressor of stress-dependent cardiac remodeling. Phosphorylated HDAC4 dissociates from transcription factor MEF2 and translocates to the cytoplasm, where it associates with 14-3-3 protein (Fig. 5.2).

In rat ventriculomyocytes, endothelin-1 has a positive inotropic effect via ET_A receptor,[115] Ca_V1 channel, and PKC-CamK2 pathway [397]. Endothelin-1 causes

---

[115]Receptor ET_A is dominant in cardiomyocytes. This GPCR is coupled to Gq subunit, like AT_1 and α1a adrenergic receptors.

phosphoinositide hydrolysis and activation of protein kinase-C. Consequently, ET1 also increases intracellular pH by PKC-induced activation of the Na^+–H^+ exchanger.

5.4.1.3 Calcium–Calmodulin-Dependent Protein Kinase-4

Calcium–calmodulin-dependent protein kinase-4 is a potent stimulator of Ca^{++}-dependent gene expression. Its activation requires both Ca^{++}–calmodulin binding and phosphorylation by Cam2K. This kinase can associate with serine/threonine protein phosphatase-2. Binding of Ca^{++}–calmodulin and PP2 is mutually exclusive [398]. Phosphatase PP2 impedes CamK4-mediated gene transcription. Kinase CamK4 is a potent activator of Ca^{++}-dependent transcription mediated by cAMP-responsive element-binding protein, CREB-related factor ATF1, MADS-box family members serum response factor and myocyte enhancer factor MEF2d, and nuclear receptors RORα, RORγ, and NR2f1. Whereas CamK1 and some CamK2 are ubiquitously expressed, CamK4 is detected in a limited number of cell types, such as thymocytes (CD4+ T lymphocytes), testis cells, and some cells of the central nervous system.

5.4.2 Myosin Light-Chain Kinase Family

Myosin light-chain kinases (MLCK) constitute a family of soluble protein kinases with 4 MLCK isoforms. In addition to MLCK4 (a.k.a. caMLCK-like and Sugen kinase SgK085) encoded by the MYLK4 gene, caMLCK (MLCK3), skMLCK (MLCK2), and smMLCK (MLCK1) subtypes reside in cardiomyocytes, skeletal myocytes, and smooth muscle cells, respectively; they are encoded by the MYLK3, MYLK2, and MYLK1 genes, respectively

Myosin light-chain kinases are protein Ser/Thr kinases that primarily phosphorylate 20-kDa regulatory myosin light chain (MLC2 or MyRL2), or myosin, light chain-9, regulatory (MyL9). In most cell types, MLCKs intervenes in response to Ca^{++} influx and binding to calmodulin. Calcium–calmodulin then associates with MLCK enzyme.

The MLCK enzymes thus contribute to muscle contraction, migration, and endo- and exocytosis. In endo- and epithelia, MLCKs control barrier function. Stress fiber and focal adhesion formation rely on myosin-2 activation. Actomyosin contractility is mainly regulated by myosin light-chain kinase (MLCK) and Rho kinase (RoCK). Phosphorylation of the 20-kDa myosin light chain is counteracted by myosin light-chain phosphatase (MLCP). Enzyme MLCK directly activates myosin-2, whereas small GTPase RhoA stimulates its effector RoCK that, in turn, phosphorylates (inactivates) MLCP, hence promoting myosin light-chain phosphorylation (Fig. 5.3). Kinases MLCK and RoCK have complementary roles.

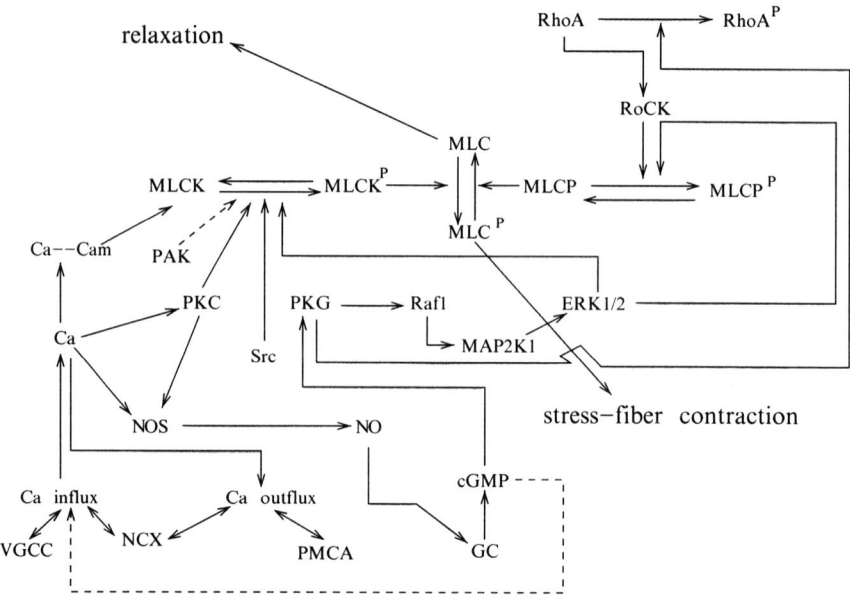

Fig. 5.3 Myosin light-chain kinase and its regulators.

5.4.2.1 MLCK Structure

MLCK contains a C2 immunoglobin domain that binds to unphosphorylated MLC, a catalytic site, and a calmodulin-binding motif. The latter operates as an auto-inhibitory domain that blocks any MLCK constitutive activity in the absence of calmodulin. This repression is relieved upon Ca^{++}–calmodulin binding.

The MLCK structure required for its binding to MLC and activity are conserved between cell type-specific isoforms. Nonetheless, these isoforms experience different types of regulation. In smooth muscle cells, Ca^{++} binding to calmodulin suffices to activate MLCK (necessary and sufficient condition). In other cell types, Ca^{++} binding to calmodulin is necessary for MLCK activity, but not sufficient.

5.4.2.2 Muscular MLCK Isoforms

Muscular MLCK isoforms encompass smooth (smcMLCK or MyLK1), skeletal (skMLCK or MyLK2), and cardiac (caMLCK or MyLK3) products. In striated myocytes, Ca^{++}-binding troponin triggers actomyosin contraction. Activation of MLCK serves to heighten the contractile strength. On the other hand, in smooth muscle cells, MLCK primes the contraction. Isotype smcMLCK has a greater structural similarity with non-muscle MLCK isoforms.

Unlike skMLCK and caMLCK, smcMLCK possesses an additional C-terminal insert and a long N-terminus with a fibronectin domain, 2 immunoglobin domains, and numerous potential phosphorylation sites [399].

5.4.2.3 Non-Muscle MLCK Isoforms

In non-muscle cells, a single gene – the MYLK1 gene – of the human genome encodes 4 high-molecular-weight MLCK isoforms (MLCK1–MLCK4). Non-muscle MLCK isoforms differ from smcMLCK mainly by their additional element of the N-terminus that results from alternative splicing. This region contains multiple sites for between-protein interactions [399].

Isoform MLCK1 was originally called endothelial cell MLCK (ecMLCK), as it was discovered in human endothelial cells. However, it is also expressed in other cell types, such as intestinal epithelial cells and neutrophils [399]. Nevertheless, MLCK1 as well as MLCK2 reach their highest levels in endothelium.

Isoform MLCK2 lacks a sequence necessary for phosphorylation and activation by Src kinase [399]. Subtypes MLCK3a and MLCK3b result from an additional distal deletion w.r.t. MLCK1 and MLCK2, respectively [399]. Isoform MLCK4 has a tissue distribution similar to other non-muscle cell MLCKs: brain, lung, kidney, liver, in addition to endothelia.

In non-muscle cells, phosphorylation of myosin-2 (Ser19) elicits stress-fiber formation. Phosphorylation of MLC enables myosin cross-bridging with actin filaments. Ca^{++}–Calmodulin-dependent MLCK catalyzes the transfer of phosphate from ATP^{Mg} to serine residue of regulatory myosin light chain to trigger actomyosin contraction. This specialized kinase of the Ca^{++}–calmodulin-regulated kinase superfamily exclusively phosphorylates regulatory light chain of myosin-2.

5.4.2.4 MLCK and the Endothelial Barrier

The microvascular endothelium consists of a monolayer of closely apposed endothelial cells (Vol. 5 – Chap. 9. Endothelium). It forms a semipermeable barrier between blood and tissue that controls exchange of water, electrolytes, and proteins. Transcellular transport of molecules offers a first route for the transfer of materials. The paracellular route is limited by intercellular junctions, especially tight and adherens junctions (Vol. 1 – Chap. 7. Plasma Membrane).

Tight junctions are connected to the actin cytoskeleton by zona occludens proteins ZO1 and ZO2. Adherens junctions are linked to the actin cytoskeleton via various catenin types (α, β, γ, and δ). Therefore, the tone of stress fibers influences the paracellular permeability.

5.4.2.5 MLCK Regulation

In vertebrates, the smooth and non-muscle MLCK gene locus produces 2 MLCK isoforms: $MLCK_L$ (long, high molecular mass) and $MLCK_S$ (short). The long MLCK variant is the predominant isoform in cultured cells. The short MLCK is a substrate for several kinases, such as kinases PKA, PKC, PKG, CamK2, PAK, CK2, ERK1, and ERK2 [400]. Enzymes ERK1 and ERK2 can phosphorylate (activate) MLCK and (inactivate) MLCP phosphatase.

In endothelial cells, stress fiber contraction influence vascular barrier regulation and leukocyte diapedesis. Two MLCK splice variants – ecMLCK1 and ecMLCK2 – exist. Isoform ecMLCK1 can be phosphorylated by Src kinase [401].

Myosin light-chain kinase acts as a contractile signal for tonic airway smooth muscle cells [402]. Agent MLCK is needed in phasic smooth muscle contraction. Enzyme MLCK is required for the initial contraction, whereas the sustained contraction may rely on Ca^{++} sensitization and Ca^{++}-independent kinases with myosin light-chain phosphatase inhibition. Nonetheless, MLCK is strongly involved in bronchoconstriction [402].

Vasodilation results from relaxation of vascular smooth muscle cells. Nitric oxide activates soluble guanylate cyclase that synthesizes cGMP messenger. The latter activates cGMP-dependent protein kinase-G that phosphorylates MLCK, but also cAMP-dependent protein kinase-A, thereby reducing the Ca^{++}-sensitivity of myosin [403]. Protein kinase-G phosphorylates (inhibits) RhoA [404]. Upon phospholipase-C stimulation, nitric oxide synthase produces NO that targets guanylate cyclase to synthesize cGMP messenger. The latter activates PKG that can work via the cRaf–MAP2K1–ERK1/2 pathway [399]. In addition, PLC elicits Ca^{++} ingress and PKC activation.

Intracellular Ca^{++} concentration that depends on in- and outflux through calcium channels, exchangers, and pumps, such as voltage-gated Ca^{++} channels, transient receptor potential Ca^{++} channels, Na^+–Ca^{++} exchanger, and Ca^{++} ATPase, is required for smooth muscle contraction.

5.4.3 AMP-Activated Protein Kinases

Adenosine monophosphate (AMP)-activated protein kinase (AMPK) is a heterotrimer with a catalytic α and 2 regulatory β and γ subunits (Table 5.16).

This sensor of the energy state of the cell is activated by a decreasing ATP concentration and an increasing AMP concentration.[116] Once activated, AMPK

---

[116] When energy demand increases, ATP concentration decays, whereas concentrations of ADP and AMP rise. Sensors monitor the cellular energy status and trigger mechanisms that restore energy balance and avoid metabolic stress caused by elevated demand or deprivation (hypoxia).

Table 5.16. Structure of AMPK. The heterotrimeric protein AMPK is consituted by a catalytic α and 2 regulatory β and γ subunits. The most common subunit isoforms synthesized in most cells are α1, β1, and γ1 chains; nevertheless, the remaining isoforms are also expressed in cardiac and skeletal muscle cells.

Subunit type	Subtypes
α	Catalytic subunit $AMPK_{c\alpha 1}$–$AMPK_{c\alpha 2}$ (encoded by the PRKAA1 to PRKAA2 genes)
β	Regulatory subunits $AMPK_{r\beta 1}$–$AMPK_{r\beta 2}$ (encoded by the PRKAB1 to PRKAB2 genes)
γ	$AMPK_{r\gamma 1}$–$AMPK_{r\gamma 3}$ (encoded by the PRKAG1 to PRKAG3 genes)

phosphorylates enzymes in all types of metabolism as well as transcription factors to regulate gene expression and redirect cellular metabolism from ATP-consuming anabolism to energy-generating catabolism.

Under normal conditions, the intracellular concentration of adenosine triphosphate is on the order of 1 mmol. Enzyme AMPK belongs to the main enzyme set that regulates ATP levels, discriminating between the mono- and triphosphate compounds. Nucleotides ATP and AMP competitively bind to the same AMPK site. Kinase AMPK then senses the relative ratio of ATP to AMP and controls the cell metabolism according to energy availability.

The γ subunit contains 3 binding domains for regulatory nucleotides: AMP and ATP bind to 2 sites in a mutually exclusive manner, whereas another site tethers AMP in a non-exchangeable manner [405]. Adenosine monophosphate binds to AMPK not only to stimulate it via allosteric regulation, but also to modify AMPK configuration, thereby enhancing its affinity for its kinase (AMPKK or AMP2K).

The AMPK activity is actually modulated not only by AMP-to-ATP ratio, but also by various types of AMP2K enzymes. Activation of AMPK depends on phosphorylation of the catalytic α subunit (Thr172) by STK11 (or LKb1), Ca^{++}–calmodulin-dependent protein kinase kinase (CamKK or Cam2K), and MAP3K7 enzyme.[117]

Enzyme Cam2K phosphorylates AMPK in response to elevated intracellular protein Ser/Thr kinase STK11 and AMPK acts as a cellular energy sensor that regulates cell metabolism, and hence, cell fate. The STK11–AMPK pathway particularly phosphorylates acetyl coenzyme-A carboxylase (ACC), which catalyzes the rate-limiting step in lipogenesis, and tumor suppressors.

---

[117] A.k.a. transforming growth factor-β-activated kinase TAK1.

Enzyme Cam2K phosphorylates AMPK in response to elevated calcium concentration, in the absence of change in cellular adenine nucleotide level, especially Cam2Kβ that activates AMPK much more rapidly than Cam2Kα [406].

Phosphorylation (activation) of AMPK is promoted by AMP binding to the γ subunit. This signal sustains AMPK activity, as it inhibits dephosphorylation of α subunit, whereas ATP fosters dephosphorylation.[118] Inactivation of AMPK occurs in response to a rising concentration of ATP.[119] Phosphorylation stimulates AMPK activity several hundred-fold, whereas AMP excites AMPK severalfold, the degree of activation depending on the nature of AMPK isoforms. Like AMP, ADP binds to γ subunit and supports α-subunit phosphorylation. Unlike AMP, ADP does not directly prime AMPK activation [405].[120] Kinase AMPK is also activated by leptin and adiponectin.

Activated AMPK promotes ATP-producing catabolism (fatty acid oxidation, glucose uptake, and glycolysis; Vol. 1 – Chap. 4. Cell Structure and Function) and impedes ATP-consuming metabolism (synthesis of fatty acid, cholesterol, glycogen, and proteins).

Ubiquitous AMPK, a sensor of cellular energy, increases sensitivity of cells to insulin [407]. In the liver, AMPK phosphorylates (inhibits): (1) acetylCoA carboxylase, which converts acetylCoA to malonylCoA for fatty acid synthesis in the liver; (2) 3-hydroxy 3-methylglutarylCoA reductase, thus decreasing cholesterol synthesis in the liver; and (3) CREB-regulated transcription coactivator-2 (TORC2), which activates gluconeogenesis and upregulates expression of insulin receptor substrate-2, a mediator of insulin signaling. In contrast, in muscles, activated AMPK promotes fatty acid oxidation and activates PGC1α promoter, thus increasing metabolism.

Enzyme AMPK regulates tight junction assembly and disassembly [408]. It is phosphorylated by STK11 that controls cell polarity. Kinase STK11, indeed, activates Par1 implicated in cell polarity and microtubule affinity-regulating kinase (MARK). It phosphorylates AMPK during tight junction assembly stimulated by calcium. Moreover, AMPK enhances the stability of tight junction from disassembly induced by calcium depletion. Enzyme AMPK also phosphorylates the regulatory site of non-muscle myosin regulatory light chain involved in mitosis and cell polarity [409]. Regulation by STK11 of epithelial cell polarity and mitosis is due to MRLC phosphorylation by AMPK.

In central neurons, $K_V2.1$ channel regulates action potential frequency. Firing of action potentials in excitable cells accelerates ATP turnover. Enzyme AMPK

---

[118]Binding of AMP stimulates phosphorylation of α subunit. This stimulation depends on the myristoylated β subunit. Once phosphorylated, AMPK is further activated 2- to 5-fold by AMP. Moreover, AMP binding also precludes inactivation (dephosphorylation) of α subunit by protein phosphatases PP2 and PPM1 [405].

[119]Exchange of AMP for ATP in the 2 cognate sites of γ subunit prevents direct activation and favors dephosphorylation of α subunit.

[120]In other words, phosphorylation of AMPK (Thr172) is regulated by both ADP and AMP, whereas only AMP directly activates AMPK [405].

activated by ATP depletion phosphorylates $K_V2.1$ channel, thereby causing hyperpolarizing shifts in the current–voltage relationship and, then, modifying channel activation and inactivation [410]. As AMPK facilitates activation of $K_V2.1$ channel, it reduces neuronal excitability and conserves energy. On the other hand, AMPK phosphorylates (inactivates) $K_{Ca}1.1$ and $K_{Ca}3.1$ as well as $K_{IR}6.2$ channels.

5.4.4 AMPK-Related Kinases

Twelve AMPK-related kinases — brain-specific kinases (BrSK1–BrSK2), microtubule affinity-regulating kinases (MARK1–MARK4), maternal embryonic leucine-zipper kinase (MELK), nuclear AMPK-related kinases (NuAK1–NuAK2), and members of the salt-induced kinase family (SIK1–SIK3; Sect. 3.1.3) constitute with AMP-activated protein kinase the AMPK family. These 12 enzymes are closely related to the $AMPK_{c\alpha1}$ and $AMPK_{c\alpha2}$ catalytic subunits.

Some members are involved in the regulation of cell polarity; others control cell differentiation. Brain-specific kinases are neuron-polarity regulators. Microtubule affinity-regulating kinases regulate microtubule dynamics, as they phosphorylate microtubule-associated proteins; they are also involved in the establishment of cell polarity. Maternal embryonic Leu-zipper kinase contributes to stem cell renewal, cell cycle progression, and pre-mRNA splicing. Enzyme NuAK1 impedes apoptosis and controls cellular senescence. Kinase NuAK2 is activated in response to glucose deprivation, ATP depletion, and hyperosmotic stress. Salt-induced kinases participate in steroido- and adipogenesis, particularly SIK2[121] that intervenes during early stage of adipogenesis.

Kinase STK11 (LKb1) phosphorylates (activates) AMPK family members [411]. In addition, STK11 phosphorylates a single member of the family of SNRK kinases (SNF1 [plant AMPK]-related protein Ser/Thr kinase) that are distant relatives of AMPK. Kinase STK11 forms a complex with regulator pseudokinase StRAd and Mo25 to phosphorylate AMPK (Sect. 3.4.6).

Except MELK kinase, all AMPKRKs are phosphorylated by STK11, but not other AMP2Ks, at a site equivalent to Thr172 in AMPK.[122] Kinase Cam2Kα phosphorylates BrSK1 kinase (Thr189). Other AMPKR2Ks (or AMPKRKKs) exist such as MAP3K16, a MAR2K for microtubule-associated protein (MARK) [411]. Glycogen synthase kinase-3β phosphorylates (inactivates) MARK2. Protein kinase-A targets salt-inducible kinase SIK1 (Ser577) to cause translocation from the nucleus, relieve cAMP-response element-binding protein (CREB) inhibition, and lower phosphorylation of CREB-regulated transcription coactivator (CRTC or transducer of regulated CREB). Kinase SIK1 can also be phosphorylated (activated) by Ca^{++}–

---

[121] A.k.a. Qin-induced kinase (QIK).
[122] Kinases BrSK1 and BrSK2 are phosphorylated by STK11 at Thr189 and Thr174, respectively.

calmodulin-dependent kinase (Thr322) in response to an increase in intracellular sodium concentration. Protein kinase-B2 phosphorylates SIK2 (Ser358) to disrupt and degrade target of rapamycin TORC2 complex.

The AMPK-related kinases undergo ubiquitination.[123] Their ubiquitin-associated domain interact swith the catalytic sequence to render kinase conformation suitable for STK11-induced phosphorylation [411]. On the other hand, ubiquitin-associated domains of MARK1 and MARK2 bind to and inhibit the catalytic motif. Conversely, MARK4 and NuAK2 interact with the deubiquitinase X-linked ubiquitin-specific USP9X peptidase.

Phosphorylated AMPK-related kinases can be sequestered in the cytosol by ubiquitous 14-3-3 proteins. For example, MARK phosphorylation by atypical PKC favors 14-3-3 binding and causes relocalization from the plasma membrane to the cytoplasm [411]. Once phosphorylated by STK11, SIK1 and SIK3 interact with 14-3-3 proteins.

5.4.4.1 Brain-Specific Kinases

Two brain-specific kinase isoforms exist: BrSK1[124] and BrSK2.[125] They are involved in neuronal polarity and synapse development. Isoform BrSK1 localizes to and associates with synaptic vesicles [411]. In addition, BrSKs participate in neuron polarity to produce axons and dendrites. Both BrSK1 and BrSK2 phosphorylate microtubule-associated protein Tau that regulates microtubule stability. Furthermore, BrSK1 acts as a checkpoint kinase, as it phosphorylates Wee1 (Ser642) and CDC25c (Ser216). It can be phosphorylated by ATMK and ATRK that are implicated in the DNA-damage response.

5.4.4.2 Maternal Embryonic Leucine Zipper Kinase

Maternal embryonic leucine zipper kinase (MELK) is maximally active during mitosis. It is counteracted by CDC25b phosphatase. It interacts with the splicing factor nuclear inhibitor of PP1 (NIPP1) to block pre-mRNA splicing during mitosis [412]. It is strongly expressed during early embryogenesis. It contributes to the self-renewal of multipotent neural progenitors. Two splice variants (MELK1 and MELK2) have been identified.

---

[123]The AMPKRK enzymes are the only kinases in the human genome that contain a ubiquitin-associated domain.

[124]A.k.a. SADb, or SAD2, as it regulates several aspects of presynaptic differentiation in Caenorhabditis elegans, as well as STK29.

[125]A.k.a. SADa or SAD1. Activated LKB1 phosphorylates (activates) both BrSK1 and BrSK2 kinases that are required for neuronal polarization with axon and dendrites in the mammalian cerebral cortex. Activity of LKB1 relies on its connection to STRADα pseudokinase and its phosphorylation by kinases (e.g., PKA and RSK).

Kinase MELK possesses many autophosphorylation sites, but its activity is not affected by phosphorylation by the STK11–STRAD–CaB39 complex that is required to activate all other AMPK-related protein kinases. It interacts with ZPR9 transcription factor [412].

5.4.4.3 Microtubule Affinity-Regulating Kinase Subfamily

Four microtubule affinity-regulating kinases constitute the MARK subfamily of the family of AMPK-related protein kinases: MARK1, MARK2 (or EMK), MARK3 (or cTAK), and MARK4 (or MARKL1). These 4 members are encoded by 4 Mark genes.

The Mark paralogs differ in tissue distribution, subcellular location, and expression regulation [413]. Expression of MARK1 is high in the brain, heart, kidney, skeletal muscle, spleen, and pancreas, and low in the lung and liver [413].

Two isoforms of MARK4 exist: MARK4$_S$ and MARK4$_L$ that are produced in neural progenitor cells and neurons, respectively [411].

Microtubule affinity-regulating kinases contribute to the regulation of cell polarity and microtubule stability as well as possibly cell cycle control and Wnt signaling. They can dissociate microtubule-associated proteins from microtubules, hence destabilizing the microtubule network, and affecting intracellular transport [411]. Overexpression of MARK kinase in cultured cells actually causes removal of Tau and related microtubule-associated proteins from the microtubule surface, hence transforming microtubules into bundles and reducing the microtubule network.

Other MARK activities can be specific to MARK isoforms. Isoform MARK1 can be involved in the regulation of synaptic remodeling, MARK2 in the regulation of metabolism, and MARK3 in cell cycle control.

Enzyme MARK phosphorylates doublecortin, thereby regulating doublecortin binding to microtubule. On the other hand, MARK2 that impedes axon formation and favors dendrite development is inhibited by P21-activated kinase PAK5. All human MARKs are phosphorylated (activated) by Ser/Thr kinase STK11 (or LKb1) [413].

MARK1

Enzymes MARK1 and MARK2[126] phosphorylate microtubule-associated protein Tau. In addition, MARK1 interacts with ubiquitin-C and 14-3-3σ (stratifin) [251].

---

[126] A.k.a. ELKL motif Ser/Thr protein kinases EMK1 and EMK3, respectively.

MARK2

In addition to STK11, MARK2 kinase can also be activated by MAP3K16 and GSK3β kinases. Conversely, MARK2 is inhibited by its association with PAK5 kinase (without phosphorylation). Kinase MARK2 interacts with PKB, MAP3K3, and 14-3-3γ, 14-3-3σ, and 14-3-3θ sequestrators [251]. Kinases PKCλ and PKCζ can phosphorylate (inactivate) MARK2 and MARK3, as they provoke binding to 14-3-3 proteins.

MARK3

Kinase MARK3 phosphorylates CDC25 phosphatase, hence preventing CDC25 activation of the CcnB–CDK1 complex. It also phosphorylates (inactivates) PIM1 kinase to promote cell cycle progression at the G2–M transition [411].

Kinase MARK3 interacts with Ras guanine nucleotide-releasing factor Ras-GRF1 vacuolar protein-sorting protein VSP11, mitochondrial processing peptidase β subunit, neuropeptide secretoneurin (or chromogranin-B), lysosomal H^+ ATPase (ATP6V0C), serum amyloid-A-like-1, patatin-like phospholipase domain-containing protein PnPLa2, transcription elongation factor TcEA2, carrier of thyroxine (T4) and retinol transthyretin, ubiquitin-C, and 14-3-3β, 14-3-3γ, 14-3-3σ, and 14-3-3θ [251].

MARK4

Kinase MARK4 interacts with microtubule-associated protein-2 and -4, and MAPT (Tau), tubulin-α1a, -β, and -γ1, myosin heavy chains-9 and -10, actin-α1, actin-dependent regulator of chromatin SMARCA4, RhoGEF2, CDC42, PP2 regulatory $PP2_{r1\alpha}$ and catalytic $PP2_{c\beta}$ subunits, hepatocyte growth factor-regulated tyrosine kinase substrate, heat shock 70-kDa protein-4, ubiquitin-C, ubiquitin ligase RNF41, ubiquitin-specific peptidase-7 and -9, and 14-3-3η [251, 413]. Two MARK4 splice isoforms exist: long ($MARK4_L$) and short ($MARK4_S$) form that has a truncated C-terminus. In the brain, $MARK4_S$ is predominant.

5.4.4.4 NuAK Kinases

Two isoforms of NuAK kinases exist: NuAK1[127] and NuAK2.[128] Both NuAK1 and NuAK2 are activated by STK11 (LKb1) kinase.

---

[127] A.k.a. AMPK-related protein kinase ARK5.
[128] A.k.a. SNF1/AMP activated-related protein kinase (SNARK).

NuAK1

Kinase NuAK1 interacts with several myosin phosphatases, such as the dimer made of protein phosphatase-1 regulatory subunit-12A ($PP1_{r12a}$), or myosin phosphatase target subunit-1 (MyPT1), and catalytic $PP1_{c\beta}$ subunit. Phosphorylation of $PP1_{r12a}$ (Ser445, Ser472, and Ser910) by NuAK1 promotes the interaction of $PP1_{r12a}$ with 14-3-3 adaptors, thereby suppressing phosphatase activity [414]. Cell detachment is characterized by phosphorylation of $PP1_{r12a}$ by NuAK1 to enhance phosphorylation of myosin light chain-2.

NuAK2

Anti-apoptotic NuAK2 is regulated by glucose or glutamine deprivation, elevated intracellular AMP concentration, as well as endoplasmic reticulum, hyperosmotic, salt, and oxidative stress [411]. Isoform NuAK2 phosphorylates regulatory $PP1_{r12a}$ subunit.

5.4.4.5 Salt-Inducible Kinases

Salt-inducible kinase SIK1 contributes to the regulation of steroidogeneis following ACTH stimulation. STK11-activated SIK1 phosphorylates transducer of regulated CREB, thereby causing its nuclear export and inhibiting CREB activity [411]. Kinase SIK1 is also activated by elevated intracellular sodium concentration and stimulates Na^+–K^+ ATPase to maintain cell volume and composition [411].

Salt-inducible kinase SIK2 (a.k.a. Qin-induced kinase [QIK]) is ubiquitous with its highest levels in adipose tissue. Kinase SIK2 phosphorylates insulin receptor substrate-1 [411]. Kinase SIK2 also inhibits CREB-mediated gene expression, as it phosphorylates target of rapamycin complex TORC2, a coactivator of CREB, that then connects to 14-3-3 for cytoplasmic sequestration.

Salt-inducible kinase SIK3 (a.k.a. QSK) can intervene in cell proliferation, as its suppression causes spindle and chromosome abnormalities [411].

5.4.4.6 Other Types of AMPK-Related Kinases

Eight other kinases — *hormonally upregulated Neu-associated kinase* (HUNK), *non-inducible immunity gene product* NIM1, *sucrose non-fermenting protein* (SNF1 [plant AMPK])-*related kinase* (SNRK), *testis-specific* Ser/Thr *kinases* (TSSK1–TSSK4 and TSSK6) — belong to the AMPK branch of the human kinome tree. Kinase SNRK is highly expressed in the testis. Kinase STK11 preferentially phosphorylates Thr residues with a Leu residue at the -2 position, hence not kinases of the HUNK and TSSK subfamilies that do not possess this Leu residue.

5.4.5 Checkpoint Kinase Family

Protein Ser/Thr kinases checkpoint kinases ChK1 and ChK2 are encoded by the CHEK1 and CHEK2 genes, respectively. These protein kinases are activated in response to DNA damage and provoke cell cycle arrest.

5.4.5.1 ChK1 Kinase

Kinase ChK1 phosphorylates phosphatases CDC25 that operate during the cell cycle, particularly for entry into mitosis. Phosphatase CDC25 (Sect. 8.3.15), when it is phosphorylated (Ser216) by ChK1 binds to an adaptor in the cytoplasm for sequestration. Hence, it cannot relieve the inhibition of the mitotic CcnB–CDK1 complex caused by Wee1 kinase. Consequently, cell is prevented from entering into mitosis.

5.4.5.2 ChK2 Kinase

Kinase ChK2 operates in DNA-damage response and mitotic progression, during which it can play a role distinct from that in DNA damage, independently of P53 factor. Kinase ChK2 is rapidly phosphorylated by ATMK in response to DNA damage. Once activated, ChK2 inhibits CDC25c phosphatase, thereby preventing entry into mitosis. It also stabilizes P53, leading to cell cycle arrest in G1 phase. In addition, ChK2 interacts with and phosphorylates another tumor suppressor with ubiquitin ligase activity, the breast cancer-related protein BrCa1 for accurate spindle formation and stable attachment of microtubules to kinetochores that aims at maintaining genome integrity [415]. It eventually allows BrCa1 to restore cell survival after DNA damage.

5.4.6 Death-Associated Protein Kinase Family

Death-associated protein kinase DAPK1 is a Ca^{++}–calmodulin-regulated Ser/Thr protein kinase that participates in apoptosis launched by factors, such as interferon-γ, tumor-necrosis factor-α, transforming growth factor-β, TNFSF6, and lipid ceramide [416]. It suppresses integrin signaling, particularly during ischemic brain injury. Autophosphorylation (Ser308) inhibits Ca^{++}–calmodulin interaction, thus attenuating its basal activity. Activity of DAPK1 rises upon phosphorylation by extracellular signal-related kinase (Ser735).

Kinase DAPK1 is expressed in the nervous system, particularly in hippocampus. It has been detected in mouse lung, bladder, liver, kidney, heart, and uterus [416]. In the mouse, DAPK has 2 α and β isoforms. Isoform DAPKβ has an 11-amino acid extension at its C terminus. The latter has a slight pro-apoptotic activity.

Substrates of DAPK1 include myosin light chain, DAPK3 kinase that regulates stress fiber–focal adhesion assembly, syntaxin-1A, and calmodulin-dependent kinase kinase (Cam2K) [416]. Kinase DAPK1 interacts with ubiquitin ligase DAPK-interacting protein DIP1,[129] Fas (TNFRSF6a)-associating death domain-containing protein FADD, ankyrin repeat-containing protein Krit1, TNFR-binding protein TNFSF1a, and Tau protein [251].

Members of the CAMK superfamily tether to calmodulin. Ability of DAPK to bind calmodulin correlates with its catalytic activity [417].

5.4.7 Death Receptor-Associated Protein-Related Apoptotic Kinase Family

Two protein Ser/Thr kinases, death receptor-associated protein-related apoptotic kinases DRAK1 and DRAK2, have a catalytic domain related to that of death-associated protein (DAP) kinase (DAPK) involved in apoptosis initiated by interferon-γ. Like death-associated protein kinase DAPK3,[130] the kinase domain of both DRAK subtypes are homologous to that of DAPK enzyme [418].[131] Both DRAK isoforms triggers cell apoptosis. However, the sensitivity to DRAK activity may depend on the cell type. Both DRAK subtypes localize to the nucleus and cytoplasm, unlike DAPK, which is tightly associated with the cytoskeleton, and DAPK3, which resides in the nucleus [418].

5.4.7.1 DRAK1

DAPK-related apoptosis-inducing protein kinase DRAK1 is also called protein Ser/Thr kinase STK17a. This calmodulin-dependent kinase interacts with calcineurin homologous protein (CHP),[132] an inhibitor of DRAK2 and PP3. It is highly expressed in the lung, pancreas, and placenta, as well as, at lower levels, in the brain, heart, liver, kidney, and skeletal muscle.

---

[129] A.k.a. Mindbomb homolog Mib1.

[130] A.k.a. DAP-like kinase (DLK) and ZIP kinase.

[131] The C-termini of DPAK, DAPK3, DRAK1, and DRAK2, which mediate between-protein interactions, are structurally different. Kinase DAPK3 is activated upon homodimerization. Kinase DAPK may be activated upon homodimerization or binding to regulators [418].

[132] A.k.a. calcium-binding protein P22. This cofactor supports activity of sodium–hydrogen exchangers NHE1 (SLC9a1) and NHE3 (SLC9a3).

5.4.7.2 DRAK2

Enzyme DRAK2, or STK17b, resides in the cytoplasm and nucleus, particularly in lymphocytes. It is expressed in many cells, with its highest expression in lymphoid organs. It is synthesized at high levels in adipose tissue and low levels in lungs and small intestine [419].

Kinase DRAK2 undergoes autophosphorylation (Ser10 and Ser12) upon antigen receptor ligation in B lymphocytes and thymocytes. It phosphorylates myosin light chain in vitro. It inhibits calcineurin homologous protein-B (CHP). In particular, DRAK2 is also able to cause apoptosis of cardiomyocytes [419].

5.4.8 PIM Kinase Family

Proto-oncogene PIM1 has been identified as an activated gene in T-cell lymphoma induced by provirus insertion of Moloney murine leukemia virus. Its product is a protein Ser/Thr kinase that is highly expressed in hematopoietic cells as well as testis. The PIM family contains 3 known members (PIM1–PIM3). Kinase PIM1 and other family members are also produced in cells other than blood cells in a cell cycle-dependent manner. Related isoforms, PIM2 and PIM3, have overlapping as well as distinct functions in cell growth and survival.

5.4.8.1 PIM1

Many PIM1 target molecules regulate cell cycle progression and apoptosis. Furthermore, in heart, PIM1 has positive inotropic effects and impedes maladaptive hypertrophy. Moreover, mitochondrial PIM1 preserves mitochondrial integrity. Kinase PIM1 enhances mitochondrial resistance to Ca^{++}-induced opening of permeability transition pores and inhibits cytochrome-C release primed by pro-apoptotic set of the BCL2 family (BAX and BAK) that elicits mitochondrial outer membrane permeabilization in response to apoptotic stimuli.

Kinase PIM1 interacts with chromobox homologs CBx1 and CBx3 (a.k.a. heterochromatin proteins HP1β and HP1γ), cyclin-dependent kinase inhibitor CKI1a, nuclear mitotic apparatus protein NuMA1, nuclear factor of activated T cells NFAT1, CDC25a phosphatase, sorting nexin SNx6, and heat shock protein HSP90a1 [251].

5.4.8.2 PIM2

Kinase PIM2 interacts with BCL2-antagonist of cell death (BAD), copper transporter CuTc, heterogeneous nuclear ribonucleoprotein HNRPh3, and NADH dehydrogenase (ubiquinone)-1β subcomplex-8 (NDUFb8) [251].

5.4.9 Protein kinase-D Family

Protein kinases of the PKD family (PKD1–PKD3) belongs, in fact, to the CAMK superfamily. Its members are activated by diacylglycerol. Protein kinases-D are activated via a PKC-dependent pathway by: (1) various regulator petides (bombesin, bradykinin, endothelin, and vasopressin), lysophosphatidic acid, and thrombin that act via Gq, G12, Gi (Vol. 3 – Chap. 7. G-Protein-Coupled Receptors) and Rho GTPases (Sect. 9.3.12); (2) growth factors (e.g., platelet-derived- and insulin-like growth factor); (3) crosslinking of B- and T-cell receptors on B and T lymphocytes, respectively, and (4) oxidative stress [420]. PKC-dependent PKD activation has been detected in many normal cell types, including fibroblasts, smooth muscle cells, cardiomyocytes, neurons, lymphocytes, mastocytes, and platelets, among others. Kinase PKCϵ particularly activates PKD.

5.4.9.1 Protein kinase-D1

Protein kinase-D1 (or protein kinase-Cμ) participates in the regulation of many processes, such as signal transduction, membrane trafficking, and cell survival, differentiation, proliferation, and migration. They targets JNK and ERK kinases. Both PKD1 and PKD2 undergo a reversible translocation from the cytosol to the plasma membrane in response to GPCR stimulation (Gq and G12/13 pathways), which requires PKC activity.

Upon activation of small RhoA GTPase, protein kinase-D1 regulates cofilin-mediated ^{F}actin reorganization and cell motility via slingshot phosphatase [422]. Actin remodeling at the leading edge of migrating cells that generates cell protrusions, defines migration direction, and initiates lamellipodium growth, is indeed controlled by a temporospatial coordination of Rho GTPases, kinases, and phosphatases.

Protein kinase-D1 transduces stress stimuli involved in maladaptive cardiac remodeling generated by pressure overload, angiotensin-2, and adrenergic signaling [421]. Protein kinase-D is a stress-responsive kinase that phosphorylates class-2 histone deacetylases, thereby dissociating their association with transcription factor myocyte enhancer factor MEF2, which, upon its liberation, stimulates prohypertrophic transcription. Class-2 histone deacetylases that silence MEF2 target genes are also targeted by Ca^{++}–calmodulin-dependent kinases to relieve MEF2 repression. Isoforms PKD2 and PKD3 cannot fully compensate for PKD1 loss. PKD1 and PKA phosphorylate cardiac troponin-I on the same sites and hence reduce myofilament Ca^{++} sensitivity.

5.4.9.2 Protein Kinase-D2

Protein kinase-D2 functions in diacylglycerol-mediated signaling cascades initiated by the activated G-protein-coupled receptors. It targets phosphatidylinositol-4 kinase-3 and regulates nuclear factor-κB. It also controls the formation of exocytic carriers from the trans-Golgi network and vesicular transport by regulation of PI(4)K3β. Activity of PKD2 is regulated by PLCγ and novel PKC isoforms (PKCδ, PKCε, PKCη, and PKCθ).

5.4.9.3 Protein Kinase-D3

Protein kinase-D3, or PKCν, intervenes in diacylglycerol-mediated signaling cascades initiated by activated B-cell receptors and G-protein-coupled receptors. Isozyme PKD3 resides in the nucleus and cytoplasm of unstimulated cells. PKD3 Activity is regulated by PLCγ and novel PKC isoforms (PKCδ, PKCε, PKCη, and PKCθ), as well as Rac GTPase and G12 and G13 subunits of G proteins [424]. Ubiquitous PKD3 operates in the formation of exocytic carriers from the trans-Golgi network and glucose transport.

5.5 PIKK Superfamily

Members of the phosphatidylinositol 3-kinase-related kinase (PIKK) superfamily include ataxia telangiectasia mutated (ATMK), ATM and Rad3-related (ATRK), DNA-dependent protein kinase (DNAPK), suppressor with morphological effect on genitalia SMG1,[133] and target of rapamycin, as well as transformation/transcription domain-associated protein (TRRAP).[134]

5.5.1 Ataxia Telangiectasia Mutated and ATM and Rad3-Related Kinases

Ataxia telangiectasia mutated (ATMK) and ataxia telangiectasia and Rad3-related (ATRK) kinases are recruited and activated by DNA damage.[135] These kinases serves as a cellular damage sensor that coordinates the cell cycle with damage-response checkpoints and DNA repair to preserve genomic integrity. These cell

---

[133] A.k.a. λ/ι (lambda)-protein kinase-C-interacting protein (LIP).

[134] Adaptor TRRAP is a component of various chromatin complexes with histone acetyltransferase activity. It contributes to transcription activation launched by MyC, P53, E2F1, and E2F4 factors.

[135] The Atmk gene is mutated in the cancer-prone ataxia telangiectasia.

cycle checkpoint kinase indeed phosphorylate several proteins that activate of the DNA-damage checkpoint to elicit cell cycle arrest and DNA repair or apoptosis. Genotoxic stresses repress the target of rapamycin complex TORC1 upon P53 activation and via P53-regulated sestrin-1 and -2.[136] Sestrin-1 and -2 activate AMPK and repress TORC1 signaling (negative feedback loop) [425]. Sestrin concentration rises when reactive oxygen species accumulate and cause activation of Jun N-terminal kinase and transcription factor Forkhead box O. Sestrins prevent excessive TOR activation.

Ataxia-telangiectasia mutated protein kinase is a component of DNA-damage processing (Vol. 1 – Chaps. 4. Cell Structure and Function and 5. Protein Synthesis). Enzyme ATMK becomes activated within a complex made of meiotic recombination-11 homolog (MRE11), radiation sensitivity DNA-repair Rad50 homolog (Rad50), and Nijmegen breakage syndrome-1 homolog or nibrin (NBS1), i.e., the MRN (MRE11–Rad50–NBS1) DNA-repair complex at sites of double-stranded DNA breaks to coordinate the diverse components of the DNA-damage response.

Kinase ATMK phosphorylates transcriptional regulators P53, BrCa1, and BrCa2, DNA repair protein RAD51, checkpoint kinases ChK1 and ChK2, checkpoint proteins RAD9 and RAD17, DNA-repair nibrin (Nbs1), double-strand break repair protein MRE11a, telomeric repeat-binding factor TeRF1, segregation of mitotic chromosome-1-like protein SMC1l1, minichromosome maintenance complex component MCM2, DNA-mismatch repair protein Mlh1, mutS homologs MSH2 and MSH6 (or G/T mismatch-binding protein GTBP), replication factor-C (RFC1), DNA polymerase-θ, ligase-4, telomerase RNA component, retinoblastoma-binding protein RbBP8, histone H2A, adaptors AP1β1, AP2α2, and AP3β2, Abl1 kinase, small RHEB GTPase, eukaryotic translation initiation factor 4E-binding protein eIF4eBP1, and cAMP responsive element binding protein CREB1 [251].

Like ATMK, ATRK targets transcriptional regulators P53, DNA-damage repair proteins BrCa1 and BrCa2, double-strand break repair protein MRE11a, mutS (mismatch repair protein) homologs MSH2, checkpoint protein RAD17, checkpoint kinase ChK1, adaptors AP1β1 and AP2α2, Abl1 kinase, and small GTPase RHEB [251]. Kinase ATRK also interacts with ATR-interacting protein (ATRIP) to recognize single-stranded DNA coated with replication protein RPA, replication protein RPA1, histone deacetylases HDac1 and HDac2, chromodomain helicase DNA-binding protein CHD4, metastasis-associated proteins Mta1 and Mta2, PTP synthase, E4F transcription factor E4F1, claspin, DNA-directed polymeraseδ1, HER3-binding protein EBP1 or proliferation-associated protein PA2G4, PI3K, protein phosphatase PP2, and RhoGEF1 [251].

Ataxia telangiectasia mutated kinase also participates in the cellular damage response to reactive oxygen species owing to a P53-independent pathway [426].[137]

---

[136]Sestrin-1 and -2 are encoded by the SESN1 and SESN2 genes.

[137]Reactive oxygen species act as signaling intermediates in many cellular processes, but elevated ROS concentration leads to diseases.

Kinase ATMK activates the tumor suppressor tuberous sclerosis protein TSC2 via the STK11 (LKb1)–AMPK metabolic pathway in the cytoplasm to repress target of rapamycin complex TORC1 and induce autophagy.[138] Moreover, ATMK is also directly activated (oxidized and dimerized) by reactive oxygen species [427]. It thus serves as a sensor of an increased production of reactive oxygen species in cells.

5.5.2 DNA-Dependent Protein Kinase

Nuclear DNA-dependent Ser/Thr protein kinase (DNAPK) is a proteic complex. The catalytic subunit of DNA-dependent protein kinase (DNAPK$_c$)[139] is inactive, but is able to undergo autophosphorylation. This subunit forms DNAPK with Ku70–Ku80 heterodimer[140] to trigger its kinase activity. Enzyme DNAPK intervenes positively or negatively in the regulation of DNA end access to repair via the homologous recombination pathway as well as in non-homologous end joining pathway of DNA repair to rejoin double-strand breaks [428]. It is also required for V(D)J recombination that uses the same NHEJ pathway to promote immune system diversity. Enzyme DNAPK also cooperates with ATRK and ATMK to phosphorylate proteins involved in the DNA-damage checkpoint. Checkpoint kinase ChK1 that regulates mitotic progression in response to DNA damage by blocking the activation of CcnB–CDK1 stimulates the kinase activity of DNAPK to phosphorylate P53 (Ser15 and Ser37) [429]

5.5.3 Target of Rapamycin

Protein Ser/Thr kinase target of rapamycin (TOR)[141] is a controller of cell growth, proliferation, survival, and metabolism (Vol. 2 – Chap. 2. Cell Growth and Proliferation). It reacts to 4 major inputs: (1) nutrients such as amino acids; (2) hormones and growth factors such as insulin; (3) cellular energy level, i.e., AMP/ATP ratio; and (4) stress such as hypoxia. It controls several anabolic and catabolic processes, as it regulates protein translation, ribosome genesis, nutrient transport, autophagy,

---

[138] Tuberous sclerosis complex TSC2 participates in energy sensing and growth factor signaling to repress target of rapamycin kinase in the TORC1 complex that regulates protein synthesis and cell growth. The TSC2 complex is itself controlled by the AMPK metabolic pathway. Activated TSC2 inhibits TORC1 under energy stress.

[139] A.k.a. PrKD, DNPK1, HYRC1, XRCC7, and P350.

[140] A.k.a. X-ray repair crosscomplementing XRCC6–XRCC5.

[141] Kinase TOR has preferred target proline, hydrophobic, and aromatic residues at the +1 position [430].

5.5 PIKK Superfamily

and activation of members of the AGC superfamily,[142] such as S6K, PKB, PKCα, and SGK kinases.[143]

The primary cilium contributes to the regulation of cell size. Cilia that act as signaling platforms control TOR signaling. In ciliated cells, bending by flow of the primary cilium[144] downregulates TOR enzyme (mechanotransduction) [432]. Cilia are actually mechanosensors that can cause calcium fluxes. The basal body of cilia serves as a restricted subcellular compartment for activation of LKb1 kinase (located at the axoneme and basal body) upon primary cilium bending that phosphorylates at the basal body the energy sensor and kinase AMPK, thereby inhibiting the TORC1 complex.[145]

5.5.3.1 TOR Complexes

Target of rapamycin kinase acts in protein complexes in a nutrient-sensing system. It forms 2 functional TOR complexes (Table 5.17): (1) mitogen-, nutrient-, and rapamycin-sensitive complex TORC1 and (2) mitogen-sensitive and nutrient- and rapamycin-insensitive complex TORC2. Protein Ser/Thr kinase target of rapamycin thus serves as the catalytic subunit of TORC1 and TORC2 complexes.

Kinase TOR also complexes with BCLxL protein of the mitochondrial outer membrane, a regulator of apoptosis and mitochondrial permeability, and voltage-dependent anion-selective channel VDAC1 [433]. In fact, it phosphorylates BCLxL agent. The mitochondrial TOR complex may regulates the balance between glycolytic and respiratory metabolism. Kinase TOR reduces aerobic glycolysis.

---

[142] Acronym AGC stands for cAMP-dependent protein kinase (PKA), protein kinase-G (PKG), and protein kinase-C (PKC).

[143] Target of rapamycin regulates most members of the AGC superfamily, which, in particular, includes kinases involved in insulin signaling, such as PKB and S6K1 enzymes.

[144] In renal tubules, bending of the primary cilium by urine flow activates polycystin-1 and -2 (TRPP1 and TRPP2), i.e., an adhesion G-protein-coupled receptor and a mechanosensitive, non-selective cation channel located at the primary cilium that causes calcium transients at the cilium. They are encoded by the genes mutated in autosomal dominant polycystic kidney disease. (Mutations in the polycystin-1 gene and polycystin-2 gene account for about 85% and 15% of autosomal dominant polycystic kidney disease, respectively.) Polycystin-1 is a multidomain glycoprotein of 4,303 amino acids. It is made of a large extracellular N-terminus, a membrane spanning domain with 11 transmembrane segments, and a cytosolic C-terminus. The last 6 transmembrane segments are homologous to polycystin-2 and voltage-activated calcium channels. It can couple with and activate several heterotrimeric G protein subtypes, such as Gi, Gq, G12/13, and stimulate Jun N-terminal kinase and AP1 transcription factor [431]. In addition, polycystin-1 binds to and stabilizes regulator of G-protein signaling RGS7. Polycystin-1 can be cleaved at the G-protein-coupled receptor proteolytic site (GPS). Following GPS proteolysis, both full length and cleaved fragment of polycystin-1 can be involved in cell signaling. Polycystin-2 activity may be regulated via a direct interaction with the C-terminus of polycystin-1.

[145] Kinase AMPK phosphorylates TSC2, thereby activating the TSC1–TSC2 heterodimer. The latter integrates environmental signals into TOR signaling. It inactivates RHEB GTPase, an activator of the TORC1 complex.

Table 5.17. Activity of target of rapamycin complexes (mitogen- and nutrient-sensitive complex TORC1 and mitogen-sensitive and nutrient-insensitive complex TORC2).

Complex	Effects
TORC1	DNA replication, RNA processing, protein synthesis cell growth (S6K1–S6K2 substrates), cell division (4eBP1–4eBP3 substrates), cell autophagy, intracellular transport, cell metabolism, lipogenesis, inhibitory feedback of growth factor signaling
TORC2	Cell survival, actin polymerization, chemotaxis (PKC–AC9–cAMP–RhoA–RoCK–MLC axis)

Table 5.18. Negative feedback loops triggered by the insulin and insulin-like growth factor pathways.

IR/IGFR–IRS–TORC1–GRB10	Inhibition of RTKs (Tier 0)
IR/IGFR–IRS–TORC1–S6K1	Inhibition of IRS (Tier 1)

The TORC1 and TORC2 complexes possess numerous substrates. Among the latter, adaptor growth factor receptor-bound protein GRB10 operates in the formation of signaling complexes linked to growth factor receptors. In particular, the cytoplasmic protein GRB10[146] suppresses signaling by insulin and insulin-like growth factors, thereby adding a complementary negative feedback loop of the PI3K and ERK pathways, in addition to the S6K1–IRS control [430, 434] (Table 5.18).

Small guanosine triphosphatase Rac1 (much more than Rac2) connects to TOR as well as with Raptor and Rictor, independently of its activity status (i.e., Rac1^{GTP} or Rac1^{GDP}) [435]. Agent Rac1 regulates both TORC1 and TORC2 in response to growth factor stimulation. Once bound to TOR, Rac1 recruits TORC1 and TORC2 at specific membrane sites.

TORC1 Complex

The TORC1 complex is composed of TOR, regulatory associated protein of TOR (Raptor), and Lethal with Sec13 (thirteen) LST8[147] In addition, TORC1 associates

---

[146] Phosphorylation of GRB10 by TORC1 stabilizes GRB10 adaptor.

[147] A.k.a. G-protein β subunit-like protein.

with 40-kDa, 14-3-3-binding, proline-rich Akt (PKB) substrate PRAS40,[148] apoptosis inhibitor FK506-binding protein FKBP38, and Rag GTPases [436]. Furthermore, TORC1 promotes the synthesis of ribosomes and transfer RNAs.

TORC1 Substrates in Cell Growth and Proliferation

Among many targets, TORC1 phosphorylates eukaryotic translation initiation factor eIF4e-binding proteins (4eBP1–4eBP3), thereby controlling protein synthesis. The 4eBP proteins are necessary for TORC1 activation of cell proliferation, but are dispensable for TORC1 effect on cell growth [437]. The TORC1 complex promotes cap-dependent translation initiation through phosphorylation and inhibition of 4eBP proteins.[149]

On the other hand, cell growth requires another TORC1 substrate, the ribosomal S6K1 and S6K2 kinases. Whereas TORC2 phosphorylates protein kinase-B to promote cell proliferation and survival, TORC1 targets S6K and 4eBP1 proteins to activate translation initiation and elongation. The TORC1 complex indeed phosphorylates S6K kinases.[150]

TORC1 in Cell Metabolism

The liver is an important site for glucose and lipid metabolism. In hepatocytes, insulin activates insulin receptor that recruits insulin receptor substrate and activates phosphoinositide 3-kinase and protein kinase-B, hence lipo- and gluconeogenesis [438].

The PI3K–PKB axis promotes glucose uptake via the translocation of glucose transporter GluT4 to the plasma membrane as well as Forkhead box FoxO1 phosphorylation (inactivation), thereby impeding gluconeogenesis.

Insulin stimulates lipogenesis via the transcription factor sterol and regulatory element-binding protein SREBP1c via the TORC1 complex. Factor SREBP1c controls the transcription of genes for cholesterol, fatty acid, triglyceride, and phospholipid synthesis. Insulin promotes SREBP1c activity, as it: (1) increases the transcription of the Srebp1c gene and (2) favors SREBP1c cleavage from its membrane-bound precursor.

---

[148] A.k.a. Akt1S1. Protein PRAS40 inhibits TOR kinase. Once phosphorylation by PKB, PRAS40 links to 14-3-3 anchor protein. This connection allows insulin to stimulate TOR enzyme.

[149] Phosphorylated 4eBP1 (4eBP1^P) releases eIF4e that then associates with eIF4g to stimulate translation initiation.

[150] Phosphorylation of S6K by TORC1 as well as phosphoinositide-dependent kinase PDK1 for full activation promotes translation initiation, as, in turn, it phosphorylates eIF4b, programmed cell death protein PDCD4, and eEF2 kinase (eEF2K). Phosphorylated eIF4b and PDCD4 activate translation initiation, whereas phosphorylated eEF2K upregulates translation elongation.

The PI3K–PKB pathway uses 2 different routes for lipogenesis and gluconeogenesis. The TORC1 complex is a component in insulin-regulated hepatic lipogenesis, but not gluconeogenesis [438].[151]

TORC2 Complex

The TORC2 complex consists of TOR, rapamycin-insensitive companion of TOR (Rictor), stress-activated protein kinase (SAPK)-interacting protein MAPKAP1 (or SIN1),[152] LST8, and the proteins observed with Rictor Protor-1 and -2.

TORC2 in Chemotaxis

Adenylate cyclase AC9 impedes the activity of RhoA and, hence, RoCK that phosphorylates myosin light chain and increases the activity of myosin-2, thereby reducing the persistence of the uropod.[153] Both Rictor and PKC (but not PKB) are required for cAMP generation at the rear of the cell and uropod retraction. Myosin-2 is regulated by TORC2 via the PKC–AC9–cAMP–RhoA–RoCK axis, independently of actin reorganization during neutrophil chemotaxis [440]. Subunit Gi controls cell polarity via TORC2, as it regulates actin polymerization at the front and impedes myosin-2 assembly at the rear.

5.5.3.2 Other Effects of TOR Kinase

Molecular Tranfer

Kinase TOR precludes turnover of amino acid, glucose, iron, and lipoprotein transporters and promotes exocytosis of transporters to the plasma membrane. The TORC1 complex is activated by amino acids via RHEB and Rag GTPases, the latter ditermining the localization of TORC1 to RHEB-containing compartments.

---

[151] Insulin resistance (Vol. 6 – Chap. 7. Vascular Diseases), i.e., insulin inability to promote efficient glucose uptake by peripheral tissues, is coupled to dyslipidemia. Ectopic fat accumulation in non-adipose tissues and inflammation reduce the efficiency of insulin action. In the liver of insulin-resistant mouse models, insulin fails to suppress gluconeogenesis, but still promotes lipogenesis. Factor FoxO1 can localize in the nucleus, hence priming the expression of gluconeogenic genes. On the other hand, high hepatic levels of SREBP1c promote synthesis of the TORC1 complex. Its nuclear accumulation, observed in type-2 diabetic mouse models, contributes to augmented hepatic lipogenesis.

[152] Alternative splicing generates at least 5 MAPKAP1 isoforms, among which 3 assemble into TORC2 to generate 3 distinct TORC2 complexes [439].

[153] Upon exposure to a chemoattractant, cells polarize to form a myosin-2-dependent rear uropod and an ^{F}-rich front pseudopod. The TORC2 complex controls cell polarity and chemotaxis in migrating neutrophils, as it is implicated in polymerization of actin at the cell front and myosin-2 at the cell sides and rear to promote contraction of the trailing edge.

Cholesterol is a constituent of the plasma membrane that confers impermeability to cellular membranes and can participate in signal transduction via membrane rafts. Two sources of cholesterol exist: intra- and extracellular. Whatever its origin, cholesterol needs proper membrane insertion. Intracellular transfer of cholesterol, particularly cholesterol egress from endosomes to the plasma membrane, requires TOR in endothelial cells [441].

Cell Autophagy

Kinase TOR inhibits autophagy. It senses cellular energy level via the STK11–AMPK–TSC1/2 pathway. It is inhibited by cellular stresses such as hypoxia via HIF1 that stimulates Regulated in development and DNA-damage response gene products REDD1 and REDD2 that activate TSC1 and TSC2 proteins.

5.6 CK1 Superfamily

Casein kinase activity can be detected in most cell types and is associated with multiple enzymes. Two families of casein kinases exist: family-1 and -2 casein kinases. Casein kinases of the CK1 and CK2 families contribute to signaling specificity in the β-catenin-independent Wnt axis (Vol. 3 – Chap. 10. Morphogen Receptors).

5.6.1 Casein Kinase-1 Family

Casein kinases of the CK1 family regulate signal transduction in most cell types, in association with multiple enzymes. The CK1 family includes at least 7 isoforms (CK1α–CK1ε) and their splice variants. The CK1γ subfamily comprises 7 isozymes (CK1α1–CK1α2, CK1γ1–CK1γ3, CK1δ, and CK1ε) that are encoded by 7 genes (CSNK1A1, CSNK1A1L, CSNK1G1–CSNK1G3, CSNK1D, and CSNK1E).

Distinct CK1 family members have various roles. Mammalian CK1 isoforms are involved in membrane trafficking, circadian rhythm, cell cycle progression, chromosome segregation, cell apoptosis, and cellular differentiation. Mutations and deregulation of CK1 expression and activity lead to multiple diseases (neurodegenerative disorders such as Alzheimer's and Parkinson's disease, sleeping disorders, and proliferative diseases such as cancer).

Substrates of CK1 include cytosolic proteins (glycogen synthase, acetylCoA carboxylase, and inhibitor-2 protein of protein phosphatase-1), cytoskeletal proteins (myosin, troponin, and ankyrin), membrane-associated proteins (spectrin, neural cell adhesion molecule, and insulin receptor), nuclear proteins (P53, cAMP response element modulator, and RNA polymerase-1 and -2), and proteins involved in protein synthesis (initiation factor-3, -4B, -4E, and-5, aminoacyl-tRNA synthase, and ribosomal protein-S6) [442].

Casein kinase-1α phosphorylates Disheveled of the Wnt pathway. It also controls the basal activity of Hh signaling. Casein kinase-1γ associated with the cell membrane binds to and phosphorylates low-density lipoprotein receptor-related protein LRP6 to promote binding of Axin to LRPs and prime Wnt signaling. Casein kinase-1δ is required for Four and a half LIM protein-mediated TGF-β-like responses. Protein kinase-A, -B and Cα and CDC-like kinase CLK2 phosphorylate CK1δ to modulate CK1δ-dependent processes. Casein kinase-1ε also intervenes in the Wnt and Hedgehog pathways and regulation of the circadian rhythm. Isoform CK1ε phosphorylates Disheveled and controls RhoA, Rac, and Rap1 GTPases.

5.6.2 Vaccinia-Related Kinase Family

The mammalian family of vaccinia-related kinases comprises 3 members (VRK1–VRK3). Whereas VRK1 and VRK2 can strongly and modestly autophosphorylate, respectively, as well as phosphorylate casein, VRK3 has key amino acid substitutions in its catalytic activity, so that it is enzymatically inert [443]. Enzymes VRK1 and VRK2 contain C-terminal extracatalytic sequences that mediate intracellular localization. Kinase VRK1 possesses a basic nuclear localization signal and resides in the nucleus. Protein VRK2 is highly hydrophobic and is located in membranes of the endoplasmic reticulum. N-terminal region of VRK3 contains a bipartite nuclear localization signal that determines its nuclear location.

5.6.2.1 Vaccinia-Related Kinase-1

Vaccinia-related kinase-1 (VRK1) belongs to a family of 3 members that has diverged from the casein kinase-1 branch. It can reside in the nucleus. Human VRK1 is expressed mainly in cell lines with high proliferation rates. VRK1 undergoes autophosphorylation. VRK1 is upregulated by E2F and downregulated by retinoblastoma protein and p16 [444].

Kinase VRK1 phosphorylates N-terminus of barrier to autointegration factor (BAF) that is required for assembly of the nuclear envelope, as well as ATF2 transcription factor (Ser62 and Thr73). Because JNK targets Thr69 and Thr71, JNK and VRK1 can have an additive effect on ATF2 phosphorylation and transcriptional activation [444]. Kinase VRK1 binds to the nuclear non-phosphorylated fraction of Jun and phosphorylates Jun (Ser62 and Ser73; Ser73 is targeted by JNK). Together, JNK and VRK1 have an additive effect on Jun dimerization and transcriptional activation. Moreover, VRK1 can form a stable complex with Jun in the nucleus. Lastly, VRK1 phosphorylates P53 (Thr18 that interacts with hdm2).

5.6.2.2 Vaccinia-Related Kinase-2

Vaccinia-related kinase-2 (VRK2) is ubiquitous. Two splice variants exist (VRK2a–VRK2b) [445]. Both full-length VRK2a (VRK2$_{FL}$; 508 amino acids) and VRK2b (397 amino acids) are regulated by Ran GTPase. Inactive Ran^{GDP} inhibits VRK2 kinase activity. In contrast, VRK2 bound to Ran^{GTP} is catalytically active. Both VRK2 isozymes undergo autophosphorylation. Human VRK2a protein resides in mitochondrial and endoplasmic reticulum membranes with a preferential perinuclear location. It colocalizes with calnexin and calreticulin. Isoform VRK2b is detected as granular deposits in both cytoplasm and nuclei.

Isoform VRK2a interacts with scaffold Jun N-terminal kinase-interacting protein JIP1 (or MAPK8IP1) in response to either interleukin-1β or hypoxia and represses the signal transmitted by the active complex. Once bound to MAPK8IP1, VRK2 connects to transforming growth factor-β-activated kinase TAK1 (or MAP3K7). The VRK2–MAPK8IP1–MAP3K7 complex can associate with MAP2K7 [445]. On the other hand, VRK2 and MAP3K7 can also interact independently of MAPK8IP1 agent. Both VRK2a and VRK2b phosphorylate P53 factor (Thr18). Substrates of VRK2 also include barrier to autointegration factor.

5.6.2.3 Vaccinia-Related Kinase-3

Unlike other VRKs, vaccinia-related kinase-3 (VRK3) does not possess kinase activity. Pseudokinase VRK3 cannot bind ATP. Protein VRK3 contains a degraded catalytic site, a highly conserved kinase fold, a nuclear localization signal, and a putative regulatory binding site [446]. It has a scaffolding role in the nucleus. It enhances the activity of dual specificity phosphatase DUSP3 that dephosphorylates phosphoERK1 and -2 [447].

Kinase VRK3 interacts with actin monomer-binding protein twinfilin-2.[154] Both twinfilin types bind to ADP–^{G}actin with a higher affinity than ATP–^{G}actin and prevent actin filament assembly. In mice, twinfilin-1 is the major isoform in embryos and most adult non-muscle cell types, whereas twinfilin-2 is the predominant isoform in the adult heart and skeletal muscles [448]. Whereas actin filament-binding proteins control the nucleation, assembly, disassembly, and crosslinking of actin filaments, ubiquitous actin monomer-binding (-sequestering) proteins, such as profilin, ADF/cofilin, twinfilin, adenylate cyclase-associated protein (CAP), WASP, WAVe, and WASP-interacting protein verprolin, regulate size, localization, and dynamics of the large pool of unpolymerized actin in cells. Twinfilin-2 is involved in neurite outgrowth.

---

[154] A.k.a. twinfilin-1-like protein, protein tyrosine kinase 9-like protein (PTK9L), as it is closely related to twinfilin-1 (PTK9), and A6-related protein (A6RP).

5.6.3 Tau–Tubulin Kinase Family

The microtubule-associated protein Tau, or MAPT, is a microtubule stabilizer encoded by the Mapt gene.[155] Its transcript undergoes alternative splicing, thus generating several variant species. It resides mainly at the plasma membrane. Protein Tau promotes microtubule assembly and stability. Its C-terminus binds axonal microtubules; Its N-terminus connects to plasmalemmal components. Its short isoforms (Tau$_S$) enables cytoskeleton restructuring. Its long isoforms (Tau$_L$) preferentially stabilize microtubules. This constituent of neuronal cytoskeleton binds enzymes and lipoproteins, in addition to microtubules.

Protein Tau is phosphorylated by various kinase types, such as cyclin-dependent protein kinase CDK5, glycogen synthase kinase GSK3β, and Jun N-terminal kinase. Neuronal dysfunction associated with hyperphosphorylation of Tau can result from abnormal activation of CDK5 kinase.

The family of Tau–tubulin kinases that phosphorylate Tau and tubulin includes 2 members (TTbK1–TTbK2).[156] Isoform TTbK1 is specifically synthesized in neurons, where it phosphorylates Tau protein. Widespread TTbK2 kinase is involved in spinocerebellar ataxia.

5.6.3.1 TTbK1

Subtype TTbK1 phosphorylates Tau, thereby promoting Tau oligomerization and aggregation. It can activate other Tau kinases, such as microtubule affinity-regulating kinase (MARK), cAMP-dependent protein kinase-A (PKA), and Ca^{++}–calmodulin-dependent protein kinase-2 [449, 450]. Activity of CDK5 depends on inhibitory (Ser9) and activating (Tyr216) phosphorylation. On the other hand, phosphorylation of Tau by TTbK1 (Ser198, Ser199, Ser202, possibly Ser396, and Ser422) may allow Tau phosphorylation by MARK, PKA, or CamK2 (Ser262) [450]. Kinase TTbK1 increases concentrations of CDK5$_{r1}$ activatory subunit and tubulin polymerization-promoting protein TPPP, improves calpain-1 activity (hence elevated TPPP levels), and reduces the postsynaptic density of hippocampal NMDA-type glutamate receptor GluN2b and GluN2d types.[157] Active CDK5 impedes the binding of Src kinase

---

[155] A.k.a. neurofibrillary tangle protein, paired helical filament-τ(PHFτ), pallido-ponto-nigral degeneration protein (PPND), frontotemporal dementia protein FTDP17, disinhibition dementia parkinsonism amyotrophy complex protein (DDPAC), and G-protein-β1γ2 subunit-interacting factor-1. Other aliases comprise MSTD, MTBT1, and MTBT2.

[156] Kinases TTbK1 and TTbK2 are also called brain-derived Tau kinase (BDTK) and type-11 spinocerebellar ataxia protein (SCA11), respectively.

[157] NMDA-type glutamate receptors are heteromeric, ionotropic glutamate receptors composed mainly of GluN1, GluN2, and GluN3 subunits. They are involved in neuronal development, synaptic restructuring, learning, memory, and addiction [451]. Activator CDK5$_{r1}$ acts as a scaffold in the GluN2b–CDK5$_{r1}$–CDK5–calpain complex, the formation of which is promoted by TTbK1 kinase.

to DLg4 and phosphorylation of GluN2b subunit (Tyr1472) by Src kinase, hence increasing binding of GluN2b to clathrin adaptor AP2 and endocytosis [451].[158] Conversely, the phosphorylation of NMDA-type glutamate receptors by Fyn is promoted by Disc large homolog DLg4. The latter bind Src and other SRC family kinases and may facilitate phosphorylation of NMDA-type glutamate receptors by SRC family kinases. Protein CDK5 phosphorylates DLg4, hence controlling the binding of Src to DLg4 agent [451].

Neuron-specific enzyme TTbK1 provokes dissociation of $CDK5_{r1}$ from ^{F}actin, thereby relieving sequestration by ^{F}actin [450]. Another activatory subunit, $CDK5_{r2}$, can bind to the actin cytoskeleton. Actin depolymerization increases of $CDK5_{r2}$–CDK5 activity [450]. Conversely, CDK5 regulates dendritic remodeling via the regulation of ^{F}actin formation that relies on phosphorylation of actin regulators, such as WAVe1, cyclin-dependent kinase inhibitor CKI1b, $PP1_{r9a}$, and RhoGEF27 agent.

5.6.3.2 TTbK2

Mutations in the gene that encodes Tau–tubulin kinase TTbK2 cause type-11 spinocerebellar ataxia [452]. Autosomal dominant spinocerebellar ataxias constitute a heterogeneous group of neurodegenerative disorders characterized by poor coordination, abnormal eye movements, and impaired speech and swallowing. Spinocerebellar ataxia type 11 is a pure progressive cerebellar ataxia with peripheral neuropathy. SCA11 truncating mutations promote TTbK2 production, suppress kinase activity, and enhance its nuclear localization [453]. Mutations cause premature termination of the TTbK2 protein, thereby eliminating most of the non-catalytic part of the enzyme, but conserving the kinase catalytic domain.

Widespread TTbK2 is detected in the heart, lung, kidney, liver, muscle, thymus, spleen, testis, and ovary, in addition to all brain regions [453]. Two phosphorylation sites of Tau (Ser208 and Ser210) targeted by TTbK2 are priming sites for the phosphorylation of Tau by GSK3β [449, 452].

5.7 Family-2 Casein Kinases

Kinase CK2 is a ubiquitous, constitutively active, protein Ser/Thr kinase that is constituted by tetramers formed by CK2α1, CK2α2, and CK2β subunits. The latter is encoded by the CSNK2B gene, whereas CK2α1 and CK2α2 are encoded

---

[158]The plasmalemmal density of NMDA-type glutamate receptor channels depends partly on clathrin-mediated endocytosis that regulates synapse maturation and causes long-term depression. Subunit GluN2b possesses a clathrin adaptor AP2-binding site. Subunit GluN2b is phosphorylated by SRC family kinases. Phosphorylation (Tyr1472) prevents the binding of AP2, thus enhancing the plasmalemmal density of NMDA-type glutamate receptors On the other hand, GluN2b is degraded after its processing by CDK5 and calpain [450].

by the CSNK2A1 and CSNK2A2 genes. Tetramer CK2 consists of 2 catalytic α (2 identical catalytic subunits [2 homodimer types] or CK2α1–CK2α2 heterodimer) and 2 regulatory β subunits that undergo autophosphorylation. The CK2α activation loop is kept in an active conformation without phosphorylation or interaction with regulatory β subunits via unique interactions between N-terminal segment and catalytic core.

Subunits of CK2 either form CK2 heterotetramers or complex with other proteins [454]. Some substrates can only be phosphorylated by free catalytic subunits, whereas others require formation of tetramers [454]. None of the CK2α2 partners exhibit isoform specificity.

In vitro, CK2 is inhibited by polyanionic compounds such as heparin and is stimulated by polycationic compounds, such as polyamines and polybasic peptides [454]. Inositol phosphates can also regulate CK2 enzyme. Kinase CK2 localizes in the nucleus and cytoplasm. It behaves as an ectokinase, as it operates in concert or consecutively with multiple components of cellular structures and organelles, such as plasma membrane, endoplasmic reticulum, Golgi body, cytoskeleton, nuclear matrix, nucleolus, and mitotic spindle [454].

Kinase CK2 contributes to the regulation of various cellular processes, such as transcription, protein translation and processing, cell polarity, transformation, and apoptosis [455]. Casein kinase-2 also participates in cell cycle control and DNA repair, as well as regulates the circadian rhythm. However, CK2α1, but not CK2α2, is phosphorylated in a cell cycle dependent manner. In addition, casein kinase-2 is stimulated by activated Wnt pathway.

5.7.1 Catalytic CK2α1 and CK2α2 Subunits)

Although CK2α1 and CK2α2 are catalytically active in the absence of CK2β, CK2β heightens catalytic activity. Kinase CK2 interacts with numerous molecules. Peptidyl-prolyl cis/trans isomerase Pin1 that regulates proline-directed kinase signaling (e.g., MAPK, CDK, and GSK3), pleckstrin homology domain-containing protein PlekHO1 (or casein kinase-2 interacting protein CKIP1) that connects to SMAD ubiquitination regulatory factor SMURF1 (but not SMURF2) to promote ubiquitination, and PP2 interact with CK2α1, but not CK2α2 [454].

On the other hand, many CK2α1-interacting proteins also bind to CK2α2, such as nucleolin that is involved in the synthesis and maturation of ribosomes, tubulin, ATF1 transcription factor, ubiquitin conjugase UbC3b, and receptor protein tyrosine phosphatase PTPRc [454].

Kinase CK2α2 also targets an accessory factor complex for access to DNA packaged into chromatin, the so-called chromatin-specific transcription elongation factor

FACT, a heterodimer that consists of suppressor of Ty 16 homolog SupT16H[159] and structure-specific recognition protein SSRP1 and links specifically to histones H2A and H2B for nucleosome disassembly and transcription elongation [455]. The CK2–FACT complex acts as a protein kinase that is activated by certain cell stresses that generate DNA damage.

5.7.2 Regulatory CK2β Subunit

Subunit CK2β has both positive and negative modulatory effects on the activity of the catalytic subunits. Activation is mediated by CK2β C-terminal domain, whereas inhibition is conferred by its N-terminal domain [456].

Subunit CK2β can be involved in many cellular processes, such as transcription, protein translation and processing, cell polarity, cell cycle control, cell transformation, and apoptosis. Subunit CK2β does not necessarily tethers to other subunits to form CK2 tetramers and can then interact with other proteins [456].

Subunit CK2β is phosphorylated both at autophosphorylation sites and at a CDK1 site. Autophosphorylation does not influence the catalytic activity of CK2 but can affect CK2β turnover [456]. Subunit CK2β can operate as a docking element to recruit CK2 tetramers to specific cellular sites or proteic complexes that modulate phosphorylation state of cognate proteins. It can also bind regulators, such as histones, fibroblast growth factor-2, and polyamines that adjust the activity of tetrameric CK2. Lastly, CK2β is also involved in CK2-independent interactions with other proteins [456]. The CK2β interactors include Raf kinase isoform aRaf of the MAPK module, oocyte maturation factor Mos, and ChK1 checkpoint kinase.

5.8 CMGC Superfamily

5.8.1 Cyclin-Dependent Kinase Family

Cyclin-dependent kinases (CDK) are Ser/Thr protein kinases that are activated by association with cyclins (Vol. 2 – Chap. 2. Cell Growth and Proliferation). Cyclin-dependent kinases and corresponding cyclins form heterodimeric enzymes. The Ccn–CDK complexes are regulated by kinases and phosphatases. Cyclin-dependent kinases, except CDK9, regulate the cell cycle as well as transcription and mRNA processing.

---

[159] A.k.a. 140-kDa Facilitates chromatin transcription complex subunit FACT140 and chromatin-specific transcription elongation factor subunit SPT16.

Cyclin-dependent kinase CDK5 activated by activator p35 following DNA damage phosphorylates ataxia telangiectasia mutated kinase that coordinates DNA-damage responses, such as cell cycle checkpoint control, DNA repair, and cell apoptosis [457]. The CDK5–ATMK signal regulates phosphorylation, hence function of ATMK targets P53 and histone-2AX.

The CDK9–CcnK complex, but not the CDK9–CcnT complex, functions in a DNA-damage response pathway to maintain genome integrity and promote replication fork recovery from a transient arrest [458].

5.8.2 Homeodomain-Interacting Protein Kinase Family

Homeodomain-interacting protein proline-directed Ser/Thr kinases constitute the HIPK family (HIPK1–HIPK4). They regulate cell proliferation and apoptosis, as they modulate the Wnt and P53 pathways. The HIPK kinases can also regulate gene expression by phosphorylating transcription factors and accessory components of the transcription machinery.

Members of the HIPK family localize to nuclear speckles. They phosphorylate several types of proteins, such as homeodomain transcription factor NKx1-2 and intracellular androgen receptor, in addition to P53 factor. Site-specific phosphorylation of P53 regulates its activity. In humans, HIPK2 phosphorylates P53 (Ser46) and the corepressor C-terminal-binding protein (CTBP; Ser422) to trigger apoptosis; HIPK2 also phosphorylates P53 (Ser9).

5.8.2.1 HIPK1

Four isoforms of P53-binding homeodomain-interacting protein kinase HIPK1 result from alternative splicing. In resting endothelial cells, HIPK1, Disabled homolog-2 (Dab2)-interacting protein (Dab2IP or apoptosis signal-regulating kinase ASK1 [MAP3K5]-interacting protein AIP1), and MAP3K5 lodge primarily in the nucleus, plasma membrane, and cytoplasm, respectively. Cytokine TNFα causes HIPK1 desumoylation and translocation from the nucleus to the cytoplasm [459]. In synergy with Dab2IP, it then activates MAP3K5 by provoking release of its inhibitors (thioredoxin, glutaredoxin, glutathione S-transferase-μ1, heat shock proteins, and 14-3-3 proteins) and launching MAP3K5 oligomerization and autophosphorylation. Subsequently, MAP3K5 stimulates MAP2K3 and MAP2K7 as well as MAP2K4 and MAP2K7 that operate on JNKs and P38MAPKs, respectively [459].

5.8.2.2 HIPK2

Kinase HIPK2 has a high turnover in unstressed cells, as its short half-life results from degradation after ubiquitination by ubiquitin ligases Seven in absentia

5.8 CMGC Superfamily

homolog SIAH1 and SIAH2, WD-repeat and SOCS box-containing protein WSB1, double minute DM2, and Skp1–Cul1–F-box SCF3 [460]. It is mainly a nuclear kinase that acts as both transcriptional activator and repressor, as it phosphorylates transcription factors as well as coactivators and corepressors. Isokinase HIPK2 regulates cell apoptosis in response to DNA damage, particpates in cell response to hypoxia and morphogenetic signals (Wnt, TGFβ, BMP, Notch, and Sonic hedgehog) [460]. The Hipk2 mRNA abounds in human tissues with its highest values in neurons. At protein level, HIPK2 expression is almost undetectable in unstressed cells and increases in the presence of genotoxic stress.

Under normoxia, HIPK2 phosphorylates SIAH2 and subsequently weakens mutual binding [461]. Isozyme HIPK2 is then protected from SIAH2-mediated ubiquitination and subsequent proteasomal degradation. Hypoxia favors association of SIAH2 with prolyl hydroxylating domain-containing proteins PHD1 and PHD3 that leads to proteasomal degradation of these HIF-prolyl hydroxylases, hence avoiding hypoxia-inducible factor elimination. Transcription factor HIF1α can accumulate and trigger gene expression. Furthermore, hypoxia provokes polyubiquitylation of transcription repressor HIPK2 and subsequent proteasomal degradation. Because HIPK2 impedes gene expression, oxygen deprivation causes reprogramming of gene expression pattern. However, a non-interacting pool of kinase HIPK2 escapes from degradation even under hypoxia.

Upon DNA damage that disrupts linkage between HIPK2 and its ubiquitin ligase WSB1, HIPK2 is stabilized by ataxia telangiectasia mutated kinase that phosphorylates SIAH1 and activated by activated protein kinases ATMK and ATRK to trigger apoptosis [460]. In response to severe DNA damage, HIPK2 phosphorylates pro-apoptotic P53 (Ser46) and anti-apoptotic corepressor C-terminal-binding protein CtBP (Ser422), hence priming its proteasomal degradation. It also interacts with histone acetyltransferases CREB-binding protein and P300 for P53 acetylation, thereby raising P53 transcriptional activity. Scaffold Axin forms the Axin–HIPK2–P53 complex to stimulate phosphorylation (activation) by HIPK2 of P53 factor. Axin also bridges Fas (TNFRSF6a)-binding Death-associated protein DAP6 (or Daxx)[160] and P53 to augment HIPK2-mediated P53 activation and cause cell death. Transcriptional regulator promyelocytic leukemia (PML) stimulates HIPK2 phosphorylation of P53 factor. Kinase HIPK2 phosphorylates PML (Ser8 and Ser38) to trigger PML sumoylation and its cooperation with HIPK2 for cell death. Moreover, PML stimulates stabilizes HIPK2 and P300, as it inhibits their SCF3-induced ubiquitination and subsequent proteasomal degradation. In addition, HIPK2 activates the pro-apoptotic JNK pathway. During the recovery phase of weak DNA damage, HIPK2, which accumulates upon repression of SIAH1- and WSB1-mediated elimination of HIPK2, is degraded via a P53–SIAH1-dependent mechanism.

---

[160]Protein DAP6 connects TNFRSF6a to the JNK pathway. It tethers to TβR2 receptor. Upon TGFβ stimulation, nuclear kinase HIPK2 translocated into the cytosol phosphorylates DAP6 that, in turn, activates the JNK pathway.

According to the molecular context, HIPK2 activates or represses transcription in response to differentiation and development stimuli. Kinase HIPK2 is a corepressor for NK1 and NK3 transcription factors in cooperation with corepressor Groucho [460]. It phosphorylates Paired box Pax6 transcription factor and enhances Pax6–p300 interaction to promote the recruitment of p300 to specific promoters. It also phosphorylates p300 transcriptional regulator. In addition, HIPK2 is involved in Wnt1-dependent phosphorylation and subsequent degradation of MyB transcriptional activator in immature hematopoietic cells. Upon BMP signaling, HIPK2 binds both transcriptional regulators Ski and SMAD1 to inhibit SMAD1–SMAD4-induced transcriptional activation. Kinase HIPK2 also intervenes in TGFβ signaling. Furthermore, HIPK2 promotes hematopoietic gene transcription, as it phosphorylates RUNX1 transcription factor and then p300 histone acetyltransferase. It also forms a corepressor complex with Groucho and HDAC1 histone deacetylase. Kinase HIPK2 also colocalizes with Four and a half LIM domains FHL2 and can form a HIPK2-FHL2–P53 complex.

5.8.2.3 HIPK3

Kinase HIPK3 is involved in basal and cAMP-stimulated (upon binding of the pituitary peptide adrenocorticotropic hormone to its cognate plasmalemmal receptor) production of the steroidogenic CYP11A1 gene that encodes cytochrome-P450-11a1 [462]. Enzyme HIPK3 operates via Jun N-terminal kinase, which phosphorylates Jun transcription factor. In cooperation with Jun^P, it launches activity of steroidogenic factor-1, or NR5a1. In addition, HIPK3 interacts with another nuclear receptor, androgen receptor, or NR3c4.

5.8.2.4 HIPK4

In humans, HIPK4, which is produced in the lung and white adipose tissue, can phosphorylate P53 (Ser9) for P53-mediated transcriptional repression [463]. Kinase HIPK4 prevents P53-dependent activity of survivin gene promoter, survivin being also called apoptosis inhibitor-4 and baculoviral IAP repeat-containing protein BIRC5.

5.9 TKL Superfamily

5.9.1 Integrin-Linked Kinase

Integrin-linked kinase (ILK or ILK1) is a ubiquitous Ser/Thr kinase that participates, rather as a pseudokinase [168, 465] (Sect. 3.4.3) and a regulator of integrin-

Table 5.19. Interactome of ILK (Source: [465]; ELMO: engulfment and cell motility; ILKAP: ILK-associated Ser/Thr phosphatase [of the PPM1 family]; kAE: kidney anion exchanger [chloride–bicarbonate exchange in basolateral surface of acid-secreting α intercalated cells in the collecting duct of the nephron]).

Partner	Effect
β1-integrin	Growth
β3-integrin	Platelet aggregation
kAE1	AE1–Actin linkage
ELMO2	Cell polarity and migration
EPHa1	Cell shape and motility
ILKAP	Regulation of GSK3β signaling, Wnt signaling
Kindlin-2	Recruitment of ILK to focal adhesion
Parvins	Cell adhesion, mural cell coating of blood vessels
Paxillin	Recruitment of ILK to focal adhesion
PINCH-1/2	Cell spreading and migration
PKB1	PKB1 phosphorylation, cell survival
Rictor	PKB phosphorylation, cell survival
RuvBL1	Spindle assembly
Src	Cofilin phosphorylation, actin organization
Thymosin-β4	Actin polymerization, PKB phosphorylation

mediated signaling,[161] in multiple cellular functions, such as cell proliferation, adhesion, and migration, as well as cadherin expression and pericellular fibronectin matrix assembly. It interacts mainly with the cytoplasmic domain of β1-integrins to regulate integrin-mediated signal transduction, especially signaling from growth factors and Wnt morphogens.

Integrin-linked kinase contributes to cell adhesion-dependent cell cycle progression, as it regulates the level or activity of several components of the cell cycle, such as cyclin-A and -D1 and CDK4 [466]. It localizes to the centrosome and can regulate mitotic spindle organization.

Integrin-linked kinase can interact with several focal adhesion proteins, such as β1 integrins, PINCH,[162] parvins, and paxillin (Table 5.19). Together with PINCH and NCK2, ILK couples and uncouples regulatory networks of cellular processes, such as cell morphology, migration, and survival [467]. Focal adhesion adaptor PINCH-1 (a.k.a. LIM and senescent cell antigen-like-containing domain proteins LIMS1) thus operates as a coupling–uncoupling platform for diverse cellular processes via its distinct LIM domains.

---

[161] Kinase ILK does not possess catalytic activity in mammalian development or adult life [465].
[162] Focal adhesion protein Particularly interesting new Cys–His protein PINCH, or PINCH-1, is a widely expressed LIM-only protein that possesses 5 LIM domains and a short C-terminus.

Integrin-linked kinase interacts with PINCH adaptors (PINCH1–PINCH2)[163] to be connected to receptor Tyr kinases and lipid kinases, especially during cell migration [468].[164] The ternary NCK2–PINCH–ILK complex is involved in signaling pathways from growth factors via their cognate receptor Tyr kinase and small GTPases.[165] Adaptor PINCH interacts with NCK2 adaptor protein that participates with small GTPases and growth factor receptors in signaling pathways, as it tethers to IRS1, EGFR, and PDGFR [468]. It is activated by phosphatidylinositol (3,4,5)-phosphate synthesized by PI3K kinase. Conversely, it is inhibited by PIP_3 dephosphorylation by PTen phosphatase. Integrin-linked kinase can inactivate myosin phosphatase and phosphorylate myosin light chain.

Members of the parvin family, i.e., α- (or actopaxin),[166] β- (or affixin) that is primarily expressed in myocytes, especially in the heart and skeletal muscle, and γ-parvin produced in hematopoietic cells, are actin-binding proteins that are associated with integrin clusters, the so-called focal adhesions. Parvin-α is involved in stabilization of vessel walls during angiogenesis as well as in mature vessels by vascular smooth muscle cells. Parvin-α limits the activity of the RhoA–RoCK axis that causes fast, random motility[167] to avoid hypercontractility and a loss of directed migration toward developing vessel wall or suitable coverage and permeability of mature vessels [469].

The ternary ILK–PINCH–parvin complex (IPP) couples integrins to the actin cytoskeleton. The assembly of the IPP complex precedes integrin-mediated cell adhesion [465]. The IPP complex contributes to adhesion strengthening and organization of the actin cytoskeleton downstream from integrins. Distinct IPP complexes formed by PINCH and parvin isoforms have specific functions in modulation of integrin signaling. Parvin-α forms a ternary complex with ILK and PINCH adaptor in adhesion sites between cells and the extracellular matrix [470]. The formation of the PINCH–ILK–α-parvin complex, an effector of protein kinase-C, is necessary, but not sufficient for ILK localization to cell–matrix adhesion sites.

---

[163] Particularly interesting new Cys-His proteins PINCH1 and PINCH2 are also called LIM and senescent cell antigen-like domain-containing proteins LIMS1 and LIMS2.

[164] Adaptor PINCH1 is ubiquitously expressed throughout mammalian development and adult life, whereas PINCH2 is produced during the second half of embryonic development with a slightly more restricted expression pattern.

[165] Cytoplasmic protein non-catalytic region of tyrosine kinase adaptor NCK2 is also named GRB4.

[166] A.k.a. calponin-like integrin-linked kinase-binding protein, calponin homology-containing ILK-binding protein (CHILKBP), and matrix remodeling-associated protein-2.

[167] On the other hand, activated Rac GTPase promotes persistent cell migration. Parvin-α activity as RhoA inhibitor is specific to cell type, as its absence in fibroblasts or endothelial cells does not induce a hypercontractile phenotype [469].

5.9.2 Interleukin-1 Receptor-Associated Kinase Family

Interleukin-1 receptor-associated kinases (IRAK1–IRAK4) were first described as signal transducers for interleukin-1. However, they are effectors of other members of the TIR (Toll-like [TLR] and interleukin-1 [IL1R]) receptor family that couple signaling events from plasmalemmal and endosomal receptor complexes to cytosolic signalosomes (Tables 5.20 and 5.21). Toll-like receptors are pattern recognition receptors of the innate immunity involved in pathogen sensing; IL1 cytokine triggers inflammation. Intracellular kinases of the IRAK family participate in innate immune responses via various signaling axes that regulate inflammation, antiviral response, and subsequent activation of the adaptive immunity, as well as in the inhibitory control of autoimmunity and chronic inflammation.

5.9.2.1 IRAK Interactors

Members of the IRAK family interact with many molecules involved in signal transduction downstream from TLR–IL1R (TIR) receptors [472] (Table 5.21). Type-1 interleukin-1 receptor (IL1RI) is a master regulator of inflammation and innate immunity. Once it is bound to IL1β, IL1R1 complexes with IL1R-associated protein and forms a *plasmalemmal signalosome* that potently activates signaling cascades. Once IL1R1 is internalized, activated receptors are often sorted via endosomes and delivered to lysosomes for degradation due to labeling ensured by ubiquitination of cargos as well as protein-sorting complexes.

Toll-like receptor-1, -2, and -4 to -6 lodge on the cell surface; their ectodomains bind pathogen-associated molecular patterns, such as lipopolysaccharide and flagellin, which are exposed on the exterior of invading microbes. Other TLRs (TLR3 and TLR7 to TLR9) reside in endosomes, on which they detect nucleic acids from pathogens. Stimulation of TLR by pathogen-associated molecular patterns permits the recruitment of the IRAK1–IRAK4 complex as well as inhibitory IRAKs to the signaling complex. Inhibitory IRAKs preclude the release of IRAK1–IRAK4 subcomplex from the TLR signaling complex. The engagement of TLR ectodomains by cognate ligands also enables the recruitment of TIR domain-containing adaptors to the cytoplasmic TIR domains of TLRs. This process triggers coordinated activation of transcription factors, such as nuclear factor-κB and interferon regulatory factors (IRF), as well as mitogen-activated protein kinase (MAPK) modules for efficient elimination of invading microorganisms.

Myd88

Both myeloid differentiation primary response protein MyD88 and its splice variant MyD88$_S$ are ubiquitous. Adaptor MyD88 impedes IRAK4 recruitment to the receptor complex. In TLR signaling, MyD88 supports cell death together with IRAK2, whereas IRAK1 and TRAF6 are involved in the survival pathway that leads to NFκB activation (Sect. 10.9.1) via the TRAF6–IKK complex [473].

Table 5.20. Interleukin-1 receptor-associated kinases (IRAK) downstream from activated interleukin (ILR) and Toll-like (TLR) receptors (Source: [471]). Toll-like receptors localize either to the plasma membrane, such as TLR1, TLR2, and TLR6, which recognize lipoproteins, and TLR4, which targets lipopolysaccharide (LPS), or to endosomes, such as TLR3, TLR7, TLR8, and TLR9 that bind to double stranded RNA (dsRNA [TLR3]), single stranded RNA (ssRNA [TLR7 and TLR8]), and cytidine^p–guanosine (CpG) motifs in DNA (TLR9). Liganded receptors dimerize and recruit TLR–IL1R (TIR) domain-containing adaptors (MAL: MyD88 adaptor-like protein; MyD88: myeloid differentiation primary response gene product-88; SArm: sterile-α- and armadillo motif-containing protein; TRAM: TRIF-related adaptor molecule; TRIF: TIR domain-containing adaptor inducing Ifnβ). Isoform IRAK4 interacts with MyD88 using death domain and may phosphorylate IRAK1 and IRAK2 that then autophosphorylate. Once IRAK1 and IRAK2 are released from the receptor complex, they connect to TNFR-associated factor-6 (TRAF6). Subunit IKKγ binds to ubiquitinated IRAK1 protein. Isoform IRAK2 causes TRAF6 polyubiquitination that enables the recruitment of the MAP3K7–MAP3K7IP2 complex; MAP3K7 activates the IKK complex. It also stimulates MAP2K3 and MAP2K6 that target P38MAPKs as well as MAP2K4 and MAP2K4 that identify JNKs, respectively. Scaffolding proteins TANK (TRAF-associated NFκB activator), NAP1 (NFκB-activating kinase-associated protein), and Sintbad (similar to NAP1 TBK1 [TANK-binding kinase-1] adaptor), assemble TBK1 and IKK complex. The IKK complex phosphorylates NFκB1, a MAP3K8 inhibitor, hence activating MAP3K8 that stimulates MAP2K1 and MAP2K2, which then phosphorylates ERK1 and ERK2 kinases. Kinases TBK1 and IKK phosphorylate interferon (Ifn) regulatory factors IRF3 and IRF7. Therefore, IRAKs provoke activation of NFκB, MAPK modules, and IRF factors.

Receptor	Adaptors and partners	Mediators	Targets
		Plasmalemmal signaling	
TLR1/2/6	MyD88, IRAK1/2/4	TRAF6, MAP3K7, MAP3K7IP1/2/3 Rac, PI3K, PKB, IKK	MAP2K3/6–P38MAPK MAP2K4/7–JNK NFκB, MAP3K1/2–ERK1/2
TLR4	MyD88, IRAK1/2/4, TRIF, TRAM, BTK	TBK1, IKKε	Inflammasome, MAPK, NFκB, IRF3
TLR5	MyD88, IRAK1/2/4	TRAF6, etc.	NFκB, MAP3K3–P38MAPK
IL1R, IL18R	MyD88, IRAK1/2/4	TRAF6, etc.	NFκB, MAPK
		Endosomal signaling	
TLR3	TRIF, RIPK1, IRAK2	TANK, TBK1, TRAF3, IKKε, Sintbad	NFκB, IRF3/7
TLR7/8/9	MyD88, IRAK1/4, TRAF3/6	IKKα	NFκB, IRF5/7

BTK

TIR domain-binding kinase BTK intervenes in B-cell maturation as well as mastocyte activation via high-affinity IgE receptor. Bruton Tyr kinase connects in particular to the TIR domain of TLR4, TLR6, TLR8, and TLR9. During BCR-mediated signaling, BTK is activated by members of the SRC family, such as Fyn, Lyn, and HCK kinases [474]. It binds PIP$_3$ and subsequently phosphorylates

Table 5.21. Partners of members of the IRAK family (Source: [472]). Some endocytic adaptors enable intracellular signaling. These adaptors are involved in the assembly of signalosomes as well as receptor cargo sorting (BTK: Bruton Tyr kinase; LPS: lipopolysaccharide; MAL: MyD88 adaptor-like protein; MyD88: myeloid differentiation primary response gene product-88; SIGIRR: single immunoglobulin [Ig] domain-containing and Toll-like and interleukin-1 receptor [TIR]-related protein, or TIR receptor TIR8; SOCS: suppressor of cytokine signaling protein; Tollip: Toll-interacting protein). Subtype IRAK4, the most receptor-proximal kinase, acts downstream from MyD88 and MAL and upstream from TRAF6 in TIR signaling. It phosphorylates (activates) only IRAK1 among members of the IRAK family. Whereas IRAK1 is an IRAK4 stimulator and IRAK2a and IRAK2b enhance IRAK4-mediated signaling, IRAK2c, IRAK2d, and IRAK3 are IRAK4 inhibitors.

Partner	Role and events
BTK	Interaction with MyD88, MAL, and IRAK1 during LPS stimulation
IRAKs	IRAK2a/2b stimulates, IRAK2c/2d/3 inhibits IRAK4
MyD88	Inhibition of IRAK4 recruitment to the receptor complex Mediation of apoptosis with IRAK2
Pellino-1–3	Ubiquitin–protein ligases Component of the Pellino–IRAK1–IRAK4–TRAF6 complex NFκB activation in response to IL1
SOCS1	Suppression of TLR4 signaling Inhibition of IRAK4
Tollip	Component of the sorting machinery for ubiquitinated proteins in the endosome, IRAK1 degradation

phospholipase-C that, in turn, hydrolyzes PIP_2 into IP_3 and DAG messagers. Moreover, it interacts with MyD88, MAL, and IRAK1 during stimulation by lipopolysaccharides [473].

TRAF and Pellino Ubiquitin Ligases

Ubiquitination contributes to the regulation of mediators involved proteins in signal transduction cascades triggered by ILRs and TLRs that activate proper transcription factors and MAPK modules. In addition to the Ub ligases Tumor-necrosis factor receptor (TNFR)-associated factors (TRAF), Pellino Ub ligases catalyze polyubiquitination of IRAK molecules. Synthesis of Pellino-1 relies on interferon regulatory factor IRF3 [475].

According to the type of ubiquitin conjugase that works in combination with Pellino-1, the latter can form Lys48- or Lys11-linked polyubiquitin chains [476]. However, when Pellino proteins act with IRAK1, they only generate Lys63-linked polyubiquitination of IRAK1 kinase.

Pseudokinases of the IRAK family (e.g., IRAK1b, a splice variant of IRAK1, IRAK2, and IRAK3) may fail to interact with Pellino proteins.

Pellino-1 and -2 that interact with IRAK1 are required for NFκB activation during TIR signaling [474]. Pellino-1 and -2 are also substrates of IRAK4 [476]. Once Pellino isoforms are phosphorylated by IRAK1 and IRAK4, they act as ubiquitin ligases.

Active IRAK kinases can promote polyubiquitination of Pellino proteins, thus causing Pellino degradation. In addition, both IKKε and TBK1 phosphorylate (Ser76, Thr288, and Ser293; activate) Pellino-1. Pellino proteins can also interact with TRAF6 and MAP3K7 [476].

Stimulated TLRs or IL1R promote the association of Pellino proteins with IRAK1 and TRAF6 effectors. Kinase IRAK1 phosphorylates (activates) Pellino proteins, thereby causing Lys63-linked polyubiquitination of IRAK1, which can then binds to IKKγ [476]. Lys63-linked polyubiquitination of TRAF6 facilitates recruitment and activation of the MAP3K7–MAP3K7IP1/2/3 complex. The IKK and MAP3K7 complexes are thus bridged. Enzyme MAP3K7 activates IKK, hence stimulating NFκB and releasing MAP3K8 that launches the ERK1/2 cascade. In addition, MAP3K7 can also boost the P38MAPK and JNK pathways via corresponding MAP2Ks.

Once connection of IRAK1 and TRAF6 to the receptor complex is disrupted, TRAF6 interacts with TRAF-interacting protein with a forkhead-associated (FHA) domain (TIFA), which promotes TRAF6 oligomerization and ubiquitination, as TRAF6 also connects to a suitable ubiquitin conjugase [476]. Whereas the TRAF6–MAP3K7–MAP3K7IP complex moves to the cytosol, IRAK1 remains at the plasma membrane and is subsequently degraded.

SIGIRR

Widespread (but not ubiquitous as it lacks in macrophages), transmembrane protein Single immunoglobulin (Ig) domain-containing and Toll-like and IL1 receptor (TIR)-related protein SIGIRR, or TIR receptor TIR8, links to both IRAK1 and TRAF6 agents. It may prevent signaling from the IRAK1–TRAF6 complex [473].

SOCS1

Suppressor of cytokine signaling SOCS1 can suppress TLR4 signaling, as it interacts directly or not with IRAKs. It regulates protein turnover, as it targets proteins for polyubiquitination and degradation.

Tollip

Toll-interacting protein — Tollip — is an adaptor and substrate for IRAK1. In resting cells, Tollip forms a complex with members of the IRAK family, thereby preventing NFκB activation, as it hinders IRAK1 phosphorylation. After receptor activation, Tollip–IRAK1 complex is recruited to the receptor, thereby causing the rapid IRAK1 autophosphorylation and its dissociation from the receptor. In addition, Tollip is phosphorylated by IRAK1 upon receptor engagement. Once it is activated, Tollip precludes IRAK1 activity (negative feedback loop). Ubiquitination of IRAK1 is controlled by Tollip-interacting protein, Target of MyB1 (TOM1), which is also a ubiquitin- and clathrin-binding protein. Tollip is required for sorting of IL1R1 to late endosomes. Agent TOM1 interacts with IL1R1 to avoid accumulation of IL1R1 in late endosomes [473]. Tollip may act as an endosomal adaptor that links IL1R1 via TOM1 to the endosomal degradation machinery.

5.9.2.2 IRAK Proteins in Cell Signaling

Functional domains of IRAK proteins can comprise a death domain for MyD88 linkage, a proST motif rich in prolines, serines, and threonines, which is required for autophosphorylation and can be hyperphosphorylated, a kinase site, and a C-terminus for interaction with TRAF6 ubiquitin ligase. Subtype IRAK1 contains 3 TRAF6-binding motifs, IRAK2 2, IRAK3 1; IRAK4 lacks this motif [471].

Once interleukin-1 binds to type-1 IL1 receptor (IL1R1), the latter connects to IL1R1 accessory protein (IL1RAcP) that converts silent IL1R1 into an active receptor. The IL1R1–IL1RAcP (receptor–coreceptor) complex then recruits adaptor MyD88 that attracts IRAKs [477, 478]. Kinase IRAK is highly phosphorylated, interacts with ubiquitin ligase TRAF6, then leaves the receptor complex. Agent TRAF6 bears autoubiquitination (Lys63). The IRAK–TRAF6 complex initiates the assembly of a signalosome that includes MAP3K7 (TAK1). The MAP3K7 (TAK1)-binding proteins MAP3K7IP2 (TAB2) and/or MAP3K7IP3 (TAB3) binds ubiquitinated TRAF6 agent. Kinase MAP3K7, in turn, activates IκB and Jun N-terminal kinases that phosphorylate IκB and Jun, respectively. Receptor TLR4 also primes interactions between MyD88, IRAK, and TRAF6 proteins.

Among the 4 known members of the IRAK family, only IRAK1 has a potent kinase activity. Yet, IRAK1 can act primarily as an adaptor for TRAF6. Whereas IRAK1, IRAK2, and IRAK4 are ubiquitous, inducible IRAK3 is mainly detected in monocytes and macrophages (hence its other alias IRAKm) [477]. Kinase IRAK4 can act as an IRAK1 activator [479]. Isozyme IRAK4 actually interacts with IRAK1 and TRAF6 to activate NFκB transcription factor and MAPK pathways. Isoform IRAK4 tethers to adaptor MyD88 to form the so-called *myddosome* (IRAK4–MyD88 complex) via their death domains [481].

Phosphatidylinositol 3-kinase controlled by membrane IP_3 and its activating effector PKB positively or negatively regulate transcription factor NFκB activity in a cell- and ligand-specific manner. Protein kinase-B phosphorylates IKKα subunit of IκB kinase (IKK; i.e., IKKα–IKKβ–IKKγ complex), but not IKKβ. However, PI3K can act without stimulating IKK. Tumor-necrosis factor-α, interleukin-1, insulin growth factor-2, platelet-derived growth factor, bradykinin, interferon-α and -β, and N-formylmethionyl-leucyl-phenylalanine[168] provoke activation of IκB kinase that phosphorylates inhibitor of NFκB (IκB) for degradation, hence eliciting NFκB activity via the PI3K–PKB axis together with other mediators. On the other hand, lipopolysaccharide- or cytokine-induced stimulation of production of inducible nitric oxide synthase (NOS2) that is controlled by NFκB is hampered by PI3K–PKB signaling. Protein kinase-B can be phosphorylated (activated) by calcium–calmodulin-dependent Cam2Kα to phosphorylate IRAK1, thereby reducing its association with MyD88 and precluding IL1β activation of NFκB via IκB kinase-β excitation [478].

Upon stimulation of interleukin-1 and some Toll-like receptors, IRAK1 and TRAF6 are ubiquitinated. Ubiquitin ligase TRAF6 causes Lys63-linked polyubiquitination of IRAK1 that is required for IKK recruitment and activation [480]. Only polyubiquitinated IRAK1 (Lys134 and Lys180) actually binds the regulatory IKKγ subunit of IκB kinase to activate NFκB.

5.9.2.3 IRAK1

In humans, ubiquitous IRAK1 has 3 splice variants. It interacts with MyD88 adaptor and can homodimerizes [471]. It is phosphorylated by IRAK4 kinase. Once it is phosphorylated and autophosphorylated, IRAK1 is released from the receptor complex and binds to TRAF6, thereby enabling NFκB activation. It is also ubiquitinated and sumoylated. Its Lys48-linked polyubiquitination causes its rapid degradation. On the other hand, Lys63-linked polyubiquitination allows interaction with IKKγ subunit.

Subtype IRAK1 regulates ubiquitination by TRAF6 of IRF5 [471]. It cooperates with STAT3 and tethers to the IL10 promoter. It also contributes to STAT1 phosphorylation during IL1 stimulation.

5.9.2.4 IRAK2

Isoform IRAK2 interacts with MyD88 and MAL adaptors as well as TRAF6 Ub ligase. It functions redundantly with IRAK1 in early signaling, but plays an important role in late, sustained NFκB and MAPK activation [471]. Subtype IRAK4 phosphorylates IRAK2 to prime its kinase activity.

5.9.2.5 IRAK3

In humans, IRAK3 is produced in monocytes and macrophages. It contributes to the regulation of the NFκB axis. In addition, IRAK3 prevents P38MAPK activation by TLR2, but not JNK or ERK activation [471]. It operates independently of IRAK1

---

[168] This formylated tripeptide is a chemoattractant for polymorphonuclear leukocytes. It elicits release of lysosomal enzymes from these activated cells.

via dual-specificity phosphatase DUSP1, or MAPK phosphatase MKP1. In addition, IRAK3 represses the non-canonical, MAP3K14-dpendent NFκB pathway [471].

5.9.2.6 IRAK4

Isoform IRAK4 also interacts with MyD88 to form the *myddosome* (7–8 MyD88 complexed to 4 IRAK4 proteins), which may recruit IRAK2 or IRAK3 [471]. Subtype IRAK4 phosphorylates IRAK1, thereby enabling IRAK1 autophosphorylation (activation). However, IRAK4 also primes IRAK1 degradation to regulate the MyD88-mediated signaling (negative feedback loop) [471]. In addition, IRAK4 interacts with IRAK2 isoform. Activity of IRAK4 kinase is necessary for activation of the MAPK module and JNK enzymes downstream from IL1R, TLR2, TLR4, and TLR7 [471].

5.9.3 LIM Domain-Containing Kinase Subfamily (LISK [LIMK–TESK] Family)

The LIMK subfamily includes 2 paralogs (LIMK1–LIMK2). These LIM domain-containing kinase isoforms regulate actin polymerization via phosphorylation (inactivation) of actin-depolymerizing factors cofilin-1 and -2 and cofilin-related destrin to accumulate actin filaments.

Ubiquitous LIMK1 and LIMK2 can form heterodimers with potential transphosphorylation. Both paralogs have splice variants for augmented diversity.

5.9.3.1 LIMK1

The activity of LIMK1 is regulated by small GTPases Rho, Rac, and CDC42 via their effector kinases RoCK and P21-activated kinases PAK1 and PAK4 that phosphorylate LIMK1 (Thr508 of kinase domain) [482]. In addition to its action on actin dynamics, LIMK1 participates in the regulation of microtubule dynamics, as it phosphorylates (inhibits) tubulin polymerization-promoting protein (TPPP), hence provoking microtubule disassembly.

Stability of LIMK1 is enhanced by transphosphorylation. Binding of heat shock protein Hsp90 to LIMK1 promotes its homodimerization and transphosphorylation, thereby heightening its stability and activity [482]. Paralog LIMK1 is also phosphorylated (Ser323; activated) by mitogen-activated protein kinase-activated protein kinase MAPKAPK2 during VEGF stimulation of endothelial cells. Last, but not least, LIMK1 phosphorylates transcription factor CREB1 and interacts with nuclear receptor-related factor NR4a2 (NuRR1).

Many factors depress LIMK1 ability to phosphorylate cofilin [482]: slingshot phosphatase Ssh1, Large tumor suppressor Lats1, a Ser/Thr kinase, interaction with unbound bone morphogenetic protein receptor BMPR2, nischarin, LIMK1-short splice variant, and ubiquitin ligases RNF6 and Parkin.

5.9.3.2 LIMK2

Paralog LIMK2 is activated by Rho GTPase effector kinases RoCK1, PAK1, and myotonic dystrophy kinase-related CDC42-binding kinase MRCKα once they are actuated by CDC42, Rac, and Rho [482]. Activated LIMK2 phosphorylates (inactivates) the 3 members of the cofilin family, like its paralog LIMK1, hence stabilizing accumulated actin filaments (single known LIMK2 function).

Whereas LIMK1 colocalizes with actin stress fibers and focal adhesions, LIMK2 is found in punctae that resemble endosomes [482]. Isoform LIMK2, but not LIMK1, is inhibited by protein partitioning defective Par3 to impede cofilin phosphorylation and tight junction assembly. Like LIMK1, heat shock protein HSP90 provokes LIMK2 dimerization.

Nuclear import of LIMK2 is suppressed after phosphorylation by PKCδ (Ser283 and Thr494). Transforming growth factor-β1 induces LIMK2 phosphorylation by RoCK kinase. Slingshot phosphatase Ssh1 dephosphorylates (inactivates) LIMK2.

5.9.4 Testis-Specific Kinase Subfamily (LISK [LIMK–TESK] Family)

Testicular protein kinases (TesK) form a subfamily that phosphorylate (deactivate) cofilin, like members of the LIMK subfamily. Kinases LIMK1 and LIMK2 are ubiquitous, whereas TesK family members are produced most abundantly in the testis.

The LIMK kinases are phosphorylated (activated; LIMK1: Thr508; LIMK2: Thr505) by distinct Rho GTPase pathways [483]: (1) CDC42 and Rac act via P21-activated kinases PAK1 and PAK4; (2) CDC42 via myotonic dystrophy-related CDC42-binding kinase; and (3) RhoA via RoCK kinase. On the other hand, TesK activation depends on integrin engagement upon cell attachment. It occurs independently of PAK1 and RoCK activation.

5.9.4.1 Testis-Specific Kinase-1

Testis-specific kinase-1 (TesK1) is a protein Ser/Thr kinase that possesses an N-terminal protein kinase region and a C-terminal proline-rich domain. Its kinase domain is closely related to those of the LIM (Lin11/Isl1/Mec3) motif-containing protein kinases. Therefore, kinases of the LIMK and TESK subfamilies constitute the LIMK/TESK family of protein Ser/Thr kinases. It is encoded by the TESK1 gene. It can phosphorylate myelin and histone.

5.9.4.2 Testis-Specific Kinase-2

Testis-specific kinase-2 (TesK2) is a protein Ser/Thr kinase that contains an N-terminal kinase domain structurally similar to that of TesK1 and LIMK kinases. Its encoded by the TESK2 gene that is predominantly expressed in the testis and prostate.

5.9.5 Receptor-Interacting Protein Kinase Family

Receptor (TNRFSF)-interacting protein Ser/Thr kinases (RIPK) constitute a family of 7 members that regulate cell survival and death [484] (Vol. 2 – Chap. 4. Cell Survival and Death). They interact with members of the superfamily of tumor-necrosis factor receptors (Vol. 3 – Chap. 11. Receptors of the Immune System). Pleiotropic tumor-necrosis factor intervenes in cellular stress and inflammation. Its signaling can lead to 3 outcomes via different working modes: survival, apoptosis, and necrosis, as TNF can cause the formation of a *necrosome*.[169] Kinases RIPK1 and RIPK3 participate in necrosis [485]. They are regulated by caspases and ubiquitination.

5.9.5.1 RIPK1

Receptor-interacting protein kinase RIPK1 (or RIP1) serves as an adaptor and integrator kinase in interactions between stress-induced signaling pathways triggered by binding of inflammatory cytokines to their receptors, stimulation of pathogen recognition receptors by pathogen-associated molecular patterns, and DNA damage and cell fate, i.e., resulting in cell survival or death. Its C-terminus contains a death domain and a caspase recruitment domain (CARD) that can recruit large proteic complexes for different signaling pathways. It is constitutively expressed in many tissues and inducible upon TNFα or T-cell activation.

Kinase RIPK1 activates NFκB as well as mitogen-activated protein kinases, such as ERK, JNK, and P38MAPK [484]. Upon binding of TNFα to TNFR1, RIPK1 and

---

[169] The necrosome provokes the production of reactive oxygen species via mitochondrial complex-I. In addition, NADPH oxidases are recruited to synthesize ROS at the plasma membrane-associated TNF receptor complex 1 and for calcium mobilization, activation of phospholipase-A2, lipoxygenases, and acid sphingomyelinases, as well as lysosomal destabilization.

TRAF2 are recruited and form a complex that primes TNFα-induced NFκB and MAPK activation. Upon TLR3 and TLR4 activation, RIPK1 is recruited and binds to TRIF to be then phosphorylated and polyubiquitinated [484].

5.9.5.2 RIPK2

Receptor-interacting protein kinase RIPK2[170] is synthesized in the brain, heart, lung, kidney, spleen, pancreas, prostate, testis, placenta, and lymph node, as well as peripheral leukocytes. It is involved in signal transduction associated with apoptosis and immune response. It actually activates NFκB as well as procaspase-1 and -8. It can be a component of tumor-necrosis factor receptor TNFRSF1a and TNRFSF5 complexes.

Kinase RIPK2 can signal by associating with other CARD domain-contaning signaling adaptors and initiator caspases. It binds to apoptosis inhibitors IAp1 (baculoviral IAP repeat-containing protein BIRC3) and IAp2 (BIRC2), tumor-necrosis factor receptor-associated factors TRAF1, TRAF2, TRAF5, and TRAF6, FADD-like apoptosis regulator CFLAR, prohibitin-2, 40S ribosomal protein-S14, NFκB essential modulator IKKγ, ubiquitin-C, and E2L ubiquitin-conjugase [251].

5.9.5.3 RIPK3

Receptor-interacting protein kinase RIPK3 (or RIP3) possesses a unique C-terminus with a RIP homotypic interaction motif (RHIM) that allows interaction with RIPK1. It is predominantly located in the cytoplasm, but can travel to the nucleus. It is a component of the tumor-necrosis factor receptor-1 complex. It weakly activates NFκB transcription factor. It can provoke apoptosis. It interacts with TNFRSF1A, TRIF, and TRAF2 [251].

5.9.5.4 RIPK4

Receptor-interacting protein kinase RIPK4 (or RIP4) is also labeled PKCδ-interacting protein kinase (DIK), ankyrin repeat domain-containing protein-3 (AnkRD3), and protein kinase-C-associated kinase (PKK). Its C-terminus possesses ankyrin repeats. It interacts with protein kinases PKCβ1 and PKCδ, as well as small GTPase RhoA [251]. It activates NFκB. It operates in keratinocyte differentiation.

---

[170] A.k.a. RIP2, CARD-containing interleukin-1β-converting enzyme-associated kinase (CARDIAK) and RIP-like interacting CLARP kinase (RICK).

5.9.5.5 RIPK5

Receptor-interacting protein kinase RIPK5 (or RIP5) is also termed dusty protein kinase. Like RIPK4, its C-terminus possesses ankyrin repeats. It interacts with 40S ribosomal protein-S9, proteasome-β3 subunit, ATP synthase B chain, chaperonin 10 (heat shock 10-kDa protein-1), IKK-related kinase-ϵ, small GTPases Rab5c and Rab14, huntingtin-interacting protein-2, and mitotic arrest-deficient MAD2 homolog-like-1 [251].

5.9.5.6 RIPK6

Receptor-interacting protein kinase RIPK6 (or RIP6) is also designated as leucine-rich repeat kinase-1 (LRRK1). It contains a leucine-rich repeat (LRR) motif that can be involved in recognition of pathogen- (PAMP), damage- (DAMP), or stress-associated molecular patterns (SAMP). It also possesses Ras of complex proteins/C-terminal of Roc (RoC/CoR) domain that can bind GTP for its activation. It interacts with lysozyme, Rab3-interacting protein-2, P21-activated kinase PAK4, and S100A8 (calgranulin-A) [251].

5.9.5.7 RIPK7

Receptor-interacting protein kinase RIPK7 (or RIP7) is also called leucine-rich repeat kinase-2 (LRRK2). Like RIPK6, it contains a leucine-rich repeat (LRR) motif for recognition of pathogen-, damage-, or stress-associated molecular patterns as well as RoC/CoR domain, hence its name Roco protein found in the literature in addition to Leu-rich repeat kinase.

5.10 Aurora Kinase Family

The mitotic, protein Ser/Thr Aurora kinases regulate the cell cycle transition from G2 down to cytokinesis. They are encoded by 3 Aurora genes (AURKA–AURKC).

5.10.1 Aurora Kinase-A

Kinase Aurora-A (AurA or AurKa)[171] is highly expressed in testis and weakly in skeletal muscle, thymus, and spleen. Its activity is regulated during the cell cycle,

---

[171] A.k.a. protein Ser/Thr kinase STK6, STK7, and STK15, as well as AIK, ARK1, and BTAK.

with a low level in G1 and S phase, rising during G2 and M phase, and rapidly falling afterward. It contributes to the regulation of the cell cycle during ana- and/or telophase.

Aurora-A participates in microtubule formation and/or stabilization, especially when the centrosome–mitotic spindle pole region is involved in chromosome segregation. Aurora-A also regulates some cellular processes in normal cells during interphase, especially microtubule dynamics, thereby intervening in the control of ciliary resorption, cell differentiation, and cell polarity. It can be activated at the basal body of cell cilia in the G0 and G1 phase of the cell cycle. Various types of stimuli release Ca^{++} from the endoplasmic reticulum rapidly and transiently to activate AurA kinase upon Ca^{++}–calmodulin binding [486].

Activation of AurA during mitosis results from binding to partners, such as its regulator Ajuba, PAK, C13ORF34 (Bora), TPX2, and NEDD9 scaffold. Its phosphorylation is required for catalytic activity during mitosis. It is phosphorylated upon DNA damage, probably by ATMK or ATRK kinase.

Aurora-A phosphorylates protein phosphatase-1 catalytic $PP1_{c\alpha}$ and $PP1_{c\gamma}$ subunits, CDC20 (a regulator of the cell division implicated in microtubule-dependent processes), CDC25b phosphatase, PKA kinase, transforming, acidic coiled-coil-containing protein TACC1, cytoskeleton-associated protein CkAP5, microtubule-associated homolog TPX2 (targeting protein for Xklp2), BrCa1, RhoGEF2, and BORA [487]. It also interacts with cytoplasmic polyadenylation element-binding protein CPEB1 that regulates translation of cyclin-B1.

In addition, AurA phosphorylates Ca^{++}-permeable, non-selective, cation channel and reduces the activity of polycystin-2 (TRPP2) [488].

5.10.1.1 Aurora Kinase-B

Kinase Aurora-B (AurB or AurKb), also called protein Ser/Thr kinase STK5 or STK12,[172] is a component of the *chromosomal passenger complex* that ensures correct chromosome alignment and segregation at the centromere [489]. It participates in the regulation of cleavage of polar spindle microtubules and onset of cytokinesis during mitosis. It also contributes to spindle checkpoint. It actually localizes on chromosome arms and inner centromeres from prophase through metaphase and then moves from anaphase through cytokinesis to the spindle midzone and midbody, where it colocalizes with γ tubulin. It is expressed during S phase and G2–M transition, at especially high levels in the thymus, but also in the lung, spleen, testis, colon, placenta, and fetal liver.

Aurora-B participates in at least 2 different chromosome passenger complexes: the AurKb–INCENP–BIRC5–CDCA8[173] and AurKb–INCENP complexes [489].

---

[172] A.k.a. AIK2, AIM1, and ARK2.

[173] Cell division cycle-associated protein CDCA8 is also named borealin. Baculoviral IAP repeat-containing protein BIRC5 is also called Apoptosis inhibitor API4 and survivin.

Protein phosphatase PP1 dephosphorylates (inactivates) Aurora-B. The latter phosphorylates histone-3 during mitosis as well as borealin (CDCA8), a component of the chromosomal passenger complex required for stability of the bipolar mitotic spindle, some centromere-interacting proteins, such as transiently interacting, inner centromere protein (InCenP) of the passenger protein set[174] and constitutively binding centromere protein CENPA, TACC1, RacGAP1 during M phase, RNA methyltransferase NSUN2, and ankyrin repeat and zinc finger domain-containing AnkZF1 [487]. Moreover, it associates with components of the chromosomal passenger complex such as inhibitor of apoptosis survivin (BIRC5). It also interacts with CDCA1, Evi5, and kinetochore complex component NDC80 that is involved in spindle checkpoint signaling. Inner centromere KIF2c stimulator (ICK2S) that stimulates mitotic centromere-associated kinesin KIF2c to promote microtubule depolymerization interacts with Aurora-B and INCENP [489]. Aurora-B phosphorylates MCAK and inhibits its microtubule depolymerization activity.

Many Aurora-B substrates that localize at the midbody are involved in cytokinesis: vimentin, desmin, myosin-2 regulatory light chain, and RacGAP1 [489]. Aurora-B also serves in the delocalization of kinesin-like protein Pavarotti (also known as mitotic kinesin-like protein MKLP1, Chinese hamster ovary monoclonal antibody ChO1, and zygotic enclosure defective protein ZEn4) that is required to build the central spindle at the end of mitosis.

5.10.1.2 Aurora Kinase-C

Kinase Aurora-C, also called Ser/Thr protein kinase STK13,[175] contributes to the organization of microtubules, especially in the centrosome–spindle pole region during mitosis. Its expression reaches a maximum during M phase. Two isoforms are produced by alternative splicing.

5.11 Polo-like Kinase Family

Polo-like kinases (PLK; homologs of Drosophila Polo kinase) constitute a family of Ser/Thr kinases involved in the regulation of the progression of the cell division cycle from G2 to M phase (Vol. 2 – Chap. 2. Cell Growth and Proliferation).

---

[174]Centromere-interacting proteins constitute 2 sets: (1) constitutively binding centromere proteins and (2) passenger (transiently interacting) proteins. Proteins of the INCENP set localize along chromosomes in the early stages of mitosis, then gradually concentrate at centromeres as the cell cycle progresses into mid-metaphase, and afterward reside within the midbody in the intercellular bridge during telophase, from which they are discarded after cytokinesis.

[175]A.k.a. AIE2 and AIK3.

All the PLKs associate with the spindle poles early in mitosis. They participate in the formation and motion of the mitotic spindle as well as activation of CDK–cyclin complexes during mitosis [490].

Polo-like kinases include several cell cycle-regulated protein kinases: PLK1,[176] PLK2,[177] PLK3,[178] and PLK4.[179]

In addition to the N-terminal catalytic domain, Polo-like kinases have C-terminal Polo boxes (motifs) that enable PLK localization to specific mitotic structures, such as centrosomes in early M phase, spindle midzone in early and late anaphase, and midbody during cytokinesis.

Polo-like kinases enable the G2–M transition, activation of CDC25a phosphatase and ubiquitin ligase cyclosome (or anaphase-promoting complex), and mitotic processes, such as centrosome maturation, bipolar spindle formation, chromosome segregation, and actin-ring formation.

5.11.1 Polo-like Kinase PLK1

Polo-like kinase PLK1 is involved in the regulation of DNA-damage repair, apoptosis, and progression of the cell cycle. It regulates mitotic entry and exit, centrosome maturation, sister chromatid cohesion, chromosome segregation, cytokinesis, and activation of APC ubiquitin ligase.

This mitotic Ser/Thr kinase changes its localization during different stages of the cell division. During interphase, PLK1 localizes to centrosomes. In early mitosis, it lodges to mitotic spindle poles. It resides to the central region of the mitotic spindle in late mitosis, where it associates with KIF23 kinesin-like protein.

Polo-like kinase-1 helps in the maturation of the centrosome in late G2 phase and early prophase as well as establishment of the bipolar spindle. It phosphorylates (activates) CDC25c phosphatase (Ser216, a residue also phosphorylated by ChK1 and ChK2) [491]. Phosphatase CDC25c dephosphorylates (activates) the CcnB–CDK1 complex.

During prometaphase, PLK1 phosphorylates the cohesin subunit SA2,[180] thereby promoting dissociation of cohesin from chromosome arms. Therefore, PP1 hinders PLK1 activity to avoid a premature loss of cohesion. In addition, centromeric shugoshin-1 recruits PP2–PP2$_{r5}$ to protect SA2 against phosphorylation by PLK1, thereby maintaining a pool of persistent cohesin.

---

[176] A.k.a. STPK13.

[177] A.k.a. serum-inducible kinase (SnK).

[178] A.k.a. cytokine-inducible kinase (CnK), FGF-inducible kinase (FnK), and proliferation-related kinase) (PRK).

[179] A.k.a. STK18 and SnK/PLK-akin kinase (SAK).

[180] A.k.a. stromal antigen StAg2.

During anaphase, PLK1 translocates from kinetochores to the central spindle by binding to the microtubule-bundling protein regulator of cytokinesis PRC1. Before anaphase onset, phosphorylation of PRC1 by CDK1 inhibits PRC1 binding to PLK1 enzyme. When CDK1 activity decreases during anaphase, PLK1 can then phosphorylate PRC1 at a site adjacent to the CDK1 phosphorylation site, thereby creating its own binding site.

In addition to centrosome- and chromatin-based microtubule nucleation, microtubule-generated microtubule polymerization participates in the formation of a proper bipolar spindle during mitosis, during which the augmin complex intervenes.[181] Polo-like kinase PLK1 localizes to the mitotic spindle and promotes the microtubule-generated microtubule nucleation, as it regulates augmin activity [492]. The phosphorylation by CDK1 of the recruiter, or γ-tubulin-targeting factor, neural precursor cell-expressed, developmentally down-regulated protein NEDD1,[182] of the γ-tubulin ring complex (γTuRC), which functions as a centrosomal and chromatin-mediated microtubule nucleator,[183] permits the NEDD1-PLK1 interaction. This event enables the phosphorylation by PLK1 of the subunit HAUS8 of the Augmin complex to support the augmin-microtubule interaction and microtubule-based microtubule nucleation within the spindle.

Damaged DNA provokes the phosphorylation (activation) of checkpoint kinase-2 at multiple sites. Polo-like kinase-1 phosphorylates ChK2 (Thr68) [491]. Kinases ChK2 and PLK1 colocalize to centrosomes in early mitosis and to the midbody in late mitosis.

5.11.2 Polo-like Kinase PLK2

Kinase PLK2 plays a role beyond the cell cycle regulation. On the other hand, PLK2 regulation depends on calcium. Activity-dependent changes in synaptic efficacy underlie learning and memory. Polo-like kinases PLK2 and PLK3 contribute to stabilization of long-term synaptic adaptivity (plasticity for long-term potentiation).[184] They are targeted to the dendrites of activated neurons. They interact

---

[181] Globular tubulin subunits form the tubulin subunit constituted by α- and β-tubulin heterodimers. The third member of the tubulin family — γ-tubulin —, an element of the microtubule-organizing center, is required for microtubule nucleation as well as centrosome duplication and spindle assembly. It is a component of the γ-tubulin small complex (γTuSC) and of the larger γ-tubulin ring complex (γTuRC). Each complex consists of numerous γ-tubulin complex proteins (GCP1, or γ-tubulin, GCP2–GCP6, and NEDD1 (or GCPWD) that regulates the localization of γTuSC to spindles and centrosomes.

[182] A.k.a. γ-tubulin ring complex protein TubγCP7 and WD40 repeat-containing protein GCPWD.

[183] The centrosome is the primary microtubule-organizing center (MTOC) in most somatic cells. It is surrounded by the pericentriolar material. The latter contains γ-tubulin ring complexes.

[184] Long-term potentiation is a long-lasting enhancement in signal transmission between neurons that are synchronously stimulated.

specifically with Ca^{++}- and integrin-binding protein CIB [493]. In addition, Polo-like kinase PLK2 phosphorylates PLK1 (Ser137) to initiate a survival signal when mitochondrium-associated cell respiration is impaired [494].

Transcription of PLK2 is regulated by P53 transcription factor. Nonetheless, the P53-binding sites in the Plk2 gene contain both stimulatory and inhibitory elements. The Plk2 promoter region contains consensus sequences for both calcium-dependent cAMP (CRE) and anti-oxidant (ARE) response elements [494].

5.11.3 Polo-like Kinase PLK3

Polo-like kinase PLK3 responds to DNA damage and/or mitotic spindle disruption [495]. It strongly interacts with P53 factor upon DNA damage, hence associating DNA damage with cell cycle arrest and apoptosis. Its activity is relatively low during G1 phase, G1–S transition, and mitosis. Its level peaks during late S and G2 stages of the cell cycle. It phosphorylates (activates) CDC25c phosphatase (Ser216, a residue also phosphorylated by ChK1 and ChK2) [496]. In addition, the transition of monocytes from peripheral blood to matrix-resident macrophages is accompanied by increasing PLK3 levels [497].

5.11.4 Polo-like Kinase PLK4

Polo-like kinase PLK4 localizes to centrioles and regulates centriole duplication during the cell cycle. Kinase PLK4, γ-tubulin, spindle assembly-6 homolog SAs6, centrosomal P4.1-associated protein (CenPj or CPAP), 135-kDa centrosomal protein CeP4 (CeP135), and centrosomal protein CP110 are required at different stages of procentriole formation, in association with different centriolar structures [498]. Protein SAs6 associates only transiently with nascent procentrioles. Centrosomal proteins CeP135 and CPAP form a core structure within the proximal lumen of both parental and nascent centrioles. Centrosomal protein CP110 is recruited early to centrosomes and then connects to the growing distal tips. Its expression is strongly induced at the G1–S transition.

5.12 With-No-K Kinase Family

Member of the WNK family display a unique substitution of cysteine for lysine at a nearly invariant residue within subdomain-2 of their catalytic core, hence their name "with no K" (lysine). Lysine-deficient kinases are Ser/Thr protein kinases.

5.12 With-No-K Kinase Family

The WNK family comprises 4 members (WNK1–WNK4) that are encoded by 4 genes (PRKWNK1–PRKWNK4). The WNK kinases are located in epithelia that are involved in chloride ion flux.

WNK kinases are composed of 3 domains: a kinase domain flanked by a short N-terminus and a large C-terminus that contains an autoinhibitory domain and 2 coiled-coil domains for protein–protein interaction. They can interact with other WNKs as well as other kinases, such as mitogen-activated protein kinases (MAP3K2 and MAP3K3), serum- and glucocorticoid-induced kinase SGK1, and Ste20-related kinases (OSR1 and STK39), as well as SMAD2 of the transforming growth factor-β signaling axis[185] and synaptotagmin.[186]

They participate in the regulation of not only transepithelial ion transport, but also paracellular permeability. In particular, they contribute to the control of electrolyte homeostasis and blood volume, hence blood pressure. In kidneys, WNKs control salt homeostasis downstream of aldosterone signaling [502]. Other physiological functions of WNKs include cell volume regulation, neurotransmission, cell proliferation, and embryogenesis.

5.12.1 WNK1

Kinase WNK1 is a cytoplasmic Ser/Thr protein kinase that is expressed in epithelia involved in chloride transport, especially distal nephron. Within kidneys, WNK1 is expressed in the cytoplasm of distal convoluted tubule (DCT), connecting tubule, and cortical and medullary collecting duct cells. It resides in other epithelia with a prominent role in Cl$^-$ flux, such as polarized epithelia that line the lumen of biliary and pancreatic ducts, epididymis, sweat ducts, colonic crypts, and gallbladder, as well as basal layers of epidermis, esophageal epithelium, and specialized endothelium of the blood–brain barrier [503]. WNK1 subcellular location varies among these epithelia, as WNK1 is cytoplasmic in kidney, colon, gallbladder, sweat duct, skin, and esophagus, but resides in lateral membrane in bile and pancreatic ducts and epididymis. In fact, WNK1 is ubiquitous, with high expression not only in kidneys, but also brain, heart, and skeletal muscles.

In humans, the Wnk1 gene expresses both kinase and pseudokinase isoforms according to both transcription initiation site among the available site set and tissue-specific regulatory elements. The Wnk1 gene indeed contains multiple promoters. Two promoters generate 2 WNK1 isoforms with a complete kinase domain [504]. Further variations are achieved by polyadenylation and tissue-specific splicing.

---

[185] Both WNK1 and WNK4 directly bind to Smad2. Kinase WNK1 possesses a dual role in TGFβ–Smad2 signaling, as it favors Smad2 synthesis, but precludes phosphoSMAD2 signaling [499].

[186] WNK1 phosphorylates synaptotagmin-2, hence increasing the amount of Ca^{++} required for synaptotagmin-2-binding to phospholipidic vesicles [500].

A *kidney-specific, kinase-defective isoform* (ksWNK1) is associated with a third promoter. It is restricted to distal convoluted tubule and possesses the auto-inhibition domain.

Kinase WNK1 activates the MAP3K2/3–ERK5 pathway [505], as it is able to phosphorylate MAP3K2 and MAP3K3. It functions as a tetramer that selectively phosphorylates synaptotagmin-2, a component of the endocytic machinery, to modulate its membrane binding, hence regulating vesicle trafficking and fusion. Moreover, WNK1 interacts with intersectins Itsn1 and Itsn2 that are involved in clathrin-mediated endocytosis and indirectly coordinate endocytic membrane traffic with the actin assembly machinery. Together with WNK4, WNK1 stimulates clathrin-dependent endocytosis of ATP-dependent renal outer medullary K^+ channel ROMK1, the apical, inward rectifier $K_{IR}1.1$ that modulates K^+ secretion.

Kinase WNK1 likely modulates WNK4 action on the sodium–chloride cotransporter. It also activates serum- and glucocorticoid-inducible protein kinase SGK1 to stimulate Na^+ transport by increasing the density of epithelial Na^+ channel (ENaC) at the cell surface [506]. Epithelial sodium channel is the rate-limiting step for Na^+ transport across high-resistance epithelia. This channel is regulated by insulin and mineralocorticoid hormones via phosphatidylinositol 3-kinase. SGK1 Activation by WNK1 is also carried out by PI3K that phosphorylates WNK1 (Thr58) [507]. Activated SGK1 reduces the interaction between NEDD4-2 ubiquitin ligase and ENaC, as it phosphorylates NEDD4-2 to heighten ENaC plasmalemmal amount.

Kinase WNK1 is required for SGK1 activation by insulin-like growth factor IGF1 [507]. Phosphorylation of WNK1 (Thr60) induced by IGF1 results from the activated PI3K–PKB pathway [508]. Protein WNK1 actually is a substrate of PKB, but neither of P70 ribosomal S6 kinase nor serum- and glucocorticoid-induced protein kinase SGK1. It also interacts with E2F transcription factor E2F3 [251].

5.12.2 WNK2

Kinase WNK2 is a cell growth regulator that modulates activation of ERK1 and ERK2 enzymes. It promotes RhoA activity, but impedes Rac1 activation, thus repressing stimulation of Rac1-effector PAK1 and its substrate MAP2K1 [509].

5.12.3 WNK3

Kinase WNK3 has its highest expression in the brain. Nevertheless, it is ubiquitous. The Wnk3 transcript has at least 2 alternatively spliced forms. Kinase WNK3 colocalizes with NKCC1, KCC1, and KCC2 cotransporters in diverse Cl^--transporting epithelial cells and GABA-responsive neurons. These neurons express the ligand-gated ion channels, ionotropic $GABA_A$ receptors. They are located in the hippocampus, cerebellum, cerebral cortex, and reticular activating system [510].

Kinase WNK3 increases Cl^- influx via NKCC1 and impedes Cl^- efflux through KCC1 and KCC2 [510]. These combined effects raise intracellular Cl^- concentration. WNK3 does not necessarily directly phosphorylate its targets but regulates effector kinases and phosphatases. In kidneys, activated WNK3 increases the surface expression and activity of NKCC2 in the thick ascending limb of Henle's loop and NCC in the distal convoluted tubule.

5.12.4 WNK4

Kinase WNK4 is synthesized in epithelia that operate in Cl^- handling, as it regulates ion carriers that are responsible for Cl^- flux across apical and basolateral membranes. Expression of WNK4 is more restricted than that of WNK1. It is predominantly produced in kidneys and other epithelial tissues.

Kinase WNK4 is a potent inhibitor of diverse epithelial transporters, such as sodium chloride cotransporter and renal outer medullary potassium ion channel. Activity of WNK4 promotes paracellular chloride ion flux.

Within kidneys, WNK4 is synthesized along the distal nephron, but its production is limited to the distal convoluted tubule (DCT), connecting tubule, and cortical collecting duct. In DCT cells, WNK4 colocalizes with ZO1 in tight junction. In other nephron segments, WNK4 is cytoplasmic. In distal nephron, it inhibits Na^+ reabsorption and K^+ secretion via both trans- and paracellular ion fluxes. Kinase WNK4 coordinates activities of diverse aldosterone-sensitive mediators of ion transport in the distal nephron [502]. Mutations in Wnk4 gene cause pseudohypoaldosteronism type-2, an autosomal dominant form of hypertension with hyperkalemia.

Like WNK1, WNK4 stimulates endocytosis of ROMK1, as it recruits endocytic scaffold intersectin, thereby impeding channel function. Like WNK1, WNK4 phosphorylates Ste20-related kinases OSR1 and STK39 (Sect. 5.1.8) that, in turn, phosphorylate N^+–K^+–$2Cl^-$ cotransporters NKCC1 and NKCC2 as well as Na^+–Cl^- cotransporter (NCC) to heighten their activity.

5.12.5 Regulation of Ion Transport in Nephron

The WNK kinases are regulators of intracellular signaling cascades throughout the nephron, especially in the aldosterone-sensitive distal nephron, where they modulate both the activity and/or abundance of sodium and potassium transporters (Tables 5.22 and 5.23).

In the distal nephron, WNKs control Na^+–Cl^- cotransporter in the distal convoluted tubule (Tables 5.24 and 5.25) as well as epithelial sodium channel (ENaC) in connecting tubule and collecting duct. Phosphorylation of ENaC depends on

Table 5.22. Main carriers for Na^+ reabsorption and K^+ secretion in distal convoluted tubule (DCT), connecting tubule (CNT), and collecting duct (CD) of the nephron (Source: [511]; ENaC: epithelial Na channel; NCC: Na^+–Cl^- cotransporter; ROMK: renal outer medullary potassium channel). Distal convoluted tubule that begins few cells after the macula densa is divided into 2 segments: upstream DCT1 and downstream DCT2. Isoforms WNK1, ksWNK1, and WNK4 are strongly expressed in distal nephron.

Ion carrier	Na^+	K^+
DCT1	NCC	ROMK
DCT2	NCC, ENaC	ROMK
CNT	ENaC	ROMK
CD	ENaC	ROMK

Table 5.23. The WNK kinases and their effects on ion transport in the nephron (Source: [511]; ENaC: epithelial Na channel; KCC: K^+–Cl^- cotransporter; NCC: Na^+–Cl^- cotransporter; ROMK: renal outer medullary potassium channel).

Ion transport	WNK1	WNK4
ENac	Stimulation	Inhibition
KCC		Inhibition
NCC		Inhibition
ROMK	Stimulation	Inhibition
Claudin-4	Stimulation	Stimulation

Table 5.24. Ion carriers in cells of distal convoluted tubule. In DCT, amount and phosphorylation state of Na^+–Cl^- cotransporters are determined by the ratio between stimulator WNK3 and inhibitor WNK4 that bear mutual inhibition. In addition, WNK1 inhibits WNK4 and is inhibited by ksWNK1. Aldosterone that binds to nuclear mineralocorticoid receptor acts on the distal tubules and collecting ducts to cause sodium and water retention as well as potassium secretion.

Apical membrane	Basolateral membrane
NCC	KCC4
	Na^+–K^+ ATPase

activities WNK1 and ksWNK1 as well as WNK4 that are controlled by serum- and glucocorticoid-regulated kinase SGK1. Kinase WNK1 indirectly activates ENaC, as it activates SGK1 via PI3K. Moreover, SGK1 reverses the inhibition of ENaC by WNK4. Both WNK1 and WNK4 inhibit ROMK by clathrin-mediated endocytosis.

Aldosterone is both a sodium-retaining and potassium-secreting hormone. It supports synthesis of ksWNK1 and WNK4 kinases. Like aldosterone, hyperkalemia increase synthesis of ksWNK1 and WNK4, whereas potassium restriction decreases ksWNK1 concentration and augments WNK1 level. Hyperkalemia stimulates renal

Table 5.25. Signaling pathways involved in hyper- and hypokalemia and -natremia and resulting effects on ROMK, BK, and ENaC channels in the apical membrane of cells of distal convoluted tubule and principal cells in the cortical collecting duct (Sources: [501,512]; ATn: angiotensin; BK: Ca^{++}-activated big-conductance K^+ channel; COx: cyclooxygenase; CyPEOx: Cytochrome-P450 epoxygenase [CyP2C23 and CyP2J]; EET: (11,12)-epoxyeicosatrienoic acid; ENaC: epithelial Na channel; MAPK: mitogen-activated protein kinase; MR: mineralocorticoid receptor; NCC: Na^+–Cl^- cotransporter; ROMK: renal outer medullary potassium channel; SGK: serum- and glucocorticoid-induced kinase).

Ion level	Signaling pathways	Channel effect
	Distal convoluted tubule	
Hypokalemia	ksWNK1–WNK1–WNK4–(–)–NCC	Stimulation
	WNK1–SGK1–WNK4–(–)–NCC	Stimulation
Aldosterone	MR–SGK1–WNK4–(–)–NCC	Inhibition
	Cortical collecting duct	
Hyperkalemia	Aldosterone–SGK1–WNK4–(–)–ROMK	Inhibition
	ksWNK1–WNK1–SGK1–WNK4–(–)—ROMK	Stimulation
	ksWNK1–WNK1–(–)—ROMK	Stimulation
	Renin–ATn2–MAPK–(–)–BK/ROMK	Stimulation
	CyPEOx–EET–(+)–BK	Stimulation
Hypokalemia	ksWNK1–WNK1–(–)—ROMK	Inhibition
	Renin–ATn2–Src–(–)—ROMK	Inhibition
	Renin–ATn2–MAPK–(–)–BK/ROMK	Inhibition
	COx2–PGE2–(–)–BK	Inhibition
	ksWNK1–(–)–ENaC	Inhibition
	WNK1–(–)—ROMK	Inhibition
	WNK1–SGK1–Nedd–(–)–ENaC	Stimulation
	WNK1–SGK1–WNK4–(–)–ENaC	Inhibition
Hyponatremia	Renin–ATn2–ATR1–aldosterone– –SGK1–WNK4–(–)–ROMK	Inhibition
	Renin–ATn2–ATR1–(–)–BK/ROMK	Inhibition

K^+ secretion by both aldosterone-dependent and -independent mechanisms. It suppresses the renin–angiotensin-2 axis. Hyperkalemia activates ksWNK1 and WNK4, inhibits Na^+–Cl^- cotransporter, and favors electrogenic sodium reabsorption by ENaC, thereby increasing the transepithelial voltage and stimulating potassium secretion. Conversely, a reduction in ksWNK1 and WNK4 activates Na^+–Cl^- cotransporter and fosters electroneutral sodium reabsorption with a relative conservation of potassium. Hypernatremia significantly raises ksWNK1 production with respect to hyponatremia. Hypovolemia and hyponatremia do not markedly affect ksWNK1 and WNK4 enzymes.

Table 5.26. Ion carriers in cells of cortical collecting duct. Isoform kAE1 of anion exchanger AE1, or solute carrier family-4, anion exchanger, type 1 (SLC4a1), is specific to the basolateral surface of α-intercalated cell (or type A) in the cortical collecting duct of the kidney. Exchanger AE1 causes electroneutral exchange of chloride (Cl^-) for bicarbonate (HCO_3^-) across the plasma membrane. Two types of H^+–K^+ ATPases exist in the kidney, gastric and colonic H^+–K^+ ATPases that contain type-1 and type-3 K^+ ATPases, respectively.

Principal cell		Intercalated cell	
Apical	Basal	Apical	Basal
H^+–K^+ ATPase	Na^+–K^+ ATPase	H^+–K^+ ATPase	K^+ channel
BK	K^+ channel	BK	kAE1
ROMK	Cl^- channel		
ENaC			

5.12.5.1 Potassium Ions

Maintenance of adequate plasma K^+ level within a narrow physiological range is needed for functioning of neurons, cardiomyocytes, and skeletal myocytes. Nephron regulates K^+ excretion, as it: (1) completely filters K^+ in the glomerulus; (2) reabsorbs K^+ extensively along proximal tubule and thick ascending limb of Henle's loop; (3) secretes K^+ in the connecting tubule (CNT) and cortical collecting duct (CCD); and (4) reabsorbs K^+ in outer medullary collecting duct (OMCD). Nephron segments DCT, CNT, and CCD constitute the *aldosterone-sensitive distal nephron*.

Principal and intercalated cells in CNT and CCD are responsible for K^+ secretion and absorption, respectively [501]. Several types of K^+ channels are synthesized and incorporated in the apical membrane of connecting tubule and cortical collecting duct: (1) ROMK ($K_{IR}1.1$); (2) Ca^{++}-activated big-conductance K^+ channel (BK); and (3) double-pore K^+ channel $K_{2P}1.1$ (or TWIK1; Table 5.26).

WNK1, WNK3, and WNK4, as well as kidney-specific WNK1 splice variant (ksWNK1) are produced in cortical collecting duct cells. Isoform ksWNK1 inhibits WNK1. In kidneys, various kinases such WNKs, SGK1, PKA, and Src participate in the regulation of K^+ secretion. Protein kinase-A phosphorylates ROMK, thereby increasing ROMK channel activity, whereas SGK1 stimulates the surface expression of ROMK1 by phosphorylating WNK4 (Ser1169) [501]. WNK1 stimulates SGK1 via PI3K. In the cortical collecting duct, SRC family kinases phosphorylate ROMK1 (Tyr337) and elicits its endocytosis. A low dietary K^+ intake increases, whereas a high K^+ intake decreases Src activity.

Potassium ions are secreted into the lumen through apical K^+ channels using a favorable electrochemical gradient after influx through basolateral Na^+–K^+ ATPase [501]. Potassium absorption is achieved by K^+ entry across apical membrane through H^+–K^+ ATPase and exit across the basolateral membrane. H^+–K^+ ATPases reside mainly in intercalated cells of collecting duct, but type-3

H^+–K^+ ATPases are also located in principal cells of connecting tubules and cortical collecting ducts. Type-3 H^+–K^+ ATPase is mainly responsible for renal K^+ reabsorption and type-2 H^+–K^+ ATPase mediates K^+-dependent H^+ secretion in collecting duct. Basolateral Na^+–H^+ exchangers supply Na^+ to sustain Na^+–K^+ ATPase activity.

Aldosterone augments activity of both basolateral Na^+–K^+ ATPase and apical ENaC in principal cells. Increased ENaC activity augments the driving force for K^+ exit across the apical membrane of CNT and CCD, whereas high Na^+–K^+ ATPase activity stimulates K^+ secretion by increasing K^+ uptake across the basolateral membrane. Aldosterone acts via SGK1 that promotes renal K^+ secretion, as it enhances ROMK recruitment to the apical membrane [501]. In addition, SGK1 phosphorylates WNK4 (Ser1169) to suppress its inhibitory effect on ROMK.

Vasopressin increases K^+ secretion, as it elevates apical ROMK activity in the CCD via the V_2–cAMP pathway and BK channel via V_1 receptor [501]. In addition, vasopressin can heighten K^+ secretion via apical KCC in the distal tubule (DCT, CNT, and CCD).

High dietary K^+ intake suppresses renin and angiotensin-2 activity and, conversely, K^+ restriction stimulates renin and angiotensin-2. Kinases of the P38MAPK and ERK family repress ROMK and BK channel activity in the cortical collecting duct.

Insulin enhances sodium reabsorption by the Na^+–Cl^- cotransporter and ENaC channel. Furthermore, it reduces WNK4 expression via the PI3K–SGK1 pathway, as SGK1 inhibits WNK4 [512]. Chronic insulin treatment can thus raise blood pressure.

Uroguanylin and guanylin are small peptides (16 and 15 amino acids, respectively). They are released by enterochromaffin cells in duodenum and proximal bowel and goblet cells in colon, respectively. Guanylin is an agonist of the guanylate cyclase receptor GCc and regulates electrolyte and water transport in intestinal and renal epithelia. Uroguanylin stimulates BK-dependent K^+ secretion in rat distal nephron [501]. Guanylin and uroguanylin cause both hyperpolarization and depolarization of cultured CCD principal cells. Peptide-induced hyperpolarization and depolarization are blocked by protein kinase-G and phospholipase-A2, respectively.

Low dietary K^+ intake stimulates renin production that enhances cyclooxygenase COX2 expression and PGE2 production in the rat kidney [501]. Prostaglandin-E2 inhibits ROMK via the PKC–MAPK pathway.

5.12.5.2 Sodium Ions

Kinase WNK4 inhibits Na^+–Cl^- cotransporter in distal convoluted tubule, hence lowering Na^+ reabsorption, as it reduces the cotransporter density in the plasma membrane. On the other hand, WNK1 relieves WNK4 inhibition of NCC in DCT.

5.12.5.3 Chloride Ions

Chloride influx and efflux through cation–Cl^- cotransporters contribute to the maintenance of intracellular volume. Cl^- influx and efflux are mediated by N^+–K^+–$2Cl^-$ and K^+–Cl^- cotransporters, respectively (Table 5.6).

In response to changes in extracellular osmolarity, intracellular volume of epithelial cells is predominantly maintained by augmenting or lowering intracellular Cl^- concentration, thereby minimizing transmembrane water flux. Intracellular Cl^- concentration is modulated by Cl^- entry mainly through NKCC1 and exit chiefly through KCC1. Transcellular cation–Cl^- cotransporters (NCC, NKCC1–NKCC2, KCC1–KCC4) belong to the SLC12A family (SLC12a1–SLC12a7). They use inward Na^+ or outward K^+ gradients to move Cl^- into or out of cells, respectively. Sensors and transducers coordinate regulation of NKCC and KCC cotransporters by triggering their phosphorylation and dephosphorylation. Phosphorylation activates NCC, NKCC1, and NKCC2 (Cl^- influx), but inactivates KCCs. Conversely, dephosphorylation inhibits NCC, NKCC1, and NKCC2, but stimulates KCCs (Cl^- efflux).

Cation–Cl^- cotransporter KCC1 is distributed in proximal and distal convoluted tubules and cortical collecting ducts, as well as smaller medullary collecting ducts and epithelial cells of the larger papillary ducts of Bellini. Smooth muscle cells in the interlobular arteries and arterioles do not express KCC1. However, endothelial cells in renal medium-sized arteries, arterioles, and vasa recta, like those of the aorta, synthesize KCC1 [513]. Expression of KCC1 is also detected in glomerular mesangial cells. Isoform KCC3 that has been initially cloned from vascular endothelial cells is expressed in kidneys. The KCC4 transcripts abound in kidneys.

Transepithelial Cl^- Permeability

Kinase WNK4 not only inhibits Na^+–Cl^- cotransporter in the distal convoluted tubule, but also phosphorylates the tight-junction constituent claudins. Consequently, WNK4 augments paracellular Cl^- permeability, especially between type A and B intercalated cells. However, elevation in transepithelial Cl^- permeability remains lower than that provoked by WNK4 mutant [514]. On the other hand, paracellular Na^+ permeability decays only slightly.

WNK and Other Cl^- Carriers

Kinase WNK4 also inhibits basolateral N^+–K^+–$2Cl^-$ cotransporter that is located in respiratory epithelium as well as other polarized epithelia. Moreover, WNK4 precludes the activity of apical chloride–base exchanger SLC26A6 (or CFEX) in intestinal, pancreatic, testicular, and hepatic epithelia [515]. Nevertheless, it does not impede the activity of apical ion exchanger pendrin (Cl^-–HCO_3^-–OH^- exchanger), or solute carrier SLC26A4, that is located in inner ear, thyroid gland, and a subpopulation of intercalated cells of the cortical collecting duct of nephron.

Chloride Handling and Dynamic Modulation of Neuron Activity

Dynamical regulation of $[Cl^-]_i$ directs the type of GABA signaling. In adult central nervous system brain, GABA is the major inhibitory neurotransmitter, whereas, during the neonatal period and in the peripheral nervous system, GABA neurotransmission is predominantly excitatory. In adults, GABA signaling also varies from inhibitory to excitatory with circadian rhythm in many brain centers. Neurotransmitter GABA acts by opening Cl^- channels on postsynaptic membranes. Resulting response depends on intracellular Cl^- concentration. When $[Cl^-]_i$ is below its equilibrium potential, Cl^- enters the cell and causes hyperpolarization and inhibition. On the other hand, if $[Cl^-]_i$ is above its equilibrium potential, GABA induces Cl^- efflux, depolarization, and neuronal excitation. Change in $[Cl^-]_i$ in neurons results from influx principally through NKCC1 and efflux through neuronal-specific cotransporter KCC2.

Kinase WNK3 colocalizes with NKCC1, KCC1, and KCC2 in neurons that possess $GABA_A$ receptors [510]. Once activated, it increases Cl^- entry via NKCC1 that is activated by phosphorylation and prevents Cl^- exit through KCC2 that is inactivated by phosphorylation.

Because NKCC1 modifies $[Cl^-]_i$, it modulates neuronal activity generated by γ-aminobutyric acid toward excitation or inhibition. Neurotransmitter GABA causes depolarization in dorsal root ganglion neurons and immature neurons that have a high NKCC1 expression level, whereas it gives rise to hyperpolarization in most adult neurons. GABA-mediated excitation actually characterizes neurogenesis, but the hyperexcitable, immature brain is more susceptible to epileptic seizures.

In rats, embryonic dorsal root ganglion neurons that need high $[Cl^-]_i$ and are characterized by a high, homogeneous NKCC1 expression are able to accumulate intracellular Cl^- via NKCC1, whereas in embryonic motor neurons that experience large variations as well as long-lasting changes in $[Cl^-]_i$ during activity, the NKCC1 expression is not only smaller, but also restricted to dendrites [516]. Both dorsal root ganglion and motor neurons can be loaded with or depleted of Cl^- during $GABA_A$ receptor activation at depolarizing or hyperpolarizing membrane potentials. After loading, the recovery rate (i.e., decay rate in $[Cl^-]_i$) is similar in both neuron types, whereas, after depletion, it (i.e., elevation rate in $[Cl^-]_i$) is markedly faster in dorsal root ganglion neurons than motor neurons, except when extracellular Na^+ level is low.[187]

---

[187] Cotransporter NKCC1, a so-called secondary active ion carrier, uses gradients of Na^+ or K^+ created by the primary active Na^+–K^+ ATPase.

5.13 Other Types of Ser/Thr Kinases

5.13.1 Doublecortin-like Kinase

Doublecortin and calcium–calmodulin-dependent protein kinase-like kinase DCLK1, or DCamKL1, is a microtubule-associated Ser/Thr kinase that contains 2 major functional domains: an N-terminal doublecortin-like domain and C-terminal kinase domain homolog to CamK1 and CamK2. It is specific to the central nervous system. Multiple splice variants include full-length isoform (DCLK1), a DCX-domain-only isoform (Dcl), a kinase-domain-only isoform (CPG16 or CLICK1), and a CamK-related peptide (CARP or Ania4) [517]. It contributes to neurogenesis.

In mice, 3 CamK1/4-related kinases possess a kinase domain structurally related not only to that of CamK1 and CamK4, but also CamK2 [518]: DCamKL1,[188] DCamKL2,[189] and DCamKL3.[190] They are highly expressed in the entire central nervous system from embryon to adult. Shorter isoforms DCamKL1α and DCamKL1β lack the doublecortin-like domain. Both DCamKL2α and DCamKL2β contain an N-terminal doublecortin-like domain that links to microtubules. On the other hand, DCamKL3 has an incomplete doublecortin-like domain-like homology and cannot localize to microtubules. These kinases target CREB-regulated transcription coactivator.

5.13.2 Haspin

Kinase haspin[191] is ubiquitous, but at a lower level than that in testis. Haspin is required for completion of mitosis in human cells. Haspin is hyperphosphorylated during mitosis. Haspin phosphorylates histone-3 during mitosis to change chromatin behavior [519]. Phosphorylated H3 appears at foci on chromosome arms in late G2 phase or early prophase. In prometaphase, pH3 reaches its highest level at inner centromeric regions.

Haspin maintains cohesion between sister chromatids and chromosome alignment during metaphase. Removal of chromosome cohesion occurs in 2 steps. In prophase and prometaphase, cohesin is released from chromosome arms under the control of Polo-like kinase-1, Aurora-B, and Wings apart-like homolog Wapal, a cohesin-binding protein. Shugoshin prevents cohesin removal from centromeres until anaphase, during which separase cleaves cohesin.

---

[188] A.k.a. DCLK1, CLICK1, and CL1.

[189] A.k.a. CLICK2 and CL2.

[190] A.k.a. CLICK1/2-related kinase (CLR).

[191] A.k.a. Germ cell-specific gene product GSG2. Haspin was initially detected in germ cells of testis.

5.13.3 Liver Kinase-B1 (STK11)

Protein Ser/Thr kinase STK11, or Liver kinase LKb1, acts upstream from adenine monophosphate-activated protein kinase (AMPK) in cell metabolism when energy and nutrient are lacking. It activates many other kinases such as AMPK-related kinases to maintain cell polarity.

In mammary glands, loss of STK11 function disturbs epithelium organization (mislocalization of cell polarity markers, lateralization of tight junctions, deterioration of desmosomes and basement membrane, and hyperbranching of mammary ducts).[192] Mutations of the LKB1 gene is detected in carcinomas of the lung, among others. Combination of LKb1 deficiency responsible for structural alterations in mammary epithelium and overexpression of the cell proliferation inducer MyC accelerates tumor formation [520]. Protein Ser peptidase hepsin, released from desmosomes upon the loss of LKb1, deteriorates desmosomes and the basement membrane.[193] Conversely, normal, LKb1-dependent epithelial architecture prevents MyC from initiating the cell cycle.

5.13.4 Never in Mitosis Gene-A-Related Expressed Kinases

5.13.4.1 NEK1

Never in mitosis gene A (NIMA)-related expressed kinase-1 (NEK1) possesses an N-terminal kinase domain that is similar to the catalytic domain of NIMA, a protein kinase that controls initiation of mitosis in Aspergillus nidulans. As both NEK1 and NIMA have a long, basic C-terminal extension, they are similar in overall structure. Kinase NEK1 is highly expressed in germ cells. It is also involved in the response to DNA damage caused by radiation. It is a dual-specificity kinase that phosphorylates substrates primarily on Ser and Thr, but also on Tyr residues [521].

5.13.4.2 NEK2

Never in mitosis gene-A-related expressed kinase-2 (NEK2) is a cell cycle-regulated Ser/Thr protein kinase that is active at the onset of mitosis. It accumulates at the centrosome, where it can participate in the assembly and maintenance of centrosome

---

[192]Epithelial structure relies on between-cell and cell–matrix adhesions. Polarized epithelial cell–cell contacts are ensured by apical tight and adherens junctions, basolateral desmosomes linked to intermediate filaments (Vol. 1 – Chap. 7. Plasma Membrane), and specialized basal junctions based on integrins of epithelial cells and extracellular components (laminin, collagen-4, nidogen, and proteoglycans) of the basement membrane (Vol. 1 – Chap. 8. Cell Environment). Kinase LKb1 participates in the formation of apicobasal polarity of epithelial cells.

[193]Targets of hepsin comprise, in particular, laminin-332, prohepatocyte growth factor, and prourokinase plasminogen activator.

structure throughout the cell cycle, the regulation of the intercentriolar linkage, as well as centrosome separation at the G2–M transition for proper bipolar spindle assembly and chromosome segregation [522]. This effect needs the presence of its substrate centrosomal coiled-coil NCK-associated protein NAP1 (or NCKAP1) and inactivation of inhibitory partner protein phosphatase PP1. Kinase NEK2 also prevents centrosome separation in the presence of DNA damages.

There are at least 2 NEK2 splice variants (NEK2a–NEK2b). Subtype NEK2a, but not NEK2b, is regulated by cell cycle-dependent protein degradation. Kinase NEK2a is ubiquitinated by APC/C and destroyed in prometaphase by 26S proteasome. Both NEK2a and NEK2b are short-lived proteins [522].

5.13.4.3 NEK6

Kinase NEK6 is a cell cycle-regulated Ser/Thr protein kinase of the NIMA family that accumulates and is activated during mitosis. It lacks an extended regulatory domain. Both NEK6 and NEK7 bind to and are phosphorylated (activated) by NEK9 kinase [523].

5.13.4.4 NEK7

Kinase NEK7, like NEK6, consists almost entirely of a catalytic domain. It is also activated during mitosis. It abounds at the centrosome. It is required for proper spindle assembly and mitotic progression [523].

5.13.4.5 NEK9

Kinase NEK9[194] is a cell cycle-regulated Ser/Thr kinase composed of an N-terminal kinase domain followed by an RCC1-like domain that causes auto-inhibition, and a C-terminus for NEK9 oligomerization. It localizes to the cytoplasm and remains inactive during interphase. It is activated during mitosis at the centrosomes and spindle poles [523].

5.13.5 NEMo-like Kinase

Nuclear factor-κB essential modulator (NEMo)-like Ser/Thr kinase (NLK) precludes Notch signaling. Upon ligand binding, intracellular domain of Notch is released and translocates into the nucleus. It then forms a ternary complex with

---

[194] A.k.a. Nercc1.

the CSL (CBF1, Suppressor of hairless, LAG1) transcription factor and a member of the Mastermind family of transcriptional coactivators. Kinase NLK: (1) inhibits Notch1^{ICD}, as it prevents the formation of the ternary transcription complex between Notch1^{ICD}, CSL, and Mastermind-like protein MamL1; (2) augments Notch3^{ICD} activity; and (3) does not influence Notch2^{ICD} and Notch4^{ICD} [524].

5.13.6 PTen-Induced Kinase

PTen-induced kinase PInK1 is a mitochondrial Ser/Thr kinase that has an antiapoptotic function. It localizes exclusively to the mitochondria. It is primarily located at the inner mitochondrial space, attached to the inner mitochondrial membrane [525]. It interacts with the mitochondrial chaperone protein tumor-necrosis factor receptor-associated protein TRAP1[195] and serine peptidase High temperature requirement protein HTRa2.[196] Factor FoxO3a regulates the transcription of PINK1 [526]. It upregulates the expression of SOD and catalase. In addition, PInK1 may intervene in dopamine release and synaptic remodeling.

Ubiquitous PInK1 yields a transcriptional target of PTen signaling. It might be a mediator of the MAPK module [525]. Kinase PInK1 also interact with Parkinson disease protein Park7[197] that participates in sensing and protecting against oxidative stress as well as ubiquitin ligase parkin.

In cardiomyocytes, ubiquitous PInK1 links mitochondrial dysfunction to oxidative stress.[198] Agent PINK1 is involved in the maintenance of the mitochondrial function. Loss of PINK1 actually causes [526]: (1) increased sensitivity to ROS-dependent depolarization of the mitochondrial membrane potential; (2) decreased oxidative phosphorylation; (3) reduced mitochondrial replication; (4) mitochondrial swelling; and (5) altered expression of mitochondrial proteins.

---

[195] Mitochondrial protein TRAP1 is also called heat shock protein HSP75. Its phosphorylation by PInK1 prevents the release of cytochrome-C from the mitochondria under oxidative stress. It is thus dispensable for PInK1 anti-apoptotic function.

[196] Serine peptidase HTRa2, also called Omi stress-regulated endoprotease and Park13, can be released from mitochondria during apoptosis and tether to caspase inhibitors, such as XIAP, IAP1, and IAP2, to cleave them, thereby reducing inhibition to caspase activation. Protein HTRa2 is phosphorylated under peculiar stress conditions by activated P38MAPK.

[197] Protein Park7 is also called DJ1. It interacts with protein inhibitor of activated STAT2 and CAP-binding protein complex interacting protein-1 (EfCab6 or Flj23588).

[198] Oxidative stress results from an imbalance between the production of reactive oxygen species and the elimination of these toxic intermediates by anti-oxidant systems. Upon ROS exposure, the heart evolves toward maladaptive hypertrophy (cell enlargement, myofibrillar disarray, and reexpression of fetal genes). Oxidative stress with associated mitochondrial dysfunction causes cardiac dysfunction and heart failure.

Kinase PInk1 should not be confused with peptidylprolyl cis/trans isomerase interacting with Never in mitosis-A (NIMA) PIN1.

5.13.7 Protein Kinase-R

Double-stranded RNA-dependent protein kinase-R (PKR) is a Ser/Thr kinase. This ubiquitous kinase that is also termed eukaryotic translation initiation factor-2α kinase-2 (eIF2αK2) is encoded by the EIF2AK2 gene. It phosphorylates translation initiation factor eIF2α (Ser51) to prevent protein synthesis.

This intracellular sensor detects various stresses. In particular, PKR is a component of innate immunity that protects against viral infections. In fact, it is activated directly by double-stranded RNA during viral replication. Moreover, PKR is also activated indirectly by bacterial lipopolysaccharides, ceramide, and DNA adducts (DNA segments covalently bound to carcinogens) [527]. Kinase PKR also participates in the regulation of transcription and signal transduction as well as cell growth and differentiation.

In normal circumstances, PKR is maintained as an inactive monomer. Upon stress, it is converted into an active homodimer. Kinase PKR is inhibited by the catalytic PP$1_{c\alpha}$ subunit of protein phosphatase-1, as well as ribosomal protein L18 (a component of the large ribosome 60S subunit) and eIF2α-associated glycoprotein GP67 [527].

Other PKR substrates comprise interleukin enhancer-binding factor-3 and transcription factor P53 [527]. It also targets interferon regulatory transcription factor IRF1 and nuclear factor-κB by phosphorylating IKKα or TNFR-associated factor (TRAF), as well as signal transducer and activator of transcription STAT1.

Protein kinase-R is induced by interferons in a latent state, hence its name interferon-inducible double-stranded RNA-dependent kinase. It is activated not only by viral double-stranded RNA, but also interferon-inducible double-stranded RNA-dependent protein kinase activator (PRKRA or PAct).[199]

Activation of PKR results from its dimerization and subsequent autophosphorylation. Kinase PKR is a pro-apoptotic protein to prevent viral spread, as it inhibits protein synthesis. Virus strategy aims at disabling PKR to promote viral production by expressing K3L that mimics eIF2α [529]. Activity of PKR relies on a high degree of flexibility to recognize mimickers aimed at subverting various cell processes and discrimination of self from foreign element.

Protein kinase-R participates in signal transduction in other situations of cellular stress than pathogen invasion. Other PKR activators comprise pro-inflammatory cytokines (interferon-γ, interleukin-3, and tumor necrosis factor-α) and heparin [530].

---

[199] A.k.a. PKR-associated protein-X (RAX). Activator of PKR binds to PKR inhibitor TAR RNA-binding protein (TRBP). The TRBP–PAct interaction impedes PAct-induced PKR activation [528].

In response to double-stranded RNA stimulation, PKR activates MAP2K6, but neither MAP2K3 nor MAP2K4, to stimulate P38MAPK [531].[200] Scaffold proteins that interact with at least 2 components of the MAPK module can act as insulators to prevent crosstalk between MAPK pathways and activators of a specific MAPK signaling, such as the MAP2K6 cascade.

5.13.8 Slicing Factor Protein Kinase

Slicing factor Ser/Arg-rich (RS domain-containing) protein kinases SRPK1 (or SFrsK1) and SRPK2 that are encoded by the SRPK1 and SRPK2 genes belong to the superfamily of SR proteins, which are Ser/Arg-rich proteins. These proteins generally have 2 domains: (1) an RS domain, rich in arginine–serine repeats and (2) an RNA-recognition motif (RRM). Kinases SRPK1 and SRPK2 constitute a family within the CMGC superfamily of Ser/Thr protein kinases.

Ubiquitous SRPK1 and SRPK2 phosphorylate Ser/Arg-rich (SR) proteins, such as splicing factors SFRS1 (also called alternative splicing factor ASF or splicing factor SF2), SFRS2 (or SC35), and SFRS6 (or SRp55) [532]. Phosphorylation by SRPK1 enables regulation of between-protein interactions between RS domain-containing splicing factors. Because SR proteins regulate splice site selection and spliceosome assembly, SRPK contributes to the control of mRNA splicing. Overexpression of SRPK1 or SRPK2 induces the phosphorylation-dependent relocalization of SR proteins from nuclear speckles to the nucleoplasm.

5.13.9 TANK-Binding Kinase 1

Tumor-necrosis factor receptor-associated factor family member (TRAF)-associated NFκB (TANK)-binding kinase TBK1[201] complexes with adaptors TRAF2 and TANK to assemble signaling complexes on the intracellular region of cytokine receptors. In particular, the TBK1–TRAF2–TANK complex activates NFκB. Once it is recruited to TNF-bound TNFR1, TBK1 indeed interacts with regulatory IKKγ subunit. In addition, TBK1 can phosphorylate class-2 NFκB protein RelA. Kinase TBK1 also operates in the innate immune response to viruses via Ifn1 agent.

---

[200] Other activators of MAP2K6 include MAP3K3 to MAP3K5, MAP3K7, MAP3K8, and MAP3K11.
[201] A.k.a. NFκB-activating kinase (NAK).

Cell evolution toward survival or death upon exposure to pro-inflammatory cytokines[202] depends on the balance between pro- and anti-apoptotic factors. Protein Ser/Thr kinase TBK1 promotes survival, as it phosphorylates the prosurvival factor RelA[203] that leads to a production of anti-apoptotic plasminogen activator inhibitor PAI2, or serpin-B2.[204] The latter stabilizes the pleiotropic enzyme transglutaminase-2 that inactivates procaspase-3 [533].

---

[202]Agent TNF can trigger apoptosis via its its receptor TNFR1 that recruits complexes on its cytoplasmic death domain to activate caspase-8, thereby stimulating the executioner caspase-3. On the other hand, cytokine TGFα yields a prosurvival signal.

[203]Factor NFκB supports cell survival via synthesis of anti-apoptotic proteins that control caspase activation, such as cellular inhibitors of apoptosis, caspase-8 inhibitor FlIP, and prosurvival BCL2-related protein BCL2a1.

[204]Protein PAI2 also targets retinoblastoma protein and protects it from cleavage by calpains.

Chapter 6
Mitogen-Activated Protein Kinase Module

Mitogen-activated protein kinase (MAPK) modules are commonly used to transduce signals aimed at regulating cell fate (cell differentiation, proliferation, senescence, and death) and inflammation. They mediate responses to extracellular signals, such as hormones, growth factors, and cytokines, as well as stresses and intercellular interactions.

The variety of possible combinations of kinases engaged by the initial stimulus, the presence of parallel or converging signaling, the intervention of specific scaffold proteins,[1] and the magnitude and duration of signaling determine the final outcome.

6.1 Main Features of MAPK Signaling

Cells possess multiple MAPK modules that are constituted by at least 3 tiers. Nevertheless, signaling MAPK network architecture can change owing to the modular design of MAPK axes that can acquire new partners [534].

Final mediators include many MAPK enzymes grouped into 3 main families: (1) extracellular signal-regulated kinases with the major types ERK1 (or MAPK3) and ERK2 (MAPK1 or MAPK2) as well as other components ERK4 (a.k.a. ERK5, MAPK6, and MAPK7) and ERK7 (ERK8 or MAPK15);[2] (2) Jun N-terminal kinases (JNK1–JNK3, or MAPK8–MAPK10); and (3) P38MAPK enzymes (P38MAPKα–P38MAPKδ, or MAPK14 and MAPK11–MAPK13, respectively).

---

[1] Protein kinases are organized in hierarchical networks, the nodes of which are assembled and regulated by scaffold proteins. Scaffold proteins bind to at least 2 signaling mediators and form platforms to enable the coordinated and ordered assembly of signaling effectors, to ensure kinase specificity, and to restrict the ability of a kinase to phosphorylate to many substrates.

[2] Kinase P38MAPKγ is also called ERK3 and ERK6 enzyme. Extracellular signal-regulated kinase ERK5, or MAPK7 (a.k.a. big MAPK1 [BMK1]), can constitute a forth family [535].

6.1.1 Basic Three-Tiered MAPK Module

The prototypical mitogen-activated protein kinase module comprises, at least, 3 types of kinases: (1) upstream Ser/Thr kinase; (2) middle dual-specificity kinase; and (3) downstream Ser/Thr kinase [536]. The 3-tiered kinase modules are constituted by different members and have distinct activation mechanisms.

At each tier, a kinase activates the kinase of the next tier. Furthermore, full activation of the kinase requires double phosphorylation. The MAPK enzymes are activated by concomitant Tyr and Thr phosphorylation of a Thr–X–Tyr (X: any amino acid) phosphoacceptor motif in the activation loop of the kinase domain [267].[3]

The upstream kinase and a phosphatase control the phosphorylation state of the 2 sites of the target kinase. The upstream kinase can be dephosphorylated (deactivated) by a phosphatase and thus must be reactivated with a given time scale to process the second phosphorylation.

6.1.2 Double Phosphorylation Cycle

The double phosphorylation cycle can either rely on a processive or distributive process. In a *distributive mechanism*, the kinase is released from its substrate after it has modified the first site. Afterward, it can rebind and modify the second site. In a *processive mechanism*, the enzyme remains bound to its substrate during the entire processing, from first to second phosphorylation. A distributive mechanism can generate an ultrasensitive response because the concentration of the fully-activated kinase depends on the upstream kinase concentration [538].

6.1.3 MAPK Pathway

A typical signaling pathway involves receptor Tyr kinases that are activated by ligand binding (leading to possible receptor dimerization and autophosphorylation) and then recruit signaling proteins. Recruited proteins are phosphorylated (activated) and then interact with other signaling mediators. In addition, multiple MAPK pathways, together with the nuclear factor-κB pathway, are targeted by environmental stress and inflammatory cytokines, especially those of the tumor-necrosis factor superfamily.

Mitogen-activated protein kinases are substrates of upstream kinases of the MAPK cascades, i.e., MAPK kinases (MAPKK, MAP2K, MEK, or MKK) and

---

[3]The Thr–X–Tyr motif corresponds to Thr203-Glu-Tyr205 in ERK1, Thr185-Glu-Tyr187 in ERK2, Thr183-Pro-Tyr185 in JNKs, and Thr-Gly-Tyr in P38MAPKs.

6.1 Main Features of MAPK Signaling

Table 6.1. Mitogen-activated protein kinase (MAPK) modules and their stimuli. Each mitogen-activated protein kinase cascade consists of a main triple-decker part with a MAP kinase kinase kinase (MAP3K), a MAP kinase kinase (MAP2K), and a MAP kinase (MAPK). Enzymes of the MAP4K set comprise various enzymes (GCK: germinal center kinase; HPK: hematopoietic protein kinase; and PAK: p21-activated kinase). The MAP3K set also include many members (ASK: apoptosis signal-regulating kinase; MLK: mixed lineage kinase; TAK: transforming growth factor-β-activated kinase; and TPL1: tumor progression locus-1). Pathway effectors differ according to the stimulus type. MAP2K1 and MAP2K2 are dual-specificity kinase that phosphorylate ERK1 (Thr183) and ERK2 (Tyr185), respectively; MAP2K3 (phosphorylated at Ser189 and Thr193) targets P38MAPK (at Thr180 and Tyr182); MAP2K4 activates both P38MAPKs and Jun N-terminal kinases (JNK) that are called stress-activated protein kinase (SAPK); MAP2K5, MAP2K6, and MAP2K7 stimulate ERK5, P38MAPK, and JNK, respectively. Proteins of the MAPK set can translocate to the nucleus and phosphorylate transcription factors (ATF: activating transcription factor; CHoP: CCAAT/enhancer-binding protein homologous protein; CREB: cAMP-responsive element-binding protein; ELk: member of DNA-binding ETS [E26] domain transcription factor family; and MEF: myocyte enhancer factor).

Stimuli	Growth factors	Growth factors, stresses Pro-inflammatory cytokines, TNFα
Small GTPases	Ras, Rap	Rho, Rac, CDC42
MAP4K		GCK, HPK1, PAK1
MAP3K	Raf	MAP3K1/4, JN3K (ASK, MLK2/3, TAK)
MAP2K	MAP2K1/2/5	MAP2K
MAPK	ERK1/2/5	JNK, P38MAPK
Transcription factors	TCF, MEF2C	ATF2, CREB, CHoP, MEF2C (P38MAPK) ATF2, ELk1, Jun (JNK)

MAPK kinase kinases (MAPKKK, MAP3K, MEKK, or MKKK; Table 6.1).[4] The latter can be activated by MAPKKK kinase (MAPKKKK or MAP4K) that can be considered as the transducers of 3-tier MAPK modules (Table 6.2). They tether to the plasma membrane or other signaling mediators.

The MAPK modules that are constituted of affectors such as MAP4Ks, a set of effector kinases (MAP3Ks, MAP2Ks, and MAPKs), and scaffold proteins, target various types of proteins (Table 6.3).

6.1.4 MAPK Stimuli

Diverse stimuli operate via various types of receptors to initiate MAPK-mediated signaling using multiple activators and inhibitors upstream from MAPK modules: (1) hormones and growth factors that act via receptor Tyr kinases (Vol. 3 –

---

[4] Aliases MEK and MKK stand for mitogen–extracellular signal-regulated kinase kinase and MAP kinase kinase.

Table 6.2. Mitogen-activated protein kinase kinase kinase kinases (GCK: germinal center kinase; GCKR: GCK-related kinase; GLK: GCK-like kinase: HGK: HPK/GCK-like kinase; HPK: hematopoietic progenitor kinase; KHS: kinase homologous to Ste20 protein kinase; MinK: misshapen-like kinase; NIK: NCK-interacting kinase [not NFκB-inducing kinase (NIK or MAP3K14)]).

Type	Other aliases
MAP4K1	HPK1
MAP4K2	GCK, BL44, Rab8IP
MAP4K3	GLK
MAP4K4	NIK, HGK
MAP4K5	GCKR, KHS, KHS1
MAP4K6	MinK

Table 6.3. Target enzymes of MAPK modules (Source: [10]; cPLA2: cytoplasmic phospholipase-A2; MAPKAPK: MAPK-activated protein kinase; MSK: mitogen- and stress-activated protein kinase; PTPR: receptor-type protein Tyr phosphatase).

Type	Examples
Kinase	MAPKAPKs, MSK1/2
Phosphatase	PTPRh
Phospholipase	cPLA2

Chap. 8. Receptor Kinases) and cytokine receptors, but transforming growth factor-β-related proteins target receptor Ser/Thr kinases; (2) vasoactive peptides that binds to G-protein-coupled receptors; (3) inflammatory cytokines of the tumor-necrosis factor superfamily; and (4) environmental stresses (e.g., ischemic injury, osmotic shock, and ionizing radiation).

6.1.5 MAPK Cascade Regulation

The MAPK cascades are composed of specific tiers of signal transmission from environmental stimuli (Fig. 6.1, Table 6.4). There are up to 6 activation levels in the cytosol and nucleus that encompass the specific 3-tiered MAP3K–MAP2K–MAPK cascade. Regulation of MAP3K is achieved by membrane recruitment, homo-oligomerization, often formation of a proteic complex with regulators, and phosphorylation.

The MAPK pathway contains autoregulatory loops. In particular, the ERK axis can regulate the production of MAPK phosphatase MKP1 (Table 6.5) that trigger a negative feedback loop.

6.1 Main Features of MAPK Signaling

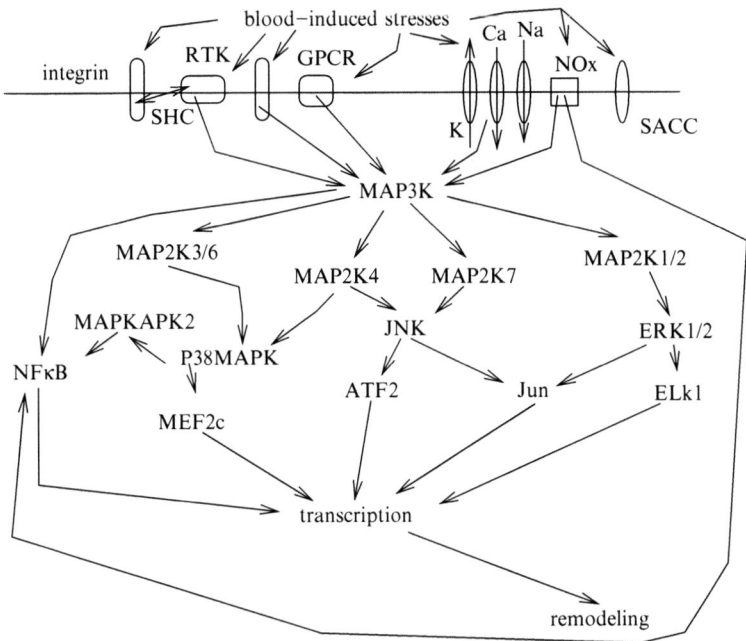

Fig. 6.1 Mitogen-activated protein kinase cascades in cells correspond to a reaction cascade with several steps that include the 3-tiered MAP3K–MAP2K–MAPK module (Source: [539]). Enzyme MAP3K is the usual entry node of MAPK cascades, although MAP3Ks can sometimes be stimulated by MAP4K proteins. Mechanical stresses, particularly in blood vessel wall, stimulate mediators of the MAPK cascade. Effectors of the MAPK cascade lead to gene transcription for adaptation and, if necessary when subjected to long-duration high-magnitude forces, for tissue remodeling.

Table 6.4. Extracellular stimuli activate MAPK pathways via small GTPases. Activated MAP3Ks (e.g., Raf, apoptosis-related kinases [ASK], mixed lineage kinases [MLK], TGFβ-activated kinase [TAK], etc.) phosphorylate MAP2Ks on 2 serine residues, which in turn phosphorylate MAPKs, extracellular signal-regulated kinase (ERK), Jun N-terminal kinase (JNK), and P38MAPK on both threonine and tyrosine residues. Activated MAPKs can translocate to the nucleus and phosphorylate transcription factors (ternary complex factor TCF, activator protein AP1 with JUN, and activating transcription factor ATF2). TCF forms a complex with serum response factor (SRF).

	Cascade 1	Cascade 2	Cascade 3
Tier 1	Ras	MAP3K	TAK
Tier 2	MAP2K1/2	MAP2K4/7	MAP2K3/6
Tier 3	ERK	JNK	P38MAPK
Transcription factor	SRF–TCF	SRF–TCF AP1	SRF–TCF ATF2

Table 6.5. Phosphatases of the MAPK modules (Source: [10]; DUSP: dual-specificity phosphatase; PTPn: protein Tyr phosphatase, non-receptor).

Category	Types
DUSP	DUSP2/3/5/7/8
MAPK phosphatases	MKP1–MKP5
PTPn	PTPn5

6.1.6 Auto-Inhibition

Auto-inhibition is a functioning mode of protein kinase. The catalytic domain binds ATP and substrates for phosphoryl transfer. They change their activation region into an auto-inhibitory module [540]. The activation region must unlock from another molecular segment to tether ATP.

6.1.7 Scaffold Proteins

Signaling pathway organization is mediated by scaffold proteins (Table 6.6). Scaffold proteins couple MAPK components to upstream activators. Signaling efficiency of MAPK modules results, at least partly, from scaffolds that organize MAPK signaling components into functional elements to activate specific MAPK pathways. Scaffold proteins of MAPK signaling also allow specific signal transduction, as they bind to selected MAPK constituents. They are able to segregate adequate axes among MAPK modules and maintain signal transduction specificity. Furthermore, they protect signaling components from undesirable activation by other signaling molecules. They also enable a coordinated activation of MAPK modules in response to given stimuli. Some MAPK interactors can couple constituents of different MAPK modules. Elements of MAPK modules can indeed be targeted by several activators and, in turn, connect to several effectors.

Stress-activated protein kinase-associated proteins, or Jun N-terminal kinase-interacting proteins, operate as scaffold proteins for JNK cascades. Jun N-terminal kinase-associated leucine zipper protein (JLP; JIP4 or MAPK8IP4) is a scaffold for the JNK and P38MAPK signaling cascades that also acts as a binding partner for MAX transcription factor. It interacts with several proteins, such as $G\alpha_{13}$ protein, kinesin light chain-1, and cell-surface receptor-like glycoprotein Cell adhesion molecule (CAM)-related downregulated by oncogenes CDO of the immunoglobulin–fibronectin-3 repeat-containing protein family in myocyte precursors.

6.1 Main Features of MAPK Signaling

Table 6.6. Scaffold proteins of the mitogen-activated protein kinases (CNK2: CNK homolog protein; CnKSR: connector enhancer of kinase suppressor of Ras; JIP: JNK-interacting protein; KSR: kinase suppressor of Ras; LAMTOR: late endosomal/lysosomal adaptor, MAPK and TOR activator; maguin: membrane-associated guanylate kinase-interacting protein; MAPKAP1: MAPK-associated protein; MP: MEK-binding partner; SIN: stress-activated MAPK (SAPK)-interacting protein [a.k.a. target of rapamycin complex TORC2 subunit]; SKRP: stress-activated protein kinase (SAPK) pathway-regulating phosphatase; TAB: TAK1-binding protein (MAP3K7IP1–MAP3K7IP3; MAP3K7IP1 is a MAP3K7 activator, whereas MAP3K7IP2 and -3 are adaptors). The dual-specificity phosphatase DUSP7 can also function as a scaffold. The MAP3K12 binding inhibitory protein MAP3K12IP1 is the MAPK upstream kinase (MUK)-binding inhibitory protein (MBIP).

Type	Other aliases
β-Arrestin-2	
CnKSR2	CNK2, maguin
DUSP7	MKPX
DUSP19	SKRP1, DSP3$_{IMW}$
KSR1	RSU2
KSR2	
MAP2K1IP1	LAMTOR3, MAPKSP1, MAPBP, MP1, ragulator-3
MAP3K7IP1	TAB1
MAP3K7IP2	TAB2
MAP3K7IP3	TAB3
MAPK8IP1	JIP1
MAPK8IP2	JIP2
MAPK8IP3	JIP3
MAPK8IP4	JIP4
MAPKAP1	MIP1, SIN1

Connector enhancer of kinase suppressor of Ras CnKSR2[5] is an enzyme that also operates as a scaffold protein to mediate MAPK signals downstream from Ras GTPase. It is stimulated by vitamin-D. The CnKSR2–KSR2 complex may enable crosstalk between the MAPK, PI3K, and insulin pathways [541]. It can complex with cRaf, RalA, MAP2K1, MAP2K2, PI3K, and CDK4 kinases, as well as PP2 phosphatase.

6.1.8 Transcription Factor Substrates of MAPK Modules

Activated MAPKs can translocate to the nucleus to phosphorylate a number of transcription factors, such as members of the ternary complex factor (TCF)

---

[5] A.k.a. CNK homolog protein 2 (CNK2) and maguin (a portmanteau for membrane-associated guanylate kinase-interacting protein).

Table 6.7. Target transcription factors of MAPK modules (ATF: activating transcription factor; ELk: ETS-like transcription factor; ETS: E26 (E-twenty six) transcription factor family; MEF: myocyte-enhancer factor; SRF: serum response factor; TCF: ternary complex factor [ELk1, ELk3, ELk4]). Transcription ETS factors, or ternary transcription factors, complex with DNA and SRF factor.

Type	Components (versions)
Activator protein-1	ATF2, Jun, JunB, JunD
ETS SRF–TCF	ETS1, ELk1
MEF2	MEF2a–MEF2d versions

family, which associate with serum response factor and target the FOS gene, and components of the Activator protein-1 complexes, such as Jun and activating transcription factor ATF2 (Table 6.7).

The Ras–MAPK signaling activates transcription factors of the ETS family. Phosphorylation improves the affinity (34-fold) of ETS1 for transcriptional coactivators CREB-binding protein and P300 that also serve as histone acetyltransferases, polyubiquitin ligases, and adaptors [542]. Ras-responsive elements include the serum-response-element of promoter of the FOS gene, tandem ETS, or ETS–AP1 composite sites associated with ETS1 and ETS2.

6.1.9 Subcellular Compartmentation

Mitogen-activated protein kinases are linked to various organelles, such as microtubules, endosomes, endoplasmic reticulum, and actin cytoskeleton. Active MAPKs often translocate from the cytosol to the nucleus to phosphorylate nuclear targets. Enzymes of the MAP2K set can carry MAPK from the cytoplasm to the nucleus and inversely.

6.2 MAPK Modules in Cell Metabolism

Signaling based on MAPK modules is involved in almost all of the changes in the extra- or intracellular medium that affect the cell metabolism. Metabolic adaptation actually relies on MAPK-dependent signal transduction. On the other hand, inappropriate MAPK signaling contributes to the development of metabolic syndrome [535].

6.2 MAPK Modules in Cell Metabolism

Table 6.8. Regulation of cell metabolism by MAPK modules and their effector kinases triggered by epithelial growth factor (EGFR), glucagon (GcgR), insulin (IR) and insulin growth factor (IGF1R), and tumor-necrosis factor (TNFR) receptors (Source: [535]) ATF: activating transcription factor; CREB: cAMP response element-binding protein; C/EBP: CCAAT/enhancer binding protein; eIF: eukaryotic translation initiation factor; Fox: forkhead box; GR: glucocorticoid receptor; GSK: glycogen synthase kinase; HIF: hypoxia-inducible factor; HSL: hormone-sensitive lipase; MAPKAPK: MAPK-activated protein kinase; MKnK: MAPK-interacting kinase (or MnK); MSK, mitogen- and stress-activated kinase (or 90-kDa ribosomal protein S6 kinase); PGC: PPARγ coactivator; PPAR: peroxisome proliferator-activated receptor; SREBP: sterol regulatory element-binding protein; TORC: target of rapamycin complex; TSC: tuberous sclerosis complex).

Effect	Effectors (and their activators)
Gluconeogenesis	C/EBPα (ERK, P38MAPK), CREB (MSK), FoxO1 (ERK, P38MAPK), GR (ERK, JNK, P38MAPK), PPARα (ERK, P38MAPK), PGC1α (P38MAPK)
Glycogen synthesis	GSK3 (MAPKAPK1, MSK)
Lipid homeostasis	C/EBPα (ERK, P38MAPK), HIF1α (ERK, P38MAPK), SREBP1a/2 (ERK)
Lipolysis	HSL (ERK), perilipin (ERK, JNK)
Protein synthesis	eIF4b (MAPKAPK1), eIF4e (MKnK), TORC1 (ERK), TSC1/2 (ERK, MAPKAPK1)
Adipocyte differentiation	ATF2 (P38MAPK), C/EBPα (ERK, P38MAPK), C/EBPβ (P38MAPK), PPARγ (ERK, JNK)

Hormones (e.g., glucagon and insulin), growth factors (e.g., IGF1 and EGF), cytokines (e.g., TNFα), and environmental stress prime MAPK signaling (Table 6.8). Activated MAPKs directly or indirectly via MAPKAPK1 proteins (or ^{P90}RSK; MAPKAPK1a–MAPKAPK1c),[6] MAPKAPK2 (or MK2), ^{P90}RSK subfamily enzymes MSK1 and MSK2),[7] MAPK-interacting protein Ser/Thr kinases MKnK1[8] and MKnK2,[9] regulate numerous metabolic agents.

---

[6] A.k.a. RSK1–RSK3 (or RPS6Kα1, RPS6Kα3, and RPS6Kα2, respectively).

[7] A.k.a. RPS6Kα5 as well as RLPK and RPS6Kα4 as well as RSKb, respectively.

[8] A.k.a. MAPK signal-integrating kinase-1 (MnK1).

[9] A.k.a. MnK2 and G-protein-coupled receptor kinase GPRK7. MAPK-interacting kinase MnK1 and MnK2 bind to the ERK1 and ERK2 [543]. Enzyme MnK1 complexes more strongly with inactive than active ERK proteins. Enzyme MnK1, but not MnK2, also links to P38MAPK enzymes. Once they are phosphorylated, MnK1 and MnK2 target eukaryotic initiation factor-4E (eIF4e), a regulatory phosphoprotein, the phosphorylation of which is increased by insulin.

6.2.1 Insulin and Glucagon Signaling

Insulin activates not only the PI3K–PKB pathway, but also MAPK modules [535]. Upon binding of the insulin receptor substrate (IRS) or SHC adaptor to the activated insulin receptor and subsequent phosphorylation, GRB2 adaptor is recruited and, then, associates with SOS guanine nucleotide-exchange factor to activate the Ras–Raf–ERK pathway. In addition, insulin can activate P38MAPK and JNK, involving in the latter case, the small CDC42 GTPase and PI3K regulatory subunit P85 ($P85_{PI3K}$) to activate MAP2K4 enzyme.

Upon insulin stimulation, inhibitory phosphorylation of glycogen synthase kinase GSK3 by ^{P90}RSK enzymes, especially those of the MAPKAPK1 group, which are substrates of ERK1 and ERK2, as well as by PKB, and activation of PP1 stimulate glycogen synthase [535].

Once it is activated by insulin, PKB phosphorylates FoxO1 transcription factor, thereby leading to its nuclear export and stopping the transcription of enzymes of gluconeogenesis, lipid catabolism, and mitochondrial function [535]. Moreover, FoxO1 is phosphorylated by ERK1, ERK2, and p38MAPKs.

Kinases P38MAPKα and P38MAPKβ may be involved in glucose uptake using glucose transporter GluT4 in response to insulin. Insulin also stimulates protein synthesis and may act via MAPK modules.

Transcription factors of the SREBP family (SREBP1a, SREBP1c, and SREBP2) are major regulators of insulin-primed lipogenesis.[10] Insulin-activated ERK1 and ERK2 phosphorylates SREBP2 and possibly SREBP1a factor [535].

On the other hand, insulin suppresses gluconeogenesis and lipid catabolism in the liver, whereas glucagon and glucocorticoids have the opposite effect during fasting.[11] Glucagon enhances the production of PPARγ and PGC1 via the cAMP pathway and CREB transcription factor [535].[12] Enzymes of the P38MAPK set stimulate the glucagon-mediated fasting response in the liver, as they phosphorylates various key agents.[13]

---

[10] Factor SREBP1a controls both lipogenic and cholesterogenic enzymes; SREBP1c mainly regulates synthesis of fatty acids; and SREBP2 controls cholesterol homeostasis.

[11] Glucocorticoid receptor is phosphorylated by P38MPAKs to enhance its transcriptional activity. On the other hand, phosphorylation by JNKs, ERK1, and ERK2 impedes its action. In addition, phosphorylation (activation) of C/EBPα by P38MAPKs regulates hepatic gluconeogenesis. Enzymes of the P38MAPK set are involved in suppression of hepatic lipogenesis initiated by glucagon and fasting.

[12] Factor PPARα has an enhanced activity after phosphorylation by ERK1, ERK2, and P38MAPKs. Factor PGC1, a substrate of P38MAPK, coactivates many transcription factors, such as glucocorticoid receptor (or NR3c1), FoxO1, and PPARα. Factor CREB is a direct target of MSK kinases (^{P90}RSK), which are activated by ERK1, ERK2, and P38MAPKs.

[13] Insulin also activates P38MAPKs, but mainly in adipocytes and skeletal myocytes.

6.2.2 MAPKs in Adipocytes

Adipose tissue participates in nutrient and energy homeostasis (Vol. 2 – Chap. 1. Remote Control Cells). In white adipose tissue, when energy is needed, triacylglycerides are hydrolyzed into free fatty acids and glycerol by 3 lipase types. Upon TNFα stimulation, activated JNKs, ERK1 and ERK2 increase basal lipolysis.

These members of the MAPK superfamily indeed repress perilipin-1 synthesis.[14] Moreover, ERK1 and ERK2 phosphorylate (activate) hormone-sensitive lipase.

Both ERK1 and ERK2 contribute to adipocyte differentiation. Temporal control of ERK activity is necessary during adipogenesis, ERK1 and ERK2 being activated early in cellular differentiation. Factor PPARγ (or nuclear receptor NR1c3), the master regulator of adipocyte differentiation, is phosphorylated (inhibited) by ERK and JNK kinases. Adipocyte differentiation is also supported by P38MAPKs that phosphorylate C/EBPβ and target activating transcription factor ATF2 that promotes PPARγ transcription.

6.2.3 MAPKs in Metabolic Syndrome

Excessive diet intake causes lipid accumulation and various metabolic imbalances that lead to insulin resistance, dyslipidemia, inflammation, and hypertension (Vol. 6 – Chap. 7. Vascular Diseases). Metabolic syndrome is a major risk factor for type-2 diabetes mellitus and atherosclerosis.

Enzymes of the MAPK set are involved in the development of insulin resistance. In particular, JNK1 contributes to increase adipose mass under hypercaloric conditions [535]. Kinases of the JNK subset can inhibit insulin signaling, as they phosphorylate IRS1 (Ser307) in insulin-sensitive cell types other than hepatocytes.[15] In hepatocytes, insulin stimulation indeed leads to IRS1 phosphorylation (Ser307).

---

[14] Triacylglycerols are mainly stored in neutral lipid droplets enveloped by lipid droplet-associating proteins. Perilipin-1, or lipid droplet-associated protein, coats lipid droplets in adipocytes. It thus restrains the action of triacylglyceride lipases under basal conditions. All perilipin subtypes (PLin1a–PLin1c) protect against lipases, such as hormone-sensitive lipase. Perilipin pertains to the PAT (perilipin, adipophilin, and TIP47) family that encompasses: (1) perilipin-2, also called adipophilin and adipose differentiation-related protein (ADRP); (2) perilipin-3, also named mannose 6-phosphate receptor-binding protein, 47-kDa tail-interacting protein TIP47 (that binds the cytoplasmic domain of cation-dependent and -independent mannose 6-phosphate receptors for their transport from endosomes to the Golgi body), and cargo selection protein; and (3) perilipin-5, also termed lipid storage droplet protein LSDP5, myocardial lipid droplet protein (MLDP), and PAT family member expressed in highly oxidative tissues (OxPAT, which is mainly produced in cells with high lipid oxidative capacity, such as cardiac and skeletal myocytes and brown adipocytes). Perilipin is hyperphosphorylated by PKA after β-adrenergic receptor activation. Phosphorylation changes perilipin conformation, thereby exposing stored lipids to hormone-sensitive lipase, and hence enabling lipolysis.

[15] Muscle-specific JNK1-deficient mice have an improved insulin sensitivity with respect to wild-type mice [544].

Kinase ERK1 interacts with and is inhibited by the adaptor IGF2-binding protein IGF2BP2 [535]. Protein ERK1 is required in the early stage of adipogenesis, but needs to be inactivated to enable PPARγ activity in late stage of adipogenesis.

In the pancreas, P38MAPKδ prevents PKD1 activity that regulates insulin secretion and β-cell survival. In addition, under stress, JNK activation that relies on the scaffold JNK-interacting protein MAPK8IP1 can provoke failure of β-cell functioning [535].

6.3 Mitogen-Activated Protein Kinase Kinase Kinase Kinases

Many mitogen-activated protein kinase kinase kinase kinases (MAP4K) belong to the Sterile-20-like kinase superfamily (Sect. 5.1), more precisely to the germinal center kinase (GCK) family.[16] Members of the GCK family (MAP4Ks) indeed include: MAP4K1 (or hematopoietic progenitor kinase HPK1); MAP4K2 (or germinal center kinase [GCK]); MAP4K3 (or GCK-like kinase [GLK]); MAP4K4 (or non-catalytic region of Tyr kinase adaptor [NCK]-interacting kinase [NIK]); MAP4K5 (or GCK-related kinase [GCKR]); and MAP4K6 (or misshapen-like [MinK or MinK1] NIK-related kinase; Sect. 5.1.2).

The GCK subfamily 1 comprises MAP4K1 to MAP4K3 and MAP4K5, whereas the GCK subfamily 4 contains MAP4K4 and MAP4K6 (MinK), as well as TRAF2- and NCK-interacting kinase (TNIK) and NIK-related kinase (NRK). Members of the GCK subfamily 4 within the GCK family are slightly divergent from subfamily-1 kinases.

Enzyme MAP4K2 is a specific activator of Jun N-terminal kinase, as it does not activate ERKs, P38MAPKs, and nuclear factor-κB. This stress-activated kinase that has a basal activity is activated in vivo by signaling affectors.

Enzyme MAP4K5 is also a specific activator of Jun N-terminal kinase that has a much smaller basal activity than MAP4K2. It is another effector of tumor-necrosis factor. Enzyme MAP4K3 is a third GCK subfamily-1 member that is also a selective JNK activator stimulated by tumor-necrosis factor.

Enzyme MAP4K1 is more distantly related to MAP4K2 than MAP4K5 and MAP4K3 enzymes. It interacts with MAP3K11 (or mixed lineage kinase MLK3). It stimulates strongly JNK kinases and weakly P38MAPK enzymes. It is recruited by adaptors GRB2, CRK, and CRKL to the plasma membrane and plasmalemmal receptors such as EGFR. Its activation of JNKs is indeed primed by mitogens such as EGF via their cognate receptor Tyr kinases.

Enzyme MAP4K4, like its close relatives NRK and TNIK, is the kinase most distantly related to MAP4K2 protein. However, like MAP4K2, it is activated by TNF and strongly and specifically activates JNKs. Adaptor NCK couples MAP4K4 to the EPH family of receptor Tyr kinases.

---

[16]Distribution of germinal center kinase is restricted to B-lymphocyte maturation germinal centers of lymphoid tissues (not cortex). B-lymphocyte maturation such as immunoglobulin class switching and selection are primed by ligands of the tumor-necrosis factor receptor superfamily.

Enzyme MAP4K6, like MAP4K4, is an effector for TNFR-associated factors (TRAF). Enzyme TNIK, another MAP4K, provokes a potent and specific JNK activation that also can interact with NCK adaptor and TRAF2 protein.

Among the group-2 GCKs (Sect. 5.1.4), high levels of STK4 (MST1 or KRS2) can activate JNK and P38MAPKs [267]. Members of another group of GCKs (GCK subfamily 8; Sect. 5.1.10), thousand and one kinases (TAOK1–TAOK3) serve as MAP3Ks (MAP3K16–MAP3K18), and not upstream affectors such as MAP4Ks, that can activate certain MAPK pathways (Sect. 6.5.7).[17]

6.4 Other Regulators of the MAPK Module

6.4.1 Small GTPases of the Ras Hyperfamily

Monomeric GTPases of the RAS hyperfamily are especially involved in the activation of the ERK pathway mainly via cRaf kinase. They are stimulated by guanine nucleotide-exchange factors (GEF) and inactivated by GTPase-activating proteins (GAP). Activation of cRaf by Ras follows its translocation to the plasma membrane and leads to its phosphorylation by recruited cytosolic Tyr protein kinases such as Src or Ser/Thr protein kinases such as p21-activated kinase PAK3 and homo-oligomerization.

In addition, small GTPase Rac1 (but not CDC42, Ras, or Rho) interacts with SH3 domain-containing ring finger protein SH3RF1 (or plenty of SH3s [POSH]) that triggers nuclear translocation of NFκB and activates Jun N-terminal kinases. Small GTPase CDC42 (but not Rac1) can target MAP3K10 (or MLK2) and MAP3K11 (or MLK3) via adaptor SH3RF1. Furthermore, PAKs as well as MAP3K1 and MAP3K4 can interact with both Rac1 and CDC42 [267].

6.4.2 Heterotrimeric G Proteins

Heterotrimeric G proteins can couple to Jun N-terminal kinases via monomeric Ras GTPase or cytosolic Tyr kinases. Bruton's Tyr kinase is activated by G12 subunit to stimulate Jun N-terminal kinases [267]. Subunits Gq and free Gβγ can activate phospholipase-Cβ, thereby causing calcium ion influx that stimulates FAK2 kinase. The latter, together with Src, transmits signals from G proteins to extracellular signal-regulated protein kinases [267].

---

[17]Isoforms TAOK1 (MAP3K16) and TAOK2 (MAP3K17) target JNKs and P38MAPKs. On the other hand, TAOK3 (MAP3K18 or JIK) inhibits JNKs, but has no effect on ERKs, ERK5, and P38MAPKs.

6.5 Mitogen-Activated Protein Kinase Kinase Kinases

Mitogen-activated protein kinase kinase kinases (MAP3K; Table 6.9) constitute an epifamily that comprises diverse families of protein kinases [545, 546] (Tables 6.10 to 6.13): (1) ASK (apoptosis signal-regulating kinase) with ASK1 and ASK2 members (i.e., MAP3K5–MAP3K6); (2) MEKK (MAP/ERK kinase kinase, i.e., MEKK1–MEKK4, or MAP3K1–MAP3K4); (3) MLK (mixed lineage kinase, i.e., MAP3K9–MAP3K13); (4) Mos (uterus Moloney sarcoma viral v-Mos protooncogene product); (5) Raf (the first RAF gene corresponds to acute transforming retrovirus oncogene homologous to the SRC family of sarcoma virus-related protooncogenes [v-Raf]); (6) TAK1 (transforming growth factor-β-activated kinase-1 or MAP3K7); (7) TAOK (thousand and one amino acid kinases, i.e., TAOK1, MAP3K16, MARKK, PSK2, or KFCb; TAOK2, MAP3K17, PSK1, or KFCc; and TAOK3, MAP3K18, DPK, JIK, or KFCa); and (8) TPL2 (tumor progression locus-2 or MAP3K8). Many MAP3Ks belong to the Sterile-20-like kinase superfamily. In fact, the GCK subfamily 8 (Sect. 5.1.10) of the Sterile-20-like kinase superfamily is constituted by members of the TAOK family.

Table 6.9. Mitogen-activated protein kinase kinase kinases (ASK: apoptosis-signal-regulating kinase; DLK: dual leucine zipper-bearing kinase; MEKK: mitogen–extracellular signal-regulated kinase kinase; MLK: mixed lineage kinase; MTK: MAP three kinase; TAK: TGFβ-activated kinase; TAOK: thousand and one amino acid kinase; TPL: tumor progression locus-2).

Type	Other aliases
aRaf	
bRaf	
cRaf	Raf1
MAP3K1	MEKK1
MAP3K2	MEKK2
MAP3K3	MEKK3
MAP3K4	MEKK4, MTK1
MAP3K5	MEKK5, ASK1
MAP3K6	MEKK6, ASK2
MAP3K7	MEKK7, TAK1
MAP3K8	MEKK8, Cot, TPL2
MAP3K9	MEKK9, MLK1
MAP3K10	MEKK10, MLK2, MST
MAP3K11	MEKK11, MLK3, SPRK, PTK1
MAP3K12	MEKK12, DLK, MUK, ZPK
MAP3K13	MEKK13, LZK
MAP3K14	NIK, HS, HSNIK
MAP3K15	MLK7, ZAK, MRK, MLTK
MAP3K16	TAOK1, MARKK, PSK2, KFCb
MAP3K17	TAOK2, PSK, PSK1, TAO1, TAO2, KFCc
MAP3K18	TAOK3, DPK, JIK, KFCa

6.5 Mitogen-Activated Protein Kinase Kinase Kinases

Table 6.10. Features of MAP3Ks. (**Part 1**) Raf and Mos (Source: Sigma-RBI eHandbook; BAG: BCL2-associated athanogene; CNKSR: connector enhancer of kinase suppressor of Ras (or CNK); KSR: kinase suppressor of Ras; MyoD: Myogenic differentiation transcription factor; MyT: myelin transcription factor; PEBP: phosphatidylethanolamine-binding protein [or RKIP]; RIP: receptor-interacting protein; SHOC: suppressor of clear homolog protein [SOC2 or SUR8]). All MAPK signal transduction axes correspond to 3-kinase (MAP3K–MAP2K–MAPK) cascades. MAPK signaling modules mediate cellular responses to hormones, growth factors, cytokines, and environmental stresses.

Distribution Location	Binding partners	Activators	Affectors	Effectors	Function
		Raf MAP3Ks: a-, b-, and cRaf			
Ubiquitous Cytosol	Pax, KSR1/2, BAG1, CNKSR, PEBP1, SHOC2, 14-3-3	Growth factors	Ras, Rap, PAK	MAP2K1/2	Cell growth
		Mos MAP3K			
Germ cell Cytosol		Progesterone Growth factors	CDK1	MAP2K1/2, MyT1, MyoD1	Oocyte maturation

Table 6.11. Features of MAP3Ks. (**Part 2**) MEKK (Source: Sigma-RBI eHandbook; GADD: growth arrest and DNA-damage protein; JLP: JNK-associated leucine zipper protein (or SpAg9); JNKBP: JNK-binding protein; JSAP: JNK/SAPK-associated protein; LAd: LCK-associated adaptor (Rlk- and Itk-binding protein RIBP); MIP1: MAP3K2-interacting protein; OsM: oncostatin-M; TRAF: tumor-necrosis factor receptor-associated factor). All MAPK signal transduction axes correspond to 3-kinase (MAP3K–MAP2K–MAPK) cascades. MAPK signaling modules mediate cellular responses to hormones, growth factors, cytokines, and environmental stresses.

Distribution Location	Binding partners	Activators	Affectors	Effectors	Function
		MEKKs: MAP3K1–MAP3K4			
Ubiquitous Cytosol	JNKBP, JSAP, JLP, GADD45, LAd, MIP1, OsM, TRAF, 14-3-3	Growth factors, stress	Rac1, Ras, CDC42, MAP4K1/2	MAP2K1–7 IKK	Stress response, development

Table 6.12. Features of MAP3Ks. (**Part 3**) ASK and MLK (Sources: Sigma-RBI eHandbook and BioGRID; Develop.: development; CKI: cyclin-dependent kinase inhibitor [CKI1a; a.k.a. p21, Cip1, and Waf1]; DAXX: death-associated protein [DAP6 or BING2]; GADD: growth arrest and DNA-damage protein; GlnRS: glutaminyl-tRNA synthase; GST: glutathione S-transferase; IRE1: inositol-requiring endoplasmic reticulum-resident transmembrane kinase/endoribonuclease (or endoplasmic reticulum-to-nucleus signaling ERN1); MBIP: MUK-binding inhibitory protein (MAP3K12IP1); SH3RF: SH3 domain-containing ring finger protein (SH3RF1 or plenty of SH3s [POSH]); TRAF: tumor-necrosis factor receptor-associated factor).

Distribution Location	Binding partners	Activators	Affectors	Effectors	Function
ASKs: ASK1/MAP3K5 and ASK2/MAP3K6					
Ubiquitous Cytosol	β-Arrestin-2, Thioredoxin, MAP3K5–7, GlnRS, PP5, eIF2αK2, CDC25a, DUSP19, DAXX, GSTμ1, TRAF1–3/5/6, Hsp70-1a, CKI1a, cRaf, JIP3, GADD45b, 14-3-3	Stress Calcium	IRE1, CamK2	MAP2K3/4, MAP2K6/7	Stress response Neuron apoptosis Heart function
MLKs: MLK1–MLK4, MLK7, DLK, LZK					
Ubiquitous Cytosol	JIP1–3, MBIP, SH3RF1, 14-3-3	Stress GF Cytokines	MAP4K1, CDC42, Rac1, PAK	MAP2K3/4, MAP2K6/7	Stress response Develop.

Most MAP3K enzymes are regulated, at least partly, by 4 processes: oligomerization (e.g., cRaf, MAP3K1, MAP3K5, and MAP3K11), adaptor binding, membrane translocation, and phosphorylation and trans-autophosphorylation. Regulatory domains of MAP3Ks interact with upstream regulators, such as Rho and Ras GTPases. Activated MAP3K phosphorylates (activates) one or several MAP2Ks. Many MAP3Ks also activate the IKK–NFκB pathway.

The MAPK pathways can be subdivided into those that predominantly respond to either mitogens to regulate cell growth and differentiation (e.g., extracellular signal-regulated kinases ERK1 and ERK2) or stress (Jun N-terminal kinases and P38MAPKs) to control cell apoptosis and immune response (Table 6.14). The ERK5 pathway responds to both growth stimuli and some stresses. Activity of MAP3K

Table 6.13. Features of MAP3Ks. (**Part 4**) TAK1, TAOK, and TPL2 (Sources: Sigma-RBI eHandbook and BioGRID; Inflam.: inflammation; IKK: inhibitor kinase of nuclear factor-κB; HGS: hepatocyte growth factor-regulated tyrosine kinase substrate; HIPK: homeodomain-interacting protein kinase; IRE1: inositol-requiring endoplasmic reticulum-resident transmembrane kinase/endoribonuclease [or endoplasmic reticulum-to-nucleus signaling ERN1]; RKIP: Raf kinase inhibitor protein [phosphatidylethanolamine-binding protein (PBP) or osteoclast differentiation factor receptor (ODFR)]; IL17Rd: interleukin-17 receptor-D [or similar expression to FGF (SEF1)]; TAB: TAK1-binding protein [MAP3K7IP1–MAP3K7IP3]; TNFAIP: tumor-necrosis factor-α-interacting protein [or A20-binding inhibitor of nuclear factor-κB (ABIN)]; TIRAP: Toll-interleukin-1 receptor domain-containing adaptor protein; TRAF: tumor-necrosis factor receptor-associated factor).

Distribution Location	Binding partners	Activators	Affectors	Effectors	Function
TAK1					
Ubiquitous Cytosol	SMAD3/7, RKIP, HGS, IL17Rd, MAP3K5/7/14, TGFβR1, TAB1–3 MyD88, TIRAP, TRAF2/6,	TGFβ Wnt Stress Cytokines	MAP4K1/4, Ubiquitin	MAP2K3/4, MAP2K6, IKK, HIPK2	Stress response
TAOK MAP3Ks: TAOK1/MAP3K16–TAOK3/MAP3K18					
Ubiquitous Cytosol	Tubulin, TRAF2	Osmotic stress	IRE1	MAP2K3/4, MAP2K6/7	Stress response
TPL2/MAP3K8					
Ubiquitous Cytosol	NFκB, PKB, STK4, TRAF2, TNFAIP2, MAP3K14, KSR2	Cytokines, LPS	MAP4K4	MAP2K1/2, MAP2K4–6 IKK	Inflam.

is regulated by various mechanisms, such as phosphorylation and proteolysis as well as interactions with small GTPases and other regulators and linkage to scaffold proteins.

Kinases Mos and Raf (aRaf–cRaf) are activated by growth factors. Members of the RAF family are components of ERK signaling modules, as they phosphorylate MAP2K1 and MAP2K2. Kinases of the RAF family are recruited to the plasma membrane, where they links to small GTPase Ras and are phosphorylated (activated). Enzyme MAP3K8 also activates the ERK pathway, especially in the immune response.

Table 6.14. Types of MAP2Ks and MAPKs targeted by MAP3Ks.

MAP3K	Alias (main)	MAP2K	ERK1/2	ERK4	JNK	P38MAPK
MAP3K1	MEKK1	MAP2K4/7	+		+	+
MAP3K2	MEKK2	MAP2K1/4	+	+	+	+
MAP3K3	MEKK3	MAP2K1/4	+	+	+	+
MAP3K4	MEKK4	MAP2K3/4/6			+	+
MAP3K5	ASK1	MAP2K3/4/6			+	+
MAP3K6	ASK2	MAP2K3/4/6			+	+
MAP3K7	TAK1	MAP2K3/4/6			+	+
MAP3K8	TPL2	MAP2K1/4	+		+	+
MAP3K9	MLK1		+	+	+	+
MAP3K10	MLK2	MAP2K4/7	+	+	+	+
MAP3K11	MLK3	MAP2K4/7	+	+	+	+
MAP3K12	DLK	MAP2K4/7	+		+	+
MAP3K13	LZK		+		+	+
MAP3K14	HSNIK					
MAP3K15	MLK7				+	+
MAP3K16	TAOK1	MAP2K3				+
MAP3K17	TAOK2	MAP2K3				+
MAP3K18	TAOK3	MAP2K3				+
MLK4					+	+
Mos			+			
aRaf		MAP2K1/2	+			
bRaf		MAP2K1/2	+			
cRaf	cRaf	MAP2K1/2	+	+		

The MEKK and MLK families can be activated by Rho GTPases and Ste20-like protein kinases (Sect. 5.1). Upon apoptotic cues, MAP3K1 (MEKK1) can bear proteolysis (activation) by caspases that cleave MAP3K1 and release its pro-apoptotic catalytic C-terminus. Although they are mainly involved in stress signaling, some members of the MAP3K and MLK families regulate the ERK pathways. Enzymes MAP3K1 and MAP3K11 (MLK3) target ERK1 and ERK2 as well as MAP3K2 and MAP3K3 targets ERK5.

Enzymes MAP3K5 (ASK1) and MAP3K7 (TAK1) are components of JNK and P38MAPK signaling pathways. Enzyme MAP3K5 participates in apoptosis caused by oxidative stress. It is activated by dissociation from a redox-sensitive inhibitor thioredoxin. Enzyme MAP3K7 is activated by cytokines and connects to tumor-necrosis factor-receptor associated factor TRAF6 via adaptor TAK1- and TRAF6-binding protein TAB2 (MAP3K7IP2; MAP3K7–MAP3K7IP2–TRAF6 complex) to activate transcription factor NFκB and JNK. TAK1(MAP3K7)-binding protein TAB2 promotes TRAF6 ubiquitination. Like MAP3K7, cRaf and MEKKs are able to activate NFκB (Sect. 10.9.1), independently of MAPK enzymes.

6.5.1 Family of Raf Kinases

The RAF family (aRaf–cRaf) of proto-oncogene products are cytoplasmic Ser/Thr kinases that operate in cell growth and tissue development. Each Raf has a distinct expression profile. Isotype cRaf, or Raf1, is widespread, whereas aRaf and bRaf are restricted to some tissues: aRaf is mainly expressed in urogenital and gastrointestinal tracts and bRaf in neural, hematopoietic, splenic, and testicular tissues. The B-RAF gene is the most frequently mutated oncogene. The common substrates for all 3 Raf kinases are MAP3K1 and MAP3K2, dual-specificity kinases that target extracellular signal-regulated protein kinases ERK1 and ERK2.

Members of the RAF family share 3 conserved regions (CR1–CR3). The CR1 domain contains the Ras-binding motif. Both CR1 and CR2 have a regulatory function that mediates binding to Ras and other regulators. The CR3 sequence in the C-terminus possesses the Ser/Thr kinase domain [547].

Activation of Raf kinase includes translocation to the plasma membrane, induction of a conformational change by Ras GTPase, and phosphorylation. The subcellular compartmentation and activity of Raf kinase is regulated by cognate kinase and phosphatase, as well as scaffold proteins and other interacting proteins (Sect. 7.7).[18] The catalytic function of Raf results from a specific mode of dimerization of its kinase domain, the side-to-side dimerization [549]. Pseudokinase related to Raf, kinase suppressor of Ras (KSR), participates in the formation of the Raf side-to-side heterodimers and can trigger Raf activation.

The Ras–Raf–MA2PK–ERK module is activated by the guanine nucleotide-exchange factor SOS that stimulates Ras GTPase. Small Ras GTPase recruits Raf to the plasma membrane to phosphorylate MAP2K, which phosphorylates (activates) ERK enzyme. Kinases aRaf to cRaf have different phosphorylation sites, so that they can be independently regulated.

Signaling based on MAPK modules requires scaffold[19] IQ motif-containing GTPase activating protein IQGAP1 to modulate the activation of bRaf by EGF [550].[20]

The Ras–Raf–PI3K signaling pathway leading to protein kinase-B can act either synergistically with or in opposition to the Ras–Raf–ERK pathway. PKB

---

[18] Kinase Raf trapping to Golgi body (RKTG) is a transmembrane protein located in the Golgi body [548]. Protein RKTG sequesters cRaf at the Golgi body. Sequestrated cRaf then cannot interact with Ras GTPase at the inner leaflet of the plasma membrane. Protein RKTG thus blocks the extracellular signal-regulated kinase pathway. Once bound to growth factors, activated receptor tyrosine kinases initiate GTP-loading of small GTPase Ras that stimulates the sequential cascade of phosphorylation of components of the MAPK module down to ERK kinases.

[19] Several scaffold proteins for cRaf, such as kinase suppression of Ras KSR1, β-arrestin, and maguin participate in MAPK signaling.

[20] Protein IQGAP1 binds to diverse partners, such as CDC42 and Rac1 (but not RhoA or hRas), actin, calmodulin, E-cadherin, β-catenin, CLIP170, and adenomatous polyposis coli (APC). Protein IQGAP1 also binds to protein kinases (protein kinase-A and -C) and protein phosphatases (PP2 and PTPRm).

phosphorylates Raf, which then can inhibit activation of the Ras–Raf–ERK pathway. Cross-action between the Ras–Raf–ERK and Ras–Raf–PI3K–PKB pathways modulates cell life modes [551] (Sect. 5.2).

Some molecules can have 2 opposing mechanisms of action according to the cellular context and genotype, as they act as inhibitors or activators of Raf dimers [552]. Drug inhibitors that bind to Raf isoforms can provoke Raf dimerization or Ras–Raf interaction and membrane localization. Drug inhibitors that connect to one member of Raf homo- (cRaf–cRaf) or heterodimers (cRaf–bRaf) inhibits one subunit, but transactivate the drug-free protomer [553].

6.5.1.1 aRaf

V-Raf murine sarcoma 3611 viral oncogene homolog aRaf[21] interacts with small GTPase rRas, MAP2K2, MAP2K-like protein PDZ-binding kinase (PBK), casein kinase-2β, phosphoinositide 3-kinase regulatory subunit $PI3K_{r1}$, as well as 14-3-3γ, 14-3-3η, and 14-3-3σ, among others [251]. Hence, whereas bRaf and cRaf bind to all the 14-3-3 isoforms, aRaf links to some isoforms (14-3-3ε, -σ, and -τ [547]) in vitro and to a lesser degree. Enzyme aRaf transduces signals from mitogens and growth factors and activates MAP2K1.

Like other RAF family members, aRaf binds lipids to enable membrane connection, thereby regulating the kinase activity. Like cRaf, aRaf tethers to monophosphorylated phosphoinositides (PI(3)P, PI(4)P, and PI(5)P) and phosphatidylinositol (3,5)-bisphosphate. In addition, aRaf also links specifically to $PI(4,5)P_2$ and $PI(3,4)P_2$ and to phosphatidic acid [547].

Like cRaf, 2 ARAF genes exist in the human genome, one of which is functional (ARAF1). The ARAF2 gene is a pseudogene [547]. Two alternative splice aRaf variants have been detected without kinase domain (but with the Ras-binding domain): daRaf1 and daRaf2. They bind to activated Ras, but act as antagonists of the Ras–ERK pathway.

Activity of aRaf is regulated by phosphorylation as well as lipidic and proteic interactions. Protein aRaf is able to form homo- and heterodimers as well as trimers with cRaf and bRaf. Heterodimer Raf has properties distinct from those of Raf monomer and homodimer, such as elevated kinase activity toward MAP2K.

Enzyme aRaf exhibits different effects according to the type of receptor Tyr kinase. It is recruited to EGFR upon EGF stimulation. However, it is constitutively associated with PDGFR.

Enzyme aRaf is weakly activated by small GTPase hRas and Src kinase with respect to bRaf and cRaf.[22] Full aRaf activation needs Ras GTPase and other tyrosine kinases. Kinase aRaf may have the lowest kinase activity among Raf kinases for both MAP2K1 and MAP2K2 [547].

---

[21] A.k.a. A-Raf1, RafA1, or presumably for kinase sequence PKS2.

[22] Kinase bRaf is strongly activated by Ras GTPase alone, cRaf by both Ras GTPase and Src kinase.

Casein kinase CK2β activates aRaf. Interleukin-3 and phosphoinositide 3-kinase contribute to aRaf activity. Kinase aRaf interacts with P85 regulatory subunit of PI3K ($P85_{PI3K}$) in the presence or not of growth factor [547]. It also regulates the transition from the inactive dimeric form to the active tetrameric form of pyruvate kinase-M2. Enzyme aRaf interacts with TOM and TIM proteins involved in the mitochondrial transport system. It also targets trihydrophobin-1, a component of the negative elongation factor complex that hinders transcriptional elongation by RNA polymerase-2. In addition, aRaf acts as a dynamic interactor of scaffold Kinase suppressor of Ras KSR2 [547]. Protein aRaf binds constitutively to pro-apoptotic kinase mammalian Sterile 20-like kinase STK3 (MST2) [547].

Other interactors include [251, 547]: argininosuccinate synthase ASS1,[23] Rab geranylgeranyltransferase β subunit, negative elongation factor proteins NElFc[24] that precludes transcriptional elongation by RNA polymerase-2, translocase of inner mitochondrial membrane homolog TIMM44, PRP6 pre-mRNA processing factor PRPF6 (or U5 small nuclear ribonucleoprotein particle-binding protein) that is involved in pre-mRNA splicing, nucleoside uridine diphosphate-linked moiety X-type motif NUDT14, lymphokine-activated killer T-cell-originated protein kinase (TOPK or PBK), COP9 signalosome complex subunit-3, mitochondrial carbamoyl-phosphate synthase CPS1, and EGF-containing fibulin-like extracellular matrix protein EFEMP1 (or fibulin-3).

6.5.1.2 bRaf

V-Raf murine sarcoma viral oncogene homolog-B1 bRaf is also known as BRaf1, and RafB1. The Ras–Raf–ERK signal transduction pathway controls multiple processes, such as proliferation, differentiation, senescence, and apoptosis, depending on the duration and strength of the external stimulus and the cell type. The Ras–Raf–ERK pathway is activated in most human tumors, especially tumor cells with bRaf[25] or Ras gene mutations. When the MAPK cascade is inhibited, tumor growth is completely hindered in bRaf mutant and only partly in Ras-mutant tumors [554]. In addition, bRaf, but neither aRaf nor cRaf, contributes to vascular patterning in the placenta, although Raf kinases play redundant roles during initial stages of embryogenesis [556].

Kinase bRaf interacts with kinases PKB, ERK1, and cRaf, small GTases hRas and mRas, Rap1 GTPase-activating protein, Ras-related protein RaP1A, and 14-3-3β, -γ, -ε, -ζ, -σ, and -τ [251]. It mobilizes ERK1 and ERK2 via MAP2K1 and MAP2K2 and potentially ERK3.

---

[23] A.k.a. citrulline–aspartate ligase Ctln1.

[24] A.k.a. NElFd and TH1L homolog.

[25] Kinase bRaf is a protein implicated in the growth and survival of cancer cells. It is mutated in the majority of melanoma and a minority of colon, breast, and lung cancers.

6.5.1.3 cRaf

Protein Ser/Thr kinase (MAP3K) cRaf, or Raf1 (v-Raf1 murine leukemia viral oncogene homolog), binds and operates downstream from membrane-associated Ras GTPase. Activated cRaf phosphorylates (activates) dual-specificity protein kinases MAP2K1 and MAP2K2 that, in turn, phosphorylate (activate) extracellular signal-regulated kinases ERK1 (MAPK3) and ERK2 (MAPK1 or MAPK2). The kinetics (sustained or transient activation) and intensity of Raf–ERK pathway stimulation by growth factors specify the signaling type. Kinase cRaf is a direct effector of Ras GTPase. Its activation involves plasmalemmal recruitment and binding to RasGTP. It is inactivated by protein phosphatase-5 (Sect. 8.3.5) [555].

Pleiotropic cRaf interacts with receptor PDGFRB, dihydrotestosterone receptor, nuclear receptor NR3c1, transcription factor Myc, G-protein subunits Gβ2 and Gγ4, monomeric GTPases Rap1a, eRas, hRas, kRas, mRas, nRas, rRas, RasD2, and RHEB, guanine nucleotide-exchange factor Vav1, arrestin-β2, kinases MAP3K5 (ASK1), MAP2K1, MAPK1 (ERK2), MAPK3 (ERK1), MAPK7 (ERK4), Mst4 (or MASK), bRaf, Src, Fyn, LCK, LTK, JaK1, JaK2, PAK1, PKG1, and PKCϵ, phosphatase CDC25A, phosphorylase kinase-α2, lipoprotein receptor-related protein associated protein LRPAP1, TNFR2–TRAF-signaling complex proteins Inhibitor of apoptosis proteins IAP2 and IAP3,[26] adaptor GRB10, interactor MAPK8IP3 (or JIP3), connectors enhancers of kinase suppressor of Ras CnKSR1 and CnKSR1, phosphatidylethanolamine-binding protein PEBP4, sprouty-2, transcriptional modulator prohibitin, transcriptional regulator TSC22-related protein TSC22D3, Soc2 suppressor of clear homolog SHOC2, chaperonin-containing TCP1 subunit CCT3, heat shock protein Hsp90α, heat shock protein-A-binding protein-2, voltage-dependent anion channel VDAC1 (porin), ubiquitin-conjugating enzyme UbE2d2, ubiquitin ligase CHIP (or STUB1), and 14-3-3β, -γ, -ζ, -η, and -τ [251].

Kinase cRaf is a regulator of apoptosis, as it can phosphorylate (inactivate) pro-apoptotic protein BCL2 antagonist of cell death (BAD) in addition to potentiating BCL2-mediated resistance to apoptosis. It connects to retinoblastoma proteins Rb1 and Rb2, B-cell lymphoma protein BCL2, BCL2-associated athanogene BAg1 that enhances BCL2 anti-apoptotic effect, BCL2-like protein BCL2l1, and caspase-like apoptosis regulatory protein ClARP (Caspase homolog [CasH], Casp8 and FADD-like apoptosis regulator [CFLAR], or usurpin-β) [251].

---

[26] A.k.a. baculoviral IAP repeat-containing proteins BIRC3 and BIRC4, respectively. The latter is also called XIAP (X-linked inhibitor of apoptosis protein). The former is also termed TNFR2–TRAF-signaling complex protein-1. The TNFR2–TRAF-signaling complex protein-2 corresponds to IAP1, or BIRC2 protein. Molecule BIRC1 is also named NLR family, apoptosis inhibitory protein (NLRb1) and neuronal apoptosis inhibitory protein (NAIP).

6.5.2 Moloney Sarcoma Viral Mos Proto-Oncogene Kinase (Mos)

Kinase Mos is also a MAP3K specific to the ERK pathway, but it operates mainly in the reproductive system, as it is primarily expressed in germ line cells. It is expressed early during oocyte maturation and disappears immediately after fertilization. It actually contributes to oocyte maturation, as it activates ERK1 and ERK2 and maturation-promoting factor, a heterodimeric CDK1–cyclin-B complex that triggers oocyte maturation. As a cytostatic factor (CsF), it controls meiosis like maturation-promoting factor (MPF) for the cell division cycle.

6.5.3 Family of Mixed Lineage Kinases

The family of mixed lineage kinases (MLK) include MLK1 (MAP3K9), MLK2 (MAP3K10), MLK3 (MAP3K11), dual leucine zipper-bearing kinase (DLK or MAP3K12), LZK (MAP3K13), MLK4, MLK7α (MRKα, MLTKα, or ZAKα) and MLK7β (MRKβ, MLTKβ, or ZAKβ). These Ser/Thr kinases also belongs to Tyr kinase-like (TKL) superfamily of protein kinases. They activate stress-activated protein kinases Jun N-terminal kinases via MAP2K4 (JNKK1) and MAP2K7 and P38MAPKs via MAP2K3 and MAP2K6, as well as ERK1 and ERK2 via MAP2K1 and MAP2K2. Mixed lineage kinases also cause mitochondrial cytochrome-C release and caspase activation.

Members of the MLK family — MLK3 and MLK7 — are detected in adult hearts. In cardiomyocytes, MLK7 is activated in response to β-adrenergic stimulation and targets the P38MAPK and JNK axes. Saturated free fatty acid causes metabolic stress and activates JNK signaling via MLK3, MLK4, and MLK7 [556].

6.5.3.1 Mixed Lineage Kinase MLK1 – MAP3K9

Mitogen-activated protein kinase kinase kinase-9 (MAP3K9)[27] contains an N-terminal glycine-rich region, an SH3 domain, a kinase domain similar to both Tyr and Ser/Thr kinases, 2 Leu/Iso (or Ile) zipper motifs, and a polybasic sequence similar to nuclear localization signals near the C-terminus [557, 558]. Kinase MLK1 requires autophosphorylation of its activation loop (Thr304, Thr305, Ser308, and Thr312) for full activity. It is predominantly produced in the central nervous system, kidney, spleen, stomach, and testis [559].

---

[27] A.k.a. Src-homology-3 domain-containing proline-rich kinase SPRK and PRKE1, in addition to MLK1 mixed lineage kinase.

Kinase MLK1 phosphorylates dual-specificity kinase MAP2K4 to activate Jun N-terminal kinases in coordination with pro-apoptotic scaffold proteins such as Plenty of SH3 domains POSH1, especially for JNK-mediated neuronal apoptosis [559]. Scaffold POSH interacts with activated Rac to stimulate a subset of MLK family members, notably CDC42–Rac-interactive binding domain-containing MLK1 to MLK3. It forms a complex with MAP3K kinase (MLK1–MLK3 or MAP3K12), MAP2K (MAP2K4 or MAP2K7), and JNK1 or JNK2. It is inhibited by PKB. It operates as a pro-apoptotic protein in fibroblasts and mature neurons.

6.5.3.2 Mixed Lineage Kinase MLK2 – MAP3K10

Enzyme MAP3K10[28] possesses a structure similar to that of MAP3K9, i.e., an SH3 sequence, a kinase motif that is associated with both Tyr and Ser/Thr specificity, a double Leu/Iso zipper region, and a basic domain similar to nuclear localization signals, as well as a large C-terminus. The production of MAP3K10 is relatively restricted, as it is predominantly detected in the central nervous system, kidney, skeletal muscle, and testis [560]. Highest expression level is observed in brain and skeletal muscles.

In neurons, huntingtin interacts with MAP3K9 and huntingtin-associated protein HAP1 that connect to neuron-specific basic helix–loop–helix transcription factor neurogenic differentiation protein NeuroD1 (or β-cell E-box transactivator β2).[29] Enzyme MAP3K9 phosphorylates (activates) NeuroD [561]. This effect is facilitated by huntingtin and huntingtin-associated HAP1 protein.

Other MAP3K10 interactors include MAP2K4, MAPK8IP1 and MAPK8IP2, CDC42, dynamin-1, kinesin-3A, kinesin-associated protein KiFAP3, clathrin heavy chain-1, Disc large homolog DLg4, keratin-associated protein-3 pseudogene product KrtAP3P1, human growth factor-regulated tyrosine kinase substrate (HGS or HRS), hippocalcin, and 14-3-3ε [251].

Active small GTPases CDC42 and Rac1 tether MAP3K10 and might then relieve its auto-inhibition. P21 (CKI1a)-activated kinase PAK1 binds MAP3K10 to partly control its capacity to activate JNK [560]. The Rho target and scaffold Connector enhancer of kinase suppressor of Ras CnKSR1 connects to MAP3K10 possibly for Rho activation of JNK kinase. Ubiquitin conjugase UbE2d3.2 expressed in a tissue-restricted fashion during development interacts with MAP3K10 kinase [560].

Although it can phosphorylate MAP2K3 and MAP2K6, MAP3K10 kinase is a weak activator of the P38MAPK and ERK pathways [560]. Its major substrates include MAP2K4 and MAP2K7. It is rather specific to the JNK pathway, operating as a JN3K kinase. Members of scaffold POSH and JIP families that can dimerize bind directly to MAP3K10 to facilitate the activation of the JNK pathway. Conversely, JNK2 phosphorylates MAP3K10 kinase.

---

[28] A.k.a. Ser/Thr kinase MST, MKN28 kinase, and mixed lineage kinase MLK2.
[29] Huntingtin is a ubiquitous protein. Transcription factor NeuroD1 also regulates expression of the insulin gene.

MAP3K10 participates in the regulation of vesicular transfer. It localizes to clathrin-coated vesicles [560]. It binds to dynamin-1. In addition, like MAP3K9, MAP3K10 is involved in neuronal apoptosis.

6.5.3.3 Mixed Lineage Kinase MLK3 – MAP3K11

Enzyme MAP3K11[30] is widespread. It has a structure identical to MLK1 and MLK2 enzymes. Enzyme MAP3K11 colocalizes with prenylated, activated monomeric GTPase CDC42 at the plasma membrane [562]. Its activation requires phosphorylation of multiple sites in the activation loop.

Enzyme MAP3K11 activates the JNK pathway in response to gonadotropin-releasing hormone, epidermal and platelet-derived growth factors, tumor-necrosis factor-α, transforming growth factor-β, T-cell receptor costimulation, and interferon-γ [562]. It contributes to cell proliferation, especially to microtubule organization during the cell division cycle.

Enzyme MAP3K11 can be phosphorylated (activated; Ser281) by MAP4K1 (or hematopoietic progenitor kinase HPK1) [560]. In addition to MAP3K1, MAP3K11 is, indeed, another effector for MAP4K1 that also interacts with adaptors GRB2, CRK, and CRKL for signaling, as they recruit MAP4K1 to autophosphorylated receptor Tyr kinases at the plasma membrane. Enzyme MAP3K11 can also tether directly to CDC42 [267]. Enzyme MAP4K2 (GCK) interacts directly with and activates MAP3K11 kinase [560].

Substrates of MAP3K11 include MAP2K1, MAP2K3, as well as MAP2K4 and MAP2K7 that activate ERK, P38MAPK, and JNK, respectively [562]. Type of scaffold proteins directs MAP3K11 signaling mode. MAPK-interacting proteins, such as JNK-interacting proteins JIP1 and JIP3 as well as SH3 domain-containing ring finger protein SH3RF1,[31] promote MAP3K11-mediated JNK activation, whereas JIP2 enhances MAP3K11-induced P38MAPK stimulation [562]. In addition, MAP3K11 phosphorylates IκB kinases IKKα and -β.

Substrates of PKB comprise glycogen synthase kinase-α and -β, cRaf, bRaf, MAP3K5, MAP3K11, and MAP2K4, as well as Fox transcription factors. In addition, PKB1 isoform (but not PKB2) interacts with scaffold JNK-interacting protein JIP1 and prevents formation of JNK signaling complex. On the other hand, PKB2 targets SH3RF1 scaffold for the JNK signaling axis. This scaffold binds to activated Rac and is able to activate the JNK pathway and promote

---

[30] A.k.a. mixed lineage kinase-3 (MLK3), protein Tyr kinase-1 (PTK1), and SH3 domain-containing proline-rich kinase (SPRK).
[31] Protein SH3RF1, or "plenty of SH3s" (POSH), acts as a ubiquitin–protein ligase in protein sorting at the trans-Golgi network and as a scaffold for the JNK signaling pathway that facilitates the formation of functional signaling complexes such as that composed of calcium-binding apoptosis-linked gene ALG2 and ALG2-interacting protein-X (ALIX, a.k.a. programmed cell death-6-interacting protein [PDCD6IP] and ALG2-interacting protein AIP1) [563].

nuclear translocation of NFκB factor. Protein kinase-B2 binds to the SH3RF1–MAP3K11–MAP2K–JNK complex and phosphorylates MLK3, thereby causing complex disassembly and impeding the JNK pathway [564].[32]

Small Rho GTPases, such as CDC42 and Rac, as well as MAP4Ks, such as MAP4K1 and MAP4K2, provoke MAP3K11 activity. Binding of a GTPase to the CDC42–Rac-interactive binding site (CRIB) disrupts auto-inhibition generated by the SH3 domain of MAP3K11 on its kinase motif. Ceramide also activates MAP3K11. Dimerization or oligomerization enhances MAP3K11 function. Conversely, MAP3K11 phosphorylation (Ser674) by PKB lowers its catalytic activity. Kinase-specific cochaperone Hsp90 regulates MAP3K11 stability.

6.5.3.4 Mixed Lineage Kinase DLK – MAP3K12

Enzyme MAP3K12[33] activates the JNK (SAPK) pathway via MAP2K7 enzyme. It is involved in the regulation of apoptosis, cAMP response element-binding protein transcriptional activity, and terminal differentiation of keratinocytes [565]. It can form homodimers. Oligomerization-dependent autophosphorylation is required for its activation.

Unlike most of the MLKs that are widely expressed, MAP3K12 exhibits a more restricted pattern of expression. It activates JNK for neuronal migration during cerebral cortex development [556]. Scaffolds MAPK-binding inhibitory proteins MAPK8IP1 (or JNK-interacting protein JIP1) and MAP3K12IP1 (or MUK-binding inhibitory protein [MBIP]) regulate MAP3K12 activity by preventing its oligomerization and activation. Protein phosphatases PP1 and PP2A dephosphorylate MAP3K12.

6.5.3.5 Mixed Lineage Kinase LZK – MAP3K13

Leucine zipper-bearing kinase (LZK), or MAP3K13, activates the Jun N-terminal kinase pathway, but not the ERK axis. It is detected in the human brain.

Enzyme MAP3K13 interacts with itself and MAP2K7, as well as mitogen-activated protein kinase-8 interacting protein MAPK8IP1 (or JNK-interacting protein JIP1) and peroxiredoxin-3 [251]. Enzyme MAP3K13 stimulates inhibitor of nuclear factor-κB kinase IKKβ, hence transcription factor NFκB. Anti-oxidant protein AOP1 that localizes to mitochondria binds to MAP3K13 to modulate its activity. Protein AOP1 enhances MAP3K13-induced NFκB activation [566].

---

[32] Scaffold SH3RF1 (or POSH) can bind MLK1 to MLK3 and DLK and hence complex with MAP2K4, MAP2K7, JNK1 and JNK2, as well as PKB2 preferentially to PKB1 in mammalian cells.

[33] A.k.a. dual leucine zipper-bearing kinase (DLK), mitogen-activated protein kinase upstream kinase (MUK), and leucine-zipper protein kinase (ZPK).

6.5 Mitogen-Activated Protein Kinase Kinase Kinases

Table 6.15. Main signaling pathways that are activated by MAP3K1 to MAP3K3 (Source: [556]).

MAP3K	MAP2K	MAPK	Effect
MAP3K1	MAP2K1	ERK1/2	Cell proliferation and differentiation
	MAP2K4/7	JNK	Cell migration and apoptosis inhibition
MAP3K2	MAP2K5	ERK5	Cell survival and proliferation
	MAP2K4/7	JNK	Cell migration and apoptosis inhibition
MAP3K3	MAP2K1	ERK1/2	Cell proliferation and differentiation
	MAP2K5	ERK5	Cell survival and proliferation
	MAP2K4/7	JNK	Cell migration and apoptosis inhibition
	MAP2K3/6	P38MAPK	Cell differentiation and apoptosis inhibition

6.5.3.6 Mixed Lineage Kinase MLK4

Mixed lineage kinase-4 (MLK4) has several splice variants. Isoform MLK4α contains a different, shorter C-terminus than MLK4β subtype. Its subcellular compartment comprise the plasma membrane and cytoplasm.

6.5.3.7 Mixed Lineage Kinase MLK7

Mixed lineage kinase-7 (MLK7)[34] is ubiquitous. In cells, it is located in the cytosol and nucleus [567]. Two splice variants exist: MLK7α and MLK7β.

In response to stress, both isoforms phosphorylate MAP2K4 and MAP2K7 to activate the JNK pathway as well as MAP2K3 and MAP2K6 to excite the P38MAPK axis [567]. Overexpression of MLK7α causes cell proliferation and transformation as well as degradation of stress fibers. In the heart, MLK7β induces cardiac hypertrophy and fibrosis. Enzyme MLK7α, but not MLK7β, phosphorylates histone-3 in response to epidermal growth factor cue. Enzyme MLK7β contributes to cell cycle arrest after DNA damage.

Enzyme MLK7α is phosphorylated by PAK1. Enzyme MLK7β phosphorylates checkpoint kinase ChK2 [567]. Osmotic shock and ionizing radiation cause MLK7β autophosphorylation (activation; Thr161 and Ser165).

6.5.4 Family of MAP/ERK Kinase Kinases (MEKK)

The main signaling pathways that are activated by MAP3K1 to MAP3K3 include ERK1 and ERK2, ERK5, JNK, and P38MAPK (Table 6.15). Enzymes MAP3K1,

---

[34] A.k.a. MLK-like mitogen-activated protein triple kinase (MLTK), MLK-related kinase (MRK), and sterile-α motif- and leucine zipper-containing kinase (AZK or ZAK).

MAP3K2, and MAP3K3 participate in cell proliferation and differentiation, and tissue development. Enzyme MAP3K1 is particularly involved in epithelial cell migration; MAP3K2 in T-cell response; MAP3K3 in angio- and cardiogenesis [556].

Two MAP3K axes determine vascular patterning: (1) bRaf–MAP2K1–ERK2 cascade that regulates blood vessel formation in the placenta and (2) MAP3K3–MAP2K5–ERK5 axis that controls vessel development in the embryo. Several MAP3Ks, such as MAP3K1, MAP3K5 (ASK1), and MLK7, can be involved in mature heart homeostasis.

6.5.4.1 MAP3K1 (MEKK1)

Mitogen-activated protein kinase kinase kinase MAP3K1 is a ubiquitous Ser/Thr kinase that transduces many mitogenic and metabolic stimuli, such as insulin and growth factors, as well as pro-inflammatory cues. Enzyme MAP3K1 regulates diverse functions in a tissue- and cell type-specific manner. In most cases, MAP3K1 preferentially activates MAP2K4 and MAP2K7 that, in turn, stimulate Jun N-terminal kinases and/or P38MAPKs [568].[35] It can also activate extracellular signal-regulated kinases ERK1 and ERK2 via MAP2K1 and MA2PK2 enzymes. Besides, it can phosphorylate IκB kinases.

Protein MAP3K1 not only phosphorylates (activates) its effector kinases, but also has a ubiquitin ligase activity for ERK1 and ERK2 kinases. It also acts as a scaffold protein that binds to components (e.g., cRaf, MAP2K1, and ERK2) of the 3-tier ERK module as well as to the module activator Ras. It also uses additional scaffold proteins such as MAPK8IP3 to regulate the signaling kinetics and threshold.

Numerous input signals trigger MAP3K1 activation, such as osmotic stress, growth factors, and TNFSF proteins. Context-specific signaling of MAP3K1-based modules requires specific scaffold proteins to yield signaling routes with milestones made of proteic complexes.

WD repeat domain-containing protein WDR68[36] couples MAP3K1 to 2 protein Ser/Thr kinases: dual-specificity Tyr phosphorylation-regulated kinase DYRK1a and DYRK1b and homeodomain-interacting protein kinase HIPK2 [569].[37] Proteins MAP3K1 and DYRK1a suppress NFAT signaling, as NFAT is sequestered into

---

[35] Autophosphorylation and phosphorylation of threonine residues in the MAP3K1 activation loop of the kinase domain are needed to phosphorylate its primary substrate MAP2K4 [568]. Enzyme MAP4K2 (GCK) and TNFR-associated factor TRAF2 can activate MAP3K1 by priming oligomerization and subsequent autophosphorylation. Enzyme MAP3K1 can also be phosphorylated in regions outside of the kinase domain by PAK1 and protein kinase-G.

[36] A.k.a. anthocyanin (An11) and DDB1 and CUL4-associated factor DCAF7. Motifs WD40 repeats are used in between-protein interactions.

[37] Kinase HIPK2 of the CMGC superfamily of protein kinases is activated in response to morphogenic signals or DNA damage. It then triggers cell differentiation and tissue development or, conversely, apoptosis. Its substrates comprise P53 and STAT3 factors. The DYRK kinase family encompasses various members. Kinase DYRK1a targets transcription and splicing factors as well

the cytoplasm, thereby counteracting T-cell activation. Both MAP2K1 and HIPK2 target P300 acetyltransferase.[38] In addition, both MAP3K1 and HIPK2 can bind to axin adaptor.[39]

It is a key MAP3K in the regulation of cell migration, especially that involved in vascular remodeling as well as epithelial cell movement primed by developmental factors FGF2, TGFβ, and activin B.

Enzyme MAP3K1 contributes to heart hypertrophy triggered by Gq subunit that is activated by angiotensin-2, endothelin, and adrenergic signals. Furthermore, it protects cardiomyocytes from oxidative stress [556].

Enzyme MAP3K1 participates in the modulation of immune responses. It favors B-cell proliferation via TNFRSF5-dependent stimulation of JNK and P38MAPK.[40] In T lymphocytes, MAP3K1 precludes T-cell-mediated autoimmune diseases.

Enzyme MAP3K1 is a prosurvival and anti-apoptotic factor in a cell-type- and stimulus-dependent manner, but when overexpressed, it can be pro-apoptotic [568]. In addition, MAP3K1 can phosphorylate transcription factors such as STAT3 and cofactors, such as transducer of regulated CREB activity TORC1 (but not TORC2) and P300. Enzyme MAP3K1 stimulates P300-dependent transcription; P300 mediates apoptosis.

Enzyme MAP3K1 interacts with MAP4K1 (HPK1), MAP4K2 (GCK), MAP4K4 (HGK or NIK), cRaf, MAP2K1, MAP2K4 (JNKK1), MAPK1 (ERK2), and MAPK8 (JNK1), as well as MAPK8IP3 [251]. Enzyme MAP3K1 is an effector for MAP4K1 that also interacts with adaptors GRB2, CRK, and CRKL for signaling, as they recruit MAP4K1 to autophosphorylated receptor Tyr kinases at the plasma membrane. Active GTPase CDC42^{GTP} binds and promotes MAP3K1 homodimerization and activation [267]. Other interactors include nuclear factor-κB inhibitor kinase IKKβ, morphogen disheveled-2, Wnt inhibitor axin-1, adaptor GRB2, RhoGAP4, NCK-interacting kinase (NIK), focal adhesion kinase, actin-crosslinking protein α-actinin, TNFR-associated factors TRAF2 and TRAF6, FK506-binding protein FKBP5, ribosomal proteins RPL4, RPL30, and RPS13, cell division cycle protein CDC37, a chaperone that complexes with HSP90 and many kinases (CDK4, CDK6, Src, cRaf, MOK,[41] and eIF2αK), heat shock protein HSP90α, ubiquitin conjugase UbE2l, and 14-3-3ε [251]. Various intracellular molecules function as activators of

---

as mediators of apoptosis. Kinase DYRK1b is mainly produced in skeletal muscles, where it is involved in myocyte differentiation.

[38] Kinase HIPK2 phosphorylates P300 at multiple sites.

[39] Protein axin interacts with itself as well as adenomatosis polyposis coli protein, catenin, glycogen synthase kinase GSK3β, and PP2 phosphatase.

[40] Inflammatory cytokine and costimulatory protein on antigen-presenting cells TNFRSF5 activate MAP3K1 in B lymphocytes. Ligation of TNFRSF5 generates the formation on the plasma membrane of a complex with TRAF2 and TRAF3 adaptors, ubiquitin conjugase UbC13, cellular inhibitor of apoptosis proteins IAP1 and IAP2, IκB kinase regulatory subunit, and MAP3K1 [568].

[41] Kinase MOK (MAPK–MAK–MRK overlapping kinase), or renal tumor antigen RAge1, that is activated in response to tissue plasminogen activator and undergoes autophosphorylation is structurally similar to members of the MAPK superfamily as well as MAK (male germ

MAP3K1, such as RhoA GTPase, filamin-B, which organizes interferon signaling, TRAF2, and glycogen synthase kinase GSK3β, whereas protein Ser/Thr kinase STK38 and multifunctional Parkinson disease protein Park7 of the peptidase C56 family sequester MAP3K1 and hinder MAP3K1 activity [568]. Receptor-interacting protein Ser/Thr kinase RIPK1 phosphorylates MAP3K1 enzyme.

Enzyme MAP3K1 can also interact with small GTPases CDC42, Rac1, and Ras. Activation of MAP3K1 by protein Tyr kinases requires the coordinated recruitment of MAP3K1 and its adaptors and activators to the plasma membrane. Activated EGFR bound to phosphorylated adaptor SHC can recruit: (1) the GRB2–MAP3K1 complex; (2) adaptors CRK or CRKL and MAP4K1; or (3) the Ras-activating GRB2–SoS complex [267]. On the other hand, EPH receptor recruits and activates MAP4K4 and NCK.

6.5.4.2 MAP3K2 (MEKK2)

Enzyme MAP3K2 selectively activates JNK and ERK5 kinases. It also phosphorylates (activates) IκB kinases, thus initiating NFκB signaling. Enzyme MAP3K2 interacts with MAP2K4 (JNKK1), MAP2K5, MAP2K7 (JNKK2), and MAPK8 (JNK1), as well as T-lymphocyte specific adaptor SH2 domain-containing protein SH2D2a, apoptosis inhibitor IAP3,[42] and 14-3-3ε and 14-3-3σ [251].

Enzyme MAP3K2 contributes to the maturation and function of cellular immunity, as it modulates T-cell receptor stimulation of the JNK pathway and favors cytokine production in response to IgE [556].

6.5.4.3 MAP3K3 (MEKK3)

Ubiquitous MAP3K3 regulates ERK1 and ERK2, ERK5, JNK, and P38MAPK signaling as well as NFκB transcription factor. It intervenes in cardiovascular development via MAP2K5 and ERK5, especially in the maturation of primary vasculature and endocardium. In addition, activation of ERK1 and ERK2, JNK, P38MAPK, and NFκB signaling pathways is necessary for epithelial-to-mesenchymal transition, particularly during the formation of the cardiovascular and nervous system [556].

Enzyme MAP3K3 interacts with MAP2K5 as well as microtubule affinity-regulating kinase-2 (MARK2). Other interactors include GRB2-associated binding protein GAB1, transcription factor zinc finger and BTB domain-containing ZBTB16, TNFR-associated factor TRAF7, P53- and DNA-damage-regulated protein PDRG1, cell division cycle homolog CDC37, DNA-damage repair mediator

---

cell-associated kinase) and MRK (MAK-related kinase). In humans, it is detected in heart, brain, lung, kidney, and pancreas [570].

[42] A.k.a. baculoviral IAP repeat-containing, homologous protein BIRC4 and X chromosome-linked inhibitor of apoptosis (XIAp).

breast cancer susceptibility protein BrCa1, JNK-associated scaffold protein sperm-associated antigen SpAg9, prefoldin-2, heat shock proteins HSP90α (HSPCal3) and HSPa4, and 14-3-3γ and 14-3-3ε [251].

Hypertonicity (e.g., high NaCl concentration)[43] activates the osmolarity-responsive NFAT5 transcription factor.[44] Small Rac1 GTPase indeed complexes with cerebral cavernous malformation-2 homolog (CCM2)[45] MAP3K3, and MAP2K3 that activates P38MAPK in response to hypertonicity. However, NFAT5 activation results from signaling via phospholipase-Cγ1 and the Rac1-CCM2 axis, but not the Rac1-CCM2-MAP3K3-MAP2K3-P38MAPK pathway [571].

6.5.4.4 MAP3K4 (MEKK4)

Enzyme MAP3K4[46] activates JNKs and P38MAPKs, but not MAP2K1, in response to environmental stresses, such as osmotic shock, UV irradiation, wound stress, and inflammatory factors. It is a major activator of the P38MAPK pathway, but a minor effector of the JNK pathway. The JNK signalosome is a complex constituted by Jun N-terminal kinase and dual-specificity kinases MAP2K4 and MAP2K7 that can activate Jun transcription factor, a AP1 component.

Enzyme MAP3K4 contributes to the regulation of cell proliferation, apoptosis, and migration. It participates in neural, cardiac (valvulogenesis, chamber partitioning, and formation of the coronary vasculature), and skeletal development [556]. Wnt signaling can operate via MAP3K4. Angiotensin-2 provokes calcium-dependent phosphorylation by FAK2 of MAP3K4 that, in turn, activates MAP2K6 in smooth muscle cells. In addition, both MAP2K6 and calcium-regulated phosphatase calcineurin (PP2B or PP3) are involved in T-cell signaling and activation.

Enzyme MAP3K4 interacts with MAP2K1, MAP2K3, MAP2K4 (JNKK1), and MAP2K6, small GTPase CDC42 and Rac1, and growth arrest and DNA-damage-inducible GADD45α, -β, and -γ [251]. These 3 GADD45 polypeptides may contribute to activation of both P38MAPKs and JNKs via MAP3K4 enzyme. However, GADD45 homologs remains at low level when JNKs are fully activated during DNA damage [267].

---

[43] Tonicity measures the osmotic pressure gradient in a 2-solution medium, such as a biological tissue with its extra- and intracellular fluids, separated by a semipermeable membrane such as the plasma membrane. Hypertonicity increases NFAT5 activity, as it augments its abundance, transactivating activity, nuclear location, and phosphorylation due to both elevated kinase activity and lowered phosphatase activity.

[44] A.k.a. tonicity-responsive enhancer-binding protein (TonEBP) and osmotic response element-binding protein (OREBP). Kinase P38MAPKα increases NFAT5 activity, whereas P38MAPKδ decreases it. Hypertonicity activates P38MAPKα via MAP3K3 and MAP2K3 or MAP2K6 (MAP3K3–MAP2K3/6–P38MAPKα axis).

[45] A.k.a. osmosensing scaffold for MAP3K3 (Osm).

[46] A.k.a. MEKK4 and MAP three kinase MTK1.

6.5.5 Family of Mitogen-Activated Protein Kinase Kinase Kinases-5/6

Mitogen-activated protein kinase kinase kinases MAP3K5 and MAP3K6 are also tagged apoptosis signal-regulating kinases ASK1 and ASK2 as well as MAP/ERK kinase kinase MEKK5 and MEKK6, respectively. They indeed regulate apoptosis downstream from cellular stresses.

6.5.5.1 MAP3K5 or ASK1

Enzyme MAP3K5, or apoptosis signal-regulating kinase ASK1, functions in both the MAP2K4–JNK (or JN2K1–JNK) and MAP2K7–JNK (or JN2K2–JNK), as well as MAP2K3–P38MAPK (or SKK2–P38MAPK) and MAP2K6–P38MAPK (or SKK3–P38MAPK) cascades. It intervenes in apoptosis under various stress types, such as oxidative and endoplasmic reticulum stresses. These stresses causes MAP3K5 homo-oligomerization and autophosphorylation (Thr838). Apoptosis that depends on MAP3K5 is mainly mediated by mitochondria-dependent caspase activation (caspase-9 and -3) [572].

Enzyme MAP3K5 also mediates signaling for cell differentiation and survival, in particular, in neurons and keratinocytes [572]. It is also involved in inflammation and innate immunity. Production of inflammatory cytokines (e.g., TNFα, IL6, and IL1β) relies on MAP3K5 in splenocytes and bone marrow-derived dendritic cells.

Oligomerization of MAP3K5 is required for its activation by trans-autophosphorylation. Enzyme MAP3K5 is phosphorylated (activated; Thr838) [572]. In addition, MAP3K6 hetero-oligomerizes with MAP3K5 (ASK1–ASK2 complex) and directly phosphorylates MAP3K5 (Thr838). The MAP3K5–MAP3K6 heteromer stabilizes MAP3K6 and improves MAP3K6 efficiency for its subtrates in the JNK and P38MAPK modules. The MAP3K5–MAP3K6 complex causes apoptosis.

Enzyme MAP3K5 interacts not only with itself and MAP3K6, but also with MAP3K7 (TAK1). It then disrupts TRAF6–MAP3K7 interaction and inhibits IL1-induced NFκB activity [572]. It connects to MAP2K6 and MAP2K7 [251]. It is also linked to MAPK8IP3 (or JNK-interacting protein JIP3), thioredoxin, glutaredoxin, TNFR-associated factors TRAF1 to TRAF3, TRAF5 and TRAF6, arrestin-β2, glutaminyl-tRNA synthase, glutathione S-transferase-M1, dual-specificity phosphatase DuSP19, cell division cycle CDC25a phosphatase, cRaf, growth arrest and DNA-damage-inducible protein GADD45β, cyclin-dependent kinase inhibitor CKI1a, programmed cell death PCD6,[47] endoplasmic reticulum-to-nucleus signaling ERN1, and 14-3-3ζ protein, among others (Table 6.11).

---

[47] A.k.a. apoptosis-linked gene product ALG2.

Various factors regulate MAP3K5 oligomerization. Both TRAF2 and TRAF6 as well as adaptor death-associated protein DAP6,[48] facilitate MAP3K5 oligomerization and activation. On the other hand, Vacuolar protein sorting-associated protein-28 homolog (VPS28)[49] is an anti-apoptotic protein that suppresses MAP3K5 oligomerization [572].

Conversely, MAP3K5 is inactivated by phosphorylation at Ser83, Ser966, and Ser1033 in humans [572]. Protein kinase-B phosphorylates (inactivates) MAP3K5 (Ser83). Phosphoinositide-dependent protein kinase PDK1 also phosphorylates (inhibits) MAP3K5 (Ser966). The receptor Tyr kinase insulin-like growth factor receptor also phosphorylates (inactivates) MAP3K5. In addition, 14-3-3 protein prevents MAP3K5 activity, as it binds to a phosphorylated Ser residue (Ser966) in the 14-3-3-binding motif of MAP3K5 agent [572].

On the other hand, protein kinase-D and Disabled homolog 2-interacting protein (Dab2IP)[50] foster the release of 14-3-3 inhibitor from MAP3K5 enzyme. Homeodomain-interacting protein kinase HIPK1 is an MAP3K5–Dab2IP complex-associated protein that activates MAP3K5, as it dissociates the inhibitors 14-3-3 and thioredoxin from MAP3K5. Kinase HIPK1 assists activation of the MAP3K5–JNK axis by TNFα and ROS agents [459].

In addition, Ca^{++}–calmodulin-dependent protein kinase CamK2 phosphorylates (activates) MAP3K5 in neurons [573]. Ste20-like kinase (SLK) interacts with and activates MAP3K5 and provokes apoptosis [572].

Phosphatase PP2 supports MAP3K5 activity, as it dephosphorylates an inactivating phosphorylation site (Ser966) and dissociates 14-3-3 from MAP3K5 enzyme [572]. Calcium–calmodulin-activated protein phosphatase PP3 also activates MAP3K5, as it promotes its release from 14-3-3 proteins in cardiomyocytes. Moreover, PP3 dephosphorylates the inactivating phosphorylation site of MAP3K5, thereby exciting its activity. On the other hand, protein Ser/Thr phosphatase PP5 inhibits MAP3K5, as it dephosphorylates an activating phosphorylation site (negative feedback).

Phosphatase CDC25a inhibits MAP3K5 independently of its phosphatase activity, as it impedes MAP3K5 oligomerization [572]. Dual-specificity phosphatase DuSP19[51] enhances MAP3K5 signaling, as it acts as a scaffold for this kinase.

Activation of the MAP3K5–P38MAPK pathway upon stimulation by reactive oxygen species is required for extracellular ATP-induced apoptosis in macrophages, which is initiated by the ATP-ligated $P2X_7$ receptor, an ATP-gated ionotropic channel [572].

---

[48] A.k.a. Daxx. It does not possess a death domain.

[49] A.k.a. caspase-activated DNase (CAD) inhibitor that interacts with ASK1 (CIIA).

[50] A.k.a. MAP3K5 (ASK1)-interacting protein-1 (AIP1). This member of RasGAP family binds to a sequence surrounding the 14-3-3-binding site on MAP3K5, but preferentially to a dephosphorylated active MAP3K5 form.

[51] A.k.a. stress-activated protein kinase (SAPK) pathway-regulating phosphatase SKRP1.

Adaptors TRAF2 and TRAF6, which link plasmalemmal receptors and kinases, are recruited to the ROS-stimulated MAP3K5 signalosome and activate MAP3K5 (full MAP3K5 activation), after dissociation of thioredoxin from inactivated MAP3K5, the MAP3K5 signalosome enlarging with respect to its resting state [572].

Redox sensing oxidoreductase thioredoxin, only in its reduced state, inhibits MAP3K5, as it sequesters MAP3K5 away from TRAF2 until TRAF2-dependent production of reactive oxygen species provokes the conversion of thioredoxin to its oxidative form, hence the dissociation of the MAP3K5–thioredoxin complex [267]. Interaction between MAP3K5 and TRAF6 is also prevented by thioredoxin.

Calcium- and integrin-binding protein-1 (CIB1)[52] that interacts with α_{2B} integrin in particular, binds to MAP3K5, interferes with the recruitment of TRAF2 to MAP3K5, and inhibits MAP3K5 autophosphorylation, thereby repressing MAP3K5 activation [573]. Inhibitor CIB1 thus competes with TRAF2 to connect to MAP3K5 kinase. Calcium influx relieves inhibition of CIB1 on MAP3K5 kinase.

Retinoblastoma RB1-inducible coiled-coil protein RB1CC1,[53] a widespread inhibitor of FAK1 and FAK2 and regulator of the retinoblastoma gene[54] contributes to the regulation of cell proliferation and migration as well as regulates TRAF2–MAP3K5 interaction [575].[55] Eukaryotic translation initiation factor-2α kinase-2,[56] an element of the antiviral response, links to MAP3K5 to activate the JNK and P38MAPK pathways [576].

Enzyme MAP3K5 is also regulated by ubiquitination. Suppressor of cytokine signaling SOCS1 constitutively binds to MAP3K5 and causes its ubiquitination and degradation in resting cells [576]. Cellular inhibitor of apoptosis protein IAP1[57] is a ubiquitin–protein ligase for MAP3K5 that can terminate TNFα-induced MAPK activation (negative feedback).

Balance between cell survival and death relies on activation of the PI3K–PKB and MAP3K5–JNK pathways. Activated MAP3K5 triggers cell death in response to diverse apoptotic stimuli, such as TNFα or reactive oxygen species. Scaffold protein Disabled homolog Dab2-interacting protein (Dab2IP), a Ras-GTPase activating protein (RasGAP)[58] causes cell cycle arrest and promotes apoptosis in response

---

[52] A.k.a. calmyrin and kinase-interacting protein KIP1.

[53] A.k.a. 200-kDa FAK family kinase-interacting protein (FIP200).

[54] Retinoblastoma protein impedes cell cycle progression until DNA damage has been repaired (Vol. 2 – Chap. 2. Cell Growth and Proliferation).

[55] Protein RB1CC1 interacts with tuberous sclerosis complex TSC1 to inhibit the TSC1–TSC2 complex that precludes target of rapamycin activity, hence cell growth.

[56] A.k.a. double-stranded RNA-activated protein kinase (PKR).

[57] A.k.a. baculoviral IAP repeat-containing protein BIRC2.

[58] A.k.a. ASK1-interacting protein AIP1 (not actin-interacting protein; Sect. 9.4.2.5).

to stress. It actually suppresses the PI3K–PKB pathway and enhances MAP3K5 activation to impede cell survival and provoke cell death [577].[59]

Enzyme MAP3K5 interacts with cell cycle regulators. Cyclin-dependent kinase inhibitor CKI1a links to MAP3K5 and inhibits the MAP3K5 signaling [572].

Several scaffold proteins, such as MAPK8IP3 and β-arrestin, interact with MAP3K5 to facilitate its signaling [572]. The scaffold Gem (nuclear Gemini of Cajal bodies)-associated protein Gemin-5, a component of the survival of motor neurons (SMN) complex,[60] interacts with ASK1 and potentiates activation of the MAP3K5–MAP2K4–JNK1 pathway. Parkinson disease protein Park7 tethers to and inhibits MAP3K5 enzyme.

Enzyme MAP3K5 phosphorylates Programmed cell death PDCD6,[61] a member of the pentaEF hand Ca^{++}-binding protein family; this phosphorylation is counteracted by cRaf [572].

Enzyme MAP3K5 contributes to cardiogenesis, as cardiac chamber partitioning and valve formation require appropriate timing of apoptosis, as well as left ventricular remodeling by inducing apoptosis after myocardial infarct via the P38MAPK and JNK pathways [556].

6.5.5.2 MAP3K6 or ASK2

Enzyme MAP3K6 interacts with MAP3K5 (ASK1). Alone, MAP3K6 weakly activates JNKs and does not stimulate ERKs and P38MAPKs [578]. Its kinase activity happens only in association with MAP3K5 enzyme. The latter may confer kinase activity to MAP3K6; it may allosterically activate MAP3K6, hence causing its autophosphorylation (Thr806). Activation of MAP3K6 does not depend on MAP3K5 kinase activity. In the presence of MAP2K5, MAP3K6 phosphorylates (activates) MAP2K4 and MAP2K6, thereby priming the MAP2K4–JNK and MAP2K6–P38MAPK pathways, to the same extent as MAP3K5 alone [578]. Reactive oxygen species leads to MAP3K6 autophosphorylation (Thr806) [578].

In addition to its heteromerization with MAP3K5, MAP3K6 can homodimerize. This homodimer may enable MAP3K6 trans-autophosphorylation. Protein 14-3-3 is a MAP3K6-interacting partner; this connection is reversibly regulated by phosphorylation.

Enzyme MAP3K6 regulates the expression of vascular endothelial growth factor under both normoxia and hypoxia in vitro [579]. Consequently, MAP3K6 favors tumor angiogenesis and tissue response to ischemia.

---

[59] It coordinates inactivation of PI3K–PKB axis via its C-terminal proline-rich (PR) and PERIOD-like (PER) domains for PI3K and PKB, respectively, and MAP3K5 activation.

[60] The SMN complex is necessary for pre-mRNA splicing in the nucleus and spliceosomal snRNP assembly in the cytoplasm.

[61] A.k.a. Apoptosis-linked gene product ALG2.

6.5.6 Mitogen-Activated Protein Kinase Kinase Kinase-7 (TAK1)

Mitogen-activated protein kinase kinase kinase-7 (MAP3K7) is also called transforming growth factor-β-activated kinase TAK1. This signaling mediator of innate immunity is activated by pro-inflammatory cytokines, such as interleukin-1 and tumor-necrosis factor-α, as well as transforming growth factor-β, bone morphogenetic protein BMP4, and Wnt family ligands [580]. Enzyme MAP3K7 contributes to vasculature development. It is also involved in the redox regulation.

Ubiquitous MAP3K7 is mainly located in the cytoplasm. Ligand Wnt1 causes MAP3K7 accumulation in the nucleus. Proteasomal degradation provokes a negative feedback. In humans, 4 splicing variants exist (MAP3K7a–MAP3K7d). Isoform MAP3K7a (i.e., MAP3K7) is the predominantly expressed form.

In mammalian cells, MAP3K7 activates NEMo-like kinase and downregulates transcriptional activation by β-catenin and T-cell factor/lymphoid enhancer factor [580]. Enzyme MAP3K7 interacts with upstream kinases MAP4K1 and MAP4K4 on the one hand and downstream enzymes MAP2K6 on the other [251]. It also connects to components of the same tier of MAPK module MAP3K5 and MAP3K14. In addition, MAP3K7 links to IκB kinases IKKα and IKKβ.

Enzyme MAP3K7 is required for the activation of JNK and nuclear factor-κB upon signaling from IL1-, TNF-, and Toll-like receptor. Upon IL1 signaling, MAP3K7 is recruited to TRAF6 complex. Mediator TRAF6 bears Lys63-linked ubiquitination that facilitates MAP3K7 activation. Activated MAP3K7 then stimulates the JNK and P38MAPK axes as well as IκB kinases to activate nuclear factor-κB [580].

Proteins MAP3K7 and TRAF6 also attract receptor activator of nuclear factor-κB, or tumor-necrosis factor-related activation-induced cytokine (TRANCE), that is expressed on activated T cells to stimulate mature dendritic cells as well as osteoblasts to prime osteoclastogenesis and osteoclast activation [580].

Two major MAP3K7-binding partners correspond to MAP3K7-interacting proteins MAP3K7IP1 and -2, or TAK1-binding proteins TAB1 and TAB2. Protein MAP3K7IP1 is an activator of MAP3K7, whereas MAP3K7IP2 is an adaptor that links MAP3K7 to TRAF6 [580]. Proteins MAP3K7IP1 and -2 participate in TRAF6-induced IKK activation. Enzyme MAP3K7 is constitutively associated with its activator MAP3K7IP1, but remains inactive in the absence of stimuli. Once stimulated by IL1, MAP3K7 autophosphorylates (activates; Thr184, Thr187, and Ser192).

Basal activity of MAP3K7 relies on autophosphorylation and its MAP3K7IP1 partner [581]. On the other hand, stimulus-dependent activity of MAP3K7 needs MAP3K7IP2. In particular, MAP3K7IP2 mediates polyubiquitin chain-dependent MAP3K7 activation in innate immunity. Autophosphorylation of MAP3K7 results from conformational changes following MAP3K7 association with MAP3K7IP2-polyubiquitin chains or MAP3K7IP1-dependent oligomerization. On the other hand, MAP3K7 is dephosphorylated by protein phosphatases PP2, PP6, and PPM1.

Inhibitor of apoptosis protein IAP3 binds MAP3K7IP1 and receptor Ser/Thr kinases BMPR1a (ALK3 or CD292) and BMPR1b (ALK6) that are targeted by BMP2, BMP4, BMP6, and BMP7 bone morphogenetic proteins, as well as GDF6 growth differentiation factor. It can then form a trimeric IAP3–MAP3K7IP1–MAP3K7 complex. Other members of the family of inhibitors of apoptosis can be recruited to TNFRSF complexes. Apoptosis inhibitors IAP1 and IAP2 interact with TRADD and TNFRSf1a (TNFR1) and can recruit MAP3K7.

Tumor-necrosis factor receptor-associated factor TRAF2 mediates TNF signaling, as it acts as an adaptor that relays signals in the NFκB and AP1 pathways. Upon TNF stimulation, TRAF2 is phosphorylated by PKC and, then, recruits IKKα and IKKβ to TNFR receptor. Phosphorylated TRAF2 also undergoes polyubiquitination that allows its connection to TAB2 and -3 (MAP3K7IP2 and -3) and activation of IKK and JNK [582].

Interleukin-17 receptor-D (IL17Rd)[62] has both inhibitory and stimulatory activities. On the one hand, this transmembrane protein is a feedback inhibitor of fibroblast growth factor signaling [583].[63] It can indeed interact with FGFR1 and prevent its phosphorylation as well as suppress Ras function, hence blocking FGF-mediated ERK activation. Its splice variant that gives rise to a cytoplasmic form of Sef likewise associates with FGFR1 and inhibits ERK activation. In addition, IL17Rd inhibits PKB. On the other hand, IL17Rd interacts with MAP3K7 to activate a MAP3K7–MAP2K4–JNK pathway [584].

Other MAP3K7 interactors include TGFβ receptor TβR1 (or activin receptor-like kinase ALK5), inhibitory SMAD6 and SMAD7, TNFRSF11a, magnesium-dependent protein phosphatase-1B, receptor (TNFRSF)-interacting Ser/Thr kinase RIPK1, cRaf kinase inhibitor protein (RKIP or phosphatidylethanolamine-binding protein PEBP1), hepatocyte growth factor-regulated tyrosine kinase substrate (HRS or HGS), myeloid differentiation primary response gene product MyD88 and Toll–IL1 receptor domain-containing adaptor protein (TIRAP), as well as cylindromatosis (Cyld) deubiquitinase [251].

Dysregulation in cellular redox systems (i.e., anti-oxidant enzymes, such as superoxide dismutase and catalase, and scavengers such as glutathione) causes accumulation of reactive oxygen species. Among molecules involved in redox regulation, transcription factors NFκB, AP1, and NRF2 regulate production of anti-oxidant enzymes. Enzyme MAP3K7 activates NFκB and AP1, especially after stimulation by cytokines and Toll-like receptors, and upregulates NRF2 [581].

---

[62] A.k.a. Similar expression to FGF (SEF).

[63] Activity of FGF is also inhibited by members of the Sprouty family and Sprouty-related EVH1 domain-containing proteins (SPRED) that antagonize Ras signaling induced by receptor Tyr kinase FGFR.

6.5.7 Family of Thousand and One Kinases (TAOK)

Thousand and one kinases (TAOK1–TAOK3 or MAP3K16–MAP3K18; Sect. 5.1.10) that constitute the GCK subfamily 8 primarily activate the P38MAPK pathway in response to cellular stress and DNA damage.

Upon DNA damage, ataxia telangiectasia mutated kinase (ATMK) is activated and phosphorylates TAOK kinase. Phosphorylated TAOKs then activate P38MAPK and cells enter the G2–M damage checkpoint for DNA repair [556].

Kinase MAP3K16 activates MAP2K3 that targets P38MAPKs. It also targets JNKs. Kinase MAP3K17 prevents activation by MAP3K7 of the NFκB pathway after osmotic stress. It also associates with MAP2K3 and MAP2K6 that both targets P38MAPKs. It can also activate JNK kinases.

6.5.8 Mitogen-Activated Protein Kinase Kinase Kinase-8 (TPL2)

Mitogen-activated protein kinase kinase kinase-8 (MAP3K8)[64] activates the ERK and JNK pathways in immunocytes stimulated by lipopolysaccharides and other inflammatory mediators. However, development of immunocytes (T and B lymphocytes, dendritic cells, natural killer cells, and macrophages) does not need MAP3K8 protein. Enzyme MAP3K8 is required for activation of ERK1 and ERK2 in myeloid cells after TLR (TLR2 and TLR9) and TNFRSF receptor stimulation. In resting cells, MAP3K8 complexes with inhibitory protein NFκB1 (P105$_{NFκB}$ monomer), thereby being unabled to connect to its MAP2K substrate. In stimulated cells, the IKK complex phosphorylates NFκB1, thereby triggering its ubiquitination by SCF ligase for proteasomal degradation and releasing MAP3K8 as well as Rel subunits for nuclear translocation [585]. During inflammation, MAP3K8 enables the production of pro-inflammatory cytokine TNFα. Enzyme MAP3K8 can also activate nuclear factor of activated T cells and prime interleukin-2 production.

The Tpl2 gene encodes full-length MAP3K8$_{FL}$[65] and MAP3K8$_S$ variant[66] by alternative translational initiation at methionine 1 (M1) or 30 (M30). C-terminal truncation increases protein stability and kinase activity; moreover, MAP3K8$_S$ is expressed at higher levels [585].

Transcript Tpl2 mRNA is expressed at highest levels in the lung, spleen, and thymus, and at low levels in the brain, liver and testis [585]. As Tpl2 mRNA is also detected in the kidney, salivary glands, stomach, intestine, and skeletal muscles, MAP3K8 is a widespread molecule. Expression of MAP3K8 is greater in the ventricular myocardium during body's development than in adults [556].

---

[64] A.k.a. MEKK8, tumor progression locus-2 (TPL2), cancer Osaka thyroid proto-oncogene product (COT), and Ewing sarcoma transformant factor (ESTF).
[65] A.k.a. ^{M1}TPL2 and 58-kDa isoform P58$_{TPL2}$.
[66] A.k.a. ^{M30}TPL2 and 52-kDa isoform P52$_{TPL2}$.

The signaling function of MAP3K8 is regulated by phosphorylation. Autophosphorylation targets Thr290 residue. Additional phosphorylation (Ser400) is needed to activate MAP2K effector [585].

Enzyme MAP3K8 interacts with MAP3K14 as well as MAP2K1 and MAP2K4 to MAP2K6 effectors [251,585]. It also links to IκB kinases IKKα[67] and IKKβ, as well as nuclear factor-κB subunits (NFκB1, NFκB2, Rel, and RelA) [251]. Other MAP3K7 interactors comprise protein kinase-B, kinase suppressor of Ras KSR2, TNFR-associated factor TRAF2, and ubiquitin-binding TNFα-induced protein-3-interacting protein TNIP2.[68] Enzyme MAP3K8 can be confined to a cytoplasmic complex with $P105_{NF\kappa B}$[69] and ubiquitin-binding TNFαIP3 protein.[70]

6.5.9 Mitogen-Activated Protein Kinase Kinase Kinase MAP3K14

Mitogen-activated protein kinase kinase kinase-14 (MAP3K14) is a TRAF2-interacting Ser/Thr kinase that shares sequence similarity with several other MAP3Ks and stimulates nuclear factor-κB. It actually participates in NFκB-mediated signaling triggered by receptors of the TNFRSF–NGFR set (Vol. 3 – Chap. 11. Receptors of the Immune System)[71] family that activates or inhibits cell death and interleukin-1 type-1 receptor. It is thus branded nuclear factor-κB-inducing kinase (NIK).[72] Besides, MAP3K14 seems to be exclusively involved in the activation of NFκB. Other MAP3K14 aliases comprise HS and HSNIK.

---

[67] A.k.a. conserved helix–loop–helix ubiquitous kinase (CHUK). The IκB kinase (IKK) complex intervenes in innate and adaptive immunity.

[68] A.k.a. A20-binding inhibitor of nuclear factor-κB ABIN2. The ABIN proteins (ABIN1–ABIN3) bind the ubiquitin-editing nuclear factor-κB inhibitor A20, or tumor-necrosis factor-α-induced protein TNFαIP3. Protein TNIP2 tethers to $P105_{NF\kappa B}$ and directly to MAP3K8, but preferentially forms a ternary complex with these 2 partners.

[69] Protein $P105_{NF\kappa B}$ serves as a precursor for $P50_{NF\kappa B}$ subunit (NFκB1) after limited proteolysis of $P105_{NF\kappa B}$ monomer by the proteasome. On the other hand, K48-linked ubiquitination of $P105_{NF\kappa B}$ by SCF triggers its complete degradation by the 26S proteasome. Protein $P105_{NF\kappa B}$ also acts as a IκB that sequester associated NFκB subunits in the cytoplasm.

[70] Proteasomal degradation of $P105_{NF\kappa B}$ also releases TNFαIP3 from the MAP3K8–$P105_{NF\kappa B}$–TNFαIP3 complex.

[71] Several receptors of the TNFR–NGFR family activate transcription factor NFκB via common TRAF2 adaptor. The TNFRSF–NGFR family includes TNFR1 (TNFRSF1a) and TNFR2 (TNFRSF1b), as well as TNFRSF3 (lymphotoxin-β receptor), TNFRSF4 (CD134), TNFRSF5 (CD40), TNFRSF6a (CD95), TNFRSF7 (CD27), TNFRSF8 (CD30), TNFRSF9 (CD137), and TNFRSF12 (TNFRSF25), and nerve-growth factor receptor (NGFR), among others. Receptors TNFR and NGFR regulate cell proliferation. Receptors TNFRSF4 and TNFRSF7 to TNFRSF9 serves as accessory regulators in lymphocyte activation, proliferation, and differentiation. Receptors TNFR and TNFRSF6 prime cell apoptosis, whereas NGFR and TNFRSF5 inhibit cell death.

[72] Alias NIK also stands for NCK-interacting kinase that corresponds to MAP4K4. Therefore, this alias should be eliminated to avoid confusion.

Upon stimulation from TNFRSF11a, TNFR1 and TNFR2, Toll-like receptors, T-cell receptors, CD3 and CD28 coreceptors, interleukin-1 receptor, and lipopolysaccharides, MAP3K14 activates IκB kinase, as it phosphorylates both NFκB inhibitors IKKα and IKKβ. Activated IKK then phosphorylates IκBα (Ser32 and Ser36) that leads to IκBα ubiquitination and degradation, thereby allowing NFκB translocation to the nucleus.

Enzyme MAP3K14 interacts not only with TRAF2, but also with TRAF1, TRAF3, and TRAF6 agents. It also connects not only to α and β subunits of inhibitor of nuclear factor-κB kinase, but also inhibitor of κlight polypeptide gene enhancer in B cells, kinase complex-associated protein (IKBKAP), and kinase-γ (IKBKγ), as well as nuclear factor-κlight polypeptide gene enhancer in B-cell inhibitor-α (NFKBIα). In addition, MAP3K14 associates with itself, MAP3K7 (or TAK1), and MAP3K8 (TPL2), as well as epidermal growth factor receptors (heregulins), TNFR-interacting protein kinase RIPK1, phosphatidylethanolamine-binding protein PeBP1 (or Raf1 kinase inhibitor protein [RKIP]), caspase-8 and -10, and caspase-8 and FADD-like apoptosis regulator (CFLAR) [251].

6.6 Mitogen-Activated Protein Kinase Kinases

Activated MAP2K phosphorylates a specific MAPK (Table 6.16). Activated MAPK phosphorylates cytosolic and nuclear effectors. MAPK members have, indeed, many substrates: (1) other protein kinases belonging to either downstream components (MAPK-activating protein kinases MAPKAPK1 and MAPKAPK2, ribosomal S6-kinase (RSK1–RSK3) that phosphorylates glycogen synthase kinase GSK3) or upstream elements (cRaf and SOS); (2) enzymes (phospholipases PLA2 and PLCγ); (3) plasmalemmal proteins, such as growth factor receptors (EGFR); and (4) transcription regulators, such as the serum response factor, Fos, Jun, MyC, and other transcription factors (ATF2, ELk1, and MEF2), as well as members of the steroid/thyroid hormone receptors. Enzymes of the MAPK modules are inhibited by MAPK phosphatases (tyrosine phosphatases and serine/threonine phosphatases).

6.6.1 MAP2K1

Enzyme MAP2K1[73] can be activated by MAP3K1 to MAP3K4, MAP3K8, and cRaf enzymes. It also interacts with itself as well as MAPK1 (ERK2) and MAPK3 (ERK1), kinase suppressor of Ras KSR2, MAP2K-interacting protein-1 and -3, mitogen-activated protein-binding protein-interacting protein (MAPBPIP), adaptor GRB10, phosphatidyl ethanolamine-binding protein PBP (or Raf kinase inhibitory

---

[73] A.k.a. MEK1, MKK1, and protein kinase mitogen-activated kinase PrKMK1.

Table 6.16. Types of MAPK enzymes targeted by MAP2K proteins (MEK:mitogen–extracellular signal-regulated kinase; MKK: MAP kinase kinase). Activators of P38MAPKs comprise MAP2K3, MAP2K4, and MAP2K6, but MAP2K3 and MAP2K6 are the specific kinases of P38MAPK. MAP2K6 phosphorylates all P38MAPK isoforms (P38MAPKα–P38MAPKδ), whereas MAP2K3 preferentially activates P38MAPKα, P38MAPKγ, and P38MAPKδ (MEK: mitogen–extracellular signal-regulated kinase kinase; MKK: MAP kinase kinase; SEK: SAPK–ERK kinase SEK; SKK: SAPK kinase [SAPK: stress-activated protein kinase]).

MAP3K	Aliases	ERK1/2	ERK5	JNK	P38MAPK
MAP2K1	MEK1, MKK1	+			
MAP2K2	MEK2, MKK2	+			
MAP2K3	MEK3, MKK3, SKK2				+
MAP2K4	MEK4, MKK4, SKK1, JNKK1, SEK1			+	+
MAP2K5	MEK5, MKK5		+		
MAP2K6	MEK6, MKK6, SKK3				+
MAP2K7	MEK7, MKK7, SKK4, JNKK2			+	

protein) and PEBP4 (or cousin of RKIP1 protein CORK1), small GTPase hRas, guanine nucleotide-exchange factor Vav1, actin-binding protein filamin-α, and urokinase-type plasminogen activator (urokinase) [251].

6.6.2 MAP2K2

Enzyme MAP2K2[74] interacts with itself as well as kinases MAPK1 (ERK2), MAPK3 (ERK1), and aRaf, and filamin-α that crosslinks actin filaments in cortical networks, which anchor membrane proteins (e.g., integrins and transmembrane receptor complexes) to the actin cytoskeleton [251].

6.6.3 MAP2K3

Enzyme MAP2K3[75] is highly selective for P38MAPKs, but preferentially activates P38MAPKα (MAPK14), P38MAPKγ, and P38MAPKδ [587]. It is more heavily activated by physical and chemical stresses. Phosphorylation sites of MAP2K3 are Ser189 and Thr193 [267]. Enzymes MAP3K1, MAP3K3 to MAP3K5, MAP3K7, MAP3K11, as well as MAP3K16 to MAP3K18 (TAOK1–TAOK3) can activate MAP2K3. Stimulated MAP2K3 phosphorylates (activates) P38MAPK.

---

[74] A.k.a. MEK2 and MKK2.
[75] A.k.a. MEK3, MKK3, and SAPK kinase SKK2.

Enzyme MAP2K3 also interacts with MAP2K6, MAPK3 (ERK1), and MAPK-interacting protein MAPK8IP2, as well as phospholipase-Cβ2, transcription inhibitor SMAD7, and nephrin[76] of the slit diaphragm of fenestrated endothelium of renal filtration barrier [251]. Interactor MAPK8IP2 that binds to RacGEFs TIAM1 and RasGRF1 activates Rac GTPase that can then stimulate MAP3K11 and MAP2K3 and P38MAPK kinases [587].

Enzymes MAP2K3 and P38MAPK associate with hepatocyte growth factor receptor (HGFR) and its partner Src kinase that stimulates Rac1 GTPase [587]. In addition, phosphorylated MAP2K3 activates minibrain-related kinase (MiRK or dual-specificity Tyr phosphorylation-regulated kinase DYRK1b) that phosphorylates (activates) the transcriptional regulator hepatocyte nuclear factor HNF1α, hence promoting cell proliferation. Conversely, P38MAPKα and P38MAPKβ bind to and disrupt the DYRK1b–MAP2K3 complex (negative feedback).

Activation of MAP2K3 is related to the stress-mediated remodeling of cytoskeletal proteins [587]. Osmosensing scaffold binds to actin, Rac GTPase, and MAP3K3 and MAP2K3 enzymes. Stress-mediated activation of MAP3K3 triggers the expression of several cytokines, such as interleukins IL1α, IL1β, IL6, and IL12 [587]. Lipopolysaccharides trigger the formation of a complex between IRAK2, TRAF6, MAPKAPK2, and MAP2K3 and MAP2K6 enzymes.

Activated MAP2K3 contributes to apoptosis [587]. Inhibitor SMAD7 interacts with the MAP3K7 (TAK1)–MAP2K3–P38MAPK complex in TGFβ1-induced apoptosis. Conversely, phosphorylated MAP2K3 can associate with anti-apoptotic factors immunoglobulin-binding protein IgBP1 and PP2 phosphatase. The latter dephosphorylates MAP2K3 and can suppress apoptosis [587].

Activated MAP2K3 can provoke clathrin-mediated endocytosis [587]. Platelet-activating factor induces the endocytosis of its receptor PAFR via the activation of the MAP2K3–P38MAPK cascade. Activation of PAFR actually triggers the formation of the PAFR–β-arrestin-1–clathrin complex, the simultaneous association of MAP2K3 with β-arrestin-1, and activation of the MAP2K3–P38MAPK signalosome.

6.6.4 MAP2K4

Enzyme MAP2K4[77] with MAP2K7 are crucial transducers that prime JNK signaling. It phosphorylates specific Tyr and Thr residues in the activation loop

---

[76] A.k.a. renal glomerulus-specific cell adhesion receptor and congenital, Finnish type, nephrosis protein-1 (Nphs1). Nephrin is a transmembrane protein that serves as a structural component of the slit diaphragm of the renal filtration barrier that consists of fenestrated endothelial cells, the glomerular basement membrane, and podocytes of epithelial cells.

[77] A.k.a. JNK-activating kinase and Jun N-terminal kinase kinase JNKK1, MEK4, MKK4, protein kinase mitogen-activated kinase PRKMK4, SAPK/ERK kinase SEK1 or SERK1, stress-activated protein kinase kinase SAP2K, and SAPK kinase SKK1.

of JNK proteins,[78] thereby activating these kinases in response to environmental stress (e.g., UV and γ-irradiation, heat shock, hyperosmolarity), peroxide exposure, pro-inflammatory cytokines, T-cell receptor stimulation, or developmental cues. In particular, it acts in dorsoventral patterning, convergent extension, and somitogenesis [588]. It targets P38MAPKs and JNKs. Phosphorylation sites of MAP2K4 are Ser257 and Thr261 [267].[79] Enzymes MAP3K1 to MAP3K5, MAP3K7, MAP3K8, and MAP3K10 to MAP3K12 can activate MAP2K4 enzyme.

Enzyme MAP2K4 also interacts with MAP4K2, MAP2K7 (or JN2K2), MAPK8 (or JNK1), MAPK-interacting proteins MAPK8IP3 and MAPK8IP4,[80] PKB, arrestin-β2, actin-binding filamins FlnA and FlnC,[81] and nephrin [251]. Anti-apoptotic MAP2K4 protects from excessive TNFSF6-stimulated apoptosis, although this process eliminates autoreactive cells that can appear during early stage of T- and B-lymphocyte development [267].

6.6.5 MAP2K5

Enzyme MAP2K5[82] is a dual-specificity protein kinase. It is the single identified component of the MAPK set that targets MAPK6 (a.k.a. MAPK7, ERK4, and ERK5).

Some splice variants of MAP2K5 generated by reading frame shift can have different subcellular locations: the long isoform (MAP2K5$_L$) localizes to the cytoskeleton, whereas the short isoform (MAP2K5$_S$) remains in the cytosol [267]. A second splicing event (in- or exclusion of the 5' alternative splice site) yields MAP2K5α (448 amino acids), which is restricted mostly to the brain and liver, and ubiquitous MAP2K5β (359 amino acids) isoforms. Both MAP2K5α and MAP2K5β have 2 isoforms according to in- or exclusion the 3' alternative splicing site, i.e., without or with a deletion of 10 amino acids in the catalytic domain. Full-length MAP2K5 contains the PB1 domain[83] that enables interaction with its partners, such as MAP3K2, MAP3K3, and protein kinase-Cζ [589].

---

[78] Activated MAP2K4 stimulates JNK by dual phosphorylation of the Thr–Pro–Tyr motif of JNK activation loop. Although MAP2K4 is a dual-specificity kinase (targeting both Thr and Tyr residues), it preferentially phosphorylates Tyr residue. Enzymes MAP2K4 and MAP2K7 act synergistically; phosphorylation of Tyr residue by MAP2K4 followed by phosphorylation of Thr residue by MAP2K7 leads to an optimal JNK activation [588].

[79] Activation of MAP2K4 results from phosphorylation at Ser and Thr residues within a Ser–X–Ala–Lys–Thr motif of the activation loop [588].

[80] A.k.a. sperm-associated antigen SpAg9, JNK-interacting protein JIP4, and JNK/SAPK-associated protein JSAP2.

[81] A.k.a. filamin-α and -γ, respectively. They tether filamentous actin to the plasma membrane.

[82] A.k.a. MEK5.

[83] The PB1 domain can also be observed in the structure of sequestosome-1, PKCζ, MAP2K2, MAP2K3, and Partitioning defective-6 homolog Par6α. These molecules can associate with MEK5 using the PB1 domain.

Phosphorylation sites of MAP2K5$_L$ are Ser311 and Thr315. Enzymes MAP3K2, MAP3K3, and MAP3K8 can activate MAP2K5 [251, 267]. Enzyme MAP2K5 also interacts with itself and protein kinase-Cζ [251]. Muscle-specific A-kinase anchoring protein AKAP6 connects to both MAP2K5 and ERK5 enzymes, thereby inhibiting PDE4d3 phosphodiesterase. In addition, RapGEF3 can be recruited by the AKAP6–PDE4d3 inhibits ERK5 activation.

6.6.6 MAP2K6

Enzyme MAP2K6[84] is a ubiquitous dual-specificity protein kinase[85] highly selective for P38MAPK enzymes in a cell- and context-specific manner. It strongly activates all known P38MAPK isoforms. It is activated by all identified P38MAPK stimuli. In particular, upon DNA damage and cell stresses, MAP2K6 phosphorylates P38MAPKα and provokes its nuclear translocation. The ability to phosphorylate and activate P38MAPKs is shared with MAP2K3 and MAP2K4, but with a non-redundant and selective manner in response to specific stimuli [590].

The Map2k6 transcript generates alternatively spliced variants in humans.[86] Its phosphorylation sites are Ser207 and Thr211 [267]. Enzymes MAP3K1, MAP3K3 to MAP3K5 (ASK1), MAP3K7 (TAK1), MAP3K8, and MAP3K17 (TAOK2) can activate MAP2K6 enzyme.[87] Enzyme MAP2K6 abounds in the skeletal muscle, where it promotes myogenic differentiation using the P38MAPK pathway [590].

Protein MAP2K6 also interacts with MAP2K3, phospholipase-Cβ2, inhibitor SMAD7, and nephrin [251]. It also associates with microtubules and proteins from the transport machinery such as dynactin [590].

Protein Ser/Thr kinase C-related kinase PKN1 (PKNα, or PAK1) phosphorylates (activates) MAP3K15 that phosphorylates (activates) MAP2K6 enzyme. This fatty acid- and Rho-activated kinase may also act as a scaffold for MAP2K6–P38MAPKγ signaling [590].

---

[84] A.k.a. MEK6, MKK6, and SKK3 (SAPK kinase).

[85] Enzyme MAP2K6 targets a threonine and a tyrosine within a tripeptide motif (Thr–Gly–Tyr) of the activation loop of P38MAPKs. Ubiquitous MAP3K6 is mainly expressed in the heart, skeletal muscle, liver, and pancreas.

[86] The Map2k6 gene encodes a 278-amino acid protein (MAP2K6a). The longer isoform (MAP2K6b) encodes a 334-amino acid protein that abounds in the heart, skeletal muscle, liver, and pancreas.

[87] Enzyme MAP3K5, or apoptosis signal-regulating kinase-1 (ASK1) activates the MAP2K6–P38MAPK module in response to pro-inflammatory cytokines and stress to induce apoptosis [590]. Upon MAP2K6 activation, transforming growth factor-β-activated protein kinase-1 (TAK1, or MAP3K7) and its binding partner MAP3K7IP1 control P38MAPK location [590].

6.6.7 MAP2K7

Enzyme MAP2K7[88] is a crucial transducer upstream from JNK signaling, in addition to MAP2K4 enzyme. It has a larger affinity for JNK enzymes.[89] It phosphorylates specific Tyr and Thr residues in the activation loop of JNK proteins, thereby activating these kinases in response to environmental stress (e.g., UV and γ-irradiation, heat shock, hyperosmolarity), peroxide exposure, pro-inflammatory cytokines, T-cell receptor stimulation, or developmental cues. In particular, it acts in dorsoventral patterning, convergent extension, and somitogenesis [588].

Enzyme MAP2K7 is strongly activated by cytokines TNF and IL1.[90] Enzyme MAP3K1 can activate MAP2K7 isoforms (MAP2K7α1–MAP2K7α2, MAP2K7β1 –MAP2K7β2, and MAP2K7γ1).[91] Enzyme MAP3K3 can activate MAP2K7α1 and, to a much lesser extent, MAP2K7α2, MAP2K7β1, and MAP2K7β2. Enzymes MAP3K11 and MAP3K12 can stimulate MAP2K7 (strongly MAP2K7β1 and MAP2K7β2, moderately MAP2K7α isoforms). Enzyme MAP3K10 can activate MAP2K7α1.

Enzyme MAP2K7 also interacts with affectors MAP3K2, MAP3K5, MAP3K11, MAP3K12, MAP3K15, and MAP2K4 (or JN2K2); effectors MAPK8 (or JNK1) and MAPK14 (or P38MAPKα); MAPK-interacting proteins MAPK8IP1 to MAPK8IP3; dual-specificity phosphatase DuSP22; and growth arrest and DNA-damage-inducible GADD45β [251].

6.7 Mitogen-Activated Protein Kinases

The MAPK cascades include in their downstream tier several intracellular signaling networks. The 3 main MAPK cascades involved in signal transduction include (Table 6.17): (1) extracellular signal-regulated protein kinases, with its 2 isoforms ERK1 and ERK2; (2) Jun N-terminal kinases (JNK), with its 3 components JNK1

---

[88] A.k.a. Jun N-terminal kinase kinase JNKK2, MKK7, MAPK/ERK kinase MEK7, and SAPK kinase SKK4.

[89] Activated MAP2K7 stimulates JNK by dual phosphorylation of the Thr–Pro–Tyr motif of JNK activation loop. Although MAP2K7 is a dual-specificity kinases (targeting both Thr and Tyr residues), it preferentially phosphorylates Thr residue. Enzymes MAP2K4 and MAP2K7 act synergistically; phosphorylation of Tyr residue by MAP2K4 followed by phosphorylation of Thr residue by MAP2K7 leads to an optimal JNK activation [588].

[90] Phosphorylation sites of MAP2K7α are Ser206 and Thr210 [267]. Activation of MAP2K7 results from phosphorylation at Ser and Thr residues within the Ser–X–Ala–Lys–Thr motif of the activation loop [588].

[91] In mice, the Map2k7 transcript contains 14 exons that can be alternatively spliced to generate protein kinases with 3 different N-termini (α-, β-, and γ-isoforms) and 2 different C-termini (type-1 and type-2 isoforms) [588]. Isoform MAP2K7α that lacks the N-terminal extension has a lower basal activity on JNKs than the MAP2K7β and MAP2K7γ isoforms.

Table 6.17. Types of mitogen-activated protein kinases (ERK: extracellular-signal-regulated kinase; JNK: Jun N-terminal kinase; SAPK: stress-activated protein kinase).

MAPK type	Other aliases
MAPK1, MAPK2	ERK2
MAPK3	ERK1
MAPK4, MAPK5, MAPK12	ERK3, ERK6, P38MAPKγ
MAPK6, MAPK7	ERK4, ERK5
MAPK8	JNK1, SAPKγ
MAPK9	JNK2, SAPKα
MAPK10	JNK3, SAPKβ
MAPK11	P38MAPKβ
MAPK13	P38MAPKδ
MAPK14	P38MAPKα
MAPK15	ERK7, ERK8

Table 6.18. Transcription factors targeted by MAPKs (Source: [267]; NFAT: nuclear factor of activated T cells). Phosphorylation of Jun is exclusively catalyzed by JNKs.

MAPK	Stimulus	Transcription factors
ERK1/2	Mitogens	ELk1/4, Fos
ERK4	Mitogens	MEF2a/c, Jun
(ERK5)	Stress, cytokines	
JNK	Mitogens	Jun
	Stress, cytokines	ATF2
		ELk1/4, Fos
		NFAT2, NFAT4 (inhibition)
P38MAPK	Stress, cytokines	ATF2
		ELk1/4, Fos
		NFAT2, NFAT4
		MEF2a/c, Jun
		GADD153
		Max

to JNK3; and (3) P38MAPK isozymes. The 2 other MAPK families are ERK3 and ERK4. The former is mainly regulated by autophosphorylation. The latter is activated by growth factors; it is required for cardiovascular development.

Enzymes of the ERK, JNK, and P38MAPK subfamilies phosphorylate both transcription factors (Table 6.18) and other protein kinases. Whereas certain substrates, such as MAPK-activated protein kinases MAPKAPK2, MAPKAPK3, and MAPKAPK5 (or PRAK), are selectively recruited by stress-activated MAPKs, MAPK-interacting Ser/Thr kinases MNK1 and MNK2 as well as mitogen- and stress-activated protein kinases MSK1 and MSK2 are activated by both stress- and mitogen-regulated MAPKs and, like transcription factor AP1, integrate both stress and mitogenic signalings (Table 6.19) [267].

6.7 Mitogen-Activated Protein Kinases

Table 6.19. MAPK targets of the RSK subfamily (RSKi: ribosomal protein S6 kinase, 90-kDa polypeptide-i, $i = 1,\ldots,5$).

Type	Gene	Alias in the MAPK superfamily
RSK1	RPS6KA1	MAPKAPK1a
RSK2	RPS6KA3	MAPKAPK1b
RSK3	RPS6KA2	MAPKAPK1c
RSKb	RPS6KA4	MSK2
RLPK	RPS6KA5	MSK1

The MAPK enzymes target some transcription factors. All the MAPKs — ERKs (mainly in response to mitogens), ERK4 (primarily in response to stress), JNKs, and P38MAPKs — contribute to activity of AP1 activator protein by stimulating the transcription of genes that encode AP1 components.[92] In resting cells, Jun is phosphorylated by glycogen synthase kinase GSK3 or CK2 casein kinase. Once cells are stimulated, GSK3 is inactivated via the PI3K axis. Conversely, phosphorylated Jun or ATF2 is active. Members of the JUN subfamily can be activated by JNKs (Jun: Ser63 and Ser73; JunD: Ser90 and Ser100). On the other hand, members of the ATF category are phosphorylated by both JNKs and P38MAPKs (ATF2: Thr69 and Ser71). Promoter of the FOS gene possesses a serum response element (SRE) that links to a heterodimer made of serum response factor (SRF) and a member of the ternary complex factor (TCF) family (e.g., ELk1 and ELk4). Enzymes of the JNK and ERK subfamilies, but not P38MAPKs, phosphorylate ELk1 (Ser383 and Ser389), whereas P38MAPKs phosphorylate nuclear transcriptional stimulator SAP1a (Ser381 and Ser387), or ELk4, to enhance the SRF–TCF connection, thereby initiating SRE activation. Enzymes ERK4 and P38MAPKs also phosphorylate some transcription factors of the myocyte enhancer factor MEF2 group (MEF2a–MEF2d) that target the JUN promoter.[93]

Transcriptional factor DNA-damage-inducible transcript DDIT3[94] represses certain cAMP-regulated genes and activates some stress-induced genes. Nuclear factor of activated T cells, such as NFAT2 (NFATc1) and NFAT4 (NFATc3), that are phosphorylated by JNKs are sequestered in the cytosol. Dephosphorylation

---

[92] Transcription factor AP1 (of the superfamily-2B3A of nucleus-resident nuclear factors in the functional classification and family-1A1 of the subclass-1A bHLH leucine zipper transcription factors [bHLHzip] in the structural classification) is a homo- or heterodimer formed by a member of the JUN subfamily (e.g., Jun or JunD) and a member of the FOS subfamily (e.g., Fos or ATF2). Transcription factors ATFs are members of the cAMP-responsive element-binding protein (CREB) family. Factor AP1 targets Il1 and Il2, Cd30 and Cd40, Tnf, and JUN genes as well as genes of proteases and cell-adhesion molecules [267].

[93] Factor MEF2a is phosphorylated by P38MAPKα (Thr312 and Thr319) and MEF2c by P38MAPKs (Thr293 and Thr300) and by ERK5 (Ser387).

[94] A.k.a. growth arrest and DNA-damage protein GADD153 and CREB homologous protein (CHoP or CHoP10).

Table 6.20. MAPK Enzymes and their kinase substrates (Source: [267]).

Type	Substrates
ERK1	MAPKAPK1, MNKs, MSKs
ERK2	MAPKAPK1, MNKs, MSKs
P38MAPKα	MAPKAPK2/3, MSKs
P38MAPKβ	MAPKAPK2/3, MSKs

(activation) of NFAT by PP3 allows nuclear translocation and DNA binding. Enzyme MAP3K1 can prevent NFAT4 dephosphorylation [267]. The class-D basic helix–loop–helix peptide MYC associated factor-X (MAX)[95] that can complex with P38MAPKα interacts with Myc transcription factor that regulates cell proliferation, differentiation, and apoptosis.

The family of calcium–calmodulin-dependent protein kinases, which act downstream from mitogen-activated protein kinases, includes: (1) members of the ribosomal S6-kinase family with the mitogen- and stress-activated kinase infrafamily (MSK1 and MSK2; i.e., RLPK and RSKb) and MAPKAPK1 infrafamily (RSK1–RSK3); (2) MAPK-activated protein kinases (MAPKAPK or MK; i.e., MK2, MK3, and MK5); and (3) MAPK-interacting kinases (MNK1–MNK2) (Table 6.20). Enzymes of the RSK set are exclusively activated by extracellular signal-regulated kinases. Enzymes of the MSK and MNK categories are downstream effectors of both ERK and P38MAPK proteins. Enzymes of the RSK and MSK sets regulate gene expression, as they both phosphorylate Fos on the one hand, and cAMP-responsive element–binding proteins or histone-3, respectively, on the other. Stabilization of mRNA induced by P38MAPKs is mediated by MAPKAPK2 [267]. The JNK enzymes can also stabilize mRNA via MAP3K1 and MAP2K7 enzymes.

Enzymes of the MNK set phosphorylate eukaryotic translation-initiation factor eIF4e and factors that bind to certain mRNAs, to regulate the expression of a specific category of proteins. These kinases are particularly involved in growth control and inflammation. Kinase MK2, activated by P38MAPK, is required for cytokine synthesis in inflammation and the cell cycle. Enzyme MK3, also activated by P38MAPK, regulates chromatin remodeling. Protein MK5, which particularly interacts with ERK3, is activated by ERK and P38MAPK, but not JNK kinases.

6.7.1 JNKs and P38MAPKs in Cardiovascular Diseases

Enzymes of the JNK and P38MAPK subfamilies, via MAP3K6 and MAP3K7, contribute to maladaptive cardiomyocyte hypertrophy in response to pressure

---

[95] A.k.a. bHLHd4 to bHLHd8.

overload. In addition, vascular wall hypertension elicits the release of the vasoactive peptides angiotensin-2 and endothelin-1 that also cause cardiomyocyte hypertrophy (Vol. 5 – Chap. 6. Heart Wall). Enzymes of the JNK and P38MAPK subfamilies participates in ischemia–reperfusion injury (Vol. 6 – Chap. 6. Heart Pathologies), in particular via MAP3K3 that preferentially recruits P38MAPKα and P38MAPKβ. Enzyme MAP3K3 also regulates the development of the cardiovascular apparatus via the P38MAPK pathway and MEF2c transcription factor.

6.7.2 Family of Extracellular Signal-Regulated Kinases

6.7.2.1 ERK Signaling Duration and Spatial Organization

ERK Signaling Duration and Cell Fate

Signaling duration contributes to the control of cell responses to stimulated signaling pathways that share signaling modules. Transient activation with negative feedback of extracellular signal-regulated kinases by epidermal (EGF), keratinocyte (KGF1, or FGF7, and KGF2, or FGF10), and insulin-like (IGF1) growth factors can cause cell proliferation [591]. On the other hand, sustained ERK activation with positive feedback by nerve growth factor and heregulin can provoke cell differentiation, whereas that by epidermal and hepatocyte growth factor can induce cell migration.

ERK Subcellular Localization and Scaffolding

A scaffold can stabilize preformed assemblies of signaling effectors as well as regulate the specificity and magnitude of signaling. Stimulation by epidermal- and nerve growth factor releases ERK from preformed complexes. In addition, nerve growth factor provokes ERK nuclear translocation. In the nucleus, ERK phosphorylates transcription factors and primes the export of transcriptional inhibitors. In the cytosol, ERK controls compartmentation of signaling mediators. Nerve growth factor induces the sustained dissociation of RasGAP neurofibromin NF1 from a Ras–ERK complex, thereby avoiding rapid Ras and subsequent ERK inactivation. Furthermore, NGF causes a sustained liberation of ERK from 15-kDa phosphoprotein enriched in astrocytes (PEA15) that sequesters ERK in the cytosol. Signaling based on ERK localizes to different subcellular compartments according to distribution of ERK partners that also serves as substrates.

Fully vs. Partially Scaffolded Signaling Complexes

Protein MAPK organizer MOrg1[96] enables the assembly of various signaling complexes with distinct outcomes. Scaffold MOrg1 can actually interact with Raf, MAP2K, and ERK (fully scaffolded complexes) as well as MAPK scaffold protein MAPKSP1[97] that recruits both MAP2K and ERK (partly scaffolded complexes) [591].

Scaffold Types and Subcellular Localization

Different scaffold proteins are used in different intracellular sites to target ERK substrates. Rapid and transient ERK activation and nuclear translocation is triggered from the plasma membrane by GPCRs, protein kinase-C, and Src kinase, as well as receptor Tyr kinases such as EGFR. On the other hand, sustained ERK signaling that is restricted to the cytosol is primed from internalized GPCRs in endosomes by β-arrestin and Raf kinase. β-Arrestin associates with entire MAPK modules, such as Raf–MAP2K–ERK and MAP3K5–MAP2K4–JNK3 modules. Discrimination among MAPK modules can be supported by cell locus (e.g., distinct Ras nanoclusters at the plasma membrane), context, and affector and partner types, such as kinase suppressor of Ras that serves as a scaffold for the Raf–MAP2K–ERK module.[98] Therefore, ERK signaling triggered by GPCRs evolves into 2 phases: (1) an early signaling from the plasma membrane with ERK nuclear translocation followed by (2) a signal emission during endocytosis with ERK signaling confined to the cytosol. Signal integration from dually phosphorylated ERK results from combination of nuclear and cytosolic ERK^{PP} concentrations.

The specificity of ERK substrate depends on the location of upstream mediators. Scaffolds participate in the determination of ERK substrates in the cytosol or nucleus according to Ras activation foci. At the plasma membrane, kinase suppressor of Ras facilitates the phosphorylation of cytoplasmic phospholipase-A2 by ERK [591]. On the cytoskeleton, the negative ERK feedback that is mediated by scaffold IQ motif-containing GTPase-activating protein IQGAP1 phosphorylates epidermal growth factor receptor at the plasma membrane. On the Golgi body membrane, interleukin-17 receptor-D also promotes phosphorylation by ERK of cytoplasmic PLA2 that generates arachidonic acid, a precursor to leukotrienes and prostaglandins.

---

[96] A.k.a. WD repeat domain-containing protein WDR83.

[97] A.k.a. MAP2K1IP1, MP1, and MAP2K-binding partner-1.

[98] Scaffolds can act as allosteric regulators of their target kinases. Kinase suppressor of Ras can activate bRaf kinase.

Spatiotemporal Dynamics of ERK Signaling

Spatiotemporal network dynamics govern cell fate decisions. Computational models exhibit distinct modes of ERK spatiotemporal dynamics according to feedback type from active, dually phosphorylated ERK to Raf kinase or its regulators, such as Raf kinase inhibitor protein (RKIP) and Ras-activating guanine nucleotide-exchange factor Son of sevenless, and kinetic parameters [591]: (1) monotone, sustained response; (2) transient, adaptive output; (3) damped and (4) sustained oscillations; and (5) bistable outcome, in which 2 switch-like, stable steady states coexist. Different ERK temporal responses can also be observed experimentally.

Signaling Termination and Phosphatases

Signaling from ERK kinases causes negative feedback using Ser/Thr, Tyr, and dual-specificity phosphatases (Sect. 8.3.13.3). The latter dephosphorylate ERK1, ERK2, JNKs, and P38MAPKs. They are encoded by *immediate-early genes*, i.e., genes that are stimulated rapidly and do not require new protein synthesis for their transcription. Nevertheless, the time scale for DUSP-mediated feedback is longer than that of negative feedback caused by already synthesized inhibitors. Dual-specificity phosphatase isoforms localize to the cytoplasm and nucleus, hence regulating both cytoplasmic and nuclear ERK pools.

6.7.2.2 ERK1 and ERK2 Infrafamily

Extracellular signal-regulated kinases ERK1 and ERK2 are expressed in many tissues. They are activated by growth factors to regulate cell adhesion, growth, proliferation, differentiation, migration (during both chemotaxis and haptotaxis), and apoptosis. In particular, activated ERK1 and ERK2 regulates membrane protrusions and focal adhesion turnover by phosphorylating (activating) myosin light-chain kinase and calpain. They also regulate focal adhesion dynamics by influencing the interaction between paxillin and focal adhesion kinase. Unlike adhesion signaling, growth factors activate both MAP2K1 and -2.

The corresponding multipotent pathway achieves response specificity especially owing to interactions with scaffold proteins. Scaffolds include kinase repressor of Ras, β-arrestin, IL17Rd, and IQGAP1 proteins. Molecule KSR produces a docking platform at the plasma membrane. Agent IQGAP1 links the MAP2K–ERK pathway to the cytoskeleton.

In unstimulated cells, scaffold protein KSR1 prevents improper ERK cascade activation by sequestering MAP2K away from Raf kinase. In response to growth factors, KSR1 that localizes to the plasma membrane contributes to the spatiotemporal control of the Raf–MAP2K–ERK signaling. Both MAP2K1 and MAP2K2 kinases possess a KSR-binding motif that allows the formation of the bRaf–MAP2K–KSR ternary complex [592]. Docking of active ERK to KSR1 scaffold allows ERK to phosphorylate KSR1 and bRaf, hence ensuring a feedback control. Phosphorylation of KSR1 and bRaf actually promotes their dissociation and causes KSR1 release from the plasma membrane. Therefore, KSR1 first potentiates signal transmission, as it colocalizes cascade components to facilitate Raf–MAP2K interaction, thereby increasing efficiency and specificity of signaling, and then attenuates ERK activation, thereby controlling intensity and duration of ERK signaling.

The Ras–Raf–MAP2K–ERK signaling cascade triggered by growth factors and intercellular contacts controlled by IQGAP family proteins controls cell proliferation. Protein IQGAP3 is specifically expressed by proliferating cells to cooperate with Ras GTPase in the cell cycle [593].

Phosphoprotein-enriched in astrocytes PEA15 binds ERK and RSK2 for sequestration in the cytoplasm. Protein RSK2 is an ERK substrate kinase that can activate CREB-mediated transcription. Protein PEA15 independently binds ERK and RSK2 and enhances ERK binding to and phosphorylation (activation) of RSK2 in a concentration-dependent manner [594].

Endosome location of the MAP2K1–ERK pathway is required for full activation of ERK. Adaptor LAMTOR1[99] anchored to membrane rafts of late endosome connects late endosomes to the MAP2K1–ERK pathway via the LAMTOR2–MP1 complex that serves as a scaffold for MAP2K1 [595].[100] LAMTOR1 is required for vesicle dynamics that comprise Rab11-mediated endosome recycling and lysosome processing, thus participating in the regulation of vesicular interactions and material transport along the cytoskeleton.

Enzymes of the MAP3K set in the ERK1 and ERK2 cascades[101] are usually members of the RAF family (aRaf–cRaf) that bind to Ras GTPase to be activated. Upstream activators bind to receptor Tyr kinases or G-protein-coupled receptors. Activation of the ERK pathway can result from stimulation of plasmalemmal

---

[99]LAMTOR stands for late endosomal and lysosomal adaptor, MAPK and TOR activator. This membrane raft adaptor protein of 18 kDa (alias P18) is also called CKIIb (p27, or Kip1)-releasing factor from RhoA (P27RFRho, or PdRo) and Ragulator. Alias P18 should be avoided as it can also designate in Homo sapiens [76]: cofilin-1, ubiquitin conjugase-E2I, cyclin-dependent kinase inhibitor-2C, zinc finger protein-197, chromosome-2 open reading frame-28, tubulin polymerization-promoting protein family member-2, v-maf musculoaponeurotic fibrosarcoma proto-oncogene product homolog, and eukaryotic translation elongation factor-1ϵ1.

[100]Adaptor LAMTOR2 is also named endosomal adaptor protein EndAP, P14, late endosomal and lysosomal MAPK scaffold protein MP1-interacting protein MAPKSP1AP, mitogen-activated protein-binding protein-interacting protein (MAPBPIP), P14, Ragulator-2, Roadblock domain-containing protein RoblD3. Protein MP1 is a MAPK scaffold in the Raf–MAP2K–ERK pathway used for the spatial organization of LAMTOR2–MP1–MAP2K1 signaling. It heterodimerizes with LAMTOR2 to recruit MAP2K to late endosomes [596]. Adaptor LAMTOR1 is a binding partner of the MP1–LAMTOR2 dimer. Protein LAMTOR1 is attached to the endosomal surface and tethers the signaling complex to the cytoplasmic surface of late endosomes. Like P18, alias P14 should be avoided as it also refers in Homo sapiens to [76]: cyclin-dependent kinase inhibitor-2A, S100 calcium-binding protein-A9, 14-kDa replication protein-A3, catenin-β-like protein-1, CDK2-associated protein-2, SUB1 homolog, and 14-kDa subunit of splicing factor-3B.

[101]Enzymes ERK1 and ERK2, final mediators of the MAPK module, are activated by MAP2K1 and MAP2K2 enzymes.

proteins Ras via receptor Tyr kinases in association with GRB2 adaptor and Son of sevenless, a guanine nucleotide-exchange factor. Activation of Ras triggers Raf recruitment to the plasma membrane and stimulation, phosphorylation (activation) of MAP2K1 and/or MAP2K2, and then activation of ERK1 and ERK2 kinases. Enzymes ERK1 and ERK2 have many targets, such as transcription factors (NFκB, Jun, and ELk1), kinases (MAPKAPK1 and MAPKAPK2), plasmalemmal receptors (EGFR), upstream MAPK tier activators (cRaf), SOS affector, and paxillin.

Recruitment of scaffolds and adaptors, phosphorylation, and dephosphorylation yields temporospatial regulation of the ERK pathway. Upon activation by growth factors and other stimulators such as urokinase plasminogen activator of the Ras–Raf–MAP2K1/2–ERK1/2 cascade, Enzymes MAP2K1 and MAP2K2 are phosphorylated by kinase Raf (Ser218 and Ser222 of MAP2K1 and Ser222 and Ser226 of MAP2K2) and phosphorylate substrates ERK1 and ERK2 (on specific tyrosine and threonine residues). The Ras–Raf–MAP2K1/2–ERK1/2 cascade initiates not only a rapid signal amplification, but also a negative feedback loop via MAP2K1/2 heterodimer formation and ERK-mediated phosphorylation of MAP2K1 [597]. Agent MAP2K1 restricts Raf-induced MAP2K2 phosphorylation. Enzyme ERK can also exert negative feedback on several upstream components of the pathway, such as SOS, PAK, cRaf, and bRaf.

Homo- and heterodimerization are common events in ERK signaling. In particular, bRaf and cRaf can form heterodimers with stronger kinase activity than that of monomers or homodimers [598]. Heterodimerization is enhanced by proteins 14-3-3 and by mitogens independently of ERK. Phosphorylation by ERK of bRaf promotes Raf heterodimer disassembly.

Nearly half of ERK1 and ERK2 are bound to microtubules with a role in polymerization dynamics. Kinases ERK1 and ERK2 also localize to adherens junctions (intercellular contact) and focal adhesions (cell–matrix contact), thereby operating in cell motility. Enzymes ERK1 and ERK2 also reside in the nucleus. In the cytoplasm, they can phosphorylate certain transcription factors prior to nuclear entry. Proteins that alter the subcellular localization of ERK1 and ERK2 can generate diseases [599].

Protein kinase-C can activate cRaf, and hence plays a role at the MAP4K stage. In endothelial cells, activation of ERK1 and ERK2 that depends on shear stress[102] relies on PKCε [600].[103] Affector cRaf, an ERK1/2 activator, is recruited to the plasma membrane, and activated by Ras, possibly phosphorylated by CSK on the one hand and by diacylglycerol-regulated PKCα and PKCε on the other hand [601].

---

[102]Wall shear stress applied on endothelial cells activates extracellular signal-regulated kinases ERK1 and ERK2 in endothelial cells in a time- and force-dependent manner.
[103]PKCε seems to be specific for shear-dependent ERK1/2 activation, as PKCα and PKCζ do not influence ERK1/2 activity.

Protein kinases contains a site aimed at selectively binding nucleotides and inhibitors. Inhibitors prevent autophosphorylation. A gatekeeper residue in ERK2 impedes auto-activation [603].[104]

Urocortin (Ucn), a member of the corticotropin-releasing factor family and its homologs urocortin-3, or stresscopin (Scp), and urocortin-2, or stresscopin-related peptide (SRP), are potent cardioprotectors in cells exposed to hypoxia–reoxygenation events. Whereas urocortin binds to both types of CRF G-protein-coupled receptors, CRFR1 and CRFR2, urocortin-2 and -3 tether exclusively and with high affinity to CRFR2. However, all 3 peptides activate the ERK1/2 and PKB pathways [604].

The signaling adaptor sequestosome-1 regulates osteoclastogenesis via ubiquitin ligase TRAF6 and activation of nuclear factor-κB. It also promotes Ras-induced cell transformation. Moreover, sequestosome-1 impedes adipogenesis and insulin resistance, without alterations in feeding or locomotion, but favors energy combustion [605]. Sequestosome-1 sequesters extracellular signal-regulated kinase ERK1 into cytoplasm in adipose tissue. Its synthesis is induced during adipocyte differentiation (negative feedback) to prevent excessive adipogenesis. Ribosomal S6 kinase is an ERK substrate in adipogenesis.

6.7.2.3 ERK3 Subfamily: ERK3 (or ERK6) and ERK4 (or ERK5)

Extracellular signal-regulated kinase-3 (ERK3)[105] and -4 (ERK4)[106] define a distinct subfamily of MAPKs, the ERK3 subfamily, that exists exclusively in vertebrates [606].

ERK3(6)

Extracellular signal-regulated kinase-3 interacts with syntrophin-$\alpha 1$,[107] a cytoskeletal protein of the cortex of myocytes, which itself connects to dystrophin and dystrophin-related proteins as well as subunits $Na_V 1.1$ and $Na_V 1.5$ of voltage-gated sodium channels [251].

---

[104] ERK2, activated by phosphorylations on 2 residues due to MAP2K1 and MAP2K2, changes its conformation and phosphorylates cytoplasmic and nuclear proteins.

[105] A.k.a. ERK6, MAPK4, MAPK5, MAPK12, p97MAPK, P38MAPKγ, protein kinase mitogen-activated PrKM6 and -12, and stress-activated protein kinase SAPK3.

[106] A.k.a. ERK3-related kinase, ERK3β, ERK5, MAPK6, MAPK7, p63MAPK, big mitogen-activated protein kinase BMK1, and PrKM4 and PrKM7.

[107] Protein Sntα1 is also termed dystrophin-associated protein-A1.

6.7 Mitogen-Activated Protein Kinases

Table 6.21. Affectors and effectors of ERK5 compared to those of ERK1 and ERK2 (Source: [607]; AKAP6: mAKAP [muscle-selective A-kinase anchoring protein]; PEA15: 15-kDa phosphoprotein enriched in astrocytes; SGK: serum- and glucocorticoid-regulated kinase). The MAP2K5–ERK5 axis regulates cell proliferation, apoptosis, and inflammation.

ERK5 (ERK4)	ERK1 and ERK2
Stimuli	
Nerve growth factor	Nerve growth factor
Oxidative stress	Platelet-derived growth factor
Hyperosmolarity	
Transcription factors	
ELk1	Myocyte enhancer factor MEF2
MyC	MyC
Fos, Fra1, Jun	Fos, Jun
PPARγ1	
Kinases	
Ribosomal S6 kinase	Ribosomal S6 kinase
SGK	
Miscellaneous	
SAP1a	SAP1a
Cyclin-D1	Cyclin-D1
AKAP6	PEA15
	Sef
Effects	
Cardiovascular development	Embryogenesis (MAP2K1–ERK2)
Neural differentiation	Mesoderm formation
Nucleocytoplasmic transport	
Nuclear localization signal	Passive diffusion (monomer)
Nuclear export signal	Active transport (dimer)
	Interaction with nuclear-pore complex

ERK4(5)

Extracellular signal-regulated kinase ERK4, or ERK5, like ERK1 and ERK2, is activated by growth factors. Hence, it participates in the regulation of cell proliferation, differentiation, and migration. especially during the development of the cardiovascular system and neural differentiation (Table 6.21). It is also stimulated by environmental stresses, such as oxidative stress (peroxide) and hyperosmolarity (sorbitol). In adults, ERK4 is mostly expressed in heart, brain, and lung [606].

Enzyme ERK4 is required in endothelial cells, as its absence in endothelial cells, but not cardiomyocytes, causes cardiovascular defects. The N-terminus of ERK4 contains the kinase domain that is similar to that of ERK1 and ERK2 kinases.

The C-terminus of ERK4 enables its coactivator-like action aimed at increasing the transcriptional activity of some of its targets. Protein ERK4 actually is a kinase that has transcriptional activity due to its long C-terminus, independently of its kinase activity. Therefore, ERK4 can transmit signals from MAP2K5 to its effectors by phosphorylation or transactivating specific transcription factors, such as peroxisome proliferator-activated receptors PPARγ and PPARδ and myocyte enhancer factor MEF2 (MEF2a, MEF2c, and MEF2d, but not MEF2b) [589]. The latter activity is precluded by reactive oxygen species and advanced glycation end product (AGE)-dependent sumoylation.[108]

Enzymes MAP3K2 and MAP3K3 activate MAP2K5, a specific kinase for ERK4 kinase. The ERK4 signaling leads to ERK4 translocation into the cell nucleus. Activated ERK4, like ERK1 and ERK2, stimulates Fos and Jun factors. It promotes the expression of cyclin-D1, which regulates the G1–S transition of the cell cycle, implicating a cAMP response element [607].

Kinase ERK4 phosphorylates (activates) serum response element via ETS domain-containing transcriptional factor ELk4,[109] but not via the related ELk1 transcription factor [589]. It phosphorylates Fos at its nuclear export signal (Thr232), hence preventing Fos egress, and binding motif (Ser32), thereby precluding the interaction with ubiquitin ligase component N-recognin-1 (UbR1).[110] Kinase ERK4 can support transactivation of Fra1, a member of the FOS subfamily (with Fos, FosB, and Fra2). In endothelial cells, ERK5 operates via Krüppel-like factor KLF2 to counteract endothelial cell activation that occurs in response to pro-inflammatory stimuli [589].

Enzyme ERK4 also interacts with serum- and glucocorticoid-regulated kinase SGK1, the ERK1/2 substrate ^{P90}RSK, protein Tyr phosphatase receptor PTPRr, cRaf, connexin-43,[111] and 14-3-3β and 14-3-3ε [251].

---

[108] Advanced glycation end products are sugar-derived substances formed at a constant, slow rate in normal conditions. They result from a chain of chemical reactions after an initial glycation or non-enzymatic glycosylation of reducing sugars with free amino groups of proteins, lipids, and nucleic acids. The set of subsequent reactions consists of successions of dehydrations, oxidation–reduction reactions, and other arrangements that lead to the formation of advanced glycation end products. Initial reactions are reversible at low concentrations of reactants. A reduced glucose concentration unhook glucids from amino groups to which they are attached. Conversely, an elevated, persistent glucose concentration favors the series of reactions. Formation of AGEs, such as $^{\varepsilon N}$carboxymethyl-lysine, pentosidine, and methylglyoxal derivatives, markedly rises in diabetes because of the increased glucose availability.

[109] A.k.a. serum response factor (SRF) accessory protein-1a (SAP1a).

[110] Factor Fos is predominantly activated by ERK1 and ERK2 enzymes.

[111] Activated ERK5 enables uncoupling of connexin-43 from gap junction upon stimulation by epidermal growth factor [589].

Table 6.22. Scaffold proteins, especially JNK-interacting proteins (JIP), or MAPK8-interacting proteins (MAPK8IP), confer specificity to MAP2K activity (Source: [588]; MAX: MyC-associated factor-X). Scaffold proteins JIP1, JIP2, and JIP3 can form homo- and hetero-oligomers. They can thus connect 2 distinct sets of signaling modules, one containing MAP2K4, and the other incorporating MAP2K7 protein. In addition, MAP3K enzymes can act as scaffolds in addition to their intrinsic kinase activities.

Type	Other alias	Interactors
JIP1	MAPK8IP1	Various MLKs (e.g., MAP3K9–MAP3K11), MAP2K7, JNK
JIP2	MAPK8IP2	Various MLKs (e.g., MAP3K9–MAP3K11), MAP2K7, JNK
JIP3	MAPK8IP3	MAP3K1, various MLKs, MAP2K4/7, JNK
JIP4	MAPK8IP4	MyC–JIP4–MAX, JNK–JIP4–MAP3K3, JNK–JIP4–MAP2K4, JNK–JIP4–P38MAPK complex

6.7.3 Family of Jun N-Terminal Kinases

The family of Jun N-terminal kinase (JNK), or stress-activated protein kinase (SAPK) regulates stress responses, neural development, inflammation, and apoptosis. In mammals, the JNK family comprises 3 proteins encoded by 3 distinct genes (Jnk1–Jnk3), transcripts of which generate 10 splicing variants. Two isoforms JNK1 and JNK2 are ubiquitous, whereas tissue-specific isoform JNK3 lodges in the brain, heart, and testis.

The MAP3K enzymes that target the JNK pathway include mixed lineage kinases (MLK1–MLK3 and DLK, i.e., MAP3K9–MAP3K12), MAP3K1, and MAP3K5. Many involved MAP3Ks are activated by Rho GTPases. These MAP3Ks activate MAP2K4 and MAP2K7 enzymes.

The specificity of JNK signaling results, partly, by the formation of distinct JNK signaling complexes with JNK and JNK interactors, such as MAP2K4 and MAP2K7 enzymes and various scaffold proteins. The latter can assemble functional multienzyme modules that encompass a MAP3K, a MAP2K, and a MAPK. These scaffold proteins bind specifically to different JNK isoforms and different MAPKs and MAP2Ks for signaling insulation (Vol. 3 – Chap. 1. Signal Transduction) and proper spatiotemporal sites.

The JNK enzymes are regulated by multiple scaffolds, JNK-interacting proteins (JIP1–JIP4; Table 6.22), or MAPK8-interacting proteins (MAPK8IP1–MAPK8IP4), as well as β-arrestin-2, filamin,[112] and CRK2, for kinase activation and/or substrate selection. Other scaffold proteins involved in JNK signaling modules comprise SH3 domain-containing ring finger protein SH3RF1.[113]

---

[112] Filamin-A interacts with MAP2K4 and MAP2K7 enzymes. Filamin-A binds to MAP2K7β and MAP2K7γ splice variants, but not to MAP2K7α that lacks the adequate binding site.

[113] A.k.a. plenty of Src homology-3 (SH3) protein (POSH), RING finger protein RNF142, and SH3 multiple domain protein SH3MD2.

The JNK enzymes phosphorylate numerous transcription factors, such as Jun, ATF2, ELk1, P53, and Myc, as well as other proteins, such as BCL2, BCLxL, paxillin and microtubule-associated proteins (e.g., MAP1b and MAP2). Diverse splice variants of the 2 ubiquitous JNK1 and JNK2 isoforms and tissue-specific JNK3 isoform display different selectivity and affinity for target transcription factors. Subtypes of JNK2 have higher affinity for Jun than JNK1 and JNK3 isozymes. Subtype JNK2α1 binds more strongly to Jun than JNK2β2 kinase. The JNK2α isoforms bind more tightly to Jun than ATF2, whereas JNK2βs bind more strongly to ATF2 than Jun [267].

Genotoxic stresses promote apoptosis, as they recruit JNKs and activate NFκB that, in turn, stimulates AP1 transcription factors. The JNK enzymes are activated in response to extracellular stimuli, such as pro-inflammatory cytokines TNF-α and IL1β.

6.7.3.1 JNK1 or MAPK8

Jun N-terminal kinase JNK1[114] has various splice variants [267]: JNK1α1,[115] JNK1α2,[116] JNK1β1[117] and JNK1β2.[118]

This major Jun N-terminal kinase is activated by most known JNK inducers and is responsible for most known JNK effects. It contributes to cell differentiation and proliferation. It also intervenes in TNFα-regulated activities, such as cell death and inflammation.[119]

Enzyme JNK1 activates Ub-ligase Itch of the HECT (NEDD4) family[120] that specifically ubiquitinates the caspase-8 inhibitor Casp8 and FADD-like apoptosis regulator (CFLAR)[121] to trigger its proteasomal degradation [608].

The pro-inflammatory cytokine tumor-necrosis factor-α exerts its pleiotropic activities via multiple effectors such as JNK1 kinase. Transcription factor Myc-interacting zinc finger protein-1 (MIZ1)[122] that operates in transcriptional activation

---

[114] A.k.a. JNK1a2, JNK21b1/2, and MAPK8, P46MAPK, PrKM8, and stress-activated protein kinase SAPKγ, SAPK1c, and SAPK1.

[115] A.k.a. SAPKp46γ2.

[116] A.k.a. SAPKp54γ2.

[117] A.k.a. SAPKp46γ1.

[118] A.k.a. SAPKp54γ1.

[119] Biological outcome of TNFα is determined by the balance between nuclear factor NFκB and Jun kinase signaling, as NFκB and JNK promote cell survival and death, respectively.

[120] A.k.a. atrophin-1 interacting protein AIP4.

[121] A.k.a. cellular FLICE-like inhibitory protein cFLIP, cFLIP$_L$, and cFLIP$_R$, as well as inhibitor of FLICE (IFLICE), usurpin, casper, Casp8AP1, caspase homolog (CASH), caspase-like apoptosis regulatory protein (CLARP), FADD-like anti-apoptotic molecule FLAMe1, and MACH-related inducer of toxicity (MRIT).

[122] A.k.a. zinc finger and BTB domain-containing protein ZBTB17, zinc finger protein ZNF151, ZNF60, and pHZ67. Alias MIZ1 is also used to designate protein inhibitor of activated STAT PIAS2 or PIASX.

in the nucleus in response to cytoskeletal changes and impedes cell proliferation (transcriptional activator and repressor) also acts in the cytoplasm as a pathway-specific regulator, as it selectively suppresses TNFα-induced JNK1 activation and cell death [609]. Protein MIZ1 specifically regulates TNFα-induced TRAF2 Lys63-linked polyubiquitination. It, indeed, affects neither JNK1 activation by IL1β, nor TNFα-mediated activation of P38MAPK, ERK, and IκB kinase. Upon TNFα stimulation, MIZ1 is degraded rapidly by the proteasome to relieve its inhibition on JNK1 kinase.

Antigen-presenting cells, such as macrophages and dendritic cells, present antigen to naive T lymphocytes via T-cell receptors, CD3 subunit of which, in conjunction with CD28, initiates the TCR–CD28 costimulatory pathway. The latter promotes: (1) proliferation and maturation of CD4+CD8+ cells to CD4+ T_H lymphocytes that then become interferon-synthesizing T_{H1} and cytokine-producing T_{H2} effectors due to IL12 and IL4 as well as (2) production of some cytokines such as IL2. Lymphocytes T_{H1} produce interferon-γ and promote Ig switching in B lymphocytes, hence stimulating phagocytosis. On the other hand, T_{H2} cells produce IL4 and IL5 that induce switching to IgE and IgG4 in B lymphocytes that both bind to Fc receptor, either on mastocytes and basophils (IgE) or phagocytic cells (IgG4), and activate eosinophils, respectively. The TCR–CD28 costimulation generates a strong JNK1–AP1 activation [267]. Transcription factor AP1 is required for T-cell maturation and development of acquired immunity. However, JNK1 suppresses the differentiation of T_H lymphocytes into T_{H2} cells, as it impedes NFAT1 activation and TCR-stimulated IL4 production, hence favoring T_{H1} cell differentiation.

6.7.3.2 JNK2 or MAPK9

The Jun N-terminal kinase JNK2[123] has diverse splice variants [267]: JNK2α1,[124] JNK2α2,[125] JNK2β1,[126] and JNK2β2.[127] An additional isoform JNK2γ has been detected in human B lymphocytes.

Dimerization of JNK2 can support JNK2 activation. Nitrosylation and ubiquitination modify JNK2 activity. Extensive S-nitrosylation inhibits JNK2 phosphorylation [610].

Like all JNKs, JNK2 promotes cell death in various cell types. However, under some circumstances, JNK2 acts as an anti-apoptotic agent. Unlike ERK1 and ERK2, JNK2 is not strongly activated by mitogens, but vigorously by: (1) environmental stresses (heat shock, ionizing radiation, oxidative stress, DNA-damaging chemicals,

---

[123] A.k.a. JNK55, MAPK9, p54MAPK, SAPKα, SAPK1a, p54aSAPK, and PrKM9.
[124] A.k.a. SAPKp46α2.
[125] A.k.a. SAPKp54α2.
[126] A.k.a. SAPKp46α1.
[127] A.k.a. SAPKp54α1.

reperfusion injury, shear stress, and protein synthesis inhibitors); (2) inflammatory cytokines of the TNF superfamily, such as TNF, interleukin-1, TNFSF5, TNFSF6, TNFSF7, TNFSF11a, etc.; and (3) vasoactive peptides, such as endothelin and angiotensin-2 [267]. According to the cell type and context, JNK2 is either preferentially activated by MAP2K4 or MAP2K7 [610].

Among JNK2 kinases, JNK2α1 and JNK2α2 have a very high affinity for Jun, whereas JNK2β1 and JNK2β2 link preferentially to Activating transcription factor ATF2. In addition, JNK2 phosphorylates P53 factor. It also activates nuclear factor of activated T-cells NFAT3 [610].

Enzyme JNK2 is involved in T-cell activation, as it enhances TCR–CD28 costimulation of T-cell proliferation and TCR-stimulated production of IL2, IL4, and Ifnγ by activated T lymphocytes [267]. Enzyme JNK2 also contributes to B-cell activation. It is also involved in the control of the helper T_{H1} and T_{H2} cells [610].

6.7.3.3 JNK3 or MAPK10

The Jun N-terminal kinase JNK3[128] has several splice variants [267]: JNK3α1,[129] JNK3α2,[130] JNK3β1,[131] and JNK3β2.[132] It is predominantly expressed in the central nervous system. It is found, to a lesser extent, in the heart and testis. According to the context, JNK3 acts as a pro- or anti-apoptotic regulator; in particular, it preserves the pancreatic β-cell pool [611]. Unphosphorylated (inactive) JNK3 lodges mostly in the cytoplasm.

Protein JNK3 is phosphorylated (activated) by 2 dual-specificity kinases — MAP2K4 and MAP2K7 — (Thr221 and Tyr223). These activating kinases are themselves activated by MAP3K1 to MAP3K4 (MEKK family), MAP3K5 and MAP3K6 (ASK group), MAP3K7, MAP3K8, and MAP3K9 to MAP3K13 (MLK family). It is dephosphorylated (inactivated) by dual-specificity phosphatase MKP7 (or DuSP16) [611].

Scaffold proteins can facilitate the phosphorylation and dephosphorylation of JNK3 by linking to JNK3 and appropriate kinase(s) or phosphatase(s). These scaffolds include MAPK8-interacting proteins (MAPK8IP1–MAPK8IP3) [611]. Protein MAPK8IP1 aggregates MAP3K11, MAP2K7, and JNK3 enzymes to form a complete MAPK module. Scaffold β-arrestin-2 can selectively bind to JNK3 and assemble MAP3K5, MAP2K4, and JNK3 proteins.[133] Both MAPK8IP1 and

---

[128] A.k.a. MAPK10, PrKM10, stress-activated protein kinase SAPKβ, SAPK1b, p54bSAPK, and somatic embryogenesis receptor-like kinase SERK2.

[129] A.k.a. SAPKp46β2.

[130] A.k.a. SAPKp54β2.

[131] A.k.a. SAPKp46β1.

[132] A.k.a. SAPKp54β1.

[133] Kinase JNK3 interacts with 4 types of arrestins. β-Arrestin-2 can relocalizes JNK3 into the nucleus.

β-arrestin-2 can bind MKP7 (DuSP16) phosphatase. Phosphorylated MAPK8IP3 may interact with MAP2K4, MAP2K7, and JNK3 enzymes.[134] Moreover, MAPK-binding protein MAPKBP1 (or JNKBP1) bind to JNK3, in addition to JNK1 and JNK2.

In addition, cyclin-dependent kinase CDK5 may phosphorylate (inactivate) JNK3 protein [611]. Cyclin-dependent kinase inhibitor CKI2a can also bind to JNK3, thereby suppressing its activity without affecting JNK3 phosphorylation. On the other hand, S-nitrosylation of JNK3 by NO can improve the phosphorylation level and activity of JNK3 kinase.

Activated JNK3 can phosphorylate transcriptional factors, such as Jun, ELk1, and ATF2 to increase the transcriptional activity [611]. Other substrates of JNK3 include P53, kinesin-1,[135] microtubule-associated protein Tau (MAPT), stathmin-like 2,[136] mitochondrial protein Second mitochondria-derived activator of caspases (SMAC),[137] and Rab3 GDP–GTP-exchange factor (Rab3GEP).[138]

6.7.4 Family of P38 Mitogen-Activated Protein Kinases

The 38-kDa polypeptides — P38MAPKs — constitute a second group of stress-activated MAPK enzymes. Kinases of the P38MAPK family are activated by numerous factors: hormones, UV beams, ischemia, cytokines (e.g., interleukin-1 and tumor-necrosis factor), osmotic pressure changes, and heat stress.[139] Almost all of these stimuli are able to recruit both JNK and P38MAPK kinases. During ischemia–reperfusion events, JNKs are activated during reperfusion, but not ischemia, whereas P38MAPKs are activated during ischemia, but not reperfusion. In addition, MAP2K3 and P38MAPK also contribute to T_{H1}-cell development, as they provoke IL12 production by macrophages and dendritic cells.

Monomeric CDC42, Rac1, and RhoA GTPases are required for the re-entry of quiescent cells into the cell cycle G1 phase. On the other hand, in cells already committed to the cell cycle, CDC42, but not Rac1 or RhoA, targets P38MAPKs that can trigger cell cycle arrest at G1–S transition via MAP3K3 and MAP3K6 [267].

---

[134] Different isoforms of MAPK8IP3 have distinct binding affinities for JNKs.

[135] Kinase JNK3 reduces kinesin-1 binding to microtubules.

[136] A.k.a. neuronal growth-associated protein Superior cervical ganglion portein SCG10. This neuron-specific, membrane-associated regulator of microtubule dynamics influence microtubule polymerization in growth cones of developing neurons.

[137] A.k.a. Diablo homolog.

[138] A.k.a. MAPK-activating death domain protein (MADD) and Differentially expressed in normal and neoplastic cells (DENN). It serves as a Rab27aGEF activator. It regulates the recycling of Rab3 and is involved in Ca^{++}-dependent neurotransmitter release.

[139] The P38MAPK enzymes are implicated in asthma and autoimmunity.

Four P38MAPK isoforms (P38MAPKα–P38MAPKδ) that are encoded by 4 distinct genes act in a cascade that involves MAP3Ks (MAP3K1–MAP3K4, ASKs, MLKs, DLK, TAK1, TAOKs, and TPL2, i.e., MAP3K5–MAP3K12 and MAP3K16–MAP3K18). These MAP3Ks activate MAP2K3, MAP2K4, and MAP2K6, which phosphorylate P38MAPK [612]. In the classic pathway (MAP3K–MAP2K cascade), P38MAPK is phosphorylated (activated) downstream from MAP2K3 and MAP2K6 enzymes. However, in some cells such as activated T cells, an alternative pathway exists for P38MAPK activation. Enzyme P38MAPK is phosphorylated by ZAP70, itself activated by SRC family kinase LCK (MAPK-independent P38MAPK activation) [613]. The P38MAPK enzymes are linked to scaffold proteins (JIP2, JIP4, and Osm).[140]

6.7.4.1 P38MAPKα or MAPK14

Enzyme P38MAPKα[141] resides in both the cytoplasm and nucleus. In humans, 3 splice variants have been detected in addition to full-length P38MAPKα [614]: (1) cytokine suppressive anti-inflammatory drug-binding protein CSBP1; (2) exon skip Exip (as it skips exon 10); and (3) MAX-interacting protein MxI2, a shorter splice variant and a kinase that may interact with heterodimeric partner MyC-associated factor-X (MAX) of MyC transcription factor.

Enzyme P38MAPKα is required in erythropoietin synthesis and angiogenesis. It transmits chemotactic signals in neurons, endothelial and epithelial cells, neutrophils, mastocytes, monocytes, and vascular smooth muscle cells [614].

Its activation results from dual phosphorylation of the Thr180–Gly181–Tyr182 motif on the activation loop by MAP2K3, MAP2K4, and MAP2K6 enzymes (in vivo, mainly by MAP2K3 and MAP2K6, and under certain circumstances, by MAP2K4, an activator of JNKs). In addition to the these phosphorylation sites (Thr180 and Tyr182), P38MAPKα can be phosphorylated on Ser2, Thr16, and Thr241 [614].

Enzyme P38MAPKα is dephosphorylated (Thr180) by magnesium-depen-dent PPM1d protein phosphatase. Phosphatase PPM1a dephosphorylates MAP2K6 activator and P38MAPKα [614]. MAPK phosphatases, such as MKP1 (or DuSP1), MKP2 (or DuSP4), MKP5 (or DuSP10), and MKP7 (or DuSP16), and other dual-specificity phosphatases, such as DuSP2 and DuSP8, also dephosphorylate (inactivate) P38MAPKα.

Scaffold MAPK8IP2 binds to MAP3K11, MAP2K3, P38MAPKα, and TIAM1 (RacGEF) and RasGRF1 [614]. * In brown adipocytes, β-adrenergic receptor and protein kinase-A stimulate P38MAPKα via JIP2 that links to MAP2K3 activator.

---

[140]Protein Osm interacts with the actin cytoskeleton.

[141]A.k.a. PrKM14, PrKM15, Crk1, MAPK14, and SAPK2a. It was initially identified as a 38-kDa polypeptide that undergoes tyrosine phosphorylation in response to hyperosmolar shock. It is the mammalian MAPK ortholog of HOG1 osmosensor in Saccharomyces cerevisiae [614].

6.7 Mitogen-Activated Protein Kinases

In addition, JIP4 (MAPK8IP4), which binds to, but does not activate JNK, supports MAP2K3- and MAP2K6-dependent activation of P38MAPKα, but does not bind MAP2K3 and MAP2K6, whereas it connects to MAP2K4 activator [614].

Cerebral cavernous malformation CCM2 homolog, or osmosensing scaffold for MAP3K3 (Osm), tethers to Rac, MAP3K3, and MAP2K3, and indirectly enables P38MAPKα activation in response to osmotic stress. In addition, MAP3K7IP1 (or TAB1) serves as a P38MAPKα binding partner. Last, but not least, P38MAPKα interacts with P38MAPK-interacting protein (P38IP) that regulates its activity.

Once it is activated by stress, it re-activates MAPKAPK2 kinase that phosphorylates heat shock protein HSP27 in response to interleukin-1β and LIMK1 kinase in response to VEGF in endothelial cells [614]. In addition, P38MAPKα targets caldesmon, an actin- and myosin-binding protein involved in the assembly of actin filaments, in smooth muscle cells, as well as paxillin, a phosphoprotein of focal adhesions [614].

Enzyme P38MAPKα participates in cell adaptation to stress, cell cycle regulation, inducing cell division arrest at various checkpoints. and control of cellular death (promotion or protection against cell death according to circumstances) [614].

Enzyme P38MAPKα is implicated in inflammation and its associated increase in vasculature permeability to enable influx of molecular and cellular mediators. It controls the production of COx2 cyclooxygenase and promotes prostaglandin synthesis, in particular in fibroblasts and endothelial cells [614]. It also regulates the synthesis of pro-inflammatory cytokines IL1 and TNFα in monocytes and neutrophils. It mediates the activation by distinct pattern-recognition receptors on airway epithelial cells.

Enzyme P38MAPKα is involved in viral and bacterial infections [614]. It provokes degranulation of neutrophils; in these cells, it also phosphorylates NADPH oxidase, thereby priming oxidative burst. In macrophages, P38MAPKα upregulates scavenger receptors for phagocytosis of bacteria in response to Toll-like receptor activation.

Enzyme P38MAPKα is involved in the activation of the transcription regulators of the superfamily-2B3A of nucleus-resident nuclear factors Activator protein AP1 and transcription factors of the ATF–CREB family,[142] in conjunction with other MAPKs [614].[143] In particular, G-protein-coupled receptor activates the Jun promoter via P38MAPKα and AP1 activation in monocytes depends on JNK and P38MAPKα. Moreover, in some circumstances, P38MAPKα enables activation of nuclear factor-κB.

---

[142] ATF: Activating transcription factor; CREB: cAMP response element-binding protein.

[143] Enzyme P38MAPKα phosphorylates ATF2 (Thr69 and Thr71), MEF2a (Thr312 and Thr319), MEF2c (Thr293 and Thr300; but neither MEF2b nor MEF2d), ELk1 (Ser383), ELk4 (Ser381 and Ser387), C/EBP homologous protein (CHoP; Ser78 and Ser81), TCFe2a (Ser140), and STAT1 (Ser 727).

6.7.4.2 P38MAPKβ or MAPK11

Signaling by P38MAPKβ[144] depends on HDAC3, a class-1 histone deacetylase.[145] During inflammation, the induction of immediate-early genes such as that of TNFα is regulated by signaling cascades, such as the JaK–STAT, NFκB, and P38MAPK pathways. Enzyme P38MAPKβ interacts directly and selectively with HDAC3 enzyme. This deacetylase decreases the P38MAPKβ phosphorylation state and inhibits the activity of the P38MAPKβ-dependent transcription factor Activating transcription factor ATF2 [615].[146] Consequently, HDAC3 represses TNF gene expression, especially in monocytes and macrophages.

6.7.4.3 P38MAPKγ or MAPK12

Enzyme P38MAPKγ[147] behaves as a stress kinase that can function independently of phosphorylation [616]. It intervenes in cell differentiation. In skeletal muscles, where it is highly expressed, it supports fusion capacity of myoblasts, thereby promoting formation from myoblasts of myotubes. Alone or together with P38MAPKα, it also acts in erythroid differentiation.

In addition, P38MAPKγ contributes to the regulation of the stress response and cell cycle. In response to hypoxia, P38MAPKγ is phosphorylated (Thr183 and Tyr185) [616]. It is activated by dual-specificity kinases MAP2K3 and MAP2K6 enzymes, as well as MAP3K8 and RhoA and Rit GTPases. Moreover, MAP3K15 (or MLK7) stimulates the MAP2K6– P38MAPKγ axis. On the other hand, it is dephosphorylated by MAPK phosphatase MKP1 (DuSP1) and cytoplasmic PTPn3 phosphatase [616].

Enzyme P38MAPKγ is activated by various cytokines, such as IL1 and TNFα, and stress types (e.g., hypoxia and osmotic stress) [616].

Enzyme P38MAPKγ efficiently phosphorylates ATF2, ELk1, Elk4, Fos, and P53 (Ser33), but weakly MEF2a, MEF2c, and MEF2d transcription factors [616].

It also phosphorylates microtubule-associated protein Tau and BTK-associated SH3 domain-binding protein SH3BP5, a mitochondrial JNK-interacting protein. Protein P38MAPKγ can also interact with estrogen receptor ERα (or nuclear receptor NR3a1) [616].

---

[144] A.k.a. MAPK11, PrKM11, SAPK2b, and P38MAPK2.

[145] The chromatin state that regulates gene transcription is regulated by 2 antogonist enzyme types: histone acetyltransferases (HAT) and deacetylases (HDAC; Vol. 1 – Chaps. 4. Cell Structure and Function and 5. Protein Synthesis). They control the acetylation state of histones and other promoter-bound transcription factors. Histone deacetylases are classified into 3 subclasses.

[146] Transcription factor ATF2 binds a cognate response element (5′-TGACGTCA-3′). It homo- and heterodimerizes with other members of the ATF family or with members of the AP1 (Jun–Fos) family of transcription factors. In particular, the ATF2–Jun heterodimer recognizes the AP1–CRE target sequence. Upon its phosphorylation (Ser121), ATF2 associates with P300 and CBP to form a transcriptional complex.

[147] A.k.a. MAPK12, PrKM12, ERK6, or SAPK3.

6.7 Mitogen-Activated Protein Kinases

Table 6.23. Stress-activated protein kinases.

Type	Other alias
SAPK1a, SAPKα	JNK2
SAPK-P46α1	JNK2β1
SAPK-P46α2	JNK2α1
SAPK-P54α1	JNK2β2
SAPK-P54α2	JNK2α2
SAPK1b, SAPKβ	JNK3
SAPK-P46β1	JNK3β1
SAPK-P46β2	JNK3α1
SAPK-P54β1	JNK3β2
SAPK-P54β2	JNK3α2
SAPK1c, SAPKγ	JNK1
SAPK-P46γ1	JNK1β1
SAPK-P46γ2	JNK1α1
SAPK-P54γ1	JNK1β2
SAPK-P54γ2	JNK1α2
SAPK2a	P38MAPKα
SAPK2b	P38MAPKβ
SAPK3	P38MAPKγ
SAPK4	P38MAPKδ

6.7.4.4 P38MAPKδ or MAPK13

Enzyme P38MAPKδ[148] is involved in cell migration, in particular in invasion of carcinoma cells. In addition, P38MAPKδ pertains to the set of 13 lung-prominent gene products with vitamin-D-dependent calcium-binding protein S100g, solute carrier family transporter SLC29a1, corticotropin-releasing hormone receptor CRHR1, and lipocalin-2, a growth factor and component of the innate immune response to bacterial infection [617].

6.7.5 Stress-Activated Protein Kinases

Stress-activated[149] protein kinases (SAPK) include JNKs and P38MAPKs activated by MAP3K1–MAP2K4/7 and MAP3K7–MAP2K3/6 cascades, respectively (Table 6.23). The MAP3K5 module stimulates both JNKs and P38MAPKs. Activated JNKs and P38MAPKs induce gene expression via different transcription factors, such as ATF2, ELk1, Jun, and MEF2.

---

[148] A.k.a. MAPK13, somatic embryogenesis receptor-like kinase SERK4, PrKM13, and SAPK4.
[149] Various stresses (chemical, physical, and metabolic) are involved.

The SAPK molecules can be more or less implicated in cardiac remodeling associated with overload or ischemia–reperfusion injury, modulating signaling pathways associated with cardiac hypertrophy and cell apoptosis [618]. Enzyme P38MAPK can promote remodeling with interstitial fibrosis, leading to a loss in contractility, but without significant hypertrophy [619]. Cells can withstand relatively high mechanical stresses and react quickly and adapt using, at least partly, their dynamical cytoskeleton. Enzymes P38MAPK (but not ERK or JNK) are responsible for stretch-induced activation of an eicosanoid pathway that involves E-prostanoid receptor EP4 and cyclooxygenase COx2 (but not EP1 and COx1) via MAPKAPK2 [620]. However, prostaglandin-E2 induces actin filament depolymerization in cells stretched during a long time (6 h, 5% elongation, 0.5 Hz).

Downstream of mitogen-activated protein kinase pathways, MAPK-activated protein kinases regulate gene expression at the transcriptional and post-transcriptional levels and control the cell cycle. Moreover, MAPKAPKs are involved in actin remodeling, and consequently, in cell shape and cell motility [621]. Hsp27, an inhibitor of Rho-dependent actin stress fiber formation, competes with cofilin for binding to 14-3-3 protein. Increased release of cofilin from the 14-3-3 protein is followed by cofilin dephosphorylation (inactivation) and actin binding. Rho activation causes phosphorylation of cofilin by LIMK kinase and its release from the barbed ends of the actin filaments.

6.8 Kinase Substrates of MAPKs

All types of MAPKs, i.e., ERKs, JNKs, and P38MAPKs, phosphorylate various types of protein kinases: (1) MAPK-activated protein kinases (MAPKAPK); (2) MAPK-interacting Ser/Thr kinases (MNK); and (3) mitogen- and stress-activated protein kinases (MSK). Their stimulators, affectors, and effectors are exhibited in Table 6.24.

6.8.1 MAPK-Activated Protein Kinases

MAPK-Activated protein kinases correspond to members of the subfamily of P90 ribosomal S6 kinases (RSK1–RSK3; Table 6.25) that are Ser/Thr kinases activated by the ERK pathway. Other MAPKAPKs that are unrelated to the MAPKAPK1 infrafamily members form a small group of Ser/Thr kinases that include MAPKAPK2, MAPKAPK3 (also called three pathway-regulated kinase [3PK]), and MAPKAPK5 (or P38MAPK-regulated and activated protein kinase PRAK). In response to cellular stress and inflammatory cytokines, MAPKAPKs are phosphorylated (activated) by MAPKs.

6.8 Kinase Substrates of MAPKs

Table 6.24. Kinases as substrates of MAPKs (Source: [267]; MAPKAPK: MAPK-activated protein kinase; MNK: MAPK-interacting kinase; MSK: mitogen- and stress-activated protein kinase; CREB: cAMP-responsive element-binding protein; eIF: eukaryotic translation initiation factor; HSP: heat shock protein). Enzyme MSK1 can phosphorylate histone 3 and high-mobility group nucleosome-binding domain protein HMGN (or HMG14).

Kinase	Stimuli	Affector	Effector	Effect
MAPKAPK	Stresses	P38MAPK	HSP25/27	Cytoskeleton organization
	Cytokines	P38MAPK		
MNK	Stresses	P38MAPK	eIF4e	Protein synthesis
	Cytokines	P38MAPK		
	Mitogens	ERK		
MSK	Stresses	P38MAPK	CREB, H3,	Protein synthesis,
	Cytokines	P38MAPK	HMGN1	chromatin remodeling
	Mitogens	ERK		
RSK	Mitogens	ERK	H3	Chromatin remodeling

Table 6.25. Group of type-1 MAPK-activated protein kinases that correspond to members of the ^{P90}RSK subfamily of ribosomal S6 kinases (RSK1–RSK3), in opposition to S6K enzymes of the ^{P70}RSK subfamily.

Type	Gene	Aliases
MAPKAPK1a	RPS6KA1	RSK1, ^{P90}RSK1
MAPKAPK1b	RPS6KA3	RSK2, ^{P90}RSK2, S6Kα3, MRX19, CLS, ISPK1, HU3
MAPKAPK1c	RPS6KA2	RSK3, ^{P90}RSK3, S6Kα2, S6Kα, HU2

Enzyme MAPKAPK2 is phosphorylated (activated; Thr25, Thr222, and Ser272) by P38MAPKα and P38MAPKβ, but not P38MAPKγ or P38MAPKδ. Additional autophosphorylation (Thr334) follows. Enzymes MAPKAPK3 and MAPKAPK5 are also mainly phosphorylated by P38MAPKα and P38MAPKβ. Once they are activated, MAPKAPK2, MAPKAPK3, and MAPKAPK5 phosphorylate heat shock protein HSP27 (Ser15, Ser78, Ser82, and Ser90) to dissociate molecular chaperones HSP27 multimers into monomers and dimers and redistribute HSP27 to the actin cytoskeleton for its reorganization into stress fibers [267].

Like protein kinase-A, MAPKAPK1s and MAPKAPK2 target transcription factor cAMP-response element-binding protein (CREB). Yet, nuclear MSK1, but neither MAPKAPK1s nor MAPKAPK2, is the relevant stress- and mitogen-activated kinase for CREB factor.

6.8.2 MAPK-Interacting Ser/Thr Kinases

The MAPK-interacting Ser/Thr kinases MNK1 and MNK2 are phosphorylated (activated) by ERK1 and ERK2 in response to insulin and mitogens and P38MAPKs in response to stress. Kinases MNK1 and MNK2 phosphorylate eIF4e (Ser209). The eIF4f complex, i.e., the eIF4a–eIF4b–eIF4g–eIF4e complex, is involved in the recognition of the mRNA 5′-cap structure and recruitment of mRNA to ribosome. N7-methylguanosine-binding protein eIF4e recruits mRNAs onto scaffold eIF4g that also binds RNA helicase eIF4a. The latter, together with RNA-binding protein eIF4b, unwinds secondary structure in the mRNA 5′-untranslated segment to facilitate scanning of the mRNA by the 40S ribosomal complex to the ATG translational initiation site. Translational repressor 4E-binding protein 4eBP1[150] is phosphorylated in response to insulin and mitogens to dissociate it from eIF4e that can then be incorporated into the eIF4f complex. Protein eIF4e is phosphorylated by target of rapamycin and other effector kinases of PI3K to raise its affinity for mRNA 5′-cap.

6.8.3 Mitogen- and Stress-Activated Protein Kinases

Mitogen- and stress-activated protein kinases MSK1,[151] and MSK2[152] constitute an infrafamily of Ser/Thr protein kinases. Kinase MSK1, a substrate of P38MAPKs, localizes to the cell nucleus. It is activated by mitogens and environmental stresses.

Enzyme MSK1 interacts with kinases MAPK1, MAPK11, and MAPK14, epidermal growth factor, transcription factor CREB1, and BCL2 antagonist of cell death (BAD) [251]. Enzyme MSK2 interacts with kinases MAPK3 and MAPK14 and transcription factors CREB1 and Fos [251].

Kinases MSK1 and MSK2 phosphorylate transcription factors cAMP response element (CRE)-binding protein CREB and cAMP-dependent ATF1 in fibroblasts in response to mitogens and cellular stress [622].[153]

Phosphorylation of CREB (Ser133) by MSK1 (with a much lower Michaelis constant than that of MAPKAPK1 isoforms) and MAPKAPK1 that are activated by ERK1 and ERK2 is required for full activation of transcription factors encoded by immediate-early genes, such as Fos and JunB, but not early growth response protein EGR1, another immediate-early gene product [622].

---

[150] A.k.a. phosphorylated heat and acid stable protein regulated by insulin (PHASI).

[151] A.k.a. 90-kDa ribosomal protein S6 kinase polypeptide 5 and 90-kDa ribosomal protein S6 kinase-α5 (RLPK).

[152] A.k.a. RSKb.

[153] Phosphorylation induced by cAMP of CREB is catalyzed by protein kinase-A; mitogen-induced phosphorylation by MAP2K1, MAPKAPK1a to MAPKAPK1c (or RSK1–RSK3), MAPKAPK2 to MAPKAPK5, MNK1 and MNK2, and MSK1 and MSK2; and stress-induced phosphorylation by P38MAPK and possibly MSK1 and MSK2 [622].

Chapter 7
Dual-Specificity Protein Kinases

Dual-specificity kinases (DSK; or Ser/Thr/Tyr kinases) phosphorylate their substrates on serine, threonine, and/or tyrosine residues. Dual-specificity kinases intervene in the regulation of cell growth, differentiation, and apoptosis. Dual-specificity kinases include, at least, members of the superfamily of mitogen-activated protein kinases as well as those of the family of glycogen synthase kinases.

Certain dual-specificity kinases can target Tyr residues for their own activation (autophosphorylation), but not on their substrates, on which they act as protein Ser/Thr kinases. Therefore, DSKs can be kinases that phosphorylate substrates at Tyr and Thr residues or kinases that exhibit dual specificity only via autophosphorylation [536].

7.1 Mitogen-Activated Protein Kinase Kinase

Dual-specificity MAPK kinase corresponds to middle tier of mitogen-activated protein kinase module that also comprises upstream and downstream Ser/Thr kinases (Chap. 6).

7.2 Glycogen Synthase Kinase

Glycogen synthase kinase-3 (GSK3) is involved in the control of several regulatory proteins. It mainly phosphorylates serine and threonine residues of its substrates. It is one among several kinases that phosphorylate glycogen synthase, the rate-limiting enzyme of glycogen formation. Moreover, it targets transcription factor cJun to prevent DNA binding [623]. It also activates MgATP-dependent form of protein Ser/Thr phosphatase PP1.

In mammals, isozymes GSK3α and GSK3β are encoded by 2 distinct genes (Gsk3A and Gsk3B). Kinase GSK3 intervenes in protein synthesis, controls cellular response to damaged DNA, and participates in Wnt signaling.

Enzyme GSK3 can autophosphorylate Ser, Thr, and Tyr residues. Autophosphorylations at Ser and Thr residues induce inactivation, whereas Tyr autophosphorylation causes activation [624]. In particular, autophosphorylation of GSK3α (Tyr279) and GSK3β (Tyr216) activates GSK3 kinase.

Phosphorylation of GSK3α (Ser21) and GSK3β (Ser9) by several types of protein kinases occurs in response to signals that activate the cAMP, PI3K, or MAPK cascade to inhibit GSK3 enzyme. Furthermore, FAK2 and Fyn kinases can phosphorylate GSK3 at Tyr residues. Protein GSK3 is also inhibited by secreted Wnt glycoproteins (Vol. 3 – Chap. 10. Morphogen Receptors).

Phosphorylation by glycogen synthase kinase-3 usually requires a priming kinase, as efficient phosphorylation only occurs if another Ser^P or Thr^P residue already exists in GSK3 substrate. The priming kinase phosphorylates a substrate that is afterward additionally phosphorylated by GSK3. Phosphorylation by GSK3 usually precludes the activity of target protein (e.g., glycogen synthase and nuclear factor of activated T cells).

Glycogen synthase kinase-3 is phosphorylated (inhibited) by PKB during insulin signaling. Protein kinase-B stimulates many signaling axes blocked by GSK3 protein. Protein kinase-C phosphorylates GSK3 upon stimulation by cytokines, such as interleukins IL3 and IL4 as well as granulocyte–macrophage colony-stimulating factor (CSF2) [625].

In addition, lithium that inhibits GSK3 is used to stabilize mood in bipolar disorder with alternating mania and depression. Lithium induces a delay in Per2 transcription and corrects circadian disturbances of bipolar disorder [626]. Kinase GSK3 thus intervenes in circadian rhythm.

7.3 Dual-Specificity Tyrosine Phosphorylation-Regulated Kinases

Dual-specificity Tyr (Y) phosphorylation-regulated kinases (DYRK1a–DYRK1b and DYRK2–DYRK4) are Ser/Thr kinases that autophosphorylate on Tyr residues. They possess a nuclear localization signal. They can shuttle between the nucleoplasm and cytoplasm. Inside the nucleus, they can localize to the splicing-factor compartment (nuclear speckles).

Among GSK3 substrates, ε subunit of eukaryotic protein-synthesis initiation factor eIF2bε is phosphorylated (inhibited) by GSK3. Priming kinases are isoforms DYRK1a and DYRK2 [627]. The DYRK isoforms also phosphorylate microtubule-associated protein Tau, or MAPT,[1] (Thr212) for phosphorylation by GSK3 (Ser208). In addition, some members of the DYRK family are involved in gene expression, as they phosphorylate certain transcription factors.

Members of the DYRK family either abound in the testis or have a testis-restricted expression pattern.

7.3.1 DYRK1a

Ubiquitous DYRK1a is encoded by the Dyrk1A gene on human chromosome 21. It autophosphorylates (Tyr321) to prime its activation. It phosphorylates or interacts with many proteins, such as transcription factors (CREB1, FoxO1, Gli1, NFAT, and STAT3), chromatin-remodeling factor ARIP4, protein-synthesis initiation factor eIF2bϵ, splicing factors (cyclin-L2, SF2, and SF3b1), inhibitor of receptor Tyr kinase Sprouty homolog Spry2, proteins involved in synapse functioning (amphiphysin-1, dynamin-1, synaptojanin-1, α-synuclein, and huntingtin-interacting protein-1), 14-3-3 proteins, in addition to GSK3 and Tau [628,629].

Phytanoyl CoAαhydroxylase-associated protein PAHxAP1 is a DYRK1a-interacting protein that can control the subcellular DYRK1a location. Two splice variants of DYRK1a are expressed at comparable level without functional differences [629].

7.3.2 DYRK1b

Isoform DYRK1b is strongly expressed in skeletal muscles and testis [630]. In muscles, it regulates MEF2-dependent transcription, as it phosphorylates class-2 histone deacetylases. It reinforces G0 arrest of myoblasts, as it phosphorylates cell cycle regulators cyclin-D1 and -D3 and CKI1b inhibitor. Splice variants include ubiquitous, predominant ^{P69}DYRK1b and ^{P75}DYRK1b, which is restricted to skeletal muscle, as well as catalytically inactive ^{P65}DYRK1b and ^{P66}DYRK1b.

Small RhoA GTPase activates transcription factor MyoD and elicits DYRK1b synthesis. Kinases MAP2K3 and MAP2K6 phosphorylate (activate) DYRK1b. Pterin-4 α-carbinolamine dehydratase/dimerization cofactor PCBD2 of HNF1α[2] that stabilizes HNF1α and enhances its transcriptional activity is a DYRK1b-binding protein [630]. In addition, P38MAPKα and P38MAPKβ (but neither P38MAPKγ nor P38MAPKδ) inhibit DYRK1b enzyme.

---

[1] Protein Tau is a component of neurofibrillary tangles that is involved in Alzheimer's disease. Six Tau isoforms exist in brain that differ by number of binding domains. Protein Tau interacts with tubulin to stabilize microtubules and promotes tubulin assembly into microtubules.
[2] A.k.a. dimerization cofactor of hepatocyte nuclear factor HNF1α from muscle (DCoHM).

7.3.3 DYRK2

Dual-specificity Tyrosine (Y)-phosphorylation-regulated kinase DYRK2 is able to autophosphorylate on Tyr residues. It also phosphorylates histones H2B and H3 as well as transcription factor P53 (Ser46) to induce apoptosis in response to DNA damage [631]. Two isoforms of DYRK2 exist.

7.3.4 DYRK3

Isoform DYRK3 is restricted to erythroid progenitor cells and testes. It inhibits nuclear factor of activated T cells. It attenuates erythropoiesis selectively during anemia [632].

7.3.5 DYRK4

Subtype DYRK4 is a testis-specific kinase that localizes in the cytoplasm [633]. It is dispensable for male fertility, as a functional redundancy exist among DYRK isoforms during spermiogenesis.

7.4 Cell Division Cycle-like Kinases

Cell division cycle (CDC)-like Kinase-1 (CLK1) has a catalytic domain similar to that of cyclin-dependent- and mitogen-activated protein kinases [634].[3] It autophosphorylates on Ser, Thr, and Tyr residues. It phosphorylates substrates essentially on Ser residues.

In humans, ubiquitous isoforms (CLK1–CLK4) phosphorylate in the nucleus Ser–Arg-rich proteins involved in pre-mRNA processing such as the spliceosome. In humans, CLK1 to CLK3 isoforms have splice variant transcripts [634]. Both CLK1 and CLK2 activate PTPn1 protein Tyr phosphatase.

---

[3] CDC-like kinase-1 is also called LAMMER, as it possesses a sequence of LAMMER motif.

7.4.1 CLK1

Isoform CLK1 can interact with Ser–Arg-rich RNA-binding proteins, such as heterogeneous nuclear RNA ribonucleoprotein hnRNPG, RNA-binding protein RNPS1, alternative splicing factor, and splicing factors SFRS3 and SFRS4 [634]. Subtype CLK1 phosphorylates cyclophilins. It also activates PRP4 pre-mRNA processing factor PRPF4 of small nuclear ribonucleoprotein complexes of spliceosomes and targets histone H1.

7.4.2 CLK2

Isoform CLK2 phosphorylates scaffold attachment factor SAFb and alternative splicing factor. It also binds to splicing factor TRA2β1. Alternative splicing factor (ASF; a.k.a. splicing factor SF2) that is encoded by the SFRS1 gene participates not only in splicing reactions, but also in post-splicing activities, such as mRNA nuclear export and translation. It is phosphorylated by SFRS protein kinase-1 as well as CDC-like kinases CLK1 and CLK2, and, specifically, CLK3 subtype.

7.4.3 CLK3

This nuclear dual-specificity kinase regulates the intranuclear distribution of the Ser/Arg-rich (SR) family of splicing factors, as it phosphorylates SR proteins of the spliceosomal complex.

Two transcript variants encode different isoforms; the long (CLK3) and short isoform (CLK3$_{152}$) that result from alternative splicing coexist in different tissue types. Isoform CLK3$_{152}$, like CLK2$_{139}$ splice variant, lacks the kinase domain [635].

7.4.4 CLK4

Human CLK4 interacts with an Ser–Arg-rich-like protein CLASP that is a post-transcriptional regulator of CLK1. The long isoform CLASP$_L$ induces the inclusion of an exon containing a premature termination codon in CLK1 messenger RNA [634].

7.5 Wee1 Kinase

Dual-specificity protein kinase Wee1[4] hampers entry into mitosis. The Wee1 family includes several members. Ubiquitous Wee1 is expressed in somatic cells, but Wee1b is synthesized only in embryonic cells.

7.6 Membrane-Associated Tyr/Thr Protein Kinase

Membrane-associated dual-specificity protein Tyr/Thr kinase-1 (PKMYT1) that is encoded by the Pkmyt1 gene is also a member of the protein Ser/Thr kinase class.[5] Enzyme PKMYT1 is a member of the Wee kinase family. It is produced in both somatic and embryonic cells. Alternatively spliced transcript variants encode distinct isoforms.

Kinase PKMYT1 preferentially phosphorylates (inactivates) CDK1, hence impeding the G2–M transition during the cell division cycle. It causes an inhibitory phosphorylation of cyclin–CDK complexes by targeting adjacent Thr14 and Tyr15 (hence its name PKMYT) located near the ATP-binding pocket of the CDK subunit [636]. However, PKMYT1 shows preferential Thr14 phosphorylation. This kinase is anchored to the membrane throughout the cell cycle [636].

Activity of PKMYT1 is regulated through the cell cycle, at least partly, via phosphorylation. During mitosis, PKMYT1 activity decays 2-fold to 5-fold [636]. Hyperphosphorylated PKMYT1 loses its ability to bind to cyclin-B1. Protein kinase-B, Polo-like kinase, P90 ribosomal S6-kinase RSK, Raf of the MAPK module, and cyclin-B–CDK complex phosphorylate PKMYT1, in addition to PKMYT1 autophosphorylation.

The primary PKMYT1 partner is the cyclin–CDK1 complex that serves both as a substrate of PKMYT1 and kinase for PKMYT1. Yet, the cyclin-B–CDK complex is not the main repressor of PKMYT1 during M phase.

---

[4] Scottish: small. Kinase Wee1 was discovered in fission yeast Schizosaccharomyces pombe. Schizosaccharomyces pombe is a unicellular, rod-shaped organism used as a model organism in biology. Fission and budding yeasts (Saccharomyces cerevisiae) remains in phases G2 and G1 of the cell cycle for an extended period, respectively, as G2–M and G1–S transitions are tightly controlled, respectively.

[5] Kinase PKMYT1 is also known as MYT1. However, the MYT1 alias is also used to designate myelin transcription factor-1 (MyT1 according to the rules defined in this book).

7.7 Raf Kinase Inhibitory Protein

Raf kinase inhibitory protein (RKIP)[6] is a member of the phosphatidyl ethanolamine binder (PEBP) family. Protein RKIP was characterized as a phospholipid-binding protein. However, it is also able to bind nucleotides and opioids [637].

Protein RKIP can inhibit serine peptidases (e.g., thrombin, neuropsin, and chymotrypsin). It promotes differentiation of keratinocytes, dendritic cells, and macrophages. It modulates signaling cascades by impeding the MAPK, GRK, and NFκB pathways. Protein RKIP indeed acts as a sensor and integrator of several pathways. In its non-phosphorylated form, it downregulates the MAPK cascade, but potentiates GPCR signaling. In its phosphorylated state, $RKIP^P$ dissociates from cRaf and inhibits GRK2, an inhibitor of G-protein-coupled receptors.

In humans, the RKIP family includes RKIP1 and PEBP4 [638]. Protein RKIP can be the precursor of *hippocampal cholinergic neurostimulatory peptide* (HCNP) that participates in acetylcholine synthesis and secretion in the brain. Endocrine factors PEBP and HCNP are secreted with catecholamines into the blood circulation. In heart, HCNP has a negative inotropic effect [639].

Activated RKIP interacts with cRaf kinase [640], but RKIP does not directly inhibit cRaf. Protein RKIP blocks activation of cRaf in cells by preventing access to kinases and phosphorylation [637]. Furthermore, RKIP binds to locostatin and is then unable to inhibit cRaf, hindering cell migration. In tumors, RKIP is thus able to hinder angiogenesis and vascular invasion. In addition, RKIP phosphorylation by protein kinase-C prevents RKIP interaction with cRaf.

Raf kinase inhibitor protein impedes the interaction between cRaf or bRaf and MAP2K. In particular, it disrupts the cRaf–ERK1/2 pathways. However, RKIP operates as a modulator rather than an off switch for MAPK signaling, as RKIP decreases signaling output amplitude in response to stimuli. RKIP may indirectly affect bRaf signaling via the cRaf–bRaf dimer. Nevertheless, RKIP can regulate bRaf independently of cRaf kinase.

Stimulated G-protein-coupled receptor dissociates RKIP from cRaf kinase. RKIP phosphorylation by protein kinase-C enhances its ability to bind G-protein-coupled receptor kinase-2. Protein RKIP phosphorylated (activated) by PKC then inhibits G-protein-coupled receptor kinase-2 that phosphorylates and impedes the activity of many G protein-coupled receptors. Protein Ser/Thr kinase GRK2 indeed downregulates various G-protein-coupled receptors, particularly those hampering cell locomotion.

Activated Raf kinase inhibitor disturbs the nuclear factor-κB pathways, as it interacts with associated kinases. Agent RKIP in fact binds to several proteins that activate NFκB, such as transforming growth factor-β-activated kinase-1, IκB kinase IKKα and IKKβ, and NFκB-inducing kinase (MAP3K14). After degradation of IκB, NFκB translocates to the nucleus and can then bind to target gene promoters.

---

[6] A.k.a. phosphatidyl ethanolamine-binding protein-2 (PEBP2).

Phosphorylated RKIP localizes to centrosomes and kinetochores during the cell division cycle [637]. It participates in the regulation of the cell spindle checkpoint. It promotes the activity of Aurora-B kinase that ensures proper chromosome segregation during anaphase. However, Aurora-B contributes to a negative feedback loop by phosphorylating RKIP that then dissociates from cRaf and relieves its inhibition on MAPK enzyme.

Chapter 8
Cytosolic Protein Phosphatases

Protein phosphorylation is a common, reversible, post-translational modification of proteins (Vols.1 – Chap. 5. Protein Synthesis and 3 – Chap. 1. Signal Transduction). Except pseudophosphatases, phosphatases removes a phosphate group from their substrates. They thus antagonize kinases that attach phosphate groups to Ser, Thr, and Tyr residues in the same substrates using ATP. Phosphorylated residues Ser^P, Thr^P, and Tyr^{P1} account for about 86, 12, and 2% of the phosphoproteome, respectively. Mammalian genomes encode about 100 protein Tyr kinases and phosphatases, and about 100 and about 25 protein Ser/Thr kinases and phosphatases.

Protein phosphatases can be grouped into 3 main categories according to their structure and catalytic function. The largest category of protein phosphatases corresponds to the superfamily of classical, cytosolic, large, phosphoprotein Ser^P/Thr^P phosphatase (PP1–PP7)[2] that includes the families of phosphoprotein phosphatase (PPP) and Mg^{++}- and/or Mn^{++}-dependent protein phosphatases (PPM1α–PPM1ε, among others).[3] Protein Tyr^P phosphatases and aspartate-based protein phosphatases constitute a second and a third category (Tables 8.1 and 8.2).

Classical protein Tyr^P phosphatases are subdivided in receptor-like, plasmalemmal PTPs and cytosolic (non-receptor) PTPs. Both agonists and antagonists bind to, at least, some PTPRs (e.g., PTPRf, PTPRj, and PTPRt). In addition, some PTPRs (e.g., RPTPa) are inhibited by dimerization.

---

[1] Phosphorylated residues are also denoted pSer, pThr, and pTyr, respectively.

[2] Protein phosphatase-2 and -3 are also termed PP2a and PP2b, or calcineurin, respectively.

[3] A.k.a. PP2cα to PP2cε. The human genome contains, at least, 36 protein Tyr phosphatases (PTP), 39 protein Ser/Thr phosphatases (PP), and 16 dual-specific protein phosphatases (DuSP). The PP superfamily of cytosolic protein Ser/Thr phosphatases (cPSTP) can be further divided into the PPM, PPP, and FCP/SCP families.

Table 8.1. Categories of human protein phosphatases and examples of functions (**Part 1**) Subclass of protein Ser/Thr phosphatases (Source: [641]; NFκB: nuclear factor-κB; PP: protein phosphatase; PPM: protein phosphatase, Mg^{++} or Mn^{++} dependent; TGFβ: transforming growth factor-β). The PPP family can be subdivided into PP1, PP2 (including PP4 and PP6), PP3, and PP5 subfamilies.

Type	Role
Subclass of protein Ser/Thr phosphatases	
PPP superfamily	
PPP family	
PP1	Chromosome condensation
PP2	Chromatid cohesion
PP3	Immune response
PP4	DNA repair
PP5	Cell stress
PP6	NFκB signaling
PP7	
PPM family	
PPM1	TGFβ signaling

8.1 Pseudophosphatases

Protein MAP3K7IP1, or transforming growth factor-β-activated kinase-1 (TAK1)-binding protein-1 (TAB1), is one of the regulatory subunits of MAP3K7 (TAK1), an effector kinase of pro-inflammatory cascades. Molecule MAP3K7IP1 possesses a fold closely related to that of Mg^{++}- or Mn^{++}-dependent protein phosphatase family member PPM1α (Sect. 8.3.8), but without phosphatase activity, as several key residues required for dual metal-binding and catalysis are absent [642]. It serves as an adaptor that binds to and regulates accessibility of phosphorylated residues on substrates downstream of MAP3K7 or on the MAP3K7–MAP3K7IP1 complex.

The myotubularin family of dual-specificity phosphatases (Sect. 8.3.13.7) comprises many pseudophosphatases, such as MTMR5 and MTMR13. The latter indeed contain alterations in their PTP homology domains that abrogate phosphatase activity. Phosphatases that are catalytically inactive are able to bind phosphorylated substrates and prevent their dephosphorylation. Therefore, myotubularin-related pseudophosphatases can trap mutants or regulate PI3P levels by opposing the actions of myotubularin phosphatases. Truncated forms of MTMR5 are oncogenic, as they deal with cellular growth and oncogenic transformation.

8.1 Pseudophosphatases

Table 8.2. Categories of human protein phosphatases and examples of functions (**Part 2**) Subclass of protein Tyr phosphatases (PTP; Source: [641]; AcP1: acid phosphatase-1, soluble [or lmwPTP]; DuSP: dual specificity phosphatase; EyA: Eye absent; FCP: TF2F-associating C-terminal domain phosphatase; HAD, haloacid dehalogenase; lmwPTP: low-molecular-weight protein phosphotyrosine phosphatase; MAPK: mitogen-activated protein kinase; MKP: MAPK phosphatase; MTM: myotubularin; NRPTP: non-receptor [cytosolic] protein Tyr phosphatase; PI: phosphoinositide (phosphorylated phosphatidylinositol); PIP_3: phosphatidylinositol trisphosphate; PRL: phosphatase of regenerating liver; PTen: phosphatase and tensin homolog; RPTP: receptor protein Tyr phosphatase; SCP: small C-terminal domain-containing phosphatase; Ssh: Slingshot; TGFβ: transforming growth factor-β).

Type	Role
Subclass of protein Tyr phosphatases	
Infraclass-1 of Cys-based PTPs	
Hyperfamily of classical PTPs	
	RPTP superfamily
RPTP	Cell adhesion, cytoskeleton organization
	NRPTP superfamily
PTPn	Insulin signaling
Hyperfamily of DuSPs	
Atypical DuSP	
CDC14	Cytokinesis
MKP	MAPK signaling
MTM	PI metabolism
PTP14a (PRL)	
PTEN	PIP_3 dephosphorylation
Ssh	Actin dynamics
Infraclass-2 of Cys-based PTPs	
CDC25	Mitosis
Infraclass-3 of Cys-based PTPs	
AcP1 (lmwPTP)	
Infraclass of Asp-based PTPs – FCP Category	
FCP1	Transcription
SCP	TGFβ signaling
FCPL	
HAD	Actin dynamics

8.2 Catalytically Active Protein Phosphatases

Protein phosphatases form a group of signal transduction enzymes that, together with protein kinases, control the level of cellular protein phosphorylation. Kinase–phosphatase cascades are characterized by protein–phosphoprotein cycles that allow negative and positive commands and feedback in a signaling pathway.

Protein phosphatases dephosphorylate: (1) phospho-serine (Ser^P) and phospho-threonine (Thr^P) in the cell or (2) phospho-tyrosine (Tyr^P), either exclusively or with Thr^P at the plasma membrane or in the cytoplasm. Both protein Tyr and Ser/Thr dephosphorylations contribute to cell growth, proliferation, and differentiation, as they turn off or switch on actuating signaling.

Protein phosphatases mainly act on proteins, but certain protein phosphatases are able to remove phosphate from either phospholipids or mRNA. Inositide phosphatases SHIP and synaptojanins dephosphorylate inositol phospholipids (Chap. 2). SH2 domain-containing inositol 5-phosphatase SHIP1 (or IP(5)Pd) binds to Tyr residues of numerous substrates. It participates in cytoskeleton rearrangement for cell mobilization.

Dual-specificity protein phosphatases (DSP or DuSP) can dephosphorylate proteins on Tyr^P, Ser^P, and Thr^P, in addition to signaling lipids and complex carbohydrates.

Protein Tyr phosphatases (PTP) and kinases (PTK) act in partnership, as they coordinate their activities to control cell signaling. Protein Tyr phosphatases dephosphorylate Tyr^P residues selectively on their substrates to initiate, sustain, or terminate signal transduction. Both PTP catalytic and non-catalytic domains contribute to substrate specificity. In general, protein phosphatases have less substrate specificity than protein kinases.

The pathway from the platelet-derived growth factor receptor (PDGFR, Vol. 3 – Chap. 8. Receptor Kinases) to mitogen-activated protein kinases ERK1 and ERK2 that involves Ras, Raf, and MAP2K, illustrates the kinase–phosphatase control of signaling (Fig. 8.1).

8.3 Cytosolic Protein Serine/Threonine Phosphatases

Protein Ser/Thr phosphatases constitute many subfamilies associated with PP1 (α, β, $\gamma1$, and $\gamma2$), PP2 (PP2a), PP3 (PP2b), and PP4 to PP7 enzymes. Protein phosphatases PP1 and PP2 account for more than 90% of all Ser/Thr phosphatase activities in most cells (Table 8.3).

Among Ser/Thr phosphatases, *chronophin* regulates phosphorylation of *cofilin*, an actin regulator (Sect. 8.3.16.3). Actin monomers and filamentous actin are cyclically exchanged. Incorporation of $actin^{ATP}$ into filaments (^{F}actin) results from ATP hydrolysis. Recycling factors, such as cofilin and related actin-depolymerizing factors (ADF), bind $actin^{ADP}$ of filaments to promote filament depolymerization.

8.3 Cytosolic Protein Serine/Threonine Phosphatases

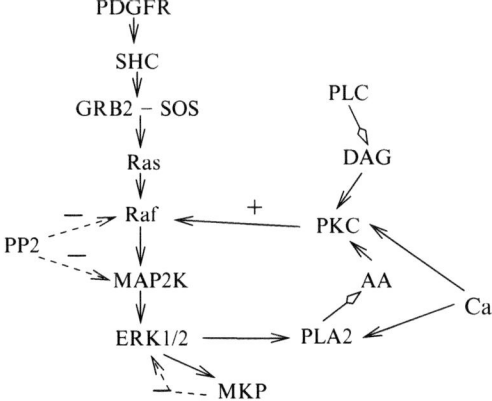

Fig. 8.1 Kinase–phosphatase cascade induced by stimulation of the platelet-derived growth factor receptor (PDGFR) associated with a balance between phosphorylation (activation) and dephosphorylation (inhibition; Source: [643]). MKP is a flexibility node of the pathway, modulating the cascade effect. Control of the Ras–ERK cascade needs the plasmalemmal recruitment of proteins serving as adaptors, anchors, and scaffolds (GRB, and SHC), and as regulators (SOS, and RasGAP). These complexes can be activated at the plasmalemma or in the endosomes, where they are able to recruit signaling molecules to trigger biological responses. ERK1/2 activates cytoplasmic phospholipase-A2 (PLA2), which produces arachidonic acid (AA). The latter stimulates protein kinase-C (PKC) via diacylglycerol (DAG). PKC in turn activates Raf (positive feedback loop). Receptor tyrosine kinases activate other effectors such as phospholipase Cγ, which stimulates PKC. Protein phosphatase-2 (PP2) dephosphorylates both Raf and mitogen-activated protein kinase kinase (MAP2K). Mitogen-activated protein kinase phosphatase (MKP) dephosphorylates ERK1/2. MAPK phosphorylates MKP, reducing MKP degradation (negative feedback).

Filament severing simultaneously stimulates actin filament disassembly and elicits actin polymerization on newly-severed barbed ends, thereby enhancing the dynamic exchange of between ^{G}actin and ^{F}actin. Interactions between cofilin and ADF on the one hand and actin on the other are improved by dephosphorylation by ubiquitous chronophin that thus assists cofilin-mediated actin organization. Chronophin counteracts LIMK kinase that inactivates actin-depolymerizing protein cofilin. Chronophin, or pyridoxal phosphatase, also dephosphorylates pyridoxal 5′-phosphate (vitamin-B6).

During mitotic entry, the regulation of phosphatases governs CDK1 activation. Conversely, after satisfactory spindle assembly checkpoint,[4] i.e., proper chromosome segregation, cytokinesis, and reassembly of interphase cell structures, mitotic exit is achieved, at least partly, by dephosphorylations of diverse substrates, hence inactivation of mitotic kinases, particularly CDK1, PLK1, and Aurora kinases, and activation of counteracting phosphatases, especially PP1 and PP2 phosphatases

---

[4]Distribution of all chromosomes to opposite spindle poles satisfies the spindle assembly checkpoint and initiates mitotic exit.

Table 8.3. Protein Ser/Thr phosphatases PP1, PP2 (PP2a), PP3 (PP2b), and PPM1 (PP2c), are the most abundant phosphatases. The number of genes that encode catalytic subunits is rather small (~40 genes). These subunits have much lower specificity for their substrates than Ser/Thr kinases, but they associate with numerous regulatory subunits that determine their substrate specificity, subcellular location, and overall activity. In addition to association with diverse types of regulatory subunits, phosphatases can be further regulated by phosphorylation, methylation, and ubiquitination.

Type	Features
PP1	Magnesium-dependent
	Numerous regulatory subunits
	Intervention in glygogen metabolism,
	Ca^{++} channel activity, muscle constraction,
	protein synthesis, cell division
PP2	Metal ion-independent
	Numerous regulatory subunits
	Heterodimer or heterotrimer
	Regulation of cell division, apoptosis
	Regulation of TOR and Wnt signaling
	Stabilization of β catenin–E-cadherin complex
	Regulation of CamK2, CDK, ERK, IKK, PKA/B/C, S6K
	Action in cytoskeleton restructuring
PP3	Active as heterodimer
	Calcium-dependent
	Stimulation by calmodulin
	Regulation of NFκB, TNFα, TGFβ
	Regulation of CamK2, MAPK
	T-lymphocyte activation
PPM1	Monomer
	Magnesium- and manganese-dependent
	Encoded by at least 10 genes

[644] (Table 8.4). A regulatory axis that involves Greatwall kinase and its substrates, the PP2-inhibitors α-endosulfine and cAMP-regulated phosphoprotein ARPP19, generates a mutual inhibition between cyclin-dependent kinase CDK1 and PP2 phosphatase.[5]

---

[5]Once the conditions related to the spindle assembly checkpoint are satisfied, activation of the anaphase-promoting complex, or cyclosome, and its coactivator CDC20 inactivate CDK1, thereby decreasing the activity of Greatwall kinase. Resulting decreased phosphorylation of Greatwall substrates (i.e., 2 small regulators α-endosulfine and cAMP-regulated phosphoprotein-19) relieves their inhibition of the $PP2_c$–$PP2_{r2}$ dimer. Therefore, Greatwall regulates $PP2_c$–$PP2_{r2}$ indirectly, via phosphorylation of α-endosulfine and cAMP-regulated phosphoprotein ARPP19, which then inhibits $PP2_c$–$PP2_{r2}$ dimer. Inhibition of this PP2 dimer increases the phosphorylation of various CDK1 substrates. In addition, Greatwall activates CDK1 (regulatory feedback loop).

8.3 Cytosolic Protein Serine/Threonine Phosphatases

Table 8.4. Major mitotic kinases and phosphatases and their phases of activity during the cell division cycle, especially mitotic entry, progression, and exit (Source: [644]; CDC: cell division cycle phosphatase; CDK: cyclin-dependent kinase; CenPe: centromere-associated protein-E; PKMYT: membrane-associated Tyr–Thr protein kinase; PLK: Polo-like kinase; PNuTS: phosphatase-1 nuclear targeting subunit).

Enzyme	Cell cycle phase of activity (location)
	Kinases
CDK1–CcnB	Prophase (nucleus), prometaphase, metaphase (cytoplasm, spindle)
Wee1, PKMYT1	G1, G2
Greatwall	Prophase, prometaphase, metaphase
Aurora-B	Prophase, prometaphase, metaphase (centromeres), anaphase, telophase (spindle)
PLK1	G2, prophase, prometaphase, metaphase (kinetochores, centromeres), anaphase, telophase (spindle)
	Phosphatases
CDC14a	G1/2 (cytoplasm, centrosome), prophase ⟶ telophase (cytoplasm)
CDC14b	G1/2 (nucleus, centrosome), prophase ⟶ metaphase (cytoplasm) anaphase, telophase (spindle)
CDC25b/c	Prophase ⟶ metaphase
PP1–CenPe	Prophase ⟶ metaphase (kinetochores)
PP1–PNuTS	Anaphase, telophase (chromosomes), G1
PP2–PP2$_{r2}$	G1/2, anaphase, telophase
PP2–PP2$_{r5}$	G1/2, prophase ⟶ metaphase (kinetochores) anaphase, telophase (spindle)

8.3.1 Protein Phosphatase-1

Ubiquitous cytosolic (non-receptor) Ser/Thr protein phosphatase-1 (PP1) regulates various cellular processes, in particular the glycogen metabolism, transcription, cell polarity, vesicle transfer, the DNA damage response, and cell cycle progression. It dephosphorylates multiple substrates and connects to numerous proteins.

In this autoregulatory loop, inhibition of PP2–PP2$_{r2}$ may enable stimulatory phosphorylation on CDC25 phosphatase and inhibitory phosphorylation on Wee1 and PKMYT1. During mitosis, PP1 activity is restrained by bound PP1$_{r1b}$, which has been phosphorylated by protein kinase-A. High CDK1 activity leads to inhibition of PP1 by phosphorylation. Inactivation of CDK1 promotes autodephosphorylation of PP1 and dephosphorylation of PP1$_{r1b}$ inhibitory subunit.

It promotes more energy-efficient fuels when nutrients abound and stimulates the storage of energy in the form of glycogen. In addition, it also enables the relaxation of actomyosin fibers as well as the return to a basal pattern of protein synthesis, and the recycling of transcription and splicing factors [645]. It also intervenes in the recovery from stress, promotes the exit from mitosis and apoptosis when cells are damaged beyond repair, and downregulates ion pumps and transporters in various cell types as well as ion channels in neurons.[6]

The PP1 catalytic subunits are encoded by 3 genes (PP1A, PP1B, or PP1D, and PP1G). They form stable dimers with numerous regulatory subunits that determine substrate specificity, subcellular location, and phosphatase activity. In addition to association with diverse types of regulatory subunits, these phosphatases can be further regulated by post-translational modifications, such as phosphorylation, methylation, and ubiquitination.

The diverse forms of protein phosphatase-1 result from association of a catalytic subunit ($PP1_c$) and different regulatory subunits (Table 8.5). The catalytic subunits of PP1 actually do not exist freely in the cell, but associate with regulatory polypeptidic interactors to form various holoenzymes. These regulatory subunits allow targeting of various subcellular locations and modulating connection to stimulators and effectors. The $PP1_c$ target depends on the interaction between the regulatory and catalytic subunits that form the PP1 complex. Three isoforms of PP1 catalytic subunit exist.

Interactors of PP1 include regulatory subunits and possible other regulators as well as other partners that may be regulators and/or substrates, such as phosphofructokinase, retinoblastoma protein, and Src-like adaptor SLA1 (Tables 8.6 to 8.8). Phosphatase PP1 associates with about 200 regulatory proteins to form highly specific holoenzymes (active complex with apoenzyme bound to its cofactor), thereby achieving target specificity. These regulatory proteins direct PP1 to its action locus within the cell as well as its enzymatic specificity for particular substrates. Spinophilin, or neuronal PP1 regulatory (inhibitory) subunit $PP1_{r9b}$, binds PP1 and blocks a substrate binding site without altering its active site [646]. In fact, Ser/Thr protein phosphatase-1 dephosphorylates hundreds of substrates.

In addition to cytosolic forms of PP1, such as those that target glycogen and myosin, nuclear PP1 exist [641].

Protein phosphatase-1 contains binding domains for hepatic and muscle glycogen and smooth muscle myosin that assign the catalytic subunit to these substrates [647]. The hepatic glycogen-binding domain has 2 binding subdomains, one that suppresses the dephosphorylation of glycogen phosphorylase, another that enhances the dephosphorylation of glycogen synthase.

---

[6]The conductance of ryanodine and inositol trisphosphate IP_3R1 receptors increases after phosphorylation by PKA, thus decreasing after dephosphorylation by PP1 phosphatase. The latter also dephosphorylates phospholamban, thereby reducing the activity of SERCA pump. Phosphatase PP1 contributes to the regulation of ionotropic AMPA-type and NMDA-type glutamate receptors [645]. It inhibits K_V7 and ClC2 channels. Phosphatase PP1 binds to $Na^+-K^+-Cl^-$ cotransporter NKCC1 and counteracts its activation by kinases.

8.3 Cytosolic Protein Serine/Threonine Phosphatases

Table 8.5. Subunits of protein phosphatase-1 (PPP: phosphoprotein phosphatase; Source: [76]). In humans, about 180 PP1-interacting proteins have been identified.

Subunit(s)	Gene(s)
Catalytic subunits	
$PP1_{c\alpha}$	PPP1CA
$PP1_{c\beta}$ ($PP1_{c\delta}$)	PPP1CB
$PP1_{c\gamma}$	PPP1CG (PPP1CC)
$PP1_{c\gamma1}$–$PP1_{c\gamma2}$ (alternatively spliced variants)	
Regulatory subunits (inhibitors)	
$PP1_{r1a}$–$PP1_{r1b}$	PPP1R1A, PPP1R1B
$PP1_{r2}$	PPP1R2
$PP1_{r3a}$–$PP1_{r3f}$	PPP1R3A–PPP1R3F
$PP1_{r7}$	PPP1R7
$PP1_{r8}$	PPP1R8
$PP1_{r9a}$–$PP1_{r9b}$	PPP1R9A–PPP1R9B
$PP1_{r10}$	PPP1R10
$PP1_{r11}$	PPP1R11
$PP1_{r12a}$–$PP1_{r12c}$	PPP1R12A–PPP1R12C
$PP1_{r13b}$, $PP1_{r13L}$	PPP1R13B, PPP1R13L
$PP1_{r14a}$–$PP1_{r14d}$	PPP1R14A–PPP1R14D
$PP1_{r15a}$–$PP1_{r15b}$	PPP1R15A–PPP1R15B
$PP1_{r16a}$–$PP1_{r16b}$	PPP1R16A, PPP1R16B

Protein phosphatase-1 modulates functioning of some membrane proteins, such as ion carriers and receptors, and participates in various cellular processes (Tables 8.9 and 8.10). Subtype PP1γ dephosphorylates histone-3 (Thr11) after DNA damage for transcriptional repression of cell cycle regulatory genes [648].

The rate-limiting enzymes of glycogen anabolism and catabolism[7] are glycogen synthase and phosphorylase, respectively. Phosphorylation of glycogen synthase by several kinases and of glycogen phosphorylase by phosphorylase kinase causes inactivation and activation, respectively. Phosphorylase kinase is activated by phosphorylation of its regulatory α and β subunits and by Ca^{++} binding to the regulatory δ subunit. Glycogen synthase and phosphorylase as well as phosphorylase kinase are dephosphorylated by PP1 anchored to glycogen via glycogen-targeting (G) subunits [645].

Phosphorylation of cAMP-responsive element-binding protein (CREB) by PKA promotes the recruitment of histone acetyltransferase CBP to facilitate access of the promoter by the transcriptional machinery. Conversely, CREB action is reduced upon CREB dephosphorylation by PP1 enzyme [645].

---

[7] αναβαλλω: throw up; καταβαλλω: throw down.

Table 8.6. Interactors of PP1 (**Part 1**; Source: [645]; AKAI: A-kinase-activated inhibitor; AnkRD: Ankyrin repeat domain-containing protein; ARD: activator of RNA decay; ASPP: apoptosis-stimulating protein of P53; CPI17: 17-kDa (PK)C-kinase-dependent phosphatase inhibitor; DARPP32: 32-kDa dopamine and cAMP-regulated protein; GADD: growth arrest and DNA damage-inducible protein; HCG5: hemochromatosis candidate gene-5 product; KEPI: kinase-enhanced PP1 inhibitor; MyPT: myosin phosphatase target subunit; NIPP1: nuclear inhibitor of PP1; PHI: phosphatase holoenzyme inhibitor; PNuTS: phosphatase-1 nuclear targeting subunit; TIMAP: TGFβ-inhibited membrane-associated protein; TP53BP: transcription factor P53-binding protein). G Subunits of PP1 have binding sites for glycogen and both PP1 and its substrates.

Interactors	Alias(es)
$PP1_{r1b}$	PP1 inhibitor-1, DARPP32, AKAI
$PP1_{r7}$	SDS22
$PP1_{r8}$	NIPP1, ARD1
$PP1_{r9a/b}$	Neurabin-1/2, spinophilin (subunit-9B)
$PP1_{r10}$	PNuTS
$PP1_{r11}$	HCG5
$PP1_{r12a/b}$	MyPT1/2 (MLCP)
$PP1_{r13}$	ASPP1
TP53BP2	ASPP2
$PP1_{r14a}$	CPI17L
$PP1_{r14b}$	PHI1
$PP1_{r14c}$	KEPI
$PP1_{r15a}$	GADD34
$PP1_{r16a}$	MyPT3
$PP1_{r16b}$	AnkRD4, TIMAP
	G (glycogen) targeting subunits
$PP1_{r3a}$	G_M (striated muscle)
$PP1_{r3b}$	G_L (liver)
$PP1_{r3c}$	Ubiquitous G_C
$PP1_{r3d}$	Ubiquitous G_D
$PP1_{r3e}$	Ubiquitous G_E
$PP1_{r3f}$	Ubiquitous G_F
$PP1_{r3g}$	Ubiquitous G_G

In human cells, during mitosis, PP1 controls the attachment of spindle microtubules to kinetochores during metaphase, as they counteract Aurora-B-mediated phosphorylation of outer kinetochore components and microtubule-destabilizing factors [644]. The kinetochore protein KNL1 (kinetochore-null protein-1),[8] $PP1_{r7}$ regulatory subunit, and kinetochore nanomotor CenPe bind PP1 and target PP1 to kinetochores. Conversely, active Aurora-B impedes PP1 linkage to kinetochores, as it phosphorylates the PP1-binding motif of KNL1 and CenPe proteins. The mutual control between Aurora-B and PP1 can establish a switch in substrate

---

[8] A.k.a. cancer susceptibility candidate CaSC5, Bub-linking kinetochore protein (blinkin), and cancer–testis antigen CT29.

8.3 Cytosolic Protein Serine/Threonine Phosphatases

Table 8.7. Interactors of PP1 (**Part 2**; Source: [645]; AKAP: A-kinase-anchoring protein; BCL: B-cell lymphoma; CGNAP: centrosome and Golgi-localized PKN-associated protein; NeK: NIMA (never in mitosis gene-A)-related protein kinase; NKCC: Na^+–K^+–Cl^- cotransporter; NLK: NIMA-like protein kinase).

Interactors	Alias(es)
AKAP1	AKAP149
AKAP9	AKAP350, AKAP450, CGNAP, Yotiao
AKAP11	AKAP220
NKCC1	SLC12a2
Vitamin-D receptor	NR1i1
Aurora-A and -B	STK6/7/15 and STK5/12
FAK	PTK2
NeK2	NLK1
Phosphofructokinase	
Protein kinase-R	eIF2αK2
BCL2, BCL2L2, BCLxL	
Retinoblastoma protein	

phosphorylation state at the outer kinetochore, thereby enabling a fast response to errors in chromosome attachment. Enzyme PP1 also opposes Aurora-B phosphorylation on chromosome arms (Ser10 on histone-3).

Protein phosphatase PP1 interacts with PP3 [649]. In addition, PP3 can dephosphorylate PP1 inhibitors such as DARPP32, hence activating PP1.

8.3.1.1 Examples of Protein Phosphatase-1 Regulatory Subunits

Protein Phosphatase-1 Inhibitory Subunit-1B

Protein phosphatase-1 regulatory (inhibitory) subunit $PP1_{r1b}$[9] that is a mediator of dopaminergic signaling. It is involved in cell differentiation as well as morphogenesis and functioning of several organs. Its effect depends on phosphorylation state of 4 regulatory residues (Thr34, Thr75, Ser102, and Ser137) [650].

Phosphorylation by protein kinase-A (Thr34) converts $PP1_{r1b}$ into a potent, high-affinity inhibitor of PP1 phosphatase. Residue Thr34 is dephosphorylated by protein phosphatase PP3 (or calcineurin). Residue Thr75 that can inhibit PKA effect is phosphorylated by cyclin-dependent kinase CDK5 and dephosphorylated by PP2 phosphatase. Residue Ser102 that is phosphorylated by casein kinase CK2 increases efficiency of Thr34 phosphorylation. Residue Ser137, once phosphorylated

---

[9] A.k.a. 32-kDa dopamine- and cAMP-regulated phosphoprotein DARPP32.

Table 8.8. Interactors of PP1 (**Part 3**; Source: [645]; CMT: Charcot-Marie-Tooth disease; FAK: focal adhesion kinase; GRP78: 78-kDa, endoplasmic reticulum luminal Ca^{++}-binding glucose-regulated protein; HCFc1: host cell factor-C1; Hox: homeodomain-containing (homeobox) protein; HSPa5: heat shock 70-kDa protein-5; InI: integrase interactor; NCoR, nuclear receptor corepressor; NPWBP: 38-kDa nuclear protein containing a WW domain-binding protein; PSF: polypyrimidine tract-binding protein-associated splicing factor; SARA: SMAD anchor for receptor activation; SNF: sucrose non-fermenting; SFPQ proline and glutamine-rich splicing factor; SIPP1: splicing factor that interacts with PQBP1 and PP1; SMARC: SWI/SNF-related, matrix-associated, actin-dependent regulator of chromatin; SNP70: 70-kDa SH3 domain-binding protein; Stau: Double-stranded RNA-binding protein Staufen homolog; TLX: T-cell leukemia homeobox; VCAF: VP16 accessory protein; WBP: WW domain-binding protein; ZFYVE: zinc finger, FYVE domain-containing protein).

Interactors	Alias(es)
PCdh7	BH (brain–heart)-protocadherin
HSPa5	GRP78
HCFc1	VCAF
TLX1	Hox11
NCoR	
Neurofilament-L	CMT1f, CMT2e
Ribosomal protein-L5	
SFPQ	PSF
SMARCb1	INI1, SNF5
Stau1	
Tau	
Trithorax	
WBP11	NPWBP, SNP70, SIPP1
ZFYVE16	Endofin, SARA

by casein kinase CK1, is able to decrease the activity of protein phosphatase PP3 [650]. Residue Ser137 is dephosphorylated by protein phosphatases PPM1 (PP2c). In summary, $PP1_{r1b}$ is inhibited when phosphorylated at Thr34, whereas PKA phosphorylation of Thr34 is prevented by Thr75 phosphorylation.

Transcription of the PPP1R1B gene is controlled by cAMP response element-binding protein. When PP1 is inhibited, cAMP response element-binding protein remains phosphorylated until $PP1_{r1b}$ is inactivated. Several agents (cocaine, morphine, ethanol, nicotine, methamphetamine, estradiol, etc.) as well as dopamine, other endogenous neurotransmitters, and hormones (serotonin, glutamate, GABA, adenosine, nitric oxide, opioids, neurotensin, cholecystokinin, and human chorionic gonadotropin) modulate $PP1_{r1b}$ phosphorylation state. Brain-derived neurotrophic factor binds to neurotrophic Tyr receptor kinase NTRK2 to activate protein kinase-B that phosphorylates CREB (Ser133).

8.3 Cytosolic Protein Serine/Threonine Phosphatases

Table 8.9. Role of protein phosphatase-1 (Source: [645]; **Part 1**; ↓: decrease). Phosphatase PP1 operates as an economizer, reducing energy expenditure, and a reset element (protein recycling, actomyosin relaxation, ion carrier inactivation, and exit from mitosis).

Cell process	Antagonist kinases, targets, and effects
Ion transport	IP_3R1, RyR2; conductance ↓; phospholamban; SERCA activity ↓; Na^+–K^+ ATPase ↓; $K_V7.1$; conductance ↓; ClC2 activation; NKCC1 inactivation
Cell signaling	Recruitment of PP1 to TβR1 by ZFYVE9 (antagonizes TβR2)
Cell metabolism	AMPK; glucose import; gluconeogenesis; use of alternative sugar and other carbon sources (fatty acids, glycerol, pyruvate, and lactate); repression of anabolism
Glycogen metabolism	Storage of glycogen
Protein synthesis	CREB inhibition; SMARCb1 — chromatin remodeling; eIF2αK2, RPL5 — translation; SFPQ — mRNA processing; recycling of transcription and splicing factors; return to basal state of protein synthesis

Table 8.10. Role of protein phosphatase-1 (Source: [645]; **Part 2**).

Cell process	Antagonist kinases, targets, and effects
Actin remodeling	$PP1_{r9}$–PP1, $PP1_{r12}$–PP1; actomyosin cytoskeleton dephosphorylation; stress fiber relaxation
Cell division	Aurora — histone-3 dephosphorylation, chromosome decondensation NeK2 — delay of centrosome splitting Lamin-B — reassembly of the nuclear envelope
Cell cycle arrest	Retinoblastoma protein activation
Cell survival or apoptosis	BCL2, BCL2L2, BCLxL PP1–BCL2/2L2/xL–BAD complex; supports apoptosis of damaged cells

Protein Phosphatase-1 Inhibitory Subunit-8

Protein phosphatase-1 regulatory (inhibitory) subunit-8 ($PP1_{r8}$)[10] is encoded by the PPP1R8 gene. It contributes to adaptation to hypoxia. This ubiquitous RNA-binding protein is required in pre-mRNA processing during a late step of spliceosome assembly [651]. In addition, it operates as a transcriptional repressor.

The $PP1_{r8}$–PP1 holoenzyme can be phosphorylated (activated) by protein kinase-A (Ser178 and Ser199) and, less effectively, by casein kinase-2 (Thr161 and Ser204) as well as Lyn (Tyr264 and Tyr335).

Protein Phosphatase-1 Inhibitory Subunit-12 – MLCP

Phosphorylation of myosin regulatory light chain is a component of the regulation of cell contraction. Indeed, MLC phosphorylation status results from the balance of activity of MLC kinase (activation) and phosphatase (inactivation). Dephosphorylation causes Ca^{++} desensitization, especially in smooth muscle cells, and reduces ability of myosin to crosslink actin filaments.

Myosin light-chain phosphatase (MLCP) is composed of 3 subunits: (1) a catalytic motif ($PP1_{c\delta}$); (2) a large myosin-binding sequence (MBS), i.e., either the protein phosphatase-1 regulatory (inhibitory) subunit-12A ($PP1_{r12a}$), also called myosin light-chain phosphatase target subunit-1 (MyPT1), or -12B ($PP1_{r12b}$), also termed MyPT2; and (3) a small subunit such as (20-kDa) smM20 protein, which elevates the Ca^{++} sensitivity of contractile fibers in vascular smooth muscle cells.

Subunit $PP1_{r12a}$ is ubiquitous, whereas $PP1_{r12b}$ mainly lodges in the brain and striated muscles. Both $PP1_{r12a}$ and $PP1_{r12b}$ can complex with $PP1_{c\delta}$ as well as other molecules, such as moesin and microtubule-associated MAP2 and Tau (MAPT) proteins. Moreover, $PP1_{r12a}$ and $PP1_{r12b}$ act as scaffolds of substrates for several kinases and signaling proteins such as RhoA GTPase [652]. Four alternatively spliced transcript variants can be described; 2 associated with the MBS subunit and 2 with smM20 subunit.

In fact, the PPP1R12B gene encodes both the large MBS and small smM20 subunit. The MBS subunit is also encoded by the PPP1R12A gene. Small smM20 subunit has a higher binding affinity for $PP1_{r12a}$ than for $PP1_{r12b}$ subunit.

In humans, a heart-specific small subunit hsM21 increases the sensitivity to Ca^{++} during the cardiomyocyte contraction. Two isoforms of hsM21 encoded by the PPP1R12B gene — hsM21a and hsM21b — preferentially bind to $PP1_{r12a}$ and $PP1_{r12b}$ subunit, respectively [652]. Subunit hsM21 heightens the contraction of renal artery smooth muscle cells and rat cardiomyocytes at a given Ca^{++} concentration.

Activity of MLCP is regulated by phosphorylation of $PP1_{r12a}$ and $PP1_{r12b}$ subunits. In the heart, hsM21 subunit supports $PP1_{r12a}$ phosphorylation (Thr696)

---

[10] A.k.a. nuclear inhibitor of protein phosphatase NIPP1.

8.3 Cytosolic Protein Serine/Threonine Phosphatases

Table 8.11. Inhibition of myosin light-chain phosphatase. This trimeric holoenzyme can undergo inhibitory phosphorylation or bind to inhibitors. Inhibition of MLCP augments MLC phosphorylation state and thus contractile force of smooth muscle cells independently of Ca^{++} ions, thereby enhancing Ca^{++} sensitivity.

Mechanism	Agents
Phosphorylation of $PP1_{r12}$	DAPK3, MRCK, cRaf, RoCK
Phosphorylated inhibitors (involved kinases)	$PP1^P_{r14a}$, $PP1^P_{r14b}$, $PP1^P_{r14c}$ (DAPK3, ILK, PAK, PKC, RoCK)

by RoCK kinase. The latter indeed phosphorylates (inhibits) human $PP1_{r12a}$ at both Thr696 and Thr853 residues (Table 8.11).[11] In addition, other kinases, such as death-associated protein kinase DAPK3[12] and P21-activated kinase, can phosphorylate $PP1_{r12a}$ (Thr696), hence causing Ca^{++} sensitization.[13]

8.3.2 Protein Phosphatase-2

In the cell, PP2 (or PP2a)[14] exists as either a heterodimer or heterotrimer. Heterotrimeric protein phosphatase-2 (PP2) is a ubiquitous, strongly regulated phosphatase that fulfills numerous functions. Two isoforms of PP2 have been identified (with >97% identity).

Protein phosphatase-2 is composed of: (1) a catalytic (C subunit; $PP2_c$); (2) scaffold (structural A subunit); and (3) regulatory (B subunit) subunits (Tables 8.12 and 8.13).[15]

---

[11] Subunit $PP1_{r12b}$ contains equivalent phosphorylation sites (Thr646 and Thr808).

[12] A.k.a. zipper-interacting protein kinase (ZIPK).

[13] Enzyme MLCP, but not MLCK, functions without Ca^{++} ion. Its inhibition thus supports action of Ca^{++} ion and MLCK enzyme.

[14] Type-2 protein phosphatases (PP2) were split into 3 categories whether they do not require metal ions, (PP2a), they are stimulated by Ca^{++} (PP2b, PP3, or calcineurin), or they depend on Mg^{++} (PP2c). According to new notation rules, PP2a is replaced by PP2, PP2b by PP3, and PP2c by PPM1 alias.

[15] Scaffold A subunit undergoes structural rearrangements upon interaction with catalytic and regulatory subunits. In mammals, both catalytic and scaffolding subunits are encoded by 2 genes (α and β), whereas several gene families (B, $B^{(1)}$ [or B′], $B^{(2)}$ [or B″], and $B^{(3)}$ [or B‴]) encode regulatory subunits [653]. Regulatory subunits of the B subset are encoded by 4 related genes that encode α, β, γ, and δ subtypes; subunits of the $B^{(1)}$ subset by 5 related genes that encode α, β, γ, δ, and ϵ subtypes, some of which generate alternatively spliced products; subunits of the $B^{(2)}$ subset by 3 related genes that encode G5PR, PR48, and splice variants PR72 and PR130 as

Table 8.12. Subunits of protein phosphatase-2 (**Part 1**; PPP: phosphoprotein phosphatase; RPc: replication protein-C; TAP42: 42-kDa 2A (two) phosphatase-associated protein; TIPRL: TIP41 [TAP42-interacting protein], TOR signaling pathway regulator-like protein). Potentially, more than 70 PP2 heterotrimers can be generated by combining various types of subunits. In mammals, both A and C subunits are encoded by 2 genes.

Subunit	Gene	Other aliases
Catalytic C subunits		
$PP2_{c\alpha}$	PPP2CA	RPc
$PP2_{c\beta}$	PPP2CB	
Regulatory A subunits		
$PP2_{r1\alpha}$	PPP2R1A	$PR65\alpha$
$PP2_{r1\beta}$	PPP2R1B	$PR65\beta$
Miscellaneous		
$PP2_{r4}$	PPP2R4	PR53
IGBP1	Igbp1	$\alpha 4$, CD79aBP1, IBP1
		TAP42 (ortholog)
TIPRL	Tiprl	TIP41

The catalytic C subunit can also bind to immunoglobulin-binding protein-1,[16] the mammalian ortholog of yeast Two (2A) phosphatase-associated protein of 42 kDa (TAP42), and TAP42-interacting protein (TIP41) to control TOR signaling.

Phosphatase PP2 is regulated by one regulatory subunit among the set of regulatory subunits that control PP2 substrate specificity, cellular location, and enzymatic activity. Phosphatase PP2 thus forms a large collection of oligomeric enzymes. In fact, A subunit links $PP2_c$ with many B-subunits to form more than 60 different heterotrimeric PP2 holoenzymes that can dephosphorylate many phosphoproteins [653].

Subunits $PP2_{r2\alpha}$, $PP2_{r5\alpha}$, $PP2_{r5\beta}$, $PP2_{r5\varepsilon}$ are cytosolic; other $PP2_{r2}$ subunits, $PP2_{r5\delta}$, and $PP2_{r5\gamma}$ abound in the nucleus; $PP2_{r3\beta}$ resides exclusively in the nucleus [641].

Post-translational modification of PP2 subunits (e.g., cysteine oxidation, tyrosine phosphorylation, and leucine methylation of $PP2_c$) regulates PP2 complex formation and its activity. Stimulation by epidermal growth factor, insulin, interleukin-1, or tumor-necrosis factor-α promotes transient Tyr phosphorylation (inactivation) of $PP2_c$. Phosphorylation targets not only $PP2_c$, but also regulatory B subunits, notably those of the $B^{(1)}$ subset, in particular by protein kinase-A. Double-stranded

---

well as ortholog PR59 (Mus musculus); and subunits of the $B^{(3)}$ subset by PR93 (or SG2NA) and PR110 (or striatin).

[16] A.k.a. CD79a-binding protein-1 and protein α-4.

Table 8.13. Subunits of protein phosphatase-2 (**Part 2**; PPP: phosphoprotein phosphatase). Four groups of B subunits are encoded by multiple genes in each group: the B group (B55 or PR55), the $B^{(1)}$ (or B') group (B56 or PR61), the $B^{(2)}$ (or B'') group (PR48, PR59, and PR72/PR130), and $B^{(3)}$ (or B''') group (PR93, or SG2NA, and PR110, or striatin). Striatin, SG2NA, and zinedin, the 3 mammalian members of the striatin family, are calmodulin-binding proteins that operate as signaling and scaffold proteins mainly expressed in the central nervous system, where they localize to the soma and dendrites, but not in the same neurons.

Subunit	Gene	Other aliases
Regulatory B subunits		
B group		
$PP2_{r2\alpha}$	PPP2R2A	B55α, PR55α
$PP2_{r2\beta}$	PPP2R2B	B55β, PR55β
$PP2_{r2\gamma}$	PPP2R2C	B55γ, PR55γ
$PP2_{r2\delta}$	PPP2R2D	B55δ, PR55δ
$B^{(1)}$ group		
$PP2_{r5\alpha}$	PPP2R5A	B56α, PR61α
$PP2_{r5\beta}$	PPP2R5B	B56β, PR61β
$PP2_{r5\gamma}$	PPP2R5G	B56γ, PR61γ
$PP2_{r5\delta}$	PPP2R5D	B56δ, PR61δ
$PP2_{r5\epsilon}$	PPP2R5E	B56ε, PR61ε
$B^{(2)}$ group		
$PP2_{r3\alpha}$	PPP2R3A	B72, B130, PR72, PR130, PR59 (ortholog)
$PP2_{r3\beta}$	PPP2R3B	PP2R3L, PR48
$PP2_{r3\gamma}$	PPP2R3G	G5PR
$B^{(3)}$ group – striatin family		
Striatin	STRN	PR110
SG2NA	STRN3	PR93
Zinedin	STRN4	

RNA-dependent protein kinase PKR can also intervene [653]. Low PP2 activity correlates with Tyr phosphorylation of $PP2_c$. Subunit $PP2_c$ is phosphorylated not only by activated growth factor receptors (e.g., epidermal growth factor and insulin receptor), but also Src kinases [654]. Demethylation of PP2 is carried out by PP2-specific methylesterase PME1 that inactivates PP2 by evicting manganese ions necessary for PP2 activity [655].

Midline-1 (Mid1)[17] is a ubiquitin ligase member of the tripartite motif-containing protein TRIM17 family that targets the microtubule-associated pool of $PP2_c$ for

---

[17] A.k.a. RING finger protein RNF59 and tripartite motif-containing protein TRIM18.

Table 8.14. Substrates, regulators, and partners of PP2 phosphatase.

Kinases (inactivation)	AMPK, CK2, CamK1/2/4, CDK1/2, ERK, JNK1, IKK, KSR1, MAP2K1/4, MAPKAPK2, P38MAPK, PAK1, PKA, PKB, PKC, PKG1, PLK, S6K, RSK1/3, Src
Kinases (activation)	CamK4, CK1/2a, GSK3, IKK, JaK2, KSR1, PAK1, PKCε, cRaf, S6K, STK4, Wee1
Phosphatases	PTPRa
Peptidases	Caspase-3
Helicases	UPF1
Receptors, ion channels	NMDAR
Other substrates	ANP32a, BCL2, CcnG1, CDC6, GADD45a, MLL, RBL1, RelA (NFκB), SET, Tau, Tlx1, paxillin, vimentin, importin-β, axin
PP2 regulators	Mid1

degradation (Table 8.14). When Mid1 concentration decays, PP2 level rises and target of rapamycin–TORC1 signaling lowers [657].[18]

Heterotrimer PP2 favors cell survival [658]. Protein phosphatase-2 can both positively and negatively influence the Ras–Raf–MAP2K–ERK signaling pathway (Fig. 8.1). Moreover, PP2 regulates G2–M transition of the cell cycle [653]. Phosphatase PP2 is required to keep the CDK1–CcnB complex[19] in its inactive precursor form. Phosphatase PP2 inhibits complete CDC25 phosphorylation (activation). It is also positively implicated in the exit from mitosis.

Phosphatase PP2 operates in the regulation of various signaling cascades, often in association with other phosphatases and kinases. It modulates the activity of several kinases, such as mitogen-activated protein kinases ERKs, calmodulin-dependent kinases, protein kinases-A, -B, and -C, S6K kinase, inhibitor of nuclear factor-κB IκB kinase, and cyclin-dependent kinases [653]. Agent $PP2_c\alpha$ stabilizes the β-catenin–E-cadherin complex that mediates interactions with the actin cytoskeleton. Phosphatase PP2 participates in the axin–APC–GSK3β complex that is involved in Wnt signaling (Vol. 3 – Chap. 10. Morphogen Receptors).

---

[18] Mutations in the MID1 gene cause the X-linked Opitz syndrome (XLOS) characterized by facial anomalies (hypertelorism [wide-spaced eyes], prominent forehead, broad nasal bridge, anteverted nares, cleft lip and palate, etc.), laryngotracheoesophageal defects, hence dysphagia, heart defects (especially tetralogy of Fallot), thymic aplasia, genitourinary abnormalities (hypospadias [urethra opening on the underside of the penis], cryptorchidism, and hypoplastic/bifid scrotum), hypocalcemia, developmental delay and intellectual disability.

[19] The complex formed by CDK1 and cyclin-B prevents genome re-replication before mitosis. Protein Tyr phosphatase CDC25 dephosphorylates (activates) the CDK1–CcnB complex.

In addition to its substrates, PP2 interacts with other proteins that influence PP2 subcellular location and/or activity. PP2-interacting proteins are involved in translation initiation (with the TOR pathway) and termination, as well as stress response down to cell apoptosis.

In the sarcoplasmic reticulum of cardiomyocytes, PP2 forms a complex with ryanodine receptor, FK506-binding protein FKBP12.6, protein kinase-A, PP1, and A-kinase anchoring protein to regulate ryanodine receptor activity.

Overexpression of PP2 precludes or strongly attenuates thrombin-induced F actin-made stress fiber formation and endothelial barrier protection [659]. Moreover, PP2 regulates the cytoskeleton structure in endothelial cells. Protein phosphatase-2 is associated with tight junction proteins zona occludens-1 and occludin and modulates the phosphorylation state of these junction proteins. Dephosphorylation of tight junction proteins induced by PP2 leads to their redistribution and epi- and endothelial barrier dysfunction.

In pancreatic β cells, glucose with other nutrients stimulates insulin release and triggers Ca^{++} influx to activate calmodulin-dependent protein kinases as well as, via bRaf, ERK1 and ERK2 that participate in nutrient sensing [661].[20] Calcium import also activates Ca^{++}-dependent phosphatase PP3, which allows maximal ERK activation by glucose. In addition, PP3 controls insulin production by ERK-dependent and -independent mechanisms. Calcineurin binds to bRaf and target, in particular, Thr401 of bRaf that is used for negative feedback phosphorylation by ERK kinases, thereby counteracting negative feedback regulation of bRaf and promoting not only bRaf activation, but also bRaf–cRaf heterodimerization [661].[21]

8.3.3 Protein Phosphatase-3

Protein phosphatase-3 (PP3 or PP2b), or calcineurin, is a dimer composed of a catalytic (calcineurin-A) and a regulatory (calcineurin-B) subunit (Table 8.15). Calcineurin-A and -B subunits are encoded by 3 (CNAα, CNAβ, and CNAγ) and 2 (CNB1 and CNB2) genes, respectively. CNAα, CNAβ, and CNB1 are ubiquitous (particularly observed in bone marrow-derived blood cell lineages), but CNAγ and CNB2 are restricted to the germ line. Three PP3 isoforms exist (with >80% identity).

---

[20] Insulin secretagogues and depolarizing stimuli regulate ERK kinases to maintain euglycemia (normal concentration of blood sugars) in the organism.

[21] Among Raf proteins, which homo- or heterodimerize, bRaf is the major activator, whereas aRaf and cRaf fine tune the signal to control the intensity and duration of ERK activity. The bRaf–cRaf heterodimer has a stronger activity than bRaf or cRaf homodimers. Phosphoprotein phosphatase PP2 not only elevates Raf activity, but also assists in the assembling of the cRaf–KSR (kinase suppressor of Ras) complex in response to hormone and growth factor signaling. Phosphatase PP2 elicits Raf activation, as it dephosphorylates its inhibitory sites.

Table 8.15. Subunits of protein phosphatase-3 (CNA, CalnA: calcineurin-A; CNB1, CalnB: calcineurin-B; CBLP: calcineurin-B-like protein; PPP: phosphoprotein phosphatase).

Subunit	Gene	Other alias
Catalytic A subunits		
$PP3_{c\alpha}$	PPP3CA	CNAα, CalnA1
$PP3_{c\beta}$	PPP3CB	CNAβ, CalnA2
$PP3_{c\gamma}$	PPP3CC	CNAγ, CalnA3
Regulatory B subunits		
$PP3_{r1}$	PPP3R1	CNB1, CalnB1
$PP3_{r2}$	PPP3R2	CNB2, CalnB2, CBLP, PP3RL

Phosphatase PP3 is activated by the Ca^{++}–calmodulin complex. Calcium–calmodulin complex activates PP3 by binding to its catalytic subunit and displacing an inhibitory regulatory subunit. Activated PP3 dephosphorylates nuclear factor of activated T cells (NFAT). This transcription factor can then move into the nucleus to launch gene expression.

The PP3–NFAT axis regulates various developmental processes (development of heart valves, nervous and vascular systems, and bone) in addition to control of T-lymphocyte activity and fiber-type switching in skeletal muscle. In particular, PP3 activates the transcription of interleukin-2 that stimulates growth and differentiation of T lymphocytes when the intracellular concentration of calcium rises.

Furthermore, calcium–calmodulin-activated PP3 participates in the regulation of cardiomyocyte growth (Vol. 5 – Chap. 5. Cardiomyocytes). Enzyme PP3 is controlled by regulator of calcineurin (RCan).[22] Inhibitor RCan1 impedes NFAT activation by binding to PP3 active site. Adaptor MAP3K7IP2[23] interacts with RCan1 to recruit MAP3K7, MAP3K7IP1, and PP3, thus forming a signaling complex [660]. Kinase MAP3K7 phosphorylates RCan1, thereby converting RCan1 from an inhibitor to a facilitator of PP3–NFAT signaling and enhancing NFAT1 nuclear translocation to promote hypertrophy of cultured cardiomyocytes. In addition, activated PP3 dephosphorylates (inactivates) MAP3K7 and MAP3K7IP1. The MAP3K7–MAP3K7IP1–MAP3K7IP2–RCan1–PP3 complex thus acts as a stimulator and feedback node, as MAP3K7–MAP3K7IP1–MAP3K7IP2-induced activation of PP3 thereafter dephosphorylates MAP3K7 and impedes MAP3K7 signaling.

---

[22] Regulator of calcineurin was previously named MCIP, DSCR, and calcipressin. The corresponding gene family includes Rcan1 to Rcan3.

[23] A.k.a. TAK1-binding protein TAB2. Adaptor TAB2 links signals from transforming growth factor-β receptor and TRAF proteins to TGFβ-activated kinase TAK1 (MAP3K7) of the mitogen-activated protein kinase family. The latter complexes with TAK1-binding proteins TAB1 and -2 to activate MAP2K3, MAP2K6, and MAP2K4, P38MAPK and JNK, and the NFκB pathway.

8.3.4 Protein Phosphatase-4

Protein phosphatase-4[24] resides in the cytoplasm and nucleus, although it is mostly located at centrosomes [662]. It is involved in microtubule organization at centrosomes during cell division and meiosis.

Protein phosphatase-4 contains regulatory subunits ($PP4_r$) that interact with the catalytic subunit ($PP4_c$) [656]. The regulatory subunit target the centrosomal microtubule organizing centers. Three PP4 regulatory subunits have been identified: immunoglobulin-binding protein IgBP1,[25] $PP4_{r1}$, and $PP4_{r2}$. The $PP4_c$–$PP4_r1$ holoenzyme may dephosphorylate HDAC3.

In response to DNA double-strand breaks, members of the phosphatidylinositol 3-kinase-like family, such as ataxia telangiectasia mutated (ATMK) and AT- and Rad3-related (ATRK) kinases, phosphorylate numerous proteins. Subsequent deactivation of these proteins can result from dephosphorylation. Phosphatase PP2 or PP4 can dephosphorylate the DNA-repair protein H2AX. Phosphatase PP4 dephosphorylates replication protein-A2 (RPa2) [663].[26] Upon DNA repair and dephosphorylation of RPa2, DNA synthesis can resume and the cell cycle starts again.

Phosphatase PP4 has a pro-apoptotic role in T lymphocytes. It associates with T-complex protein TCP1 and chaperonin-containing TCP1 subunits CCT2 (CCTβ), CCT3 (CCTγ), CCT4 (CCTδ), CCT5 (CCTε), CCT6A (CCTζ), CCT7 (CCTη), and CCT8 (CCTθ), as well as TRAF6, and TLR4 [251].

In addition, PP4 interacts with components of NFκB (Rel, P50, and RelA), dephosphorylates RelA, stimulates Rel binding to DNA, and activates NFκB-mediated transcription [277] (Sect. 10.9.1).

Phosphatase PP4 also relays TNFα signal to the JNK pathway. It downregulates insulin receptor substrate-4. Lastly, PP4 interacts with, stabilizes, and activates MAP4K1 (Sect. 5.1.3). Stimulation of TCR enhances PP4–MAP4K1 interaction.

Regulatory subunits-3a and -3b of protein phosphatase-4 ($PP4_{r3a}$ and $PP4_{r3b}$)[27] are involved in the regulation of hepatic glucose metabolism. They are upregulated during fasting that promotes hepatic gluconeogenesis to maintain glucose homeostasis [664]. Their synthesis depends on peroxisome proliferator-activated receptor-γ coactivator PGC1α. The catalytic subunit of PP4 ($PP4_c$) dephosphorylates cAMP-response element binding protein (CREB)-regulated transcriptional coactivator CRTC2 that enables transcriptional activation of gluconeogenic genes in the liver.

---

[24] Formerly called PPX.
[25] A.k.a. B-cell signal transduction molecule-α4 or mammalian protein-α4 (yeast ortholog Tap42). It associates with PP2α and PP2β catalytic subunits, as well as PP4 and PP6.
[26] Protein RPa2 helps in DNA repair via homologous recombination.
[27] A.k.a. suppressor of MEK null SMEK1 and SMEK2.

8.3.5 Protein Phosphatase-5

Protein phosphatase-5[28] localizes to the cytoplasm, centrosomes, and microtubules, or the nucleus. Phosphatase PP5, provided with auto-inhibition from its targeting and regulatory domain, acts in microtubule organization and controls the cell cycle. It regulates steroid receptor signaling [665]. It can be activated in vitro by arachidonic acid.

Its unique feature corresponds to an N-terminal extension that harbors 3 tandemly arranged tetratricopeptide repeat (TPR) motifs [666]. It is predominantly found in brain and, at a lower level, in testis, but is nearly undetectable in spleen, lung, skeletal muscle, kidney, and liver.

Protein phosphatase-5 binds to and inhibits the activated form of MAP3K5 in response to oxidative stress. It can thus prevent oxidative stress-induced, MAP3K5-dependent apoptosis. In fact, PP5 dephosphorylates the activating or inactivating phosphorylation sites of MAP3K5 enzyme.

8.3.6 Protein Phosphatase-6

Protein phosphatase-6[29] resides in both the nucleus and cytoplasm. It operates in cell cycle regulation. Three forms of PP6 mRNA can be detected with high levels in the heart, skeletal muscle, and testis [667].

Protein phosphatase-6 interacts with IGBP1 and TIP41. It has several binding partners, such as Sit4 (PP6)-associated proteins (SAP).

8.3.7 Protein Phosphatase-7

Photoreceptor Ser/Thr protein phosphatase-7 is produced from human retina. It has multiple Ca^{++}-binding sites. Like PP1 to PP6, it possesses a phosphatase catalytic core domain, but has unique N- and C-termini [668]. Human PP7 is directly activated by Ca^{++} and its activity depends on Mg^{++}.

---

[28] A.k.a. PPT.
[29] A.k.a. PPV and Sit4 in yeast.

Table 8.16. Family of Mg^{++}-dependent protein phosphatase-1 (Source: [76]). At least 11 Ppm1 genes have been detected in humans (CamKP: Ca^{++}–calmodulin-dependent protein kinase phosphatase; CamKPn: Ca^{++}–calmodulin-dependent protein kinase phosphatase, nuclear subtype; Fem2: feminization of XX and XO animals protein-2; Fln: fibroblast growth factor-inducible protein phosphatase; HFeM: hereditary hemochromatosis modifier; NerPP2c: neuronal protein phosphatase-2C-related subtype, or neurite extension-related protein phosphatase-2C; POPX: partner of PAK-interacting exchange factor (PIX, or RhoGEF6/7); PP2cm: Mg^{++}- and Mn^{++}-dependent, PP2c-like mitochondrial protein phosphatase; PPM1L: PPM1-like subtype; WiP1: wild-type P53-induced phosphatase-1, or wound-induced protein-1).

Type	Gene	Aliases
PPM1a	Ppm1A	PP2cα
PPM1b	Ppm1B	PP2cβ
PPM1c, PPM1g	Ppm1G	PP2cγ, Fln13
PPM1d	Ppm1D	PP2cδ, WiP1
PPM1e	Ppm1E	CamKPn, POPX1
PPM1f	Ppm1F	CamKP, Fem2, HFeM2, POPX2
PPM1h	Ppm1H	NerPP2c
PPM1j	Ppm1J	PP2cζ
PPM1k	Ppm1K	PP2cκ, PP2cm
PPM1l	Ppm1L	PP2cε
PPM1m	Ppm1M	PP2cη, PP2cη1

8.3.8 Magnesium-Dependent Protein Phosphatase-1

In humans, Mg^{++}- and/or Mn^{++}-dependent protein Ser/Thr phosphatases are encoded by 11 genes (Table 8.16). In addition, 2 splice variants of PPM1a and 6 splice variants of PPM1b have been identified. The PPM1 family include PPM1a to PPM1m, i.e., PP2cα to PP2cη and mitochondrial PP2cκ, as well as PP2c domain-containing protein phosphatase-1F (PPM1f).

Among PPM1 family members, PPM1h, PPM1j, and PPM1m constitute a subfamily. In humans, the PPM1 family has been enlarged to: (1) protein phosphatase PPTC7 homolog;[30] (2) integrin-linked kinase-associated Ser/Thr phosphatase-2C (ILKAP, also PP2cδ); (3) PH domain and leucine-rich repeat protein phosphatases PHLPP1 and PHLPP2; and (4) pyruvate dehydrogenase phosphatases PDP1 and PDP2, hence to 17 members.

Although the PPM and PPP families do not share any sequence homology, PPM1α possesses an active site that is similar to that of the PPP family members.

In humans, several PPM1 phosphatases localize almost exclusively to the nucleus (PPM1a, PPM1d, PPM1e, and PPM1l) [641]. In addition, PPM1b, PPM1g, PPM1k, and PPM1m resides mainly in the nucleus. Isozyme PPM1g is a component of the spliceosome.

---

[30] A.k.a. T-cell activation protein phosphatase-2C (TAPP2C).

Table 8.17. Members of the PPM1 family and their effects ($\oplus \longrightarrow$: stimulation; $\ominus \longrightarrow$: inhibition).

Type	Substrates and partners
PPM1a	$\ominus \longrightarrow$ PI3K, MAP3K3/7, JNK, P38MAPK, IKKβ
	$\ominus \longrightarrow$ CDK9, $\oplus \longrightarrow$ P53
	GluR3
PPM1b	$\ominus \longrightarrow$ PI3K, MAP3K7, JNK, P38MAPK, IKKβ
	$\ominus \longrightarrow$ CDK2/6/9
PPM1c, PPM1g	$\ominus \longrightarrow$ Histone-2
PPM1d	$\ominus \longrightarrow$ ATMK, ATRK, ChK1/2, P53
PPM1e	$\ominus \longrightarrow$ AMPK, CamK1/4
	RhoGEF6/7
PPM1f	RhoGEF6/7
PPM1h	$\ominus \longrightarrow$ CKI1b, exportin-2
PPM1j	UbC9
PPM1k	membrane permeability transition pore
PPM1l	$\ominus \longrightarrow$ MAP3K7
PPM1m	$\ominus \longrightarrow$ MAP3K7
PDP1/2	$\oplus \longrightarrow$ pyruvate dehydrogenase
PHLPP1	$\ominus \longrightarrow$ PKB2/3, kRas
PHLPP2	$\ominus \longrightarrow$ PKB1/3

The catalytic subunits of PPM1a, PPM1b, and PPM1g are stimulated by Mg^{++} ions. On the other hand, PPM1d is inhibited by Mg^{++} ions.

Members of the PPM1 family are involved in many signaling axes that control cell differentiation, proliferation, growth, survival, and metabolism (Table 8.17).

8.3.8.1 PPM1a

Magnesium-dependent protein phosphatase-1A[31] is encoded by the PPM1A gene. Three alternatively spliced transcript variants encode 2 isoforms.

Phosphatase PPM1a impedes response to cellular stresses. It dephosphorylates (inactivates) MAP2K and MAPK enzymes, such as P38MAPK and JNK enzymes.[32]

---

[31] Formely called PP2cα.

[32] Alternatively spliced variant PPM1a2 is a potent inhibitor of P38MAPK and JNK kinases activated by tumor-necrosis factor-α. Isoform PPM1a2 can also dephosphorylate (inactivate) MAP2K3, MAP2K4, MAP2K6, and MAP2K7 enzymes. In fact, both PPM1a2 and PPM1b can hinder the activation of P38MAPK and JNK enzymes.

Phosphatase PPM1a can also dephosphorylate cyclin-dependent kinases, in particular CDK9, the catalytic subunit of a general RNA polymerase-2 elongation factor [669].[33]

Upon stimulation by neurotransmitters, hormones, growth factors, cytokines, and antigens, various plasmalemmal receptors activate phosphatidylinositol 3-kinase that generates the lipid second messenger phosphoinositide 3-phosphate. In particular, the PI3K–PKB pathway supports cell growth, proliferation, and survival. Once activated, class-1A PI3K catalytic subunit ($PI3K_{c1}$ or $^{P110}PI3K$) phosphorylates $^{P85}PI3K$ regulatory subunit, thereby impeding its lipid kinase activity (negative feedback). Like PP1 and PP2, PPM1a regulates P85 phosphorylation state. In the differentiation of fibroblasts into adipocytes, PPM1a expression rises, whereas those of PP1, PP2, and PP3 decay [670].

Overexpression of PPM1α activates P53 transcription factor to provoke G2–M arrest and apoptosis. Phosphatase PPM1α dephosphorylates metabotropic glutamate GluR3 receptor [251]. It can also interact with inhibitor latexin (Lxn), or endogenous carboxypeptidase inhibitor (ECI), that can complex with carboxypeptidase-A4.

8.3.8.2 PPM1b

Magnesium-dependent Ser/Thr protein phosphatase-1B (PPM1b)[34] interacts with numerous proteins [251]: inhibitor of nuclear factor-κB kinases IKKα (or IKK1), IKKβ (or IKK2), and IKKγ, catalytic subunit-α of Ser/Thr protein phosphatase-2, cyclin-dependent kinases CDK2 and CDK6, MAP3K7,[35] coregulators annexin-A1 and -A2, actin-binding proteins villin-1 and gelsolin, intermediate filament component keratin-18, S100 calcium-binding protein-A8 (S100a8, or calgranulin-A), inducible polyadenylate-binding protein PABPc4 that binds to the poly(A) tail at the 3' ends of mRNAs,[36] ATP-binding platform for protein assembly valosin-containing protein (VCP), and Ras guanine nucleotide-releasing protein RasGRP1.

Phosphatase PPM1b is widely distributed with its highest expression in cardiac and skeletal myocytes. Unlike PPM1b1 variant, PPM1b2 is not widespread, but specifically lodges in the brain and heart.

---

[33] A.k.a. positive transcription elongation factor pTEFb. This kinase complex regulates elongation of most protein-coding genes transcribed by RNA polymerase-2. It enhances the processivity of RNA polymerase-2, as it phosphorylates the C-terminus of the polymerase as well as antagonizes the action of negative elongation factor. It is composed of CDK9 and its cyclin partner, either cyclin-T1, cyclin-T2, or cyclin-K.

[34] Formely called PP2cβ.

[35] A.k.a. TGFβ-activated kinase TAK1.

[36] The poly(A) tail consists of multiple adenosine monophosphates.

Isozyme PPM1b dephosphorylates (inactivates) IKKβ stimulator. Like PPM1a, PPM1b acts as an IKKβ phosphatase that terminates NFκB activation generated by tumor-necrosis factor-α [671]. Furthermore, PPM1b and PPM11 dephosphorylate (inactivate) MAP3K7, thereby suppressing IL1-induced activation of JNK kinase.

8.3.8.3 PPM1c (PPM1g)

Subtype PPM1c, also called PPM1g and PPM1γ, mostly resides in the heart, skeletal muscle, and testis. It helps to form the spliceosome; this splicing factor dephosphorylates specific substrates required for the spliceosome formation.

Phosphatase PPM1c also regulates phosphorylation of H2A and H2B histones.[37] Together with PP2, PPM1c intervenes in the recovery phase of the DNA-damage response [670].

8.3.8.4 PPM1d

Overexpression of PPM1d subtype can stop apoptosis. It is overexpressed in many cancers. Phosphatase PPM1d can deactivate both P53 and checkpoint kinase ChK1 that activates P53 in response to DNA damage. Although PPM1d is stimulated by P53, it impedes the activity of P38MAPK, thereby preventing P53 phosphorylation (negative feedback). In addition, PPM1d represses ATMK signaling by eliminating ATMK phosphorylation (Ser1981) as well as removes the phosphorylation of ChK1 (Ser345) and ChK2 (Thr68) checkpoint kinases by ATMK and ATRK enzymes [670].

8.3.8.5 PPM1e

Among its substrates, Mg^{++}- and Mn^{++}-dependent protein Ser/Thr phosphatase PPM1e targets AMP-activated protein kinase (AMPK) [672]. This phosphatase is also an interactor of guanine nucleotide-exchange factors RhoGEF6 and RhoGEF7 that regulate P21-, CDC42-, and Rac1-activated kinase PAK1, thereby repressing PAK1 effects. Glyceraldehyde 3-phosphate dehydrogenase (GAPDH) and fructose bisphosphate aldolase as major binding partners of PPM1e phosphatase, the latter counteracting GAPDH phosphorylation by CamK1 or CamK4 [673]. Calmodulin-dependent PPM1e phosphatase indeed dephosphorylates both CamK1 or CamK4 phosphorylated by Cam2K, but not GAPDH enzyme.

---

[37]Repair of DNA damage occurs when DNA is wrapped by an octamer made of H2a, H2b, H3, and H4 histones.

8.3.8.6 PPM1f

Calmodulin-, Mg^{++}-, and Mn^{++}-dependent protein Ser/Thr phosphatase PPM1f, like PPM1e, dephosphorylates Ca^{++}–calmodulin-dependent protein kinases, in particular CamK2γ.[38] Whereas PPM1e localizes to the nucleus, PPM1f resides in the cytosol. Like PPM1e, PPM1f interacts with guanine nucleotide-exchange factors RhoGEF6 and RhoGEF7, thus preventing the action of P21-, CDC42-, and Rac1-activated kinase PAK1. The latter is a substrate of PPM1f phosphatase.

8.3.8.7 PPM1h

The Ser/Thr phosphatase PPM1h dephosphorylates cyclin-dependent kinase inhibitor CKI1b [674]. It controls cell cycle and proliferation of cancer cells, as it dephosphorylates exportin-2.[39] It has a greater affinity for Mn^{++} than Mg^{++} ions. In particular, PPM1h prefers Mn^{++} when nitrophenyl phosphate or phosphopeptide is the substrate and Mg^{++} when casein is the substrate [675].

8.3.8.8 PPM1j

Subtype PPM1j is specifically synthesized in testicular germ cells. It can associate with ubiquitin conjugase UbC9. This association is enhanced by coexpression of small ubiquitin-related modifier-1 (SUMo1).

8.3.8.9 PPM1k

Mitochondria are involved in cellular metabolism and survival. They are integrative sites of cell signaling, as phosphorylation and dephosphorylation cycles occur in mitochondria. Many Ser/Thr protein kinases (PKA, PKB, PKC, and JNK) localize to mitochondria, where they contribute to the regulation of cell metabolism and apoptosis, as they target matrix and inner and outer membrane components (e.g., ATP synthase, voltage-dependent anion channel [porin], BCL2 antagonist of cell death [BAD], and BCL2-associated X protein [BAX]).

Conversely, mitochondrial proteins are dephosphorylated by resident mitochondrial protein phosphatases, such as protein Ser/Thr phosphatases PP1 and PP2, as well as mitochondrial protein Tyr phosphatase PTPM1, or dual-specificity phosphatase DuSP23 [676]. The latter participates in ATP production and insulin secretion.

---

[38] In response to Ca^{++} stimuli, CamK2 kinase autophosphorylates (Thr286 for CamK2α), thereby leading to an autonomous (Ca^{++}-independent) activity. Dephosphorylation can be caused by protein PP1, PP2, and PPM1 phosphatases.

[39] A.k.a. importin-α re-exporter and chromosome segregation 1-like protein (CSE1L).

In addition, PPM1k localizes exclusively to the mitochondrial matrix [676]. Loss of PPM1k does not affect mitochondrial oxygen consumption and ATP production, but renders mitochondria hypersensitive to membrane permeability transition pore opening. Hence, PPM1k contributes to cell death regulation [670].

8.3.8.10 PPM1l

Subtype PPM1l inhibits the MAP3K7- and IL1-induced activation of the MAP2K3–P38MAPK and MAP2K4–JNK pathways. Therefore, like PPM1a and PPM1b, PPM1l interacts directly or indirectly with several components of the MAPK modules to operate in stress responses and regulate cell death and survival [670].

8.3.8.11 PPM1m

Protein Ser/Thr phosphatase PPM1m, or PPM1m1, can localize mainly in the nucleus. An alternative splice variant — PPM1m2 — resides in the cytoplasm. This isoform inhibits interleukin-1 signaling mediated by nuclear factor-κB that triggers inflammation, as it inhibits MAP3K7, hence IKKβ [671].[40] Both PP2 and PP3 also hamper the NFκB pathway. The former dephosphorylates P65$_{NF\kappa B}$ (RelA), the latter dephosphorylates IκB.

8.3.8.12 PH Domain and Leucine-Rich Repeat Protein Phosphatases

PH domain and Leu-rich repeat protein phosphatases (PHLPP) constitute a family with 2 identified members, PHLPP1[41] and PHLPP2.[42] They dephosphorylate (inactivate) all protein kinase-B isoforms (PKB1–PKB3; Ser473) as well as conventional and novel protein kinase-C isoforms. Both PHLPP1 and PHLPP2 selectively dephosphorylate PKB at Ser473, but not Thr308, which is targeted by PP2

---

[40]Interleukin-1 is a pro-inflammatory cytokine that initiates a cascade characterized by the activation of NFκB and JNK that upregulate several inflammatory genes. Once IL1 is bound to its cognate receptor, the adaptor myeloid differentiation factor MyD88, IL1R-associated kinases IRAK1 and IRAK4, and tumor-necrosis factor receptor-associated factor TRAF6 are recruited. Phosphorylation and degradation of IRAK1 induces the ubiquitination of TRAF6 that then complexes with transforming-growth-factor-β-activated kinase (TAK1)-binding protein TAB2 or TAB3 for subsequent activation of TAK1 (MAP3K7). The latter activates IKKβ that phosphorylates IκBα degradation, thereby allowing NFκB translocation to the nucleus and transcription of several IL1-regulated genes. Enzyme MAP3K7 also activates JNK via MAP2K4 and MAP2K7. Once activated, JNK translocates to the nucleus to boost the activity of the transcriptional activator AP1 complex.

[41]A.k.a. suprachiasmatic nucleus circadian oscillatory protein (SCOP).

[42]A.k.a. PH domain leucine-rich repeat-containing protein phosphatase-like protein (PHLPPL).

phosphatase [670]. Phosphatase PHLPP1 dephosphorylates PKB2 that induces glucose transport upon insulin stimulation and PKB3 predominantly expressed in brain, whereas PHLPP2 targets PKB1 involved in growth and PKB3 enzyme.

Both PHLPP1 and PHLPP2 contain a PH domain for membrane recruitment and a leucine-rich repeat (LRR) domain to impede kRas activation. In particular, PHLPP1 is constitutively expressed in neurons and colocalizes in membrane rafts with kRas^{GDP}, prevents GTP binding to kRas, thereby precluding kRas–MAPK signaling in membrane rafts [677].

8.3.8.13 Pyruvate Dehydrogenase Phosphatases

The first irreversible reaction in glucose oxidation is catalyzed by the pyruvate dehydrogenase complex inside mitochondria. Pyruvate dehydrogenase (PDC^{E1}) is a component of the large pyruvate dehydrogenase complex (PDC).[43] Pyruvate dehydrogenase kinases phosphorylates (inactivates) Ser residues of pyruvate dehydrogenase component, thereby inhibiting the enzymatic complex. Conversely, pyruvate dehydrogenase phosphatases dephosphorylate (activate) pyruvate dehydrogenase component. All these enzymes localize in the mitochondrial matrix.

Magnesium-dependent, calcium-stimulated, pyruvate dehydrogenase phosphatase PDP1, which belongs to the protein phosphatase PPM1 category, is a heterodimer with a catalytic (PDP1$_c$) and a regulatory subunit (PDP1$_r$). On the other hand, PDP2 consists of a single catalytic subunit (PDP2$_c$) [678]. Two catalytic subunit types abounds in the skeletal muscle and liver, respectively.

In skeletal myocytes and hepatocytes, insulin elevates the activity of the pyruvate dehydrogenase complex, as it causes activation and mitochondrial translocation of PKCδ that phosphorylates (activates) PDP1 and PDP2 phosphatases [679].

8.3.9 Cytosolic Protein Tyrosine Phosphatases

Classical protein Tyr phosphatases can be classified as: (1) plasmalemmal, receptor or receptor-like, transmembrane enzymes (rPTP or PTPR; Vol. 3 – Chap. 9. Receptor Tyrosine Phosphatases) and (2) cytoplasmic (cPTP or non-receptor PTPn) proteins. Protein Tyr phosphatases can dephosphorylate (inactivate) receptor Tyr kinases (RTK), such as EPH receptors [680].

---

[43] Pyruvate dehydrogenase tetramer catalyzes the oxidative decarboxylation of pyruvic acid and reductive acetylation of lipoic acid residues of dihydrolipoamide acetyltransferase (PDC^{E2}). The latter transfers acetyl moieties to CoA, and lipoic residues are reoxidized by dihydrolipoamide dehydrogenase (PDC^{E3}) with the formation of NADH.

Table 8.18. Examples of signaling substances targeted by protein tyrosine phosphatases – strictly tyrosine or dual-specificity phosphatases – (PTPn11: protein Tyr phosphatase non-receptor type 6 [a.k.a. SHP2: Src homology protein Tyr phosphatase-2]; PTen: dual-specificity phosphatase and tensin homolog; MKP: dual-specificity mitogen-activated protein kinase phosphatase; Source: [680]).

Tyr phosphatase	Signaling molecules
PTPn11	PDGF, endothelin-1
PTen	PDGF, EGF
MKP (JNK phosphatase)	TNFα

Table 8.19. Classification of protein Tyr phosphatases (PTP; **Part 1**; NRPTP: non-receptor protein Tyr phosphatases [superfamily]; PTPR: protein Tyr phosphatase receptor; RPTP: receptor protein Tyr phosphatases [superfamily]).

	Infraclass-1 Cys-based PTPs (PTPIC1)
	Hyperfamily of classical PTPs
RPTP	PTPRa–PTPRh, PTPRj–PTPRk, PTPRm, PTPRn–PTPRn2, PTPRo, PTPRq–PTPRv, PTPRz1, PTPRz2(?)
NRPTP	PTPn1–PTPn7, PTPn9, PTPn11–PTPn14, PTPn18, PTPn20a–PTPn20c, PTPn21–PTPn23

In humans, among (at least) 107 known protein Tyr phosphatases, 11 are catalytically inactive, 2 dephosphorylate mRNAs, and 13 dephosphorylate inositol phospholipids. The remaining 81 PTPs dephosphorylate Tyr^P residue.

Phosphatases that are able to dephosphorylate Tyr^P residues are classified into various superfamilies (superfamilies of plasmalemmal and cytosolic protein Tyr phosphatases and dual-specificity phosphatases). Among them, dual-specificity phosphatases target not only Tyr, but also Ser and Thr residues. For example, MAPK dephosphorylation (inactivation) is catalyzed by cell cycle regulatory phosphatase CDC25 and PTen (phosphatase and tensin homolog deleted on chromosome 10), an antagonist of the PI3K–PKB pathway, hence its designation as a tumor-suppressor phosphatase (Table 8.18).

Protein Tyr phosphatases are classified according to the key amino acid involved in the catalysis: Cys-based and Asp-based PTPs. The Cys-based PTPs share a canonical P loop in the phosphatase domain that contains the cysteine required to form the thiol phosphate intermediate. In fact, protein Tyr phosphatases can be grouped into 4 infraclasses depending on the amino acid sequence of their catalytic domains (Tables 8.19 to 8.21). The Cys-based PTPs are indeed subdivided based on the characteristics of the catalytic domain.

Protein Tyr phosphatases of the 3 infraclasses display a highly conserved catalytic domain (~ 240 residues). Phosphatases of infraclass 1 contain a catalytically inactive CDC25 homology (CH2) domain. These phosphatases possess an Asp distant from the catalytic Cys that acts in the catalysis via protonation.

8.3 Cytosolic Protein Serine/Threonine Phosphatases

Table 8.20. Classification of protein Tyr phosphatases (PTP; **Part 2**; ADUSP: atypical dual-specificity phosphatase hyperfamily [aDuSP proteins]; CDC14: cell division cycle phosphatase-14; DuSP: dual-specificity phosphatase; MKP: mitogen-activated protein kinase phosphatase; MTM: myotubularin (myotubular myopathy-associated gene product); PRL: phosphatase of regenerating liver; PTEN: superfamily of phosphatase and tensin homolog [PTen] deleted on chromosome 10 [ten] and its related proteins; PTP4A: family of type-4A protein tyrosine phosphatases [PTP4a1–PTP4a3]; SBF: SET-binding factor; SSH: family of slingshot homolog phosphatases [Ssh]).

	Infraclass-1 Cys-based PTPs (PTPIC1)
	Hyperfamily of dual-specificity phosphatases
MKP	DuSP1–DuSP2, DuSP4, DuSP10, STYXL1
ADUSP	DuSP3, DuSP11–DuSP12, DuSP13a–DuSP13b, DuSP14–DuSP15, DuSP18–DuSP19, DuSP21–DuSP23, DuSP26–DuSP28, Laforin (EPM2a), PTPmt1, RNGTT, STYX
SSH	Ssh1–Ssh3
PTP4A	PTP4a1 (PRL1)–PTP4a3 (PRL3)
CDC14	CDC14a–CDC14b, CDKn3, PTPDC1
PTEN	PTen, TenC1 (Tns2), Tns1, TPTe1–TPTe2
MTM	MTM1, MTMR1–MTMR5 (SBF1), MTMR6–MTMR13 (SBF2), MTMR14

Infraclass-1 cysteine-based PTPs constitute the largest PTP subfamily with at least 102 proteins. It comprises 2 hyperfamilies: (1) classical PTPs that are strictly Tyr-specific (at least 41 members) and constitute the 2 superfamilies of transmembrane and cytoplasmic PTPs and (2) dual-specific protein phosphatases with at least 61 members that form diverse families with numerous types of substrates, from 5' cap end nucleotide structure of primary RNA transcripts such as precursor messenger RNAs, to inositol phospholipids and Ser^P, Thr^P, and Tyr^P proteins. Among these dual-specificity protein Tyr phosphatases, some have non-proteic substrates, such as phosphatase and tensin homolog and myotubularins that dephosphorylate phosphoinositides. In addition, laforin dephosphorylates phosphorylated glycogen; RNA guanylatetransferase and 5'-phosphatase (RNGTT) can remove a phosphate from the 5' end of nascent mRNA; mitochondrial PTPmt1 selectively processes mainly phosphatidylglycerol phosphate.[44]

Infraclass-2 cysteine-based PTP in humans corresponds to a single phosphatase, the Tyr-specific low-molecular-weight phosphatase, that has isoforms. The phosphatase domain has a structure that differs from that of Cys-based PTPs of 2 other

---

[44] The resulting product phosphatidylglycerol is a component of pulmonary surfactant and a precursor of cardiolipin.

Table 8.21. Classification of protein Tyr phosphatases (PTP). **(Part 3)** Infraclasses 2 to 4 (CTDP1: carboxy (C)-terminal domain (CTD), RNA polymerase-2, polypeptide-A phosphatase subunit-1 [a.k.a. RNA polymerase-2 subunit-A C-terminal domain phosphatase and TF2F-associating CTD phosphatase; CTDSP: CTD-containing small phosphatase [CTDSP1 is a.k.a. small CTD-containing phosphatase SCP1 and nuclear LIM interactor (NLI)-interacting factor NIF3; CTDSP2 as NIF2 and SCP2]; CTDSPL: CTD-containing small phosphatase-like protein [CTDSPL1 is a.k.a. NIFL (NLI-interacting factor-like protein) and SCP3]).

Set	Members
Infraclass-2 of Cys-based PTPs (PTPIC2)	
Acid phosphatase-1	AcP1
Infraclass-3 of Cys-based PTPs (PTPIC3)	
Cell division cycle-25 homologs	CDC25a–CDC25c
Infraclass of Asp-based PTPs (PTPIC4)	
Eye absent homologs	EyA1–EyA4
Haloacid dehalogenases	HADs
CTD phosphatase	CTDP1 (FCP1)
CTDSP phosphatases	CTDSP1–CTDSP1 (SCP1–SCP2), CTDSPL1–CTDSPL2

infraclasses. The Asp involved in the catalysis, which is N-terminally located w.r.t. the catalytic Cys in infraclass-1 PTPs, lodges in the C-terminus of AcP1 (or lmwPTP).

Infraclass-3 cysteine-based PTPs are rhodanese-related Tyr/Thr-specific phosphatases represented by 3 cell cycle regulatory CDC25 phosphatases (CDC25a–CDC25c). The phosphatase domain contains a rhodanese fold. The catalytic residue Asp is absent in CDC25 enzymes. The CDC25 phosphatases lack any sequence similarity to other DuSPs, except the canonical P-loop Cys–X_5–Arg (CX_5R) motif in the phosphatase domain with the Cys residue common to all Cys-based PTPs. Infraclass-3 phosphatases dephosphorylate cyclin-dependent kinases.

The last infraclass includes *aspartate-based PTPs*, for which the aspartic acid acts as a key element in the catalysis. Activity of these phosphatases depends on the presence of divalent cations. Members of the Asp-based phosphatase infraclass correspond to: (1) Eyes absent (EyA) proteins that are Tyr-specific protein phosphatases and operate as transcriptional coactivators and (2) bacterial haloacid dehalogenases (HAD) that are either Tyr-, Ser-, or dual Ser/Tyr-specific protein phosphatases and have various substrates (proteins, phospholipids, carbohydrates, and nucleotides).[45] Other members of this infraclass dephosphorylate Ser^P and

---

[45] Humans possess haloacid dehalogenase-like hydrolase domain-containing proteins HDHD1 to HDHD3. Haloacid dehalogenase-like hydrolase domain-containing protein-1 (HDH1 or HDHD1a) is also named pseudouridine 5′-monophosphatase and GS1 protein. In humans, phos-

Thr^P substrates. *Chronophin* regulates the cytoskeleton by processing cofilin. Carboxy-terminal domain *(CTD), RNA polymerase-2, polypeptide-A phosphatase subunit-1* (CTDP1)[46] dephosphorylates the C-terminal domain of RNA polymerase-2, thereby regulating transcription. Carboxy-terminal domain *(CTD)-containing nuclear envelope phosphatase* CTDNEP1, or Dullard homolog, participates in nuclear membrane genesis, as it dephosphorylates the phosphatidic acid phosphatase lipin, a phosphatidate phosphatase and transcriptional coactivators.[47]

8.3.10 Protein Tyrosine Phosphatase Regulation by Reversible Oxidation

Oxidation of the catalytic cysteine of protein Tyr phosphatases causes a reversible inactivation. Different types of oxidative modification of the PTP catalytic cysteine exist.[48] This mechanism ensures signaling by reactive oxygen or nitrogen species (Tables 8.22 and 8.23). Both the general cellular redox state and extracellular ligand-stimulated ROS production can prime PTP oxidation [682] (Sect. 10.6). However, protein Tyr phosphatases differ in their intrinsic susceptibility to oxidation.[49] Many types of oxidants interact with PTP active sites (Table 8.24).

Growth factors cause oxidation of many classical PTPs, in addition to other cell surface triggers (Table 8.25). In addition to classical PTPs, dual-specificity phosphatases (DuSP) and PTen that catalyze dephosphorylation by a cysteine-based mechanism undergo oxidation (inactivation; Table 8.26).

---

phoserine phosphatase (PSP), an enzyme involved in serine synthesis, shares 3 motifs with P-type ATPases and haloacid dehalogenases.

[46] A.k.a. TF2F-associating C-terminal domain-containing phosphatase FCP1 and RNA polymerase-2 subunit-A C-terminal domain phosphatase.

[47] Lipins have 2 substrates — phosphatidate and water — and 2 products — (1,2)-diacylglycerol and phosphate —. Cytosolic phosphatidate phosphatases (PAP) dephosphorylate phosphatidate, thereby contributing to glycerolipid synthesis. They control the balance between phosphatidate used for the synthesis of acidic phospholipids, and diacylglycerol used to synthesize triacylglycerol, phosphatidylcholine, and phosphatidylethanolamine [681]. Lipins function as both homo- and hetero-oligomers at the endoplasmic reticulum. These cytosolic enzymes translocate from the cytosol to membranes, particularly that of the endoplasmic reticulum, the major site of glycerolipid synthesis. Haloacid dehalogenase (HAD)-like domain-containing lipin-1 is the major phosphatidate phosphatase in adipose tissue, myocardium, and skeletal muscle. Two lipin-1 splice variants, full-length lipin-1b and short lipin-1a, operate as both phosphatases and transcriptional coactivators. A third lipin-1 isoform abounds in the brain (lipin-1γ). Lipin-2 has high levels in the liver. Lipin-3 is highly expressed in the intestine.

[48] The catalytic site of classical PTPs contains a cysteine, the SH group of which exists in the thiolate state (S$^-$) to enable catalysis.

[49] The sensitivity of the catalytic D2 domain of RPTPa is higher than that of the catalytic D1 domain.

Table 8.22. Protein Tyr phosphatases and reactive oxygen species (Source: [682]; AA; arachidonic acid; COx: cyclooxygenase; GF: growth factor; Gpx: glutathione peroxidase; LOx: lipoxygenase; NOx: NADPH oxidase; PLA2: phospholipase A2; Prx: peroxiredoxin; RTK: receptor Tyr kinase; SHC: Src homology-2 domain-containing transforming protein). A reducing enviroment, i.e., low activity of ROS producers and high activity of ROS scavengers, lowers PTP oxidation, thereby enhancing their activation and attenuating RTK signaling. Conversely, an oxidizing enviroment elevates PTP oxidation and improves RTK activity. Phosphorylation (activation) of SHC1 causes mitochondrial production of hydrogen peroxide (H_2O_2). Inhibitory phosphorylation of ROS scavenger Prx1 is another mechanism for growth factor-induced PTP oxidation.

Producers	Inhibitors
ROS	
NOx	Catalase
SHC1 (mitochondrion)	Prx, Gpx
Peroxidized lipids	
COx, LOx	Gpx
PTP oxidation (inactivation) pathways	
GF–RTK–PI3K–Rac–NOx	
GF–RTK–SHC1	
GF–RTK⊖ ⟶ Prx1	
GF–RTK–cPLA$_2$–AA–peroxidized lipids	

Table 8.23. Redox enzymes in PTP oxidation (Source: [682]).

	ROS producers
DuOx1	PTPn11 oxidation for enhanced and sustained TCR signaling
NOx	PTPn12 oxidation in focal adhesions for cell migration
NOx1	PTPn1 oxidation in IL-4 signaling
NOx4	PTPn1 oxidation at the endoplamoic reticumul in EGF signaling
SOD1	PTP oxidation in growth factor signaling
	ROS scavengers
Gpx1	Scavenges H2O2 to protect PTen from oxidation, desensitization of insulin signaling
Gpx4	Protection of PTPs from oxidation by peroxided lipids in PDGF signaling
Prx2	Protection of PTPs from oxidation in PDGF signaling

8.3.11 Classical Cytosolic Protein Tyrosine Phosphatases

In humans, protein Tyr phosphatases can be categorized into 20 sets according to the presence, number, and position of each detected domain [683]. For example, PTPn5, PTPn7, and PTPn20 contain a single PTPase domain at their C-terminus.

8.3 Cytosolic Protein Serine/Threonine Phosphatases

Table 8.24. Examples of PTPs inhibited by oxidants, such as ROS, RNS, peroxidized lipids such as 15-hydroperoxy eicosatetraenoic acid, and toxic constituents of ambient particulate matter, such as quinones and aldehydes (Source: [682]). Reaction with H_2O_2 oxidizes PTP active-site thiolates initially to sulfenic acid (SOH), and then, at high concentrations, sulfinic (SO2H) and sulfonic acid (SO3H) derivatives.

Oxidant	Target PTPs
Hydrogen peroxide	RPTPa, PTPn1/6/11, DuSP3/6, CDC25, PTen
Superoxide	PTPRc, PTPn1
Lipid peroxides	PTPn2/3/6
Oxidized glutathione	PTPRm, PTPn1/6, DuSP6, PTen
S-nitrosothiols	PTPn1, PTen
Polyaromatic quinones	PTPRc, PTPn1
Amino acid and peptide hydroperoxides	PTPn1
Peroxymonophosphate	PTPn1
Hypothiocyanous acid	PTPn1

Table 8.25. Oxidation of PTPs by activated plasmalemmal receptors (Source: [682]; EGFR: epidermal growth factor receptor; InsR: insulin receptor; PDGFR: platelet-derived growth factor receptor; VEGFR: vascular endothelial growth factor receptor).

Plasmalemmal receptors	Target PTPs
G-protein-coupled receptors	
Adiponectin	PTPn1
Angiotensin-2 receptor-1A	PTPn11
Endothelin-1 receptor	PTPn11
S1P$_2$	PTPn1
Urotensin-2 receptor	PTPn11
Receptor Tyr kinases	
EGFR	PTPn1, PTen
InsR	PTPn1, PTPn2, PTen
PDGFR	PTPn11, lmwPTP, PTen
VEGFR2	PTPRj, PTPn1
Cytokine receptors	
Interleukin-4 receptor	PTPn1
TNFR1	MKP1/3/5/7
Immune system receptors	
BCR	PTPn6
TCR	PTPn6/11
Integrins	
Integrins	PTPn11, lmwPTP

Table 8.26. Regulation of protein Tyr phosphatases (PTP) by reversible oxidation of the active site cysteine (Source: [682]; DuSP: dual-specificity phosphatase; MKP: MAPK phosphatase; PRL: phosphatase of regenerating liver; PTen: phosphatase and tensin homolog).

Category	Members
	Infraclass-1 Cys-based PTPs
Transmembrane PTPs (receptor-like)	PTPRa, PTPRc, PTPRf, PTPRj, PTPRk, PTPRm, PTPRz
Cytosolic PTPs	PTPn1, PTPn2, PTPn3, PTPn6, PTPn11, PTPn12, PTPn13, PTPn22
DSPs	DuSP1 (MKP1), DuSP6 (MKP3), DuSP10 (MKP5), DuSP16 (MKP7), DuSP3, DuSP12
PRLs	PRL1, PTP4a1
	Infraclass-2 Cys-based PTPs
	lmwPTP
	APC1
	Infraclass-3 Cys-based PTPs
CDC25	CDC25b, CDC25c
PTEN	PTEN PTen, Tns1/2, TPTe1/2

Subtype PTPn1 and PTPn2 also have a single PTPase domain, but at their N-terminus.[50] The B41 domain responsible for targeting of cytoskeletal proteins to the membrane–cytoskeleton interfaces is found in PTPn3, PTPn4, PTPn13, PTPn14, and PTPn21 phosphatases.

Protein Tyr phosphatases counteract the reversible protein Tyr phosphorylation on many substrates carried out by protein Tyr kinases to regulate cell differentiation and proliferation (Tables 8.27 to 8.30).

8.3.11.1 Non-Receptor Protein Tyrosine Phosphatase-1

Non-receptor protein Tyr phosphatase-1 (PTPn1 or PTPn1b) is an abundant enzyme with a broad tissue distribution. It is associated with the outer leaflet of the endoplasmic reticulum membrane [684]. Growth factor stimulation provokes production of a PTPn1 splice variant with a modified C-terminus. Phosphatase PTPn1 is encoded by the PTPN1 gene (N: non-receptor).

---

[50]Similarly, plasmalemmal phosphatases PTPRb and PTPRj have a single C-terminal PTPase domain preceded by a transmembrane domain. However, each protein can be classified differently as the former has 17 fibronectin type-3 repeats, whereas the latter has 9 fibronectin type-3 repeats.

8.3 Cytosolic Protein Serine/Threonine Phosphatases 423

Table 8.27. Partners and substrates of cytoplasmic protein Tyr phosphatases (**Part 1**; Sources: Wikipedia and BioGRID; ADAM: a disintegrin and metallopeptidase ATRX: α-thalassemia/mental retardation syndrome RAD54 homolog X-linked [DNA-dependent ATPase and helicase]; eEF: eukaryotic translation elongation factor; EGFR: epidermal growth factor receptor; GP: glycoprotein; GluRδ2: ionotropic glutamate receptor-δ subunit [GRID2]; GluN2a: ionotropic glutamate receptor, N-methyl D-aspartate subunit [GRIN2A]; IGF1R: insulin-like growth factor-1 receptor; InsR: insulin receptor; LPase: lipoprotein lipase; YWHAB: Tyr 3-monooxygenase/Trp 5-monooxygenase activation protein of the 14-3-3 protein family (14-3-3β/α); ZZEF1: zinc finger, ZZ-type with EF-hand domain-1 protein).

Gene	Expression pattern	Partners and substrates
PTPN1	Ubiquitous	EGFR, InsR, IGF1R, IRS1 NTRK1–3, LTK, JaK2, TyK2, GSK3β GRB2, BCAR1 (CAS), CRK, STAT5a/b Cav1, Catenin-β1
PTPN2	Hematopoietic cell	JaK1/3, STAT1/3 14-3-3σ
PTPN3	Neuron	CDC48, YWHAB, ADAM15, 14-3-3σ
PTPN4	Neuron	GluRδ2, GRIN2A eEF1α and γ, ATRX, ZZEF1, LPase
PTPN5	Neuron	Fyn, ERK1/2, P38MAPK

Phosphatase PTPn1 particularly inhibits signaling downstream of insulin and leptin receptors [684]. Phosphatase PTPn1 indeed dephosphorylates (inactivates) insulin receptor Tyr kinase (InsR) and possibly insulin receptor substrate proteins as well as both mediators JaK2 and STAT3 of the leptin pathway [685] (Tables 8.27 to 8.29). It also inhibits prolactin signaling by preventing nuclear translocation of dephosphorylated effectors STAT5a and STAT5b [686]. On the other hand, PTPn1 dephosphorylates (activates) Src kinase, particularly in platelets.

Phosphatase PTPn1 also regulates the insulin response element-mediated unfolded protein response in the endoplasmic reticulum. It reduces caveolin-1 phosphorylation (Tyr14), thus its docking capacity. It forms a complex with α-actinin and dephosphorylates (Tyr397; inactivates) FAK [684]. Among hyper(Tyr) phosphorylated proteins that link to PTPn1, there are regulators of cell motility and adhesion, such as cortactin, lipoma-preferred partner, tight junction protein zona occludens ZO1, and catenin-δ1, as well as docking protein Dok1 and p120RasGAP to lower cell motility and heighten proliferation [687].

Phosphatase PTPn1 also interacts with adaptor GRB2 and cytoskeletal component adaptor BCAR1 (CAS or p130CAS) as well as SUMo ligase Protein inhibitor of activated STAT PIAS1. It can associate with and dephosphorylate many receptor (PDGFRa and PDGFRb, InsR, IGFR1, and EGFR) and non-receptor (JaK2, TyK2, and Src) protein Tyr kinases [684]. It also binds to N- and E-cadherins, then dephosphorylates β-catenin to increase the strength of cadhesin–catenin linkage, thus intercellular adhesion.

Table 8.28. Partners and substrates of cytoplasmic protein Tyr phosphatases (**Part 2**; Sources: Wikipedia and BioGRID; Arfaptin: adpribosylation factor interacting protein; BCR: B-cell receptor; EGFR: epidermal growth factor receptor; EpoR: erythropoietin receptor; FCRL3: Fc receptor-like-3; Hox: homeobox transcription factor; InsR: insulin receptor; LAIR: leukocyte-associated immunoglobulin-like receptor; LCP: lymphocyte cytosolic protein (LCP2 alias: SLP76); LILRB: leukocyte immunoglobulin-like receptor-B; M6PRBP: mannose-6-phosphate receptor-binding protein [or perilipin-3]; PAG: phosphoprotein associated with glycosphingolipid microdomains [or CBP: CSK-binding protein]; PILRα: paired immunoglobin-like type 2 receptor-α; SIRP: signal regulatory protein; SLAMF: signaling lymphocytic activation molecule family protein).

Gene	Expression pattern	Partners and substrates
PTPN6	Hematopoietic cell Ubiquitous	EGFR, SstR2, InsR, EpoR, SCFR, CSF2Rb, IL4R, BCR, CD5, CD22, CD72, SIRPA, Siglec2/3/11, FcRL3, PILRα, LIFR, LAIR1, LILRB1/2/4, SLAMF6, BLnk, IL6ST, TREML1, C6orf25, CXCR4, Ros1, Abl1, Lck, Lyn, SYK, ZAP70, FAK2, JaK1/2, TyK2, PKCδ, PI3KR1, NOS1, PTPNS1, SOS, Vav1, GRB2, GAB2, LCP2, CBL, PAG1, Cav1, Catenin-δ1, PECAM1, STAT5b, HoxA10
PTPN7	Hematopoietic cell	ERK1/2, P38MAPK
PTPN8(22)	B and T cells, Myeloid cell lines	CSK, LCK, Fyn, ZAP70 GRB2, CBL
PTPN9	Hematopoietic cell	M6PRBP, Arfaptin (ArfIP1), reticulon-3

Small RhoA GTPase inhibits PTPn1 enzyme. Transcription factor Yb1 contributes to the regulation of PTPn1 synthesis. Activity of PTPn1 is regulated by proteolysis, oxidation, phosphorylation, glutathionylation, and sumoylation [684]. Phosphatase PTPn1 is phosphorylated by PKB (Ser50) to impede its phosphatase activity, thus creating a positive feedback loop for insulin signaling [688].

8.3.11.2 Non-Receptor Protein Tyrosine Phosphatase-2

Non-receptor protein Tyr phosphatase-2 (PTPn2) that is encoded by the PTPN2 gene is also called T-cell protein Tyr phosphatase (TC-PTP). Three alternatively spliced variants of mRNA from the PTPN2 gene encode isoforms that differ at their non-catalytic C-termini: 48-kDa, long (PTPn2$_L$; also called TC48 and TCPTPβ) and 45-kDa, short (PTPn2$_S$; also termed TC45 and TCPTPα) form of PTPn2. Different C-termini determine substrate specificity and cellular location of the isoforms. PTPn2$_L$ localizes to the endoplasmic reticulum, whereas under basal conditions PTPn2$_S$ resides in the nucleus [689]. Phosphatase PTPn2 is mainly produced by hematopoietic cells.

Table 8.29. Partners and substrates of cytoplasmic protein Tyr phosphatases (**Part 3**; Sources: Wikipedia, BioGRID; CEACAM: carcinoembryonic antigen-related cell adhesion molecule; EFS: embryonal Fyn-associated substrate; EpoR: erythropoietin receptor; FCRL3: Fc receptor-like-3; FRS2: fibroblast growth factor receptor substrate; GHR: growth hormone receptor; GP: glycoprotein; IGF1R: insulin-like growth factor-1 receptor; InsR: insulin receptor; LAIR: leukocyte-associated immunoglobulin-like receptor; LILRB: leukocyte immunoglobulin-like receptor-B; PSTPIP: Pro–Ser–Thr phosphatase-interacting protein; SLAMF: signaling lymphocytic activation molecule family protein; SorbS2: sorbin and SH3 domain-containing protein-2 [a.k.a. Arg–Abl-interacting protein ArgBP2]; TGFβ1I1: transforming growth factor-β1-induced transcript-1 [a.k.a. 55-kDa androgen receptor coactivator ARA55 and hydrogen peroxide-inducible clone-5 protein HIC5]).

Gene	Expression pattern	Partners and substrates
PTPN11	Hematopoietic cell Ubiquitous	PDGFRβ, HER1/2/4, GHR, PrlR, IGF1R, InsR, EPHa2, SCFR, CSF2Rβ, CSF3R, EpoR, IL4R GP130, CD33, PIIRα, LILRB4, MPZL1 SLAMF1/6, NK-cell CD244, FcGR2B, FcRL3 IRS1/2, FRS2, IL6ST, Siglec11, Cav1, C6orf25 Ros1, JaK1/2, FAK1/2, Fyn, SYK, TyK2 ERK, PI3KR1, PTPn6, PLCγ2 GAB1/2, GRB2, SHC, LAIR1, SIT1, CRKL, CBL SOS1, STAT3, STAT5a/b, SOCS3 SynCRIP, catenin-β1 PECAM1, CEACAM1, VE-cadherin, selectin-E
PTPN12	Ubiquitous	Abl, CSK, FAK1/2, LCK, JaK2 BCAR1 (CAS), CASL, GRB2, SHC, Sin, PSTPIP1, TGFβ1I1 Paxillin, leupaxin, gelsolin, WASP EFS, p190RhoGAP, ArgBP2

Among PTPn2 substrates, there are epidermal growth factor receptor and adaptor SHC [689, 690]. Both PTPn2$_L$ and PTPn2$_S$ isoforms recognize phosphorylated EGFR. Long subtype PTPn2$_L$ can prevent inappropriate signaling by nascent receptor, whereas PTPn2$_S$ exits the nucleus in response to EGF and is recruited to plasmalemmal, EGFR-containing complexes. Splice variant PTPn2$_S$ hampers EGFR signaling after EGFR or integrin stimulation toward the PI3K–PKB axis [690]. Splice variant PTPn2$_S$ also precludes EGF-induced association of SHC with GRB2 [689].

In addition to STAT1, PTPn2$_S$ binds to signal transducers and activators of transcription STAT3 and STAT5a/b to repress IL6 and prolactin pathways, respectively [691]. Both endoplasmic reticulum and nuclear isoforms dephosphorylate (inactivate) insulin receptor [692]. Phosphatase PTPn2 targets JaK1 and JaK3, whereas PTPn1 dephosphorylates (inactivates) JaK2 kinase.

Table 8.30. Partners and substrates of cytoplasmic protein Tyr phosphatases (**Part 4**; Sources: Wikipedia and BioGRID; APC: adenomatous polyposis coli Ub ligase; KIF: kinesin-like protein; PSTPIP: Pro–Ser–Thr phosphatase-interacting protein).

Gene	Expression pattern	Partners and substrates
PTPN13	Ubiquitous	TNFRSF6a, TNFSF6, PAK2, IκBα Catenin-β1, ephrin-B1, RhoGAP29, APC
PTPN14	Ubiquitous	Catenin-α1, -β1, and -γ Zona occludens-1
PTPN18	Hematopoietic progenitors Brain, colon	CSK, TEC, Fyn PSTPIP
PTPN21	Hematopoietic cell	BMX, Src KIF1C
PTPN22: see PTPN8		
PTPN23	Endothelial cell	

8.3.11.3 Non-Receptor Protein Tyrosine Phosphatase-3

Non-receptor protein Tyr phosphatase-3 (PTPn3)[51] is encoded by the PTPN3 gene. It is characterized by an N-terminus that is homologous to cytoskeletal-associated proteins of the band 4.1 superfamily (band 4.1, ezrin, radixin, moesin, merlin, and talin) [693]. Band 4.1 Domain and phosphatase segment in the C-terminus are separated by a central motif with one PDZ binding domain. Phosphatase PTPn3 is involved in thalamocortical connections.

Phosphatase PTPn3 interacts with 14-3-3β adaptor to regulate its activity [694] as well as metallopeptidase ADAM17 to prevent its action [695].[52] Moreover, PTPn3 shifts activity of voltage-gated sodium channel Na$_V$1.5 toward hyperpolarized potentials in cardiomyocytes [696]. Phosphatase PTPn3 also dephosphorylates T-cell receptor TCRζ subunit [697].

---

[51] A.k.a. protein Tyr phosphatase PTP-H1 (cloned from the HeLa cell cDNA library).

[52] Metallopeptidase ADAM17, also called tumor-necrosis factor-α-convertase (TACe), is a member of the ADAM (a disintegrin and metallopeptidase) or MDC (metallopeptidase, disintegrin, cysteine-rich) family of metallopeptidases. This sheddase processes TNFα, transforming growth factor-α, L-selectin, Notch, TNF receptor-1 and -2, interleukin-1 receptor-2, and HER4. It is also an amyloid precursor protein α-secretase.

8.3.11.4 Non-Receptor Protein Tyrosine Phosphatase-4

Non-receptor protein Tyr phosphatase-4 (PTPn4)[53] is encoded by the PTPN4 gene. Like PTPn3, this testis-enriched phosphatase contains a C-terminal phosphatase domain and N-terminal motif homologous to band 4.1 superfamily of cytoskeletal-associated proteins. It is prominently expressed in cerebellar Purkinje and thalamus cells [698].

Phosphatase PTPn4 interacts with δ2 and ε subunits of the heteromeric glutamate receptor, the predominant excitatory neurotransmitter receptors in the brain [698]. Glutamate receptor GluRδ2 is selectively synthesized in cerebellar Purkinje cells. Phosphatase PTPn4 in thalamus cells enhances Fyn-mediated Tyr phosphorylation of N-methyl-D-aspartate receptor GluN2a (or GluRε1).

8.3.11.5 Non-Receptor Protein Tyrosine Phosphatase-5

Brain-specific non-receptor protein Tyr phosphatase-5 (PTPn5) is highly enriched within dopaminergic neurons (soma, axons, and dendrites) in the striatum, neocortex, amygdala and other related regions relatively to other areas of the central nervous system [699, 700].[54] It operates in the rapid integration of adrenergic, dopaminergic, and glutamatergic signaling. Cytosolic PTPn5 is related to transmembrane PTPRr phosphatase. Both are mainly expressed in neurons.

Four major PTPn5 isoforms exist (20-, 38-, 46-, and 61-kDa PTPn5). However, only two isoforms — $PTPn5_{46}$ and $PTPn5_{61}$ — are active, as they contain a phosphatase domain [700].

Phosphatase PTPn5 localizes to postsynaptic densities, which are specialized parts of the cytoskeleton at neuronal synapses in striatal neurons, where it binds to and dephosphorylates non-receptor protein Tyr kinase Fyn (Tyr420) [701]. In addition, PTPn5 is a substrate and inactivator of extracellular signal-regulated kinases ERK1 and ERK2 [702].

Several post-translational modifications regulate PTPn5 activity, such as phosphorylation, ubiquitination and proteasomal degradation, and proteolysis.[55] Phosphatase PTPn5 contributes to the regulation of endocytosis of NMDA-type glutamate receptors.

Phosphatase PTPn5 is phosphorylated by protein kinase-A and dephosphorylated by protein phosphatases PP1 and PP3 [700]. Phosphatase PTPn5 dephosphorylates NMDA-type glutamate receptors GluN2b, thereby causing NMDAR internalization. It also targets GluR2 subunit of AMPA-type glutamate receptors, MAPKs (ERK1, ERK2, P38MAPK, and JNK), in addtion to Fyn kinase [700].

---

[53] A.k.a. cytosolic megakaryocyte protein Tyr phosphatase PTP-Meg1 (or Meg1).

[54] A.k.a. protein Tyr phosphatase striatum-enriched phosphatase (STEP or PTP-STEP).

[55] Stimulated dopamine D_1 receptor activates protein kinase-A that phosphorylates PTPn5. On the other hand, stimulated NMDA-type glutamate receptors cause PTPn5 dephosphorylation (activation). During hypoxia, amount of PTPN5 decays, $PTPn5_{61}$ is cleaved by calpain into $PTPn5_{33}$ that does not binds to any of PTPn5 substrates.

8.3.11.6 Non-Receptor Protein Tyrosine Phosphatase-6

SH2 domain-containing, non-receptor protein Tyr phosphatase-6 (PTPn6)[56] is encoded by the PTPN6 gene. Its N-terminus contains 2 tandem Src homolog-2 (SH2) binding domains. Multiple alternatively spliced transcript variants generate distinct isoforms.

Phosphatase PTPn6 is expressed primarily in hematopoietic cells, where it participates in multiple signaling pathways that regulate cell growth, differentiation, and proliferation, especially during hematopoiesis. However, it is ubiquitous and can be particularly detected in neurons and epithelial cells.

In cells, it resides in the cytoplasm and nucleus. In particular, in neurons, it translocates into the nucleus upon angiotensin-2 stimulation. Its post-translational modifications include phosphorylations on Ser and Tyr residues.

Phosphatase PTPn6 impedes signaling from receptor Tyr kinases, adhesion receptors, cytokine receptors, and immunoreceptors, in particular receptors Ros1 in the proximal segment of the epididymis [703], EGFR [704], SCFR [705], and EpoR [706], as well as Fc receptor-like-3, inhibitory leukocyte-associated immunoglobulin-like receptor LAIR1 [708],[57] and leukocyte immunoglobulin-like receptors LILRb2 and LILRb4, adhesion molecules PECAM-1 and catenin-δ1, and sialic acid-binding Ig-like lectin Siglec-2.[58]

Phosphatase PTPn6 controls TCR signaling thresholds in both developing and mature T lymphocytes. It interacts with receptor and cytoplasmic Tyr kinases (Abl, FAK2, JaK2, LCK, Lyn, SYK, and TyK2) for activation or, most often, inactivation of protein kinase-Cδ [707], adaptor GRB2, Homeobox-A10 transcription factor, and scaffold signal-regulatory protein-α.

In cooperation with mitochondrial (or microtubule-associated) tumor suppressor MTuS1[59] that interacts with angiotensin-2 receptor AT_2, PTPn6 upregulates ubiquitin conjugase variant UbE2V2 upon angiotensin-2 stimulation.

8.3.11.7 Non-Receptor Protein Tyrosine Phosphatase-7

Non-receptor protein Tyr phosphatase-7 (PTPn7),[60] encoded by the PTPN7 gene, is predominantly expressed in hematopoietic cells. Its N-terminus contains a kinase interaction motif (KIM) used by PKA to phosphorylate PTPn7 and to tether ERK kinase. Two alternatively spliced transcript variants encode 2 isoforms.

---

[56] A.k.a. hematopoietic cell phosphatase (HCP), PTP1c, SHPTP1, and SHP1.

[57] Transmembrane glycoprotein LAIR1 recruits phosphatases PTPn6 and PTPn11 upon activation. Crosslinking of LAIR1 antigen on natural killer cells inhibits NK cell-mediated cytotoxicity.

[58] A.k.a. B-cell receptor or lectin CD22.

[59] A.k.a. angiotensin-2 type-2 receptor-interacting or -binding protein (ATIP or ATBP).

[60] A.k.a. B-cell protein Tyr phosphatase BPTP4, hematopoietic PTP (HePTP), leukocyte PTP (LC-PTP or L-PTP), and PTP-NI.

Phosphatase PTPn7 can form a cholesterol-regulated complex with PP2 that has dual-specificity due to the combined activities of Ser/Thr phosphatase PP2 and Tyr phosphatase PTPn7, hence being able to dephosphorylate ERK1 and ERK2 [709]. Acute depletion of cholesterol causes the disassembly of the PP2–PTPn7 complex.[61]

Phosphatase PTPn7 is closely related to receptor PTPRr and cytoplasmic PTPn5 PTPs that target ERK1/2 and P38MAPKs. Its non-catalytic N-terminus can interact with MAPKs to suppress their activities. In addition to Ser/Thr phosphatases PP1 and PP2, dual-specificity MAPK phosphatases, PTPn7 also binds and dephosphorylates P38MAPK and ERK enzymes.[62] Phosphatase PTPn7 regulates ERK1 and ERK2 nuclear translocation, rather than ERK activity [710].

8.3.11.8 Non-Receptor Protein Tyrosine Phosphatase-8

Non-receptor protein Tyr phosphatase-8 (PTPn8) that is encoded by the PTPN8 gene is also identified as cytoplasmic protein Tyr phosphatase-22 (PTPn22). Alternative splicing transcript variants generate 2 isoforms.

8.3.11.9 Non-Receptor Protein Tyrosine Phosphatase-9

Non-receptor protein Tyr phosphatase-9 (PTPn9)[63] is encoded by the PTPN9 gene. It is expressed in some hematopoietic cell lineages such as neutrophils, where it is mainly associated with the cytoplasmic face of granules and secretory vesicles [711]. Its non-catalytic N-terminus negatively regulates the enzymatic activity of the C-terminal phosphatase domain.

The N-terminus of PTPn9 phosphatase is homologous to lipid-binding proteins involved in intracellular transport such as retinaldehyde-binding protein R1BP1

---

[61] Cholesterol depletion of non-caveolar rafts activates the Ras–ERK pathway. Plasmalemmal rafts include caveolae and other types of nanodomains that all participate in the spatial organization of signal transduction at the cell surface. Cholesterol depletion attenuates signal transduction from endothelin receptor-A that then relocalizes to clathrin-coated pits, where it is internalized. Insulin and platelet-derived growth factor receptors also reside in caveolae. Cholesterol sensors in each membrane compartment enable quick responses to local changes in cholesterol and regulate cholesterol transport to the membrane. A cholesterol-sensitive feedback loop links activation to deactivation of ERK1 and ERK2 by kinases (MAP2K) and phosphatases (PTPRr, PTPn5, PTPn7, PP2, and MKP3), respectively.

[62] In resting cells, extracellular signal-regulated kinases ERK1 and ERK2 are predominantly cytosolic. They are stimulated by growth and differentiation factors that bind to receptor Tyr kinases and G-protein-coupled and cytokine receptors. A fraction of ERK1 and ERK2 then translocates to the nucleus to phosphorylate transcription factors and trigger gene expression. According to cell type and stimulus duration and magnitude, activated ERK initiates cell proliferation or differentiation.

[63] A.k.a. cytosolic megakaryocyte protein Tyr phosphatase PTPMeg2 or Meg2.

(also known as cellular retinaldehyde binding protein [CRalBP]) and members of the SEC14 cytosolic factor family. This domain can serve as a secretory vesicle targeting signal. The Sec14 domain is responsible for PTPn9 localization to the perinuclear region.

Activity of PTPn9 is enhanced by PI(4)P, PI(4,5)P$_2$, and PI(3,4,5)P$_3$. It contributes to vesicle fusion and phagocytosis. Furthermore, PTPn9 precludes hepatic insulin signaling [712].[64]

8.3.11.10 Non-Receptor Protein Tyrosine Phosphatase-11

Ubiquitous, Src homology-2 (SH2) domain-containing, non-receptor protein Tyr phosphatase-11 (PTPn11)[65] is encoded by the gene PTPN11 in all hematopoietic cells as well as other cell types. Its N-terminus contains 2 tandem SH2 domains (^{N}SH2 and ^{C}SH2) that are Tyr^P-binding sites. In its inactive state, its N-terminus attaches to PTP domain and blocks access of substrates to the active site (auto-inhibition).

Phosphatase PTPn11 interacts with receptors of hormones, such as insulin and growth hormone [713], growth factors, such as EGF, PDGF, and IGF1, cytokines (SCFR and IL6Rβ subunit [a.k.a. glycoprotein GP130 and CD130]) [714], some immunoreceptors (e.g., leukocyte-associated immunoglobulin-like receptor LAIR1, or CD305), as well as insulin receptor substrate IRS1, Janus kinases JaK1 and JaK2 (but not JaK3) [715], transcription factors STAT5a and STAT5b [716], suppressor of cytokine signaling SOCS3, kinase FAK2, phospholipase-Cγ2 [717], adaptors GRB2-associated binders GAB1 and GAB2, CBL, and adhesion molecule PECAM1.

Mutations of the PTPN11 gene are the most frequent cause of juvenile myelomonocytic leukemia characterized by a selective hypersensitivity of hematopoietic progenitors to granulocyte–macrophage colony-stimulating factor (CSF2). In more than 75% of cases, mutually exclusive gain-of-function mutations appear in N-RAS, K-RAS, or PTPN11 or homozygous loss-of-function mutations in NF1 (RASGAP), i.e., mutations in genes that encode components of the Ras–ERK cascade. Protein PTPn11 also operates in signal transduction of many receptors involved in aging, such as GHR.

Phosphatase PTPn11 intervenes in embryogenesis, as it can regulate cell survival, differentiation, proliferation ,and migration using the ERK1/2 pathway that is transiently activated during early embryogenesis. Mutations in PTPN11 gene also cause Noonan syndrome often characterized by cardiac disease, craniofacial

---

[64] In liver, insulin inhibits glucose production, as it causes nuclear exclusion of gluconeogenic transcription factor FoxO1 via PKB kinase. Phosphatase PTPn9 modulates insulin-dependent FoxO1 subcellular localization.

[65] A.k.a. SHP2, SHPTP2, PTP1d, PTP2c, and BPTP3.

abnormalities, growth impairment, and a disturbance in IGF1 signaling.[66] Whereas mutations associated with Noonan syndrome often generate gain of ERK1/2 function, Leopard syndrome also associated with PTPN11 mutations leads to loss of ERK1/2 function.

Neural crest cells (the fourth germ layer) participate in both heart and skull formation. Neural crest cell determination, proliferation, migration, and differentiation are controlled by sonic Hedgehog, Wnts, bone morphogenic proteins and fibroblast growth factors. Furthermore, PTPn11 is essential for normal migration of neural crest cells and differentiation into the diverse lineages of heart and skull [718].[67]

8.3.11.11 Non-Receptor Protein Tyrosine Phosphatase-12

Non-receptor protein Tyr phosphatase-12 (PTPn12),[68] encoded by the PTPN12 gene, belongs to the *PEST family* of protein Tyr phosphatases with PTPn18, PTPn22, and PTPn23 phosphatases. Phosphatase PTPn12 resides in nonimmunocytes, but abounds in hematopoietic cells.

Phosphorylated focal adhesion-associated proteins and dephosphorylation of corresponding phosphoproteins contribute to the regulation of the cytoskeleton dynamics and control of signals for cell adhesion, migration, growth, and survival. In non-hematopoietic cells, PTPn12 is a potent regulator of adhesion and migration.

Phosphatase PTPn12 prominently associates with cytoskeletal proteins (CAS and CAS-related proteins CAS family member-3 (CAS3) and CASL, paxillin and paxillin-related polypeptides TGFβ1-induced transcript-1 TGFβ1-1[69] and leupaxin) as well as kinases and adaptors (CSK, SHC, and GRB2).

Phosphatase PTPn12 dephosphorylates Abl kinase, as well as various cytoskeleton and cell-adhesion molecules, such as focal adhesion-associated adaptor BCAR1 (a.k.a. CAS and p130CAS), WASP, kinases FAK1 and FAK2 (Sect. 4.6), Pro–Ser–Thr phosphatase-interacting protein PSTPIP1, Vav GEF, RhoGAP35, and paxillin [719].

---

[66]Other mutations responsible for Noonan syndrome imply K-RAS, C-RAF, SOS. Neurocardiofacial cutaneous syndromes include not only Noonan syndrome, but also Costello syndrome due to H-RAS mutations and cardiofaciocutaneous syndrome caused by mutations of the B-RAF, K-RAS, MEK1, or MEK2 gene. All these pathologies are associated with a defect in ERK1/2 signaling.

[67]Neural crest cells are first localized in the neural folds at the dorsal aspect of the developing spinal cord and delaminate from the neural tube. They then migrate toward the organ primordia, where they give rise to and influence the development of diverse tissues, such as numerous craniofacial structures, cardiac outflow tract, endocrine glands, and neurons.

[68]A.k.a. PTP-PEST, as its C-terminus contains a PEST motif (proline-, glutamic acid-, serine-, and threonine-rich) for between-protein interaction, and PTP-G1.

[69]A.k.a. 55-kDa androgen receptor-associated protein ARA55 and hydrogen peroxide-inducible clone-5 protein (HIC5).

Phosphatase PTPn12 participates not only in cytoskeletal reorganization for cell adhesion and migration, but also inhibition of lymphocyte activation [720]. It dephosphorylates a subset of BCR- and TCR-regulated substrates, such as adaptors (e.g., SHC and BCAR1 [CAS]) and kinases (e.g., FAK and PYK2 that intervene with Lyn, SYK, and BTK in B-cell receptor signaling; Lck, Fyn, and ZAP70 for T-cell receptor signaling).

8.3.11.12 Non-Receptor Protein Tyrosine Phosphatase-13

Ubiquitous non-receptor protein Tyr phosphatase-13 (PTPn13),[70] encoded by the PTPN13 gene, dephosphorylates TNFRSF6a receptor. Four alternatively spliced transcript variants exist. Like PTPn3 and PTPn4, PTPn13 belongs to the group of PTPs characterized by an N-terminal segment homologous to proteins of the band 4.1 superfamily, PDZ (PSD95, Disc large, zonula occludens) domains, and a PTPase catalytic motif.

Phosphatase PTPn13 interacts via its PDZ domain with PAK2[71] that is regulated by small GTPase Rho [721]. Both PTPn13 and PAK2 colocalize in lamellipodium-like structures, where they can contribute to the regulation of the actin cytoskeleton. As it interacts with GTPase-activating proteins, it can operate as a regulator of Rho signaling. Phosphatase PTPn13 also dephosphorylates IκBα agent.

8.3.11.13 Non-Receptor Protein Tyrosine Phosphatase-14

Non-receptor cytoskeletal-associated protein Tyr phosphatase-14 (PTPn14)[72] is encoded by the PTPN14 gene. Like PTPn3, PTPn4, and PTPn13, it belongs to the group of PTPs characterized by an N-terminal segment homologous to proteins of the band 4.1 superfamily, PDZ domains, and a PTPase catalytic domain. Hence it is a member of the FERM (four-point-one, ezrin, radixin, moesin) family of PTPs. It localizes to adherens junctions, where it dephosphorylates β-catenin [722]. It is synthesized in many human organs, such as lung, kidney, skeletal muscle, and placenta. It is phosphorylated upon DNA damage, probably by ATMK or ATRK kinase.

---

[70] A.k.a. Fas (TNFSF6)-associated phosphatase FAP1, protein Tyr phosphatase-basophil-like PTP-BAS or PTP-BL, PNP1, PTP1E, protein Tyr phosphatase-like protein-1 (PTP-L1), and PTP-LE.
[71] A.k.a. protein kinase-N2 (PKN2), protein kinase-C-like-2, and protein kinase-C-related kinase-2.
[72] A.k.a. PEZ, PTP-D2, and PTP36.

8.3.11.14 Non-Receptor Protein Tyrosine Phosphatase-18

Non-receptor protein Tyr phosphatase-18 (PTPn18)[73] is encoded by the PTPN18 gene. Like PTPn12 and PTPn22, it contains a PEST (proline-, glutamic acid-, serine-, and threonine-rich domain) motif. It is expressed in hematopoietic cells as well as other cell types, especially in brain and colon.

Like PTPn12, PTPn18 links to cytoskeleton-associated Pro–Ser–Thr phosphatase-interacting proteins PSTPIP1 and PSTPIP2 [723].[74] It can be a positive and negative regulator of cytoskeletal rearrangement and inflammation, respectively. It can be phosphorylated by PKA or PKC as well as Fyn or TEC. It can associate with C-terminal Src kinase that inactivates Src kinases. This association can enable its recruitment in the vicinity of CSK targets, in particular Src kinases [720].

8.3.11.15 Non-Receptor Protein Tyrosine Phosphatase-20

Non-receptor protein Tyr phosphatase-20 (PTPn20)[75] contains a single catalytic domain at its C-terminal half and an N-terminal half with a region rich in PEST motifs [724]. It is specifically expressed in mouse testicular germ cells that undergo meiosis. It is detected between week 2 and 3 after birth with the onset of meiosis.

In mammals, spermatogenesis happens in the seminiferous tubule, in which mitotic division of spermatogonia, meiosis of spermatocytes, and differentiation of spermatid to sperm occurs. Many enzymes are specifically produced before or at meiosis, such as Tyr kinases SCFR (Vol. 3 – Chap. 8. Receptor Kinases), Abl (Sect. 4.2), and Fer (Sect. 4.7), Ser/Thr kinases male germ cell-associated kinase (MAK), v-mos Moloney murine sarcoma viral oncogene homolog Mos (a MAP3K; Sect. 6.5.2), and TesK (Sect. 5.9.4), dual-specific Never in mitosis gene-A-related kinase Nek1, receptor protein Tyr phosphatase PTPRv, Ser/Thr phosphatase calcineurin-B subunit isoformβ1 (PP3$_{r1}$), and dual-specific phosphatase CDC25, in addition to cytoplasmic protein Tyr phosphatase PTPn20 [724].

8.3.11.16 Non-Receptor Protein Tyrosine Phosphatase-21

Non-receptor protein Tyr phosphatase-21 (PTPn21)[76] is encoded by the PTPN21 gene. Like PTPn3, PTPn4, and PTPn13, PTPn21 belongs to the group of PTPs characterized by an N-terminal segment homologous to proteins of the band 4.1 superfamily, PDZ domains, and a PTPase catalytic domain.

---

[73] A.k.a. brain-derived phosphatase BDP1, PTP hematopoietic stem cell factor (PTP-HSCF), PTP20, PTP-K1, and fetal liver phosphatase FLP1.
[74] Proteins PSTPIP1 and PSTPIP2 homodimerize, but do not heterodimerize.
[75] A.k.a. testis-specific Tyr phosphatase PTP-Typ.
[76] A.k.a. PTP-D1, PTP-RL10, and PTP2E.

Phosphatase PTPn21 interacts with BMX kinase of the TEC kinase family. It localizes along actin filaments and to adhesion plaques. It can form distinct complexes with actin, Src kinase, and focal adhesion kinase to cause cell adhesion and migration [725].

8.3.11.17 Non-Receptor Protein Tyrosine Phosphatase-22

Non-receptor protein Tyr phosphatase-22 (PTPn22)[77] is encoded by the PTPN22 gene (or PTPN8 gene). In humans, PTPN22 polymorphism is a significant risk factor for autoimmune diseases, such as type-1 diabetes, rheumatoid arthritis, juvenile idiopathic arthritis, myasthenia gravis, and systemic lupus erythematosus.

Phosphatase PTPn22 interacts with adaptors CBL and GRB2, as well as kinases CSK, Fyn, LCK, and ZAP70 [720]. It prevents T-cell activation, mainly by suppressing the activity of SRC family PTKs via CSK kinase.

8.3.11.18 Non-Receptor Protein Tyrosine Phosphatase-23

Non-receptor protein Tyr phosphatase-23 (PTPn23)[78] is encoded by the PTPN23 gene. It possesses an N-terminal domain, a Tyr phosphatase motif, and a PEST sequence at its C-terminus. Phosphatase PTPn23 participates in sorting cargo proteins to endosomes.

Endothelial cells synthesize various transmembrane and cytoplasmic PTPs, such as PTPRb (VE-PTP), PTPRe, PTPRj, PTPRm, and PTPn23, in addition to RPTKs, such as FGFR, PDGFR, and VEGFR1 and -2, and NRTRKs, such as Src, JaK, and FAK. Endothelial cell migration during angiogenesis (Vol. 5 – Chap. 10. Vasculature Growth) is controlled via phosphorylation and dephosphorylation of different substrates by coordinated kinases and phosphatases. Phosphatase PTPn23 binds focal adhesion kinase [726]. Fibroblast growth factor FGF2 that stimulates endothelial cell migration precludes PTPn23–FAK attachment. On the other hand, Src kinase favors cell migration, as it phosphorylates FAK, BCAR1, and E-cadherin. In addition, Src phosphorylates (inactivates) PTPn23 [727]. Fibroblast growth factor controls Src–PTPn23 interaction, as it activates Src kinase.

---

[77] A.k.a. PTPn8, lymphoid Tyr phosphatase (LyP), as it is expressed primarily in lymphoid tissues, and PEP, as it is a PEST domain-containing phosphatase.

[78] A.k.a. His domain-containing protein Tyr phosphatase HD-PTP and PTP-TD14 (rat homolog).

8.3.12 PEST Family Phosphatases in Inflammation and Immunity

The PEST (proline-, glutamic acid-, serine- and threonine-rich) family of cytoplasmic protein Tyr phosphatases includes PTPn12, PTPn18, PTPn22, and PTPn23. T-cell activation results from binding of an antigen with a major histocompatibility complex molecule that is displayed to a T-cell antigen receptor. Class-1 and -2 MHCs interacts with CD8 and CD4, respectively, thereby building receptor complexes. These associations trigger phosphorylation initiated by Src family PTKs and amplified by SYK family PTKs. Molecules of the major histocompatibility complex are dispensable for B-cell antigen receptors and receptors recognizing Fc-receptor-binding part of immunoglobulins, but these receptors prime equivalent signalings.

Phosphatase PTPn12 is implicated with kinases of the Src family members, FAK1/2, and PTPn1 in cytoskeletal reorganization for leukocyte adhesion and migration as well as granule exocytosis and phagocytosis [720]. In addition, it hampers lymphocyte activation, as it can dephosphorylate TCR- and BCR-regulated substrates. Conversely, PTPn12 is downregulated during T-cell activation. Phosphorylation of PTPn12 by PKC and PKA impedes PTPn12 activity.

Phosphatase PTPn18 can contribute to inhibition of inflammation. It promotes cytoskeleton rearrangement. Kinases of the SRC and TEC families can phosphorylate PTPn18, thereby inactivating PTPn18 [720].

Phosphatase PTPn22 links to CSK to cooperatively inhibit activation of both CD4+ and CD8+ mature (effector and memory) T lymphocytes, but not naive T lymphocytes [720]. Phosphatase PTPn22 can also regulate BCR-triggered responses in immature B lymphocytes. Phosphorylation of PTPn22 by PKC reduces PTPn22 activity.

8.3.13 Dual-Specificity Protein Phosphatases

Dual-specificity phosphatases (DuSP) form a heterogeneous hyperfamily of protein phosphatases that can dephosphorylate Ser^P, Thr^P, and Tyr^P residues within a given substrate.

These phosphatases constitute 7 families on the basis of sequence similarity: (1) mitogen-activated protein kinase (MAPK) phosphatases (MKP) that dephosphorylate (inactivate) MAPKs; (2) Slingshots; (3) protein Tyr phosphatases-4A that are also termed phosphatases of regenerating liver; (4) cell-division cycle-14 homolog phosphatases; (5) phosphatase and tensin homologs; (6) myotubularins; and (7) atypical dual-specific phosphatases.

8.3.13.1 Lipid and Protein Phosphatases – PTen and PTen-like Phosphatases

Plasmalemmal phosphatase and tensin homolog deleted on chromosome 10 (PTen; Sect. 2.10.1)[79] possesses a tensin-like domain as well as a catalytic domain similar to that of dual-specificity protein phosphatases. However, cytoplasmic PTen preferentially dephosphorylates phosphoinositide substrates. It is called a tumor suppressor, as this phosphoinositide phosphatase restricts the activity of phosphatidylinositol 3-kinase by dephosphorylating $PI(3,4,5)P_3$ to $PI(4,5)P_2$, thereby impeding the PI3K–PKB pathway.

Phosphatase PTen is both a phosphatidylinositol 3-phosphatase and a protein phosphatase that is able to dephosphorylate Ser^P, Thr^P, and Tyr^P, residues. It indeed dephosphorylates Tyr-phosphorylated focal adhesion kinases. Phosphatidylinositol 3-kinase interacts with FAK in PKB phosphorylation. Protein phosphatase-1δ is also located in focal adhesions and associates with focal adhesion kinases.

Phosphatase PTen interacts with adaptor NHERF[80] and platelet-derived growth factor receptors [728]. Ubiquination regulates PTen stability and its nuclear location, where PTen favors chromosome integrity [729]. Polyubiquination degrades PTen, whereas nuclear monoubiquitination stabilizes PTen. PTen acts on chromatin and regulates expression of small GTPase Rad51 that reduces the incidence of spontaneous DNA double-strand breaks.

Phosphorylation represents a mode of regulation of PTen activity. Phosphatase PTen is a substrate of Src family protein Tyr kinases. It modulates the activity of the PI3K–PKB axis, whereas SH2 domain-containing protein Tyr phosphatase PTPn6 binds and dephosphorylates PTen. Casein kinase CK2 also targets PTen [730].

Five *PTen-like phosphatases* dephosphorylate D3-phosphorylated inositol phospholipids, hence contributing to reversible phosphorylation of inositol lipids that serves in cell signaling. Some myotubularin phosphatases also dephosphorylate D3-phosphorylated inositol phospholipids.

8.3.13.2 PTEN Superfamily

Members of the PTEN superfamily, PTen, tensins, and transmembrane phosphatases with tensin homology, are encoded by 5 genes (Table 8.31).

---

[79] A.k.a. MMAC1 or TEP1.

[80] Sodium–hydrogen exchanger regulatory factor (NHERF) is also called Na^+–H^+ antiporter-3 regulator-1 and ezrin–radixin–moesin-binding protein EBP50.

8.3 Cytosolic Protein Serine/Threonine Phosphatases

Table 8.31. Members of the PTEN superfamily (C1Ten: C1 domain-containing phosphatase and tensin homolog; MMAC: mutated in multiple advanced cancers; MxRA: matrix-remodeling associated protein; TenC: tensin-like C1 domain-containing phosphatase; Tns: tensin; TPIP: TPTE and PTen homologous inositol lipid phosphatase; TPTe: transmembrane phosphatase with tensin homology, or transmembrane phosphoinositide 3-phosphatase and tensin homolog).

Type	Other aliases
PTen	MMAC1, TEP1
Tns1	MST091, MSTP091, MST122, MSTP122, MST127, MSTP127, MxRA6
Tns2	C1Ten, TenC1
TPTe1	TPTe, PTen2
TPTe2	TPIP

Tensin-1

Tensin-1 encoded by the TNS1 gene localizes to focal adhesions,[81] more precisely *fibrillar adhesions*.[82] Tensin-1 is involved in integrin removal from focal adhesions and transfer to fibrillar adhesions as well as in fibronectin fibrillogenesis.[83] This protein binds to and crosslinks actin filaments. However, tensin-1 can reduce the actin polymerization rate by interacting with the barbed end of actin. Tensin has a protein Tyr phosphatase domain related to PTen, an Src homology-2 (SH2) motif, and a Tyr^P-binding sequence (PTB).

Platelet-derived growth factor, thrombin, angiotensin, among others, provoke Tyr phosphorylation of tensin-1. Both Abl and Src kinases phosphorylate tensin-1 [731]. It is a substrate of calpain-2, a focal adhesion peptidase involved in disassembly of focal adhesions. Therefore, tensin-1 links the extracellular matrix and the actin cytoskeleton to signal transduction.

Protein Ser/Thr phosphatase PP1 resides also in focal adhesions. Among the 3 predominant, widespread PP1 isoforms, PP1β (or PP1δ) can link to focal adhesion kinase. On the other hand, PP1α tethers to tensin [732].

---

[81] These attachment zones between the cell and the extracellular matrix transmit forces generated by the actin cytoskeleton to the extracellular matrix. These adhesions are assembled initially upon integrin engagement. Upon integrin ligation, tensin-1 rapidly translocates to the assembling focal adhesions. These sites also contain focal adhesion kinase, paxillin, vinculin, among others. These constituents are phosphorylated.

[82] Fibrillar adhesions are characterized by actomyosin-dependent displacement of fibronectin receptors. Binding of $\alpha_5\beta_1$-integrin to fibronectin activates the formation of fibrillar adhesions and actin filaments. These adhesions are relatively enriched in tensin. In human fibroblasts, these structures constituted by $\alpha_5\beta_1$-integrin binds to fibrils of fibronectin parallel to actin bundles and tensin-1, but do not contain other focal adhesion molecules, such as vinculin and paxillin.

[83] Soluble fibronectins form fibrillar matrices.

Tensin-2

Tensin-2 is strongly produced in the heart, kidney, skeletal muscle, and liver. Three transcript variants generate 3 isoforms. Tensin-2 also localizes to focal adhesions. Both tensin-1 and -2 are able to promote cell migration [731]. Tensin-2 binds to Axl receptor Tyr kinase [733]. This intracellular phosphatase precludes PKB signaling.

Transmembrane Phosphatases with Tensin Homology

The human transmembrane phosphatase with tensin homology genes (TPTE1–TPTE2) encode PTen-related Tyr phosphatases TPTe1 and TPTe2. These 2 phosphatases are observed in secondary spermatocytes and/or prespermatids [734]. Several alternatively spliced isoforms of these 2 proteins possess a variable number of transmembrane domains among the 4 potential transmembrane domains. Phosphatase TPTe2 is a membrane-associated phosphatases with substrate specificity for the 3-position phosphate of inositol phospholipids.

8.3.13.3 Mitogen-Activated Protein Kinase Phosphatases

Non-dual-specificity mitogen-activated protein kinase phosphatases include protein Ser/Thr phosphatases PP2 and PPM1 as well as receptor-like PTPRr and cytoplasmic PTPn5 and PTPn7 protein Tyr phosphatases.

Dual-specificity MAPK phosphatases (DuSP/MKP) dephosphorylate (inactivate) both Thr^P and Tyr^P on activated MAPKs, i.e., extracellular-signal-regulated kinases, Jun N-terminal kinases, and P38MAPKs (Table 8.32; Chap. 6), thereby antagonizing MAPK signaling and contributing to the setting of the magnitude and duration of MAPK activation. They are involved in development, metabolism, immunity, and cellular stress responses.

According to sequence similarity, gene structure, substrate specificity, subcellular location, and transcriptional regulation, the MKP superfamily can be subdivided into 3 families [738, 739]: (1) family 1 that comprises 4 inducible, nuclear MKPs (DuSP1, DuSP2, DuSP4, and DuSP5); (2) family 2 with 3 related, cytoplasmic, ERK-specific MKPs (DuSP6, DuSP7, and DuSP9); and (3) family 3 also with 3 MKPs (DuSP8, DuSP10, and DuSP16). that preferentially inactivate stress-activated MAPKs, i.e., P38MAPKs and JNKs.

Except DuSP5 that targets ERK1 and ERK2, subfamily-1 MKPs display a rather broad specificity for ERK, JNK, and P38MAPKs. All of these DuSPs are encoded by inducible genes that are rapidly upregulated in response to mitogenes and cell stresses.

Pseudophosphatase mitogen-activated protein kinase (MAPK) Ser^P (S)/Thr^P (T)/Tyr^P (Y)-interacting (X) protein (MKSTYX), or STYX-like protein (STYXL1), that is characterized by a substitution of Ser for Cys in the catalytic domain is included in MKP subfamily on the basis of domain structure [736].

Table 8.32. Types of mitogen-activated protein kinase phosphatases and their specific substrate with order of potency (Sources: [735–737]; KAP: kinesin-associated protein). Substrate specificity of MKPs can differ in vivo from that in vitro. Dual-specificity MKPs constitute a structural superfamily of 11 proteins within the hyperfamily of dual-specificity protein phosphatases. This subset encompasses 10 enzymes and DuSP24 pseudophosphatase.

Type	Substrates (potency order)
MKP1 (DuSP1)	ERK < P38MAPK ~ JNK
MKP2 (DuSP4)	P38MAPK < ERK ~ JNK
MKP3 (DuSP6)	P38MAPK ~ JNK < ERK
MKP4 (DuSP9)	JNK < P38MAPK < ERK
MKP5 (DuSP10)	ERK < P38MAPK ~ JNK
MKP6 (DuSP14)	ERK, JNK
MKP7 (DuSP16)	ERK < P38MAPK ~ JNK
MKP8 (DuSP26)	ERK, JNK, P38MAPK, PKB, KAP3
DuSP2	JNK < P38MAPK ~ ERK (in vivo, preferentially JNK)
DuSP3	ERK, JNK, P38MAPK, STAT5
DuSP5	ERK
DuSP7	P38MAPK ~ JNK < ERK
DuSP8	ERK < P38MAPK ~ JNK
DuSP18	JNK
DuSP19	JNK
DuSP22	ERK, JNK, P38MAPK, ERα, STAT3
DuSP23	ERK, JNK, P38MAPK

Numerous MKPs have overlapping substrate specificities. These enzymes interact specifically with their substrates via a kinase-interaction motif (KIM) located within the non-catalytic, N-terminal domain of the protein. Binding of MAPKs to MKPs often primes activation of the C-terminal catalytic domain, thereby ensuring substrate specificity.

Four types of DuSP/MKP proteins can be defined according to their structure [741]. Classical MKPs possess a DuSP domain and a MAPK-binding domain (MKB), whereas atypical DuSPs that still function as MKP (e.g., DuSP3, DuSP14, and DuSP22) are much smaller proteins that lack a MAPK-binding domain. Type-1 DuSP/MKPs are characterized by a single PTP domain between their C- and N-termini. They include DuSP3, DuSP14, DuSP15, and DuSP22. Type-2 DuSP/MKPs contain a PTP and a MKB domain. They encompass DuSP1, DuSP2, DuSP4 to DuSP7, and DuSP9. Type-3 DuSP/MKP represented by a single element DuSP10 possesses a unique N-terminal NT domain in addition to MKB and PTP motifs. Type-4 DuSP/MKPs, such as DuSP8 and DuSP16, are constituted by NT, MKB, PTP, and PEST (proline-, glutamine-, serine-, threonine-rich) sequences.

Therefore, type-2 DuSP/MKPs are split into MKP subfamilies 1 and 2, whereas type-3 and -4 DuSP/MKPs constitute the MKP subfamily 3.

Table 8.33. The DuSP hyperfamily of phosphatases with its members of the MKP superfamily (**Part 1**; Sources: [736, 739, 740]; VH1: virus-encoded protein homolog, i.e., vaccinia virus VH1 gene product). The set of MAPK phosphatases also include protein Ser/Thr phosphatases PP2 and PPM1 as well as receptor-like protein Tyr phosphatase PTPRr and cytoplasmic types PTPn5 and PTPn7. Various synonyms exist for some of the DuSP genes or proteins. The MKP list takes into account Human Genome Organisation (HUGO) nomenclature. Mitogen-activated protein kinase phosphatases are classified into 2 main families according to intracellular location and transcriptional regulation. Family-1 MKPs are primarily located in the nucleus and encoded by immediate-early genes. They include MKP1, MKP2, DuSP2, and DuSP5. The expression of the nuclear MKPs is induced shortly after cell stimulation by growth factors and stresses. Family-2 MKPs are located either primarily in the cytoplasm or in both the cytosol and nucleus. They are characterized by slower kinetics. Mitogen-activated protein kinase phosphatases, which are specific for MAPKs, could also serve as anchors for MAPKs and control their intracellular location.

Name	Alias	Subcellular localization	Substrates	Stimuli
DuSP1	PTPn10, MKP1, CL100, VH1	Nuclear	ERK, JNK, P38MAPK	Hypoxia, LPS, heat shock Oxidative stress
DuSP2	PAC1	Nuclear	ERK, P38MAPK >JNK	Growth factors, heat shock, LPS
DuSP4	MKP2, VH2, TYP, STY8	Nuclear	ERK>JNK >P38MAPK	Growth factors, Other mitogens
DuSP5	DuSP, VH3, B23	Nuclear	ERK	Growth factors, IL2, heat shock
DuSP6	MKP3, PYST1, VH6	Cytosolic	ERK>JNK =P38MAPK	
DuSP7	MKPx, PYST2, B59	Cytosolic	ERK>P38MAPK	
DuSP8	VH5, VH8, HB5, C11ORF81	Nuclear, cytosolic	JNK>P38MAPK >ERK	
DuSP9	MKP4, PYST3	Nuclear, cytosolic	ERK>P38MAPK =JNK	

The MKP phosphatases possess an N-terminal rhodanese or CDC25 homology-2 domain (CH2) with kinase-interacting motifs that confer MAPK substrate specificity for one or more MAPKs among ERKs, JNKs, and P38MAPKs. Members of the MKP subfamily are listed in Tables 8.33 and 8.34. Many MKPs are inducible and have a low expression level in unstressed cells. Their expression rapidly heightens upon stimulation by growth factors or cell stresses. Their production kinetics and magnitude can depend on cell type and context.

The MKP enzymes bear post-translational modification. As the catalytic cleft possesses a cysteine residue, MKPs can undergo reversible oxidation (inactivation). Certain MKPs are inducible, others are stabilized or destabilized by phosphorylation. Phosphorylation of MKPs can lead to their stabilization by attenuating ubiquitination and subsequent degradation. Isozymes MKP1 to MKP3 and MKP7 are regulated by phosphorylation.

8.3 Cytosolic Protein Serine/Threonine Phosphatases

Table 8.34. The DuSP family of phosphatases with its members of the MKP subfamily (**Part 2**; Sources: [736, 739, 740]; MKPL: MKP1-like protein Tyr phosphatase; SKRP: stress-activated protein kinase pathway-regulating phosphatase; VHx: Vaccinia virus VH1 related (VHR)-related MKPx).

Name	Alias	Subcellular localization	Substrates	Stimuli
DuSP10	MKP5	Nuclear, cytosolic	JNK=P38MAPK >ERK	Peptidoglycans, LPS
DuSP14	MKP6, MKPL	Nuclear, cytosolic	ERK, JNK, P38MAPK	
DuSP16	MKP7, MKPm	Cytosolic	JNK2/3, P38MAPK$\alpha\beta$	
DuSP18		Nuclear, cytosolic	JNK	
DuSP22	JKAP, JSP1, VHx		JNK	
DuSP26	MKP8, SKRP3, NEAP	Nuclear, cytosolic	ERK, P38MAPK	
STYXL1	DuSP24, MKSTYX			

In fact, MKPs can be phosphorylated by their substrates: DuSP1 is phosphorylated by ERK (Ser359 and Ser364) and DuSP16 by P38MAPK (Ser446) [736]. Phosphorylation by ERK stabilizes MKP1 and increases its half-life, whereas it elicits MKP3 degradation [735]. However, the interaction of MKP3 with its specific substrate ERK2 enhances its activity. Furthermore, MKPs can be epigenetically regulated via promoter methylation or chromatin modification. Upon Toll-like receptor stimulation, histones associated with DuSP1 are acetylated.

The MAPK enzymes target numerous substrates, such as transcription factors of the ATF–CREB and AP1 family, kinases MAPKAPK2 and RSK, and controllers of mRNA stability and translation.

MKPs in Cell Metabolism

Upon insulin stimulation, liganded insulin receptor triggers the PI3K–PKB pathway for glucose uptake. This pathway possesses upstream mediators, insulin receptor substrates. The latter serves as a negative feedback node.

Phosphatase DuSP1 (MKP1) regulates multiple MAPK pathways in the nucleus, thereby controlling the expression of genes involved in fatty acid metabolism and energy expenditure. As cytosolic MAPKs are not subjected to inducible, nuclear MKP1, negative feedback exerted by phosphorylation of IRS by MAPK kinases does not operate.

Table 8.35. DuSPs in immune cells (Source: [741]).

DuSP type	Cell types
DuSP1	Macrophages, mastocytes, microglial cells, fibroblasts
DuSP2	B and T cells, neutrophils, mastocytes, macrophages
DuSP5	T cells, fibroblasts
DuSP10	Macrophages, T cells

MKPs in Development

Members of the subfamily-2, cytoplasmic, ERK-specific MKPs (DuSP6, DuSP7, and DuSP9) inactivates mainly ERK1 and ERK2. Phosphatase DuSP6 is an inhibitor of fibroblast growth factor signaling. Phosphatase DuSP9 intervenes in placental development and function [738].

MKPs in Stress Response

Members of the MKP subfamily 3 (DuSP8, DuSP10, and DuSP16) target stress-activated MAPKs (JNKs and P38MAPKs). In addition, DuSP1 promotes cell survival, as it attenuates effects of stress-activated MAPK pathways. In response to oxidative stress, the P53 pathway upregulates DuSP2 [738].

MKPs in Immunity

The MKP phosphatases can elicit synthesis of cytokines and chemokines. Activation of ERK results from TLR2 stimulation and leads to the production of anti-inflammatory IL10, whereas P38MAPK activation is caused by excited TLR4, TLR5, or TLR9 to provoke IL12 synthesis.

Phosphatases DuSP1, DuSP2, and DuSP10 regulate immune function. Activated B and T lymphocytes, mastocytes, eosinophils, macrophages, and dendritic cells, are characterized by high concentrations of nuclear MKPs (DuSP1, DuSP2, DuSP4, and DuSP5 [MKP subfamily 1]). Upon activation by a set of TLR ligands, DuSP1, DuSP2, and DuSP16 are the most strongly induced MKPs in macrophages [741] (Table 8.35).

Phosphatase DuSP1 controls macrophage function, inflammatory response to TLR signaling, and both innate and adaptive immune responses by inactivating P38MAPK and JNK [736]. In macrophages, lipopolysaccharides signal via Toll-like receptors that recruit MyD88 and TRIF adaptors to launch an early response using P38MAPKs and JNKs characterized by the production of pro-inflammatory TNF cytokine. This reaction is followed by synthesis of MKP1 that then prevents P38MAPK and JNK activities to avoid overproduction of cytokines, hence endo-toxic shock. Later, when MKP1 level declines, P38MAPKs and JNKs promote the synthesis of anti-inflammatory IL10 [738].

Among inducible, nuclear MKPs of the subfamily 1, DuSP2 is the most closely associated with immunocytes. Phosphatase DuSP2 not only impedes directly ERK and P38MAPK signaling, but also operates in JNK activity. In fact, it favors immune response via crosstalk between JNK and ERK pathways [738].

Phosphatase DuSP10 modulates gene expression in innate immunocytes. It precludes inflammation by inactivating JNK kinase. Whereas DuSP10 mainly operates on JNK signaling in immune cells, DuSP1 that influences the activity of all 3 MAPK types predominantly acts via the P38MAPK pathway [738].

DuSP1 – MKP1

Dual-specificity phosphatase-1 (DuSP1)[84] links to ERK1, ERK2, JNK1 to JNK3, and P38MAPKα (i.e., MAPK1, MAPK3, MAPK8–MAPK10, and MAPK14 [251]). Its other partners include CDC28 protein kinase regulatory subunit CKS1b, S-phase kinase-associated protein SKP2, and ubiquitin-C.

Phosphatase DuSP1 inhibits immune function, participates in metabolic homeostasis, and mediates resistance to cellular stress [739]. Activity of MKP1 is elicited by TLR stimulation via both myeloid differentiation factor MyD88, TIR domain-containing adaptor inducing Ifnβ(TRIF), and P38MAPK, to avoid excessive response to lipopolysaccharides of bacteria, especially by macrophages [742]. In macrophages, DuSP1 expression is synergistically induced by TLR and IL10 or glucocorticoid receptor signaling [741]. Phosphatase DuSP1 acts primarily on P38MAPK, thereby limiting the activity of CREB and AP1 transcription factors and controlling the expression of a set of LPS target genes. In microglia, endocannabinoid anandamide upregulates DuSP1 via CB_2, hence dampening MAPK activation.

DuSP2

Dual-specificity phosphatase-2 (DuSP2) is also named PAC1.[85] It is predominantly expressed in hematopoietic cells. It is located in the nucleus. It inactivates ERK1 and ERK2. Phosphatase DuSP2 is a positive regulator of inflammatory response [739].

Phosphatase DuSP2 is highly inducible in macrophages, mastocytes, eosinophils, and B and T lymphocytes by activated receptors. In mastocytes and macrophages,

---

[84] A.k.a. PTP10, HVH1, and MKP1.

[85] Acronym PAC1 also stands for platelet activation complex-1. Resting platelets do not bind coagulation factor VIII, but activated platelets can tether unactivated FVIII even in the absence of von Willebrand factor. Platelet activation complex-1-binding is enhanced by FVIII. Molecule PAC1 can also designate first procaspase-activating compound that induces apoptosis.

DuSP2 controlc JNK–ERK crosstalk [741]. Expression of DuSP2 is induced in macrophages by TLR ligands and in mastocytes by FcϵR1 ligation. Protein DuSP2 associates with MAPKs and blocks JNK-mediated inhibition of ERK activation.

DuSP4 – MKP2

Dual-specificity phosphatase-4 (DuSP4)[86] is widespread. It localizes to the nucleus. Two alternatively spliced transcript variants encode distinct isoforms.

Phosphatase DuSP4 interacts with MAPK3 (or ERK1), MAPK8 (or JNK1), and MAPK14 (or P38MAPKα) as well as cytokine suppressive anti-inflammatory drug-binding protein CSBP2 that itself connects to growth arrest and DNA-damage-inducible molecule GADD45α, MAP2K3, and MAP2K6 [251].

DuSP5

Dual-specificity phosphatase-5 (DuSP5)[87] is produced in various tissues, but with high level in pancreas and brain. It lodges in the nucleus. It inactivates ERK1 [251].

Phosphatase DuSP5 is encoded by the vascular-specific Dusp1 gene. It is expressed in angioblasts and in the vasculature, where it counteracts Ser/Thr sucrose non-fermenting-related kinase SNRK1. Kinase SNRK1 operates in migration, maintenance, and differentiation of angioblasts and favors endothelial cell migration. It tags arterial vasculature in association with Notch signaling [743], whereas the Rho pathway favors venous differentiation of microvasculature. Enzymes DuSP5 and SNRK1 function together to control angioblast population density [744]. In addition, DuSP5 impedes apoptosis of endothelial cells.

DuSP6 – MKP3

Dual-specificity phosphatase-6 (DuSP6) targets ERK1, ERK2, and P38MAPKα, in addition to MyoD family inhibitor and testis expressed protein-11 (TEx11) [251]. It is also identified as MKP3 and PYST1. The Dusp6 gene is expressed in various tissues with the highest levels in heart and pancreas. Unlike most other members of the group of typical DuSPs, it resides in the cytoplasm. Two transcript variants encode different isoforms.

Phosphatase DuSP6 is negative feedback regulator of ERK2 downstream from FGFR signaling. Fibroblast growth factors are secreted glycoproteins that contribute to morphogenesis by initiating the FGFR–Ras–MAPK cascade. Dual-specificity phosphatase DuSP6, sonic Hedgehog, Sproutys, and FGF-binding protein, the

---

[86] A.k.a. HVH2, TYP, and MKP2.
[87] A.k.a. VH1-like phosphatase HVH3.

Similar expression to FGFs (Sef), operate in coordination as feedback attenuators of the Ras–MAPK pathway to adjust FGF signaling to optimal level in embryogenesis. Whereas Sef and Sproutys impedes the Ras–MAPK axis at multiple nodes, DuSP6 restrains this signaling by dephosphorylating a single component, ERK.

At least in some animal species, DuSP6 is synthesized in cardiac progenitor cells and attenuates the FGF–ERK signaling by dephosphorylating ERK1/2. Inhibitor of DusP1 and DusP6[88] reversibly represses these phosphatases that are directly activated by ERK binding. As this inhibitor interferes with ERK-mediated activation, it does not disturb DuSP6 basal activity, but precludes ERK-stimulated DuSP6 enzymatic activity that thus remains at a low level and unresponsive to additional stimulus. Transient DuSP6 inhibition in zebrafish embryos elevates FGF signaling and expands the pool of cardiac progenitor cells at the expense of hematopoietic or endothelial cell lineages [745].

DuSP7 (MKPx)

Dual-specificity phosphatase-7 (DuSP7)[89] abounds in leukocytes of patients with acute myeloid or lymphoid leukemia. There is 2 splice variants: $DuSP7_L$ and $DuSP7_S$. Isoform $DuSP7_S$ may function as an inhibitor of $DuSP7_L$ [746].

Enzyme DuSP7 possesses a nuclear export sequence that confines the protein to the cytoplasm [747]. In vivo, DuSP7 binds with high efficiency and dephosphorylates extracellular signal-regulated kinases ERK1 and ERK2.

Protein DuSP7 has a broad tissue distribution (brain, heart, lung, kidney, liver, skeletal muscle, pancreas, and placenta) [747]. Two splice isoforms have been described: $DuSP7_L$ and $DuSP7_S$ that lacks the catalytic domain.

DuSP8

Dual-specificity phosphatase-8 (DuSP8)[90] is expressed predominantly in the adult brain, heart, and skeletal muscle. It abounds in the central and peripheral nervous system. It inactivates mitogen-activated protein kinase [748].

Olfactory stimuli primes expression of several immediate early genes, among which the DUSP8 gene in certain brain regions of the set of odor-processing nervous structures [749].

---

[88]Similarly, dorsomorphin is an inhibitor of bone morphogenetic protein that regulates iron metabolism and prostaglandin-E2 regulates hematopoietic stem cell. A small compound BCI (benzylidene cyclohexylamino-dihydro-1H-inden-1) blocks DuSP6 activity and hyperactivates FGF [745].

[89]A.k.a. MKPx and PYST2.

[90]A.k.a. HB5, homolog of vaccinia virus H1 phosphatase gene clone VH5, and VH8.

DuSP9 – MKP4

Dual-specificity phosphatase-9 (DuSP9)[91] is highly synthesized in the developing liver and placenta, but at low level in other organs. It thus has a developmentally regulated expression. It is indeed essential for placental development and function [739]. It is also produced in the lung, kidney, and adipose tissue [750].

Phosphatase MKP4 has a high specificity for extracellular signal-regulated kinases ERK1 and ERK2, but a low one for P38MPAKα and Jun N-terminal kinases [750]. It possesses a nuclear export sequence that confines the protein to the cytoplasm.

DuSP10 – MKP5

Dual-specificity phosphatase-10 (DuSP10), or MAPK phosphatase MKP5, interacts with MAPK8, MAPK9, and MAPK10 (JNK1–JNK3) as well as MAPK14 (P38MAPKα) [251].

It functions in innate and adaptive immunity, primarily by regulating the JNK pathway in innate and adaptive immune effector cells [751]. It limits the cytokine production in CD4+ and CD8+ effector T cells. On the other hand, it promotes T-cell proliferation upon activation. In macrophages, TLR ligands induce DuSP10 production to impede JNK activity, thereby constraining the production of cytokines and reducing costimulation of T cells. In T lymphocytes, DuSP10 is constitutively expressed and inhibits early JNK activation after TCR ligation [741].

DuSP14

Phosphatase DuSP14 interacts with T-cell costimulatory factor CD28 [736].[92] It can also target ERK and JNK and, via these kinases, interleukin-2. It may then inhibit costimulatory signaling in T lymphocytes.

In pancreatic β cells, DuSP14 production rises in response to glucagon-like peptide GLP1, a growth and differentiation factor for these cells, to launch a negative feedback [737]. Activated dopamine D_1 receptor can provoke the synthesis of DuSP14 phosphatase.

DuSP16 – MKP7

Dual-specificity phosphatase-16 (DuSP16), or MKP7, interacts not only with ERK2 (MAPK1), MAPK8 to MAPK10 (JNK1–JNK3), and MAPK14 (P38MAPKα), but also with scaffold proteins MAPK8IP1 and MAPK8IP2 (a.k.a. JNK-interacting proteins JIP1 and JIP2).

---

[91] A.k.a. MKP4.
[92] Protein DuSP14 was initially identified as a CD28-interacting protein.

Table 8.36. Slingshot properties (Source: [483]). Isoform Ssh3 is less effective than the others in dephosphorylating phosphoADF and phospho-cofilin.

Ssh Type	Distribution	Localization	Regulators	Substrates
Ssh1	Ubiquitous	Lamellipodia	14-3-3	Cofilin, ADF, LIMK1
Ssh2	Ubiquitous	Lamellipodia	Calcineurin	Cofilin, ADF
Ssh3	Ubiquitous	Membrane protrusion	PI3K	Cofilin, ADF

DuSP24 – STYXL1

Dual-specificity phosphatase-24, or MAPK phosphatase-like inhibitor Ser/Thr/Tyr (STY)-interacting-like protein STYXL1,[93] is a catalytically inactive dual-specificity phosphatase as a result of a naturally occurring replacement of the phosphatase catalytic cysteine by serine.

Among 5 potential isoforms, DuSP24 is the single isoform that encodes both the CDC25-like domain and dual-specificity phosphatase segment. It may antagonize cell survival [752]. It is widespread.

DuSP26 – MKP8

Dual-specificity phosphatase-26 (DuSP26 or DuSP24), or mitogen-activated protein kinase phosphatase MKP8,[94] targets P38MAPK and prevents P38MAPK-mediated apoptosis.

8.3.13.4 Slingshot Phosphatases

The slingshot (Ssh) family of phosphatases contains 3 members (Ssh1–Ssh3) encoded by 3 genes (SSH1L–SSH3L). Each Ssh has long and short variants. They have distinct subcellular location [483] (Table 8.36).

They dephosphorylate proteins of the actin-depolymerization factor–cofilin family. Mammalian slingshots can dephosphorylate Ser^P and Thr^P residues. They possess the PTP catalytic domain as well as 14-3-3-binding motifs, a C-terminal filamentous actin-binding site, and a SH3 binding sequence. Phosphorylated (Ser) slingshot phosphatases bind to 14-3-3 proteins for cytoplasmic sequestration (inactivation).

---

[93] A.k.a. MKSTYX.
[94] A.k.a. low-molecular-mass dual-specificity phosphatase LDP4 and stress-activated protein kinase pathway-regulating phosphatase SKRP3.

The 3 slingshots are widespread. However, they have distinct subcellular locations, and hence related but distinct functions in the regulation of actin polymerization. Slingshot phosphatases dephosphorylate ADF^P and cofilin^P to impede actin filament assembly caused LIM domain-containing kinase LIMK1 and testis-specific kinase TesK1. Moreover, slingshot Ssh1L dephosphorylates LIMK1 [736].

Cofilin is a ubiquitous actin-binding molecule required for the reorganization of actin filaments. Cofilin dephosphorylation enables its depolymerizing activity during cell migration. In fact, 2 distinct families of phosphatases dephosphorylate cofilin (Ser3): slingshot phosphatases and haloacid dehalogenase phosphatase chronophin (Sect. 8.3.16.3).

Slingshot phosphatases are regulated by pathways that involve Ca^{++} influx (e.g., Ca^{++}-mobilizing agents ATP and histamine), cAMP and PI3K [483]. Calcium-regulated PP3 dephosphorylates (activates) Ssh1, whereas PAK4 inhibits Ssh1. Insulin-primed actin reorganization occurs via PI3K. Slingshot activity is also directly stimulated by ^{F}actin binding.

During healing of wounded skin, keratinocytes migrate. Integrins β_4 regulate the laminin matrix assembly through which keratinocytes move. Whereas in neurons, small GTPase Rac1 signals via p21-activated kinase to phosphorylate (inactivate) cofilin, in keratinocytes, $\alpha_6\beta_4$ integrin works via Rac1 to maintain active cofilin via slingshot phosphatases [753].

As they control cofilin activity, slingshot phosphatases are implicated in mitosis, growth cone motility and morphology, and neurite extension, as well as regulation of membrane protrusion [483]. During metaphase of the cell division, they participate in cortical actin regulation; during anaphase and telophase in the cleavage furrow; and during late telophase, in the midbody (a set of bundles of microtubules derived from the mitotic spindle at the end of cytokinesis, just prior to the complete separation of the daughter cells).

8.3.13.5 Protein Tyrosine Phosphatases-4A

Members of the protein Tyr phosphatase-4A family (PTP4a1–PTP4a3) are also called phosphatases of regenerating liver (PRL1–PRL3). They possess the conserved PTP catalytic domain and a C-terminal CAAX box that allows post-translational farnesylation and localization to membranes. They have a specific tissue distribution. Like other PTPs, PTP4as can be inactivated by reversible oxidation.

8.3.13.6 Cell-Division Cycle-14 Phosphatase Homologs

Phosphatases of the CDC14 family specifically dephosphorylate cyclin-dependent kinases. Four CDC14 phosphatases exist: CDC14a and CDC14b as well as kinase-associated phosphatase (KAP), or cyclin-dependent kinase inhibitor CIP3, and protein Tyr phosphatase domain-containing protein-1 (PTPDC1 or PTP9Q22). They are all related to class-3 cysteine-based CDC25 Tyr phosphatases that regulate the initiation of mitosis and DNA-damage checkpoint control.

Phosphatase CIP3 dephosphorylates cyclin-dependent kinase CDK2, hence hindering cell cycle progression. Mammalian CDC14 phosphatases contribute to centrosome maturation, spindle stability, cytokinesis, and cell cycle progression. They are primarily regulated by sequestration in the nucleolus during interphase [736].

8.3.13.7 Myotubularins

Myotubularins (myotubular myopathy-associated gene product [MTM])[95] and myotubularin-related proteins (Sect. 2.10.2; Table 8.37) form a family of protein Tyr phosphatases. They use inositol phospholipids rather than phosphoproteins as substrates. These lipid phosphatases dephosphorylate phosphatidylinositol 3-phosphate and phosphatidylinositol (3,5)-bisphosphate. Myotubularin can hydrolyze Ser^P and Tyr^P substrates due to the presence of the DSP catalytic motif, but its activity is relatively inefficient.

Myotubularins commonly contain motifs assigned to between-protein or protein–lipid interactions. Most myotubularins indeed possess PH and FYVE (Fab1p, YOTB, Vac1p, and EEA1) domains that link to phosphoinositide lipids and/or PDZ domains that connect to protein. In any case, they have an N-terminal PH–GRAM (pleckstrin homology–glucosyl transferase, Rab-like GTPase activator, and myotubularin domain) region that binds particular lipids and recruits proteins to the plasma membrane and a C-terminal coiled-coil domain. The PTP domain contains a SET-interacting domain (SID sequence) that is conserved in myotubularins [754]. The PH-G domain is a phosphoinositide-binding site for all 3 monophosphoinositides as well as $PI(3,5)P_2$.

Among the 12 identified members of the myotubularin family, a subgroup of catalytically inactive members serve as adaptors for the active members. Nearly half the myotubularins are indeed catalytically inactive (*pseudophosphatase*).[96]

The prototypical family member MTM1 is encoded by a gene on chromosome-X.[97] Other members include myotubularin-related proteins MTMR1 to MTMR9 as well as MTMR13.[98] Members of the MTM family can heteromerize. Both MTM1

---

[95] The myotubularin gene is implicated in X-linked myotubular myopathy, a disorder characterized by failure of normal development of muscle fibers..

[96] Myotubularins contain a PTP catalytic domain, but catalytically inactive myotubularins lack critical cysteine, arginine, and aspartic acid residues that render the phosphatase inactive.

[97] Mutations of the MTM1 gene cause myotubular myopathy. Mutations of the MTMR2 gene are associated with the Charcot-Marie-Tooth syndrome.

[98] A.k.a. SET-binding factor SBF2; SBF1 is an alias for MTMR5 phosphatase.

Table 8.37. Superfamily of myotubularins (myotubular myopathy-associated gene products [MTM] and myotubularin-related phosphatases [MTMR]; SBF: SET binding factor). Both SBF1 and SBF2 are approved symbol for MTMR5 and MTMR13 (3PAP: 3-phosphatase adaptor protein; CNM: X-linked centronuclear myopathy [XLCNM; a.k.a. myotubular myopathy] protein; MTMx: X-linked recessive myotubular myopathy [XLMTM] protein; CMT4b: Charcot-Marie-Tooth disease type-4B protein; CRA: cisplatin resistance-associated protein; EDTP: egg-derived tyrosine phosphatase; FAN: Fanconi-associated nuclease; FYVEDuSP: FYVE domain-containing dual-specificity protein phosphatase; PIP3AP: phosphatidylinositol 3-phosphate 3-phosphatase adaptor; lipSTYX: lipid-selective Ser/Thr/Tyr-interacting (STYX) protein-like protein; ZFYVE: zinc finger FYVE domain-containing protein). The abnormal Mtmr14 gene produces the JUMPY phenotype in flies (a muscle defect and progressive loss of muscle control with shaky and slower motions, hence the name JUMPY).

Type	Other alias(es)
MTM1	CNM, MTMx
MTMR1	
MTMR2	CMT4b1
MTMR3	ZFYVE10, FYVEDuSP1
MTMR4	ZFYVE11, FYVEDuSP2
MTMR5	SBF1
MTMR6	
MTMR7	
MTMR8	
MTMR9	MTMR8, lipSTYX, C8ORF9
MTMR10	
MTMR11	CRA
MTMR12	PIP3AP, 3PAP
MTMR13	SBF2, CMT4b2
MTMR14	C3ORF29, EDTP, Jumpy
MTMR15	FAN1

and MTMR2 can interact with MTMR12; MTMR6 and MTMR7 with MTMR9; MTMR2 with MTMR5 and MTMR13 [754]. Hetero-oligomerization enhances enzymatic activity.

Myotubularin-related proteins participate in lipid-mediated signaling. Several myotubularins, such as MTMR1, MTMR6 to MTMR8, and MTMR12, promote cell survival. Lipid PI(5)P actually regulates a PI(3,4,5)P$_3$-specific 5-phosphatase that controls PKB activation. Protein MTMR2 promotes the formation of adherens junctions. Protein MTMR6 binds to and inhibits Ca^{++}-activated $K_{Ca}3.1$ channels.

Myotubularins and Endocytosis

Myotubularins regulate endocytosis. Endocytic membrane trafficking depends on the synthesis of phosphatidylinositol 3-phosphate and phosphatidylinositol (3,5)-phosphate. Phosphoinositide PI3P resides in sorting endosomes, early and late endosome. Class-3 PI3K corresponds to a single kinase (PI3KC3 with catalytic PI3K$_{c3}$ subunit) that is responsible for the synthesis of its specific substrate phosphatidylinositol 3-phosphate.[99] In humans, it is encoded by the PIK3C3 gene. Endosomal synthesis of PI3P is initiated with the activation of Rab5 and Rab7 GTPases on early and late endosomes, respectively. Phosphoinositide PI3P governs the endosomal recruitment of membrane-trafficking FYVE- and PX-domain-containing regulators, such as early endosome antigen EEA1 and hepatocyte growth factor-regulated Tyr kinase substrate Hrs. Kinase PIP5K3 (or PIKfyve) generates PI(3,5)P$_2$ from PtdIns3P. The conversion of PI3P into PI(3,5)P$_2$ occurs on multivesicular late endosomes.

Myotubularin substrates PI3P and PI(3,5)P$_2$ are regulators of the endocytic pathway. Myotubularins antagonize PIK3C3 and PIP5K3, as MTM1 and MTMR2 utilize both PI3P and PI(3,5)P$_2$ as substrates. Hydrolysis of PI(3,5)P$_2$ produces PI5P that activates myotubularins. Myotubularins can repress accumulation of substrate lipids at inappropriate compartments. Under certain circumstances, the cell needs to limit PI3P concentration on the secretory pathway to avoid recruitment of endosomal proteins [754].

Myotubularins MTM1 and MTMR2 are associated with early and late endosomes, respectively, although MMT1 is recruited at low level in late endosomes. Enzyme MTM1 lowers the endosomal pool of PI3P, but not its overall cellular mass that resides elsewhere in the cell, presumably at lower concentrations [754]. On the other hand, myotubularins do not extract EEA1 from endosomes.

As myotubularins impede endosome formation, it prevents receptor internalization. MTMR2 translocates to EGFR-containing endosomes after EGF stimulation to block receptor degradation via late endosomes [755].

8.3.13.8 Atypical Dual-Specific Phosphatases

Atypical dual-specificity phosphatases have a low molecular mass, as they lack the N-terminal CDC25 homology-2 domain common to mitogen-activated protein kinase phosphatases. However, some atypical DuSPs dephosphorylate MAPK enzymes (Sect. 8.3.13.3). Sixteen atypical DuSPs have been identified (Table 8.38).[100] All DuSPs contain the conserved catalytic domain. Among these atypical dual-specificity phosphatases, some are pseudophosphatases (Table 8.39)

---

[99] Kinase PI3KC3 (PI3K$_{c3}$) is labeled VPS34 in yeast and PI3K59F in fruit fly.

[100] They are similar to VH1 phosphatase detected in vaccinia virus (VH1 stands for vaccinia virus open reading frame H1).

Table 8.38. Atypical dual-specificity phosphatases (Source: [736]; HUGO: Human Genome Organisation; BEDP: branching enzyme-interacting DuSP; Cap: mRNA-capping enzyme; DuPD: DuSP and proisomerase domain-containing protein; EPM; epilepsy, progressive myoclonus; GKAP: glucokinase-associated phosphatase; HCE: human corneal epithelial cell phosphatase; JKAP: JNK pathway-associated phosphatase; JSP: JNK-stimulating phosphatase; LD: Lafora disease protein [or laforin]; LDP: low-molecular-mass DuSP; lmwDSP: low-molecular-weight DuSP; mDSP: muscle-restricted DuSP; MKPL: MKP1-like protein Tyr phosphatase; MOSP: myelin- and oligodendrocyte-specific protein phosphatase; PTPmt: protein Tyr phosphatase localized to the mitochondrion; PIR: phosphatase that interacts with RNA–ribonucleoprotein complex; PLIP: PTen-like phosphoinositide [PI(5)P] phosphatase; RNGTT: RNA guanylatetransferase and 5′-phosphatase; SKRP: stress-activated protein kinase pathway-regulating phosphatase; STYX: Ser/Thr/Tyr-interacting protein; tmDP: testis- and skeletal muscle-specific DuSP; VHR: vaccinia virus open reading frame H1 (VH1)-related; VHX: VH1-related phosphatase-X; VHY: VH1-like member-Y; VHZ: VH1-related phosphatase-Z).

Type	Alias(es)	Subcellular localization
DuSP3	VHR	Nucleolus
DuSP11	PIR1	Nucleus
DuSP12	DuSP1, yVH1, GKAP	Cytoplasm, nucleus
DuSP13a	LDP1, tmDP	
DuSP13b	BEDP, mDSP, SKRP4, tsDSP6	Cytoplasm
DuSP14	MKP6, MKPL	Cytoplasm (T lymphocytes)
DuSP15	VHY	Cytoplasm (plasma membrane)
DuSP18	DuSP20, lmwDSP20	Cytoplasm, nucleus, mitochondrion
DuSP19	DuSP17, SKRP1, LDP2, lmwDSP1, lmwDSP3, tsDSP1	Cytoplasm
DuSP21	lmwDSP21	Cytoplasm, nucleus, mitochondrion
DuSP22	lmwDSP2, JKAP, JSP1, VHX, tsDSP2, VHX	Nucleus and cytoplasm (enriched in Golgi body)
DuSP23	DuSP25, LDP3, VHZ	Cytoplasm
DuSP26	DuSP24, LDP4, MKP8, NATA1, SKRP3	Cytoplasm (perinuclear region, Golgi body), nucleus
DuSP27	DuPD1, FMDSP, STYXL2	Cytoplasm
DuSP28	DuSP26, VHP	
EPM2a	Laforin, LD	Cytoplasm
PTPmt1	DuSP23, PLIP, MOSP	Mitochondria
RNGTT	Cap1a, HCE1	Nucleus
STYX		

8.3 Cytosolic Protein Serine/Threonine Phosphatases

Table 8.39. Subtrates of atypical dual-specificity phosphatases (Source: [737]; GSK: glycogen synthase kinase; PI(5)P: phosphatidylinositol 5-phosphate). Substrates of atypical DuSPs include not only proteins phosphorylated on Ser, Thr, and Tyr, but also phosphatidylglycerol phosphate (PGP), mRNA capping structure, and glycogen.

Type	Substrates
DuSP11	RNA 5'triphosphate
DuSP12	Glucokinase
DuSP27	Pseudophosphatase
Laforin	Glycogen, GSK3β
PTPmt1	PGP, PI(5)P
RNGTT	5'Cap (RNA phosphatase)
STYX	Pseudophosphatase

Laforin

The unique dual-specificity phosphatase laforin with a carbohydrate-binding domain (CBM20) is encoded by the EPM2A gene.[101] The EPM2B gene encodes ubiquitin ligase malin, or NCL1,[102] HT2a,[103] and Lin41 (NHL) repeat-containing protein-1 (NHLRC1), that ubiquitinates laforin for proteasomal degradation. Laforin phosphatase dephosphorylates glycogen to maintain a normal pool of cellular glycogen.[104]

In addition to a DuSP catalytic domain, laforin possesses a carbohydrate-binding module-20 (or starch-binding domain) in its N-terminus that is a glycogen-binding site. Laforin binds glycogen-targeting regulatory subunit-5 of protein phosphatase-1 ($PP1_{r5}$), or protein targeting to glycogen (PTG), that recruits glycogen synthase [736]. Protein phosphatase-1 subunit $PP1_{r5}$ also impedes the activity of glycogen degradation enzymes phosphorylase and phosphorylase kinase [737]. Laforin and malin form a complex that ubiquitinates $PP1_{r5}$ for proteasomal degradation, thereby precluding glycogen accumulation, especially in neurons that do not synthesize phosphorylase and phosphorylase kinase involved in glycogen hydrolysis. Laforin may serve as a substrate adaptor that recruits proteins in glycogen particles for ubiquitination by malin and subsequent degradation by the proteasome. In addition, malin ubiquitinates laforin (autoregulation). Interaction

---

[101] Mutations in the EPM2A gene generate progressive myoclonus epilepsy (EPM: epilepsy, progressive myoclonus) or Lafora's disease. The Spanish neuropathologist G.R. Lafora (1886–1971) recognized small inclusions in Lafora patients. Lafora disease is indeed characterized by the presence of Lafora bodies composed of abnormal glycogen (insoluble polyglucosans).

[102] I.e., new calpain-1, muscle-specific calcium-activated neutral peptidase-3, and calpain-3.

[103] I.e., HIV1 Tat activation domain-binding protein HT2a, 72-kDa, zinc finger-containing Tat-interacting protein (TatIP), or ubiquitin ligase tripartite motif-containing protein TriM32.

[104] Glycogen, a branched polymer of glucose, contains a small amount of covalently linked phosphate.

between laforin and malin relies on laforin phosphorylation by AMPK. Moreover, AMPK phosphorylates PP1$_{r5}$ (Ser8 and Ser268) to accelerate its ubiquitination by the laforin–malin complex [737].

STYX

The Ser (S)/Thr (T)/Tyr (Y)-interacting (X) protein (STYX)[105] is a pseudophosphatase.[106] Like its related pseudophosphatases, MAPK Ser (S)/Thr (T)/Tyr (Y)-interacting protein (MKSTYX or STYXL1), Dusp27 (or STYXL2), and MTMR9 (or lipid-selective STYX-like protein [lipSTYX]), STYX functions as an adaptor for phosphorylated targets that operate in the differentiation of male germ cells [756]. Protein STYX is predominantly produced not only in the testis, but also in the myocardium and skeletal muscle; it is widespread.

Protein STYX interacts with the phosphorylated, 24-kDa, RNA-binding, Ca^{++}-regulated, heat-stable protein CRHSP24. The latter is substrate of PP3 phosphatase [737].

RNA Phosphatase RNGTT

Like DuSP11, RNGTT phosphatase possesses an N-terminal phosphatase domain that contains the P-loop motif, but lacks the catalytically involved Asp residue of the Cys-based PTPs. This RNA triphosphatase removes the γ-phosphate from the 5' end of nascent mRNA to promote the formation of unmethylated 5' cap structure, which facilitates transcript translation and subsequent mRNA processing.

Mitochondrial Protein Tyr Phosphatase PTPmt1

Mitochondrial atypical dual-specificity phosphatases include DuSP18, DuSP21, and PTPmt1. Mitochondrial protein Tyr phosphatase PTPmt1 localizes to the mitochondrion inner membrane; more precisely, it is anchored to the matrix face of the inner membrane. It colocalizes with components of the respiratory chain.

Its preferred substrate is PI(5)P in vitro. It dephosphorylates phosphatidylglycerol phosphate, which is similar PI(5)P that exists only in very low concentrations in mitochondria, to phosphatidylglycerol, an intermediate in the cardiolipin synthesis [757].

---

[105] More precisely phosphoserine/phosphothreonine/phosphotyrosine-interacting protein.

[106] This inactive aDuSP presents a Gly residue instead of the catalytic Cys in the P loop of the phosphatase domain.

It is also called myelin- and oligodendrocyte-specific protein (MOSP), as it is detected in oligodendrocytes and myelin of the central nervous system, where it can serve as a late marker.[107]

In pancreatic β cells, this mitochondrial atypical dual-specificity phosphatase attenuates ATP production and insulin secretion under both basal and glucose-stimulated conditions [737].

DuSP3

Phosphatase DuSP3 preferentially targets diphosphorylated Tyr^P and Ser^P–Thr^P peptides rather than monophosphorylated compounds and Tyr^P in monophosphorylated substrates rather than Ser^P or Thr^P. Unlike some other DuSPs, the DUSP3 gene is not an early-response gene.

Ubiquitous DuSP3 intervenes during the cell cycle as well as several signaling pathways involved in the immune response. Regulation of DuSP3 function is achieved by phosphorylation. In T lymphocytes, DuSP3 is phosphorylated by ZAP70 to enable antigen signaling from T-cell receptors.

Kinases ERK1, ERK2, and JNK, as well as EGFR and STAT5 are DuSP3 substrates [736]. Once phosphorylated (Tyr138) by ZAP70 or TyK2 kinase, DuSP3 targets STAT5 especially in cells stimulated by Ifnα and Ifnβ [737].[108]

When DuSP3 expression is reduced, JNK and ERK kinases are hyperactivated and the cell cycle stops during the G1–S and G2–M transitions. In addition, cells undergo senescence [737].

DuSP11

Phosphatase DuSP11 serves as a mRNA triphosphatase. It hydrolizes RNA 5′-triphosphate to 5′-diphosphate and 5′-monophosphate [737]. It can associate with splicing factors SFRS7 and SFRS9 as well as glutamine tRNA ligase (or glutaminyl-tRNA synthase [251].

---

[107] Myelin is produced by oligodendrocytes in the central nervous system and by Schwann cells in the peripheral nervous system. During demyelination, immunoreactivity for MOSP, like that of the early marker myelin basic protein (MBP), is lost.

[108] Kinase TyK2 also phosphorylates STAT5 factor.

DuSP12

Dual-specificity phosphatase-12 (DuSP12)[109] dephosphorylates glucokinase, thereby enhancing its activity. It also interacts with pleiotrophin, methylCpG-binding domain protein MBD6, and prenylated Rab acceptor RabAC1 [736]. It possesses a N-terminal DUSP domain and a Cys-rich C-terminus that can bind zinc.

Phosphatase DuSP12 acts as a ribosome assembly factor that regulates ribosome stalk assembly, in particular in humans [758]. It interacts with heat shock protein HSP70 using its Zn-binding domain together with the HSP70 ATPase domain to enhance cell survival.

DuSP13

Two isoforms — DuSP13a and DuSP13b — are encoded by alternative open reading frames of the single DUSP13 gene [736]. Isoform DuSP13a is highly expressed in the testis (hence the alias tmDP for testis and skeletal muscle-specific dual-specificity phosphatase); DuSP13b is only produced in the muscle (hence the alias mDSP for muscle-restricted dual-specificity phosphatase). Protein DuSP13 interacts with coiled-coil domain-containing protein CCDC85b [251].[110]

DuSP15

Phosphatase DuSP15 is highly expressed in testis, but is also synthesized at low levels in other organs, such as the brain, spinal cord, and thyroid [736]. This myristoylated atypical dual-specificity phosphatase is characterized by the addition of myristic acid to Gly residue that enables its recruitment to the plasma membrane [737].

DuSP18

Mitochondrial DuSP18 (or lmwDuSP20) is strongly produced in the brain, liver, testis, and ovary, but is also synthesized in most tissues. In cells, it can be restricted to mitochondria, from which it is released during apoptosis. It may then dephosphorylate JNK, but not ERK and P38MAPK [736].

---

[109] The ortholog of yeast VH1 (yVH1). The mouse ortholog mVH1 has a splice variant that contains only the phosphatase domain: low-molecular-weight, dual-specificity phosphatase-4 (lmwDSP4, or DSP4$_{lMW}$). The rat ortholog was cloned with glucokinase, hence named glucokinase-associated phosphatase (GKAP).

[110] A.k.a. hepatitis-δ antigen-interacting protein-A (DIPA).

Unlike PTPmt1 that possesses an N-terminal mitochondrial localization signal, DuSP18 contains an internal mitochondrial localization signal that enables its localization in the mitochondrial inner membrane. Unlike PTPmt1 that is anchored to the matrix face of the inner membrane, it is located at the external edge that faces the intermembrane space [737].

DuSP19

Protein DuSP19[111] localizes to the cytoplasm of most cell types, where it regulates JNK2 signaling by targeting upstream kinases, such as MAP2K7 (JN2K2) and apoptosis signal-regulating kinase MAP3K5 (or ASK1), or acting as a scaffold [736].

DuSP21

Mitochondrial DuSP21 is restricted to testis. In cells, it localizes to mitochondria [736]. Like PTPmt1 and DuSP18, it also has a mitochondrial localization signal, but lodges to the mitochondrial matrix [737].

DuSP22

Two DuSP22 variants can be found, a ubiquitous and a testis- and liver-restricted isoform. DuSP22 is required for full JNK1 activation in response to cytokines via MAP2K4 and MAP2K7 [736]. It also interacts with ERK2 [251]. Like DuSP15, DuSP22 is myristoylated to be recruited to the plasma membrane [737].

Protein DuSP22 is also involved in the regulation of certain transcription factors, such as STAT3 and estrogen receptor-α. It indeed dephosphorylates ERα (Ser118) [737]. It also dephosphorylates STAT3, especially during interleukin-6 stimulation.

DuSP23

Phosphatase DuSP23 is detected in the cytosol. It is widespread with high expression in the heart, spleen, liver, and testis [736]. It is produced in different hematopoietic cell lines.

---

[111]A.k.a. DuSP17, stress-activated protein kinase pathway-regulating phosphatase SKRP1, low-molecular-weight dual-specificity phosphatase lmwDSP3, and tsDSP1.

DuSP26

Phosphatase DuSP26, also called DuSP24 and DuSP28,[112] It has its highest expression in the brain, heart, adrenal glands, and skeletal muscles [736]. It localizes mainly in the nucleus. Its synthesis rises upon nerve growth factor stimulation.

Substrates of DuSP26 include MAPKs and PKB (Ser473) [737]. Phosphatase DuSP26 impedes cell growth in response to epidermal growth factor, as it lowers EGFR production. In addition, DuSP26 participates in intracellular transport and cell adhesion, since it binds microtubule-associated nanomotor KIF3 that carries β-catenin and N-cadherin to the plasma membrane [737].

DuSP27

Protein DuSP27 is also labeled dual-specificity phosphatase and proisomerase domain-containing DUPD1, but refers to another open reading frame that encodes a pseudophosphatase [737]. It is produced in adipose tissue, skeletal muscles, and liver [736].

Protein DuSP27 can dephosphorylate Ser^P and Thr^P peptides, but is more efficient on Tyr^P peptides. It processes particularly phosphorylated substrates in which 2 phosphorylated residues are separated by 2 amino acids instead of 1 such as MAPK enzymes.

DuPD1

Atypical dual-specificity phosphatase and proisomerase domain-containing protein-1, in fact, does not contain the proisomerase sequence [737].

8.3.14 Infraclass-2 Protein Tyrosine Phosphatase

Soluble acid phosphatase-1 (AcP1), or low-molecular-weight protein phosphotyrosine phosphatases (lmwPTP; $\sim 20\,kDa$)[113] generated by the highly polymorphic ACP1 gene have a sequence homology with other classes of PTPs, the active site sequence that contains Cys and Arg residues involved in enzyme catalysis found in all PTPs [759].

---

[112] A.k.a. low-molecular-mass dual-specificity phosphatase-4 and stress-activated protein kinase pathway-regulating phosphatase SKRP3.

[113] A.k.a. adipocyte acid phosphatase (AAP) and erythrocyte acid phosphatase-1.

8.3 Cytosolic Protein Serine/Threonine Phosphatases

In both vascular smooth muscle and endothelial cells, low-molecular-weight protein Tyr phosphatases include 2 active isoforms – AcP1^{IF1} and AcP1^{IF2} – that operate in cell growth and migration [761].[114]

Both isoforms AcP1^{IF1} and AcP1^{IF2} inhibit PDGF-induced protein synthesis and cell migration in smooth myocytes. On the other hand, both isoforms enhance lysophosphatidic acid-stimulated migration of endothelial cells, without change in DNA synthesis. In addtion, high concentrations of glucose suppress AcP1 production.

Phosphatase AcP1 targets phosphorylated platelet-derived growth factor receptor [760]. It also dephosphorylates insulin receptor. Isoform IF1 is more active than isoform IF2 on the insulin receptor.

8.3.15 Infraclass-3 PTP Cell-Division Cycle-25 Phosphatases

Cell-division cycle dual-specificity phosphatases CDC25 remove inhibitory phosphate on specific Thr and Tyr residues in cyclin-dependent kinases, hence controlling progression through different phases of the cell cycle. Phosphatases of the CDC25 infraclass include 3 isoforms (CDC25a–CDC25c) that are encoded by 3 genes (Cdc25A–Cdc25C).

The CDC25 isozymes have both different and redundant specificities and regulations. Phosphatase CDC25a intervenes at the G1–S transition. Paralog CDC25b is activated during S phase and activates the CDK1–CcnB complex in the cytoplasm. Active CDK1–CcnB complex then phosphorylates (activates) CDC25c, thereby generating a positive feedback for entry into mitosis [762]. Moreover, CDC25 contributes to G2 arrest caused by DNA damage or unreplicated DNA. All isoforms are also targets and regulators of the G2–M checkpoint mechanism that is initiated in response to DNA injury.

The expression and activity of CDC25 phosphatases is finely regulated by multiple processes, such as post-translational modifications, interactions with regulatory partners, control of their intracellular localization, and cell cycle-regulated degradation [763].

The CDC25 phosphatases interact with many substances, such as kinases CDK1, Src, Polo-like kinases PlK1 and PlK3, checkpoint kinase ChK1, P38α MAPK, cell cycle mediators, such as cyclin-B1, peptidyl-isomerase Pin1 that regulates CKI1b expression, and proliferating cell nuclear antigen (PCNA), an auxiliary processivity factor for DNA polymerase-δ, as well as 14-3-3 protein family members tyrosine

---

[114]The ACP1 gene is composed of 7 exons and 6 introns. Alternative splicing of ACP1 transcript leads to mature mRNAs that encode 2 isoforms, fast and slow AcP1 according to their electophoretic mobility), a.k.a. ACP1a and ACP1b as well as isoform AcP1^{IF1} and AcP1^{IF2}, respectively. In AcP1^{IF1}, exon 4 is excised and exon 3 is retained (exons 1–2–3–5–6–7) and conversely in AcP1^{IF2} isoform (exons 1–2–4–5–6–7).

3-monooxygenase/tryptophan 5-monooxygenase activation protein YWHAα (or 14-3-3α; encoded by the YWHAA gene), YWHAε (14-3-3ε; encoded by the YWHAE gene), and YWHAη (14-3-3η; encoded by the YWHAH gene), and ubiquitin ligase NEDD4 [251].

Proteolysis of cyclin-B is a major mechanism that inactivates cyclin-dependent kinase CDK1 to promote exit from mitosis. Inhibitory phosphorylation of CDK1 (Thr14 and Tyr15) by Wee1 and PKMYT1 that are removed by CDC25 phosphatases not only repress CDK1 during S- and G2 phases, but also in G1 phase to avoid reversal to M phase. Downregulation of CDK1 activates Wee1 and PKMYT1 kinases and simultaneously inhibits CDC25 phosphatase during the M–G1 transition [764]. Therefore, a reciprocal action of CDK1 exist on its kinases and phosphatases (feedback loop). Conversely, during the G2–M transition, Wee1 and PKMYT1 are phosphorylated (inactivated) when CDK1 activity rises, as CDK activity stimulates its activator CDC25 and inhibits its inhibitors Wee1 and PKMYT1 kinases.

The CDC25 phosphatases are not only rate-limiting activators of cyclin-dependent kinases, but also targets of the ChK1–ChK2-mediated checkpoint pathway. Each CDC25 isoform exerts specific functions. Isoform CDC25a regulates both G1–S and G2–M transitions. During embryogenesis, it intervenes in cell proliferation after the blastocyst stage. Isoform CDC25b is dispensable in embryogenesis, but indispensable in meiotic progression of oocytes. In addition, CDC25a and CDC25b regulate different stages of mitosis. In fact, CDC25b activates CDK1–cyclin-B at the centrosome during prophase, whereas CDC25a operates for subsequent full activation of the nuclear CDK1–cyclin-B complex [765]. Isoform CDC25c is dispensable for both mitosis and meiosis, although it is highly regulated during these processes.

8.3.15.1 CDC25a Phosphatase

Dual-specificity protein phosphatase CDC25a regulates not only the cell cycle progression, but also mitogen-activated protein kinase pathway. Isoform CDC25a actually receives signals from mitogens that influences its stability, localization, and activity.

Phosphatase CDC25a has a non-redundant activity during the cell cycle. It indeed acts during the G1 phase and G1–S transition, with a peak activity that precedes that of CDC25b in prophase, as it activates the CDK2–cyclin-E and CDK2–cyclin-A complexes, a function that has not been observed with other members [766]. Furthermore, in G2 phase, CDC25a activates both CDK1/2–cyclin-A and CDK1–cyclin-B complexes, whereas CDC25b only activates CDK1–cyclin-B [767].

Phosphatase CDC25a and CDK6 are effectors of transcription factor Nanog[115] during the G1–S transition in human embryonic stem cells [768]. Factor Nanog binds to the regulatory regions of Cdk6 and Cdc25A genes.

8.3.15.2 CDC25b Phosphatase

Phosphatase CDC25b intervenes in the control of the activity of the CDK1–CcnB complex at the entry into mitosis and together with Polo-like kinase PLK1 in the regulation of the resumption of cell cycle progression after DNA damage-dependent checkpoint arrest in G2. Activity of PLK1 is necessary for the relocation of CDC25b from the cytoplasm to the nucleus at the G2–M transition [769].

8.3.15.3 CDC25c Phosphatase

Activity of CDC25c is precluded by sequestration by 14-3-3ϵ and 14-3-3γ, especially to complete S phase and G2 checkpoint [770].

8.3.16 Infraclass Aspartate-Based Protein Tyrosine Phosphatases

8.3.16.1 Eye Absent Phosphatases

Eye absent phosphatases (EyA) regulate organogenesis, especially developing kidney, branchial arches, eye, and ear. They are implicated in morphogenesis of thymus, parathyroid, and thyroid that are derived from the pharyngeal region, i.e., cranial neural crest that yields mesenchymal cells of the pharyngeal region [771].

Four mammalian EYA genes encode proteins that contain a N-terminal transactivation domain and a highly conserved C-terminal Eya domain for protein–protein interactions (Table 8.40). The EYA genes are coexpressed with Pax and Six genes. Four EyA1 transcript variants encode 3 distinct isoforms. The EyA phosphatases can act as transcriptional activator.

8.3.16.2 Haloacid Dehalogenases

Haloacid dehalogenases (HAD) possess a highly conserved catalytic domain and 3 characteristic conserved sequence motifs. They use an unconventional catalytic mechanism whereby a phospho-aspartate intermediate transiently forms following nucleophilic processing on the substrate phosphate group.

---

[115]Celtic: na nÓg. In old Irish, "Tr inna n-Óc" means Land of the Young, a place of eternal youth, beauty, and happiness in Irish mythology. The greatest Irish tales include "Oisn in Tr na nÓg".

Table 8.40. Family of eye absent homologs (EyA; BOR: autosomal dominant branchiootorenal syndrome protein; CMD: dilated cardiomyopathy; DfNA10: autosomal dominant, progressive, sensorineural, non-syndromic deafness [hearing loss]).

Type	Aliases
EyA1	BOR
EyA2	EAb1
EyA3	
EyA4	CMD1J, DFNA10

Soluble nucleotidases belong to the haloacid dehalogenase superfamily. Nucleotidase catalyzes the hydrolysis of a nucleotide into a nucleoside (e.g., AMP to adenosine and GMP to guanosine).

P-type ATPases constitute a superclass of ubiquitous ATP-driven pumps involved in active transport across membranes of charged substrates. They can be grouped into 5 major families according to substrate specificity: type-1 ATPases are heavy metal pumps that are decomposed into 2 subtypes, the type 1A for K^+ import and type 1B for transport of cations (Cu^+, Ag^+, Cu^{++}, Zn^{++}, Cd^{++}, Pb^{++}, and Co^{++}; type-2 ATPases Ca^{++} pumps (types 2A and -2B) as well as Na^+–K^+- and H^+–K^+ pumps (type 2C); type-3 ATPases H^+- and Mg^{++} pumps; type-4 ATPases are phospholipid transporting pumps, and type-5 ubiquitous ATPases that have no allocated substrate specificity.

P-type ATPases have several functions: action potentials in heart and nervous system, secretion and reabsorption of solutes in kidneys, acidification in stomach, nutrient absorption in intestine, relaxation of muscles, and Ca^{++}-dependent signal transduction.

The family of haloacid dehalogenase-like hydrolase domain-containing enzymes includes HDHD1a, HDHD2, and HDHD3 that interact with thyroid receptor-interacting protein TRIP13. Subtype HDHD1A[116] that can bind to RNA. It can then function as an RNA-binding factor involved in RNA processing or transport. It can also represent an atypical transcription regulatory factor.

8.3.16.3 Chronophin

Chronophin is a phosphatase of the haloacid dehalogenase superfamily that carries out 2 tasks. Chronophin works with LIMK kinase on actin regulators cofilin and ADF to control cofilin^P and ADF^P turnover, hence the dynamics of the actin cytoskeleton. Chronophin also acts as a pyridoxal phosphatase, as it removes a phosphate from pyridoxal phosphate (also known as vitamin-B6). Chronophin is an unusual phosphatase, it has a large domain that covers the active site with its magnesium ion to help phosphate removal [772].

---

[116] A.k.a. GS1. However, acronym GS1 serves also to designate glutamine synthase-1.

8.3 Cytosolic Protein Serine/Threonine Phosphatases

Chronophin is widespread. It is particulalrly abundant in brain, heart, skeletal muscle, liver, and kidney. It localizes to actin-rich ruffles and membrane protrusions. Chronophin possesses several interaction motifs to link to PI3K and PLC enzymes.

Cofilin is a key regulator of actin cytoskeletal dynamics, the activity of which is controlled by phosphorylation of a single serine residue (Ser3). Cofilin and related actin depolymerizing factor family proteins are phosphorylated (inactivated). Dephosphorylation re-activates cofilin and ADF. Cofilin phosphorylation is regulated by several phosphatases, such as slingshot (Sect. 8.3.13.4) and chronophin. Elevated chronophin activity is associated with lowered cellular ^{F}actin content.

Reorganization of actin at cell leading edge during cell displacement requires the coordinated action of both actin-polymerizing and -depolymerizing (severing) factors. Branched actin filaments are created at the leading edge by the actin nucleator ARP2–ARP3 complex and actin filament crosslinkers filamin-A1 and -A2. Actin depolymerizing proteins, such as members of the cofilin–actin depolymerizing factor family, disassemble ^{F}actin in the cell rear to recycle actin monomers to the leading edge. Cofilin-dependent ^{F}actin depolymerizing activity then contributes to actin polymerization. Cofilin-depolymerized ^{F}actin fragments are preferred substrates for the ARP2–ARP3 complex [483]. Cofilin bears phosphorylation-dephosphorylation cycles by LIMK (Sect. 5.9.3) and TesK kinases (Sect. 5.9.4) that prevent cofilin binding to ^{F}actin, hence stabilizing ^{F}actin, and phosphatases Ssh and chronophin that stimulate ^{F}actin depolymerization and subsequent actin nucleation for directional motility.

During mitosis, chronophin relocalizes from cytoplasm, where it resides in prophase and metaphase to cell poles in anaphase and telophase and to cleavage furrow, actomyosin contractile ring, and midbody in late telophase [483].

Chapter 9
Guanosine Triphosphatases and Their Regulators

Guanosine triphosphatases intervene in: (1) signal transduction from the intracellular edge of the plasma membrane and intracellular domain of transmembrane receptors; (2) protein synthesis at the ribosome (Vol. 1 – Chap. 5. Protein Synthesis); (3) control of cell division (Vol. 2 – Chap. 2. Cell Growth and Proliferation); (4) proper protein folding; (5) translocation of proteins through the membrane of the endoplasmic reticulum; and (6) vesicular transport within the cell (Vol. 1 – Chap. 9. Intracellular Transport).

Guanosine triphosphatases include: (1) translation factors of the TRAFAC family that control initiation, elongation, and termination of protein synthesis (or translation at ribosomes; they act as ribosome-assembly factors); (2) translocation ribonucleoproteic factors;[1] (3) large GTPases, such as *dynamin*[2] and members

---

[1] The signal recognition particle (SRP) binds to newly synthesized peptides that emerge from ribosomes, thereby coupling protein translation to protein translocation through the protein-conducting channel, the so-called translocon, in the endoplasmic reticulum membrane. Once SRP is connected to the SRP receptor (SRPR), the nascent peptide chain is inserted into the translocon and, then, enters into the endoplasmic reticulum. The SRP-SRPR complex dissociates via GTP hydrolysis.

[2] The GTP hydrolysis rate of dynamins accelerates in the presence of microtubules, liposomes, and certain SH3 domain-containing proteins, to which dynamin can link. Its monomers self-associate into homodimers that connect each other to form homotetramers for optimal GTPase activity [773]. Dynamin is able to build homo-oligomers and polymers. Its membrane binding and self-association occur independently from its nucleotide-bound state. Dynamin operates in the scission of newly formed, coated vesicles from the membrane of native compartment for their release and fusion with recipient compartment. Dynamin also intervenes in cytokinesis and division of organelles. The dynamin family includes classical dynamins, dynamin-like proteins, interferon-induced GTP-binding protein Myxovirus resistance proteins (Mx), and mitochondrial membrane proteins mitofusins. In mammals, 3 known dynamin types are encoded by 3 genes: dynamin-1 in neurons and neuroendocrine cells; dynamin-2 in most cell types; and dynamin-3 strongly produced in testes and, to a lesser extent, in the brain, heart, and lung. Dynamins are founding member of a family of dynamin-like GTPases. Dynamin-like proteins include atlastin, or GTP-binding protein GBP3, dynamin-related protein DRP1, optic atrophy 1 homolog (OpA1), and mitofusin-1 and -2.

of the *septin* family[3] of polymerizing GTPases; (4) heterotrimeric, relatively large, guanine nucleotide-binding proteins, the so-called *G proteins*; and (5) monomeric, *small guanosine triphosphatases* (small GTPases).[4]

In addition, proteins require assistance of molecular chaperones to get a functional conformation or for refolding of unfolded or misfolded proteins that underwent a stress. Among these chaperones, chaperonins are large complexes that enhance the efficiency of protein folding.[5] The reaction cycle consists of a nucleotide-regulated conformational change between an open substrate-acceptor state and a closed folding-active state. Cytosolic, hetero-oligomeric chaperonin with ATPase and GTPase activity (at similar hydrolysis rate) chaperonin containing TCP1 (CCT)[6] assists in protein folding. Its 8 different subunits have an ATPase domain and distinct substrate-binding domain. Its substrates include actin and tubulin. Nucleotide GTP induces an open-to-close conformational change in the CCT structure, hence regulating protein folding [776].

Receptor activation primes chemical events that can involve heterotrimeric and monomeric guanosine triphosphatases. These 2 types of proteins — G proteins and small GTPases — cycle between 2 states, as they bind either guanosine di- (GDP) or triphosphate (GTP). These proteic switches are flicked off (inactive, GDP-bound state [GTPase^{GDP}]) and on (active, GTP-bound state [GTPase^{GTP}]; Fig. 9.1).[7] This hydrolytic activation–deactivation cycle is regulated to avoid pathological states. Once activated, the switch activates the pathway effector immediately downstream from it.

They localize to sites of membrane fission and fusion in the endoplasmic reticulum, mitochondria, and peroxisomes.

[3] Septins constitute a group of GTP-binding, hetero-oligomeric proteins that can form filaments. They are involved in different stages of the cell cycle. Septins also participate in secretion, membrane remodeling, and cytoskeleton dynamics. Post-translational modifications and partners regulate septin filament function. Binding of GTP induces conformational changes. Binding of GTP and hydrolysis influence the stability of septin interfaces and contribute to septin filament assembly and disassembly [774].

[4] Heterotrimeric guanine nucleotide-binding proteins, monomeric small guanosine triphosphatases, and related ATP-binding proteins are P-loop proteins. Guanosine triphosphatases can be subdivided into conventional guanosine triphosphatases and G proteins activated by nucleotide-dependent dimerization. Conventional guanosine triphosphatases include small GTPases of the RAS superfamily that form heterodimeric complexes with their cognate GTPase-activating proteins. G proteins activated by dimerization are not regulated by guanine nucleotide-exchange factors and GTPase-activating proteins [775].

[5] Chaperonins has been classified into 2 groups: group-1 chaperonins that possess 14 subunits and localize to organelles such as mitochondria (e.g., HSP60) and cytoplasm and group-2 chaperonins that have 16 or 18 subunits and reside in the cytosol.

[6] A.k.a. TCP1 ring complex (TRiC).

[7] Two motifs of small GTPases — switch-1 and -2 — recognize the nature of the bound nucleotide and change their conformation accordingly. These 2 switch domains signal the nucleotide status of the GTPase to effectors or regulators, as they influence between-protein interactions with these binding partners.

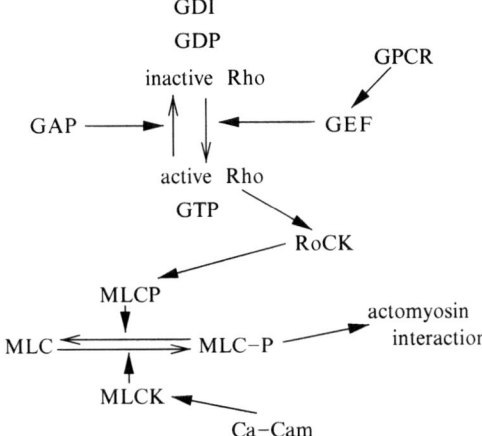

Fig. 9.1 Small GTPase states and effectors. They cycle between active, guanosine triphosphate (GTP)-bound and inactive, guanosine diphosphate (GDP)-loaded states. This cycling is controlled by 3 types of regulatory proteins: (1) guanine nucleotide-dissociation inhibitors (GDI) that stabilize inactive form; (2) guanine nucleotide-exchange factors (GEF) that catalyze the exchange of GDP for GTP; and (3) GTPase-activating proteins (GAP) that enhance the low intrinsic GTPase activity of GTPases.

9.1 Heterotrimeric G Proteins (Gαβγ GTPases)

Heterotrimeric guanine nucleotide-binding proteins (G proteins) are signal transducers attached to the inner leaflet of the plasma membrane. These transducers respond to various extracellular signals, such as hormones, neurotransmitters, chemokines, photons, odorants, tastants, nucleotides, and ions. They are controlled by a set of regulators. Conversely, they regulate the production or influx of second messengers, such as cAMP and calcium. The activation of G proteins is induced by ligand-bound G-protein-coupled receptors (Fig. 9.2; Vol. 3 – Chap. 7. G-Protein-Coupled Receptors).[8]

Heterotrimeric G protein is composed of a Gα^{GDP} subunit and a Gβ–Gγ dimer (Fig. 9.3). Many isoforms of each of the 3 subunits exist. Genes have been identified for 16 Gα, grouped in 4 main subclasses (Tables 9.1 and 9.2), 5 Gβ, and 15 Gγ subunits, with possible splice variants and post-translational modifications [778]. The G-protein subunit genes encode for [778]: (1) subclass-1 of Gα$_s$

---

[8]Each member of the GPCR superclass has a similar structure with 7 transmembrane helices, extracellular amino (N)- and intracellular carboxy (C)-termini, and 3 interhelical loops on each side of the plasma membrane. G-protein-coupled receptors form both homo- and heterodimers. When a single element of the dimer bind to its ligand, the dimer cannot fully activate the linked G protein. Besides, when 2 receptors dimerize to form a signaling unit, the signal can differ from the response given by either receptor alone. Furthermore, the heterodimerization of a GPCR with another affects the functioning of the bound GPCR, modulating its activity.

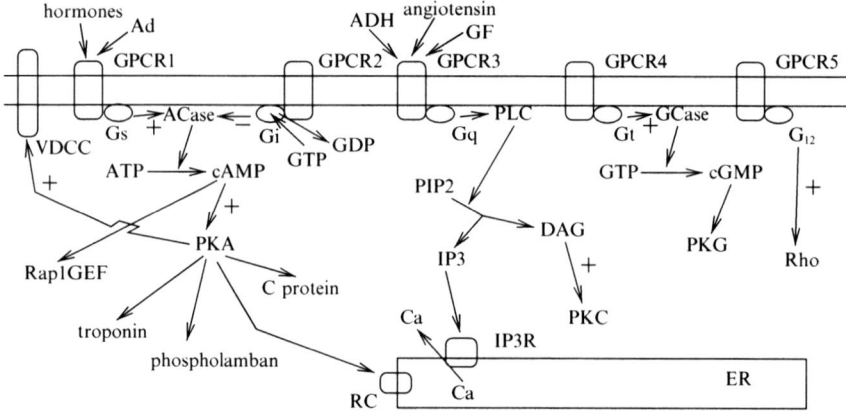

Fig. 9.2 G-protein-coupled receptors (GPCR) are associated with 3 G-protein subunits: Gα that binds GDP (inactive state) or GTP after ligand binding (stimulated form), Gβ, and Gγ. Subunit Gα shares homology with small monomeric GTPase of the RAS superfamily. It binds guanine nucleotides in a signal-dependent manner. Heterodimer Gβγ has high affinity for Gα^{GDP} form. It is released from active Gα^{GTP} after receptor activation. Activated Gα activates an effector. Types of Gα subunits include Gα$_s$ (stimulatory), Gα$_i$ (inhibitory), Gα$_q$, Gα$_t$, and Gα$_{12}$. Subunit Gα$_s$ stimulates adenylate cyclase (ACase) that produces cAMP messenger. On the other hand, Gα$_i$ inhibits adenylate cyclase. Gα$_q$-coupled receptor activates phospholipase-C (PLC) that generates second messengers inositol trisphosphate (IP$_3$) and diacylglycerol (DAG). Subunit Gα$_t$ stimulates guanylate cyclase (GCase) that forms cGMP. Subunit Gα$_{12}$ activates small RhoA GTPase. Protein kinase-A (PKA) regulates myocardial contraction, acting on contractile proteins, activating ion channels, such as ryanodine receptors (RR) of the endoplasmic reticulum (ER), voltage-dependent calcium channels (VDCC), and sodium channels of the plasma membrane (Gβγ inhibits these voltage-gated channels), and carriers. Protein kinase-A also increases metabolic energy, stimulating phosphorylase kinases and hormone-sensitive lipases, and activates gene transcription. Messenger cAMP exerts direct effects on other effectors, such as Rap1GEF. Mediator IP$_3$ binds to its endoplasmic reticulum receptor IP$_3$R and causes the release of stored Ca^{++} which rises transiently in the cytosol. Phospholipase-C also activates plasmalemmal transient receptor potential (TRP) Ca^{++} channels (TRPC), hence Ca^{++} influx (adapted from [777]).

subunits with Gs and Gα$_{olf}$; (2) subclass-2 of Gα$_i$ subunits with Gi1 to Gi3, Go1, Go2, Gt1, Gt2, Ggust (gustducin), and Gz; (3) subclass-3 of Gα$_q$ subunits with G11, Gq, and G14 to G16; (4) subclass-4 of Gα$_{12}$ subunits with G12 and G13; (5) subclass of Gβ subunits with Gβ1 to Gβ5; and (6) subclass of Gγ subunits with farnesylated isoforms Gγ1, Gγ11, and Gγ9 (Gγ8$_{cone}$ or Gγ14), and geranyl-geranylated isoforms Gγ2 to Gγ5, Gγ7, Gγ8 (a.k.a. Gγ8$_{olf}$), Gγ10, and Gγ12.

A single receptor can stimulate a single G protein (linear signaling). Some receptors promiscuously interact with G proteins, whereas others specifically interact. A given G protein can be regulated by several receptors, either positively or negatively (convergent receptor signaling). A given receptor can activate several G proteins (divergent receptor signaling), such as the α-adrenergic receptor targeting both Gs and Gi subunits.

9.1 Heterotrimeric G Proteins (Gαβγ GTPases)

Table 9.1. The 4 main subclasses of Gα and corresponding targets. Activated GPCRs stimulate G proteins, as they catalyze guanine nucleotide-exchange on Gα subunit. G Proteins are classified according to Gα subunit into Gs, Gi/o, Gq/11, and G12/13. Members of the Gs and Gi/o subclasses converge to the second messenger cAMP. Second messenger activation occurs quickly due to the presence of complexes between Gs and adenylate cyclase as well as Gq and phospholipase-Cβ, but substantial change in second messenger concentration needs much longer time in comparison with the activation kinetics of initial signaling stages.

Gα	Effector
Gs	Adenylate cyclase (stimulation)
Gi/o	Adenylate cyclase (inhibition)
Gq/11	Phospholipase-Cβ
G12/13	Small guanosine triphosphatase

Table 9.2. G-protein subunits and their signaling in the cardiovascular system (Source: [779]).

Subunit	Pathway	Effect
Gs	ACase–cAMP–PKA	Positive inotropy and lusitropy
Gi/o	ACase	Inhibition
	NO	Vasodilation
	PLC–IP$_3$–Ca^{++}	Vasoconstriction
	PLC–DAG–PKC	
Gq/G11	PLC–IP$_3$–Ca^{++}	Vasoconstriction
	PLC–DAG–PKC	Positive inotropy
G12/13	RhoGEF–Rho–RoCK	Vasoconstriction
Gβγ	K$_{ACh}$	Negative chronotropy

Activated G protein is dissociated into Gα^{GTP} and Gβ–Gγ (or Gβγ; Fig. 9.3). Isolated Gα^{GTP} acts on a second messenger. However, although certain G proteins dissociate after receptor activation in vivo, other G proteins undergo simple rearrangements, possibly modulating G-protein effector activity [780].

Once the target has been stimulated, Gα fills its GTPase activity. When GTP is hydrolyzed, Gα returns to the inactivated GDP-bound conformation, dissociates from its target, and reassembles with Gβγ. The duration of G-protein activation is thus controlled by the intrinsic GTPase activity of Gα subunit.

Subunits Gα^{GTP} and Gβγ can propagate signals, as they interact with adenylate cyclases, phospholipase-C isoforms, potassium and calcium ion channels, guanine nucleotide-exchange factors for RhoA GTPase, and other effectors.

However, certain ligands modify the conformation of the receptor such that it interacts with either heterotrimeric G proteins or another effector, such as GPCR kinases (GRK) that phosphorylate activated GPCRs and arrestins that bind phosphorylated GPCRs (β-*arrestin-biased ligands*) and support clathrin-mediated GPCR endocytosis.

Certain G proteins are regulated by homodimerization. These G proteins, activated by nucleotide-dependent dimerization, belong to the set of proteins that also includes signal recognition particles, dynamins, septins, and receptor-interacting protein kinase RIPK7 [775].

In addition to GPCR types and effectors, accessory proteins influence specificity, magnitude, and duration of GPCR signaling. GPCR signaling accessory proteins include scaffold proteins and molecules that directly affect the basal activation state of effectors, such as regulators of G-protein signaling and activators of G-protein signaling (Sect. 9.2).

In dual signaling pathways activated by a single G protein, both activated Gα and G$\beta\gamma$ stimulate effectors. In other pathways, the major regulator is G$\beta\gamma$, the activity of which can be suppressed by excess Gα liberated by other activated G proteins.

9.1.1 Set of Gα Subunits

Four categories of Gα subunits include: (1) subclass of Gs that stimulates adenylate cyclase and activates calcium channels (Sect. 9.1.1.1); (2) subclass of Gi that inhibits adenylate cyclase and potassium channels (Sect. 9.1.1.2); (3) subclass of Gq that activates phospholipase-C, and protein kinase-C and -D (Sect. 9.1.1.3); and (4) subclass of G12 that regulate cell activity, especially actin cytoskeleton remodeling, via guanine nucleotide-exchange factors (Sect. 9.1.1.4).

9.1.1.1 Gα_s Subclass

Ubiquitous Gα_s is an activator of adenylate cyclase that produces messenger cAMP from ATP. In its quiescent state, Gα_s is associated with Gβ/γ dimer. The intrinsic GTPase activity of Gs causes its deactivation and reassociation with Gβ/γ dimer. In its inactive form, Gs binds to G$\beta 1/\gamma 1$, G$\beta 1/\gamma 2$, G$\beta 1/\gamma 5$, G$\beta 1/\gamma 7$, G$\beta 2/\gamma 5$, and G$\beta 2/\gamma 7$ [781].

Subunit Gα_s is activated by G-protein-coupled receptors, such as β-adrenergic, glucagon, luteinizing hormone, calcitonin, and dopamine D_1 and D_5 receptors [781]. Subunit Gs^{GTP} activates adenylate cyclase isoforms. Affinity of Gs for both Gβ/γ and adenylate cyclase is increased by N-terminal palmitoylation.

9.1.1.2 Gα_i Subclass

Several types of Gα_i exist (Gα_{i1}–Gα_{i3}) that mainly inhibit the cAMP-dependent pathway, but activate the phospholipase-C pathway. In fact, the Gi/o subclass is composed not only of Gα_i, but also Gα_o, Gα_{gust}, and Gα_t (transducin) proteins (Table 9.3).

9.1 Heterotrimeric G Proteins (Gαβγ GTPases)

Table 9.3. Examples of G-protein-coupled receptors (GPCRs) preferentially coupled to members of the Gi/o subclass (Source: [785]).

GPCR type	Gα subunit type
A_1	$G\alpha_{i1}$–$G\alpha_{i2}$, $G\alpha_o$
AT_2	$G\alpha_{i1}$–$G\alpha_{i2}$, $G\alpha_o$
APJ	$G\alpha_{i1}$–$G\alpha_{i2}$
CB_1	$G\alpha_{i1}$–$G\alpha_{i3}$, $G\alpha_o$
D_2	$G\alpha_{i1}$
$GABA_B$	$G\alpha_{i1}$, $G\alpha_o$
mGluR2/3	Gi/o
LPA_1	$G\alpha_{i1}$, $G\alpha_{o1}$
M_2	$G\alpha_{i1}$–$G\alpha_{i3}$, $G\alpha_o$
M_4	$G\alpha_{i2}$, $G\alpha_o$
Op_3	$G\alpha_{i1}$, $G\alpha_{i3}$, $G\alpha_o$
$P2Y_{12}$	Gi
$S1P_1$	$G\alpha_{i1}$, $G\alpha_{i3}$
$5HT_{1A}$	$G\alpha_{i1}$–$G\alpha_{i3}$
α2a	$G\alpha_{i1}$–$G\alpha_{i2}$

Cells respond to growth factors by either migrating or proliferating. Cell movement and proliferation are indeed mutually exclusive. Growth factor receptors, such as EGFR, VEGFR, and PDGFR receptors, can actually trigger cell motion or division according to the type and concentration of the signaling ligand on the one hand and the density and distribution of receptors on the other. During cell migration, phospholipase-Cγ1 and phosphatidylinositol 3-kinase and its effector protein kinase-B are stimulated and coupled to actin remodeling at the leading edge. In proliferating cells, another set of signals (ERK1 and ERK2, Src kinase, and STAT5b) is involved for the activation of nuclear transcription factors and DNA synthesis. Girdin (girders of actin filaments)[9] is an actin-binding PKB substrate[10] that operates in actin organization and PKB-dependent cell motility, at least, in fibroblasts. It is phosphorylated upon stimulation by insulin-like growth factor IGF1

---

[9] A.k.a. Akt (PKB) phosphorylation enhancer (APE), Gα-interacting vesicle-associated protein (GIV), Hook-related protein HkRP1, and coiled-coil domain-containing protein CCDC88a.

[10] Protein kinase-B phosphorylates girdin at Ser1416 in lamellipodia of migrating cells.

that fosters tumor cell movement [782].[11] Girdin is a guanine nucleotide-exchange factor for G proteins involved in the regulation of the PI3K–PKB axis, actin cytoskeleton remodeling, and cell migration.[12] Proteic subunit $G\alpha_i$ and its ubiquitous guanine nucleotide-exchange factor girdin orchestrate the migration–proliferation dichotomy downstream from EGFR signaling [784]. Girdin directly interacts with EGFR and associates $G\alpha_i$ to the receptor. A $G\alpha_i$–girdin–EGFR complex indeed assembles, EGFR autophosphorylates (a process necessary for cell migration, but not for mitosis), and EGFR residence in the plasma membrane is prolonged (signaling from the cell surface rather than from endosomes). Plasma membrane-based motogenic signals (PLCγ1 and the PI3K–PKB axis) are triggered and cell migration starts. Activation of $G\alpha_i$ by girdin is required in cell migration primed by EGF and insulin during epithelial wound healing, macrophage chemotaxis, and tumor cell migration. In addition, cancer angiogenesis, invasion, and metastasis also rely on girdin mediator following VEGF and IGF stimulation. On the other hand, a GEF-deficient girdin splice variant promotes mitogenic signals and cell proliferation occurs.[13]

$G\alpha_{i1}$

Ubiquitous $G\alpha_{i1}$ inhibits adenylate cyclases AC1, AC5, and AC6 [785]. On the other hand, associated Gβγ primes the activation of acetylcholine-activated inwardly rectifying K+ GIRK channels ($K_{IR}3$). In addition, $G\alpha_{i1}$ can also activate $K_{IR}3$ activity.

In platelets, adrenaline stimulates the association of Src with $G\alpha_{i1}$ Protein $G\alpha_{i1}$ can then trigger the PI3Kγ–PKB pathway. In endothelial cells, stimulation of β2-adrenoceptor activates the $G\alpha_{i1}$–Src–PKB axis, whereas in cardiomyocytes, this receptor activates predominately the Gβγ–PI3K–PKB pathway [785].

Activated Gi (Gi1–Gi3)-coupled receptors can signal via Jun N-terminal kinase. Protein $G\alpha_{i1}$ mediates not only GPCR signaling, but also that of receptor Tyr kinases. Both $G\alpha_{i1}$ and $GG\alpha_{i3}$ downstream from epidermal growth factor receptor use GAB1 to activate the PKB–TORC1 pathway [785].

---

[11]Protein kinase-B contributes to tumor growth and metastasis, as it activates nuclear factor-κB, target of rapamycin, double minute-2, BCLxL/BCL2-associated death promoter homolog, and matrix metallopeptidases on the one hand and inactivates tuberous sclerosis complex-2, cyclin-dependent kinase inhibitor CKI1b, and FoxO transcription factor on the other.

[12]Girdin acts as a signal amplifier for PKB signaling. A threshold of GEF activity exists in the PKB activation by girdin in different cell types and by various stimuli [783]. Signaling from PKB is minimal at low GEF activity. It abruptly rises to reach a maximum above a threshold of GEF activity (switch-like behavior or all-or-none response). Girdin aims at amplifying signaling to initiate cell migration. The regulated amplification of the input signal is associated with slight changes in girdin's GEF activity close to a certain threshold that allows a rapid cell decision toward motion.

[13]Receptor EGFR is internalized mitogenic signals are then executed.

9.1 Heterotrimeric G Proteins ($G\alpha\beta\gamma$ GTPases)

Table 9.4. Examples of Go-coupled receptors (Source: [786]).

Chemokine receptors	CXCR4
Morphogen receptors	Fz
Neurotransmitter GPCRs	$\alpha 2a$, D_2, $5HT_{1P}$, M_2, M_4 (Go1), $GABA_B$
Nucleoside receptors Nucleotide receptors	A_1 $P2Y_2$
Opioid receptors	Op_1, Op_3, Op_4
Peptide receptors	Galanin, neuropeptide-Y, somatostatin (Sst_2; Go2), corticotropin-releasing factor receptor (CRF_2)

Protein $G\alpha_{i1}$ can be N-myristoylated and palmitoylated to enhance membrane insertion and affinity for $G\beta\gamma$ subunit on the one hand, and to reduce the affinity for the GAP and RGS4 proteins [785].

Protein $G\alpha_{i1}$ intervenes in synaptic transmission and platelet aggregation. It is linked to serotonin $5HT_{1A}$, dopamine D_2, cannabinoid CB_1, metabotropic glutamate mGluR2 and mGluR3, muscarinic acetylcholine M_2, and μ-opioid, as well as adenosine A_1, angiotensin-2 AT_2, apelin, sphingosine-1-phosphate $S1P_1$, lysophosphatidic acid LPA_1, and metabotropic $P2Y_{12}$ receptors [785]. Gi1-coupled $\alpha 2a$-adrenoreceptor activates voltage-dependent Ca_V channels in rat portal vein myocytes.

$G\alpha_o$

The "other" G-protein subunit $G\alpha_o$ that constitute the Go protein with $G\beta\gamma$ dimer is the most abundant G protein in the central and peripheral nervous systems. Two splice variants — $G\alpha_{o1}$ and $G\alpha_{o2}$ — exist. It is activated not only by many G-protein-coupled receptors bound to neurotransmitters and hormones (Table 9.4), but also by 43-kDa growth cone-associated protein GAP43,[14] and amyloid-β (A4) precursor protein (APP) [786].

Protein Go also plays a role in visual signal transduction, olfactory reception, and neurotransmitters monoamine and glutamate uptake after exocytosis by their cognate vesicle membrane transporters [786]. Frizzled receptor that transmits developmental signals relies on $G\alpha_o$ subunit.

In the heart, Go protein is preferentially synthesized in endocrine atriomyocytes and peripheral neurons. It enhances calcium cycling and contractility [786]. In addition, it supports the cardiac control by the parasympathetic activity mediated by muscarinic acetylcholine receptors.

---

[14] A.k.a. axonal membrane protein B50, neural phosphoprotein PP46, and neuromodulin.

Activated $G\alpha_o$ can interact directly with many types of ion channels [786]. It impedes transmitter release, as it precludes the activity of $Ca_V2.2$ channel in presynaptic terminals. Moreover, opioid receptors inhibit $Ca_V2.1$ channel in dorsal root ganglion neurons. In ventriculomyocytes, muscarinic GPCRs prevents the activation of $Ca_V1.2a$ channel. On the other hand, Go protein activate $K_{IR}3$ (GIRK) channel. It also modulates Na^+ channel gating and prevents the activaty of transient receptor potential cation channel $TRPM1_L$ in retinal bipolar cells.

Protein $G\alpha_o$ directly binds to Rap1GAP protein, then targeting it for proteasomal degradation. It also interacts with Rab5 and Rit1 GTPases. It activates the GTPase activity of tubulin. Activated $G\alpha_o$ protein interacts directly with zinc finger and BTB domain-containing transcription factor ZBTB16, thereby enhancing its transcriptional repression.

Galanin[15] localizes to autonomic nerve terminals in the endocrine pancreas. It impedes insulin release via Gi/o-coupled galanin receptors (GalR1–GalR3)[16] and Go2 proteins in pancreatic β cells to prevent oversecretion [787].

$G\alpha_z$

The Gz subunit is activated by GPCRs (adenosine [A_1], adrenergic [α2a], complement C5a, corticotropin-releasing hormone, dopaminergic receptors [D_2–D_5], endothelin, fractalkine, lysophosphatidic acid, macrophage inflammatory protein-3α, melatonin [MT_1 and MT_2], muscarinic [M_2], neurokinin, nociceptin, opioids, serotonin receptors [$5HT_{1A}$], and thrombin) that activate other members of the Gi subclass. It is unique among the Gi subclass, with very low rates of GDP–GTP exchange (0.02/mn) and GTP hydrolysis (0.05/mn).

Because Gz is recognized by several RGS proteins (RGS4, RGS10, RGS17, RGS19, RGS20, and retinal RGS1 [RetRGS1]), sustained state of activation due to slow GTP hydrolysis can be shortened [788]. Myristoylation and palmitoylation are needed for stable anchorage to the plasma membrane. Palmitoylation prevents GTP hydrolysis by RGS proteins. Protein Gz is phosphorylated by protein kinase-C and P21-activated protein kinase. Phosphorylation disrupts $G\alpha_z$–Gβγ interaction. Phosphorylation (Ser16) blocks binding to RGS17, RGS19, and RGS20 proteins.

Protein Gz directly binds and inhibits adenylate cyclases AC1, AC5, and AC6, but indirectly stimulates adenylate cyclase AC2 [788]. Its other effects include: (1) inhibition of $Ca_V2.2$ channels; (2) stimulation of inwardly rectifying K^+

---

[15]The neuropeptide galanin resides in the central and peripheral nervous systems and digestive neuroendocrine system. It is coexpressed with many neurotransmitters. It functions as an inhibitor. It operates in metabolism, feeding, learning and memory, nociception, spinal reflexes, neuron regeneration, and anxiety. In particular, galanin precludes cAMP production and activates G-protein-regulated inwardly rectifying K^+ channels.

[16]Proteins Gi1 to Gi3 as well as Go1 and Go2 are synthesized in pancreatic islets. However, Go2 is the major transducer that mediates inhibition of insulin release. It intervenes in galanin effects on ATP-sensitive $K_{IR}6.2$ (activation) and Ca_V1 (inhibition) channels, thereby regulating exocytosis.

9.1 Heterotrimeric G Proteins (Gαβγ GTPases)

Table 9.5. Examples of plamalemmal, heptahelical $G\alpha_{q/11}$-protein-coupled receptors (**Part 1**; Source: [789]).

Type	Name
Receptors for amino acids and dicarboxylic acids	
Metabotropic glutamate receptor	mGluR1, mGluR5 (class C)
α-Ketoglutarate receptor or oxoglutarate receptor	GPR99/OXGR1 (subfamily A11)
Succinate receptor	GPR91 (subfamily A11)
^{L}L-amino acid receptor (^{L}arginine and ^{L}lysine)	GPRC6A (class C)
Receptors for biogenic amines	
Muscarinic acetylcholine receptor	M1, M3, M5 (subfamily A18)
Adrenergic receptor	α1A, α1B, α1D (subfamily A18)
Histamine receptor	H1 (subfamily A18)
Serotonin receptor	5HT2A, 5HT2B, 5HT2C (subfamily A17)

Table 9.6. Examples of plamalemmal, heptahelical $G\alpha_{q/11}$ protein-coupled receptors (**Part 2**; Source: [789]).

Type	Name
Receptors for ions	
Calcium-sensing receptor	CaSR (class C)
Receptors for nucleotides and nucleosides	
Purinergic receptor (subfamily A11)	P2Y1 (ADP/ATP), P2Y2 (UTP/ATP), P2Y4 (UTP/ATP), P2Y6 (UDP), P2Y11 (ATP)

channels; (3) repression of Rap1 signaling by interacting with Rap1 GTPase-activating protein; (4) release of tumor-necrosis factor-α from immunoglobulin-E-stimulated mastocytes; (5) natural killer cell-mediated lysis of allogeneic and tumor cells; and (6) reduction in insulin secretion.

9.1.1.3 Gα$_q$ Subclass

Subunit Gα$_q$ links certain types of activated G-protein-coupled receptors to intracellular signaling cascades. Various types of GPCRs couple to Gα$_q$, such as acetylcholine, angiotensin, catecholamine, endothelin, glutamate, histamine, lysophospholipid, and serotonin receptors [789] (Tables 9.5 to 9.8).

The subclass of ubiquitous Gα$_q$ subunits include 4 families: Gq, G11, G14, and G16 (G15 and G16 are murine and human orthologs, respectively; Table 9.9).

Table 9.7. Examples of plamalemmal, heptahelical $G\alpha_{q/11}$ protein-coupled receptors (**Part 3**; Source: [789]).

Type	Name
Receptors for lipids	
Fatty acid receptor	GPR40, GPR43, GPR120 (subfamily A11)
Leukotriene C4/D4 receptor	CysLT1, CysLT2 (subfamily A5)
Lysophosphatidic acid receptor	LPA1/2/3/5 (subfamily A13)
Platelet-activating factor receptor	PAF (subfamily A12)
Prostaglandin F2a receptor	PF2R (subfamily A14)
Prostaglandin E2 receptor	EP1, EP3 (subfamily A14)
Sphingosine-1-phosphate receptor	S1P2, S1P3 (subfamily A13)
Thromboxane A2 receptor	TP (subfamily A14)

$G\alpha_q$

Once G protein-coupled receptors are activated by ligand binding, $G\alpha_q$ releases GDP, binds GTP, and dissociates from $G\beta\gamma$. Subunits $G\alpha_q^{GTP}$ and $G\beta\gamma$ target phospholipase-Cβ and phosphatidylinositol 3-kinase, respectively. Phospholipase-Cβ catalyzes the hydrolysis of phosphatidylinositol bisphosphate to release inositol trisphosphate and diacylglycerol. Second messengers IP_3 and DAG in turn cause calcium influx from intracellular stores via IP_3Rs and stimulate protein kinase-C, respectively. Signaling pathways independent of PLCβ are also primed, as Gq-based cue activates Rho guanine nucleotide-exchange factors and interacts with G-protein-coupled receptor kinases [789].

Therefore, owing to Ca^{++} entry into the cytosol, $G\alpha_q$ contributes to the regulation of cell contractility and secretion, especially the control of smooth muscle cell tone in blood vessel and bronchus walls via calcium-dependent activation of myosin light chain kinase that phosphorylates (activates) myosin light chain to heighten actomyosin (stress fiber)-based contractility with $G\alpha_{11}$ as well as membrane fusion of secretory vesicles, particularly in endocrine cells and platelets. In neurons, $G\alpha_q$ modulates the synaptic transmission. In addition, $G\alpha_q$ is involved in pre- and postnatal cardiomyocyte growth, craniofacial development, and phototransduction [789].

Intrinsic GTPase activity of the $G\alpha_q$ subunit hydrolyzes $G\alpha_q^{GTP}$ to $G\alpha_q^{GDP}$, thereby terminating the signal and allowing the heterotrimeric G-protein to reassemble and take its quiescent state. However, the GTPase activity of isolated G proteins is much lower than that of G proteins in the pool of cellular molecules.

Phospholipase-Cβ (negative feedback) and regulators of G-protein signaling act as GTPase-activating proteins for subunits of the $G\alpha_q$ subclass [791]. Inhibitors of $G\alpha_q$ signaling include RGS1 to RGS5, RGS8, RGS13, RGS16 to RGS19 [789]. In addition to GTPase activation, these RGSs act as effector antagonists, as they are able to reduce PLCβ activity induced by $G\alpha_q$ subunit.

Table 9.8. Examples of plamalemmal, heptahelical $G\alpha_{q/11}$ protein-coupled receptors (**Part 4**; Source: [789]).

Type	Name
Receptors for peptides and proteins	
Angiotensin-2 receptor	AT1A, AT1B (subfamily A3)
Bradykinin receptor	B1, B2 (subfamily A3)
Calcitonin receptor	CT (class B)
Calcitonin gene-related peptide receptor	CGRP1 (CALCRL; class B)
Cholecystokinin-8 receptor	CCK1, CCK2 (subfamily A6)
Endothelin-1, -2, -3 receptors	ETA, ETB (subfamily A7)
Galanin receptor	GAL2 (subfamily A5)
Gastrin-releasing peptide/bombesin receptor	BB2 (subfamily A7)
Ghrelin receptor	GHSR (subfamily A7)
Gonadotropin-releasing hormone receptor	GnRHR (subfamily A6)
Kisspeptin and metastin receptor	GPR54 (subfamily A5)
Melanin-concentrating hormone receptor	MCHR2 (subfamily A5)
Orexin A/B receptor	OX1R, OX2R (subfamily A6)
Oxytocin receptor	OxTR (class A)
Parathyroid hormone receptor	PTH (class B)
Parathyroid-related peptide receptor	PTHrP1, PTHrP2 (class B)
Prokineticin-1/2 receptor	PKR1, PKR2 (subfamily A9)
Substance-P	NK1R (subfamily A9)
Neurokinin-A/B receptor	NK2R (subfamily A9)
Thyrotropin receptor	TSH (subfamily A10)
Thyrotropin-releasing hormone receptor	TRH1, TRH2 (subfamily A10)
Urotensin-2 receptor	Uts2R (GPR14) (subfamily A5)
Vasopressin receptor	V1a, V1b (subfamily A6)
Receptors for peptidases	
Thrombin receptors	PAR1, PAR3, PAR4 (subfamily A15)
Trypsin receptor	PAR2 (subfamily A15)

The inhibition by regulators of G-protein signaling on G-protein-mediated signaling can actually be exerted: (1) by serving as GTPase-activating proteins (GAP) for Gα subunits of heterotrimeric G proteins, thereby accelerating G-protein inactivation (GAP mechanism of action); or (2) by blocking Gα-mediated signaling, as they compete by binding to Gα effectors (effector antagonism). Among the members of the R4 RGS subfamily that have the shortest N- and C-terminal flanking regions among the RGS family members and interact with Gq and Gi/o subunits, RGS2 and RGS3 operate mainly by effector antagonism, whereas RGS5 and RGS16 inactivate G proteins [792].

Ubiquitously expressed Gα_q undergoes reversible palmitoylation at adjacent cysteines in its N-terminus (Cys9 and Cys10) that enables its localization to the cytoplasmic face of the plasma membrane. In addition, receptor activation promotes palmitate turnover on Gα subunits [789].

Table 9.9. Properties of the Gq family members (Source: [790]; RIC8: resistance to inhibitors of cholinesterase-8 homolog that serve as Gα protein guanine nucleotide-exchange factor). Regulator of G-protein signaling associated with $G\alpha_q$ belongs to the B/R4 subfamily (RGS1–RGS5, RGS8, RGS13, RGS16, and RGS18). Among them, RGS2, RGS3, RGS4, and RGS18 exhibit GAP activity for $G\alpha_q$ subunits. Members of the Gq/11 and G12/13 subclasses activate RhoA, but via distinct transmission pathways from RhoGEFs to RhoA. Whereas members of the G12/13 subclass use components of the RhoGEF1 subfamily (RhoGEF1, RhoGEF11, and RhoGEF12), members of the Gq/11 subclass target RhoGEF11, RhoGEF12, RhoGEF13 and RhoGEF25 proteins. In addition, Resistance to inhibitors of cholinesterase-8 homolog RIC8a has GEF activity for Gi1, Go, Gq, and G12, but not Gs, whereas RIC8b homolog interacts with Gs and Gq, but not Gi and G12. Scaffold caveolin-1 in calveolae allows efficient Gq–receptor coupling. Lipid raft resident proteins, flotillin-1 and -2, also interact with Gq proteins.

Type	Distribution	Effectors	Partners
Gq	Ubiquitous	PLCβ1/3/4 RGS, AKAP13, RhoGEF11/12/25	RIC8a/b, caveolin-1, flotillin-1/2
G11	Ubiquitous	PLCβ1/3/4 RGS, RhoGEF25	RIC8a/b, caveolin-1, flotillin-1/2
G14	Lung, kidney, liver	PLCβ1/3/4	
G16	Hematopoietic cells	PLCβ1/2/3	

Stimulated G-protein-coupled receptors are able to activate Ras–MAPK cascades, albeit with smaller efficacy than that of receptor Tyr kinases. Monomeric GTPases of the RHO superfamily can transmit signaling from certain activated GPCRs. Subunit $G\alpha_q$ can induce activation of small GTPase RhoA, though with a lower potency than that of members of the $G\alpha_{12}$ family. Most of the GPCRs that prime Rho-dependent responses, such as GPCRs of bombesin, endothelin, thrombin, lysophosphatidic acid, and thromboxane-A2, as well as calcium-sensing receptor that stimulates stress fibers, focal adhesions, and cell rounding, activate Gq or G12/13 subclass protein. Activation of RhoA mediated by $G\alpha_q$ relies on member of the Rho-specific GEF family, such as RhoGEF11 to RhoGEF13 and RhoGEF25 [793].[17]

Another group of $G\alpha_q$ interactors corresponds to G-protein-coupled receptor kinases, such as GRK2 and GRK3, that phosphorylate activated $G\alpha_q$, hence blocking $G\alpha_q$-mediated signaling and recruit arrestins and other proteins that can desensitize GPCRs [789].

---

[17]RhoGEF11 is also identified as PDZRhoGEF; RhoGEF12 as leukemia-associated RhoGEF (LARG); RhoGEF25 as GEFT and p63RhoGEF; RhoGEF13 as AKAP13, AKAP-LBC, and LBC-RhoGEF. The latter functions as an A-kinase-anchoring protein (AKAP) and a Rho-selective guanine nucleotide-exchange factor. Expression of AKAP13 is restricted to human hematopoietic cells as well as lung, heart, and skeletal muscle.

Once bound to angiotensin-2, AT_1 receptor triggers multiple signaling axes, particularly those mediated by G proteins ($G\alpha_i$, $G\alpha_q$ [mainly], and $G\alpha_{12/13}$) and by β-arrestin, in addition to receptor Tyr kinase transactivation.[18] Different types of GPCR agonists can engage distinct signaling pathways (*ligand type-dependent signaling*), such as: (1) G-protein- and β-arrestin-dependent and (2) G-protein-independent and β-arrestin-dependent activation of ERK1 and ERK2 kinases.[19] Both Tyr4 and Phe8 of octapeptide angiotensin-2 are essential for both AT_1–ligand binding and G-protein-dependent signaling, i.e., ensure the structure–function relationship for angiotensin-2 and its analogs. Substitution of these residues at position 4 and/or 8 in angiotensin-2 indeed produces agonists with different functional outcomes. These β-arrestin-biased ligands actually trigger β-arrestin-selective signaling downstream from AT_1, instead of supporting IP_3 production and PKC activation [795]. Different βArr–AT_1 complexes resulting from distinct angiotensin-2 analogs, i.e., different changes in AT_1 conformation initiated by the analogs and in β-arrestin conformation upon recruitment to AT_1 receptor as well as possible selective actions of GRKs (GRK2 and GRK6),[20] which can cause β-arrestin-dependent, selective engagement of adaptors and/or signaling effectors, create different activated states and, hence, cellular outcomes (vascular smooth muscle cell growth or migration) [795]. Efficiency of ERK1 and ERK2 activation is correlated to the stability of endosomal βArr–AT_1 complex, which depends on

---

[18]Activation of phospholipase-C by $G\alpha_q$ causes IP_3 production that triggers activation of calcium-dependent protein kinase-C. On the other hand, AT_1 phosphorylation by G-protein-coupled receptor kinase increases AT_1 avidity for β-arrestin and uncouples AT_1 from G-protein. Both signaling from AT_1 based on the MAPK module and AT_1-mediated transactivation of receptor Tyr kinases can be initiated by both AT_1-primed activation of G-protein and the βArr–AT_1 complex.

[19]G-protein-coupled receptors can activate ERK1 and ERK2 via both G-protein-dependent and -independent mechanisms and different pools of β-arrestins. For example, stimulation of $G\alpha_s$-coupled V_2 vasopressin receptor in rat renal medulla collecting ducts provokes translocation of aquaporin-2 to apical membranes to foster water reabsorption (Gs–ACase–cAMP pathway). Vasopressin V_2 receptor can also activate ERK1 and ERK2 upon stimulation of G proteins as well as upon RTK (EGFR, FGFR, IGFR, NTRK1, PDGFR, and VEGFR) transactivation (independently of Gs, Gi, Gq, or Gβγ subunit). β-Arrestins not only act as GPCR regulators, but also as signal transducers of the MAPK module and as activators of Src kinase. Receptor Tyr kinases can recruit β-arrestins once bound to their respective growth factors. In particular, β-arrestins contribute to the activation of ERK1, ERK2, and PI3K by the IGFR–IGF1 complex [794]. Receptor IGFR can be transactivated by thrombin, μ-opioid, $GABA_B$, and AT_1 receptors. After Src-dependent shedding by a membrane-associated metallopeptidase of an activator of the insulin-like growth factor receptor for auto- and paracrine signaling, β-arrestins that promotes the GPCR-mediated ERK1/2 activation are engaged by the transactivated IGFR, but not by V_2 [794]. β-Arrestins operate in other transactivation types, as they are also involved in transactivation from platelet-activating factor receptor [794].

[20]Kinase GRK6 is more efficient than GRK2 on the recruitment of β-arrestin to AT_1 and β-arrestin conformational changes [795]. The effect of GRK6 relies on the agonist type. It remains mild for angiotensin-2.

the nature of the ligand–receptor complex, rather than to the extent of β-arrestin recruitment to AT_1 receptor [795].[21] The propensity of different analogs to promote distinct β-arrestin recruitment and conformational changes depends or not on phosphorylation of AT_1 by GRK kinases. Once β2-adrenoceptor is stimulated, ligand promotes phosphorylation of distinct sites by different GRKs and distinctively influences the conformation of β-arrestin. Therefore, different β-arrestin conformations generate distinct signaling types and cell responses.

$G\alpha_{11}$

Widespread $G\alpha_{11}$ subunit interacts with multiple G-protein-coupled receptors, such as muscarinic, adrenergic, endothelin, bombesin, cholecystokinin receptors, among others. $G\alpha_{11}^{GTP}$ binds to its effectors and regulators, such as phospholipase-Cβ, G-protein-coupled receptor kinases, and regulators of G-protein signaling.

In cardiomyocytes, RGS2 precludes Gq/11-mediated phospholipase-C activation, but not Gq/11-mediated ERK1, ERK2, P38MAPK, and JNK activation downstream from endothelin receptor [796]. In addition, RGS4 inhibits Gq-mediated activation of mitogen-activated protein kinase and phospholipase-C downstream from bombesin receptor [797].

Subunit $G\alpha_{11}$ activates phospholipase-Cβ to generate the following effects [791]: (1) stimulation of transcription factor tubby-like protein, as PIP_2 tethers tubby-like protein to the plasma membrane; (2) hydrolysis of PIP_2 into inositol trisphosphate and diacylglycerol; (3) IP_3-triggered Ca^{++} release from its intracellular store; (4) DAG activation of Unc13 (or Munc13) in hyperglycemia that causes apoptosis; and (5) DAG activation of protein kinase-C.

Subunits of the Gq/11 subclass can also stimulate the mitogen-activated protein kinase cascade. Vascular endothelial growth factor binds to its cognate receptors on vascular endothelial cells (VEGFR1–VEGFR2). Receptor VEGFR2 triggers signaling, whereas VEGFR1 represses VEGFR2-initiated effect. Both receptors stimulate G proteins. Receptor VEGFR2 activates Gq/11 that target small GTPase RhoA and the MAPK module and provokes Ca^{++} influx via phospholipase-C [798]. On the other hand, VEGFR1 activates small GTPase CDC42. Constitutively active G11 mutant (but not constitutively active Gq mutant) can cause phosphorylation of both VEGFR2 and MAPK. In addition, Gβ/γ subunit intervenes in MAPK phosphorylation and intracellular Ca^{++} mobilization, but does not influence VEGFR2 phosphorylation.

---

[21] For some agonist types, the magnitude of conformational rearrangement of β-arrestin is well correlated with the extent of recruitment to AT_1 receptor. For other ligand types, the magnitude of recruitment and of conformational change are not correlated. The extent of β-arrestin conformational change can be indeed lower for an equivalent degree of recruitment to AT_1 receptor [795].

$G\alpha_{16}$

Protein $G\alpha_{16}$ is a subunit of the Gq subclass that activates phospholipase-Cβ (PLCβ1–PLCβ4) [799]. This effector enzyme hydrolyzes PIP$_2$, thereby releasing inositol trisphosphate that provoke intracellular Ca^{++} flux and diacylglycerol messenger that activate protein kinase-C.

Protein $G\alpha_{16}$ is expressed predominately in hematopoietic cell types. It is coupled to most GPCRs. It is a substrate of 2 types of GTPase-activating proteins ($G\alpha_{16}$GAP): PLCβ and many regulators of G-protein signaling [799].

9.1.1.4 $G\alpha_{12}$ Subclass

Members of the subclass 4 of $G\alpha$ subunits, i.e., G12 and G13, participate in various cellular functions. They activate Rho GTPases. Both $G\alpha_{12}$ and $G\alpha_{13}$ trigger similar cellular effects that lead to gene transcription, reorganization of the actin cytoskeleton, formation of actin stress fibers, and assembly of focal adhesions, events that are associated with RhoA activation [800]. Nonetheless, $G\alpha_{12}$ and $G\alpha_{13}$ do not have only redundant activities. Subunit $G\alpha_{13}$, but not $G\alpha_{12}$, contributes to the development of the vasculature. Factors RhoGEF1 and RhoGEF12 are more effective for G13 than G12. Whereas $G\alpha_{13}$ is able to stimulate RhoGEF1 and unphosphorylated RhoGEF12 in vitro, $G\alpha_{12}$ only activates phosphorylated RhoGEF12 [800]. Whereas $G\alpha_{12}$ activates RhoA for stress-fiber formation upon activation by G-protein-coupled receptors bound to endothelin, thrombin, and vasopressin, $G\alpha_{13}$ primes RhoA signaling as a mediator of GPCRs ligated by bradykinin, serotonin, and lysophosphatidic acid. Consequently, $G\alpha_{12}$ and $G\alpha_{13}$ can be targeted by distinct receptors. They not only possess different affectors, but also different effectors. Furthermore, they can have opposite effect on a given mediator. Subunit G13 activates the transcription factor NRF2, whereas G12 antagonizes it [800]. Besides, G13 is involved in the regulation of NOS2 expression.

Subunits G12 and G13 not only target small GTPase Rho via RhoGEF, but also phospholipase-Cϵ and -D, MAPK, and Na$^+$–H$^+$ exchanger, as G12 and G13 interact with several proteins (Table 9.10). Ezrin, radixin, and moesin connect various receptors, ion channels, and integrins to the cytoskeleton. Four Rho GTPase guanine nucleotide-exchange factors (Sect. 9.4.1.9) are regulated by G12/13 (Table 9.11): (1) RhoGEF1,[22] which is produced in blood cells; (2) RhoGEF11,[23] which is widespread, but at low level, except in the central nervous system; (3) ubiquitous RhoGEF12;[24] and (4) widespread RhoGEF13.[25]

---

[22] A.k.a. p115RhoGEF.

[23] A.k.a. PDZRhoGEF.

[24] A.k.a. leukemia-associated RhoGEF (LARG).

[25] A.k.a. lymphoid blast crisis (LBC)-RhoGEF, AKAP13, and AKAP-LBC.

Table 9.10. Interactors of G12 and G13 proteins (Source: [801]; AKAP: A-kinase anchoring protein; CTK: cytosolic Tyr kinase; HSP: heat shock protein; PP5: protein phosphatase; RasGAP: GTPase-activating protein of Ras; RhoGEF: guanine nucleotide-exchange factor of Rho).

Protein	Function
Cadherin	G12 and G13-mediated release of β-catenin
Radixin	G13-induced linkage between cytokeletal and plasmalemmal proteins
HSP90	G12-induced actin reorganization (stress fiber formation)
AKAP	G12-mediated stimulation of AKAP13 (especially in heart) G13 interaction with AKAP3 in testis and PKA stimulation
CTK	G12-mediated activation of BTK G13-mediated activation of FAK2 Gi- and Gs-mediated activation of Src
PP5	G12 and G13-mediated activation
RasGAP	G12-mediated stimulation
RhoGEF	G13-mediated stimulation of RhoGEF1 (p115RhoGEF) G12-mediated stimulation of RhoGEF12 (LARG)

Table 9.11. G12/13-regulated Rho guanine nucleotide-exchange factors (Source: [802]; AKAP-LBC: A-kinase anchoring protein-lymphoid blast crisis, a cardiac splice variant of LBC-RhoGEF [AKAP13 or RhoGEF13]; FAK: focal adhesion kinase; PAK: P21-activated kinase; PKA(C): protein kinase-A(C); TEC: Tyr kinase expressed in hepatocellular carcinoma). G12 and G13 prime the RhoGEF–RhoA–RoCK pathway. They cause RhoGEF translocation from the cytosol to the plasma membrane and stimulate their GEF activity. Whereas RhoGEF oligomers are inactive, active monomers bind to activated G12/13 at the plasma membrane. Conversely, these RhoGEFs serve as GTPase-activating proteins for G12/13 (negative feedback). Targets of RhoA include Rho kinase (RoCK) that: (1) provokes cell contraction, as it inhibits myosin light chain phosphatase, and (2) initiates serum response factor-dependent gene transcription.

RhoGEF	Effect	GEF Activity regulators
RhoGEF1 (p115RhoGEF)	G13-induced GEF activation	PKC (+) PAK1 (− via Rac)
RhoGEF11 (PDZ-RhoGEF)	G13-induced GEF activation	TEC, FAK (+)
RhoGEF12 (LARG)	G12-induced GEF activation G13-induced GEF activation	TEC, FAK (+)
RhoGEF13 (LBC-RhoGEF) (AKAP13)	G12-induced GEF activation G13-induced GEF activation	PAK1 (−)

Gα_{12}

Ubiquitous Gα_{12} protein transmits signals from G-protein-coupled receptors to RhoGEFs to stimulate small Rho GTPase. Unlike G13, G12 does not activate Rho via RhoGEF1, despite its interaction with RhoGEF1 [803]. Protein Gα_{12} also connects to regulators of G-protein signaling with GTPase-activating protein activity that regulate its functioning. It mainly targets small GTPase RhoA, thereby stimulating its effectors in a Rho-dependent manner. However, it also directly binds and activates certain signaling effectors, such as Bruton's Tyr kinase, protein phosphatase-2, small Ras GTPase, and N- and E-cadherin [803]. In addition, it also links to Ras(Rap)GAP RasA2[26] Furthermore, G12 synergistically regulates cell responses with other Gα proteins. Both G12 and G13 cooperate with Gs to integrate cAMP synthesis via adenylate cyclase AC7 [803].

Protein Gα_{12} can stimulate the activity of Na^+–H^+ exchangers via the Ras–PKC pathway [803]. It contributes to cell proliferation, particularly the G1–S transition via both Ras and Rac small GTPases. It activates the JNK pathways also via Ras, Rac, or CDC42 GTPase, as well as MAP3K1 and MAP3K5 for cell apoptosis.

An active form of G12 binds to the cytoplasmic domain of cadherins to release β-catenin that has an inhibitory effect on intercellular interactions and can then stimulate cell migration [803]. Subunit Gα_{12} is required for the S1P-induced migration of vascular smooth muscle cells that depends on phospholipase-C It also connects to tight junction protein zonula occludens-1, together with heat shock protein HSP90 and Src kinase [803]. It then disrupts tight junctions. Moreover, G12 or G13 heightens the activity of Rho–Rac-dependent activator protein AP1 that provoke synthesis of transforming growth factor-β1.

In migrating cells, G12 generates backward signals to activate myosin-2 of contractile actin–myosin complexes formed at cell trailing region via Rho small GTPase for a forward motion [803]. Conversely, G12 can impede cell migration, as it can inhibit $\alpha_2\beta_1$-integrin.

Subunit Gα_{12} binds GDP very tightly. The estimated rate of guanine nucleotide-exchange for G12 is very slow, i.e., 10- to 20-fold slower than that for Gs and Go, but similar to that of Gz and G13 [803]. The GTP hydrolysis rate of Gα_{12} is very slow, i.e., comparable to that of Gz and G13, but is 5 to 40 times slower than that of other Gα subunits. Kinetics of GTP–GDP exchange are accelerated by GEFs. The duration of Gα_{12} signaling is controlled by the GAP activity of RGS domain of RhoGEFs, such as RhoGEF1 and RhoGEF12 [803].

Protein Gα_{12} undergoes post-translational modifications. Palmitoylation determines its subcellular localization, especially to membrane rafts via HSP90, and interaction with other proteins [803]. Phosphorylation by PKCα, PKCβ, PKCδ, PKCϵ, and PKCζ reduces its affinity for G$\beta\gamma$ subunit.

---

[26]A.k.a. GAP1m, as it is related to GTPase-activating protein GAP1, or inositol (1,3,4,5)-tetrakisphosphate-binding protein IP$_4$BP1 with a different subcellular distribution. Whereas GAP1 is located solely at the plasma membrane, RasA2 (or GAP1m) lodges in the perinuclear region.

Subunit $G\alpha_{12}$ couples to cognate GPCRs of angiotensin-2 (AT_1), cholecystokinin (CCK_1), endothelin (ET_A and ET_B), galanin, acetylcholine (M_1 and M_3), peptidase-activated or thrombin (PAR_1, PAR_2, and PAR_4), lysophosphatidic acid and sphingosine 1-phosphate ($S1P_2$ to $S1P_5$), thromboxane-A2, thyroid-stimulating hormone, vasopressin, lysophosphatidylcholine, and formylpeptide, as well as calcium-sensing receptor [803].

$G\alpha_{13}$

Activity of $G\alpha_{13}$ is modulated by Rho-specific GEFs (RhoGEF1, RhoGEF11, and RhoGEF12). Subunit G13 also bears post-translational modification, such as acylation and phosphorylation. Protein kinase-A phosphorylates G13 (Thr203) [800]. In response to thrombin or thromboxane-A2, protein kinase-C also phosphorylates G13 subunit.

Subunit $G\alpha_{13}$ couples to endothelin ET_A and ET_B, lysophosphatidic acid, peptidase-activated, sphingosine 1-phosphate, thromboxane-A2, and thyrotropin receptors, as well as angiotensin receptor-1A, cholecystokinin receptor-A, dopamine receptor-D5, galanin receptor-2, neurokinin (substance P) receptor NK_1, and serotonin receptor-4 [800].

Subunit $G\alpha_{13}$ interacts with cytoplasmic tails of N- and E-cadherins, radixin, and A-kinase anchor protein AKAP3 [800]. It also binds to protein phosphatase PP5 and kinases FAK2 and Tec in addition to PKA and PKC. In addition, it targets regulator of G-protein signaling RGS16 [251].

9.1.2 Set of Gβ Subunits

Heterotrimeric guanine nucleotide-binding proteins that transmit signals from G-protein-coupled receptors to effectors comprise a Gβ subunit. Subunits Gβ1 to Gβ4 are expressed ubiquitously, whereas Gβ5 subunit is produced only in the brain.

The combinatorial linkage of individual subtypes of Gα, Gβ, and Gγ subunits contributes to G-protein specificity for GPCRs. In a cell that contains many different types of Gα, Gβ, and Gγ subunits, only some types are able to heterotrimerize because of differences in affinity of these subunit types for one another.

In general, Gβγ inhibits GTP binding to Gα. Dimer Gβγ can interact with protein kinase-D, G-protein-coupled receptor kinases, Bruton Tyr kinase, several components of the Raf–Ras–MAPK cascade, RhoGEF18 for monomeric GTPases RhoA and Rac1, adaptors Src homology-2 domain-containing protein SHC and kinase suppressor of Ras KSR1, retinal phosducin, calmodulin, membrane integrated qSNARE protein syntaxin-1B and synaptosome-associated protein SNAP25B, tubulin, and dynamin-1, as well as voltage-dependent calcium channels [804]. The G-protein-coupled receptor kinase β-adrenoceptor kinase impedes binding of Gβγ to cRaf [805].

9.1.3 Set of Gγ Subunits

Heterotrimeric G proteins contains a Gγ subunit that is irreversibly tethered to a Gβ subunit with variable affinity. Twelve known Gγ subunit types exist. Many, but not all, Gγ subtypes can associate with Gβ protein.

Like G-protein-α subunit that undergoes myristoylation and palmitoylation, Gβγ dimers are covalently modified by lipids for membrane anchorage and specific proteic interactions. Members of Gγ subclass experience prenylation (farnesylation or geranylgeranylation) [806]. The activity of Gβγ is increased by reversible carboxymethylation of Gγ, whereas the nature of isoprenylation (farnesyl or geranylgeranyl group) determines the membrane affinity of Gβγ permanently (irreversible post-translational modification) [804].

Subunits Gγ show greater variation in tissue distribution than Gβ subunits. Subunit Gγ1 is specific to rod photoreceptors; Gγ2 and Gγ7 are present in several tissues, yet they are enriched in brain; Gγ3 and Gγ4 are brain-specific; and Gγ5 is ubiquitous [807]. Subunit Gγ8 could function in chemosensory transduction, both in olfactory and vomeronasal neurons [804]. Subunit Gγ1 interacts poorly with Gβ2 and Gβ3, but fails to bind to Gβ4 and Gβ5; Gγ2 links poorly to Gβ3; Gγ3 binds well with all Gβsubunits; Gγ4 tethers especially well with Gβ5 (preferential interaction with Gβ1, Gβ2, and Gβ5); Gγ5 and Gγ7 favor interactions with Gβ1 and Gβ2 subunits [807]. Subunit Gγ8 is coexpressed with Gβ1 and forms a stable dimer [804].

Gβγ (especially Gβ1γ2 and Gβ2γ2) effectors include adenylate cyclases, phospholipase-Cβ1, -Cβ2, and -Cβ3, cGMP phosphodiesterases, and ion channels.

9.1.4 G Protein and Myogenic Response

Local blood flow is regulated to match the metabolic demand of peripheral tissues. Small-resistance arteries possess the intrinsic property to constrict in response to a rise in intraluminal pressure (*Bayliss effect*). Vascular tone determines local vascular resistance and hence organ perfusion. This pressure-induced myogenic vasoconstriction does not depend on endothelium functioning. Myogenic responsiveness actually is an inherent property of vascular smooth muscle cell that can be further fine-tuned by endothelial and neurohumoral factors. Although increased intravascular pressure causes depolarization of arterial myocyte membrane that activates voltage-dependent Ca_V1 channel, the latter does not influence pressure-induced depolarization, which thus implies another stretch-activated ion channel.

The myogenic response triggered by mechanosensors in vascular smooth muscle cells involves ligand-independent activation of Gq/11-protein-coupled receptors rather than membrane stretch-induced gating of mechanosensitive transient receptor potential ion channels [808]. Mechanically activated receptors adopt an active conformation that allows G-protein coupling and recruitment of β-arrestin. Activated

Gq/11 protein signals hastens phospholipase-C that stimulates transient receptor potential channel via DAG, thus leading to membrane depolarization and finally to pressure-induced myogenic vasoconstriction.

9.2 Regulators of Heterotrimeric G Proteins

Heterotrimeric G proteins are activated by a GPCR-induced GDP–GTP exchange at the Gα subunit. This activation is followed by regulation of specific effectors, such as adenylate cyclase, phospholipase-Cβ, kinases, and ion channels. Such a process is controlled by various regulators.

9.2.1 Nucleoside Diphosphate Kinases

Nucleoside diphosphate kinases (NDPK) form proteic complexes, supply nucleoside triphosphates, and catalyze the transfer of phosphate between nucleoside triphosphates and nucleoside diphosphates. In humans, the NDPK family includes 9 known members (mainly NDPKa–NDPKc). They are receptor-independent activators of G-protein. Phosphate conjugation mediated by NDPK acts as an alternative mechanism to the GPCR-induced GDP–GTP exchange. Phosphate is transferred via a plasma membrane-associated complex of NDPKb (but not NDPKa) and Gβγ dimers.

Receptor-independent activation of G proteins via the NDPKb–Gβγ complex requires the intermediate phosphorylation of Gβ subunits [809]. Enzyme NDPKb hence acts firstly as a histidine kinase for Gβ subunits. The phosphate can then be transferred onto GDP. The greater the number of $Gα_s$–Gβγ–NDPKb complexes, the stronger the NDPKb–Gβγ-mediated phosphotransfer to adenylate cyclase-regulating $Gα_s$ and $Gα_i$ subunits. Conversely, NDPKb depletion leads to strongly reduced amounts of Gβ1γ2 dimers as well as caveolin-1 and -3 [810]. Therefore, caveolins that connect to NDPKb and Gβγ subunits contribute to the regulation of plasmalemmal G-protein content.

In cardiomyocytes, activation of Gs and Gi via GPCRs regulates intracellular cAMP concentration that controls myocardial contractility via stimulation of protein kinase-A and changes in Ca^{++} transients via L-type Ca^{++} channels. β-Adrenoceptor activation can regulate plasmalemmal NDPK content, hence $Gα_s$ activity that causes cAMP synthesis [809]. Conversely, the higher the NDPK and Gi levels and density of $Gα_i$βγ–NDPKb complexes, the lower the cAMP concentration and contractility.

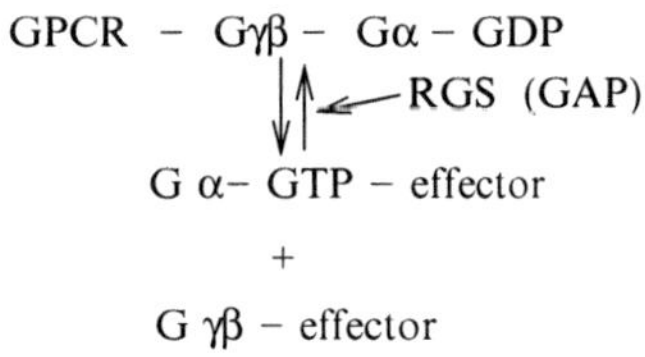

Fig. 9.3 G-protein, its components, its activation–deactivation, and regulators of G-protein signaling proteins (RGS)

$$\text{GPCR} - \text{G}\gamma\beta - \text{G}\alpha - \text{GDP}$$
$$\Updownarrow \leftarrow \text{RGS (GAP)}$$
$$\text{G}\alpha - \text{GTP} - \text{effector}$$
$$+$$
$$\text{G}\gamma\beta - \text{effector}$$

Table 9.12. G-protein-coupled receptors in the cardiovascular system, associated G proteins, regulators (inhibitors) of G-protein signaling, and targets that most often experience a reduced activity, as RGSs terminate G-protein activation by enhancing GTP hydrolysis (**Part 1**; Source: [779]; AT_1: angiotensin-2 receptor; ET_X: any endothelin receptor type (X: A or B); IP: inositol phosphate; MAPK: mitogen-activated protein kinase; $S1P_i$: type-i sphingosine 1-phosphate receptor). Receptors ET_A and ET_B augment and reduce heart contractility, respectively. Receptors ET_A and NOS-coupled ET_B cause a vasoconstriction and -dilation, respectively. The RGS proteins discriminate not only Gα types, but also GPCRs coupled to the same Gα type to selectively regulate signaling.

GPCR	Gα	RGS	Targets
AT_1	Gq/11 (Gi)	RGS1/2/3I/4	MAPK
		RGS5	MAPK
		RGS5	IP
		RGS2	cAMP
		RGS2	IP
		RGS4/5	Ca^{++}
ET_A	Gq/11 (Gi)	RGS3I/4	MAPK
		RGS4/5	Ca^{++}
ET_X		RGS2/4	MAPK
		RGS2/3/4/5	IP
$S1P_1$	Gi	RGS1/3I/4	MAPK
$S1P_2$	Gi, Gq/11, G12/13	RGS1/3I/4	MAPK
$S1P_3$	Gi, Gq/11, G12/13	RGS1/2/3I	MAPK

9.2.2 Regulators of G-Protein Signaling

The interactions of GPCR, G protein, and effectors also involve fine-tuned regulators of G-protein signaling proteins (RGS), which accelerate GTP hydrolysis by Gα subunits. Regulators of G-protein signaling act as GTPase-activating proteins (Fig. 9.3). They thus inhibit signal transduction mediated by Gi/o and Gq/11 (Tables 9.12 and 9.13).

Members of the family of small RGSs are encoded by different genes and share an RGS homology domain. However, their specific activities in GPCR signaling is associated with another structural domain. The smaller RGS proteins likely function as GαGAPs, whereas the larger RGS proteins and RGS-like proteins are Gα effectors. Moreover, certain RGS proteins are involved in the assembly of signaling complexes.

Table 9.13. G-protein-coupled receptors in the cardiovascular system, associated G proteins, regulators (inhibitors) of G-protein signaling, and targets that most often experience a reduced activity, as RGSs terminate G-protein activation by enhancing GTP hydrolysis (**Part 2**; Source: [779]; AR: adrenergic receptor; GIRK: G-protein inwardly rectifying K^+ channel; IP: inositol phosphate; M_i: type-i muscarinic receptor; MAPK: mitogen-activated protein kinase).

GPCR	Gα	RGS	Targets
α1aAR	Gq/11	RGS2	IP
α1bAR	Gq/11		IP
α1xAR		RGS2/4	MAPK
		RGS2	MAPK; JNK (+)
		RGS2/3/4/5	IP
		RGS2	MAPK
β 2AR	Gs (Gi)	RGS2	cAMP
M_1	Gq/11	RGS2/3/8	Ca^{++}
		RGS2/8	IP
		RGS2/16	IP
		RGS2/5	IP, Ca^{++}
M_2	Gi	RGS1/3/4	GIRK
		RGS3/3s/4/5/16	MAPK
		RGS2/3/3s/4/5/16	PKB
		RGS3/4/5	cAMP
M_3	Gq/11	RGS3	MAPK
		RGS2/3/8	Ca^{++}
		RGS2/8	IP
		RGS2/3/5/16	IP
		RGS2/3	IP
		RGS2/3/4	Ca^{++}
		RGS2/3/3s/5	MAPK
		RGS3/3s	PKB
M_5	Gq/11	RGS2/3/8	Ca^{++}
		RGS2/8	IP

9.2.2.1 Superfamily of RGS-Box-Containing Proteins

In humans, the RGS superfamily contains at least 37 identified proteins. It can be decomposed into many families [811]: (1) A or RZ family with members (RGS17, RGS19, and RGS20) that are characterized by an N-terminus with many cysteine residues, which can be reversibly palmitoylated; (2) B or R4 family (RGS1–RGS5, RGS8, RGS13, RGS16, RGS18, and RGS21); (3) C or R7 family, members (RGS6, RGS7, RGS9, and RGS11) of which can couple Gα subunits to GPCRs in the absence of Gβγ dimers. They indeed bind Gβ5 via Gγ-like (GGl) domain to form heterodimers;[27] (4) D or R12 family (RGS10, RGS12, RGS14);

---

[27]C-class RGS proteins act as conventional Gβγ dimers, as they couple Gα subunits to GPCRs.

(5) E or RA family that is composed of RGS box-containing axin1 and axin2, negative regulators of the Wnt pathway, that interact with adenomatous polyposis coli protein; (6) F or GEF family that comprises RhoA-specific guanine nucleotide exchange factors (RhoGEF1, RhoGEF11, and RhoGEF12); (7) G or GRK family that encompasses RGS box-containing G-protein-coupled receptor kinases (GRK1–GRK7; Sect. 5.2.11); and (8) H or Snx family that contains RGS box-containing sorting nexins (Snx13, Snx14, Snx25). Other RGS groups include [811]: (1) dual-specificity A-kinase anchor proteins AKAP10a and AKAP10b that bind PKA regulatory subunits and (2) RGS22 isoforms (RGS22a–RGS22c).

Most RGS proteins are GPCR inhibitors, either by their GAP activity for Gα or by effector antagonism, as they can bind activated Gα^{GTP} in competition with effectors. On the other hand, members of the F class of RGS proteins are positive regulators; RhoGEF1, RhoGEF11, and RhoGEF12 (Sect. 9.4.1.9) couple Gα_q, Gα_{12}, and/or Gα_{13} subunits to monomeric RhoA GTPase (Sect. 9.3.13.1). Owing to their RGS box, they convert inactive RhoA^{GDP} into active RhoA^{GTP}. These 3 RGS box-containing, RhoA-specific guanine nucleotide-exchange factors (RGS-RhoGEFs) serve as Gα effectors that couple not only GPCRs, but also semaphorin receptors to RhoA GTPases [811].[28] The RGS box of RhoGEF1 serves as a GTPase-activating protein motif Gα_{12} and Gα_{13} subunits. Protein RhoGEF12 is a Gα-responsive RhoGEF for Gq, G12 and G13 [811]. However, RhoGEF12 stimulation by Gα_{12}^{GTP} depends on RhoGEF12 phosphorylation by TEC family kinases or focal adhesion kinase. In addition, these 3 RGS-RhoGEFs couple distinct receptors to RhoA activation (RGS-RhoGEF signaling specificity).[29]

The RGS proteins with a G$\alpha_{i/o}$–Loco interaction (GoLoco or G-protein regulatory [GPR]) motif, such as RGS12 and RGS14, have a guanine nucleotide-dissociation inhibitor activity, as they slow spontaneous exchange of GDP for GTP and inhibit association with G$\beta\gamma$ subunits (sequestration) [811]. GoLoco motif-containing proteins generally bind to Gα^{GDP} subunits of the Gi/o subclass. Other GoLoco motif-containing proteins include Rap1GAP (Sect. 9.4.2.3) that accelerates GTP hydrolysis by Rap1 and Rap2-interacting protein (Sect. 9.3.9), as well as G-protein signaling modulators GPSM1,[30] GPSM2,[31] GPSM3,[32] which is able to simultaneously bind more than one Gα_{i1} subunit, and GPSM4.[33] The GPSM proteins do not modify kinetics or magnitude of effector activation, but can reduce agonist binding affinity, hence signaling intensity [811].

---

[28] Both RhoGEF11 and RhoGEF12 bind plexin-B1, a transmembrane receptor for semaphorin Sema4D.

[29] For example, RhoGEF12 and RhoGEF11 couple peptidase-activated receptor PAR$_1$ and lysophosphatidic acid receptor, respectively [811]. Protein RhoGEF12 also couples insulin-like growth factor receptor IGF1R for possible crosstalk between GPCRs and RTKs to converge toward RhoA-mediated cytoskeletal rearrangement.

[30] A.k.a. AGS3.

[31] A.k.a. AGS3-like and Leu(L)–Gly(G)–Asn(N) repeat-containing protein (LGN).

[32] A.k.a. G18.

[33] A.k.a. Purkinje cell protein PCP2.

In summary, regulators of G-protein signaling serve as: (1) guanosine triphosphate-accelerating protein for Gα proteins (GαGAP), hence as signaling inhibitors; (2) Gα effectors, such as RGS-RhoGEFs that couple receptors to monomeric GTPase RhoA; (3) signaling scaffolds between signaling mediators, such as RGS12 that operates as a nexus between Gα, small guanosine triphosphatase Ras, and protein kinases; (4) coupling factors between Gα and GPCRs, such as dimers composed of C-class RGSs and Gβ5; and (5) guanine nucleotide-dissociation inhibitors (e.g., RGS12 and RGS14).

9.2.2.2 Regulation of Activity

Activity of RGS proteins is regulated within a cell. Phosphatidylinositol (1,4,5)-trisphosphate inhibits RGS proteins [812]. Ca^{++}–calmodulin restores the GAP activity. Both PIP_3 and Ca^{++}–calmodulin bind to the same RGS site. Inhibition and disinhibition of GAP activity of RGS4 by these molecules explain oscillations in intracellular calcium concentration.[34] Furthermore, RGS proteins are targeted by different protein kinases.[35]

The RGS proteins are implicated in interactions involving various signaling molecules associated with G-protein signaling pathways and ion carriers at the plasma membrane, such as Ca^{++}, phospholipids (especially phosphoinositides), and Tyr kinases.[36] Besides, lipopolysaccharides and angiotensin-2 increase the expression of RGS proteins in vascular cells.

9.2.2.3 RGS Isoforms

RGS1

In humans, RGS1 is identified in heart, aorta, lung, and olfactory bulb, as well as B lymphoblasts, T lymphocytes, dentritic cells, and peripheral blood monocytes. Isoform RGS1 is a GTPase-activating protein for Gi/o and Gq subunits [813]. It interacts with spinophilin and neurabin to lower GPCR-mediated signaling and counteract RGS inhibition, respectively.

---

[34]At low $[Ca^{++}]_i$, GAP activity of RGS4 is hampered by PIP_3. Hence, Gq/11 stimulates phospholipase-C that raises $[Ca^{++}]_i$. Ca^{++} then binds to calmodulin. Ca^{++}–calmodulin reduces Gq/11 activation and $[Ca^{++}]_i$ decays. When Ca^{++}–calmodulin complex is dissociated, PIP_3 inhibition of RGS4 is restored.

[35]Protein kinase-C phosphorylates RGS2, thereby decreasing its Gq/11 inhibition. The cGMP-dependent protein kinase PKG phosphorylates RGS4, inducing its translocation to the plasma membrane. Protein kinase-A phosphorylates RGS9, thus inhibiting the GAP activity on Gt.

[36]Receptor Tyr kinases regulate phosphoinositide activity.

RGS3

Ubiquitous RGS3 is a GTPase-activating protein for Gi and Gq subclass members. Alternative splicing generates long ($RGS3_L$) and short ($RGS3_S$) forms, in addition to $_{PDZ}$RGS3 and $_{C2PA}$RGS3 isoforms [813].

Binding of 14-3-3ζ and 14-3-3τ proteins to RGS3 hampers RGS3 activity, whereas Ca^{++} influx through voltage-gated channels activates RGS3 [813]. It interacts with: (1) SMAD2 to SMAD4 transcription factors and interferes with their heteromerization; (2) Na^+-dependent inorganic phosphate cotransporters SLC17a1 and SLC34a1; (3) type-1 membrane neuroligins of the postsynaptic membrane, ligands of presynaptical β-neurexins; and (4) EPHb2 and $mGlu_{1A}$ receptors.

RGS8

Protein RGS8 is a GTPase-activating protein that binds Gα subunits of the Gi and Gq subclasses [814].[37] In particular, RGS8 inhibits Gq-coupled M_1 muscarinic acetylcholine (mAChR) or substance-P receptors, but not M_3 or cholecystokinin CCK_2 receptors.[38]

Agent RGS8 is recruited to the plasma membrane via its interaction with $PP1_{r9b}$ regulatory subunit [814].[39] In humans, RGS8 is primarily expressed in the brain.

9.2.2.4 Cardiac RGSs

Cardiac RGSs hinder phospholipase-C activity via Gq/11, especially phospholipase-C stimulation of endothelin-1.[40] Hence, RGS4 has an antihypertrophic effect. Furthermore, cardiac RGSs regulates the activation and deactivation kinetics of G$\beta\gamma$-gated K^+ channels, thus the inward rectifier K^+ channel regulated by acetylcholine via Gi/o (RGSs accelerate GTP hydrolysis rate of G$\alpha_{i/o}$ and regulate ACh-dependent relaxation).[41]

---

[37] In rat, the majority of RGS8 binds to Go and Gi3. Protein RGS8 can also interact weakly with Gi1, Gi2, Gz, and Gq/11.

[38] Protein RGS8 binds strongly to M_1 mAChR, but weakly to M_3 mAChR receptor. It also tethers to melanin-concentrating hormone receptor (MCH1R) [814]. It can also form a quarternary complex with liganded $GABA_{B1b}$ and $GABA_{B2}$ receptors and Gi2 or GoA.

[39] A.k.a. spinophilin and neurabin-2 (neural tissue-specific ^{F}actin-binding protein).

[40] Endothelin-1 has positive inotropic effects and stimulates heart wall growth via Gq activation.

[41] Binding of ACh to cardiac muscarinic M_2 receptors enables GTP to replace GDP at the Gα-subunit of Gi/o-proteins. The subunits dissociate and G$\beta\gamma$ subunit activates K^+ channels. Phosphatidylinositol (1,4,5)-trisphosphate inhibits RGS activity on G$\alpha_{i/o}$ subunits. The rising intracellular Ca^{++} concentration leads to the formation of Ca^{++}–calmodulin complexes. Calcium–calmodulin binds to RGS, then inhibits the effect of phosphatidylinositol trisphosphate and activates RGS. Consequently, K^+ channels are inactivated. With decaying intracellular Ca^{++} concentration, Ca^{++}–calmodulin dissociates and phosphatidylinositol-trisphosphate inhibits RGS again.

Membrane-attached RGS3, a $G\beta\gamma$-binding protein [815] that is strongly expressed in the heart [816], attenuates signaling not only via $G\alpha_i$ and $G\alpha_{q/11}$, but also via $G\beta\gamma$-mediated signaling (via phospholipase-C, mitogen-activated protein kinase, and phosphatidylinositol 3-kinase). Agent RGS6 is relatively abundant in atriomyocytes [817]. Sorting nexin SNx13,[42] a $G\alpha_s$-specific GAP, inhibits adenylate cyclase stimulation induced by the α-adrenoceptor–$G\alpha_s$ complex. It may then modulate the activity of cardiac Ca^{++} channels. It binds to membrane phosphatidylinositol 3-phosphate, thus participating in early endosome structure. The RhoGEF proteins have Dbl- (DH) and pleckstrin (PH) homology domains. The DH domain is responsible for exchange activity and the PH domain is likely involved in subcellular localization. The RGS-like RhoGEFs act on $G\alpha_{12}$ and $G\alpha_{13}$ [815]. Coupling of GPCRs to G12/13 in cardiomyocytes is associated with contractility [818], and protein kinase-C–mediated activation of sarcolemmal Na^+–H^+ exchangers [819]. GPCR kinases phosphorylate activated GPCR receptors. Agent GRK2, highly expressed in the human heart (as well as GRK5 and GRK6), interacts with Gq/11 (GRK2 sequesters activated $G\alpha_q$).

9.2.3 Gα Guanine Nucleotide-Exchange Factors

Resistance to inhibitors of cholinesterase-8 homolog-A (RIC8a)[43] acts as a guanine nucleotide-exchange factor for a subset of $G\alpha$ proteins activated by ligand-bound G-protein-coupled receptors that include $G\alpha_{i1}$ to $_{i3}$, $G\alpha_o$, $G\alpha_q$, and $G\alpha_{13}$ [820]. It also associates with $G\alpha_s$ [251]. Protein RIC8a connects to $G\alpha^{GDP}$ proteins, stimulates GDP release, and forms a nucleotide-free, transition-state complex with $G\alpha$ that dissociates upon GTP binding to $G\alpha$. In addition, RIC8a binds adenylate cyclase-5 that is highly expressed in the brain striatum and heart. It suppresses AC5 activity upon Gi stimulation [821].[44] Therefore, RIC8a potentiates $G\alpha$-mediated signaling. During Gq-mediated signal transduction, RIC8a enhances ERK activation [822].

9.2.4 Activators of G-Protein Signaling

Activators of G-protein signaling (AGS) exert distinct effects depending on the G-protein activation–deactivation cycle. They can be classified into 3 distinct functional sets: (1) AGSs that function as direct $G\alpha$ activators, as they operate

---

[42] A.k.a. RGS and PHOX domain-containing protein RGSPX1.

[43] A.k.a. RIC8 and synembryn.

[44] Adenylate cyclase-5 can also be inhibited by Gi, Gz, and RGS2, as well as protein kinase-A.

Table 9.14. Receptor-independent activators of G-protein signaling (Source: [823]). Activated G-protein-coupled receptors intervene as guanine nucleotide exchange factors (GEF) that trigger the transformation of $G\alpha^{GDP}$ into $G\alpha^{GTP}$ and the dissociation of $G\alpha$ from $G\beta\gamma$ for signaling. Upon GTP hydrolysis, the heterotrimer then reforms. The stages of G-protein activation–deactivation cycle are regulated to optimize signal magnitude and duration, as well as to keep signal specificity. Inhibitors of G-protein activation are either regulators of G-protein signaling, which act as GTPase-activating proteins (GAP) by enhancing GTP hydrolysis, or guanine dissociation inhibitors (GDI), which inhibit GDP dissociation. Stimulators of G proteins can operate as GEFs for $G\alpha$. The AGS subfamily 1 includes AGS1, subfamily-2 AGS3 to AGS6 (AGS5 and AGS6 targeting only Gi3), and subfamily-3 AGS2, AGS7, and AGS8.

Type	Target	Effect
AGS1	Gi2, Gi3	GEF
AGS2	Gi2, Gi3, Gs, G16	$G\beta\gamma$ binding
AGS3	Gi2, Gi3	GDI

as guanine nucleotide-exchange factors; (2) AGSs that bind to $G\alpha$, serving as guanine nucleotide dissociation inhibitors; and (3) AGSs that bind to $G\beta\gamma$ that either dissociate the inactive heterotrimer or sequester released $G\beta\gamma$ upon G-protein activation [823] (Table 9.14). Accessory AGS proteins affect the signal features and propagation. In addition, AGSs can facilitate crosstalk between signaling pathways.

9.2.4.1 AGS1

Activator of G-protein signaling AGS1 selectively activates Gi/o independently of G-protein-coupled receptors. Protein AGS1 is a member of the RAS hyperfamily of small GTPases that plays the role of guanine nucleotide-exchange factor for Gi subunit. Protein AGS1 can antagonize GPCR by reducing the pool of G-proteins available for GPCR coupling.

In addition, AGS1 impedes increase in activity of $G\beta\gamma$-regulated inwardly rectifying K^+ channel (GIRK) activated by muscarinic M_2 receptors [824]. In neurons, AGS1 interacts with adaptors, such as NOS-associated CaPON and NCK2, as well as nitric oxide synthase NOS1. Adaptor CaPON competes with adaptor DLg4 involved in the ^{NMDA}Glu–NOS1 pathway to form a ^{NMDA}Glu–CaPON–NOS1–AGS1 complex [825]. Moreover, AGS1 is expressed in the suprachiasmatic nuclei according to circadian rhythm [826].

9.2.4.2 AGS2

Activator of G-protein signaling AGS2 is similar to a light chain component of cytoskeletal and ciliary dynein. AGS2 interacts with numerous signaling effectors, especially GPCRs. Dynein could organize signaling complexes to regulate organelle displacement. AGS8 is produced by ventricular cardiomyocytes (but not in cardiac fibroblasts, aortic smooth muscle cells, and endothelial cells) in response to hypoxia (but not by tachycardia, hypertrophy, or failure) [827]. AGS8 interacts directly with G$\beta\gamma$ without disturbing the regulation of PLCβ2 by G$\beta\gamma$.

9.2.4.3 AGS3

Activator of G-protein signaling AGS3, which is widely expressed (with tissue-specific splicing),[45] acts on plasmalemmal concentrations of certain receptors and channels by modulating cellular transport of receptors and channels such as K$_{IR}$2.1. Moreover, AGS3 activates G protein (Gi inhibition and G$\beta\gamma$ stimulation) in a receptor-independent fashion [828]. AGS3 interacts only with members of the G$\alpha_{i/o}$ subclass (except Gz), i.e., with both Gi and Go, but it is a guanine dissociation inhibitor for Gi3 only (not for Go) [829]. Agent AGS3 binds and stabilizes Gα_i^{GDP}, and then blocks reassociation of Gα_i with G$\beta\gamma$ dimer. Therefore, AGS3 inhibits Gα_i, but favors G$\beta\gamma$ signaling.

9.2.5 GPCR-Interacting Proteins

Many GPCRs interact with GPCR-interacting proteins (GIP). These proteins are involved in [779]: (1) GPCR compartmentation in membrane rafts and/or caveolae; (2) assembling of large signaling complexes to direct signaling specificity, intensity, and duration; (3) transfer to and from the plasma membrane; and (4) fine-tuning of GPCR signaling.

Some GIPs are transmembrane proteins, whereas others are intracellular. Certain GIPs can interact with several receptors, such as arrestin and G-protein-coupled receptor kinases, whereas other interact specifically with a single type of GPCR, such as AT$_1$ receptor-associated protein (ATRAP).[46]

Several GPCR-interacting proteins, such as Homer proteins, Gα-interacting protein (GAIP)-interacting protein C-terminus GIPC1,[47] 14-3-3 proteins, calmodulin,

---

[45] A splice variant of AGS3 lacking TPR domains is expressed in the heart.

[46] Angiotensin-2 receptor AT$_1$ tethers ATRAP that can then impede signaling (cell proliferation and vascular remodeling) by enhancing AT$_1$ endocytosis.

[47] Protein GIPC1 interacts with β1-adrenoceptor, dopamine receptor D$_3$, neurotrophic Tyr receptor kinases NTRK$_1$ and NTRK$_2$, RGS19, α_5- and α_6-integrins, myosin-1C and -6, α-actinin-1, GluT1, low-density lipoprotein receptor-related proteins LRP1, LRP2, and LRP8, as well as KIF1b kinesin [251].

and spinophilin,[48] link GPCRs to receptors, ion channels, cytoskeletal proteins, protein kinases, and regulators of G-protein signaling [779].

Several GPCR–GIP interactions are enhanced by ligand stimulation, such as connections between GPCRs and arrestins, GRKs, Na^+–H^+ exchanger regulatory factor NHERF1 [779].[49]

Among GPCR-interacting proteins, arrestin causes GPCR desensitization via endocytosis. In addition, arrestins are scaffold proteins for Src kinase and several components of MAPK modules, such as cRaf, extracellular signal-regulated kinases, and Jun N-terminal kinase JNK3 [779].

G-protein-coupled receptor kinases also interact with numerous proteins involved in signaling and trafficking, such as Gq/11, Gβγ, PI3K, PKB, MAP2K1, calmodulin, clathrin, caveolin, and actin.

A-kinase anchoring proteins AKAP5 and AKAP12 that link to adrenoceptors also interact with kinases PKA and PKC as well as PP3 phosphatase. Upon stimulation of adrenoceptors, PKA anchored to AKAP5 phosphorylates the receptor that then switches from Gs to Gi coupling and promotes MAPK signaling [779]. It also phosphorylates GRK2 to heighten adrenoceptor desensitization.

9.2.6 G-Protein-Coupled Receptor Kinases

G-protein-coupled receptor kinases (GRK) can complex with heterotrimeric G protein on the inner surface of the plasma membrane. Interactions of G proteins, in particular Gβ1γ2 subunit, with the lipid bilayer facilitate GPCR-catalyzed GTP exchange on the Gα subunit. Ubiquitous G-protein-coupled receptor kinases constitute a 7-member family (GRK1–GRK7; Vol. 3 – Chap. 7. G-Protein-Coupled Receptors and Sect. 5.2.11). These kinases have different distribution patterns among the body's tissues as well as distinct binding preferences for some receptors.

Subunit Gβ1γ2 links to phospholipase-Cβ and G-protein-regulated inwardly rectifying potassium channels (GIRK or $K_{IR}3$), as well as GRK2 kinase [830]. Kinase GRK2 can simultaneously interact with activated $Gα_q$ and Gβγ, GPCRs, and the plasma membrane. Therefore, GRK2 enables the assembly and organization of signaling complexes at sites of GPCRs. Subunit Gβ1γ2 changes its orientation after binding to GRK2 protein recruited to the plasma membrane by G proteins to phosphorylate activated GPCRs [830].

---

[48] Spinophilin is a regulatory subunit of protein phosphatase-1.
[49] Sodium–hydrogen exchanger-regulating factor NHERF1 interacts with β2-adrenoceptor, once the latter is phosphorylated by GRK5 kinase. Protein NHERF1 also mediates recycling of internalized β2-adrenoceptors.

9.3 Monomeric (Small) GTPases

Guanosine triphosphatases form a set of hydrolases that process guanosine triphosphate. This set comprises both large and small GTPases that operate in signal transduction, protein synthesis, intracellular transport of vesicles, translocation of proteins across cellular membranes, and organization of the cytoskeleton and cell adhesion plaques, hence cell growth, division, differentiation, spreading, and migration. Many small GTPases are dynamically acylated to modify their membrane affinity.

Monomeric, cytosolic, regulatory G proteins constitute a class of approximately 150 members that can be split to form the RAS hyperfamily ($\sim$ 36 members). The RAS hyperfamily can be decomposed into the ARF, RAB, and RAS superfamilies, the latter being subdivided into the DIRAS, NKIRAS, RAL, RAN, RAP, RAS, RHO ($\sim$ 22 members), RGK, RHEB, and RIT families (Table 9.15).

Among large GTPases, dynamins act in endocytosis, organelle division, and cytokinesis. Small GTPases share common features with G proteins, but have different structure and mechanism of action. Monomeric, small GTPases can be categorized according to their functions (Tables 9.16 and 9.17).

The *RAS hyperfamily* of small GTPases includes 5 major superfamilies — ARF, RAB, RAN, RAS, and RHO — based on sequence homology and specific functions. Members of the RAS superfamily regulate cytoskeletal rearrangement and contraction and are activated upon stimulation of G-protein-coupled and growth factor receptor Tyr kinases.

The *ARF superfamily* of adpribosylation factors is implicated in the regulation of intracellular vesicle genesis and motion.

The *RAB superfamily* is composed of about 70 types in humans that regulate vesicle formation, displacement along actin and tubulin filaments, and fusion.

The *RAN superfamily* of Ras-related nuclear GTPases is involved in nuclear export and import.

The *RAS superfamily* of small GTPases comprises the DIRAS, RAL, RAP, RAN, RAS, RASD, RAS-like, RGK, RHEB, NKIRAS, and RIT families.

The *RAS family* embodies 2 subfamilies: the *P21RAS subfamily* with hRas, kRas, and nRas, and the *RRAS subfamily* with rRas1, rRas2 (or TC21), and mRas (or rRas3).

The *RASD family* is constituted by RasD1[50] and RasD2.[51]

The *RAS-like family* contains the RASL10 (with RasL10a, or RRP22, and RasL10b); RASL11 (with RasL11a and RasL11b); RASL12 (with RasL12);[52] and

---

[50] A.k.a. activator of G-protein signaling AGS1. Protein RasD1 interacts with nitric oxide synthase NOS1 and its adaptor NOS1AP (or CaPON), as well as $G\alpha_i$ [251].

[51] A.k.a. Ras homolog enriched in striatum (RHES). The RasD2 protein links to cRaf and PI3K [251].

[52] A.k.a. Ris.

9.3 Monomeric (Small) GTPases

Table 9.15. Superfamilies of the class of monomeric GTPases.

Category	Subcategories Members
	ARF superfamily
ARF family	Arf1–Arf6
ARL family	
ARFRP family	
SAR family	
	RAB superfamily
14 families	Rab1–Rab35
	RAS superfamily
DIRAS family	DIRas1–DIRas3
NKIRAS family	
RAL family	RalA–RalB
RAN family	
RAP family	Rap1–Rap2
RAS family	P21RAS subfamily (hRas, kRas, nRas)
	RRAS subfamily (rRas1, rRas2, mRas/rRas3)
	RASD subfamily (RasD1–RasD2)
	RAS-like subfamily
	RASL10 subfamily (RasL10a–RasL10b)
	RASL11 subfamily (RasL11a–RasL11b)
	RASL12 subfamily
	RASL13 subfamily
	RERG subfamily
RGK family	
RHEB family	
RIN/RIT family	RIN subfamily
	RIT subfamily
RHO family	CDC42 subfamily
	MIRO subfamily (Miro1–Miro2 [RhoT1–RhoT2])
	RAC subfamily (Rac1–Rac3, and RhoG)
	RHO subfamily (RhoA–RhoC)
	RHOBTB subfamily (RhoBTB1–RhoBTB2)
	RHOD/RIF subfamily
	RHOF subfamily
	RHOH/TTF subfamily
	RND subfamily (Rnd1, Rnd2, Rnd3/RhoE)
	TC10 subfamily (RhoJ–RhoQ)
	WRCH subfamily (RhoU–RhoV)

Table 9.16. Main functions of small GTPases (**Part 1**; ARF: adpribosylation factors; CDC42: cell-division cycle-42; PIKE: phosphatidylinositol 3-kinase enhancer). PIKE is a nuclear GTPase with PLCγ1 as guanine nucleotide-exchange factor.

Type	Role
ARF	Control of cell transfer of molecules (vesicular transport — vesicular budding and maturation), phosphoinositide metabolism
CDC42	Control of cytoskeleton dynamics, formation of filopodia and adhesive-like complexes, activation of the MAPK module and gene expression, initiation of DNA synthesis, regulation of G1 phase progression
Gem	Nucleocytoplasmic transport
Miro	Mitochondrial fusion and transport
PIKE	Activation of nuclear PI3K activity

RASL13 (with RasL13)[53] subfamilies, as well as Ras-like, estrogen-regulated, growth inhibitor RERG.

The *DIRAS family* of RAS-like GTPases groups DIRas1 (a.k.a. Rig) to DIRas3 that operate in intracellular protein transport, including nucleocytoplasmic transfer.

The *family of NFκB inhibitor-interacting Ras-like proteins* comprises NκIRas1 and NκIRas2.

The *RGK family* consists of Rad, Gem/Kir, and Rem.

The *RHO superfamily* of small Ras homolog GTPases encompasses the 3 main families, i.e., CDC42, Rac, and Rho families. In fact, it is made up of many families [831]: family **1** with CDC42; **2** with Miro1 and Miro2; **3** with Rac1,[54] Rac2,[55] Rac3, and RhoG; **4** with RhoA, RhoB, and RhoC; **5** with RhoD; **6** with RhoF;[56] **7** with RhoH; **8** with RhoJ[57] and RhoQ;[58] **9** with RhoU[59] and RhoV;[60] **10** with RhoBTB1 and RhoBTB2; and **11** with Rnd1 to Rnd3.[61]

Among the 22 identified members of the RHO superfamily, most are classically activated (i.e., cycle between active GTP-bound and inactive GDP-loaded forms),

---

[53] A.k.a. Rab44.
[54] A.k.a. cell migration-inducing protein MIg5, teratocarcinoma protein TC25, and p21Rac1.
[55] A.k.a. RacB.
[56] A.k.a. Rap1-interacting factor-1 (RIF1).
[57] A.k.a. TC10-like protein (TCL).
[58] A.k.a. TC10.
[59] A.k.a. Wnt1-responsive CDC42 homolog WRCH1.
[60] A.k.a. WRCH2 and Chp.
[61] A.k.a. RhoE.

9.3 Monomeric (Small) GTPases

Table 9.17. Main functions of small GTPases (**Part 2**; Rab: Ras from brain; Rad: Ras associated with diabetes; Ran: Ras-related nuclear proteins; Rap: Ras-related proteins; Ras: rat sarcoma viral proto-oncogene product homolog; RHEB: Ras homolog enriched in brain; Rho: Ras homology; Rit: Ras-like expressed in many tissues).

Type	Role
Rab	Control of vesicle transport, membrane trafficking (clathrin-coated vesicle formation, endosomal motility)
Rac	Control of cytoskeleton, formation of filopodia and adhesive-like complexes, activation of the MAPK module and gene expression, initiation of DNA synthesis, regulation of G1 phase progression
Rad	Nucleocytoplasmic shuttling, cardiomyocyte growth cardiac excitation-contraction coupling
Ran	Control of nucleocytoplasmic transport, especially during phases G1, S, and G2 of the cell cycle, and microtubule organization during M phase
Rap	Vesicular transport
Ras	Control of cell differentiation, adhesion, growth, proliferation, migration, and apoptosis
Rem	Nucleocytoplasmic shuttling, muscular excitation–contraction coupling, interaction with ion channels and 14-3-3 proteins
RHEB	Cell growth (TOR pathway)
Rho	Control of cytoskeleton dynamics and integrin activity, formation of stress fibers and focal adhesion complexes, initiation of DNA synthesis, activation of gene expression, regulation of progression of cell cycle G1 phase
RIN	Neuron development and trophicity
RIT	Neuron differentiation, axonal and dendritic growth

but 8 are atypical Rho GTPases. *Classical Rho GTPases* correspond to CDC42, RAC, RHO, RHOD, and RHOF families. *Atypical Rho GTPases* comprise RhoBTB, RhoH, RhoJ, RhoQ, RhoU, RhoV, and Rnd GTPases. These GTPases are either predominantly GTP-bound (e.g., Rnd and RhoH) or have an increased nucleotide exchange (e.g., RhoU2).

Proteins of the RHO, RAC, and CDC42 families of the RHO superfamily intervene in the reorganization of the actin cytoskeleton as mediators of extracellular signals for formation of stress bundles and focal adhesions (RhoA), membrane ruffles and lamellipodia (Rac1), and filopodia or microspikes (CDC42) [832].

Table 9.18. Some effectors of small GTPases CDC42, Rac1, and RhoA and corresponding effects (Source: [833]; Dia: diaphanous formin; IQGAP: IQ motif-containing GTPase-activating protein; MRCK: myotonic dystrophy kinase-related CDC42-binding kinase nWASP: neural Wiskott-Aldrich syndrome protein P67PhOx: P67 phagocyte oxidase protein; PAK: P21-activated kinase; Par6: partitioning defective-6; PKN: protein kinase-N; PLC: phospholipase-C; RoCK: Rho-associated coiled-coil-containing protein kinase SRA1: specifically Rac1-associated protein-1; WAVe: WASP family verprolin-homologous protein). CDC42 is involved in organelle positioning; Rac1 in cytoskeleton polarization, microtubule stabilization, formation of tight junctions and cell protrusions; RhoA in formation of actin stress fibers and focal adhesions, membrane retraction, and microfilament contraction.

Effector	Effect
	Small GTPase CDC42
IQGAP	Intercellular adhesion, capture of microtubule
MRCK	Actomyosin-based contraction, kinase
nWASP	Actin polymerization
PAK	Actin polymerization, microtubule stabilization, kinase
Par6	Cell polarity
	Small GTPase Rac1
IQGAP	Intercellular adhesion, capture of microtubule
P67PhOx	NADPH oxidation
PAK	Actin polymerization, microtubule stabilization, kinase
Par6	Cell polarity
SRA1	Actin polymerization
WAVe	Actin polymerization
	Small GTPase RhoA
Citron	Cytokinesis, kinase
PKN	Cell cycle, kinase
PLC	Cell signaling
RoCK	F-actin stabilization, kinase
Dia	Actin polymerization, microtubule stabilization

They have additional roles. Members of the RHO family operate during progression through G1 phase of the cell division cycle. Transcription activation by serum response factor in response to lysophosphatic acid requires RhoA GTPase. Proteins CDC42 and Rac1 are able to induce sequential phosphorylations (activation) of Jun N-terminal kinases. Small Rac GTPases participate in the activation of NADPH oxidases in neutrophils and macrophages.

Because small GTPases of the RHO superfamily of the RAS hyperfamily activate various mediators (actin nucleators, protein kinases, phospholipases, and scaffold proteins), they can regulate the cytoskeleton dynamics and thus act in various cell activities that require changes in the cell cytoskeleton, such as intracellular vesicle transport and cell polarity, shape, adhesion, motion, division, and differentiation. About 70 Rho GTPase effectors have been identified in addition to the high number of regulators (Table 9.18).

9.3 Monomeric (Small) GTPases

During cell migration, Rho GTPases drive membrane protrusion at the leading edge and contractility of the cell body. Rho GTPases indeed regulate the assembly of filamentous actin (^{F}actin) in response to signaling. Their effectors induce the assembly of contractile actin–myosin filaments (stress fibers in particular) and integrin-containing focal adhesions. Consequently, small Rho GTPases act in vascular processes, such as smooth muscle cell (SMC) contraction, cell adhesion, endothelial permeability, leukocyte extravasation, platelet activation, and migration of smooth muscle cells and endothelial cells involved in angiogenesis and wall remodeling [834].

Small Rho GTPases can be activated by: (1) G-protein-coupled receptors,[62] (2) Tyr kinase receptors, and (3) cytokine receptors. In particular, growth factors recruit Rho, Rac, and ERM.[63] Activation of Rho GTPases increases their membrane-associated level and decreases their cytosolic concentration.

Members of the RHO superfamily, such as Rac and CDC42, activate the JNK pathway. Activated JNKs, in turn, phosphorylate (activate) transcription factors Jun and ELk1 that boost transcription from genes with AP1 and SRE reponse elements, respectively. Activated RhoA does not act on JNK, but targets the transcription factor serum response factor that cooperates with ELk1 and Fos to stimulate transcription from promoters containing SRE elements. In addition, Rho, Rac, and CDC42 target the transcription factor nuclear factor-κB.

Small Rho GTPases are involved in vascular disorders associated with pathological remodeling and altered cell contractility. Kinase RoCK, an effector of the small Rho GTPase, is involved in atherosclerosis as well as in post-stenting restenosis. Ezrin, radixin, and moesin of the ERM family are phosphorylated by RoCK [836]. Small Rho GTPase also controls other cellular activities [837]. Small Rho GTPase regulates several enzymes involved in phospholipid metabolism (phospholipase-D and phosphatidylinositol kinase). It controls delayed rectifier K^+ channels.

The activation–inactivation cycle of Rho GTPases is a regulated process. Activation of GTPases into a GTP-bound conformation is controlled by specific *guanine nucleotide-exchange factors* (GEF; Sect. 9.4.1). Conversely, GTP is hydrolyzed to GDP by GTPase in combination with *GTPase-activating proteins* (GAP; Sect. 9.4.2). In the absence of signaling, the major fraction of Rho^{GDP} GTPases is sequestered in the cytosol by *guanine nucleotide-dissociation inhibitors* (GDI; Sect. 9.4.3). In mammals, about 70 RhoGEFs, 60 RhoGAPs, and 3 RhoGDIs have been detected. Certain members of the RHO superfamily are able to regulate activity of other members. Small CDC42 GTPase can activate Rac1; Rac1 can inhibit RhoA GTPase.

Activity of small GTPases Ras and Rho is not only regulated by GEFs, GAPs, and GDIs for Rho GTPases, but also by post-translational modification such as

---

[62]GPCR ligands, such as thrombin, endothelin, prostaglandin-E2, angiotensin, α-adrenergics, sphingolipids, etc., activate Rho GTPases.

[63]Ezrin, radixin, and moesin are involved in membrane recruitment of Rho GTPases. In endothelial cells, RhoA colocalizes with ERM proteins [835].

isoprenylation by farnesyl and geranylgeranyl transferases to promote membrane anchorage and subsequent effector association.

Except Ran GTPases, all of small GTPases of the RAS superfamily are post-translationally modified. Small ARF GTPases are myristoylated, Rab GTPases and members of the RAS and RHO superfamilies attach 1 or 2 farnesyl or geranylgeranyl groups. Conversely, Rab and Rho GTPases are sequestered in the cytosol by GDIs that extract GTPases^{GDP} from the membrane by binding to their prenylated motifs [832].

9.3.1 Superfamily of ARF GTPases

Adpribosylation factors[64] (ARF) constitute a superfamily of ubiquitous proteins of the RAS hyperfamily of small GTPases. These myristoylated GTP–GDP switch proteins are engaged in the regulation of vesicular transport in cells. Their activity is controlled by a cycle of successive GTP binding and hydrolysis, i.e., a cycle of activation and inactivation, using ARF guanine nucleotide-exchange factors and GTPase-activating proteins, respectively. They localize to cellular membranes (plasma membrane as well as membranes of the secretory vesicles, endosomes, and lysosomes).

Small ARF GTPases recruit coat proteins for cargo sorting, enzymes to adapt the lipid composition of membranes (e.g., phosphatidylinositol kinases), and cytoskeletal components for motion (Tables 9.19 and 9.20). They interact with regulators of other guanine nucleotide-binding proteins and form molecular platforms constituted by ARF^{GDP}, GEFs, GAPs, and effectors [838]. They can act simultaneously or successively at membranes of the endoplasmic reticulum and Golgi body and at the plasma membrane. Regulators of the ARF network also integrate ARF activities with other GTPase signaling axes. Both ArfGEFs and ArfGAPs can serve as scaffolding effectors, recruiting signaling mediators and promoting conformational changes for appropriate binding and activation.

9.3.1.1 Families of the ARF Superfamily

The superfamily of ARF GTPases includes several families constitued by: (1) ARF isoforms (ARF1–ARF6);[65] (2) secretion-associated and Ras-related protein SAR1;

---

[64]The name of these monomeric GTPases comes from their ability to act as cofactors for cholera toxin-catalyzed adpribosylation of heterotrimeric G-protein subunit Gα_s. Six ARF proteins (ARF1–ARF6) operate as cofactors that stimulate the adpribosylating activity of cholera toxin. Yet, the function of ARFs does not involve adpribosylation.

[65]Subtypes ARF1 to ARF5 localize to the Golgi body, whereas ARF6 resides at the cell surface.

Table 9.19. Partners and interactors of ARF GTPases. The ARFs regulators recruit: (1) coat proteins that support sorting of cargo into vesicles; (2) enzymes that change lipid composition of membranes; and cytoskeletal factors (**Part 1**; Source: [838] ERGIC: endoplasmic reticulum–Golgi intermediate compartment; AP: adaptor proteic complex; BBSome: Bardet-Biedl syndrome coat complex (transport of membrane proteins into cilium); CerT: ceramide-transfer protein; CoP: coatomer protein; GGA: Golgi-localized, γ-ear-containing, ARF-binding protein; GCC: GRIP and coiled-coil domain-containing protein; GMAP: Golgi-associated microtubule-binding protein; PLD, phospholipase D; PI4K: phosphatidylinositol 4-kinase; PI4P5K: phosphatidylinositol 4-phosphate 5-kinase). The Golgi-associated retrograde protein (GARP) complex, or vacuolar protein sorting (VPS)-53 (fifty-three [VFT]) tetramer (VPS51–VPS54) is required for the fusion of early and late endodomes with the trans-Golgi network. Pleckstrin homology (PH) domain-containing family-A member PlekHa8, a phosphoinositol 4-phosphate adaptor, also called four-phosphate adaptor protein FAPP2, binds phosphatidylinositol 4-phosphate and ARF1 GTPase and serves as glucosylceramide-transfer protein.

Effector	ARF	Location
Coat complexes		
AP1	ARF1/3	Endosomes, trans-Golgi network
AP3	ARF1/3	Endosomes, trans-Golgi network
AP4	ARF1/3	Trans-Golgi network
BBSome	ARL6	Plasma membrane
CoP1	ARF1/3	Golgi body, ERGIC
CoP2	SAR1	Endoplasmic reticulum exit
GGA1/2/3	ARF1/3	Endosomes, trans-Golgi network
Lipid-processing enzymes		
CerT	ARF1	Golgi body
PlekHa3/8	ARF1	Golgi body
PI4K	ARF1	Golgi body
PI4P5K	ARF1	Plasma membrane
PLD	ARF1–6, ARL1	Plasma membrane
Tethers		
Exocyst	ARF6	Plasma membrane
GARP	ARL1	Endosomes, trans-Golgi network
GMAP210	ARF1	Cis-Golgi network
Golgin-A1/4, GCC1/2	ARL1	trans-Golgi network

(3) ARF-related protein ArfRP1;[66] (4) ARF-like GTPases (ARL1–ARL22);[67] and (5) adpribosylation factor domain-containing protein-1 (ARD1 or ARFD1) that

---

[66] Alias ARP1 used to designate ARF-related protein should be avoided, as ARP means actin-related protein.

[67] Small ARL GTPases are related to ARFs structurally, but not functionally or phylogenetically.

Table 9.20. Partners and interactors of ARF GTPases (**Part 2**; Source: [838]; arfaptin: ARF-interacting protein; BART: binding partner binder of ARF-like protein-2 (two; ARL2); FIP: family of Rab11-interacting protein; MAPK8IP: MAPK8-interacting protein; NDPKa: nucleoside diphosphate kinase-A; PDE: phosphodiesterase; RhoGAP: Rho GTPase-activating protein; SCoC: short coiled-coil protein; Unc: uncoordinated).

Effector	ARF	Location
Regulators		
Cytohesin	ARF6, ARL4	Plasma membrane
RhoGAP21	ARF1/6	Golgi body, plasma membrane
Scaffolds		
FIP3/4	ARF5/6	Recycling endosomes
MAPK8IP3/4	ARF6	Endosomes
Tubulin folding chaperone		
β-Tubulin cofactor-D	ARL2	Cytosol
Miscellaneous		
Arfaptin1/2	ARF1, ARL1	Golgi body, trans-Golgi network
BART2	ARL2	Mitochondria, nucleus
NDPKa	ARF6	Plasma membrane, cell junctions
PDE6δ	ARL2/3	
SCoC	ARL1	Golgi body
Unc119	ARL2/3	

corresponds to tripartite motif-containing protein TriM23,[68] a ubiquitin-protein ligase and a GTP-binding protein.[69]

Small ARF GTPases can be classified into 3 subfamilies: subfamily 1 with ARF1 to ARF3 that regulate the assembly of coat complexes onto budding vesicles and activate enzymes that target lipids; subfamily 2 with ARF4 and ARF5 that may operate in Golgi transport and recruitment of coat components to trans-Golgi membranes; and subfamily 3 with ARF6 that controls endosomal-membrane trafficking. Isoform ARF2 is absent in humans [839].

---

[68] A.k.a. RING finger protein-46 (RNF46).

[69] This protein should not be confused with N-terminal acetyltransferase complex ArD1 (cell division-arrest defective) catalytic subunit homologs (ArD1a and ArD1b, also called $^N\alpha$-acetyltransferase-10 and -11).

9.3.1.2 ARF Compartmentation and Function

Adpribosylation factors are characterized by their subcellular compartmentation and binding partners that dictates the function of each member.[70] However, the majority of ARF effectors can interact with several ARF GTPases. All ARF GTPases are myristoylated (second N-terminus Gly), thus allowing ARF tethering to membranes. Activation–deactivation cycle of ARFs, hence association–dissociation of transport vesicle coat proteins, are regulated by ArfGEFs (Sect. 9.4.1.1) and ArfGAPs (Sect. 9.4.2.1).

The primary role of ARF GTPases is the regulation of vesicular membrane transfer, especially the formation of coated vesicles. The ARF proteins actually regulate structure, budding from donor membrane, motion, and fusion with acceptor membrane of vesicles devoted to intracellular transport of cargos, as they recruit coat proteins, prime proteic complex assembly, modulate phospholipid metabolism, contribute to actin remodeling, especially at the cell cortex, and participate in some signaling pathways. Therefore, ARF GTPases are involved in endo- and exocytosis, phagocytosis, cytokinesis, and cell adhesion and migration. Small ARF GTPases not only regulate vesicular motions, but also organelle structure.

Small ARF GTPases execute their function by anchoring to membrane surfaces, where they interact with other proteins to initiate budding and maturation of transport vesicles. They are mainly involved in the vesicular transport of lipids and proteins between the endoplasmic reticulum, where cargos are synthesized, and the Golgi body, where cargos undergo post-translational modification (Table 9.21 and 9.22).

Small ARF GTPases contribute to morphology and location of the Golgi body that depends on the cell cytoskeleton as well as actin cytoskeleton-dependent changes in cell morphology. Golgi-associated ARF1 is involved in the formation of focal adhesions via the delivery of adaptor paxillin to the plasma membrane and Rho activation. Small ARF6 GTPase facilitates recruitment of active Rac to the plasma membrane. Small Rac GTPase alone can promote the migration of epithelial cell sheets during wound healing, but not cell scattering that requires cooperation of ARF6 [841].

---

[70] ARF1 regulates the early secretory pathway, from the Golgi body to the endoplasmic reticulum and between Golgi cisternae, recruiting coat protein complex-1, clathrin, and adaptor proteins (AP1, AP3, and AP4) [840]. ARF1 also interacts with membrin and SNAREs. ARF6 is located at the plasma membrane and endosomal compartments, where it regulates endocytic membrane trafficking and actin remodeling. It recruits clathrin and AP2. ARF6 acts on phosphatidylinositol 4-phosphate 5-kinase and phospholipase-D, for production of phosphatidylinositol (4,5)-bisphosphate.

Table 9.21. Monomeric ARF and ARL GTPases in the secretory pathway (Source: [838]; ERGIC: endoplasmic reticulum–Golgi intermediate compartment; ADRP: adipose differentiation-related protein, or adipophilin; AP: adaptor protein complex; ATGL: adipose triglyceride lipase; CAPS: calcium-dependent activator protein for secretion; CoP: coat protein of the coatomer complex; GGA: Golgi-localized, γ-ear-containing, ARF-binding protein PI4K: phosphatidylinositol 4-kinase).

Site	Involved molecules
Endoplasmic reticulum	SAR1, Sec12 (SAR1GEF), Sec23–Sec24 (SAR1GAP), CoP2
ERGIC	ARF1, GBF1 (ARF1GEF), CoP1, CoP2, ADRP, ATGL (formation of lipid droplets)
Cis-Golgi network	ARF1, ARF4, GBF1 (recruited by Rab1 and PI4P), Cert, PlekHa8 (lipid transport) PI4K
Medial-Golgi network	
Trans-Golgi network	ARF3
Trans-Golgi network	ARF3 (constitutive exocytosis), ARF4, ARF5 (regulated secretion), ARL1 (recruitment of golgins and arfaptin), BIG1, BIG2 (ArfGEF; recruited by PDE3a), AP1, CoP1, GGA, (endosome and lysosome), CAPS (regulated secretion)

9.3.1.3 ARF Structure

The ARF proteins possess a 2-domain structure [842]. The myristoylated N-terminal helix used for membrane binding[71] is separated from the C-terminus by a flexible linker. This linker may yield a certain degree of adaptability in binding modes for the huge quantity of ARF-interacting proteins, in addition to specific binding sites for lipids on some of these small GTPases.

Exchange of GDP by GTP occurs at the ARF myristoyl binding site. Active ARFs promote and stabilize curved surfaces of budding vesicles [842]. C-terminus positioned on the membrane surface provides an interaction structure for activators, adaptors, effectors, and inhibitory GAP proteins. At least certain ARF types (e.g., ARF4 to ARF6) remain bound to membranes in their GDP-bound conformation.

---

[71] The myristoylated N-terminal helix is inserted into membrane upon GTP binding. Unlike other small GTPases of the RAS superfamily, which have a long C-terminal linker with a lipid membrane anchor, thereby being located at some distance from the membrane, ARFs are very close to the membrane surface.

Table 9.22. Small ARF and ARL GTPases at the plasma membrane and in endocytosis (Source: [838]; IPCEF: interaction protein for cytohesin exchange factor).

Type	Involved partners and role
ARF1	Clathrin-independent endocytosis
ARF6	Cortical actin cytoskeleton structuring, endosomal membrane recycling, Rac, DOCK1, IPCEF (protrusion), AP2, clathrin (endocytosis), PI4P5K, PLD (clathrin-coated pit formation, sorting endodome), MAPK8IP4 (microtubule motor adaptor for rapid recycling), exocyst RhoGEF24, Rac, RhoG (actin dynamics), cytohesin (cell adhesion assembling), BRAG2, Rac, SOS1 (adherens junction disassembly)
ARL2	Regulation of microtubule-based motion, ELMOD2
ARL3	Ciliogenesis, intraciliary transport RP2
ARL4	Clathrin-independent endocytosis, cytohesin
ARL6 (BBS3)	Ciliogenesis, intraciliary transport BBSome
ARL8	Late endosome (fusion with lysosome)
ARL13	Ciliogenesis, intraciliary transport

9.3.1.4 ARF Partners

To function, ARFs must interact sequentially or simultaneously with: (1) guanine nucleotide-exchange factors and lipids that catalyze their activation; (2) proteic adaptors that modulate recruitment of cargo into nascent buds; (3) lipid-modifying enzymes; and (4) GTPase-activating proteins, the 2 latter categories being required for budding and maturation of transport carriers.

Partners of ARFs can be grouped into 3 categories: (1) ArfGEFs that have a Sec7 domain (Sect. 9.4.1.1);[72] (2) ArfGAPs that possess a cysteine-rich ArfGAP domain (Sect. 9.4.2.1);[73] and (3) effectors and ARF-binding lipids, either non-specific partners that promote nucleotide-exchange, or specific partners, such as PI(4,5)P$_2$ and phosphatidic acid [839]. Several GEFs and GAPs interact with ARF proteins. Proteins of the ArfGEF and ArfGAP sets, like ARF, can translocate to

---

[72] In humans, 15 ArfGEFs have been identified. They are classified into 6 categories.

[73] In mammals, 31 ArfGAPs have been detected. They are categorized into 9 sets.

membranes in a regulated fashion. Nucleotide-sensitive partners comprise mitotic kinesin-like protein MKlP1, arfaptin-1 and -2, and arfophilins.[74] In addition to arfophilins, JNK-interacting proteins JIP3 and JIP4 (MAPK8IP3 and MAPK8IP4) that also selectively bind ARF6 mediate postendocytic recycling during cytokinesis. All ARF GTPases share several effectors, such as phospholipase-D and several phosphoinositide kinases (PI4P5Kα–PI4P5Kγ).

9.3.1.5 ARF GTPase Isoforms

Soluble ARFs (ARF1–ARF5) are regulators of vesicular transport, particularly to and from the Golgi body, and endosomes, as well as activators of phospholipase-D. Active ^{GTP}ARF1 to ^{GTP}ARF5 translocate onto membranes where they recruit: (1) ARF-dependent coat proteins AP1, AP3, AP4, GGA13, and CoP1; (2) scaffolding and trafficking Munc18-interactors MInt1 (a.k.a. amyloid-β A4 precursor-binding protein APBa1 and neuron-specific adaptor X11α), MInt2 (a.k.a. APBa2, X11β, and X11-like protein [X11l]), and MInt3 (a.k.a. APBa3, X11γ, and X11-like-2 protein X11l2);[75] and (3) lipid enzymes, such as phospholipase-D, phosphatidylinositol 4-kinase-3, and phosphatidylinositol 4-phosphate 5-kinase [844].

ARF1

Protein ARF1 regulates the membrane association of clathrin-associated adaptor proteic complex AP1 and AP3 used for trans-Golgi network (TGN)-to-endosome and endosome-to-lysosome transport. In its active state, it recruits coat and adaptor proteins, such as clathrin, AP1, and AP3 to the membrane and induces the assembly and nucleation of coated vesicles.

In addition to its myristoylated N-terminus that promotes membrane tethering, inactive, cytosolic ARF1^{GDP} attaches to membranes of the Golgi body and

---

[74]Dual Rab11–ARF-interacting proteins arfophilin-1 and -2 are also called members of family of Rab11-interacting proteins FIP3 and FIP4. They function in the delivery of recycling endosomes to the cleavage furrow. They selectively bind ARF6 GTPAse.

[75]Munc18 (mammalian homolog of uncoordinated mutant Unc18) that is also called syntaxin-binding protein StxBP1 is a regulator of Ca^{++}-regulated exocytosis in neurons and neuroendocrine cells as well as platelets, adipocytes, etc. Mammalian MInt proteins are homologs of Lin10 in Caenorhabditis elegans. In conjunction with Ca^{++}–calmodulin-dependent Ser/Thr protein kinase-3, or membrane-associated guanylate kinase-2 (CaSK, CaMGuK2, or Lin2 homolog) of the MAGUK family, and Lin7 (a.k.a. vertebrate Lin7 homolog [VeLi] and mammalian Lin-Seven homolog MaLS1), MInt proteins that form MInt–CaSK–Lin7 complexes are required for a precise targeting and localization of certain membrane proteins. The MInt proteins bind to the Munc18–syntaxin complex involved in synaptic vesicle fusion. In the central nervous system, MInt substances are scaffolding and trafficking proteins that associate via Disc large homolog DLg1 (or synapse-associated protein SAP97) with inward rectifier potassium channels K$_{IR}$2.1 to K$_{IR}$2.3, which control cell excitability [843].

9.3 Monomeric (Small) GTPases

endoplasmic reticulum owing to transmembrane trafficking proteins, transmembrane EMP24 domain-containing trafficking proteins TMED2, TMED3, and TMED10[76] on the one hand and *membrin*, a vesicular fusion mediator SNARE as well as a component of CoP1-mediated coats on the other [840].

Proteins GBF1 and BIG2 (Sect. 9.4.1.1), GEFs for ARF1, stimulate ARF1 in the cis-Golgi network. Upon activation, ARF1^{GTP} dissociates from transmembrane Golgi cargo receptors TMEDs, thereby yielding binding sites for coatomers that can then prime vesicle coat assembly [840]. Active ARF1^{GTP} indeed recruits pre-assembled heptameric CoP1 coatomer complex and controls, together with coatomers, vesicle budding. CoP1-coated vesicles mediate: (1) retrograde transport from the cis-Golgi network to the endoplasmic reticulum[77] and (2) retrograde and possibly anterograde, transport between Golgi cisternae. Active ARF1 is also involved in clathrin-coated vesicle formation at the trans-Golgi network for exocytosis to the plasma membrane or cell organelle membranes. Clathrin-coated vesicles transport cargos from the trans-Golgi network. Active ARF1^{GTP} promotes recruitment of heterotetrameric adaptor proteic complexes AP1, AP3, and AP4, as well as monomeric Golgi-localized γ-ear-containing ARF-binding proteins (GGA) from the cytosol onto membranes. Active ARF1 also favors assembly of spectrin and actin on Golgi membranes.

Conversely, ArfGAP1 bound to ligand-coupled KDEL (Lys–Asp–Glu–Leu)-containing receptors and packaged into budding vesicles is activated by coatomers and membrane curvature. It then hydrolyzes GTP on ARF1, thereby dissociating CoP1 from vesicles and eliciting cargo packaging, as CoP1-coated vesicles are depleted of cargo. Whereas GTP hydrolysis leads to CoP1-coated vesicle uncoating, it is not sufficient to induce clathrin-mediated coat disassembly [840].

Activated ARF1 can control Golgi membrane–spectrin connection that contributes to the maintenance of Golgi body organization via phosphatidylinositol

---

[76]CoP1-coated vesicles that bud off the Golgi body contain 2 major transmembrane proteins, i.e., Golgi cargo receptors, that travel between the intermediate compartment and the Golgi body: transmembrane EMP24 domain-containing proteins (TMED). In Saccharomyces cerevisiae, 2 molecules encoded by the Emp24 (EMP24: endomembrane protein of 24 kDa) and Erv25 (ERV25: endoplasmic reticulum vesicle protein of 25 kDa) genes, members of the budding yeast P24 family (EMP24, ERV25, and ERP1–ERP6), heterodimerize for an efficient transport of target proteins from the endoplasmic reticulum to the Golgi body. They localize to both CoP1+ and CoP2+ vesicles as well as endoplasmic reticulum and Golgi membranes. Proteins Emp24 and Erv25 complex. Furthermore, Emp24 and Erv25 depend on each other for stability and incorporation into CoP2-coated vesicles. Type-1 transmembrane protein Emp24 is identical to 24-kDa protein (P24). Transmembrane trafficking proteins TMED2 and TMED3 correspond to membrane proteins P24α and P24β, respectively. Protein P24α is also called transmembrane Emp24-domain-containing trafficking protein TMEδ2. Proteins TMED3 and TMED10 complex. Protein TMED10 is also named P23, P24δ, and transmembrane protein TMP21. In humans, 2 TMP21 variants exist — TMP21-1 and TMP21-2; TMP21-2 is transcribed, but not translated [846]. Protein TMED10 thus corresponds to TMP21-1.

[77]Newly synthesized membrane proteins and lipids are conveyed from the endoplasmic reticulum to the Golgi body.

(4,5)-bisphosphate levels on Golgi membranes, as it causes activation of type-3 phosphatidylinositol 4-kinase[78] and type-1 phosphatidylinositol 4-phosphate 5-kinase [840].[79] Protein kinase-D is also recruited to the trans-Golgi network, where it causes the fission of vesicles. All PKD isoforms (PKD1–PKD3) phosphorylate PI4K3β.

Protein ARF1 promotes actin assembly on Golgi membranes that favors vesicular budding, as it facilitates ARP2/3-dependent actin polymerization via CDC42 and its effector nWASP [840].[80] In addition, ARF1 regulates exocytosis from the Golgi body via the cortactin–dynamin-2 complex. Cortactin and dynamin-2 are involved in actin remodeling and stabilization via the ARP2–ARP3 actin-nucleating complex and vesicle scission, respectively.

ARF3

Adpribosylation factor-3 interacts with adpribosylation factor-interacting proteins-1 (ARFIP1) and -2 (ARFIP2); Golgi-associated, γ-adaptin ear-containing, ARF-binding proteins GCA1 to GCA3; and kinesin-like protein-5 (a.k.a. KiF23 and mitotic kinesin-like protein MKLP1) [251].

ARF4 or ARF2

Adpribosylation factor-4 that belongs to the ARF subfamily-2 interacts with epidermal growth factor receptor [251].

ARF5

Adpribosylation factor-5 interacts with adpribosylation factor-interacting proteins-1 (ARFIP1) and -2 (ARFIP2) as well as β1 subunit of the adaptor-related protein complex-3 (AP3β1) [251].

ARF6

Whereas ARF1 is involved in Golgi membrane dynamics, ARF6 localizes to the plasma membrane and endosomes, where it regulates endocytic membrane

---

[78] In mammals, PI4K3α isoform is predominantly expressed in the brain, whereas PI4K3β is more spread.

[79] All isoforms PI(4)P5K1α, PI(4)P5K1β, and PI(4)P5K1γ synthesize PI(4,5)P$_2$.

[80] Protein ARF1 facilitates the recruitment of CDC42^{GTP} via the γ-subunit of the coatomer. The CDC42 effector nWASP activates the ARP2–ARP3 actin-nucleating complex.

trafficking and molecule sorting in the absence of adaptor complex AP2 and clathrin, as well as actin remodeling at the cell cortex via Rac1 GTPase and phospholipids [840]. In vitro, ARF6 binds directly to the β1 subunit of the AP1 complex and both the β3 and δ subunits of the AP3 complex, but not to the AP2 complex [847]. However, ARF6 stimulates the local activity of PI(4)P5K1γ, thereby promoting AP2 assembly at the membrane. In vitro, ARF6 may recruit the AP2 (but not AP1) complex onto liposomal membranes.

Protein ARF6 regulates clathrin-dependent and -independent endocytosis, endosome recycling, and actin reorganization. It operates in highly dynamical events at the cell surface, such as phagocytosis, between-cell adhesion, and cell migration.

In cell migration, ARF6 promotes the activation of Rac necessary for the formation of lamellipodia at the leading edge and mediates the recycling of integrins [847].

Protein ARF6 activates phosphatidylinositol 4-phosphate 5-kinase-1γ and phospholipase-D. The former synthesizes $PI(4,5)P_2$ that regulates clathrin-mediated endocytosis. The latter produces phosphatidic acid, a cofactor in PI(4)P5K activation [840].

Protein ARF6 also regulates clathrin- and caveola-independent endocytosis of various subtances such as major histocompatibility complex class-1 proteins and peripheral myelin-membrane protein PMP22 [840]. Activated ARF6 also favors arrestin dissociation to facilitate internalization of G-protein-coupled receptors. All these molecules are delivered to sorting endosomes with possible plasmalemmal recycling that needs phospholipase-D.

Among transfer of G-protein-coupled receptors, ARF6 controls endocytosis of β-adrenoceptor, angiotensin AT_1, vasopressin V_2, and endothelin receptors. It also supports neurotransmitter-triggered internalization of AMPA-type glutamate receptors [847].

Small ARF6 GTPase also regulates postendocytic transport (e.g., transferrin recycling) [847]. Several transmembrane and glycosyl-phosphatidylinositol (GPI)-anchored cargos enter an ARF6+ endosomes (e.g., β1-integrins, MHC class-1 molecules, complement regulator CD59,[81] syndecans, SNAP25, mucolipin-2, GluT1 and GluT4, T-cell surface glycoprotein CD1, epican (CD44; Indian blood group), CD55 (decay accelerating factor for complement; Cromer blood group), basigin (Ok blood group; or CD147), SLC3a (activators of dibasic and neutral amino acid transport; CD98), and ICAM1 adhesion molecule [847]. Membrane raft components such as GM1 ganglioside are also recycled to the plasma membrane using ARF6 GTPase.

Several effectors bound by active ARF6 intervene in the postendocytic transfer, such as vacuolar ATPase and a component of the exocyst complex Sec10 that mediates docking and fusion of carrier vesicles with the plasma membrane [847].

Protein ARF6 promotes actin remodeling, as it influences lipid metabolism, links to partner of Rac1 POR1 and arfaptin-2, and modulates Rac1 activity. In epithelial

---

[81] A.k.a. membrane attack complex inhibition factor (MACIP), and protectin.

Table 9.23. Members of the ARL family.

Protein	Human genes
ARL1	Arl1
ARL2	Arl2
ARL3	Arl3
ARL4a, ARL4c–ARL4d	Arl4A, Arl4C, Arl4D
ARL7	Arl4C
ARL5, ARL5a–ARL5b	Arl5A–Arl5B
ARL6	Arl6
ARL8a–ARL8b	Arl8A–Arl8B
ARL9	Arl9
ARL10	Arl10
ARL11	Arl11
ARL13a–ARL13b	Arl13A–Arl13B
ARL2L1	Arl13b
ARL14	Arl14
ARL15	Arl15
ARL16	Arl16
ARL17	Arl17 (Arl17B)
ARL18	ARFRP1

cells, ARF6 promotes the internalization of E-cadherin, hence causing disassembly of adherens junctions [840]. Protein ARF6 may promote endocytosis of E-cadherin via nucleotide-diphosphate kinase (NDK) [847]. During mitosis, activated ARF6 localizes to the cleavage furrow and acts via arfophilins and the exocyst complex.[82]

Among 15 guanine nucleotide-exchange factors ArfGEFs, 12 can activate ARF6, such as 4 cytohesins[83] that may act on both ARF6 and ARF1, 3 BRAGs, and 3 EFA6s [847]. Among ArfGAPs that regulate ARF6 activity, ACAP1, ACAP2, and ARAP3 may be specific, whereas others, such as APAP1 and APAP2 act on both ARF6 and other ARFs [847].

9.3.1.6 Family of ARL Regulatory GTPases (ARF Superfamily)

Adpribosylation factor-like proteins are members of the ARF superfamily. The ARL family include 24 members (Table 9.23).

Generally, ARL proteins bind to cell membranes using an N-terminal helix that is inserted into the lipid bilayer once activated. In addition, this N-terminal helix contains a myristoyl or an acetyl group.

---

[82] Arfophilin-1 and -2 form complexes with Rab11 and ARF6 GTPases.

[83] I.e., ARNO, or PSCD2, ARNO2, or PSCD1, ARNO3, or PSCD3, and ARNO4, or PSCD4.

ARL1

Ubiquitous ARF-like protein-1 (ARL1) is enriched on vesiculotubular structures on the side of the Golgi body [848]. Binding of ARL1 to membranes depends on co-translational N-myristoylation. Agent ARL1^{GDP} is cytosolic, whereas ARL1^{GTP} is anchored to membranes, especially those of the trans-Golgi network.

ARL2

Ubiquitous ARF-like protein-2 (ARL2) is involved in microtubule dynamics and stability [849]. Isoform ARL2 participates in folding of tubulin and/or tubulin heterodimer assembly by binding to tubulin-specific cochaperone cofactor-D and protein phosphatase-2. Protein ARL2 associates with binder of ARL2 (BART) and adenine nucleotide transporter ANT1 at the inner mitochondrial membrane.

ARL4

Three ARL4 proteins exist. Proteins ARL4a, ARL4c, and ARL4d differ from other members of the ARF superfamily, as they possess a basic C-terminal bipartite nuclear localization signal [850]. Small GTPase ARL4d is similar to ARL4a and ARL4c (60% and 58% identity, respectively). All the ARL4 GTPases recruit cytohesin ArfGEFs to the plasma membrane.

Adpribosylation factor-like protein-4d (ARL4d)[84] resides in the nucleus and cytoplasm. Localization of ARL4d at the plasma membrane depends on both GTP binding and N-terminal myristoylation. Human Arl4D mRNA is predominantly detected in the kidney, esophagus, testis, and uterus.

ARL8

The ARL8 subfamily contains 2 isoforms (ARL8a–ARL8b). These proteins contain an N-terminal helix and GTP-binding domains (G1-G5). Both ARL8a and ARL8b localize to lysosomes. Both ARL8a and ARL8b cannot be myristoylated. However, they can be acetylated.

Ubiquitous ARL8a and ARL8b acts mainly in the delivery of endocytosed macromolecules to lysosomes, lysosome motility, chromosome segregation, and axonal transport of presynaptic cargos [851]. In addition, ARL8b associates with β-tubulin.

---

[84]Previously named ARF4L, ARL4L, ARL5, ARL6, and ARL9 [850].

ARL13

Ubiquitous ARF-like protein-13b (ARL13b) is required for the formation and/or maintenance of cilia, hence associated with sonic Hedgehog signaling [845]. Isotype ARL13b is able to self-associate. Two alternatively spliced transcripts produce long and short isoforms.

9.3.1.7 ArfRP GTPase (ARF Superfamily)

Adpribosylation factor-related protein ArfRP1[85] is another Ras-related, membrane-associated, monomeric GTPase with some degree of similarity with ARF molecules. Protein ArfRP1 has an unusual feature for an ARF superfamily member: it has a rather high intrinsic GTPase activity. It is mainly associated with the trans-Golgi network [852]. This regulator works synergistically with ARL1 as well as golgin-97 and -245 on Golgi membranes.

In addition, ArfRP1 associates with the plasma membrane. Protein ArfRP1 connects to ArfGEF cytohesin-1 and -2 (Sect. 9.4.1.1. Protein ArfRP1^{GTP} stimulates phospholipase-D to produce phosphatidic acid, a messenger for vesicle formation and trafficking [853]. Phospholipase-D1a and -D1b reside in the perinuclear region, whereas PLD2 is attached to the plasma membrane and is less responsive to ARFs than PLD1 enzyme. Phospholipase-D2 can be stimulated from activated receptor Tyr kinase and G-protein-coupled receptors that prime ARF-, Rho-, and/or PKC-dependent signaling.

9.3.1.8 ARF Domain-Containing Protein ARD1 (ARF Superfamily)

Adpribosylation factor domain-containing protein-1 (ARD1) that pertains to the ARF superfamily of small GTPases corresponds to ubiquitin-protein ligase TriM23 of the tripartite motif (TRIM) family. Members of the TRIM family are implicated in gene transcription, signal transduction, vesicular transport, antiviral defense,[86] phospholipase-D activation, and protein degradation via ubiquitination. Three alternatively spliced transcript variants have been detected in humans (ARD1α–ARD1γ).

---

[85]One should avoid alias ARP1 because ARP is used for actin-related protein.

[86]Ubiquitination by ARD1 of NFκB essential modulator (NEMo), or IKKγ regulatory subunit, is used during antiviral innate and inflammatory responses launched by Toll-like receptor TLR3.

The TRIM motif includes 3 zinc-binding domains, a RING, B-box type 1 and type 2, and a coiled-coil region. Its C-terminus contains an adpribosylation factor (ARF or P3) domain and a guanine nucleotide-binding site. Its N-terminus possesses a GTPase-activating protein (GAP or P5) domain. Protein ARD1 cycles between active (GTP-bound) and inactive (GDP-bound) forms. Unlike ARFs, ARD1 lacks the myristoylation site in its ARF domain. The GAP activity of ARD1 is restricted to its ARF domain, without GAP activity for any other ARFs [854].

Protein ARD1 localizes to lysosomes and the Golgi body. It operates in the formation of intracellular transport vesicles and their displacement between cellular compartments, as well as phospholipase-D activation [854].

Cytohesin-1 (or ARNO2) serves not only as ArfGEF, but also supports GTP binding to ARD1, at least in vitro [854]. Free Mg^{++} at concentration in the micromolar range favors guanine nucleotide release from ARD1 protein. In addition, ARD1 interacts with TriM29[87] and TriM31 Ub ligases.[88]

9.3.1.9 SAR GTPase (ARF Superfamily)

Small GTPase secretion-associated and Ras-related protein SAR1, a member of the RAS superfamily, operates as a molecular switch to control protein–protein and protein–lipid interactions during vesicle budding from the endoplasmic reticulum. Unlike all Ras GTPases that use either myristoyl or prenyl groups to direct membrane association and function, SAR1 lacks such modifications, but contains a SAR1-NH2-terminal activation recruitment (STAR) motif [855]. The STAR motif mediates the recruitment of SAR1 to endoplasmic reticulum membranes and facilitates its interaction with membrane-associated guanine nucleotide exchange factor Sec12. In addition, an N-terminal motif assists interaction with the GTPase-activating protein complex Sec23/24. Small GTPase SAR1 coordinates CoP2 coat assembly and disassembly and cargo selection with the recruitment of CoP2 coat to initiate export from the endoplasmic reticulum.

9.3.2 CDC42 GTPase (RHO Superfamily)

CDC42 regulates the actin cytoskeleton and thus cell polarity and migration. CDC42 actually controls the formation of actin-rich filopodia (Vol. 2 – Chap. 6. Cell

---

[87] A.k.a. ataxia telangiectasia group D-associated protein (ATDC).

[88] The stability and activity of P53 transcription factor is regulated by several ubiquitin ligases, not only DM2, but also caspase recruitment domain-containing protein CaRD16, ring finger and CHY zinc finger domain-containing protein RCHY1, as well as Tripartite motif (TriM) proteins TriM24, whereas TriM29 inhibits P53 via its repression of KAT5 acetyltransferase.

Motility),[89] as well as Rac, RhoD, RhoF, RhoQ, and RhoU [831]. CDC42 induces actin polymerization by binding to Wiskott-Aldrich syndrome protein (WASP), related nWASP, or via insulin-receptor substrate IRSp53 Tyr kinase to induce actin bundles and branched actin filaments owing to actin-related protein ARP2–ARP3 complex. In addition, CDC42-mediated activation of P21-activated kinase phosphorylates LIMK kinase that phosphorylates (inhibits) cofilin.

CDC42 is involved in chemotaxis of several cell types, such as macrophages, neutrophils, T lymphocytes, and fibroblasts. CDC42 contribution is cell-type specific, as CDC42 is not required for migration of some cell types. In addition, CDC42 regulates cell polarity via partitioning-defective proteins Par6 and Par3 and atypical protein kinase-C. Protein CDC42 also operates via myotonic-dystrophy-kinase-related CDC42-binding kinase (MRCK) to move the nucleus behind the microtubule-organizing center. CDC42 also influences hematopoietic cell differentiation and cell cycle progression.

9.3.3 Superfamily of Rab GTPases

Small Rab (Ras-related proteins in brain) GTPases constitute the largest superfamily of the RAS hyperfamily, with more than 70 members encoded by the human genome.

The Rab proteins regulate all aspects of intracellular transfer of materials between different membrane-enclosed organelles or between organelles and the plasma membrane. They ensure that cargos are delivered to their correct destinations and control membrane identity. Transport of molecules (Vol. 1 – Chap. 9. Intracellular Transport) starts from budding of vesicular or tubular carriers from donor membranes and culminates in their docking and fusion to specific acceptor membranes. Small Rab GTPases of the plasma membrane and organelle membranes behave as membrane-associated switches that regulate vesicle budding, motility, uncoating, docking, and fusion, via recruitment of effectors, such as sorting adaptors, tethering factors, kinases, and phosphatases, and their interactions with coat components, SNARE mediators of membrane fusion, and nanomotors that propel vesicles along microtubules and actin filaments, such as microtubule minus-end-directed dynein, plus-end-directed kinesins,[90] and myosins.

---

[89] Filopodia are thin protrusions that contain parallel bundles of actin filaments that extend from the leading edge of migratory cells.

[90] As microtubules are organized by the centrosome, microtubule-dependent transfer toward the cell cortex requires plus-end-directed nanomotors of the kinesin superfamily.

9.3 Monomeric (Small) GTPases

Small Rab GTPases control transport fluxes, as they cycle between active and inactive forms. This cycle is regulated by GTPase-activating proteins (RabGAP; Sect. 9.4.2.2) and guanine nucleotide-exchange factors (RabGEF; Sect. 9.4.1.2), such as guanine-exchange factor complex TraPP (transport protein particle).[91] The RabGAP proteins inactivate Rab GTPases that are re-activated by RabGEF factors. Active Rab^{GTP} GTPases bind their effectors, although some Rab effectors prefer the GDP-bound form [856].

Multiple Rab GTPases interact as they share effectors or via recruitment of selective Rab activators to warrant the spatiotemporal regulation of vesicle transfer. Different Rab GTPases associated with distinct subdomains of the same membrane or distinct types of cellular membranes can actually interact via effectors that are coupled to specific guanine nucleotide exchange factors and GTPase-activating proteins.

A Rab effector complex can contain guanine nucleotide-exchange factors for the same Rab GTPase, thereby generating a positive feedback loop, or for another Rab that is then activated (*effector-mediated Rab activation*). Effectors of Rab proteins frequently contain separate binding sites for 2 types of Rab GTPases that enable coordination of nanodomains in a given membrane or tethering between 2 membranes (*effector-mediated Rab coupling*). The effector complex of the secondarily activated Rab GTPase that contains a GTPase-activating protein for the first Rab GTPase triggers a *negative feedback* loop.

Newly synthesized Rab^{GDP} GTPases link to *Rab escort protein* (REP) [856]. The latter presents Rab GTPases to a geranylgeranyl transferase (GGT). Geranylgeranylated, Rab^{GDP} are targeted by Rab dissociation inhibitor (RabGDI). The RabGDI sequestrators preclude GDP release from Rab, thereby stabilizing the inactive form. Yet, the Rab–RabGDI complex that chaperones geranylgeranylated Rab GTPases in the cytosol mediates their delivery to their recipient membranes and recycles them back to the cytosol. The Rab–RabGDI complex, indeed, identifies specific membranes via a membrane-bound *GDI displacement factor* (GDF) [856].

The superfamily of Rab GTPases is the largest set within the hyperfamily of Ras-like monomeric GTPases. In humans, about 70 Rab types have been identified, each with a specific subcellular location, many with a specific tissue distribution (Table 9.24).

Small GTPases Rab1 and Rab2 localize to the endoplasmic reticulum, especially the pre-Golgi intermediate compartment, and Golgi stack;[92] Rab3a in secretory

---

[91] Intrinsic GTP hydrolysis by Rab GTPases is slow, but is significantly heightened by RabGAPs that accelerate Rab conversion to inactive, GDP-bound form.

[92] Small Rab1 and Rab2 GTPases mediate endoplasmic reticulum–Golgi body and Golgi body–endoplasmic reticulum transfer, respectively.

Table 9.24. Locations of selected Rab GTPases in epithelial cells (Sources: [856, 858]). Different Rab GTPases localize to different organelles as well as to distinct membrane nanodomains of the same organelle.

Membrane type	Rab Types
Apical recycling endosome	Rab15, Rab17, Rab25
Apical tubule (kidney)	Rab18, Rab20
Autophagosome	Rab7, Rab24, Rab33
Caveosome	Rab5
Centrosome	Rab12
Cilium	Rab8, Rab17, Rab23
Early endosome	Rab4, Rab5, Rab15, Rab21, Rab22
Early phagosome	Rab5, Rab14, Rab22
Endoplasmic reticulum	Rab1, Rab2
Endosomes	Rab8, Rab13, Rab35
Golgi body	Rab1, Rab2, Rab6, Rab8, Rab10, Rab33, Rab40
GluT4 vesicle	Rab8, Rab10, Rab13, Rab14
Late endosome	Rab7, Rab9
Late phagosome	Rab5, Rab7, Rab14, Rab22
Liquid droplet	Rab18
Lysosome	Rab7, Rab34
Macropinosome	Rab34, Rab35
Melanosome	Rab27, Rab32, Rab38
Mitochondrium	Rab32
Nuclear inclusion	Rab24
Plasma membrane	Rab3, Rab4, Rab5, Rab8, Rab10, Rab11, Rab14, Rab15, Rab17, Rab22, Rab23, Rab25, Rab26, Rab27, Rab34, Rab35, Rab37
Recycling endosome	Rab4, Rab11, Rab15, Rab35
Secretory granule	
in neuroendocrine cells,	Rab18
in mastocyte	Rab37
Insulin secretory vesicle	Rab3, Rab26, Rab27, Rab37
Synaptic vesicle	Rab3
Tight junction	Rab13
Trans-Golgi network	Rab8, Rab9, Rab22, Rab31

vesicles; Rab4 and Rab5 in early endosomes;[93] Rab6 in the Golgi body;[94] Rab7 and Rab9 in late endosomes; and Rab11 in recycling endosomes.

Proteins of the RAB superfamily have been classified into 8 functional groups according to sequence similarity, subcellular location, and function [859]: Group 1

---

[93] Protein Rab5 mediates endocytosis and endosome fusion of clathrin-coated vesicles, macropinocytosis with Rab34, and maturation of early phagosomes with Rab14 and Rab22 GTPases.

[94] Three known isoforms of Rab6 exist (Rab6a–Rab6c) [857]. Isoforms Rab6a and Rab6b are encoded by separate genes, whereas Rab6a and Rab6c are splice variants. Isoforms Rab6a and Rab6c are ubiquitous, whereas Rab6b is mainly expressed in the brain. Molecule Rab6a localizes

9.3 Monomeric (Small) GTPases

Table 9.25. Families of Rab GTPases (Source: [858]; RasEF: RAS and EF-hand domain-containing protein, or Ras-related protein Rab45, an atypical GTPase with a coiled-coil motif at the mid region, an N-terminal EF-hand domain, and a C-terminal Rab-homology domain). Some members of the RAB superfamily that signal to the nucleus (Rab5, Rab8, Rab24, and possibly others) may cooperate with other RAB superfamily members — Ran GTPases —, which control nucleocytoplasmic shuttling. Small GTPase Rab32 regulates mitochondrial fission and may participate in adaptation to changing energy requirements during growth. Cell growth and differentiation can be modulated by the coordinated actions of Rab GTPases that regulate cell–matrix and cell–cell adhesion (Rab4a, Rab8b, Rab13, and Rab21) and those involved in growth regulation and cell proliferation and apoptosis (Rab6a, Rab11, Rab12, Rab23, Rab25, Rab35, as well as Ran and likely others).

family	Members
1	Rab23
2	Rab29, Rab32, Rab38, Rab7L1
3	RabL2, RabL3, RabL5
4	Ran
5	Rab7, Rab7b, Rab9a/b/c
6	Rab28, RabL4
7	Rab34, Rab36
8	Rab6a/a'/b/c, Rab41
9	Rab5a/b/c, Rab17, Rab20, Rab21, Rab22a/b (or Rab31)/c, Rab24
10	Rab18
11	Rab2a/b, Rab4a/b/c, Rab11a/b, Rab14, Rab25, Rab39, Rab42
12	Rab19, Rab30, Rab33a/b, Rab43
13	Rab1a/b, Rab3a/b/c/d, Rab8a/b, Rab10, Rab12, Rab13, Rab15, Rab35, Rab40
14	Rab26, Rab27a/b, Rab37, Rab44, RasEF (Rab45)

Rab with Rab1a, Rab1b, and Rab35; 2 with Rab2a, Rab2b, Rab4a, Rab4b, Rab11a, Rab11b, Rab14, and Rab25; 3 with Rab3a to Rab3d, Rab26, Rab27a, Rab27b, and Rab37; 4 with Rab19, Rab30, and Rab43; 5 with Rab5a to Rab5c, Rab21, and Rab22a to Rab22c; 6 with Rab6a to Rab6c and Rab41; 7 with Rab7, Rab9a, and Rab9b; 8 with Rab8a, Rab8b, Rab10, and Rab13. Other components of the dendogram[95] comprises the Rab12–Rab15 branch between groups 1 and 3; Rab18, Rab33a–Rab33b, and Rab39a–Rab39b branches between groups 2 and 4; Rab20 and Rab17–Rab24–Rab34–Rab36 branches between groups 2 and 5; Rab28 between groups 5 and 6, the Rab29–Rab32–Rab38 branch between groups 6 and 7; and Rab23 and the Rab40a-to-Rab40c branches between groups 7 and 8. Dendrogram clustering suggests 14 RAB families, but uncharacterized members exist [858] (Table 9.25).

---

to the Golgi body. Protein Rab6a inhibits anterograde and hastens retrograde intra-Golgi transport. In addition to Rab6, Rab33 and Rab40 mediate intra-Golgi transfer.

[95] A dendogram is a tree for visual classification based on structural and/or functional similarity of molecular species.

Small Rab GTPases are characterized by their steady-state location at cytosolic surface of particular cell membranes. Each Rab protein indeed has a unique subcellular membrane distribution, although they travel between the cytosol and their target membrane(s).[96] In addition, Rab effectors can target at least some members of certain Rab functional groups, without interacting with constituents of other groups. Ten major sets of Rab include Rab1, Rab3, Rab4, Rab5, Rab6, Rab7, Rab8, Rab11, Rab28, and Rab38 groups [860].

Sequential activation of a series of Rabs by a RabGEF module in which RabGEF operates successively is required for intracellular trafficking. Rabs can recruit GEFs that activate subsequently acting Rabs to initiate next transport stage, such as vesicle formation to release or vesicle targeting to docking and then fusion.[97] Complexes formed by Rab GTPases and their effectors may exclude other Rab types. Rapid Rab re-activation by specific guanine nucleotide-exchange factors can be prevented by RabGDI that extracts from the membrane and sequesters Rab GTPases into the cytosol.

Recruitment of Rab-specific effectors defines functional nanocompartments that form transport vesicles, link vesicles to motor proteins, and dock and fuse vesicles with their targets. Rab GTPases avoid mixing with one another on a membrane surface, hence disturbances in transport sorting. A given Rab protein can inactivate a previously acting Rab, as it can recruit its cognate GTPase-activating protein to delineate boundaries between Rab GTPases on membranes and determine directionality to membrane traffic. Similar to RabGEF cascade, RabGAP cascade restricts the spatial and temporal overlap of Rab GTPases [862].

9.3.3.1 Group-1 Rab GTPases

Rab1

The Rab1 isoforms (Rab1a–Rab1b) reside at the endoplasmic reticulum and Golgi body as well as intermediate tubules. They regulate the anterograde transport of cargos between the endoplasmic reticulum and Golgi body. They interact with vesicle docking protein, golgin-A2, -A5, and -B1 (giantin), Rab-binding effector protein RabBP1, Rab escort protein-1 and -2, Rab acceptor-1, Rab-interacting factor, Rab1-interacting iporin,[98] actin-bundling fascin Fscn1, SLC16a8, cyclin-dependent kinase inhibitor-1A, riboflavin kinase, MiCaL1,[99] etc. [251].

---

[96]Protein RabGDI forms a cytosolic heterodimer with Rab GTPase for sequestration. This complex is targeted for recruitment of Rabs to cognate membrane.

[97]On endosomes, Rab5 recruits GEF for Rab7 that allows maturation of early endosomes into late endosomes [861]. Activation of Rab7 then associates with inactivation of Rab5 agent. Therefore, predominant resident Rab isoform determines the compartment identity.

[98]Iporin also binds to golgin-A2.

[99]microtubule-associated monoxygenase, calponin, and LIM domain-containing proteins (MiCaL1–MiCaL3) are predominantly cytosolic Rab1-interacting proteins that bind cytoskeletal components, such as intermediate filament vimentin, and actin.

Rab1a

Protein Rab1a assists in transferring secretory vesicles from the endoplasmic reticulum to the endoplasmic reticulum–Golgi intermediate compartment (ERGIC) and cis-Golgi network [863]. Three Rab1a splice variants can be observed, isoform Rab1a1 being the reference sequence.

During cardiogenesis, Rab1a is involved in plasmalemmal localization of angiotensin-2 and adrenergic receptors, except α2b-adrenoceptor [863].

In addition, GTPase Rab1a binds to epithelial Ca^{++} channel. It interacts with Rab escort proteins (REP1 and REP2) to facilitate the geranylgeranylation and membrane delivery of Rab proteins [863].

Rab35

Ubiquitous Rab35 regulates fast recycling of endocytic membranes, especially during cytokinesis as it forms a protein complex to stabilize the bridge between daughter cells prior to abscission. It is also involved in fast recycling of transferrin receptors to the plasma membrane and transport of class-2 major histocompatibility complexes from the plasma membrane to early endosomes and back to plasma membrane in antigen-presenting cells, as well as immunological synapse between antigen-presenting cells and T lymphocytes [864].

9.3.3.2 Group-2 Rab GTPases

Rab4

Small GTPase Rab4 mediates fast endocytic recycling directly from early endosomes [856]. Early and recycling endosomes contain 3 types of nanodomains with: (1) only Rab5, (2) Rab4 and Rab5, and (3) Rab4 and Rab11 GTPases. Small GTPases Rab4 and Rab5 are involved in endocytic recycling and endosome fusion, hence in the formation of recycling and late endosomes, respectively. Whereas Rab4 GTPase is involved in fast endocytic recycling, Rab11 and Rab35 in slow endocytic recycling via recycling endosomes; Rab21 in integrin endocytosis; Rab15 in transfer from early endosomes to recycling endosomes and from apical recycling endosomes to the basolateral plasma membrane [856]. The Rab effector Rabenosyn-5 contains separate binding sites for Rab4 and Rab5 GTPases.

Rab11

The subfamily of Rab11 GTPases includes Rab11a, Rab11b, and Rab25. Ubiquitous Rab11a and Rab11b have redundant functions, whereas Rab25 expression is restricted to polarized epithelial cells. Among Rab11 effectors, Rab11-interacting proteins target all members of the Rab11 group. Small GTPases Rab11 and Rab35 mediate slow endocytic recycling through recycling endosomes [856].

Small GTPase Rab11a regulates the recycling of transferrin receptor, β-integrin, TRPV5 and TRPV6 channels, glucose transporter GluT4, and chemokine CXCR receptor [865]. It binds to epithelial Ca^{++} channel. Active Rab11b interacts with Rab11-interacting proteins. Like Rab11a, Rab11b associates with recycling endosomes. In polarized epithelial cells, Rab11b regulates the apical recycling compartment. It also contributes to localization of recycling endosome to the cleavage furrow during mitosis.

Rab11 family-interacting proteins (Rab11FIP) that are classified into groups 1 (Rab11FIP1–Rab11FIP2 and Rab11FIP5) and -2 (Rab11FIP3–Rab11FIP4) bind to Rab11^{GTP}. They cycle between the cytosol and recycling endosomes. Group-1 Rab11FIPs are implicated in the regulation of endocytic protein sorting and recycling, whereas group-2 Rab11FIPs intervene in endosomal transport.[100] Adaptor Rab11FIP2 links Rab11a+ endocytic recycling vesicles to myosin-5B for vesicle motility [856]. Myosin-5B is also a direct effector of Rab11a GTPase.

9.3.3.3 Group-3 Rab GTPases

Rab3

Isoforms of Rab3 molecule (Rab3a–Rab3d) operate in synaptic vesicle and calcium-dependent regulated exocytosis. They interact and complex with Rab escort proteins. Small Rab3a GTPase participates in the regulation of neurotransmitter release. Paralogs Rab3a and Rab3b can bind Ca^{++}-binding rabphilin-3, a regulator of neurotransmitter release. Rabphilin, brain-specific scaffolds regulating synaptic membrane exocytosis proteins Rim1 and Rim2, and Rab effector rabphilin-3a-like protein (Rph3aL or Noc2) involved in exocytosis in endocrine cells are Rab3-specific effectors that control vesicle exocytosis in neurons and some endocrine cells.[101]

Rab3-interacting molecule (RIM) is a scaffold protein that participates in docking and fusion of secretory vesicles at release sites. It tethers to synaptosomal-associated protein SNAP25 and membrane-trafficking synaptotagmin-1. It also binds to pore-forming subunit-α1B of Ca$_V$2.2 channels. It also weakly connects to subunit-α1c of Ca$_V$1 channels. In the presence of Ca^{++}, RIM interaction with synaptotagmin-1 rises, but that with SNAP25 decays [867].

---

[100]Member Rab11FIP3 involved in localization of recycling endosomes binds to Rab11a and Rab11b, Rab25, ARF6, RhoAGAP Cyk4, and ARF1GAP ASAP1 [866].

[101]These effectors, in fact, target members of both group-3 and -8 Rab GTPases.

Rab26

Rab26 is tethered to membranes of secretory granules and synaptic vesicles. It thus mainly resides in cells with regulated exocytosis or high secretory activity. Scaffold Rim1, a member of the synaptotagmin-like protein family, is a known effector of Rab26 that also interacts with Rab3a to Rab3d, Rab10, and Rab37 [868]. Rab26 constitutes a subfamily with Rab3 isoforms, Rab27 isoforms, Rab37, Rab44, and Rab45.

Rab27

Small GTPase Rab27 regulates exocytosis of granules in cytotoxic T lymphocytes and melanosomes in melanocytes. Members of the *synaptotagmin-like protein* group (SLP1–SLP4 or SytL1–SytL4)^{102} of the C2 domain-containing protein family and SLP homologs lacking the C2 domain SLaC2a, or melanophilin, and SLaC2b interact specifically with Rab27a^{GTP} and Rab27b^{GTP} [869].^{103}

Small Rab27 GTPases coordinate vesicle motility, as adaptor SLaC2a also attaches to myosin-5a and -7a and actin [869]. Myosin-5a then transfers Rab27a+ vesicles toward the cell cortex. Synaptotagmin-like proteins SLP1 and SLP2 colocalize with Rab27a in melanosomes.^{104}

---

^{102}Synaptotagmin-like protein SytL4 is also called granuphilin-a.

^{103}The C2 domain is a motif of various mediators involved in vesicular trafficking. It serves as a binding site both Ca^{++} and membrane phospholipids. Among various C2 domain-containing proteins, 4 families of C-terminal tandem C2 domain-containing proteins (C2a and C2b) exist: double C2-like proteins (DoC2), rabphilin, synaptotagmin (Syt), and Syt-like protein (SLP). The C2a domain regulates fusion of synaptic vesicles for neurotransmitter release. The C2b domain is a binding motif for phosphatidylinositol trisphosphate and bisphosphate in the absence and presence of calcium ions, respectively. Subtype DoC2a that is mainly expressed in the central nervous system and ubiquitous DoC2b are involved in Ca^{++}-dependent neurotransmitter release and intracellular vesicle trafficking, respectively. Rabphilin-3 also regulates neurotransmitter release in hippocampal neurons. Synaptotagmins are calcium sensors that intervene in the regulation of neurotransmitter release and hormone secretion. Only 8 synaptotagmins among 15 synaptotagmins bind to calcium (synaptotagmins-1 to -3, -5 to -7, -9, and -10). C2 domain-containing protein families that are related to synaptotagmins include: (1) transmembrane proteins, such as: (1.1) *ferlins* that are involved in Ca^{++}-mediated membrane fusion; (1.2) extended synaptotagmin-like proteins (intracellular membrane-associated eSyt1 and plasmalemmal eSyt2 and eSyt3); and (1.3) Ca^{++}-binding, membrane-anchored, multiple C2-domain and transmembrane region proteins MCTP1 and MCTP2 that are encoded by 2 genes; as well as (2) soluble proteins, such as: (2.1) members of the Rim family of regulating synaptic membrane exocytosis proteins, scaffold proteins that contribute to synaptic vesicle exocytosis during short-term remodeling; (2.2) brain-specific peripheral membrane protein Munc13 that interacts with syntaxin, a component of the exocytic synaptic core complex, and RIM1; (2.3) synaptotagmin-related proteins; and (2.4) trans-Golgi network membrane-associated B/K proteins (or SytB/K) that are expressed in the brain and kidney.

^{104}The Rab27a effector SLP2 is recruited after melanophilin for the correct peripheral distribution of melanosomes in melanocytes. In melanocytes, Rab27 binds to melanosomes and recruits

In addition, small GTPase Rab27 regulates dense core granule secretion in platelets with Rab27-binding protein Munc13-4 [871]. Isoform Rab27a complexes with granuphilin in pancreatic β cells to regulate exocytosis of insulin-containing dense-core granules [872]. Small GTPase Rab27b is most abundantly produced in the pituitary gland. Yet, Rab27b localizes to zymogen granules and regulates pancreatic acinar exocytosis.

Small GTPase Rab27a not only mediates vesicle–nanomotor attachment, but also controls docking of exocytic dense-core vesicles to the plasma membrane. Granuphilin, or SLP4, an effector of Rab27a, interacts directly with Sec1-related syntaxin-binding protein StxBP1 (a.k.a. Munc18-1) [856]. Release of granuphilin precedes SNARE complex assembly and membrane fusion.

Rab37

Small GTPase Rab37 that is closely related to Rab26 contributes to masocyte degranulation.

9.3.3.4 Group-4 Rab GTPases

Golgins are long coiled-coil proteins that localize to particular Golgi subdomains via their C-termini. GRIP domain-containing golgins (golgin-97 and -245 and GRIP and coiled-coil domain-containing proteins GCC1 or GCC88 and GCC2 or GCC185) bind via their C-termini Arf-like protein ArL1 on the trans-Golgi network as well as Rab2, Rab6, Rab19, and Rab30. These molecules of the trans-Golgi network as well as other compartments of the Golgi body such as cis-Golgi network capture Rab-containing vesicles and exclude ribosomes (sise ∼25 nm) [873]. Actin and spectrin form a mesh around the Golgi body that reorganizes for vesicle arrival and departure.

Rab19

Small GTPase Rab19 can be detected at high levels in lung, intestine, and spleen and at a lower level in kidney, whereas the brain, heart, and liver contain only very little or no detectable amounts [874].

---

melanophilin and myosin-5a to transfer melanosomes from microtubules to actin filaments. Whereas C-terminal tandem C2 domain-containing protein rabphilin as well as rabphilin-3A-like without C2 domain protein NoC2 (or Rph3aL) that is involved in exocytosis in endo- and exocrine cells interact with all Rab27 isoforms as well as members of group-3 (Rab3 set) and group-8 (Rab8a set) Rab GTPases, Rim1 interacts with Rab3 isoforms, Rab26, and Rab37, as well as Rab10 (group-8 Rab GTPases), and Rim2 with Rab3 isoforms and Rab8a [870].

Rab30

Small GTPase Rab30 can be detected in vascular endothelial cells. Jun N-terminal kinase can influence intracellular transport via small GTPase Rab30 [875].

Rab43

Small GTPase Rab43 (a.k.a. Rab11b, and Rab41) is expressed in the brain, lung, heart, kidney, spleen, colon, uterus, ovary, and testis, but not in liver. It is involved in trafficking between the Golgi body and endoplasmic reticulum [876].

9.3.3.5 Group-5 Rab GTPases

Rab5

Protein Rab5 is involved in budding of clathrin-coated vesicles from the plasma membrane and their transport to early endosomes as well as fusion of early endosomes. It localizes not only to early endosomes, but also phagosomes and caveosomes, in addition to plasma membrane [856]. It mediates endocytosis and endosome fusion of clathrin-coated vesicles, macropinocytosis with Rab34, and maturation of early phagosomes with Rab14 and Rab22 GTPases [856].

The Rab GTPases cooperate with components of the vesicle docking and fusion machinery such as SNARE proteins. Small Rab5 GTPase links to tethering factors early endosome antigen EEA1 and rabenosyn-5 to mediate fusion of endosomal membranes [856]. During membrane fusion of homotypic early endosomes, the complex of RabGEF1 (or RabEx5) and Rab-binding effector protein RabEP1 (or rabaptin-5) stimulates and stabilizes Rab5 protein. The latter can then recruit PI3K to produce PI(3)P as well as Rab5 effectors such as early endosome antigen EEA1 and rabenosyn-5 that assemble with VPS45 vacuolar protein sorting, a member of the Sec1 family of SNARE regulators. In addition, EEA1 and rabenosyn-5 can directly interact with endosomal SNAREs syntaxin-6 and -13 and syntaxin-7, respectively.

During membrane fission, Rab5-binding protein RIn3 interacts with amphiphysin-2 that mediates receptor-induced endocytosis from plasma membranes to early endosomes [877].

Vesicle coat complexes required for cargo sequestration and membrane budding must be shed prior to acceptor membrane fusion. Endocytic vesicles with clathrin coat can recruit cargo adaptor complex AP2 via phosphatidylinositol (4,5)-bisphosphate and adaptin-associated kinase AAK1 that phosphorylates AP2 μ2 subunit. Clathrin-coated vesicle-associated Rab5 and its GEF, GAP, and VPS9 domain-containing protein GAPVD1 coordinate AP2 uncoating, as they promote dephosphorylation of AP2 subunit μ2 and increase PI(4,5)P$_2$ turnover [856].

Small Rab GTPases specify membrane identity, as they control local levels of phosphoinositides and recruit specific effectors to restricted membrane nanodomains. The Rab5 effectors EEA1, rabenosyn-5, and rabankyrin-5 (or AnkFy1) contain PI(3)P-binding FYVE domains. Phosphoinositide PI(3)P, in turn, recruits the retromer subunits — sorting nexins SNx1 and SNx2 — that control the plasmalemmal density of transmembrane receptors. Subunits SNx1 and SNx2 associate with the cargo-interacting retromer VPS26–VPS29–VPS35 subcomplex, an effector of Rab7 GTPase (*Rab crosstalk*) [856].

Endocytic vesicles experience a Rab5-to-Rab7 conversion. Once it is activated by Rab5GEFs on donor membranes, Rab5^{GTP} is recruited to the endocytic vesicular membrane. Accumulation of PI(3)P[105] in the vesicular membrane enables the recruitment of vacuolar fusion protein Mon1b, thereby attracting and activating Rab7 and inhibiting the Rab5–Rab5GEF *positive feedback*.[106] The maturation of Rab5+ early endocytic structures into later Rab7+ vesicles involves Rab5-mediated recruitment of the tether homotypic fusion and vacuole protein sorting complex (HoPS), in which VPS39 subunit is a Rab7GEF.[107] Small GTPase Rab7 then recruits a Rab5GAP and causes Rab5–Mon1b–Rab7 *negative feedback loop*). Then, Rab5GAPs inactivate Rab5 and Rab5GDIs sequester inactivated Rab5 into the cytosol, avoiding its degradation and allowing its transfer to the plasma membrane, where it may once more be activated on newly formed vesicles. Similarly, as the vesicles reach the stage of late endosomes, Rab7GAPs on these membranes inactivate Rab7, and Rab7GDIs sequester it into the cytosol and transporting it to vesicles for Rab5–Rab7 conversion.

Class-1 PI3K kinases that phosphorylate PI(4,5)P$_2$ to PI(3,4,5)P$_3$ at the plasma membrane as well as class-3 PI3K kinases that phosphorylate PI to PI(3)P on endosome membranes are effectors of Rab5 [856]. Phosphoinositide 4- and 5-phosphatases that dephosphorylate PI(3,4)P$_2$ to PI(3)P and PI(3,4,5)P$_3$ to PI(3,4)P$_2$, respectively, are also effectors of Rab5 GTPase.

Small GTPase Rab5 then controls the formation of PI(3)P that recruits KIF16b kinesin [856]. Kinesins can hence be indirectly regulated by Rab GTPases for vesicle motility. Once their destination is reached, membrane-tethering complexes enter into action. They often contain RabGEFs. The Rab5 effector rabaptin-5 complexes with RabGEF1 (or rabex-5) to amplify Rab5 activation in nanodomains of endosomal membrane [856].

---

[105] Numerous adaptors support activation of PI3K using in particular signals emanating from cargos that concentrate into the vesicle.

[106] As the PI(3)P concentration reaches a certain threshold, Mon1b is recruited to the vesicular membrane, where it removes Rab5GEF agent.

[107] Protein Mon1b also interacts with the HoPS complex, hence recruiting and activating Rab7 GTPase.

Endocytosis-associated Rab5 serves as a hub for cooperation between cell signaling and trafficking. Activation of the epidermal growth factor receptor activates Rab5 via RasGEF Son of sevenless SOS1, small GTPase Ras, and Rab5GEF Ras and Rab interactor RIn1 to stimulate EGFR transfer to early endosomes [856]. Small GTPase Rab5 also stimulates macropinocytosis via Rab5 effector USP6 N-terminal-like protein (USP6NL or Related to the N-terminus of Tre [RNTRE]) that interacts directly with actin or via actin-binding actinin-4. Protein USP6NL, a Rab5GAP, is recruited to EGFR via EFGR kinase substrate EPS8 that activates Rac1 GTPase via adaptor Abl interactor ABI1 and RasGEF SOS1. In addition, Rac1 interacts with Rab5 on endosomes via the recruitment of Rab5GEF alsin. The Rab5 effectors — adaptor phosphotyrosine interaction, PH domain, and leucine zipper-containing proteins APPL1 and APPL2 — reside in a subpopulation of endosomes and translocate to the nucleus, where they connect to the chromatin remodeling complex NuRD, an histone deacetylase complex, in response to EGF stimulation.

Rab21

Ubiquitous Rab21 GTPase is located in early endosomes [878]. Small Rab21 GTPase mediates integrin endocytosis [856].

Rab22

Small GTPase Rab22 mediates trafficking between the trans-Golgi network and early endosomes and conversely [856].

9.3.3.6 Group-6 Rab GTPases

Rab6

Golgi body-localized Rab6, Rab33, and Rab40 mediate intra-Golgi trafficking [856]. Kinesins are directly or indirectly regulated by Rab6 GTPase [856]. Protein Rab6 of the Golgi body directly regulates kinesins for vesicle motility, as cytokinesis regulator KiF20a, or rabkinesin-6, is its direct effector. The Rab6 effector Bicaudal-D1 mediates attachment of Golgi vesicles to dynein–dynactin complex in transport from the Golgi body to the endoplasmic reticulum.

The Rab effector Golgi coiled-coil protein GCC2 that contains several Rab-binding sites localizes to Golgi membranes via interactions with Rab6 and adpribosylation factor-like protein ARL1 [856]. It also binds to Rab1, Rab2, Rab9, Rab15, Rab27b, Rab30, and Rab33b.

9.3.3.7 Group-7 Rab GTPases

Rab7

Late endosome-associated GTPase Rab7 mediates maturation from early endosome of late endosomes and phagosomes, and their fusion with lysosomes [856]. Rubicon, a component of the endosomal class-3 PI3K complex (or vacuolar protein sorting VPS34) prevents endosome maturation, as it impedes Rab7 activation [879]. Molecule UV radiation resistance-associated gene product (UVRAG, or VPS38) activates PI3K$_{c3}$ and class-C VPS, a Rab7GEF protein. Rubicon sequesters UVRAG from class-C VPS [880]. Active Rab7 competes for Rubicon binding and releases UVRAG to associate with C-VPS, which, in turn, further activates Rab7 GTPase. This feedforward loop ensures rapid amplification of Rab7 activation and subsequent endosome maturation.

Microtubule minus-end-directed nanomotor dynein is an indirect effector for late endosome-associated Rab7 GTPase. The Rab7 effector Rab-interacting lysosomal protein (RILP) recruits a subunit of the dynactin complex that connects to dynein.

Rab9

Late endosome-associated Rab9 GTPase mediates transfer from late endosomes to the trans-Golgi network [856]. Widespread Rab9 GTPase intervenes in cargo-specific coat assembly.

Small Rab9 GTPase operates in the recycling of mannose 6-phosphate receptors (M6PR) from late endosomes to the trans-Golgi network.[108] The Rab9 effector — sorting adaptor M6PRBP1 —[109] recognizes the cytosolic tail of the 2 mannose 6-phosphate receptors.[110]

Protein Rab9 activates ubiquitous Rho-related BTB (Broad complex, Tramtrack, and Bric-à-brac) domain-containing GTPase RhoBTB3 for the docking of transport vesicles at the trans-Golgi network [881].

Protein Rab9 links to Biogenesis of lysosome-related organelles complex BLOC3 and may participate in the membrane recruitment of this complex [881, 883].[111]

---

[108] Mannose 6-phosphate binds 2 transmembrane proteins: cation-dependent (cdM6PR) and -independent (ciM6PR or IGF2R).

[109] A.k.a. 47-kDa tail-interacting protein (TIP47).

[110] The retromer also participates in mannose 6-phosphate receptor transport to the trans-Golgi network. It involves Rab7+ late endosomes.

[111] Three cytosolic complexes — the Biogenesis of lysosome-related organelles complexes — (BLOC1–BLOC3), like clathrin-associated adaptor proteic complex AP3 and homotypic fusion and vacuole protein sorting (HoPS) complex, mediate vesicular transport during the genesis of lysosome-related organelles. The AP3 and HoPS complexes participate in vesicular transport in the endosomal–lysosomal compartment. Active HoPS complex that binds phosphoinositides

9.3.3.8 Group-8 Rab GTPases

Skeletal muscle is a major site of dietary glucose storage, using insulin-mediated translocation of GluT4 at the plasma membrane via the PKB–TBC1D4 pathway. Agent TBC1 domain-containing protein TBC1D4,[112] a RabGAP does not target the same Rab GTPase in skeletal myocytes and adipocytes; Rab8a is involved in GluT4 exocytosis in myocytes and Rab10 in adipocytes. Insulin promotes activation of both Rab8a and Rab13 of the group-8 Rab GTPases, but not Rab10, in rat myocytes, Rab8a activation preceding that of Rab13 [884]. The latter is characterized by a wider intracellular distribution than that of Rab8a GTPase, restricted to the perinuclear region.

Rab8

Rab8a regulates export of vesicles from the trans-Golgi network, transport along actin filaments and microtubules, and fusion with the plasma membrane [885]. Numerous regulators, such as Rab8a-specific GEFs, GAPs, and kinases, modulate Rab8a activity, whereas cytoskeletal nanomotors or motor-binding proteins and effectors coordinate Rab8a function during membrane trafficking.

Small Rab8a GTPase contributes to cell morphogenesis, signaling, and development. Small GTPase Rab8 also participates in translocation of glucose transporter GluT4 vesicle with Rab10 and Rab14 and ciliogenesis with Rab17 and Rab23 GTPases [856]. Myosin-5b is a direct effector of Rab8a GTPase that coordinate vesicle motility [856].

---

and vacuolar SNARE proteins couples Rab GTPase activation and SNARE complex assembly during membrane fusion. Biogenesis of lysosome-related organelles complex BLOC1 is a ubiquitous, multisubunit, proteic complex required for the formation of specialized organelles of the endosomal–lysosomal network, such as melanosomes and platelet dense granules. The complex includes Cappuccino, dysbindin, Muted homolog, pallidin, snapin, and BLOC1 subunits BLOC1S1 to BLOC1S3 (or BLOS1–BLOS3). The Hermansky-Pudlak syndrome (HPS) is a set of autosomal recessive disorders characterized by deficiencies in lysosome-related organelles, such as melanosomes and platelet-dense granules, hence oculocutaneous albinism, prolonged bleeding, and mild ceroid lipofuscinosis. This disease results from gene mutations. Eight genes can be involved, which encode subunits of vesicular transport complexes, such as AP3 and HoPS complexes. In addition to β3a subunit of the AP3 complex that mediates signal-dependent transfer of integral membrane proteins to lysosomes and related organelles, other implicated genes encode HPS1, HPS3, and HPS4 proteins. These proteins exist in soluble and membrane-associated forms. Subtype HPS4, but not HPS3, associates with HPS1 in the BLOC3 complex, the simplest of the above-mentioned complexes [882].

[112] A.k.a. 160-kDa Akt (PKB) substrate (AS160).

Rab10

Insulin stimulates the translocation of glucose transporter GluT4 from intracellular vesicles to the plasma membrane. Three Rab GTPases — Rab10, Rab11, and Rab14 — lodge on GluT4+ vesicles.

Monomeric Rab10 GTPase can localize to the inner face of the plasma membrane using a lipid anchor. In polarized cells such as epithelial cells, Rab10 mediates transport from basolateral sorting endosomes to common endosomes [886].

Rab13

Small GTPase Rab13 regulates the assembly of tight junctions between epithelial cells [856]. In addition, Rab13, which is synthesized in skeletal myocytes, participates in insulin-stimulated translocation of the GluT4 glucose transporter [884].

9.3.3.9 Other Rab GTPases

Most cells have members of the hyperfamily of small GTPases that are structurally classified into, at least, 5 superfamilies (ARF, Rab, Ran, Ras, and Rho), although expression levels of these members vary. Yet, a few members have a tissue-specific expression. Small GTPase Rab17 is detected only in epithelial cells.

Rab15

Small Rab15 GTPase is involved in the trafficking from early endosomes to recycling endosomes and from apical recycling endosomes to the basolateral plasma membrane [856].

Rab17

In epithelial cells, Rab17 and Rab25 control transfer through the apical recycling endosomes to the apical plasma membrane [856].

Rab18

Small Rab18 GTPase controls the vesicular transport by defining organelle identity and organizing functional membrane nanodomains. It controls the formation of lipid droplets [856].

Rab23

Small Rab23 GTPase is produced in most tissues at a low level. Yet, it is highly expressed in developing brain and spinal cord. It acts as an inhibitor of sonic Hedgehog [887]. Endosomal Rab23 actually causes receptor endocytosis to endosomes, thereby attenuating sonic Hedgehog signaling. It prevents formation of Gli2 activator. It could regulate the subcellular location of Hh signaling components, downstream from Smoothened effector and upstream from Gli mediator.

Rab24

Small Rab24 GTPase that has a very low intrinsic GTPase activity is associated with the autophagosome [888]. It may be implicated in degradation of misfolded proteins as well as in nucleocytoplasmic transport. Protein Rab24 is not efficiently geranylgeranylated. Consequently, Rab24 does not form an observable complex with Rab guanine nucleotide-dissociation inhibitor, as geranylgeranylated Rab^{GDP} favors linkage with GDI agents.

Rab32

Small GTPases Rab32 and Rab38 are involved in the genesis of melanosomes. Small GTPase Rab32 also controls mitochondrial fission [856].

Rab33

Small GTPase Rab33, together with Rab24, regulates the formation of autophagosomes [856].

Rab44 and Rab45

Small GTPases Rab44 and Rab45 represent unconventional members of the Rab26 family.

9.3.4 Rac GTPases

The Rac family of small GTPases, a subset of the RHO superfamily of the RAS hyperfamily, include 3 isoforms (Rac1–Rac3; Table 9.26) and RhoG [831]. Protein Rac1 is ubiquitous; Rac2 is mainly restricted to hematopoietic cells; Rac3 abounds

Table 9.26. Rac GTPases and their effectors (Source: [889]; BAIAP: brain-specific angiogenesis inhibitor-1-associated insulin receptor substrate; ElMo: engulfment and cell motility adaptor; Ktn: endoplasmic reticulum membrane and microtubule-associated kinectin; p67phox: NADPH oxidase subunit; SH3RF: SH3 domain-containing ring finger ubiquitin-ligase; SRA: steroid receptor RNA activator).

Type	Effectors
Rac1	PAK1–PAK3, MAP2K
	SH3RF1, SRA1, BAIAP, p67phox
Rac2	PAK1–PAK3
Rac3	PAK1–PAK3
RhoG	ElMo, Ktn1

in brain cells and fibroblasts.[113] Small GTPase RhoG is widely expressed albeit at varying levels according to the cell type (Sect. 9.3.17).

Formation and turnover of cell adhesions with adjacent cells and the extracellular matrix is a necessary function of epithelial cells to regulate epithelial barrier integrity and circulating leukocytes that are recruited in tissues to clean and repair. Activated Rac GTPase weakens tight junctions, but stimulates adherens junction formation.

Protein Rac activates actin polymerization during lamellipodium formation via Wiskott-Aldrich syndrome protein family verprolin-homologous protein complex (WAVe) that activates actin-related protein ARP2–ARP3 complex, which generates actin filament branches, and formin diaphanous-2 that nucleates unbranched actin filaments. Moreover, Rac-mediated activation of P21-activated kinase phosphorylates LIMK kinase that phosphorylates (inhibits) actin-severing cofilin. Small Rac GTPases not only stimulate the formation of lamellipodia and membrane ruffles, but also induce membrane protrusions during phagocytosis.[114]

All components of the renin–angiotensin system, renin, angiotensinogen (or serpin-A8), angiotensin-converting enzyme (ACE), angiotensin-2, angiotensin-2 receptors) localize in the ventricular myocardium, as they are produced by cardiac fibroblasts. Activated angiotensin-2 fosters cardiac remodeling (maladaptive hypertrophy) and fibrosis in overloaded myocardium. In both cardiac myocytes and fibroblasts, integrin-$\beta 1$, a major mechanosensor, contributes to the regulation of cell growth and gene expression via its effectors of the RHO superfamily and stress-activated protein kinases (JNKs and P38MAPKs). Transcription of the ATG (angiotensinogen) gene is inhibited by Rac1 via both JNK-dependent and -independent mechanisms, and stimulated by RhoA via a P38MAPK-dependent mechanism [890].[115]

---

[113] For example, cardiomyocytes express neither Rac2 nor Rac3.

[114] Agent Rac2 is involved in both enzyme release from granules and NADPH oxidase activation during phagocytosis to kill bacteria.

[115] In stretched neonatal rat cardiac fibroblasts, Rac1 activity returns to its original level after 4 h, whereas RhoA remains at a high level of activity until the end of the prolonged stretch period

9.3.4.1 Rac1

Phosphatidylinositol (4,5)-bisphosphate is produced at sites of N-cadherin-mediated intercellular adhesion due to activated Rac1 GTPase. The latter mediates insertion of transient receptor potential channel TRPC5 into the plasma membrane via its stimulation of PI(4)P5K enzyme.

Isoform Rac1 elicits migration of multiple cell types (e.g., macrophages, T lymphocytes, Schwann cells, epithelial cells, and fibroblasts). Response generated by Rac1 relies particularly on chemotactic fMLP peptide. Both Rac1 and Rac2 regulate cell spreading. Both Rac1 and Rac2 intervene in the actin mesh and spectrin scaffold of erythrocytes. Proteins Rac1, Rac3, and RhoG promote axon growth and guidance.

Sumoylation of Rac1 promotes its activity, hence cell migration. Protein inhibitor of activated STAT PIAS3, a small ubiquitin like modifier (SUMo) ligase, interacts with Rac1, particularly Rac1^{GTP}, and primes its sumoylation [891].

9.3.4.2 Rac2

Isoform Rac2 is required for migration and phagocytosis of neutrophils. Protein Rac2 interacts with phagocyte NADPH oxidase NOx2 [892]. Inactive Rac2^{GDP} is bound to RhoGDI that sequesters Rac2 and prevents Rac2 from interacting with membranes as well as GEFs, GAPs, and effectors.

9.3.4.3 Rac3

Ubiquitous Rac3 has an association rate of GDP similar to Rac1, but 13-fold higher than that of Rac2. The intrinsic GTP hydrolysis rate of Rac3 is similar to those of Rac1 or Rac2, but is about 5-fold faster than that of Ras [893]. GTPase-activating protein Bcr can stimulate GTP hydrolysis of Rac3 to a similar degree to that of Rac1. Like other Rac GTPases, guanine nucleotide-exchange factors such as Tiam1 increases the dissociation rate of GDP on Rac3. Protein Rac3 also interacts with guanine nucleotide-dissociation inhibitors.

Its specific effectors include calcium–integrin-binding protein (CIB), nuclear DNA-binding protein C1D, and nuclear receptor-binding protein (NRBP). Constitutively active Rac3 localizes to internal membranes of the endoplasmatic reticulum and Golgi body. Active ^{GTP}Rac3 can also be detected at the inner leaflet of the plasma membrane and membrane protrusions.

(24 h) [890]. Mechanical stretch initially causes a moderate decrease in AGT gene expression and a secondary increase from 8 h. Small RhoA GTPase mediates both the stretch-induced inhibition of ATG gene at 4 h and subsequent upregulation of ATG gene expression at 24 h. Integrin-β1 enables acute (2 and 15 min) stretch-induced Rac1 activation, but represses RhoA activity. Unlike Rac1, RhoA mediates hypertrophy in the myocardium via its effector RoCK kinase.

9.3.5 Rad GTPase

Protein Ras associated with diabetes, (Rad; Sect. 9.3.25),[116] a member of the RGK (Rad, Gem, and Kir) family of Ras-related GTPases, shares GTP-binding domains with Ras, but lacks the C-terminal CAAX motif involved in prenylation of RAS superfamily members.

Small Rad GTPase interacts with calmodulin and calmodulin-dependent protein kinase CamK2 to preclude its signaling.

In the myocardium, Rad hinders the activity of β-adrenoceptor and of $Ca_V1.2a$ channel, thereby reducing cardiomyocyte contractility. Protein Rad then operates as a regulator of cardiac electromechanical coupling and βAR signaling.

Protein Rad contributes to the attenuation of cardiac hypertrophy. It indeed represses connective tissue growth factor (CTGF).[117] Cardiomyocytes release Rad that targets cardiac fibroblasts, which produce extracellular matrix, to prevent cardiac fibrosis. In cardiomyocytes, CCAAT/enhancer-binding protein-δ activates CTGF production [894].

In addition, Rad lowers intimal hyperplasia after balloon injury, as it prevents migration of vascular smooth muscle cells activated by the Rho–RoCK pathway in rat carotid arteries [895]. Protein Rad indeed reduces the formation of both focal contacts and stress fibers in this cell type by blocking RoCK signaling.

9.3.6 Rag GTPases

Small Ras-related GTPases Rag include 4 known members (RagA–RagD). Proteins Rag function as heterodimers that consist of RagA or RagB and RagC or RagD.

Subtype RagC associates with the nutrient–energy–redox sensor TOR complex-1 (TORC1) that acts as protein kinase to influence its location within endomembranes of the cell [896]. Small Rag GTPase binds to raptor that mediates Rag–TORC1 interaction.

---

[116]Concentration of Rad GPTase rises in the skeletal muscle of a subset of patients with type-2 diabetes mellitus.

[117]A.k.a. CCN (CTGF, CyR61 [cysteine-rich angiogenic inducer 61], NOv [nephroblastoma overexpressed]) family member-2 (CCN2). In cultured neonatal cardiomyocytes, Rad overexpression suppresses both basal and transforming growth factor-β1-induced CTGF expression. Factor TGFβ1 actually strongly fosters CTGF expression via SMAD3 in many cell types, particularly cardiac myocytes and fibroblasts. However, SMAD3 phosphorylation and SMAD reporter activity is not disturbed when Rad activity is altered.

Table 9.27. Ral GTPases and their effectors (Source: [889]).

Type	Effectors
RalA	PLD1, PLCδ1, RalBP1
RalB	

9.3.7 Ral GTPases

Closely related Ral GTPases — RalA and RalB —[118] constitute a family within the RAS hyperfamily (Table 9.27). Small Ral GTPases are activated by a unique nucleotide-exchange factor RalGEF (a.k.a. RalGDS) and inactivated by a distinct GTPase-activating protein RalGAP. The Ral proteins are prenylated to be associated with cellular membranes. Whereas Ras is found almost exclusively in the cell cortex (intracellular egde of the plasma membrane), Ral primarily localizes in cytoplasmic vesicles, in addtion to plasma membrane. Unlike Ras GTPases, Ral does not produce transformed cells when it bears oncogenic mutations that lock it in its active GTP state.

Both RalA and RalB have a calcium-dependent calmodulin-binding domain. Small GTPase Ral and calcium–calmodulin activate phospholipase-Cδ1 [897].

Protein RalA mediates signaling initiated by plasmalemmal, ligand-bound receptors. Active ^{GTP}RalA binds filamin, an actin filament crosslinker and recruiter of membrane and intracellular proteins to actin [898]. It is a CDC42 effector to recruit filamin and generate actin-rich filopodia. In addition, RalA GTPase associates with ARF-responsive, phosphatidylinositol (4,5)-bisphosphate-dependent phospholipase-D1 that hydrolyzes phosphatidylcholine into choline and messenger phosphatidic acid [899]. Moreover, RalA synergistically enhances activation of phospholipase-D1 by ARF1 [900].[119]

The RalB effectors comprise exocyst complex component Exoc2 (or Sec5)[120] The RalB–Exoc2 complex recruits and activates atypical IκB kinase family member TRAF-associated NFκB activator (TANK)-binding kinase TBK1 to stimulate transcription factor NFκB [901].

---

[118]Ras-related Ral simian leukemia viral proto-oncogene product homolog.
[119]Phospholipase-D1 is also activated by small GTPase CDC42, Rac1, and RhoA as well as PKC-α.
[120]Exocyst complex connects exocytic vesicles to specific docking sites on the plasma membrane.

Table 9.28. Small Rap GTPases and their effectors (Source: [889]; RPIP: Rap2-interacting protein).

Type	Effectors
Rap1a	bRaf, cRaf, RalGEF, RGS14
Rap1b	bRaf, cRaf, RalGEF, RGS14
Rap2a	Rif GTPase, RPIP8, RalGEF, RGS14
Rap2b	Rif GTPase, RPIP8, RalGEF, RGS14
Rap2c	

9.3.8 Ran GTPases

Small Ran GTPase intervenes in the transport of proteins between the nucleus and cytoplasm and conversely. Protein Ran can diffuse between the 2 compartments.

The transfer from the cytoplasm to the nucleus is allowed for cytosolic proteins that contain a nuclear localization signal (NLS; protein^{NLS}). These proteins bind to the importin-α–importin-β complex. The resulting ternary complex crosses the nuclear pore (Vol. 1 – Chaps. 4. Cell Structure and Function and 9. Intracellular Transport). Binding of Ran^{GTP} to importin-β dissociates the ternary complex, thereby liberating the NLS-bearing protein.

The transfer from the nucleus to the cytoplasm is allowed for proteins that possess a nuclear export signal (NES). The latter bind to the karyopherin exportin. The resulting complex is exported when bound to Ran^{GTP} GTPase.

The basic selectivity associated with structural motifs is complemented by the distributions of proteins affecting the activation state of Ran GTPase. The single known RanGEF — regulator of chromosome condensation RCC1 — is bound to chromatin, whatever the phase of the cell cycle [902]. This regulator continuously generates Ran^{GTP} in the nucleus. On the other hand, RanGAPs localize to the cytosol and cytoplasmic face of the nuclear envelope. It dissociating the Ran–exportin–protein^{NES} complex arriving from the nucleus to the cytoplasm.

9.3.9 Rap GTPases

Monomeric Rap GTPases that are similar in structure to Ras constitute a family of small cytosolic GTPases. These Ras-like small GTPases participate in the control of between-cell and cell–matrix adhesion, especially epithelial and endothelial cell junctions [903].

Small Ras-related Rap GTPases can be subdivided into 2 subfamilies by sequence homology. Subfamily-1 Rap GTPases (Rap1a–Rap1b) are ubiquitous; subfamily-2 Rap GTPases (Rap2) are mainly found in the central nervous system and platelets (Table 9.28).

9.3.9.1 RAP1 Subfamily

Members of the Rap1 subfamily are able to bind Ras effectors, with an affinity similar to or sometimes higher than Ras, such as cRaf, PI3K, and RalGEFs. They promote activation of certain component of the mitogen-activated protein kinase module, such as bRaf kinase. In addition to their effect on MAPK signaling, Rap1 GTPases participate in cytoskeletal rearrangement and cadherin- and integrin-mediated cell adhesions, as well as the regulation of cell proliferation and differentiation. Both Rap1a and Rap1b contribute to vesicular delivery of E-cadherins with Ral and exocyst complex components ExoC2 and ExoC8, hence controlling plasmalemmal E-cadherin density, as well as cytoskeletal linkage to E-cadherins with catenin-δ1 and Rap1 effector, afadin,[121] and thus cell adhesion stabilization. Afadin also binds to Rap1GAP signal-induced proliferation-associated protein SPA1 (or SIPA1). It can then control integrin-mediated cell adhesion via the recruitment of SPA1 and Rap1^{GTP}.

Small GTPase Rap1 is also able to block cell transformation owing to its capacity to form non-productive complexes with Ras effectors. Anti-oncogenic effect of Rap1 indeed relies on sequestration of RasA1 (or p120RasGAP) and cRaf kinase. Conversely, Rap1 can cause Rac activation and mediate relocalization of RacGEFs Vav2 and TIAM1 during intercellular adhesion maturation and cell spreading [903].

In hematopoietic cells, Rap1 provokes lymphocyte aggregation, T-cell anergy, and platelet activation. In the nervous system, Rap1 contributes to the regulation of axonal differentiation, dendritic development, dendritic and spine morphology, and synapse remodeling, in particular synaptic depression by removing AMPARs from synapses via P38MAPK [904]. In addition, Rap1 couples cAMP signaling to ERK1 and ERK2 kinases.

Several Rap guanine nucleotide-exchange factors (RapGEF; Sect. 9.4.1.6)) activate Rap GTPases. They include RapGEF1 (or C3G), RapGEF2 (PDZ-GEF), RapGEF3 and -4 (EPAC1 and -2), Rap(Ras)GRPs, and dedicator of cytokinesis DOCK4.

Ras-related GTPases Rap1 and Rap2 are able to interact with RalGEFs, such as Ral GDP-dissociation stimulator (RalGDS or RalGEF), RalGDS-like (RGL GEF), and RalGDS-like factor (RLF) that, all, are effectors of Ras [905]. In the central nervous system, Rap2 specifically interacts with Rap2-interacting protein RpIP8 (or Rundc3a) [906].

Activation of Rap1 follows its interaction with Rap1-specific RapGEF1 protein. Under basal conditions, RapGEF1 is associated with CRK2 adaptor. Stretch of the extracellular matrix is transduced into intracellular biochemical signals. Small Rap1 GTPase provokes integrin-mediated adhesion and then influences actin dynamics. Protein Rap1 activates integrins via Rap1^{GTP}-interacting adaptor (RIAM) [907].[122]

---

[121] A.k.a. AF6 and cytoskeleton-anchoring mixed lineage leukemia translocated protein MLLT4.

[122] Adaptor RIAM binds to activated Rap1 to cause adhesion of T lymphocytes to the extracellular matrix. It also connects to VASP proteins and actin-elongation factor profilin.

Cell stretch activates Rap1 using RapGEF1 and CRK2 (Rap1–CRK2–RapGEF1 pathway) [908]. Stretch-induced phosphorylation of CRK-associated substrate (CAS or BCAR1) of the cytoskeleton is due to SRC family kinases. Adaptor CRK2 binds directly to phosphorylated CAS, especially in cell–matrix adhesion sites.

Small Rap1 GTPase is a member of the *shelterin* complex, or *telosome* complex, that regulates the length of telomeres and protects chromosome ends [909].[123] Shelterin protects chromosome ends from the activity of DNA-repair molecules.[124] Repressor and activator Rap1 protein[125] is not involved in the maintenance of telomere length, the organization of telomeric chromatin, and inhibition of non-homologous end joining (NHEJ), a repair process of double-strand DNA breaks at telomeres, but is required for the inhibition of homology-directed repair (HDR) [910].[126]

Small Rap1 GTPase not only protects telomeres from sister chromatid exchange, but also intervenes in transcriptional regulation. Protein Rap1 may bind to telomeric and extratelomeric $(TTAGGG)_2$ DNA consensus motifs.[127] It operates in transcriptional regulation and silences genes in proximal and subtelomeric regions [911].

Small Rap1 GTPase also controls the NFκB pathway that mediates the transcriptional response to various types of cellular and developmental signals and stresses [912]. A significant proportion of cytoplasmic Rap1 is constitutively bound to inhibitor of NFκB kinase (IκB kinases [IKK]). Protein Rap1 enables efficient recruitment of IKKs and phosphorylation of $P65_{NF\kappa B}$.[128] Moreover, NFκB may, in turn, regulate Rap1 production via the NFκB-binding site in the RAP1 promoter region.

---

[123]Telomere maintenance by telomerase, tankyrase, and telomeric proteins is necessary for normal cell life, as its disregulation can accelerate aging and cause cancer. Tankyrase-1, a member of the poly^{ADP}ribose polymerase (PARP) family of enzymes is also called TRF1-interacting ankyrin-related ^{ADP}ribose polymerase TIN1 and PARP5a. It binds to and adpribosylates TRF1 protein. The shelterin hexamer is composed of 2 double-stranded telomeric DNA-binding repeat-binding factors TRF1 and TRF2, a single-stranded telomeric DNA-binding protein Protection of telomeres POT1, TRF1-interacting nuclear factors TIN2, adrenocortical dysplasia homolog (ACD), a binding partner of POT1 (a.k.a. TIN2 (two)-interacting protein TINT1, POT1-interacting protein PIP1, POT1 and TIN2 organizing protein [PTOP], and TINT1/PTOP/PIP1 protein TPP1), and TRF2-interacting factor Rap1. Component TRF2 recruits and stabilizes the shelterin complex to telomeres. Component ACD regulates both POT1 telomere localization and telosome assembly via TIN2 binding. It prevents telomerase-mediated telomere elongation.

[124]In the absence of shelterin, factors of DNA-damage checkpoints and repair would process chromosome ends as double-strand breaks.

[125]A.k.a. telomeric repeat-binding factor TRF2-interacting protein TRF2IP1 and telomeric repeat-binding factor TeRF2-interacting protein (TeRF2IP).

[126]The HDR process can create undesirable telomeric sister chromatid exchange when NHEJ process is abrogated. Protein TRF2 inhibits HDR, as it tethers Rap1 to telomeres [910].

[127]Telomerase adds DNA sequence repeats (TTAGGG) to the 3' end of DNA strands in telomeres. Telomere, a region of repeated nucleotides, contains non-coding DNA material and prevents DNA loss from chromosome ends.

[128]In addition to their IκB substrates, IKK1 and IKK2 phosphorylate $P65_{NF\kappa B}$, P53, β-catenin, and 23-kDa synaptosomal-associated protein SNAP23.

Rap1a

Small GTPase Rap1a counteracts Ras mitogenic action, because it competes with Ras to connect to Raf and binds RasGAPs. Protein Rap1a can be phosphorylated by protein kinase-A and calmodulin-dependent kinase.

Rap1b

Small GTPase Rap1b is particularly detected in platelets. It is implicated in cAMP-associated signaling. Like Rap1a, Rap1b can be phosphorylated by cAMP-dependent protein kinase PKA. Phosphorylation of Rap1b by PKA (Ser179) is required for cAMP-dependent cell proliferation and PKB inhibition [913].

9.3.9.2 RAP2 Subfamily

Proteins Rap2a and Rap2b have a low intrinsic GTPase activity. Like Rap1, Rap2 GTPases contribute to synapse remodeling, particularly removal of AMPA receptors via Jun N-terminal kinases. Besides, neither Rap2a nor Rap2b antagonize Ras GTPases. The Rap2 proteins can also be involved in cell signaling, as they can mediate activation of phospholipase-Cε.

Rap2a

Small Rap2a GTPase can be farnesylated. It is ubiquitously expressed, with a higher level in brain and hematopoietic tissues. Kinase MAP4K4 interacts with Rap2 to enhance activation by MAP4K4 of Jun N-terminal kinase [914].

Rap2b

Small Rap2b GTPase undergoes post-translational modifications. It can indeed be polyisoprenylated and palmitoylated. Small GTPase Rap2b is involved in platelet activation.[129] Like Rap1b that is activated upon thrombin and ADP stimulation mainly via Ca^{++}-dependent signaling and the Gi–PI3K pathway (Gi-coupled $P2Y_{12}$ receptor), respectively, Rap2b is stimulated by Ca^{++}, PKC, PI3K upon excitation of G-protein-coupled receptors by ADP and receptor tyrosine kinases by the collagen receptor platelet glycoprotein-6 (GP6) [915]. However, Rap1b rather than Rap2b seems to be directly regulated by thrombin receptors, whereas glycoprotein-6 stimulation preferentially leads to activation of Rap2b rather than Rap1b.

---

[129] Human platelets produce at significant levels Rap1b and Rap2b, the amount of Rap1b being about 10-fold higher than that of Rap2b [915].

Rap2c

Small GTPase Rap2c is widely expressed (mainly in skeletal muscle, liver, digestive tract, bladder, prostate, uterus, and, to a lesser extent, in brain, kidney, pancreas, and bone marrow) [916]. Small Rap2c GTPase is the predominant type in circulating mononuclear leukocytes. It is also expressed in human megakaryocytes, but not in platelets [917].

Small GTPase Rap2c can be involved in gene transcription via the serum response element in the promoter region of target genes [916]. It binds to GTP in the presence of Mg^{++}, but less efficiently than homologous Rap2b GTPase. It also has a slower rate of GDP release than that of Rap2b GTPase [917].

9.3.10 Ras GTPases

Membrane-associated Ras[130] and Rho GTPases activate intracellular pathways in response to extracellular signals for multiple functions. Activation of Ras by receptor Tyr kinases involves the binding of Ras to Ras-specific GEF Son of sevenless, which has been recruited to the plasma membrane. The binding is followed by nucleotide exchange with SOS agent. G-protein-coupled receptors can also initiate Ras signaling, using a SOS homolog, Ras guanine nucleotide-releasing factor-1 (RasGRF1), which, like SOS, catalyzes Ras nucleotide exchange.

Signal transduction by growth factor receptors proceeds via recruitment to the cell cortex of cytosolic signaling effectors such as adaptor growth factor receptor-bound protein GRB2, which then attracts nucleotide-exchange factor Son of sevenless. Colocalization at the cell membrane increases the effect of effectors. The kinetics of Ras activation by SOS indeed changes according to Ras location. Signaling is more efficient when Ras is tethered to phospholipid membranes instead of being in cytosol, as the activity of SOS catalytic unit is about 500-fold higher when Ras is on membranes [918]. Protein SOS has 2 Ras-binding sites that can both be simultaneously occupied by membrane-bound Ras. Small GTPase Ras transiently binds to the active site for nucleotide exchange. Occupation of the second Ras-binding site of SOS stimulates nucleotide exchange of SOS allosterically, by causing conformational changes of active site and favoring Ras binding. Access to the allosteric site that anchors SOS catalytic unit to the membrane is controlled by SOS regulatory domain. Binding of Ras^{GTP} to the allosteric site leads to a positive feedback loop for Ras activation. The N-terminal segment of SOS can occlude the allosteric site, hence blocking membrane anchoring and preventing unchecked activation of Ras. On the other hand, the large increase in activity of

---

[130]Proteins of the RAS superfamily consist of 6 β-sheets and 5 α-helices interconnected by a set of 10 loops. They have RasGEF and RasGAP-binding interfaces. Inactive (Ras^{GDP}) and active (Ras^{GTP}) conformations differ by switch-1 and -2 regions.

SOS catalytic unit requires Ras binding to SOS allosteric site and conversion of Ras^{GDP} to Ras^{GTP}. Protein SOS responds depending on Ras membrane density, GTP loading, and membrane concentration of phosphatidylinositol (4,5)-bisphosphate. Agent SOS integrates these data to relieve its auto-inhibition with greater efficiency.

9.3.10.1 RAS Hyperfamily

Small Ras GTPases include Ras, Rap, Ral, and others [919]. Small GTPases Rho, Rac, and CDC42 are the 3 best known families of the RHO superfamily. Kinases of the RoCK family are effectors of Rho GTPases. Small Rac and CDC42 GTPases act via P21-activated kinases. Small Rho GTPases regulate cytoskeletal activity during cell motility, shape change, and contraction.[131] Each RHO family is characterized by specific effects on the actin cytoskeleton. Agent Rho is involved in the formation of stress fibers, Rac of membrane ruffles and lamellipodia, and CDC42 of filopodia (radial unipolar bundles). Members of all RHO families also regulate the assembly of integrin-containing adhesion complexes, and thus cell–matrix interactions and cell adhesion.

9.3.10.2 Members of the RAS Family

Members of the Ras family constitute 2 subfamilies. A first subfamily includes Harvey hRas, Kirsten kRas,[132] and neuroblastoma nRas, whereas related rRas isoforms are grouped in a second subfamily. In mammals, 3 RAS genes generate subfamily-1 Ras GTPases (hRAS, kRAS, and nRAS). The KRAS transcript can be alternatively spliced, thereby producing kRas4a and kRas4b subtypes.

9.3.10.3 Post-Translational Modifications

Post-translational modifications are aimed at enabling proper protein folding and localization, signal transmission, as well as protein degradation. Post-translational modifications of Ras GTPases direct them to various cellular membranes and, in some cases, modulate GTP–GDP exchange [920]. These modifications are either *constitutive*, such as irreversible farnesylation of the C-terminal CAAX

---

[131] The regulation of the non-muscle and smooth muscle F-actin cytoskeleton involves phosphorylation of the regulatory myosin light chains, leading to the formation of actin filaments.

[132] Aliases hRas and kRas stand for proto-oncogene products associated with Harvey and Kirsten sarcoma virus-associated oncogenes, respectively.

motif,[133] reversible palmitoylation,[134] methylation, and proteolysis of a C-terminal propeptide,[135] or *conditional*, such as reversible phosphorylation,[136] peptidyl-prolyl isomerisation,[137] mono- and diubiquitination,[138] nitrosylation,[139] adpribosylation, and glucosylation.

Membrane tethering and trafficking of small GTPases of the P21RAS subfamily other than kRas4b are regulated by reversible palmitoylation of Cys residues in their C-terminus [920].[140] The covalent attachment of the acyl chain of a fatty acid to a protein is called protein acylation. Unlike farnesylation, palmitoylation is reversible. Reversibility is ensured by one or more thioesterases, such as acyl-protein thioesterase APT1. Both hRas and nRas undergo palmitoylation–depalmitoylation (acylation–deacylation) cycles for Ras transfer from (anterograde) and to (retrograde transport) the Golgi body. Palmitoylation is required for the transfer of hRas and nRas from endomembranes to the plasma membrane. Farnesylated Ras has only a slight affinity for membranes, but farnesylated and palmitoylated Ras has a much higher affinity [920]. Mono- and dipalmitoylation control Ras localization.

Cis–trans isomerization of a peptidyl-prolyl bond adjacent to a palmitate in hRAS acts as a molecular timer that regulates depalmitoylation and retrograde transfer.

---

[133] In a CAAX sequence, C means Cys, A is usually, but not always, an aliphatic amino acid, and X is any amino acid. Farnesylpyrophosphate is added to Cys of the CAAX motif of Ras by farnesyltransferase via a stable thioether linkage [920]. The CAAX motif is sequentially processed to undergo prenylation, proteolysis, and then methylation by 3 enzymes: rate-limiting, cytosolic farnesyltransferase, endoplasmic reticulum RAS-converting enzyme-1, and endoplasmic reticulum isoprenylcysteine carboxylmethyltransferase. The 2 first reactions are irreversible, whereas the third is reversible.

[134] Palmitate, a saturated (acyl) fatty acid, is added to 1 or 2 Cys residues immediately upstream of the CAAX sequence of Ras GTPases by one or more transmembrane palmitoyl acyltransferases, such as the complex made of DHHC domain-containing DHHC9 and 16-kDa Golgi body-associated protein GPC16 and other Ras-processing, DHHC motif-containing palmitoyl acyltransferases, via a labile thioester bond [920].

[135] The AAX amino acids of the CAAX sequence are substrates for the peptidase Ras-converting enzyme RCE1. After proteolysis, the carboxyl group of the new C-terminal prenylcysteine of Ras is methylesterified by isoprenylcysteine carboxylmethyltransferase (ICMT) [920].

[136] In particular, protein kinase-C phosphorylates kRas4b (Ser181) [920]. Small RalA GTPase is phosphorylated (Ser194) by Aurora-A kinase.

[137] The Gly–Pro peptidyl-prolyl bond of hRas (at position 178–179) undergoes cis–trans isomerization processed by 12-kDa FK506-binding protein (FKBP12) [920].

[138] Agents hRas, nRas, and kRas4b can be mono- and diubiquitinated on Lys residues by the ligase rabaptin-5-associated exchange factor for Rab5 (RabEx5), or RabGEF1 [920].

[139] Residue Cys118 in all Ras isoforms operates as a redox indicator that can be nitrosylated when exposed to nitric oxide [920].

[140] Agent kRas4b is not palmitoylated. It possesses a Lys-rich domain that complements farnesylation.

Phosphorylation of kRas4b in its polybasic region allows kRAS4b dissociation from the plasma membrane. S-nitrosylation of Ras promotes guanine nucleotide exchange.

Mono- and diubiquitination of hRAS regulate its association with endosomes; kRas4b monoubiquitination enhances its activation. Ubiquitination regulates hRas transfer to and from endosomes. Effector of Ras GTPase Ras and Rab interactor RIn1 is required for RabEx5-dependent Ras ubiquitination.

9.3.10.4 Ras Signaling

Small Ras GTPases are involved in the regulation of cell growth, differentiation, and survival. They are encoded by genes that experience mutations in about 30% of cancers in humans.

The components of the P21RAS subfamily have a similar effector domain, which interacts with downstream pathway targets. Both Rap and Ras[141] can bind the same effectors to regulate intracellular signaling events.

In the GTP-bound conformation, Ras binds to and activates its effectors, members of the RAF family, phosphatidylinositol 3-kinase, and members of the RalGEF family [922] (Table 9.29). Effector of Ras for small Ras-like GTPases RalA and RalB are the guanine nucleotide-exchange factors Ral guanine nucleotide-dissociation stimulator (RalGDS or RalGEF) and RalGDS-like protein. Other effectors include phospholipase-Cϵ,[142] T-cell lymphoma invasion and metastasis GEF TIAM1,[143] and Ras interaction/interference protein RIn1, afadin, and Ras association domain-containing family proteins (RASSF) [923].[144]

---

[141] The Ras GTPases – hRas, nRas, and kRas – have isoform-specific effects, in particular due to their different location at the inner surface of the plasma membrane and possibly in the Golgi body membranes, both sites that are involved in Ras signaling. Small GTPases hRas and nRas are palmitoylated and can be located in the plasma membrane and Golgi body membranes. Small kRas GTPase localizes to the plasma membrane. A deacylation and reacylation (depalmitoylation and repalmitoylation) cycle maintains the specific intracellular distribution [921]. Moreover, the kinetics of hRas and nRas GTPase trafficking are different. Small hRas GTPase, stimulated by growth factors, is rapidly and transiently activated at the plasma membrane, whereas it has a delayed and sustained activation at the Golgi body membranes. Small GTPase nRas is activated sooner than hRas at the Golgi body. Inhibition of palmitoylation blocks Ras activation. Depalmitoylation redistributes farnesylated Ras in required membranes. Repalmitoylation, which occurs at the Golgi body, enables Golgi membrane anchorage. Protein Ras is then redirected to the plasma membrane by exocytosis.

[142] Enzyme PLCϵ cleaves PIP$_2$ into IP$_3$ and DAG that release Ca^{++} and activates PKC, respectively.

[143] Agent TIAM1 is involved in Ras-associated tumorigenesis.

[144] Proteins RIN1 and RASSF are Ras-interacting proteins that act as tumor suppressors. The RASSF family includes 8 identified members (RASSF1–RASSF8) that have pro-apoptotic function.

Table 9.29. Interactors of small Ras GTPases (Source: [920]).

Set	Agents
GEFs	RasGRF, RasGRP, SOS
GAPS	RasA1/2/3/4, RasAL, neurofibromin-1
Effectors	RalGEF, RASSF, RIn1, TIAM1 PI3K, PLCε, cRaf

Table 9.30. Small Ras GTPases and their effectors (Sources: [889,926]; PM: plasma membrane). Residence in the Golgi body results from a Golgi-localized palmitoylation reaction. Both hRas and nRas, but not kRas, are palmitoylatable.

Type	Location	Effectors
	P21RAS subfamily	
hRas	Golgi, PM	cRaf, PI3K, RalGEF
kRas	PM	cRaf, PI3K, RalGEF
nRas	Golgi, PM	cRaf, RasGRP2
cRaf activation order:		kRas > nRas > hRas
	RRAS subfamily	
rRas1		bRaf, cRaf, PI3K, PLCε, RalGEF
rRas2		PI3K
rRas3		Raf, PI3K
cRaf activation		Weak

Small GTPases Ras, hRas and kRas (Table 9.30) in particular, are activated by most growth factors as well as integrins. Activated receptor Tyr kinases dock adaptors such as GRB2 that attract RasGEFs. The latter recruit and activate Ras GTPases. Other receptors can stimulate Ras GTPases via SRC family kinases.

Members of the P21RAS subfamily of small GTPases — hRas, nRas, and kRas (kRas4a–kRas4b) isoforms — mediate mitogenic signaling triggered by growth factor receptors, hence, when mutated, cancer. These Ras proteins participate in cell proliferation and migration via the Raf–MAP2K–ERK pathway, but not survival, at least in mouse embryonic fibroblasts [924].[145] The Ras proteins mainly

---

[145] Selective ablation of Ras loci in mouse embryonic fibroblasts has shown that each Ras is capable of sustaining proliferation of these cells in culture in the absence of the other two [924]. However, kRas are more efficient to cause cell proliferation than hRas or nRas. Other Ras-like small GTPases, even constitutively active types, cannot compensate for the absence of Ras proteins. The PI3K–PTen–PKB and RalGEF–Ral pathways, either alone or combined, do not launch cell proliferation or migration of Ras-less cells, although they are able to cooperate with the Raf–MAP2K–ERK axis [924]. Moreover, Ras signaling does not induce cell proliferation via the expression of cyclin-D, as CcnD1–CDK4 and CcnE–CDK2 complexes have normal concentrations

signal via the Raf–MAP2K–ERK, PI3K–PTen–PDK1–PKB, and RalGEF–Ral pathways [924].

In addition to RalGEF, Ras effectors involve other guanine nucleotide-exchange factors, such as TIAM1 and RalGDS. Agent TIAM1 activates small GTPases of the RHO and RAC families to regulate cell polarity, motility, and adhesion. Molecule RalGDS stimulates Ral GTPases that are primarily involved in membrane trafficking.

Additional effectors include components of the PLCε–PKC–Ca^{++} cascade, afadin (or AF6 GEF) that regulates intercellular adhesions and can interact with cytoskeletal structures, Ras and Rab interactor RIn1, a GEF, and Ras association domain family (RASSF) proteins.

On the other hand, constitutively active GTPases, such as members of the RRAS subfamily and embryonic stem cell-expressed Ras (eRas or hRAS2), activate the Raf–ERK pathway.

Small Ras GTPases can influence ion channel activity (Sect. 11.4). Small GT-Pases Rap and Ras have opposite effects on atrial M_2 muscarinic receptor-coupled K^+ channels (GIRK). Small GTPase Rap1 also antagonizes Ras action on voltage-gated Na^+ channels, ^{NMDA}Glu receptors and ^{AMPA}Glu receptors [925].

Tumor suppression by Ras results from cell cycle arrest and/or senescence that can exceed unlimited proliferation and/or transformation [923]. Yet, deregulated Ras signaling causes developmental disorders (cardiofaciocutaneous syndromes) characterized by accumulation of tumors and skeletal, cardiac, and visual abnormalities.

The RRAS subfamily includes rRas isoforms (rRas1–rRas3). Protein rRas[146] antagonizes hRas signaling. In vitro, rRas enhances integrin-mediated cell adhesion, whereas hRas inhibits integrin activities. rRas promotes the differentiation of myoblasts, whereas hRas inhibits it. rRas regulates cell survival and integrin activity, particularly in the remodeling of blood vessels. In vivo rRas is mainly expressed by smooth muscle and endothelial cells [927]. In the absence of rRas, neointimal thickening in response to injury and tumor angiogenesis are increased, whereas rRas expression is greatly reduced in hyperplastic neointimal smooth muscle cells and angiogenic endothelial cells.

Specific interactions between isoforms of membrane-anchored Ras and their guanine nucleotide-exchange factors, GTPase-activating proteins, and effectors depends not only on distinct molecular structure (e.g., C-terminal hypervariable region that is modified by lipids, anchors the G domain to the membrane, and allows distinct exocytosis sorting, subcellular localization, including segregation

---

in Ras-less mouse embryonic fibroblasts, but remain inactive. Signaling from Ras initiates the cell cycle by activating pre-existing CcnD–CDK4, CcnD–CDK6, or CcnE–CDK2 complexes rather than by inducing expression of cyclins-D [924].

[146]Protein rRas differs from the other members of the Ras family. It contains a proline-rich SH3 domain binding site. It can be phosphorylated by EPH receptors and Src. Both the SH3 binding site and phosphorylation regulate rRas activity.

into specific nanodomain in the plasma membrane),[147] but also recognition of orientation of the G domain of small GTPase Ras isoforms with respect to the plane of the plasma membrane [928]. Orientation of the G domain of small hRas GTPase that differs from that of kRas, mRas (or rRas3), and nRas, is recognized by effectors cRaf and PI3Kα as well as scaffold galectin-1 (that create nanoclusters of active hRas).

9.3.10.5 Ras Localization, Post-Translational Modifications, and Signaling

Small Ras GTPases lodge in various intracellular membranes, in addition to the plasma membrane. Golgi-associated hRas can be activated after growth factor stimulation with kinetics that differ from that of hRas activation at the plasma membrane. Furthermore, hRas tethered to the endoplasmic reticulum is able to activate the extracellular signal-regulated kinase and protein kinase-B preferentially, whereas a Golgi-tethered hRas can activate predominantly the Jun N-terminal kinase [929]. Therefore, the subcellular localization of Ras influences the type of engaged effectors. The activation of hRas localized at given subcellular compartments may be mediated by specific guanine nucleotide-exchange factors with a restricted residence. The subcellular localization depends on adequate post-translational modifications, especially lipidation (Sect. 9.3.10.9).[148]

9.3.10.6 Membrane Ras Nanoclusters

Plasmalemmal nanodomains with specialized protein and lipid composition regulate not only cell transport (membrane rafts and caveolae), but also cell signaling. Small Ras GTPases — hRas, kRas, and nRas — either reside in plasmalemmal nanoclusters (bore of ~ 9 nm; lifetime 0.5–1.0 s) or correspond to freely diffusing monomers. Small GTPases hRas and nRas are farnesylated and acylated, whereas kRas is only farnesylated for stable anchoring to the plasma membrane.

Plasmalemmal Ras clusters are more or less associated with cholesterol and cortical components of the cytoskeleton. Plasmalemmal kRas nanoclusters are cholesterol-independent and actin-dependent, whereas hRas nanoclusters are cholesterol-dependent and actin-independent. Yet, the link with elements of the cell membrane and cortex depends not only on Ras isoform, but also on its bound nucleotide. Plasmalemmal Ras-containing nanoclusters can actually be segregated

---

[147] Differences in nanocompartmentation between hRras and kRras may partly explain isoform specific differences in MAPK signaling. Kinase cRaf is, indeed, selectively retained in kRas, but not in hRas nanoclusters.

[148] Prenylation is the attachment of lipid chains to proteins to enable their attachment to cell membranes. Bound fatty acids, such as palmitate or myristate, serve to anchor proteins to inner or outer face of a given membrane.

9.3 Monomeric (Small) GTPases

into GTP- and GDP-loaded protein assemblies. Nanoclusters that contain hRas^{GDP} depend on cholesterol and actin, but not hRas^{GTP} nanoclusters. Nanoclusters that possess kRas^{GTP} depend weakly on actin and do not depend on cholesterol. Assembling of hRas^{GTP} and kRas^{GTP} nanoclusters rely on scaffolds galectin-1 or -3, respectively. The higher the Ras^{GTP} concentration, the greater the number of nanoclusters in the plasma membrane.

In particular, the mitogen-activated protein kinase pathway is underpinned on nanoclusters of small GTPase Ras on the inner leaflet of the plasma membrane.[149] Ras-containing nanoclusters are required for MAPK-based signaling; indeed, Ras^{GTP} nanoclusters recruit Raf and KSR–MAP2K–ERK complexes from the cytosol [591]. Signaling from ERK terminates by spontaneous disassembly of nanoclusters.

Plasmalemmal nanodomains aimed at signaling yield a local concentration of signaling components. They also allow segregation of signaling components inside and outside nanoclusters. These sites act as switches by recruiting and activating pathway effectors causing maximal output above a given threshold rather than eliciting a graded signaling. However, a graded response directly proportional to input can be provided by the formation and activation of nanoclusters.

Signal transmission deeply depends on Ras spatial organization, i.e., on transient Ras cluster tethered to the inner leaflet of the plasma membrane. The plasmalemmal concentration of specific signaling effectors increase the efficiency and specificity of signaling using the Ras–Raf–MAP2K–ERK module. A set of plasmalemmal nanodomains of Ras GTPases convert graded ligand inputs into adapted outputs of activated extracellular signal-regulated kinase [930]. The higher the ligand concentration, the larger the number of transient Ras clusters (linear relationship).

Activation of Ras in association with scaffold proteins such as galectins triggers the formation of signaling clusters that can then activate the MAPK module. The regulated motion of scaffold proteins can control the extent of Ras clusters and hence modulate MAPK signaling magnitude. The recruitment of MAPK components to nanoclusters during cluster lifetime not only locally increases the kinase concentration and favors activation, but also protects them from degradation by phosphatases.

Post-translational lipid processing (palmitoylation) of Ras allows its attachment to the cell membrane, a prerequisite to its activation. The addition of a farnesyl isoprenoid lipid is catalyzed by farnesyl transferase. This reaction is followed by proteolytic cleavage processed by Ras-converting enzyme-1 and carboxymethylation by isoprenylcysteine carboxymethyltransferase-1 [923]. Farnesylated kRas4b is transported directly to the plasma membrane, whereas hRas, nRas, and kRas4a undergo palmitoylation in the Golgi body.

---

[149]Unlike hRas and nRas, kRas localizes exclusively to the plasma membrane, where it recruits cRaf.

9.3.10.7 hRas

Small hRas GTPase (Harvey rat sarcoma virus oncogene homolog) is encoded by the HRAS gene. It relays growth regulatory signals that are initiated from plasmalemmal receptors, as, upon activation by inducers, hRas^{GTP} can interact with effectors. Ubiquitous hRas is predominantly localized to the inner leaflet of the plasma membrane.

The weak intrinsic GTPase activity of hRas is enhanced by association with GTPase-activating proteins, such as RasA1 (or p120RasGAP) to RasA4 and neurofibromin-1 [931]. Conversely, hRas^{GDP} is converted into active hRas^{GTP} when hRas links to guanine nucleotide-exchange factors, such as SOS1, SOS2, RasGRF1, RasGRF2, and RasGRP4.

Effectors of hRas include Raf isoforms (aRaf–cRaf), Ral guanine nucleotide-dissociation stimulator RalGDS (or RalGEF), RalGDS-like proteins RGL1 and RGL2, which bind hRas^{GTP}, but not hRas^{GDP}, Ras-responsive Ub ligase Impedes mitogenic signal propagation (IMP) that inactivates KSR scaffold, which couples activated Raf to MAP2K [932], RalGEFs AF6 and Ras and Rab interactor RIn1,[150] Ras-association (RalGDS–AF6) domain-containing proteins RASSF1 and RASSF5, PI3K$_{c1\alpha}$, PI3K$_{c1\delta}$, and PI3K$_{c1\gamma}$ subunits, and PLCϵ [931].

Small hRas GTPase decreases the activity of inward rectifier K$_{IR}$2.1 channel (IRK1), as it promotes its internalization [925]. On the other hand, it elevates Ca$_V$3 channel functioning.

9.3.10.8 kRas

Small kRas GTPase (Kirsten rat sarcoma viral oncogene homolog) is encoded by the KRAS gene. (Mutations of the KRAS gene produce potent oncogenes.) Protein kRas is usually tethered to cell membranes via a C-terminal isoprenyl group. Like hRas and nRas, kRas is farnesylated, but does not contain a palmitoylatable cysteine. Phosphorylation of a serine residue causes kRas to target mitochondria. It can also reside in the endoplasmic reticulum and late endosomes [902].

Small GTPase kRas interacts with cRaf of the MAPK module, Ras-association domain-containing protein RASSF2, Ral guanine nucleotide dissociation stimulator (RalGDS or RalGEF), and PI3Kγ. In addition, small GTPase kRas increases activity of epithelial Na$^+$ channel (ENaC) via PI3K and PI(4)P5K enzymes.[151]

---

[150] Guanine nucleotide-exchange factors of Ral — AF6 and RIn1 — are junctional adhesion molecule-binding partner [933]. Protein Ras and Rab interactor is an effector for membrane-associated Ras that is recruited to the Golgi body by activated Ras [934].

[151] Aldosterone augments ENaC density and activity in kidneys [925]. ENaC channels connect to PI(3,4,5)P$_3$ that raises the channel activity.

Nucleolar CKI2A-locus alternate reading frame tumor suppressor ARFTS (or P19^{ARF})[152] controls the activity of kRas GTPase [935]. It also indirectly activate P53 transcription factor, as it inhibits DM2 ubiquitin ligase.

9.3.10.9 nRas

Small nRas GTPase (neuroblastoma rat sarcoma viral oncogene homolog) is encoded by the NRAS gene that generates to 2 main transcripts. Both hRas and nRas are continuously exchanged between 2 cellular pools, the Golgi body and plasma membrane, at a relatively fast rate; this *spatial cycle* that depends on reversible palmitoylation is superimposed on the *temporal GTPase cycle*. Palmitoylatable hRas and nRas proteins are farnesylated, hence weakly attached to membranes and able to rapidly diffuse throughout the cell. However, because hRas and nRas are reversibly palmitoylated (at cysteine residues close to the farnesylation site), the resulting additional hydrophobicity confers higher stability on cellular membranes [902]. Switching between high- and low-membrane-affinity associated with the palmitoylation–depalmitoylation cycle due to the action of palmitoyltransferases (for palmitoylation and repalmitoylation) and thioesterases (for depalmitoylation) together with localized acylation–deacylation events creates the spatial cycle. Depalmitoylation occurs throughout the cytoplasm by acyl protein thioesterase-1 (APT1). Due to its ubiquitous distribution within the cell, APT1 enzyme converts by depalmitoylation mislocalized palmitoylated Ras to the fast-diffusing solely farnesylated form.

9.3.10.10 rRas

Small rRas GTPases (related rat sarcoma viral oncogene homolog) are encoded by genes of the RRAS group (RRAS1–RRAS3). Proteins rRas2 and rRas3 are also termed TC21 and mRas (muscle rat sarcoma viral oncogene product homolog), respectively.

Small rRas GTPase interacts with Ras-association domain-containing protein RASSF5, RalGEF, adaptor NCK1 that links receptor Tyr kinases to Ras, anti-apoptotic molecule BCL2, and kinase aRaf [251]. Protein aRaf binds to MAP2K2, translocase of inner mitochondrial membrane TIMM44, EGF-containing fibulin-like extracellular matrix protein EFEMP1, PRP6 pre-mRNA processing factor PRPF6, a bridging factor between snRNPs in the spliceosome, and TH1-like of the NELF complex that interacts with the DSIF complex to repress transcriptional elongation by RNA polymerase-2.

---

[152] Alias ARFTS is here used rather than the usual abbreviation ARF to avoid confusion with small GTPase ARF (adpribosylation factor).

9.3.10.11 Ras-Association Domain Family Members

Molecular switches Ras control cell proliferation, differentiation, motility, and apoptosis according to extracellular signals. Their effectors specifically bind to Ras^{GTP} GTPases. These effectors include not only Raf kinases of the MAPK module and PI3K, but also Ras-association (RA; RalGDS/AF6) domain[153] family proteins (RASSF1–RASSF8) that act as tumor suppressors [936].

RASSF1

Seven different transcripts are generated (RASSF1a-RASSF1g) by differential promoter use and alternative splicings [936]. Ubiquitous RASSF1a and RASSF1c are the major isoforms. Subtype RASSF1b is expressed predominantly in hematopoietic cells. Four isoforms RASSF1d to RASSF1g are splice variants of RASSF1a protein. Transcripts RASSF1d and RASSF1e are expressed specifically in cardiac and pancreatic cells, respectively. Protein RASSF1 interacts with plasma membrane calmodulin-dependent calcium ATPase PMCA4b [936].

Isotype RASSF1a is a microtubule-binding and -stabilizing molecule that control genome stability during the cell division cycle, as it impedes activated Ras to cause genomic instability. Subtype RASSF1a localizes to microtubules in interphase, relocalizes to separated centrosomes during prophase, then to spindle fibers and poles during metaphase and anaphase, and finally to the midbody during cytokinesis [936]. Isoform RASSF1c has identical cellular locations as those of RASSF1a. Protein RASSF1a primes cell cycle arrest by engaging the checkpoint retinoblastoma proteins that regulates entry into S phase.

In addition, RASSF1a as well as RASSF5 may serve as detectors of pro-apoptotic signals initiated via the Ras pathway. Effectors RASSF1a or RASSF1c can bind scaffold Connector enhancer of kinase suppressor of Ras CnK1 that allows Ras to activate Raf kinase. Isoform RASSF1a can also connect to kinases STK3 and STK4 (Sect. 10.7), hence favoring CnK1-induced apoptosis [936]. Moreover, binding of RASSF1a to Modulator of apoptosis MAP1 provokes MAP1 association with BAX protein. Activation of BAX leads to cell death. Besides, RASSF1c is a partner of Death-associated protein DAP6 (DAXX).

---

[153] The RA domain is a conserved motif defined by sequence homologies between 2 Ras effectors: (1) Ral guanosine nucleotide-exchange factor (RalGDS) involved in Ras-induced transformation and (2) Afadin (Aggressive acute lymphocytic leukemia ALL1 fusion partner from chromosome-6 [AF6] or myeloid–lymphoid or mixed lineage leukemia translocated to 4 [MLLT4]) implicated in intercellular adhesion.

RASSF2

Protein RASSF2, originally termed Rasfadin, possesses 3 isoforms (RASSF2a–RASSF2c), but only the RASSF2a transcript is translated [936]. Overexpression of RASSF2 causes apoptosis via caspase-3 activation and cell cycle arrest.

RASSF3 and RASSF4

The RASSF3 gene can produce 3 transcripts (RASSF3A–RASSF3C) by alternative splicing. Protein RASSF3a corresponds to RASSF3 that is longer than RASSF3b and RASSF3c isoforms [936]. Protein RASSF4 is broadly expressed in human tissues (heart, brain, lung, liver, skeletal muscle, pancreas, and placenta) [936].

RASSF5

Protein RASSF5[154] possesses 2 isoforms RASSF5a and RASSF5b [936]. It associates with hRas and kRas as well as, with a comparable affinity, Ras-like GTPases, such as Rap1 and Rap2, as well as rRas, rRas2, and rRas3 (mRas). Like RASSF1, it impedes cell proliferation. It also modulates Ras signaling triggered by stimulated TCR via recruitment of active Ras to the plasma membrane and control of localization of caspase recruitment domain family member CARD11, a membrane raft-associated regulator that supports TCR-induced NFκB activation.

RASSF6 to RASSF8

RASSF6 Protein exists in 2 forms (RASSF6a–RASSF6b). It may interact with kRas GTPase. It has pro-apoptotic effects [936]. Protein RASSF7[155] possesses 3 splice variants (RASSF7a–RASSF7c) [936]. Protein RASSF8 also exist in several isoforms [936].

9.3.10.12 Small GTPases Ras and Cancer

Many cancers are associated with mutations in RAS genes that remain in the active (GTP-bound), oncogenic state. Mutations of the KRAS gene initiate and maintain tumor growth by stimulating the PI3K–PKB pathway that phosphorylates (activates) nitric oxide synthase NOS3 [937]. The latter synthesizes nitric oxide that

---

[154] A.k.a. regulator for cell adhesion and polarization enriched in lymphoocytes (RAPL) and novel Ras effector NoRE1.

[155] Originally termed hRAS1 cluster HRC1.

promotes Ras nitrosylation and GTP binding of hRas and nRas needed for growth and maintenance of kRas-mutant tumors.

9.3.10.13 Small GTPases Ras and Immunity

Many GTPases of the RAS hyperfamily (Ras, Rap1a, CDC42, Rac1, Rac2, and RhoA) contribute to signal transduction primed by antigen receptors, costimulators, and cytokine and chemokine receptors to regulate the immune response [938]. Small Ras GTPase is activated by T-cell antigen and cytokine receptors (particularly IL2R and IL15R). It mediates some of preTCR signals that select thymocytes.[156] In mature T lymphocytes, Ras targets transcription factors that regulate cytokine genes, such as ELk1, SRF, AP1, and NFAT factors.

In lymphocytes, Ras GTPases are regulated by RasA1 (or p120RasGAP) and phosphorylated adaptors such as docking proteins DOK1 and DOK2 on the one hand and the single RasGEF in lymphocytes SOS that tethers to GRB2 on the other [938].[157] In addition, lymphocytes express Ras guanine nucleotide-releasing protein (RasGRP) that associates with diacylglycerol.

Among members of the RHO superfamily of the RAS hyperfamily, CDC42 regulates interactions between T lymphocytes and antigen-presenting cells and chemokine-stimulated lymphocyte migration [938]. Other member RhoA controls thymocyte development as well as integrin-mediated cell adhesion. Small Rac1 GTPase is involved in T-cell activation. The RacGEF Vav1 is phosphorylated in response to engagement of antigen receptor or costimulator CD28 with their ligands.[158] Both Rac1 and Rac2 modulate distinct T-cell functions. Only Rac1 can direct thymocytes from positive to negative selection, whereas Rac2 causes apoptosis and depletion of CD4+, CD8+ double-positive lymphocytes [938].

Small Rap1a GTPase is rapidly activated after antigen receptor ligation in both B and T lymphocytes via Fyn kinase that promotes formation of a complex with adaptors CBL and CRKL and RapGEF1 [938]. Small GTPase Rap1a can also be activated via increase in intracellular calcium concentration, diacylglycerol release, and cAMP synthesis. It can antagonize Ras, and, conversely, can be antagonized

---

[156] Early thymocyte progenitors initiate rearrangements of the TCR locus that enables production of functional preTCR complexes. The latter primes thymocyte proliferation and differentiation. T lymphocytes then express mature TCR complexes.

[157] Adaptor GRB2 then binds phosphorylated receptors and adaptors, such as plasmalemmal linker for activated T cells (LAT). Adaptor LAT is a substrate for the kinase 70-kDa ζ-associated protein (ZAP70). It acts also as a scaffold for various signaling complexes, as it interacts with phospholipase-Cγ1 and lymphocyte cytosolic protein LCP2 (or SH2 domain-containing phosphoprotein of 76 kDa [SLP76]) via GRB2-like adaptors such as GRB2-related adaptor protein GRAP2.

[158] Like LAT, Vav1 forms proteic complexes with ZAP70 kinase and LCP2 adaptor in activated T lymphocytes. In T lymphocytes, Vav1 is the major type, whereas in B lymphocytes, Vav1 and Vav2 have redundant functions.

by CD28 agent. However, Rap1A and Ras may be involved in different signaling pathways, so that active Rap1A does not interfere with Ras signaling and T-cell activation [938].

9.3.11 RHEB GTPase

Small GTPase Ras homolog enriched in brain RHEB is involved in cell growth and nutrient uptake. Tuberous sclerosis complex TSC2 is a GTPase-activating protein that deactivates RHEB GTPase.[159] Proteins TSC1 and TSC2 form a complex that inhibits phosphorylation of S6K and 4eBP1 proteins. The PI3K–PKB axis activates RHEB by inhibiting TSC1 or TSC2. In response to nutrients and cellular energy status, RHEB then activates Ser/Thr protein kinase target of rapamycin that phosphorylates kinase S6K and translational repressor initiation factor-4E for eukaryotic translation (eIF4e)-binding protein 4eBP1 [939]. In addition, RHEB promotes the formation of late endosomes [940]. It regulates endocytic trafficking pathway independently from the mTOR pathway.

9.3.12 Superfamily of RHO GTPases (CDC42, Rac, and Rho)

The RHO superfamily of small GTPases encompasses 3 major families: RHO (RhoA–RhoC), RAC (Rac1–Rac3) and CDC42 families. Each of these monomeric GTPases controls the formation of a distinct cytoskeletal constituent.[160] In fact, families of the RHO superfamily include: (1) RHO family with its 3 isoforms; (2) RAC family also with 3 isoforms and RhoG; (3) CDC42 family with 2 isoforms; (4) RND family with 3 isoforms (Rnd1, or Rho6, Rnd2, or Rho7, and Rnd3, or RhoE; (5) MIRO family with 2 isoforms (Miro1–Miro2, or RhoT1–RhoT2); as well as families that contain (6) RhoJ and RhoQ; (7) RhoU and RhoV; (8) RhoD, (9) RhoF, and (10) RhoH GTPase.

Major activators of members of the RHO superfamily comprise G-protein-coupled receptors, receptor Tyr kinases, adhesion molecules, and mechanical stresses (tension, compression, and shear). These signaling mediators are, indeed,

---

[159]Tuberous sclerosis is a genetic disease caused by mutations in either the Tsc1 or Tsc2 genes.

[160]Small Rho GTPases regulate bundling of actin filaments into stress fibers and formation of focal adhesion complexes. Activation of Rac GTPases causes actin polymerization to form lamellipodia (broad cytoplasmic extensions) during cell migration (Vol. 2 – Chap. 6. Cell Motility), whereas activated CDC42 stimulates actin polymerization in filopodia (long and thin cytoplasmic extensions).

activated upon stimulation by hormones, growth factors, cytokines,[161] adhesion molecules, especially integrins[162] and cadherins, to reorganize the actin cytoskeleton and, hence, regulate vesicle release, transfer, and uptake, activate NADPH oxidase, contribute to gene transcription, execute cell division cycle or apoptosis, control morphogenesis, and participate in platelet activation and aggregation during blood coagulation as well as neutrophil activation and phagocytosis in immune response, among other biological events. Small GTPases CDC42, Rac1, and RhoA modulate dynamics of contractile actin–myosin-2 filaments via their associated kinases and other target proteins (Tables 9.31 to 9.33).

Multiple Rho-associated Ser/Thr kinases interact with and are regulated by their partner GTPases: citron Rho-interacting kinase (CRIK), mixed lineage kinases (MLK) of the MAPK module (MAP3K9–MAP3K11 and MAP3K15, in addition to MLK4), myotonin-related CCD42-binding kinase (MRCK), P21-activated kinase (PAK), protein kinase novel (PKN), and RoCK kinase. All of these kinases can dimerize [943].

Myosin phosphorylation depends mainly on the balance of 2 enzymes: Ca^{++}-dependent regulatory myosin light-chain kinase and phosphatase. Phosphorylation of myosin-2 light chain provokes its interaction with actin and subsequent activation of myosin ATPase to cause cell contraction. Kinase RoCK phosphorylates MBS subunit (Thr697, Ser854, and Thr855) to prevent dephosphorylation and also phosphorylates regulatory myosin light chain (Ser19). In addition, it phosphorylates LIMK2 (Thr505) that phosphorylates (inactivates) cofilin, an actin-depolymerizing agent, to promote actin–myosin-2 assembly.

Among Rho interactors and indirect effectors, activated proteins ezrin, radixin, and moesin (ERM) connect directly to adhesion molecules, such as intercellular adhesion molecules ICAM1 to ICAM3 and indirectly with other integral membrane proteins such as Na^+–H^+ exchanger NHE3 (SLC9a3) via NHE regulatory factor (NHERF) [941].[163] Activated ERM proteins also bind to RhoGDI to activate Rho GTPase that, in turn, can stimulate ERM proteins beneath the plasma membrane (positive feedback). Kinectin (Ktn1), a Dia-related protein, interacts with RhoA and CDC42 GTPases. Citron Rho-interacting kinase (CRIK or STK21) is a RoCK-related kinase that operates in cytokinesis and other aspects of cell cycle progression. Diaphanous proteins (Dia1–Dia3) mediate both actin polymerization and stabilization of microtubule plus-ends during cell migration.

---

[161] Growth factors and cytokines activate Rho GTPases via interactions with guanine nucleotide-exchange factors, GTPase-activating proteins, guanine nucleotide-dissociation inhibitors, and GDI dissociation molecules.

[162] Integrins can both directly activate and enhance growth factor activation of small GTPases Rac and CDC42 via their recruitment to the plasma membrane. Integrins contribute to activation of phosphatidylinositol 3-kinase, possibly via focal adhesion kinase and CDC42, and promote polymerization and organization of actin filaments.

[163] A.k.a. ERM-binding phosphoprotein EBP50 and SLC9a3r1.

Table 9.31. Targets of small GTPases of the Rho superfamily (**Part 1**; Sources: [941–943]; ACK: activated CDC42-associated kinase; ARP: actin-related protein; ERK: extracellular signal-regulated kinase; ERM: ezrin, radixin, and moesin; JNK: Jun N-terminal kinase; IQGAP: IQ motif-containing GTPase-activating protein; MLC: myosin light chain; MLK: mixed lineage kinase; MRCK: myotonic dystrophy kinase-related CDC42 binding kinase; Par: partitioning defective protein; Par6–Par3–aPKC: Par6–Par3–atypical PKC complex; PIP_2: phosphatidylinositol (4,5)-bisphosphate; PIP5K: phosphatidylinositol 4-phosphate 5-kinase; PKB/C: Protein kinase-B/C; WASP: Wiskott-Aldrich syndrome protein; WIP: WASP-interacting protein).

Effector	Effect
	CDC42
ACK1/2	Focal adhesion formation
Binders of Rho GTPases	Cytokinesis via septins
Coatamers	Membrane trafficking
IQGAP	Intercellular adhesion, microtubule orientation
MRCK	Promotion of cell contraction and focal adhesion
	LIMK2 phosphorylation (activation)
P21-activated kinases	LIMK phosphorylation (cofilin inactivation)
(PAK1–PAK4)	Stathmin phosphorylation (microtubule stability)
	Triggering of cRaf–MAP2K1–ERK1 axis
	Stimulation of MLK–JNK pathway
	(activation of transcription factors Jun, Fos, Elk1)
	LIMK1 activation (PAK1/4)
	MLCK inhibition (PAK1)
	MRCK phosphorylation
Par6–Par3–aPKC	Cell polarity (tight junction)
	Reorientation of the microtubule organizing center during cell migration
PI3K	Stimulation of RacGEFs,
	WASP (actin polymerization); PKB activation
PIP5K	PIP_2 production
	(ERM processing for subsequent phosphorylation)
WASP, WIP	Formation of filopodia via $PI(4,5)P_2$ and ARP2/3

Protein kinases PKN1 and PKN2[164] are Rho effectors involved in endosomal trafficking. Enzyme PKN also phosphorylates (inactivates) intermediate filament proteins, such as neurofilament, vimentin, and glial fibrillary acidic protein, as well as microtubule-associated protein Tau [944]. Widespread PKN has a catalytic domain homologous to that of members of the PKC family and a unique regulatory region. It can bind to small GTPase RhoA to be activated [941] as well as RhoB [943].

Ubiquitous PKNα (PKN1) is expressed especially in central neurons. It is concentrated in a region of the endoplasmic reticulum and its derived vesicles

---

[164] At least 3 different isoforms of Ser/Thr protein kinase PKN exist: PKNα (a.k.a. PKN1, PAK1, and PRK1), PKNβ (a.k.a. PKN3 and PAK3), and PKNγ (a.k.a. PKN2, PAK2, and PRK2).

Table 9.32. Targets of small GTPases of the RHO superfamily (**Part 2**; Sources: [941–943]; ARP: actin-related protein; ERK: extracellular signal-regulated kinase; JNK: Jun N-terminal kinase; IQGAP: IQ motif-containing GTPase-activating protein; MLK: mixed lineage kinase; p140SRA1: specifically Rac1-associated protein; Par: partitioning defective protein; Par6–Par3–aPKC: Par6–Par3–atypical PKC complex; PIP$_2$: phosphatidylinositol (4,5)-bisphosphate; PIP5K: phosphatidylinositol 4-phosphate 5-kinase; PKB/C: Protein kinase-B/C; WASP: Wiskott-Aldrich syndrome protein; WAVe: WASP family verprolin homology domain-containing protein).

Effector	Effect
	Rac
IQGAP	Intercellular adhesion, microtubule orientation
NADPH Oxidase	Synthesis of reactive oxygen species (activation of nuclear factor-κB)
p140SRA1	Membrane ruffling
P21-activated kinases (PAK1–PAK4)	LIMK phosphorylation (cofilin inactivation)
	Stathmin phosphorylation (microtubule stability)
	Triggering of cRaf–MAP2K1–ERK1 axis
	Stimulation of MLK–JNK pathway (stimulation of AP1-dependent gene expression)
Par6–Par3–aPKC	Cell polarity (tight junction)
Partner of Rac1 (POR1)	Membrane ruffling
Phospholipase-D	Phosphatidic acid production
PI3K	Activation of WASP and PKB
PIP5K	PIP$_2$ production
WAVe	Activation of the ARP2–ARP3 complex (actin polymerization)

localized to the apical compartment of the juxtanuclear cytoplasm, as well as Golgi body, late endosomes, multivesicular bodies, and secretory vesicles, in addition to cell nucleus [944]. It phosphorylates myristoylated alanine-rich C kinase (MARCK) and vimentin. It can be stimulated by phosphoinositide-dependent protein kinase PDK1 as well as arachidonic and linoleic acids, cardiolipin, phospholipids, such as phosphatidylinositol (4,5)-bisphosphate and (3,4,5)-trisphosphate, and lysophospholipids, such as lysophosphatidic acid and lysophosphatidylinositol. Isoform PKNα interacts with RhoA, RhoB, RhoC, and Rac1 GTPases [941,943]. It sustains the ^{GTP}RhoA form [944]. In addition, PKNα is able to bind to actin crosslinking protein α-actinin as well as actin-cytoskeleton proteins, such as caldesmon and ^{G}actin.

Isoform PKNβ (PKN3) attaches to GTPase-activating proteins for RhoA GRAFs (GTPase regulator associated with focal adhesion kinase) and CDC42 [944].

Isozyme PKNγ (PKN2) tethers to NCK and GRB4 adaptors [944]. It can associate with MAP3K2 kinase. However, MAP3K2 may act as a scaffold to regulate PKNγ activity, similarly to PKCα that interacts with and activates PLD independently of its kinase activity. It can also phosphorylate PDK1 for maximal PKB activation, but can inhibit PDK1 autophosphorylation, hence impeding the ability of PDK1 to phosphorylate PKCζ and PKCδ, as well as S6K and SGK [944].

9.3 Monomeric (Small) GTPases

Table 9.33. Targets of small GTPases of the RHO superfamily (**Part 3**; Sources: [941–943]; ARP: actin-related protein; CRIK: citron Rho-interacting kinase; ERM: ezrin, radixin, and moesin; MBS: myosin-binding subunit; MLC: myosin light chain; NHE1: sodium–hydrogen exchanger; PIP$_2$: phosphatidylinositol (4,5)-bisphosphate; PIP5K: phosphatidylinositol 4-phosphate 5-kinase; WASP: Wiskott-Aldrich syndrome protein).

Effector	Effect
	Rho
Binders of Rho GTPases	Cytokinesis via septins
CRIK	Cytokinesis, cell division cycle progression
Diaphanous	Microtubule stability, actin polymerization
GDIA	
Kinectin	
LIMK1/2	Cofilin phosphorylation (inactivation)
MLC kinase	Actin–myosin assembly, cell contraction
MLC phosphatase (MBS PP1δ)	Actin–myosin disassembly, cell relaxation
NET1	Guanine nucleotide-exchange factor
Phospholipase-D	Phosphatidylcholine hydrolysis
PIP5K	PIP$_2$ production (ERM processing for phosphorylation)
Protein kinase-N (PKN1 [PRK1]–PKN2)	Endosome-mediated transfer
RoCK	Phosphorylation of MLC and MBS (inactivation of myosin phosphatase)
	LIMK2 phosphorylation (cofilin inactivation)
	ERM and NHE1 phosphorylation (actin–membrane link)
	Phosphorylation of vimentin and desmin (reorganization of intermediate filaments)
Rhophilins-1/2	Rhophilin-2-induced disassembly of stress fibers
Rhotekins-1/2	Lymphocyte function (rhotekin-2)
WASP	Activation of the ARP2–ARP3 complex (promotion of actin polymerization)

In addition, it interacts with protein Tyr phosphatase PTPn13 to modulate activity of the actin cytoskeleton.

Alias PKN can be used to designate a protein kinase novel (PKN1–PKN3 or PKNα–PKNγ), some being also called PKC-related kinase (PRK1/2), and P21-activated kinases. Some authors describe both types of kinases in a review on Rho-associated kinases [943].

The P21-activated kinases (PAK1–PAK3; i.e., group-1 PAKs) interact with activated Rac1 to Rac3 and CDC42, as well as other small GTPases, such as RhoQ, RhoU, and RhoV proteins, but neither other Rho types (RhoA to RhoC, RhoE, and RhoG) nor other RAS superfamily members [943]. They require autophosphorylation to become active. They are also implicated in transcription via mitogen-activated protein kinase cascades. Membrane recruitment of PAK1 via NCK and GRB2 adaptors stimulates its kinase activity, possibly by PDK1 kinase. Kinases PAK1 to PAK3 complex with focal adhesion-associated CDC42–Rac1GEF Rho(Arh)GEF6 protein.

Non-conventional members of group-2 P21-activated kinases include ubiquitous PAK4, brain-enriched PAK5, and androgen receptor-interacting PAK6. Isoform PAK4 binds to $CDC42^{GTP}$ and, to a much lesser extent, Rac^{GTP}. Binding of $CDC42^{GTP}$ to PAK4 does not stimulate kinase activity, but instead causes PAK4 translocation to the Golgi body [943]. Similarly, linkage between CDC42 or Rac and PAK5 or PAK6 does not enhance PAK activity.

Mixed lineage kinases (MAP3K9–MAP3K11, MLK4, and MAP3K15) influence cytoskeletal organization (Sect. 6.5.3) [943]. Both MLK1 to MLK4 are auto-inhibited. Kinase MAP3K11 binds activated CDC42 and Rac1 activators to dimerize and autophosphorylate. Ubiquitous myotonin-related CDC42-binding kinase (MRCK), an effector of CDC42 and Rac GTPases, targets both RoCK and LIMK2 kinases. Myotonic dystrophy-associated protein kinase DMPK1 is not regulated by small GTPases [943].

9.3.13 Family of RHO GTPases (RhoA, RhoB, and RhoC)

The family of Ras homolog (Rho) GTPases include 3 isoforms (RhoA–RhoC; Table 9.34).[165]

Members of the RHO superfamily are prenylated and methylated at their C-termini and thus can reside in cell membrane. Similarly to other Ras-related GTPases, members of the RHO superfamily are inefficient GTPases, as they tightly bind both GTP and GDP and slowly catalyze GTP hydrolysis. The rates of GTP hydrolysis and guanine nucleotide release are greatly accelerated by GTPase-activating proteins and guanine nucleotide-exchange factors, respectively. Therefore, Rho GTPase functions primarily as a molecular switch. Specific GAPs favor GTP hydrolysis, as they transfer the switch into its inactive GDP-coupled form. Specific GEFs assist GTP binding and displace the switch into its active GTP-charged form that can bind effectors to generate and/or transmit signals.

---

[165] Small Rho GTPases isoforms RhoX are also called ArhX as they were called low-density lipoprotein receptor adaptor proteins. The ARH2 gene encodes low-density lipoprotein receptor adaptor protein-1. Acronym ARH stands for autosomal recessive hypercholesterolemia. In fact, low-density lipoprotein receptor (LDLR) plays a pivotal role in cholesterol metabolism. Defects in LDLR or apolipoprotein-B, the proteic component of LDL particles that binds the LDL receptor, elevate circulating LDL–cholesterol levels. Genetic mutations in ARH gene on chromosome 1 cause autosomal recessive hypercholesterolemia (ARH).

9.3 Monomeric (Small) GTPases

Table 9.34. Small Rho GTPases and their effectors (Source: [889]; CRIK: citron Rho-interacting, serine/threonine kinase STK21; Dia: diaphanous formin; FHoD1: formin homology-2 domain-containing protein). RhoA is more closely related to RhoC than RhoB. Whereas RhoA and RhoC promote cell growth, RhoB most often inhibits cell growth. Rho GTPases link to a large number of proteins, such as members of the Rho-associated kinase family and mammalian Diaphanous proteins. They participate in gene expression, cell morphology control, smooth muscle contraction, formation of stress fibers and focal adhesions, cytokinesis, cell cycle progression, axon guidance and extension, vesicular transfer, tissue development, and cell transformation.

Type	Effectors
RhoA	RoCK1/2, PAK1, CRIK, Dia1/2
RhoB	RoCK1/2, PAK1, CRIK, Dia1/2
RhoC	RoCK1/2, PAK1, Dia1/2, FHoD1
RhoD	Dia3
RhoF	Dia2

9.3.13.1 RhoA

Small RhoA GTPase is the prototypical member of the RHO family that is involved in multiple signal transduction pathways, as it associates with a large number of proteins. It undergoes adpribosylation by C3 ^{ADP}ribosyltransferase [945]. Cytoskeleton assembly and actin and myosin dynamics are regulated by the balance between RhoA, Rac1, and CDC42 activities.

Small RhoA GTPase binds to RoCK1 and RoCK2 kinases that phosphorylate and activate LIMK1 kinase that, in turn, phosphorylates (inactivates) cofilin, hence inhibiting depolymerization of filamentous actin. The RoCK kinases also enhance myosin activation, as they phosphorylate regulatory myosin light chain and inhibit MLC phosphatase (Sect. 5.2.14). In mammals, Diaphanous-related formins (DRF) that include Dia1 to Dia3, Disheveled-associated activators of morphogenesis DAAM1 and DAAM2, and FRL1 to FRL3 that bind and bundle actin filaments are direct effectors of RHO family proteins. They are auto-inhibited by intramolecular interactions. Binding of GTP to Rho relieves auto-inhibition. Isoform Dia1 specifically binds to and is activated by RhoA to RhoC, whereas Dia2 and Dia3 are activated by RhoA to RhoC, CDC42, and RhoD [945]. Formins initiates actin filament assembly, as they bind and stabilize actin dimers or trimers. Diaphanous initiates the assembly of new actin filaments from actin monomers, extension of non-branching actin filaments, and formation of thin actin stress fibers. Effectors RoCK kinases and Diaphanous cooperate in Rho-induced actin reorganization to create actin fibers of various thicknesses and densities according to the balance of RoCK and Dia activities. On the other hand, RoCK antagonizes Dia in Rho-dependent Rac stimulation via Src activation and formation of the CAS–CRK–DOCK180 complex [945].

RhoA and Cell Migration

In migrating cells, CDC42 and Rac1 operate at the leading edge, in filopodia and lamellipodia, respectively, whereas RhoA stimulates contractile actin–myosin filaments in the cell body and at the cell rear to induce tail retraction as well as in lamellipodia and membrane ruffles, in which RhoA associates with Dia1 to control actin assembly (Vol. 2 – Chap. 6. Cell Motility). The Rho–Dia pathway regulates cell polarity and focal adhesion turnover, as it localizes adenomatous polyposis coli protein and Src kinase at their respective sites [945].

RhoA and Phospholipids

Small RhoA GTPase associates with phospholipase-D that generates phosphatidic acid. The latter activates PI(4)P 5-kinase that locally produces $PI(4,5)P_2$. Small RhoA GTPase also connects to PI(4)P 5-kinase. Synthesized $PI(4,5)P_2$ regulates actin-binding proteins, such as cofilin and gelsolin [945]. Phosphatidylinositol 4-phosphate 5-kinase-1γ associates with N-cadherin-mediated intercellular adhesions owing to activated RhoA GTPase. Agent $PI(4,5)P_2$ regulates intercellular adhesion strength via actin-binding gelsolin [946]. In addition, RhoA links to $K_V1.2$ channel subunit. It also heightens the activity of ENaC channels.

RhoA and the Cell Cycle

Small GTPase contributes to the control of cell cycle progression, as it regulates cyclin-dependent kinase inhibitor and cyclin-D1 concentrations [945]. During G1 phase, the RhoA-RoCK pathway prevents the activity of CDK inhibitors CKI1a and CKI1b via adhesion-dependent signaling and causes sustained extracellular signal-regulated kinase stimulation that controls expression of cyclin-D1 and repression on Rac and CDC42 signals. In addition, the RhoA–Dia pathway promotes ubiquitin-mediated degradation of CKI1b, thereby stimulating the G1–S transition. Last, but not least, RhoA and its effectors CRIK kinase and Dia2 actin nucleator colocalize to the cleavage furrow. Small GTPase RhoA is activated by RhoA-GEF epithelial-cell transforming gene ECT2 at the cell equator to form the contractile ring of actin–myosin-2 filaments that cleaves the cell into 2 daughter cells. Simultaneously, at the midbody during cytokinesis, Aurora-B colocalizes with RhoA and RacGAP1 and phosphorylates RacGAP1 (Ser387) to focus RhoA activity only to the cleavage furrow.

9.3 Monomeric (Small) GTPases

RhoA and Vesicular Transfer

Both RhoA and RhoB are involved in the regulation of exo- and endocytosis, as they stimulate actin dynamics via Wiskott-Aldrich syndrome protein and WASP family verprolin homologous proteins [945].

RhoA and Epi- and Endothelium Integrity

Cadherins suppress RhoA activity partly by increased RhoGAP35 activity. Catenin-δ1 not only interacts with cadherins to prevent their endocytosis, but also inhibits RhoA to avoid disruption of apical junctions and maintain epithelial architecture [945]. Ezrin–radixin–moesin proteins promote cortical actin assembly and cell polarity. They link transmembrane proteins to apical actin cytoskeleton of epithelial cells. A negative feedback exists between activated ERM and RhoA GTPase [945]. Small RhoA GTPase causes formation and contraction of actin stress fibers, hence endothelial hyperpermeability once stimulated by vascular epidermal growth factor, angiopoietin-1 and -2, and thrombin.

Regulation of Activity

Activity of RhoA is governed by (1) activation by GEFs and deactivation by GAPs; (2) adpribosylation (inactivation); (3) prenylation (geranylgeranylation) for interactions with membranes as well as kinesin-associated protein KIFAP3, RhoGDIs, and GEFs; (4) phosphorylation by protein kinase-A[166] and dephosphorylation[167] that regulates RhoA activity and location; (5) methylation; (6) ubiquitination for degradation; and (7) crosstalk with Rac or Ras GTPase pathways.[168]

Activity of RhoA can be regulated by angiopoietin-2, bombesin, chemoattractants, lysophosphatidic acid, sphingosine-1-phosphate, thrombin, Wnt morphogens, as well as hepatocyte, platelet-derived, transforming- (TGFβ), and vascular endothelial growth factors, and plexin-B and EPHa ligands via particular GEFs linked to G-protein-coupled receptors and receptor Tyr or Ser/Thr kinases [945].

---

[166]Phosphorylation prevents its association with the plasma membrane, enhances its affinity for RhoGDIs, and protects it from degradation [945]. Other kinases, such as FAK, Src, and PKG, indirectly regulate RhoA activity.

[167]Phosphatase PTPn11 dephosphorylates FAK for focal adhesion turnover during migration and RoCK2 to stimulate its binding to RhoA.

[168]Crosstalk between RhoA and other small GTPases is mediated via RhoGAPs and GEFs. Protein RasA1 interacts with Deleted in liver cancer protein DLC1 and then inhibits GAP activity of DLC1 in focal adhesions to increase RhoA activity. Small GTPases Rac or CDC42 can interfere with Rho via P21-activated-kinases and RhoGEFs [945]. Kinase PAK1 binds to RhoGEF1 to block the signaling from thrombin receptors to RhoA. Kinase PAK4 interacts with and phosphorylates RhoGEF2 to inhibit stress fiber formation. Protein BCR serves both as a RhoGEF and RacGAP.

Cadherins, integrins, fibronectins, and T-cell receptors also influence RhoA activity possibly via focal adhesion kinases that phosphorylate various GEFs.

9.3.13.2 RhoB

Small RhoB GTPase localizes to endocytic vesicles and regulates endocytosis, especially receptor internalization. It influences nuclear transport of protein kinase-B in endothelial cells.

9.3.13.3 RhoC

Small RhoC GTPase favors cell migration, particularly metastasis of tumor cells.[169] It indeed contributes to the regulation of actin stress-fiber assembly and focal adhesion formation. It interacts with numerous affectors and effectors: (1) RhoC-GEFs include Scambio and RhoGEF5 (Tim); (2) RhoC-GAPs p50RhoGAP, p190RhoGAP, Myr5, and RhoGAP26; (3) RhoC-GDIs RhoGDIα, RhoGDIβ, and RhoGDIγ. Its effectors comprise phospholipase-Cε, scaffold rhotekin (inhibitor of GTPase activity of Rho), rhophilin-1 (Rho GTPase-binding protein), Rho-associated kinase RoCK1/2, Rho-interacting Ser/Thr kinase-21 Citron (STK21), Diaphanous Dia1, and PAK1 [947].

9.3.14 Atypical RhoBTB GTPases

Small RhoBTB GTPases (BTB: Broad complex, Tramtrack, and Bric-à-brac) constitute a family of atypical GTPases within the RHO superfamily. In addition to the GTPase domain (most often non-functional), RhoBTB is constituted by a tandem of 2 Bric-à-brac domains.

In humans, the RhoBTB family includes 3 isoforms (RhoBTB1–RhoBTB3). The RhoBTB1 and RhoBTB3 isoforms are ubiquitous, but with different distribution patterns [948]. Subtype RhoBTB1 is produced at high levels in the kidney, skeletal muscle, stomach, placenta, and testis, and RhoBTB3 in the nervous system, heart, pancreas, placenta and testis. On the other hand, RhoBTB2 is expressed at much lower levels than the other 2 isoforms.

---

[169]The activity of many proteins is implicated in metastasis: (1) constituents of the extracellular matrix and regulators of the matrix assembly and (2) components of the cytoskeleton and regulators of its dynamics: collagen-1α2 and -3α1, fibronectin, a ligand for integrins, calcium-binding matrix Gla proteins, biglycans, keratan sulfate proteoglycan fibromodulin, a collagen fibril formation regulator, tPA, angiopoietin-1, lysozyme-M, cathepsin-S, ERK1, α-catenin, α-actinin, calmodulin, actin-sequestering thymosin-β4, and other cytoplasmic and nuclear actin-related proteins, as well as eukaryotic initiation and elongation factors, endoplasmic reticulum-membrane translocators, etc.

Small GTPases RhoBTB can recruit cullin-3, a scaffold of ubiquitin ligases [949]. Once RhoBTB2 is bound to cullin-3, the latter regulates RhoBTB2 level by ubiquitinating RhoBTB2 directly for degradation [950]. The RhoBTB proteins may participate in cell cycle regulation and vesicle transport, as they tag substrates for degradation by 26S proteasome.

9.3.15 Classical RhoD GTPase

Small RhoD GTPase regulates early endosome movement between the cell cortex and central region. A splice variant of human Diaphanous Dia2c specifically binds to GTPase RhoD and is recruited onto early endosomes. Together with Dia2c, RhoD aligns early endosomes along actin filaments and reduces their motility, once Src kinase has been recruited and activated [951]. Small RhoD GTPase antagonizes RhoA, as RhoD causes loss of stress fibers and focal adhesions, whereas RhoA favors formation of actin stress fibers and associated focal adhesions [952].

9.3.16 Classical RhoF GTPase

Small RhoF GTPase (Ras homolog gene family, member F [in filopodia], also named RIF) is expressed in lymphocyte subpopulations. Cells of B-cell origin expressed higher RhoF levels than their T-cell counterparts [953]. It is upregulated in transformed follicular lymphoma, the most common form of non-Hodgkin lymphoma.

In addition to filopodium formation caused by GTPase CDC42,[170] GTPase RhoF induces actin-rich filopodia that contain parallel actin filaments bundled by actin crosslinking proteins and help cells to sense their environment. RhoF-mediated filopodium formation occurs via Diaphanous-related formin Dia2 [954].

9.3.17 Atypical RhoG GTPase

Small GTPase RhoG shares a significant degree of sequence identity with Rac1 and CDC42 [955]. Protein RhoG not only accumulates in the perinuclear region, but

---

[170]GTPase CDC42 binds and activates WASP proteins that, in turn, activate the actin-nucleating ARP2–ARP3 complex. It also binds and activates Brain-specific angiogenesis inhibitor 1-associated protein BAIAP2 (or 53-kDa insulin receptor substrate protein [IRSP53]) that recruits WASP family protein Enabled homolog EnaH to the filopodial tip and protects elongating actin filaments from capping.

concentrates in plasmalemmal spots. Constitutively active RhoG causes cytoskeletal changes similar to those generated by a simultaneous activation of Rac1 and CDC42 (formation of microvilli, ruffles, lamellipodia, filopodia, and partial loss of stress fibers). Small GTPase RhoG controls a pathway that requires microtubules and activates Rac1 and CDC42 independently of their stimuli, such as PDGF or bradykinin [955]. Monomeric GTPase RhoG is activated at the transcriptional level in the mid-G1 phase in fibroblasts [956].

9.3.18 Atypical RhoH GTPase

Small RhoH GTPase is predominantly expressed in hematopoietic cells. Monomeric RhoH has low or no GTPase activity and is thus constitutively bound to GTP [831]. Small GTPase RhoH hinders Rac recruitment to the plasma membrane. Protein RhoH precludes proliferation and homing of murine hematopoietic progenitor cells. It is needed to keep leukocytes in a resting, non-adhesive state [957]. It also intervenes in signal transduction during T-cell development.

Activity of RhoH is regulated by tyrosine phosphorylation (ITAM-like motif). In addition, RhoH C-terminal CAAX box is targeted by farnesyl transferase and geranylgeranyl transferase. Isoprenylation of RhoH and translocation yield further elements for fine-tuned signaling [957]. Small GTPase RhoH is used by the NFκB, PI3K, and MAPK pathways.

9.3.19 Atypical RhoJ GTPase

Small RhoJ GTPase[171] is related to CDC42 and RhoQ (or TC10). It localizes to the plasma membrane and early endosomes [958].

Receptor-dependent endocytosis of transferrin remains unaffected in the absence of endogenous RhoJ, but transferrin accumulates in Rab5+ uncoated endocytic vesicles and fails to reach early endosome and pericentriolar recycling endosomes. Monomeric GTPase RhoJ is thus needed for clathrin-dependent endocytosis of receptors via early endosomes and direct recycling without accumulating in perinuclear recycling endosomes.

Small RhoJ GTPase participates in the early stage of adipocyte differentiation, probably linked to the peroxisome proliferator-activated receptor PPARγ pathway [959]. It regulates the expression of specific genes.

---

[171] A.k.a. TC10-like (TCL) GTPase, TC10βL (TC10β long), ArhJ, RasL7B, and RhoI.

9.3.20 Atypical RhoQ GTPase

Small Ras-like RhoQ GTPase[172] is a unusual member of the RHO superfamily. Monomeric RhoQ GTPase with 2 RhoQ homologous transcripts belongs to Ras-related teratocarcinoma (TC)-associated proteins, such as Ran (TC4), rRas2 (TC21), Rac1 (TC25), and Rab11. Small RhoQ GTPase is predominantly located at the plasma membrane and endosomes due to post-translational prenylation [960].[173] It colocalizes with caveolin and flotillin at the plasma membrane.

Protein RhoQ regulates various cellular processes, such as actin cytoskeletal organization and cell shape, intracellular protein transport, signal transduction associated with the mitogen-activated protein kinase module, and cell growth and proliferation. Ubiquitous RhoQ [960]: (1) promotes formation of filopodia; (2) activates Jun N-terminal kinase; (3) stimulates nuclear factor-κB- and serum response factor-dependent transcription; and (4) synergizes with activated Raf kinase and contributes to full hRas action in cell transformation. Small RhoQ GTPase interacts with actin-binding and filament-forming profilin.

Small RhoQ GTPase binds to several signaling effectors, such as mixed lineage kinase MLK2 (MAP3K10), CDC42-binding protein kinase-α (a.k.a. myotonic dystrophy-related CDC42 kinase [MRCK]), P21-activated protein kinases, the Borg family of CDC42 effector proteins (CDC42EP), partition-defective homolog Par6, and nWASP [961].

Small RhoQ GTPase intervenes in intracellular protein transport. Upon insulin stimulation, RhoQ regulates assembly and fusion of vesicles responsible for GluT4 transporter exocytosis to the plasma membrane [962]. Protein RhoQα^{GTP} (but not RhoQβ) promotes insulin-stimulated GluT4 translocation. Active RhoQ^{GTP} fastens to Exo70, a subunit of a proteic complex that tethers vesicles to sites of secretion. Protein Exo70 is involved in GluT4 transfer. Whereas CDC42-interacting protein CIP4 mediates the association of WASP with microtubules [963], RhoQ-interacting protein CIP4 is recruited to the plasma membrane in response to insulin to enhance GluT4 translocation [964]. Phosphoinositides contribute to GluT4 exocytosis. In fact, GluT4 translocation requires activation not only of RhoQ, but also phosphatidylinositol 3-kinase, as phosphatidylinositol (3,4,5)-trisphosphate serves in GluT4 translocation to the plasma membrane. On the other hand, phosphatidylinositol 3-phosphate is produced at plasmalemmal rafts upon stimulation by insulin via RhoQ to be incorporated into endosomes [965].

Protein RhoQ inhibits a cystic fibrosis transmembrane conductance regulator-interacting protein, CFTR-associated ligand (CAL), that mediates CFTR degradation in the lysosome, hence raising plasmalemmal CFTR density [966]. Active RhoQ^{GTP} redistributes CAL from the Golgi body to the plasma membrane, where it

---

[172] A.k.a. teratocarcinoma-associated protein TC10 or RasL7a.

[173] Most RHO superfamily members undergo geranylgeranylation. However, RhoQ is farnesylated and palmitoylated. These post-translational modifications allow RhoQ to connect to membrane rafts.

cannot influence CFTR transport to lysosomes for degradation. Monomeric RhoQ recruits effectors, such as CDC42-interacting protein, to the plasma membrane.

Small RhoQ GTPase is activated by RapGEF1 and inactivated by RhoGAP1 and RhoGAP35 [967]. It interacts with α, β, and γ isoforms of P21-activated kinases. It is also able to bind to mixed lineage kinase MLK2 (MAP3K10) that links to huntingtin as well as carrier IQGAP1 involved in neurite outgrowth. It tethers to nWASP that activates the ARP2–ARP3 complex as well as profilin that converts globular into filamentous actin and myotonic dystrophy kinase-related CDC42-binding kinase (MRCK; also known as CDC42-binding protein kinase CDC42BPα), a Ser/Thr kinase. Small GTPase RhoQ can recruit PKCζ and -λ to the plasma membrane in adipocytes. Activated RhoQ stimulates transcription factors or promoters, such as that of the JNK–Jun axis, as well as SRF, cyclin-D1, and NFκB promoters [967].

Binder of Rho GTPases (BORG) are CDC42 and RhoQ GTPase-interacting proteins that behave as negative regulators [968]. Whereas BORG3 binds only to CDC42, BORG1, BORG2, BORG4, and BORG5 tether to both CDC42 and RhoQ in a GTP-dependent manner. Both full-length BORG1 and BORG2 proteins as well as fragments of BORG1, BORG4, and BORG5 are able to interact with RhoQ.

9.3.21 RhoT GTPase

Small RhoT GTPase is closely related to RhoQ agent. Protein RhoT is predominantly expressed in heart and uterus. It is also induced during neuronal differentiation. Whereas RhoQ generates actin filament-containing peripheral processes longer than CDC42-launched filopodia, RhoT produces much longer and thicker processes that are also composed of actin filaments. Both RhoQ and RhoT assist neurite outgrowth [969]. They as well as CDC42 bind to and activate nWASP to cause ARP2/3-mediated actin polymerization.

9.3.22 Atypical RhoT1/2 GTPases (Miro-1/2)

Monomeric GTPases Ras homolog gene products RhoT1[174] and RhoT2[175] have a C-terminal transmembrane domain that serves as a label for mitochondria. Protein RhoT2 induces mitochondrium aggregation. In addition, Miro interacts with kinesin-binding proteins, thereby linking mitochondria to microtubules [970].

---

[174] A.k.a. Miro-1 or ArhT1.
[175] A.k.a. Miro-2.

9.3.23 Atypical RhoU GTPase

Monomeric RhoU GTPase[176] actuates filopodium formation and stress fiber dissolution. In addition, it localizes to focal adhesions and Src-induced podosomes to enhance cell migration [971].

Small RhoU GTPase is targeted by transcription factor signal transducer and activator of transcription STAT3 that is stimulated by growth factors and morphogen Wnt1 using the non-canonical planar cell polarity pathway with activation of Jun N-terminal kinase [972].

9.3.24 Atypical RhoV GTPase

Small RhoV GTPase[177] has been initially identified as a protein interacting with the regulatory domain of PAK2 kinase. Activated RhoV stimulates the JNK pathway and provokes the formation of small lamellipodia and focal adhesions [973]. Activity of RhoV depends on palmitoylation to associate with cellular membranes. It is insensitive to RhoGDI agents.

9.3.25 RGK (Rad–Gem/Kir–Rem) Family of GTPases

The RGK family of small GTPases includes Rad, Rem1,[178] Rem2, and Gem[179] that are involved in the nucleocytoplasmic transport. Among the small GTPases of the RGK family, Rem2 is the single member that abounds in neurons.

Rem-interacting proteins include 14-3-3 isoforms (ϵ, η, θ, and ζ). Protein Rem is phosphorylated and then binds to 14-3-3ζ, as the presence of protein phosphatase-1 abolishes this association [974]. Protein 14-3-3 may recruit mediators of Rem-dependent signaling.

Small Ras GTPase associated with diabetes (Rad; Sect. 9.3.5),[180] is highly expressed in human skeletal and cardiac muscles and lung. It inhibits glucose uptake. It is overexpressed in skeletal myocytes in insulin-independent type-2 diabetes [975]. In addition, Rad GTPase is an inhibitor of cardiac hypertrophy via calmodulin-dependent kinase-2δ [976].

---

[176] A.k.a. WRCH1, as it is a Wnt-responsive CDC42 homolog.

[177] A.k.a. CHP (CDC42 homologous protein), ArhV, and WRCH2 (Wnt-responsive CDC42 homolog-2).

[178] A.k.a. Rem and Ges.

[179] Mouse ortholog Kir.

[180] A.k.a. Ras-related associated with diabetes (RRad) and Rem3.

Table 9.35. Rnd GTPases and their effectors (Source: [889]). Formin-binding protein FnBP1 (a.k.a. Rnd2 apostle rapostlin) that binds to microtubules favors neurite branching. Pragmin (pragma of Rnd2) causes cell contraction via the RhoA–RoCK pathway. Socius (a.k.a. UBX domain-containing protein UBXD5) is a RndGAP involved in disassembly of actin stress fibers. It connects to Gα subunits.

Type	Effectors
Rnd1	Socius
Rnd2	FnBP1, pragmin
Rnd3	RocK1, socius, p190RhoGAP

Small GTPase Rem alters the excitation–contraction coupling by reducing the number of functional plasmalemmal Ca_V1 channels [977]. In excitable cells, inhibition of voltage-gated Ca^{++} channels by RGK GTPases participates in the regulation of Ca^{++} influx, in addition to channel expression, among other mechanisms. In fact, RGK GTPases are potent inhibitors of Ca^{++} influx via high-threshold voltage-gated Ca^{++} channels $Ca_V1.2$ and $Ca_V2.2$, as small RGK GTPases acts on channel auxiliary $Ca_V\beta1$ and -2 subunits [978, 979]. Effect of RGK corresponds to a decrease in amplitude without modification in voltage dependence and ion flux kinetics [979]. In neurons, Rem2 nearly abolishes calcium currents arising from high-voltage-gated Ca^{++} channels without affecting low-voltage-gated Ca^{++} channels. Small Rem2 GTPase localizes to the plasma membrane and interacts with calcium channel β subunits in the pre-assembled $Ca_V2.2$ channel, thereby forming a non-conducting carrier [980].

Insulin secretion by pancreatic β cells also requires Ca^{++} ingress. Monomeric Rem2 GTPase that is synthesized upon exposure to high glucose level associates with $Ca_V1.2$ and $Ca_V1.3$, thereby preventing Ca^{++} import and glucose-stimulated insulin secretion [981].

Small Gem GTPase interacts with the microtubule network through kinesin-like protein KiF9. Moreover, it regulates actin dynamics downstream from RhoA by inhibiting kinase RoCKβ and interacts with Rho GTPase-activating protein Gmip [982].

9.3.26 RND Family of Atypical GTPases

The family of Rnd GTPases, a subset of the RHO superfamily, includes 3 members (Table 9.35): Rnd1 or Rho6, Rnd2 or Rho7, and Rnd3 or RhoE. They regulate actin cytoskeleton dynamics. The Rnd proteins antagonize small GTPase Rho, as they hinder the formation of and even disrupt contractile actomyosin stress fibers. Constitutively active Rnd^{GTP} is regulated by its synthesis level, location, phosphorylation, and degradation [983].

The Rnd proteins operate in fibroblast-growth-factor-receptor-1 signaling. Small Rnd GTPases are involved in axon guidance. They interact with plexins. Plexins (plexin-A–plexin-D) are semaphorin receptors that govern cell adhesion, migration, and axon guidance. Among plexins, plexin-A1 and -B1 operate as rRasGAP. Activity of rRasGAP requires Rnd1 binding. Plexin-D1 also acts as rRasGAP and inhibits cell migration with the help of Rnd2 [984]. Like other plexins, Plexin-C1 exhibits rRasGAP activity. Nevertheless, it can prevent cell migration without Rnd GTPase.

9.3.26.1 Rnd1 GTPase

Isotypes Rnd1 and Rnd3 interact with RhoGAP35, and recruit this protein at sites where Rho should be inhibited. Both RhoGAP35 isoforms, RhoGAP35a and RhoGAP35b, stimulate Rnd1 GTPase. Isotype Rnd1 interacts with plexin-A1 and -B1, the semaphorin-3A and -4D receptors, UBX domain-containing protein UBXD5 (or Socius; a RndGAP involved in disassembly of stress fibers), GRB7 adaptor, and rRas [985]. In addition, Rnd1 interacts with membrane-anchored, PTB domain-containing fibroblast growth factor receptor substrates FRS2a and FRS2b that are docking proteins, which recruit signaling proteins, such as PKCλ and NTRK1 kinases, PTPn11 phosphatase, CBL and GRB2 adaptors, and SOS guanine nucleotide-exchange factor to the plasma membrane during fibroblast growth factor stimulation. Interaction of Rnd1 with FRS2b prevents Rnd1 ability to downregulate RhoA GTPase.

9.3.26.2 Rnd2 GTPase

Subtype Rnd2 is predominantly synthesized in testes as well as, to a lesser extent, the brain and liver. In neurons, it regulates the actin cytoskeleton [983]. Small GTPase Rnd2 interacts with RhoGAP5, albeit less strongly than Rnd1 and Rnd3 GTPases. In the testis, Rnd2 also binds to RacGAP1. In neurons and testis cells, it works with UBXD5, or UBX domain-containing protein UBXn11, a RndGAP molecule.

Small GTPase Rnd2 is less efficient at eliminating stress fibers in fibroblasts than other Rnd isoforms [985]. In neurons, it induces neurite branching together with rapostlin (apostle of Rnd2) and nWASP protein.

Unlike Rnd1 and Rnd3, Rnd2 can have a GTPase activity. Like Rnd1, but not Rnd3, Rnd2 interacts with FRS2a and FRS2b substrates. These molecules sequester Rnd2 in an inactive complex. In addition, Rnd2 interacts with vacuolar protein sorting VPS4a in early endosomes. Isoform Rnd2 can link to the plasma membrane as well as other membranes.

Table 9.36. Small RIN and RIT GTPases (Source: [889]). Whereas RIT is ubiquitous, RIN is detected only in neurons.

Type	Aliases
RIN	RIBa, RIT2, Roc2
RIT	RIBb, RIT1, Roc1

9.3.26.3 Rnd3 GTPase

Isotype Rnd3, or RhoE, is ubiquitous, but at very low levels. Its activity is regulated by RoCK1 [985]. Small Rnd3 GTPase contributes to the regulation of P53-mediated stress response that triggers actin depolymerization [986]. It inhibits RoCK1 during genotoxic stress, thereby suppressing apoptosis. Monomeric Rnd3 GTPase also participates in the regulation of cell cycle progression, eventually by decreasing adhesion and inducing dissociation of integrin-based focal adhesions that hamper cell cycle progression, and above all by inhibiting induction of cyclin-D1 [983]. Protein Rnd modulates smooth muscle contractility. In the brain, they are implicated in neurite extension (growth cone). Amphetamine and cocaine upregulate expression of Rnd3 in mice brain by distinct pathways [987].

9.3.27 Family of RIN and RIT GTPases

Small GTPases RIN (Ras-like protein expressed in neurons), a neuron-specific and calmodulin-binding Ras-related protein, and RIT (Ras-like protein expressed in many tissues; Table 9.36) are 25-kDa Ras-like GTPases that are 64% identical [988]. They lack a binding motif (Ras-like proteins without a terminal cysteine–aliphatic amino acid–aliphatic amino acid–any amino acid [CAAX] motif) for C-terminal lipid anchor that enables plasma membrane association that is essential for their activity.[181] Nonetheless, transiently produced RIN and RIT localize at the plasma membrane.

---

[181] Membrane tethering of Ras GTPases generally requires a C-terminal isoprenyl group that is post-translationally added on cysteine–alanine–alanine–any amino acid (CAAX) motif. In some RAS hyperfamily members, this CAAX motif is replaced by a CXC or CC motif. Internal palmitoylation or a C-terminal cluster of basic amino acids yields another source [988]. Certain unusual Ras-related GTPases (Rad, Kir, and Gem) that lack a CAAX or similar site contain a cysteine residue that can be targeted for isoprenylation.

9.3.27.1 RIN1 (RIT2)

Small RIN GTPase (or RIN1), a Ras-related protein as well as a Ras effector, is a calmodulin-binding protein expressed predominantly (if not solely) in neurons. It is also called Ras-like without CAAX motif type-2 protein (RIT2).[182] The Rin1 gene encodes a Ras effector that regulates epithelial cell behavior.[183] The transcriptional repressor Snai1 (Snail homolog) that silences the CDH1 (E-cadherin) gene as well as other proepithelial genes hampers RIN1 synthesis [989]. Moreover, DNA methylation within the Rin1 promoter and first exon can also contribute to gene silencing.

Monomeric RIN GTPase forms a family with RIT agent. Protein RIN localizes to both the plasma membrane and nucleus. The GTP dissociation rate is 5-to 10-fold faster than most Ras-like GTPases. Small GTPase RIN is activated by Src kinase following stimulation of both Gs and Gi subunits.

Protein RIN1 activates Abl kinases and Rab5 GTPase to regulate cytoskeletal remodeling and endocytic pathways. It thus reduces motility of epithelial cells, as activated Abl kinases precludes cytoskeletal rearrangements needed for cell dissociation and migration [989]. Molecule RIN operates via HSP27 and the cAMP–PKA pathway. Neuropeptide pituitary adenylate cyclase-activating polypeptide PACAP38 acts via G-protein-coupled receptors and RIN for its neurotrophic and neurodevelopmental effects [990]. Signaling based on RIN appears to contribute to nerve growth factor-dependent neuronal differentiation in cooperation with both Rho and P38MAPK mediators.

Protein RIN can associate with Ca^{++}–calmodulin [991]. Active RIN^{GTP} is able to interact with Par6 scaffold and bRaf kinase, whereas RIN^{GDP} can associate with transcription factor POU (pituitary-specific Pit1/octamers Oct/Unc) class-4 homeobox gene product POU4F1, or brain-specific homeodomain-containing protein Brn3a.

9.3.27.2 RIT1

Protein RIT1 (Ras-like protein expressed in many tissues)[184] is ubiquitous, but transiently expressed at the plasma membrane. The GTP dissociation rate is 5- to 10-

---

[182] The CAAX motif corresponds to the Cys–Ala–Ala-any amino acid sequence that attaches prenyl groups to anchor selected cellular membranes.

[183] Among common features in human tumors, elevated Ras signaling can originate from overexpression of upstream receptor tyrosine kinases or mutations in RAS genes. Signaling pathways downstream from small GTPases Ras prime many Ras effectors, such as bRaf and PI3K kinases that stimulate cell proliferation and repress apoptosis, respectively. Members of the Ras-association domain family (RASSF) operate as tumor suppressors that enhance apoptosis; silencing of RASSF genes may promote tumor cell survival. Mutations in genes that encode Ras effectors, such as bRaf and PI3K, also contribute to cancer.

[184] A.k.a. RIC-like expressed in many tissues and type-1 Ras-like without CAAX motif.

fold faster than most Ras-like GTPases. Small GTPase RIT intervenes in neuronal differentiation. It selectively activates neuronal bRaf isoform. Protein RIT1 can also complex with Par6 and CDC42 [992].

9.3.28 Phosphoinositide 3-Kinase Enhancer: GTPase and GAP

Phosphoinositide 3-kinase enhancer (PIKE) is a GTPase that binds to and stimulates PI3K kinase. Protein PIKE exists in 3 types: 2 brain-specific isoforms, a long form $PIKE_L$ (or PIKE1) and a short form $PIKE_S$ (or PIKE2) and a ubiquitous PIKEa subtype. Whereas $PIKE_S$ is exclusively nuclear, $PIKE_L$ resides in both the nucleus and cytoplasm. The latter contains a C-terminal extension with an adpribosylation GTPase-activating protein domain and 2 ankyrin repeats, in addition to the N-terminal GTPase domain. Therefore, it is also termed ArfGAP with GTP-binding protein-like, ankyrin repeat and pleckstrin homology domain-containing **AGAP2** protein.

Isoform PIKEa contains the domains of $PIKE_L$, except the N-terminal proline-rich domain that binds PI3K and PLCγ1 enzymes. Yet, PIKEa specifically binds to active protein kinase-B.

Small PIKE GTPase is also known as centaurin-γ1 (encoded by the CentG1 gene) and GTP-binding and GTPase-activating protein GGAP2. It has a strong GTPase activity. Moreover, it actually acts as a GTPase-activating protein for ARF1 and ARF5 GTPases.

Cytoplasmic PI3K and its products phosphatidylinositol (4,5)-bisphosphate and (3,4,5)-trisphosphate regulate membrane translocation of many signaling molecules, as they bind to and activate these mediators. Small PIKE GTPases are regulated by $PI(4,5)P_2$ and, to a lesser extent, by $PI(3,4,5)P_3$. Phosphatidic acid potentiates PIP_2 stimulation.

Nerve growth factor activates $PIKE_S$, as it triggers nuclear translocation of phospholipase PLCγ1 that acts as a guanine nucleotide-exchange factor for $PIKE_S$ [993].

Isoform $PIKE_L$ associates with Homer-1, a metabotropic glutamate receptor mGluR1-binding adaptor. The Homer1–$PIKE_L$ complex couples PI3K to mGluR1 to prevent neuronal apoptosis. In addition, $PIKE_L$ may regulate postsynaptic signaling by metabotropic glutamate receptors. Isoform $PIKE_S$ also binds to and enhances PKB activity to prevent apoptosis. Besides, it also regulates adaptor protein-1-dependent transport of proteins in the endosomal system.

9.4 Regulators of Monomeric Guanosine Triphosphatases

Monomeric GTPases exist in 3 states: (1) a transient, active, GTP-bound state, in which they connect to signaling effectors; (2) a transient, free, inactive, GDP-loaded state that results from the action of GTPase-activating proteins (GAP); and

(3) a sequestered, GDP-bound, guanine nucleotide-dissociation inhibitor (GDI)-complexed state.

The bimodal functioning of small GTPases during transmission of intracellular signals involves a cascade of between-protein interactions modulated by chemical modifications, structural rearrangements, and intracellular relocalizations. The GDP–GTP binding cycle enables monomeric GTPases to filter, amplify, or temporize receiving signals.

Switching between an inactive and active, effector-binding conformation is regulated by: (1) guanine nucleotide-exchange factors (GEF) that act as activators, as they promote the release of GDP and binding of GTP; and (2) GTPase-activating proteins (GAP) that accelerate the intrinsically slow GTPase activity of monomeric GTPases. Therefore, GTP binding is primed by GEFs (that then do not improve a prerequisite condition, but trigger signal transmission), whereas GAPs enhance intrinsic catalysis rate of small GTPases (hence they do not create a new state) or cause the release of GTP of small GTPases devoid of strong enzymatic capacity.

Membrane anchoring of monomeric GTPases is often a prerequisite for their activity. Therefore, small GTPases require not only the presence of GEFs and GAPs to modulate upstream signaling, but also mediators that relieve their cytosolic sequestration by guanine nucleotide-dissociation inhibitors. The control of the cellular membrane association–dissociation cycle corresponds to an additional regulation level.

9.4.1 Guanine Nucleotide-Exchange Factors

Guanine nucleotide-exchange factors increase the activity of small GTPases, as they catalyze GDP release and cause GTP association (i.e., exchange of GDP for GTP), thereby activating small GTPases (Fig. 9.1).[185]

Multiple GEFs have been described in humans. Many GEFs can activate various Rho GTPases.[186] More than 80 RhoGEFs (Table 9.37) activate Rho GTPases that are encoded by 22 genes and regulate the actin cytoskeleton according to the type of control input. Monomeric GTPases CDC42, Rac1, and RhoA intervene in the formation of filopodia, lamellipodia, and contractile actin–myosin filaments (stress fibers), respectively. Guanine nucleotide-exchange factors link specific cytoskeletal

---

[185]Rho guanine nucleotide-exchange factors (RhoGEF) has Diffuse B-cell lymphoma (DBL) homology (DH) and pleckstrin homology (PH) domains that catalyze GDP–GTP exchange. Other domains are specific to each member. On the other hand, the regulator of G-protein signaling (RGS) domain acts as a GTPase-activating protein (GAP). Some GEFs such as 180-kDa protein downstream of CRK-related proteins contain 2 DOCK homology domains.

[186]For example, Vav1 activates RhoA, RhoG, Rac1, and CDC42. T-cell-lymphoma invasion and metastasis TIAM1 targets Rac1, Rac2, and Rac3. Triple functional domain protein (TRIO) stimulates Rac, RhoA, and RhoG [831].

Table 9.37. Examples of guanine nucleotide-exchange factors(ARNO: ARF nucleotide site opener; BRAG: brefeldin-resistant ArfGEF; DOCK: dedicator of cytokinesis; EFA6: exchange factor for ARF6; FARP: FERM, RhoGEF and pleckstrin domain protein, a.k.a. FERM domain including RhoGEF [FRG] and pleckstrin homology domain-containing family-C member PlekHc3; FBx: F-box only protein; DEF6: differentially expressed in FDCP6 homolog, a.k.a. SWAP70L and IRF4-binding protein (IBP); PREx: PIP$_3$-dependent Rac exchanger; RabIn3: Rab3a-interacting protein; RIn: Ras and Rab interactor; RGL: Ral guanine nucleotide-dissociation stimulator-like protein; SOS: Son of sevenless; SWAP: Switch-associated protein; TIAM: T-cell lymphoma invasion and metastasis;).

Category	Examples of members
ArfGEFs	ARF1GEF (GBF1), ArfGEF1–ArfGEF3 (BIG1–BIG3), cytohesin-1–cytohesin-4 (ARNO1–ARNO4), BRAG1–BRAG3 (ARF6GEF1–ARF6GEF3), EFA6a–EFA6d (ARF6GEFa–ARF6GEFd), FBx8
CDC42GEF	FARP2, intersectin-1$_L$, Tuba
RabGEFs	RIn1–RIn3, RabGEF1, Rab3IP, Rab5GEF (alsin), RabIn8, RabIn3L
RacGEF	DEF6, PREx, SwAP70, TIAM1–TIAM2
RalGEFs	RalGEF (RalGDS), RGL1–RGL4
RanGEFs	
RapGEFs	RapGEF1–RapGEF6
RasGEFs	RasGRF1–RasGRF3, RasGRP1–RasGRP4, DOCK1–DOCK4
RhoGEFs	Duo, FGD1, RhoGEF1–RhoGEF19, RhoGEF21/23–25/27, RhoGEF30–RhoGEF31, SOS1/2, Scambio, Vav1–Vav3

responses to corresponding signaling. Reprogrammed GEFs can act on unrelated processes [994].

An additional level of regulation relies on restricted subcellular distributions of GTPases and their GEF and GAP regulators. In addition, GEFs promote crosstalks between different sets of small GTPases. Activated Ras activates GEFs for small GTPases Ral [995], Rab5 [996], and Rac [997].

9.4 Regulators of Monomeric Guanosine Triphosphatases

Table 9.38. Effects of ArfGEFs (Source: [840]; ER: endoplasmic reticulum; AP: adaptor proteic heterotetramer; ARNO: ARF nucleotide site opener; BIG: brefeldin-A-inhibited GEFs for ARFs; BRAG: brefeldin-resistant ArfGEF; CoP: coating protein complex (coatomer); EFA6: exchange factor for ARF6; GBF: Golgi-associated brefeldin-A-resistant GEF; GGA: Golgi-localized γ-ear-containing ARF-binding protein).

ArfGEF	Functions
GBF1	Endoplasmic reticulum–Golgi body transport, CoP1 recruitment
BIG1	Exocytosis from Golgi body
BIG2	Exocytosis from Golgi body, AP1 and GGA recruitment
ARNO	Actin remodeling, GPCR desensitization
	Ca^{++}-regulated exocytosis, dendritic branching
	Cell migration
ARNO2	β2-Integrin adhesion
	Synaptic transmission
ARNO3	Growth-factor signaling
	(e.g., insulin, EGF, and NGF)
ARNO4	
EFA6a	Endocytosis, recycling, actin remodeling
EFA6b	Endocytosis, recycling, actin remodeling
EFA6c	
EFA6d	
BRAG1	Vesicle trafficking at synapses
	Association with NMDA receptors at excitatory synapses
BRAG2	Endocytosis and recycling of $β_1$-integrin
BRAG3	

9.4.1.1 ArfGEFs

Guanine nucleotide-exchange factors that target ARF GTPases possess a Sec7 domain responsible for GEF activity.[187] The human genome encodes 15 known Sec7 family members (Table 9.38). They contribute to the regulation of membrane remodeling and trafficking associated with vesicular transport [998]

Only 2 among 7 ArfGEF sets exist in all eukaryotes: (1) BIG set of brefeldin-A-inhibited guanine nucleotide-exchange factors for adpribosylation factors (BIG1–BIG7) and (2) GBF set of Golgi-associated brefeldin-A-resistant guanine nucleotide-exchange factors (GBF1–GBF3) that are also termed guanine nucleotide-exchange on ARF (GEA).[188] The set of eukaryotic ArfGEFs is

---

[187] The Sec7 family is based on homology of ArfGEF catalytic domains to yeast ArfGEF Sec7p protein.

[188] Brefeldin-A, a fungal toxin, blocks secretion by preventing the assembly of coat protein components onto donor membranes. It is an uncompetitive inhibitor of the exchange reaction that binds to an ARF^{GDP}–ArfGEF complex.

Table 9.39. Guanine nucleotide-exchange factors of ARF GTPases (ArfGEFs) in humans (**Part 1**; Sources: [838, 998, 999] and Wiki Professional; ArfD: ARF domain protein; ARNO: ARF nucleotide-binding site opener; CNKSR: connector enhancer of kinase suppressor of Ras; PKAR: protein kinase-A regulatory subunit; PSCD: PH, Sec7, and coiled-coil domain-containing protein). Vesicles from the endoplasmic reticulum travel via the vesicular-tubular cluster (VTC) to the cis-Golgi network (CGN), where they fuse. Materials subsequently progress through the Golgi body. Vesicles also carry substances from the trans-Golgi network (TGN) to their final destination (Endos: endosome; rEndos: recycling endosomes; N: nucleus; PM: plasma membrane). In addition to their substrates, ArfGEF bind to many partners, such as other ArfGEFs, receptors (A_{2A} adenosine receptor; THR: thyroid hormone receptor), ion channels, adaptors (CASP: cytohesin-associated scaffold protein; GRASP: GRP1 [ARNO3]-associated scaffold protein; GRSP: GRP1 [ARNO3] signaling partner; IPCEF: interaction protein for cytohesin exchange factor; VDP: vesicle docking protein [P115]), enzymes (vATPase: vacuolar adenosine triphosphatase), exocyst subunits (Exo), and signaling regulators (Arrβ: β-arrestin).

Type	Aliases	Site	Substrates	Partners
BIG1	ArfGEF1, P200ArfGEF1	TGN, N, endosome	ARF1/3	BIG2, FKBP13, Exo70, myosin-9b
BIG2	ArfGEF2, BIG5	TGN, rEndos	ARF1/3	BIG1, PKAR1/2, Exo70, GABAR
BIG3	ArfGEF3			
GBF1		CGN, VTC, ERGIC	ARF1/3/5	Rab1b, VDP
Cytohesin-1	PSCD1, ARNO2	PM, endosome	ARF1/6, ArfD1	ARP, ARL4, ArfRP, CASP, CNKSR1
Cytohesin-2	PSCD2, ARNO	PM, endosome	ARF1/3/6 ARL4	ARF6, ARL4, A_{2A}, Arrβ, CASP, GRASP, CNKSR1, HER, IPCEF1, vATPase
Cytohesin-3	PSCD3, ARNO3, GRP1	PM endosome	ARF1/6	ARF6, ARL4, CASP, CNKSR1, GrASP, GRSP1, THR
Cytohesin-4	PSCD4, ARNO4		ARF1/5	

subdivided into 5 families based on overall structure and domain organization [999] (Tables 9.39 and 9.40): (1) *BIG–GBF family* of BFA-inhibited GEF (BIG) and Golgi body brefeldin-A-resistance factor (GBF) that has representatives in all eukaryotes; (2) *ARNO/cytohesin family* of ARF nucleotide-binding site openers (ARNOs) or cytohesins;[189] (3) *EFA6 family* of exchange factors for ARF6; (4) *BRAG family* of brefeldin-resistant ArfGEF; and (5) *FBx family* of F-box only protein-8 (FBx8).

---

[189]The name cytohesin originates from observation that cytohesin-1 activates β2-integrin-mediated adhesion.

9.4 Regulators of Monomeric Guanosine Triphosphatases

Table 9.40. Guanine nucleotide-exchange factors targeting ARF GTPases (ArfGEFs) in humans (**Part 2**; Sources: [998, 999] and Wiki Professional; ArfGEP: ARF guanine nucleotide-exchange protein; BAIAP: brain-specific angiogenesis inhibitor 1-associated protein [or insulin receptor substrate P53 (IRSP53)]; BIG: brefeldin-A-inhibited GEFs for ARFs; BRAG: brefeldin-resistant ArfGEF; EFA6: exchange factor for ARF6; FBx: F-box only protein; FBS: F-box, Sec7 protein; IQSec: IQ motif and Sec7 domain-containing protein; PSD: pleckstrin- and Sec7 domain-containing protein). Vesicles from the endoplasmic reticulum travel to the cis-Golgi network (CGN), where they fuse. Cargos subsequently progress through the Golgi body. Vesicles also carry substances from the trans-Golgi network to their final destination (Endos: endosome; N: nucleus; PM: plasma membrane; PsD: postsynaptic density). In addition to their substrates, ArfGEF bind to many partners, such as other ArfGEFs, receptors, ion channels ($K_{2P}1.1$: two pore-forming K^+ channel [KCNK1]), adaptors (Homer-1; DLg: Disc large homolog), enzymes, exocyst subunits, and signaling regulators.

Type	Aliases	Site	Substrates	Partners
EFA6a	EFA6, PSD, PSD1, TYL	PM	ARF6	$K_{2P}1.1$
EFA6b	PSD4, TIC	PM	ARF6	
EFA6c	PSD2	PsD	ARF6	
EFA6d	PSD3	PsD	ARF6	
BRAG1	IQSec2, ArfGEF4	PsD	ARF6	BAIAP2
BRAG2	IQSec1, ArfGEP100	PM, Endos, N	ARF6	AMPAR
BRAG3	IQSec3, Sag, synArfGEF	PsD	ARF6	DLg1/4, Homer1 dystrophin
FBx8	FBS			

Cytohesins

In humans, the cytohesin family contains 4 known members (cytohesin-1–cytohesin-4).[190] Cytohesin-2 and -3 are ubiquitous. Cytohesin-1 lodges principally in leukocytes; cytohesin-4 is more leukocyte specific. Cytohesins reside primarily in the cell periphery. Two splice isoforms of cytohesin-1, -2, and -3 are synthesized [999].

Cytohesin-1 is a guanine nucleotide-exchange factor for membrane-associated ARF GTPases. It is highly synthesized in natural killer and peripheral T cells. It

---

[190] Cytohesin-1 is also called pleckstrin homology, Sec7, and coiled-coil domain-containing protein PSCD1 and ARF nucleotide-binding site opener ARNO2. Cytohesin-2 is also named ARF nucleotide-binding site opener (ARNO), PSCD2, Sec7-containing protein-like substance Sec7l, and CTS18. Cytohesin-3 is also termed ARNO3, PSCD3, or general receptor for phosphoinositides RasGRP1. Cytohesin-4 is also labeled as ARNO4 and PSCD4. The Sec7 domain contains a guanine nucleotide-exchange motif, a coiled-coil region involved in homodimerization, and a PH domain that interacts with phospholipids.

elicits ARF-dependent activation of phospholipase-D. This integrin-binding GEF favors adhesion of lymphocytes mediated by $\alpha_L\beta_2$-integrin, hence their attachment and migration [1000].

Cytohesin-2 targets ARF1, ARF3, and ARF6 that promotes both migration of epithelial cells and outgrowth and branching of neurites [999]. Integrin-mediated activation of rRas leads to cytohesin-2 recruitment via rRas effector RalA-binding protein RalBP1. In migrating epithelia, cytohesin-2 promotes Rac1 activation by recruitment of the DOCK1–ElMo complex, a Rac1GEF (Sect. 9.4.1.7). It is also involved in docking and fusion of secretory granules, postendocytic trafficking via vacuolar adenosine triphosphatase, as well as endocytosis of some G-protein-coupled receptors via β-arrestins.

Recruitment of cytohesins to membranes is done via phosphoinositides and proteic adaptors, such as cytohesin-interacting protein (CytIP) or cytohesin-associated scaffold protein (CASP), GRP1-associated scaffold protein (GRASP or tamalin), and interacting protein for cytohesin exchange factors (IPCEF) [999]. Among these tethering proteins, GRASP couples metabotrobic glutamate receptors and the neurotrophin receptor NTRK3 to ARF activation.

ARF1GEF

Protein ARF1GEF, or Golgi-specific resistance to brefeldin-A factor GBF1, targets both subfamily-1 and -2 ARFs. It localizes to cis-compartments of the Golgi body, where it regulates assembly of CoP1 coat. On the other hand, ArfGEF1 and ArfGEF2, or BIG1 and BIG2, that activate subfamily-1 ARFs concentrate on trans-compartments of the Golgi body, where they control the recruitment of clathrin adaptors [1001].

Protein ARF1GEF presents GEF activity specifically for ARF5 at physiological Mg^{++} concentration [1002]. It associates primarily with vesicular–tubular clusters. It is involved in transport between the endoplasmic reticulum and Golgi body as well as inside the Golgi body. In vesicular–tubular clusters, it interacts with vesicle docking protein (VDP) and actuates ARF-dependent recruitment of vesicle coat protein CoP1 [999].

Agent ARF1GEF is recruited to exit sites of the endoplasmic reticulum and Golgi membranes by interacting with Rab1b GTPase [999]. It rapidly attaches to and detaches from membranes of the inner Golgi compartment (cis-Golgi network). In fact, ARF1GEF substrates include ARF1 among others [1003]. Protein ARF1GEF not only mediates the recruitment of CoP1 coat to cis-Golgi membranes, but also stimulates ARF1 in the cis-Golgi network.

ArfGEF1–ArfGEF3 (BIGs)

Proteins ArfGEF1 to ArfGEF3 correspond to brefeldin-A-inhibited GEFs for ARFs BIG1 to BIG3, respectively. Proteins ArfGEF1 and ArfGEF2 (BIG1 and BIG2) as well as ARF1GEF (or GBF1) can homodimerize. Both ArfGEF1 and ArfGEF2 associate with outer compartments of the Golgi body.

Like ARF1GEF, ArfGEF1 rapidly connects to and is released from membranes. Whereas ARF1GEF is associated with the cis-Golgi network, ArfGEF1 is linked to the trans-Golgi network.

Protein ArfGEF2 regulates the association of heterotetrameric adaptor protein complex AP1 and Golgi-localized γ-ear-containing ARF-binding proteins (GGA) to the trans-Golgi network. Both adaptor-related protein complex AP1[191] as well as Golgi-associated, γ-adaptin ear-containing, ARF-binding adaptor complexes are components of clathrin coat complexes. Protein ArfGEF2 is also linked to perinuclear recycling endosomes.

ARF6GEF1 to ARF6GEF3 (BRAGs)

Three brefeldin-resistant ArfGEFs exist (BRAG1–BRAG3), each with splice variants [999]. All activate ARF6;[192] in the present text, they are then called ARF6GEF1 to ARF6GEF3 proteins. Agent ArfGEF4, or ARF6GEF1 (BRAG1), is a member of group 3 of large ArfGEFs; ArfGEF1 to ArfGEF3 (BIGs) and ARF1GEF (GBF1) pertain to group-1 and -2.

Like most ArfGEFs, ARF6GEFs are expressed at low levels. Proteins ArfGEF4, or ARF6GEF1 (BRAG1), and ARF6GEF3 (or BRAG3) are mainly produced in the brain; ARF6GEF2 (or BRAG2) abounds in the brain, but is also synthesized in various tissues. Protein ArfGEF4 targets ARF1 and ARF6 in neurons.

Postsynaptic densities of excitatory and inhibitory synapses contain diverse types of GEFs (ARF6GEF1, ARF6GEF2b, RapGEF4, RhoGEF9, and RhoGEF24) and GAPs (RasA1, APAP1, AGAP2$_L$, and SpAR RapGAP) for small GTPases [1005]. These regulators modulate the synaptic transmission, as they control the formation and maintenance of dendritic spines and synapses via the actin cytoskeleton structuring. At postsynaptic densities of excitatory synapses, ARF6GEF1 complexes with NMDA-type glutamate receptors. On the other hand, ARF6GEF3, like

---

[191] Vesicles with an AP1-containing coat are mainly located in the trans-Golgi network. Coat adaptor complex consists of AP1β1, -γ1, -μ1, and -σ1 subunits (adaptins). It links clathrin to the membrane of vesicles. Three classes of coated transport vesicles can be defined: (1) CoP2-coated vesicles that bud from the endoplasmic reticulum; (2) CoP1-coated vesicle coats that assemble onto cisternae of the Golgi body and intermediate compartment between the endoplasmic reticulum and Golgi body; and (3) clathrin-coated vesicles that contain AP1 and AP2 adaptor complexes that arise from the trans-Golgi network and the plasma membrane. Whereas adaptor-related proteic complex AP1 is related to Golgi processing, AP2 adaptor complex that is also formed by 4 adaptins (β2, -γ2, -μ2, and -σ2 adaptins) works on the plasma membrane to internalize cargo in clathrin-mediated endocytosis.

[192] Agent ARF6, which localizes to the plasma membrane and endosomes, regulates the endosome–plasma membrane transfer and remodeling of the actin cytoskeleton at the cell cortex. At synapses, the submembrane cytoskeleton controls the number and dynamics of neurotransmitter receptors on the postsynaptic membrane, thereby modulating synaptic transmission efficiency.

RhoGEF9, localizes preferentially to postsynaptic densities of inhibitory synapses, with gephyrin, neuroligin-2, dystrophin, syntrophin, α- and β-dystroglycan, α- and β-dystrobrevin, and membrane associated guanylate kinase with inverted domain organization MAGI2, in addition to glycine and/or $GABA_A$ receptors. Long ARF6GEF3 splice variant ($ARF6GEF3_L$) interacts with postsynaptic proteins Disc large homologs DLg1 and DLg4 as well as Homer-1.[193]

Guanine nucleotide-exchange factor ARF6GEF1 (ArfGEF4) is confined to early Golgi compartments. Unlike ARF1GEF, it displays broader ARF substrate specificity [1001].

In non-neuronal cells, ARF6GEF2 regulates endocytosis of some cargos. Agent ARF6GEF2 cycles between the cytoplasm, where it contributes to endocytosis regulation at the plasma membrane, and the nucleus, where it can regulate nucleolar architecture [1006]. Its 2 splice variants — ARF6GEF2a and ARF6GEF2b — can both enter in and exit the nucleus.

ARF6GEFa to ARF6GEFd (EFA6s)

Exchange factor for adpribosylation factor ARF6 (EFA6) localizes mainly at the apical pole of polarized epithelial cells. It contributes to the stability of the apical actin ring onto which the tight junction is anchored [1004]. In response to E-cadherin engagement, it is also involved in tight junction formation.

Protein EFA6 was the first identified set member that activates ARF6, especially in developing and mature neurons, where it regulates actin cytoskeleton reorganization associated with intracellular transfer of cargos. Additional ARF6-specific GEFs constitute a family of 4 structurally related proteins (EFA6a–EFA6d); in the present text, they are called ARF6GEFa to ARF6GEFd, as it is done for BRAG family members, called ARF6GEF1 to ARF6GEF3 for a straightforward meaning. Members of the EFA6 family are highly selective for ARF6 [999]. They also interact selectively with PI(4,5)P_2 and localize mostly to the plasma membrane. These EFA6 isoforms promote the reorganization of cortical actin into microvillus-like structures.

Protein EFA6a is synthesized predominantly in the central nervous system, but also in intestine; EFA6b is more widespread with its highest levels in the pancreas, spleen, thymus, and placenta; EFA6c is also produced primarily in the brain; EFA6d is ubiquitous [999]. In the brain, EFA6a, EFA6c, and EFA6d have distinct regional distributions.

---

[193] Adaptor DLg4 that anchors synaptic proteins directly and indirectly binds neuroligin, 2 ionotropic glutamate receptors (^{N}methyl ^{D}aspartate [^{NMDA}Glu receptors (NMDAR)] and α-amino 3-hydroxy 5-methyl 4-isoxazolepropionic acid [^{AMPA}Glu receptors (AMPAR)] receptors), and voltage-gated potassium channels $K_V1.2$, $K_V1.4$, and $K_V1.5$. Homer-1 regulates group-1 metabotropic glutamate receptors.

9.4 Regulators of Monomeric Guanosine Triphosphatases

Table 9.41. Guanine nucleotide-exchange factors for Rab (RabGEFs) and their effectors (Source: [889]; RabEx5: Rab5 GDP–GTP exchange factor; RASSF: Ras interaction/interference RIN1, afadin, and Ras association domain-containing protein family member; RIn: Ras and Rab interactor).

Type	Aliases	Effectors
RIn1		Rab5
RIn2	RASSF4	
RIn3		
RabGEF1	RabEx5	Rab5

FBx8

Molecule F-box only protein FBx8 mediates incorporation of proteins via ubiquitination, as it can interact with both ARFs and ubiquitin ligases [999].

9.4.1.2 RabGEFs

Protein *Ras and Rab interactor* RIn1 that impedes activated Ras is stimulated by Ras and acts as a guanine nucleotide-exchange factor for Rab5 [996] (Table 9.41).

Rab5GEF (Alsin)

Amyotrophic lateral sclerosis protein-2 (ALS2), also called alsin, acts as a Rab5GEF involved in endosome dynamics [1007]. Endocytosis mediated by Rab5 and appropriate rate of endosomal conversion to lysosomes, i.e., proper lysosomal degradation of internalized glutamate receptors in neurons, requires Rab5GEF [1008]. Agent Rab5GEF preferentially interacts with activated Rac1 that recruits cytosolic Rab5GEF to membranes. Activator Rab5GEF participates in Rac1-activated macropinocytosis [1009].

9.4.1.3 RacGEFs

Dendritic spines, small protrusions from a dendrite that increase the number of possible contacts between neurons, serve for memory storage and synaptic transmission. Dendritic spine morphogenesis and remodeling of excitatory synapses is needed for adequate neuronal development. Calcium–calmodulin-dependent kinases and guanine nucleotide-exchange factors RacGEFs interact to control dendritic spine morphogenesis [1010]. This connection between these 2 agents transduces Ca^{++} influx into small GTPase activity that leads to actin reorganization, hence into spatially and temporally regulated remodeling of dendritic spines.

Table 9.42. Guanine nucleotide-exchange factors for Rac GTPases (RacGEFs) — TIAM1 and TIAM2 — and their effectors (Source: [889]).

Type	Effectors
TIAM1	Rac1, Rac2, Rac3
TIAM2	Rac1

Small Rap1 GTPase modulates Rac1 activity. The latter is activated and translocates to intercellular adhesion plaques. Proteins Vav2 and TIAM1 are recruited by atrial natriuretic peptide and prostaglandin-E2 to enhance the endothelial barrier (Sect. 9.4.1.6).

TIAM1–TIAM2

Agent T-cell lymphoma invasion and metastasis-inducing protein TIAM1, a Rac-specific guanine nucleotide-exchange factor, can augment endothelial permeability, hence leukocyte transendothelial migration. Platelet-activating factor disrupts interendothelial junctions, as it provokes translocation of Rac1 and TIAM1 [1011]. It binds $PI(3,4,5)P_3$ with high affinity. In vascular endothelial cells, RapGEF3-dependent Rap1 activation causes relocalization and stimulates Rac-specific GEFs TIAM1 and Vav2 and promotes accumulation of suppressor of cytokine signaling SOCS3 (cAMP–RapGEF3–Rap1–SOCS3 pathway), thereby inhibiting signal transducer and activator of transcription [1012]. Protein TIAM1 mediates Ras activation of Rac to stimulate membrane ruffling as well as activate the JNK and NFκB pathways, hence promoting cell survival, independently of phosphatidylinositol 3-kinase [997]. Protein TIAM1 regulates cell migration, as it modulates both intercellular and cell–matrix adhesions.

Other TIAM1 partners include protein kinase-$C\alpha$, -$C\beta 1$, -$C\gamma$, -$C\delta$, -$C\epsilon$, and -$C\zeta$, protein phosphatase-1, MAPK8-interacting protein-1 and -2, and ankyrins-1 and -3 [251].

Activator TIAM2 that is also called SIF and TIAM1-like exchange factor (STEF) targets Rac1 GTPase (Table 9.42).

During cell migration, focal adhesions that attach the cytoskeleton to the extracellular matrix, assemble and disassemble. Numerous agents intervenes in this process, such as ERK, FAK, MLCK, PAK, RoCK, and Src kinases as well as Rho GTPases. Member of the spectraplakin family of cytoskeletal crosslinking proteins Microfilament and actin filament crosslinker protein MAcF1[194] binds to both actin and microtubule cytoskeletons, thereby guiding microtubules toward actin filaments connected to focal adhesions. In addition, microtubule growth activates Rac GTPase that, in turn, promotes microtubule growth into the lamellipodium

---

[194] A.k.a. actin crosslinking family member ACF7.

(positive feedback loop), as it targets the microtubule plus-tip protein Cytoplasmic linker-associated protein CLAsP2. The latter is involved in the local regulation of microtubule dynamics in response to positional signals. Agent TIAM1 regulates microtubule stability. Subtype TIAM2 is required during microtubule growth that is involved in focal adhesion disassembly for optimal cell migration [1013]. Protein TIAM2 stimulated by microtubules to activate Rac GTPase promotes the growth of microtubules toward focal adhesions.

PREx1 and PREx2

Guanine nucleotide-exchange factors of the PREX (PI(3,4,5)P$_3$-dependent Rac exchanger) family activate Rac GTPases. The PREX family comprises PREx1 and PREx2 (or PREx2a) and its splice variant PREx2b.

Isoform PREx1 is synthesized in leukocytes. Subtype PREx2 is more widely produced, but not in leukocytes. Variant PREx2b is only generated in the heart. Protein PREx1 is also widespread in the central nervous system, whereas PREx2 is specifically expressed in Purkinje neurons of the cerebellum [100].

Both PREx1 and PREx2 connect signaling from G$\beta\gamma$ subunits of heterotrimeric G proteins and phosphatidylinositol (3,4,5)-trisphosphate synthesized by phosphatidylinositol 3-kinase to Rac activation, hence linking GPCR stimulation to that of PI3K kinase. Members of the PREX family of RacGEFs differ from others in their mode of regulation, because they are synergistically activated by G$\beta\gamma$ subunit and PIP$_3$ agent.

Target of rapamycin complex TORC2 regulates the actin cytoskeleton via Rho GTPases. Both PREx1 and PREx2 interact with TOR kinase, thereby acting as effectors in the TOR signaling for Rac activation for cell migration [1014].

Regulator PREx1 regulates GPCR-dependent Rac2 activation, ROS production, neurotrophin-stimulated neuron migration, and neutrophil recruitment to inflammatory sites [100]. Isoform PREx2a is a direct regulator of phosphatase and tensin homolog on chromosome 10 (PTen) that can stimulate cell proliferation, as it inhibits PTen phosphatase, hence stimulating PI3K signaling. In endothelial cells, PREx2b governs Rac1 activation and cell migration in response to sphingosine 1-phosphate.

Rac and RacGEFs in the Nervous System

Small Rac isoforms — Rac1, Rac2, and Rac3 — control the organization of the actomyosin cytoskeleton, gene expression, and production of reactive oxygen species. In the nervous system, Rac GTPases intervene in all stages of neuronal development (neurite, axon, and dendrite formation; axon pathfinding; dendrite branching; dendritic spine formation; and neuron survival). Small Rac GTPases are activated by guanine nucleotide-exchange factors that control neuronal development [100].

Table 9.43. Guanine nucleotide-exchange factors for Ral (RalGEFs) and their effectors (Source: [889]).

Type	Effectors
RalGEF (RalGDS)	RalA, RalB
RGL1	RalA, RalB
RGL2	RalA, RalB
RGL3	RalA, RalB

Activator of Rac GTPases TIAM1 regulates neurite and axon outgrowth and dendritic spine formation; Vav2 and Vav3 axon outgrowth from retina to thalamus; RhoGEF23 neurite outgrowth, axon extension, and pathfinding; RhoGEF24 regulates neurite and axon outgrowth as well as dendritic spine formation; Dedicator of cytokinesis DOCK1 neurite outgrowth and DOCK7 axon formation; Rac–CDC42 exchange factors RhoGEF7 and RhoGEF25 dendritic spine formation; and Rab5GEF regulates neuronal survival [100].

9.4.1.4 RalGEFs

Protein Ral GDP-dissociation stimulator (RalGDS or RalGEF) is a Ras effector for RalA and RalB GTPases. In addition, RalGDS-like proteins (RGL1–RGL3) also activate both RalA and RalB (Table 9.43). Molecule RalGDS binds preferentially to active forms of Ras, Rap1a, and rRas GTPases. It competes with Ras1 for connection to hRas.

9.4.1.5 RanGEFs

Small GTPase Ras-related nuclear protein (Ran) is involved in nuclear transfer as well as chromatin condensation via its control of microtubule assembly. Regulator of chromosome condensation RCC1 (or ChC1) is a GEF for Ran GTPase [1007]. Ran-binding protein RanBP3 associates with RCC1 to facilitate its activation and to regulate the nucleocytoplasmic transport.

9.4.1.6 RapGEFs

Rap-specific guanine nucleotide-exchange factors that serve as Ras activators constitute several groups that operate in junctional control. According to activators and binding partners, RapGEFs include the subfamilies of [903]:

1. RapGEF1;
2. $_{PDZ}$GEFs, or RAGEFs (RapGEF2–RapGEF6);

9.4 Regulators of Monomeric Guanosine Triphosphatases

Table 9.44. Guanine nucleotide-exchange factors for Rap GTPase (RapGEFs) and their effectors (Source: [889]).

Type	Main alias	Effectors
RapGEF1	C3G	BCAR1, CBL, CRK, CRKL, GRB2, NEDD9
RapGEF2	$_{PDZ}$GEF1, RAGEF1	Ral, Rap1, Rap2
RapGEF3	EPAC1	Rap1, Rap2
RapGEF4	EPAC2	Rap1, Rap2
RapGEF5	GFR	Rap1, Rap2, rRas3
RapGEF6	$_{PDZ}$GEF2, RAGEF2	Rap1, Rap2

3. cAMP-activated GEF (EPAC or $_{cAMP}$GEF) with 2 members RapGEF3 and RapGEF4;
4. calcium- and diacylglycerol-regulated GEF ($_{CalDAG}$GEF; RasGRP1–RasGRP2); and
5. atypical RapGEF DOCK4; in addition to
6. phospholipase-Cε.

Small Rap GTPases are involved, in particular, in the control of between-cell and cell–matrix adhesions.[195] Four RapGEF families are implicated in junction regulation.

RapGEF1

Ubiquitous guanine nucleotide-exchange factor RapGEF1 targets Ras GTPase.[196] It localizes to endosomes. It activates Rap1 upon E-cadherin and EGFR internalization and nectin engagement [903]. Protein RapGEF1 interacts with many adaptors, such as BCAR1, CRK, CRKL, and GRB2 (Table 9.44). It also tethers to E-cadherin during the initial step of junction formation (Table 9.45).

---

[195] In epithelia and endothelia, apical tight junctions create a barrier for ions and solutes, whereas cadherin-mediated adherens junctions link actin networks of neighboring cells.

[196] A.k.a. CRK SH3-binding GEF (C3G).

Table 9.45. Activation of Rap1 by RapGEFs in adherens junctions (Source: [903]).

Pathway	Effect
Nectin–Src–RapGEF1–Rap1	Adherens junction formation
Stretch–BCAR1–RapGEF1–Rap1	Adherens junction formation
RTK—Src–RapGEF1–Rap1	Adherens junction formation
GPCR–cAMP–RapGEF3–Rap1a	Adherens junction tightening
RhoG–DOCK4–Rap1	E-Cadherin recruitment
RapGEF2/6–Rap1a	Adherens junction formation and maturation

PDZ Domain-Containing RapGEF2 and RapGEF6

Guanine nucleotide-exchange factors RapGEF2 and RapGEF6[197] localize to cell junctions.

Protein RapGEF2[198] connects to Rap1a and Ral guanine nucleotide-dissociation stimulator (RalGDS). Agent RapGEF2 associates with β-catenin both directly and via Membrane-associated guanylate kinases (MAGuK) with inverted domain organization scaffolds MAGI1 and MAGI2 [903]. Protein MAGI1[199] localizes to tight and adherens junctions in epithelial cells as well as synapses in neurons.[200] It is required for cell adhesion-induced activation of Rap1 GTPase.

Molecule RapGEF6 is closely related to RapGEF2 protein. It has 2 splice variants: RapGEF6a and RapGEF6b [1016]. Like RapGEF2, RapGEF6 is specific for Rap1 and Rap2 downstream from rRas3; it is unresponsive to cAMP and other nucleotides. Protein RapGEF6 has a different tissue distribution than that of RapGEF2 agent. It operates in the cell cortex, whereas RapGEF2 exerts its function in the perinuclear compartment. Protein RapGEF6 does not connect to hRas, nRas, RalA, RIN, RIT, and RHEB [1017]. It is an activator of Rap1 during junction formation and maturation both in epithelial and endothelial cells [903].

---

[197] A.k.a. PDZ domain-containing GEFs ($_{PDZ}$GEF1 and $_{PDZ}$GEF2). Alias PDZ combines the first letters of 3 proteins: postsynaptic density protein PSD95, or DLg4, Disc large homolog DLg1, and zonula occludens protein ZO1. ¡they are also called Ras/Rap1-association GEFs (RAGEF1 and RAGEF2).

[198] A.k.a. cAMP-dependent guanine nucleotide-exchange factor CNRasGEF in addition to $_{PDZ}$GEF1 and RAGEF1.

[199] A.k.a. atrophin-1-interacting protein-3 and brain-specific angiogenesis inhibitor (BAI1)-associated protein BAIAP1 (or BAP1).

[200] Hence the name synaptic scaffolding molecule (SSscM).

cAMP-Regulated RapGEF3 and RapGEF4

Guanine nucleotide-exchange factors for Rap GTPases RapGEF3 and RapGEF4[201] can act as RasGEFs. These cAMP-responsive guanine nucleotide exchange factors for Rap GTPase are involved in the regulation of endothelial cell adhesion modulation, myocardial contraction, and insulin secretion, among other tasks.

These proteins possess an N-terminal regulatory region with 1 (RapGEF3) or 2 (RapGEF4) cAMP-binding domains and a C-terminal catalytic sequence with a CDC25-homology domain for GEF activity and a Ras exchange motif (REM).

These 2 isoforms have different tissue-specific distribution. Isoform RapGEF4 is predominantly produced in the brain and adrenal glands, whereas RapGEF3 is ubiquitous. Agent RapGEF3 can activate ERK signaling via Rap2b, PLCε, and subsequent hRas activation.

Agent RapGEF3 binds directly to negatively charged phosphatidic acid using its Disheveled, EGL10, and pleckstrin (DEP) domain under the control of cAMP for subsequent activation of Rap GTPase at the plasma membrane [1018].[202] In addition, RapGEF3 can also be recruited to the plasma membrane independently of cAMP via activated ezrin, radixin, and moesin.

Dynamic control of between-endothelial cell adhesions by cAMP that can be stimulated by prostaglandins and atrial natriuretic peptide is mediated by both PKA and RapGEF3 [903]. Junctional Rap1 GTPase is indeed activated by RapGEF3 agent. The cAMP–RapGEF3–Rap1 axis impedes vascular permeability, as it stabilizes adherens junctions. Cerebral cavernous malformation protein CCM1[203] actually contributes to signaling between adhesion molecules and the cytoskeleton. It complexes with VE-cadherin, α-, β-, and δ-catenins, and cytoskeleton-anchoring Rap1 effector AF6[204] and transmits Rap1 signaling to cell adhesion plaques to tighten intercellular contacts. In addition, CCM1 binds to microtubules to regulate endothelial cell shape.

---

[201] A.k.a. cAMP-dependent $_{cAMP}$GEF1 and 2 and exchange protein directly activated by cAMP EPAC1 and EPAC2. Upon cAMP binding, EPACs undergo a conformational change that relieves auto-inhibition. In addition, cAMP enables translocation of EPACs from the cytosol to the plasma membrane.

[202] Phosphatidic acid also enables recruitment of DOCK2 RacGEF at the leading edge of migrating neutrophils and SOS RasGEF to cell adhesion-free membrane regions.

[203] A.k.a. Rap1-Binding protein K-Rev1 interaction trapped gene product KRIT1. Protein CCM1 forms a family with cerebral cavernous malformation proteins CCM2 and CCM3 (a.k.a. programmed cell death protein PDCD10).

[204] A.k.a. mixed lineage leukemia translocated protein MLLT4.

RapGEF5

Protein RapGEF5[205] localizes to the nucleus, as it contains a nuclear localization signal [1019]. However, it lacks a cAMP-binding motif. It is strongly expressed in the central nervous system.

Rap(Ras)GRPs

Protein RapGEF5 connects to Rap1 and, in vitro, mRas^{GTP}; RapGEF1 bind Rap1 and Rap2; RasGRP3 activates both Rap1 and Ras [1020]. All these 3 types of GEFs can promote ELk1 activation.

Guanine nucleotide-releasing proteins for Ras (RasGRPs) that are calcium- and diacylglycerol-activated GEFs[206] (Sect. 9.4.1.7) constitute a category of Rap guanine nucleotide-exchange factors that abound in the central nervous system.

Protein RasGRP2 has a substrate specificity for Rap1a GTPase. The basal ganglia of the brain is well supplied with RasGRP2 activator. It precludes the Ras-dependent activation of the ERK cascade. Like RasGRP2, RasGRP1 is synthesized in hematopoietic cells. Yet, RasGRP1 fails to activate Rap1a, but activates hRas and rRas in response to Ca^{++} and diacylglycerol.

DOCK4

Dedicator of cytokinesis DOCK4 (Sect. 9.4.1.7) is a member of atypical RhoGEFs that has GEF activity for both Rap and Rac. In particular, DOCK4 is a RapGEF in junction formation [903]. It is activated by RhoG GTPase (Table 9.45). It regulates cell migration via Rac GTPase.

Rap1GDS1

Protein Rap1 GTP–GDP dissociation stimulator-1 (Rap1GDS1)[207] catalyzes GDP exchange for GTP on Rap1a and Rap1b, as well as other small GTPases, such as kRas, Rac1, Rac2, RhoA, RalB, and CDC42 [1020, 1021]. Besides, Rap1a and kRas can be antagonist. However, kRas as well as RhoA and Rac2 are more important substrates for Rap1GDS1 than Rap1a [1021].

---

[205] A.k.a. guanine nucleotide-exchange factor for Rap1 (GFR) and mRas (rRas3)-regulated GEF (MRGEF).

[206] Proteins RasGRP1 and RasGRP2 are also named $_{CalDAG}$GEF2 and $_{CalDAG}$GEF1, respectively.

[207] A.k.a. small-molecular-weight, GTP-binding, guanine nucleotide (GDP)-dissociation stimulator (SMGGDS).

9.4.1.7 RasGEFs

Members of the RAS hyperfamily impinge on cell proliferation, differentiation, and apoptosis. Upon gene mutation, these oncogene products provoke cell transformation. They are counteracted by Rap1 GTPase. Guanine nucleotide-exchange factor for Ras GTPases is also called Ras guanine nucleotide-dissociation stimulator (RasGDS).

Families of the RASGEF Superfamily

Three main families of guanosine nucleotide-exchange factors are associated with Ras GTPases (hRas, kRasA, kRasB, and nRas) and link plasmalemmal receptors to Ras activation [1022]: (1) Son of sevenless (SOS1–SOS2); (2) Ras guanosine nucleotide-releasing factors (RasGRF1–RasGRF2); and (3) Ras guanosine nucleotide-releasing proteins (RasGRP1–RasGRP4).

In fact, the superfamily of guanine nucleotide-exchange factors for Ras GTPases also encompass (1) RapGEF1 to RapGEF6; (2) members of the Ral guanine nucleotide-dissociation stimulator family (RalGDS, Ras association [RalGDS–AF6] and pleckstrin homology domain-containing protein RAPH1 [or lamellipodin], and RalGDS-like proteins RGL1, RGL2, and RGL4); (3) RalGEFs with PH domain and SH3-binding motif (RalGPS1–RalGPS2); (4) breast cancer antiestrogen resistance BCAR3 (a CDC42GEF and RacGEF);[208] Rap1GDS1 (Rap1, GTP–GDP dissociation stimulator-1);[209] and phospholipase-Cε [1028].

The RasGEF superfamily can be subdivided into 2 epifamilies according to sequence similarities: CDC24- and CDC25 motif-containing RasGEFs. The catalytic domain of RasGEFs, RasGRFs, and RasGRPs corresponds to catalytic CDC25 homology region. Proteins of the RASGEF category that possess a CDC24 motif act on CDC42 GTPase.

Signaling Mediated by RasGEFs

Activators of the SOS and RASGRP categories link plasmalemmal or cytoplasmic Tyr kinases to Ras GTPases via GRB2 adaptor and diacylglycerol (synthesized by phospholipase-Cγ), respectively [1022].[210] Most cell types use primarily SOS

---

[208] A.k.a. SH2 domain-containing protein SH2D3b, novel SH2-containing protein NSP2, and AND34.
[209] A.k.a. small monomeric GTPases, GTP–GDP dissociation stimulator SMGGDS.
[210] Proteins of the RASGRP category have calcium-binding EF hand sequences and diacylglycerol-binding domain [1023].

proteins. In lymphocytes, the main GEF involved in the TCR-mediated activation of Ras is diacylglycerol-dependent RasGRP1.[211]

Receptor Tyr kinases can be transactivated by Gi- and Gq-coupled receptors to stimulate extracellular signal-regulated kinases (ERK1 and ERK2). In addition, G$\beta\gamma$ dimer associated with Gi- and Gq-coupled receptors can activate Ras via Src kinase, GRB2 adaptor, and SOS activator.[212] Moreover, Gi and Gq subunits can activate ERK1 and ERK2 by binding to Rap1GAP1, thereby inhibiting Rap1 that antagonizes Ras GTPase. In particular, Gq-coupled serotonin receptors 5HT$_2$ can trigger the PLC–IP$_3$–Ca^{++}–RasGRP1 and PLC–IP$_3$–Ca^{++}–RasGRF1 pathways. Similarly to RTKs, Gs-coupled receptors can launch the MAPK module via Ras GTPase. Gs-coupled serotonin receptors 5HT$_{4B}$ and 5HT$_{7A}$ excite ERK1 and ERK2 kinases, without transactivation of epidermal growth factor receptors, via Ras, but not Rap1 [1024]. In summary, RasGRP1 is stimulated not only from receptor (e.g., growth factor receptors) and cytosolic Tyr kinases, but also GPCRs (e.g., serotonin HT$_1$ and HT$_2$, dopamine, and prostaglandin receptors) and protein Tyr phosphatase receptors.

On the other hand, RasGRFs are coupled to G-protein-coupled receptors.[213] These receptors support RasGRF translocation to the plasma membrane as a result of Ca^{++} influx and their enzymatic activity on account of phosphorylation [1022]. However, RasGRFs do not work only downstream from G-protein-coupled receptors. For example, RasGRF2 also participates in T-cell signaling [1025].

The plasma membrane is not the exclusive platform from which Ras regulates signaling. Activation of Ras at distinct subcellular sites represents a mechanism for signal diversification from a given receptor. Whereas kRas resides exclusively at the plasma membrane, hRas and nRas also lodge in organelle membranes, such as that of the endoplasmic reticulum and Golgi body. Both hRas and nRas can then engage cRaf at organelle membranes during cell signaling with different outcomes. Activation of JNK and ERK is more efficient in the endoplasmic reticulum and Golgi body, respectively [1026]. In T lymphocytes, the plasma membrane remains the single site of TCR-driven nRas activation [1027]. On the oher hand, in response to Src-dependent activation of phospholipase-Cγ1, RasGRP1 can translocate to the Golgi body where it then activates Ras.

---

[211] Stimulation of TCRs causes phosphorylation of Rho- and RacGEF Vav1. The latter activates PLCγ1 that synthesizes diacylglycerol. Diacylglycerol then binds to and activates RasGRP1 in leukocytes and neurons. Protein RasGRP1 can thus be activated by both PKC-dependent and -independent axes. In addition, the Vav1–PLCγ1 pathway promotes actin-dependent translocation of RasGRP1 to the plasma membrane. Besides, PKC can activate ERK1 and ERK2 enzymes. Therefore, 3 pathways leads to the MAPK module: (1) TCR–Vav–PLC–DAG–PKCθ–MAPK; (2) TCR–Vav–PLC–DAG–PKC–RAsGRP1–Ras–MAPK; and (3) TCR–Vav–PLC–DAG–RAsGRP1–Ras–MAPK.

[212] Kinase Src is a common effector of both G-protein-coupled receptors and receptor Tyr kinases.

[213] Agents RasGRF1 and RasGRF2 possess 2 pleckstrin homology (PH) regions, a coiled-coil motif, a Ca^{++}–calmodulin binding ilimaquinone (IQ) domain, a DBL homology (DH) sequence for both homodimerization and catalysis of the GDP–GTP exchange on Rac1, and the prototypical CDC25 Ras exchange domain [1025].

9.4 Regulators of Monomeric Guanosine Triphosphatases

Table 9.46. Guanine nucleotide-exchange factors for Ras of the RASGRP category and their effectors (Sources: [889, 1067]).

Type	Effectors [signaling axis]
RasGRP1	hRas, kRas, nRas, rRas1–rRas3 (hRas>kRas, nRas) [Ras–bRaf–MAP2K3/6–P38MAPK pathway]
RasGRP2	kRas, nRas, rRas1–rRas2, Rap1a, Rap2a [Rap–cRaf–MAP2K1/2–ERK1/2 pathway]
RasGRP3	hRas, kRas, nRas, rRas1–rRas3, Rap1a, Rap2b (hRas>kRas, nRas)
RasGRP4	hRas, kRas, nRas, rRas1–rRas2

RasGRPs

Ras guanine nucleotide-releasing proteins are activated synergistically by Ca^{++} and diacylglycerol (Sect. 9.4.1.6). Four known members (RasGRP1–RasGRP4) are produced in various tissues. Activation of most RasGRPs relies on phospholipase-Cγ-dependent generation of diacylglycerol that facilitates RasGRP translocation to membranes and association with target GTPases (Table 9.46). In neurons, RasGRP2 principally activates Rap GTPases, whereas RasGRP1 activates Ras GTPases.[214]

RasGRP1

Protein RasGRP1, or $_{CalDAG}$GEF2, interacts with diacylglycerol kinase-ζ, PTPn1 phosphatase, and von Hippel-Landau protein [251]. It activates members of the RAS family, such as hRas, kRas, nRas, rRas1, and rRas2 GTPases.

RasGRP2

Protein RasGRP2, also called $_{CalDAG}$GEF1 and CDC25-like protein (CDC25L), interacts with patched domain-containing-2 (or Dispatched Disp3) [251]. It is specific for Rap1 GTPase [1067].

RasGRP3

Protein RasGRP3 activates some RAS superfamily members as well as Rap1 GTPase. It interacts with protein kinase-Cα, PKCβ1, PKCδ, and PKCϵ, as well as PI3K and 14-3-3-τ protein [251].

---

[214] Agent RasGRP1 is highly expressed in neurons of the forebrain (prosencephalon), whereas RasGRP2 is particularly highly produced in the striatum [926].

Table 9.47. Expression in hematopoietic cells of members of the RASGRP family.

Type	Expression in hematopoietic cells
RasGRP1	T lymphocytes
RasGRP2	Neutrophils, platelets
RasGRP3	B lymphocytes
RasGRP4	Mastocytes

Table 9.48. Guanine nucleotide-exchange factors for Ras of the RASGRF category and their effectors (Sources: [889, 1067]).

Type	Effectors
RasGRF1	hRas, kRas, nRas, rRas1–rRas3, Rac1
RasGRF2	hRas>kRas, nRas; Rac1

RasGRP4

Protein RasGRP4 resides in hematopoietic cells, like other group members (Table 9.47). Members RasGRP1 to RasGRP3 are located in various tissues, mainly in cells of the nervous system.

RasGRFs

The RASGRF family comprises 2 known members (RasGRF1–RasGRF2) that are mainly synthesized in cells of the nervous tissue (Table 9.48). Protein Ras-GRF1 is involved in the maintenance of population and functioning of pancreatic β cells [1067]. It indeed participates in PKB and ERK activation in response to IGF1 factor. Both RasGRF1 and RasGRF2 are stimulated by activated GPCR receptors. They are sensitive to calcium–calmodulin. They target Rac GTPases. They undergo ubiquitination.

RasGRF1

Protein RasGRF1 is able to target various small GTPases of the RAS family, such as hRas, kRas, nRas, and rRas1 to rRas3, as well as Rac1, but not Rap GTPase. It activates Ras downstream from various signaling mediators, i.e., via: (1) increase in intracellular calcium concentration that leads to calmodulin binding to RasGRF1; (2) Gβγ subunit release; (3) elevation of intracellular cAMP concentration and subsequent phosphorylation of RasGRF1 by protein kinase-A; (4) receptor Tyr kinase activation, such as nerve growth factor receptor NTRK1 that provokes RasGRF phosphorylation; (5) activation of ionotropic glutamate

9.4 Regulators of Monomeric Guanosine Triphosphatases

receptors, either NMDA-type glutamate receptor (NMDAR) that primes RasGRF1 binding to GluN2b subunit to activate ERK or P38MAPK enzyme, or AMPA-type glutamate receptor (AMPAR) to activate the Ras–ERK axis [1029]. It mediates one-third of the NMDAR-mediated activation of the Ras–Raf–ERK cascade in cultured hippocampal neurons and a significant fraction in intact brain [926].

Both RasGRF1 and RasGRF2 can form homo- and hetero-oligomers to execute their activities [1030]. Protein RasGRF1 mediates at least 2 pathways for Raf activation [1030]: (1) a constitutive Ras-dependent and (2) a calcium-induced ERK activation that cooperates with the constitutive signal without augmenting the level of active Ras, despite an increased Ras activity.

RasGRF2

Protein RasGRF2 is widespread. It is a bifunctional signaling protein that can bind and activate both Ras and Rac GTPases. It coordinates the activation of the MAPK modules in response to extracellular signals, i.e., the extracellular-signal-regulated kinase and stress-activated protein kinase (JNK and P38MAPK) pathways [1031].

Development and activity of T lymphocytes rely on Ras GTPase. The latter is stimulated by [1025]: (1) diacylglycerol-responsive RasGRP1 following Vav1 and PLCγ1 actions and (2) RasGRF2 to activate nuclear factor of activated T cells in cooperation with Vav1 and calcium ions.

DOCKs

The dedicator of cytokinesis (DOCK) family of guanine nucleotide-exchange factors include several known members (DOCK1–DOCK11). Members of the DOCK family operate in organism development as well as cell migration, T-cell activation, and removal of apoptotic cells.

DOCK1

Protein DOCK1[215] is produced in many tissues, except peripheral blood cells [1032].

Protein DOCK1 and related proteins possess DHR1 and DHR2 domains that allow binding to phospholipids and Rac activation. Protein DOCK1 also has an SH3-binding domain for engulfment and cell motility (ElMo) proteins. These adaptors recruit DOCK1 to the plasma membrane and augment its GEF efficiency [1033]. The ELMO family comprises 3 paralogs (ElMo1–ElMo3) that all interact with DOCK1 agent. In addition, ElMo1 inhibits DOCK1 ubiquitination [1034].

---

[215] A.k.a. DOCK180, a mammalian ortholog of Caenorhabditis elegans protein Ced5.

DOCK2

Protein DOCK2 is a member of the DOCK-A subfamily that specifically activates isoforms of Rac GTPases. It is expressed specifically in lymphocytes and macrophages [1032]. Whereas DOCK1 binds to CRK2 to form CRK2-DOKk1–Rac1 complex required for cell migration and phagocytosis of apoptotic cells, DOCK2 does not connect to CRK2 that transduces signals at focal adhesions.

DOCK3

Protein DOCK3[216] is a member of the DOCK-B subfamily that is specifically expressed in neurons. Like DOCK1, it tethers to CRK adaptor. Activity of DOCK3 is specific to Rac1, as it does not target Rac3, which is primarily expressed in the brain, where it antagonizes Rac1 [1035].

DOCK4

Member DOCK4 of the DOCK-B subfamily activates Rac and Rap1 GTPases. It intervenes in the Wnt pathway, in which it undergoes phosphorylation by glycogen synthase kinase GSK3β of the β-catenin degradation complex, hence enhancing β-catenin stability and axin degradation [1036].

9.4.1.8 GEF Domain-Containing NSPs

Association between members of the family of Novel Src homology 2 (SH2) domain-containing proteins (NSP) and those of the family of CRK-associated substrate (CAS) contributes to integrin and receptor Tyr kinase signaling.[217] In particular, breast cancer anti-estrogen resistance BCAR3, a member of the NSP family (Table 9.49), complexes with Neural precursor cell expressed, developmentally down-regulated protein NEDD9, or CRK-associated substrate-related protein (CASL or CAS2),[218] an element of the CAS family to form additional BCAR1-associated complex [1037]. These proteins possess a GEF domain that targets RalA, Rap1A, and rRas [1038].[219]

---

[216] A.k.a. modifier of cell adhesion (MOCA) and presenilin-binding protein (PBP).

[217] Other association with members of the CAS family, such as the CAS-CRK and CAS-Src complexes, are implicated in cell processes that involve the regulation of the actin cytoskeleton (cell adhesion, migration, proliferation, and survival).

[218] A.k.a. human enhancer of filamentation HEF1.

[219] Adaptor BCAR1, also called CRK-associated substrate (CAS or P130CAS), is a binding partner and potent enhancer of Src kinase activity. Kinases of the SRC family (e.g., Fyn, Src, and Yes) mediate integrin-dependent CAS phosphorylation.

9.4 Regulators of Monomeric Guanosine Triphosphatases

Table 9.49. BCAR1-binding GEFs novel SH2 domain-containing proteins (NSP), not neuroendocrine-specific proteins (NSP). Scaffold BCAR1, a.k.a. CRK-associated substrate (CRKAS, CAS, and P130CAS) is a binding partner and potent enhancer of Src kinase activity. Kinases of the SRC family (e.g., Src, Yes, and Fyn) mediate integrin-dependent BCAR1 phosphorylation. These proteins share a common structural organization with an N-terminal SH2 domain, a central proline-rich motif, and a C-terminal GEF sequence. They target Ra1A, Rap1a, and rRas GTPases. On the other hand, neuroendocrine-specific protein is also designated as reticulon-1; neuroendocrine-specific protein-like-1 (NSPL1), or neuroendocrine-specific protein NSP2, as reticulon-2; neuroendocrine-specific protein-like-2 (NSPL2) as reticulon-3; and neuroendocrine-specific protein-C homolog (NSPcL) as neurite outgrowth inhibitor (NOGo) and reticulon-4. All reticulon types abound in the endoplasmic reticulum. The main members NSPa and NSPc of the NSP (reticulon [RTN]) family are encoded by a single gene (ASYIP: ASY interacting protein; HAP: homolog of ASY protein; NA: not applicable).

NSP type	Reticulon type	Other aliases
Novel SH2 domain-containing proteins with RasGEF and RasGRF (CDC25) domains		
NSP1	NA	SH2D3a
NSP2	NA	SH2D3b, SHEP2, AND34, BCAR3
NSP3	NA	SH2D3c, SHEP1, CHAT
Neuroendocrine-specific proteins (reticulons		
NSP1	Rtn1	
NSP2	Rtn2	NSPL1
NSPL2	Rtn3	ASYIP, HAP
NSPcL	Rtn4, Rtn5	ASY, foocen, NOGo, NSPcL, RtnX

Family of Novel SH2 Domain-Containing Proteins

The family of novel SH2 domain-containing proteins (NSP)[220] includes 3 major members (Table 9.49): NSP1,[221] NSP2,[222] and NSP3.[223] All 3 BCAR1-binding adaptors of the NSP family bind to BCAR1 and CASL[224] owing to their C-terminal domain with homology to CDC25 motif of members of the RASGEF epifamily. The association between NSP and BCAR1 modulates GEF function of NSPs [1038].

---

[220] Neither novel Ser peptidase (alias NSP), nor nuclear structure protein (also NSP).

[221] A.k.a. SH2 domain-containing protein SH2D3a.

[222] A.k.a. breast cancer anti-estrogen resistance BCAR3, SH2D3b, SHEP2, and AND34 (from thymic "AND" T-cell receptor in transgenic mice).

[223] A.k.a. SH2D3c, CRK-associated substrate and human enhancer of filamentation (CAS–HEF1)-associated signal transducer [CHAT] and SH2 domain-containing EPH receptor-binding protein SHEP1. Protein NSP3 binds both EPH receptors such as EPHb2 as well as rRas and Rap1a, but neither hRas nor RalA, thereby linking activated EPH receptors to small Ras GTPases [1039].

[224] A.k.a. neural precursor cell expressed, developmentally down-regulated NEDD9 and human enhancer of filamentation HEF1.

Family of Neuroendocrine-Specific Proteins

Alias NSP is also used for *reticulons* (Rtn), or neuroendocrine-specific proteins (NSP), that link to the endoplasmic reticulum and are involved in secretion in neuroendocrine cells.[225]

Members of the reticulon family (Rtn1–Rtn4), i.e., protein NOGo (reticulon-4) and related reticulons and their isoforms, localize both to the cell surface and within the cell, especially to the endoplasmic reticulum and nuclear envelope. Reticulons are involved in bending endoplasmic reticulum membrane, in transfer of material from the endoplasmic reticulum to the Golgi body, and in apoptosis.[226]

Proteins NOGoA, NOGoB, and NOGoC are produced by the reticulon RTN4 (or NOGO) gene.[227] On the other hand, proteins Rtn1 to Rtn3 are encoded by 3 other reticulon genes.[228]

Reticulon-3 that can homo- or heterodimerize operates in the early secretory pathway, between the endoplasmic reticulum and Golgi body. Short, non-inhibitory NOGoB and NOGoC[229] proteins are widespread [1040].

Transmembrane neurite outgrowth inhibitor and repulsive molecule NOGoA is strongly, but not exclusively, expressed in the nervous system. It is synthesized in the developing skin, differentiating skeletal muscle, and in the heart, as well as some types of immunocytes, in particular macrophages. It abounds in oligodendrocytes and myelin in adults, as well as in neurons and precursor cells during development. It stabilizes wiring in the central nervous system (brain and spinal cord), impeding remodeling and regeneration after injury [1040]. Vascular endothelial cells that produce NOGoB and its receptor NOGbR respond to NOGoB by increased adhesion, whereas vascular smooth muscle cells react by decreased adhesion.

The NOGo proteins regulate cell motility and tissue growth, particularly in the nervous system and blood vessels. They operate in secretase regulation, endoplasmic reticulum structure, and cell survival [1040]. Protein NOGoA targets

---

[225] Neuroendocrine-specific proteins, or reticulons, are indicators of neuroendocrine tumor cells of the lung.

[226] Reticulons interact with proteins involved in vesicle formation and fusion, such as SNAREs and SNAPs; they act as regulators of Rab-controlled intracellular transport. Both Rtn1c and Rtn4a are inhibitors of BCLxL apoptosis inhibitor; Rtn1c also inhibits BCL2. All reticulons interact with membrane-bound aspartyl peptidase BACE1 (β-site APP-cleaving enzyme-1), a δ-secretase that cleaves amyloid precursor protein into β-amyloid peptide; they prevent access of BACE1 to APP substrate.

[227] These 3 isoforms share only the C-terminal reticulon homology domain. Protein NOGoC N-terminus is encoded by a transcript generated from a different promoter to those of primary transcripts of N termini of NOGoA and NOGoB subtypes [1040].

[228] Nine detected transcripts are generated from 4 reticulon genes (Rtn1a–Rtn1b, Rtn2a–Rtn2c, Rtn3a1, and Rtn4a, Rtn4b1–Rtn4b2, Rtn4c). The common C-terminus encodes the reticulon homology (RH) domain, whereas N-termini are specific for each paralog. Different isoforms of reticulon-3 exists (Rtn3a1–RTN3a4); in addition, Rtn3a3 and Rtn3a4 yields 2 subtypes (Rtn3a3a–Rtn3a3b and Rtn3a4a–Rtn3a4b).

[229] Shortest NOGo isoform (Rtn4c).

9.4 Regulators of Monomeric Guanosine Triphosphatases

Table 9.50. Interactors of Novel SH2 domain-containing proteins (Source: [1043]; CAS3: CAS scaffold family member-3, or embryonal Fyn-associated substrate EFS1; IMPDH: inosine monophosphate (IMP) dehydrogenase; PABP: polyadenylate-binding protein; SncαIP: synuclein-α-interacting protein, or synphilin-1; TARS: threonyl-tRNA synthetase; TSG101: ESCRT1 complex subunit tumor susceptibility gene-101, or VPS23).

Type	Interactors
NSP1	HER2, MAPK8, 14-3-3σ, TSG101, BCAR1
NSP2	IMPDH2, TARS, PABP1, NEDD9, BCAR1
NSP3	FAK2 (PTK2β), SncαIP, NEDD9, CAS3, BCAR1
BCAR1 (CAS)	CRK, Fyn, PTPn12, MMP14, actin-β

GPI-anchored NOGo receptor NOGR1, or reticulon-4 receptor (Rtn4R), and membrane paired immunoglobulin-like receptor PIRb.[230] G-protein-coupled receptor GPR50 can serve as a NOGoA-specific receptor. In addition, NOGoB interacts with NOGoB receptor (NOGbR) and may cause proliferation of vascular endothelial and smooth muscle cells.

The NOGo proteins act via Rho GTPase, RoCK and LIMK1 kinases, slingshot phosphatase, and the actin regulator cofilin. They can also increase intracellular Ca^{++} concentration and influence integrins, protein kinase-C, target of rapamycin, signal transducer and activator of transcription STAT3, and epidermal growth factor receptor [1040].

Activity of Novel SH2 Domain-Containing Proteins

In breast cancer cells, BCAR3 can activate Rac and CDC42 GTPases as well as PKB [1041] (Table 9.50). Protein NSP3 activates Rap1 indirectly via BCAR1 NSP1 links both receptor Tyr kinases and integrin receptors to Jun N-terminal kinase JNK1, but not ERK2, and increases the activity of AP1-containing promoter [1042]. Integrin-mediated adhesion recruits 2 GEFs to BCAR1: RapGEF1 and SOS proteins.[231]

Agents BCAR1 and BCAR3 can cooperate to stimulate Src signaling and promote cell migration [1044]. Protein BCAR3 also enhances Src activation and

---

[230] Both NOGR1 and PIRb interact with 3 NOGo isoforms. Receptor NOGR1 binds also to the growth inhibitor myelin-associated glycoprotein (MAG), or siglec-4A, and oligodendrocyte myelin glycoprotein (OMGP) [1040]. Myelin-associated glycoprotein, its soluble cleaved form, and NOGoA are glia-derived inhibitors that yield a non-permissive environment for elongating nerve fibers via RhoA activation and Rac1 repression. The RhoA effector RoCK phosphorylates LIMK1 kinase, which, in turn, phosphorylates cofilin, there leading to actin depolymerization. Protein MAG binds with higher affinity to NOGo receptor-2 (NOGR2), or reticulon-4 receptor-like protein (Rtn4RL). Other proteins compete with NOGoA for binding to NOGR1 receptor.

[231] Integrin signaling also induces formation of a complex with BCAR1, adaptor CRK, and GEF DOCK1, that is required for membrane ruffling in cell migration.

Table 9.51. Guanine nucleotide-exchange factors for small GTPases of the RHO superfamily, their known targets and regulators (**Part 1**; Source: [1047]). Guanine nucleotide-exchange factors with aliases "RhoGEFs" are usually named ArhGEF proteins. Active breakpoint cluster region (BCR)-related protein (ABR) possesses both GEF and GAP domains.

Type	Targets	Regulators
ABR	RhoA, Rac, CDC42	
Alsin (Rab5GEF)	Rac	
RhoGEF1	RhoA	G12/13
RhoGEF2	RhoA/B/C, Rac1	Microtubule
RhoGEF3	RhoA/B	
RhoGEF4	Rac, CDC42	
RhoGEF5	RhoA, Rac, CDC42	
RhoGEF6	Rac, CDC42	PI3K
RhoGEF7	Rac	
RhoGEF8	RhoA	
RhoGEF9	CDC42	
RhoGEF10		
RhoGEF11	RhoA	Gq/11, G12/13
RhoGEF12	RhoA	G12/13
RhoGEF13 (AKAP13)	RhoA	G12/13
RhoGEF14	RhoA, CDC42	
RhoGEF15	RhoA	EPHa4
RhoGEF16		
RhoGEF17	RhoA	Gβγ
RhoGEF18	RhoA	Gβγ
RhoGEF19	RhoA	

BCAR1-dependent cell migration [1044]. Protein BCAR1 indeed redistributes from focal adhesions to lamellipodia. Like Src kinase, BCAR3 is another binding partner of the focal adhesion adaptor BCAR1 (linkage at BCAR1 C-terminal binding site) that operate as a RasGEF via activated PI3K kinase. Enzyme PI3K can then activates Rac [1045]. In addition, BCAR3 can activate CDC42, at least in B lymphocytes [1046]. Protein BCAR3 also targets Ral, and, to a lesser extent, Rap1 and rRas, but not hRas.

9.4.1.9 RhoGEFs

Numerous guanine nucleotide-exchange factors possess a RhoGEF DBL homology (DH; DBL: diffuse B-cell lymphoma) domain to bind CDC42, Rac, and Rho GTPases (Tables 9.51 to 9.54). The DBL epifamily encompasses about 70 members generated from mammalian genomes. Twin PH domains localize DBL proteins to the plasma membrane and regulate their GEF activity through allosteric mechanisms.

Table 9.52. Guanine nucleotide-exchange factors for small GTPases of the RHO superfamily, their known targets and regulators (**Part 2**; Source: [1047]; ASEF: adenomatous polyposis coli (APC)-stimulated GEF; CLG: common site lymphoma/leukemia GEF; CTAge: cutaneous T-cell lymphoma-associated antigen; DBL: diffuse B-cell lymphoma; DnmBP: dynamin-binding protein; ECT2: epithelial cell transforming sequence-2 proto-oncogene product; LFDH: lung-specific F-box and DH domain-containing protein; MCF2: MCF2 cell line-derived transforming sequence proto-oncogene product; MyoGEF: myosin-interacting GEF; NFκBAP: NFκB-activating protein; PlekHg: pleckstrin homology (PH) domain-containing, family-G member; SCA: spinocerebellar ataxia protein, a.k.a. Purkinje cell atrophy-associated protein-1, or puratrophin-1; SGEF: SH3 domain-containing GEF; SeStD: Sec14 and spectrin domain-containing protein; SpatA: spermatogenesis-associated protein; TECH: transcript highly enriched in cortex and hippocampus).

Type	Other aliases	Targets	Regulators
RhoGEF21	DBL, MCF2	RhoA, Rac, CDC42	G13, Gβγ
	RhoG		
RhoGEF22	MCF2L2	CDC42, Rnd2	
RhoGEF23	Trio	RhoA, RhoG, Rac	
Trio^C domain		RhoA	
Trio^N domain		RhoG, Rac	
RhoGEF24	Duo	Rac1	
	Duet	RhoG, Rac	
RhoGEF25	GEFT, P63RhoGEF	RhoA, Rac, CDC42	Gq/11
RhoGEF26	SGEF	RhoG	
RhoGEF27	Ephexin	RhoA, Rac, CDC42	EPHa4
RhoGEF29	ASEF2, SpatA13	CDC42, Rac1	
RhoGEF30	Obscurin	RhoA, Rac, CDC42	
RhoGEF31	ECT2	RhoA, Rac, CDC42	
RhoGEF32	ECT2L, LFDH	Rho	
RhoGEF33		Rho	
RhoGEF35	CTAge4, ArhGEF5L	Rho	
RhoGEF36	Tuba, DnmBP	CDC42	
RhoGEF37		Rho	
RhoGEF38		Rho	
RhoGEF40	Solo, SeStD1	Rho	
RhoGEF41	PlekHg1	Rho	
RhoGEF42	PlekHg2, CLG	CDC42	
RhoGEF43	PlekHg3	Rho	
RhoGEF44	PlekHg4, SCA4	Rho	
PlekHg5	GEF720, TECH, NFκBAP	RhoA/B/D/G/Q, Rac2	
PlekHg6	MyoGEF	Rho	
PlekHg7		Rho	

Multiple RhoGEFs (or ArhGEFs) exist (RhoGEF1–RhoGEF19, RhoGEF21–RhoGEF27, RhoGEF29–RhoGEF33, RhoGEF35–RhoGEF38, and RhoGEF40 to RhoGEF44, and other $_{Rho}$GEFs, i.e., GEFs for Rho GTPases without known alias Rho/ArhGEF; Tables 9.55 and 9.56).

Table 9.53. Guanine nucleotide-exchange factors for small GTPases of the RHO superfamily and their known targets (**Part 3**; Source: [1047]; BCR: breakpoint cluster region protein with both GEF and GAP activities; FARP: FERM, RhoGEF, and pleckstrin domain-containing protein [FARP1 a.k.a. chondrocyte-derived ezrin-like protein (CDEP) and pleckstrin homology (PH) domain-containing, family-C member PlekHc2; FARP2 as PlekHc3]; FGD: Fyve, RhoGEF, and PH domain-containing protein; FRG: FGD1-related CDC42-guanine nucleotide-exchange factor). Intersectins Itsn1 and Itsn2 are also called SH3 domain-containing protein SH3D1a and SH3D1b, or SH3P18, respectively (activators NSP1 to NSP3 corresponds to SH2D3a to SH2D3c, respectively).

Type	Other aliases	Targets
BCR		RhoA, Rac, CDC42
FARP1	CDEP, PlekHc2	RhoA
FARP2	FRG, PlekHc3	CDC42
FGD1	ZFYVE3	CDC42
FGD2	ZFYVE4	CDC42
FGD3	ZFYVE5	CDC42
FGD4	ZFYVE6, frabin	CDC42
Itsn1	SH3D1a, SH3P17	CDC42
Itsn2	SH3D1b, SH3P18	CDC42

Table 9.54. Guanine nucleotide-exchange factors for small GTPases of the RHO superfamily, their known targets and regulators (**Part 4**; Source: [1047]; AbI1: Abelson interactor-1, a.k.a. Abl-binding protein AblBP4 and EPS8-binding SH3 domain-containing protein E3B1; EPS: epidermal growth factor receptor (EGFR) pathway substrate; PREx: Phosphatidylinositol trisphosphate-dependent Rac exchanger; RasGRF: Ras guanosine nucleotide-releasing factor; SOS: Son of sevenless; TIAM: T-lymphoma invasion and metastasis-inducing protein). Protein Trio contains 2 RhoGEF domains, an N-terminal domain active on RhoG and Rac1 and a C-terminal motif active on RhoA, in addition to a kinase domain.

Type	Targets	Regulators
PREx1	Rac	$G\beta\gamma$, $PI(3,4,5)P_3$
PREx2	Rac	
RasGRF1 (P190RhoGEF)	Rac1, RhoA	
RasGRF2	Rac1	
Scambio	RhoA, RhoC	
SOS1	Rac	AbI1–EPS8
SOS2	Rac	
TIAM1	Rac	
TIAM2	Rac	
Vav1	RhoA, Rac, CDC42	
Vav2	RhoA, Rac, CDC42	
Vav3	RhoA, Rac, CDC42	

Table 9.55. Guanine nucleotide-exchange factors named RhoGEFs in the present text and their aliases (**Part 1**).

Type	Aliases
RhoGEF1	GEF1, P115RhoGEF (115 kDa Rho-specific GEF protein), LSC homolog (LBC's second cousin), LBCL2, Sub1.5
RhoGEF2	GEF2, GEFH1, LFC homolog (LBC's first cousin), LFP40, Proliferating cell nucleolar antigen P40
RhoGEF3	GEF3, STA3, XPLN (exchange factor in platelets and leukemic and neuronal tissues)
RhoGEF4	GEF4, APC-stimulated GEF (ASEF or ASEF1), STM6
RhoGEF5	GEF5, P60, TIM (or TIM1: transforming immortalized mammary proto-oncogene product)
RhoGEF6	GEF6, PIXα (PAK-interacting exchange factor), COOL2 (Cloned out of library), MRX46 (mental retardation, X-linked)
RhoGEF7	PIXβ, COOL1, P85COOL1, P50, P50BP, P85, P85SPR, SH3 domain-containing proline-rich protein, PAK3
RhoGEF8	GEF8, NET1 (or NET1a: neuroepithelial cell transforming gene product), P65
RhoGEF9	Collybistin, hPEM2 (or PEM2: human homolog of posterior end mark-2 ascidian protein)
RhoGEF10	GEF10
RhoGEF11	GEF11, $_{PDZ}$RhoGEF, RhoGEF glutamate transport modulator, RhoA-specific GEF, glutamate transporter EAAT4-associated protein GTrAP48
RhoGEF12	GEF12, LARG (leukemia-associated Rho GEF)
RhoGEF13	AKAP13, AKAP-LBC (lymphoid blast crisis), LBC-RhoGEF
RhoGEF14	DBS (DBL's big sister), Ost, MCF2L
RhoGEF15	GEF15, vsmRhoGEF (vascular smooth muscle cell-specific GEF for Rho)
RhoGEF16	GEF16, NBR (neuroblastoma protein)
RhoGEF17	GEF17, P164RhoGEF, TEM4 (tumor endothelial marker)
RhoGEF18	GEF18, P114RhoGEF
RhoGEF19	GEF19, WGEF (weakly similar to Rho GEF)

These RhoGEFs tether Rho GTPases to trigger signaling, especially those initiated by extracellular stimuli that target G-protein-coupled receptors. Some GEFs are assigned to specific biological processes, such as RhoGEF31 (or ECT2) for cytokinesis and Vav for adaptive immunity. Proteins RhoGEF1, RhoGEF11, and RhoGEF12, also act as direct effectors or modulators of heterotrimeric G proteins, hence integrating cellular responses for combined action of monomeric and heterotrimeric GTP-binding proteins. Proteins RhoGEF11 and RhoGEF12 can enhance the intrinsic GTPase activity of G12/13 subunits and concentrate signaling output to RhoA [945]. Ephexin (RhoGEF27) and RhoGEF15 interact with EPHa4 receptor Tyr kinase to activate RhoA and attenuate the activaty of Rac1 and CDC42 GTPases.

Table 9.56. Guanine nucleotide-exchange factors named RhoGEFs in the present text and their aliases (**Part 2**). Members of the RHO superfamily include CDC42 and members of the RAC family, and various RHO families. The GEF activity of members of the DBL epifamily (RhoGEF1, RhoGEF2, RhoGEF5 to RhoGEF8, RhoGEF11 to RhoGEF14, RhoGEF21 to RhoGEF24, RhoGEF27, RhoGEF31, FGD1, intersectin-1, PREx1, TIAM1 and TIAM2, and Vav1 to Vav3) can be activated upon stimulation of numerous plasmalemmal receptors (growth factor and cytokine receptors as well as adhesion molecules).

Type	Aliases
RhoGEF21	MCF2, DBL (diffuse B-cell lymphoma)
RhoGEF23	Trio
RhoGEF24	Duo, kalirin, HAPIP (Huntingtin-associated protein-interacting protein)
RhoGEF25	P63RhoGEF, GEFT (RhoA–Rac–CDC42 exchange factor)
RhoGEF27	Ephexin, NGEF (neuronal GEF)
RhoGEF30	Obscurin
RhoGEF31	ECT2 (epithelial cell transforming sequence-2; mitotic RhoGEF)
Rho/RacGEF	FGD1 (faciogenital dysplasia); FYVE, RhoGEF, and PH domain-containing GEF, ZFYVE3
Itsn1L	(CDC42-specific GEF intersectin-1_L)
PREx1	(PI(3,4,5)P$_3$-dependent Rac exchanger)
Scambio	(RhoA and RhoC GEF)
SOS1/2	SOS1: HGF, GF1 (hereditary gingival fibromatosis)
TIAM1/2	(T-cell lymphoma invasion and metastasis)
Vav1/2/3	

RGS Homology Domain-Containing RhoGEF1, RhoGEF11, and RhoGEF12

In arterial smooth muscle cells, RhoGEF1 is the RhoA guanine nucleotide-exchange factor specifically responsible for angiotensin-2-induced activation of RhoA signaling. Angiotensin-2 activates RhoGEF1 via Janus kinase JaK2 that phosphorylates RhoGEF1 (Tyr738) [1048]. Increased activity of RhoA GTPase in arterial media characterizes angiotensin-2-dependent hypertension.

Protein RhoGEF1 activates Rho via its C-terminal DBL homology (DH) domain. It also contains a pleckstrin homology (PH) domain used to regulate its localization to the plasma membrane. Moreover, RhoGEF1 has a dual role, as it operates as both a GAP for heterotrimeric G protein subunit Gα of the G12/13 subclass and a GEF for RhoA GTPase, thereby linking these 2 mediators.[232]

Proteins RhoGEF1, RhoGEF11, and RhoGEF12 actually constitute a category of RhoGEFs that have a domain similar to that of regulators of G-protein signaling

---

[232] G12/13-coupled receptors include lysophosphatidic acid, lysophosphatidylcholine (GPR132), sphingosine 1-phosphate, thromboxane-A2, peptidase-activated (PAR$_1$), muscarinic M$_3$, serotonin, and α1-adrenergic receptors.

(RGS). These proteins function as GAPs for G proteins. This functional RGS homology domain (RH)[233] enables the connection with $G\alpha_{12}$ and $G\alpha_{13}$ that then regulate the activity of these RhoGEFs, in addition to the termination of signaling mediated by $G\alpha_{12}$ and $G\alpha_{13}$ [1049]. Binding to $G\alpha_{12}$ and $G\alpha_{13}$ indeed stimulates their GEF activity for RhoA GTPase. The DH domain-containing proteins RhoGEF1, RhoGEF11, and RhoGEF12 can transmit signals from G12- and G13-coupled receptors to Rho GTPase.

Proteins RhoGEF11 and RhoGEF12 can also be phosphorylated by protein Tyr kinases such as FAK and associate with plexin-B, a semaphorin receptor. Proteins RhoGEF11 and RhoGEF12 can homo- and hetero-oligomerize, whereas RhoGEF1 can only homo-oligomerize.

Unlike RhoGEF11 and RhoGEF12, RhoGEF1 is not regulated by Tyr phosphorylation downstream from G12 signaling. Yet, it is Tyr phosphorylated by Janus kinase JaK2 via $G\alpha_q$ after stimulation of arterial smooth muscle cells by angiotensin-2 [1050]. Moreover, thrombin stimulates Ser/Thr phosphorylation of RhoGEF1 by protein kinase PKCα for Rho activation and maintenance of the endothelial barrier function.

Ubiquitous RhoGEF1 interacts with the scaffold connector enhancer of kinase suppressor of Ras CNKSR1, hyaluronan receptor CD44 variant-3 (CD44v3), and P21-activated kinase PAK1 [1050].

Two additional isoforms — RhoGEF1$_{c237}$ and RhoGEF1$_{c249}$ lack the C-terminal coiled-coil and regulatory domains.

Unlike guanine nucleotide-exchange factors of the DBL epifamily that catalyze exchange of GDP for GTP by hampering the slight GTPase activity via their tandem DBL homology (DH) and pleckstrin homology (PH) domains,[234] the $_{RGS}$RhoGEF RhoGEF11 possesses 2 binding sites for RhoA (but not CDC42 or Rac1) [1051]. Protein RhoGEF11 indeed binds tightly to both RhoA^{GDP} and RhoA^{GTP}. Its PH domain used to tether RhoA^{GTP} differs from that serving for the exchange reaction, as RhoGEF11 may compose a ternary complex with both active and inactive forms of RhoA simultaneously (RhoA^{GDP}–RhoGEF11–RhoA^{GTP}). Interaction of RhoGEF11 with activated RhoA may yield a positive or negative feedback on RhoGEF11 activity, like the RasGEF Son of sevenless or may target RhoGEF11 to the plasma membrane.

The PDZ domain of RhoGEF11 and RhoGEF12 can also assist in localizing these RhoGEFs to membranes by interacting with lysophosphatidic acid receptors, plexins, or insulin-like growth factors, in contact with geranylgeranylated, membrane-anchored RhoA [1051]. Binding of RhoGEF11 to RhoA^{GTP} at the plasma membrane may also foster further activation of RhoA (feedforward mechanism).

---

[233] The RH domain of RhoGEF1, RhoGEF11, and RhoGEF12 that requires further 60 amino acids in addition to the conserved 120 amino acid RGS box to confer GAP activity is also named RGS-like (RGL or RGSL) and RhoGEF RGS (rgRGS) domain.

[234] The DH domain catalyzes nucleotide-exchange. The PH domain differs according to the RhoGEF type with a function that facilitates interaction with Rho or localizes the GEF via connection to specific polyphosphoinositides.

Table 9.57. Examples of small GTPases and their corresponding GEFs and regulators. GEFs act both upstream and downstream from a small GTPase. Agent TIAM1, an effector for Ras, also activates Rac for temporal and spatial coordination of Ras and Rac pathways. Activated Ras activates its upstream GEF SOS (stimulatory feedback loop). Auto-inhibition of RhoGEF7 activity is relieved by EGF stimulation via Src- and FAK-dependent phosphorylation of RhoGEF7. Src-mediated phosphorylation of RhoGEF7 and CBL promotes and inhibits RhoGEF7–CBL binding, respectively. Activated CDC42 promotes the formation of the CDC42–RhoGEF7–CBL complex, which hinders EGFR signaling by sequestering CBL away from EGFR, thus preventing CBL-catalyzed EGFR degradation. CBL negatively regulates EGFR as well as other receptor Tyr kinases for platelet-derived growth factor, colony-stimulating factor-1, stem cell factor, and fibroblast growth factor. EGFR hyperactivation is associated with various human tumors. Conversely, the RhoGEF7–CBL–EGFR complex prevents CBL-catalyzed ubiquitination of EGFR, thus EGFR downregulation. Formation of complexes with CBL balance EGFR signaling and EGFR degradation. RhoGEF6 binds activated CDC42 and promotes GEF activity for Rac; but activated Rac inhibits the GEF activity of RhoGEF6XS (Source: [1052]).

Small GTPase	GEF	Upstream regulator
CDC42	RhoGEF7–CBL	Src–FAK on RhoGEF7 (activation)
	RhoGEF7–CBL	Src on CBL (inhibition)
Rac	TIAM1	Ras^{GTP}
	RhoGEF6	$CDC42^{GTP}$
Ras	SOS	Ras^{GTP}

RhoGEF2

Protein RhoGEF2 is an unusual GEF capable of binding to active GTPases of the RHO family. It indeed binds to both activated Rac and Rho, and is able to activate them. This GEF may assist in the transport and activate Rac at microtubules.

PAK-Interacting RhoGEF6 and RhoGEF7

Proteins RhoGEF6 and RhoGEF7[235] constitute the COOL or PIX group. Agent RhoGEF7 is characterized by auto-inhibition, containing an element that prevents its GEF activity [1052].

Protein RhoGEF7 regulates degradation of epidermal growth factor receptor. It acts as a scaffold protein, linking CDC42 to CBL ubiquitin ligase. Protein RhoGEF7 is both an activator of CDC42 and an effector of activated CDC42 (Tables 9.57 and 9.58).

---

[235] A.k.a. Cloned-out of library proteins COOL2 and COOL1, respectively, or P90COOL2 and P85COOL1, with P50COOL1 splice variant. They are also called P21-activated kinase-interacting exchange factors α- and β-PIX, respectively).

Table 9.58. Guanine nucleotide-exchange factors RhoGEF6 and RhoGEF7 and their effectors (Sources: [889, 1067]).

Type	Aliases	Effectors
RhoGEF6	PIXα, COOL2	CDC42, Rac1
RhoGEF7	PIXβ, COOL1	CDC42, Rac1

RhoGEF9

Protein RhoGEF9[236] enables the translocation of gephyrin to the cell membrane. Gephyrin is a component of the postsynaptic protein set of inhibitory synapses that clusters glycine receptors as well as specific subtypes of ionotropic GABA$_A$ receptors. Gephyrin displaces GABA receptors from the GABARAP–P130 complex[237] to bring these receptors to the synapse, where gephyrin binds to RhoGEF9 agent.

Sec14-like Domain-Containing RhoGEF14 and RhoGEF21

Members of the DBL epifamily — Diffuse B-cell lymphoma protein (DBL), or RhoGEF21, and DBL's big sister (DBS), or RhoGEF14 — possess a Sec14-like domain in their N-terminus that determines the subcellular localization of these GEFs and also that of CDC42 substrate.

Protein RhoGEF14[238] can catalyze guanine nucleotide exchange on RhoA and CDC42, but does not associate with these GTPases in vivo. It interacts specifically with Rac1^{GTP} [1053].

Subtype RhoGEF14-1 and RhoGEF21-1, which lack the Sec14-like domain, translocate CDC42 to the plasma membrane, where they colocalize [1053]. On the other hand, RhoGEF14-2 and RhoGEF21-2, which contain the Sec14-like domain, colocalize with CDC42 in endomembranes.

A splice variant RhoGEF14-3 contains a unique C-terminus with a SH3 domain. This regulator of Rac1 inhibits receptor endocytosis. On the other hand, other splice variants RhoGEF14-1 and RhoGEF14-2 exert no effect on receptor endocytosis [1054]. Each RhoGEF14 splice variant directs distinct subcellular localization of synaptojanin-2, depending on Rac1 activation. Agent GABA$_A$ receptor-associated protein (GABARAP) binds to RhoGEF14-3 and potently suppresses RhoGEF14-3-dependent Rac1 activation, thereby relieving the inhibition of receptor endocytosis.[239]

---

[236] A.k.a. GDP–GTP exchange factor collybistin.

[237] GABARAP: GABA$_A$ receptor-associated protein; P130: IP$_3$-binding, PLCδ1-related, catalytically inactive protein, an inhibitor of binding of γ2 subunit of GABA$_A$ receptor to GABARAP.

[238] A.k.a. osteosarcoma protein Ost.

[239] Lipid modification of GABARAP is necessary for the suppression of RhoGEF14-3 activity.

RhoGEF23

Protein RhoGEF23, or Triple functional domain (Trio), is a Rac1- and RhoG-specific guanine nucleotide-exchange factor. It interacts with actin crosslinker filamin via its pleckstrin homology (PH) domain as well as RhoA GTPase, as it has 2 separate GEF domains: an N-terminal domain ($Trio^N$ or GEFD1) for RhoG and Rac1 and a C-terminal domain ($Trio^C$ or GEFD2) for RhoA [1055].

RhoGEF24

Protein RhoGEF24[240] is a guanine nucleotide-exchange factor for Rac-like GTPases. It is mainly produced in the central nervous system. It contributes to nerve growth and axonal development.

Several isoforms result from alternative splicing. Kalirin-7 is the predominant adult splice form that is the single RhoGEF in postsynaptic density [1056]. Kalirin-7 is needed for synapse remodeling in mature cortical neurons. It is activated by EPHb and targets Rac GTPase [926].

RhoGEF25

Protein RhoGEF25 is RhoA-specific, DBL-like guanine nucleotide-exchange factor that promotes stress fiber formation in fibroblasts and cardiac myoblasts.

RhoGEF36

Protein RhoGEF36[241] a CDC42-specific GEF, concentrates at the apical region of cell junctions in epithelia via its interaction with zonula occludens ZO1 protein. Agent DnmBP links dynamin, Rho GTPase, and the actin cytoskeleton. In fact, RhoGEF36 connects dynamin to many actin regulators. It binds directly to neuronal Wiskott-Aldrich syndrome protein (nWASP) and Enabled–vasodilator-stimulated phosphoprotein (Ena–VASP) proteins, as well as indirectly to WASP family verprolin homolog (WAVe1), Wiskott-Aldrich syndrome (WASP) and WAS-like (WASL, i.e., nWASP) protein-interacting protein family members WIPF2[242] and WIPF3,[243]

---

[240] A.k.a. Duo, kalirin, and Huntingtin-associated protein-interacting protein (HAPIP).

[241] A.k.a. dynamin-binding protein (DnmBP) and Tuba according to the tradition of naming large synaptic proteins as musical instruments.

[242] A.k.a. Wiskott-Aldrich protein-interacting protein (WIP)-related protein WIRe.

[243] A.k.a. corticosteroid and regional expression protein-16 homolog (CR16).

cytoplasmic FMR1-interacting protein CyFIP2,[244] and NCK-associated protein NCKAP1, a member of the WAVE actin-remodeling complex [1057].

Intersectin-1

Intersectin-1 is a cytoplasmic membrane-associated protein that indirectly coordinates endocytic membrane trafficking with the actin assembly machinery. In addition, it may regulate the formation of clathrin-coated vesicles. It may also be involved in synaptic vesicle recycling.

Intersectin-1 links to dynamin, CDC42, SOS1, secretory carrier-associated membrane protein SCAMP1, synaptosomal-associated proteins SNAP23 and SNAP25, NCK-interacting protein with SH3 domain (NCKIPSD),[245] epidermal growth factor receptor pathway substrate EPS15, epsin-1 and -2, and stonin-2.

This multidomain adaptor involved in endocytosis possesses 2 major splice variants: (1) a ubiquitous short variant Itsn1$_S$ with 2 EH domains and 5 SH3 domains and (2) a neuron-specific long isoform Itsn1$_L$ that additionally contains DH, PH, and C2 domains. The DH domain acts as a specific GEF for CDC42 GTPase.

P190RhoGEF (RasGRF1)

Some Rho guanine nucleotide-exchange factors bind to microtubules, such as RhoGEF2, RhoGEF31, and Ras protein-specific guanine nucleotide-releasing factor RasGRF1 (Sect. 9.4.1.3).[246] Ubiquitous RasGRF1 connects to and activates RhoA, but neither Rac1 nor CDC42 GTPase [1058].

Small GTPase RhoA can be activated by G12/13-coupled receptors. Subunits Gα_{12} and Gα_{13} can directly bind to both RhoGEF1 and RhoGEF11, but not RasGRF1 agent. On the other hand, RasGRF1 can directly bind to Jun N-terminal kinase-interacting protein JIP1 (or MAPK8IP1) [1058]. Jun N-Terminal kinase also localizes to microtubules.

PlekHg RhoGEFs

Guanine nucleotide-exchange factors of the family of pleckstrin homology (PH) domain-containing proteins with RhoGEF (or DBL-homology [DH]) domain encompass PlekHg1 to PlekHg4, or RhoGEF41 to RhoGEF44, and PlekHg5 to PlekHg7.

---

[244] A.k.a. P53-inducible protein PIR121.

[245] A.k.a. 90-kDa SH3 adaptor protein interacting with NCK (SPIN90) and Wiskott-Aldrich syndrome protein (WASP)-interacting SH3-domain protein (WISH).

[246] A.k.a. 190-kDa RhoA-binding protein P190RhoGEF and Ras-specific nucleotide-exchange factor CDC25.

PlekHg5

Member PlekHg5[247] is widespread. It specifically targets RhoA GTPase, but not RhoC, Rac1, Rac3, and CDC42 [1059]. In humans, PlekHg5 is expressed with similar expression levels in the brain, heart, skeletal muscle, lung, kidney, liver, etc.

Agent PlekHg5 is required for endothelial cell migration and angiogenesis. It is also involved in the maintenance of tight junctions. It can tether to PDZ domain-containing adaptors synectin and Multiple PDZ domain protein MPDZ (or MuPP1). Adaptor MPDZ as well as Protein associated with Lin7 (PALS1)-associated tight junction protein (PATJ)[248] bind *angiomotin* that couples PlekHg5 to members of the Par6–Par3 apicobasal polarity complex.[249] The resulting angiomotin–PlekHg5–Par6–Par3 complex includes CDC42-specific RhoGAP17 and scaffold Membrane protein palmitoylated MPP5 [1059].[250] Agent RhoGAP17, an important mediator of the polarity complex, binds scaffold angiomotin and is then attracted to a proteic complex at tight junctions that contains a PDZ domain (MPP5, PATJ, and polarity protein partitioning defective-3 homolog Par3) [1062]. In addition, PlekHg5 RhoAGEF can activate nuclear factor-κB [1059].

Neuronal RhoA guanine nucleotide-exchange factor PlekHg5 is highly produced in brain neurons, but also in spinal cord and peripheral nerves. It activates RhoA for assembly of actin stress fibers [1063]. It inhibits nerve growth factor-induced neurite outgrowth. Its concentration increases during postnatal development and remains high during adulthood to regulate RhoA signaling in developing and mature neurons.

PREx1

Phosphatidylinositol PI(3,4,5)P$_3$-dependent Rac exchanger-1 (Sect. 9.4.1.3) integrates signals from GPCRs and protein Tyr kinases. This PIP$_3$–G$\beta\gamma$-dependent RacGEF can mediate HER-induced activation of Rac1 for cell migration using the CXCR4–Gi–PI3Kγ pathway [1064].

---

[247] A.k.a. GEF720, synectin-binding RhoA exchange factor [Syx], and transcript highly enriched in cortex and hippocampus (TECH).

[248] A.k.a. InaD-like protein (InadL).

[249] Angiomotin that is expressed in the endothelial cells from capillaries to larger vessels in the placenta regulates junctions between endothelial cells and cell migration. Angiomotin binds and internalizes angiostatin, a circulating inhibitor of angiogenesis [1060]. It localizes to the leading edge of migrating endothelial cells and stimulates cell motility. Conversely, angiostatin binds angiomotin on the cell surface to preclude the migration of angiomotin-expressing cells. Angiomotin not only controls cell motility, but also intervenes in the assembly of junctions between endothelial cells. Angiomotin colocalizes with scaffold ZO1 in tight junctions [1061], the presence of ZO1 depending on PlekHg5 protein. Paracellular permeability is reduced by p80 and p130 angiomotin isoforms.

[250] A.k.a. Protein associated with Lin-7 PALS1.

Scambio

Scambio is highly expressed in the heart and skeletal muscle, and, to a lesser extent, in the lung [1065]. It binds to activated Rac and CDC42 GTPases. As it connects to $CDC42^{GTP}$ and Rac^{GTP}, Scambio does not activate these GTPases, but operates as a RhoGEF for RhoA and RhoC. Scambio operates not only as a RhoGEF, but also as a Rho effector.

In vascular smooth muscle cells, overactivation of the RhoA–RoCK pathway is involved in excessive migration and proliferation. In rat aorta and mesenteric and pulmonary arteries, Transcript of Scambio belongs to the set of the 8 most expressed RHOAGEF mRNAs, with those of RhoGEF1, RhoGEF2, RhoGEF12, RhoGEF13, RhoGEF17, RhoGEF23, and RhoGEF25 [1066].[251] Among 28 detected RhoAGEFs[252] in mesenteric arteries of normo- and hypertensive (angiotensin-2-treated) rats, 16 RhoAGEFs (e.g., RhoGEF1, RhoGEF2, RhoGEF3, RhoGEF11, RhoGEF12, RhoGEF17, RhoGEF18, RhoGEF21, and ABR) are downregulated and 5 RhoAGEFs (e.g., Vav2) are upregulated by the vasoconstrictor angiotensin-2 that acts via AT_1 receptor to activate some RhoAGEFs (e.g., RhoGEF1, RhoGEF11, RhoGEF25), and subsequently RhoA and RoCK enzyme [1066]. Stimulation of the RhoA–RoCK axis creates a negative feedback on the expression of 9 RhoAGEFs involved in the activation of RhoA by vasoconstrictors (in particular, RhoGEF1, RhoGEF2, RhoGEF11, RhoGEF12, RhoGEF17, and RhoGEF18).

Son of Sevenless

Son of sevenless (SOS) is a Ras guanine nucleotide-exchange factor.[253] It is stimulated by activated receptor Tyr kinases such as growth factor receptors. This RasGEF forms 2 signaling complexes according to adaptor types, either growth factor receptor-bound protein GRB2, or Abl-interactor AbI1 and epidermal growth factor receptor pathway substrate EPS8 [1067]: (1) Ras-specific SOS–GRB2 and (2) Rac-specific SOS–AbI1–EPS8 complexes (Table 9.59). Activation of Ras and Rac GTPases is transient and sustained, respectively.

---

[251] The human genome generates more than 70 RhoGEFs with tissue-restricted expression (but only 22 Rho GTPases). Among RhoGEFs, 25 to 30 are RhoAGEFs. Except RhoGEF30 (obscurin), the expression of which is restricted to the heart and striated muscle, RhoAGEFs do not exhibit a tissue-specific pattern. A group of 13 RhoAGEF mRNAs is characterized by a moderate and heterogeneous expression according to the artery type (e.g., RasGRF1 [P190RhoGEF] is produced at high levels in the aorta and at low levels in mesenteric and pulmonary arteries).

[252] I.e., Scambio, RhoGEF1, RhoGEF2, RhoGEF3, RhoGEF5, RhoGEF11, RhoGEF12, RhoGEF13, RhoGEF14, RhoGEF15, RhoGEF17, RhoGEF18, RhoGEF19, RhoGEF21, RhoGEF23, RhoGEF24, RhoGEF25, RhoGEF27, RhoGEF31, ABR, BCR, Vav1–Vav3, RasGRF1 (p190RhoGEF), FARP1, and PlekHg5.

[253] The name "Son of sevenless" derives from the fact that SOS operates downstream of the sevenless gene in Drosophila melanogaster.

Table 9.59. Monomeric target GTPases by SOS GEF and effect hierarchy (Sources: [889, 1067]).

Type	Alias	Effectors
SOS1	RasGEF1	hRas, kRas, nRas, rRas2/3, Rac1
		hRas>nRas>kRas
SOS2	RasGEF2	hRas, Rac1

Table 9.60. VavGEFs and their effectors (Source: [889]).

Type	Effectors
Vav1	CDC42, Rac1, RhoA, RhoG
Vav2	CDC42, Rac1, RhoA, RhoG
Vav3	CDC42, Rac1, RhoA, RhoG

Mammalian paralogs — SOS1 and SOS2 — function downstream from many growth factor and adhesion receptors. Subtype SOS1 participates in short- and long-term activation of the Ras–MAPK pathway, whereas SOS2-dependent signaling is mainly short-term.

Activator SOS1 of Ras interacts with receptor Tyr kinases, such as EGFR and HGFR, with CRK, GRB2, GRB2-related adaptors GRAP1 and GRAP2, NCK1 and NCK2, SHC1, FGFR substrate FGFRS2, Abelson interactor AbI1, epidermal growth factor receptor pathway substrate EPS8 and related protein EPS8L1, SH3-domain kinase-binding protein SH3KBP1; protein Tyr kinases Abl1, Fyn, and FAK2; protein Tyr phosphatases PTPn6 and PTPn11; phospholipase PLCγ1; RIT2 GTPase; BCR RhoGEF and GAP; transport component intersectin-1, mucin-1, and annexin-A2 [251]. Isoform SOS2 interacts with receptor Tyr kinases such as EGFR, adaptors GRB2, SHC1, and GRAP2, and phospholipase PLCγ1 [251].

Vav GEFs

The 3 members of the VAV family[254] (Vav1–Vav3) are guanine nucleotide-exchange factors for Rho GTPases (Table 9.60). The Vav proteins can be activated by numerous plasmalemmal receptors, such as G-protein-coupled, growth factor, and immune receptors, as well as integrins. Guanine nucleotide-exchange activity of all Vav proteins is triggered by Tyr phosphorylation by SRC family and SYK kinases.

Protein Vav1 targets Rac1, Rac2, and RhoG, but is significantly less active on RhoA [1068]. On the other hand, widespread Vav2 and Vav3 act on RhoA and RhoG, but are less active on Rac1 GTPase. The Vav proteins operate not only in the

---

[254] As Vav protein was the sixth discovered oncogene, it was named "Vav", the sixth letter of the Hebrew alphabet.

immune system, but also in the cardiovascular apparatus.[255] Stimulation of vascular smooth muscle cells by angiotensin-2 provokes Tyr phosphorylation of Vav proteins.

The Rho guanine nucleotide-exchange factors of the VAV family regulate angiogenesis [1069]. In endothelial cells, Vav2 and Vav3 bind to EPHa2 in response to ephrin-A1 stimulation and trigger Rac1 activity. Small Rac1 GTPase is essential in cell migration, as it is involved in the formation of lamellipodia at the cell leading edge and forward movement as well as cell retraction at the trailing edge via its effector P21-activated kinase.

The Vav proteins also attenuate the blood pressure via a catecholamine-dependent stimulation of the renin–angiotensin system [1070].

In rats, Vav3 favors the proliferation and migration of vascular smooth muscle cells [1068]. Protein Vav3 causes an enrichment of Rac1 in the membrane and PAK activation. Two distinct pathways exist downstream from Rac1: (1) Rac1–PAK–CcnD and (2) Rac1–actin–SKP2 (S-phase kinase-associated protein-2) that targets CcnA-CDK2 complex. Both cascades are required for efficient Rac1-mediated cell cycle progression.

Vav1

Protein Vav1, a hematopoietic cell-specific signaling mediator, acts as a phosphorylation-dependent GEF for Rho and Rac GTPases [1071]. It supports the development and effector functions of lymphocytes and, to a lesser extent, specific signaling responses of other hematopoietic cell types.

This DBL epifamily member needs to be phosphorylated by protein Tyr kinases to activate Rho and Rac GTPases. Receptors with intrinsic or associated Tyr kinase activity recruit Vav1 to the plasma membrane and phosphorylate (activate) it (Table 9.61). In addition to integrins, Vav1 actually works downstream from antigen receptors of T and B lymphocytes (TCRs and BCRs) and immunoreceptor tyrosine-based motif (ITAM)-containing receptors [1071]. Adaptors and coreceptors favor Vav1 translocation to the plasma membrane, such as Linker of activated T cells, B-cell linker, IL2-inducible T-cell kinase (ITK), costimulatory receptor T-cell-specific surface glycoprotein CD28, and B-cell coreceptor CD19 [1071]. Furthermore, some molecules can activate Vav1 in a phosphorylation-independent manner, such as neutrophil cytosol factor NCF2.[256]

In addition, Vav1 can operate as an adaptor. Independently of its catalytic activity, it can stimulate nuclear factor of activated T cells (NFAT), a transcription factor involved in the regulation of interleukin-2 and other cytokines in lymphoid cells [1071]. It also contributes to signal amplification in hematopoietic cells. In particular, it enables the optimal activation of phosphatidylinositol 3-kinase and

---

[255] Mice deficient in Vav3 exhibit tachycardia and hypertension.
[256] A.k.a. P67PhOx (phagocyte oxidase), a 67-kDa neutrophil oxidase used as a subunit of NADPH oxidase.

Table 9.61. Interactors of Vav1 (Source: [1071]). Interleukin-2-inducible T-cell kinase (ITK) promotes the translocation of Vav1 to the plasma membrane independently of its kinase activity. Proline-rich focal adhesion kinase FAK2 (or PYK2) may be involved in the tethering of Vav1 to specific receptors such as $\alpha_V\beta_3$ integrin in neutrophils.

Plasmalemmal kinases	EGFR, IR, IGF1R, PDGFR, SCFR, STK1 (FLT3)
Intracellular kinases	FAK2, ITK, JAK, Src, SYK, ZAP70, TEC
	PKC
	PI3K
Phosphatases	PTPn6/11, PPM1
GTPases	Rac > RhoA ≥ CDC42
Other enzyme types	PLCγ
Transmembrane coreceptors	CD19, CD28
Adaptors	BLnk, DOCK2, GRB2, LAT, LCP2, SHB, STAP2
Miscellaneous	calmodulin, dynamin-2, EZH2, RhoGDI, talin, tubulin, zyxin

phospholipase-Cγ. These enzymes cause signaling diversification, as they stimulate Ras^{GDP}-releasing protein (RasGRP), Ras, ERKs, and protein kinases PKB, PKC, and PKD [1071].

Many agents support the activity of Vav1, such as lymphocyte cytosolic protein LCP2,[257] calmodulin, histone lysine ^{N}methyltransferase enhancer of zeste homolog EZH2,[258] dynamin-2, PI3K, PLCγ, and PKC family members [1071]. On the other hand, many proteins counteract Vav1, such as Casitas B-lineage lymphoma proto-oncogene product (CBL; ubiquitination [inhibition]), as well as PTPn11 and PPM1 phosphatases [1071].

9.4.2 GTPase-Activating Proteins

Guanosine triphosphate is hydrolyzed to guanosine diphosphate by monomeric guanosine triphosphatase in combination with GTPase-activating proteins. GTPase-activating proteins regulate the inactivation of small GTPases by accelerating their slow intrinsic GTPase activity. Many (at least 80) GTPase-activating proteins comprise: (1) selective GAPs that target a single small GTPase (e.g., CDC42GAP) and (2) non-specific GAPs (Table 9.62).

---

[257] A.k.a. 76-kDa SH2 domain-containing leukocyte protein (SLP76).

[258] A.k.a. Enx1. It pertains to the Polycomb group of transcriptional repressors.

9.4 Regulators of Monomeric Guanosine Triphosphatases

Table 9.62. GTPase-activating proteins (ACAP: ArfGAP with coiled-coil, ankyrin repeat, PH domains; ADAP: ArfGAP with dual PH domains; AGAP: ArfGAP with GTPAse, ankyrin repeat, and PH domains; AGFG: ArfGAP with FG repeats; APAP: ArfGAP with PIX- and paxillin-binding domains; ARAP: ArfGAP with RhoGAP, ankyrin repeat, PH domains; ASAP: ArfGAP with SH3, ankyrin repeat, PH domains; DAB2IP: Disabled homolog-2 (Dab2)-interacting protein; IQGAP: IQ motif-containing GTPase-activating protein; NF: neurofibromin; RasAL: Ras GTPase-activating-like protein; SH3BP: SH3-domain binding protein, or RhoGAP43; SMAP: stromal membrane-associated GTPase-activating protein; TBC1D: Tre2/USP6, BUB2, CDC16 (TBC1) domain-containing family member).

Category	Examples of members
ArfGAPs	ArfGAP1–ArfGAP13, ACAP1–ACAP3, ADAP1–ADAP2, AGAP1–AGAP11, AGFG1–AGFG2, APAP1–APAP2, ARAP1–ARAP3, ASAP1–ASAP3, SMAP1–SMAP2
CDC42GAPs	RhoGAP10
RacGAPs	RabGAP1, Rab3GAP1/2, RhoGAP43, TBC1D1–TBC1D30
RapGAPs	RasA1, SpAR1–SpAR2
RasGAPs	Dab2IP, IQGAP1–IQGAP3, NF1, RasA1–RasA4, RasAL1–RasAL2
RhoGAPs	RhoGAP1–RhoGAP30, RhoGAP32/35/43

Interplays between small GTPases exist not only via GEFs, but also GAP proteins. Effector Ral-binding protein RalBP1 operates as a GAP for CDC42 [1072]. Proteins RasA1 and phosphorylated RhoGAP35 form a complex that connects the Ras and Rho signaling pathways [1073]. Epithelial cell motility that occurs in embryogenesis, wound repair, and metastasis requires coordination between Rho and ARF GTPases. Small GTPase Rac1 communicates with ARF6 via a complex that consists of Rac effector PAK1, RacGEF PIX, and ArfGAP GIT1 [841].

9.4.2.1 ARF GTPase-Activating Proteins (ArfGAPs)

Adpribosylation factors do not have detectable intrinsic GTPase activity. Therefore, hydrolysis of GTP on ARF requires GTPase-activating proteins. The human genome contains at least 31 genes that encode ArfGAPs. In addition to ArfGAP1 to ArfGAP8, ArfGAPs include the following types: (1) APAP type (ArfGAPs with PIX- and paxillin-binding domains; APAP1–APAP2); (2) AMAP type (a multidomain ArfGAP protein; AMAP1–AMPA3); (3) ACAP type (ArfGAP with coiled-coil (BAR), ankyrin repeat, PH domains; ACAP1–ACAP3); (4) AGAP type (ArfGAP with GTPAse, ankyrin repeat, and PH domains; AGAP1–AGAP3); (5) ARAP type (ArfGAP with RhoGAP, ankyrin repeat, and PH domains; ARAP1–ARAP3); (6) AFAP type (ArfGAP with phosphoinositide-binding and PH domains; AFAP1–AFAP2); and (7) ASAP type (ArfGAP with SH3, ankyrin repeat, and PH domains; ASAP1–ASAP3) that partially corresponds to the AMAP type.

Table 9.63. Adpribosylation factor GTPases and their effectors (**Part 1**; Sources: [838, 889]; ACAP: ArfGAP with coiled-coil, ankyrin repeat, and PH domains; ADAP: ArfGAP with dual PH domains; AGAP: ArfGAP with GTPase domain, ankyrin repeat, and PH domain; CtGlF: Centaurin-γ-like family member; PIKE: phosphatidylinositol-3-kinase (PI3K) enhancer).

Type	Aliases	Effectors
ArfGAP1		ARF1–ARF3, ARF5
ArfGAP2		ARF1–ARF5
ArfGAP3		ARF1–ARF5
ACAP1	Centaurin-β1	ARF6
ACAP2	Centaurin-β2	ARF6
ACAP3	Centaurin-β5	ARF6
ADAP1	Centaurin-α1	
ADAP2	Centaurin-α2	
AGAP1	Centaurin-γ2	ARF1
AGAP2	Centaurin-γ1	ARF1
	PIKE	$PI3K_{r1}$, PLCγ1, Homer-1/2
AGAP3	Centaurin-γ3	
AGAP4	CtGlF1	
AGAP5	CtGlF2	
AGAP6	CtGlF3	
AGAP7	CtGlF4	
AGAP8	CtGlF5	
AGAP9	CtGlF6	
AGAP10	CtGlF7	
AGAP11		

The superfamily of ArfGAPs can be subdivided into 10 families according to the structure of the ArfGAP domain (Tables 9.63 and 9.64) [1074]: family 1: ArfGAP1; 2: ArfGAP2/3; 3: ACAPs; 4: ADAPs (ArfGAPs with dual PH domains); 5: AGAPs; 6: AGFGs (ArfGAPs with FG repeats); 7: AMAPs; 8: ASAP; 9: GITs (G-protein-coupled receptor kinase-interacting ArfGAPs); and 10: SmAPs (small ArfGAPs [SmAP1–SmAP2]).

Members of the ARFGAP superfamily have various degrees of specificity for members of the ARF superfamily. Some ArfGAPs can distinguish ARFs from ARLs, but other ArfGAPs, as well as at least some ArlGAPs, exhibit GAP activity toward both ARFs and ARL proteins.

These ArfGAPs localize to cell–matrix adhesion sites, as they associate with focal adhesion protein paxillin. They can also bind to RhoGEF6 mediator.

ArfGAP1

Protein ArfGAP1 shuttles between cytosol and the Golgi body, where it contributes to the regulation of CoP1-associated membrane trafficking. It interacts with

9.4 Regulators of Monomeric Guanosine Triphosphatases

Table 9.64. Adpribosylation factor GTPases and their effectors (**Part 2**; Source: [838, 889]; AGFG: ArfGAP with FG repeats; ARAP: ArfGAP with Rho GAP domain, ankyrin repeats and PH domain; ASAP: ArfGAP with SH3 domain, ankyrin repeat, and PH domain; CAT: COOL-associated tyrosine phosphorylated protein; DDEF: development and differentiation-enhancing factor; ELMOD: ELMO domain-containing protein; GIT: GRK-interacting ArfGAP; PKL: paxillin-kinase linker; RP2: X-linked recessive retinitis pigmentosa protein-2 [or XRP2, lodges in plasma membrane, periciliary ridge, and cilial basal body] SmAP: Small ArfGAP protein).

Type	Aliases	Effectors
AGFG1	HRB1	
AGFG2	HRB2	
ARAP1	Centaurin-δ2	
ARAP2	Centaurin-δ1	
ARAP3	Centaurin-δ3	
ASAP1	AMAP1, DDEF1 Centaurin-β4	ARF1, ARF5, ARF6
ASAP2	AMAP2, DDEF2 Centaurin-β3	ARF1, ARF5, ARF6
ASAP3	ACAP4, DDEFL1 Centaurin-β6	ARF1, ARF5, ARF6
GIT1	APAP1, CAT1	ARF1, ARF3, ARF5, ARF6
GIT2	APAP2, CAT2, PKL	ARF1, ARF3, ARF5, ARF6
SmAP1		ARF1, ARF6
SmAP2		ARF1, ARF6
RP2	ArlGAP XRP2	ARL3
ELMOD2		ARL2, ARL3

components of clathrin-coated vesicles, such as clathrin and clathrin-coat adaptor complexes AP1 and AP2 [1074] (Table 9.65).

Agent ArfGAP1 is recruited to Golgi membranes via interaction with KDEL motif containing receptors and transmembrane Emp24 protein transport domain-containing proteins TMED2 and TMED10 transmembrane Golgi-cargo receptors. Hydrolysis of GTP by ArfGAP1 allows cargo packaging as well as vesicle uncoating [840]. Protein ArfGAP1 favors cargo packaging, as CoP1-coated vesicles are depleted of cargo.

ArfGAP2 and ArfGAP3

Proteins ArfGAP2 and ArfGAP3 are closely related, but with little similarity to ArfGAP1, except the catalytic domain. They strongly interact with the CoP1 coat that is responsible for retrograde transport from trans-Golgi network to cis-Golgi

Table 9.65. Proteins ArfGAP1, -2, and -3 and interacting proteins (Sources: [838, 1075]; CoP: coat protein; KDELR1: KDEL (Lys–Asp–Glu–Leu) endoplasmic reticulum retention receptor).

Type	Location	Effectors
ArfGAP1	Golgi body	Clathrin, adaptin-γ1, KDELR1, RGS2
ArfGAP2	Golgi body	CoP1
ArfGAP3	Golgi body	CoP1

Table 9.66. ACAP ArfGAPs and interacting proteins (Sources: [838, 1075]; NOD: nucleotide-binding oligomerization domain-containing protein; TfnR: transferrin receptor; VAMP: vesicle-associated membrane protein [VAMP3: cellubrevin]).

Type	Location	Binding partners
ACAP1	PM, endosome	NOD1, NOD2, integrin-β_1, VAMP3, TfnR
ACAP2	PM, endosome	
ACAP3	PM, endosome	Integrin-β_4-binding protein

network and endoplasmic reticulum.[259] Activity of ArfGAP2 and ArfGAP3 is stimulated by CoP1 coatomer [1074] (Table 9.65).

ACAPs

Member of the ACAP family (ACAP1–ACAP3) regulate ARF6-dependent actin remodeling that occurs during endocytosis and receptor Tyr kinase-dependent cell movement (Table 9.66). Protein ACAP1 intervenes in ARF6-regulated clathrin coat [1074].

Protein Arf6GAP ACAP1 associates with ARF6+ tubular structures and protrusions. It is involved in endosomal recycling of transferrin receptors, and β_1 integrins, although β_1-integrin recycling depends on its phosphorylation by protein kinase-B [840].

ADAPs

In humans, 2 ADAP proteins (ADAP1–ADAP2) have an ArfGAP domain followed by 2 PH domains ($_N$PH and $_C$PH), which may bind phosphatidylinositols. However, unlike ADAP1, ADAP2 lacks the N-terminal nuclear localization signal.

---

[259]Coatomer CoP2 operates in anterograde transport from endoplasmic reticulum to the cis-Golgi network.

ADAP1

Protein ADAP1[260] links with a high affinity to PI(3,4,5)P$_3$ and I(1,3,4,5)P$_4$. It acts as Arf6GAP [1074]. It resides in dendrites, spines, and synapses of developing and mature neurons, where it regulates the actin cytoskeleton, secretory vesicle transport and membrane trafficking, as well as neuronal differentiation.

ADAP2

Protein ADAP2[261] specifically binds phosphatidylinositol (3,4)-bisphosphate and (3,4,5)-trisphosphate, which are plasmalemmal second messengers produced by activated phosphatidylinositol 3-kinase [1076]. It also links to inositol (1,3,4,5)-tetrakisphosphate and phosphatidylinositol (4,5)-bisphosphate. Regulator ADAP2 is recruited to the plasma membrane by connecting to PI3K lipid products, where it can then inhibits ARF6 GTPase.

Inhibitor ADAP2 also tethers to NFκB essential modulator (NEMO, or IKKγ regulatory subunit), and, like ADAP1, Ran-binding protein in microtubule-organizing center (RanBPM), as well as nucleolin,[262] and nardilysin, a zinc metalloendopeptidase of the insulinase family [1076].[263]

AGAPs

In humans, AGAPs are encoded by 11 know genes. They can directly bind and activate PKB and other Ras effectors [1074]. Members AGAP1 and AGAP2 act in endocysis. Proteins AGAP1 and AGAP2, ARF1GAPs, interact with and regulate the function of endosomal AP3 and AP1 complexes, respectively.

Protein AGAP1 operates with heterotetrameric clathrin-coat adaptor AP3 complex[264] that is associated with the Golgi body and peripheral cell structures (Table 9.67).

Protein AGAP2 has multiple splice variants. It works with clathrin-coat adaptor AP1 complex.

AGFGs

The AGFG family is constituted of ArfGAPs that possess 10 Phe–Gly (FG) repeats [1074].

---

[260] A.k.a. centaurin-α1 and PIP$_3$BP.

[261] A.k.a. centaurin-α2.

[262] Nucleolin is involved in the synthesis and maturation of ribosomes.

[263] A.k.a. ^{N}arginine dibasic convertase (NRD).

[264] This complex is made of large adaptin-δ3 and -β3, medium adaptin-μ3, and small adaptin-σ3.

Table 9.67. APAP ArfGAPs and interacting proteins (Source: [1075]; AP: clathrin-coat adaptor protein complex; EPB4l1; erythrocyte membrane protein band 4.1-like; sGC: NO-sensitive soluble guanylate cyclase).

Type	Binding partners
AGAP1	AP3, sGC
AGAP2	AP1, PLCγ, PI3K, PKB, Homer1, EPB41L1
AGAP3	

APAPs

Members of the APAP family interact with G-protein-coupled receptor kinases. Therefore, they are also termed *G-protein-coupled receptor kinase interactors* (GIT). They constitute a group of ArfGAPs that bind to paxillin.

Protein APAP2[265] is nearly ubiquitous, whereas APAP1[266] is absent from many cell types (e.g., myocytes, hepatocytes, pneumocytes, and adipocytes), but it is prominent in endothelial cells [999].

Unlike other ArfGAPs, APAPs are tightly associated with RhoGEF6 and RhoGEF7 (PIX or COOL group) for CDC42 and Rac1 GTPases to form oligomers [999]. These APAP–RhoGEF6/7 complexes act as scaffolds for various kinases (GRKs, PAKs, FAKs, and MAPKs) and phospholipase-Cγ. They are recruited to distinct subcellular locations by specific partners: at focal adhesions via paxillin or α_4-integrin; at synapses via presynaptic cytomatrix protein piccolo (PClo) or liprin-α; and at the plasma membrane via Scribbled homolog (Scrib) [999].

Protein APAP1 connects to RhoGEF6 and RhoGEF7, APAP2, and RGS2, as well as FAK1, FAK2, and PAK2 kinases, PTPRb and PTPRf phosphatases, and PTPRf-interacting proteins liprin-α1 to -α4, among other molecules [251] (Table 9.68).

Protein APAP2 also links to PAK2 [251]. It is implicated in the regulation of cell spreading and motility, as it transiently recruits P21-activated kinase PAK1 to focal adhesions. Integrin engagement and Rac activation lead to phosphorylation by Src kinase of focal adhesion kinase that can, in turn, phosphorylate APAP2 (Tyr286, Tyr392, and Tyr592) [1077]. Phosphorylated APAP2 localizes to focal adhesions and can bind paxillin. It also associates with NCK adaptor and the NCK–PAK–RhoGEF6/7–ArfGAP2 complex is built at focal adhesions. Protein APAP2 is a substrate for PTPn12 that binds focal adhesion adaptor paxillin to cause cell migration [1078]. Interaction between APAP2 and paxillin prevents cell

---

[265] A.k.a. GIT2, paxillin kinase linker (PKL) and COOL-associated and Tyr-phosphorylated protein CAT2. Alias PKL also stands for liver (L)- and red blood cell (R)-type pyruvate kinase (other aliases PKLR, PKR, and RPK) that also interacts with paxillin, Rac-activated kinase PAK1, and RhoGEF6, in addition to kinesin KIF23 [251].

[266] A.k.a. GIT1 and COOL-associated and Tyr-phosphorylated protein CAT1.

9.4 Regulators of Monomeric Guanosine Triphosphatases

Table 9.68. APAP ArfGAPs and interacting proteins (Sources: [838, 1075]; Htt: huntingtin; TGFβ1I: transforming growth factor-β1-induced transcript).

Type	Location	Binding partners
APAP1	Plasma membrane	RhoGEF6, RhoGEF7, PLCγ, FAK, MAP2K1, GRK, paxillin, Htt, TGFβ1I1, 14-3-3ζ
APAP2	Plasma membrane	RhoGEF6, RhoGEF7, GRK, paxillin, leupaxin, TGFβ1I1

Table 9.69. ARAP ArfGAPs and interacting proteins (Sources: [251, 1075]; InPPl: inositol polyphosphate phosphatase-like protein; SH3KBP: SH3-domain kinase-binding protein).

Type	Binding partners
ARAP1	FGFR3, ARP2/3
ARAP2	RhoA
ARAP3	Rap1, InPPL1, SH3KBP1, PIP$_3$

spreading. Adaptors CRK and CRKL that bind to Tyr-phosphorylated proteins associate with APAP2, paxillin, and RhoGEF7 for cell spreading and lamellipodium formation [1079].

ARAPs

Centaurins-δ (Centδ1–Centδ3), or ARAP1 to ARAP3, are implicated in the regulation of actin cytoskeleton. The domain structures of the 3 ARAPs are similar, but these ArfGAPs have distinct subcellular locations, functions, and GTPase-binding specificities. They are involved in signaling from EGFR, focal adhesion dynamics, and lamellipodium formation (Table 9.69).

Members of the ARAP family can act on small GTPases of both ARF and RHO superfamilies. Ubiquitous ARAP3 is synthesized at high levels in leukocytes. It is a phosphatidylinositol (3,4,5)-trisphosphate-binding protein for its translocation to the plasma membrane, where it regulates the actin cytoskeleton, formation of lamellipodia, and cell spreading [1080].[267] It also tethers to Rap1 GTPase. Therefore, ARAP3 is a phosphoinositide 3-kinase- and Rap1-regulated GAP for RhoA and ARF6.

Protein ARAP3 is Tyr phosphorylated (inactivated) by members of the SRC family, Lyn and Src kinases. It also interacts with SH3 domain-containing kinase-binding protein SH3KBP1 (or CIN85) and its splice variant Cluster of

---

[267] Stimulation by growth factors such as EGF that provokes PI(3,4,5)P$_3$ formation causes ARAP3 translocation to the plasma membrane [1080]. Stimulation by PDGF that also induces PI(3,4,5)P$_3$ synthesis leads to ARAP3 translocation to lamellipodia.

Table 9.70. ASAP ArfGAPs and interacting proteins (Sources: [251, 1075]; CD2AP: CD2-associated protein; REps: RALBP1-associated Eps domain-containing protein; SH3KBP: SH3-domain kinase-binding protein; TbRg4: transforming growth factor-β regulator).

Type	Location	Binding partners
ASAP1	Plasma membrane, focal adhesion	FASrc, FAK1, FAK2, CRK, CRKL, Cortactin, REps2, CD2AP, SH3KBP1
ASAP2	Plasma membrane, focal adhesion	Paxillin, amphiphysin, FAK2, Src REps2, TbRg4
ASAP3		

differentiation CD2-associated protein (CD2AP)[268] and with SH2-containing inositol phosphatase-2 (SHIP2). Agents ARAP3, SHIP2, and SH3KBP1 or CD2AP can complex [1080].

ASAPs

Agents ASAP1, or AMAP1, and ASAP2, or AMAP2,[269] as well as ASAP3,[270] form the ASAP family of ArfGAPs. The ASAP proteins are associated with specialized sites of the plasma membrane such as focal adhesions. They regulate endocytosis and associated actin remodeling [1074].

These multidomain proteins interact with many signaling molecules. Protein ASAP1 binds to Src and FAK1 kinases, CRK and CRKL adaptors, CD2-associated protein (CD2AP) scaffold that localizes to membrane rafts and ruffles, where it interacts with filamentous actin, and SH3 domain kinase-binding protein SH3KBP1 interactor (Table 9.70). The latter binds to ASAP1 and recruits ubiquitin ligase CBL to trigger ASAP1 monoubiquitination. Protein ASAP2 connects to focal adhesion kinase FAK2. Protein ASAP3 associates with focal adhesions and regulates stress fibers.

---

[268] A.k.a. CAS ligand with multiple SH3 domains (CMS). Adaptor CD2AP binds to BCAR1, SRC family kinases, phosphatidylinositol 3-kinase, and GRB2 adaptor. Cluster of differentiation CD2 receptor (a.k.a. LFA2 and LFA3) is a cell-adhesion molecule on the surface of T lymphocytes and natural killer cells.

[269] A.k.a. Development and differentiation-enhancing factors DDEF1 and DDEF2, as well as centaurin-β4 and -β3, respectively.

[270] A.k.a. Development and differentiation-enhancing factor-like protein DDEFL1, ACAP4, and Centβ6.

Table 9.71. Small ArfGAPs and interacting proteins (Source: [1075]).

Type	Binding partners
SmAP1	Clathrin
SmAP2	Clathrin, PICAlm

SmAPs

Small ArfGAPs (SmAP) are regulators of endocytosis. In humans, the SMAP family includes 2 members that bind to clathrin heavy chain and phosphatidylinositol-binding clathrin assembly protein PICAlm (or CAlm) [1074] (Table 9.71).

Protein SmAP1 resides in the cytosol. It is recruited to membranes to regulate endocytosis. This ArfGAP directly connects to the clathrin heavy chain. It interacts preferentially with ARF6 than ARF1 GTPase.

Protein SmAP2 is more stably bound to endosomes. It is involved in the clathrin- and AP1-dependent retrograde transport from early endosomes to the trans-Golgi network. It localizes to AP1+, epsinR+ endosomes, but not on AP3+ or CoP1+ endosomes. It targets both ARF1 and ARF6 in vitro, but as an ARF1-specific GAP activity in vivo [1081]. Two alternatively spliced variants of SmAP2 — canonical SmAP2-1 and shorter SmAP2-2 — exist.

ArfGAPs and Membrane Trafficking

Both ArfGAP1 and ACAP1 bind directly to transmembrane cargo or cargo receptors [1075]. The former binds to the transmembrane trafficking protein Golgi-cargo receptor transmembrane emp24-domain trafficking protein TMED2 and KDEL (Lys-Asp-Glu-Leu) endoplasmic reticulum protein retention receptor KDELR1[271] that mediates retrograde transport of endoplasmic reticulum-resident proteins from the Golgi body to the endoplasmic reticulum. Members of the TMED family prevent ArfGAP activity. On the other hand, ACAP1 interacts with transferrin receptor, β_1-integrin, and vesicle-associated membrane protein VAMP3 (cellubrevin) and participates in molecule recycling.

Other ArfGAPs bind to vesicle coat proteins or coat protein adaptors. ArfGAP1 connects to coatomer CoP1 and clathrin-associated AP1 complexes [1082]. Members of the SMAP family bind to clathrin. Agent SMAP1 tethers to clathrin to regulate endocytosis of transferrin receptors. Molecule SMAP2 interacts with clathrin heavy chain and phosphatidylinositol-binding clathrin assembly protein PICAlm [1083]. Proteins AGAP1 and AGAP2 associate with clathrin adaptor AP3 and AP1 complexes, respectively [1084].

---

[271] A.k.a. endoplasmic reticulum retention signal-containing protein ERD2.

Proteins ASAP1 and ASAP2 as well as ARAP3 bind to adaptors involved in membrane trafficking, such as SH3 domain kinase-binding protein SH3KBP1,[272] RalBP1-associated Eps domain-containing REps2,[273] that complexes with ASAP1 and RalBP1 to regulate actin remodeling by controlling RhoA GTPase, and ubiquitous splice form of amphiphysin-2 that is highly produced in myocytes [1075].[274]

ArfGAPs and Signaling

Among ArfGAPs, APAPs, AGAPs, and ARAP3 interact with enzymes of lipid metabolism [1075]. Agents APAP1 and AGAP2 bind to phospholipase-Cγ and ARAP3 to SH-containing inositol 5-phosphatase SHIP2.

The ArfGAP proteins also connects to protein kinases. Protein APAP1 links to P21-activated kinase via Rac–CDC42 exchange RhoGEF6/7 factors. Once phosphorylated by SRC family kinase in response to angiotensin-2 and EGF, APAP1 also interacts with MAP2K1 [1075]. Both APAP1 and APAP2 tether to GRK2 and cause internalization of G-protein-coupled receptors.

Protein AGAP2 is phosphorylated by SRC family kinases to prevent apoptosis [1075]. Furthermore, AGAP2 complexes with protein kinase-B. Protein ASAP1 attaches to and is phosphorylated by Src, FAK1, and FAK2 during cell motion.

ArfGAPs, Cytoskeleton, and Cell Adhesion

The ArfGAP proteins not only target ARFs, but other monomeric GTPases to regulate remodeling of the actin cytoskeleton [1075]. Members of the ARAP family operate with Rho and Rap GTPases. Both ARAP1 and ARAP3 link to RhoA preferentially to Rac1 and CDC42 GTPases. Protein ARAP2 that has an inactive RhoGAP domain tethers RhoA^{GTP} to build focal adhesions and stress fibers.

Proteins of the APAP and ASAP families regulate focal adhesions [1075]. In addition, ASAPs contribute to the regulation of podosomes as well as invadopodia in tumor cells. Besides, APAP1, APAP2, and ASAP2 interact with focal adhesion adaptor paxillin and its homolog leupaxin that is highly expressed in hematopoietic cells and transforming growth factor-β1-induced transcript TGFβ1I1.

The ArfGAP proteins associate with scaffold proteins. Protein ASAP1 links to cortactin that can be detected in peripheral membrane ruffles, podosomes, and invadopodia. In focal adhesions as well as membrane ruffles, ASAP1 connects to CRK and CRKL adaptors. The latter binds to paxillin and breast cancer anti-

---

[272] A.k.a. 85-kDa CBL-interacting protein CIn85 for epidermal growth factor receptor internalization.

[273] A.k.a. proline-rich motif in EH-domain-containing protein POB1.

[274] Hence its name amphiphysin-2M.

estrogen resistance adaptor BCAR1 (CAS or P130CAS) in focal adhesions.[275] Both BCAR1–Src and BCAR1–CRK complexes contribute to actin cytoskeleton regulation. The BCAR1–CRK complex promotes cytoskeletal rearrangements by interacting with GEFs RapGEF1 and RasGEF DOCK1.

The ArfGAP proteins, such as AGAPs, APAPs, ARAPs, and ASAPs, can operate as scaffolds for signaling mediators, such as non-receptor Tyr kinases of the SRC family, ARFs, RalBP1, phospholipids, adaptor CRK, and cortactin.

9.4.2.2 RabGAPs

Small GTPases of the RAB superfamily that comprises at least 30 members is implicated in intracellular vesicle transfer.

Rab3GAP

The RAB3 infrafamily include 4 known members (Rab3a–Rab3d). Heterodimer Rab3 GTPase-activating protein Rab3GAP1 inactivates members of the RAB3 infrafamily that are implicated in the exocytosis of neurotransmitters and hormones. It is active only on the lipid-modified form [1085]. Proteins Rab3GAP1 and Rab3GAP2 encoded by 2 distinct genes correspond to catalytic and non-catalytic subunits of Rab3 GTPase-activating protein.

RabGAP1

Protein RabGAP1 is also named GAP and centrosome-associated protein (GAP-CenA) and TBC1D11 protein. The Tre2–BUB2–CDC16 (TBC) domain is a motif that functions as a specific RabGAP domain. More than 40 distinct TBC domain-containing proteins have been identified in humans (TBC1D1, -2a–2b, -3a–3c, -4–7, -8a–8b, -9a–9b, -10a–10c, -11–21, -22a–22b, and -23–30 [76])[276]

---

[275] Members of the BCAR–CAS family include human enhancer of filamentation HEF1 (a.k.a. CAS1 and neural precursor cell expressed, developmentally down-regulated protein NEDD9) mostly found in lymphocytes, lung, and breast, and embryonal Fyn-associated substrate (EFS or Src-interacting protein SIN) in embryonic tissues. Non-receptor protein Tyr kinases Src, FAK1, FAK2, and Abl phosphorylate BCAR1. Adaptor BCAR1 allows CRK to interact with effectors that activate Rac1 GTPase. The BCAR1–CRK–Rac1 signaling leads to JNK activation [1038]. In addition, CRK shuttles from BCAR1–CRK complexes to CRK–GAB, CRK–CBL, or CRK–IRS1 complexes in response to certain signals.

[276] Protein TBC1D6 is also called growth hormone-regulated TBC protein GRTP1; TBC1D11 RabGAP1; and TBC1D18 RabGAP1L.

9.4.2.3 RapGAPs

Small Rap GTPases (Sect. 9.3.9) are inactivated by cognate inhibitors. Gz-coupled receptors recruit Rap1GAP1 isoform from the cytosol to attenuate the cAMP–Rap–ERK pathway that primes neuron differentiation [1086]. On the other hand, activated Gi-coupled receptors recruit Rap1GAP2 isoform to the plasma membrane for Rap GTPase deactivation, hence supporting the MAPK signaling [811]. In addition, Go^{GDP} interacts with Rap1GAP to block Rap-directed GAP activity. Interactions between Gα subunit and Rap1GAP depend on cell and splice variant type.

Synaptic RapGAPs

Activation of postsynaptic EPHb receptors by ephrin-B1 induces rapid formation of mature spines. Receptors EPHb recruit RacGEF kalirin-7 into clusters in the dendritic shaft and spines to activate Rac GTPase that, in turn, stimulates CDK5 and PAK1 or PAK3 kinases. Spine-associated RapGAP SpAR1 regulates dendritic spine morphology in hippocampal neurons. It is phosphorylated by Polo-like kinase PLK2 for proteosomal degradation.

Spine-associated RapGAP SpAR2 has a GAP activity for Rap1 and Rap2 GTPases. Like SpAR1, SpAR2 interacts with synaptic scaffold protein prosynapse-associated protein-interacting protein ProSAPIP1 that, in turn, binds to postsynaptic density scaffold proteins ProSAP1[277] and ProSAP2,[278] that make a molecular interface between glutamate receptor clusters and the actin cytoskeleton [904].

Signal-Induced Proliferation-Associated RapGAPs

Signal-induced proliferation-associated GTPase-activating protein SIPA1 (or SPA1) targets Rap1 and Rap2 GTPases, but not Ran, Ras, and Rho GTPases. This mitogen-induced GAP precludes cell cycle progression when it is prematurely expressed [1087]. It is located in the perinuclear region and nucleus. Two alternatively spliced variants exist. It interacts with chromosome-associated protein CAP[279] that links to replication factor RFC1 to RFC5 and constitutive photomorphogenic protein COP1[280] that connects to transcription factor Jun [251].

---

[277] A.k.a. SH3 and multiple ankyrin repeat domain-containing protein SHAnk2.

[278] A.k.a. SHAnk3 protein.

[279] A.k.a. bromodomain-containing protein BrD4. The Bromodomain is involved in chromatin targeting.

[280] A.k.a. ring finger and WD repeat domain-containing protein RFWD2.

9.4.2.4 Dual-Specificity RasGAPs

Dual-specificity RasGAPs of the GAP1 family activate the GTP hydrolysis of both Rap and Ras GTPases that correspond to 2 distinct types of monomeric GTPases of the RAS superfamily. In general, RasGAPs and RapGAPs use different strategies to stimulate the GTPase reaction of their cognate small GTPases. Activity of RasGAP involves 2 major residues [1088]: (1) Gln61 from Ras in a highly mobile motif (switch-2) and (2) Arg (finger) yielded by the RasGAP that moves into the binding site and orients Gln61 to position a water molecule for nucleophilic attack and reduce the free activation energy. On the other hand, proteins of the RAP family (Rap1a–Rap1b and Rap2a–Rap2c) do not possess a residue corresponding to Gln61, but instead a non-essential threonine residue. Rap-specific RapGAPs that are structurally unrelated to RasGAPs do supply an asparagine (thumb) rather than an arginine finger.

Yet, several RasGAPs have dual specificity for both Ras and Rap GTPases: RasA1, or RasA5,[281] as well as 3 members of the GAP1 family, i.e., RasA3,[282] RasA4,[283] and RAS activator-like RasAL1 protein.[284]

Protein RasA1 is more active on Rap than on Ras. Full-length RasA3 is an effective GAP for both Ras and Rap, but its GAP domain is only active on Ras GTPase. Protein RasAL responds to Ca^{++} oscillations and activates the GTPase activity of Ras only when bound to membranes, but its RapGAP activity is independent of membrane binding [1088]. The adaptivity of RasA3 and RasAL is mediated by extra GAP domains [1088].

9.4.2.5 RasGAPs

Proteins of the large RasGAP set include: (1) neurofibromin; (2) Ras P21 activator RasA1;[285] (3) the GAP1 family with the group of others RAS P21 activators — RasA2, or GAP1^M, RasA3, or GAP1^{IP4BP}), and RasA4 — and the first member of the group of Ras GTPase-activating-like protein RasAL1; as well as (4) the synaptic RasGAP family (SynGAPs) with DAB2-interacting protein (DAB2IP) and RasAL2 (i.e., nGAP),[286] in addition to RasA1 (Table 9.72).

---

[281] A.k.a. synaptic Ras GTPase-activating protein SynGAP1.

[282] A.k.a. inositol (1,3,4,5)-tetrakisphosphate-binding protein (GAP1^{IP4BP}).

[283] A.k.a. calcium-promoted Ras inactivator (CaPRI).

[284] A.k.a. RasAL and GAP1-like protein.

[285] A.k.a. RasA5 and synaptic GTPase-activating protein SynGAP1.

[286] Acronym nGAP is also used for neuronal growth-associated proteins, such as stathmin, stathmin-like-2, and growth-associated protein GAP43 that are targets of neurotrophin. They are also related to small GTPase. Activated Rac and, to a lesser extent, CDC42 downstream from epidermal growth factor lead to stathmin phosphorylation, hence inhibition of stathmin-induced destabilization of microtubules.

Table 9.72. RasGAPs and their effectors (Source: [251, 889]; CaPRI: Ca^{++}-promoted Ras inactivator; DLg: Disc large homolog; GAP: Ras GTPase-activating protein; GAPl: Ras GTPase-activating protein-like; GRIF1: glucocorticoid receptor DNA binding factor; Gβ2L1: G-protein subunit-β2-like 1; G3BP: Ras-GAP SH3 domain-binding protein; HCK: hemopoietic cell kinase IP4BP: inositol (1,3,4,5)-tetrakisphosphate-binding protein; NF: neurofibromatosis protein; RasA: RAS P21 protein activator; RasAL: Ras-activator-like protein; SynGAP: synaptic Ras GTPase-activating protein; UlK: Unc51-like kinase; VRNF: von Recklinghausen neurofibromatosis protein; WSS: Watson or Wrinkly skin syndrome protein).

Type	Aliases	Effectors and interactors
RasA1	RasA5, RasGAP p120GAP, SynGAP1	hRas, nRas, rRas1–rRas3, Rab5, Rap1, HCK, LcK, FAK2, Src, CSK, ZAP70, MAP4K4, PLCγ1, NCK, GRB2, Dok1/2, Cav2, UlK1, DLg3, Htt, G3BP, Annexin-6, synapsin-1, paxillin PDGFRβ, InsR, IGF1R, EPHb2/b3, VEGFR2, GRIF1, SLC9A2
RasA2	GAP1^M	hRas, rRas1–rRas3
RasA3	GAP3, GAP1^{IP4BP}	HCK, Gβ2L1 $I(1,3,4,5)P_4$, $PI(4,5)P_2$, $PI(3,4,5)P_3$
RasA4	GAPl, CaPRI	hRas, Rap1
RasAL1		hRas, Rap1
RasAL2	nGAP	
Neurofibromin-1	NF1, NFNS, VRNF, WSS	hRas, kRas, nRas, rRas1–rRas3

Neurofibromins

Neurofibromin-1 is a cytoplasmic protein prominently synthesized in glial (Schwann cells, oligodendrocytes, and astrocytes) and neuronal cells, as well as leukocytes. Mutations of the NF1 gene cause autosomal dominant neurofibromatosis type 1 (von Recklinghausen or Watson disease).

Four alternative splicing exons are able to produce 5 human neurofibromin isoforms [1089]. Alternatively splicing can generate 2 isoforms with different neurofibromin-1 GAP-related domain (GRD): NF1^{GRD1} and NF1^{GRD2}). The latter is produced in Schwann cells.

Neurofibromin-2[287] also controls Ras activity [1090]. It uncouples both Ras and Rac from growth factor receptor Tyr kinases and counteracts ezrin–radixin–moesin-dependent activation of Ras.

In humans, neurofibromin-3 and -4 are produced mainly in the myocardium and skeletal muscles [1089]. Neurofibromin-9a has a limited neuronal expression. An additional isoform with a transmembrane domain likely serves in the maintenance of intracellular membranes.

---

[287] A.k.a. Merlin, a portmanteau for moesin-ezrin-radixin-like protein, and schwannomin.

Neurofibromin-1 increases the GTP hydrolysis rate of hRas, kRas, and nRas that are involved in cell growth and differentiation via the PI3K–PKB–RHEB–TOR and Raf–ERK pathways as well as RalGEF and Rac GTPase [1089]. In addition, it links to microtubules and participates in several signaling pathways. It actually modulates adenylate cyclase activity, hence the PKA–cAMP pathway. Neurofibromin-1 also binds to caveolin-1 that connects to protein kinase-C and growth factor receptors. It is rapidly ubiquitinated after activation of receptor Tyr kinases and G-protein-coupled receptors. Neurofibromin-1 also intervenes during wound healing and endothelial repair. It could suppress excessive proliferation and migration of smooth muscle cells in blood vessel wall [1089]. Neurofibromin-1 contributes to cell motility by regulating actin filament dynamics via the Rho–RoCK–LIMK2–cofilin pathway [1091].

RasA1

Ras GTPase-activating protein RasA1[288] is a cytosolic protein that inactivates Ras and transmits mitogenic signals. Phosphorylation of RasA1 by CamK2 increases inactivation of Ras GTPase. Protein RasA1 also inactivates Rap GTPase [926].

Protein 68-kDa Src substrate-associated during mitosis protein SAM68 links to RasA1 after its phosphorylation by insulin receptor and connection to PI3K [1092]. Protein RasA1 contains a C2 domain that mediates Ca^{++}-dependent membrane association and between-protein interactions. It indeed forms a complex with annexin-6, Fyn kinase, and focal adhesion FAK2 kinase [1093]. It also possesses a PH domain that allows it to tether scaffold RACK1, a membrane-associated protein that binds protein kinase-C [1094].

Synaptic RasA1 is an abundant component of the NMDAR signaling complex that binds to NMDAR scaffold protein Disc large homolog DLg4 (a.k.a. postsynaptic density protein PSD95). Activated NMDAR causes phosphorylation of RasA1 by CamK2 [926]. Both RasA1 and CamK2 bind to scaffold protein multiple PDZ-domain protein MPDZ (or MuPP1). Activated NMDAR also provokes dissociation of RasA1 from MPDZ, RasA1 dephosphorylation, reduction in p38MAPK activity, and elevation in AMPAR insertion into the postsynaptic membrane.

Synaptic RasA1 also interacts with DLg3 as well as uncoordinated-51 (Unc51)-like Ser/Thr kinases ULK1 (Unc51 homolog) and ULK2 [251]. Kinase ULK1 regulates axon extension [1095]. Pseudokinase ULK1 also binds to scaffold protein syntenin that connects to Rab5 GTPase. Activity of Rab5 is also lowered by RasA1.

GAP1 Family (RasA2–RasA4 and RasAL1)

Members of the GAP1 family — RasA2 and RasA3 — associate with phosphoinositides at the cytosolic leaflet of the plasma membrane, where they inactivate Ras

---

[288] A.k.a. RAS P21 protein activator-1, SynGAP1, and P120GAP.

Table 9.73. Synaptic Ras- and RapGAPs in the central nervous system (Source: [926]; SynGAP: synaptic GTPase-activating protein; SpAR: spine-associated Rap GTPase-activating protein).

GAP	Aliases	Effector	Activator	Expression
RasA1	SynGAP	Rap1, Rap2, Ras	CamK2	Neurons
SpAR1		Rap1a, Rap2a		Neurons
SpAR2				

GTPase. Members of the GAP1 family — RasA4 and RasAL1 — also translocate to the plasma membrane and bind phospholipids upon elevation in intracellular Ca^{++} concentration.

Calcium influx can trigger inhibition of Ras GTPases. Two Ca^{++}-sensitive Ras inhibitors are recruited to the plasma membrane according to temporal features of Ca^{++} signaling [1096]. Activation of RasA4 and RasAL1 results from distinct increase in cytosolic Ca^{++} concentration. (1) Calcium-promoted Ras inactivator (CaPRI or RasA4) that senses Ca^{++} signal amplitude by experiencing either a Ca^{++}-dependent, transient association with the plasma membrane or predominantly Ca^{++}-independent, sustained plasmalemmal recruitment following integration of Ca^{++} cues. Consequently, RasA4 converts different intensities of Ca^{++} stimulation into diverse durations of Ca^{++} effect on hRas activity. (2) Ras GTPase-activating-like protein (RasAL1) is a sensor of Ca^{++} signal frequency to which it responds by repetitively tethering with the plasma membrane and deactivating Ras GTPase. In other words, the frequency of cytosolic Ca^{++} oscillations is thus decoded by fluctuating connections with the plasma membrane.

rRasGAP Plexins

Membrane-associated rRasGAP has stronger GAP activity for integrin activator rRas than hRas. Plexins are receptors for axon-guidance semaphorins. The Sema4d–PlxnB1 signaling that provokes repulsion primes rRasGAP activity to inhibit cell migration by targeting $β_1$-integrin [1097]. In the nervous system, activation of plexin-B1 by semaphorin-4D causes growth cone collapse, as it precludes the PI3K pathway. It indeed leads to dephosphorylation of PKB and GSK3β (activation) that, in turn, phosphorylates collapsin response mediator protein CRMP2 [1098].[289] Plexin-A1 and -B1 that govern axon guidance function as a rRasGAP upon binding of Rnd1 GTPase, whereas in cortical neurons, plexin-D1 rRasGAP activity is exerted on Rnd2 GTPase [984]. On the other hand, plexin-C1 that also displays rRasGAP activity represses cell migration without Rnd GTPases.

---

[289] A.k.a. dihydropyrimidinase-like protein DPysL2.

Dab2IP

Disabled homolog Dab2-interacting protein (Dab2IP)[290] is a RasGAP that inhibits hRas, rRas, and rRas2, but not Rap1a GTPase. It suppresses epidermal growth factor-elicited cell growth. In addition, it causes cell cycle arrest and promotes TNFα-mediated apoptosis in response to stress, especially in endothelial cells, as it facilitates the dissociation of MAP3K5 (or ASK1; Sect. 6.5.5.1) from its inhibitor 14-3-3 protein [577].

Scaffold protein Dab2IP associates with and inactivates phosphatidylinositol 3-kinase and protein kinase-B via its C-terminal Pro-rich (PR) and Period-like (PER) binding sites. Platform Dab2IP sequesters PI3K, PKB, and MAP3K5 and modulates their phosphorylation status in response to different types of signals.

9.4.2.6 RhoGAPs

A large group of GTPase-activating proteins targets RHO superfamily members: CDC42, Rac, and Rho GTPases. Many are active on several Rho species. However, some could be specific to a single member of the RHO superfamily. They comprise RhoGAP1[291] to RhoGAP15 and RhoGAP17 to RhoGAP49 as well as other types (Tables 9.74 to 9.77). In addition, several members of the RhoGAP superfamily possess both RhoGAP and RhoGEF domains, such as ABR and BCR proteins [1100]. Others have lipid phosphatase and protein kinase domains. Lastly, ARAPs contain both RhoGAP and ArfGAP sequences.

Some GTPase-activating proteins, such as RhoGAP6, RhoGAP7, GMIP, Myo9a, and ARAP3, are specific for RhoA; RhoGAP1, -20, -26, -35, and -37 prefer RhoA to Rac1 and CDC42; ARAP1, RhoGAP4, -10, -17, and -41 also inactivate RhoA [945].

These GAP activities are regulated by phosphorylation, lipid binding, and between-protein interaction. Certain GAPs not only catalyze the intrinsic GTPase activity of RHO superfamily GTPases that then return to their inactive state, but also act as effectors and transmit signals.

RhoGAP1

Rho GTPase-activating protein-1 (RhoGAP1)[292] interacts with CDC42, RhoA, and RhoC GTPases.

---

[290] A.k.a. ASK1-interacting protein AIP1, an alias to avoid for the sake of disambiguation.
[291] A.k.a. ArhGAP1 and CDC42GAP.
[292] A.k.a. ArhGAP1, CDC42GAP, and P50RhoGAP.

Table 9.74. RhoGAPs and their effectors (**Part 1**; Sources: [889, 1043, 1100]; BPGAP: BH3-only protein family never in mitosis protein-related kinase (NeK2)-interacting protein-2 (BNIP2) and CDC42GAP Homology (BCH) domain-containing, proline-rich, and CDC42GAP-like protein; Chn: chimerin; DLC1: deleted in liver cancer protein-1, a.k.a. StARD12 [StAR-related lipid transfer (StART) domain-containing protein]; FAM7b1: family with sequence similarity-7 member-B1; GRAF: GTPase regulator associated with focal adhesion kinase; SRGAP: Slit–Robo Rho GTPase-activating protein [SRGAP is also denoted SRGAP2 and SRGAP3]; TCGAP: TC10–CDC42 GTPase-activating protein, a.k.a. sorting nexin SNx26; WRP: WAVE-associated Rac GTPase-activating protein).

Type	Aliases	Effectors, interactors
RhoGAP1	CDC42GAP, p50RhoGAP ArhGAP1	CDC42 > Rac1, RhoA
RhoGAP2	Chn1, αChn ArhGAP2, RhoGAP22	CDC42, Rac1
RhoGAP3	Chn2, βChn, ArhGAP3	Rac1
RhoGAP4	C1, RGC1, C7, Pn5, ArhGAP4, P115RhoGAP	RhoA
RhoGAP5	ArhGAP5, P190b	RhoA > CDC42, Rac1; Rnd1–Rnd3
RhoGAP6	ArhGAP6, RhoGAP-X1	RhoA
RhoGAP7	ArhGAP7, DLC1, HP, P122RhoGAP, StarD12	RhoA > CDC42
RhoGAP8	BPGAP1, PP610 ArhGAP8	CDC42 > RhoA, Rac1
RhoGAP9	ArhGAP9, RGL1, 10c	CDC42, Rac1>RhoA
RhoGAP10	GRAF2, ArhGAP10/21 PSGAP	CDC42 > Rac1, RhoA
RhoGAP11a	ArhGAP11a	14-3-3σ
RhoGAP11b	ArhGAP11b, FAM7b1	
RhoGAP12	ArhGAP12	MLC2b
RhoGAP13	ArhGAP13, SRGAP1	CDC42
RhoGAP14	SRGAP3, MEGAP, WRP, ArhGAP14	Rac1 > CDC42
RhoGAP15	ArhGAP15	Rac1

RhoGAP2 and RhoGAP3 (Chimerins)

The chimerin family of GTPase-activating proteins includes α-chimerin (Chn1), or RhoGAP2, with its isoforms chimerin-α1 and -α2 (RhoGAP2-1 and RhoGAP2-2), as well as β-chimerin (Chn2), or RhoGAP3, with chimerin-β1 and -β2 isozymes (RhoGAP3-1 and RhoGAP3-2). Chimerin-α2 and -β2 result from alternative splicing of chimerins-α1 and -β1, respectively. These variants have a distinct tissue distribution in the central nervous system.

Table 9.75. RhoGAPs and their effectors (**Part 2**; Sources: [889,1043,1100]; FilGAP: filamin-A-binding RhoGTPase-activating protein; GRAF: GTPase regulator associated with focal adhesion kinase; GRLF: glucocorticoid receptor repression DNA-binding factor; OphnL1 (Ophn1L1): oligophrenin-1-like protein; RARhoGAP (RAGAP): Rap-activated Rho GTPase-activating protein; RICH1: RhoGAP interacting with CIP4 homolog-1 [a.k.a. nadrin: neuron-associated developmentally regulated protein]).

Type	Aliases	Effectors, interactors
RhoGAP17	ArhGAP17, RICH1, nadrin, WBP15	CDC42, Rac1
RhoGAP18	ArhGAP18, MacGAP	Rac2, RhoB/D
RhoGAP19	ArhGAP19	Rac2, RhoB/D/F
RhoGAP20	ArhGAP20, RAGAP RARhoGAP	Rho (Rap1 affector)
RhoGAP21	ArhGAP21, ArhGAP10	RhoA
RhoGAP22	ArhGAP22, RhoGAP2	Rac2, RhoB/D/F/G
RhoGAP23	ArhGAP23	hnRNPa1, MePCE, RPAP1
RhoGAP24	FilGAP, RCGAP72, ArhGAP24, p73RhoGAP	CDC42, Rac1 RoCK1
RhoGAP25	ArhGAP25	
RhoGAP26	ArhGAP26, OphnL1, GRAF	CDC42 ≫ RhoA
RhoGAP27	ArhGAP27, CAMGAP1	CDC42, Rac1
RhoGAP28	ArhGAP28	
RhoGAP29	ArhGAP29, PARG1	CDC42, Rac1, RhoA, Rap2a SIRT1, PKB1, PTPn13
RhoGAP30	ArhGAP30	Rac2, RhoB/D, RhoBTB1/2

Proteins RhoGAP2 and RhoGAP3 target Rac GTPase mostly in nervous tissues. Agent RhoGAP2 is specific for CDC42 and Rac GTPases, hence contributing to the regulation of cytoskeleton dynamics in neurons. Diacylglycerol-responsive RhoGAP3 operates in proliferation and migration of smooth muscle cells.

In myocytes, cyclin-dependent kinase CDK5 controls differentiation, organization, and signal transduction. In neurons, CDK5 is a regulator of neuronal development and function. It contributes to the control of cytoskeletal dynamics with its neuron-specific activators via chimerin-1 [1101]. Agent RhoGAP2 binds cyclin-dependent kinase CDK5 regulatory subunit $CDK5_{r1}$.[293]

---

[293] Activators of CDK5, short-lived $CDK5_{r1}$ (or 35-kDa protein P35) and the regulatory subunit-2 ($CDK5_{r2}$ (or P39), complex CDK5 similarly to cyclins. The $CDK5_{r1}$–CDK5 complex phosphorylates: (1) structural proteins, such as constituents of intermediate neurofilaments (nestin and neurofilament heavy and middle chains) and actin filaments, as well as cytoskeleton-regulatory proteins such as microtubule-associated proteins Tau, and (2) proteins involved in neurotransmitter release, such as synapsin-1 and syntaxin-binding protein-1.

Table 9.76. RhoGAPs and their effectors (**Part 3**; Sources: [889, 1043, 1100]; CDGAP: CDC42 GTPase-activating protein; CrGAP: crossGAP homolog; FAM13a1: family with sequence similarity-13 member-A1; FnBP: formin-binding protein; GCGAP: GAB-associated CDC42/Rac GTPase-activating protein; GMIP: GTP-binding protein overexpressed in skeletal muscle (GEM)-interacting protein; GRAF: GTPase regulator associated with focal adhesion kinase; GRIT: GTPase regulator-interacting with TRKa; GRLF: glucocorticoid receptor DNA-binding factor; HA1: minor histocompatibility antigen-1 [HLA-HA1]; Ophn [Opn]: oligophrenin, a.k.a. MRX60 [X-linked mental retardation]; SRGAP: Slit-Robo Rho GTPase-activating protein; TAGAP: T-cell activation RhoGAP; TCGAP: TC10–CDC42 GTPase-activating protein, a.k.a. sorting nexin SNx26).

Type	Aliases	Effectors, interactors
RhoGAP31	CDGAP	CDC42, Rac1
RhoGAP32	ArhGAP32, RICS, GCGAP, GRIT,	CDC42, RhoA > Rac1
RhoGAP33	ArhGAP33, TCGAP, SNx26	CDC42, RhoQ, Rac1, RhoJ
RhoGAP34	FnBP2, SRGAP2	TNFSF6, WASP, 14-3-3γ/ζ
RhoGAP35	ArhGAP35, GRLF1, P190RhoGAP, P190a	RhoA
RhoGAP36	ArhGAP36	
RhoGAP37	ArhGAP37, DLC2, StARD13, GT650	RhoA > CDC42
RhoGAP38	ArhGAP38, DLC3, StARD8, StARTGAP3	Rac2, RhoB/D/F/G
RhoGAP39	ArhGAP39, CrGAP	Rac1/2, RhoB
RhoGAP40	ArhGAP40, C20orf95	Rac2, RhoB/D, RhoBTB1/2
RhoGAP41	ArhGAP41, MRX60, Ophn1, Opn1	CDC42, Rac1, RhoA
RhoGAP42	ArhGAP42, GRAF3	
RhoGAP43	ArhGAP43, SH3BP1, 3BP1	RhoA Abl1, HCK, InPP5g1
RhoGAP44	ArhGAP44, RICH2	CDC42, Rac1, Rac2
RhoGAP45	ArhGAP45, HA1	Rac2, RhoB/D, RhoBTB1/2
RhoGAP46	ArhGAP46, GMIP	Rac2, RhoA/B, RhoBTB1, GEM (KIR)
RhoGAP47	ArhGAP47, TAGAP1	
RhoGAP48	ArhGAP48, FAM13a1	RapGEF2, MAP3K10, ATF2, Jun, MEF2c
RhoGAP49	ArhGAP49, FAM13b1	Jun, WNK1, 14-3-3-$\gamma/\epsilon/\zeta$

Inactivation of Rac by RhoGAP3 precludes ERK activity that results from heregulin-β1 stimulation [1102]. Heregulin-β1 belongs to the category of EGF-like ligands (a.k.a. neuregulins) of the family of human epidermal growth factor receptor HER3 and HER4 kinases. Heregulins activate the PI3K–PKB pathway and extracellular signal-regulated protein kinase, at least, in some cancer cell types.

9.4 Regulators of Monomeric Guanosine Triphosphatases

Table 9.77. RhoGAPs and their effectors (**Part 4**; Sources: [889, 1043, 1100]; ABR: active BCR-related gene; ARAP: ArfGAP, RhoGAP, ankyrin repeat, and pleckstrin homology domain-containing protein, a.k.a. centaurin-δ; BCR: breakpoint cluster region protein, a.k.a. ALL (acute lymphoblastic leukemia), CML (chronic myeloid leukemia), and Phl (Philadelphia chromosome translocation); DRAG: dual-specificity Rho and ARF GTPase-activating protein; MgcRacGAP: male germ cell RacGAP; Myo9b: myosin-9b, a.k.a celiac disease-4 (Cceliac4); OCRL: oculocerebrorenal syndrome of Lowe protein, a.k.a. LOCR [Lowe oculocerebrorenal syndrom] and InPP5f [inositol polyphosphate 5-phosphatase, a PIP2 5-phosphatase]).

Type	Aliases	Effectors
ARAP1	Centδ2	CDC42, Rac1, RhoA
ARAP2	Centδ1	ARF6, Rac2, RhoB, RhoBTB1/2
ARAP3	DRAG1, Centδ3	RhoA > CDC42, Rac1
ABR		CDC42, Rac1, Rac2
BCR	ALL, CML, Phl	Rac1 > CDC42, RhoA
InPP5b		Rab6a
InPP5f	OCRL1, LOCR, NPHL2	RhoA > CDC42, Rac1
		Rab1a/5a/6a/14
Myo9a	Myr7	RhoA
Myo9b	Myr5, Celiac4	RhoA
RacGAP1	MgcRacGAP, HsCYK-4, IDGAP	CDC42, Rac1, RhoA
RalBP1	RlIP76, RIP1, P200RhoGAP, P250GAP	CDC42 > Rac1

RhoGAP4

Protein RhoGAP4, or P115RhoGAP, is predominantly produced in hematopoietic cells. It specifically targets RhoA GTPase [1103]. It is connected to MAP3K1 enzyme that binds to MAP2K4 enzyme to stimulate the ERK and JNK cascades. It thus reduces MAP3K1-induced signaling to AP1 transcription factor.

RhoGAP5 and RhoGAP35 (P190RhoGAPs)

Two ubiquitous 190-kDa RhoGAPs are encoded by 2 genes: RhoGAP5[294] and RhoGAP35.[295] Both are involved in cortical actin assembly and disassembly, actin–myosin filament contraction, and cell motility and differentiation [1104]. However, RhoGAP5 regulates cell and organism size via cAMP response element-binding protein.

---

[294] A.k.a. P190bRhoGAP, P190b, ArhGAP5, and growth factor independent GFI2.

[295] A.k.a. P190aRhoGAP (or simply P190RhoGAP), P190a, ArhGAP35, and glucocorticoid receptor DNA-binding factor GRF1.

Table 9.78. Interactors of RhoGAP5 (Source: [1105]; DIP: Diaphanous interacting protein; ECT2: epithelial cell transforming sequence-2 oncogene protein; GAP: GTPase-activating protein; GEF: guanine nucleotide-exchange factor).

Partner type	Proteins
Small GTPase	CDC42, Rac1, RhoA, Rnd1, Rnd2, Rnd3 (RhoE)
Phosphatase	PTPn11

RhoGAP5

Stable interactions between RhoGAP5 and Rac1^{GTP} modulate its subcellular localization and activity, thus yielding crosstalk between the Rac and Rho signaling that are up- and downregulated, respectively [1105].

Phosphorylation of RhoGAP5 (Tyr306) may follow insulin or insulin-like growth factor IGF1 stimulation. Protein RhoGAP5 binds to RhoE GTPase that can then activates RhoGAP5, thereby lowering Rho^{GTP} level. It intervenes in the regulation of the synthesis of 2 collaborative peptidases — matrix metallopeptidases MMP2 and MMP14 — that contribute to matrix remodeling and angiogenesis [1105]. Several proteins are binding partners of RhoGAP5 (Table 9.78).

RhoGAP35

Various integrins, together with soluble fibronectin, collagen-1, or laminin favor RhoGAP35 activation [1104]. Protein RhoGAP35 is also stimulated upon cadherin engagement, plexin activation, and filamin accumulation. It binds RhoE GTPase that, in turn, activates the GAP activity of RhoGAP35 protein. In vitro, it functions with the following hierarchy on RHO superfamily members: Rho > Rac > CDC42. Activity of RhoGAP35 is regulated by Rho effectors, such as RoCK, CRIK, Diaphanous, etc. Using Rho-dependent and -independent mechanisms, RhoGAP35 regulates the dynamics of both actin filaments and microtubules.

Kinases of the SRC family and other protein Tyr kinases (Abl, BrK, FAK, and Fyn) phosphorylate (activate) RhoGAP35 (Tyr1105 of the RasA1 [P120RasGAP]-binding site). This modification promotes RasA1 binding, hence RasGAP activity of bound RasA1 [1104]. Epidermal, platelet-derived, and vascular endothelial growth factor increase RhoGAP35 activity via Src kinase. In addition, NADPH oxidase NOx1, angiotensin-2, Gα_{13} subunit, protein kinase-C, and phosphatidylinositol 3-kinase (PI3K$_{c1\delta}$, or P110δ isoform) indirectly promote RhoGAP35 activation [1104] (Table 9.79).

Focal adhesion kinase helps to maintain the integrity of vascular endothelial barrier and supports barrier function restoration in response to permeability-raising mediators. Focal adhesion kinase phosphorylates RhoGAP35, hence reducing RhoA activity [1106]. Plexin activated by semaphorin associates with RhoGAP35 to preclude cell migration [1107].

9.4 Regulators of Monomeric Guanosine Triphosphatases

Table 9.79. Interactors of RhoGAP35 (Source: [1104]; DIP: Diaphanous interacting protein; ECT2: epithelial cell transforming sequence-2 oncogene protein; GAP: GTPase-activating protein; GEF: guanine nucleotide-exchange factor).

Partner type	Proteins
	Stable associations
Small GTPase	Rac1, RhoA, RhoQ, Rnd1, Rnd2, Rnd3 (RhoE)
GEF	ECT2 (RhoGEF)
GAP	p120RasGAP (RasA1)
Scaffold	Paxillin, filamin, cateninδ1, DIP
Kinase	FAK1/2, FGR, Fyn
Transcription factor	GATA2, GTF2I
Receptor	Plexins-A1/B1
	Transient interactions
Kinase	Src, Abl2, BrK, LcK, GSK3β
Phosphatase	PTPRz; PTPn12/18; LMW-PTP
	Stable complexes
FAK–p120RasGAP	
GRB2–p120RasGAP–SHC1	
Paxillin–RasA1	
Paxillin–RasA1–FAK2–Abl–CAS–HER2	
PKCδ–RasA1	

The Rnd proteins that lack intrinsic GTPase activity antagonize RhoA inactivation by connecting to RhoGAP35 [1108]. Agent RhoGAP35 interacts with Rnd1 to Rnd3 GTPases and RhoA as well as paxillin [251].

On the other hand, glycogen synthase kinase GSK3β phosphorylates RhoGAP35 and decreases its activity [1104]. Different phospholipids inhibit RhoGAP35 activity. Dephosphorylation of RhoGAP35 by protein Tyr phosphatases PTPRz1 as well as PTPn12 and PTPn18 also reduces its activity. Infraclass-2 Cys-based phosphoTyr protein phosphatase, low-molecular weight PTP (lmwPTP), or acid phosphatase-1, also dephosphorylates RhoGAP35 agent.

In atriomyocytes, TGFβ modulates RhoGAP35 concentration, hence the expression of muscarinic M$_2$ receptor by interfering with the promoter activity and the response of atriomyocytes to parasympathetic stimulation [1104].

RhoGAP7, RhoGAP37, and RhoGAP38 (DLCs)

In mammals, 3 genes encode GAPs for both CDC42 and RhoA GTPases: RhoGAP7, RhoGAP37, and RhoGAP38 [1109].[296]

---

[296] Agents RhoGAP7, RhoGAP37, and RhoGAP38 are also called StAR-related lipid transfer domain-containing proteins StARD12, StARD13, and StARD8, respectively, steroidogenic

Inhibitor RhoGAP7,[297] localizes to focal adhesions. It enhances the GTPase activity of CDC42 and RhoA GTPases. It is involved in phophoinositide and insulin signalings, as it interacts with phospholipase-Cδ1 and is phosphorylated by protein kinase-B [1109]. Protein RhoGAP7 regulates cell motility at sites of cell contact, at least partly, as it controls activation–inactivation cycle of Rho GTPases. Inhibitor RhoGAP37[298] also precludes RhoA activity.

RhoGAP10, RhoGAP26, and RhoGAP42 (GRAFs)

The family of GTPase regulators associated with focal adhesion kinase (GRAF) are characterized by an N-terminus that possesses a BAR (Bin/amphiphysin/Rvs) domain. The BAR domain interacts directly with the GAP domain and inhibits its activity, thus maintaining these GAPs (GRAFGAPs) in an auto-inhibited state [1110].

GTPase-activating protein RhoGAP26[299] is involved in clathrin-independent endocytosis. The GRAF family also includes (or oligophrenin), RhoGAP10 (or GRAF2), and RhoGAP42 (or GRAF3) members.

Inhibitor RhoGAP41 is expressed in the developing spinal cord and later in brain areas with high synaptic remodeling. It is produced in both glial and neuronal cells, in which it colocalizes with actin. Agent RhoGAP41 inhibits Rho GTPases.

RhoGAP13, RhoGAP14, and RhoGAP34 (SRGAPs)

Protein Slit is a guidance molecule for migrating neurons and leukocytes that acts essentially via its cognate receptor Roundabout (Robo).[300] The group of Slit–Robo Rho GTPase-activating proteins include SRGAP1, or RhoGAP13, SRGAP2, or RhoGAP34,[301] and SRGAP3, or RhoGAP14 agent.[302] Protein RhoGAP13 inactivates CDC42; RhoGAP34 is a Rac-selective GTPase-activating protein.

---

acute regulator (StAR)-related lipid transfer (StART) domain-containing proteins (StARTGAP1–StARTGAP3), and Deleted in liver cancer proteins (DLC1–DLC3).

[297] A.k.a. ArhGAP7, P122RhoGAP, and adaptor for Rho and PLC P122ARP. Alias DLC1 is also used for dynein light chain, which is better termed dynein light chain LC8-type-1 DynLL1 for disambiguation.

[298] Not dynein light chain-2 that should be termed dynein light chain LC8-type DynLL2 for the sake of disambiguation.

[299] A.k.a. GTPase regulator associated with focal adhesion kinase (GRAF) and oligophrenin1-like protein (OphnL1).

[300] This repellant molecule prevents axon crossing through the midline of the brain and spinal cord.

[301] A.k.a. formin-binding protein FnBP2.

[302] A.k.a. mental disorder-associated GAP (MeGAP) and WAVE-associated RacGAP protein (WRP). Members of the WASP family of scaffold proteins such as WAVE participate in actin reorganization, especially during cell migration.

RhoGAP17 and RhoGAP44 (RICHs)

Agent RhoGAP17, or RhoGAP interacting with CDC42-interacting protein CIP4 homolog RICH1, is a CDC42GAP that localizes to tight and adherens junctions. It associates with components of epithelial polarity structure and maintains the integrity of tight junctions in epithelial cells by regulating CDC42 [1111].[303] Protein RhoGAP17 complexes with angiomotin (AMot),[304] a scaffold of the motin family of angiostatin-binding proteins, and polarity components such as membrane-associated guanylate kinase (MAGuK) P55 subfamily adaptor MPP5[305] and subapical, cell-polarity Par3 protein. Scaffolding MAGuKs as well as Par3 contain PDZ sequences. Adaptor MPP5 thus also heterodimerizes with Inactivation no after-potential D (INAD)-like protein (INADL),[306] a protein with 10 PDZ domains. Both MPP5 and PATJ tether apical transmembrane protein Crumbs homolog Crb3. Both Par3 and Crb3 complexes participate in establishment and maintenance of apical junctions. In addition, PATJ binds to tight junction components zona occludens protein ZO3 and claudin.

The RhoGAP domains of RhoGAP17 and RhoGAP44 specifically activate GTP hydrolysis of both Rac1 and CDC42, but not of RhoA [1112]. Protein RhoGAP44 (or RICH2) as well as ezrin of the ERM family of linkers between integral membrane and cytoskeletal proteins and ERM-binding protein EBP50[307] connect tetherin[308] to the apical actin network [1113]. Tetherin is a raft-associated, transmembrane protein expressed at the apical surface of polarized epithelial cells. Protein RhoGAP44 thus links membrane trafficking regions to apical actin network.

RhoGAP31

Protein RhoGAP31, or CDC42 GTPase-activating protein (CDGAP), targets Rac1 and CDC42 GTPases. Two isoforms are produced in distinct tissues: long (CDGAP$_L$) and short (CDGAP$_S$) subtypes. In mice, CDGAP$_L$ resides in the brain, heart, and lung; CDGAP$_S$ in the liver and kidney [1114]. Protein CDGAP is phosphorylated by extracellular signal-regulated kinase in response to platelet-derived growth factor [1114]. It thus links the Ras–MAPK pathway to Rac1 GTPase.

---

[303] The plasma membrane of epithelial cells is asymmetrically organized into apical and basolateral regions. These 2 regions are separated by tight junctions (Vol. 1 – Chap. 7. Plasma Membrane) that encircle cells and yield lateral contact with neighboring cells. Cell polarity structures are regulated via phosphorylation and Rho GTPases.

[304] Angiomotin supports migration of endothelial cells triggered by growth factors. The motin family of proteins comprises 3 members: angiomotin (AMot), angiomotin-like-1 (AMotL1), and angiomotin-like-2 (AMotL2). Alternative splicing in endothelial cells and other cell types raises the molecular diversity within this protein family.

[305] I.e., membrane palmitoylated protein-5; a.k.a. protein associated with Lin-7 (PALS1).

[306] A.k.a. PALS1-associated tight junction protein (PATJ).

[307] A.k.a. sodium–hydrogen antiporter-3 regulator NHERF1 and SLC9a3r1.

[308] A.k.a. bone marrow stromal cell antigen BSt2 and CD317.

RhoGAP33

Protein RhoGAP33, or TC10 and CDC42 GTPase-activating protein (TCGAP), specifically interacts with CDC42 and RhoQ (or TC10). It translocates to the plasma membrane in response to insulin in adipocytes [1115]. Insulin stimulates glucose uptake especially in adipocytes and myocytes by priming GluT4 translocation from intracellular storage vesicles to the plasma membrane via RhoQ GTPase. Agent RhoGAP33 inhibits insulin-stimulated GluT4 translocation and thus glucose uptake.

RhoGAP43

Agent RhoGAP43[309] stimulates the GTPase activity of RhoA, thereby reducing signaling from protein Tyr kinases to RhoA GTPase. It also precludes formation of CDC42- and Rac-induced actin structures [1116].

RhoGAP46

Agent RhoGAP46[310] connects to the small GTPase GTP-binding protein overexpressed in skeletal muscle (GEM).[311] The latter links to GTP-binding adpribosylation factor-domain protein ARD1, or tripartite motif protein TRIM23, and PDZ and LIM domain-containing protein PDLIM7 that is involved in assembly of actin filament-associated complex [251].

BCR1

Breakpoint cluster region protein (BCR or BCR1),[312] like SH3BP1, interacts with activated RhoA with a much higher Michaelis kinetic constant than RhoGAP1 and RhoGAP35 [1117]. It also links to CDC42 and Rac1, SOS GEF, mast and stem cell growth factor receptor (SCFR), GRB2, GRB10, GAB2, CBL, and CRKL adaptors, Abl1, Fes, HCK, and PI3Kγ kinases, cytosolic PTPn6 phosphatase, RB1 retinoblastoma protein, and talin-1 [251].

---

[309] A.k.a. SH3 domain-binding protein SH3BP1 (or 3BP1).

[310] A.k.a. GTP-binding protein overexpressed in skeletal muscle (GEM)-interacting protein (GMIP).

[311] A.k.a. GTP-binding mitogen-induced T-cell protein and kinase-inducible Ras-like protein (KIR).

[312] The Breakpoint cluster region (BCR) gene is involved in chromosomal translocations that cause the development of chronic myeloid leukemia and a subset of acute lymphoblastic leukemia.

ABR

Active breakpoint cluster region-related protein (ABR or MDB) is homologous to BCR RhoGEF and RhoGAP protein. Both target Rac and CDC42 GTPases. Proteins ABR and BCR specifically prevent Rac activity in macrophages, as small GTPases of the RHO superfamily are regulators of phagocytic function [1118].

RacGAP1

Rac GTPase-activating protein RacGAP1[313] corresponds to a heterotetrameric centralspindlin subunit. It interacts with RhoGEF31 for equatorial recruitment of RhoA GTPase that controls actomyosin dynamics during cytokinesis (mitosis) [1119].[314] Agent RacGAP1 also interacts with Rnd2 GTPase (or Rho7), Vav1 GEF, as well as GRB2 adaptor and 14-3-3-β, -γ, and -σ [251].

RalBP1

RalA-binding protein RalBP1[315] serves as CDC42 and Rac1 GTPase-activating protein [1120]. It enables a crosstalk between the Rho and Ras pathways via the Ras–RalGDS–Ral–RalBP1–CDC42/Rac1 cascade.[316]

Myosin-9

In mammals, subclass-9 myosins include 2 members (Myo9a and Myo9b) that both possess numerous differentially spliced variants. Unconventional myosin-9a is an actin-dependent nanomotor with a Rho GTPase-activating domain. It is expressed during development and in many adult tissues, with highest levels in brain and testis. It inhibits RoCK kinase. In vitro, Myo9a inactivates RhoA, RhoB, and RhoC GTPases.

In humans, myosin-9b is a nanomotor that is also a GTPase-activating protein for Rho GTPases. It is indeed a homolog of rat Myr5, an unconventional myosin that binds actin filaments in an ATP-dependent manner and has GAP activity mainly on

---

[313] A.k.a. male germ-cell RacGAP (MGCRacGAP), IDGAP, and Homo sapiens cytokinesis defect protein HSCYK4.

[314] Cytokinesis is related to the assembly and constriction of an actomyosin ring at the cell equator. It requires regulated changes in protein phosphorylation. Cyclin-dependent kinase CDK1 on the one hand and Polo-like kinase PlK1 and Aurora-B on the other regulate cytokinesis negatively and positively, respectively. Polo-like kinase PlK1 phosphorylates RacGAP1 that can then assemble with ECT2.

[315] A.k.a. Ral-interacting protein (RIP or RIP1) and 76-kDa Ral-interacting protein (RlIP76).

[316] Agent RalGDS and RalGDS-like protein are activators of both RalA and RalB GTPases.

Rho and CDC42 GTPases. (Myosin-9a is a homolog of Myr7.) Myosin-9b interacts with Arf1GEF BIG1 (ArfGEF1) [1121]. This linkage prevents RhoA binding to myosin-9b. Conversely, RhoA GTPase impedes ArfGEF1 tethering to myosin-9b.

InPP5f (OCRL1)

Oculocerebrorenal Lowe syndrome protein OCRL1[317] is an inositol polyphosphate 5-phosphatase (InPP5f) with phosphatidylinositol (4,5)-bisphosphate as a preferred substrate. It primarily localizes to the trans-Golgi network and possibly to lysosomes and endosomes [1122]. It avoids accumulation of PIP_2 that is involved in cell signaling, vesicle trafficking, and actin polymerization. OCRL1 forms a stable complex with Rac^{GTP} via its RhoGAP domain within the cell, especially in the trans-Golgi network. Upon PDGF and EGF-induced Rac activation, OCRL1 translocates from the trans-Golgi network to the plasma membrane and concentrates in membrane ruffles.

9.4.3 Guanine Nucleotide-Dissociation Inhibitors

In the absence of cell signaling, the major fraction of GDP-bound Rho GTPases localizes to the cytosol, bound to guanine nucleotide-dissociation inhibitors. Guanine nucleotide-dissociation inhibitors slow the rate of GDP dissociation from Rho GTPases, which remain inactive. They sequester Rho GTPases in the cytoplasm away from their regulators and targets, thereby preventing Rho GTPase translocation from the cytosol to the plasma membrane. Dissociation of monomeric GTPases from GDIs is required for membrane translocation and GEF-mediated activation.

9.4.3.1 RabGDIs

The cytosolic distribution, consequently membrane extraction and delivery, of Rab GTPases depend on their interaction with GDP-dissociation inhibitor (RabGDI) that binds to cytosolic, prenylated, GDP-bound (inactive) Rab proteins. The Rab–RabGDI complexes represent a pool of recycling Rab GTPases that can deliver Rab proteins to specific intracellular membranes on release from GDI to regulate intracellular vesicle transfer. The Rab^{GDP}–GDI linkage is characterized by a high affinity (dissociation constant K_d : $O[1 \text{ nmol}]$), only when Rab is both prenylated and in its inactive state. When active and/or unprenylated, this affinity decays by 3 to 4 orders of magnitude (K_d : $O[1 \text{ μmol}]$). In mammalian cells, 2 subtypes of RabGDI proteins are capable of binding the entire set of prenylated Rab GTPases, thereby extracting prenylated Rab GTPases from membranes and incorporating them in soluble Rab–GDI complexes.

---

[317] A.k.a. Lowe oculocerebrorenal syndrom (LOCR).

Membrane-associated GDI-displacement factor (GDF) can detach Rabs from GDIs. It especially targets Rab9–GDI complexes [1123]. To tether to the cytosolic surface of intracellular membranes, most Rab proteins possess 2 hydrophobic prenyl (geranylgeranyl) moieties at their C-terminus. These motifs are attached via the concerted action of Rab geranylgeranyl transferase and Rab escort protein (REP), a neccessary cofactor of Rab geranylgeranyl transferase that prenylates Rab, using geranylgeranyl pyrophosphate as a cosubstrate. Whereas Rab escort protein binds monogeranylgeranylated Rab7, GDI tethers both mono- and diprenylated Rab7 GTPases (dissociation constant $K_d = 1--5$ nmol) [1124].

Post-translational modifications of Rab proteins modulate the affinity for GDI, thereby causing disruption from GDI agents and precluding reformation of Rab–GDI complexes [1125]. Membrane association of endosomal Rab5 and Rab9 is accompanied by nucleotide exchange shortly after membrane association.

Phosphorylation modifies the membrane–cytosol partitioning of Rab proteins. In particular, Rab1a, Rab1b, Rab4, Rab7b, Rab8a, and Rab40 are phosphorylated by CDK1 during mitosis, thereby hindering intracellular transport at the onset of mitosis [1125]. Upon phosphorylation, the cytosolic concentration of Rab1 and Rab4 falls and rises, respectively. Phosphorylation of Rab6 by PKC also lowers the cytosolic concentration. Moreover, GDI molecules are also phosphorylated. For example, GDP-dissociation inhibitor GDI1 is phosphorylated by serum- and glucocorticoid-inducible kinase SGK1; Serine phosphorylation of GDI1 and tyrosine phosphorylation of GDI2 elevates the amount of cytoplasmic Rab4–GDI complex [1125].

Insulin stimulates glucose transporter GluT4 translocation in adipocytes and myocytes. Conversely, both RhoGDI1 and RhoGDI2 prevents GluT4 translocation, but RhoGDI1 connects to Rab10 with higher affinity [1126].

9.4.3.2 RacGDIs

In phagocytes, NADPH oxidase is activated by interaction of membrane-associated cytochrome-B559 (flavocytochrome) with its subunits P47PhOx[318] and P67PhOx as well as RacGDI1 that corresponds to RhoGDI1 [832]. In mastocytes, the Rac1–RhoGDI1 complex regulates secretion [832].

Angiogenesis requires polarized activation of Rac1 GTPase that must then locally dissociate from RhoGDI1 and associate with the plasma membrane. Polarized activation of Rac1 for endothelial cell migration relies on its interaction with its adaptor synectin that is also a binding partner for syndecan-4, a transmembrane heparan sulfate proteoglycan, as well as RhoG activation [1127]. Inactive RhoG

---

[318] Subunit Neutrophil P47PhOx (phagocytic oxidase) is also called neutrophil cytosol factor-1 and neutrophil NADPH oxidase factor-1. The NADPH oxidase complex of phagocytes is constituted by membrane-associated flavocytochrome-B559 and 4 cytosolic components: P40PhOx, P47PhOx, P67PhOx, and Rac GTPase.

is sequestered by RhoGDI1 that complexes with synectin and syndecan-4.[319] On the other hand, syndecan-4 intervenes in FGF2 signaling. Fibroblast growth factor-2 regulates angiogenesis via Rac1 activation. Clustering of syndecan-4 activates PKCα that phosphorylates RhoGDI1 (Ser96), thereby priming RhoG release, which can activate Rac1 GTPase.

9.4.3.3 RhoGDIs

Rho guanine nucleotide-dissociation inhibitors (RhoGDI) extract Rho GTPases from membranes,[320] prevent their activation, maintain a stable pool of soluble, inactive Rho GTPases in the cytoplasm, and protect them from degradation [1128].[321] In addition to transport vesicles, RhoGDIs transfer Rho GTPases from a given cellular membrane to another, thereby regulating the removal and delivery of Rho GTPases from and to their sites of action.[322]

In particular, Rho guanine nucleotide-dissociation inhibitors can bind both GDP- and GTP-bound forms of geranylgeranylated RhoA GTPase. They mask the prenyl group of RhoA, prevent RhoA to bind downstream effectors, retrieve RhoA from membranes, and stabilize the cytosolic pool of inactive RhoA GTPase [945]. Furthermore, RhoGDIs may escort translocating RhoA to specific membrane compartments. Active RhoA can also be labeled for degradation by the ubiquitin ligase SMAD ubiquitination regulatory factor SMURF1.

RhoGDI Family

At steady state, most Rho GTPases are bound in the cytosol to Rho-specific guanine nucleotide-dissociation inhibitors. In mammals, despite the numerous types of Rho GTPases, only 3 genes encode RhoGDIs. Hence, RhoGDIs constitute a family of 3 known abundant members (RhoGDI1–RhoGDI3).

---

[319] Syndecan-4 tethering to synectin increases synectin binding to RhoGDI1 and synectin enhances RhoGDI1 affinity for RhoG.

[320] Monomeric Rho GTPases that are prenylated act at membranes. Newly synthesized small GTPases of the RHO superfamily are geranylgeranylated and then post-translationally modified by the peptidase RAS-converting enzyme RCE1 and by isoprenylcysteine carboxyl methyltransferase (ICMT) at the cytoplasmic face of the endoplasmic reticulum. After geranylgeranylation by geranylgeranyl transferase, Rho proteins associate with RhoGDIs [1128].

[321] Small Rho GTPases that are neither bound to RhoGDIs nor associated with membranes are destroyed. Free prenylated cytosolic Rho GTPases are unstable; they are rapidly degraded by the proteasome.

[322] The rate of cycling of the RhoGDI–Rho complex between the cytosol and target membrane can be regulated by post-translational modifications on both Rho GTPase and RhoGDI protein.

Proteins RhoGDI1 (the most abundant) and RhoGDI2 are cytosolic proteins, whereas RhoGDI3 is linked to the actin cytoskeleton and vesicles in the perinuclear region [832].

Ubiquitous RhoGDI1 binds RhoA, RhoC, and RhoG, as well as Rac1, Rac2, and CDC42. Protein RhoGDI2, expressed in high levels in hematopoietic cells, but also in other cell types, binds RhoA.[323] Protein RhoGDI3, usually expressed at low levels, exclusively binds RhoB and RhoG at the Golgi body and other cellular membranes [945].

Regulation of RhoGDI–Rho Binding

The interaction between RhoGDIs and Rho GTPases can be regulated by several types of mechanisms. Some processes control the displacement of a single Rho GTPase from RhoGDI, whereas others remove all types of Rho GTPases.

In the presence of cellular membranes, RhoGDI has a lower affinity for Rho^{GTP} than Rho^{GDP} GTPases. Nevertheless, nucleotide-exchange may happen in the Rho–RhoGDI complex, as, at least, some GEFs may act on RHO superfamily members such as Rac1 complexed with RhoGDI proteins [1128]. In fact, RhoGEFs can assist in the release of Rho GTPases from RhoGDI agents. Both RhoGEFs and RhoGDIs interact with the switch-1 and -2 domains of Rho GTPases, hence connecting to RHO superfamily proteins in a mutually exclusive fashion.

The RhoGDI–Rho complex may be also dissociated by acidic lipids and phosphoinositides [1128]. Release of Rho GTPases from RhoGDIs can also be supported by proteins. Activated ezrin, radixin, and moesin proteins as well as TNFRSF16 interact with RhoGDIs and facilitate RhoA release [1128].[324]

In addition, phosphorylation of RhoGDI (Ser, Thr, and Tyr) promotes the liberation of Rho GTPases (Table 9.80). Enzyme P21-activated kinase (PAK) phosphorylates RhoGDI to remove Rac1 GTPase, but neither RhoA nor CDC42 GTPase [1128].

Upon stimulation by hepatocyte growth factor, diacylglycerol kinase-α promotes the formation of phosphatidic acid that recruits PKCζ and PKCι to the RhoGDI–Rac1 complex [1128]. Diacylglycerol kinase-ζ complexes with PAK, Rac1, and RhoGDI protein. In response to platelet-derived growth factor, diacylglycerol kinase-ζ augments the production of phosphatidic acid that stimulates PAK activity. Activated PAK then phosphorylates RhoGDI, thereby releasing Rac1 GTPase [1128].

---

[323] Protein RhoGDI2 associates with several Rho GTPases in vitro, but with significantly lower affinity than RhoGDI1. Many of these interactions are not detected in vivo [1128].

[324] The interaction of TNFRSF16 with RhoGDIs is enhanced by myelin-derived proteins, such as myelin-associated glycoprotein (MAG), or siglec-4a, and reticulon-4, also called reticulon-5, neuroendocrine-specific protein-C homolog (NSP or NSPcL), and neurite outgrowth inhibitor (NOGo).

Table 9.80. Phosphorylation of RhoGDI modulates the RhoGDI–Rho GTPase interaction (Source: [1128]; Fer: feline sarcoma kinase (Fes)-related Tyr kinase; PAK: p21-activated kinase; PKA/B/C/G: protein kinase-A/B/C/G; SLK, Ste20-like kinase; Src: sarcoma-associated kinase).

Complex type	Phosphorylation site	Kinase	Effect
RhoGDI–CDC42	Ser71	PKB	Dissociation
	Ser185	PKA	Increased association
	Tyr64	Src	Increased association
RhoGDI1–CDC42	Ser101, Ser174	PAK1	Dissociation
	Thr	PKCζ	Dissociation
	Tyr27, Tyr156	Src	Dissociation
RhoGDI2–CDC42	Tyr24, Tyr153	Src	Dissociation
RhoGDI–Rac1	Ser71	PKB	Dissociation
RhoGDI1–Rac1	Ser101, Ser174	PAK1	Dissociation
	Thr	PKCζ	Dissociation
	Tyr27, Tyr156	Src	Dissociation
	Tyr156	Fer	Dissociation
RhoGDI2–Rac1	Tyr24, Tyr153	Src	Dissociation
RhoGDI–RhoA	Ser188	PKA/G, SLK	Increased association
RhoGDI1–RhoA	Ser34, Ser96	PKCα	Dissociation
	Ser174	PKA	Inhibition of interaction
	Thr	PKCζ	Dissociation
	Tyr27, Tyr156	Src	Dissociation
RhoGDI2–RhoA	Tyr24, Tyr153	Src	Dissociation
RhoGDI1–RhoG	Ser96	PKCα	Dissociation
	Ser187	PKA	Increased association

Syndecan-4 (a heparan sulfate proteoglycan) and its intracellular adaptor synectin complex with RhoGDI that increases RhoGDI affinity for RhoG GTPase. Binding of syndecan-4 to its ligand, fibroblast growth factor FGF2, fosters the activation of PKCα that then phosphorylates RhoGDI, thereby releasing RhoG protein [1128].

RhoGDIs and Rho GTPase Crosstalk

In steady state, RHO superfamily GTPases reach an equilibrium between their ability to bind RhoGDIs, competition between different members of the RHO superfamily (CDC42, Rac1, and RhoA being the major members), and their turnover by degradation. Crosstalk between Rho GTPases via RhoGDIs determines the affinity and binding level of a given Rho GTPase for a given RhoGDI [1128].

For example, in smooth muscle cells, an augmented PKG-mediated phosphorylation (Ser188) of RhoA increases the binding of RhoA to RhoGDI, thereby preventing Rac1 binding; Rac1 then is activated to foster cell migration [1128].

RhoGDI1

Ubiquitous Rho guanine nucleotide-dissociation inhibitor RhoGDI1[325] sequesters in the cytoplasm CDC42, Rac1, Rac2, RhoA, RhoB, and RhoC [832, 1116]. Geranylgeranylation of C-terminal CAAX motif of RHO superfamily members is necessary for GDI interaction.

Protein RhoGDI1 promotes association of cytosolic RhoA with HSP70 chaperone. Protein RhoGDI1 not only keeps these Rho GTPases in an inactive state, but also contributes to proper folding of these cytosolic Rho GTPases, hence acting as chaperone, and protects them from degradation [1129]. Members of the RHO superfamily indeed compete to bind to RhoGDI, thereby avoiding proteasomal destruction.

In numerous mammalian cell types, RhoGDI1 depletion provokes misfolding and degradation of the cytosolic prenylated (geranylgeranylated) pool of Rho GTPases, but activates the remaining membrane-bound fraction. In addition, RhoGDI1 regulates Rho activity, as it mediates shuttling of Rho GTPases between membranes. Deficiency of RhoGDI1 sequesters Rho GTPases to the endoplasmic reticulum, hence reducing localization at the plasma membrane. Isotype RhoGDI2 does not compensate the loss of RhoGDI1 protein. Therefore, RhoGDI1 contributes to the regulation of crosstalk between RHO superfamily members, as well as their localization and degradation.

The RhoA–RhoGDI complex stability is regulated by RhoGDI phosphorylation. RhoGDI1 is phosphorylated (Ser34 and Ser101) by active P21-activated kinase PAK2 to dissociate Rac1 and CDC42 in response to FGF2 stimulation that primes neurite outgrowth [1130].

Cytoplasmic bone marrow Tyr kinase with its gene in chromosome X (BMX)[326] of the Bruton Tyr kinase (BTK) family, which participates in cytoskeletal reorganization and cell motility, promotes the removal of RhoA (but not Rac1 and CDC42) from RhoGDI1 in a kinase-independent manner [1128]. Kinase BMX interacts with RhoA and competes for its binding to RhoGDI agents.

Protein RhoGDI1 may connect to $CDC42^{GTP}$ or $RhoA^{GTP}$ GTPases [832]. It can associate with a complex composed of Rac1, PI(4)P5K1, and diacylglycerol kinase [832].[327] Protein RhoGDI1 can also associate with a complex that contains ezrin, radixin, and moesin, which link actin filaments to the plasma membrane, as well as cell-surface glycoproteic ERM partner CD44 [832]. Hence, RhoGDI1 release from RHO superfamily GTPases can involve phosphatidylinositols and members of the ERM family.

---

[325] A.k.a. RhoGDIα and aplysia Ras-related homolog GDI ArhGDIα.

[326] A.k.a. epithelial and endothelial Tyr kinase (ETK).

[327] Diacylglycerol kinase phosphorylates diacylglycerol to produce phosphatidic acid that stimulates PI(4)P5K1. Lipid $PI(4,5)P_2$ may cause the release of Rac from RhoGDI1 protein. Protein Rac can then be activated by GEFs.

RhoGDI2

Protein RhoGDI2[328] is specifically synthesized in hematopoietic cells, predominately in B and T lymphocytes [832, 1116]. Cleavage of RhoGDI2 by interleukin-1β-converting enzyme-like peptidases prevents its activity on Rho GTPases.

Protein RhoGDI2 targets RhoA, Rac1, and CDC42, but only forms transient complexes. It is less efficient (by a factor of 10–20) than RhoGDI1 in inhibiting GDP–GTP exchange and causes membrane dissociation of these GTPases [832].

RhoGDI3

Protein RhoGDI3[329] is restricted to the brain, lung, and testis [832]. It targets RhoB and RhoG, but not Rac [832, 1116]. In humans, RhoGDI3 is also able to interact with RhoA and CDC42, but with a lower affinity than that of RhoGDI1.

---

[328] A.k.a. RhoGDIβ, developmentally regulated GDI (D4GDI), and lymphocyte GDI (LyGDI). Expression of RhoGDI2 in hematopoietic cell lines is modulated by cell fate. Cell division and differentiation indeed augment RhoGDI2 production.

[329] A.k.a. RhoGDIγ.

Chapter 10
Other Major Types of Signaling Mediators

Other important signaling mediators include: (1) endogenous, gaseous, diffusible messengers, or *gasotransmitters*, such as *carbon monoxide* (CO; Sect. 10.2), *nitric oxide* (NO; Sect. 10.3), and *hydrogen sulfide* (H_2S; Sect. 10.4), in addition to *oxygen* and *carbon dioxide*, as well as several *reactive oxygen* (ROS) and *nitrogen species* (RNS; Sect. 10.6), that act as intra- and intercellular regulators; (2) membrane-bound enzymatic complexes such as the reduced form of *nicotinamide adenine dinucleotide phosphate oxidase* (NADPH oxidase or NOx; Sect. 10.5); (3) some *coregulators*, such as (protein kinase) *A-kinase-anchoring proteins* (AKAP; Sect. 10.8.4) and *annexins* (Sect. 10.8.5); and (4) *transcription factors*[1] involved in stress response, such as the proteic complex nuclear factor κ light-chain-enhancer of activated B cells (NFκB; Sect. 10.9.1), the heterodimer hypoxia-inducible factors (HIF; Sect. 10.9.2), and members of the Forkhead box (Fox; Sect. 10.9.3) and P53 families (Sect. 10.9.4; Table 10.1).

More than 2000 transcription factors are encoded in the human genome. Some transcription factors reside permanently in the nucleus and are constitutively active (functional class **1**). Regulatory, conditionally-active transcription factors (class **2**) constitute 2 sets: developmentally regulated transcription factors (class **2.A**), often cell type-specific, and signal-dependent transcription factors (class **2.B**) that are subdivided into extracellular ligand-dependent nuclear receptors, or nuclear factors (class **2.B.1**), among which some are activated by certain types of hormones; internal signal-regulated (class **2.B.2**); and cell-surface receptor-regulated (class **2.B.3**) transcription factors. The latter category is split into nuclear resident (class **2.B.3.a**),

---

[1] Transcription factors that contain at least 1 DNA-binding domain regulate gene transcription, during which messenger RNA is synthesized from DNA by RNA polymerase (Vol. 1 – Chap. 5. Protein Synthesis). Coactivators, corepressors, chromatin remodelers, and histone acetylases, deacetylases, kinases, and methylases, participate in the regulation of gene transcription, but lack DNA-binding domains.

Table 10.1. Classification of some transcription factors based on their function and mode of activation (Sources: [1131, 1132]; bHLH: basic helix–loop–helix; bZip: basic leucine zipper; AP1: Activating protein-1 (Fos and Jun components); ATF: activating transcription factor; β-Ctn: β-catenin; C/EBP: bZip CCAAT [(cytidine)$_2$–(adenosine)$_2$–thymidine) box motif]/enhancer-binding protein; CREB: cAMP response element-binding protein; DREAM: Downstream regulatory element antagonistic modulator; ER: estrogen receptor; ETS: E twenty-six; Fox: Forkhead box; GATA: DNA GATA sequence-binding transcription factor; GR: glucocorticoid receptor; GTF: general transcription factor; HIF: hypoxia-inducible factor; HNF: hepatocyte nuclear factor; MeCP2: Methyl-CpG-binding protein-2 transcriptional repressor; MEF: myocyte enhancer factor; MRF: bHLH myogenic regulatory factor; MyF: myogenic factor [MRF set]; MyoD: myocyte differentiation transcription factor [MRF set]; NF1: nuclear factor-1; NFAT: nuclear factor of activated T cells; NFκB: nuclear factor-κB; ONR: orphan nuclear receptor; PiT1: pituitary-specific positive transcription factor-1, or POU domain-containing class-1 transcription factor 1 [POU1F1]; PPAR: peroxisome-proliferator-activated receptor; RAR: retinoic acid receptor; RXR: retinoid X receptor; SMAD: small (son of, similar to) mothers against decapentaplegia homolog; SP1: Specificity protein-1; SREBP: sterol regulatory element-binding protein; SRF: serum response factor: STAT: signal transducer and activator of transcription; TR: thyroid hormone receptor; TuLP: Tubby-like protein).

Constitutively active, nuclear transcription factors	Developmentally regulated	Conditionally active transcription factors				
			Signal-dependent			
		Nuclear factors	Internal signal-regulated	Cell-surface receptor-dependent		
				Resident nuclear	Shuttle	Latent cytoplasmic
C/EBP	GATA	ER	ATF6	AP1	DREAM	βCtn
GTF2	HNF	GR	HIF	ATMs	FoxO	Gli
NF1	Hox	PR	P53	CREB		MITF
SP1	MyF1/5/6	TR	SREBP	E2F1		NFAT
	MyoD (MyF3)	PPAR	ONRs	ETS		NFκB
	PiT1	RAR		MeCP2		SMAD
	Winged-helix (FoxA-FoxQ)	RXR		MEF2		STAT
				MyC		TuLP
				SRF		Notch^{ICD}

shuttle (between the nucleus and cytoplasm; class **2.B.3.b**), and latent cytoplasmic (class **2.B.3.c**) transcription factors that translocate into the nucleus upon receptor activation.

10.1 Gaseous Neurotransmitters

Different categories of chemical species constitute a set of 50 to 100 neuro- and gliotransmitters, neuromodulators, and cotransmitters, in addition to acetylcholine and adenosine [1133]: (1) biogenic amines (e.g., adrenaline, noradrenaline, dopamine, histamine, serotonin, and tryptamine); (2) amino acids (aspartate, glutamate, and glycine, as well as ^{D}aspartate and ^{D}serine, which is released by protoplasmic [grey matter] type-2 astrocytes close to NMDA-type glutamate receptor synapses), in addition to GABA synthesized by glutamic acid decarboxylase; (3) peptides (enkephalins, substance-P, and other neuropeptides, especially those of neurons of the digestive tract, such as vasoactive intestinal polypeptide, cholecystokinin, gastrin, and insulin, the enteric nervous system using nearly all transmitters of the central nervous system); (4) gases (CO, NO, and H_2S).

Most often, neurotransmitters are stored in nerve terminals and signal via both fast and slow mechanisms, using ionotropic and metabotropic receptors, respectively. Signaling is terminated by specific reuptake carriers and/or degradation.

Nitric oxide functions as a neuromodulator and/or -transmitter in the central, peripheral, and enteric nervous systems. In the enteric nervous system, NO is produced in enteric neurons by NOS1 subtype. It acts as an inhibitory neurotransmitter. It diffuses from neurons to adjacent smooth muscle cells, in which it activates soluble guanylate cyclase, thereby relaxating smooth muscle cells.

In cardiac nodal cells, nitric oxide supports acetylcholine release to augment vagal neurotransmission and trigger bradycardia [1134]. Isoform NOS1 is detected in parasympathetic ganglion that innervates the sinoatrial node. Nitric oxide acts presynaptically to facilitate vagal neurotransmission via the cGMP–PDE3 pathway that promotes cAMP–PKA-dependent phosphorylation of presynaptic $Ca_V2.2$ channels to increase presynaptic calcium influx and acetylcholine release. In addition, NO can function postsynaptically via a cGMP-dependent increase in the hyperpolarization-activated current in sinoatrial cells. Moreover, NO can be generated upon M_2 muscarinic receptor stimulation of NOS3 and preclude the adrenergic stimulation of $Ca_V1.2$ channels via the PDE2–cAMP axis [1134].

Once glutamate is bound to NMDA-type receptor ion channel, calcium enters the neuron, binds calmodulin to activate NOS1 synthase. Freely diffusible NO enters adjacent neurons or other cell types, links to iron in heme of soluble guanylate cyclase at the postsynaptic membrane near NMDA-type glutamate receptors to form cGMP messenger [1133].

Hydrogen sulfide (H_2S) acts on serotoninergic neurons in the dorsal raphe nucleus [1133].

Like other gasotransmitters, CO is neither stored in synaptic vesicles nor released by exocytosis, and do not act at postsynaptic membrane receptors. In specific neurons of the brain such as olfactory neurons, CO is synthesized by constitutive heme oxygenase (HO2) and can activate soluble guanylate cyclase. Carbon monoxide is also produced by heme oxygenase in collaboration with cytochrome-P450 reductase and biliverdin reductase. Heme oxygenase has inducible (HO1) and constitutive (HO2) isoforms. Heme oxygenase-1 is involved in cellular responses to stress. Heme oxygenase-2 resides in neurons of the brain, enteric neurons, and interstitial pacemaker cells of Cajal in the the gastrointestinal tract. Interstitial cells of Cajal produce low NOS1 levels. Interstitial cells of Cajal, enteric neurons, and intestinal smooth muscle cells interact to cuase smooth muscle relaxation.

Nitric oxide and carbon monoxide operate as coneurotransmitters produced by the same neurons not only in the brain, but also in the enteric nervous system. Their respective synthases, neuronal NO synthase (NOS1) and heme oxygenase-2 (HO2), colocalize in enteric neurons [1135]. In small intestinal (jejunal and ileal) smooth muscle cells, NO signaling requires CO generation for muscle relaxation and inhibitory non-adrenergic non-cholinergic neurotransmission that controls intestinal motility.

10.2 Carbon Monoxide and Heme Oxygenases

Carbon monoxide (CO), a stable, lipid-soluble gas, is produced by heme oxygenase (HO), the rate-limiting enzyme of the catabolism of heme [1136].[2] Although CO can bind to heme-containing proteins, or *hemeproteins*, (but not vesicles) for transient storage and later release,[3] the predominant source is de novo synthesis [1137]. It then diffuses according to the local partial pressure gradient. Carbon monoxide binds to the ferrous iron of heme and, then, influences the activity of all hemeproteins.

10.2.1 Carbon Monoxide Synthesis and its Regulation

Two CO sources — HO-dependent (major) and HO-independent — exist. The latter results from oxidation of organic molecules, phenols, and flavenoids and peroxidation of lipids during severe stress [1136].

---

[2]The heme oxygenase system includes HO and biliverdin reductase, as well as products of heme degradation: CO, iron, *biliverdin*, and *bilirubin*. The catabolism of heme, especially of hemoglobin of senescent or damaged erythrocytes in the spleen by macrophages. leads to the formation of equimolar amounts of biliverdin, free iron, and carbon monoxide. Biliverdin, a green (Latin viridis, French vert: green) tetrapyrrolic bile pigment, is rapidly converted by *biliverdin reductase* to bilirubin, a yellow (Latin ruber = red, ruddy) tetrapyrrolic catabolite and another bile pigment. (80 to 85% of the bilirubin derives from hemoglobin. Preferred cofactor of biliverdin reductase, which undergoes autophosphorylation, is NADPH (rather than NADH). Bilirubin is excreted in bile and, then, feces, as well as urine. When it is oxidized, it reverts to biliverdin. In fact, in the liver, bilirubin is conjugated with glucuronic acid by ^{UDP}glucuronyl transferase, to be soluble in water. Once excreted into bile, it goes into the small intestine. Conjugated bilirubin can be processed in the large intestine by colonic bacteria to *urobilinogen*, which is further metabolized to *stercobilinogen*, and finally oxidized to *stercobilin*, which is eliminated in feces. Yet, a fraction of urobilinogen is reabsorbed and excreted in the urine together with its oxidized form, *urobilin*.

[3]Hemeproteins, or hemoproteins, are metalloproteins that have diverse biological functions, such as oxygen transport (hemoglobin, myoglobin, neuroglobin, and cytoglobin), metabolism (peroxidases, nitric oxide synthases, cyclooxygenases, cytochromes-P450, and cytochrome-C oxidase), membrane transport (cytochromes), electron transfer (cytochrome-C and catalase), and sensing (nitric oxide sensor and soluble guanylate cyclase receptor).

Heme oxygenase catalyzes heme degradation to carbon monoxide, iron, and biliverdin with the oxidation of reducing agents NADH and NADPH.[4] Heme oxygenase is responsible for the recycling of iron from senescent and damaged erythrocytes and other cell types, such as hepatocytes. Heme oxygenase allows to limit the overall production of reactive oxygen species, as excess free heme catalyzes the formation of these species. Among other agents, atrial natriretic peptide and donors of nitric oxide are important modulators of the heme–heme oxygenase couple [1138].

Heme oxygenase has 2 active membrane-bound isoforms encoded by 2 distinct genes: inducible HO1 and constitutive HO2 subtypes. Therefore, HO2 participates in the immediate response and HO1 in long-term regulation.[5] Isozyme HO1 acts as a heat shock and stress protein induced by several agents that cause oxidative damage. Induction of HO1 coupled with ferritin synthesis yields a rapid, protective, anti-oxidant response. In skin fibroblasts, it protects against UV-induced oxidative stress [1136]. Heme oxygenases are anchored to the endoplasmic reticulum membrane. They also localize to the nuclear envelope and perinuclear region.

10.2.1.1 Heme Oxygenase-1

Heme oxygenase-1 is expressed basally only in the liver and spleen, but can be produced in most tissues. It participates in the degradation of erythrocytes and elimination of toxic heme.

In pulmonary artery endothelial cells, HO1 resides in plasma membrane caveolae. Caveolin can bind to HO1 and regulate its activity.

Oxidative stress upregulates HO1 in various tissues.[6] Cell-specific inducers of HO1 includes heme, cobalt protoporphyrin, hemin (oxidized heme),[7] ROS and RNS (e.g., peroxynitrite), transition metal complexes (iron $FeCl_2$ and $FeCl_3$ and cobalt $CoCl_2$), NO, and oxidized lipids. Unlike in other cell types, in cerebral

---

[4]Both intracellular heme and iron contribute to the regulation of many cellular functions. The concentration of heme is regulated by the synthesis by rate-limiting enzymes δ-aminolevulinic acid (ALA) synthase and porphobilinogen deaminase on the one hand and degradation by heme oxygenase on the other. Heme released in excess from mitochondria into the cytoplasm suppresses ALA synthesis, impedes ALA transport to mitochondria, and enhances the heme degradation rate [1136]. However, in erythroid cells, excess heme improves hemoglobin synthesis.

[5]Production of HO1 primed by pharmacological agents or gene transfer of human HO1 into endothelial cells in vitro fosters the cell cycle progression and attenuates the effect of angiotensin-2 and DNA damage caused by TNF and heme; in vivo administration corrects blood pressure elevation triggered by angiotensin-2 [1138].

[6]The Ho1 gene pertains to the category of early response genes regulated by stress (hyperthermia and hypoxia) and the anti-oxidant response in conjunction with the redox-sensitive transcription factor NFE2-related factor NRF2, nuclear factor-κB, and cAMP-responsive element.

[7]Cellular heme is produced in the reduced state.

vascular endothelial cells, hydrogen peroxide, arachidonic acid, NO donors (3-morpholinosydnonimine and ^{S}nitroso ^{N}acetyl dl-penicillamine), transition metals, TNFα, and glutamate fail to upregulate HO1 enzyme.

Acute and chronic productions of HO1 reduces concentrations of vasoconstrictors such as 20HETE. In particular, HO1 impedes the activity of heme-containing thromboxane synthase and cyclooxygenase-2 [1136]. Chronic, modest expression of HO1 increases adiponectin, hence, resistance to oxidative stress.

10.2.1.2 Heme Oxygenase-2

Heme oxygenase-2 has its highest expression in the central nervous system and testes. Consequently, activity of heme oxygenase in the brain exceeds that in other organs. In the brain, HO2 is highly produced in vascular endothelial and smooth muscle cells as well as perivascular astrocytes.[8] Carbon monoxide is also synthesized in neurons. In the brain, HO2 associates with guanylate cyclase, ALA synthase, cytochrome-P450 reductase, and nitric oxide synthase [1136]. In addition, HO2 concentration is relatively high in keratinocytes, but low in human dermal fibroblasts.

Like other gaseous mediators, CO causes a vasodilation of cerebral arterioles. Certain vasodilatory stimuli, such as hypoxia, reduced arterial pressure in the autoregulation range, and ADP, raise CO concentration owing to a rapid increase in HO2 activity, but not HO2 synthesis. Only glucocorticoid hormones augment HO2 production, as a glucocorticoid response element exist in the Ho2 gene promoter [1137]. Glucocorticoids are slow-acting, weak inducers of HO2 enzyme.

Elevated heme concentration, Ca^{++} influx, and Ca^{++}–calmodulin-dependent processes also elevate HO2 activity [1137]. Post-translational phosphorylation (activation) of HO2 indeed depends on Ca^{++}–calmodulin. On the other hand, bilirubin may yield a negative feedback on HO2 function. In hippocampal and olfactory neurons, heme oxygenase-2 is activated by casein kinase-2 and protein kinase-C.

In endothelial cells, HO2 is stimulated by protein Tyr kinases as effectors of the Glu–iGluR axis. In fact, CO synthesized by HO2 in astrocytes is responsible for glutamate-induced dilation of pial arterioles. The CO production in endothelial cells and astrocytes rises in response not only to glutamate, but also tumor-necrosis factor-α.

Brain production of CO causes an accumulation in the cerebrospinal fluid, especially in the cortical, periarachnoid space (basal CO concentration 50–80 nmol).

In cerebral vascular endothelial cells, TNFα rapidly activates HO2 via ROS possibly via a thiol–disulfide redox switch (Sect. 10.6) [1137]. Agents ROS are mainly produced by NADPH oxidase NOx4 isoform.

---

[8]In cerebral vessels, HO2 is synthesized in endothelial cells, adjacent astrocytes, and, to a lesser extent, vascular smooth muscle cells.

Table 10.2. Effects of carbon monoxide and other products of heme oxygenase activity (Source: [1136]; ⊕: stimulation; ⊖: inhibition ↑: increase; ↓: decrease; ND: not determined; AP1: activator protein-1; BK: high-conductance, Ca^{++}-activated, voltage-gated K^+ channel; CO: carbon monoxide; COx: cyclooxygenase; CyP: cytochrome-P450; GSH: anti-oxidant glutathione (reduced form); HETE: hydroxyeicosatetraenoic acid; HO: heme oxygenase; MAPK: mitogen-activated protein kinase; ^{Mn}SOD: manganese-containing superoxide dismutase; PKC: protein kinase-C; ROS: reactive oxygen species; RSK, ribosomal S6 kinase; sGC: soluble guanylate cyclase; SOD_{ec}: extracellular superoxide dismutase; TxaS1: thromboxane-A synthase-1 [or CyP5a1]). Uncoupling of nitric oxide synthase (NOS) can result from glucose that enhances superoxide (O_2^-) production. The latter promotes the formation of peroxynitrite ($ONOO^-$), which oxidizes NOS cofactor tetrahydrobiopterin, thereby uncoupling NOS, i.e., favoring O_2^- production instead of NO synthesis. Prostaglandin-A is a potent inducer of HO synthesis to protect against stress and hypoxia, particularly in cardiomyocytes. Hypoxia fosters the production of HO in vascular smooth muscle cells but limits that of endothelin-1, platelet-derived growth factor-B, and vascular endothelial growth factor in endothelial cells.

Target	Effect
BK channel	Vasodilation
SOD_{ec}	↑
GSH	⊕, especially in endothelial cells
20HETE	⊖ (hence, vasodilation)
P38MAPK	⊖ cell proliferation,
	⊖ inflammation, apoptosis
^{Mn}SOD	Anti-oxidant
RSK	⊖ apoptosis
sGC	cGMP synthesis, vasodilation
ND	Angiogenesis
	Heme–HO system
HO	⊖ CyP (TxaS1), COx (heme degradation, resulting vasodilation)
Biliverdin	↓ oxidative and nitrosative stress
Bilirubin	ROS scavenger, inhibition of NADPH oxidase and PKC, anti-inflammatory effect
Biliverdin reductase	AP1 complex activation

10.2.2 Vasodilation and Other Effects

Carbon monoxide not only causes vasodilation, but also supports action of vasodilators and attenuates that of vasoconstrictors such as platelet-activating factor [1137]. Its vasorelaxant effect results not only from a cGMP-dependent, but also cGMP-independent, stimulation of some K^+ channels and an increase in adiponectin (NO-dependent vasodilation) [1136] (Table 10.2).

Carbon monoxide relaxes vascular smooth muscle cells via BK channels. It indeed binds to BK channel and increases Ca^{++} sparks that activate BK channels. In the arterial wall, the transient BK current provokes a membrane hyperpolarization that reduces the activity of voltage-dependent Ca^{++} channels. Both CO and CO-releasing molecule CORM3 stimulate BK channels [1136].

Both NO and CO activate BK channels on vascular smooth muscle cells, but targeting different channel subunits. Heme possesses a high affinity for the heme-binding domain (Cys-Lys-Ala-Cys-His) of the α subunit of BK channel. Binding of heme to the heme-binding domain inhibits the BK channel. Carbon monoxide tethers to channel-bound reduced heme, thereby relieving heme-induced inhibition of the BK channel and leading to channel activation [1137]. In addition, CO binding increases Ca^{++} sensitivity of the BK channel. Furthermore, HO enzyme and BK channel colocalize; HO1 and HO2 can reduce inhibition caused by heme, as they destroy heme.

On the abluminal side of the neurovascular unit, glutamate binds to ionotropic (iGluR) and metabotropic (mGluR) glutamate receptors of astrocytes, causing a Ca^{++} influx and increasing HO2 activity. On the luminal side, endothelial nitric oxide and prostacyclin contribute to the response. The former activates guanylate cyclase, thereby producing cGMP messenger. The latter stimulates the PGI2 receptor (IP) that triggers the ACase–cAMP pathway. Carbon monoxide is much less efficient in activating guanylate cyclase than nitric oxide. Nevertheless, cGMP acts as a mediator of CO-induced vasodilation. Activated protein kinase-G can phosphorylate ryanodin receptors of the sarcoplasmic reticulum, thereby increasing Ca^{++} spark frequency and BK channel activity.

In the carotid body, HO2 colocalizes with BK channels. Carbon monoxide generated by HO2 may mediate O_2 sensitiviy of BK channels in glomus cells of the carotid body [1137]. Carbon monoxide may also serve as a neurotransmitter in the carotid body [1136]. However, HO2 is not necessary for O_2 sensing by cells of the carotid body and adrenal medulla.

During hypoxia, CO reduces the proliferation of endothelial cells, as it impedes VEGF production by adjacent vascular smooth muscle cells, as well as that of vascular smooth muscle cells via cGMP [1137]. Cytosolic messenger cGMP activates P38MAPK, thus upregulating caveolin-1 that prevents cell proliferation. In a different context, CO can promote the proliferation of microvascular endothelial cells.

Heme oxygenase-1 and its product CO participate in the maintenance of the cell homeostasis, including DNA stability and repair to ensure cell survival [1139]. DNA repair pathways encompass repair of single- and double-strand DNA breaks (Vol. 1 – Chaps. 4. Cell Structure and Function and 5. Protein Synthesis). Double-strand DNA breaks are detected initially by damage sensors that trigger the recruitment to altered DNA of proper kinases and their activation.[9] The HO1–CO axis contributes to DNA repair of double-strand DNA breaks via ATMK kinase.

Endogenous CO at nanomolar levels primes the production of anti-oxidant enzymes, such as heme oxygenase-1 and mitochondrial superoxide dismutase SOD2, to protect against oxidative stress [1140]. Hence, pro-oxidant and cytotoxic CO primes a protective response.

Gaseous signaling molecules carbon monoxide and nitric oxide also stimulates mitochondrial genesis via cytochrome-C oxidase. In the heart, carbon monoxide works via transcriptional activation of peroxisome proliferator-activated receptor-γ coactivator PGC1α, nuclear respiratory factor-1 and -2, and mitochondrial transcription factor-A (TFaM) [1140].

10.2.3 Carbon Monoxide and Nitric Oxide

Endothelial nitric oxide and prostacyclin cooperate with the HO–CO axis to regulate the blood circulation. Nitric oxide heightens HO2 activity in cerebral microvessels using a cGMP-dependent mechanism. In aortic endothelial cells, NO stimulates the CO production. However, CO inhibits NO-mediated relaxation when the NO concentration rises. On the other hand, CO stimulates soluble guanylate cyclase and cGMP when the NO concentration falls [1136]. Vasodilatory effect of CO thus depends on BK channel activity as well as NO concentration and soluble guanylate cyclase.

In addition, in cerebral arterioles, whereas acute production of CO causes a vasodilation, prolonged exposure of cerebral arterioles to elevated CO produces progressive constriction, as it inhibits nitric oxide synthase [1137]. Moreover, NO can hinder HO2 activity.

The HO–CO axis protects the vasculature. Carbon monoxide precludes apoptosis and generation of endogenous oxidants [1137].[10] Carbon monoxide binds to a heme prosthetic group and regulates components of cell signaling, such as BK channels and guanylate cyclase. It inhibits NADPH oxidase and cytochrome-C oxidase of the mitochondrial respiratory chain, thereby hampering ROS formation. In addition, bilirubin is a potent scavenger of reactive oxygen species.

Whereas NO can simulate cystathionine γ-lyase, this enzyme can be inhibited by CO [1137]. Gasotransmitter H_2S can inhibit HO2 in aortic endothelial cells.

---

[9] Both ataxia telangiectasia mutated kinase and ataxia telangiectasia and Rad3-related kinase can phosphorylate multiple targets (e.g., P53-binding protein-1, breast cancer protein-1 (BrCa1), checkpoint kinases ChK1 and ChK2, and the MRN mediator complex) as well as histone-2A to enable the recruitment of additional DNA damage-responsive proteins.

[10] On the other hand, high, toxic levels of exogenous CO release superoxide anions from mitochondria. In addition, excessive CO competitively removes NO from hemoproteins and can lead to the production of $ONOO^-$ and tissue damage [1136].

10.3 Nitric Oxide, Nitric Oxide Synthases, Nitrite, and Nitrate

Nitric oxide (NO) a highly diffusible gas that runs very rapidly from its site of synthesis and crosses quickly cell membranes. Nitric oxide can act as an intra- and autocrine second messenger as well as a paracrine regulator on neighboring cells.

However, NO is a labile vasodilator with a limited half-life ($O[1\,s]$); its effects are thus localized close to the site of release. This free radical is highly reactive with other chemical species, notably oxygen, superoxide, and iron-containing hemes that act as NO scavengers.

This stable radical possesses an unpaired electron (hence the notation NO•).[11] Extracellular nitric oxide was first designated as endothelium-derived relaxing factor, as this messenger is involved in the regulation of vascular tone and blood flow (Vol. 5 – Chap. 9. Endothelium).[12]

The primary source for NO synthesis in the blood circulation is endothelial nitric oxide synthase. Nitric oxide is generated from the enzymatic oxidation of ^{L}arginine by nitric oxide synthase. In particular, synthesized endothelial NO contributes to the control of the vasomotor tone and, hence, blood pressure. However, NO can also be produced from the reduction of nitrite and nitrate. Inorganic nitrate derives either from NO oxidation or from diet. It can serve as a storage form of reactive nitrogen oxides that can be reduced back to nitrite and nitric oxide. Release of NO from stored nitrite and nitrate compensate the very short half-life of NO [1141].

Nitric oxide can complex with all transition metals, the so-called *metal nitrosyl complexes* (M^{NO}), or simply metal nitrosyls (e.g., Cu^{++}, Mn^{++}, and Fe^{3+}).[13] Nitric oxide interacts with metalloproteins, especially those that contain hemes, and with thiols. Nitric oxide can act in the form of free radical (NO or NO•) or as chemical species, such as nitrosyl (NO^+) or nitroxyl (NO^-) ions or ^{S}nitrosothiols (SNO). Hemes either transform NO into nitrate or promote its transformation into ^{S}nitrosothiols.

---

[11] The chemical structure of nitric oxide (NO) is intermediate between molecular oxygen (O_2) and nitrogen (N_2). A single unpaired electron exists in nitric oxide (•N–O), 2 in oxygen (•O–O•), and 4 in nitrogen (•̇N–Ṅ•). In nitrogen, one of the most inert molecules known, unpaired electrons in antibonding orbitals are repulsive, tend to destabilize the molecule, and counteract the 3 bonding orbitals that are more stable and promote the bonding of the 2 N atoms into N_2 molecule. Nitric oxide has only one unpaired electron, which allows it to bind strongly to the iron in heme groups, which serves in the activation of guanylate cyclase and slowing mitochondrial respiration by binding to cytochrome-C oxidase.

[12] Flow-mediated dilation refers to the link between an increase in local blood flow, mechanical stress, and NOS3 and NO activity.

[13] Metal nitrosyls follows the same bonding rules like metal carbonyl complexes (M^{CO}), or metal carbonyls, formed by the linkage between transition metals and carbon monoxide, such as hydrogenases that contain CO bound to iron.

Nitric oxide controls: (1) at the tissue level, neurotransmission and vascular tone; (2) at the cell level, gene transcription (as it binds to iron-responsive elements) and mRNA translation; and (3) at the molecule level, post-translational modifications of proteins (e.g., adpribosylation).

10.3.1 Nitric Oxide Synthases

Nitric oxide is synthesized by nitric oxide synthases (NOS). Three nitric oxide synthases (NOS1–NOS3) differs according to subcellular location, gene expression, protein interactions, post-translational modifications, and catalytic behavior for specific roles. Yet, these different isoforms share a similar synthesis mode that uses ^{L}arginine and molecular oxygen (O_2), reduced nicotinamide adenine dinucleotide phosphate (NADPH) as substrates and flavin adenine dinucleotide (FAD), flavin mononucleotide (FMN), and (5,6,7,8)-tetrahydrobiopterin (BH4) as cofactors to form nitric oxide. All NOS isoforms bind to calmodulin. They contain heme. Moreover, these widespread isozymes can coexist in many cell types. In particular, NO is continuously synthesized and released from the vascular endothelium.

Nitric oxide synthases require oxygen and other cofactors, especially tetrahydrobiopterin and Ca^{++}–calmodulin for efficient production of nitric oxide.[14] Both NADPH and O_2 serve as cosubstrates. In addition to pteridine tetrahydrobiopterin, NO synthesis also relies on other cofactors, such as flavin mononucleotide and adenine dinucleotide, and reduced thiols. Synthesis of NO depends on substrate and oxygen availability, hence on blood supply. It is altered by endogenous NOS inhibitor asymmetric dimethylarginine.

In the absence of pteridine tetrahydrobiopterin, NOS3 dimer is uncoupled from ^{L}arginine oxidation and oxygen is reduced to form superoxide anion (O_2^-) instead of nitric oxide.

Active nitric oxide synthases are homodimers, whatever the isoform. Enzymes NOS are hemeproteins; heme irons can bind NO in both ferric or ferrous states within seconds from the start of NO synthesis. Nitric oxide binding to heme reversibly inhibits NOS catalysis (negative feedback by the inactive NOS–NO complex). Displacement of NOS away from the cell cortex can decrease its activity.

Nitric oxide synthases transfer electrons from reduced NADPH to FAD and FMN in the C-terminal reductase domain to heme in the N-terminal oxygenase domain. On the other hand, NOS has a limited capacity to reduce O_2 to superoxide.

---

[14]Tetrahydrobiopterin, synthesized from guanosine triphosphate, is a cofactor of 3 hydroxylases — phenylalanine [PAH], tryptophan [TPH], and tyrosine hydroxylase [TH] — that convert ^{L}Trp into 5-hydroxytryptophan, ^{L}Phe into ^{L}Tyr, and ^{L}Tyr into ^{L}(3,4)-dihydroxyphenylalanine (^{L}DOPA), respectively. They then operate in the synthesis of neurotransmitters adrenaline, noradrenaline, dopamine, melatonin, and serotonin, in addition to be a cofactor for the production of nitric oxide by nitric oxide synthases.

The oxygenase domain also binds BH4 cofactor, O_2, and ^{L}arginine [1142]. At the heme site, electrons are used to reduce O_2 and oxidize ^{L}arginine, thereby producing ^{L}citrulline and nitric oxide.[15]

All NOS isoforms bind calmodulin. Both NOS1 and NOS3 connect to calmodulin upon intracellular Ca^{++} influx (half-maximal activity 200–400 nmol) [1142]. Calmodulin facilitates the electron transfer from NADPH in the reductase domain to the heme in the oxygenase domain. Calmodulin links to NOS2 at very low intracellular Ca^{++} concentrations (~ 40 nmol), i.e., with high affinity even in the absence of Ca^{++} ion, because of a different amino acid structure of the calmodulin-binding site [1142].

All NOS proteins contain a zinc–thiolate cluster formed by a zinc ion with 2 CXXXXC (CysXXXXCys) motifs (one motif from each monomer) at the NOS dimer interface [1142]. Zinc has a structural rather than catalytic function. This site actually tethers to BH4 and ^{L}arginine.

Electron transfer from the reductase domain enables NOS ferric (Fe^{3+}) heme to bind O_2 and form a ferrous (Fe^{++})–dioxy species. This species may receive a second electron from BH4 or the reductase domain [1142]. Oxidized BH4 is the trihydrobiopterin radical (BH3$^\bullet$) or trihydropterin radical cation (protonated at N5; BH3$^{\bullet^{H^+}}$). Trihydrobiopterin radical or radical cation can be recycled to BH4 by NOS. Alternatively, reducing agents such as ascorbic acid (AscH; concentration $O[1\,\text{mmol}]$ in cells) can reduce BH3$^\bullet$ back to BH4, hence producing ascorbate radical (Asc$^\bullet$) [1142].

10.3.1.1 Constitutive NOS1 and NOS3

Two Ca^{++}-dependent constitutive NOS isoforms include neuronal NOS (nNOS or NOS1) and endothelial NOS (eNOS or NOS3).[16] Calcium-activated calmodulin binds to and transiently activates constitutive NOS dimers.

Transiently activated constitutive NOS synthesizes NO in response to: (1) increased intracellular calcium concentration by different activators followed by calmodulin binding and (2) other stimuli such as fluid flow stress exerted at the wetted surface and within the vessel wall.

NOS1

Nitric oxide synthase-1 is, in particular, located in synapses of central (brain and spinal cord) and peripheral (sympathetic ganglia, adrenal glands, and nitrergic

---

[15]Oxidation occurs in 2 steps. In a first step, NOS hydroxylates ^{L}arginine to $^{N^\omega}$hydroxy ^{L}arginine. In a second step, NOS oxidizes $^{N^\omega}$hydroxy ^{L}arginine to ^{L}citrulline and NO [1142].

[16]Constitutive NOSs are designated after the cell types in which they were originally discovered.

nerves) neurons. It resides also in other cell types such as skeletal muscle cells, in which an alternatively spliced variant exist. It is detected in epithelial, kidney macula densa, pancreatic islet, and vascular smooth muscle cells.

Nitric oxide synthase-1 operates in synaptic remodeling in the central nervous system, relaxation of corpus cavernosum and penile erection, central regulation of blood pressure, and smooth muscle relaxation, especially vasodilatation primed by peripheral nitrergic nerves, as NOS1-derived NO acts as a neurotransmitter that stimulates NO-sensitive guanylate cyclase in effector cells [1142]. Neuronal NOS1 is involved in neurogenesis, learning, memory, and long-term regulation of synaptic transmission (long-term potentiation or inhibition).

Enzymes NOS1 and NOS3 have distinct roles in the vascular tone regulation. In particular, NOS1 is implicated in the blood flow rate control in the human forearm and coronary circulation, independently of its effects in the central nervous system [1142]. On the other hand, NOS3 provokes vasodilation in response to acetylcholine, substance-P, or mechanical stress.

Synthase NOS1 contains a PDZ domain used for between-protein interactions that determine NOS subcellular distribution and activity. Isozyme NOS1 interacts [1143]: (1) in the nervous system with anchoring adaptors Disc large homologs DLg2 and DLg4,[17] as well as (2) in the striated muscle with α1-syntrophin. Caveolin-3 binds to NOS1 and inhibits NO synthesis.

NOS3

Nitric oxide synthase NOS3 has the weakest activity with respect to NOS1 and NOS2 subtypes. It is expressed not only by vascular endothelia, but also by airway epithelia as well as other cell types, such as neurons, cardiomyocytes, platelets, kidney tubular epithelial cells, and syncytiotrophoblasts of the human placenta).

In endothelial cells, NOS3 synthesizes NO markedly when intracellular Ca^{++} rises. Calcium ion provokes the binding of calmodulin to NOS3 enzyme. Several other proteins also interact with NOS3 and regulate its activity. Heat shock protein HSP90 links to NOS3 and serves as an allosteric activator that promotes NOS3 recoupling [1142].

In endothelial cells, NOS3 generates NO in response to mechanical stress as well as acetylcholine and bradykinin. Nitric oxide released from endothelial cells diffuses in the neighborhood to regulate smooth muscle cell tone and proliferation, platelet aggregation, and leucocyte adhesion to the endothelium. Endothelial NOS-derived NO is a vasodilator and an inhibitor of platelet aggregation and adhesion to the vascular wall as well as smooth muscle cell proliferation (as it precludes PDGF secretion) and production of matrix molecules and leucocyte adhesion to vascular endothelium and subsequent diapedesis (as it impedes the production of many cell adhesion molecules).

---

[17] A.k.a. postsynaptic density proteins PSD93 and PSD95, respectively.

Endothelial NOS also intervenes in lung morphogenesis and postnatal angiogenesis [1142]. Enzyme NOS3 produced by bone marrow stromal cells supports via NO the mobilization of endothelial progenitor cells by VEGF factor, hence neovascularization.

Post-translational processing (primarily by acylation) incorporates NOS3 to plasmalemmal caveolae. This compartmentation facilitates between-protein interactions and signal transduction. Within caveolae, NOS3 is targeted by G-protein-coupled receptors as well as other receptors (estrogen receptor and high-density lipoprotein receptor ScaRb1) [1144]. Binding to caveolin-1 inhibits NO synthesis by NOS3 [1145].

Enzyme NOS3 can be myristoylated, palmitoylated, farnesylated, acetylated, and phosphorylated. These modifications assist in recruiting the enzyme to various cell compartments. Except phosphorylation, these changes do not significantly affect NOS3 activity.

Nitric oxide synthase NOS3 can generate both nitric oxide and superoxide.[18] In the presence of Ca^{++}–calmodulin, NOS3 produces NO from ^{L}Arg by means of electron transfer from NADPH via a flavin-containing reductase domain to oxygen-bound at the heme of an oxygenase domain, which also contains binding sites for tetrahydrobiopterin and ^{L}Arg [1146]. In the absence of tetrahydrobiopterin (BH_4), NO synthesis is abrogated and instead superoxide is generated. The NOS3 uncoupling caused by S-glutathiolation may be triggered by other uncoupling processes such as BH_4 depletion and may further enhance BH_4 depletion.

Thiols potentiate NOS3 activity. Subtype NOS3 possesses specific redox-sensitive thiols. These thiols can be S-glutathiolated.[19] This oxidative modification switches NOS3 from a NO synthase function to an NADPH-dependent oxidase generation of O_2^-. Under oxidative stress, S-glutathiolation occurs via a thiol–disulfide exchange with oxidized glutathione or reaction of oxidant-induced protein thiyl radicals with reduced glutathione. S-glutathiolation of NOS3 reversibly decreases NO production and increases in O_2^- generation.[20] In endothelial cells, S-glutathiolation of NOS3 is associated with an impaired endothelium-dependent vasodilation [1146].

Overexpressed cardiac NOS3 increases concentrations of nitrite, nitrate, and nitrosothiol not only in the heart, but also plasma and liver, and ensures cytoprotection of remote organs after blood transport of nitrite and ^{S}nitrosothiol [1147].

Protein kinases PKA, PKB, and AMPK phosphorylate (Ser1177; activate) NOS3 enzyme. On the other hand, PKC phosphorylates (Thr495; inactivate)[21] NOS3

---

[18] Messengers NO and O_2^- have many opposing roles in cell signaling.

[19] S-Glutathiolation is a reversible protein modification involved in cellular signaling and adaptation.

[20] Oxidized glutathione induces a dose-dependent S-glutathiolation of NOS3. This process is reversed by reducing agents.

[21] Residue Thr495 corresponds to a calmodulin-binding site.

10.3 Nitric Oxide, Nitric Oxide Synthases, Nitrite, and Nitrate

Table 10.3. NOS3 phosphorylation (Phosphoryl.) and dephosphorylation (Dephosphoryl.). Ser1177 phosphorylation leads to NOS3 activation, and Thr495 phosphorylation to NOS3 inactivation (Source: [1148]).

	Ser1177		Thr495	
	Phosphoryl. (activation)	Dephosphoryl. (inhibition)	Phosphoryl. (inhibition)	Dephosphoryl. (activation)
Phosphatases		PP2		PP1
Kinases	PKA PKB PKG AMPK CamK2	PKC (via PP2)	PKC AMPK	PKA (via PP1)

Table 10.4. Stimulators of multisite NOS3 phosphorylation and dephosphorylation of specific serine and threonine residues (Source: [1149]).

Residue site	Phosphorylation stimulators	Dephosphorylation stimulators
Ser114	Mechanical stress, HDL	ATP, VEGF
Ser615	ATP, bradykinin, VEGF	
Ser633	Mechanical stress, ATP, bradykinin, VEGF	
Ser1177	Mechanical stress, ATP, bradykinin, histamine, thrombin, hydrogen peroxide, insulin, estrogen, adiponectin, leptin, sphingosine 1-phosphate, VEGF, IGF1	
Thr495		Bradykinin, hydrogen peroxide, VEGF

as well as promotes dephosphorylation (Ser1177), hence inhibiting NOS3 [1148] (Tables 10.3 and 10.4).[22] Phosphatases PP1 and PP2 dephosphorylate NOS3 (Thr495 and Ser1177, respectively).

---

[22] Factor VEGF stimulates PKB and PKC, IGF1 stimulates PKB.

10.3.1.2 Inducible NOS2

Cytokine-inducible nitric oxide synthase (iNOS or NOS2), or immunocyte NOS,[23] produces NO only in selected tissues such as lung epithelium, but in many cell types, such as cardiac, endothelial, vascular smooth muscle, and glial cells, as well as dermal fibroblasts, macrophages, and CD8+ T lymphocytes.

Nitric oxide intervenes in host defense. It is typically synthesized in response to inflammation, i.e., in response to lipopolysaccharide and cytokines, among other agents. It is implicated in vascular diseases and transplant rejection.

The NOS2 activity appears slowly after exposure to cytokines, is sustained, and can function independently of calcium and calmodulin. However, calmodulin can be tightly bound to NOS2 enzyme.[24] Cells regulate NOS1 via aggresome [1150]. The dynein–dynactin complex associates NOS2 to the aggresome at the microtubule-organizing center.

Inducible NOS2 in non-stimulated T lymphocytes is implicated in atherosclerosis and graft-induced intimal thickening. The JaK–STAT signaling primed during T-cell activation inhibits NOS2 expression. Endothelial cells and other stromal cells induce NOS2 expression in CD8+ T cells, at a greater extent than in CD4+ T lymphocytes. Inducible NOS2 in resting T cells and low NO concentrations increase T-cell proliferation in response to allogeneic endothelial cells of grafted vessels [1151]. Induction of NOS2 depends on NFκB that is inhibited by STAT factor.

Both soluble and membrane-associated chloride intracellular channel ClIC4 are sensitive to the redox state. This P53- and MyC-responsive pro-apoptotic protein intervenes in innate immunity as well as in PKC-dependent differentiation of keratinocytes [1152]. In keratinocytes, ClIC4 channel also supports TGFβ signaling, as it hinders interaction of SMAD2^P and SMAD3^P with PPM1a phosphatase [1152]. In macrophages, ClIC4 is encoded by an early response gene, which is a transcriptional target of NFκB and interferon regulatory IRF3 factor [1152]. Its synthesis is hampered by anti-inflammatory glucocorticoid receptor. Many of ClIC4 functions depend on its nuclear localization. Its S-nitrosylation enables connection to nuclear import proteins. Nuclear translocation of ClIC4 channel indeed relies on NOS2 activation in pro-inflammatory peritoneal macrophages [1152]. Nuclear ClIC4 deactivates the pro-inflammatory program in macrophages to allow the phenotype transition from microorganism struggle to tissue repair.

---

[23]Enzyme NOS2 was first identified in macrophages. Nitric oxide produced in large amounts becomes cytotoxic. It can inhibit iron–sulfur cluster-dependent enzymes (complex-1 and -2) of the mitochondrial electron transport chain, ribonucleotide reductase (the rate-limiting enzyme in DNA replication), and cis-aconitase of the tricarboxylic acid cycle [1142].

[24]Affinity of NOS for calmodulin obeys the following order:
$$NOS1 < NOS3 \ll NOS2.$$

Asymmetric dimethylarginine and monomethyl arginine inhibit all NOS isoforms.[25] Asymmetric dimethylarginine is a cardiovascular risk factor because it reduces NO signaling, hence eliciting endothelial dysfunction and augmenting systemic and pulmonary blood pressure [1153]. Its plasmatic concentration rises particularly in pulmonary hypertension.

Short-lived nitric oxide is rapidly oxidized to nitrite in aqueous solutions; It can also be directly oxidized to nitrate in the presence of superoxide or oxyhemoglobin.

10.3.2 Nitric Oxide Production from Nitrite and Nitrate

Additional sources of nitric oxide than nitric oxide synthases arise from the cycling of nitrate, nitrite, and nitric oxide. Several pathways modulate the NO_3^-–NO_2^-–NO cycling, such as oxidation or reduction by hemoglobin, myoglobin, neuroglobin, xanthine oxidoreductase (XOR), nitric oxide synthase, carbonic anhydrase, cytochrome-C oxidase, cyclooxygenase, microsomal cytochrome-P450, mitochondrial aldehyde oxidase, and cytochrome-C, as well as bacteria, in different tissues under different conditions [1141, 1154]. Hemoglobin operates at normal pH and oxygen saturation in the vasculature.

Free radical gases are actually synthesized by the body's cells and transported nearby or remotely to influence cell function and signaling. Nitric oxide has a short half-life (<2 ms) in blood. Nitric oxide freely diffuses to adjacent cells, but the diffusion distance in tissues is relatively limited. Nitric oxide half-life within normoxic tissues has been estimated less than 0.1 s with a diffusion distance dependent on oxygen concentration, NO being metabolized by heart mainly to nitrite [1204]. Moreover, NO short half-life in blood due to hemoglobin consumption limits remote NO action [1205].

Nevertheless, NO can be transported as nitrite (NO_2^-) and ^{S}nitrosothiol. Nitrites can indeed serve as both a stored NO precursor used for intravascular endocrine NO transport, and an oxidation product of NO metabolism [1206]. Plasma contains approximately 7 μmol of ^{S}nitrosothiols, of which 96% are ^{S}nitrosoproteins (mainly relatively long-lived serum ^{S}nitroso-albumin) that have endothelium-derived relaxing factor-like properties [1207]. Nitric oxide activates specific soluble guanylate cyclase (Sect. 11.2). It also acts via cGMP-independent mechanisms.

---

[25] Methylarginines are formed by arginine methyltransferases. After proteolysis, methylarginines are released. These endogenous amino acids are degraded by dimethylarginine dimethylaminohydrolase.

10.3.2.1 Nitrite

Nitrite (concentration 0.3–1.0 μmol in plasma and 1–20 μmol in tissues) is an oxidative product of nitric oxide and reduction element of nitrate (NO_3^-) by commensal bacteria, in addition to dietary sources. Nitrites represent a major storage form of nitric oxide in blood and tissues, as NO_2^- is much more stable than NO or ^{S}nitrosothiols.

Nitrite-Mediated Vasodilation

Nitrites (10 μmol–2 mmol) cause vasorelaxation of isolated aortic rings and decrease both systolic and diastolic blood pressure. Nitrite can also prime vasodilation in isolated vessels at therapeutic (≤ 200 μmol) and physiological concentrations (100–200 nmol) [1154].

Nitrite concentration may be higher in the erythrocyte (~ 290 nmol) than in the plasma (~ 120 nmol) [1154]. It depends on the rate of nitrite import in and export out of the erythrocyte.

Nitrite forms a storage pool of nitric oxide that can be mobilized to trigger vasodilation during hypoxia, but at relatively large time scale. In fact, the conversion of free NO to nitrate is much faster than NO_2^- reduction to nitric oxide.

Hemoglobin (Hb) is an important NO buffer and a modulator of NO bioavailability. Free NO produced in an erythrocyte is very quickly scavenged by both oxy- or deoxyhemoglobin [1154]. Nitrite may cross membranes by simple diffusion coupled with protonation–deprotonation according to concentration gradients. Nitrite transport across the erythrocyte is regulated by O_2 content and deoxyHb-mediated nitrite reduction, thus preventing nitrite export from the erythrocyte by inhibiting anion exchanger-1 (or SLC4a1).[26]

Nitrite anion is reduced to nitric oxide as oxygen concentration decays by several mechanisms that rely on xanthine oxidoreductase,[27] deoxyhemoglobin, and deoxymyoglobin, among others. Members of the heme globin family (cyto-, hemo-, neuro-, and myoglobin) can act as nitrite reductases that regulate the hypoxia-induced NO generation [1156].

Hemoglobin-mediated nitrite reduction is much faster under partially oxygenated than fully oxygenated conditions. Nitrite reduction kinetics by erythrocytic or cell-free Hb are directly regulated by oxygen fractional saturation; the bell-shaped relation reaches a maximal nitrite reduction rate around the oxygen partial pressure

---

[26] Nitrite binds to and is reduced by ferrous heme to liberate NO, which can then bind to a vacant deoxygenated heme.

[27] In erythrocytes, xanthine oxidoreductase reduces nitrite (NO_2^-) to NO only when P_{O_2} is very low and NO_2^- concentration very high. The formation of Hb^{NO} increases when oxygen saturation of hemoglobin drops. Hence, circulating deoxyhemoglobin acts as a NO_2^- reductase in hypoxia [1155].

10.3 Nitric Oxide, Nitric Oxide Synthases, Nitrite, and Nitrate

at which Hb is 50% oxygenated (Hb^{O_2} P_{50}) [1154]. Nitrite reduction thus becomes faster as erythrocytes are deoxygenated. The combination of deoxygenated erythrocytes and nitrite primes the NO-mediated signaling.

Nitrites, which are not vasoactive at physiological concentrations, can be reduced to nitric oxide in particular by hemoglobin [1157]. Hemoglobin acts as a reductase that then releases NO under allosteric control during its transition from the high-oxygen affinity R (relaxed) to low-oxygen affinity T (tense) state [1158].[28] When hemoglobin is exposed to NO_2^-, ferrous nitrosyl hemoglobin ($^{Fe^{++}}Hb^{NO}$) is formed. Then, free NO can be possibly produced according to the following reaction [1158]:[29]

$$NO_2^- + {}^{Fe^{++}}Hb + H^+ \to {}^{Fe^{3+}}Hb + NO + OH^-.$$

Afterward, released NO can rapidly bind to deoxygenated hemoglobin to produce nitrosylhemoglobin ($^{Fe^{++}}Hb-NO$) or react with any oxygenated hemoglobin chains to form methemoglobin and nitrate [1155]:

$$^{Fe^{++}}Hb + NO \to {}^{Fe^{++}}Hb^{NO};$$
$$^{Fe^{++}}Hb^{O_2} + NO \to {}^{Fe^{3+}}Hb + NO_3^-.$$

Many chemical reactions associate NO, nitrite, and various forms of hemoglobin (oxyHb, deoxyHb, metHb, Hb^{NO}, and Hb^{SNO}) [1154]. Reaction 1 is the initial step in the deoxyHb–nitrite reaction that may create the transient $^{Fe^{3+}}Hb$ intermediate. In the reversible reaction 2, $^{Fe^{3+}}Hb$ generates metHb and NO. However, in the erythrocyte, NO unlikely rebinds to metHb. Reaction 3 related to nitrite and oxyHb is characterized by a slow initiation phase and a fast autocatalytic phase. Yet, the autocatalytic phase is unlikely in the erythrocyte because of the competition of intermediate species. Reaction 4 corresponds to the reversible binding of nitrite to metHb.[30] Reaction 5 of nitrite-bound metHb ($^{Fe^{3+}}Hb^{NO_2^-}$) with NO forms reduced deoxyHb and N_2O_3. The latter may diffuse out of the erythrocyte and subsequently generate NO and nitrite. Alternatively it can react with a thiol (e.g., glutathione or cysteine; reaction 6) followed by export of the nitrosothiol, hence of NO activity. In reaction 7, Hb^{SNO} results from the $^{Fe^{++}}Hb^{NO}$ intermediate. In reaction 8 (intramolecular transfer), NO^+ from $^{Fe^{++}}Hb^{NO^+}$ is transferred within Hb to Cysβ93. Alternatively (as in reductive nitrosylation), NO^+ from $^{Fe^{++}}Hb^{NO^+}$ can also reacts with OH^- to generate nitrite.[31] In reaction 9, N_2O_3, a strong

---

[28] NEthylmaleimide alkylates cysteinyl residues of hemoglobin, in particular the Cysβ93, thereby stabilizing Hb in its high-oxygen affinity R state. It accelerates nitrite-dependent deoxyHb decay. On the other hand, inositol hexakisphosphate (IP_6) binds to and stabilizes Hb in the low-oxygen affinity T state [1154]. It decelerates nitrite-dependent deoxyHb decay.

[29] Ferric hemoglobin ($^{Fe^{3+}}Hb$) is used for the production of ^{S}nitrosohemoglobin (or ^{S}nitrosohemoglobin [Hb^{SNO}]), and ferrous nitrosyl hemoglobin ($^{Fe^{++}}Hb^{NO}$) may also undergo an oxidation-promoted transfer of NO to Cysβ93.

[30] The bound form may be either the N-bound (nitro) or O-bound (nitrito) form.

[31] In reductive nitrosylation, $^{Fe^{3+}}Hb^{NO}$ reacts with OH^- or water to make nitrite, leaving a reduced iron (Fe^{++}).

nitrosating agent, results from the reaction of nitrite with the $^{Fe^{3+}}Hb^{NO}$ intermediate. Reaction 10 represents NO binding to deoxyHb. In reaction 11 (oxidative denitrosylation), intermediates in the oxyHb–nitrite reaction oxidize the ferrous nitrosyl heme, thus producing NO and a ferric heme. Reaction 12 deals with the NO dioxygenation reaction that limits the export of NO from the erythrocyte, in which NO reacts with oxyHb to form metHb and nitrate. In addition, nitrite-dependent ATP export out of the erythrocyte can contribute to vasodilation. Hemoglobin that has reacted with nitrite releases membrane-bound glycolytic enzymes, which then produce intracellular ATP for subsequent secretion under hypoxia.

The reduction of nitrite back to NO by deoxyhemoglobin in erythrocytes can be followed by NO transport when its scavenging by hemoglobin is bypassed. Nitric oxide can be transferred to the Cysβ93 thiol group of hemoglobin to form S-nitrosylated hemoglobin when heme nitrosylated hemoglobin ($^{Fe^{++}}Hb++^{NO}$) is oxygenated [1158]. Subsequent to the formation of ^{S}nitrosohemoglobin,[32] S-transnitrosylation, the transfer of NO groups to thiols on the erythrocyte membrane and then in the plasma, enables NO delivery to the vascular wall.

Oxyhemoglobin is an efficient NO scavenger. Because nitrite can be reduced to NO by deoxyhemoglobin, an *arteriovenous gradient* of blood of nitrite exists. Nitrite is consumed between artery (176 ± 10 nmol) and vein (143 ± 7 nmol); this consumption rises during exercise.[33]

The short lifespan of nitric oxide in blood (< 2 ms) requires a mechanism to retain NO activity in the circulation, especially in the microcirculation, where the oxygen partial pressure is lower and less oxygen is available for NO synthesis by endothelial nitric oxide synthase. Nitrite can deliver active NO to the vasculature after its reduction by deoxyhemoglobin in erythrocytes.[34]

10.3.2.2 Nitrate

Both nitrate and nitrite are generated from endogenous and dietary sources. The main dietary source of nitrate is leafy green and root vegetables. The primary endogenous source of nitrate and nitrite is NO that is rapidly oxidized in biological tissues, including blood, into these anions.

---

[32] A nitric oxide group is covalently attached to Cysβ93 of hemoglobin β chain.

[33] A significant decrease in plasma nitrite level and erythrocyte nitrosyl hemoglobin (Hb^{NO}) level, but not Hb^{SNO} level, was observed during the transit of blood from the artery to the vein in the forearm circulation after NO inhalation [1154].

[34] Two intermediates can be identified [1155]. (1) Heme–nitrite binding ($^{Fe^{++}}$deoxyHb–NO_2^-) forms a first intermediate ($^{Fe^{++}}$Hb–ONOH) that releases NO in acid solution in the presence of ascorbate. (2) A second, much more stable intermediate corresponds to the interconversion of $^{Fe^{++}}$Hb–NO$^+$ and $^{Fe^{3+}}$Hb–NO molecule. This intermediate liberates NO, especially when nitrites are in excess, at neutral pH in the presence of ferricyanide when reacted with an Fe^{3+} ligand like azide. The methemoglobin–nitric oxide molecule ($^{Fe^{3+}}$Hb–NO) is transiently formed.

Nitric oxide reacts with oxyhemoglobin in erythrocytes to form nitrate and ceruloplasmin and oxygen in plasma to form nitrite. Approximately half of plasma nitrite comes from dietary nitrate and half from oxidation of endogenous NO [1159].

Inorganic nitrate is serially reduced to nitrite and, then, NO, nitrosothiols, and other bioactive nitrogen oxides via numerous pathways in biological tissues, including blood.

Exogenous Nitrate

Ingestion of nitrate amounts from 1 to 2 mmols per day (western diet) [1141]. A diet rich in fruit and vegetables is an important source of nitrate.[35] Dietary nitrate is rapidly and completely absorbed in the upper digestive tract. About 60% ingested nitrate is excreted in the urine within 48 h. The half-life of an oral dose of inorganic nitrate is estimated to range from 5 to 8 h, due to reabsorption into the proximal renal tubule. The urinary clearance of nitrate equals about 26 ml/mn.

Dietary supplementation with sodium nitrate, in amounts similar to those derived from NOS3 under normal conditions (0.1 mmol/(kg·d), reverses features of the metabolic syndrome (Vol. 6 – Chap. 7. Vascular Diseases) in NOS3-deficient mice [1160].[36]

Endogenous Nitrate

Nitric oxide synthesized by NOS is rapidly oxidized to nitrate in the presence of oxyhemoglobin, which yields approximately 1 mmol nitrate per day [1141]. Methemoglobin is a by-product of this oxidation process that cannot bind oxygen. It needs to be reduced back to hemoglobin by methemoglobin reductase.

Nitrite has a short half-life in plasma, because it is oxidized to nitrate. In addition, the concentration of nitrite is much higher in vascular wall than in plasma.[37] Oxidation of nitrite to nitrate in hepatocytes utilizes tetrahydrobiopterin [1141].

---

[35]Green leafy and cruciferous vegetables have a high nitrate content. Drinking 250 ml of beetroot juice elevates plasma nitrite concentration from 380 ± 70 nmol at baseline to 580 ± 90 nmol after 2.5 h [1154].

[36]The metabolic syndrome, a set of risk factors of metabolic origin that augments the probability of occurrence of cardiovascular disease and type-2 diabetes, is characterized by a decay of nitric oxide synthesis by NOS3 enzyme (endothelial dysfunction).

[37]Nitrite may serve as a messenger independently of NO and be used in post-translational modification (protein nitrosylation).

Distribution and Excretion

Whatever the source, circulating nitrate is concentrated in the salivary glands and subsequently secreted into the mouth. The salivary nitrate concentration is at least 10 times that of plasma [1141]. Local anaerobic bacteria can reduce nitrate to nitrite. The resulting salivary nitrite concentration is more than 1000 times greater than that of plasma [1141].

Swallowed nitrite can be reduced to NO in the acid solution of the stomach. Intragastric NO regulates gastric blood flow, mucus production, and host defense. In conditions of very low pH, such as in the stomach, nitrous acid (HNO_2) spontaneously generates nitric oxide [1154].

In some organs, nitrate can also be reduced into nitrite. In the liver, xanthine oxidoreductase serves as nitrate reductase. In erythrocytes and vascular endothelial cells, in normal conditions, the activity of xanthine oxidoreductase is minimal, but rises during acidosis and hypoxia [1141].

Nitrate released from skin stores has a substantial concentration in sweat. It is then reduced to nitrite by skin commensals and, then, nitric oxide.

10.3.3 Nitric Oxide and Hypoxic Vasodilation

When an overall regulation is required, both central and local mechanisms provide a coordinated adjustment of blood flow to metabolic demand. When an increased demand is confined to a restricted region, i.e., when a change in metabolic activity concerns a limited number of adjacent groups of working cells, a local vasodilation of the microvascular bed (arterioles), the so-called *hypoxic vasodilation* (Table 10.5),[38] occurs in response to decreased oxygen saturation of hemoglobin to promptly raise the local blood perfusion, hence supplying more oxygen, without changes in the upstream arterial condition, such as vascular resistance at a distance from the site of interest, as well as without modifications in the control by the central nervous system and activity of circulating hormones.

Hypoxic vasodilation relies on a pathway, in which the signal is the decreased ATP production that results from a local lack in oxygen. The sensor localizes to hemoglobin in erythrocytes, as the response is correlated to $Hb-O_2$ saturation, but not tissue partial pressure in oxygen.

In addition to hemoglobin, other factors, such as ATP, adenosine, vasoactive peptides, CO_2, as well as H^+ and K^+ cations can contribute to the hypoxic vasodilation [1158].

---

[38]Hypoxic vasodilation that is restricted to a local microvasculature due to local changes in metabolic conditions differs from a response to a global hypoxemia as well as hypoxic potentiation of vasoactivity mediated by some agents that can depend on the partial pressure of oxygen, but not changes in $Hb-O_2$ saturation.

Table 10.5. Types of mechanotransduction in different compartments of the arterial bed. Autoregulation associated with the myogenic tone results from paradoxical vasoconstriction (smooth mucle cell contraction) when the local arterial pressure heightens and vasodilation when the flow rate rises.

Compartment	Effect
Large arteries	Vasodilation in response to prolonged increase of intraluminal pressure (direct and endothelium-mediated response of VSMCs)
Resistive arteries	Autoregulation (maintenance of constant flow rate) Calcium sparklets and membrane potential change in both endothelial and smooth muscle cells
Arterioles	Hypoxia-triggered vasodilation on increased local metabolism (^{S}nitrosothiol-based process)

The distribution of blood flow in arteriolar networks according to metabolism of active cells and regulation of capillary perfusion by arterioles is also controlled by sympathetic nerve activity [1161]. Sympathetic constriction of proximal arterioles and feed arteries can restrict functional hyperemia on resting territories, whereas dilation prevails in distal arterioles that irrigate working tissues to promote oxygen extraction. In addition, venules yield feedback from metabolic state of drained tissues to nearby arterioles via the production of vasodilators.

10.3.3.1 Role of Erythrocyte and Hemoglobin

Hemoglobin tetramer, during its transformation from the relaxed state (R state) to tense state (T state), releases NO bound to a cysteine in its β chain (Cysβ93). The latter is subsequently carried as low-molecular-weight ^{S}nitrosothiols, such as ^{S}nitrosocysteine (Cys^{NO}) and ^{S}nitrosoglutathione (G^{SNO}). These agents have longer life duration than free NO and are protected from scavenging by hemoglobin hemes.

Owing to erythrocytic Hb, erythrocytes consume and produce NO using nitrite. Nitric oxide reacts with oxyHb slightly faster than with deoxyhemoglobin. At relatively low oxygen partial pressures, NO consumption is thus slightly less efficient, but NO production rises due to the reaction of nitrite and deoxyhemoglobin. Therefore, relatively more NO is produced than consumed as erythrocytes deoxygenate [1154].

Erythrocytes actually release ^{S}nitrosothiols into the blood stream during hypoxia [1162].[39] ^{S}Nitrosohemoglobin (Hb^{SNO}) binds to solute carrier superclass

---

[39] Hemoglobin desaturation in blood vessels that perfuse tissues causes the release from erythrocytes of O_2 and vasodilation. Nitric oxide cannot escape red blood cells, as it is sequestered by both

Table 10.6. Reduction and oxidation of nitric oxide into hyponitrite (or nitroxyl) anion (NO$^-$) and nitrosonium (or nitrosyl) cation (NO$^+$), respectively.

Reduction	NO + e$^-$	→	NO$^-$
Oxidation	NO	→	NO$^+$ + e$^-$

anion exchanger SLC4a1[40] that facilitates electroneutral anion exchange. By transnitrosylation, the NO group then transfers to Cys in the cytoplasmic N-terminus of SLC4a1. On the other hand, ATP-binding cassette protein can transport ^{S}nitrosoglutathione.

^{S}Nitrosohemoglobin in erythrocytes can elicit a relaxation of vascular smooth muscle cells via cGMP and Ca^{++}-dependent K$^+$ channels and Ca^{++} ATPase, independently of endothelium-derived NO and nitrite [1162]. In addition, oxyhemoglobin (HbO$_2$) can be oxidized by NO to yield methemoglobin (metHb, i.e., $^{Fe^{3+}}$Hb that cannot carry oxygen, unlike normal $^{Fe^{++}}$Hb, and NO$_3^-$ [1163]. Afterward, NADH-dependent methemoglobin reductase converts methemoglobin into hemoglobin. In systemic vessels,[41] ^{S}nitrosohemoglobin intervenes in hypoxic vasodilation with a time scale of second.[42]

10.3.4 Reduction–Oxidation Reaction Products of Nitric Oxide

Reduction (gain of electron) and oxidation (loss of electron) of nitric oxide give redox forms: (1) *hyponitrite* anion (NO$^-$), or *nitroxyl* anion and (2) *nitrosonium* cation (NO$^+$), or nitrosyl cation, respectively (Table 10.6). The uncharged *free radical form* of nitric oxide (NO$^\bullet$) that activates Ca^{++}-activated K$^+$ channels (K$_{Ca}$) repolarizes smooth muscle cells at a lower extent than nitroxyl [1165].

Nitric oxide yields several different redox forms. Its 1-electron reduction product, protonated nitroxyl (HNO) and nitroxyl anion (NO$^-$),[43] causes a hyperpolarization of vascular smooth muscle cells and repolarization as well as vasorelaxation of contracted cells in resistance arteries, as it acts on voltage-dependent K$^+$ channels (K$_V$) via the sGC–cGMP pathway [1165].

---

oxygenated and deoxygenated hemoglobin [1162]. Moreover, erythrocytes and hemoglobin block nitrite-mediated relaxations. On the other hand, erythrocytes that contain ^{S}nitrosohemoglobin (Hb^{SNO}) and isolated Hb^{SNO} can replicate hypoxic vasodilation.

[40] A.k.a. anion exchanger AE1.

[41] In pulmonary arteries, hypoxemia causes a vasoconstriction.

[42] Substitution of Cys93 of hemoglobin β chain by alanine impedes ^{S}nitrosohemoglobin formation, but does not change total erythrocytic ^{S}nitrosothiol amount, as it shifts ^{S}nitrosothiol distribution to lower molecular weight species [1164].

[43] Nitroxyl (HNO) deprotonates to NO$^-$ nitroxyl anion.

10.3 Nitric Oxide, Nitric Oxide Syntheses, Nitrite, and Nitrate

Table 10.7. Metabolites of nitric oxide (Source: [1163]). Reaction of NO with O_2^- (reactions 1a and 1b) and O_2 (reactions 2 and 3). Reaction 1 is irrelevant in situations where a high local concentration of O_2^- occurs, such as in the immediate vicinity of activated macrophages and neutrophils. Reaction 4 is analogous to the normal reaction of SOD, in which $^{Cu^{++}}$SOD accepts an electron from O_2^- to produce O_2 and converts into $^{Cu^+}$SOD. A proton donation from SOD may facilitate a reduction of NO to HNO (reaction 5). The reduction (reverse reaction from NO to NO^- in reaction 4) is analogous to another normal reaction of SOD with O_2^- that yields H_2O_2 (reaction 6).

$NO + O_2^-$	$\rightleftarrows$	$ONOO^-$	(1a)
$ONOO^-$	$\longrightarrow$	NO_3^-	(1b)
$2\,NO + O_2 \longrightarrow 2\,NO_2$	$\xrightarrow{H_2O}$	$NO_2^- + NO_3^- + 2\,H^+$	(2)
$NO_2 + NO \rightleftarrows N_2O_3$	$\xrightarrow{H_2O}$	$2\,NO_2^- + 2\,H^+$	(3)
$NO^- + {}^{Cu^{++}}SOD$	$\rightleftarrows$	$NO + {}^{Cu^+}SOD$	(4)
$NO + {}^{Cu^+}SOD^{H^+}$	$\rightleftarrows$	$HNO + {}^{Cu^{++}}SOD$	(5)
$O_2^- + {}^{Cu^+}SOD^{H^+} + H^+$	$\rightleftarrows$	$H_2O_2 + {}^{Cu^{++}}SOD$	(6)

The endogenous production of HNO may occur via various pathways. Nitroxyl can be synthesized by NOS, particularly in the absence of the cofactor tetrahydrobiopterin or after oxidation of the NOS intermediates $^{N^\omega}$hydroxy ^{L}arginine and hydroxylamine [1166].[44] Moreover, HNO can be produced from the reduction of NO• by mitochondrial cytochrome-C, xanthine oxidase, and hemoglobin.

Nitroxyl anion induces vasodilation in association with the formation of iron–nitrosyl complexes and the conversion of NO^- to NO• [1167]. Protonated nitroxyl is an endothelium-derived relaxing and hyperpolarizing factor in resistance mesenteric arteries of rats and mice, but with different potency order in these 2 rodent species with respect to other vasodilatory agents (NO• $\sim$ HNO > EDHF in mice and EDHF > HNO $\sim$ NO• in rats; EDHF: endothelium-derived hyperpolarizing factor) [1166].

Superoxide dismutase (SOD) rapidly scavenges superoxide (O_2^-), thereby prolonging the vasorelaxant effects of nitric oxide. Enzyme SOD supports a reversible reduction of NO to NO^- [1163] (Table 10.7). Although peroxynitrite ($ONOO^-$) formed by dismutation of NO with O_2^- can decompose into NO_2 and OH• in the absence of SOD or may convert to NO_2^+ in its presence, nitrate (NO_3^-) results from peroxynitrite.

In addition, ^{S}nitrosothiols (RSNO; R denotes an organic group; R–N=O: nitroso functional group),[45] or thionitrites, that serve as donors of the nitrosonium ion NO^+

---

[44] N^ω-hydroxylation of ^{L}arginine depends on NADPH.

[45] Nitroso and curved nitrosyl are synonyms. A thiol is a compound that contains a functional group composed of a sulfur–hydrogen bond (R–S–H). The amino acid cysteine possesses a thiol group. Peptides and proteins, such as antioxidant glutathione, cysteine peptidases, and coenzyme-A involved in synthesis and oxidation of fatty acids as well as oxidation of pyruvate in the citric acid cycle, are thiols.

can react with other thiol species to yield HNO at physiological pH [1166]. Activity of NO actually results from NO and ^{S}nitrosothiol derivatives.

Under hypoxia, nitrite ions (NO_2^-) may release nitric oxide that causes potent vasodilation. Nitrite may also react with deoxyhemoglobin in venous blood to generate nitrosylated iron ($^{Fe^{3+}}$NO) [1155]. The subsequent conversion to SNO results from oxygenation at neutral pH in the lung vasculature and acidification in the gut. Endogenous ^{S}nitrosothiols are much more potent vasodilators than nitrites.

In platelets, both HNO and NO• exert an anti-aggregating effect.

In cardiomyocytes, HNO enhances Ca^{++} cycling and sensitivity of Ca^{++} handling proteins. Nitroxyl anion has positive inotropic and lusitropic effects. Cardiac inotropic effect of nitroxyl anion is mediated by the neuromodulator calcitonin gene-related peptide via the release of CGRP from non-adrenergic, non-cholinergic fibers (NANC) [1167]. Improved left ventricular contractility is associated with a concomitant selective venodilation (without any change in arterial resistance). Therefore, HNO is called an *inodilator*.[46]

10.3.5 NO Transfer

Nitric oxide is a small hydrophobic molecule that can cross cell membranes without necessarily using carriers, unlike the gaseous molecules oxygen and carbon dioxide that use gas transporter (Vol. 3 – Chap. 4. Membrane Compound Carriers). Nevertheless, aquaporin Aqp1 carries hydrophobic NO and O_2 molecules.

Nitric oxide can diffuse from its synthesis site to surrounding cells, where it activates soluble guanylate cyclase (sGC; Vol. 3 – Chap. 6. Receptors). Erythrocytes yield a major sink for NO in the flowing blood that creates a diffusion gradient between irrigated tissues and blood.

The diffusion coefficient of NO in water at 37°C is slightly faster than that of oxygen and carbon dioxide [1168]. This short-lived molecule (half-life of $\sim$ 1 s) diffuse rapidly with an average molecular velocity of $O[100 \text{m/s}]$, following a trajectory in solution determined by $O[10^9]$ collisions per time unit (1 s). Due to its short life,[47] Agent NO travels a relatively short averaged distance. Its action is thus limited to cells close to the source of production.

Nitric oxide mainly signals via soluble guanylate cyclase that produce cGMP, a second messenger.[48] Like hemoglobin, soluble guanylate cyclase contains heme protoporphyrin-9 with iron in the ferrous form that binds NO with high affinity.[49] In turn, cGMP activates cGMP-dependent protein kinase-G. The greater the local NO concentration, the larger the amount of cGMP synthesized.

---

[46] An inodilator is a substance with both positive inotropic and vasorelaxant effects.

[47] Nevertheless, the NO lifespan is relatively long compared with signals carried by action potential in nerves.

[48] Only 5 to 10 nmol NO suffice to activate guanylate cyclase [1168].

[49] Deoxyhemoglobin binds NO with a 10,000-fold higher affinity than molecular oxygen [1168].

10.3.6 NO Effects

Nitric oxide transmits information in different modes. The first triggered pathway relies on soluble guanylate cyclase (sGC)[50] that synthesizes cyclic guanosine monophosphate. Soluble guanylate cyclases that pertain to the heme nitric oxide- and oxygen-binding (HNOX) proteins are heterodimeric hemeprotein.[51] Two isoforms exist ($\alpha 1\beta 1$ and $\alpha 2\beta 1$) that have similar enzymatic properties. At least in some tissues, these isoforms can compensate for each other. Endogenous modulators of sGC include ATP, GTP, Ca^{++}, Mg^{++}, and protein kinases and phosphatases, among other interactors: [1169].[52] Nitric oxide exerts its effects only partly by binding to guanylate cyclase receptors, thereby priming cGMP accumulation in target cells (Vol. 3 – Chap. 6. Receptors).[53]

Second messenger cGMP can modify the activity of ion channels (e.g., cyclic nucleotide-gated channels), protein kinase-G, protein phosphatases, and phosphodiesterases,[54] especially cGMP-specific PDE5 isoform. Substrates of protein kinase-G include inositol trisphosphate receptor-associated PKG1 substrate (IRAG),[55] BK channels, PDE5, cerebellar G substrate,[56] vesicle-associated membrane protein (VASP), and telokin [10]. The NO–GCase–cGMP pathways (Table 10.8) are involved in smooth muscle relaxation, synaptic plasticity, and cardiac hypertrophy. Nitric oxide thus participates in the regulation of the vasomotor tone (Vol. 5 – Chaps. 8. Smooth Muscle Cells and 9. Endothelium), as well as ion conductance, glycolysis, immunity, neurotransmission, and cell apoptosis.

Maximal relaxation of airway smooth muscle by nitric oxide that stimulates soluble guanylate cyclase is greater than that produced in response to atrial natriuretic peptide that activates the particulate guanylate cyclase (pGC). Stimulation

---

[50] Plasmalemmal particulate guanylate cyclase (pGC) is a single membrane-spanning receptor activated by various peptides, such as atrial (ANP), brain (BNP), and C-type (CNP) natriuretic peptides, and guanylin (Vol. 3 – Chap. 6. Receptors). Particulate guanylate cyclases constitute a family of 7 known plasmalemmal receptors (GCa–GCg). They have an extracellular ligand-binding domain, a single transmembrane region, and an intracellular cyclase domain.

[51] In contrast to many other hemeproteins that bind both NO and O_2, the HNox domain of sGC is selective for NO.

[52] The heat shock proteins — HSP70 and HSP90 chaperones — may be involved in sGC maturation and transfer to intracellular compartments and plasma membrane [1169].

[53] Nitric oxide receptors are constituted by a prosthetic heme that serves as NO-binding site and guanylate cyclase domain. Phosphodiesterases that convert cGMP to GMP impede NO signal transmission. However, the PDE activity can be weak. Furthermore, PDE5 subtype has a low activation rate.

[54] The PDE class includes 11-gene families, i.e., at least 18 human genes (Pde1a–Pde1c, Pde2a, Pde3a–Pde3b, Pde4a–Pde4d, Pde5a–Pde5b, Pde7a–Pde7b, Pde8a–Pde8b, Pde9a, Pde10a, Pde11a, and Pde12) that may encode more than 100 protein variants. Among PDEs, some are cAMP-selective (PDE4, PDE7, and PDE8) and cGMP-selective hydrolases (PDE5, PDE6, and PDE9), whereas others can hydrolyze both cAMP and cGMP (PDE1–PDE3, and PDE10–PDE11).

[55] Agent IRAG localizes to the endoplasmic reticulum in association with IP_3 receptors.

[56] Cerebellar G substrate may be involved in long-term depression. It inhibits protein phosphatase-2 more efficiently than protein phosphatase-1.

Table 10.8. Signaling by the NO–sGC–cGMP pathway (Source: [1169]; CNG: cyclic nucleotide-gated channel; IRAG: IP$_3$R-associated PKG1β substrate; MLCP: myosin light-chain phosphatase; PDE: phosphodiesterase; pGC: particulate guanylate cyclase; PKG: cGMP-dependent protein kinase-G (or cGK); sGC: soluble guanylate cyclase; VASP: vasodilator-stimulated phosphoprotein). Three types of cGMP-binding proteins transduce the cGMP signal: (1) cGMP-modulated cation channels; (2) cGMP-dependent protein kinases; and (3) cGMP-regulated phosphodiesterases that degrade cGMP and/or cAMP messengers. Membrane-bound PKG2 mediates effects of pGC-derived cGMP on intestinal electrolyte transport and bone formation. Cytosolic PKG1, which has 2 isoforms (PKG1α–PKG1β) transduces many effects of sGC-derived cGMP messenger. Enzyme PKG1 stimulates sumoylation of the ETS-like transcription factor ELk1, thereby derepressing smooth muscle-specific gene promoters.

Target	Pathway Effect
Ion channel	NO–sGC–cGMP–CNG–Ca^{++}
Phosphodiesterases	NO–sGC–cGMP–PDE–cAMP/cGMP Cell growth and differentiation Inhibition of platelet aggregation
Protein kinase-G	NO–sGC–cGMP–PKG1α Growth of vascular smooth muscle cell NO–sGC–cGMP–PKG1α–MLCP Smooth muscle relaxation NO–sGC–cGMP–PKG1β–IRAG–Ca^{++} Inhibition of intracellular Ca^{++} release Smooth muscle relaxation NO–sGC–cGMP–PKG1α/β–VASP Inhibition of platelet adhesion to endothelium NO–sGC–cGMP–PKG1–ELk1 Smooth muscle gene transcription

of pGC causes a bronchodilation exclusively by decreasing the intracellular Ca^{++} concentration, whereas stimulation of sGC decreases both [Ca^{++}]$_i$ and Ca^{++} sensitivity, i.e., the force developed for a given Ca^{++} level [1170].

In rat renal glomeruli, inhibitors of inducible nitric oxide synthase NOS2 may also repress the activity of soluble guanylate cyclase, but increase the activity of particulate guanylate cyclase to compensate that of sGC [1171].

Nitric oxide also contributes to the increase in muscle blood flow during exercise after acute administration of ascorbic acid [1172].

Biological effects of NO can also be mediated by modifications of proteins. Nitric oxide can indeed operate via reactive oxygen and nitrogen species, thereby modifying the activity of target proteins via nitrosylation. Nitric oxide actually interacts with various acceptors, such as superoxide radical (O$_2^{-\bullet}$), cysteine, glutathione, and transition metal ions:

$$NO + O_2^- \rightarrow ONO-NO$$
$$NO + Mn^{++} \rightarrow {}^{Mn^{++}}NO$$

10.3 Nitric Oxide, Nitric Oxide Synthases, Nitrite, and Nitrate

$$NO + GSH \rightarrow GS^{NO}$$
$$NO + Cys \rightarrow Cys^{NO}$$

The S-nitrosylation reaction relies on the transfer of NO from one of the reactive nitrogen species to a thiol group of target proteins (R–S) [10]:

$$RS + ONO\text{–}NO \rightarrow R\text{–}S\text{–}NO$$
$$RS + Mn^{++}NO \rightarrow R\text{–}S\text{–}NO$$
$$RS + GS^{NO} \rightarrow R\text{–}S\text{–}NO$$
$$RS + Cys^{NO} \rightarrow R\text{–}S\text{–}NO$$

Nitric oxide may reversibly inhibit enzymes with transition metals or with free radical intermediates in their catalytic cycle. In micromolar concentrations, NO reversibly impedes the activity of catalase and cytochrome-P450 [1168]. It can also hinder the action of ribonucleotide reductase used in DNA synthesis that contains a tyrosine radical.

Cells are very sensitive NO detectors, as they are able to respond to tiny concentration of NO (1 pmol) by rising their cGMP content (≥ 30 nmol) [1173]. Moreover, they can sense brief NO puffs (100 ms) that yield a peak intracellular NO concentration of about 20 pmol. Therefore, NO is able to signal at extremely low concentrations from a tiny proportion ($\sim 1\%$) of cognate activated receptors.

Concentration of nitric oxide influences its effects. At 100 nmol or less, NO activates the cGMP synthesis as well as the protein kinase-G[57] and extracellular signal-regulated protein kinase pathways [1174]. At higher concentrations (300–800 nmol), it stimulates the protein kinase-B pathway and hypoxia inducible factor-1α, and stabilizes P53 transcription factor.

Nitric oxide and its messenger cGMP control Ca^{++} homeostasis in a cell-specific manner. In pancreatic acinar cells, nitric oxide potentiates Ca^{++} entry. In hepatocytes, NO causes oscillatory increase in intracellular Ca^{++} level.

In platelets and vascular smooth muscle cells, NO precludes Ca^{++} entry and intracellular store refilling. In cardiomyocytes, NO impedes the activity of $Ca_V 1.2a$ channel.

Most of NO effects result from its regulation of gene expression.[58] The NO pathway supports the activation of the telomerase reverse transcriptase (TERT; i.e., catalytic subunit) during angiogenesis [1174]. In addition, S-nitrosylation of cysteines represses HDAC2 histone deacetylase complex.

---

[57]Two homodimeric cGMP-dependent protein kinases-G exist (PKG1 and PKG2). In addition, PKG1 has 2 alternatively spliced forms (PKG1α and PKG1β).

[58]Epigenetic processes also regulate NOS isoforms. The proximal promoter of the Nos3 gene is non-methylated in endothelial cells, but heavily methylated in non-endothelial cell types [1174]. Transcription factors — E26 factor ETS1 and specificity proteins SP1 (stimulatory) and SP3 (inhibitory) — are recruited to the Nos3 promoter in endothelial cells, but not in vascular smooth muscle cells. On the other hand, methyl-CpG-binding protein MeCP2 is preferentially recruited to the Nos3 promoter in vascular smooth muscle cells. In chondrocytes, the synthesis of NOS2 is initiated upon methylation of Lys4 of histone-3. In neurons, nuclear factor-κB activates the NOS1 promoter, once chromatin is remodeled [1174].

Table 10.9. NO targets in the cardiovascular system (Source: [1179]).

Target	Function
Guanylate cyclase	cGMP formation
NADPH oxidase	Enzyme inhibition
Tissue plasminogen activator	Vasodilation, platelet inhibition
Ca^{++}-activated K^+ channel	Vasodilation
Cyclooxygenase-2	Prostaglandin synthesis
Cytoskeleton proteins	Dysfunction

The ubiquitin-dependent N-end rule pathway recognizes degradation signals based on a destabilizing N-terminal residue. Oxidation of N-terminal cysteine (before its arginylation) of regulatory proteins requires nitric oxide. In the heart, NO regulates particularly proteolysis of regulator of G-protein signaling RGS4, RGS5, and RGS16 [1175].

Perturbation of the PI3K–PKB–NOS3 cascade pertains to the set of mechanisms responsible for the reduction and dysfunction of bone marrow-derived endothelial progenitor cells in type-2 diabetes mellitus [1176].[59]

10.3.6.1 NO Activity in Blood and Lymph Vessels

Once bound to the heme iron of soluble guanylate cyclase, nitric oxide ultimately causes smooth muscle relaxation [1177, 1178] (Table 10.9). Subtype NOS3 is phosphorylated (activated) by Src kinase, protein kinases PKA, PKB, and PKC, and AMP-activated protein kinase. In endothelial cells, NOS3 is also activated via angiotensin-2 AT_1 and bradykinin B_2 receptors. Phosphorylation of NOS3 by the Src–PI3K–PKB axis causes AT_1–NOS3 dissociation and subsequent increase in NO synthesis [1185].

The NOS–NO signaling axis is vasoprotective, preventing the occurrence of pulmonary arterial hypertension. In addition, nitrite anion, a complementary source of NO, yields another adaptive, vasoregulatory pathway in the pulmonary and systemic vasculatures. Multiple nitrite reductases convert nitrite to NO, such as deoxyhemoglobin and myoglobin in the circulation and heart, respectively, and xanthine oxidoreductase in the lung parenchyma [1159].

---

[59] Hyperglycemia increases oxidative stress, thereby causing dysfunction in endothelial progenitor cells. The combination of an elevated plasma oxLDL level and hyperglycemia further aggravates the impaired migration of endothelial progenitor cells and NO production.

In the heart, abundant myoglobin act as a nitrite reductase.[60] At least in the murine vasculature, myoglobin (at very low concentrations in vascular smooth muscle cells) contributes significantly to nitrite-induced vasodilation [1180].

Angiogenesis after injury and ischemia allows tissue repair. It relies, at least partly, on pro-angiogenic nitric oxide synthesized by NOS3 synthase activated by PKB kinase. Nitric oxide represses anti-angiogenic thrombospondin-2 [1181]. Thrombospondins enhance clearance of matrix metallopeptidases MMP2 and MMP9 and interact with plasmalemmal receptors, such as $\alpha_V\beta_3$-integrin, very low density lipoprotein receptor (VLDLR), thrombospondin receptor CD36 (or scavenger receptor ScaRb3), and integrin-associated protein CD47, another thrombospondin receptor, to inhibit angiogenesis. Conversely, thrombospondin-1 impedes the ability of NO to activate soluble guanylate cyclase in vascular smooth muscle cells and NOS3 activity by blocking phosphorylation (Ser1176), hence endothelial-dependent arterial relaxation [1182].

NO in Lymphatics

Immune surveillance relies on antigen-presenting cell transport in lymphatics to bring antigens in draining lymph nodes. *Initial lymphatics* are endothelium-lined conduits without perivascular cells (pericytes and smooth muscle cells). They are endowed of overlapping cell junctions forming primary valves in addition to traditional secondary lymph valves [1183]. Fluid transport inside lymphatics relies on surrounding tissue motions that generate lymphatic expansion and compression. On the other hand, *collecting contractile lymphatics* are equipped with smooth muscle cells and bileaflet valves to carry fluid from initial lymphatics to lymph nodes. Lymphatic transport thus depends on smooth muscle cell contraction, hence on regulators of cell contraction and relaxation. In particular, it is influenced by nitric oxide. Under physiological conditions, NO produced by NOS3 in lymphatic endothelial cells is required for autonomous, periodic contractions of lymphatics [1184]. During inflammation, NO synthesized by NOS2 in bone marrow-derived CD11b+ myeloid cells that infiltrate surrounding tissue of contractile lymphatics attenuates lymphatic contraction [1184].

Nitric Oxide, Hemoglobin, and Oxygen Content

Nitric oxide targets heme-containing molecules to form NO–hemoprotein complexes (NO–hemoglobin, NO–myoglobin, NO–sGC, etc.). Nitric oxide binds to

---

[60] Myoglobin is a more potent reductase at low pH and low oxygen partial pressures [1180]. Nitrite reduction by myoglobin yields NO and metmyoglobin. Metmyoglobin is a relatively inefficient NO scavenger that allows NO to be liberated, whereas methemoglobin is a more abundant and more efficient scavenger that can impede NO release.

hemoglobin about 10^6 times more tightly than oxygen. Extracellular, but not intracellular, hemoglobin interferes with NO-dependent vasoactivity. A cycle of reversible NO binding to hemoglobin that sequesters NO in erythrocytes and releases it from these cells participates in respiratory gas transport in the tissues. Hypoxia drives more or less simultaneous unloading of NO and oxygen. The relative number of hemoglobin molecules in erythrocytes that carry out the vasoactive function is very small ($\sim 0.1\%$ of the total amount).

Oxyhemoglobin lowers cGMP levels, thereby impeding effects of endothelium-dependent relaxants acetylcholine[61] and bradykinin on isolated rings of bovine intrapulmonary artery and vein, as well as of nitric oxide radical on endothelium-denuded rings.

Nitric Oxide and Natriuretic Peptides

Nitric oxide and atrial natriuretic peptide provoke cGMP synthesis and influence cell functioning via cGMP-dependent protein kinases (PKG) and cGMP-specific phosphodiesterases (for signaling termination), as well as cyclic nucleotide-activated ion channels. They prime 2 different pathways in vascular smooth muscle cells and cardiomyocytes owing to spatial segregation of cGMP signaling [1186].

Natriuretic peptides ANP, BNP, and CNP target their specific plasmalemmal receptor guanylate cyclases NPR1 (or pGCa) and NPR2 (or pGCb) to cause sustained, compartmentalized cGMP synthesis.[62] Nitric oxide activates cytosolic soluble guanylate cyclases sGCα1 and sGCβ1 to transiently produce cGMP messenger.

In aortic vascular smooth muscle cells, nitric oxide lowers the expression of Gi-coupled natriuretic peptide receptor NPR3 via a cGMP-independent, MAPK-dependent pathway [1187]. Decreased expression of Gα_i subunit of G protein is delayed with respect to that of NPR3 receptor. Phosphorylated extracellular signal-regulated kinases ERK1 and ERK2 increases and then decreases.

Vascular Cell Fate

Nitric oxide inhibits growth and migration of vascular smooth muscle cells. On the other hand, NO stimulates endothelial cell proliferation and prevents endothelial cell apoptosis. Moreover, nitric oxide provokes apoptosis of vascular smooth muscle cells via reactive nitrogen species. Factor P53 protects vascular smooth muscle cells from NO-induced apoptosis, as it promotes synthesis of anti-oxidant

---

[61] Acetylcholine acts on muscarinic receptors of endothelial cells to prime NO release, thereby relaxing vascular smooth muscle cells.

[62] Cyclic guanosine monophosphate production is spatially confined in vascular smooth muscle cells and cardiomyocytes due to PDE5 phosphodiesterase.

proteins such as peroxiredoxin-3 (PRx3) [1188]. Anti-oxidant PRx3 operates with its specific electron acceptor thioredoxin-2. Transcription factor P53 can also exert its protection via mitogen-activated protein kinase modules and heme oxygenase-2.

Matrix Metallopeptidases

Remodeling of the extracellular matrix requires matrix metallopeptidases. Nitric oxide regulates MMP1, MMP9, and MMP13 [1189]. It controls MMP9 secretion from macrophages via protein modification mediated by reactive nitrogen species, soluble guanylate-cyclase-dependent modulation of the MMP9–TIMP1 balance, as well as MMP1- and MMP13-dependent cleavage of MMP9 enzyme.

Sirtuin

During caloric restriction, sirtuin-1 deacetylase[63] decreases arterial blood pressure and provokes vasodilation. Sirtuin-1 deacetylates NOS3 [1190]. Among other possible factors, it reduces both systolic and diastolic blood pressure during long-term reduced caloric intake not only in overweight individuals, but also in healthy subjects.

Tumor vasculature has abnormal structure, organization, and function that compromise both tumor oxygenation and delivery of antitumor agents. Nitric oxide mediates the effects of many angiogenic factors, such as vascular endothelial growth factor, angiopoietin-1, and sphingosine 1-phosphate. It can also induce the expression of angiogenic factors, such as VEGF and FGF2. Elimination of nitric oxide production from tumor cells by inhibiting NOS1 creates a perivascular NO gradient that normalizes the tumor vasculature [1191].

10.3.6.2 NO Activity in the Heart

In the heart, NO is an important regulator of coronary perfusion and cardiac contractility [1192]. Both NOS1 and NOS3 are constitutively expressed in cardiomyocytes. In addition, NOS3 abounds in endothelial and endocardial cells and NOS1 in both adrenergic and cholinergic nervous fibers.[64] In fact, NO is synthetized by all cardiac cell types.

---

[63] Sirtuin-1 targets many substances, such as transcription factors FoxO, P53, PGC1α, as well as transcriptional coactivators P300 and CBP.
[64] Subtype NOS3 mainly localizes to both coronary endothelium and cardiac endocardium, as well as, to a lesser extent, to cardiomyocytes, platelets, and monocytes.

In cardiomyocytes, NOS lodges in distinct subcellular compartments.[65] Synthase NOS3 resides in sarcolemmal and T-tubular caveolae, associated with myocyte-specific caveolin-3.[66] It is also observed in the sarcoplasmic reticulum, associated with ryanodine receptors, and mitochondria. Enzyme NOS1 localizes to the sarcolemma and sarcoplasmic reticulum;[67] NOS1α in mitochondria.

Nitric oxide synthesized in cardiomyocyte exerts intracrine effects as well as paracrine action on adjacent cardiomyocytes. Nitric oxide generated from non-cardiomyocyte sources (endocardial cells, coronary vascular cells, autonomic neurons, fibroblasts, and blood cells) operates on cardiomyocytes.

In the heart, nitric oxide produced by endothelial cells favors the perfusion via vasodilation and prevention of platelet aggregation. Nitric oxide in endothelial cells subjected to mechanical and chemical stimulators increases the intracellular calcium level that activates NOS3 enzyme. It indeed promotes the binding of Ca^{++}–calmodulin to NOS3 and activation of PI3K and, subsequently, PKB kinase. Nitric oxide produced by cardiomyocytes regulates the force and rate of contraction. Moreover, noradrenaline and acetylcholine released from autonomic nerve endings stimulate their respective receptors to activate NOS3 in cardiomyocytes (Fig. 10.1).

Effects of NO on Cardiac Ion Carriers

Nitric oxide modulates the activity of cardiac ion channels involved in the genesis of the cardiac action potential and can exert anti-arrhythmic effects via cGMP-dependent (protein kinase-G and phosphodiesterases) and cGMP-independent mechanisms (S-nitrosylation and direct effects on G proteins) [1193].

In the sarcolemma, NOS1 interacts with Na^+–K^+ ATPase and plasma membrane Ca^{++}–calmodulin-dependent Ca^{++} ATPase PMCA4b. In the sarcoplasmic reticulum, NOS1 is associated with ryanodine receptor RyR2 and Ca^{++} ATPase SERCA2 to regulate intracellular calcium cycling and excitation–contraction coupling.

Sodium Carriers

In atrio- and ventriculomyocytes, NO either inhibits inward Na^+ current without altering channel conductance and kinetics via both PKG and PKA, or causes a late Na^+ flux via S-nitrosylation of cardiac $Na_V1.5$ channel encoded by the

---

[65]Whereas constitutive NOS1 and NOS3 depend on Ca^{++}–calmodulin, inducible NOS2 does not depends on Ca^{++} and produces much higher NO levels. Inducible NOS2 is produced in cardiomyocytes, vascular endothelial and smooth muscle cells, fibroblasts, and macrophages.

[66]Enzyme NOS3 colocalizes with β-adrenoceptors, muscarinic M_2 receptors, and voltage-gated ion channels in sarcolemmal caveolae.

[67]On the sarcolemma, NOS1 links to Na^+–K^+ ATPase and PMCA pump; on the sarcoplasmic reticulum to SERCA2 pump and RyR channels.

SCN5A gene [1193]. In the sarcolemma, $Na_V1.5$ complexes with NOS1 and its inhibitor PMCA4b, as well as α1-syntrophin, a scaffold for NOS1, PMCA4b, and α1-syntrophin.

On the other hand, NO does not seem to influence sarcolemmal Na^+–Ca^{++} exchanger. Both NOS1 and NOS3 colocalize with caveolin-3 and Na^+–K^+ ATPase in the sarcolemma of cardiomyocytes. Nitric oxide increases the activity of these pumps.

Nitric oxide exerts a cGMP-dependent stimulation on pacemaker, hyperpolarization-activated, cyclic nucleotide-gated, HCN1 to HCN4 channels that convey an inward, mixed Na^+ and K^+ flux.[68]

Calcium Channels

Nitric oxide can increase, decrease, or produce a biphasic effect on $Ca_V1.2$ channel, according to its concentration and the context [1193]. S-nitrosylation reduces its functioning. Nitric oxide regulates β-adrenoceptors that activates $Ca_V1.2$ via the ACase–cAMP–PKA pathway. Adaptor NOS1AP inhibits $Ca_V1.2$ channel.

Cardiac excitation–contraction coupling relies on ryanodine receptors. S-nitrosylation supports channel activation.[69] In addition, NOS3-derived NO enhances the open probability of RyRs and frequency of Ca^{++} sparks in response to myocardium stretch via the PI3K–PKB–NOS3 pathway [1193]. In rat ventriculomyocytes, NO influences RyR function according to the level of β-adrenergic stimulation and the state of PKA activation [1193].[70]

Potassium Channels

In human atrial and ventricular myocytes, $K_V4.3$ channel is responsible for early rapid repolarization (phase 1 of the action potential), thereby influencing the activation of other ion carriers that control repolarization ($Ca_V1.2$ and delayed rectifier K^+ channels). Nitric oxide inhibits $K_V4.2$ and $K_V4.3$ channels via PP2 phosphatase and PKA kinase [1193].

Ultrarapid delayed rectifier $K_V1.5$ in human atriomyocytes is repressed by NO via activation of the GC–cGMP–PKG pathway and S-nitrosylation [1193]. Rapid delayed rectifier $K_V11.1$ that contributes largely to phases 2 and 3 of the repolarization in most nodal cells and cardiomyocytes can be inhibited, in particular via radical oxygen species. Slow delayed rectifier $K_V7.1$ that activates slowly during depolarization and deactivates slowly during repolarization and are involved in the late repolarization phase also can be S-nitrosylated.

---

[68]These channels activates slowly during phases 3 and 4 and inactivates slowly during depolarization.
[69]Colocalized NOS1 causes S-nitrosylation of ryanodine receptors.
[70]At high level of PKA activation, NO decreases the frequency of Ca^{++} sparks.

Inward rectifiers $K_{IR}2.1$ to $K_{IR}2.3$ works over a limited range of membrane potentials (-30 to -40 mV), thereby acting in the late phase 3. In human atriomyocytes, NO augments the activity of these channels, thereby shortening the final phase of repolarization.[71]

Muscarinic M_2 receptor stimulates acetylcholine-activated $K_{IR}3.1$ to $K_{IR}3.4$ channels in the sinoatrial and atrioventricular nodes and atriomyocytes. Nitric oxide may potentiate the increase in $i_{K_{ACh}}$ current primed by acetylcholine [1193].

ATP-sensitive $K_{IR}6.1$ and $K_{IR}6.2$ in the sarcolemma and mitochondria couple cell metabolism to membrane potential and exert a cardioprotection during ischemia. In ventriculomyocytes, NO enhances the $i_{K_{ATP}}$ current via the GCase–cGMP–PKG pathway.

In guinea pig ventriculomyocytes, 2-pore domain K^+ channels TALK1 ($K_{2P}16.1$) and TALK2 ($K_{2P}17.1$) generate an outward current during ischemic preconditioning. Nitric oxide may assist in the protection against prolonged ischemia [1193].

Effects of NO on Heart Intrinsic Properties

In the cardiomyocyte, the regulation of intracrine nitric oxide signaling depends on the location (sarcolemma or sarcoplasmic reticulum) of NO synthase isoforms (NOS1 and NOS3) [1194]. Synthases NOS1 and NOS3 have opposite effects on Ca^{++} influx. Nitric oxide inhibits β-adrenergic-induced inotropy after compartmentation of NOS3 in caveolae, hampering Ca_V1 channels.[72] On the other hand, NOS1 associates with ryanodine receptors of the sarcoplasmic reticulum and then stimulates Ca^{++} release.

Isoform NOS3 facilitates the electromechanical coupling in response to sarcomere stretch. It actually sustains length-dependent slow increase in calcium transient and force generation in stretched fibers (*Anrep effect*). Isozyme NOS1 hinders calcium current through Ca_V1 and enhances calcium reuptake into the sarcoplasmic reticulum by SERCA pumps (Fig. 10.1). Hence, NOS1 and possibly paracrine NO from adjoining endothelial cells promote cardiomyocyte relaxation and thereby ventricular filling that increases stretch. Moreover, NOS1 and NOS3 attenuate β1- and β2-adrenergic positive inotropy and chronotropy. The NOS3 enzyme also potentiates acetylcholine activity. The NOS synthases reinforce pre- and postsynaptic vagal control of the cardiac contraction. They thus protect the heart against excessive stimulation by catecholamines. In an ischemic myocardium, NOS1 augments the effect of constituve NOS. Last but not least, NOS modulates oxygen consumption, promotes free fatty acids rather than glucose oxidation, struggles against oxidative stresses, prevents apoptosis at low concentrations, and acts in adaptive and regenerative processes.

---

[71] Effect of NO results from a selective S-nitrosylation of the major ventricular isoform $K_{IR}2.1$ channel.

[72] Isoform NOS3 binds to caveolin-3 in caveolae that also incorporate β-adrenergic receptors and Ca_V1 channels.

10.3 Nitric Oxide, Nitric Oxide Synthases, Nitrite, and Nitrate

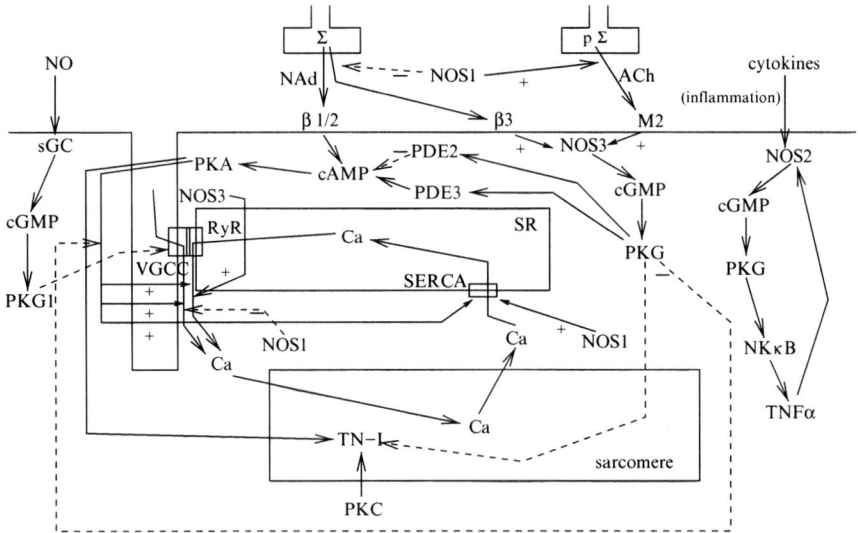

Fig. 10.1 Enzymes NOS1 and NOS3 cooperate to regulate the sympathovagal balance in the heart (Sources: [1179, 1192, 1198, 1199]). Isoform NOS1 acts in the parasympathetic and sympathetic endings. It potentiates acetylcholine release and hampers noradrenaline. release in the synaptic cleft. Isoform NOS3 (eNOS) is activated in cardiomyocytes by both stimulated muscarinic cholinergic and β3-adrenergic receptors. Activated NOS3 in the cardiomyocyte opposes adrenergic activity, but reinforces vagal input. Catecholamine stimulation of β3-adrenoceptors of cardiomyocytes activates NOS3 via $G\alpha_i$, which increases cGMP level and leads to negative inotropy. Colocalized NOS3 and ryanodine channels RyR2 (RC2) in the T-tubule–sarcoplasmic reticulum junction favor calcium influx. Messenger cGMP opposes cAMP-associated positive inotropy because it activates phosphodiesterase PDE2 for cAMP degradation, although cGMP can potentiate cAMP effects via inhibition of PDE3 enzyme. Activation of PKG reduces the $Ca_V1.2$ activity. Protein kinase-G also decreases myofilament sensitivity to calcium, thereby promoting relaxation. Enzyme NOS1 hinders $Ca_V1.2$ channels (VGCC), but promotes sarcoplasmic reticulum calcium ATPase (SERCA). Interstitial NO acts on soluble guanylate cyclase (sGC) for cGMP production, and activates PKG1 kinase. Activated PKG1 inhibits $Ca_V1.2a$ channels. Enzyme NOS2 triggered by cytokines and other inflammatory mediators in the cardiomyocyte activates the cGMP–PKG pathway, and hence NFκB, leading to TNFα synthesis. The latter upregulates NOS2 in cardiomyocytes.

Cyclic adenosine (cAMP) and guanosine (cGMP) monophosphate have opposing effects in cardiomyocytes.[73] A high NO concentration increases the cGMP level leading to negative inotropy by a PKG-dependent reduction in myofilament responsiveness to calcium. (Calcium ions remain available.) Low NO concentration

---

[73] Partners of cAMP and cGMP include: (1) cAMP- and cGMP-dependent protein kinases PKA and PKG; (2) several phosphodiesterases, and (3) A- and G-kinase-anchoring proteins. Phosphodiesterases PDE4 and PDE5 selectively degrade cAMP and cGMP, respectively. Both PDE1 and PDE2 target both cAMP or cGMP. Compartmentalized signaling events involve scaffold proteins AKAPs and GKAPs for PKA and PKG, respectively. Sphingosine kinase-1-interacting protein (SKIP), an anchor for sphingosine kinase-1, can serve as AKAP agent.

increases cAMP level by adenylate cyclase activation, leading to positive inotropy [1195]. The negative cGMP effects in cardiomyocytes are mediated by cGMP-gated ion channels, protein kinase-G, and phosphodiesterases. The cGMP-stimulated cAMP phosphodiesterase PDE2 and cGMP-inhibited cAMP phosphodiesterase PDE3 are regulated by the intracellular cGMP concentration. Enzyme PDE3 is inhibited by an increase in cGMP level that raises the cAMP level. Protein kinase-A then activates calcium channels, countering PKG effects. On the other hand, cGMP can stimulate PDE2, thereby reducing the cAMP level and PKA activity. Interaction between cGMP and cAMP signaling pathways is impaired in failing cardiomyocytes [1196].

Protein kinase-A phosphorylates Ca_V1 channels to increase calcium influx using calcium-induced calcium release, thereby enhancing cardiomyocyte contraction [1179]. Messenger cGMP stimulates PKG that inhibits Ca_V1 channels and activates BK channel and myosin light-chain phosphatase.

Adaptor NOS1AP[74] is a NOS1 regulator that interacts with NOS1, but not NOS3 subtype. It accelerates cardiac repolarization (short duration of action potential), as it inhibits $Ca_V1.2$ channels and enhances the activity of rapid delayed rectifier potassium channels ($K_V11.1$; i_{Kr} current) by cGMP-dependent or -independent pathways in guinea pig ventriculomyocytes [1197]. Adaptor NOS1AP directs NOS1 to specific target proteins.

Nitric oxide acts on soluble guanylate cyclase and triggers the cGMP–PKG pathway. Protein kinase-G1 then inhibits $Ca_V1.2a$ channels [1198]. Muscarinic receptor stimulation does not act on this cascade but stimulates calmodulin-dependent cardiac nitric oxide synthase. Muscarinic inhibition occurs rather via inhibition of cAMP signaling via $G\alpha_{i2}$-dependent inhibition of adenylate cyclase. Stimulation of β1- and β2-adrenergic receptor produces positive inotropy via $G\alpha_s$-mediated activation of adenylate cyclase that augments cAMP levels (Fig. 10.1).

Nitric oxide produced by inducible NOS2 in all cardiac cell types, including interstitial macrophages, under exposure to cytokines and other inflammatory mediators, exerts para- and autocrine effects. Large NO concentrations decrease the contraction of cardiomyocytes and vascular smooth muscle cells. Nitric oxide produced by NOS2 also causes apoptosis of various cell types. In the cardiomyocyte, NO provokes TNFα synthesis using the cGMP–PKG pathway. Agent TNFα induces NOS2 expression and yields a positive feedback inflammatory loop [1199].

Low myocardial NO availability during coronary vascular inflammation, ^{L}Arg methylation that produces NOS inhibitors, and low concentration of NO metabolites in coronary veins reduces NO lusitropic effect and restrains left ventricle filling [1200].

Nitric oxide operates in differentiation of embryonic stem cells into myocardial cells [1201]. Nitric oxide and its receptor, soluble guanylate cyclase, activate expression of cardiac genes (e.g., myosin light chain MLC2 and Nkx2-5 factor).

---

[74] A.k.a. carboxy-terminal PDZ ligand of NOS1 (CaPON).

10.3 Nitric Oxide, Nitric Oxide Synthases, Nitrite, and Nitrate

Table 10.10. Insulin-triggered pathways for cardioprotection (Source: [1202]; ↑: increase; ⊖: inhibition). Insulin binds to insulin receptor and activates 2 main signaling cascades: PI3K–PKB–NOS3 and the mitogen-activated protein kinase (MAPK) module. Activated insulin receptor (IR) phosphorylates insulin receptor substrate IRS1, leading to the binding and activation of phosphatidylinositol 3-kinase (PI3K), subsequently of phosphoinositide-dependent protein kinase PDK1 and protein kinase-B (PKB) that phosphorylates NOS3. Production of NO then rises within mn. In addition, PKB inhibits GSK3β and activates target of rapamycin (TOR) and P70 ribosomal S6 kinase. Insulin stimulates K^+ reuptake through Na^+–K^+ ATPase and glucose uptake via GluT4 transporter for glycolytic energy production. The insulin–MAPK pathway regulates the secretion of the vasoconstrictor endothelin-1 from endothelium.

Pathway	Effect
IR–IRS1–PI3K–PDK1–PKB–NOS3–NO	Vasodilation, coronary perfusion ↑
	Cell survival
	⊖ oxidative stress
	⊖ inflammation
IR–IRS1–PI3K–PDK1–PKB–GluT4	Glucose endocytosis
IR–IRS1–PI3K–PDK1–PKB–TOR	Protein synthesis
IR–Ras–MAPK	Cell proliferation
	Endothelin-1-mediated vasoconstriction

Insulin exerts a cardiovascular protection via the PI3K–PKB–NOS3 pathway in vascular endothelial cells and cardiomyocytes [1202]. Insulin then supports NO production by NOS3 enzyme. Nitric oxide causes vasodilation on the one hand and exerts anti-apoptotic and prosurvival effects on the other (Table 10.10).

10.3.6.3 Nitric Oxide and Nephron Transport

Nitric oxide influences several functions of the kidney: salt and fluid reabsorption, renal hemodynamics, and renin secretion. Three nitric oxide synthase isoforms (NOS1–NOS3) are expressed by nephron segments, in addition to renal vessels.

In the kidney, nitric oxide has natriuretic and diuretic effects, as NO modulates electrolyte and water transport in different segments of the nephron (Table 10.11). The natriuretic and diuretic effects of NO are not actually accompanied by proportional changes in glomerular filtration rate or renal blood flow. In the proximal tubule, NO inhibits Na^+–H^+ exchanger that is responsible for HCO_3^- reabsorption and basolateral Na^+–K^+ ATPase that yields energy required for Na^+-coupled transport [1203]. In the thick ascending limb, NO impedes the activity of Na^+–K^+–$2Cl^-$ cotransporter that reabsorbs NaCl and Na^+–H^+ exchanger that uptakes most of HCO_3^-, which escapes proximal reabsorption, whereas it stimulates apical K^+ channels that recycle K^+ across the apical membrane necessary for Na^+–K^+–$2Cl^-$ cotransport and positive luminal potential to provide driving force for paracellular transport of Ca^{++} and Mg^{++} ions. In the collecting duct, NO hinders H^+ ATPase. In addition, NO reduces vasopressin-stimulated water reabsorption.

Table 10.11. Nitric oxide effect on nephron transport (Source: [1203]). Proximal tubule reabsorbs 50 to 60% of filtered inorganic solutes and water, whereas organic solutes (glucids, amino acids, and other metabolites) are completely reabsorbed. Solutes are transported by apical Na^+-coupled cotransporters (e.g., Na^+–glucose, Na^+–PO_4^{--}, and Na^+–amino acid cotransporters) or exchangers (Na^+–H^+ exchanger). Water is reabsorbed by diffusion due to the osmotic gradient generated by active solute transport. Thick ascending limb reabsorbs 25 to 30% of filtered salt, but is water impermeable. Cortical collecting duct controls Na^+ absorption by epithelial (apical) Na^+ channels (ENaC) and basolateral Na^+–K^+ ATPase, as well as K^+ excretion by luminal and basolateral K^+ channels. Water transport that is low is stimulated by arginine vasopressin that augments apical aquaporin number. Cortical collecting ducts also secrete either H^+ or HCO_3^- whether α- or β-intercalated cells predominate. NO does not seem to influence outer medullary collecting duct transport. Inner medullary collecting duct is the last site that regulates salt and water excretion in urine. Na^+ reabsorption in this segment occurs mainly by apical epithelial Na^+ channels and basolateral Na^+–K^+ ATPase.

Nephron segment	Ion and water flux
Proximal tubule	Decrease in water and HCO_3^- flux
	Inhibition of Na^+–K^+ ATPase
	Inhibition of type-3 Na^+–H^+ exchanger
Thick ascending limb	Decrease in Cl^- and HCO_3^- flux
	Inhibition of type-3 Na^+–H^+ exchanger
	Inhibition of Na^+–K^+–$2Cl^-$ cotransporter
	Stimulation of K^+ channel
Cortical collecting duct	Decrease in water and Na^+ flux
	Inhibition or stimulation of ENaCs
	Inhibition or stimulation of K^+ channel
Inner medullary collecting duct	Decrease in Na^+ flux
	Inhibition of H^+ ATPase

Nitric oxide action on nephron transport depends not only on direct effect, but also indirect effect via renal blood flow. Moreover, both nephron cells and neighboring endothelial cells produce NO that operates on nephron with a delay that depends on diffusion time necessary to reach NO targets. Nitric oxide synthesis is regulated. Nitric oxide release can be triggered by many substances such as acetylcholine.

10.3.7 Interaction between Nitric Oxide and Hydrogen Sulfide

Nitric oxide cooperates with another gaseous mediator — hydrogen sulfide (H_2S) — in various organs, in particular the heart. Nitric oxide donors can enhance H_2S production in smooth muscle cells. Hydrogen sulfide exerts its cardioprotective effect, at least partially, by activating NOS1 and NOS3 [1208].

Owing to its strong reducing capability, H_2S may reduce NO to form nitroxyl (HNO), a thiol-sensitive molecule.[75] The latter has positive inotropic and lusitropic effects [1208].[76] Nitroxyl can operate independently of the cAMP–PKA and cGMP–PKG pathways, as well as β-adrenoceptor. Nitroxyl enhances calcium release from the sarcoplasmic reticulum, as it augments the open probability of ryanodine receptors. It also activates SERCA pumps by S-glutathiolation (Cys674) [1208].

10.4 Hydrogen Sulfide

Endogenous gaseous signaling molecules that are produced at micromolar levels include carbon monoxide, nitric oxide,[77] and hydrogen sulfide (H_2S).[78] They have vasculoprotective effects.

10.4.1 Hydrogen Sulfide Production

Hydrogen sulfide is synthesized from homocysteine metabolism by cystathionine β-synthase and cystathionine γ-lyase[79] in vascular smooth muscle cells as well as

---

[75] Inactive nitrosothiol may be formed from the reaction between NO and H_2S. The nitrosothiol intermediate may further react with thiols to form nitroxyl.

[76] In cardiomyocytes, NO produced by constitutive NOS1 and NOS3 modulates contractility via activated guanylate cyclase or nitrotyrosylation of proteins involved in calcium handling and sarcomere functioning (Sect. 10.3.6.2). Endogenous H_2S is cardioprotective. Nitric oxide donors produce negative inotropic effect in cardiomyocytes. On the other hand, simultaneous administration of NO and H_2S donors increases cardiomyocyte contractility. Nitroxyl donor mimics the cooperative effect of NO and H_2S donors. Thiols that are effective HNO scavenger suppress the effect of HNO donors as well that of simultaneous administration of NO and H_2S donors.

[77] Nitric oxide was the first identified gasotransmitter. It relaxes smooth muscle of blood vessels, reduces inflammation, stimulates hypothalamic hormone release, and transmits signals between neurons.

[78] Hydrogen sulfide, the so-called rotten egg gas, is known for its very high toxicity. Exposure threshold is set at 10 ppm. Breathing 500 ppm of exogenous H_2S causes death.

[79] Cystathionine β-synthase and γ-lyase catalyze the synthesis of cystathionine from ^{L}serine and ^{L}homocysteine (β-replacement reaction) and the degradation of cystathionine into ^{L}cysteine, α-ketobutyrate, and ammonia (γ-elimination reaction), respectively [1209]. In addition to these main reactions, cystathionine β-synthase and γ-lyase can also catalyze the synthesis of cystathionine from ^{L}cysteine and ^{L}homocysteine to generate H_2S. Yet, cystathionine γ-lyase prefers ^{L}cystine to ^{L}cysteine and the degradion leads to pyruvate, ammonia, and thiocysteine.

in many tissues, such as the brain, liver, kidney, ileum, uterus, and placenta [1209].[80] Both enzymes contain heme. Their activity depends on cofactor pyridoxal 5′-phosphate. Calcium–calmodulin-activated cystathionine γ-lyase is predominantly located in the vascular endothelium.

Therefore, the trans-sulfuration pathway convert the risk factor homocysteine to the protectant hydrogen sulfide. The extracellular trans-sulfuration enzymes — cystathionine β-synthase and cystathionine γ-lyase — are secreted by microvascular endothelial cells and hepatocytes, circulate in blood, and produce hydrogen sulfide to avoid damages caused by free homocysteine [1210].

An additional enzyme — 3-mercaptopyruvate sulfurtransferase — coupled with cysteine (aspartate) aminotransferase also produces H_2S in the presence of α-ketoglutarate [1209].[81] This enzyme localizes to the brain and vascular endothelium (e.g., thoracic aorta).

When supplied with elemental sulfur or inorganic polysulfides, erythrocytes produce H_2S. In addition, garlic-derived organic polysulfides are converted by erythrocytes and vascular cells into hydrogen sulfide gas [1211].[82]

Like for other gaseous mediators NO and CO, H_2S production is improved by muscarinic cholinergic stimulation of the vascular endothelium. Muscarinic receptors activates cystathionine γ-lyase via the binding to calcium–calmodulin [1209]. However, the mechanism of action of H_2S differ from that of NO and CO that activate soluble guanylate cyclase and increase the intracellular cGMP concentration.

In addition, feedbacks exist among different gasotransmitters that enable the regulation of the action of a given gas by others gaseous mediators [1209].

10.4.2 Hydrogen Sulfide Effects

Endogenous hydrogen sulfide modulates neuronal transmission, participates in the endothelium-dependent vasorelaxation, stimulates angiogenesis, and regulates

---

[80]The rate of H_2S production in tissue homogenates ranges about 1 to 10 pmol/s/mg protein [1208]. Human plasma content of H_2S may equal approximately 50μmol (but some investigations show that free H_2S gas is undetectable).

[81]Cysteine aminotransferase is similar to aspartate aminotransferase. It catalyzes the transamination between ^{l}-cysteine and α-ketoglutarate to produce 3-mercaptopyruvate and ^{l}-glutamate. 3-Mercaptopyruvate sulfurtransferase then transfers sulfur from 3-mercaptopyruvate to sulfurous acid to produce pyruvate and thiosulfate, which is finally reduced to H_2S by another sulfurtransferase in the presence of reduced glutathione. Cysteine aminotransferase prefers ^{l}-aspartate than to ^{l}-cysteine.

[82]Mutant mice lacking cystathionine γ-lyase have reduced H_2S levels in blood, heart, aorta, and other tissues and develop hypertension [1212]. In addition, vasorelaxation in response to muscarinic stimulation (that triggers calcium influx to form calcium–calmodulin complex) is deeply attenuated.

insulin release.[83] It relaxes vascular smooth muscle cells via a possible activation of ion channels, in particular K_{ATP} channels, thereby hyperpolarizing the membrane potential. Hydrogen sulfide acts as a neuromodulator and metabolic regulator that lowers metabolism. At normal concentrations, it transiently inhibits mitochondrial respiration in a dose-dependent manner. However, its main function is to preserve mitochondrial function. Many of its effects are mediated by potassium channels that cause K^+ outflux.

Moreover, H_2S has anti-inflammatory and anti-oxidant effects. This endogenous gasotransmitter inhibits leukocyte–endothelial cell interactions. Anti-oxidant H_2S gas counteracts oxidative and proteolytic stresses, as it opposes activation of matrix metallopeptidases by reactive oxygen and nitrogen species, increases concentrations of tissue inhibitors of metallopeptidases, and lowers apoptosis and fibrosis in chronic heart failure [1213]. Cytoprotectant H_2S reduces intimal hyperplasia (restenosis) [1214] and favors angiogenesis [1215].

10.4.3 Hydrogen Sulfide in Mitochondria

Endogenous H_2S can regulate energy production in cells during cell stress to cope with energy demand when oxygen supply is insufficient. In vascular smooth muscle cells under normal conditions, cystathionine γ-lyase localizes to the cytosol. Under stress conditions, this enzyme translocates from the cytosol to mitochondria via 20-kDa translocase of the outer mitochondrial membrane TOM20 [1216].[84] Cysteine concentration in mitochondria is approximately 3 times that in the cytosol. Mitochondria can use sulfide as an energetic substrate at low micromolar concentrations to maintain ATP production under stress conditions. Translocation of cystathionine γ-lyase to mitochondria processes cysteine, produces H_2S inside

---

[83]In the brain, hydrogen sulfide acts as a neuromodulator that facilitates hippocampal long-term potentiation, as it enhances the activity of NMDA receptors in neurons and increases Ca^{++} influx into astrocytes [1209]. In the thoracic aorta and portal vein, as well as ileum, H_2S is a smooth muscle relaxant in synergy with nitric oxide.

[84]Except proteins synthesized within mitochondria, mitochondrial proteic constituents are produced in the cytosol as preproteins. These preproteins are then imported into mitochondrial compartments (outer membrane, intermembrane space, inner membrane, and matrix) through translocases of the outer and inner mitochondrial membranes. Preproteins are then processed to their mature forms. In addition, translocases are used by tRNA and heme oxygenase-1 that produces carbon monoxide. Upon stimulation with the calcium ionophore (a lipid-soluble carrier of ions across the lipid bilayer of the cell membrane, usually synthesized by microorganisms) A23187 (a carboxylic acid antibiotic derived from Streptomyces chartreusensis that transfers calcium, magnesium, and other divalent cations across biological membranes) and increased intracellular calcium concentration, cystathionine γ-lyase translocates from the cytosol into mitochondria, and H_2S is then produced inside mitochondria [1216].

mitochondria, and improves ATP production under hypoxia. Hence, at physiological concentrations,[85] H_2S preserves structural and functional integrity of mitochondria. In the cardiovascular system, H_2 can then mitigate ischemia injury.

10.5 NADPH and Dual Oxidases

Nicotinamide adenine dinucleotide phosphate (NADPH) oxidase is a membrane-bound enzyme complex. These enzymes comprise many subtypes with different cellular locations. The NADPH oxidase family actually includes 7 known members, i.e., 5 types of *NADPH oxidases* (NOx1–NOx5) and 2 types of *dual oxidases* (DuOx1–DuOx2). Every cell type of the vessel wall contains the reduced form of nicotinamide adenine dinucleotide phosphate (NAD[P]H) oxidases (NOx).

Dual oxidases are sensitive to Ca^{++} ions and intervene in Ca^{++} signaling.

The NADPH oxidases belong to the small set of superoxide-producing enzymes of the vasculature.[86] Therefore, they produce reactive oxygen species (ROS) [1217] (Fig. 10.2). Isoforms NOx2 and NOx4 are responsible for basal ROS production in vascular endothelial and smooth muscle cells.

Reactive oxygen species are oxygen-derived small molecules, i.e., oxygen radicals, such as superoxide ($O_2^{\bullet -}$, hydroxyl ($^\bullet OH$), peroxyl ($RO_2^\bullet$), alkoxyl ($RO^\bullet$), and certain non-radicals, such as hypochlorous acid (HOCl), ozone (O_3), singlet oxygen ($^1O_2$), and hydrogen peroxide (H_2O_2) [1218]. Reactive oxygen species act on cell growth and apoptosis, and cause vasodilation, as they stimulate protein kinases and inhibit protein Tyr phosphatases to activate various transcription factors.

In phagocytes, external signals stimulate their specific plasmalemmal receptors and then trigger the PI3K–PI(3,4,5)P_3 pathway. The latter stimulates NADPH oxidase via Rac GTPase. NADPH oxidase uses NADPH as an electron donor to carry out the first step of ROS formation, i.e., the reduction of oxygen to superoxide radical (O_2^-):

$$NADPH + 2O_2 \rightarrow NADP^+ + H^+ + O_2^{-\bullet}.$$

In addition to NOx activation via Rac, the second messengers diacylglycerol and calcium may act via protein kinase-C. The latter phosphorylates P47PhOx, a cytoplasmic component of NADPH oxidase, to assemble the NADPH complex [10].

---

[85] At toxic concentrations, H_2S inhibits cytochrome-C oxidase of the mitochondrial respiratory complex-IV.

[86] The set of superoxide-producing enzymes of the vasculature includes xanthine oxidase, cytochrome-P450, and uncoupled nitric oxide synthase.

10.5 NADPH and Dual Oxidases

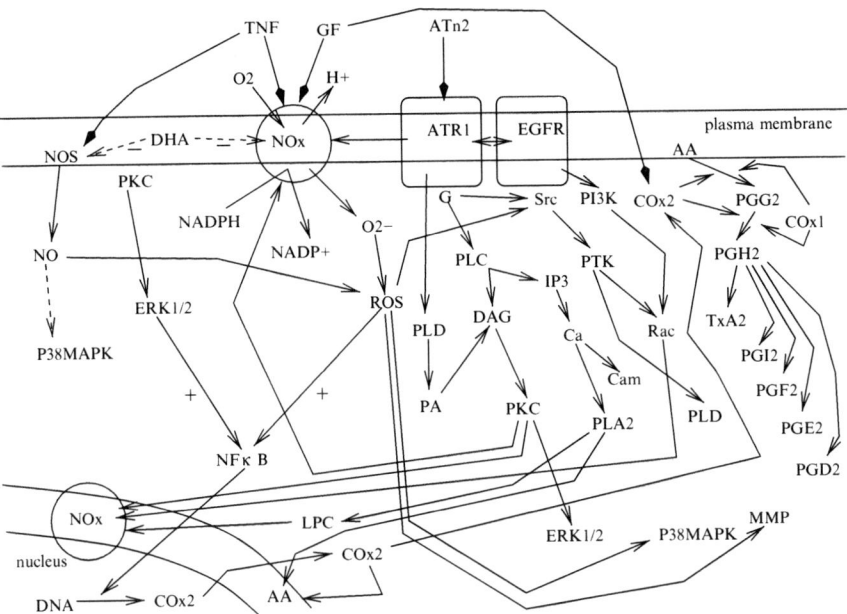

Fig. 10.2 NAD(P)H oxidase (NOx) in vascular cells (Sources: [1217, 1219]. This enzyme type is upregulated in vascular smooth muscle cells by platelet-derived growth factor (PDGF), transforming growth factor-β (TGF), and tumor-necrosis factor-α (TNF), and, in endothelial cells, by angiotensin-2 (ATn2). Stimulation of AT_1 receptor (AT1R) by angiotensin-2 activates phospholipase-C (PLC) via G protein. Phospholipase-C produces inositol trisphosphate (IP_3) and diacylglycerol (DAG), thereby releasing calcium from intracellular stores and activating protein kinase-C (PKC), respectively. The latter phosphorylates the catalytic NOx subunit. Calcium ions activate phospholipase-A2 (PLA2), which produces lysophosphatidylcholine and arachidonic acid (AA). Arachidonic acid is metabolized into prostaglandins by constitutive (COx1) and inducible (COx2) cyclooxygenases. Angiotensin-2 also activates Src kinase, which participates to the activation of EGF receptor (EGFR) and activates protein tyrosine kinases. Activated phospholipase-D produces phosphatidic acid (PA), which is converted to DAG, subsequently enhancing PKC activity. Reactive oxygen species (ROS) activate Src associated with EGFR receptor. Long-duration exposure of vascular endothelial cells to ω3 fatty acid docosahexaenoate (DHA) causes its incorporation into the plasma membrane. Docosahexaenoate inhibits nuclear factor-κB (NFκB) and, subsequently, COx2 enzyme. Furthermore, it decreases the production of reactive oxygen species by inhibiting NADPH oxidase (NOx), and reduces PKC association to the plasma membrane. Last but not least, NOx is stimulated by pulsatile hemodynamic stress.

10.5.1 Structure

Nicotinamide adenine dinucleotide phosphate oxidases comprise a set of transmembrane proteins with 6 transmembrane domains that have binding sites for FAD and NADPH and 4 heme-binding histidines in the third and fifth transmembrane domains to produce ROS by transferring an electron to molecular oxygen. Members of the NOX family represent an important source of reactive oxygen species that oxidize nucleic acids (e.g., DNA), carbohydrates, proteins, and lipids.

Table 10.12. Subunits of NADPH oxidases that connect to NOx1 to NOx5 catalytic components (Sources: [76, 1043]; CyB: cytochrome-B245, or cytochrome-B558; CyBL(H)C: CyB light (heavy) chain; NCF: neutrophil cytosol factor; NOxA1: NOx activator; NOxO1: NOx organizer; SH3PXD: SH3 and PX domain-containing protein; SNx: sorting nexin). Phagocyte NADPH oxidase consists of a dimer of transmembrane subunits (GP91PhOx and P22PhOx) associated with 4 cytosolic subunits (P40PhOx, P47PhOx, P67PhOx, and Rac2).

Subunits	Other aliases	Interactors
CyBα	CyBLC, P22PhOx	NOxA2, NOxO2, P40PhOx, NOx1/4
CyBβ	CyBHC, NOx2, GP91PhOx, GP91-1	CyBα, NOxA2, Rac1, IQGAP1, actin-γ1
NOxA1	NCF2L, P51NOx, P67PhOxL	NOxO1, NOx1, Rac2, PKCγ, 14-3-3ζ
NOxA2	NCF2, P67PhOx	NOxO2, P40PhOx, Rac1/2, CDC42
NOxO1	P41NOx, P41NOxA/B/C, SH3PXD5, SNx28	CyBα/β, NOxA1, NoxO1, TβR1,
NOxO2	NCF1, NCF1a, P47PhOx, SH3PXD1a	
P40PhOx	NCF4, SH3PXD4	CyBα, NOxA2, NOxO2, PKCδ, coronin-1A
Rac1		
Rac2		

NADPH oxidases are composed of several subunits. Nine regulatory components exist, including 2 NADPH oxidase organizer subunits — NOxO1 and NOxO2 (P47PhOx) — 2 NADPH oxidase activator subunits — NOxA1 and NOxA2 (P67PhOx) — as well as 2 Rac isozymes (Rac1 and Rac2). The other subunits encompass P22PhOx, P40PhOx, and GP91PhOx[87] (Tables 10.12 and 10.13).

Membrane-bound *cytochrome-B245* is a heterodimer formed by a 22-kDa polypeptide (P22PhOx, or light α chain encoded by the CYBA gene [cytochrome-B245 and -B558 (CyB) α-polypeptide]) and a 91-kDa glycoprotein (GP91PhOx, or heavy β chain encoded by the CYBB gene [CyB β-polypeptide]) and a component of NADPH oxidase, which synthesizes superoxide (O_2^-).

The catalytic component of *cytochrome-B558* localizes to the membrane; it is constituted by the GP91PhOx–P22PhOx heterodimer (CyBβ–CyBβ or CyBHC–CyBLC). Other regulatory components, such as P47PhOx and P67PhOx, as well as Rap1A and Rac regulators, reside in the cytoplasm.

---

[87] A.k.a. NADPH oxidase-2 (NOx2). Alias PhOx stands for phagocytic oxidase.

Table 10.13. Composition of NADPH oxidase isoforms (Source: [1218]). Component P22PhOx is essential for NOx activity; it complexes with other subunits to form NOx oxidases (NOx1–NOx3). In general, the organizer subunit of NOx1 and NOx3 is NOxO1, and that of NOx2 is NOxO2 (P47PhOx); the activator subunit for NOx1 is NOxA1 and that for NOx2 is NOxA2 (P67PhOx). Isotype NOx3 may not need an activator subunit. Isozyme NOx4 does not require cytosolic subunits. Isozymes NOx5, DuOx1, and DuOx2 are activated by Ca^{++} ions; they do not seem to need subunits. Production of P22PhOx subunit is controlled by AP1 and NFκB transcription factors.

Catalytic subunit	Associated regulatory subunits
NOx1	P22PhOx, NOxO1, NOxA1, Rac1
NOx2	P22PhOx, NOxO2, NOxA2, Rac, possibly CyBβ
NOx3	P22PhOx, NOxO1, Rac1
NOx4	P22PhOx
NOx5	
DuOx1	
DuOx2	

10.5.2 Subtypes

A large number of potential combinations of subunits exists. Therefore, NADPH oxidases are categorized according to NOx constituents. NADPH oxidase isoforms are tissue specific.

NADPH oxidase isoforms NOx1 to NOx3 (Tables 10.14 to 10.16) are complexes with an activator P22PhOx subunit [1220]. However, NOx2 requires 2 additional activator subunits (P47PhOx and P67PhOx) in addition to P22PhOx to be active. Subunit P47PhOx can be phosphorylated by different kinases, such as PKCζ and P38MAPK. Regulatory subunits, such as NOxO1 and -2 and NOxA1 and -2, are components of different oxidase complexes.

All members of the NOX family share common structural characteristics, such as 6 hydrophobic transmembrane domains, NADPH- and FAD-binding motifs in the cytoplasmic region, and 2 heme moieties in the intramembranous sequence.

In addition, NOx5 contains an N-terminal extension with 4 Ca^{++}-binding EF hand domains [1222].[88] Whereas NOx1, NOx2, and NOx3 require cytosolic subunits and cofactors for full activity, NOx5 can be activated by Ca^{++} ion alone.

---

[88] The EF-hand domains of NOx5 form 2 pairs: N-terminal and C-terminal pairs that bind Ca^{++} with low and high affinity, respectively.

Table 10.14. Subtypes of NADPH oxidases (catalytic subunits and corresponding [possibly heteromeric] oxidase; Sources: [76, 1043]; KOx: kidney oxidase; LNOx: long (large) NOx; MOx: mitogenic oxidase; NOD: nucleotide-binding oligomerization domain-containing protein; NOH: NADPH oxidase homolog; NOxEF: NADH/NADPH oxidase/peroxidase EF hand domain-containing protein; ReNOx: renal NADPH-oxidase; ThOx: thyroid oxidase; TPO: thyroid peroxidase; TxnDC: thioredoxin domain-containing protein).

Type	Other aliases	Interactors
NOx1	MOx1, NOH1, GP91-2, P65MOx	NOxA1, CyBα/β, MAPK11, RelA
NOx2	CyBβ, GP91-1, GP91PhOx	CyBα, NOxA2, Rac1, IQGAP1, actin-γ1
NOx3	MOx2, GP91-3	MAPK11–MAPK13, MMP2/9
NOx4	KOx, KOx1, ReNOx	CyBα, TLR4, RelA, PTPn1
NOx5	NOx5a/b	Abl
DuOx1	ThOx1, LNOx1, NOxEF1	TPO, TxnDC11
DuOx2	ThOx2, LNOx2, NOxEF2	NOD2, TxnDC11

However, NADPH oxidases without EF hand domains associates with partners to sense and react to elevated intracellular Ca^{++} concentration. For example, S100A8 and S100A9 proteins with 2 EF-hands that are expressed mainly in phagocytes heterodimerize upon increased Ca^{++} concentration; this dimer then connects directly to the P22PhOx–GP91PhOx complex [1222].

In addition, NOx5 is activated by Abl kinase via a Ca^{++}-mediated, redox-dependent signaling pathway [1223]. Therefore, a positive feedback occurs, as H_2O_2 can activate NOx5 via cytosolic protein Tyr kinase Abl.

Four NOx5 isoforms (NOx5α–NOx5δ) possess a long intracellular N-terminus that has Ca^{++}-binding EF hand domains. They are produced in human aortic smooth muscle cells. A fifth NOx5 isoform (NOx5ε or NOX5$_S$) lacks the EF-hand region, thereby resembling NOx1 to NOx4 isoforms [1218]. Unlike other NOx subtypes, NOx5 does need for its activity neither P22PhOx, nor cytosolic organizer or activator subunits, nor any cytosolic proteins.

NADPH oxidase-1 (NOx1) is a pyridine nucleotide-dependent, superoxide-generating oxidase homolog of the catalytic subunit of the NADPH oxidase of phagocytes (GP91PhOx). NADPH oxidase NOx1 contains regulatory cytosolic subunits P47PhOx (NOxO1), P67PhOx (NOxA2), and Rac1 GTPase. Phosphorylation by protein kinase-A of NoxA2 (Ser172 and Ser461) provokes its interaction with 14-3-3ζ protein and subsequent inhibition of NOx1 activity.

Subtype NOx2 possesses several regulatory components (P22PhOx, P4OPhOx, P47PhOx, P67PhOx, and Rac1 GTPase), in addition to NOx2 catalytic subunit. In other words, NOx2 oxidase, a heterodimer made up of NOx2 subunit

Table 10.15. NADPH oxidase isoforms (**Part 1**) NOx proteins (Sources: [1218, 1221]). Component P22PhOx is essential for NOx activity; it complexes with other subunits to form NOx oxidases (NOx1–NOx3). In general, the organizer subunit of NOx1 and NOx3 is NOxO1, and that of NOx2 is NOxO2 (P47PhOx); the activator subunit for NOx1 is NOxA1 and that for NOX2 is NOxA2 (P67PhOx). Isotype NOx3 may not need an activator subunit. Isozyme NOx4 does not require cytosolic subunits. Production of P22PhOx subunit is controlled by AP1 and NFκB transcription factors.

Isoform	Distribution Expression stimuli Cell site
NOx1	Vasculature, kidney, retina, colon, uterus, prostate, placenta Smooth muscle and endothelial cells, pericytes, osteoclasts Upregulation by Infγ Plasma membrane, caveolin
NOx2	Vasculature, heart, lung, skeletal muscle Phagocytes, B lymphocytes, neurons, endothelial and smooth muscle cells, cardiomyocytes, hematopoietic stem cells, hepatocytes Upregulation by Infγ Intracellular
NOx3	Fetal kidney, lung, liver, spleen, skull bone, brain, inner ear Plasma membrane, vinculin
NOx4	Vasculature, heart, kidney, skeletal muscle, bone, eye, ovary, placenta Stimulation by ATn2, TGFβ, TNFα Inhibition by BMP4 Intracellular
NOx5	Brain, vasculature, lymphoid tissue, testis, pancreas, stomach, prostate, breast, placenta, ovary, uterus Endothelium and smooth muscle cells

(i.e., GP91PhOx or CyBβ) and P22PhOx, is activated when the P22PhOx–GP91PhOx heterodimer associates with P40PhOx, P47PhOx (NOxO2), P67PhOx (NOxA2), and Rac1 or Rac2 GTPase.

Unlike NOx1 and NOx2, NOx4 activation does not require P47PhOx, P67PhOx, or Rac GTPase.

At least 4 NOx homologs reside in vascular cells (Table 10.17). NADPH oxidase-1 is expressed in vascular smooth muscle cells, particularly upon stimulation by platelet-derived growth factor, but not in blood leukocytes. In vascular smooth

Table 10.16. NADPH oxidase isoforms (**Part 1**): DuOx proteins (Sources: [1218, 1221]). Airway epithelia contain DuOx1 and DuOx2 as well as possibly NOx2 enzymes.

Isoform	Distribution Expression stimuli Cell site
DuOx1	Lung, thyroid, tongue, cerebellum, testis Upregulation by IL1, IL3, IL4 and Infγ Plasma membrane Calcium sensitive
DuOx2	Gastrointestinal tract, salivary glands, thyroid, airway epithelia, uterus, gall bladder, pancreatic islets Calcium sensitive

Table 10.17. NADPH oxidases in vascular cells (Source: [1225, 1228]; EC: endothelial cell; VSMC: vascular smooth muscle cell). Vascular NOx comprises 3 main catalytic subunits (NOx1, NOx2, NOx4) and 4 cytosolic regulatory components (NOxO1, NOxA1, Rac1, and Rac2). Endothelial cells possess NOx1, NOx2, NOx4, and NOx5, but only NOx2 and NOx4 abound. Isozyme NOx5 is extensively expressed in lymphatic cells. In vascular smooth muscle cells, NADPH oxidases can be activated by angiotensin-2, lysophosphatidylcholine, thrombin, and tumor-necrosis factor-α; redox-sensitive protein kinases include, in particular, JaK2, PKB, and P38MAPK.

Cell type	NOx subunit types
EC	P22PhOx, P47PhOx (NOxO2), Rac, NOx1/2/4/5
VSMC	P22PhOx, P47PhOx, Rac, NOx1/2/4/5
Fibroblast	P22PhOx, P47PhOx, Rac, NOx2/4
Cardiomyocyte	NOx2/4

muscle cells, prostaglandin F2α and angiotensin-2 also provoke NOx1 production. Subtype NOx1 serves as voltage-gated proton (hydrogen) channels involved in cell defense against acidic stress [1224].

Whereas NOx1 and NOx2 that lodge mainly in the plasma membrane generate O_2^-, NOx4 that resides in the endoplasmic reticulum and focal adhesions produces primarily H_2O_2 agent. Isotype NOx1 colocalizes with caveolin in patches of the plasma membrane, whereas NOx4 resides with vinculin in focal adhesions [1226]. Endothelial cells and adventitial fibroblasts synthesize mainly NOx2 and NOx4 and vascular smooth muscle cells NOx1, NOx2, and NOx4 isozymes [1225].

Vascular NOx5 is implicated in: (1) endothelial cell proliferation and angiogenesis; (2) PDGF-induced proliferation of vascular smooth muscle cells; and (3) atherosclerosis [1225]. Vascular NOx5 is activated by thrombin and platelet-derived growth factor.

10.5 NADPH and Dual Oxidases

In vascular smooth muscle cells, reactive oxygen species act as signaling molecules that contribute to cell growth, hypertrophy, and migration. Rodent aortic smooth muscle cells mainly possess NOx1 and NOx4 subtypes; NOx2 is expressed at very low levels. Whereas NOx1 is inactive under basal conditions, NOx4 is constitutively active. In human aortic smooth muscle cells, synthesis of NOx1 and NOx4 isoforms as well as P22PhOx, P47PhOx, and P67PhOx subunits is regulated by the JaK–STAT signaling cascade, especially STAT1 and STAT3 transcription factors [1227].

In endothelial cells, NOx2 and NOx4 are predominantly ($\sim 90\%$) located in the endoplasmic reticulum and nuclear membranes.

Members of the DUOX set possess a NOx1–NOx4 homology domain, an EF-hand region, and a seventh transmembrane domain at the N-terminus with an ecto-facing peroxidase like domain [1218].[89] These enzymes are glycosylated. Immature, partly glycosylated form of DuOx2 generates superoxide, whereas mature form produces hydrogen peroxide [1218]. The DuOx enzymes do not require activator or organizer subunits. Enzyme DuOx1 is synthesized in response to interleukin-4 and -13 and DuOx2 to interferon-γ in respiratory epithelia [1218].

10.5.3 Activity

NADPH oxidases have constitutive and inducible activities as well as substrate specificity. Stimulation leads to subunit assembling. Activated neutrophils, monocytes, and macrophages produce large quantities of superoxide for host defense via NOx2 enzyme. Fibroblasts, vascular smooth muscle and endothelial cells, and cardiomyocytes contain several NADPH oxidase family members to regulate signaling cascades. Different NOx isozymes activate different kinases: (1) in NOx2-overexpressing cells, P38MAPKs are phosphorylated in response to angiotensin-2; (2) in NOx4-overexpressing cells, ERK1, ERK2, and JNKs are activated [1225].

NADPH oxidases interact with small Rac1 or Rac2 GTPases, according to the cell type. Subtypes NOx1 and NOx2 increase Rac GTPase activity.

Isoforms NOx1 and NOx2 are directly and indirectly activated by TNFα, respectively [1220]. Vascular NADPH oxidases are activated by protein kinase-C. Phospholipase-A2 and -D may also stimulate NOx via effectors. Thrombin, several growth factors, and angiotensin-2 stimulate NOx activity in vascular smooth muscle and endothelial cells (Table 10.18). Besides, angiotensin-2 transactivates EGF receptor with ROS via Src kinase.

Isozymes NOx2 and NOx4 that are coexpressed in many cell types respond distinctively to stimulations by angiotensin-2, TNFα, and insulin. They regulate the activation of different protein kinases. In particular, insulin raises phosphorylation of

---

[89] However, the DuOx peroxidase homology domain lacks many amino acid residues that are essential for peroxidase function.

Table 10.18. Expression of NAD(P)H oxidase in vascular smooth muscle cells depends on numerous compounds that either favor or prevent its activity.

Upregulation	Downregulation
ATn2	Leptin
ET1	Adiponectin
Thrombin	Estradiol
PDGF	Statins
TGFβ	
TNFα	
PGF2α	
oxLDL	
WSS	
Fatty acids	

P38MAPK, PKB, and GSK3β specifically in NOx4-overexpressing cells and JNK specifically in NOx2-overexpressing cells [1229]. In addition, activation caused by angiotensin-2 of ERK1, ERK2, and P38MAPK depends on NOx2 isoform.

NADPH oxidase is able to sense mechanical stress. It can particularly react to directional changes of flow. Activated NOx generates superoxide (O_2^-) in a dose-dependent fashion during flow reversal [1230]. Superoxide can eliminate nitric oxide and provoke endothelial dysfunction.

Vascular NOxs are regulated via synthesis and post-translational modifications such as phosphorylation by mechanical stresses, chemical agents (pH, O_2), and vasoactive factors (angiotensin-2, endothelin-1, and aldosterone) and growth factors. In endothelial cells, NOx5 is activated by vasoactive peptides angiotensin-2 and endothelin-1 via a Ca^{++}–calmodulin-dependent, Rac1-independent mechanisms. Subsequently generated ROS lead to ERK1/2 phosphorylation under stimulation by angiotensin-2, but not endothelin-1 [1225].

In hyperglycemia, NADPH oxidase excessively generates reactive oxygen species that impair anti-oxidant defense, especially redox-sensitive transcription factors involved in adaptive responses to oxidative stress [1221].[90] Whereas NADPH contributes to the anti-oxidant potential by regenerating reduced glutathione from oxidized glutathione, high glucose concentration (10–25 mmol/l) activates protein kinase-A that phosphorylates (inhibits) glucose-6 phosphate

---

[90] NADPH oxidase catalyzes the transfer of electrons from NADPH to oxygen. It hence generates superoxide (O_2^-) and possibly hydrogen peroxide (H_2O_2). In phagocytic cells, NADPH oxidase produces reactive oxygen species against pathogens. In endothelial cells, NOx isoforms, such as NOx2 and NOx4, generates ROS that accumulate, although expression of anti-oxidant enzymes (superoxide dismutase, catalase, and glutathione peroxidase) can sometimes rise in diabetes. Furthermore, plasmalemmal anion channels such as Cl^- channel can facilitate O_2^- flux to the extracellular medium [1221]. Hydrogen peroxide that is able to cross the cell membrane through aquaporins can elicit vascular contraction or relaxation according to animal species, vascular bed, contractile state, and context (possible disease).

Table 10.19. Vascular NADPH oxidase effects (Source: [1221]).

Basal activity	Sustained stimulation
Anti-oxidant gene expression	ROS overproduction
Oxygen sensing (HIF1α)	Mitochondrial dysfunction
NOS3 activity	NOS3 uncoupling

dehydrogenase, thereby leading to decreased NADPH level.[91] Therefore, sustained NOx activation in diabetes reduces intracellular concentration of NADPH that is a cofactor for NOS3 and several anti-oxidant enzymes. Prolonged NOx excitation thus causes oxidative stress with NOS3 uncoupling,[92] mitochondrial dysfunction, attenuated anti-oxidant gene expression, and eventually endothelial dysfunction (Table 10.19). However, NADPH oxidase can provoke a transient anti-oxidant response via receptor Tyr kinases and redox-sensitive transcription factors, such as NF-E2-related factor NRF2 [1221].[93]

10.5.4 NOx–ROS Axis

In vascular cells, NADPH oxidase isoforms — NOx1, NOx2, NOx4, and NOx5 — differ in their response to stimuli, activity, and released ROS type [1231]. In particular, NOx activator NoxA1 regulates redox signaling and VSMC phenotype [1232]. Subtype NOx1 is involved in responses to stimuli of vascular smooth muscle cells (hypertrophy upon angiotensin-2 stimulation, serum-induced proliferation, and migration primed by fibroblast growth factor-2).

In cultured smooth muscle cells from thoracic aorta, NADPH oxidase operates via Src-mediated phosphorylation (transactivation) of epidermal growth factor

---

[91] Overexpression of glucose-6 phosphate dehydrogenase in vascular endothelial cells decreases ROS accumulation in response to oxidant stress and improves NO availability [1221].

[92] Superoxide can interact with NO to form peroxynitrite (ONOO$^-$) that can oxidize tetrahydrobiopterin BH$_4$, an NOS3 cofactor, thereby causing NOS3 uncoupling. Uncoupling of NOS3 results from depletion of ^{L}-arginine or BH$_4$ cofactor. Uncoupling causes NOS3 to produce O$_2^-$ rather than NO. In addition to NOS, BH$_4$ is a cofactor for: (1) tryptophan hydroxylase that converts tryptophan into 5-hydroxytryptophan; (2) phenylalanine hydroxylase that transforms phenylalanine into tyrosine; and (3) tyrosine hydroxylase that synthesizes dihydroxyphenylalanine (DOPA) from tyrosine.

[93] Reactive oxygen and nitrogen species are able to activate many transcription factors, such as NFκB, AP1, P53, Ets1, HIF1α, and NRF2. Upregulation of NOx4 is actually associated with elevated levels of HIF1α and VEGF [1221]. Transcription factor NRF2 binds to anti-oxidant response element (ARE) of the promoter region of genes of detoxifying and anti-oxidant enzymes, such as heme oxygenase HO1 and peroxiredoxin-1. Whereas short-term exposure of endothelial cells to elevated glucose level (25 mmol/l) moderately increases NOS3 expression and activity, chronic exposure (25 mmol/l during 7 days) reduces NOS3 expression [1221].

receptor to activate matrix metallopeptidase MMP9 and promote cell migration [1233].[94] Alteration in N-cadherin-mediated intercellular adhesion that results from shedding of the extracellular portion of N-cadherin by MMP9 enables VSMC migration from the media to the subendothelial space, where they can proliferate. This migration that relies on elevation of β-catenin signaling and cyclin-D1 production support intimal thickening [1231].

In atherosclerosis, heightened thrombin synthesis and activity raise NOx activity and then ROS production [1233].

Migration of vascular smooth muscle cells that depends on NOx1 relies on JNK and the Src–PDK1–PAK axis under stimulation by FGF2 and PDGF, respectively [1231]. On the other hand, NOx1-dependent, thrombin-induced VSMC migration depends on P38MAPK activation upon ROS stimulation.

10.6 Redox Signaling and Reactive Oxygen and Nitrogen Species

During redox signaling, free radicals (atoms, molecules, or ions with unpaired electrons), reactive oxygen species, and other electronically activated species act as intra- and extracellular messengers (Table 10.20). Reactive oxygen species modify subtrates by oxidizing thiol groups (-SH),[95] thereby forming disulfide bonds (-S–S-)[96] that reversibly influence protein structure and function:

$$\text{-SH} + \text{-SH} \xrightarrow{\text{oxidation}} \text{-S–S-} + 2\,H^+ + 2\,e^-.$$

Target proteins with reduction–oxidation (redox) sensitive thiol groups include: (1) signaling mediators, such as mitogen-activated protein kinases and protein Tyr phosphatases [1234]; (2) transcription factors; (3) RNA-binding proteins used in DNA methylation; and (4) mediators of histone acetylation, deacetylation, or methylation [1235]. Protein S-glutathiolation and disulfide formation contribute to the redox regulation of signaling proteins and protein regulation during intracellular oxidative stress.

---

[94]Signaling steps comprise [1233]: (1) activation by NOx1 of Src kinase; (2) phosphorylation (activation) of EGFR at multiple tyrosine residues by Src kinase; (3) phosphorylation of ERK1 and ERK2 by activated EGFR; (4) MMP9 phosphorylation (activation) by ERK1 and ERK2; and (5) N-cadherin shedding. Single-span, transmembrane N-cadherin is the predominant type in smooth muscle cells that afford calcium-dependent, homophilic intercellular adhesion. The cytoplasmic domain of N-cadherin contains binding sites for catenin-δ1 and for β- or γ-catenin.
[95]θειον: sulfur.
[96]I.e., covalent bonds derived by the coupling of 2 thiol groups.

10.6 Redox Signaling and Reactive Oxygen and Nitrogen Species

Table 10.20. Major components of the redox signaling pathway (Source: [10]; ROS: reactive oxygen species). Free radicals are implicated in cell signaling and phagocytosis, but their accumulation causes oxidative stress, damages DNA, proteins, carbohydrates, and lipids, thereby provoking cell injury. Consequently, both enzymes that synthesize and catabolize free radicals are required for normal cell function and minimization of induced damage. Sulfiredoxin catalyzes the reduction of hyperperoxidized proteins. Vitamin-A, -C, and -E, bilirubin, uric acid, dietary polyphenol anti-oxidants help neutralize free radicals.

Component	Members
ROS	Superoxide ($O_2^{\bullet-}$)
	Hydrogen peroxide (H_2O_2)
	Hydroxyl radical ($OH^{\bullet}$)
Enzymes of the ROS synthesis	NADPH oxidases
	Dual oxidases
Enzymes of the ROS metabolism	Superoxide dismutase
	Catalase (peroxisome)
	Glutathione peroxidase (cytosol and mitochondria)
	Peroxiredoxins
Thiol-containing proteins	Glutathione
	Glutaredoxin
	Thioredoxin
Reductases	Glutathione reductase
	Glutaredoxin reductase
	Thioredoxin reductase
	Sulfiredoxin

Redox signaling participates in many cellular processes: gene transcription,[97] modulation of Ca^{++} signaling,[98] DNA damage, cell proliferation,[99] apoptosis,[100] and vascular tone [10].[101]

---

[97] Multiple transcription factors are redox sensitive, such as nuclear factor-κB, Activating protein AP1, specificity protein SP1, MyB, P53, and EGR1 factor. In the nucleus, Redox factor ReF1, or transcriptional enhancer factor TEF1 or TEAD1, promotes gene transcription and interacts with P53 to support DNA repair.

[98] Calcium ions enhance redox signaling and inhibits thioredoxin reductase. Conversely, redox signaling influences Ca^{++} signaling. In particular, both IP3R and RyR Ca^{++} channels can be activated by oxidation of key cysteine residues. In addition, hydrogen peroxide can inhibit PTPn6 phosphatase to support Ca^{++} signaling.

[99] When many growth factors signal, the production of hydrogen peroxide increases and facilitates growth factor signaling, as it impedes the activity of protein Tyr phosphatases and PTen phosphatase that precludes the hydrolysis of PIP_3. The latter favors the H_2O_2 production.

[100] Redox signaling operates in mitochondria, where it contributes to Ca^{++}-induced apoptosis via opening of mitochondrial permeability transition pore. However, cells can struggle against

The intracellular medium is a reducing environment. The cell has a large redox buffer capacity using glutathione, which maintains the intracellular redox balance. The redox system signals as well as modulates the activity of other signaling pathways. Conversely, Ca^{++} can stimulate redox signaling, particularly in mitochondria. On the other hand, oxidative stress can cause apoptosis.

The redox balance of the cell is maintained by the energy metabolism, primarily the pentose phosphate pathway that creates NADPH. The latter maintains redox buffers such as glutathione.

The *pentose phosphate pathway*, also called the pentose phosphate shunt, the phosphogluconate pathway, and the hexose monophosphate shunt, generates NADPH and pentoses (5-carbon carbohydrates).[102]

Non-phagocytic cells use reactive oxygen species (ROS) as intracellular messengers to activate signaling cascades. On the other hand, phagocytes such as neutrophils rapidly generate superoxide radical (O_2^-) and hydrogen peroxide (H_2O_2) that can be released in the extracellular space to kill invading microorganisms and cells.

Free radicals and anti-oxidants are involved in cell signaling and activation or inhibition of transcription factors. In addition, the synthesis of growth factors and cytokines, such as platelet-derived (PDGF) and epidermal (EGF) growth factor, transforming growth factor-β, and tumor-necrosis factor-α, may be influenced by free radicals.

Two main types of redox signaling exist: signaling mediated by reactive oxygen species and that primed by reactive nitrogen species (RNS), associated with the NO–cGMP pathway.

10.6.1 Signaling by Reactive Oxygen Species

Reactive oxygen species can operate as second messengers produced at appropriate concentrations by various enzymes, such as nicotinamide adenine dinucleotide phosphate (NADPH) oxidase (NOx) and Dual oxidase (DuOx; Sect. 10.5), especially for cell proliferation. At abnormal concentrations, they function as noxious molecules that target various cellular components, thereby causing oxidative damage and thus aging, cell death, and human diseases.

ROS-induced apoptosis. The PI3K–PKB axis operates via FoxO3a transcription factor to raise the amount of ^{Mn}SOD superoxide dismutase.

[101] Hydrogen peroxide may act as an endothelium-derived hyperpolarizing factor that relaxes neighboring smooth muscle cells, especially in cerebral arteries.

[102] The pentose phosphate pathway is subdivided into 2 phases: an oxidative phase, during which the reducing factor NADPH is generated, and a non-oxidative synthesis of 5-carbon sugars such as ribose 5-phosphate used in the synthesis of nucleotides and nucleic acids. This pathway offers an alternative to glycolysis.

10.6 Redox Signaling and Reactive Oxygen and Nitrogen Species

Reactive oxygen species refer to oxygen species — superoxide, hydrogen peroxide, and hydroxyl radical — that are more reactive than oxygen. Several proteins, such as protein Tyr phosphatases and members of the thioredoxin (TRX) and peroxiredoxin (PRX) families, that commonly possess a highly reactive cysteine residue function as ROS effectors; they are reversibly oxidized by ROS agents.

Oxygen is a strong oxidizing agent that possesses 2 unpaired electrons (one by oxygen atom), which have parallel spins and occupy separate π-antibonding orbitals. These unpaired electrons confer to oxygen the feature of a free radical. However, oxygen is relatively inert, because it has to accept a pair of electrons with antiparallel spins to fit into the empty spaces in the π orbitals to react with another molecule. Oxygen accepts electrons sequentially, thereby forming successively different ROS agents. Superoxide ($O_2^{\bullet -}$ [$\bullet$: radical] or simply O_2^-) is transformed by the addition of electrons into hydrogen peroxide (H_2O_2). The generation of hydroxyl radical ($OH^{\bullet}$) then follows. Therefore, ROS are not all free radicals, as free radicals are atoms and molecules with at least one unpaired electrons.

Cells of respiratory epithelia are exposed to higher oxygen concentrations than most other tissues, hence with a greater probability to free radicals. Rapid catabolism of reactive oxygen species ensures a short half-life.

The sequential ROS synthesis can be triggered by an external messenger that activates its cognate receptors to generate superoxide radical. Numerous receptors activated by external messengers stimulate the formation of reactive oxygen species, such as receptors of cytokines (tumor-necrosis factor-α, interleukin-1, and interferon-γ), receptor Tyr kinases of growth factors (platelet-derived, epidermal, and fibroblast growth factor) and G-protein-coupled receptors of various regulators (angiotensin-2, bradykinin, endothelin, serotonin, and thrombin). The above-mentioned regulators activate receptors coupled to PI3K, thereby synthesizing PIP3 that then stimulates plasmalemmal NADPH oxidases via Rac GTPase. NADPH oxidase removes an electron from NADPH and transfers it to oxygen to create superoxide radical.

In addition to the plasma membrane, mitochondria also generate reactive oxygen species. In mitochondria, most electrons that enter the electron transport chain are transferred to oxygen, oxygen being reduced to water by accepting 4 electrons from cytochrome-C oxidase:

$$O_2 + 4\,e^- + 4\,H^+ \longrightarrow 2\,H_2O.$$

Nevertheless, a small leakage is responsible for a direct transfer of electrons to oxygen molecules to produce superoxide. Mitochondrial ROS may contribute to apoptosis. They may act synergistically with Ca^{++} ions to stimulate the formation of the mitochondrial permeability transition pore.

Oxidized proteins can prime gene transcription, in addition to the modulation of the activity of ion channels, in particular of the Ca^{++} flux, and activated mitogen-activated protein kinase modules.

Signaling molecules, the activity of which is reduced by oxidation, include protein Tyr phosphatases and PTen phosphatase, a protein $Tyr^P/Ser^P/Thr^P$ phosphatase that preferentially dephosphorylates phosphoinositides. Other proteins that possess cysteine residues hypersensitive to oxidation (Cys^{S-}) include the cell cycle regulatory phosphatase CDC25c and Ca^{++}-release channels, ryanodine receptors and inositol trisphosphate receptors.

10.6.1.1 Superoxide

The superoxide radical (O_2^-) that results from the one-electron reduction of O_2 is short-lived (half-life $O[1\,\mu s]$) due to the rapid transformation by superoxide dismutate (SOD) into hydrogen peroxide. Superoxide anion is rapidly converted by adding further electrons into hydrogen peroxide by superoxide dismutase. Hydrogen peroxide, one of the main messengers used by the redox signaling pathway, provokes oxidation of target proteins.

10.6.1.2 Superoxide Dismutases

The most important anti-oxidant enzymes that scavenge superoxide are intra- and extracellular (SOD_{ec}) superoxide dismutases. Superoxide dismutases (SOD) constitute a family of metallopeptidases that converts superoxide radical into hydrogen peroxide:

$$2\,O_2^- + 2\,H^+ \to H_2O_2 + O_2.$$

Subtypes of superoxide dismutases encompass: (1) iron-containing superoxide dismutases (^{Fe}SOD); (2) copper-containing superoxide dismutases (^{Cu}SOD); (3) copper and zinc-containing superoxide dismutases ($^{(Cu,Zn)}SOD$); and (4) manganese-containing superoxide dismutases (^{Mn}SOD).

In the lung, the ^{Mn}SOD subtype is weakly expressed in alveolar macrophages and airway epithelial cells. It is induced by cytokines [1237]. $^{(Cu,Zn)}SOD$ is constitutively expressed in human bronchial and alveolar epithelium. Extracellular SOD_{ec} is mainly released from alveolar macrophages, in regions that contain high amounts of collagen-1 fibers and near lung capillaries.

10.6.1.3 Hydrogen Peroxide

Like superoxide (O_2^-), hydrogen peroxide (H_2O_2) has a short half-life due to its rapid catabolism. Its action depends upon its ability to react with cysteine residues of target proteins that have reactive thiol groups, which are rapidly oxidized to form a disulfide bond. Conversely, glutaredoxin and/or thioredoxin allow the recovery from the oxidized state back to a fully reduced thiol group.

Upon receptor activation and subsequent stimulation of NADPH oxidase, H_2O_2 is generated near the plasma membrane, once PI(3,4,5)P$_3$ is synthesized. Hydrogen peroxide inhibits phosphatase and tensin homolog deleted on chromosome 10 (PTen) that hydrolyzes PIP$_3$, thereby initiating a positive feedback loop.

Oxidation utilizes numerous modes, but is specific, as H_2O_2 selectively modifies only a subset of proteins. Proteins can indeed be characterized by their sensitivity to mild oxidizing agents such as H_2O_2. Most of the cysteine residues in proteins resists to oxidation by H_2O_2 agent. However, some of the cysteine residues, particularly those located next to positively charged amino acids, exist as *thiolate anions* (Cys^{S-}) that are very vulnerable to oxidation (*hyperreactive cysteine residues*).

Hydrogen peroxide targets, in particular, the reduced cysteine residues in N-termini ($_NCys^{SH}$) of thioredoxin dimers to create 2 oxidized *sulfenic* cysteine residues ($_NCys^{SOH}$). These residues then interact with $_CCys^{SH}$ of C-termini of neighboring dimers to form 2 intermolecular disulfide bonds, structural units composed of a linked pair of sulfur atoms. Hydrogen peroxide is transformed by a set of enzymes: catalase, glutathione peroxidase (GPx), and peroxiredoxin (Prx).

In fact, the unstable sulfenic acid group (-SOH) can be processed using different modes [10]. (1) The sulfenic acid residue can be converted into an intramolecular disulfide bond with the elimination of water. (2) The sulfenic acid residue can interact with the reduced form of glutathione (GSH) to form an intermolecular disulfide bond. (3) The sulfenic acid residue can interact with the main-chain nitrogen atom of an adjacent serine residue to form an intramolecular cyclic *sulfenyl amide* (-SN), as during the oxidation of protein Tyr phosphatases (e.g., PTPn1). (4) The sulfenic acid residue can undergo hyperperoxidation by interacting with another molecule of H_2O_2 to form a *sulfinic acid intermediate* (-SO_2H; Cys^{SO_2H}). (5) The sulfinic acid intermediate undergoes further hyperperoxidation to form the *sulfonic acid intermediate* (-SO_3H; Cys^{SO_3H}). (6) The sulfinic acid group can be reduced by sulfiredoxin with ATP.

The irreversible oxidation may be avoided by an internal reaction. The sulfenic acid is rapidly converted into a sulfenyl-amide species, as it reacts with the neighboring amide nitrogen to yield a cyclic sulfenyl amide (Cys–S–N). This sulfenyl-amide intermediate protects against further oxidation, and, upon enzyme reactivation, can be converted back into a thiol group.[103]

Hydrogen peroxide can be catabolized by 3 main types of enzymes: catalase, glutathione peroxidase, and peroxiredoxins.

10.6.1.4 Catalase

Heme-containing catalase decomposes hydrogen peroxide to water and oxygen:

$$2 H_2O_2 \rightarrow 2 H_2 + O_2.$$

Most of the catalases in the cell lodge in peroxisomes. Their action is thus restricted to H_2O_2 produced at the plasma membrane.

---

[103] When protection ensured by disulfide formation or sulfenyl-amide species against irreversible overoxidation fails, thiosulfinate (Cys–S–S=O) and sulfinyl-amide (Cys–SON) and sulfonyl-amide residue (Cys–SO_2N) can appear.

10.6.1.5 Glutathione Redox Couple (GSH–GSSG)

Glutathione is a Glu–CysH–Gly tripeptide. Two reduced forms of GSH molecules tether together via a disulfide bond to create the oxidized form, glutathione disulfide (GSSG). The 2GSH–GSSG redox couple is the most abundant redox couple in the cell (cytoplasmic GSH concentration 1–10 mmol) [10]. The oxidized form of glutathione (GSSG) is converted back into GSH by glutathione reductase.

The oxidized form of glutathione interacts with reduced glutaredoxin $(GRx^{(SH)_2})$ to generate oxidized glutaredoxin (GRx^{S_2}); this disulfide bond can be transferred to oxidize target proteins.

Under normal reducing conditions, the cell reduction potential is high and cell proliferation is favored. Cell differentiation seems to occur at lower potentials. Low potentials supports apoptosis [10].

10.6.1.6 Glutathione Reductase

Glutathione reductase converts oxidized form of glutathione (GSSG) back into the reduced form (GSH) [10]:

$$GSSG + NADPH + H^+ \rightarrow 2\,GSH + NADP^+.$$

10.6.1.7 Glutathione Peroxidase

Glutathione peroxidases (GPx) use the reducing capacity of glutathione to convert H_2O_2 into water:

$$H_2O_2 + 2\,GSH \xrightarrow{GPx} 2\,H_2O + GSSG.$$

Human bronchial epithelial cells, alveolar type-2 cells, and macrophages are characterized by a constitutive activity of catalase and glutathione peroxidase. The epithelial lining fluid of airways has a glutathione level more than 100 times higher than that in circulating blood.

10.6.1.8 Peroxiredoxins

Peroxiredoxins (PRx) are thiol-specific anti-oxidant enzymes that catabolize hydrogen peroxide into water:

$$H_2O_2 + PRx^{(SH)_2} \xrightarrow{PRx} 2\,H_2O + PRx^{S_2}.$$

They also exert a protective role using their peroxidase activity against peroxynitrite and phospholipid hydroperoxides. They homo-oligomerize under strong oxidative stress.

Table 10.21. The peroxiredoxin family (Sources: [10, 1237]). Two-Cys PRx1 to PRx4 contain 2 cysteines in the catalytic motif; atypical 2-Cys Prdx5 uses thioredoxin as an electron donor; 1-Cys Prdx6 utilizes glutathione as an electron donor to catalyze the reduction of peroxides.

Isoform	Features
	$_{2-Cys}$Prx
PRx1	Cytosol, inducible, 22 kDa
PRx2	Cytosol, inducible, 22 kDa
PRx3	Mitochondria, 28 kDa
PRx4	Endoplasmic reticulum, 31 kDa, extracellular space (heparin binding)
PRx5	Atypical $_{2-Cys}$Prx, 17 kDa, cytosol, peroxisomes, mitochondria
PRx6	$_{1-Cys}$Prx, 25 kDa

Table 10.22. Distribution and expression degree (weak +, mid ++ and strong +++) of peroxiredoxins in the normal human lung (Source: [1237]).

Isoform	Bronchial epithelium	Alveolar epithelium	Alveolar macrophages	Vascular endothelial cells
PRx1	++	+	++	+
PRx2	+	+	+	
PRx3	+++	++	+++	+
PRx4	+		+	
PRx5	+++	+++	++	
PRx6	+++	+++	++	+

The active cysteine site of peroxiredoxins can be selectively oxidized to cysteine sulfinic acid ($Cys^{SO(OH)}$), thereby inactivating the peroxidase activity. Nonetheless, the sulfinic form of peroxiredoxin-1 produced during the cell exposure to H_2O_2 is rapidly reduced to the catalytically active thiol form [1236].

The PRX family contains 6 ubiquitous members, especially in human lung cells [1237] (Tables 10.21 and 10.22). These proteins are classified into 3 subfamilies: 2-cysteine ($_{2-Cys}$Prx), atypical $_{2-Cys}$Prx, and $_{1-Cys}$Prx according to the number and position of cysteine residues that participate in the catalytic activity.

Peroxiredoxin-6 is detected in bronchial Clara cells, alveolar epithelial type-2 cells, and alveolar macrophages [1238]. It regulates cellular signaling and protects against membrane lipid peroxidation and apoptosis caused by oxidative stress. It can indeed reduce phospholipid hydroperoxides, thereby being capable of repairing membrane damage caused by oxidative stress.

Peroxiredoxins contain catalytic cysteine residues. They use thioredoxin as an electron donor. The reducing equivalents derived from thioredoxin (Trx) are used to regenerate Prx^{SH_2}:

$$Prx^{S_2} + Trx^{SH_2} \to Prx^{SH_2} + Trx^{S_2}.$$

10.6.1.9 Glutaredoxin

Recovery from oxidation is based on thioredoxin (TRx) and glutaredoxin (GRx). These 2 proteins have common and distinct features. In particular, they are characterized by their substrate specificity and types of target disulfide bonds. For example, GRx is a less efficient reducing agent on protein Tyr phosphatase PTP1b than TRx agent.

In a reduced state, both TRx and GRx reduce their substrates and become oxidized. Oxidized glutaredoxin is converts back into a reduced state using glutathione, which is regenerated by a glutathione reductase.

10.6.1.10 Thioredoxin

The various oxidized intermediates can be converted back into the initial reduced state by either thioredoxin or glutaredoxin. In particular, the anti-oxidant defense of the human lung include both small cysteine-containing proteins thioredoxin and peroxiredoxins.

Ubiquitous thioredoxins act as anti-oxidants that foster the reduction of proteins by cysteine thiol–disulfide exchange. The redox regulator thioredoxin possess multiple substrates, such as ribonuclease, chorionic gonadotropin (a glycoprotein hormone produced during pregnancy), coagulation factors, glucocorticoid receptor, and insulin [10].

The reducing ability of the regulator thioredoxin is repressed by Ca^{++} ions that support the oxidized form of reduced thioredoxin (TRx^{S_2}) [10].

Reduced $TRx^{(SH)_2}$ bind to MAP3K5 enzyme.[104] When it is oxidized (TRx^{S_2}), MAP3K5 is released and causes apoptosis [10].

Thioredoxin may also increase the expression of hypoxia-inducible factor HIF1α, thereby supporting the activity of vascular endothelial growth factor and angiogenesis [10].

10.6.1.11 Thioredoxin Reductase

Thioredoxin reductase converts back oxidized thioredoxin into a reduced state. Hence, thioredoxin reductase and thioredoxin compose an important redox regulator. Three proteins constitute the TRX family: TRx1, mitochondria-specific TRx2, and sperm spTRx. Thioredoxin-2 regulate the mitochondrial membrane potential and contributes to the inhibition of apoptosis.

Thioredoxin reductase contains selenocysteine (^{Se}Cys) in its C-terminal active site as well as flavin adenine dinucleotide (FAD) in its N-terminal region. This enzyme operates by transferring electrons from NADPH to FAD and then on to the C-terminal active site.

---

[104] A.k.a. apoptosis signal-regulating kinase ASK1.

10.6 Redox Signaling and Reactive Oxygen and Nitrogen Species

The thioredoxin redox regulator controls numerous transcription factors involved in cell proliferation and death. In human adipose tissue-derived mesenchymal stem cells, overexpression of TRx1 and TRx2 increases ERK1 and ERK2 phosphorylation, nuclear factor-κB activation, and βCtn–TCF–promoter activity, but prevents leucine zipper tumor suppressor LZTS2 expression [1239].

10.6.1.12 Nucleoredoxin

Nucleoredoxin (NRx) that mainly localizes to the nucleus, but also in the cytoplasm. This oxidoreductase[105] contains a pair of Cys residues in its catalytic sequence. It inhibits Disheveled, thereby suppressing Wnt signaling. However, NRx impedes Dvl ubiquitination and subsequent degradation, hence maintaining a pool of inactive Dvl for robust signaling under Wnt stimulation.

10.6.1.13 Peroxidases

Peroxidases catalyze the oxidation of substrates by hydrogen peroxide. Thyroid peroxidase (TPO) are used in the oxidation of iodine and formation of thyroid hormone. Myeloperoxidase (MPO) of neutrophils, monocytes, and macrophages serve in the production of the weak acid and bactericidal oxidizer hypochlorous acid (HOCl) from hydrogen peroxide (H_2O_2) and chloride anion (Cl^-) .[106] Hypochlorous acid reacts with various types of molecules, such as DNA, RNA, fatty acids, cholesterol, and proteins.

The set of peroxidases also includes lactoperoxidase, eosinophil peroxidase, and salivary peroxidase, as well as vascular peroxidase (VPO1; or insect peroxidasin homolog) that have a very low enzymatic activity in comparison with other peroxidases [1240].

Heme-containing VPO1 can be produced and secreted into the extracellular medium upon stimulation by transforming growth factor-β1 [1240]. It is mainly detected in the vascular wall, lung, liver, among other organs [1241].

Increased synthesis of reactive oxygen species can foster smooth muscle cell proliferation and hypertrophy as well as matrix formation. In the cardiovascular system, myeloperoxidase promotes lipid peroxidation and scavenges nitric oxide. Its products activate matrix metallopeptidases, thereby favoring tissue remodeling. In addition, vascular peroxidase contributes to the proliferation of vascular smooth

---

[105] Enzyme that catalyzes the transfer of electrons from one substrate — reductant — (hydrogen or electron donor) to another — oxidant — (hydrogen or electron acceptor). Oxidoreductase usually uses NADP or NAD cofactors.

[106] Immunocytes use NADPH oxidase to synthesize H_2O_2 and, then, myeloperoxidase to combine H_2O_2 with Cl^- to produce cytotoxic hypochlorous acid and its salt (hypochlorite anion ClO^-) to destroy bacteria.

muscle cells caused by angiotensin-2 via the NOx–H_2O_2–VPO1–HOCl–ERK1/2 pathway [1241]. Hypochlorous acid is synthesized upon angiotensin-2 stimulation. Angiotensin-2, like another profibrotic agent TGFβ, activates NADPH oxidase that increases VPO1 expression. Vascular peroxidase-1 uses H_2O_2 produced by NADPH oxidase to generate peroxides such as hypochlorous acid. The latter modifies intracellular proteins and, once it is secreted, matrix proteins.

10.6.1.14 Hydroxyl Radical

Hydroxyl radical (OH•) is the highly reactive, short-lived (half life O[1 ns]), neutral form of the hydroxide ion [(OH)$^-$]. Hydrogen peroxide can be converted into highly toxic hydroxyl radical, as it undergoes a reduction catalyzed by transition metals (Fe^{3+} or Cu^{++}).

10.6.1.15 Cysteine Oxidation by Reactive Oxygen Species

Among 20 amino acids that build proteins, cysteine has a thiol moiety (R–SH functional group, R denoting an attached hydrogen, a hydrocarbon side chain, or any group of atoms) in the side chain of Cys is very sensitive to oxidation. This thiol moiety can form disulfide bonds with another thiol moiety (R–S–S–R) [1242]. These disulfide bonds can be degraded to reform free thiol moiety.

The thiol moiety (R–SH) is oxidized to become the *sulfenyl* moiety (R–SOH). This process is reversible, as the sulfenyl moiety can be reduced to the thiol moiety by various anti-oxidants. The sulfenyl moiety can form reversible disulfide bond with another thiol moiety. Disulfide bond can be reduced by various anti-oxidants. In turn, the sulfenyl moiety can be further oxidized to the *sulfinyl* (R–SO_2H) and *sulfonyl* moiety (R–SO_3H) that cannot be reduced under normal intracellular conditions.

Protein Tyr phosphatases possess a Cys residue in the catalytic domain that acts as a transient acceptor for the phosphate during dephosphorylation. Several PTPs are inhibited by oxidation of their catalytic Cys residue [1218]. This process can be observed in endothelial cells stimulated with PDGF factor [1242]. Phosphatase and tensin homolog deleted on chromosome 10 is a PTP domain-containing protein that can experience oxidation; the catalytic Cys residue forms a disulfide bond with another Cys that protects itself from further irreversible oxidation. Protein Tyr phosphatase PTPn1 contributes to the regulation of insulin signaling. Its oxidized Cys in the catalytic region forms a reversible sulfenyl-amide bond (R–S–N–R) with nitrogen atom in the polypeptidic chain that also protects it from further oxidation [1242]. Oxidation of MAPK phosphatases (MKP) prevents their phosphatase activity. Oxidized MKPs form proteic complexes labeled for proteasomal degradation.

On the other hand, Lyn kinase in leukocytes is activated by wound-derived H_2O_2 and triggers movement of leukocytes to wound sites [1242]. Ataxia-telangiectasia

10.6 Redox Signaling and Reactive Oxygen and Nitrogen Species

mutated protein kinase can be oxidized (activated) under oxidative stress, even in the absence of DNA double-strand breaks [427]. The oxidized ATMK form is a disulfide-crosslinked dimer.

Pyruvate kinase muscle isozyme PKM2 normally highly expressed only in undifferentiated tissues catalyzes the transfer of a high-energy phosphate group from phosphoenol pyruvate to ATP, i.e., the rate-limiting step of glycolysis.[107] Oxidative stress causes PKM2 oxidation (Cys358), thereby provoking reversible enzyme inactivation [1245]. The diversion of glucose flux into the pentose phosphate pathway enables to generate sufficient reducing potential for ROS detoxification.

10.6.1.16 Summary of ROS Effects

Inhibition of Phosphatases and Activation of Kinases

Reactive oxygen species inhibit protein Tyr phosphatases that control cell proliferation, differentiation, survival, metabolism, and motility. On the other hand, they activate MAPK modules.

Regulation of Ion Channels

Reactive oxygen species can regulate intracellular and plasmalemmal ion channels (e.g., K^+ channels [inactivation], especially in the aging brain, Ca_V1 [activation], store-operated Ca^{++} channels [inhibition], Ca^{++} ATPases [activation at low ROS concentrations; inhibition during oxidative stress], and ryanodine receptors [stimulation]) directly or via effectors [1218]. Reactive oxygen species can regulate ion channel activity via either post-translational modifications (cysteine oxidation and S-glutathiolation) or cell depolarization that results from NOx-dependent electron transport.

protein kinase-G phosphorylates (inactivates) cardiac ATP-sensitive, hetero-octameric potassium channel (K_{ATP},[108] more precisely cardiac $K_{IR}6.2$–SUR2a carrier used in metabolic stress adaptation in the heart, or some closely associated proteins. Kinase PKG also stimulates K_{ATP} channel via ROS (H_2O_2) in particular) and the calmodulin–CamK2 axis [1246].

---

[107] Alternatively spliced isoform PKM1 is ubiquitous. Tumors express PKM2 subtype, allowing these cells to withstand oxidative stress [1244].

[108] This channel is composed of 4 inwardly rectifying subunits ($K_{IR}6.1$ or $K_{IR}6.2$) and 4 sulphonylurea receptors (SUR1, SUR2a, or SUR2b). In neurons and* pancreatic β-cells, where it serves in neurotransmitter release and glucose-stimulated insulin secretion, K_{ATP} channel is made up of $K_{IR}6.2$ and SUR1 subunits. K_{ATP} channel is also phosphorylated by protein kinases PKA and PKC, as well as extracellular signal-regulated kinase. In cultured vascular smooth muscle cells, vasorelaxant agents that increase intracellular cGMP concentrations modulate the activity of K_{ATP} and K_{Ca} channels. Atrial natriuretic peptide targets particulate guanylate cyclase and activate K_{ATP} and K_{Ca} channels [1247].

Gene Expression

Generation of ROS stimulates the production of angiotensin-2, TNFα, TGFβ1, CCL2 chemokine, and plasminogen activator inhibitor-1 [1218]. Transcriptional upregulation results either from redox-sensitive second messengers, such as MAPK, or transcription factors (AP1, P53, and NFκB) with redox-sensitive cysteine residues in their DNA-binding domain.

Oxygen Sensing

Some organs, such as the kidney cortex, carotid bodies, and pulmonary hypoxia-sensitive neuroepithelial bodies, are specialized in oxygen sensing. Hypoxia may lower concentrations of by-products of O_2 reduction by NADPH oxidase (ROS), thereby altering the cellular redox potential and priming a conformational change and inactivation of the redox-sensitive K^+ channels. The resulting drop in outward flux of K^+ ions leads to membrane depolarization, calcium influx, and neurotransmitter secretion. In addition, increased ROS generation under hypoxia can also contribute to HIF stabilization [1218].

Regulation of Matrix Metallopeptidases

Expression and/or activation of matrix metallopeptidases may be, at least partly, regulated by ROS agents.

Cell Fate

Under some circumstances, ROS may support cell proliferation. Agents ROS can accelerate cell senescence due to oxidative stress. Cell apoptosis can follow damage of DNA, lipids, and proteins or be primed by activated signaling mediators (JNKs, ERKs, and P38MAPKs) [1218]. Under certain circumstances, ROS may act as anti-apoptotic signals via NFκB and the PKB–MAP3K5 pathway.

10.6.2 Signaling by Reactive Nitrogen Species

Reactive nitrogen species (RNS) are antimicrobial molecules derived from free radical nitric oxide (NO•) and another free radical, superoxide ($O_2^{•-}$), produced by inducible nitric oxide synthase NOS2 and NADPH oxidase, respectively. The former enzyme is expressed primarily in macrophages after stimulation by cytokines (e.g., interferon-γ) and microbial products (e.g., lipopolysaccharides). Reactive nitrogen species act synergistically with reactive oxygen species to cause nitrosative stress and damage their targets.

However, neither nitric oxide nor superoxide is particularly toxic when these substances do not accumulate. High concentrations of superoxide are avoided by superoxide dismutases located in mitochondria and cytosol, as well as in the extracellular matrix. Nitric oxide, after a quick intratissular transfer, is rapidly eliminated by erythrocytes, where it is rapidly converted to nitrate, as it reacts with oxyhemoglobin.[109]

The production of reactive nitrogen species begins with the very rapid reaction of free radical nitric oxide with superoxide, when they are synthesized simultaneously close to each other (but not necessarily in the same cell) to form peroxynitrite ($ONOO^-$), a strong oxidant:

$$NO^\bullet + (O_2^{-\bullet} \to ONOO^-$$

Peroxynitrite is able to cross cell membranes, especially that of erythrocytes, either in the anionic form through anion channel or in the protonated form by passive diffusion [1248]. It then rapidly reacts with oxyhemoglobin to yield methemoglobin.

Peroxynitrite can react with other molecules to form additional types of reactive nitrogen species, such as nitrogen dioxide ($NO_2^\bullet$) and dinitrogen trioxide (N_2O_3), as well as other types of free radicals (Table 10.23):

$$NO^\bullet + NO_2^\bullet \rightleftharpoons N_2O_3;$$
$$ONOO^- + H^+ \to ONOOH \text{ (peroxynitrous acid)} \to NO_2^\bullet + OH^\bullet;$$
$$ONOO^- + CO_2 \to ONOOCO_2^- \text{ (nitrosoperoxycarbonate)} \to NO_2^\bullet + O=C(O^\bullet)O^-$$
$$\text{(carbonate radical)}.$$

The terms "nitrosative stress" and "nitrative stress" are used to denote damage from reactive nitrogen species without refering specifically to nitrosation and nitration. The expression "nitroxidative stress" enables to distinguish an oxidative from a purely nitrosative or nitrative cause [1249]. Major agents are $O_2^{\bullet-}$, $^1O_2$, H_2O_2, and $^\bullet OH$ for oxidative stress and $ONOO^-$, $CO_3^{\bullet-}$, and $^\bullet NO_2$ for nitroxidative stress.

10.6.2.1 Reactants of Peroxynitrite

Peroxynitrite, a strong oxidant, directly reacts at a relatively slow rate with thiols,[110] lipids, amino acids, DNA bases, and low-molecular-weight anti-oxidants [1168].

---

[109] Nitric oxide reacts with oxyhemoglobin to form nitrate. This reaction prevents the interaction of NO with oxygen to form nitrogen dioxide (NO_2). However, when the NO synthesis rate is elevated, the formation of nitrogen dioxide is mainly precluded by the reaction of NO with O_2^- to produce peroxynitrite.

[110] Compounds with a carbon-bonded sulfhydryl, or thiol, group (-C–SH or R–SH; R: alkane, alkene, or other carbon-containing group of atoms). Thiol is a portmanteau of "thio" (SH) and "alcohol" (reference to hydroxyl group [OH] of an alcohol).

Table 10.23. Relative contributions of oxidation, nitrosation, and nitration to products of reactive nitrogen species for a steady-state concentration $[^{\bullet}NO]_{SS} = 20$ nmol (Source: [1249]). Reactions occur in an aqueous solution within the complex environment of the cell; some reaction types dominate because of competition for reactants and intermediates. The dominant reactions in a set of interacting rapid reactions are those involving reactants with the highest concentrations. Nitration introductes a nitro group (-NO$_2$) into a chemical compound. Nitrosylation adds a nitrosyl ion (NO$^-$) to a metal or a thiol. Nitrosation connects a nitrosonium ion (NO$^+$) to an amine (-NH$_2$) leading to a nitrosamine. Nitrosation occurs mainly by radical combination rather than using the potent nitrosating species nitrous anhydride (N$_2$O$_3$). Nitrosation predominates over nitration. Nitrosation is eliminated when $[O_2^{\bullet-}] \geq [^{\bullet}NO]$. Nitration is relatively independent of excess $^{\bullet}$NO vs. excess $O_2^{\bullet-}$. Intermediate (not endproduct) nitrosothiol (GSNO) is formed when $^{\bullet}$NO is in excess over $O_2^{\bullet-}$. The extremely rapid reaction between $^{\bullet}$NO and $O_2^{\bullet-}$ produces peroxynitrite (ONOO$^-$), a potent oxidizing agent, which also react very rapidly with either thiol or CO$_2$. The reaction of either $^{\bullet}$NO or $O_2^{\bullet-}$ with ONOO$^-$ forms the potent nitrosating species nitrous anhydride (N$_2$O$_3$) or nitrogen dioxide ($^{\bullet}$NO$_2$, respectively. The products of reaction of CO$_2$ with ONOO$^-$ and homolysis of peroxynitrous acid, $^{\bullet}$NO$_2$ and $^{\bullet}$OH, rather than ONOO$^-$ or ONOOH. Both $^{\bullet}$NO$_2$ and $^{\bullet}$OH can react with excess $^{\bullet}$NO and $O_2^{\bullet-}$. The intermediate reactive products of CO$_2$ reaction, CO$_3^{\bullet-}$ and $^{\bullet}$NO$_2$, disappear rapidly. Nevertheless, competition of cellular targets (mainly thiol) for reactive intermediate oxidants (CO$_3^{\bullet-}$ and $^{\bullet}$NO$_2$) is much more effective than that for excess $^{\bullet}$NO and $O_2^{\bullet-}$.

Reaction	flux of $^{\bullet}$NO w.r.t. that of $O_2^{\bullet-}$		
	$vO_2^{\bullet-} < v^{\bullet}NO$	$vO_2^{\bullet-} = v^{\bullet}NO$	$vO_2^{\bullet-} < v^{\bullet}NO$
Oxidation	98.6%	99.3%	99.6%
Nitration	0.2%	0.1%	0.1%
Nitrosation	1.2%	0.6%	0.4%

Proteins

Peroxynitrite interacts with proteins that contain transition metals. Peroxynitrite rapidly reacts with (inactivates) iron–sulfur domain-containing enzymes (e.g., mitochondrial aconitase Aco2) as well as zinc–sulfur motif-containing enzymes (e.g., NOS3 and alcohol dehydrogenase) [1168].

The direct reaction of peroxynitrite with transition metal centers pertains to the fastest reaction with peroxynitrite. Peroxynitrite thus modifies proteins that contain a heme prosthetic group, such as hemoglobin, myoglobin, and cytochrome-C by oxidizing ferrous heme into its ferric form.

Peroxynitrite can also inactivate NOS2 by oxidative modification of its heme group, thereby priming a negative feedback of peroxynitrite generation.

Amino Acids

Peroxynitrite may also change protein structure by reacting with amino acids, such as cysteine oxidation and tyrosine nitration [1168]. Peroxynitrite reacts directly with cysteine, methionine, and tryptophan residues. It also targets transition metal centers and selenium-containing amino acids.

In addition, free radicals arising from peroxynitrite homolysis[111] such as hydroxyl, nitrogen dioxide, and carbonate radical (O=C(O$^\bullet$)O$^-$) formed in the presence of carbon dioxide, also react with protein moieties.

Cysteine Oxidation

Peroxynitrite reacts with thiols, particularly with the anion form (R–S$^-$) to generate an intermediate *sulfenic acid* (R–SOH), which then reacts with another thiol, forming a *disulfide* (R–S–S–R).

Thiols may also be oxidized by radicals formed from peroxynitrite, thereby producing *thiyl radicals* (R–S$^\bullet$). Thiyl radicals may react with oxygen and promote oxidative stress. They also react with NO to form *nitrosothiols*, or *thionitrites* (R–S–NO).

Enzyme Inhibition

The oxidation of certain cysteine residues by peroxynitrite inactivates many enzymes involved in cellular energetic metabolism, such as glyceraldehyde-3-phosphate deshydrogenase, creatine kinase, and complex-I (NADH dehydrogenase), -II (succinate dehydrogenase), -III (cytochrome-C reductase), and -V (ATP synthase) of the mitochondrial respiratory chain [1168]. In addition, tyrosine nitration also inactivates these enzymes.

Cysteine oxidation by peroxynitrite also leads to the inactivation of several protein Tyr phosphatases. In addition to protein-bound thiols, peroxynitrite can directly oxidize low-molecular-weight thiols, especially reduced glutathione (GSH). The latter scavenges peroxynitrite.

Enzyme Activation

Cysteine oxidation by peroxynitrite can also activate certain types of enzymes such as matrix metallopeptidases. Peroxynitrite activates the proenzymes (proMMP) via thiol oxidation and S-glutathiolation of the auto-inhibitory domain.

Cysteine oxidation by peroxynitrite stimulates the SRC family kinase hematopoietic cell kinase (HCK) in erythrocytes.

---

[111]homolysis (ομοιος: like, similar, resembling; λυσις: loosing, releasing), or homolytic fission, is a chemical bond dissociation of a neutral molecule that generates 2 free radicals. Two electrons involved in the interatomic bond are distributed to the 2 species S_A and S_B. Each of the 2 covalently shared electrons are withdrawn by the bonded atoms:

$$S_A\text{–}S_B \rightarrow S_A^\bullet + S_B^\bullet.$$

Tyrosine Nitration

Protein tyrosine nitration is a covalent protein modification mediated by reactive nitrogen species such as peroxynitrite anion and nitrogen dioxide, i.e., introduction of a nitro group (-NO2) into a target molecule. Tyrosine nitration is a selective process limited to specific tyrosine residues on a relatively small number of proteins.[112]

A nitro group ($-NO_2$) is added to Tyr residues to form nitrotyrosines (Tyr^{NO_2}). Tyrosine nitration influences protein structure and function. Tyrosine does not react directly with peroxynitrite. During tyrosine nitration, a hydrogen atom (Tyr^H is first abstracted from tyrosine to form a tyrosyl radical ($Tyr^\bullet$) that quickly combines with $NO_2^\bullet$ to produce 3-nitrotyrosine. A second combination with another tyrosyl radical forms dityrosine [1168].

Radicals involved in the reaction may come from peroxynitrite homolysis ($OH^\bullet$ and $NO_2^\bullet$). However, radicals can be $CO_3^{-\bullet}$ and $NO_2^\bullet$ radicals produced by the reaction between $ONOO^-$ and CO_2.

Tyrosine nitration is enhanced in the presence of transition metals due to the formation of secondary radicals at the metal center plus $NO_2^\bullet$. Metalloproteins, such as heme-containing proteins (e.g., prostacyclin synthase) or $^{Cu,Zn}SOD$ and ^{Mn}SOD may catalyze peroxynitrite-mediated tyrosine nitration.

Another mechanism of tyrosine nitration relies on the generation of the $NO_2^\bullet$ radical by various heme-peroxidases (mainly myeloperoxidase and eosinophil peroxidase) in the presence of hydrogen peroxide [1168].

Tryptophan, Methionine, and Histidine Oxidation

Peroxynitrite can directly oxidize methionine, thus forming methionine sulfoxide (Met^{SO}), and to a lesser extent, ethylene and dimethyldisulfide. Conversely, ubiquitous peptide methionine sulfoxide reductase carries out the enzymatic reduction of methionine sulfoxide to methionine, thereby repairing oxidative damage and restoring normal activity.

---

[112] Mitochondrial ^{Mn}SOD, the first protein found to be nitrated in vivo, loses its catalytic activity upon nitration of a single Tyr residue. Prostacyclin synthase is rapidly nitrated during inflammation of arterial walls [1168]. In the heart, Tyr nitration rapidly inactivates creatine kinase, an energetic controller of cardiomyocyte contractility (Vol. 1 – Chap. 4. Cell Structure and Function), and sarcoplasmic reticulum Ca^{++} ATPase SERCA2a. Peroxynitrite also nitrates (inactivates) voltage-gated K^+ channels on the coronary endothelial cells, thereby imparing the coronary flow reserve and causing cardiac dysfunction. Moreover, peroxynitrite also nitrates several structural proteins in cardiomyocytes, such as desmin, myosin heavy chain, and α-actinin, thus altering myocardium contractility. Many cytoskeletal proteins can undergo nitration, such as actin, profilin, and tubulin, hence disturbing many processes, such as platelet function and migration and phagocytosis of activated leukocytes, and disrupting endothelial barrier [1168].

10.6 Redox Signaling and Reactive Oxygen and Nitrogen Species

Methionine oxidation can participate in immunity, via inactivation of glutamine synthase and chaperonin GroEL in bacteria [1168].[113] Methionine oxidation also inhibits serpin-A1.[114]

Peroxynitrite can also oxidize tryptophan to yield ^{N}formylkynurenine,[115] oxindole, hydropyrroloindole, and nitrotryptophan [1168].

Peroxynitrite modifies histidine to create a histidinyl radical, which inactivates $^{(Cu,Zn)}$SOD type of superoxide dismutase [1168].

Lipids

Peroxynitrite can cause lipid peroxidation in cell membranes, liposomes, and lipoproteins, as it extracts a hydrogen atom from polyunsaturated fatty acids [1168]. Resulting products include lipid hydroperoxyradicals, conjugated dienes (C_nH_{2n-2}), and aldehydes (R–CHO).

These radicals attack neighboring polyunsaturated fatty acids, thereby generating additional radicals that amplify free radical reactions and subsequent degradations.

Peroxynitrite acts as a potent oxidizing agent for low-density lipoproteins. Peroxynitrite-modified LDLs bind with high affinity to scavenger receptors, thereby leading to the accumulation of oxidized cholesteryl esters and formation of foam cells, an early event in atherogenesis.

Last, but not least, peroxynitrite interacts with membrane lipids and can generate various nitrated lipids that can serve as signaling mediators, as well as intermediate products, such as isoprostanes[116] and 4-hydroxynonenal (4HNE [$C_9H_{16}O_2$]),[117] which can further trigger secondary oxidative insults [1168].

Nucleic Acids

Peroxynitrite can cause oxidative modifications in both nucleobases and carbohydrate–phosphate backbone, thereby damaging DNA [1168]. Among the 4 nucleobases, guanine is the most reactive with peroxynitrite. The major product of guanine oxidation is 8-oxoguanine, which further reacts with peroxynitrite to form cyanuric acid and oxazolone. Guanine oxidation by peroxynitrite provokes guanine fragmentation, thereby leading to carcinogenesis.

---

[113]In eukaryotes, HSP60 is structurally and functionally nearly identical to GroEL chaperonin.

[114]A.k.a. α1-antitrypsin and α1-antipeptidase.

[115]A formylated derivative of kynurenine and catabolite of tryptophan.

[116]Prostaglandin-like compounds generated from peroxidation are catalyzed by free radicals of fatty acids, mainly arachidonic acid. These non-classical eicosanoids serve as inflammatory mediators.

[117]This (α,β)-unsaturated hydroxyalkenal has 3 reactive groups: an aldehyde, a double-bond at carbon 2, and a hydroxy group at carbon 4. It may intervene in cell signaling.

Peroxynitrite can nitrate guanine to form 8-nitroguanine. Resulting abasic sites can then be cleaved by endonucleases and generate DNA single-strand breaks [1168].

Peroxynitrite may also degrade the sugar phosphate backbone by removing a hydrogen atom from the deoxyribose moiety, hence opening the sugar ring and creating DNA strand breaks.

10.6.2.2 Peroxynitrite in Mitochondria

Peroxynitrite may enter into mitochondria or be produced in mitochondria. Mitochondria can actually synthesize both nitric oxide using Ca^{++}-sensitive mitochondrial NOS (mtNOS) and superoxide upon partial reduction of oxygen within the mitochondrial matrix due to the natural leak of electron from the respiratory chain. In mitochondria, nitric oxide reversibly inhibits cytochrome-C oxidase (complex IV of the electron transport chain), as it competes with oxygen for the binding site, thereby regulating oxygen consumption [1168]. Moreover, peroxynitrite nitrates (inhibits) ^{Mn}SOD, thus preventing the breakdown of locally produced superoxide and amplifying the formation of peroxynitrite.

Peroxynitrite also inhibits most components of the electron transport chain: NADH dehydrogenase (complex I), succinate dehydrogenase (complex II), cytochrome-C reductase (complex III), and ATP synthase (complex V) using cysteine oxidation, tyrosine nitration, and damage of iron–sulfur centers [1168].

Furthermore, peroxynitrite impedes the activity of aconitase of the tricarboxylic acid cycle in the mitochondrial matrix as well as mitochondrial creatine kinase in the intermembrane space.

Peroxynitrite targets nicotinamide nucleotide transhydrogenase that generates NADPH from NADH and $NADP^+$, thereby reducing the mitochondrial ability to regenerate the reduced form of glutathione (GSH) [1168].

10.6.2.3 Peroxynitrite and Cell Signaling

Peroxynitrite nitrates tyrosine residues and can impair Tyr^P-dependent signaling. Phosphorylation of critical Tyr residues can be prevented by peroxynitrite-mediated Tyr nitration, thereby blocking signaling [1168].

On the other hand, Tyr^P-dependent signaling can be activated by peroxynitrite [1168]. In particular, peroxynitrite stimulates MAPK superfamily members (Table 10.24). Phosphotyrosine signaling stimulation results either from the inhibition of protein Tyr^P phosphatases (PTP) or activation of protein Tyr kinases (PTK).

Table 10.24. Main signaling pathways influenced by peroxynitrite in certain explored cell types (Source: [1168]; EGFR: epidermal growth factor receptor protein Tyr kinase; ERK: extracellular-regulated kinase; JNK: Janus kinase; MAPK: mitogen-activated protein kinase; NFκB: nuclear factor-κB; NTRK: neurotrophic protein Tyr receptor kinase; PDGFR: platelet-derived growth factor receptor protein Tyr kinase; PI3K: phosphatidylinositol 3-kinase; PKB(C): protein kinase-B(C); TNFRSF: tumor-necrosis factor receptor superfamily member).

Mediators	Effect of peroxynitrite	Explored cell types
EGFR	Activation	Fibroblast, myofibroblast
	Inhibition	Colonic (Caco2B) cell
PDGFR	Activation	Fibroblast
NTRKb	Activation	Fibroblast
TNFRSF6a	Inhibition	Hepatocyte
Src	Activation	Erythrocyte, endothelial cell, astrocyte
ERK	Activation	Cardiomyocyte, neuron, fibroblast, neutrophil, vascular smooth muscle cell, endothelial cell
JNK	Activation	Alveolar and bronchial cells, endothelial cell, cardiomyocyte, fibroblast
P38MAPK	Activation	Neurons, endothelial and bronchial cells, vascular smooth muscle cell, cardiomyocyte, hepatocyte
PKC	Activation	Cardiomyocyte, endothelial cell
	Inhibition	Neuron
PI3K–PKB	Activation	Endothelial cells, hepatocyte, fibroblast
	Inhibition	Endothelial cell, macrophage, adipocyte, retinal cell
NFκB	Activation	Neutrophil, monocyte
	Inhibition	Cardiomyocyte

Plasmalemmal and Cytosolic PTKs

Peroxynitrite anion ($ONOO^-$) is structurally similar to phosphate anion (PO_4^{3-}). Therefore, protein Tyr phosphatases are vulnerable to peroxynitrite. On the other hand, protein Tyr kinases can be directly activated by peroxynitrite. Receptor protein Tyr kinases, particularly growth factor receptors EGFR and PDGFR, undergo Tyr phosphorylation upon exposure to oxidants. However, peroxynitrite-mediated nitration of EGFR can prevent its phosphorylation in a certain type (Caco2) intestinal cells [1168].

Cytosolic protein Tyr kinases are also targeted by peroxynitrite. In particular, SRC family kinases can be activated by peroxynitrite at least in erythrocytes and endothelial cells. In erythrocytes, hematopoietic cell kinase (HCK) is activated by peroxynitrite via cysteine oxidation [1168]. Another SRC family kinase, Lyn kinase, is activated via the inhibition of the binding of auto-inhibitory Tyr527 C-terminal domain to the interaction SH2 domain.

MAPK Module

Components of the MAPK module (ERKs, JNKs, and P38MAPKs) can be activated by oxidants and free radicals. Peroxynitrite can potently stimulate ERK in endothelial and vascular smooth muscle cells, fibroblasts, cardiomyocytes, neutrophils, and neurons, using cell-specific mechanisms [1168]. Activation of JNK kinases by peroxynitrite has been observed in endothelial cells subjected to flow. Peroxynitrite is very efficient in activating P38MAPK in cardiomyocytes, endothelial and vascular smooth muscle cells, bronchial epithelial cells, and neurons [1168]. In particular, peroxynitrite stimulates P38MAPK: (1) via MAP3K3 and MAP3K6 following ERK-dependent activation of cytosolic phosholipase-A2 in human bronchial epithelial cells; (2) via activated Ca^{++}–calmodulin kinase CamK2 and Src in PC12 cells; and (3) via the release of Zn^{++} in neurons.

The PI3K–PKB Axis

In human skin fibroblasts, peroxynitrite can support PDGFR–PI3K–PKB signaling. On the other hand, peroxynitrite can preclude the PI3K signaling and PKB activation in several cell types, in particular endothelial cells. The chemical microenvironment may affect the response to peroxynitrite [1168].

PKC Isoforms

In cardiomyocytes, peroxynitrite can activate PKCϵ during ischemic preconditioning. In endothelial cells, stimulation of PKC by peroxynitrite is associated with the activation of cytosolic phospholipase-A2 and an enhanced release of vasoactive mediators. On the other hand, peroxynitrite reduces the activity of PKCα, PKCβ, PKCϵ, and PKCζ in neurons [1168].

NFκB

Nuclear factor-κB is a transcriptional activator of inflammation and anti-apoptotic genes. In addition to reactive oxygen species, peroxynitrite can activate or repress NFκB [1168].

10.7 Hippo Signaling (STK3–STK4 Pathway)

The Hippo signaling pathway participates in the control of organ size, during the body's development as well as tissue regeneration, as it restricts cell proliferation

10.7 Hippo Signaling (STK3–STK4 Pathway)

and promotes cell elimination.[118] The appropriate number of cells, hence size of the body's tissues and organs, is actually determined by coordinated cell growth, proliferation, and apoptosis. Disruption of these processes can cause cancers.

The Hippo kinase is the Drosophila homolog of mammalian protein Ser/Thr kinases STK3 and STK4[119] (Sect. 5.1) that controls both cell proliferation and death. Kinases STK3 and STK4 are named tumor suppressors, as they promote apoptosis in particular in hepatocytes [1250].

Another Hippo signaling component, the Salvador ortholog Sav1 (or WW45) facilitates activation of STK3 and STK4 kinases. In mammals, each component of the Hippo cascade regulates cytosolic or nuclear location of mammalian yorkie orthologs Yes-associated protein YAP1 and WW domain-containing transcription regulator WWTR1 (or transcriptional coactivator with PDZ-binding motif [TAZ]) to control cell proliferation and survival, as well as self-renewal and differentiation of stem and progenitor cells. In addition, the Hippo pathway crosstalks with other signaling axes, such as those of Notch, sonic Hedgehog, and Wnt morphogens.

10.7.1 Hippo Signaling Components

Known components of the Hippo (STK3–STK4) pathway include (Fig. 10.3 and Tables 10.25 to 10.27): (1) receptor Fat1 protocadherin and the other non-classical cadherin Dachsous from adjacent cells that can modulate the Hippo axis;[120] (2) adaptors neurofibromatosis NF2 (mammalian homolog of Merlin) and FERM domain-containing protein-6 (FRMD6), or Expanded-1 (Ex1; human ortholog of Expanded) that operates in parallel with NF2; (3) scaffold protein WWC1 (or KiBra) that tethers to both FRMD6 and NF2 to form an apical scaffold that promotes Hippo axis activity; (4) kinases STK3 and STK4 (human ortholog of Hippo); (5) scaffold protein WW45 (or Salvador-1 [Sav1]; ortholog of Drosophila Salvador) that binds to kinases STK3/4 and LaTS (active STK3/4–WW45 platform); (6) large tumor suppressor kinases LaTS1 and LaTS2 (mammalian homologs of Warts) that are phosphorylated (activated) by STK3 and STK4 complexed with WW45; (7) transcriptional coactivator YAP1 (human ortholog of Yorkie), which is phosphorylated (stimulated) by LaTS kinases complexed with MOB1, to activate genes responsible for cell proliferation and survival, as well as its paralog WWTR1; and (8) other components, such as Ras and Ras association domain-containing

---

[118] Protein Ser/Thr kinase Hippo is a tumor suppressor in Drosophila melanogaster, in which it was originally discovered. The Hippo pathway regulates organ size in Drosophila melanogaster. Mutations of its gene causes tumorigenesis. In Drosophila, 2 main organ-size control pathways include bone morphogenetic protein (BMP) and Hippo axes.

[119] A.k.a. mammalian Ste20-like kinases MST2 and MST1, respectively.

[120] Cadherin Fat is the largest cadherin with 34 cadherin repeats.

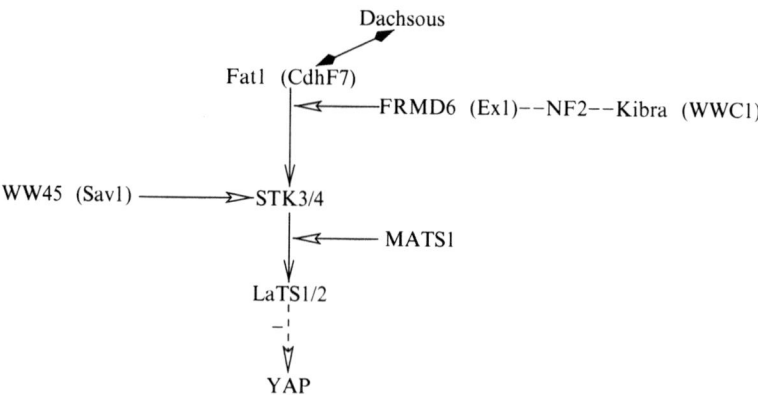

Fig. 10.3 Signaling by Hippo regulator (STK3–STK4 pathway; Source: [1251]). The Hippo cascade controls tissue growth, as it impedes cell proliferation and promotes apoptosis. The Hippo signaling requires 2 related cortical band-4.1 proteins, the tumor suppressor neurofibromatosis NF2 (Merlin homolog), a membrane-cytoskeleton scaffold, and FERM domain-containing protein-6 (FRMD6 or Expanded-1 [Ex1]; human ortholog of Expanded), the upstream effectors of the STK3/4 pathway targeted by protocadherin Fat1. Activity of NF2 depends on its phosphorylation by P21-activated kinase and activation by Rac GTPase. Expanded-1 homolog is a potent inhibitor of Yes-associated protein (YAP1) in parallel to YAP1 phosphorylation and cytosolic sequestration by 14-3-3 protein. Proteins NF2 and Expanded-1 promote the phosphorylation (activation) of large tumor suppressor kinase (LaTS) by activated protein Ser/Thr kinases STK3 and STK4 (a.k.a. mammalian sterile-20-like kinases MST2/1, i.e., mammalian homolog of Hippo). The scaffold protein WW domain-containing protein WW45 (mammalian homolog of Salvador) binds to LaTS and STK3/4 to favor LaTS phosphorylation. Activation of LaTS also requires plasmalemmal localization of Mps one-binder enzyme (MOB) as tumor suppressor (MATS1 or MOB kinase activator-like protein MOBKL1b). Once activated by STK3 and/or STK4 kinases, LaTS phosphorylates (inactivates) YAP1 transcriptional coactivator. The latter regulates the activity of various transcriptional regulators, such as P73 and P53-binding protein-2. Protocadherin Fat1 acts as a transmembrane receptor for protocadherin Dachsous to trigger the Hippo pathway.

family proteins (RASSF) and microRNA Bantam. In addition to this canonical pathway (PCdh–NF2–STK3/4–LaTS1/2–YAP1/WWTR1), Hippo signaling relies on crosstalks with other signaling cascades.

Cell proliferation during embryogenesis and childhood becomes restricted in adulthood. Cell division then serve for tissue maintenance. Kinases STK3 and STK4 particularly promote apoptosis when they are primed by irradiation and activated by P53 [1251]. Activation of P53 by these kinases is modulated by proteins of the Ras and Ras association domain-containing family (RASSF).[121] Hippo signaling restrains cell proliferation, as it prevents activation of YAP1 transcriptional coactivator.

---

[121] Ras association domain-containing family protein RASSF1a causes apoptosis, as it stimulates kinases STK3 and STK4 [1252]. Once bound to STK3, RASSF6 hinders STK3 activity to repress Hippo signaling. When released from activated STK3, RASSF6 causes apoptosis. This RASSF6-mediated pathway works in parallel to the canonical Hippo axis.

10.7 Hippo Signaling (STK3–STK4 Pathway)

Table 10.25. Components of the Hippo (STK3–STK4) pathway. (**Part 1**) Cell-surface mediators (protocadherins). Some mediators of the Hippo pathway in Drosophila melanogaster, in which it was originally discovered, can have several homologs in mammalian cells. These paralogs sometimes play redundant roles, but in most cases have distinct properties. . The Hippo pathway regulates organ size. It is composed of: (1) plasmalemmal regulators, such as cell adhesion molecules and cell polarity complexes; (2) a cytoplasmic cascade with 2 protein Ser/Thr kinases and their regulators and adaptors; and (3) a transcription coactivator (CDHF: cadherin family member; CDHF7 is also called cadherin ME5 and cadherin-related homolog CdhR8; MGEF1: multiple EGF-like domain-containing protein; PCdh: protocadherin).

	Mammals	Drosophila
Type	Other aliases	
Dchs1	Cdh19 (25), CdhR6, PCdh16	Dachsous
Dchs2	Cdh27, CdhR7, PCdh23, PCdhJ	Dachsous
Fat1	CDHF7, CdhR8, ME5	Fat
Fat2	CDHF8, CdhR9, MEGF1	Fat
Fat3	CDHF15, CdhR10,	Fat
Fat4	CDHF14, CdhR11, FatJ	Fat

The Hippo pathway comprises Fat1 cadherin and microRNA Bantam. Protocadherin Fat1 stabilizes Large tumor suppressor kinases (LaTS).[122]

The Hippo signaling may repress microRNA Bantam that elicits tissue growth and patterning and inhibits apoptosis. On the other hand, miR372 and miR373 target transcripts of large tumor suppressor kinases. The latter then cannot phosphorylate (inactivate) the 2 transcription coactivators YAP1 and WWTR1 [1253].

10.7.1.1 Signaling Crosstalk

Signaling efficiency in cells compels many proteins to function in various modes and possible locations. Proteins WWTR1 and YAP1 acts with other mediators to regulate transcription in response to activation of the Hippo pathway. Factors WWTR1 and YAP1 can move between the nucleus and cytoplasm.

Moreover, WWTR1 participates in the Wnt pathway that controls cell fate in developing and remodeling tissues. It interacts directly with Disheveled. It impedes the phosphorylation of Disheveled by casein kinases CK1δ and CK1ε. Consequently, it restricts the canonical Wnt–β-catenin pathway [1254].

Hippo signaling restrains cardiomyocyte proliferation, thereby controlling the heart size in developing mice. In cardiomyocytes, Hippo regulates the genes that encode transcription factors Sox2, Snai2, and lung-derived MycL1 (and Myc in

---

[122] Protein Ser/Thr kinase LaTS1 complexes with the cell cycle controller kinase CDK1 in early mitosis. Protein Ser/Thr kinase LaTS2 localizes to centrosomes during interphase as well as metaphase, during which it interacts with Aurora-A kinase. It also favors P53 activity.

Table 10.26. Components of the Hippo (STK3–STK4) pathway. (**Part 2**) Cytoplasmic mediators (CDHF: cadherin family member, CDHF7 is also called cadherin ME5 and cadherin-related homolog CdhR8; DTEF: divergent TEF; ETFR: enhancer of transcriptional factor-related; FRMD: FERM domain-containing protein, or Expanded-1 [Ex1]; KiBra: kidney and brain protein; KPM: kinase phosphorylated during mitosis; LaTS: large tumor suppressor kinase; MATS: Mps one-binder (MOB) as tumor suppressor; MATS1 is also named MOB1a and MOB kinase activator-like protein MOBK1Lb; MST: mammalian Ste20-like kinase; NF2: neurofibromatosis-2; RTEF: related to TEF; STK: protein Ser/Thr kinase; TAZ: transcriptional coactivator with PDZ-binding motif; TCF: transcription factor; TCF13L: transcription factor-13-like; TEAD: TEA domain-containing family member; TEF: transcriptional enhancer factor; VgL: Vestigial-like protein; ViTo: vestigial- and Tondu-related protein; WW45: 45-kDa WW domain-containing protein; WWC: WW and C2 domain-containing protein; WWTR: WW domain-containing transcription regulator; YAP: Yes Tyr kinase-associated protein).

Type	Mammals Other aliases	Drosophila
Kinases		
LaTS1		Warts (Wts)
LaTS2	KPM	Warts
STK4	MST1	Hippo
STK3	MST2	Hippo
Adaptors, scaffolds, and activators		
FRMD6	Ex1	Expanded
MATS1	MOB1a (4b), MOBK1Lb	MATS
MATS2	MOB1b (4a), MOBK1La	MATS
NF2		Merlin
WW45	WWP4, Sav1	Salvador
WWC1	KiBra	
Transcriptional coactivators		
TEAD1	TEF1, TCF13	Scalloped
TEAD2	TEF4	Scalloped
TEAD3	TEAD5, TEF5, DTEF1, ETFR1	Scalloped
TEAD4	TEF3, EFTR2, RTEF1b, TCF13L1	Scalloped
VgL1	Tondu (Tdu)	Vestigial
VgL2	ViTo1	Vestigial
VgL3		Vestigial
WWTR1	TAZ	Yorkie (Yki)
		Taffazin (*Saccharomyces cerevisiae*)
YAP1	YAP2, YAP65	Yorkie

hepatocytes) as well as those that encode the cell cycle regulators cyclin-D1 and CDC20 [1256]. Effector YAP1 interacts with β-catenin on Sox2 and Snai2 genes [1256]. However, phosphorylation of YAP1 triggered by Hippo signaling provokes YAP1 sequestration into the cytoplasm. Activity of Hippo- and Wnt-regulated genes is thus repressed.

10.7 Hippo Signaling (STK3–STK4 Pathway)

Table 10.27. Effector complexes of Hippo signaling. Upon stimulation, several molecular coupling platforms are built along the Hippo axis in different cellular compartment in the canonical pathway. In addition, the Hippo pathway crosstalks with other signaling axes, especially those involved in morphogenesis (TGF, Notch, Hedgehog, and Wnt).

Site	Effectors
Cell surface and cortex	Fat–Dachsous–NF2
Cytoplasm	STK3–STK4–WW45–LaTS1/2
	LaTS1–LaTS2–MOB1–YAP1/WWTR1
Nucleus	YAP1–VgL–TEAD
	WWTR1–TEAD

Agent WWTR1 is also implicated in signaling by transforming growth factor-β [1255]. In human embryonic stem cells, WWTR1 interacts with SMAD complexes. In particular, YAP1 binds SMAD1 and supports subsequent gene transcription.

Furthermore, WWTR1 is a component of the ubiquitin ligase complex SCF$^{\beta TRCP}$ that degrades calcium-permeable cation channel polycystin-2 [1257].

In Drosophila, Hippo signaling can activate Notch signaling. Sonic Hedgehog increases YAP1 expression and promotes its nuclear localization; subsequently, YAP1 primes expression of Gli2 transcription factor [1258].

Moreover, YAP1 targets the amphiregulin gene; this ligand of epidermal growth factor receptor activates EGFR signaling [1259].

In Drosophila melanogaster, the growth suppressor Yorkie, a transcriptional regulator, binds to the MYC gene [1260]. Conversely, transcription factor MyC can regulate Yorkie expression, i.e., its own cellular level. Growth-promoting MyC represses Yorkie expression using both transcriptional and post-transcriptional mechanisms.

10.7.1.2 Intercellular Contacts – Protocadherins

The Hippo pathway is activated by intercellular contacts and engagement of protocadherin Fat1, the largest cadherin.[123] Protocadherin Fat1 acts as a receptor to regulate cell growth and polarity [1262]. Protocadherin Dachsous can extracellularly interact with Fat1 protocadherin.

---

[123] In Drosophila, Fat cadherin regulates cell proliferation and planar cell polarity. Its mammalian homolog Fat1 binds to vasodilator-stimulated phosphoprotein (VASP) to regulate actin dynamics and intercellular interactions [1261]. This member of the ENA-VASP family contains an N-terminal EVH1 domain that enables its recruitment to focal adhesions, a proline-rich region that binds SH3 and WW domain-containing proteins, and a C-terminal EVH2 domain that allows tetramerization and binds globular and filamentous actin.

Table 10.28. Mediators of the Hippo pathway and their partners and substrates. (Source: [1043, 1258]; HTRa2: mitochondrial high temperature requirement serine peptidase-A2; NeK: never in mitosis gene-A (NIMA)-related kinase; RCC: regulator of chromosome condensation). Mediators of the Hippo pathway interact with various proteins, such as Na^+–H^+ exchanger regulatory factor NHERF1, or SLC9a3R1, 14-3-3 protein, HER4 receptor, as well as P73, Runx1, Runx2, SMAD7, and TEAD1 factors.

Components	Interactors
STK3/4	LaTS, WW45, MOBKL1a/1b, YAP1, WWTR1
	NeK2, JNK, RASSF6, RCC1, FoxO, histone-2AX/2B
STK3	histone-1H4A–histone-1H4E
STK4	PKB1, Aurora-B, STK38/38L, RASSF5
LaTS1	CDK1, LIMK1, STK3/4, HTRA2, zyxin
LaTS2	STK3/4, YAP1, MOBKL1a, $PP1_{r13b}$, ajuba (Jub)
YAP1	Abl, PKB, RASSF1a, 14-3-3, SLC9a3R1,
	cyclin-E (in Drosophila), cyclin-D1 in vertebrates,
	P73, PML, Runx, TEAD
WWTR1	LaTS2, GSK3β, TEADs
WWC1	NF2, FRMD6
NF2	PAK, $PP1_{r12}$, CRL4–DCAF1 ligase

Protocadherins Fat1 and Dachsous indeed form heterodimeric bridges between adjacent cells.[124] Kinases STK3 and STK4 transfer signals from these atypical cadherins of the Fat and Dachsous categories at the plasma membrane to intracellular kinase LaTS1 to regulate transcriptional coactivators.

10.7.1.3 STK3 and STK4 Kinases

Substrates of STK3 (Sect. 5.1.4.1) and STK4 (Sect. 5.1.4.2) kinases include not only components of the Hippo pathway, such as LaTS kinases, WW45, and Mps one binder kinase (MOB) activator-like proteins, but also STK38, NeK2 (never in mitosis gene-A [NIMA]-related kinase-2), and JNK (activation) kinases, histone-2AX and -2B, and FoxO factor [1258] (Table 10.28).

Kinases STK3 is cleaved by caspase-3 to then undergo an irreversible autophosphorylation. Its full activation indeed requires both caspase-mediated cleavage and phosphorylation. Kinase STK4 mediates apoptosis upon cleavage, autophosphorylation and nuclear translocation [1264].

---

[124]Amounts of Fat1 and Dachsous on the cell surface of a given cell affect the distribution of Fat1 and Dachsous on its neighbors and participate in polarization of these cells. In Drosophila melanogaster, protein Ser/Thr kinase Four-jointed of the Golgi body phosphorylates Fat1 extracellular domain and its transmembrane ligand Dachsous before exocytosis [1263].

10.7 Hippo Signaling (STK3–STK4 Pathway)

Kinase PKB phosphorylates STK3 (Thr117 and Thr384) and STK4 (Thr120 and Thr387), hence impeding their cleavage. Kinase STK4 is dephosphorylated by PHLPP1 and PHLPP2 phosphatases [1258]. Agent RASSF1 promotes STK3 and STK4 autophosphorylation.

Activity of STK3 and STK4 kinases is controlled by cRaf and phosphatases. Protein phosphatase-2 activates STK4 kinase. Connection between STK3 and cRaf is promoted by PKB [1258]. On the other hand, RASSF1a (RASSF1a and RASSF1c being major RASSF1 isoforms) relieves STK3 inhibition by cRaf kinase. In addition, STK3 and STK4 kinases and RASSF6 mutually inhibit each other. Kinase JNK, a substrate of STK4, phosphorylates STK4 (Ser82) to enhance its activity [1258].

Kinase STK4 phosphorylates (inactivates) Aurora-B as well as P53 transcription factor; it also inhibits PKB1 enzyme.

Members of the RASSF family (RASSF1–RASSF10), are components of the Hippo pathway.[125] Agent RASSF1a complexes with STK3, WW45, and LaTS1 components [1258].[126]

The other family member RASSF6 interacts with STK3 and STK4, but it can support apoptosis independently of the Hippo pathway. Both RASSF1a and RASSF6 participate in apoptosis partly via modulator of apoptosis protein MOAP1. Kinase STK3 precludes RASSF6–MOAP1 interaction, thereby hindering RASSF6-mediated apoptosis. Conversely, RASSF6 suppresses STK3 activity, possibly by inhibiting homo-oligomerization [1258].

10.7.1.4 LaTS Kinases

Mammalian LaTS kinases pertain to the nuclear DBF2-related (NDR) family. Member NDR1, or STK38, is involved in the regulation of centrosome and chromosome alignment. Both NDR1 and NDR2 (STK38L) are activated by STK4 to mediate apoptosis downstream from the RASSF1a–STK4 module [1258].

Both LaTS1 and LaTS2 localize to the centrosome. The centrosomal location of LaTS2 is controlled by Aurora-A [1258]. During mitosis, LaTS1 interacts with zyxin, a zinc-binding phosphoprotein that accumulates in focal adhesions and along the actin cytoskeleton, to ensure a normal mitosis progression.

Kinase LaTS2 associates with ajuba, a regulator of cell proliferation and differentiation, and recruits γ-tubulin to centrosomes during mitosis. It also binds to DM2 and inhibits its ubiquitin ligase activity, thereby stabilizing P53 factor. Enzyme

---

[125] Two subfamilies can be defined whether RASSF proteins possess the Salvador/RASSF/Hippo (SARAH) domain in their C-termini (RASSF1–RASSF6) or not (RASSF7–RASSF10). The SARAH domain, which mediates RASSF heterodimerization with STK3 and STK4 kinases is shared by STK3, STK4, WW45, and NeK2 proteins.

[126] In response to DNA damage, RASSF1a is phosphorylated by ATMK to activate STK3 and LaTS1 kinases. In addition, RASSF1a interacts with CDC20, an activator of the anaphase-promoting complex (APC), as well as Aurora-A and Aurora-B kinases.

LaTS2 phosphorylates inhibitory subunit of protein phosphatase-1 ($PP1_{r13b}$) to induce its nuclear translocation. In the nucleus, P53 then targets promoters of cell cycle regulator genes rather than those of pro-apoptotic genes [1258]. On the other hand, cytoplasmic $PP1_{r13b}$ prevents the phosphorylation by LaTS1 of YAP1 and WWTR1 and increases their nuclear levels to promote cell survival.

In addition, LaTS1 operates with enzyme Mps one-binder (MOB) as tumor suppressor MATS1 (a.k.a. MOB1a and MOB kinase activator-like protein MOBK1Lb) that mediates growth inhibition and tumor suppression [1265].[127]

10.7.1.5 YAP1 and WWTR1 Transcriptional Coregulators

Transcriptional coregulators YAP1 and WWTR1 share interactors. Yes-associated protein-1 binds to the Src homology SH3 domain of Yes kinase proto-oncogene product. Molecule YAP1 is characterized by domains with 2 tryptophans (**WW**) that enables binding of proteins with a PPXY sequence, among other between-protein interaction domains; YAP1 is a founding member of the family of WW domain-containing proteins. Protein YAP1 links to members of the P53 family (P73α–P73β and P63), SMAD7, and Runx2 factors, as well as HER4 receptor [1267]. Proteins that interact with YAP1 N-terminus include heterogeneous nuclear ribonucleoprotein hnRNPu, TEA domain-containing transcription factors, and LaTS kinases. Moreover, YAP1 C-terminus contains a PDZ-binding motif for binding to PDZ domain-containing proteins.

Transcriptional coactivator YAP1 is the nuclear effector of STK3 and STK4 that targets cyclin-E (in Drosophila), among other mediators [1268]. Transcriptional coactivator YAP1 promotes epithelial–mesenchymal transition, hence malignant transformation. Regulator YAP also operates as a secondary coactivator of steroid hormone receptors to enhance their transcriptional activity.

Phosphorylation of YAP1 by Abl kinase switches YAP1 activity from a proliferative to an apoptotic mode [1269]. Upon DNA damage, phosphorylated YAP ceases its interaction with Runx transcriptional factor, thus precluding the expression of Itch Ub ligase to favor P73 accumulation. Furthermore, its interaction with P73 via RASSF1a initiates expression of pro-apoptotic genes. Mediator YAP contributes to 2 distinct tumor suppressor pathways: via promyelocytic leukemia (PML)[128] and P73 transcription factors.

---

[127] Two MOB proteins (MOB1–MOB2) exist in yeasts. They serve as cell cycle regulators. Seven human homologs of yeast enzyme MOB exist with different expression patterns (MOB1a–MOB1b, MOB2a–MOB2c, and MOB3–MOB4). Only MOB1a and MOB1b interact with both LaTS1 and LaTS2 agents [1266]. Agent MOB1a and MOB1b interact with LaTS1 and LaTS2 and support their phosphorylation (activation) by STK3 and ST4 kinases.

[128] A.k.a. RNF71 and TRIM19.

10.7 Hippo Signaling (STK3–STK4 Pathway)

YAP1 Subcellular Localization

YAP1 Cytoplasmic Sequestration

In some cell types, YAP1 localizes mainly to the cytosol, as it is phosphorylated by PKB kinase. Once YAP1 is phosphorylated by PKB, YAP binds to 14-3-3 protein, thereby being sequestered in the cytoplasm. Kinase LaTS phosphorylates YAP1, thus labelling it for degradation. The Crumbs homolog complex[129] (MPP5–InaDL–Lin7c–Angiomotin) assembled at tight junctions interacts with YAP1 and also sequesters it in the cytosol. In addition, YAP1 connects to SLC9a3R1 regulator and can then be anchored to the plasma membrane. Tight junction proteins ZO1 and ZO2 bind WWTR1; the latter, which can lodge in the nucleus, also tethers to YAP1 factor.

When LaTS kinases phosphorylate YAP1, the latter is further phosphorylated by casein kinases CK1δ and CK1ε for ubiquitination and subsequent degradation.

YAP1 Nuclear Accumulation

In other cell types and context, YAP1 nuclear accumulation promotes P73-dependent transcription of pro-apoptotic genes. Kinase Abl phosphorylates YAP1, thereby supporting the association of YAP1 with P73 factor [1258]. Factor YAP1 competes with ubiquitin ligase Itch for P73 binding. Promyelocytic leukemia transcription factor (PML) recruits YAP1 and stabilizes P73 factor via sumoylation. As PML is a target of P73, a positive feedback loop exists between these 2 molecules.

Role of YAP1 and WWTR1 in Mechanotransduction

Activity of YAP1 and WWTR1 on cognate genes, such as those that encode connective tissue growth factor (CTGF or CCN2) and cardiac ankyrin repeat domain-containing protein AnkRD1, depends on the rheology of the matrix. Expression of the Ctgf and ANKRD1 genes is observed in human cells cultured on a stiff material, but not on soft matrices [1270]. In the latter case, YAP1 and WWTR1 lodge predominantly in the cytoplasm. Tension of the actin cytoskeleton transmitted by stress fibers enables localization and retention of YAP1 and WWTR1 in the nucleus. This action of YAP1 and WWTR1 is mediated by Rho GTPase that acts on the actomyosin cytoskeleton, but not by the NF2–STK3/4–LaTS axis [1270]. Therefore, YAP1 and WWTR1 operate as mediators of mechanical signals exerted by the cell environment.

---

[129] The Crumbs complex is a signaling platform of the cell apex that keeps the aPKC (atypical protein kinase-C or Par) complex (aPKC–Par3–Par6) apical during cell shape changes.

10.7.1.6 Scaffold WW45 – Salvador-1

Scaffold protein WW45 (or Salvador-1 homolog) binds to STK3, STK4, and LaTS1 kinases. These kinases synergistically phosphorylate YAP1 and WWTR1, respectively, hence causing their cytoplasmic sequestration to suppress tissue growth.[130] Both YAP1 and WWTR1 regulate gene expression via their cognate transcription factor TEA domain transcription factors TEAD1, also called transcription enhancer factor TEF1 (homolog of Drosophila Scalloped) [1269].[131]

Activity of WWTR1 is not only regulated by LaTS2 kinase, but also by glycogen synthase kinase GSK3β that primes its ubiquitination for proteolysis [1269]. Yes-associated protein forms a transactivation complex with TEF1 and its cofactors Vestigial-like proteins (VgL1–VgL3; homologs of Drosophila Vestigial).[132] Vestigial-like proteins can induce the expression of transcription factor E2F1 [1273].

10.7.1.7 WWC1

Human WW domain-containing protein WWC1 modulates the activation of Hippo signaling. This early component of the Hippo pathway links to Hippo-related, cortical proteins FRMD6 and NF2 and promotes FRMD6–NF2 binding [1274–1276]. The FRMD6–NF2–WWC1 complex localizes to the apical region of epithelial cells to regulate the Hippo cascade. In addition, in cell cultures, WWC1 complexes with LaTS1 and causes a marked reduction in YAP1 phosphorylation without affecting the LaTS–YAP1 interaction [1275]. Protein WWC1 can also stimulate the phosphorylation of LaTS1 and LaTS2 enzymes.

10.8 Coregulators

Signaling networks create transient and focal points of enzyme activity that disseminate in a spatially and temporally controlled manner within the cell signals brought by neurotransmitters, hormones, and growth factors. Compartmentation is dictated by certain mediators, themself supervised by intracellular messengers.

Manifold cell functions can be specifically and efficiently controlled by a limited number of proteins over various distinct enzymes. Cell activities, particularly effec-

---

[130] Taffazin is a paralog of Yes-associated protein (YAP). Localization of YAP and Taffazin in the nucleus or cytoplasm depends on whether cells are sparse or confluent.

[131] Casein kinase CK2 phosphorylates TEAD1 to impede DNA binding. Overexpression of TEAD1 promotes neoplastic transformation of cells.

[132] Mammalian Vestigial-like protein VgL2, a cofactor of TEF1 and MEF2 transcription factors, promotes skeletal muscle differentiation [1271]. Scalloped interaction domain-containing protein VgL2 intervenes in the development of skeletal muscles and Rathke's pouch (pharyngeal clefts)-derived structures such as adenohypophysis [1272].

10.8 Coregulators

Table 10.29. Specific anchoring proteins for protein kinase-A, -C, and -G. These anchoring proteins are aimed at ensuring the accessibility and phosphorylation of specific substrates at given subcellular compartments.

Type of protein kinase	Specific anchoring proteins
Protein kinase-A	A-kinase-anchoring proteins (AKAP1–AKAP14)
Protein kinase-C	Receptors for activated protein kinase-C (RACK1–RACK2)
Protein kinase-G	G-kinase-anchoring protein (GKAP1)

tor responses to receptor stimulations, are processed by between-protein interactions required in signaling molecules. Regulators and targets must be appropriately located to achieve the cell functions, especially in the heart for the spatiotemporal regulation of firing action potentials by nodal cells or secretion of hormones by cardiomyocytes.

Signaling networks are particularly associated with cytosolic foci of enzyme activity that transmit the action of messengers in the cytosol. Anchoring proteins (or anchor proteins) yield a framework that orients these enzymes toward selected substrates to elicit a specific response.

Coregulators participate in signaling by organizing molecular interactions and determining pathway dynamics.

A-kinase-anchoring proteins (AKAP) control the location and substrate specificity of protein kinase-A. Protein kinase-A1 resides mainly free in the cytoplasm, where it has a high affinity for cAMP, whereas protein kinase-A2 localizes more precisely, as it couples to A-kinase-anchoring proteins. The latter are scaffold proteins that determine the spatial organization of signaling pathways, as it associates PKA with its substrates. A-kinase-anchoring proteins possess various binding domains that enable the identification and anchoring of specific targets in given subcellular regions. On the other hand, inhibition of PKA–AKAP interactions modulates the PKA signaling.

Annexins have many partners implicated in signaling cascades and operate in interactions between the cell membrane and cytoskeleton.

Like AKAPs that connect to protein kinase-A (Sect. 10.8.4), a G-kinase-anchoring protein (GKAP1) tethers to protein kinase-G. Alias GKAP1 avoid confusion with guanylate kinase-associated protein (GKAP) implicated in the formation and maintenance of molecular platforms in synapses of the nervous system.

Specific anchoring proteins exist also for protein kinase-C, receptors for activated protein kinase-C (RACK; Table 10.29).

10.8.1 Receptors for Activated C-Kinase

Isozymes of the PKC family (Sect. 5.2.7) are split into 3 subfamilies: (1) the subfamily of *conventional* enzymes that includes PKCα, PKCβ1, PKCβ2, and PKCγ, which require diacylglycerol, calcium, and phospholipid for activation;[133] (2) the subfamily of *novel* enzymes that encompasses PKCδ, PKCε, PKCη, and PKCθ, which necessitate diacylglycerol for activation; and (3) the subfamily of *atypical* enzymes that comprises PKCζ, PKCι, PKN1, and PKN2, which involve neither diacylglycerol nor calcium for activation.

Different receptor for activated C-kinases target different PKC isoforms. The RACK proteins bind activated PKC in a selective manner, but are not substrates of PKC isozymes. Among the RACK family members, RACK1 is selective for activated PKCβ2, RACK2 for PKCε [1277]. The multifunctional chaperone glycoprotein C1QBP (complement component-1, Q-subcomponent-binding protein)[134] is a PKC-binding protein that interacts with PKCα, PKCδ, PKCζ, and protein kinase-D1 (PKD1 or PKCμ).

Receptors for activated C-kinase support the translocation of each PKC isoform to a specific subcellular compartment. For example, RACK activates mitochondrial PKCε that interacts directly with mitochondrial ATP-sensitive K^+ channel at the inner mitochondria membrane [1278]. The activity of this functional signaling module is impeded by protein phosphatase-2. Receptors for activated C-kinase also enhance the duration of PKC activation, thereby increasing substrate phosphorylation.

On the other hand, pseudo-rack regions on PKC, which are homologous to RACK protein, interacts with the RACK-binding site and stabilizes the enzyme in its inactive state. Activators that match this sequence interfere with the intramolecular interaction and, thus, activate PKC enzyme.

Receptors for activated protein kinase-C constitute a family (Table 10.30). Receptor for activated C-kinase RACK1 is homologous to the Gβ subunit guanine nucleotide-binding (G) proteins. The RACK proteins serve as adaptors not only for PKCβ1 and PKCε, but also for several other signaling mediators. In particular, RACK1 also binds to integrin-β1, dynamin-1, RasA1 GAP, Src and Fyn kinases, PDE4d5 phosphodiesterase, and phospholipase-Cγ [1279], as well as PTPRm phosphatases, cyclin-A1, STAT1 and NR3c4 transcription factors, and AT_1, CSF2Rβ, and IfnαR2 receptors[251]. In addition, RACK2 is a coated vesicle component involved in vesicular transfer.

---

[133] The main mechanism of activation originates from stimulated G-protein coupled receptors and activation of phospholipase-C. The latter hydrolyzes phosphatidylinositol bisphosphate into diacylglycerol and inositol trisphosphate that causes a Ca^{++} influx.

[134] A.k.a. mitochondrial matrix protein P32.

Table 10.30. Family of receptors for activated protein kinase-C (bHLHb(e): class-B(E) basic helix–loop–helix protein; CoP: coatomer protein; GNβ2L1: guanine nucleotide-binding (G) protein-β polypeptide-2-like-1; Olig2: oligodendrocyte lineage transcription factor-2; PrKCBP: protein kinase C-binding protein ZMYND8: zinc finger, MYND-type domain-containing protein-8).

Type	Other aliases
RACK1	GNβ2L1
RACK2	β'-CoP, RACKε
RACK7	PrKCBP1, ZMYND8
RACK17	PrKCBP2, Olig2, bHLHb1, bHLHe19

10.8.2 G-Kinase-Anchoring Protein

Second messenger cGMP is produced in response to nitric oxide and natriuretic peptides. The NO–cGMP and NP–cGMP pathways intervene in numerous processes in the nervous, cardiovascular, and immune systems. Agent cGMP is involved in the regulation of smooth muscle relaxation, platelet aggregation, intestinal secretion, and endochondral ossification via activated cGMP-dependent protein kinases-G (Sect. 5.2.9). The latter possesses 2 subtypes — PKG1 and PKG2 — that are encoded by distinct genes. In addition, PKG1 has 2 isoforms (PKG1α and PKG1β) that are generated by alternative splicing.

The cGMP-dependent signal transduction involves not only protein kinase-G, but also its substrates and regulators. Among regulators, G-kinase-anchoring protein-1 (GKAP1)[135] interacts with PKG1α, but neither with PKG1β nor with PKG2 isozyme [1280].[136]

G-kinase-anchoring protein GKAP1 is specifically expressed in testis, more precisely spermatocytes and early round spermatids [1280]. Inside the cell, GKAP1 localizes to the Golgi body. In fact, PKG1α directly interacts with and phosphorylates GKAP1 to facilitate the translocation of PKG1α to the Golgi body in response to intracellular cGMP accumulation that causes a conformational change and releases PKG1α from GKAP1.[137]

---

[135] A.k.a. 42-kDa cGMP-dependent protein kinase-anchoring protein (GKAP42).

[136] In vitro, but not in cells, both PKG1α and PKG1β, but not cAMP-dependent protein kinase-A, phosphorylate GKAP1 protein. Troponin-T is another anchoring protein for PKG1, but not PKG2; PKG1 may participate in the regulation of muscle contraction via phosphorylated troponin-I [1280].

[137] Similarly, scaffold AKAPs bind PKA via its regulatory subunits and release the catalytic subunits from regulatory subunits in response to an elevation of intracellular cAMP concentration.

Table 10.31. Scaffold proteins — Disc large homologs — members of the PSD95 family of the MAGUK superfamily. Members of the MAGUK superfamily intervene in the maintenance of the structure of submembrane domains, especially cell junctions, and can thus participate in signaling at these membrane regions.

Member	Aliases
DLg1	SAP97
DLg2	PSD93
DLg3	SAP102
DLg4	PSD95, SAP90

10.8.3 Scaffolds of Synapses of the Central Nervous System

The GKAP–SAPAP/DLGAP (guanylate kinase-associated protein–SAP90/PSD95-associated protein/Disc large homolog-associated protein) family comprises scaffold proteins for assembling and maintenance of proteic signaling platforms associated with synaptic membranes.[138]

Members of the membrane-associated guanylate kinase (MAGUK) superfamily are organizers of cellular junctions.[139] In particular, these molecular scaffolds structure ion channels and signaling molecules at the synaptic junction, especially the assembly and function of signaling complex at postsynaptic densities of excitatory synapses.

Among the MAGUK superfamily members, members of the PSD95 family, i.e., scaffold Disc large homologs (DLg1–DLg4; Table 10.31) anchor glutamate receptors to neuron terminals. NMDA (^{N}methyl ^{D}aspartate)-type glutamate receptors interact directly with DLg4 that organizes a cytoskeletal–signaling complex at the postsynaptic membrane. Disc large homolog-4 interacts not only with GluN2a and GluN2b, but also K_V1 channels, and nitric oxide synthase to cluster these molecules at synaptic junctions.

Guanylate kinase-associated protein (GKAP), or Disc large homolog-associated protein-1 (DLgAP1), binds directly to the 4 Disc large homologs, as well as SH3 and multiple ankyrin repeat domains protein SHAnk2,[140] 8-kDa cytoplasmic dynein

---

[138] In this context, GKAP alias stands for guanylate kinase (GuK)-associated protein, but not G-kinase-anchoring protein.

[139] Proteins of the MAGUK superfamily contain PDZ, SH3, and guanylate kinase domains. Between-protein interactions relies on PDZ and SH3 domains; the guanylate kinase domain binds GMP and GDP, but does not have kinase activity. The MAGUK superfamily includes Disc large homologs, zona occludens (tight junction) protein-1 and -2, MAGUK P55 subfamily members (55-kDa membrane proteins, palmitoylated MPP1–MPP7), and calcium–calmodulin-dependent serine protein kinase (CASK), among others.

[140] Homer and SHAnk proteins are among the most abundant scaffolds in the postsynaptic density, working synergistically for maturation and structural integrity of dendritic spines. They recruit proteins to synapses.

10.8 Coregulators

Table 10.32. Disc large homolog-associated proteins (GKAP: guanylate kinase-associated protein; SAPAP: SAP90/PSD95 (DLg4)-associated protein).

Type	Other Aliases
DLgAP1	SAPAP1, DAP1, GKAP
DLgAP2	SAPAP2, DAP2
DLgAP3	SAPAP3, DAP3
DLgAP4	SAPAP4, DAP4, DLP4

light chain LC8 type 1 (DynLL1),[141] and LC8 type 2 (DynLL2)[142] to constitute molecular scaffolds in postsynaptic densities of neurons.[143]

Disc large homolog-associated proteins (Table 10.32) function in the transfer of DLg4 from the cytosol to the plasma membrane and may anchor molecules to facilitate interaction between postsynaptic density components, thereby being involved in signaling at neuronal postsynaptic densities [1282].

The SH3 and multiple ankyrin repeat domain-containing protein SHAnk2[144] abounds in postsynaptic densities. Members of the SHANK family that function as scaffolds in postsynaptic densities interact with synaptic proteins of the SAPAP–GKAP family. In particular, they can link DLg4 to plasmalemmal receptors and the cytoskeleton at glutamatergic synapses of the central nervous system [1283]. The SHAnk proteins connects to GKAP glutamate receptor-binding protein (GRIP), and Homer, thereby bridging the SHAnk–GKAP–DLg4–NMDAR, SHAnk–GRIP–AMDAR, and SHAnk–Homer–mGluR complexes in synapses [1284, 1285]. Alternative splicing of SHANK family members may regulate the spectrum of their interactors.

Synaptic cell adhesion molecules are implicated in the regulation of initial contacts between dendrites and axons, synapse formation and maturation, maintenance, and remodeling of established synapses. Adhesion molecules, such as neuroligins, neurexins, cell adhesion molecules (CAdM1–CAdM4; Table 10.34),[145] netrin-G

---

[141] A.k.a. DLC1 (ambiguous alias) and protein inhibitor of neuronal nitric oxide synthase NOS1 (PIN) and SAP90/PSD95-associated protein SAPAP1 (SAP: synapse-associated protein; PSD: postsynaptic density protein).

[142] A.k.a. DLC2 (ambiguous alias).

[143] Disc large homolog-4 is expressed at various submembrane domains. It localizes at postsynaptic densities. Other components of postsynaptic densities include actin, calmodulin, Ca^{++}-calmodulin-dependent protein kinase-2, fodrin, and tubulin. Dynein light chain colocalizes with DLg4 and ^{F}actin in dendritic spines. A molecular complex made up of DLg4, GKAP, DynLL, and myosin-5 may be involved in the intracellular transfer of the DLg4–GKAP complex [1281].

[144] A.k.a. proline-rich synapse-associated protein ProSAP1.

[145] A.k.a. synaptic cell adhesion molecules (SynCAM).

Table 10.33. The SHANK family of scaffold proteins that serve as assembly platforms for other postsynaptic density proteins (CortBP: cortactin-binding protein ProSAP: Proline-rich synapse-associated protein SHAnk: SH3 and multiple ankyrin repeat-containing protein; SstRIP: somatostatin receptor-interacting protein).

Subtype	Other aliases
SHAnk1	SPAnk1, SstRIP, synamon
SHAnk2	SPAnk3, ProSAP1, CortBP1
SHAnk3	SPAnk2, ProSAP2

Table 10.34. Synaptic cell adhesion molecules (BIgR: brain immunoglobulin receptor; CadM: cell adhesion molecule; IGSF: immunoglobulin superfamily member; LRRC4: leucine rich repeat-containing protein-4; LRRTM: leucine rich repeat-containing, transmembrane, neuronal protein; NecL: nectin-like protein; NGLi: netrin-Gi ligand; SynCAM: synaptic cell adhesion molecule).

Subtype	Other aliases
CadM1	SynCAM1, NecL2, IGSF4a
CadM2	SynCAM2, NecL3, IGSF4d
CadM3	SynCAM3, NecL1, IGSF4b, BIgR
CadM4	SynCAM4, NecL4, IGSF4c
LRRC4a	NGL2
LRRC4b	NGL3
LRRC4c	NGL1
LRRTM1	
LRRTM2	LRRN2
LRRTM3	
LRRTM4	

ligands (NGL1–NGL3),[146] PTPRf, leucine rich repeat-containing transmembrane neuronal proteins (LRRTM1–LRRTM4), and EPHb receptors, can induce pre- and postsynaptic differentiation in contacting axons and dendrites, respectively. Synaptic cell adhesion molecules regulate synapse formation and structure via trans-synaptic and heterophilic adhesions.

---

[146] A.k.a. leucine rich repeat-containing proteins LRRC4a to LRRC4c. The NGL family of synaptic adhesion molecules contains 3 detected members that mainly lodge at postsynaptic sites of excitatory synapses. The extracellular regions of NGLs interact with distinct presynaptic ligands: NGL1 and NGL2 interact with netrin-G1 and netrin-G2, respectively, which are GPI-anchored adhesion molecules; NGL3 with receptor Tyr phosphatases PTPRd, PTPRf, and PTPRs [1286]. Postsynaptic netrin-G ligand cell adhesion molecules connects to netrin-G cell adhesion molecules via their cytosolic tail in an isoform-specific manner owing to DLg4 scaffold. Netrin-G ligands regulate the formation of excitatory synapses.

10.8.4 A-Kinase-Anchoring Proteins

A-kinase-anchoring proteins[147] are signal-organizing molecules that target protein kinases and phosphatases to subcellular loci where these enzymes control the phosphorylation state of neighboring substrates [1287]. A-kinase-anchoring proteins actually bind to diverse enzymes, not only protein kinase-A[148] but also protein kinase-C, protein phosphatases, and phosphodiesterases [1288].

Signaling and scaffolding AKAP proteins coordinate different signaling enzymes according to the binding partner. In other words, AKAP proteins organize clusters of signaling mediators, thereby allowing signal segregation.

10.8.4.1 AKAP Family

A-kinase-anchoring proteins constitute a family of structurally diverse proteins (AKAP1–AKAP14; Table 10.35) that, in addition to protein kinase-A and -C, bind P21-activated kinase, phosphodiesterases, and phosphatases, such as PP1, PP2, and PP3. Alternative splicing generates transcript variants that encode different isoforms.

Some AKAPs are able to bind both PKA type-1 and -2 regulatory subunits (dual-specific AKAP1 and AKAP10). Both AKAP1 and AKAP10 are mitochondrial; AKAP3 and AKAP4 are linked to sperm; AKAP5 is predominantly expressed in the cerebral cortex, but also in T lymphocytes; AKAP6 anchors PKA to the nuclear envelope and sarcoplasmic reticulum, especially in the central nervous system, skeletal muscles, and myocardium; AKAP8 has a cycle-dependent interaction with PKA; AKAP9 localizes to the Golgi body and centrosome; AKAP11 to testis; AKAP12 to endothelial cells. Protein AKAP13 serves as guanine nucleotide-exchange factor for the RHO superfamily of small GTPases (RhoGEF13).

10.8.4.2 Spatiotemporal Regulation of Cell Signaling

A-kinase-anchoring proteins contribute to the *spatial regulation* of signaling events, targeting the enzyme to specific sites (plasma membrane, mitochondria, cytoskeleton, or centrosome), where they localize, as different AKAP types can have specific cytoplasmic distribution.

---

[147] The first described AKAP was microtubule-associated protein-2 (MAP2). All AKAPs share common properties. They contain a PKA-anchoring domain. Most mammalian AKAPs, indeed, bind to protein kinase-A2. They are able to complex with different signaling molecules.

[148] Protein kinase-A1 in lamellipodia of migrating cells phosphorylates α_4-integrins, the cytoplasmic domain of which acts as AKAPs that are specific to PKA1 in cell protrusions [1289].

Table 10.35. A-kinase-anchoring proteins (AKAP; AKAP-Lbc: AKAP-lymphoid blast crisis; BRx: breast cancer nuclear receptor-binding auxiliary protein; CG-NAP: centrosome- and Golgi-localized PKN-associated protein; D-AKAP: dual-specific AKAP; mAKAP: muscle-selective AKAP; PrKA: protein kinase A-anchoring protein; sAKAP: spermatid AKAP). A-kinase-anchoring proteins (PKA-anchoring molecules) contain a motif that binds to the N-terminal dimerization domain of the R subunit of protein kinase-A. Most AKAPs interact specifically with the PKA type-2 regulatory (R2) subunit (i.e., PKA2), but several R1-specific AKAPs exist. Dual-specificity D-AKAPs interact with both type-1 and -2 regulatory subunits (of PKA1 and PKA2, respectively).

Type	Other aliases and names
AKAP1	AKAP84, AKAP121, AKAP149, D-AKAP1, sAKAP84
AKAP2	AKAPkl
AKAP3	AKAP110
AKAP4	AKAP82, FSC1
AKAP5	AKAP75, AKAP79, AKAP150
AKAP6	AKAP100, mAKAP, PrKA6
AKAP7	AKAP15, AKAP18
AKAP8	AKAP95
AKAP9	AKAP350, AKAP450, CG-NAP, PrKA9, hyperion, yotiao
AKAP10	D-AKAP2
AKAP11	AKAP220
AKAP12	AKAP250, gravin
AKAP13	AKAP-Lbc, Ht31, BRx, ARHGEF13 (RhoGEF13)
AKAP14	AKAP28

Within a given site, a given AKAP can link to diverse substrates and different AKAPs can assemble distinct signaling complexes.[149] The displacement of enzymes into and out of these complexes contributes to the *temporal regulation* of signaling events.

Members of the AKAP family form complexes with enzymes that trigger the following step of signal transduction or terminate signaling. These site then regulate the forward and backward steps of a given signaling process.

The location of an AKAP complex can be modulated by competition between binding partners, for example, between an enzyme and a component of the cytoskeleton. It can also be modulated by AKAP phosphorylation. The recruitment or release of AKAP-binding partners can alter the response to signals or change the location of a signaling complex to yield a dynamic localization and reorganization of AKAP complexes.

Compartmentation of AKAP–PKA complexes relies on specialized targeting domains of each anchoring protein, the so-called *localization signals*. Several AKAPs can lodge at the same subcellular compartment. On the other hand, alternatively spliced variants can be assigned to different regions [1290].

---

[149] Protein kinase-C interacts with AKAP5, AKAP9, AKAP13, etc., PKN with AKAP5, and PKD with AKAP13 [1290].

10.8 Coregulators

Table 10.36. Examples of AKAP-based complexes that create focal points for signal transduction (Sources: [1290, 1291]). Numerous AKAPs target PKA to the plasma membrane, cytoskeleton, endoplasmic reticulum, Golgi body, mitochondria, perinuclear region, and nuclear matrix. The AKAP scaffolds not only anchor PKA, but also assemble multienzyme signaling complexes that ensure the integration and processing of diverse signaling pathways. They indeed coordinate the activity of many kinases, phosphatases, phosphodiesterases, adenylate cyclases, GTPases, and other regulators.

Type	Partners
	Plasma membrane
AKAP5	PKA, PKC, PP3, AMPAR, NMDAR, DLg1/4 (neuron); PKA, PKC, PP3, AC5, AC6, Cav3, β2AR, $Ca_V1.2$ (cardiomyocyte $Ca_V1.2$, $K_V7.2$, Aqp
AKAP7	PKA, $Ca_V1.2$, Na_V
AKAP9	PKA, CK1, PP1, NMDAR, $K_V7.1$, IP_3R
AKAP12	PKA, PKC, β2AR
	Perinuclear region
AKAP6	PKA, PDE4d3
	Endo(sarco)plasmic reticulum
AKAP6	PKA, RyR2
	Centrosome, Golgi body
AKAP9	PKA, PKA, PKN, PP1, PP3, PDE4d3, GCP2/3 ClIC
	Actin cytoskeleton
AKAP12	PKA, PKC
AKAP13	PKA, PKC, PKD, Rho, 14-3-3
	Mitochondrion
AKAP1	PKA, PP1
	Peroxisomes
AKAP11	PKA, PP1, GSK3β

Protein AKAP5 anchors to plasmalemmal phospholipids; AKAP12 also tethers to the plasma membrane using an N-terminal myristoyl group and phospholipid-binding sequences (Tables 10.36 to 10.38). Recruitment to membranes of AKAP7α and AKAP7β relies on myristoyl and palmitoyl groups. Protein AKAP9 targets the centrosome using the pericentrin-AKAP9 centrosomal-targeting domain (PACT).[150]

---

[150] Both pericentrin and AKAP9 contain the PACT domain. Pericentrin, or kendrin, is a component of the pericentriolar material (PCM) that binds to calmodulin. It also interacts with the microtubule nucleation component γ-tubulin.

Table 10.37. Cardiac AKAPs (**Part 1**; Source: [1291]; AC: adenylate cyclase; AR, adrenoceptor; ERK: extracellular signal-regulated kinase; HIF: hypoxia-inducible factor; NCX: sodium–calcium exchanger; MyCBP: Myc-binding protein; NFAT: nuclear factor of activated T cells; PDE: phosphodiesterase; PDK: phosphoinositide-dependent kinase; PKA/C/D/N: protein kinase-A/C/D/N; PP: protein phosphatase; PTP: protein Tyr phosphatase; RyR: ryanodine receptor; SIAH: seven in absentia homolog [Ub ligase]; VHL: von Hippel-Lindau ubiquitine ligase). In cardiomyocytes, cAMP messenger transmits signals sent by β2-adrenergic receptors of sympathetic nerves via catecholamines (noradrenaline and adrenaline). It activates PKA to enhance cardiac inotropy, chronotropy, and lusitropy. The cAMP–PKA axis leads to phosphorylation (opening) of $Ca_V1.2$ and RyR2 channels. In addition, phosphorylation of phospholamban that inhibits SERCA2 causes the dissociation of PLb from SERCA2, thereby activating SERCA2 for relaxation. Phosphorylation of sarcomeric troponin-I and myosin-binding protein-C also promotes relaxation by decreasing Ca^{++} responsiveness of myofilaments. Under prolonged catecholaminergic stimulation, cAMP contributes to cardiomyocyte hypertrophy and apoptosis.

Type	Location	Binding partners	Function
AKAP1	Mitochondria, nuclear envelope, endoplasmic reticulum	PKA1/2, PKCα, Src, RSK1, PP1/3, PTPn21, PDE7a, RhoGEF2, MyCBP	Regulation of cardiomyocyte hypertrophy
AKAP5	Plasma membrane, T tubules	PKA2, PKC, PP3, AC5/6 $K_V7.2$, $Ca_V1.2$, βAR, DLg1, Cav3	Calcium signaling
AKAP6	Nuclear envelope	PKA2, PDK1, RSK3, MAP2K5, ERK5, PP2/3, PDE4d3, AC5, RyR2, NCX1, NFAT, Rap1, RapGEF3, HIF1α, VHL, SIAH2, nesprin-1α, synaptopodin-2	Regulation of CMC hypertrophy, HIF1α stability, calcium signaling
AKAP7	Plasma membrane, endoplasmic reticulum	PKA2, PP1, $Ca_V1.2$, phospholamban, inhibitor-1	Calcium signaling

10.8.4.3 AKAP Post-Translational Modifications

The location of AKAP complexes can reorganize. The recruitment or release of AKAP-binding partners influences the response to transmitted signals. Competition between binding partners results from mutually exclusive binding to AKAP proteins [1290]. In addition, modification (e.g., phosphorylation, myristoylation, and palmitoylation) of the AKAP localization signal affects the location of AKAP scaffolds.

10.8 Coregulators

Table 10.38. Cardiac AKAPs (**Part 2**; Source: [1291]; BCR, breakpoint cluster region protein [GAP]; ClIC: chloride intracellular channel; ERK: extracellular signal-regulated kinase; IP$_3$R: inositol trisphosphate receptor; KSR: kinase suppressor of Ras; NHERF3: sodium–proton exchange regulatory factor-3 [a.k.a. CFTR-associated protein CAP70 and PDZ domain-containing protein PDZK1]; PDE: phosphodiesterase; PI3K: phosphatidylinositol 3-kinase; PKA/C/D/N: protein kinase-A/C/D/N).

Type	Location	Binding partners	Function
AKAP9	Plasma membrane, Golgi body, centrosome	PKA2, PKCε, PKN1, CK1, ACase, PDE4d3, IP$_3$R, K$_V$7.1, ClIC	Cardiac repolarization
AKAP10	Outer mitochondrial membrane	PKA1/2, NHERF3, Rab4/11	Cardiac rhythm
AKAP13	Cytoskeleton	PKA2, PKCη, PKD, Raf, MAP3K15, MAP2K1–3, ERK1/2, P38MAPKα, PKNα, PI3K, Ras, KSR1, PDE3b, G12, Gβγ, RhoA, BCR, 14-3-3	Regulation of CMC hypertrophy

Regulatory phosphorylation can constitute feedback loops. Dynamic regulation of PKA activity relies on antagonist component of AKAP-based complexes, such as cAMP activator and its inhibitor phosphodiesterase that can both anchor to AKAP protein. Whereas hormonal stimulation increases the cAMP concentration, hence enabling PKA activation, recruited phosphodiesterase prevents cAMP action, thereby repressing PKA activation (negative feedback).

Recruitment and release of AKAP-binding partners is regulated by phosphorylation. For example, phosphorylation-dependent recruitment of 14-3-3 protein suppresses the RhoGEF activity of AKAP13 agent. On the other hand, phosphorylation of AKAP13 at another site precludes its association with protein kinase-D [1290]. Recruitment and release of binding partners to and from AKAP12 are also controlled by phosphorylation.

AKAP1

Protein AKAP1 and its shorter splice variant AKAP1s are widespread mitochondrial proteins. In ventriculomyocytes, AKAP1 precludes maladaptive hypertrophy [1291]. It actually hinders dephosphorylation and nuclear translocation of the prohypertrophic NFAT3 transcription factor.

AKAP2

Protein AKAP2 lodges in the lung, kidney, and cerebellum [1292]. Different isoforms predominate in different tissues. Anchor AKAP2 has a polarized distribution, as it accumulates near the apical surface of the epithelium in nephron tubules. It also abounds in alveolar epithelial cells, below the plasma membrane. It interacts with and modulates the structure of the actin cytoskeleton.

AKAP3 and AKAP4

Protein AKAP3 is expressed in the testis, but not in other tissues. Protein AKAP4 resides in the fibrous sheath of the sperm flagellum, a unique cytoskeletal structure. The AKAP4 mature form can bind AKAP3 protein. The latter is involved in the organization of the basic structure of the fibrous sheath, whereas AKAP4 intervenes in completing fibrous sheath assembly [1293].

AKAP5

Neuronal AKAP5 anchors not only the cAMP-dependent kinase PKA, but also Ca^{++}- and lipid-regulated PKC, and calmodulin-stimulated protein phosphatase PP3.

In cardiomyocytes, AKAP5 assembles a signaling complex that contains β-adrenoceptors, adenylate cyclases AC5 and AC6, PKA, PP3, caveolin-3, and $Ca_V1.2$ channels that are phosphorylated upon sympathetic stimulation [1291]. In fact, the AKAP5 complex recruits AC5 and AC6 and produces cAMP in nanodomains, where PKA processes not only $Ca_V1.2$ channels, phospholamban, and RyR2 receptors located in caveolin-3-associated junctional regions of the sarcoplasmic reticulum adjacent to T tubules (Vol. 5 – Chap. 5. Cardiomyocytes).

In dendrites of excitatory synapses, AKAP5 forms a postsynaptic signaling complex with PKA and PP3 enzymes as well as GluR1 subunit of the AMPA-type glutamate receptor via DLg1 or DLg4 adaptor [1290]. Anchored PKA phosphorylates (activates) GluR1 (Ser845), whereas anchored PP3 dephosphorylates this site, thereby activating DLg4-associated double minute-2 ubiquitin ligase. The latter ubiquitinates DLg4 for proteasomal degradation and enables internalization of AMPA receptors.

In fact, neuronal AKAP5 can bind to different types of ion channels: (1) ionotropic AMPA-type glutamate receptors (iGluR) that propagate excitatory stimuli via interaction with membrane-associated guanylate kinase (MAGuK) scaffold protein Disc large homolog DLg1 and (2) M-type $K_V7.2$ and $K_V7.3$ channels. Ligand-gated ion channel iGluR requires anchoring of PKA and PP3 for full activity. Protein kinase-A is necessary to maintain ^{AMPA}Glu channels at the plasma membrane. Protein phosphatase-3 is needed to lower AMPA activity in response to tonic stimulation. Protein AKAP5 regulates activities of these different enzymes and channels in hippocampal (PKA and PP3 are active; PKC is inactive in this subset of excitatory synapses) and superior cervical ganglion neurons of the sympathetic nervous system (PKA and PP3 remain inactive; PKC is active), respectively [1294]. The AKAP5–GluR complex brings PKA kinase and PP3 phosphatase in proximity to the channel to control its phosphorylation state. (PKA activates, whereas PP3 inactivates). On the other hand, the AKAP5–K_V7.2 (3)–PKC complex inhibits K_V7.2 and K_V7.3 channels, thereby attenuating the hyperpolarizing K^+ flux in response to exciting signals from M_1 muscarinic receptors. Muscarinic suppression of M currents proceeds via Gq/11-coupled signaling requires Ca^{++}, phosphatidylinositol (4,5)-bisphosphate, and diacylglycerol, as well as phosphorylation by AKAP5-bound PKC enzyme.

AKAP6

Muscle-specific A-kinase-anchoring protein AKAP6 complexes with PKA, phosphodiesterase PDE4d3 and the guanine nucleotide-exchange factors for GTPase Rap1 RapGEF3 [1295].

In cardiomyocytes, assembling occurs at the perinuclear membrane. Phosphodiesterase-D3 serves as an adaptor for RapGEF3 and ERK5 proteins. Phosphorylation of PDE4d3 by AKAP6-associated ERK favors cAMP production and subsequent PKA and RapGEF3 activation. However, subsequent phosphorylation of PDE4d3 by PKA reduces cAMP concentration, thereby causing PKA deactivation and suppressing RapGEF3 excitation of AKAP6-associated ERK5 activity. The latter provokes cardiomyocyte hypertrophy.

Scaffold AKAP6 directly binds RyR2 channels. It may favor RyR2 phosphorylation by PKA and subsequent release of the inhibitor 12.6-kDa FK506-binding protein FKBP1b to increase RyR2 open probability. The RyR2–AKAP6 complex may localize either to the entire sarcoplasmic reticulum or only to a perinuclear compartment [1291]. The AKAP6 complex only permits a transient activation of PKA in response to βAR stimulation.

In cardiomyocytes, AKAP6β localizes to the outer nuclear membrane, as it interacts with nesprin-1α, which binds to actin filaments [1291].[151] Scaffold AKAP6β assembles a complex with PKA, ERK5, PP3, AC5, PDE4d3, RyR2, NFAT3, and RapGEF3 proteins. Activated PKA phosphorylates (inactivates) adenylate cyclase AC5 and (activates) phosphodiesterase PDE4d3, thereby decreasing cAMP synthesis and increasing cAMP degradation (negative feedback loop).

Anchored PKA also fosters PP2 activity that dephosphorylates PDE4d3 and RyR2, thereby enhancing Ca^{++} influx from intracellular stores. In addition, PDE4d3 can also act as a scaffold that recruits the cAMP-activated guanine nucleotide-exchange factor RapGEF3, prohypertrophic mitogen-activated protein kinase kinase MAP2K5, and ERK5 to the AKAP6 complex. Upon elevation of

---

[151] Nesprin is a portmanteau for nuclear envelope spectrin repeat-containing protein.

local cAMP concentration that stimulates RapGEF3, activated Rap1 inhibits the ERK5 axis. The latter kinase impedes PDE4d3 action, thereby repressing cAMP degradation.

The AKAP6 complex integrates hypertrophic signals initiated by α1- and β-adrenergic receptors, endothelin-1 receptors, leukemia inhibitor factor receptors, phosphatase PP3 catalytic subunit isoform $PP3_{c\beta}$, and NFAT3 transcription factor. In response to adrenoceptor activation, anchored $PP3_{c\beta}$ dephosphorylates (activates) NFAT3 that supports the transcription of hypertrophic genes. Scaffold AKAP6β also binds phospholipase-Cϵ that assists endothelin-1-mediated hypertrophy [1291].

Protein AKAP6 assembles a signaling complex with hypoxia-inducible factor HIF1α, prolyl hydroxylase domain-containing proteins (PHD),[152] and ubiquitin–protein ligases von Hippel-Lindau protein and Seven in absentia homolog SIAH2 that ubiquitinate HIF and PHD, respectively, for proteasomal degradation [1291]. This AKAP6 complex thus serves as a regulator of cell response to hypoxia.

AKAP7

Four isoforms (AKAP18α–AKAP18δ) have been identified. Subtype AKAP18δ anchors and controls β-adrenergic regulation of SERCA2 activity [1291]. It indeed favors PKA-mediated phosphorylation of phospholamban, thereby dissociating phospholamban from SERCA2 and fostering Ca^{++} reuptake into the sarcoplasmic reticulum and cardiomyocyte relaxation.[153]

In addition, AKAP18δ recruits protein phosphatase PP1 and $PP1_{r1c}$ inhibitory subunit. Phosphorylation by PKA of $PP1_{r1c}$ prevents PP1 activity [1291].

AKAP8

In mammalian cells, AKAP8 localizes exclusively to the nuclear matrix. DEAD (Asp-Glu-Ala-Asp) box protein DDx5, or RNA helicase P68, is a binding partner of AKAP8 [1296]. Protein AKAP8 strongly interacts with the 3 cyclins-D, but

---

[152] Prolyl hydroxylase domain-containing protein-2 (PHD2) is also called Egg Laying defective nine (EgL9) homolog-1 (EgLN1) and HIF-prolyl hydroxylase HPH2. Isoforms PHD1 and PHD3 are also named EgLN2 and EgLN3, respectively.

[153] β-Adrenoceptors promote phosphorylation by PKA of 2 sarcomeric regulators of actomyosin interactions: troponin-I and slow-type muscle myosin-binding protein-C (MyBPc1). The phosphorylation of these 2 proteins by PKA reduces Ca^{++} responsiveness of myofilaments, thereby lowering contraction force and promoting relaxation. In fact, several AKAPs associate with sarcomeric proteins: desmuslin, or synemin, an intermediate filament protein and mechanical linker, which can be recruited to the Z disc; cardiomyopathy-associated protein CMyA5, or myospryn, which colocalizes with desmin at the peripheral Z disc (costamere); cardiac troponin-T; and phosphodiesterase-4D-interacting protein, or myomegalin, which may regulate the phosphorylation of MyBPc1.

not with cyclin-dependent kinase CDK4 [1297]. It also directly interacts with minichromosome maintenance protein MCM2, a component of the hexameric MCM2–MCM7 DNA helicase prereplication complex.

AKAP9

In cardiomyocytes, AKAP9 recruits an enzymatic complex made of a kinase (PKA), a phosphatase, and a phosphodiesterase (PDE4d3), to fine tune slow-activating, delayed rectifier K^+ current (i_{Ks}) through K_V7 channels that contributes to the repolarization of the plasma membrane of cardiomyocytes [1291]. In response to β-AR stimulation, PKA phosphorylates $K_V7.1$ (Ser27) to augment K^+ efflux, accelerates repolarization, and shortens action potential duration. Maximal activation of K_V7 by the sympathetic system occurs when anchored PKA also phosphorylates AKAP9 (Ser43). In addition, AKAP9 links to chloride intracellular channel (ClIC).

AKAP10

Scaffold AKAP10 is involved in the regulation of the cardiac rhythm [1291]. This dual-specific AKAP that targets both PKA1 and PKA2 abounds in mitochondria. It contains regulator of G-protein signaling domains, in addition to a PKAR2 subunit-binding domain.

In the kidney, many plasmalemmal transporters of the apical brush border of proximal tubular cells ensure the maintenance of electrolyte concentrations. Sodium–phosphate cotransporter-2, or SLC34a1, interacts with 7 types of PDZ domain-containing proteins, among which sodium–hydrogen exchanger regulatory factors NHERF1 (SLC9a3R1) and NHERF3 (PDZK1) [1298]. Factor NHERF3 also links to solute carrier SLC17a1, or sodium–phosphate cotransporter-1, Na^+–H^+ exchanger NHE3 (SLC9a3), organic cation transporter OCTN1 (SLC22a4), chloride–formate exchanger (CFEx or SLC26a6), and urate–anion exchanger Urat1 (SLC22a12), in addition to Ste20-like kinase SLK and AKAP10, in the brush border of proximal tubular cells.

AKAP11

Glycogen synthase kinase GSK3β binds to AKAP11, in addition to cAMP-dependent protein kinase-A and protein phosphatase PP1 [1299]. Kinase PKA phosphorylates (inactivates) AKAP11-anchored GSK3β, more efficiently than free GSK3β. On the other hand, PP1 dephosphorylates GSK3β.

AKAP12

Two major isoforms (AKAP12α–AKAP12β) are created by independent promoters. They are synthesized in most mesechymal cells, some epithelial cell types, endothelial cells, and some specialized cells, such as podocytes, Purkinje cells, and astrocytes [1300]. A shorter third isoform, AKAP12γ, is specific to the testis.

Scaffold AKAP12 binds to PKC, Src kinase, calmodulin, cyclins, ^{F}actin, a membrane-associated form of (1,4)-galactosyltransferase associated with adhesion signaling, and phosphoinositols [1300].

Protein AKAP12 abounds at inner plasma membrane sites, along the cortical cytoskeleton, and in the perinuclear region.

At the neuromuscular junction, recruitment and release of binding partners to AKAP12 depend on phosphorylation by PKA and PKC kinases. The AKAP12 complex is recruited to the plasma membrane by interacting with β2-adrenoceptor. Agonist stimulation initially strengthens the association of AKAP12 with β2AR, upon AKAP12 phosphorylation by PKA kinase [1290]. Prolonged agonist stimulation dismantles the AKAP12 complex, upon phosphorylation by PKC that primes ubiquitination by DM2 ligase, which is recruited by adaptor β-arrestin, for proteasomal degradation.

AKAP13

Regulator AKAP13 is a guanine nucleotide (GDP-to-GTP)-exchange factor that links G12/13 subclass subunits to Rho activation [1290]. Dimerizing 14-3-3 polypeptides bind to specific motifs with phosphorylated Ser or Thr residues. Phosphorylation by PKA of AKAP13 enables the recruitment of 14-3-3 proteins, thereby preventing the AKAP13 RhoGEF activity.

In cardiomyocytes, AKAP13 acts as a PKA2 anchor and RhoAGEF protein. Its activity is fostered by Gq-coupled α1-adrenoceptors and G12-coupled endothelin-1 receptor. On the other hand, AKAP13-anchored PKA limits AKAP13 activity, thereby lowering RhoA activation via the recruitment of 14-3-3 protein [1291].

Scaffold AKAP13 can form a signaling complex with the scaffold kinase suppressor of Ras KSR1, and Raf, MAP2K1, and MAP2K2 kinase, that excites ERK1 and ERK2 kinases [1291].

Protein AKAP13 mediates the hypertrophic response induced by several G-protein-coupled receptors, such as α1-adrenoceptors, AT_1 angiotensin-2 receptor, and endothelin-1 receptor [1291]. Scaffold AKAP13 recruits protein kinase-D and its upstream activator protein kinase-Cη. Stimulation of α1-adrenoceptors or endothelin-1 receptors excites anchored PKCη that phosphorylates (activates) AKAP13-bound PKD enzyme [1291]. Nuclear PKD activity that is enhanced by AKAP13 leads to phosphorylation and nuclear export of class-2 histone deacetylase HDAC5, thereby supporting myocyte-specific enhancer-binding factor-2-dependent transcriptional activation of hypertrophic genes. On the other hand, active PKD is released from AKAP13 when this scaffold is phosphorylated by anchored PKA kinase.

AKAP14

Mucociliary clearance relies on the coordinated beating of cilia that line the wetted surface of the airway epithelium in the respiratory tract. Ciliary beat results from the coupling between ATP hydrolysis by dynein, the axonemal nanomotor, to microtubule sliding. Activated axonemal protein kinase-A increases the ciliary beat frequency in humans. In ciliary axonemes, AKAP14 anchors PKA holoenzyme in the axoneme, close to its substrates [1301].

10.8.5 Annexins

Annexins[154] are Ca^{++}-regulated proteins and Ca^{++} sensors.[155] These Ca^{++} effectors are organizers of membrane domains and membrane-recruitment platforms for proteins with which they interact. Annexins mediate cellular responses to changes in the intracellular Ca^{++} concentration.

10.8.5.1 Annexin Family

Annexins constitute a family of proteins that belongs to the set of EF-hand proteins.[156] Proteins of the annexin family thus share the ability to link to both Ca^{++} and membrane lipids.[157]

Five major annexin subclasses (A–E) are defined with vertebrate (human) annexins (A1–A13) encoded by the genes ANXA1 to ANXA11 and ANXA13. Animals, but not humans, possess members of the annexin-B set [1302]. Annexins-C, -D, and -E are observed in fungi, plants, and protists, respectively.

---

[154] ανεχω: to hold up, to support. Annexins are scaffolding and bridging proteins that are capable of binding to and possibly holding together cell membranes.

[155] Calcium-regulated proteins include Ca^{++} channels and pumps, Ca^{++} transport and buffers, and Ca^{++} effectors. These proteins ensure a low Ca^{++} concentration at rest as well as signaling upon stimulation (Sect. 11.5).

[156] This set comprises annexins, C2 domain-containing proteins, calmodulin, troponin-C, and S100 proteins. The EF hand is a Ca^{++}- and phospholipid-binding motif. The C2 domain, another Ca^{++}- and phospholipid-binding motif, is a membrane targeting domain for a wide range of lipids.

[157] Annexins were differently named according to their tethering properties: *synexin* (granule aggregators [Greek συν: with, συναξις: gathering, Latin nexus: fastening, joining, interlacing, entwining]) and *lipocortins* (steroid-inducible lipase inhibitors), as well as portmanteaux *chromobindins* (chromaffin granule binders), *calcimedins* (Ca^{++} signaling mediators), and *calpactins* (binders of Ca^{++}, phospholipid, and actin).

10.8.5.2 Subcellular Localization

Soluble annexins are distributed in the cytoplasm at a resting Ca^{++} concentration. Several factors influence annexin transport and localization in cells.

In fibroblasts, annexin-A1, -A4, and -A5 are observed in the nucleus with higher concentrations than in the cytoplasm [1303]. Extranuclear annexin-A4 partly localizes to the endoplasmic reticulum. Annexin-A1 and -A2 lodge in the plasma membrane. When the intracellular calcium concentration rises, intranuclear annexin-A4 and -A5 relocalize to the nuclear envelope, whereas the cytosolic pool of these annexins is situated at the plasma membrane; plasmalemmal annexin-A2 follows a punctate pattern, although it has a homogeneous distribution before calcium stimulation; annexin-A6 relocalizes to the plasma membrane with a more homogeneous distribution.

10.8.5.3 Structural and Functional Features

Annexins operate on actin polymerization at cellular membranes, structuring of endosomal compartments, intracellular Ca^{++}-regulated transfer, and midbody formation during cytokinesis [1304]. Annexins are then involved in membrane–cytoskeleton attachments, endo- and exocytosis, and regulation of ion fluxes across membranes.

Structure–Function Relationship

Annexins contain the *annexin core domain*, a Ca^{++}- and membrane-binding module. This core domain enables Ca^{++}-bound annexins to dock onto membranes that contain negatively charged phospholipids [1304]. Most annexins can then interact with cellular membranes in a reversible and controlled fashion.

Phosphatidylserine, phosphatidylinositol, and phosphatidic acid are among the preferred binding acidic phospholipids. Some annexin types interact more specifically with certain membrane lipids: annexin-A2 binds to phosphatidylinositol (4,5)-bisphosphate; annexin-A3, -A4, -A5, and -A6 to phosphatidylethanolamine; and annexin-A5 to phosphatidylcholine [1304].

Annexins can tether to 2 membranes. This feature is influenced by phosphorylation of their N-terminus. Simultaneous linkage to 2 cell membrane of annexin-A1, -A2, and -A4 is inhibited by phosphorylation, whereas that of annexin-A7 is promoted [1304]. On the other hand, phosphorylation do not strongly affects the ability of annexins to bind to a single membrane.

Calcium concentration that initiate phospholipid binding differ markedly between annexin types. It ranges from 20 µmol for the connection between annexin-A5 and phosphatidylserine of liposomes to submicromolar Ca^{++} concentrations for annexin-A1 and -A2 to phosphatidylserine and phosphatidic acid and less than 100 nmol for the annexin-A2–S100a10 complex to phosphatidylserine [1304].

10.8 Coregulators

Calcium-independent membrane binding depends on the pH value. For example, annexin-A5 binds to and apparently penetrates the bilayer of phosphatidylserine vesicles at pH 4 in the absence of Ca^{++} ions, whereas, at neutral pH, Ca^{++} binding enables lipid interaction [1302].

Annexin also possess a N-terminal interaction domain with binding sites for cytoplasmic proteic ligands, which can then be recruited to membranes via the annexin core-mediated phospholipid interaction. Once annexin is attached to a membrane, its cytoplasmic face can connect to proteins, thereby assembling these interactors at membrane sites in a Ca^{++}-regulated manner.

Cell Export and Import

Membrane-bound annexins can form self-assemblies that influence the mobility and organization of membrane lipids. They can then participate in the regulation of organization of membrane domain, in particular, those involved in material tranfer.

Exocytosis

Externalization of proteins, polypeptide hormones, and amine neurotransmitters terminates with the fusion of secretory vesicle membranes with the plasma membrane. Mature secretory vesicles can accumulate within secretory cells, waiting for a stimulus-triggered Ca^{++} influx. Upon Ca^{++} entry, vSNAREs and tSNAREs intertwine with one another and rearrange the lipid bilayers for fusion (Vol. 1 – Chap. 9. Intracellular Transport). Calcium-binding proteins lodge on membranes and in the cytoplasm of secretory cells to participate in exocytosis, such as EF-hand proteins (calmodulin and S100 proteins) and C2 domain-containing proteins (synaptotagmin, rabphilin, Rab3-interacting molecule, and MUnc13), in addition to annexins.

Annexin-A2 is a component of filamentous actin that carry newly formed endocytic vesicles from the plasma membrane into the cytoplasm. The activity of annexin-A2 in exocytosis depends on its phosphorylation of its N-terminus by protein kinase-C [1304]. Annexin-A2 and -A7 mediates aggregation and apparent fusion of hormone-containing chromaffin granules of neuroendocrine cells in the medulla of the adrenal gland [1305]. Annexin-A1 promotes the calcium-dependent fusion of liposomes as well as fusion of phospholipid vesicles with plasma membranes in neutrophils.

Exocytosis

Several annexins, such as annexin-A1, -A2, and -A6, reside in endosomal compartments. In particular, annexin-A2 intervenes in the lysosomal transport and endocytosis of EGF receptors. Two sequential steps of multivesicular endosome genesis exist. The first step require hepatocyte growth factor-regulated Tyr kinase substrate (HRS) and endosomal sorting complex required for transport proteins (ESCRT) for receptor sorting and inward vesiculation. The second step relies on annexin-A2 on early endosomes that facilitates the detachment of multivesicular regions of early endosomes [1304].

The annexin-A2–S100a10 complex also contributes to the morphology of perinuclear Rab11+ recycling endosomes [1304]. In addition, annexin-A2 composes the chaperone complex with caveolin-1 and cholesteryl esters that carries cholesteryl esters from caveolae to internal membranes.

Annexin-A1 is involved in the inward vesiculation in multivesicular endosomes Annexin-A6 is required in cysteine peptidase-dependent budding of clathrin-coated pits.

Ion Flux

Annexin-A2, -A4, and -A6 modulate plasmalemmal Cl^- channels and sarcoplasmic reticulum Ca^{++} channels. In association with S100a10, annexin-A2 interacts with Na^+ channel, 2-pore acid-sensitive K^+ channel-1 (TASK1), and Ca^{++} TRPV channels (TRPV5 and TRPV6) [1304].

Extracellular Activity

Certain annexins not only act inside the cell, but also have extracellular functions, once they are released from the cell. Annexin-A1 inhibits leukocyte extravasation. On the vascular endothelium, annexin-A2 is a coreceptor for tissue plasminogen activator and plasminogen, inducing fibrinolysis. Annexin-A5 may mask the membrane phospholipids used by coagulation factors in the clotting cascade.

10.8.5.4 Partners

Once bound to certain cell membrane phospholipids via membrane-binding domain owing to calcium ions (Ca^{++}-dependent linkage),[158] annexins interact with proteins (Table 10.39) and provide Ca^{++} signaling. Many cytosolic ligands pertain to the EF-hand Ca^{++}-binding proteins, especially S100 proteins.[159]

---

[158] Annexin-A2 binds directly to phosphatidylinositol (4,5)-bisphosphate. This association stabilizes actin assembly sites on the cell membrane.

[159] Annexin-A1 and -A2 interact with 2 consecutive EF-hand-containing S100a10 and S100a11 proteins. Annexin-A7 binds to 4 EF-hand-containing sorcin, annexin-A11 to 2 EF-hand-containing S100a6 (calcyclin).

10.8 Coregulators

Table 10.39. Annexins (Anx) and their partners (CAP: calcyclin-associated protein; Cal: calpactin; Cal1H: calpactin-1 heavy chain; Cal1L: calpactin-1 light chain; Cpb: calphobindin; Cbd: chromobindin; ClCa: Ca^{++}-activated Cl^- channel; Enx: endonexin; GAG: glycosaminoglycan; IP_2: inositol (1,2)-bisphosphate; ISA: intestine-specific annexin: Lpc: lipocortin; Maxi: Maxi chloride channel; PAP: placental anticoagulant protein; Ppx: pemphaxin; Snx: synexin; VAC: vascular anticoagulant).

Type	Aliases	Partners
AnxA1	Lpc1, Cal2, Cbd9	$PI(4,5)P_2$, cPLA2, actin, S100a10 (Cal1L), S100a11, EGFR, ATP, purine-rich RNA, pyrimidine-rich DNA
AnxA2	Lpc2, Cal1H, Cbd8, PAP4	$PI(4,5)P_2$, cholesteryl esters, caveolin-1, S100a10, Cl^- channel, TASK1, TRPV5/6, GAGs, RNA
AnxA3	Lpc3, PAP3	IP_2
AnxA4	Lpc4, Enx1, Cbd4, PAP2	ClCa, GAGs
AnxA5	Lpc5, Enx2, Cbp1 Cbd4, PAP1	CFTR, VEGFR2, GAGs
AnxA6	Lpc6, Cpb2, Cbd20	Maxi, GAGs, ATP, dynamin, RasA1, Fyn, PYK2
AnxA7	Snx	Sorcin, GAGs, GTP
AnxA9	Anx31, Ppx	
AnxA10	Anx14	
AnxA11	CAP50	S100a6
AnxA13	ISA	NEDD4

Annexins respond to an increase in Ca^{++} concentration by translocating to membranes (both the plasma membrane and organelle membranes). Membrane-bound annexins acts as membrane scaffolds, forming protein clusters.

10.8.5.5 Annexin-A1

Annexin-A1[160] may merge opposing membranes of the invaginating bud to close the neck of the nascent vesicle and enable scission from the plasma membrane. Moreover, annexin-A1 can act on vesicle fusion in the presence of Ca^{++}. Stimulation

---

[160] A.k.a. calpactin-2, chromobindin-9, lipocortin-1 (Lpc1), and phospholipase-A2 inhibitory protein.

by EGF of EGF receptor Tyr kinase leads to formation of a class of endosomes,[161] EGFR-containing multivesicular endosomes, which requires phosphatidyl inositol 3-kinase and annexin-A1 [1306].[162]

Phospholipid-binding protein annexin-A1 can connect to cytosolic phospholipase-A2α (cPLA2α) at the Golgi body in endothelial cells [1307]. Cytosolic phospholipase-A2 cleaves phospholipids in a Ca^{++}-dependent manner into arachidonic acid and lysophospholipid. The former is then converted into prostaglandins by cyclooxygenases and proper synthases.

Annexin-A1 can be phosphorylated by Tyr kinases to alter its Ca^{++} sensitivity. Annexin-A1 can segregate lipids within the plasma membrane. It has a high affinity for $PI(4,5)P_2$ to form raft-like domains rich in this lipid. Annexin-A1 binds to actin, thereby yielding attachment points where the cytoskeleton links to the plasma membrane. In addition, annexin-A1 can heteromerize with calcium-binding S100a10 and S100a11 proteins.[163]

Annexin-A1 can be exported from cells and intervene in inflammation. It hinders extravasation of neutrophils and monocytes. Extracellular annexin-A1 and/or its proteolytically removed N-terminus ($_N$AnxA1) can bind to G-protein-coupled chemoattractant receptors of the formyl peptide receptor family (FPR and FPR-like receptors FPRL1 and FPRL2; Vol. 3 – Chap. 7. G-Protein-Coupled Receptors) [1304].

Annexin-A1 is also involved in apoptosis. Plasmalemmal annexin-A1 may be required for the clearance of apoptotic cells.

10.8.5.6 Annexin-A2

Annexin-A2[164] is involved in cell motility, linkage of membrane-anchored protein complexes to the actin cytoskeleton, endo- and exocytosis, fibrinolysis, ion channel formation, and cell–matrix interactions. Annexin-A2 is a component of the ^{F}actin end, which propels newly formed endocytic vesicles from the membrane to the cytoplasm (actin-dependent transport). Annexin-A2 is also involved in membrane raft formation, especially in the smooth muscle cells. Annexin-A2 forms a complex with caveolin-1 and cholesteryl esters.

---

[161] Multivesicular endosomes that are composed of activated EGFR are distinct from endosomes labeled by lysobisphosphatidic acid.

[162] Annexin-A1 is phosphorylated within EGFR-containing endosomes. Annexin-A1 mediates vesicle formation after EGF stimulation (and not protein selection and sorting). Vesicle inclusion and degradation are selected by EGF or EGFR via endosomal sorting complexes required for transport. Annexin-A1 can reduce EGF signaling because it can remove the catalytic domain of EGF receptor.

[163] Calcium-binding protein S100a10 is also called calpactin-1 light chain (Cal1L or Clp1L) and cellular ligand of annexin-A2 (Anx2Lg); S100a11 is also named S100c and calgizzarin.

[164] A.k.a. calpactin-1 heavy chain, chromobindin-8, lipocortin-2 (Lip2 or Lpc2), placental anticoagulant protein-4 (PAP4).

10.8 Coregulators

Like annexin-A1, annexin-A2 binds to $PI(4,5)P_2$ and can segregate lipids within the plasma membrane and provide membrane attachment points for the cytoskeleton. The complex between annexin-A2 and S100a10 is implicated in the control of plasmalemmal Cl^- channels, TRPV5 and TRPV6 channels, and acid-sensitive K^+ channel TASK1 ($K_{2P}3.1$).

Annexin-A2 can be phosphorylated and, then, translocates into the nucleus, where it can bind to RNA.

The annexin-A2–S100a10 complex can localize to the surface of vascular endothelial cells, where it functions as a receptor for tissue plasminogen activator and plasminogen [1304]. Therefore, it supports the synthesis of plasmin that degrades fibrin. Furthermore, annexin-A2 controls Ca^{++}-regulated exocytosis of von Willebrand factor and P-selectin.

10.8.5.7 Annexin-A3

Annexin-A3[165] impedes the activity of phospholipase-A2. It is associated with acidic phospholipid-stimulated, Ca^{++}- and Mg^{++}-dependent inositol (1,2)-cyclic phosphate 2-phosphohydrolase that cleaves inositol (1,2)-bisphosphate into inositol 1-phosphate. It also operate as an anticoagulant.

Annexin-A3 abounds in neutrophils, whereas it is produced at low levels or is undetectable in many cell types. Annexin-A3 is associated with cytoplasmic granules in neutrophils and monocytes and translocates to the plasma membrane in activated cells [1305].

10.8.5.8 Annexin-A4

Annexin-A4,[166] a Ca^{++}-dependent phospholipid-binding protein, is involved in the control of Cl^- flux in the plasma membrane. Annexin-A4 indeed inhibits Ca^{++}-activated Cl^- channels [1309]. Annexin-A4 impedes the interaction between this ion channel and Ca^{++}–calmodulin-dependent kinase CamK2 [1310].

---

[165] A.k.a. lipocortin-3, placental anticoagulant protein-3 (PAP3), inositol (1,2)-cyclic phosphate 2-phosphohydrolase, and calcimedin-35α.

[166] A.k.a. calcimedin-35β, carbohydrate-binding protein P33 and P41, chromobindin-4, endonexin-1 (Enx1), lipocortin-4 (Lpc4), and placental anticoagulant protein-2 (PAP2).

10.8.5.9 Annexin-A5

Like annexin-A2, annexin-A5[167] can also enter into the nucleus upon tyrosine phosphorylation.

Annexin-A5 forms a shield around negatively charged phospholipids, thereby blocking the entry of phospholipids into blood coagulation.[168]

Annexin-A5 tethers to the Ca^{++}- and cAMP-activated cystic fibrosis transmembrane conductance regulator (CFTR), a chloride channel [1311]. The function and localization of CFTR in normal epithelial cells of human bronchi depend on annexin-A5.

Apoptosis is characterized by an early change in cell volume, the so-called *apoptotic volume decrease*, especially in cardiomyocytes. The cell shrinkage results from the activity of volume-sensitive chloride channels [1313]. Annexin-A5 serves as an early markers of apoptosis. Agents PI3K, PKB, and ERK1 and ERK2 repress the activity of volume-sensitive chloride channels and protect against apoptosis [1314].

10.8.5.10 Annexin-A6

Annexin-A6[169] possesses 2 Ca^{++}-binding core domains that enables it to bind to different membrane regions. Alternative splicing gives rise to 2 isoforms. Annexin-A6 is involved in endosome aggregation and vesicle fusion during exocytosis in secreting epithelia.

Chloride is the main anion of extracellular fluid in the fetus and adult, but, at all gestational stages, fetal Cl^- ion concentration is 5 to 6 mmol higher than in maternal blood. Maxi chloride channels lodge in secreting and absorbing epithelia as well as non-epithelial cell types, and in the apical membrane from human placenta. The human placental syncytiotrophoblast is the main barrier for maternofetal exchange. Annexin-A6 regulate the Cl^- conductance of Maxi-chloride channel of apical membrane in human placenta [1312].

10.8.5.11 Annexin-A7

Annexin-A7[170] participates in the control of Ca^{++} release from internal stores by both inositol trisphosphate (IP_3R) and ryanodine (RyR) receptors. Annexin-A7

---

[167] A.k.a. anchorin-C2, calphobindin-1 (Cbp1), endonexin-2 (Enx2), lipocortin-5, placental anticoagulant protein-1 (PAP1), thromboplastin inhibitor, and vascular anticoagulant-α(VACα).

[168] Once they are activated, platelets expose phosphatidylserine on their surface that serves as binding site for various coagulation factors, thereby producing a procoagulant surface.

[169] A.k.a. 67-kDa calelectrin, calphobindin-2 (Cbp2), chromobindin-20, and lipocortin-4.

[170] A.k.a. synexin.

10.8 Coregulators

associates with sorcin, an inhibitor of the coupling between of Ca_V1 and RyR2 channels in cardiomyocytes [1132]. Alternative splicing, which is tissue specific (brain, heart, and skeletal muscle) generates 2 isoforms.

Annexin-A7 promote the fusion of chromaffin granules to the plasma membrane in neuroendocrine cells of the adrenal medulla. Arachidonic acid and SNARE proteins promotes the fusion of membranes that come into contact via annexin-A7 [1304].

10.8.5.12 Annexin-A9

Annexin-A9,[171] an atypical member of the annexin family, does not involve a membrane scaffold used by other annexins. It has 4 homologous calcium-binding sites in its tetrad core that also contains amino acid substitutions that ablate their function.

10.8.5.13 Annexin-A10

Annexin-A10[172] has a restricted expression. It is synthesized in the adult liver, but not in multiple adult and fetal tissues.

10.8.5.14 Annexin-A11

Annexin-A11[173] is a midbody protein involved in the cytokinesis stage of the cell division. It may intervene in the trafficking and insertion of vesicles during cytokinesis. It enters the nucleus at prophase, before localizing to the midbody. Annexin-A11 binds to S100a6 protein (calcyclin).

Annexin-A11 may be a target of S100a6 protein. It can be phosphorylated by mitogen-activated protein kinase; once phosphorylated by MAPK, it can still bind to phosphatidylserine-containing vesicles.

10.8.5.15 Annexin-A13

Annexin-A13[174] can associate with membranes owing to an N-terminal myristoylation in a Ca^{++}-independent manner. In particular, it is associated with the plasma membrane of proliferating endothelial cells. Alternative splicing generates different isoforms.

---

[171] A.k.a. annexin-31, and pemphaxin.
[172] A.k.a. annexin-14.
[173] A.k.a. calcyclin-associated protein-50 (CAP50).
[174] A.k.a. intestine-specific annexin (ISA).

In epithelial cells, annexin-A13b splice variant links specifically to sphingolipid- and cholesterol-rich membrane domains of the trans-Golgi network [1304]. N-terminally myristoylated annexin-A13b is required for the budding of these domains, which are subsequently delivered to the apical plasma membrane.

10.8.5.16 Annexin Role in Ca^{++} Handling in Cardiac Cells

Heart cells contains annexin-A1, -A2, and -A4 to -A7. Endothelial cells express annexin-A1, -A2, -A5, and -A6; smooth muscle cells mainly annexin-A2 and -A6, and cardiomyocytes mostly annexin-A4 to -A7.

These annexin species modulate the cytosolic concentration of free Ca^{++} ions with other Ca^{++} handlers and Ca^{++} sensors. Whereas Ca^{++}-binding sorcin and AHNAK nucleoprotein, or desmoyokin, regulate $Ca_V1.2$ channels and calsequestrin, histidine-rich Ca^{++}-binding protein and sorcin control sarco(endo)plasmic reticulum Ca^{++}-release channels, and annexin-A2 and -A6 modulate Na^+–Ca^{++} exchanger activity [1315].

Annexin-A2

Annexin-A2 localizes to extracellular matrix and endothelial cells of coronary arteries. This cytoskeleton-binding protein below the plasma membrane can form heterotetramers with S100a10, thereby reducing Ca^{++} requirement for phospholipid binding. Activated annexin-A2 operates as a plasmalemmal coreceptor for plasminogen and tissue-plasminogen activator [1315]. Annexin-A2 thus controls plasmin-mediated processes, hence acting as an anti-thrombolytic.[175] In addition, annexin-A2 associates with HSP90a.

Annexin-A4

In fibroblasts, annexin-A4 impedes Ca^{++}-dependent Cl^- flux by precluding interaction between Ca^{++}–calmodulin kinase-2 and ion channel and channel phosphorylation [1315]. In cardiomyocytes, cytosolic annexin-A4 level increases during heart failure.

Annexin-A5

Annexin-A5 abounds in human myocardium; it resides mainly in cardiomyocytes. In atriomyocytes, it may mediate release of atrial natriuretic peptide. Annexin-A5 is also present in vascular endothelial cells. It may have anticoagulant, anti-apoptotic, and anti-inflammatory effects [1315].

---

[175] Prothrombotic homocysteine inhibits tPA binding to annexin-A2.

Annexin-A6

Annexin-A6 also abounds in the heart. In cardiomyocytes, it resides in the cytosol, sarcolemma, T tubules, and intercalated discs. It is also detected in various cell types, such as smooth muscle and endothelial cells. It is involved in secretion of atrial natriuretic peptide. Annexin-A6 relocalizes from the cytoplasm to plasma membrane in smooth muscle cells and cardiomyocytes during contraction. It has an indirect negative inotropic effect [1315].

Annexin-A7

Annexin-A7 is expressed in cardiomyocytes. It has a Ca^{++}-dependent GTPase activity regulated by protein kinase-C. The annexin-A7–sorcin complex interacts with ryanodin (RyR2) and Ca_V1 channels and may act as a mediator of interchannel communication during excitation–contraction coupling.

10.9 Transcription Factors Involved in Stress Responses

Main participants of gene transcription include RNA polymerase, transcription factors, coactivators, and corepressors. Because DNA exists in a condensed state wrapped by histones that complex to form chromatin, chromatin remodelers with nucleosome disassembly associated with histone acetylases, deacetylases, kinases, and methylases must operate to free the access to RNA polymerase and transcription factors.[176] Chromatin-restructuring complexes can be classified into 2 main categories: (1) ATP-dependent complexes, such as members of the SWI/SNF and ISWI (imitation SWI) remodeler families (Table 10.40), and (2) histone acetyltransferases and deacetylases.

The gene expression is controlled by transcription factors that depend on cell signaling, in addition to chromatin modifications, RNA splicing, and RNA control mechanisms. More than 2000 transcription factors are encoded in the human genome.

---

[176] Chromatin remodeler ATPases constitute 5 families with specialized roles. Remodeler ATPases attach to proteins that bind to histones and nucleosomal DNA. This fastening is influenced by the histone modification state. Histone modification can open chromatin, thus permitting selective binding of transcription factors that, in turn, recruit RNA polymerase-2.

Table 10.40. Examples of ATP-dependent chromatin-remodeling complexes (Source: [1316]; ActL: actin-like protein; ARID: AT rich-interactive domain [SWI1-like]; CHD: chromodomain helicase DNA-binding protein; MBD: methyl-CpG (cytosine–phosphate–guanine)-binding domain protein; RBBP: nucleosome-remodeling retinoblastoma-binding protein; RSF: remodeling and spacing factor; SMARC: SWI/SNF-related, matrix associated, actin-dependent regulator of chromatin).

Type	Subunits
SWI2/SNF2	ActL6a/b, ARID1a/b, SMARCa2/4, SMARCb1, SMARCc1/2 SMARCd1–SMARCd3, SMARCe1
ISWI	RSF1, SMARCa5
CHD	CHD3/4, HDAC1/2, MBD3, RBBP4/7

10.9.1 Nuclear Factor-κB Signaling Module

Nuclear factor κ light chain enhancer of activated B cells, or simply nuclear factor-κB (NFκB), was originally discovered as a transcription factor that regulates immunoglobulin gene expression. In fact, genes regulated by nuclear factor-κB in coordination with different signaling pathways control cell adhesion, survival, proliferation, apoptosis, and stress response, as well as tissue remodeling, inflammation, and innate and adaptive immunity. This ubiquitous, pleiotropic factor that constitute a family of dimeric transcription factors is thus involved in angiogenesis. Deregulated NFκB signaling is involved in malignant transformation, autoimmunity, chronic inflammation, metabolic disorders, and neurodegenerative diseases.

Factor NFκB corresponds to any dimeric transcriptional factor of the *REL family*. Five members are categorized into 2 structural groups (Table 10.41): group-1 NFκB factors, i.e., precursors NFκB1 (or $P105_{NFκB}$ monomer), which is processed into $P50_{NFκB}$ subunit by the 26S proteasome, and NFκB2 (or $P100_{NFκB}$ monomer), which produces $P52_{NFκB}$ subunit; and group-2 factors with RelA (P65), RelB, and Rel (or cRel). Subunits NFκB1 and NFκB2 actually give rise to $P50_{NFκB}$ and $P52_{NFκB}$ homodimers, respectively. Subunit $P100_{NFκB}$ selectively stabilizes RelB-containing dimers,[177] although it can associate with and inhibits several NFκB family members [1317]. The prototypical nuclear factor-κB is the P50–RelA heterodimer.

In resting cells, NFκB is inactive due to its sequestration in the cytosol by one of its inhibitors. Dimeric NFκB factors indeed complex with inhibitors of NFκB: typical IκBα, IκBβ, and IκBε, as well as atypical IκBδ (or IκBNS), IκBζ,

---

[177] Protein RelB links mainly to $P100_{NFκB}$ and $P52_{NFκB}$ monomers.

10.9 Transcription Factors Involved in Stress Responses

Table 10.41. The nuclear factor-κB (or REL) family of dimeric transcription factors and its structural categories. All family members contain an N-terminal Rel homology domain (RHD) that enables dimerization, nuclear localization, and DNA binding. Category-2 members possess a transcription activation domain (TAD) to activate gene expression. Both $P50_{NF\kappa B}$ and $P52_{NF\kappa B}$ are generated from precursors $P105_{NF\kappa B}$ and $P100_{NF\kappa B}$, respectively, by proteasomal degradation of the C-terminal IκB-like domain. They then bind to a TAD-containing NFκB category-2 member to create a functional NFκB heterodimer.

Category	Types
1	NFκB1 ($P105_{NF\kappa B}$ precursor monomer processed to generate $P50_{NF\kappa B}$ dimer) NFκB2 ($P100_{NF\kappa B}$ precursor monomer processed to generate $P52_{NF\kappa B}$ dimer)
2	Rel RelA ($P65_{NF\kappa B}$ or NFκB3) RelB

Table 10.42. Functional categories of the IκB family of inhibitors of NFκB transcription factor (BCL3: B-cell lymphoma-3). Protein $P105_{NF\kappa B}$ serves as a precursor for $P50_{NF\kappa B}$ subunit (NFκB1) after limited proteolysis of $P105_{NF\kappa B}$ monomer by the proteasome. On the other hand, K48-linked ubiquitination of $P105_{NF\kappa B}$ by ubiquitin ligase SCF triggers its complete degradation by the 26S proteasome.

Category	Members
Typical IκB proteins	IκBα, IκBβ, IκBε
Atypical IκB proteins	IκBδ, IκBζ, BCL3
NFκB precursors	$P100_{NF\kappa B}$, $P105_{NF\kappa B}$

and BCL3 (Tables 10.42 and 10.43). In addition, C-termini of $P100_{NF\kappa B}$ and $P105_{NF\kappa B}$ contain IκB-like ankyrin repeats that mimic the inhibition of IκB family members (with its 2 functional groups).

Typical IκB proteins sequester NFκB in the cytoplasm, whereas atypical IκB proteins act in the nucleus as NFκB coregulators. Ubiquitination regulates NFκB activity via proteasomal degradation of inhibitors upon activation of IκB kinase that phosphorylates IκB proteins.

Ubiquitination and subsequent degradation of inhibitors activate both canonical and non-canonical NFκB pathways. These 2 basic pathways are indeed triggered by the release of NFκB dimers from their tethered inhibitors. Activation of NFκB depends on its induction pattern. For example, NFκB regulation of Jun N-terminal kinase can have opposite effects according to the cell type.

In summary, the nuclear factor-κB signaling module is composed of: (1) homo- and heterodimeric transcription factors; (2) many regulatory subunits, NFκB-bound

Table 10.43. Inhibitors of NFκB that sequester the inactive transcription factor in the cytoplasm form a family (Source: [1317]). The IκB family includes: (1) typical IκB proteins (IκBα, IκBβ, and IκBε); (2) precursor proteins P100$_{NFκB}$ and P105$_{NFκB}$; and (3) atypical IκB proteins (IκBδ, IκBζ, and B-cell lymphoma-3 (BCL3) that are expressed after cell stimulation, but generally not in unstimulated cells (inducible IκBs). Atypical IκB proteins reside in the nucleus rather than cytoplasm and have a more restricted cell expression pattern.

Protein	Function
IκBα	NFκB cytosolic sequestration, binding to RelA and Rel
IκBβ	NFκB cytosolic sequestration
IκBε	NFκB cytosolic sequestration
IκBδ	Homodimer P50–DNA binding (repression)
IκBζ	Homodimer P50–DNA binding (transcription)
P100$_{NFκB}$	Release of P52$_{NFκB}$ (transcription)
	Formation of P52–RelB complex (repression)
	Binding to RelB for P52$_{NFκB}$ activation
	Binding to free NFκB dimers (repression)
P105$_{NFκB}$	Release of P50$_{NFκB}$ (transcription)
	Binding to free NFκB dimers (repression)
BCL3	Binding of P50–P50 and P52–P52 homodimers to DNA upon phosphorylation (transcription)
	P50$_{NFκB}$ cytosolic sequestration upon deubiquitination

inhibitors of NFκB (IκB) that cause cytosolic sequestration in unstimulated cells (IκBα, IκBβ, IκBγ, IκBε, IκBζ, as well as nucleus-resident TLR-inducible IκBδ and BCL3), in addition to precursors P100$_{NFκB}$ and P105$_{NFκB}$); and (3) inhibitor of NFκB kinase complexes such as the IKKα–IKKβ–IKKγ complex made of 3 IκB kinase subunits that comprise 2 catalytic components – IκB kinase subunits IKKα (IKK1) and IKKβ (IKK2) – and a regulatory protomer IKKγ[178] In addition, the IκB kinase family contains another member, IKKε, that forms another IκB kinase complex [1318].

10.9.1.1 IκB Regulators

Proteins IκB are not simple inhibitors of NFκB activity, but pleiotropic NFκB cofactors and regulators of gene expression. The IκB family comprises 8 members

---

[178] A.k.a. NFκB essential modulator (NEMo), inhibitor of κ light polypeptide gene enhancer in B-cell kinase γ (IκBKγ), IκB kinase-associated protein (IKKAP) and adenovirus protein Ad E3-14.7K (fourteen)-interacting protein (FIP3).

that are classified into 3 functional groups [1319]: (1) precursor proteins (P100$_{NF\kappa B}$ and P105$_{NF\kappa B}$); (2) typical IκB proteins (IκBα, IκBβ, and IκBε); and (3) atypical IκB proteins (IκBδ, IκBζ, and B-cell lymphoma protein BCL3).

Typical IκB Proteins

Typical IκB regulators IκBα and IκBβ have different preferences for NFκB dimers and experience distinct histories. The former associates with the Rel–P50 and RelA–P50 dimer, whereas the latter, in an hypophosphorylated form, enters into the nucleus and links to the Rel–RelA dimer [1319]. The linkage between IκBβ and Rel does not prevent DNA binding of Rel protein. Furthermore, the binding of the IκBβ–Rel–RelA complex to DNA resists to IκBα and may enhance the expression of genes. In addition, IκBβ is subjected to slow phosphorylation in the cytosol and degradation in a different manner than IκBα. Members of the typical IκB group can operate as both positive and negative NFκB regulators. Agent IκBβ specifically promotes TNFα transcription.

In cultured rat vascular smooth muscle cells, persistent activation of NFκB by IL1β relies on extracellular signal-regulated kinase. Interleukin-1β activates ERK then P90 ribosomal S6 kinase RSK1. The latter is a component of the RSK1–IκBβ–NFκB complex, whatever its phosphorylation status [1320]. Both active ERK2 and RSK1 phosphorylate IκBβ, but only active RSK1 phosphorylates IκBβ on Ser19 and Ser23, 2 sites that mediate the subsequent ubiquitination and degradation.

Atypical IκB Proteins

Members of the atypical IκB group act also as activators and inhibitors of gene expression. Atypical IκB proteins of the NFκB inhibitor family complex with inhibitory P50$_{NF\kappa B}$ or P52$_{NF\kappa B}$ homodimers [1317].

IκBζ

The IκBζ protein complexes with P50 homodimers and RelA-containing dimers for transcription activation and repression, respectively.[179] Agent IκBζ is not expressed constitutively, but rapidly produced after stimulations by lipopolysaccharide and

---

[179]Protein IκBζ is upregulated in response to activation of myeloid differentiation primary-response gene MyD88 by interleukin-1 and lipopolysaccharides in cooperation with Toll-like or IL1 receptors, or costimulation by IL17 and TNF cytokines. Agent IκBζ associates with P50 and P52 homodimers.

interleukin-1 via specific NFκB signaling mediated by MyD88 adaptor. The IκBζ regulator is an activator or inhibitor of NFκB according to the context. It can associate with P50 homodimers bound to specific sites on the IL6 promoter to prime gene expression [1319].

IκBδ or IκBNS

Expression of IκBδ, or IκBNS, is associated with apoptosis of immature thymocytes. Agent IκBδ is rapidly expressed upon TCR-triggered thymocyte death that eliminates thymocytes with autoreactive T-cell receptors [1321]. Protein IκBδ activated by anti-inflammatory cytokines such as interleukin-10 can repress transcription by stabilizing P50 homodimers. Agent IκBδ inhibits the expression of NFκB target genes such as the interleukin-6 gene, but promotes that of a small set of genes [1319].

BCL3

B-cell lymphoma protein-3 of the IκB family possesses a typical transactivating domain (TAD) in its C-terminus. It thus yields a transactivating activity to P50 and P52 homodimers that lack a TAD domain, once it has experienced adequate post-translational modifications, such as phosphorylation and ubiquitination, especially following pro-inflammatory signaling.[180]

On the other hand, in the absence of proper post-translational modifications, BCL3 impedes NFκB-induced gene expression. Agent BCL3 stabilizes P50 homodimer occupancy of target gene promoters, as it prevents its ubiquitination and subsequent degradation, thereby precluding their replacement with active NFκB dimers. Upon stimulation by anti-inflammatory cytokines, ubiquitinated, non-phosphorylated BCL3 may also stabilize P50 and P52 homodimers, hence functioning as an inhibitor. On the other hand, cylindromatosis deubiquinase (Cyld) causes cytosolic sequestration of the BCL3–(P52)$_2$ and BCL3–(P50)$_2$ trimers.

Activation of B and T lymphocytes by a large number of extracellular stimuli, such as cytokines IL1 and TNFα, bacterial lipopolysaccharides, and virus, causes the phosphorylation of 2 serines near the N-terminus of IκBα (Ser32 and Ser36).

---

[180]Protein BCL3 was identified as a proto-oncogene overexpressed in certain B-cell chronic lymphocytic leukemias. According to the context, BCL3 acts as a transactivating protein or NFκB inhibitor. Protein BCL3 is mainly located in the nucleus. It binds to P50 and P52 homodimers that lack transcriptional activation domain and thus acts as a repressor. Binding of BCL3 to P50 and P52 homodimers may confer transactivating ability. Protein BCL3 may also remove P50 and P52 homodimers from inhibited sites, thereby allowing other activator NFκB dimers to take their place. In any case, BCL3 transcriptional activity depends on post-translational modifications and nuclear location.

10.9 Transcription Factors Involved in Stress Responses

Table 10.44. Family of IκB kinases.

Type	Effect
IKKα	Phosphorylation of IκB (repression)
IKKβ	Phosphorylation of IκB
IKKγ	Regulatory subunit (NFκB essential modulator NEMo)
IKKε	NFκB stimulation by T-cell receptor

This phosphorylation leads to the ubiquitination of 2 lysines on IκBα (Lys21 and Lys22) by the ubiquitin ligase complex SKP1–cullin–F-box β-transducin repeat-containing protein (SCF$^{\beta TRCP}$) and its degradation by the 26S proteasome.

10.9.1.2 IκB Kinase

Both IKKα and IKKβ isozymes phosphorylate IκB proteins, thereby releasing NFκB dimers (Table 10.44). Repressor IKKα limits NFκB stimulation. The regulatory IKKγ subunit is required for NFκB activation.

In addition, IKKα and IKKβ phosphorylate NFκB proteins, in particular RelA to regulate the recruitment of transcriptional coactivators and corepressors [1317]. Phosphorylated cAMP-responsive element-binding (CREB) protein-binding protein (CBP) binds preferentially to RelA w.r.t. other transcription factors such as P53 (P53–NFκB cross-regulation).

Activation of the IKK complex in response to 12°tetradecanoylphorbol 13-acetate (TPA or phorbol 12-myristate 13-acetate [PMA]) is mediated by protein kinase-Cα [1322].

IKKα

Enzyme IKKα also targets corepressor complex silencing mediator of retinoic acid and thyroid hormone receptor (SMRT) that links repressive NFκB complexes such as P50 homodimers and histone deacetylase-3 to remove HDAC3 and replace P50 homodimers with NFκB activators such as the P50–RelA complex.

In the nucleus, IKKα can phosphorylate histone-3 for greater κB-site accessibility [1317]. Furthermore, IKKα can activate TNFSF11 to inhibit the transcription of serpin-B5, which is regulated by P53, without intervention of NFκB dimers. Therefore, IKKα can associate with serpin-B5 gene promoter and then prevent gene expression independently of NFκB.

The IKKα kinase is not required for cytokine-dependent activation of NFκB, but does intervene in keratinocyte differentiation [1318].

IKKβ

Kinase IKKβ is required for activation of NFκB by tumor-necrosis factor-α (TNFSF1), but not interleukin-1 in embryonic fibroblasts.

IKKε

Kinase IKKε is involved in NFκB stimulation by T-cell receptor, but neither by TNFα nor IL1 cytokines.

Trimeric inhibitor of nuclear factor-κB kinase intervenes not only in transcription by NFκB, but also in signaling mediated by MAPK modules, Toll-like and T- and B-cell receptors, insulin, and adipocytokines. Both IKKε and TRAF family member-associated NFκB activator (TANK)-binding Ser/Thr protein kinase TBK1 transmit signals that emanate from pattern recognition receptors leading to the phosphorylation of IRF3 transcription factor.

Kinase IKKε also acts in cell apoptosis and substance transport [1323]. Endocytosis as well as cytokinesis and cell migration rely on the spatiotemporal regulation of transmembrane transfer events (localized recruitment and activation of molecules and coordination of their activities). Kinase IKKε phosphorylates Rab11 effectors. Localized activation of IKKε can regulate the delivery of Rab11+ vesicles derived from recycling endosomes.[181]

10.9.1.3 Types of NFκB Signalings

Two NFκB signaling pathways exist: classical or canonical and alternative or non-canonical (Tables 10.45 and 10.46). Canonical and non-canonical NFκB signaling are responsible for the release of NFκB dimers from their inhibitors.

Canonical NFκB Pathway

The canonical NFκB pathway is related to P50–RelA and P50–Rel heterodimers. In resting cells, inhibitory IκB binds to NFκB dimers, thereby sequestering these dimers in the cytoplasm. Subunit IKKβ is necessary and sufficient to phosphorylate IκBα in cooperation with IKKγ. The IκB kinase complex phosphorylates IκB that is then polyubiquitinated by SCF ubiquitin ligase for proteasomal degradation, thereby activating NFκB factor. Multiple negative feedbacks control the canonical NFκB signaling.

---

[181] Extracellular signal-regulated kinase also phosphorylates a Rab11 effector to modulate the distribution of Rab11+ vesicles and the efficiency of transcytosis.

10.9 Transcription Factors Involved in Stress Responses

Table 10.45. Features of the canonical and non-canonical NFκB pathways (**Part 1**; Sources: [1324, 1325]; IKK: IκB kinase; MAP3K: mitogen-activated protein kinase kinase kinase). Upon canonical IKK activation, IκBα, IκBβ, and IκBε that sequester NFκB components into the cytoplasm are phosphorylated on specific N-terminal residues that are docking sites for the SCF$^{\beta TRCP}$ ubiquitin ligase complex, hence labeling for proteasomal degradation. Four transcriptional activators (RelA–RelA, P50–RelA, Rel–Rel, P50–Rel) are then formed. Factor NFκB provokes production of IκBα and IκBε, thereby yielding a negative feedback. The non-canonical pathway depends on the protein–ubiquitin ligase complex composed of inhibitors of apoptosis (IAP1 and IAP2) as well as tumor-necrosis factor receptor (TNFR)-associated factors (TRAF2 and TRAF3). Disruption of the TRAF2–TRAF3–IAP complex triggers the non-canonical NFκB signaling. Persistent degradation of this complex enables a sustained signaling. Kinase MAP3K14 phosphorylates P100$_{NFκB}$ (Ser866 and Ser870), thus activating IKKα. Activated IKKα-containing IKK complex further phosphorylates P100$_{NFκB}$ (Ser99, Ser108, Ser115, Ser123, and Ser872) that is then recognized by the SCF$^{\beta TRCP}$ ubiquitin ligase complex for partial degradation by the 26S proteosome, hence allowing nuclear translocation of the major P52–RelB and minor P52–RelA dimers that act in the late sustained phase. Degradation of IκBδ enables release of P50–RelB and P50–RelA dimers in the early phase.

Feature	Canonical axis	Non-canonical axis
Occurrence	Rapid Independence of protein synthesis	Slow Dependence of protein synthesis
Duration	Transient	Sustained
Regulated functions	Inflammation, Immunity	Lymphoid organogenesis, B-cell survival and maturation, dendritic cell activation, bone metabolism
Components (kinases)	MAP3K7 IKKβ/γ	MAP3K14 IKKα
Targeted inhibitor	IκBα (predominantly)	P100$_{NFκB}$
Transcription effector	P50–RelA (predominantly)	P52–RelB (prototypical heterodimer)

The canonical NFκB pathway is active in few minutes and does not require additional protein synthesis, whereas activation of the non-canonical NFκB pathway takes several hours and needs the production of new proteins, such as NFκB-inducing kinase (NIK [or MAP3K14]; Table 10.47)).

Both Rel- and RelA-containing NFκB dimers intervene in the immune and inflammatory response in various cell types. In addition, RelA prevents TNF-mediated apoptosis in many cell types, such as macrophages, T lymphocytes, and hepatocytes [1325].

The NFκB dimer specificity for target genes may not depend on affinity differences in dimer–κB site interactions, but rather on DNA-triggered conformations of the NFκB dimer that operate with coactivators or other transcriptional activators, such as BCL3, CBP, and IRF3 factors [1325].

Table 10.46. Features of the canonical and non-canonical NFκB pathways (**Part 2**; Sources: [1324, 1325]; DAMP: danger-associated molecular pattern; NLRP: NLR family, pyrin domain-containing protein; PAMP: pathogen-associated molecular pattern; TNAP: TRAF- and NIK-associated protein; TNFRSF: tumor-necrosis factor (TNF) receptor superfamily member; ZFP: zinc finger protein).

Feature	Canonical axis	Non-canonical axis
Activating substances	Cytokines, PAMPs, DAMPs	Agonists of some TNFRSFs
Triggering receptors	Various	Subset of TNFRSFs (TNFRSF1b/3/5/7/8, TNFRSF11a/12a/13C)
Ubiquitinases	SCF$^{\beta TRCP}$ (IκB degradation) TRAF2/5 (TNFR) IAP1–UbE2d [RIPK1] LUbAC [IKKγ] TRAF6–UbC13 (TLR/IL1R) [IRAK1]	SCF$^{\beta TRCP}$ (P100 cleavage) TRAF2/3, IAP1/2
Regulators		NLRP12, TNAP (MAP3K14 inhibitors) ZFP91 (MAP3K14 activator)
Feedback	NFκB⊕⟶IκBα/ε	IKKα→MAP3K14

Several deubiquitinases inhibit MAP3K7 and IKK, as they cleave polyubiquitin chains and prevent their formation, such as cylindromatosis tumor suppressor protein (Cyld) and TNFαIP3, which disassembles polyubiquitin chains and impedes TRAF6–UbC13 interaction [1326].

Alternative NFκB Pathway

Activation of the non-canonical NFκB pathway is required for the organization of secondary lymphoid tissue and innate and adaptive immune responses. This pathway actually regulates immune cell maturation and differentiation, in addition to lymphoid organogenesis.

The non-canonical NFκB pathway is triggered by specific members of the TNFSF superfamily, such as TNFSF3, TNFSF5, and TNFSF13b. It relies on IKKα subunit. It is indeed independent of IKKβ and IKKγ subunits. Subunit IKKα specifically phosphorylates NFκB2 associated with RelB component.

The non-canonical NFκB pathway actually depends on NFκB2 component. In unstimulated cells, NFκB2 is mainly associated with RelB subunit. Kinase IKKα phosphorylates NFκB2 and leads to polyubiquitination that causes limited NFκB2

10.9 Transcription Factors Involved in Stress Responses

Table 10.47. Canonical (classical) and non-canonical (alternative) NFκB signaling (Source: [1326]). Canonical NFκB pathways include signal transduction axis triggered by cytokines, such as tumor-necrosis factor-α (TNF) and interleukin-1β (IL1), and bacterial products such as lipopolysaccharides. Activated cytokine receptors such as IL1R and pathogen recognition receptors such as TLR receptors recruit myeloid differentiation gene product MyD88 adaptor and subsequently IL1R-associated kinases IRAK1 and IRAK4, and TRAF6 ubiquitin ligase. The latter then oligomerizes and ubiquitinates, in cooperation with UbC13 conjugase, MAP3K7IP2 and MAP3K7IP3 of the MAP3K7 complex as well as IKKγ. Upon MAP3K7 autophosphorylation that results from partner ubiquitination, MAP3K7 phosphorylates (activates) IKKβ in an IKKγ-dependent manner. In the non-canonical axis, stimulated TNFRSF5 and TNFRSF13c receptors on the B-lymphocyte surface recruit TRAF2, TRAF3, and inhibitor of apoptosis protein (IAP) ligases. Ligase TRAF2 ubiquitinates IAPs that, in turn, target TRAF3 for degradation. Kinase MAP3K14 is then stabilized and hence activates IKKα that phosphorylates P100$_{NFκB}$ precursor. This phosphorylation causes polyubiquitination of P100$_{NFκB}$ by the SCF^{βTRCP} ubiquitin ligase complex for partial proteolysis. Mature P52$_{NFκB}$ then complexes with RelB and the dimer enters the nucleus for gene transcription (B-lymphocyte maturation and activation).

Signaling type	Pathway example
Canonical	IL1R/TLR–MyD88–IRAK1/4–TRAF6–MAP3K7IP1/2/3– –MAP3K7–IKK–IκB–(P50–RelA)–genes TnfAip3–TNFαIP3⊖ ⟶ TRAF6 (feedback)
Non-canonical	(TNFSF5–TNFRSF5)/(TNFSF13b–TNFRSF13c)–TRAF2–IAP– –TRAF3–MAP3K14–IKKα–(P100–RelB)–(P52–RelB)

proteolysis and P52$_{NFκB}$ release. Subunit P52$_{NFκB}$ then heterodimerizes with RelB subunit. The P52–RelB dimer slowly translocates to the nucleus.

A low MAP3K14 concentration results from its degradation by the ubiquitin ligase complex made of TRAF2, TRAF3, IAP1 and IAP2 proteins. Liganded TNFRSF3, TNFRSF5, and TNFRSF13c trigger the degradation of TRAF3, thus inactivating the TRAF–IAP complex and stabilizing MAP3K14 enzyme. The latter phosphorylates IKKα that phosphorylates NFκB2 to prime proteasomal processing.

A negative feedback control of the non-canonical NFκB signaling attenuates the stabilization of MAP3K14 kinase. The non-canonical NFκB signaling is terminated when IKKα phosphorylates (inactivates) MAP3K14 [1327]. Therefore, in addition to the regulation of MAP3K14 concentration in unstimulated cells by the TRAF–IAP ligase complex,[182] IKKα prevents the uncontrolled activity of the non-canonical NFκB pathway after receptor stimulation.

In stimulated cells, accumulated NFκB-inducing kinase activates IKKα that leads to processing of NFκB2 into P52 dimer.

---

[182] In normal conditions, MAP3K14 undergoes a constitutive degradation.

Nuclear-Initiated DNA-Damage NFκB pathway

Both canonical and non-canonical NFκB pathways result from the engagement of plasmalemmal receptors that trigger the release of NFκB dimers from inactive complexes sequestered in the cytoplasm followed by their nuclear translocation to launch programs of gene transcription.

A third NFκB pathway is initiated in the nucleus in response to genotoxic agents that are responsible for DNA double-strand breaks and replication stress. This additional pathway is characterized by activation of ataxia telangiectasia mutated kinase (ATMK; Vol. 1 – Chap. 4. Cell Structure and Function and Sect. 5.5.1) and post-translational modifications of IKKγ in the nucleus, as well as stimulation of cytoplasmic IKK complex [1328].

Nuclear DNA double-strand breaks can also prime the canonical and non-canonical NFκB pathway in response to many types of genotoxic agents [1328].

The DNA-damage response is coordinated by members of the phosphoinositide 3-kinase-related protein kinase (PIKK) family, i.e., ATMK, ataxia telangiectasia and Rad3-related kinase (ATRK), and DNA-dependent protein kinase (DNAPK; Sect. 5.5). In general, ATMK and DNAPK are activated in response to DNA double-strand breaks, whereas ATRK responds to replication stress. Once they are activated, these kinases phosphorylate numerous effectors, such as histone-2AX, checkpoint kinase-1 or -2, P53, and its binding TP53BP1 partner.

Upon DNA damage, P53-induced protein with a death domain (PIDD) translocates to the nucleus and forms a complex with receptor-interacting protein kinase RIPK1 and IKKγ. Ataxia telangiectasia mutated kinase phosphorylates (activates) IKKγ. Once IKKγ is phosphorylated, sumoylated, and ubiquitinated, it translocates to the cytosol to activate the IKK complex (Table 10.48).

Under genotoxic stresses, IKKγ is phosphorylated (Ser85) by ATMK, sumoylated (Lys277 and Lys309) by the SUMO-protein ligase Protein inhibitor of activated STAT PIAS4, and monoubiquitinated (also Lys277 and Lys309) by IAP1 [1328]. The nuclear PIDD-RIPK1 complex supports IKKγ sumoylation.

Poly^{ADP}ribose polymerase PARP1 recruited to DNA-damage sites can aid assembling of a nuclear complex with PIAS4, IKKγ, and ATMK [1328]. Agent PARP1 rapidly (within s) binds to DNA-strand breaks and synthesizes poly^{ADP}ribose attached to itself or other acceptor proteins.

A small fraction of activated nuclear ATMK translocates to the cytosol and plasma membrane in a calcium-dependent manner using RanGEF RCC1 and Ran GTPase [1330]. Kinase ATMK activates MAP3K7 with the help of IKKγ and ERC2 (ELKS [protein rich in glutamic acid (E), leucine (L), lysine (K), and serine (S)], Rab6-interacting, CAST [cytomatrix at the active zone-associated structural protein] family member).[183] K63-linked polyubiquitination of ERC2 that depends on ATMK and IKKγ by ubiquitin ligase XIAP and conjugase UbC13

---

[183]Cytomatrix at the active zone-associated structural protein (CAST) and protein rich in glutamic acid (E), leucine (L), lysine (K), and serine (S; ELKS) regulate the assembly and function of the

10.9 Transcription Factors Involved in Stress Responses

Table 10.48. Three types of NFκB pathways triggered by TLR4, TCR, TNFRSF1a (TNFR1), TNFRSF5, and genotoxic stress (Source: [1332]). Canonical NFκB signaling starts with the formation of complexes that undergo non-degradative polyubiquitination (K63-linked or linear ubiquitin chains) and are recognized by the MAP3K7 complex and IKK kinase (PIDD: P53-induced protein with a death domain; TRAM: TRIF-related adaptor molecule; TRIF: Toll–IL1R domain-containing adaptor inducing Ifnβ). Stimulation of B-cell (BCR) and T-cell (TCR) receptors leads to activation of phospholipase-C (PLC) and protein kinase-C (PKC). The latter phosphorylates CARD11 that can then link to BCL10 and MALT1 to form the CBM complex to activate the IKK kinase multimer. In addition, TNFR1 receptor can also prime the formation of a caspase-8-containing complex that initiates apoptosis. In addition to tumor-necrosis factor receptor-associated death domain adaptor (TRADD), TNFRSF6a (Fas)-associated protein with death domain (FADD) and receptor-interacting protein kinase RIPK1 forms a platform for caspase-8 to launch apoptosis. Pairs of adaptors connect to TLRs to activate signaling. In particular, a sorting adaptor (TIRAP or TRAM) helps to localize TLRs to a cell region that promotes signaling. Once sorted, a signaling adaptor (MyD88 or TRIF) is then recruited to initiate signaling that triggers gene transcription.

Receptor	Effectors	Involved NFκB components
	Canonical pathway	
TCR	MAP3K7, IKKβ, MyD88, IRAK1/4, TRAF6, TRAM, TRIF, RIPK1	P50–RelA
BCR	PLCγ, PKCθ, CARD11, BCL10, MALT1, TRAF6	P50–RelA
TLR4	MAP3K7, IKKβ, PKCθ, CARD11, BCL10, MALT1, TRAF6,	P50–RelA
TNFR1	MAP3K7, IKKβ, TRADD, RIPK1, TRAF2, IAP1/2, RNFR31/54	P50–RelA
	Non-canonical pathway	
TNFRSF3/5 TNFRSF11a/13c	MAP3K14, IKKα	P52–RelB
	Genotoxic stress	
DNA damage	PIDD, RIPK1, ATMK, IKKγ	P50–RelA

enables ERC2 connection to MAP3K7 via its ubiquitin-binding MAP3K7IP2 and MAP3K7IP3 partners [1329]. Ligases IAP1 and XIAP are also needed for MAP3K7 activation. Ubiquitinated ERC2 elicits ubiquitin-dependent assembly of

---

presynaptic active zone of nerve terminals during synapse maturation. Protein ELKS may also intervene in the canonical NFκB pathway initiated by TNFα cytokine.

MAP3K7IP2–MAP3K7IP3–MAP3K7 and IKK complexes mediated by ubiquitin-binding subunits (MAP3K7IP2 and MAP3K7IP3 on the one hand and IKKγ on the other) to activate NFκB upon genotoxic stresses.

Enzyme ATMK links to IKKβ with ERC2 help [1330]. Kinase ATMK links to and activates TRAF6 ubiquitin ligase to synthezise K63-linked polyubiquitin using UbC13 and recruit IAP1 ligase. Phosphorylation by ATMK of exported IKKγ and subsequent ubiquitination by IAP1 enable activation of the cytoplasmic IKK complex. In addition, the ATMK–TRAF6–IAP1 module stimulates MAP3K7IP2-dependent MAP3K7 phosphorylation.

In summary, activation of the IKK complex upon exposure to inducers of DNA double-strand breaks thus needs: (1) assembling of a short-lived nucleoplasmic signalosome supported by automodified PARP1 — the PARP1–ATMK–PIAS4–IKKγ complex — that triggers phosphorylation by ATMK, monoubiquitination by IAP1, and sumoylation by PIAS4 of nuclear IKKγ that is subsequently exported to the cytoplasm; (2) Export of ATMK from the nucleus and subsequent formation of K63-linked polyubiquitin chain by TRAF6, K63-linked polyubiquitination of ERC2 and TRAF6, and building of a cytoplasmic complex composed of ATMK, TRAF6, polyUb chain, IAP1, ERC2, MAP3K7IP2, MAP3K7, and IKKγ; (3) monoubiquitination of IKKγ (Lys285) by the ATMK–TRAF6–IAP1 complex; and (4) activation of MAP3K7 that activates IKK in the cytoplasm.

10.9.1.4 NFκB Pathways

Canonical NFκB pathway interferes with non-canonical NFκB pathway. The former is mainly triggered by pro-inflammatory cytokine and pattern recognition receptors to launch the transcription of numerous genes, which encode regulators of immune and inflammatory responses (cytokines, chemokines, adhesion molecules, and other immunoregulators) as well as proproliferative, anti-apoptotic, and anti-oxidant proteins, thereby priming inflammation and protecting cells from damaging effects of inflammation.

Triggering Receptors

Signaling mediated by NFκB can be triggered by numerous types of receptors using specific adaptors and scaffold proteins. Receptors of TNFRSF superfamily and interleukin receptors initiate the NFκB conventional axis (Sect. 10.9.1.3).

Various scaffold proteins operate in signaling cascades upon stimulation of immunoglubulin Fc, antigen (T- and B-cell receptors), G-protein-coupled, and C-type lectin receptors to activate NFκB, in particular members of the caspase activation and recruitment domain (CARD)-containing protein family, or CARD and membrane-associated guanylate kinase-like (MAGuK) domain-containing protein family (CARMA) [1331]. The CARD family includes 3 members

10.9 Transcription Factors Involved in Stress Responses

Table 10.49. Scaffold proteins of the caspase activation and recruitment domain (CARD)-containing protein family (Source: [1331]; BIMP: B-cell lymphoma adaptor BCL10-interacting membrane-associated guanylate kinase-like [MAGuK] domain-containing protein; CARMA: CARD and MAGuK domain-containing protein family member). Protein CARD10 links G-protein-coupled receptors to the NFκB pathway. In airway epithelial cells, CARD10 enables NFκB activity via polyubiquitination of IKKγ triggered by lysophosphatidic acid. Protein CARD9 that lacks the C-terminal MAGuK domain may also couple BCL10 and MALT1 to activate the NFκB axis. Oligomerized BCL10 can act as scaffold for the IKK and JNK axes.

Type	Other aliases	Location
CARD11	CARMA1, BIMP3	Hematopoietic cells
CARD14	CARMA2, BIMP2	Mucosa
CARD10	CARMA3, BIMP1	Widespread (not in hematopoietic cells)

encoded by 3 genes (CARD10, CARD11, and CARD14, or CARMA1–CARMA3; Table 10.49). They contain an N-terminal CARD domain, a coiled-coil sequence, and a C-terminal MAGuK motif (or PDZ-SH3-GUK region). They have distinct tissue distribution pattern.

The canonical NFκB pathway is activated not only by inflammatory cytokines, but also antigen receptors (BCRs and TCRs). T-cell receptors activated by antigens presented by major histocompatibility complex molecules on the surface of antigen-presenting cells stimulate PKCθ kinase [1326]. The latter recruits membrane-associated adaptor CARD11, B-cell lymphoma adaptor BCL10, and mucosa-associated lymphoid tissue lymphoma translocation peptidase MALT1 to constitute the CBM (CARD11–BCL10–MALT1) complex that activates IKK enzyme. Proteins BCL10 and MALT1 promote TRAF6 oligomerization. Activated TRAF6 operates with UbC13 as well as MAP3K7 and IKK kinases.

In addition, the CBM complex mediates NFκB activation by immunoreceptor tyrosine-based activation motif (ITAM)-containing receptors on natural killer cells, whereas the BCL10–MALT1 complex ensures signaling from FcεR1 receptor on mastocytes [1331].

On the other hand, CBL ubiquitin ligase monoubiquitinates CARD11, thereby precluding the CBM complex and NFκB action [1326]. Deubiquitinase Cyld also hinders MAP3K7 and IKK activation. Another deubiquitinase TNFαIP3 inhibits NFκB in T lymphocytes, as it targets MALT1 paracaspase. Conversely, MALT1 cleaves TNFαIP3 and BCL10 proteins [1326].

Protein MALT1 serves as both an adaptor and peptidase to initiate a full NFκB response in antigen-stimulated lymphocytes [1332].

In the cytosol, viral RNAs are recognized by RNA helicase domain-containing members of the RLR family, such as retinoic acid-inducible gene product RIG1, melanoma differentiation-associated gene product MDA5, and laboratory of genetics and physiology protein LGP2 [1326]. The signaling cascade targets type-1 interferons via mitochondrial antiviral signaling adaptor (MAVS)[184] that activates IKK enzyme.

Nucleotide-binding domain (NBD), leucine-rich repeat (LRR)-containing proteins (NLR), or NOD-like receptors (Vol. 3 – Chap. 11. Receptors of the Immune System), are cytosolic proteins that detect pathogen (PAMP)- or damage (DAMP)-associated molecular patterns. Among about 20 NLR proteins, NOD1 and NOD2 recognize bacterial peptidoglycans and activate NFκB factor [1326].

Proteolysis

Signaling mediated by NFκB is regulated by proteolysis. In addition to the proteasome, caspases and MALT1 paracaspase participate in the NFκB signaling downstream from specific receptors (Table 10.50 and 10.51) [1332].

Non-Degradative Ubiquitination and Sumoylation

Both degradative and non-degradative ubiquitination of proteins are involved in activation of NFκB transcription factor. Activation of NFκB triggers degradation of IκBα after assembly of *non-degradative Lys63-linked polyubiquitin chains* that enable association with IκBα kinase complex and *degradative Lys48-linked polyubiquitin chains* on IκBα for destruction by the 26S proteasome.[185]

Two Ub ligases, RING finger protein RNF54[186] and RNF31,[187] that assemble linear polyubiquitin chains participate in NFκB regulation [1334]. An alternative mechanism exists that involves non-degradative, linear polyubiquitination mediated by the ubiquitin ligase Linear ubiquitin chain assembly complex (LUbAC) composed of SH3 and multiple ankyrin repeat domain (SHAnk)-associated RH domain interactor Sharpin, RNF31, and RNF54. This ligase complex thus contains 2 zinc

---

[184] A.k.a. IPS1, VISA or Cardif. It localizes predominantly to the mitochondrial outer membrane.

[185] Lysine-63-linked polyubiquitin chains are made by adding ubiquitin directly on the target, whereas Lys48-linked polyubiquitin chains are assembled before being appended to the target. Inhibitor IκBα is polyubiquitinated via Lys48 by Skp1–cullin–F-box ligase β-transducin repeat-containing protein SCF$^{\beta TRCP}$ for recognition by the 26S proteasome. On the other hand, Lys63-linked polyubiquitin chains that are not recognized by the 26S proteasome serve as scaffolds to recruit ubiquitin-binding proteins and activate IKK kinase. Excitation of IKK leads to IκBα phosphorylation that is required for SCF$^{\beta TRCP}$ activity.

[186] A.k.a. RanBP-type and C3HC4-type zinc finger-containing RBCK1, heme-oxidized iron regulatory protein IRP2 ubiquitin ligase HOIL1, hepatitis-B virus X-associated protein XAP4, and ubiquitin-conjugating enzyme-7-interacting protein UbCe7IP3. Iron regulatory protein-2 is a modulator of iron metabolism that is regulated by iron-induced ubiquitination and degradation [1333].

[187] A.k.a. HOIL1L-interacting protein (HOIP) and Zinc in-between-RING finger ubiquitin-associated domain-containing protein (ZIBRA).

10.9 Transcription Factors Involved in Stress Responses

Table 10.50. Caspases (cysteine-dependent aspartate-specific proteases) in NFκB signaling (Source: [1332]; BCL10: B-cell lymphoma adaptor-10; CFLAR: cysteine-dependent aspartate-specific peptidase CASP8 and Fas (TNFRSF6a)-associated via death domain (FADD)-like apoptosis regulator; dsRNA: double-stranded RNA; IKK: IκB kinase; LPS: lipopolysaccharide; MALT1: mucosa-associated lymphoid tissue lymphoma translocation peptidase; MDP: muramyl dipeptide [nucleotide-binding oligomerization domain-containing protein NOD2 ligand]; MST: macrophage-stimulating protein, or hepatocyte growth factor-like protein [HGFL]; NLR: NOD-like receptor; RCC: regulator of chromosome condensation; RIOK: RIO kinase; RIPK: receptor-interacting protein kinase; TCR: T-cell receptor; TIRAP: Toll-like and IL1 receptor adaptor protein; TLR: Toll-like receptor; TRAF: tumor-necrosis factor receptor [TNFR]-associated factor). Killer activation (KAR) and inhibitory (KIR) receptors on the plasma membrane of natural killer cells can only function in presence of immunoreceptors that have immunoreceptor tyrosine-based activation (ITAM) or inhibition (ITIM) motif (ITAM- [ITAMIR] and ITIM-bearing immunoreceptors). Zymosan is a glucan that initiates signaling via pro-inflammatory cytokines in macrophages. According to their main functions, caspases are grouped into apoptotic (with initiator and executioner caspases) and pro-inflammatory caspases. However, pro-apoptotic caspases also intervene in inflammation and pro-inflammatory caspases contribute to apoptosis. Activation platforms include the death-inducing signaling complex (DISC) for caspase-8 and -10, apoptosome for caspase-9, the Piddosome for caspase-2, and inflammasomes for caspase-1 and -5.

Type	Triggers and partners	Role in NFκB signaling
	Apoptotic caspases	
Caspase-2	RIPK1, TRAF2	Activation
	Initiator caspases	
Caspase-8	TCR, LPS, dsRNA, TNFSF10, BCL10, MALT1, TRAF6, CFLAR, RIPK1, MAP3K14, TRAFs, IKK, IκBα	Activation
Caspase-10	TNF, dsRNA, RIPK1, MAP3K14, RIOK3	Activation
	Executioner caspases	
Caspase-3	TNF, IκBα, RelA, RCC1, MST1, caspase-8	Inhibition
	Pro-inflammatory caspases	
Caspase-1	TLR2/4 on macrophages, NLRs, RIPK2, TIRAP	Activation
caspase-4	LPS, TRAF6	Activation
Caspase-12	MDP, RIPK2, TRAF6	Inhibition

RING (really interesting new gene) finger proteins for ubiquitin polymerization: RNF54 longer form (RNF54$_L$), an alternatively spliced variant), and RNF54$_L$-interacting protein RNF31. The RNF54$_L$–RNF31 heterodimer that tethers linear ubiquitin chains catalyzes the assembly of a polyubiquitin chain on IKKγ (Lys285 or Lys309) that serves as a scaffold to recruit the activation complex [1335]. Ligase

Table 10.51. Paracaspase mucosa-associated lymphoid tissue lymphoma translocation peptidase MALT1 in NFκB signaling (Source: [1332]; BCL10: B-cell lymphoma adaptor-10; BCR: B-cell receptor; CARD: caspase recruitment domain-containing family member; CXCL12: chemokine (C–X–C motif) ligand-12, or stromal cell-derived factor SDF1α; Ig: immunoglobulin; IKK: IκB kinase; IL: interleukin; IRAK: IL1 receptor-associated kinase; LPA: lysophosphatidic acid; LPS: lipopolysaccharide; PKC: protein kinase-C; TCR: T-cell receptor; TRAF: tumor-necrosis factor receptor [TNFR]-associated factor).

Type	Triggers and partners	Role in NFκB signaling
MALT1	BCR, TCR, LPS, LPA, IL8, TNFSF13b, CXCL12α, IgE, LPA, ITAMIR, zymosan, angiotensin-2, IRAK1, BCL10, TRAF3/6, MAP3K7, IKKγ, PKC, CARD9/11, TNFαIP3	Activation

RNF54$_L$, but not Lys63-linked polyubiquitin chain formation of the canonical NFκB activation mechanism, is essential for IκBα degradation and subsequent NFκB activation.

Ubiquitin chains are involved in signal transduction from tumor-necrosis factor (TNFRSF) and Toll-like (TLR) receptors to the IKK complex, but in different ways.

In the Toll-like and interleukin-1 receptor pathways, Ub conjugase UbC13 and ligase TRAF6 synthesize Lys63-linked chains. In fact, ubiquitin ligase TRAF6 activates NFκB downstream from interleukin-1 and Toll-like receptors. Together with ubiquitin conjugase UbC13 and variant UbE2V1 (or UEV1a), TRAF6 catalyzes Lys63-linked ubiquitination that causes phosphorylation (activation) of MAP3K7[188] and subsequent phosphorylation of the IκB kinase to stimulate NFκB factor. Enzyme TRAF6 generates unanchored polyubiquitin chains that link to MAP3K7IP2[189] and IKKγ and bind to the regulatory subunits of the MAP3K7–IKK complexes [1336]. Free Lys63-linked polyubiquitin chains (that are not conjugated to any target protein) activate MAP3K7 by binding to MAP3K7IP2, thereby causing MAP3K7 dimerization or oligomerization and subsequent autophosphorylation. In addition, unanchored polyubiquitin chains synthesized by TRAF6 and ubiquitin conjugase UbE2d3 (or UbCh5c) tether to IKKγ and activate (by autophosphorylation) the IKK complex. Free Lys63-linked Ub chains synthesized by TRAF6 may then act as a second messenger that, after recognition by MAP3K7IP2, activates MAP3K7 and stimulates adjoining IKK molecules recruited via IKKγ subunit.

---

[188] A.k.a. TGFβ-activated kinase TAK1.

[189] A.k.a. TAK1-binding protein TAB2.

10.9 Transcription Factors Involved in Stress Responses

In TNFRSF1a-primed signaling, TRAF2 scaffold is recruited by TRADD adaptor to attract Receptor-interacting kinase RIPK1 (Sect. 5.9.5.1) and the Ub ligases Inhibitors of apoptosis IAP1 and IAP2.[190] Kinase RIPK1 anchors Lys63-linked Ub chains that are then bound by MAP3K7IP2 and MAP3K7IP3 and IKKγ subunit. This process brings MAP3K7 close to IKK for IKK activation.

Ubiquitin-binding proteins bind to ubiquitinated proteins for linkage to cell processes. Inhibitor of apoptosis contains a ubiquitin-binding domain that regulates NFκB, hence cell survival and tumorigenesis [1337]. Owing to this domain, IAP1 is able to maintain endothelial cell survival and to protect cells from TNFα-induced apoptosis; Ub ligases X-linked inhibitor of apoptosis protein XIAP[191] and IAP2 can activate NFκB factor.

In response to DNA damage, signaling to NFκB involves the sumoylation of nuclear IKKγ by the SUMo ligase Protein inhibitor of activated STAT PIAS4, followed by its phosphorylation by ataxia telangiectasia mutated kinase (Sect. 5.5.1), ubiquitination, and nuclear export. Poly^{ADP}ribose polymerase PARP1 generates poly^{ADP}ribose chains and, via polyadpribosylation, mediates the assembling of a the IKKγ–PIAS4–ATMK complex [1338]. Once this complex formed, ATMK translocates to the cytoplasm and activates TRAF6 that produces Lys63-linked Ub chains and activates IKK kinase.

Ubiquitination is counteracted by deubiquitinases that can terminate NFκB signaling, in addition to prevent spontaneous activation of NFκB factor.

Reactive Oxygen Species

Riboflavin kinase, a rate-limiting enzyme in the synthesis of flavine adenine dinucleotide (FAD) that is an essential prosthetic group (cofactor) of NADPH oxidase, links TNFR1 and some TLRs to NADPH oxidases [1339]. Produced reactive oxygen species are required to activate the inflammasome, but not for signaling to IKK enzyme.

Cellular ROS sources include mitochondrial electron transport chain complex-1 and -3, NAD(P)H oxidases, particularly in response to cytokine and growth factor receptors, as well as metabolic enzymes. Above a given concentration threshold, ROS are toxic.

The NFκB pathway controls the synthesis of pro- or anti-oxidants (Table 10.52). On the other hand, ROS inhibit or stimulate NFκB signaling [1340].

Anti-Oxidants Targets

Heme-containing catalases, mostly tetramers, convert hydrogen peroxide to water and molecular oxygen. Some catalases act as both catalase and peroxidase.

---

[190] A.k.a. baculoviral IAP repeat-containing protein BIRC3 and BIRC2, respectively.

[191] A.k.a. baculoviral IAP repeat-containing protein BIRC4.

Table 10.52. Transcriptional NFκB activity targets genes of anti- and pro-oxidants (Source: [1340]; AKR: Aldo-keto reductase; COx: cyclooxygenase; CyP: cytochrome-P450 superfamily member; FHC: ferritin heavy chain; GPx: glutathione peroxidase; GSTP: glutathione S-transferase-π; LOx: lipoxygenase; MT3: metallothionein-3; NOS: nitric oxide synthase; NOx: NADPH oxidase; NQo: NAD(P)H dehydrogenase [quinone]; $^{(Cu,Zn)}$SOD: copper–zinc-containing superoxide dismutase; ^{Mn}SOD: mitochondrial manganese-containing superoxide dismutase; TRx: thioredoxin; XDH(O): xanthine dehydrogenase and oxidase). Protective anti-oxidant enzymes include catalase, superoxide dismutase, and glutathione peroxidase

Anti-oxidants	Pro-oxidants
Enzymes of ROS metabolism	Enzymes of ROS synthesis
Catalase	NOx2
GPx1	COx2
GSTP1	LOx1/2
^{Mn}SOD (SOD2)	XDH(O)
$^{(Cu,Zn)}$SOD	CyP2c11/2e1/7b
Reductases	NOS1/2 (RNS: peroxynitrite)
NQo1	
Thiol-containing proteins	
TRx1/2	
Miscellaneous	
AKR1c1	
FHC	
HOx1	
MT3	

Heme oxygenase-1 degrades heme and forms carbon monoxide and biliverdin, which is subsequently reduced by biliverdin reductase to bilirubin, a potent anti-oxidant (Sect. 10.2.1).

NAD(P)H dehydrogenase [quinone]-1, or NAD(P)H:quinone oxidoreductase-1, is a cytoplasmic, 2-electron reductase that processes quinones to hydroquinones, thereby preventing the 1-electron reduction of quinones into radical species [1340].

Aldo-keto reductase AKR1c1, or dihydrodiol dehydrogenase DDH1, oxidizes transdihydrodiols of polycyclic aromatic hydrocarbons.

Metallothioneins (MT1,[192] and MT2–MT4) on the membrane of the Golgi body not only bind to many metal types (arsenic, cadmium, copper, mercury, selenium, silver, and zinc), hence regulating metal toxicity, but also scavenge superoxide ($O_2^{\bullet-}$) and $^{\bullet}$OH radicals.

---

[192]This metallothionein subtype possesses many isoforms (MT1a–MT1b, MT1e–MT1h, MT1l–MT1m, and MT1x.

10.9 Transcription Factors Involved in Stress Responses

Ferritin heavy chain is a ferroxidase enzyme and subunit of an iron storage protein that collects iron in a soluble and non-toxic form and protects cells from oxidative damage, as it precludes iron-mediated generation of highly reactive hydroxyl radicals ($^{\bullet}$OH) from hydrogen peroxide (H_2O_2) [1340].

Pro-Oxidants Targets

Xanthine dehydrogenase and oxidase (XDH[O]), or xanthine oxidoreductase (XOR), catalyzes the interconversion of xanthine and urate with NAD^+ and water cofactors. The dehydrogenase form is the most dominant form that can be converted to the oxidase form upon oxidation of its sulfhydryl (thiol) groups (SH). It can transfer electrons to O_2 instead of NAD^+, thus generating superoxide and hydrogen peroxide [1340].

Cyclooxygenase-2, or prostaglandin-G/H synthase-2, converts arachidonic acid into prostaglandin-H2 and generates superoxide. Other catalytic NFκB target that generate ROS during arachidonic acid metabolism include arachidonate 12-lipoxygenase (LOx12 or ALOx12) and arachidonate 5-lipoxygenase (LOx5 or ALOx5).

Uncoupled cytochrome-P450 enzymes CyP2e1, CyP2c11, and CyP7b produce ROS. They are encoded by genes with NFκB promoter elements [1340].

Nitric oxide synthases NOS1 and NOS2 produce reactive nitrogen species, as NO can react with superoxide to form peroxynitrite ($ONOO^-$), which is both an oxidant and nitrating agent. The latter reacts with CO_2 to form nitrosoperoxycarbonate ($ONOOCO_2^-$), which then homolyzes to form carbonate ($CO_3^{\bullet -}$) and nitrogen dioxide radicals ($NOO_2^{\bullet}$) [1340].

Action of ROS on NFκB

Reactive oxygen species often stimulates the NFκB pathway in the cytoplasm, but inhibits NFκB activity in the nucleus [1340]. They both activate in some cell types and inactivate the IKK complex. On the other hand, hydrogen peroxide affects IκBα phosphorylation, hence preventing normal ubiquitination and degradation. In addition, oxidation of NFκB by ROS hampers its DNA binding.

Signaling Interferences

Oscillations in NFκB activity result from cycles of degradation and resynthesis of IκB inhibitors. Activation of the canonical NFκB pathway generally results from inflammatory cues, whereas the non-canonical pathway is mostly activated by developmental signals. However, interactions between these 2 pathways exist. Interdependency operates at various stages of NFκB signaling, including transcriptional control.

The P50–RelA dimer is controlled by non-canonical signals, as it can be sequestered by $P100_{NFκB}$. Therefore, both canonical and non-canonical axes control P50–RelA dimer, which is inhibited by IκBα, IκBβ, IκBδ, and IκBε. Canonical signals lead to degradation of IκBα, IκBβ, and IκBε. Non-canonical developmental NFκB signals can indeed disrupt the IκBδ inhibitory complex, thus allowing entry of the canonical P50–RelA effector into the nucleus [1325].

On the other hand, expression of P100$_{NF\kappa B}$ and RelB depend on P50–RelA activity and hence on canonical signals. Signaling from TNFRSF3 activates P50–RelA and then P52–RelB via 2 distinct axis [1325].

Factor NFκB is activated by multiple plasmalemmal or intracellular receptors. Cooperative and antagonist crosstalks exist between NFκB and other signaling pathways based on JNK, P53 transcription factor, and nuclear receptors [1341]. Adaptor B-cell lymphoma–leukemia protein BCL10 is required for NFκB activation by ligand-bound GPCRs, such as that of endothelin-1 [1342].

Gene Transcription

After degradation of their cytoplasmic inhibitor IκB, homo- and heterodimers composed of members of the NFκB family (i.e., P50$_{NF\kappa B}$, P52$_{NF\kappa B}$, Rel, RelA, and RelB) translocate to the nucleus and bind to promoters or enhancers of target genes to interact with associated transcription factors (coactivators and corepressors). The RelA–P50 heterodimer (NFκB1–RelA complex) moves toward the nucleus following stimulation using dynein, whereas back and forth motion in the absence of IκB degradation is based on dynactin [1343].

Transcription factor NFκB activates transcription of numerous genes implicated in inflammation, immune response, and struggle against apoptosis. Transcriptional activation domains exist in structural group-2 NFκB proteins (Rel, RelA, and RelB), but are lacking in structural group-1 NFκB proteins (NFκB1 [P50$_{NF\kappa B}$] and -B2 [P52$_{NF\kappa B}$]).

Members of the structural group-1 NFκB proteins act in several ways. (1) They target distinct specific sites according to the type of heterodimers they make with other members that possess transactivation domain. (2) They repress transcription when bound to target sites as homodimers. (3) They promote transcription by recruiting other Rel members with transactivation domain to target sites [1317].

Binding and unbinding of NFκB on its target sites in gene is characterized by fast dynamics [1344].

Inhibitor of NFκB kinase that initiates the NFκB pathway by degrading IκB, exerts additional effects on the NFκB transcriptional activity. It targets NFκB for degradation and selectively augments coactivator recruitment to facilitate the transcription of target genes [1317]. Some members of the IκB family selectively upregulate the transcription of specific NFκB target genes by providing transactivating activity to repressive P50 and P52 homodimers. Inhibitor IκB can also influence NFκB transcriptional activity by influencing the formation, stability, and responsiveness of the NFκB complex.

Components of the NFκB pathway can both promote and impede gene expression. In addition, NFκB transcriptional activity is modulated by interactions with other signaling pathways, such as those involving P38MAPK [1317]. The latter phosphorylates histone-3 and increases NFκB accessibility to specific promoters. It also targets mitogen- and stress-activated kinase MSK1[193] of the ^{P90}RSK subfamily of the MAPKAPK family of MAPK-activated protein kinases that may regulate cAMP-response element-binding transcription factor (CREB), which cooperates with NFκB factor. Kinase MAPKAPK2 dampens NFκB activity by regulating P38MAPK enzymes. Enzyme MAPKAPK2 that phosphorylates HSP27 also decreases nuclear retention of P38MAPK and RelA.

Control and Termination of NFκB Signaling

Termination of NFκB activity can result from: (1) dephosphorylation by protein phosphatase magnesium-dependent PPM1d;[194] (2) other post-translational modifications that target substrates for degradation; (3) nuclear degradation of NFκB subunits; (4) dissociation of coactivators; and (5) subnuclear relocalization of NFκB complexes.

Negative feedback loops involve stimulatory and inhibitory phosphorylation and ubiquitination of signaling intermediates. NFκB-induced IκBα expression actually yields a negative feedback.

Negative feedback limits the duration and magnitude of NFκB activity. Inhibitor IκBε dampens IκBα oscillations, because the delayed production of its mRNA happens in opposite phase w.r.t. that of IκBα [1325].

Newly synthesized IκBδ primes a negative feedback loop to regulate P50–RelA activity to limit signaling from lipopolysaccharides, but not from tumor-necrosis factor-α [1325]. TNF-induced IKK activity both rises and decays rapidly due to TNFαIP3 action. On the other hand, LPS-induced IKK activity has a slower activation, but longer duration.

Inhibitor IκBα is able to enhance the dissociation rate of NFκB from target genes [1325]. Once it is phosphorylated (Ser536) by nuclear IKK, DNA-bound RelA is processed by protein inhibitor of activated STAT ligase PIAS1 and then degraded. In addition, the P50–RelA dimer causes the production of P100$_{NFκB}$ (slow-kinetic, negative feedback loop) [1325].

On the other hand, peptidyl-prolyl cis-trans isomerase NIMA-interacting PIN1 targets RelA to prevent the termination of NFκB activity primed by the ECS complex. The PDZ and LIM domain-containing protein PDLIM2 interacts directly with RelA and promotes its relocalization to insoluble nuclear bodies and its ubiquitination for degradation.

---

[193] A.k.a. 90-kDa ribosomal protein S6 kinase-5 (RPS6Kα5 or RSKL).

[194] A.k.a. protein phosphatase-2Cδ and wild-type P53-induced phosphatase WIP1.

In the non-canonical NFκB pathway, MAP3K14 destabilization results from its phosphorylation (Ser809, Ser812, and Ser815) by IKKα. MicroRNAs (miR223, miR15a, and miR16) target IKKα during macrophage differentiation.

NFκB Dephosphorylation

Phosphorylation of NFκB enhances its activity, whereas dephosphorylation prevents it. Phosphatase PPM1d dephosphorylates RelA subunit (Ser536; i.e., the site required for P300 recruitment) [1345].

Magnesium-dependent protein Ser/Thr phosphatase PPM1d has many substrates, such as kinases ATMK, ChK1, and P38MAPK, DM2 Ub ligase, and P53 transcription factor.

Degradative NFκB Ubiquitination

Members of the PIAS family of SUMo ligases inhibit STAT and NFκB factors. Both PIAS1 and PIAS4 inhibit binding of RelA-containing dimers to DNA, but target distinct NFκB-regulated genes [1317].

Activated NFκB induces the expression of suppressor of cytokine signaling SOCS1 that translocates to the nucleus to form the elongin-B–elongin-C–cullin-2–SOCS1 ubiquitin ligase complex (ECS) [1317]. Agent NFκB also promotes the association of copper metabolism Murr1 domain-containing COMMD1 with the ECS complex. Protein COMMD1 promotes ECS binding to RelA-containing dimers for proteasomal degradation.

Deubiquitination

In most cell types, TNFαIP3 ligase and deubiquitinase is expressed at very low levels [1334]. Inhibitor TNFαIP3 operates particularly in B and T lymphocytes as well as dendritic cells. Agent TNFαIP3 is constitutively produced in B lymphocytes. This enzyme can be rapidly synthesized upon action of pro-inflammatory cytokines or mitogens. On the other hand, thymocytes and peripheral T lymphocytes synthesize TNFαIP3 at high levels. Upon ligation of T-cell receptor by agonists, TNFαIP3 is cleaved by MALT1 paracaspase. Conversely, TNFαIP3 deubiquitinates MALT1 peptidase.

Enzyme TNFαIP3 acts as a feedback inhibitor of NFκB in multiple innate immune pathways, especially those triggered by TNFR and TLR receptors (Table 10.53) [1334]. During early stages (15–30 mn) of TNFR1 signaling, RIPK1 is ubiquitinated (adding K63-linked ubiquitin chains) to activate IKK kinase. The latter phosphorylates (inactivates) Cyld deubiquitinase. Afterward (>30 mn), TNFαIP3 and IκBαare synthesized due to NFκB action (negative feedback loop). Enzyme TNFαIP3 interacts with human T-cell leukemia virus type-1 (Tax1)-binding protein TAX1BP1 (also TRAF6-binding protein), Itch, and RNF11, as well as possibly adaptors TNFαIP3-interacting protein TNIP1 and coiled-coil

10.9 Transcription Factors Involved in Stress Responses

Table 10.53. Inactivation of NFκB in innate immune pathways (Source: [1334]; Ag: antigen; BCL10: B-cell lymphoma adaptor-10; CARD: caspase recruitment domain-containing family member; Cyld: cylindromatosis tumor suppressor protein [deubiquitinase USPL2]; IKK: IκB kinase; IRAK: IL1 receptor-associated kinase; LCK: leukocyte-specific cytosolic Tyr kinase; LPS: lipopolysaccharide; MALT1: mucosa-associated lymphoid tissue lymphoma translocation paracaspase; MHC: major histocompatibility complex; MyD88: myeloid differentiation primary response gene product-88; OTUD7b: OTU domain-containing protein-7B, or cellular zinc finger anti-NFκB protein (Cezanne); PDK: phosphoinositide-dependent kinase; PI3K: phosphatidylinositol 3-kinase; PKC: protein kinase-C; PLC: phospholipase-C; RIPK: receptor-interacting protein kinase; TLR: Toll-like receptor; TNF: tumor-necrosis factor; TNFαIP: tumor-necrosis factor-α-induced protein; TRAF: tumor-necrosis factor receptor-associated factor; UbC: ubiquitin conjugase; USP: ubiquitin specific peptidase; ZAP70: ζ-associated protein 70). Once TNFR1, TLR4 and TCR–CD28 are activated, non-degradative Lys63-linked polyubiquitin chains are attached to RIPK1, TRAF6, and MALT1, respectively. These ubiquitinated mediators can be deubiquitinated by TNFαIP3, Cyld, OTUD7b, or USP21 enzyme. Paracaspase MALT1 cleaves (inactivates) TNFαIP3 enzyme. Deubiquitinase Cyld also processes Lys63-linked ubiquitined MAP3K7 kinase.

Trigger	Mediators	Target [K63-linked ubiquitination] (deubiquitinases)
TNFα–TNFR1	TRAF2/5, IAP1/2, UbC13, MAP3K7IP1–3, MAP3K7, IKK	RIPK1 (TNFαIP3 Cyld, OTUD7b, USP21)
LPS–TLR4	MyD88, IRAK1/4, MAP3K7IP1–3, MAP3K7, IKK	TRAF6 (TNFαIP3 Cyld)
[(Ag–MHC)–TCR]–[CD28–CD80/86]	LCK, ZAP70, PI3K, PDK1, PLCγ, PKCθ, CARD11, BCL10, TRAF2/6, MAP3K7IP1–3, IKK	MALT1 (TNFαIP3) MAP3K7 (Cyld)

domain-containing protein CCDC50, to form a ubiquitin-editing complex [1334]. Enzyme TNFαIP3 then processes (inactivates) RIPK1 via successive deubiquitinase and ligase activities: TNFαIP3 removes K63-linked ubiquitin chains on RIPK1 and adds K48-linked ubiquitin chains for degradation.

Early after lipopolysaccharide stimulation, TNFαIP3 disrupts UbC13–TRAF6 and UBE2d3–TRAF6 binding [1334]. In addition, it also degrades UbC13–TRAF2 and UbC13–IAP1 connections. Later (>4 h), it triggers the proteasomal degradation of UbC13 and UbE2d3 conjugases.

In addition, TNFαIP3 impedes signaling from the cytoplasmic pathogen recognition receptor nucleotide-binding oligomerization domain-containing protein NOD2, which recognizes bacterial wall component muramyl dipeptide, and activates NFκB, as it deubiquitinates RIPK2 kinase [1334].

Cylindromatosis (turban tumor syndrome) deubiquitinase counteracts TRAF2 activity. In lymphocytes, Cyld removes K63-linked polyubiquitin chains from MAP3K7 (inactivation), hence preventing spontaneous activation of NFκB [1334]. Enzyme TNFαIP3 preferentially cleaves K48-linked polyubiquitin chains; OTUD7b (or Cezanne) preferentially hydrolyzes K11-linked polyubiquitin chains; Cyld processes K63-linked polyubiquitin chains as well as linear ubiquitin chains [1334].

10.9.1.5 NFκB in Cell Fate

Nuclear factor-κB inhibits cell apoptosis induced by TNFα, as it upregulates the expression of anti-apoptotic genes such as caspase-8 and Fas receptor (TNFRSF6a)-associated death domain (FADD)-like apoptosis regulator (CFLAR),[195] certain BCL2 family proteins, and chromosome-X-linked inhibitor of apoptosis.

The mitogen-activated protein kinase module also regulates cell death and survival, particularly the JNK pathway. Jun N-terminal kinase is activated by MAP2K4 and MAP2K7, themselves activated by MAP3Ks (MAP3K1, MAP3K4, MAP3K5 (or apoptosis signal-regulating kinase ASK1), and MAP3K7 (or TGFβ-activated kinase TAK1).

Cytokines, such as TNFα and IL1, activate MAPK relatively quickly and transiently to foster cell survival. However, in the absence of CFLAR, TNFα causes a prolonged JNK activation and ROS accumulation.

A prolonged MAPK activation promotes cell apoptosis (Fig. 10.4). In the absence of NFκB, TNFα also favors ROS accumulation. On the other hand, in the presence of NFκB and long CFLAR form (CFLAR$_L$), TNFα upregulates the expression of anti-oxidants. Long CFLAR form interacts with MAP2K7 and inhibits interactions of MAP2K7 with MAP3K1, MAP3K5, and MAP3K7 enzymes. In fact, CFLAR$_L$

---

[195] A.k.a. caspase homolog (CasH), caspase-eight-related protein (Casper), caspase-8-associated protein Casp8AP1, mediator of receptor induced toxicity (MORT1)-associated cell death-3 (CeD3) homolog (MACH)-related inducer of toxicity (MRIT), usurpin (identified as a regulator of apoptosis initiated by TNFRSF6a that involves the recruitment, oligomerization, and autocatalytic activation of caspase-8), and cellular FADD-homologous interleukin-1β-converting enzyme (ICE)-like peptidase (FLICE)-inhibitory protein (CFLIP). Protein CFLAR interacts with caspase-3, -8, -10, FADD (or MORT1), death effector domain-containing DNA-binding proteins DEDD1 and DEDD2, cRaf, MAP3K14, RIPK1 and RIPK2, TNFRSF6, TNFRSF10a and TNFRSF10b, TRAF1, TRAF2 (the TRAF1–TRAF2 heterodimer transmits TNFα-mediated activation of JNK and NFκB and interacts with IAPs in the anti-apoptotic signaling from TNF receptors), TRAF3, polyubiquitin-C, Itch Ub ligase, P53, among others[251]. Protein CFLAR acts as an attenuator or initiator in apoptosis mediated by TNFRSF1a (TNFR1) and TNFRSF6a. The Cflar gene encodes 2 splicing variants, the long (CFLAR$_L$) and short (CFLAR$_S$) forms.

10.9 Transcription Factors Involved in Stress Responses

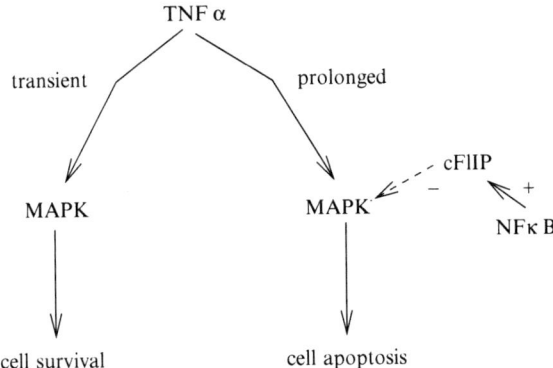

Fig. 10.4 Tumor-necrosis factor-α can support both cell survival and death. It stimulates MAP3Ks that phosphorylate MAP2K7, which, in turn, activates JNK enzyme. Nuclear factor-κB upregulates the expression of anti-oxidants, anti-apoptotic proteins, and cellular FLICE-inhibitory protein (cFLIP or CFLAR). Factor TNFα favors the interaction of CFLAR with MAP2K7, thereby inhibiting the prolonged activation of the JNK pathway.

binds to MAP2K7 and MAP2K1, thereby canceling JNK and ERK signaling [1346]. In summary, during transient activation of JNK activation, TNFα favors interaction of CFLAR with MAP2K7 enzyme; during prolonged JNK activation, TNFα promotes a quick degradation of $CFLAR_L$ via Itch ubiquitin ligase.

In disturbed endoplasmic reticulum during glucose starvation, perturbation in intracellular calcium stores, or inhibition of protein glycosylation, protein folding is compromised. An adaptive response is then triggered to avoid alteration of the processing of unfolded proteins in the endoplasmic reticulum. This reaction is mediated by 3 transmembrane proteins of the endoplasmic reticulum: (1) the protein Ser/Thr kinase and endoribonuclease Endoplasmic reticulum-to-nucleus signaling protein ERN1;[196] (2) eukaryotic translation initiation factor-2α kinase-3 (eIF2αK3);[197] and (3) ATF6 transcriptional activator. In disturbed endoplasmic reticulum, TNFR1 activates JNK downstream from IRE1 complexed to TNFR1 [1347, 1348].

Cellular inhibitors of apoptosis (IAP) promote cell survival and proliferation, especially of B lymphocytes, as they activate the non-canonical NFκB pathway. Isoforms IAP1 and IAP2 ubiquitinate tumor-necrosis factor-receptor associated factor TRAF3 in association with TRAF2 to stabilize MAP3K14 (Sect. 5.1.6) in response to cognate receptors bound to members of the tumor-necrosis factor superfamily, such as B-cell survival and maturation factors (e.g., TNFSF5 and TNFSF13b) and lymphorganogenic cytokines (e.g., TNFSF2–TNFSF3 heterotrimer).[198]

---

[196] A.k.a. inositol-requiring protein-1 (IRE1 or IRE1P).

[197] A.k.a. protein kinase-R (PKR)-like endoplasmic reticulum kinase (PERK).

[198] Members of the TNFSF superfamily signal via trimeric TNFRSF receptors and TRAF proteins.

In unstimulated cells, rapid turnover of MAP3K14 represses the non-canonical NFκB pathway, as Ub ligases IAP1 and IAP2 as well as TRAF2 and TRAF3 impede alternative NFκB signaling. Factors TRAF2 and TRAF3 play distinct roles in recruiting IAP1 and IAP2 to a regulatory complex that targets MAP3K14 for degradation. Agents TRAF2 and TRAF3 bridge MAP3K14 and IAPs for MAP3K14 ubiquitination and degradation in resting cells [1349, 1350].

On the other hand, activation of cognate receptors (e.g., TNFRSF5 and TNFRSF13c) disrupts the IAP1–IAP2–TRAF2–TRAF3–MAP3K14 complex and leads to TRAF3 degradation via TRAF2-dependent IAP-mediated ubiquitination. This process prevents ubiquitination of newly synthesized MAP3K14, thereby allowing MAP3K14 stabilization, accumulation, and activation by autophosphorylation. Stabilized MAP3K14 then activates IKKα and the non-canonical NFκB pathway.

NFκB and Angiogenesis

In vascular endothelial cells, mesodermal homeodomain-containing protein Meox2 binds to both RelA and IκBβ (but not IκBα) in the nucleus [1351]. Factor Meox2[199] stimulates and represses NFκB activity at low and high concentrations, respectively. This regulator operates not only on angiogenesis, but also inhibits responses of vascular endothelial cells to chronic inflammation.

10.9.1.6 NFκB in Inflammation and Immunity

Nuclear factor-κB regulates the development and activation of the innate and adaptive immune system. Target genes of the classical pathway encode mediators of inflammation, such as adhesion molecules, inflammatory cytokines (tumor-necrosis factor-α and interleukins IL1, IL6, and IL8), chemokines, peptidases, and apoptosis inhibitors.

In most cases, epithelial cells of infected tissue or tissue-resident hematopoietic cells (mastocytes or dendritic cells) initiate the inflammatory response by stimulating NFκB [1317]. NFκB also intervenes in the production of antimicrobial effectors and survival of leukocytes in inflammatory sites. Factor NFκB not only initiates inflammation, but can also prime the resolution of inflammation.

Nuclear factor-κB initiates both innate and adaptive immune responses. It orchestrates multiple mechanisms of inflammation, as it regulates transcriptional

---

[199] A.k.a. growth arrest-specific homeobox (GAx). It is encoded by the mesodermal homeobox MEOX2 gene both in vascular smooth muscle and endothelial cells. Its synthesis decays upon stimulation by pro-angiogenic and pro-inflammatory agents that activate NFκB factor. In quiescent vascular endothelial cells, it suppresses cell division by upregulating cyclin-dependent kinase inhibitor CKI1a and angiogenesis capacity.

10.9 Transcription Factors Involved in Stress Responses

programs in epithelial and stromal cells, vascular endothelial cells, and hematopoietic cells. Pro-inflammatory transcriptional programs regulated by NFκB differ according to stimulus and cell types that execute the inflammatory response [1317].

In endothelial cells, microbial lipopolysaccharide and tumor-necrosis factor stimulate 2 different sets of NFκB-activated genes. In addition, repeated lipopolysaccharide stimulation leads to LPS tolerance with hyporesponsive genes, as continuously expressed NFκB-regulated genes may be harmful, whereas other NFκB-regulated genes do not exhibit tolerance.

Nuclear factor-κB is activated by inflammatory cytokines (TNFα and interleukin-1) and cell stress (UV and γ-irradiation). Signaling by NFκB uses canonical and non-canonical pathways according to the stimulus type. These 2 pathways mediate different immune responses.

Both the classical and alternative pathways induce the release of NFκB from IκB. Crosstalk exists between inflammatory (canonical pathway) and developmental (non-canonical pathway) stimuli in the immune system owing to a fourth inhibitor of NFκB.

The classical pathway is particularly triggered by bacterial and viral infections and inflammatory cytokines. The alternative pathway is initiated mainly by members of the tumor-necrosis factor receptor superfamily (TNFRSF1b, TNFRSF3, TNFRSF5, and TNFRSF13c).

Tumor-necrosis factor receptor-1[200] (TNFRSF1a) trimerizes and complexes with several signaling adaptors, such as TNFR-associating factor-2 (TRAF2) and TNFR-interacting protein. Tumor-necrosis factor receptor-1 recruits receptor-interacting protein kinase-1 (RIPK1). Kinase RIPK1 is polyubiquitinated and targets IKKγ regulatory subunit of IκB kinase to activate NFκB [1352]. This complex then triggers also the MAPK pathway.

Two forms of trimeric tumor-necrosis factor-α, membrane-bound ($_m$TNFα) and soluble (TNFα^S) produced by $_m$TNFα cleavage by membrane-bound ADAM17 sheddase, bind to TNFR1 and TNFR2 receptors. Liganded trimeric TNFR1 recruits TNFR1-associated death domain-containing protein (TRADD) and triggers the formation of receptor complex-1 with TRAF2 or TRAF5, RIPK1, IAP1 and IAP2, MAP3K7 and IKK that activate NFκB [1326]. Subsequently, TRADD, TRAF2, and RIPK1 dissociate from TNFR1 and build complex-2 with FADD and procaspase-8. The latter matures and then cleaves (activates) caspase-3 that launches apoptosis. In most cells, caspase-8 activation is prevented by NFκB-mediated production of anti-apoptotic proteins such as FLIP and IAPs. On the other hand, apoptosis is primed by Itch and mitochondrial Diablo homolog[201] that link to CFLAR and IAPs, respectively, thereby labeling them for proteasomal degradation.

---

[200] Tumor-necrosis factor receptor-1 is the main member of the TNFRSF superfamily.

[201] Direct inhibitor of apoptosis protein (IAP)-binding protein with low pI (Diablo) is also named Second mitochondria-derived activator of caspase (SMAC). Inhibition of X-linked inhibitor of apoptosis (XIAP) results from binding to active Diablo once the latter is cleaved upon death stimuli and released into the cytosol.

The classical pathway triggers activation of inhibitor of NFκB kinase. The IKK complex phosphorylates IκBs for polyubiquitination and proteasomal degradation, thereby releasing NFκB components (predominantly Rel, RelA, and P50$_{NFκB}$). Released dimers (e.g., P50–RelA heterodimer) translocate to the nucleus, bind κB sites in promoters or enhancers of target genes to launch transcription.

In the presence of NFκB2, TNFRSF3 receptor can also activate the P50–RelA dimer using the canonical pathway [1353]. Inhibitor P100$_{NFκB}$ sequesters RelA–P50 dimers in the cytoplasm, thus impeding translocation to the nucleus. In addition, P100$_{NFκB}$ allows crosstalk between canonical and non-canonical signaling pathways. However, P100$_{NFκB}$ forms 2 different inhibitory complexes with RelA for pathway specificity (insulation). One pathway mediates developmental NFκB activation.

Developmental stimulus triggers the alternative pathway that involves MAP3K14 enzyme. The latter phosphorylates (activates) IKKα that processes NFκB2 subunit. The P52–RelB heterodimer is then released to enter the nucleus.

The adaptive immune response is initiated by Toll-like receptors and C-type lectins on dendritic cells that are bound to infectious agents to trigger cytokine production and stimulate T helper cell differentiation. In response to fungal infection, the C-type lectin dectin-1[202] bound to fungal components signals via spleen tyrosine kinase (SYK) and cRaf to activate RelA-associated, canonical and RelB-associated, non-canonical NFκB pathways with RelA-mediated transcription and T_{H1}- and T_{H17}-targeting cytokines, as well as formation of active P52–RelB dimers [1354]. However, cRaf kinase that phosphorylates RelA antagonizes SYK-induced RelB activation, as it supports RelB sequestration into inactive RelA–RelB dimers.

Nuclear akirins[203] are required for the transcription of a subset of NFκB-dependent genes. Two mammalian akirin genes exist: akirin-1 and -2. Akirins act downstream of Toll-like receptors and tumor-necrosis factor and IL1β pathways that lead to IL6 production in mice [1355].

Dysregulation of the transcription factor NFκB causes chronic inflammation, autoimmunity, and malignancy. Tumor-necrosis factor-α-induced protein TNFαIP3[204] that contains deubiquitinase and ubiquitin ligase domains is required with ubiquitin for NFκB clearance [1356].[205] Enzyme TNFαIP3 precludes ligase activities of TRAF2, TRAF6, and IAP1, as it antagonizes interactions with ubiquitin conjugases UbC13 and UbE2d3 (or UbCh5c). Interleukin-1 and lipopolysaccharides stimulate the association of Ub ligase TRAF6 with Ub conjugase UbC13

---

[202] Dectin-1 belongs to the family of pattern recognition receptors. It is able to recognize β-glycan carbohydrates on various fungi. This unique C-type lectin stimulates both canonical and non-canonical NFκB after ligand binding.

[203] Japanese hakkiri [saseru]: [making] things clear.

[204] A.k.a. A20 zinc finger protein.

[205] Enzyme TNFαIP3 with both ligase and deubiquitinase activities possesses a zinc finger domain for both ligase activity and recognition of Lys63 chains.

or UbE2d3. The resulting complex promotes TRAF6 autoubiquitination that then activates the IKK complex and NFκB factor. Transcription factor NFκB targets pro-inflammatory genes as well as the TNFAIP3 gene. Protein TNFαIP3 complexes with adaptor Tax1 (T-cell leukemia virus type-1)-binding protein Tax1BP1 and ubiquitin ligases Itch and RING finger protein RnF11 to remove ubiquitin chains from TRAF6, disrupt TRAF6 association with UbC13 or UbC5, thus terminating NFκB activation. In addition, TNFαIP3 modifies adaptor receptor-interacting protein RIPK1 in the TNFR pathway, as it facilitates removal of activating ubiquitin chains and addition of deactivating ubiquitin chains on another residue for proteasomal degradation, thereby preventing activation of the IKK complex.

10.9.1.7 NFκB in Hypercapnia

Many physiological gas — carbon mono- (CO) and dioxide (CO_2), hydrogen sulfide (H_2S), nitric oxide (NO), and oxygen (O_2) — are sensed by cells and triggers signaling. In particular, NO operates via guanylate cyclase and cytochrome-C oxidase and O_2 via hydroxylases.

Hypercapnia operates not only by associated acidosis, but also by disturbed carbon dioxide concentration, as CO_2 can act as an intracellular signaling messenger, especially in inflammation and immunity. Elevated arterial CO_2 levels (arterial CO_2 partial pressure p_{aO_2} > 45 mm Hg [>6 kPa]) result, in particular, from chronic obstructive pulmonary disease.

Permissive hypercapnia in patients in respiratory distress using ventilators lower their tidal volume to reduce mechanical ventilation, thereby protecting lungs against ventilator-associated mechanical damage of lungs as well as overachieved inflammation.

The immediate early transcription factor NFκB is involved in the regulation of inflammation. Agent RelB, a CO_2-sensitive REL family member, participates in CO_2-primed signaling. Signaling based on NFκB utilizes either activated IKKα–IKK/β–IKKγ heterodimer (canonical pathway) or activated IKKα homodimer (alternative pathway). In the nucleus, RelB can build various types of active and inactive homo- and heterodimers. Non-canonical pathway generates RelB–P52 heterodimer, once $P100_{NFκB}$ is processed by IKKα kinase. On the other hand, RelA–RelB heterodimer are transcriptionally inactive. Therefore, RelB cleavage by the proteasome ensures NFκB activity. Processing of RelB under stimuli other than hypercapnia can involve glycogen synthase kinase GSK3β, especially after stimulation by the T-cell coreceptor–receptor CD3–CD28 complex [1357], and mucosa-associated lymphoid tissue lymphoma translocation MALT1 peptidase. In immunocytes, proteasomal degradation of RelB can result from phosphorylation by GSK3β (Ser552) and N-terminal cleavage. Cleavage of RelB by MALT1 leads to RelB degradation and enhanced RelA- and Rel-dependent DNA binding in activated lymphocytes [1358].

In mouse embryonic fibroblasts and human pulmonary epithelial cells, hypercapnia can suppress NFκB-dependent transcription of pro-inflammatory genes, independently of changes in extracellular pH [1359]. Acidosis actually strengthens the transcription of several pro-inflammatory genes. In particular, acidosis raises synthesis of TNFα and COx2 proteins. Moreover, RelA is activated by acidosis, whereas it is inhibited by hypercapnia. On the other hand, CO_2 can provoke proteasomal-mediated cleavage of RelB to a low-molecular-weight form that translocates to the nucleus when CO_2 content rises above 0.03%, in a rapid and reversible manner [1359]. Hypercapnia also causes nuclear accumulation of IKKα that supports RelB activation.

10.9.1.8 NFκB in Hypoxia

Hypoxia activates IκB kinase-β, which degrades IκBα and releases NFκB transcription factor. IKKβ also amplifies cellular sensitivity to TNFα agent. IKKβ hydroxylation by prolyl hydroxylases,[206] mainly PHD1, withdraws the repression of NFκB [1360].

Nuclear factor-κB induces transcription of hypoxia-inducible transcription factor-1 in vivo, in both normal and hypoxic conditions [1361]. Hypoxia promotes nuclear accumulation of RelA subunit that binds to Hif1 promoter, thereby producing HIF1α in macrophages. Subunit RelA also facilitates basal Hif1 transcription in fibroblasts under normoxia. However in normoxic cells, Hif1 mRNA accumulates, but not HIF1 protein. In hypoxia, concomitant inhibition of prolyl hydroxylase domain-containing proteins that degrade HIF1 is required.

10.9.2 Hypoxia-Inducible Factor

Cells adapt to changes in oxygen availability. The cellular response to hypoxia involves 2 pathways associated with NFκB and hypoxia-inducible factor. Survival transcription factors stimulated by decreased oxygen levels are mediated by hypoxia-inducible transcription factor (HIF). Hypoxia-inducible factor binds to *hypoxia response elements* of hypoxia-inducible genes.

Hypoxia-inducible factor-1 regulates the transcription of many genes that produce proteins involved in: (1) energy metabolism, i.e., transcriptional regulation of numerous enzymes and carriers of the cell metabolism, especially glucose uptake (GluT1) and glycolysis (glycolytic enzymes); (2) pH regulation; (3) cell survival; (4) erythropoesis required for oxygen transport (erythropoietin); (5) vasodilation (nitric oxide synthase); and (6) angiogenesis (vascular endothelial growth factor).

---

[206]Three known prolyl hydroxylase domain-containing isoforms exist (PHD1–PHD3).

10.9.2.1 HIF Structure

Hypoxia-inducible factor is a heterodimer that consists of hypoxia-induced α- and constitutively expressed β subunit.[207] The concentration of HIFα subunit (but not that of HIFβ) is regulated by oxygen.

Three types of α (HIFα1–HIFα3) and β (ARNT1–ARNT3) subunits exist. Isoform HIFα1 is ubiquitous, whereas HIFα2 expression is restricted (glomerular endothelial cells, hepatocytes, and endothelial cells in the hippocampal region) [1362].[208] On the other hand, HIFα3 precludes gene transcription.

10.9.2.2 HIF Regulation

Ubiquitous HIFα1 is constitutively synthesized. Post-translational modifications of HIF comprise hydroxylation, phosphorylation, sumoylation, and ubiquitination. Proline hydroxylation (Pro564 and Pro402) regulates HIFα ubiquitination-dependent degradation using von Hippel-Lindau ubiquitin ligase complex.[209]

The activity of HIF1 is controlled by at least 2 complementary mechanisms: (1) HIF1 degradation by ubiquitination and proteasomal degradation via O_2-dependent prolyl hydroxylase domain-containing enzymes and (2) blockage of HIF1 transcriptional activity. Hypoxia prevents HIF1 activity, as it precludes these 2 mechanisms. On the other hand, during normoxia, HIFα1 is degraded.

Under hypoxia, hypoxia-inducible factor dimerizes. Dimer HIF1 is stabilized and translocates to the cell nucleus, where it triggers the transcription of hypoxia-sensitive genes.

Two oxygen-dependent repression mechanisms exist. Factor HIF1 is repressed by hydroxylation of HIFα1 catalyzed by 4 oxoglutarate dioxygenases: (1) 3 *prolyl hydroxylase domain-containing enzymes* (PHD1–PHD3) with cosubstrate oxygen to undergo ligase-dependent ubiquitination initiated by the binding of von Hipple-Lindau protein and subsequent proteasomal degradation; and (2) an *asparagine hydroxylase*, which inhibits HIF protein, hence its other name Factor inhibiting HIF1 (alias FIH1).

Prolyl hydroxylase domain-containing enzymes catalyze Pro hydroxylation. In fact, several PHD cofactors are necessary for HIF hydroxylation: 2-oxoglutarate,

---

[207] A.k.a. aryl hydrocarbon receptor nuclear translocator (ARNT) and class-E basic helix–loop–helix protein bHLHe2. Once the HIF1α–HIF1β (HIF1α–ANRT) dimer is formed, HIF becomes an active transcription factor that binds to the hypoxia response element of promoters of target genes.

[208] Some genes are specifically targeted by HIFα1 such as phosphoglycerate kinase PGK1, whereas HIFα2 specifically stimulates octamer-binding transcription factor Oct4 for stem cell maintenance [1362].

[209] The von Hippel-Lindau Ub ligase complex consists of von Hippel-Lindau protein, elongin-B and -C, cullin-2, and RING (really interesting new gene) box protein RBx1 (a.k.a. regulator of cullin-1 [ROC1] and RING finger protein RNF75).

Fe^{++}, and ascorbic acid, in addition to oxygen [1362]. Activity and cellular distribution vary according to the isozyme type. Enzyme PHD2 may have the dominant role, because the absence of PHD1 and PHD3 has no effect on HIFα1 stability. Production of PHD2 and PHD3 is upregulated by HIF1 (autoregulation). These enzymes may then rapidly terminate HIF1 action upon reoxygenation.

Ubiquitin ligase SIAH2 that possesses 2 isoforms (SIAH1–SIAH2) targets PHD3, thereby activating HIFα1, whereas FKBP8, an adaptor that mediates ubiquitin-independent proteasomal interaction, promotes PHD2 degradation [1362].

Protein asparagine (Asn) hydroxylation catalyzed by iron-dependent dioxygenase, or HIF asparagine hydroxylase, which operates on manifold proteins (HIF, IκB, etc.), prevents recruitment of coactivators and yields an oxygen-sensitive signal [1363]. A reduced hydroxylation rate during hypoxia allows HIF to activate transcription.

Oxygen-dependent regulation of the HIF1 response also targets the transactivation stage. Subunit HIFα1 can indeed interact with members of the CBP–P300 family (CBP: cAMP-responsive element-binding protein (CREB)-binding protein) of transcriptional coactivators. This interaction is also regulated by a hydroxylation-dependent switch. Asparagine hydroxylation of HIF1 actually prevents its interaction with CBP and P300 histone acetyltransferases that are necessary cofactors for HIF1-mediated gene transcription. Asparagyl hydroxylation is blocked by hypoxia.

Ubiquitous CBP–P300-interacting transactivator with ED-rich tail CITED2 also prevents the interaction of HIF1 with CBP and P300, as it competes with HIFα1 for binding to CBP and P300 cofactors. Protein CITED2 is activated by hypoxia via HIF1, thereby yielding an additional autoregulation that limits HIF1-dependent response to hypoxia and causes the rapid inactivation of the hypoxic response on reoxygenation.

10.9.2.3 HIF in Fetus and Newborn

Changes in oxygen availability at birth can trigger a transcriptional adaptation. The transition from intrauterine fetal to air-breathing newborn life, i.e., from a low to a higher oxygen environment, is indeed regulated by a transcriptional program. Synthesis of HIF1 and its activity that are indispensable in fetuses are rapidly impeded after the acute elevation in O_2 level at birth due to the new oxygen-rich environment.

In smooth muscle cells of fetal pulmonary artery, the concentration of the fetal HIF1 transcription factor does not depend on oxygen [1364]. On the other hand, in smooth muscle cells of adult pulmonary artery, hypoxia increases the production of oxygen-sensitive HIF1 transcription factor. However, hypoxia increases HIF1 mRNA production in smooth muscle cells of fetal (but not adult) pulmonary artery. Therefore, in the fetus, HIF1 concentration is O_2 insensitive, whereas HIF1 mRNA synthesis is sensitive to hypoxia. Hypoxia increases O_2-sensitive PHD2 production in smooth muscle cells of adult pulmonary artery. In smooth muscle cells of fetal pulmonary artery, neither PHD2 nor its mRNA expression change with hypoxia. Nevertheless, concentrations of PHD2 and PHD3 mRNA heigthen in late gestation fetus in comparison with those in adults to ensure sufficient quantity of PHD2 at birth. The fetal environment may be insufficiently hypoxic to elicit the hydroxylation that primes HIF1 degradation.

Other molecules control HIF1 expression and action in the fetus. Expressions of FIH1 asparagyl hydroxylase and CITED2 transactivator are developmentally regulated, as their levels are greater in the fetus than in adults.

10.9.2.4 HIF and Cell Differentiation

Factor HIF also controls cell differentiation. Isoforms HIF1 and HIF2 are required in cardiovascular and hematopoietic development. Hypoxia-inducible factor-1 is involved in many diseases, such as pulmonary hypertension, chronic obstructive lung disease, myocardial ischemia,[210] cancer, etc.

10.9.2.5 HIF and Cell Metabolism

Hypoxia-inducible factor operates in the transcriptional regulation of genes that encode enzymes of the glycolysis. In addition, HIF lowers mitochondrial oxygen consumption, as it supports pyruvate dehydrogenase kinase that inhibits the mitochondrial pyruvate dehydrogenase complex. It also regulates the expression of cytochrome-C oxidase subunit-4 isoforms to optimize the efficiency of cell respiration according to O_2 availability. Moreover, HIF can cause mitochondrial autophagy.

In humans with Chuvash polycythemia, an autosomal recessive disorder observed in Chuvashia (a region of European Russia) that results from homozygous germline mutation in exon 3 of the Vhl gene (alteration of HIFα subunit binding) is characterized by a reduced HIF degradation rate, hence augmented HIF concentration under normoxia. Symptoms comprise elevated hematocrit and hemoglobin levels, heightened pulmonary arterial blood pressures, but reduced systemic arterial pressures. Patients have an impaired adaptation to exercise and lower maximum exercise capacities [1366].

---

[210]Ischemia is caused by restricted regional flow. Factor HIF can promote cell survival during ischemia–reperfusion episodes via the regulation of cognate genes, in particular genes coding for glycolysis enzymes, anti-oxidant enzymes such as heme oxygenase-1, inducible nitric oxide synthase, cyclooxygenase-2, as well as vascular endothelial growth factor [1365]. Heme oxygenase-1 that is activated by HIFα1, degrades heme and generates carbon monoxide and bilirubin. Both can have anti-oxidant and anti-inflammatory effects, hence protecting against ischemia.

10.9.2.6 HIF and Blood Pressure

Normal development of the heart and vasculature as well as erythropoiesis depend on HIF1 transcriptional activator. Activity of HIF1α results from ischemia in the heart. It ensures ischemic preconditioning on the heart. Cardiomyocytes synthesize HIF1α to promote cardiac vascularization and contractility. However, cardiomyocyte-specific overexpression of HIF1α causes a progressive cardiomyopathy.

Factor HIF1α boosts VEGF activity to stimulate vascularization. In endothelial cells of cardiac blood vessels, HIF1 protects the heart and aorta to pressure overload, as it precludes TGFβ signaling [1367]. It avoids excessive activation of the ERK axis.

10.9.3 Forkhead Box Transcription Factors

Forkhead box[211] transcription factors (Fox) regulate cell proliferation and differentiation, from placental development to formation of the inner ear [1368].

The forkhead domain is a DNA-binding motif. In the absence of hormones and growth factors, FoxO N- and C-terminal non-phosphorylated PKB sites are masked. Under insulin stimulation, PKB phosphorylates the PKB site in the forkhead domain, thereby disrupting DNA binding and unmasking of the N- and C-terminal PKB sites. In addition, phosphorylation by PKB of the N-terminal PKB site creates a binding domain for 14-3-3 proteins, hence blocking the FoxO nuclear localization sequence and preventing DNA binding. On the other hand, phosphorylation of the C-terminal PKB site increases the export rate using Ran^{GTP} and exportin-1 due to the existence of a nuclear export sequence.

Forkhead box proteins can relieve chromatin compaction at target enhancer and/or promoter sites of DNA, thereby allowing other transcriptional activators to prime gene expression.

At least, 4 FOX genes, FOXC1, FOXC2, FOXP2, and FOXP3 cause human diseases when they experience mutations [1368].[212] The FOXO genes also contribute to cancer.[213]

---

[211] Numerous forkhead box (FOX) genes constitute the FOX subclass of Fox transcriptional regulators of the class 3 of helix–turn–helix transcription factors. This subclass is an evolutionarily ancient gene set that is named after the fork head gene in Drosophila melanogaster.

[212] Iris hypoplasia and Rieger syndrome are caused by FOXC1 gene mutations; autosomal dominant lymphedema–distichiasis syndrome by FOXC2 mutations; X-linked immunodysregulation, polyendocrinopathy, and enteropathy (IPEX) by FOXP3 mutations.

[213] Fusion proteins create artificial transcription factors. Fusion proteins of FOXO3 or FOXO4 with mixed lineage leukemia gene product are found in acute lymphocytic leukemia. Protein FoxM1 that has a pro-proliferative function, bears an augmentation of its synthesis in many cancers.

10.9.3.1 FOX Subclass

In mammals, Fox factors form a subclass of more than 40 proteins [1368]. The FOX subclass is subdivided into 19 families (FoxA–FoxS).[214] Forkhead domain-containing proteins within a family are often redundant to regulate embryo- and fetogenesis, but have distinct functions in adults to control metabolism, stress response, cell cycle, etc.

The functional diversity of Fox proteins partly relies on differences in partners, such as enzymes and cofactors [1368]. Functional diversity is also partially ensured by the spatiotemporal regulation of FOX genes.

10.9.3.2 FOX Subsets

FOXA Family

The FOXA family contains founding members of the FOX subclass. Members of the FOXA family operate in chromatin reorganization, cell metabolism, and organ development.[215] The FoxA proteins operate as pioneer transcription factors, as they interact with chromatin. In addition, they are binding partners of nuclear hormone receptors.[216]

Proteins FoxA1[217] and FoxA2 act in epithelial cell differentiation, branching morphogenesis, as well as development of the lung, liver, pancreas, and prostate, whereas FoxA3 intervenes in glucose homeostasis and response to starvation [1369]. Factors FoxA1 and FoxA2 cooperate during liver and lung morphogenesis.

FOXB Family

Factor FoxB1 contributes to development of neural tube and mammillary body nerve process [1369].

---

[214] In humans, the forkhead box gene subclass consists of 43 detected members (FOXA1–FOXA3, FOXB1, FOXC1–FOXC2, FOXD1–FOXD6, FOXE1–FOXE3, FOXF1–FOXF2, FOXG1, FOXH1, FOXI1, FOXJ1–FOXJ3, FOXK1–FOXK2, FOXL1–FOXL2, FOXM1, FOXN1–FOXN6, FOXO1–FOXO4, FOXP1–FOXP4, and FOXQ1). Members FOXN5 and FOXN6 are also called FOXR1 and FOXR2, respectively. Some phylogenetic analyses separate FOXL1 from FOXL2, FOXJ1 from FOXJ2 and FOXJ3, and FOXN1 and FOXN4 from FOXN2 and FOXN3 [1368].

[215] FoxA1 and -A2 bind to the enhancer region of the gene that encodes the pancreatic transcription factor PDX1 required in pancreas development in the embryo. In adults, endocrine pancreas needs only FoxA2 factor. Both FoxF1 and -F2 also intervene in gut development.

[216] The FoxA proteins interact with the androgen receptor. Factor FoxA2 promotes activity of glucocorticoid receptor.

[217] A.k.a. hepatocyte nuclear factor-3α.

FOXC Family

Factor FoxC1 participates in germ cell migration, mesenchymal cell differentiation, as well as development of brain, heart, kidney, skeleton, ureter, lacrimal gland, ovarian follicle, and tooth [1369].

Protein FoxC2 is expressed by epithelial cells that undergo an epithelial–mesenchymal transition. It intervenes in cell proliferation and development of heart, kidney, ureter, and skeleton [1369].

FOXD Family

Factor FoxD1 acts in axon guidance and kidney development; FoxD2 modulates cAMP sensitivity and operates in kidney development; and FoxD3 in trophectodermal cell differentiation and placenta development [1369].

FOXE Family

Protein FoxE1 and FoxE3 intervene in thyroid morphogenesis and eye development, respectively [1369].

FOXF Family

Factor FoxF1 is synthesized in endothelial and smooth muscle cells in embryonic and adult lungs. It acts in epithelial cell proliferation and development of smooth muscle, lung, colon, gall bladder, and mesenchyme. Factor FoxF2 operates in colon development [1369].

FOXG Family

Protein FoxG1 contributes to the development of the central nervous system [1369].

FOXH Family

Factor FoxH1 is a TGFβ and activin signaling mediator that operates in development of axial and prechordal mesoderm, definitive and visceral endoderm, notochord, and primitive streak [1369].

10.9 Transcription Factors Involved in Stress Responses   795

FOXI Family

Protein FoxI1 contributes to inner ear development [1369].

FOXJ Family

Protein FoxJ1 acts in T-lymphocyte proliferation [1370].

FOXK Family

Factor FoxK1 regulates cell cycle progression in myogenic progenitors [1370].

FOXL Family

Factors FoxL1 and FoxL2 are involved in proteoglycan biosynthesis and sex determination [1370]. The FOXL1 and FOXF2 genes are targeted by Hedgehog signaling.

FOXM Family

Protein FoxM1 participates in cell cycle progression (G1–S and G2–M transitions) [1370].

FOXN Family

Factor FoxN1 causes epithelial cell proliferation, keratinocyte differentiation, as well as development of hair follicle and thymus [1370]. Factor FoxN3 is involved in G2–M transition of the cell cycle.

FOXO Family

Members of the FoxO family mediate insulin signaling and glucose metabolism. They also intervene in the regulation of cell cycle arrest, DNA-damage repair, and cell death. In particular, FoxO1 and FoxO2 suppress cell proliferation and can induce apoptosis. The FoxO factors regulate cell cycle checkpoints at the G1–S and G2–M transitions. Target genes of FoxO factors that provoke cell cycle arrest include those that encode CDK inhibitors CKI1a and CKI1b in the presence of

TGFβ, Retinoblastoma-like protein RBL2, as well as cyclin-D1 and -D2. At the G2–M transition, they target cyclin-G2 and the stress sensor Growth arrest and DNA-damage-inducible GADD45 protein to trigger DNA-damage repair.

Factor FoxO1 acts in development of blood vessels and diaphragm. Factor FoxO3 operates in differentiation of erythrocytes. It suppresses spontaneous T-cell activation and autoimmunity. Protein FoxO3a participates in the elimination of damaged and old cells, hence avoiding occurrence of cancer. Factor FoxO4 intervenes in cell cycle progression [1370].

FOXP Family

Members of the FoxP family are involved in language acquisition. Factors FoxP1 and FoxP2 are also necessary for proper development of the brain and lung. Regulator FoxP1 controls differentiation of macrophages and cardiomyocytes. Agent FoxP2 operates in development of caudate nucleus, cerebellum, and putamen, as well as language acquisition. Protein FoxP3 regulates the development and function of regulatory T lymphocytes.[218] Factor FoxP4 participates in heart development [1370].

FOXQ Family

Protein FoxQ1 modulates natural killer cell function [1370].

10.9.3.3 Focus on Forkhead Box O Family

The FoxO factors (FoxO1 [FKHR], FoxO3 [FKHRL1], FoxO4, and FoxO6) exert positive and negative effects on gene expression, as they either bind to DNA target sites or interact with other transcription factors and coactivators. In particular, family-O forkhead domain-containing transcription factors regulates myocardium remodeling.

Cell Metabolism and Fate

Members of the FOXO family of forkhead box transcription factors are implicated in cell metabolism, in addition to cell fate. The FoxO proteins target genes involved in cell survival (TNFSF6, TGFβ2, etc.) and cell cycle (cyclins).

Transcription factors of the FOXO family controls insulin signaling and glucose and lipid metabolism (Sect. 11.8). They favor synthesis of enzymes of gluco-

---

[218]Protein FoxP3 serves as a marker of CD25+, CD4+ regulatory T cells.

10.9 Transcription Factors Involved in Stress Responses

neogenesis and suppress that of enzymes of glycolysis. Factor FoxO1 represses insulin-dependent signaling, but activates transcription of insulin receptor and insulin receptor substrate-2 (insulin sensitization).

Activity of FoxO family members is launched downstream from the insulin–PI3K–PKB pathway. Protein kinase-B is phosphorylated by phosphoinositide-dependent protein kinase PDK1 and target of rapamycin. Insulin and insulin-like growth factor IGF-1 inactivate FoxO via protein kinase-B, as phosphorylated FoxO in the nucleus relocalizes to the cytosol, where it interacts with 14-3-3 protein [1371]. Activator PKB and its effector FoxO are implicated in a negative feedback loop that reduces insulin sensitivity and glucose metabolism with diminished membrane translocation of GluT4, and insulin-triggered glucose uptake.

Defective PKB signaling contributes to development of insulin resistance. Sustained activation of either FoxO1 or FoxO3 in cardiomyocytes increases PKB phosphorylation [1372].[219]

Small RHEB GTPase regulates cell growth via the PI3K and TOR pathways. An increasing activity of the RHEB–TOR–S6K pathway sensitizes the cell to oxidative stresses; it is also implicated in decaying locomotor activity with aging [1373].

Both FoxOs and P53 regulate cell cycle and apoptosis. Deacetylase SIRT1 and USP7 deubiquinase affect FoxO and P53 in an opposite manner, whereas JNK stimulates both protein types under stress conditions [1371]. Enzyme SIRT1 stimulates FoxO agents, hence increasing cell life duration, and concomitantly inhibits P53, augmenting cancer risk. Conversely, USP7 inhibits FoxO proteins and stimulates P53. Factor P53 controls FoxO regulators, such as SGK, PI3K, and PTen enzymes.

Cyclin-dependent kinase CDK1 phosphorylates FoxO1, hence disrupting FoxO1 connection to 14-3-3 proteins. Factor FoxO1 then accumulates in the nucleus and activates FoxO1-dependent gene [1374]. In proliferating cells, CDK1 induces FoxO1 phosphorylation at the G2–M transition of the cell cycle, hence causing FoxO1-dependent expression of the mitotic regulator Polo-like kinase. On the other hand, cyclin-dependent kinase CDK2 phosphorylates FoxO1 for nuclear export, thereby impeding FoxO1-dependent transcription.

Cell Stress Responses

Forkhead box-O family transcription factors are also involved in stress responses, especially those primed by high concentrations of reactive oxygen species. They increase the production of several anti-oxidant enzymes. On the other hand, they provoke cell cycle arrest and entry into quiescence.

Phosphorylation by JNK, acetylation, and monoubiquitination of FoxO proteins can be caused by oxidative stress. In turn, modified FoxOs intervene in cell

---

[219]Forced expression of FoxO1 depends on the tissue type. Factor FoxO1 leads to increased phosphorylation of GSK3 in hepatocytes, but not in cardiomyocytes.

oxidative stress resistance. Oxidative stresses enhance FoxO binding to histone acetyltransferases, such as P300 and CBP, as well as lysine acetyltransferase KAT2b, and hence acetylation. Deacetylation of FoxO proteins can result from action of SIRT1 and other histone deacetylases.

Post-Translational Modifications

The FoxO transcriptional activity is regulated by reversible post-translational modifications (acetylation, phosphorylation, and ubiquitination). Phosphorylated FoxO proteins either inhibit via phosphatidylinositol 3-kinase and protein kinase-B or stimulate via Jun N-terminal kinase the transcription of target genes.

Both PI3K effectors, protein kinase-B and serum- and glucocorticoid-induced kinase (SGK), phosphorylate FoxO in the nucleus; FoxO is then exported to the cytoplasm (Fig. 10.5).[220]

FoxO signaling is modulated by protein Ser/Thr kinase CK1 and the Ras–Ral pathway [1375]. Factor FoxO is phosphorylated not only by CK1, but also by protein Ser/Thr kinase DYRK. Signaling via PI3K is counterbalanced by PTen phosphatase. Upon growth factor depletion, FoxOs reside predominantly in the cell nucleus, where their transcriptional activity can also be modulated.

Polyubiquitination leads to slow proteasomal degradation of FoxO factors; monoubiquitination stimulates, as it induces FoxO nuclear translocation. Acetylation may inhibit FoxO agents.

The FoxO factors prevent the activity of both protein phosphatases PP2 and PP3 that bind to and dephosphorylate (inactivate) PKB [1376].[221] Conversely, maintenance of a cytosolic pool of phosphorylated PKB resulting from reduced activity of phosphatases due to FoxO1 or FoxO3 enables FoxO phosphorylation by PKB (negative feedback). FoxO is thus able to bind and sequester its own PKB inhibitor. Therefore, sustained FoxO activation reduces insulin signaling in cardiomyocytes. In addition, FoxO can also affect insulin signaling via PKB-independent nodes, such as Ca^{++}–calmodulin-dependent protein kinase kinase and AMP-activated protein kinase by promoting proteasomal degradation of phosphorylated substrates due to the reduced phosphatase activity.

---

[220] The transcription factor FoxO3 is targeted by insulin and IGF1 signaling. In the absence of stimulation, FoxO localizes to the nucleus for transcription of target genes. Growth or survival signals activate PKB which phosphorylates FoxO. Moreover, PKB inhibits MAP3K5 and GSK3, which are implicated in cell apoptosis. Kinase SGK is related to PKB via PI3K and phosphoinositide-dependent kinase (PDK).

[221] Kinase PKB is also dephosphorylated by PP1 and PH domain and leucine rich repeat-containing protein phosphatase PHLPP1.

10.9 Transcription Factors Involved in Stress Responses

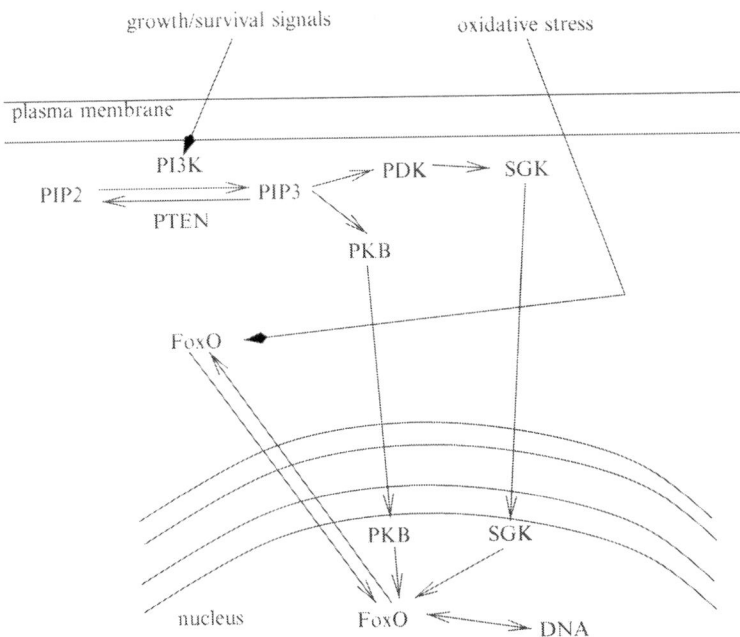

Fig. 10.5 Transcription factor FoxO and interactors. Factor FoxO is phosphorylated by several enzymes, especially protein kinase-B (PKB), and serum- and glucocorticoid-induced kinase (SGK). Both PKB and SGK depend on phosphoinositide 3-kinase (PI3K), which is counterbalanced by phosphatase and tensin homolog deleted on chromosome 10 (PTen). Phosphorylated FoxO is translocated to the cytosol. Growth and survival signals activate PKB kinase.

FoxO1

Like FoxO3a, FoxO4, and FoxO6, FoxO1 regulates cell survival, proliferation, differentiation, and metabolism.

FoxO3

In the presence of survival factors, protein kinase-B phosphorylates (inactivates) components of the apoptotic machinery (e.g., BAD and caspase-9) and regulates the activity of FoxO3 transcription factor [1377].

Activity of FoxO3 is regulated by post-translational modifications, such as acetylation, phosphorylation, and ubiquitination. Targets of FoxO3 include the genes that encode apoptosis regulators BCL2-like protein-11 (BCL2L11 or BIM) and BCL2 binding component BBC3,[222] as well as those of the cell cycle regulator CKI1b and of ROS metabolism (manganese superoxide dismutase and catalase), in addition to caveolin-1 [1378].

---

[222] A.k.a. P53-upregulated modulator of apoptosis (PUMA).

Partners

Factor FoxO3 is phosphorylated by PKB, SGK1, JNK, AMPK, and STK4 kinases. It interacts with 14-3-3 proteins, CBP and P300 acetyl transferases, sirtuins, and other transcrition factors, such as P53, MyC, and β-catenin [1378].

DNA replication

The binding partner of FoxO3 transcription factor, chromatin licensing and DNA replication factor CDT1 ensures a single DNA replication during each cell cycle. Stable maintenance of the genome relies on careful coordination of DNA replication, which is achieved by successive selection of sites of replication initiation (origin selection) and then activation of these sites to initiate replication (origin activation).

Origin selection for DNA replication relies on the assembly of a pre-replication complex in G1 phase. *Origin activation* can is triggered only in S phase by 2 kinases, Cell division cycle CDC7-related protein kinase and the activator of S-phase kinase DBF4 homolog on the one hand and cyclin-dependent kinase CDK2 targets replication factors.

During S phase, pre-replication complexes are sequentially processed by CDC7 and CDK2, thereby recruiting the DNA synthesis machinery (proteins required for helicase activation and replisome assembly) at sites of pre-replication complex formation and causing origin DNA unwinding and DNA synthesis.

Initiation factors of DNA replication are sequentially loaded on the pre-replication complex. These factors include Origin recognition complex (ORC), ORC- and origin DNA-dependent Cell division control protein-6 homolog (CDC6 or CDC18L), CDT1, and heterohexameric MCM2-7 replicative DNA helicase. Some of these factors are responsible for the correct temporospatial regulation of the formation of the pre-replication complex. Agent ORC first binds to origin DNA sites and then recruits CDC6 and CDT1 proteins. Together, they cooperate to load MCM2-7 onto origin DNA sites in an ATP-dependent manner.

At least 10 among 14 constituents of the pre-replication complex are members of the set of ATPases associated with diverse cellular activities (AAA)[1379]. Components ORC, CDC6, and all 6 MCM2-7 subunits possess ATP-binding motifs. ATP binding by ORC is required for origin DNA binding; DNA binding precludes ATP hydrolysis by ORC agent. Protein ORC^{ATP} enables initial MCM2-7 loading [1379]. However, ATP hydrolysis by ORC enables multiple rounds of MCM2-7 loading. On the other hand, earlier ATP hydrolysis by CDC6 ATPase destabilizes CDT1 link to origin DNA and allows MCM2-7 loading [1379].

10.9 Transcription Factors Involved in Stress Responses

Pre-replication complexes are dismantled after DNA replication to avoid another duplication. Component CDC6 is eliminated from the nucleus at the onset of S phase. Protein CDT1 is sequestered by the nuclear inhibitor of DNA replication geminin, which appears and accumulates in S, G2, and M phases of the cell cycle and drops at the metaphase–anaphase transition of mitosis. Kinase CDK2 phosphorylates CDT1, thereby eliciting its ubiquitination by the SCF–SKP2 complex for degradation during S phase. Moreover, CDT1 interacts with homotrimeric Proliferating cell nuclear antigen (PCNA), a processivity cofactor for DNA polymerase-δ. The PCNA–DDB1–cullin-4 complex (DDB: damage-specific DNA binding protein) also degrades CDT1 during S phase. In addition, adaptor CDC20 homolog CDH1, an activator of Ub ligase anaphase-promoting complex (or cyclosome) forms the APC–CDH1 that can label CDT1 for degradation.

Repression of re-establishment of the pre-replication complex during the S, G2, and M phases depends in particular on replication licensing factor CDT1. Factor FoxO3 complexes with CDT1, hence stabilizing CDT1 levels, as it hinders CDT1 interaction with DDB1 and PCNA agents [1380].

Cardiomyocyte Atrophy

Factor FoxO3a elicits cardiomyocyte atrophy [1381]. The PKB–FoxO pathway regulates the expression of multiple atrophy-related genes such as ubiquitin ligase atrogin-1[223] Constitutively active PKB that regulates cardiomyocyte survival and growth, as well as its contractile function, suppresses atrogin-1 expression. In contrast, FoxO3a stimulates atrogin-1 expression and thereby impedes cardiomyocyte hypertrophy. Insulin and insulin-like growth factor IGF1 prime the PI3K–PKB pathway and inhibit FoxO3a subtype.

Vitamin-D and FoxO3a Activity

The hormonal form [(1,25)-dihydroxyvitamin-D] of vitamin-D, a prohormone,[224] acts via its nuclear receptor NR1i1, or nuclear vitamin-D receptor (VDR). The latter heterodimerizes with related retinoid X receptors to connect to vitamin-D response

---

[223] A.k.a. F-box only protein FBxO32.

[224] Sources of vitamin-D encompass diet and vitamin-D3 synthesis under sun exposure, i.e., photoconversion stimulated by ultraviolet-B (315–280 nm) of 7-dehydrocholesterol in skin. In the liver, the major circulating, relatively long-lived metabolite 25-hydroxyvitamin-D (25(OH)vitamin-D) results from hydroxylation by members of the cytochrome-P450 superfamily (CyP27a1 and CyP2r1). In the kidney and other peripheral organs, 25(OH)vitamin-D is hydroxylated to produce hormonal (1,25)-dihydroxyvitamin-D [(1,25)-(OH)$_2$ vitamin-D].

elements (VDRE) of target genes (Vol. 3 – Chap. 6. Receptors). Agent (1,25)-dihydroxyvitamin-D supports VDR and FoxO3a binding to promoters of target genes [1382]. Liganded VDR complexes with FoxO, class-3 histone deacetylase sirtuin-1, and protein phosphatase-1. Agent (1,25)-(OH)$_2$vitamin-D rapidly (<4 h) induces FoxO deacetylation and dephosphorylation. It indeed promotes FoxO dephosphorylation by protein phosphatase-1 following its deacetylation by sirtuin-1. Whereas FoxO proteins do not influence regulation by (1,25)-(OH)$_2$vitamin-D of VDR-specific target genes, it reduces that of genes targeted by both VDR and FoxO transcription factors. Consequently, FoxO3a can repress cell proliferation, thereby operating as a cancer chemopreventive agent. Similarly to FoxO proteins, (1,25)-(OH)$_2$vitamin-D represses expression of the CCND2 gene that encodes cyclin-D2, but stimulates that of the CCNG2 gene, which encodes the inhibitory cyclin-G2 [1382].

FoxO4

Similarly to FoxO3a, ligand-bound VDR enhances binding of FoxO4 within 4 h to promoters of FoxO target genes and blocks mitogen-induced FoxO nuclear export [1382].

FoxO6

In the developing brain, FoxO6 is expressed with a specific temporospatial pattern [1383]. In adults, FoxO6 production is maintained in some areas of the central nervous system. Factor FoxO6 localizes mainly in the nucleus, especially after growth factor stimulation, whereas FoxO1 and FoxO3 lodge predominantly in the cytosol [1383]. Nuclear FoxO6 export results from the activity of the PI3K–PKB pathway.

Whereas FoxO1, FoxO3, and FoxO4 possess an N- and C-terminal PKB motif, FoxO6 lacks the C-terminal PKB motif, hence its defective nucleocytoplasmic shuttling [1383]. In addition, a PKB motif exists in the forkhead domain (Ser184) that regulates DNA binding. Inducible FoxO6 transcriptional activity is precluded by growth factors, The N-terminal PKB motif acts as a growth factor sensor.

Activity of FoxO6 augments hepatic gluconeogenesis, which maintains the blood glucose level within the physiological range and yields a fuel source for the brain, thereby counteracting insulin [1385]. Conversely, insulin, the action of which on hepatic gluconeogenesis partly relies on FoxO1, impedes FoxO6 activity.

10.9 Transcription Factors Involved in Stress Responses

10.9.4 P53 Family of Transcription Factors

Members of the P53 family of transcription factors (P53, P63, and P73), also called tumor suppressors (TP53, TP63, and TP73) are encoded by the tp53, tp63, and tp73 genes, respectively. They serve as hubs of numerous signaling pathways triggered by multiple cellular stresses. Transcription factor P73 is closely and more distantly related to P63 and P53 transcription factors, respectively. Tumor suppressors: (1) initiate cell cycle arrest, (2) participate in DNA repair, and (3) can cause apoptosis.

All these transcription factors have an N-terminal transcriptional transactivation domain, a DNA-binding motif, a tetramerization segment, and a C-terminal regulatory region [1386]. Members of the P53 family tetramerize to control the cell cycle and tissue development. These homotetramers correspond to dimer of primary dimers [1386].

Members of the P53 family also exist in different N- and C-terminally truncated isoforms (e.g., $_C$P53 and $_N$P53). Factors P63 and P73 have an additional sterile α motif that confers stability.

Factor P53 intervenes in somatic cells in response to many types of cellular stresses such as DNA damage. It initiates cell cycle arrest, senescence, differentiation, or apoptosis. It is inactivated by mutations of its gene in approximately 50% of cancers in humans.

Factors P63 and P73, in addition to their role in development and differentiation, can also act as tumor suppressors, like P53 paralog. According to cellular context, they exert their function either in concert with or independently of P53 factor [1386].

Factor P63 is one of the major transcription factors of epithelium development. Factor P73 participates in the development of the central nervous and immune systems. It also responds to certain stresses that stimulate P53 to prime apoptosis.

10.9.4.1 Transcription Factor P53

Transcription factor P53 regulates transcription of genes with P53-response element-containing promoters. It thus contributes to the regulation of cell growth, senescence, and apoptosis, as well as glycolysis, development, angiogenesis, and fertility.

Protein P53 is the most frequently disturbed tumor suppressor gene product in cancers. Furthermore, most of the remaining malignancies deactivate the P53 pathway, as they stimulate its inhibitors, repress its activators, or inactivate its downstream targets.

Transcription factor P53 accumulates in response to DNA damage, oncogene activation, and hypoxia, among other stresses. Upon stress exposure, P53 is activated (Table 10.54).

Table 10.54. Stages of P53 activation (Source: [1390]). Factor P53 inhibits various genes such as Myc and activates others. Activation of P53 is caused by various stresses, such as telomere shortening, hypoxia, mitotic spindle damage, heat or cold shock, unfolded proteins, improper ribosomal biogenesis, nutritional deprivation, and activation of some oncogenes by mutation. The P53 pathway is controlled by positive and negative feedback loops. Activated P53 has 2 main outcomes, either cell death via apoptosis or senescence, or cell survival via cell cycle arrest. Cell fate is selected according to multiple variables. Other cellular processes are targeted by P53: autophagy, DNA repair, regulation of protein translation, signal transduction, cell cytoskeleton, extracellular matrix, and exosome and endosome activity. Gene are activated via P53 binding and cofactor recruitment to the promoter–enhancer region of P53-regulated genes. Genes are repressed by direct and indirect mechanisms.

Step 1	P53 phosphorylation and inhibition of DM2 ubiquination
Step 2	P53 accumulation
Step 3	P53 modifications (stabilization) by acetyltransferases (CBP, P300, PCAF) and methyltransferases (SET9)
Step 4	Possible inhibition of P53 binding to DNA by deacetylase HDAC2
Step 5	P53 tetramer binding to P53 response element
Step 6	Recruitment of cofactors, such as histone acetyltransferases (HAT) and TATA-binding protein-associated factors (TAF)
Step 7	Gene activation or repression

Factor P53 is modified by: (1) phosphorylation upon DNA damage by ataxia telangiectasia mutated kinase (ATMK) and ataxia telangiectasia and Rad3-related protein complex (ATRK) [1387]; (2) lysine acetylation and methylation [1388]; and (3) arginine methylation by protein arginine methyltransferase PRMT5 that favors cell cycle arrest rather than death [1389].

In a injured cell, P53 supports: (1) cell cycle arrest via cyclin-dependent kinase inhibitor CKI1a, growth arrest and DNA-damage-induced protein GADD45, Reprimo, and 14-3-3σ or (2) cell apoptosis via BCL2-associated X protein BAX, serine peptidase inhibitor serpin-F1,[225] BCL2-binding component BBC3,[226] and leucine-rich repeats and death domain-containing protein (LRDD).[227]

Action of P53 is initiated upon repression of inhibitors of P53 (e.g., DM2 and DM4), according to the type and severity of damage as well as the cell environment [1391]. Disruption of interactions between P53 and its inhibitors on cellular stresses relies on post-translational modifications as well as sequestration and degradation of its inhibitors.

The majority of gene transcription relies on a subset of general transcription factors that construct the RNA polymerase-2 pre-initiation complex, such as

---

[225] A.k.a. pigment epithelium-derived factor PEDF and Pig3.
[226] A.k.a. P53-upregulated modulator of apoptosis (PUMA).
[227] A.k.a. P53-induced protein with a death domain (PIDD).

10.9 Transcription Factors Involved in Stress Responses

the transcriptional coactivator Mediator complex and TF2h transcription factor. The Mediator complex, TF2h, and RNA polymerase-2 works synergistically with activators to initiate gene transcription.

The Mediator complex is a major regulator of the pre-elongation complex, which includes numerous Mediator of RNA polymerase-2 transcription subunits (Med1, Med4, Med6–Med31, CcnC, CDK8, and CDK19) interacts with transcription initiation factor TF2 with its subunits (TF2a, TF2b, and TF2d–TF2h), and RNA polymerase-2. Different P53 domains tether to different Mediator subunits. Distinct resulting P53–Mediator structures differentially influence RNA polymerase-2 activity and coordinate activation of TF2h and RNA polymerase-2 in the pre-elongation complex [1392].

In addition to its nuclear activity devoted to gene transcription using its DNA-binding domain, P53 also has cytosolic, transcription-independent activities. Certain aspects of its apoptotic function indeed depend neither on transcription nor translation.

Factor P53 shuttles between the cytoplasm and nucleus. It operates in both cellular compartments (Table 10.55). Polyadpribosylation of P53 leads to its nuclear accumulation, whereas monoubiquitination by DM2 stimulates its nuclear export.

The P53–DM2 feedback loop is an important axis in the response of cells to damage that exhibits sustained oscillations upon DNA double-stranded breaks.[228] Fourier spectrum of oscillation time courses of P53 and DM2 concentrations. displays distinct low-frequency components typical of a third-order linear model with white noise [1394].

P53 Partners

Activity of transcription factor P53 is controlled by: (1) multiple interactors that cause post-translational modifications to regulate P53 stability, DNA binding, and transcription activation; (2) cofactors that participate in the recruitment of P53 to specific promoters; (3) regulators of transcription-independent P53 functions; as well as (4) its subcellular location.

Transcription factor P53 is a short-lived protein (half-life 6–20 mn), as it is quickly degraded by the proteasome under normal conditions. Numerous pathways control P53 via proteolysis, transcriptional regulation, localization, post-translational modifications, and cofactor binding.

Ubiquitin Ligases

Ubiquitination of P53 can heighten its nuclear transcriptional activity. On the other hand, P53 undergoes both multiple monoubiquitination that enhances P53 nuclear export and polyubiquitination that primes P53 proteasomal degradation in the cytoplasm in the absence of stress.

---

[228]Concentration of P53 rises and falls, with a period of 6 to 7 h; concentration of DM2 oscillates out of phase with that of P53 factor.

Table 10.55. Cytoplasmic, non-transcriptional and nuclear, transcriptional effects of transcription factor P53 (Source: [1393]; APAF: apoptotic peptidase-activating factor; BAX: BCL2-associated X protein; CDKN1A: cyclin-dependent kinase inhibitor-1A (CKI1a) gene; DRAM: damage-regulated autophagy modulator, a lysosomal protein that contributes to autophagosome accumulation in response to activated P53 as well as apoptosis in response to activated P53; DR5: death receptor-5 (a.k.a. P53-regulated DNA-damage-inducible cell death receptor Killer, tumor-necrosis factor receptor superfamily member TNFRSf10b, and TNF-related apoptosis-inducing ligand receptor TRAIL2); GADD: growth arrest and DNA-damage-induced protein; Gpx1: selenium-containing glutathione peroxidase-1 that operates in the detoxification of hydrogen peroxide; PERP: P53 apoptosis effector-related to peripheral myelin protein PMP22; PIDD: P53-induced protein with a death domain protein (a.k.a. leucine-rich repeats and death domain-containing protein [LRDD]) that interacts with adaptor Fas receptor (a.k.a. CD95 and TNFRSF6a)-associated with death domain protein (FADD), which connects Fas receptor and other death receptors to caspase-8 to form the death-inducing signaling complex (DISC) during apoptosis; PUMA: P53-upregulated modulator of apoptosis; P53AIP: P53-regulated apoptosis-inducing protein; SCO2: synthesis of cytochrome-C oxidase that pumps protons across the inner mitochondrial membrane; TIGAR: TP53-inducible glycolysis and apoptosis regulator, a fructose-2,6-bisphosphatase regulated by P53 that precludes glycolysis to favor the pentose phosphate shunt; TP53I3: tumor protein P53-inducible protein-3, a quinone oxidoreductase).

Cytoplasmic effects	Nuclear effects
Centrosome duplication Mitochondrial outer membrane permeabilization (apoptosis) Autophagy inhibition	Cell cycle arrest (CDKN1A, 14-3-3σ, Reprimo, Gadd45 genes) Apoptosis stimulation (Apaf1, Bax, Dram, Gadd45, NOXA, Perp, Pidd, Puma, P53Aip1, Tnfrsf6a, Tnfrsf10b, TP53I3 genes) Autophagy inhibition (Dram, SESTRIN1/2 genes) Anti-oxidant (GPX1, SESTRIN1/2 genes) Metabolism (Sco2, Tigar genes)

Double Minute-2

Concentration of P53 is mainly controlled by ubiquitination via Ub ligase double minute DM2 for proteasomal degradation, when P53 has already been ubiquitinated. This P53 inhibitor is upregulated by P53 itself. Activity of P53 is regulated by ubiquitin ligase DM2 in synergy with transcriptional coactivators CBP and P300 that act as histone acetyltransferases, polyubiquitin ligases, and adaptors. Cytoplasmic (but not nuclear) CBP and P300 regulate P53 via their acetylase and polyubiquitin ligase function [1395]. Enzyme DM2 associates with several subunits of the 19S proteasome regulatory particle in a ubiquitination-independent manner [1396].[229] It promotes the formation of a ternary proteasome–P53–DM2 complex.

Upon stress, DM2 increases the abundance of polyubiquitinated P53 adducts. In unstressed conditions, specific polyubiquitin ligases or ubiquitin chain-extending factors such as P300 target P53 factor. Polyubiquitination of P53 that depends on P300 occurs after multiple monoubiquitination by DM2 ligase.

Factor P53 can bind simultaneously to DM2 and to CBP or P300 histone acetyltransferases to form a ternary complex. In unstressed cells, the P53–DM2–CBP(P300) complex promotes P53 polyubiquitination and degradation. In stressed cells, DM2–P53 interaction is prevented by DM2 phosphorylation (Ser395 and Tyr394) by ATMK and Abl kinases (respectively) [1391]. Furthermore, P53 phosphorylation (e.g., Thr18) impedes binding to DM2, but favors P53 linkage to CBP [1397]. Phosphorylation of P53 (Ser15 and Ser20 in its transactivation domain) by stress-induced kinases, such as ATMK, ATRK, checkpoint kinases ChK1 and ChK2, and DNAPK, stabilizes P53. In vivo, phosphorylation may prime a series of post-translational events, as alone it does not suffice for P53 activation. Phosphorylated P53 binds preferentially to CBP or P300 to be stabilized and activated. Acetylation of DM2 by CBP or P300 disrupts the P53–DM2 interaction.

Ligase DM2 possesses inherent self-ubiquitination activity that is inhibited in the absence of cellular stress by Ubiquitin-specific peptidase USP7,[230] in synergy with death domain-associated protein DAP6[231] and Ras association (RalGDS/AF6) domain-containing family member RASSF1a [1391]. Deubiquitase USP7 deubiquitinates both DM2 and P53 in a mutually exclusive manner.

Ubiquitous transcriptional repressor or activator YY1 of the Gli-Krüppel infraclass of zinc finger proteins stabilizes DM2–P53 complexes and allows DM2 to processively elongate ubiquitin chain.

Ataxia telangiectasia mutated kinase phosphorylates DM2 (Ser395), thereby lowering P53 degradation in response to DNA damage. Conversely, magnesium-dependent protein phosphatase-1D dephosphorylates DM2 (Ser395) to heighten P53 degradation.

Protein kinase-B phosphorylates (Ser166 and Ser188) and stabilizes DM2 that can then translocate from the cytoplasm into the nucleus. Kinase Abl also phosphorylates DM2 (Tyr276 and Tyr394) after DNA damage to prevent DM2-mediated P53 degradation [1398].

---

[229] Enzyme DM2 tethers to the 19S subunit of the proteasome as well as 2 other ubiquitin ligases, Casitas B-lineage lymphoma adaptor (CBL) and Seven in absentia homolog SIAH1. Other ubiquitin ligases also link to the proteasome, such as von Hippel-Lindau protein involved in the ubiquitination and degradation of hypoxia-inducible factor, RING finger protein RNF2 that targets transcription factor TFCP2, and Parkin (or parkinson protein Park2) that degrades proteins toxic to dopaminergic neurons.

[230] A.k.a herpes virus-associated ubiquitin-specific peptidase (HAUSP).

[231] A.k.a. ETS1-associated protein EAP1 and Daxx.

Nutlins (nutlin-1–nutlin-3) inhibit DM2, as they prevent the formation of the P53–DM2 complex that inactivates P53 factor. Cell senescence caused by nutlin-3a depends on P53 protein. Moreover, nutlin-3a upregulates the expression of numerous P53-dependent genes, particularly those that encode miR34a to -34c [1399]. These microRNAs are effectors of P53 for cell senescence. Nutlin-3a has about 150 times greater affinity for DM2 than nutlin-3b. Nutlin-3b has, indeed, very low effect on cell senescence. Inhibitor of growth InG2 decreases nutlin-3a effect. On the other hand, P53 represses InG2 expression, as it binds to ING2 promoter.

Other Types of P53–Ubiquitin Ligases

Several other ubiquitin ligases, such as HECT, UBA, and WWE domain-containing ligase HUWE1, RING finger protein RNF199, which preferentially ubiquitinates transcriptionally active P53 tetramer, and male-specific lethal-2 homolog (MSL2), can ubiquitinate P53 and cause its degradation or translocation [1391]. Many ubiquitin ligases promote ubiquitin-mediated P53 degradation [1398] (Table 10.56).

RING finger proteins RNF34 and RNF34L can ubiquitinate phosphorylated P53 (Ser20) after DNA damage.

Protein RNF199 interacts with calmodulin. It is phosphorylated (inactivated) by calmodulin-dependent kinase-2 (Thr154 and Ser155) to attenuate P53 degradation.

Phosphorylation (Ser387) by ATMK of ubiquitin ligase RNF200 after DNA damage causes dissociation of RNF200 from P53.

Ubiquitin ligase HUWE1 targets the stress-induced tumor suppressor CKI2a, a regulator of P53 stabilization, as it sequesters DM2 in the nucleolus and inhibits its ubiquitin ligase activity. Protein HUWE1 also ubiquitinate DNA polymerase-β, in addition to P53 factor.

Synoviolin is a Ub ligase implicated in endoplasmic reticulum-associated degradation that also targets P53, at least in house mouse (Mus musculus).

The ubiquitin–protein ligase Topoisomerase-1-binding, arginine (R)–serine (S)-rich protein (TopoRS or simply Topors) contains 2 bipartite nuclear localization signals. It is widespread. This P53 binding partner is especially detected in the alveolar epithelium.

Protein TriM24 belongs to the tripartite motif protein family (TRIM) of transcriptional repressors that bind to chromatin. It interacts with retinoic acid receptors to regulate their transcriptional activity. It also acts as a ubiquitin ligase for P53 [1400].

Atypical ubiquitin ligase E4F1, a transcriptional coregulator, ubiquitinates chromatin-associated P53 and modulates its activity independently of degradation. Ubiquitous zinc finger protein E4F1 stimulates oligoubiquitination of P53 on lysine residues distinct from those targeted by DM2 that are acetylated by K (lysine) acetyltransferase KAT2b that also acts as a Ub ligase. Hence, E4F1 and PCAF mediate mutually exclusive post-translational modifications of P53 [1401]. Regulator E4F1 modulates P53 activity in alternative cell fates: growth arrest or apoptosis.

10.9 Transcription Factors Involved in Stress Responses

Table 10.56. Ubiquitin ligases that target P53 for degradation (Source: [1398]; NA: not applicable; ARFBP1: ARF [CKI2a] tumor suppressor-binding protein-1; ARNIP: androgen receptor N-terminus-interacting protein; CARP: cell cycle and apoptosis regulatory protein; CoP1: constitutive photomorphogenesis protein; E4F1: E4F factor-1, also a transcription coactivator and corepressor; DM: double minute; HECTh9: Homologous to E6 accessory protein (AP) C-terminus; HRD: hydroxymethylglutaryl reductase degradation protein; HUWE: HECT, UBA, and WWE domain-containing ligase; MSL: male-specific Lethal-2 homolog; MULE: Mcl1 ubiquitin ligase-E3; P53BP: P53-binding protein; PCAF: P300–CBP-associated factor; PIRH: P53-induced RING-H2 protein; RCHY: RING finger and CHY zinc finger domain-containing protein; RFWD: RING finger and WD repeat domain-containing protein; RIFF: FYVE zinc finger and RING finger domain-containing protein; RFFL: RING finger and FYVE-like domain-containing protein; RNF: RING finger protein; RP: retinitis pigmentosa; synoviolin: synovial apoptosis inhibitor; TIF: transcription intermediary factor; Topors: topoisomerase-1-binding protein; TP53BPL: tumor protein P53-binding protein-like; TriM: tripartite motif-containing protein; UREB: Upstream regulatory element-binding protein; ZNF: zinc finger protein). The HERC family encompasses ubiquitin ligases that possess a HECT domain that is involved in the transfer of ubiquitin from a conjugase to the substrate. Other ubiquitin ligases have a RING (really interesting new gene) domain that binds ubiquitin conjugase.

Ligase	Type	Other aliases	Regulator(s)
DM2	RING		Abl, ATMK, PKB, PPM1d, USP7 HUWE1
HUWE1	HECT	ARFBP1, HECTh9, MULE, UREB1	
MLS2			
RNF199	RING	ARNIP, PIRH2, RCHY1, ZNF363	CamK2
RNF200	RING	CoP1, RFWD2	ATMK
RNF34	RING	CARP2, RIFF	
RNF34L	RING	CARP1, RFFL1	
Synoviolin	RING	HRD1	
Topors	RING	P53BP3, RP31, TP53BPL	
TriM24	RING	TIF1α	
E4F1	NA		
KAT2b	NA	PCAF	

On the other hand, ubiquitin-specific peptidase USP7[232] is a deubiquitinase that cleaves ubiquitin from its substrates. It particularly antagonizes DM2 enzyme.

Tenovins

Tenovins activate the tumor-suppressive function of P53 by inhibiting sirtuin-1 and -2 [1393]. Deleted in bladder cancer protein DBC1 (FAM5A) inhibits sirtuin-1, hence acting as a P53 activator.

---

[232] A.k.a. herpesvirus-associated ubiquitin-specific peptidase (HAUSP).

JMY Protein

Transcription cofactor junction-mediating and regulatory protein JMY regulates the P53 response. In addition, JMY binds monomeric actin and facilitates actin nucleation (Vol. 1 – Chap. 6. Cell Cytoskeleton). Actin reorganizes in response to DNA damage. Nuclear actin and actin-related proteins actually contribute to chromatin remodeling and transcription by RNA polymerase-1, -2, and -3 [1402]. Upon DNA damage, JMY, which is protected against DM2 Ub ligase,[233] accumulates in the nucleus[234] and complexes with stress-responsive activator of P300, P300, and protein arginine methyltransferase PRMT5 to regulate P53 transcriptional activity [1402].

RSK2 and MSK2 Kinases (RPS6Kα3 and RPS6Kα4)

P90 ribosomal S6 kinase RSK2 (a.k.a. RPS6Kα3 and MAPKAPK1b) that colocalizes with P53 in the nucleus, phosphorylates (activates) P53 (Ser15) [1403]. On the other hand, mitogen- and stress-activated kinase MSK2 (a.k.a. RPS6Kα4 and RSKb) inhibits P53 [1404]. Kinase MSK2 actually precludes P53 coactivator P300 and associates with Noxa promoter, thereby impeding P53-dependent Noxa transactivation. On the other hand, apoptotic stimuli promote MSK2 degradation.

Upon genotoxic stress, P53 is phosphorylated and acetylated (by CBP and P300) to raise its activity and avoid its degradation. Multisite phosphorylation of the N-terminal transcriptional activation domain (TAD) of P53 yields a gradual enhancement of CBP and P300 binding according to the genotoxic stress magnitude. The TAD domain is also used to connect to DM2 ubiquitin ligase. Single-site phosphorylation (Thr18) reduces DM2 binding and slightly enhances CBP binding, whereas triple-site phosphorylation (Ser15, Thr18, and Ser20 or Ser33, Ser37, and Ser46) improves 20 to 30-fold CBP binding [1405].

BrD7 Protein

Ubiquitous Bromodomain-containing protein BrD7[235] is required for efficient P53-mediated transcription of a subset of target genes. Protein BrD7 interacts with P53 and P300 factors. It affects histone acetylation, P53 acetylation, and target gene promoter activity [1406]. It serves as a P53 cofactor for the efficient induction of P53-dependent oncogene-induced senescence.

---

[233] Ubiquitin ligase DM2 leads to JMY degradation, hence preventing P53 activation by JMY.

[234] Nuclear translocation of JMY enhances its role as a P53 cofactor. Conversely, nuclear JMY loses its influence on cell adhesion and motility. Cytoplasmic JMY indeed regulates cadherin concentration [1402].

[235] A.k.a. 75-kDa bromodomain protein (BP75), Celtix-1 (a cell-cycle regulator that connects to the C-terminus of interferon regulatory factor IRF2), and nasopharyngeal carcinoma protein NAG4.

MicroRNAs

MicroRNAs are post-transcriptional regulators of gene expression. Among numerous proteins that control P53, microRNAs of the MIR29 family (miR29a–miR29c) upregulate P53 level and induce apoptosis [1407]. They repress the P53 inhibitor PI3K regulatory PI3K$_{rl}$ subunit (P85$_{PI3K}$) that stabilizes PI3K catalytic P110$_{PI3K}$ subunit and CDC42 GTPase. They thus reduce PI3K activity and PKB phosphorylation. Activated PKB phosphorylates (inactivates) DM2 Ub ligase, thereby stabilizing P53 factor.

MicroRNAs also participate in the P53 pathway, as P53 binds to miR34 gene promoter in response to DNA damage. Members of the MIR34 set (miR34a–miR34c) that downregulate genes promoting cell cycle are targets of P53 to inhibit inappropriate cell proliferation [1408].

On the other hand, P53 enhances the post-transcriptional maturation of several growth-suppressive microRNAs, such as miR16-1, miR143, and miR145, in response to DNA damage [1409]. Transcription factor P53 can interact with the nuclear ribonuclease-3 Drosha processing complex, as it connects to ATP-dependent, DEAD (Asp–Glu–Ala–Asp motif) box-containing RNA helicase Ddx5 to facilitate processing of primary microRNAs (pri-miRs) into precursor microRNAs (pre-miRs).[236]

Factor P53 can regulate hypoxic signaling and angiogenesis via miR107 agent. The latter precludes production of Hypoxia-inducible factor-1β [1410]. In mice, miR107 impedes tumor growth, VEGF synthesis, and angiogenesis.

P53 Functions

Factor P53 activates or represses the transcription of specific genes, such as those that contribute to DNA repair, cell cycle arrest, and apoptosis.[237] In addition, it impedes autophagy. It binds to promoters and introns of genes and recruits numerous proteins, such as components of the basal transcriptional apparatus, histone acetyltransferases and deacetylases, and other transcriptional cofactors.

In the human genome, 122 genes are regulated by a set of 160 P53-binding sites [1390].[238] Cluster-site response elements devoted to P53 anchoring to DNA are listed in Table 10.57.[239]

---

[236] The Drosha complex comprises Drosha, DiGeorge syndrome critical region gene product DGCR8, and multiple RNA-associated proteins, such as DEAD box RNA helicases Ddx5 and Ddx17.

[237] MiR145 is expressed by the PI3K–PKB and P53 pathways to silence Myc expression [1411].

[238] About 50% of P53-binding sites are in the 5′ promoter–enhancer region of a gene and the remainder in exons and introns. Activator sites dedicated to P53 have a different distribution of spacer lengths with respect to repressor sites.

[239] Factor P53 can repress B-cell lymphoma BCL2-related genes. It prevents the activity of activators, such as forkhead box-A1 and regulatory factor-X1. Moreover, P53 binds to and suppresses DNA-bound and unbound activators of the CCAAT box (nuclear transcription factor-Y and

Table 10.57. Examples of P53 target genes and P53 cluster-site response elements (Source: [1390]). A cluster-site response element contains at least 3 sites, each separated by less than 16 base pairs. Cell cycle arrest and apoptosis triggered by P53 involves activation of cell proliferation genes, such as cyclin-dependent kinase inhibitor CKI1a and 14-3-3σ genes, and apoptosis genes, such as Noxa (Latin: damage; a.k.a. phorbol 12-myristate 13-acetate-induced protein PMAIP1), APAF1, and PUMA proteins.

Gene	Protein
Btg2	BTG family protein-2
CDKN1A	Cyclin-dependent kinase inhibitor-1A (CKI1a)
Ddb2	Damage-specific DNA-binding protein-2
HRAS	Harvey rat sarcoma viral oncogene homolog
Igfbp3	Insulin-like growth factor-binding protein-3
Dm2	Double minute-2
Pcna	Proliferating cell nuclear antigen
Sh2d1a	SH2 domain protein-1A, Duncan disease SH2 protein-4
Tp53i3	Tumor protein P53-inducible protein-3 (PIG3)
Tp73	Tumor protein P73
Trpm2	Transient receptor potential cation channel-M2
TYRP1	Tyrosinase-related protein-1
Vdr	Vitamin-D receptor

In addition, P53 mediates proteic interactions that repress activators of manifold genes (cyclin-B1, telomerase reverse transcriptase, insulin-like growth factor receptor-1, albumin, matrix metallopeptidase-1, etc.). P53 promotes the recruitment of histone deacetylases for gene inactivation (e.g., microtubule-associated protein-4). The P53 response element and adjacent cofactor response elements confer the potential for P53 activation, repression, or both. Furthermore, gene stimulation or squelching results from the type of combinations between P53 and cofactors.

Factor eIF5a promotes P53 expression. Once it is synthesized, P53 undergoes post-transcriptional modifications (phosphorylation, acetylation, methylation, and ubiquitination) that affect its stability and activity [1412]. Most of these modifications are reversible via phosphatases, deacetylases, and deubiquitinases. Factor P53 then can be activated by various signals using different pathways.

Two major functions of P53 are cell cycle arrest and apoptosis. Factor P53 prevents the transmission of mutations to daughter cells, as it enhances DNA repair or induces apoptosis when DNA repair is impossible. It also impedes accumulation of damages due to continuous endogenous production of hydrogen peroxide.

Known P53-interacting molecules include caspase-3 (apoptosis), 14-3-3σ (cell growth control), in addition to CKI1a, DM2, BAX, and BCL2-binding component-3 (BBC3 or PUMA). The Ras–MAPK pathway is also implicated. Certain P53 activities, but not all, involve ARF GTPase.

CCAAT/enhancer binding protein) that target numerous genes (cyclin-A2, CDK1 kinase, CDC25c phosphatase, checkpoint kinase ChK2, heat shock protein HSP70, fibronectin-1, cyclooxygenase COx2, etc.).

10.9 Transcription Factors Involved in Stress Responses

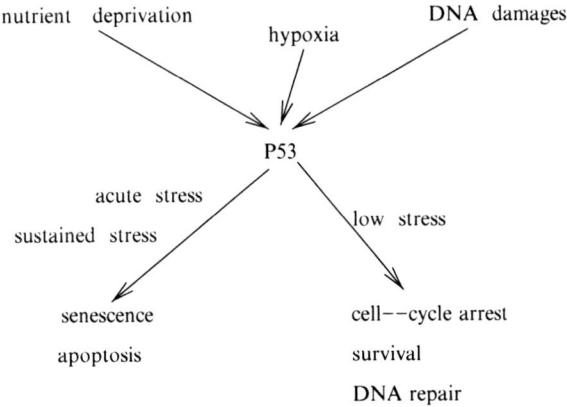

Fig. 10.6 P53 function according to stress features, intensity, occurrence, and duration.

In addition to the protection against cancer development by the regulation of DNA-damage repair, cell cycle arrest, and cell apoptosis, particularly during ischemia, P53 regulates glycolysis as well as cell survival, senescence, and autophagy. This transcription factor possesses targets that are independent of transcription.

Protein P53 primes both the extrinsic pathway via activation of cell-surface receptors and, predominantly, the intrinsic pathway associated with cell stresses (Fig. 10.6). It couples transforming growth factor-β to the RTK–Ras–MAPK cascade to stop cell growth [1413]. Kinases CK1ε and CK1δ phosphorylate P53 in response to RTK–Ras–MAPK signaling.

P53, Cellular Energy Metabolism, and Anti-Oxidant Defense

Factor P53 regulates cellular energy metabolism and anti-oxidant defense. It indeed augments the synthesis of mitochondrial glutaminase-2 (Gls2) that converts glutamine to glutamate [1414]. Glutaminase-2 also increases production of α-ketoglutarate for enhanced mitochondrial respiration and ATP generation, hence cellular energy metabolism [1414]. Furthermore, glutaminase-2 augments reduced glutathione concentration to lower ROS concentrations, thereby protecting cells against oxidative stress [1414]. Glutaminase-2 expression is induced in response to DNA damage or oxidative stress in a p53-dependent manner, as p53 associates with the GLS2 promoter [1415].

P53 and DNA Repair

Transcription factor P53 promotes the repair of minor DNA damages but suppresses the repair of severe damages and directs cell death. Concentration of P53 can evolve with a series of pulses after DNA damage caused by ionizing radiation. In a first stage, DNA repair proteins bind to double-strand breaks and form complexes that are targeted by ataxia telangiectasia mutated kinase. Activated ATMK then initiates

P53 pulses to control cell fate. Activity of ATMK is suppressed by P53 pulses, but reactivated by DNA damage according to a negative feedback loop between ATMK and P53 via PPM1d phosphatase. Cell fate (i.e., DNA-damage repair vs. cell death) is determined by the number of P53 pulses [1416]. Sustained P53 pulses trigger apoptosis by exciting P53-regulated apoptosis-inducing protein P53AIP1. Pulses of P53 thus act as nanotimers.

P53 and Cell Cycle Arrest

Agent P53 activates cyclin-dependent kinase inhibitor-1A (Vol. 2 – Chap. 2. Cell Growth and Proliferation), a mediator of G1 arrest, but CKI1a protects against apoptosis. When cell apoptosis is programmed, death factors are dominant over CKI1a-mediated protection.

Replicating precancerous cells can escape G1 blockade mediated by CKI1a inhibitor. The control mediated by CKI1a of the cell cycle at G1 phase is complemented by G2–M arrest due to P53-activated CKI1a, Growth arrest and DNA-damage protein GADD45, and 14-3-3 proteins.

Furthermore, S-phase checkpoint targeted by P53 further controls dividing cells. Factor P53 targets the gene killin on chromosome 10 that encodes DNA-binding protein Killin to impede DNA synthesis [1417]. Killin triggers S-phase arrest before apoptosis.[240]

P53 and Cell Apoptosis

Factor P53 activated by genotoxic stresses regulates numerous genes involved in cell apoptosis, such as BCL2-related genes (Bax and P53-inducible pro-apoptotic members of the BCL2 family [BH3-only protein subfamily] Puma [BBC3] and NOXA). Cell fate between growth arrest and death due to P53 activation depends on several factors.

The mitochondrial membrane is the site where pro- and anti-apoptotic factors induce or prevent mitochondrial outer membrane permeabilization. Upon stress, P53 can actually rapidly move to and associate with the outer mitochondrial membrane to trigger release of pro-apoptotic factors from the mitochondrial intermembrane space (Vol. 2 – Chap. 4. Cell Survival and Death).[241]

---

[240] In contrast to S-phase control, G1 arrest is not coupled to apoptosis. Inhibitor CKI1a prevents cell entry into S phase, thereby escaping apoptosis following S-phase arrest via Killin.

[241] In normal conditions, mitochondrial outer membrane permeabilization is inhibited by anti-apoptotic proteins of the BCL2 family (e.g., B-cell leukemia/lymphoma-2 [BCL2] and B-cell lymphoma-extra large BCLxL). Conversely, pro-apoptotic proteins of the same family (e.g., BCL2-associated X protein [BAX] and BCL2-antagonist/killer-1 [BAK1]) can homo-oligomerize within the outer mitochondrial membrane to prime mitochondrial outer membrane permeabilization. Homo-oligomerization of BAX and BAK1 is initiated by pro-apoptotic BH3-only proteins (that contain a single BCL2 homology [BH] domain) that connect to BAX and BAK1. BH3-only proteins can also neutralize anti-apoptotic proteins (hence their names sensitizers or derepressors).

Monoubiquitinated P53 (by DM2) exits from the nucleus and reaches mitochondria, where it is deubiquitinated by mitochondrial HAUSP [1393]. Non-ubiquitinated P53 then launches the apoptotic program. In addition, PUMA releases cytoplasmic P53 from its sequestered pool by anti-apototic BCLxL to activate BCL2-associated X protein BAX. Factor P53 may serve as a BH3-only protein, either as a direct activator of BAX and/or BAK1 or as a derepressor.

Inositol hexakisphosphate kinase IP_6K2 that generates inositol pyrophosphate diphosphoinositol pentakisphosphate (IP_7; Chap. 2) is required for P53-mediated apoptosis [1420]. Kinase IP_6K2 binds to P53 and decreases expression of pro-arrest genes such as the CDKN1A gene (that encodes CKI1a inhibitor).

P53 and Hypoxia

Reduced oxygen concentration primes the activity of the heterodimeric transcription factor hypoxia-inducible factor. The latter can antagonize P53-mediated apoptosis, as it can upregulate the expression of secreted tyrosinase-related protein TRP2, an inhibitor of apoptosis, produced by melanocytes [1418].[242]

Osteopontin[243] is an extracellular, structural glycoprotein that operates as an anti-apoptotic factor in many circumstances. It hampers death of macrophages, T lymphocytes, fibroblasts, and endothelial cells exposed to strong stimuli. Osteopontin is a caspase-8 substrate (Asp135 and Asp157 cleavage) that regulates cell death during hypoxia–reoxygenation events, during which its production rises rapidly [1419]. It is subsequently cleaved. Osteopontin cleavage fragment increases P53 level and induces apoptosis. Osteopontin cleavage by caspase-8 also reduces PKB-mediated survival signal. On the other hand, osteopontin-induced activation of PKB suppresses P53 accumulation for cell survival.

P53 and Cell Survival

Transcription factor P53 regulates cell growth only at physiological levels. Therefore, P53 concentration must be strongly controlled in an appropriate range to avoid unwanted effects, such as tumorigenesis in the case of a rising P53 level. The P53-associated cellular, testis-derived protein (PACT)[244] binds to P53 and inhibits it via DM2 to hamper apoptosis and favor cell growth and proliferation [1421].

On the other hand, direct activators (e.g., BCL2L11 [or BIM]) and BH3-interacting domain death agonist [BID])antagonize inhibition of mitochondrial outer membrane permeabilization.

[242]In humans, the tyrosinase family includes: (1) tyrosinase that converts ^{L}-tyrosine into ^{L}-DOPA and (2) tyrosinase-related protein-1 (TRP1 or TyRP1) and -2 (TRP2 or DCT).

[243]A.k.a. secreted phosphoprotein SPP1, bone sialoprotein BSP1, early T-lymphocyte activation protein ETA1. Osteopontin interacts with $\alpha_V\beta_3$- and $\alpha_V\beta_5$-integrins and then can activate the PI3K–PKB cascade as well as MAP4K4 and MAP3K1 kinases. In addition, an alternative isoform of osteopontin resides in cytosol, where it can link to the ezrin–radixin–moesin complex as well as in the nucleus during the cell cycle, where it can associate with Polo-like kinase-1.

[244]A.k.a. P2P-R or RBBP6.

Table 10.58. Targets of transcription factor P53 (**Part 1**; Source: [1422]). Factor P53 activates cell responses to non-repairable DNA damage caused by endogenous or exogenous sources that lead to apoptosis. It also operates in the repair of damaged DNA, as it participates in mismatch repair, non-homologous end-joining, homologous recombination, and nucleotide and base excision repair. It impedes the cell cycle progression at various checkpoints to ensure repair of damaged DNA. Cell cycle arrest hinders apoptosis, at least transiently.

Targets	Effects
DNA repair	
Mismatch repair	Pro- and anti-apoptotic effects
Nucleotide excision repair	Apoptosis inhibition
Base excision repair	Apoptosis resistance
Cell cycle control	
CKI1a	Inhibition of CDKs, caspase-3, and ASK1
14-3-3δ	Sequestration of CDKs, Bad, Bax, FoxO3, and MAP3K5
B-cell translocation genes BTG2 and BTG3	Inhibition of cyclin-D1 and -E
Polo-like kinases	Apoptosis modulation
Miscellaneous	
Checkpoint kinase ChK2	Pro-apoptotic ChK2 inhibition
TRAIL decoy receptors DcR1 and DcR2 (TNFRSF10c/d)	Inhibition of TRAILinduced apoptosis
Netrin-150	Apoptosis inhibition

The AMPK pathway activates cyclin-dependent kinase inhibitor CKI1a and P53 for short-term survival of cells subjected to transient glucose deprivation. Agent P53 regulates the expression of TP53-inducible glycolysis and apoptosis regulator (TIGAR), decreasing the activity of glycolytic 6-phospho 1-kinase. This diversion in the major glycolytic pathway into the pentose phosphate pathway increases NADPH production, eliciting decay in concentration of reactive oxygen species. However, sustained stresses switch P53 function from cell survival to apoptosis, targeting PUMA (BBC3).

Transcription factor P53 is able to activate survival pathways, hence counteracting apoptosis [1422]. Anti-apoptotic effects of P53 result from DNA repair, cell cycle control at checkpoints via CKI1a and 14-3-3δ, respectively, expression of anti-oxidant enzymes, activation of transcription factors and MAPK signaling components (Tables 10.58 to 10.60).

P53 and Autophagy

The autophagic pathway allows cells to adapt to starvation, trophic factor deprivation, hypoxia, endoplasmic reticulum and oxidative stresses, oncogenic

10.9 Transcription Factors Involved in Stress Responses

Table 10.59. Targets of transcription factor P53 (**Part 2**; Source: [1422]; TIGAR: TP53-induced glycolysis and apoptosis regulator [a.k.a. chromosome 12 open reading frame C12ORF5]). Factor P53 targets components of MAPK pathways that hamper directly or indirectly P53 (such as ATMK). Reactive oxygen species that can conduct apoptosis activate P53 that elicits expression of anti-oxidant enzymes.

Targets	Effects
MAPK signaling	
Epidermal growth factor receptor	Anti-apoptotic PKB activation
Heparin-binding EGF-like growth factor	EGFR and PKB activation
Discoidin domain receptor DDR1	NFκB and PKB activation
Cyclooxygenase COx2	PKB activation
PPM1d phosphatase	Inhibition of P53, P38MAPK, ATMK, and ChK1
MAPK phosphatase-1	Inhibition of P38MAPK and JNK
Oxidative stress response	
Aldehyde dehydrogenase-4	Anti-oxidant enzyme
Glutathione peroxidase	Anti-oxidant enzyme
Manganese superoxide dismutase	Anti-oxidant enzyme
Sestrins SESN1 and SESN2	Anti-oxidant enzyme
TIGAR	Anti-oxidant enzyme

activation, and genotoxic stress caused by DNA-damaging agents. However, strong autophagy frequently leads to cell death, as it corresponds to a failed attempt to adapt to stress and survive.

Cell autophagy is either activated or inhibited by P53 agent. Autophagy caused by P53 involves both transcription-independent (e.g., AMPK activation) and -dependent processes (upregulation of TOR inhibitors PTen and TSC1, autophagy and cell death gene DRAM, as well as the cell survival regulators sestrins Sesn1 and Sesn2). However, autophagy can be induced by P53 degradation and loss via AMPK activation and TOR repression [1423].

Dichotomous function of P53 that mediates both induction and inhibition of autophagy depends on the context. Oncogenes and genotoxic stress stimulate autophagy via TOR inhibition. Acute stress that provokes high P53 level promotes cell death, whereas low-magnitude stress favors prosurvival pathway.

Degradation of P53 and TOR inhibition primed by nutrient deprivation and toxins that alter the endoplasmic reticulum as well as P53 loss in the absence of stress-induced autophagy. Factor P53 may coordinate distinct regulations at basal level and in stress-activated contexts. Different P53 isoforms may also be involved.

P53 and Embryonic Stem Cells

In embryonic stem cells, P53 participates in stress defense and provokes apoptosis rapidly in response to genotoxic stresses. In addition, P53-mediated differentiation of embryonic stem cells may represent another mechanism to escape from

Table 10.60. Targets of transcription factor P53 (**Part 3**; Source: [1422]; CoP1: constitutively photomorphogenic Ub ligase; DM: double minute Ub ligase; DRAL: downregulated in rhabdomyosarcoma LIM domain-containing protein [a.k.a. Four and a half LIM domain-containing proteins FHL2]; GTSE1: G2- and S-phase-expressed protein; KLF: Krüppel-like factor; PIDD: P53-induced protein with a death domain [a.k.a. leucine-rich repeats and death domain-containing protein [LRDD]; PIRH2: P53-induced protein with a really interesting new gene RING-H2 domain, a Ub ligase [a.k.a. ring finger and CHY zinc finger domain-containing protein RCHY1]; PUMA: P53-upregulated modulator of apoptosis; P53R2: P53-induced R2 homolog [a.k.a. TP53-inducible ribonucleotide (ribonucleoside diphosphate) reductase M2b subunit (RRM2b)]; Snai2: snail homolog-2 [a.k.a. Slug]; ZNF385a: zinc finger protein-385A [a.k.a. hematopoietic zinc finger protein]). Variant P73$^{\Delta NT}$ is a P73 isoform that lacks the N-terminal transactivation domain and antagonizes the induction by P53 of gene expression. Factor P53 targets anti-apoptotic transcription factors (e.g., NFκB) as well as those that repress P53-induced apoptosis (e.g., KLF4 and Slug). It regulates the expression of several P53-binding proteins acting as P53 inhibitors.

Targets	Effects
Transcription factors	
Snai2	PUMA-induced repression
NFκB	Activation via DRAL and PIDD
KLF4	Activation of CKIIa; inhibition of P53 and Bax
P73$^{\Delta NT}$	Inhibition of P53
P53-Binding proteins	
DM2	P53 inhibition
RNF199 (PIRH2)	P53 inhibition
RNF200 (CoP1)	P53 inhibition
P53R2	G2–M arrest; P53 inhibition
GTSE1	G2–M arrest; P53 inhibition
ZNF385	Promotion of cell cycle arrest

DNA damage (prodifferentiation role). On the other hand, P53 regulates Wnt signaling, hence inhibiting embryonic stem cell differentiation (antidifferentiation role) [1424].

P53 and Cell Migration

Factor P53 also prevents cell migration, thus tumor cell metastasis. Both Ras and P53 control RhoA activation for cell motility [1425]. Small GTPase Ras promotes RhoA translocation to the plasma membrane, where RhoA can be activated. On the other hand, P53 restricts Ras stimulation of RhoA and stimulates RhoGAP35 to inactivate RhoA GTPase.

10.9.4.2 Transcription Factor P63

The P63 transcription factor, a 63-kDa protein with strong homology to P53,[245] pertains to the P53 family of tumor suppressor. It is involved in cell cycle arrest, growth suppression, cell differentiation and senescence, and apoptosis. Upon exposure to various stimuli, P63 concentration is regulated by P53 and other transcription factors.

Due to 2 distinct promoters, 2 major P63 variants exist: TAP63 and P63$^{\Delta NT}$. Both variants generate various isoforms via alternative splicing.

Stability of P63 is regulated by several ubiquitin ligases, such as Itch, NEDD4-like WW domain-containing protein WWP1, and SCF$^{\beta TRCP}$.

RNA-binding motif-containing protein RBM38,[246] more precisely both isoforms RBM38a and RBM38b, represses P63 expression via mRNA turnover (feedback loop between RBM38 and P63) [1426].

Factor P63 interacts with heterogeneous nuclear ribonucleoprotein-A and -B that influence pre-mRNA processing, splicing factor arginine (R)–serine (S)-rich SFRS15, homeodomain-interacting protein kinase HIPK2, Gβ2-like subunit-1, and ubiquitin-C [251].

10.9.4.3 Transcription Factor P73

The P73 transcription factor is a structural and functional homolog of P53 factor that favors cell growth, as it upregulates AP1 transcription factor. It also promotes cell survival when it functions synergistically with Jun, as it targets cyclin-D1. It enhances the binding of phosphorylated Jun and Fra1 to AP1 [1427]. The synergy between P73 and Jun activities amplifies signaling.

10.9.5 Transcription Factor p27

Transcription factor P27, which is strongly expressed in the heart, blocks cell proliferation, as it inhibits cyclin-dependent kinase CDK2.[247] Growth factors activate the CcnE–CDK2 complex, which then phosphorylates P27 factor.

Unphosphorylated P27 also inhibits CK2 kinase.[248] On the other hand, angiotensin-2 induces P27 ubiquitination and degradation via phosphorylation by CK2 in cardiomyocytes, thereby enabling cardiac hypertrophy [1428].

---

[245] A.k.a. tumor protein P73-like (TP73L).

[246] A.k.a. RNA-binding region-containing protein-1.

[247] Agent P27 suppresses adaptive hypertrophy as well as maladaptive growth.

[248] Kinase CK2 consists of 2 catalytic (α and α') and 2 regulatory (β) subunits. Factor P27 targets and inactivates CK2α' subunit. Activated CK2 primes P27 destruction.

Chapter 11
Signaling Pathways

> "..., because of the interconnection of all things with one another." (G. Leibniz, Philosophical Writings, 1670)

Multiple signaling processes are characterized by bursts of calcium ions that moves to specific locations for optimal activation of cell activity. Intracellular calcium regulates numerous protein functions in tiny cellular domains (<1 μm) and small time scales (<1 ms). For example, $Ca_V1.2$ channels form signaling clusters in plasmalemmal nanodomains. Fluorescence microscopy is aimed at imaging in real time communication within and between cells.[1] Calcium fluxes through Ca^{++} channels and gap junctions can be imaged once constitutive proteins (e.g., connexin) have been tagged with tetracysteines [1429].

Cyclic nucleotide signaling is compartmentalized. The location of most signaling complexes triggered by cyclic adenosine monophosphate (cAMP) that contain protein kinase-A is determined by A-kinase (PKA)-anchoring proteins (Sect. 10.8.4), although others contain guanine nucleotide-exchange factor activated by cAMP (EPAC1 and -2 or RapGEF3 and -4; Sect. 9.4.1).

Between-protein interactions determine the location and composition of each node of signaling pathways. A prototypical signaling pathway is the cAMP–PKA axis based on the production of cAMP that activates cAMP-dependent protein kinase-A. This signaling cascade can be insulated or can interact with other pathways. Signaling precision, specificity, and coordination rely on the multiple modes used by signaling axes to insulate or collaborate as well as cross-regulate [1430].

---

[1] Calcium sensor (1-kDa biarsenical Ca^{++} indicator or calcium green flash) bound to tetracysteines can interact with calcium signaling components to detect the activity of calcium ions. When calcium ions are trapped by these calcium sensors, these detectors becomes 10-fold more fluorescent. This sensor has high and low affinity for Ca^{++} ($K_d \sim 100$ mmol)) and Mg^{++} ($K_d > 10$ mmol), respectively, and short time response (<1 ms).

Table 11.1. Functions of cAMP signaling (Source: [10]; CREB: cAMP-responsive element-binding protein; MC_4: type-4 melanocortin receptor; PPP1R1b: protein phosphatase-1, regulatory (inhibitory) subunit-1B; SiM1: Single-minded-1 [hypothalamic transcription factor]).

Effect	Mechanism
Neurotransmission	Phosphorylation of PPP1R1b by PKA to coordinate the dopamine and glutamate signaling
Glycogenolysis	Phosphorylation of phosphorylase kinase by PKA in skeletal muscle induced by adrenaline
Glucagon synthesis	Activation of CREB
Heat	Production control by the noradrenaline–cAMP axis
Lipolysis	Stimulation of hormone-sensitive lipase
Diet	Reduction of food intake via the MC_4–SiM1 axis
Inflammation	Inhibition of macrophages and mastocytes
Oocyte maturation	Suppression of spontaneous Ca^{++} oscillations

Cyclic nucleotide-dependent kinases include cAMP-dependent protein kinase-A and cGMP-dependent protein kinase-G. Protein kinase-G-binding proteins and cognate phosphodiesterases govern the subcellular PKG distribution. Activated PKG phosphorylates (activates) PDE5 tethered to the IP_3R1–PKG complex (but not non-attached PDE5), thereby enabling the spatial and temporal regulation of cGMP signaling by hydrolyzing cGMP that inhibits IP_3R1-mediated calcium release in platelets [1431].

11.1 cAMP Signaling

Cyclic adenosine monophosphate is produced by both transmembrane and bicarbonate-sensitive, soluble adenylate cyclases. Plasma membrane-bound adenylate cyclases are regulated by both stimulatory agonists that act via the $G\alpha_s$ subunit of the trimeric guanine nucleotide-binding (G) protein and inhibitory agonists that operate via either the $G\alpha_i$ or $G\beta\gamma$ subunits.

On the other hand, cAMP signaling can terminate either via cAMP hydrolysis by phosphodiesterases or cAMP efflux from the cell through ATP-binding cassette ABCc4 transporter.

This second messenger controls numerous cellular functions (Table 11.1). The diversity of action of this diffusible messenger can be explained by its location and subsequent metabolism in subcellular compartments. The organization and function of the cAMP signaling, i.e., the spatiotemporal regulation of cAMP, is indeed associated with cortical nanodomains immediately beneath the plasma membrane. The compartmentation of second messenger cAMP allows the spatial

11.1 cAMP Signaling

segregation of cAMP signaling events.[2] The cAMP subcellular compartmentation has been particularly described in cardiomyocytes [1433]. In cardiomyocytes, the multiproteic complex formed by cAMP, PKAs, and PDEs attracts G proteins, adenylate cyclases, A-kinase anchoring proteins, and phosphoprotein phosphatases.

The cAMP signaling pathways require many components, as cAMP effectors cause divergent responses. Second messenger cAMP diffuses from its synthesis site at the plasma membrane to subcellular compartments. Targets of cAMP include: (1) cAMP-dependent protein kinase-A (PKA), which mediates most of the actions of cAMP; (2) cyclic nucleotide-gated ion channels (CNG); and (3) guanine nucleotide-exchange factors RapGEF3 and RapGEF4[3] that regulate Rap1 activity.

The cAMP–PKA pathway can [10]: (1) stimulates gene transcription via phosphorylation of cAMP response element-binding protein (CREB), thereby supporting gluconeogenesis in hepatocytes; (2) activates ion channels, such as neuronal, ionotropic, AMPA-type glutamate receptors (AMPAR) and cystic fibrosis transmembrane conductance regulator (CFTR),[4] as well as other types of carriers such as aquaporin-2 in collecting ducts of nephrons for water reabsorption; and (3) primes the activity of various enzymes that control the cell metabolism, such as fructose (2,6)-bisphosphate 2-phosphatase,[5] hormone-sensitive lipase,[6] and phosphorylase kinase.[7] Protein kinase-A phosphorylates the regulatory (inhibitory) subunit-1A of protein phosphatase-1 (PPP1R1a),[8] thereby fostering PP1 activity. In platelets, PKA phosphorylates vasodilator-stimulated phosphoprotein (VASP) associated with filamentous actin formation, thereby attenuating actin-dependent processes involved in clotting. In insulin-secreting β cells, PKA phosphorylates (inhibits) salt-inducible kinase-2 (SIK2), which phosphorylates CREB-regulated transcription coactivator, thereby preventing its nuclear import and limiting the activity of the transcriptional factor CREB.

---

[2] The subplasmalemmal complex that contains cAMP, A-kinase anchoring protein AKAP12, phosphodiesterase-4, and protein kinase-A, regulates the activity of ion channels, plasmalemmal receptors, and enzymes [1432].

[3] A.k.a. exchange protein activated by cAMP EPAC1 and EPAC2.

[4] Protein kinase-A phosphorylates the CFTR regulatory domain to activate this anion channel and, hence, fluid secretion.

[5] A.k.a. fructose (2,6)-bisphosphatase and phosphofructokinase-2 (PFK2). It degrades fructose (2,6)-bisphosphate that stimulates phosphofructokinase-1 (PFK1), or 6-phosphofructokinase, thereby repressing glycolysis and promoting gluconeogenesis.

[6] Hormone-sensitive lipase hydrolyzes triacylglycerol to free fatty acids and glycerol in both white and brown adipocytes.

[7] Phosphorylase Ser/Thr kinase activates glycogen phosphorylase to release glucose 1-phosphate from glycogen. This homotetramer of $\alpha\beta\gamma\delta$ tetramers is auto-inhibited. Phosphorylation by protein kinase-A causes a conformational shift and caonverts inactive glycogen phosphorylase-B into active phosphorylase-A in skeletal myocytes and hepatocytes.

[8] A.k.a. protein phosphatase inhibitor IPP1 and inhibitor-1 (I1). In addition, PKA phosphorylates the regulatory (inhibitory) subunit-1B of protein phosphatase-1 (PPP1R1b), or 32-kDa dopamine- and cAMP-regulated phosphoprotein DARPP32, that functions as a molecular switch to regulate the activity of protein phosphatase-1.

Other effectors are components of other signaling axes, such as cGMP-dependent phosphodiesterase PDE1a,[9] phospholamban,[10] ryanodine receptors,[11] and $Ca_V1.1$ and $Ca_V1.2$ channels.[12]

In addition, PKA phosphorylates cytoskeletal proteins and contributes to the transfer of ion carriers as well as translocation and fusion of vesicles with the apical membrane during the onset of acid secretion by gastric oxyntic cells.[13]

The cAMP concentration is balanced by adenylate cyclases (ACase) and phosphodiesterases (PDE), particularly members of the PDE4 family. In bacteria and amoebae such as Dictyostelium discoideum, cAMP also targets members of a family of 4 cAMP receptors (CAR1–CAR4).[14]

The cAMP-dependent signaling cascades are activated by G-protein-coupled receptors that can lead to opposing effects. Moreover, the coupling between cAMP pathways and other signaling pathways participates in distinct effects of cAMP signaling.

The 3 β-adrenergic receptors act via distinct cAMP signaling pathways in neonatal cardiomyocytes of mice.[15] β1- and β2-adrenergic agonists produce different effects on myocardium contractility [1434].[16]

Subcellular compartmentation of cAMP and colocalization of components of the cAMP cascade is implicated in β-adrenergic stimulation of $Ca_V1.2a$ channel [1435]. A molecular complex composed of ryanodine RyR2 channel, FK506-binding proteins FKBP12.6, PKA, protein phosphatases PP1 and PP2, and A-kinase anchorprotein AKAP6, controls the functioning of ryanodine channels of

---

[9]Phosphorylation of PDE1a by PKA decreases its sensitivity to Ca^{++}-mediated activation.

[10]Phospholamban inhibits the sarco(endo)plasmic reticulum Ca^{++} ATPase (SERCA). Activity of SERCA2a rises upon phospholamban phosphorylation by PKA.

[11]Ryanodine receptor RyR2 is phosphorylated by PKA.

[12]Cyclic nucleotide-dependent kinase PKA phosphorylates Ca_V1 channels, thereby enhancing their activity.

[13]When stimulated, parietal cells of the stomach secrete hydrochloric acid (HCl) at a concentration of about 160 mmol (equivalent to a pH 0.8). This acid is secreted into deep cannaliculi of the plasma membrane owing to a Mg^{++}-dependent H^+–K^+ ATPase of the cannalicular membrane, which are continuous with the lumen of the stomach. Bicarbonate is exported across the basolateral membrane in exchange for chloride. Chloride and potassium ions are transported into the lumen of the cannaliculus, potassium being reimported in exchange for proton. Parietal cells possess receptors for neural and endo- and paracrine stimulators of acid secretion, such as acetylcholine (muscarinic receptors) gastrin, and histamine (H_2 receptor) released from enterochromaffin-like cells.

[14]The CAR alias also designates constitutive activator of retinoid response, or nuclear receptor NR1i3, and carbonic anhydrase.

[15]The sympathetic nervous system acts via β-adrenergic receptors of cardiomyocytes and nodal cells. β-Adrenergic receptors are also involved in cardiac remodeling.

[16]Stimulation of β1-adrenergic receptors in β2-adrenoceptor-knockout cardiomyocytes produces the greatest increase in contraction rate via protein kinase-A. Activation of β2-adrenergic receptors in β1-adrenoceptor-knockout cardiomyocytes causes an initial increase in contractibility without requiring PKA followed by a decrease in contraction rate involving coupling to a G protein.

11.1 cAMP Signaling

the sarcoplasmic reticulum. In failing human hearts, these calcium channels are phosphorylated by PKA and dissociates [1436].

11.1.1 Crosstalk between Calcium and cAMP Signalings

Signalings initiated by calcium ion and cyclic adenosine monophosphate interfere. Membrane-bound, Ca^{++}-regulated adenylyl cyclases colocalize with store-operated Ca^{++} channels (SOC) and are influenced by Ca^{++} entry. Calcium-stimulated adenylyl cyclase-8 interacts with the pore component of SOC channels, Orai1 protein, to coordinate subcellular changes in both Ca^{++} and cAMP messengers [1437].

11.1.2 Crosstalk between the ERK and cAMP Signaling Pathways

Crosstalk between the cAMP and ERK pathways occurs via Raf enzyme. Elevated cAMP level activates or inhibits ERK via bRaf or cRaf, respectively. These kinases are expressed in a cell type-specific manner. Inhibition of cRaf is achieved via phosphorylation by protein kinase-A. Conversely, augmented cAMP concentration activates bRaf and hence ERK kinase. In addition, protein Tyr phosphatases dephosphorylate (inactivate) ERK enzymes. Activity of these phosphatases requires a distinct docking site that contains a serine residue, which can be phosphorylated by PKA, thereby impeding ERK binding. Uncoupling PTP from ERK thus modulates the cAMP activity via Raf on ERK kinase.

Phosphodiesterases degrade cAMP messenger. The PDE4 family is aimed at interacting with ERK kinases. Activated ERK acts on PDE4s to regulate cAMP signaling either negatively or positively depending on the expression pattern and localization of long and short PDE4 isoforms [1438].[17] Phosphorylation by ERK of PDE4 long isoform inhibits PDE4 and thus increases cAMP level. Furthermore, PDE4 long isoform is phosphorylated (inactivated) by PKA kianse. Consequently, PKA activation can relieve inhibition of phosphorylated ERK on PDE4 long isoforms. On the other hand, ERK phosphorylation of PDE4 short forms causes activation. Super-short isoform is only slightly responsive to ERK phosphorylation with a slight inhibition and without feedback from cAMP as it is not targeted by PKA enzyme.

---

[17] The multi-gene PDE4 family (Pde4A–Pde4D) encodes at least 20 different PDE4 isoforms using distinct promoters coupled to alternative mRNA splicing. Moreover, there are long, short, and super-short isoforms. Kinase ERK phosphorylates the PDE4b, PDE4c, and PDE4d (but not PDE4a) family members.

However, in smooth muscle cells, growth factors activate ERK and subsequently PKA via an autocrine process, which then stimulate PDE4d long forms as activated ERK also activates phospholipase-A2. The latter generates prostaglandin-E2 via cyclooxygenase, which activates adenylate cyclase in an autocrine fashion, leading to cAMP generation and afterward PKA phosphorylation of PDE4d long isoforms.

β2-Adrenoceptor coupling to Gs activates adenylate cyclase, but also triggers its phosphorylation by G-protein receptor kinases and recruitment of cytosolic β-arrestin, which hinders the association of β2AR with Gs subunit. Moreover, activated PKA phosphorylates β2AR, thereby not only facilitating its uncoupling from Gs, but also priming its coupling to Gi that leads to ERK activation.

The PDE4 isoforms interact with β-arrestin. The PDE4–βArr complex is recruited to β2-adrenoceptor. This process delivers active PDE4 to the site of cAMP synthesis. The PDE4–βArr complex controls a pool of membrane-anchored PKAs that phosphorylate β2ARs by lowering local cAMP levels. The recruited PDE4 can thus serve to regulate the switching of coupling of β2AR from Gs to Gi, thereby activating ERK [1438].

11.1.3 cAMP Response Element-Binding Protein and Cofactors

The cAMP signaling pathway activates cAMP response element-binding proteins. Transcription factors of the CREB family (CREB1, CREB3, and CREB5; CREB2 is ATF4; in humans, other genes encode CREB-like proteins CREBL1, CREBL2, and CREB3L1 to CREB3L4) bind to cAMP response elements of DNA, then recruit coactivators and regulate the transcription of target genes. Protein kinase-A operates in protein synthesis via CREB factors. Factor CREB is related in structure and function to cAMP response element modulator (CREM) and activating transcription factor-1 (ATF1). In neurons, CREB factors can be involved in long-term memories and neuron survival.

Transcription factor CREB recruits 2 classes of coactivators: (1) transcriptional coactivators and histone acetyltransferases (as well as polyubiquitin ligase and adaptor) CREB-binding protein (CBP) and P300 and (2) CREB-regulated transcription coactivators (CRTC1–CRTC4).[18] Under basal conditions, CRTC1 and CRTC2 are phosphorylated by SNF-related kinases (SNRK or SNFRK) and bind to 14-3-3 proteins that sequester CRTCs in the cytoplasm.

Augmented intracellular cAMP levels separate PKA catalytic subunits from PKA regulatory subunits. PKA catalytic subunits then phosphorylates SNF-related kinases, hence preventing CRTC phosphorylation. Dephosphorylated CRTC1 and CRTC2 are thus released from 14-3-3 proteins. Factors of the CRTC family

---

[18] A.k.a. transducers of regulated CREB activity (TORC1–TORC4). Agents CRTC1 and CRTC2 (TORC1 and TORC2) do not correspond to target of rapamycin complexes (TORC1 and TORC2).

move to the nucleus to bind to CREB and recruit CBP or P300 agents. Histone acetyltransferases CBP and P300 then interact with RNA polymerase-2 that is recruited by CRTC-interacting protein Nono,[19] which is required for cAMP-dependent activation of CREB target genes [1439]. Partner Nono complexes with CRTC2 on cAMP-responsive promoters.

Cytoplasmic retention of CRTC by 14-3-3 proteins allows integration of converging signalings. In hepatocytes, glucose homeostasis is regulated by hormones such as insulin and glucagon, and within the cell by energy status. Glucagon enhances glucose delivery from the liver, as it stimulates the transcription of gluconeogenic genes via CREB factor. When cellular ATP level is low, energy-sensing AMPK inhibits hepatic gluconeogenesis. Antagonist hormonal and energy-sensing pathways converge on CRTC2 coactivator [1440]. Phosphorylated CRTC2 is targeted by competitive hormone and energy-sensing pathways to modulate glucose output via CREB-mediated hepatic gene expression. Sequestered in the cytoplasm under feeding conditions, CRTC2 is dephosphorylated and transported to the nucleus where it elicits CREB-dependent transcription. Conversely, activated AMPK represses the gluconeogenic program as it promotes CRTC2 phosphorylation. In pancreatic islet cells, phosphorylated CRTC serves as a convergent node for cAMP and calcium-signaling pathways to CREB-dependent transcription [1441]. Synergy of these pathways on cellular gene expression is mediated by a module made of calcium-regulated phosphatase PP3 and Ser/Thr kinase SIK2 that leads to CRTC2. Circulating glucose and gut hormones trigger calcium and cAMP pathways that disrupt CRTC2–14-3-3 complexes. Calcium influx increases calcineurin activity and cAMP inhibits SIK2 activity.

11.1.4 Adenylate Cyclases

Nine adenylate cyclases catalyze the synthesis of the second messenger cAMP (Table 11.2). These 9 differentially regulated, membrane-bound adenylate cyclases are isoforms of type-1, G-protein-responsive, transmembrane adenylate cyclases (tmAC or $_m$AC). Adenylate cyclase-10 and its isoforms that are encoded by the ADCY10 gene constitute a distinct type of adenylate cyclases.

11.1.4.1 Soluble Adenylate Cyclase Isoforms

A distinct category (type-2) of adenylate cyclase is constituted by soluble enzymes (sAC). They depend on Mn^{++} ion. They are modulated by Ca^{++} and HCO_3^- ion in a pH-independent process. They are insensitive to heterotrimeric G protein and other regulators of membrane-bound enzymes.

---

[19]CRTC-interacting partner Non-POU domain-containing octamer-binding protein Nono also binds to thyroid hormone and retinoid X receptors, as well as other types of transcription factors.

Table 11.2. Adenylate cyclase distribution and regulators (Source: [889]; Cam: calmodulin; CamK: calmodulin-dependent kinase).

Type	Distribution	Activators	Inhibitors
AC1	Brain, adrenal glands	Gs, Ca^{++}–Cam	Gi, G$\beta\gamma$
AC2	Brain	Gs, G$\beta\gamma$, PKC	Gi
AC3	Brain, olfactory epithelium	Gs, Ca^{++}–Cam	Gi, CamK2
AC4	Ubiquitous	Gs, G$\beta\gamma$, PKC	
AC5	Brain, heart, adrenal glands	Gs	Gi, Ca^{++}
AC6	Brain, heart, kidney, liver	Gs	Gi, Ca^{++}, PKA PKA, PKC
AC7	Ubiquitous	Gs, G$\beta\gamma$, PKC	
AC8	Brain, lung, pancreas	Gs, Ca^{++}–Cam	Gi, G$\beta\gamma$
AC9	Brain, heart, cochlea, skeletal muscle	Gs	Ca^{++}–PP3

Calcium and bicarbonate synergistically activate sAC [1442]. In fact, Mg^{++}, Mn^{++}, and Ca^{++} increase adenylate cyclase activity in a dose-dependent manner [1443]. Whereas human orthologs of sAC responds minimally to HCO_3^- in the absence of divalent cations, HCO_3^- stimulates Mg^{++}-bound sAC, but inhibits Mn^{++}-bound sAC in a dose-dependent manner.

Soluble adenylate cyclase has approximately 10-fold lower affinity for its ATP^{Mg} substrate [1444]. It is a predominant source of cAMP in testis and sperm cells. Soluble adenylate cyclase is an evolutionary conserved bicarbonate sensor. In mammals, it is involved in bicarbonate-induced, cAMP-dependent processes in sperm that enable fertilization. This ubiquitous protein also participates in multiple other bicarbonate- and carbon dioxide-dependent mechanisms, e.g., diuresis, breathing, blood flow, and cerebrospinal fluid and aqueous humor formation.

Alternatively spliced isoforms include: (1) 187-kDa, full-length (sAC_{FL}) and (2) 53-kDa, truncated, highly active, of relatively low abundance ($_tsAC$) variants [1445].

Soluble adenylate cyclase is associated with various intracellular organelles (centrioles, nucleus, and mitochondria, as well as mitotic spindle and midbodies) [1442]. Cytoplasmic sAC colocalizes with microtubules, but not with microfilaments [1443]. Soluble adenylate cyclase resides throughout the nervous system (dorsal root ganglia, spinal cord, cerebellum, hypothalamus, and thalamus) [1445].[20]

---

[20] Soluble adenylate cyclase is necessary for netrin-1 induced axonal outgrowth and contributes to cell responses to nerve growth factor.

Soluble adenylate cyclase may act as a cellular sensor of pH in epididymis and kidney, a CO_2 and HCO_3^- sensor in airway cilia, a mediator of oxidative burst in response to tumor-necrosis factor in human neutrophils, and a modulator of the cystic fibrosis transmembrane conductance regulator (CFTR) in the corneal endothelium and human airway epithelium [1445].

11.1.4.2 Transmembrane Adenylate Cyclase Isoforms

Many cell types express more than one adenylate cyclase species. However, Ca^{++}–calmodulin-stimulated adenylate cyclases AC1, are restricted to neurons and secretory cells.

Adenylate cyclases are activated by Gs proteins and inhibited by Gi proteins. Adenylate cyclases are regulated by different other signaling effectors, particularly Ca^{++} ion. Like AC1, AC3 and AC8 are also activated by Ca^{++}–calmodulin [1446]. The Ca^{++}–PP3 complex inhibits AC9 enzyme. Calcium ion and protein kinase-C repress AC6 subtype. Calcium-inhibited AC5 mainly localizes to the striatum and heart. Isozymes AC2, AC4, and AC7 are insensitive to Ca^{++}, but stimulated by protein kinase-C. Adenylate cyclases AC5 and AC6 are highly synthesized in the heart.

Like G-protein-coupled receptors and glycosyl-phosphatidylinositol-anchored proteins, adenylate cyclases, especially Ca^{++}-regulated ones, can be found in membrane rafts and caveolae. Certain regulatory complexes are made of G-protein-coupled receptors, like β2-adrenergic receptors, with $K_{IR}3$ channels and adenylate cyclase [1447]. These complexes are not disturbed by receptor activation or functioning of Gα subunit. However, Gβγ interferes with the formation of the dopamine receptor–$K_{IR}3$ channel complex, but not with the maintenance of the complex.

Parathyroid hormone stimulates cAMP formation and sensitizes inositol trisphosphate receptors via cAMP second messenger. Adenylate cyclases are closely apposed to IP_3R receptors. Upon stimulation by cAMP following adenylate cyclase activation, IP_3R2 that directly links to adenylate cyclases AC6 is able to release huge amounts of calcium from intracellular stores [1448]. This massive calcium release functions as a switch (all-or-none process). A graded response results from the recruitment of activated complexes, but not modulation of a given complex. Two modes of cAMP signaling then exist: (1) binary mode that depends on local cAMP delivery, during which cAMP synthesized by AC6 activates IP_3R2; and (2) analog that is much more influenced by cAMP degradation, as local cAMP gradients target signaling effectors remote from adenylate cyclases. In addition, the IP_3R2–AC6 complex is inhibited by calcium. Therefore, local calcium concentration can fluctuate due to this negative feedback loop.

Table 11.3. Substrate specificity of cyclic nucleotide phosphodiesterases. Despite homology of their catalytic domains, slight structural differences determine the type of substrates of phosphodiesterases. Some phosphodiesterases are selective hydrolases, others have a dual substrate specificity.

Substrate(s)	PDE subtypes
cAMP	PDE4, PDE7–PDE8
cGMP	PDE5–PDE6, PDE9
cAMP, cGMP	PDE1–PDE3, PDE10–PDE11

Table 11.4. Phosphodiesterase distribution and substrates (**Part 1**; Sources: [889, 1450, 1451]; Cam: calmodulin; CamK: calmodulin-dependent kinase; SMC: smooth muscle cell; vSMC: vascular smooth muscle cell).

Type	Distribution	Substrates
PDE1a	Ubiquitous	cGMP > cAMP
	PDE1a1 in lung and heart; PDE1a2 in brain	
PDE1b	Brain	cGMP > cAMP
	PDE1b1 in neurons, lymphocytes, SMCs	
	PDE1b2 in macrophages and lymphocytes	
PDE1c	Brain, blood vessels	cGMP = cAMP
PDE2a	Brain, heart,	cGMP = cAMP
	adrenal cortex, liver	
	Platelets, macrophage subtypes, endothelial cell subsets; thymocytes	
	PDE2a1 is cytosolic, PDE2a3 and PDE2a2 are membrane-bound	
PDE3a	Heart, kidney,	cGMP = cAMP
	vSMCs, platelets	
PDE3b	Brain, kidney,	cAMP
	vSMCs, adipocytes,	
	T lymphocytes, macrophages, hepatocytes, β cells	

11.1.5 Phosphodiesterases

Cyclic nucleotide phosphodiesterases hydrolyze cAMP and cGMP cyclic nucleotides to AMP and GMP, respectively. They thus control the rate of degradation of cAMP and cGMP second messengers and terminate their signaling (Table 11.3). Phosphodiesterases bind metal ions (Mg^{++}, Mn^{++}, and Zn^{++}) [1449]. Calmodulin is a Ca^{++}-dependent regulator of some types of phosphodiesterases (Tables 11.4 to 11.7).

Cyclic nucleotide phosphodiesterases differ according to their three-dimensional structure, kinetic properties, synthesis cell types, subcellular localization, modes of regulation, and inhibitor sensitivities. Initially, the multiple forms of PDE were classified into 3 categories: cAMP-targeting PDEs, cGMP-targeting PDEs, and PDEs, according to their affinities for cyclic nucleotides. They were

Table 11.5. Phosphodiesterase distribution and substrates (**Part 2**; Sources: [889, 1450, 1451]; Cam: calmodulin; CamK: calmodulin-dependent kinase; SMC: smooth muscle cell; vSMC: vascular smooth muscle cell).

Type	Distribution	Substrates
PDE4a	Ubiquitous	cAMP
	PDE4a1 links to membrane; AKAP-bound PDE4a5 in membrane ruffles	
PDE4b	Ubiquitous	cAMP
PDE4c	Lung, testis, several cell lines mainly of neuronal origin	cAMP
PDE4d	Ubiquitous	cAMP
PDE5a	SMC, platelets, brain, lung, heart, kidney, skeletal muscle	cGMP
	PDE5a1 and PDE5a2 are widespread, PDE5a3 is specific to vSMCs	

Table 11.6. Phosphodiesterase distribution and substrates (**Part 3**; Sources: [889, 1450, 1451]; Cam: calmodulin; CamK: calmodulin-dependent kinase; SMC: smooth muscle cell; vSMC: vascular smooth muscle cell).

Type	Distribution	Substrates
PDE6a	Rods, pineal gland	cGMP
PDE6b	Rods, pineal gland	cGMP
PDE6c	Cones	cGMP
PDE6d	Rods	cGMP
PDE6g	Rods	cGMP
PDE6h	Cones	cGMP
PDE7a	Brain, heart, kidney, skeletal muscle, endothelial cells, immunocytes	cAMP
	PDE7a1 in immunocytes, PDE7a2 in heart	
PDE7b	Brain, heart, liver, pancreas, skeletal muscle	cAMP
	PDE7b1 is widespread, PDE7b2 in testis, PDE7b3 in heart	

afterward categorized into major categories, such as Ca^{++}–calmodulin-stimulated PDE, cGMP-stimulated PDE, cGMP-inhibited PDE, and cAMP-stimulated PDE, according to their regulatory and kinetic properties. Because the list of new PDE isozymes expands, a new nomenclature has been proposed based on the primary structure. In fact, cyclic nucleotide phosphodiesterases constitute 11 families, each family containing several isoforms and splice variants. In fact, 21 different transcripts generate via alternative transcriptional start sites and alternative splicing more than 100 functional PDE enzymes.

Table 11.7. Phosphodiesterase distribution and substrates (**Part 4**; Sources: [889, 1450, 1451]; Cam: calmodulin; CamK: calmodulin-dependent kinase; SMC: smooth muscle cell; vSMC: vascular smooth muscle cell).

Type	Distribution	Substrates
PDE8a	Widespread	cAMP
PDE8b	Brain, thyroid	cAMP
	PDE8b1 in thyroid, PDE8b3 in brain and thyroid	
PDE9a	Brain, kidney, spleen, digestive tract, prostate	cGMP
	PDE9a1 is nuclear, PDE9a5 cytosolic	
PDE10a	Brain, heart, thyroid, pituitary, testis	cAMP < cGMP
	(cytosolic PDE10a1–PDE10a3 variants)	
PDE11a	Pituitary, adrenal glands, thyroid, salivary glands, liver, skeletal muscle, prostate, testis	cAMP = cGMP
	PDE11a3 is specific to testis; PDE11a4 level is the highest in prostate	

Certain members of the PDE family induce vasodilation either by acting directly in vascular smooth muscle cells such as those of pulmonary arteries [1452][21] or via endothelial cells such as those of the aorta [1453][22] (Tables 11.8 and 11.9).

Protein PDE2 is stimulated by cGMP. Subtypes PDE4, PDE7, and PDE8 are specific for cAMP [1454, 1455]. Phosphodiesterases limit the diffusion of second messengers, subsequently influencing the activity of cyclic nucleotide-gated ion channels, RapGEF3 and RapGEF4 regulators (Sect. 9.4.1.6), and PKA and PKG kinases.

11.1.5.1 Phosphodiesterase-1 Family

Three genes (Pde1A–Pde1C) encode Ca^{++}–calmodulin-dependent members of the PDE1 family. The 3 resulting gene products (PDE1a–PDE1c) differ in their substrate affinities, specific activities, activation constants for calmodulin, tissue distribution, molecular weights, and regulation by Ca^{++} ion and phosphorylation. Complementary DNA segments encode cardiac PDE1 isozyme PDE1a1, cerebral 60-kDa PDE1a2 and 63-kDa PDE1b1, and 70-kDa PDE1c [1456]. In addition,

---

[21] In chronic hypoxia-induced pulmonary hypertensive rats, cAMP and cGMP PDE activities rise. Heightened cAMP–PDE activity in the first pulmonary arterial branches and intrapulmonary vessels is due to PDE3, augmented cGMP–PDE in the main pulmonary artery to PDE1. Activity of PDE5 is higher in the first pulmonary branches and intrapulmonary arteries.

[22] Cultured bovine aortic endothelial cells only contain cGMP-stimulated PDE2 and PDE4 enzymes. The ^{L}-arginine–NO–cGMP pathway leads to vasodilation.

11.1 cAMP Signaling

Table 11.8. Function of PDEs (**Part 1**; Source: [1450]: ANP: atrial natriuretic peptide).

Type	Effect
	Calcium- and calmodulin-activated PDE1 phosphodiesterases
	Phosphorylation by PKA (PDE1a and PDE1c)
	Phosphorylation by CamK2 (PDE1b)
PDE1a	Vasomotor tone regulation
PDE1b	Dopamine signaling, immunocyte activation and survival
PDE1c	Vascular smooth muscle cell proliferation
	cGMP-activated PDE2 phosphodiesterases
PDE2	Crosstalk between cGMP and cAMP pathways (inhibition), inhibition by ANP of aldosterone secretion, phosphorylation by PKA of cardiac $Ca_V1.2a$ channels, long-term memory, vascular permeability in inflammation
	cGMP-inhibited PDE3 phosphodiesterases
	Phosphorylation by PKA
PDE3a	Regulation of cardiac contractility, platelet aggregation, vascular smooth muscle contraction, oocyte maturation, renin release
PDE3b	Cell proliferation, inhibitory effects of leptin, signaling by insulin and renin
	Phosphorylation by PKA and ERK
PDE4	Cerebral function, monocyte and macrophage activation, neutrophil infiltration, vascular smooth muscle proliferation, vasodilation, cardiac contractility, fertility

Table 11.9. Function of PDEs (**Part 2**; Source: [1450]).

Type	Effect
	Phosphorylation by PKA and PKG
PDE5	Regulation of vascular smooth muscle contraction, NO–cGMP signaling in platelets (aggregation)
PDE6	Ocular phototransduction
PDE7	Activation of T lymphocytes and inflammatory cells
PDE8	T-lymphocyte activation
PDE9	Cerebral NO–cGMP signaling
PDE10	Cerebral functions
PDE11	Sperm development and function

Table 11.10. Kinetic properties of PDE1 isoforms (Source: [1457]; K_{Cam}: association constant for calmodulin). The activation of PDE1a by Ca^{++}–calmodulin relieves auto-inhibition.

Subtype	Species	Molecular weight (kDa)	$K_{M_{cAMP}}$ (μmol)	$K_{M_{cGMP}}$ (μmol)	K_{Cam} (nmol)
PDE1a1	Bovine (lung)	58	42	2.75	
	Bovine (heart)	59	40	3.2	0.1
	Dog (heart)	68	2.8	2.1	
PDE1a2	Bovine (brain)	60–61	32	2.7	1
PDE1a3	Human	61	51	3.5	
PDE1b1	Bovine (brain)	63	12	1.2	1
PDE1c1	Mouse	72	3.5	2.2	
PDE1c2	Rat		1.2	1.1	
PDE1c3	Human	72	0.57	0.33	
PDE1c4/5	Mouse	74	1.1	1.0	

alternative splicing creates diverse N- and C-termini. In humans, several splice variants of PDE1s have been identified with different Michaelis constants (K_M) for a given substrate and Ca^{++} sensitivity (PDE1a3 and PDE1c1–PDE1c5) [1457] (Table 11.10). Nonetheless, the bulk structure of PDE1 isoforms is conserved with 4 domains: 2 calmodulin-binding domains, an inhibitory motif, and a PDE1 catalytic sequence.

The calmodulin-stimulated phosphodiesterases-1 are ubiquitous. In particular, the 3 PDE1 isoforms are expressed in the central nervous system and many peripheral neurons, but to different degrees according to the region. Different PDE1 isoforms differentially reside in the heart and blood vessels as well as macrophages and T lymphocytes [1450]. Most PDE1 isoforms are cytosolic.

Stimulation of PDE1 enzymatic activity requires a physiological concentration of Ca^{++} and calmodulin, once both molecules have complexed. Inactive calmodulin and PDE1 exist separately at low Ca^{++} concentration. When Ca^{++} concentration rises, calmodulin binds to Ca^{++} and becomes active. Active calmodulin then associates with PDE1 (Ca^{++}–Cam–PDE1 complex).

The inhibitory domain of PDE1, which is conserved among all PDE1 isoforms, maintains the enzyme in a low activity state in the absence of Ca^{++} ions; full activation is restored by Ca^{++}–calmodulin [1457]. Like Ca^{++}-sensitive adenylate cyclases, phosphodiesterase-1 is activated by Ca^{++} influx.[23] Unlike Ca^{++}-

---

[23] An increase in intracellular Ca^{++} concentration is often associated with a reduction of intracellular cAMP concentration due to either Ca^{++}-inhibited adenylate cyclases AC5 and AC6 or phosphodiesterase-1.

sensitive adenylate cyclases that are exclusively regulated by the capacitative Ca^{++} entry through transient receptor potential channels in non-excitable cells,[24] PDE1 is activated almost exclusively by Ca^{++} influx from the extracellular space [1457].[25]

Phosphorylation of PDE1a and PDE1c by cAMP-dependent protein kinase PKA and of PDE1b by calmodulin-dependent kinase CamK2 reduces the affinity of PDE1 isoforms for calmodulin and, thus, their sensitivity to Ca^{++} ions [1457]. On the other hand, PDE1 isoforms are dephosphorylated (re-activated) by Ca^{++}–calmodulin-dependent protein phosphatase PP3 (or calcineurin).

PDE1a

Cardiac PDE1a1 isoform, cerebral PDE1a2, and pulmonary and ocular PDE1a isozymes share almost identical kinetic and immunological properties, but are differentially activated by calmodulin [1456]. Calmodulin concentration in mammalian brain is approximately 10 times higher than that in mammalian heart. Consequently, cardiac PDE1a1 and ocular PDE1a isozymes have a higher affinity for calmodulin than cerebral PDE1a2 subtype. Pulmonary PDE1a isoform has the highest apparent affinity for calmodulin, because it contains calmodulin as a subunit. In addition, when calmodulin concentration increases, Ca^{++} concentration required for half-maximal activation decreases. Differential affinity for Ca^{++} and calmodulin of tissue-specific isozymes may result from adaptive regulation in respective tissues as well as fine-tuned control. Moreover, several acidic phospholipids, such as lysophospholipids phosphatidylinositol, and phosphotidylserine, as well as unsaturated fatty acids and gangliosides, can activate PDE1a in a Ca^{++}-independent manner.

Isozymes PDE1a1 and PDE1a2 have a higher affinity for cGMP than cAMP (Table 11.11) [1456].

---

[24] Calcium-mobilizing messengers such as IP_3 trigger Ca^{++} import secondary to the depletion of Ca^{++} intracellular stores (capacitative Ca^{++} entry [CCE]). The duration of the resulting sustained elevation of intracellular Ca^{++} concentration is determined by the filling of intracellular Ca^{++} stores. On the other hand, the Ca^{++} release from intracellular stores is transient. Therefore, the kinetics of intracellular Ca^{++} release and capacitative Ca^{++} entry differ. Besides, Ca^{++}-sensitive adenylate cyclases colocalize with CCE channels, mainly in membrane rafts.

[25] The calcium-based regulation is preferentially mediated by Ca^{++} influx through plasmalemmal Ca^{++} carriers than Ca^{++} release from intracellular stores. Unlike transmembrane adenylate cyclases, PDE1 isoforms, to be anchored to the plasma membrane must undergo lipid modification or complex with scaffold proteins.

Table 11.11. Affinity of PDE1a1 and PDE1a2 for cGMP and cAMP (Source: [1456]).

Type	$K_{M_{cAMP}}$ (µmol)	$K_{M_{cGMP}}$ (µmol)
PDE1a1	40	3.2
PDE1a2	35	2.7

Enzyme PDE1a interacts with apolipoprotein apoA1, neurocalcin-δ,[26] neuronal calcium sensor NCS1,[27] and hippocalcin.[28] Cardiac PDE1a1 and cerebral PDE1a2 are phosphorylated by cAMP-dependent protein kinase-A (Ser120) that attenuates its affinity for calmodulin [1456]. Phosphorylation is inhibited by Ca^{++} and calmodulin. Isozyme PDE1a2 can be dephosphorylated by the calmodulin-dependent PP3 phosphatase.

The central nervous system contains the highest PDE1a activity. Yet, it is also detected in human lymphocytes and monocytes that circulate in blood. The PDE1A gene gives rise to many alternatively spliced transcripts (PDE1a1–PDE1a6 and PDE1aL42). $5'$-Splice variants of the PDE1A gene generate 59-kDa PDE1a1 and 61-kDa PDE1a2 isozymes [1456].

PDE1b

Isoform PDE1b1 is detected in the central nervous system, lung, heart, smooth and skeletal muscle, and olfactory epithelium. In addition to neurons and smooth muscle cells, among other cell types, cytosolic PDE1b1 localizes to lymphocytes, monocytes, macrophages, and spermatids [1456]; cytosolic PDE1b2 in macrophages and lymphocytes [1450].

PDE1b1

Brain PDE1b1 isoform is phosphorylated by calmodulin-dependent protein kinase CamK2 that decreases PDE1b1 affinity for calmodulin, hence an increase in Ca^{++} concentration is required for enzyme activation by calmodulin [1456]. Conversely, PDE1b1 can be dephosphorylated by calmodulin-dependent protein phosphatase

---

[26] This member (nCalδ) of the neuronal calcium sensor (NCS) family is expressed in the central nervous system, retina, and adrenal gland.

[27] A.k.a. frequenin homolog. This calcium sensor is a high-affinity, low-capacity, calcium-binding protein.

[28] This calcium-binding protein also belongs to the neuronal calcium sensor (NCS) family. It is produced especially in the hippocampus. Hippocalcin is involved in PLD1 and PLD2 as well as MAPK signaling.

PP3 that increases PDE1b1 affinity for calmodulin, thereby reducing Ca^{++} concentration for its calmodulin-mediated activation.[29] When cAMP concentration augments, cAMP-dependent protein kinase PKA phosphorylates (activates) protein phosphatase inhibitor PPI1. Phosphorylated PPI1 can inhibit protein phosphatase PP1 that reverses effect of CamK2 autophosphorylation. On the other hand, PPI1 is dephosphorylated (inactivated) by PP3 phosphatase. Consequently, PP1 is reactivated and cAMP can then exert its inhibition on PDE1b1 activity.

Isoform PDE1b1 has a basal affinity for cAMP and cGMP. In the presence of Ca^{++}–calmodulin, PDE1b1 affinity rises 2- to 3-fold for both cAMP (K_M 12 µmol) and cGMP (K_M 1.2 µmol) [1456].

PDE1b2

Granulocyte–macrophage colony-stimulating factor (CSF2) that is involved in the differentiation of the myeloid lineage upregulates PDE1b2 isoform during monocyte-to-macrophage differentiation. On the other hand, upregulation of PDE1b2 is suppressed during the differentiation of macrophages to a dendritic cell phenotype. T lymphocytes can also produce PDE1b2 [1456]. In addition, PDE1b2 is expressed strongly in the spinal cord and slightly in the putamen and caudate nucleus, thyroid, thymus, small intestine, and uterus.

PDE1c

Isoform PDE1c is expressed in the cilia of olfactory sensory neurons, where olfactory signal transduction takes place [1456]. Yet, PDE1c is elevated in pulmonary artery smooth muscle cells in both idiopathic and secondary pulmonary arterial hypertension. It also resides in cytosol of human cardiomyocytes as well as cells of the central nervous system, pancreas, bone, and testis. Several splice variants exist [1456]. Splice variants of PDE1c have different affinities for cAMP and cGMP (human native PDE1c1 $K_{M_{cAMP}}$ 0.9 ± 0.2µmol; $K_{M_{cGMP}}$ 1.2 ± 0.2µmol).

Activity of PDE1c is inhibited by protein kinase-A. Subtype PDE1c interacts with various types of adenylate cyclases (AC1, AC3, and AC5–AC9) and calmodulin-like protein-3 and atrial natriuretic peptide receptor-A precursor [1456].

11.1.5.2 Phosphodiesterase-2 Family

Phosphodiesterases-2 are encoded by a single gene (Pde2A). Three splice variants have been observed (PDE2a1–PDE2a3). Soluble PDE2a1 is cytosolic, whereas PDE2a1 and PDE2a3 bind to the plasma membrane [1450]..

---

[29]Calcium concentrations required for 50% activation of phosphorylated and non-phosphorylated PDE1b are 1.9 and 1.1 µmol, respectively [1456].

Members of the PDE2 family target both cAMP and cGMP messengers. They are produced in various cell types of the brain and heart, as well as in platelets, endothelial cells, adrenal glomerulosa cells, and macrophages [1450]. Expression of PDE2 is upregulated during the differentiation of monocyte to macrophage.

In endothelial cells, PDE2a is synthesized under basal conditions only in small vessels and capillaries, but not in large vessels. In the brain, PDE2a abounds in specific regions and cell types.

Members of the PDE2 family are responsible for mutual inhibition between the cGMP and cAMP pathways. In particular, in adrenal glomerulosa cells, PDE2 mediates inhibition exerted by atrial natriuretic peptide on aldosterone secretion.[30] In these cells, elevation of cGMP mediated by ANP activates PDE2 that, in turn, lowers cAMP stimulated by adrenocorticotropin. In platelets, NO causes cGMP accumulation that activates PDE2 and reduces cAMP agent.[31]

In human cardiomyocytes, PDE2 regulates $Ca_V1.2a$ channel. In addition, PDE2 intervenes in the cAMP response to catecholamine stimulation. It also mediate the inhibition of NO on cAMP agent. Like in platelets, the activation of PDE2 by cGMP is antagonized by the inhibition of PDE3 enzyme.

11.1.5.3 Phosphodiesterase-3 Family

Members of the PDE3 family hydrolyze both cAMP and cGMP messengers. They are encoded by 2 genes (Pde3A and Pde3B). Variants have been identified only for the PDE3a isoform (PDE3a1–PDE3a3) [1450].

Both PDE3a and PDE3b are regulated by phosphorylation in response to hormonal stimulation in several cell types. In platelets, adrenaline and prostaglandins exert a feedback via PKA to activate PDE3a isoform.

Insulin, insulin-like growth factor IGF1, and leptin use the PI3K pathway to phosphorylate PDE3b by activated PKB kinase [1451] (Table 11.12).[32]

Subtype PDE3a is relatively highly expressed in platelets, vascular smooth muscle cells, cardiomyocytes, and oocytes. In platelets, PDE3a contributes to the regulation of aggregation. On the other hand, PDE3b is a major PDE in adipose

---

[30] Atrial natriuretic peptide antagonizes the secretagogue action of K^+ ions and angiotensin-2, as it decreases appropriate Ca^{++} influx and, hence, intracellular Ca^{++} concentration. It also inhibits adenylate cyclase, but activates guanylate cyclase.

[31] Enzyme PDE3 plays a prominent role on platelet aggregation.

[32] In hepatocytes, activated PDE3b by the PI3K axis decreases the production of glucose. In basal conditions, leptin does not change glucose production in hepatocytes, but in the presence of several gluconeogenic precursors, leptin decreases the production of glucose. It also affects ATP-sensitive K^+ channels [1458]. In pancreatic β cells, cAMP synthesized by adenylate cyclase stimulated by glucose amplifies glucose-induced insulin secretion via elevated intracellular calcium concentration after opening of Ca_V channels and ryanodine receptors as well as calcium-induced Ca^{++}-release [1451]. Messenger cAMP is rapidly degraded by phosphodiesterases. Only PDE3b is produced in β cells, in which its activity is inhibited by cGMP.

11.1 cAMP Signaling

Table 11.12. Pancreatic β-cell phosphodiesterase (Source: [1451]). Membrane-bound PDE3b regulates the cAMP pool that modulates insulin secretion. Isoform PDE3b opposes insulin release. Activity of PDE3b in β cells may be regulated by insulin, insulin-like growth factor-1 (IGF1), leptin, and glucose. In adipocytes, insulin activates PDE3b to support its antilipolytic effect. Autocrine regulator insulin binds to its receptor on secreting β cell to inhibit its own release (negative feedback). Factor IGF1 potently activates PDE activity using the PI3K–PKB axis. Leptin may prevent augmentation by glucagon-like peptide GLP1 of glucose-induced insulin release due to reduced cAMP concentration.

Regulator	Target	Product Metabolism	Catabolism Effect
Ca^{++}	**PDE1**		
	ACase	cAMP	AMP (**PDE1**)
			PKA $\longrightarrow$ **PDE3b**
	Exocytosis		Insulin release
Glucose		ATP	K_{ATP},
			membrane depolarization,
			Ca^{++} influx
			Activation of **PDE1/3b**
Insulin, IGF1, leptin	**PDE3b**		
NO	GCase	cGMP	Inhibition of **PDE3b**
			GMP (**PDE5**)

tissue, liver, and pancreas, in addition to the cardiovascular apparatus [1450]. Isoform PDE3b is also synthesized in T lymphocytes and macrophages, in addition to pancreatic β cells and adipocytes.

Members of the PDE3 family are also involved in the regulation of cardiac contractility and vasomotor tone. Enzyme PI3Kγ can associate with PDE3b to control cardiac contractility. Both PDE3a and PDE3b are synthesized in vascular smooth muscle cells to modulate the vasomotor tone.

11.1.5.4 Phosphodiesterase-4 Family (PDE4)

Four cAMP-specific PDE4 subtypes (PDE4a–PDE4d), encoded by 4 genes (Pde4A-Pde4D), hydrolyze cAMP to AMP agent. More than 20 different PDE4 isoforms have been detected; they are created via alternative start sites and alternative splicing. The PDE4 isozymes have multiple promoters and transcription factors [1459].[33]

---

[33] Long variants have an N-terminus with regulatory domains and phosphorylation sites, 2 upstream conserved regions (UCR1 and UCR2), and a catalytic domain. In general, each gene also gives rise to one or more short isoforms with different N-termini. For example, 9 PDE4d variants have been identified.

Many of the PDE4 variants have a cell type-specific expression. In addition, several variants have distinct subcellular localizations [1450]. Therefore, PDE4 isoforms control spatially distinct pools of cAMP messenger.

Phosphodiesterases-4 are phosphorylated (activated) by PKA kinase. In most cells, in particular immunocytes that possess PDE4 subtypes, PDE4b and PDE4d predominate. Enzymes of the PDE4 family can homodimerize, but not heterodimerize.

Upstream conserved regions (UCR1–UCR2) of PDE4 enzymes influence outcome of phosphorylation of the catalytic unit. They regulate the catalytic activity and can interact with each other.[34] More precisely, ERK kinases phosphorylate enzymes of the PDE4B, PDE4C, and PDE4D subfamilies. Consequently, ERK kinases: (1) prevent the activity of long alternatively spliced variants that have both UCR1 and UCR2 motifs; (2) activate short variants that possess only UCR2; and (3) do not influence the action of supershort variants that have a truncated UCR2, thereby being catalytically inactive [1460]. Catalytically inactive PDE4 dead-short isoforms lack both UCR1 and UCR2 and have truncated N- and C-termini of their catalytic units.

Phosphatidic acid binds to long PDE4 isoforms ($PDE4_L$) and primes their activation [1460]. Reactive oxygen species can trigger the phosphorylation of $PDE4_L$ isoforms at 2 sites: (1) phosphorylation by ERK at the C-terminus of the catalytic unit and (2) that at the N-terminus of the catalytic unit, which does not modify the PDE4 activity, but switch the inhibitory phosphorylation by ERK to activation [1460].

Multiple other interactors include β-arrestin-1 and -2, PDE4d-interacting protein (or myomegalin), receptor for activated C-kinase RACK1,[35] several A-kinase anchoring proteins, and protein Tyr kinases such as members of the SRC family [1450].

PDE4a

The isoform-specific N-terminus controls the subcellular localization of phosphodiesterases. In the absence of its N-terminus, membrane-associated PDE4a1 isoform lodges in the cytosol.

---

[34] The UCR1 domain yields the site for stimulatory phosphorylation by protein kinase-A. The UCR2 sequence gates the access to the catalytic site. Both UCR1 and UCR2 control the outcome of phosphorylation of the catalytic unit of enzymes of the PDE4B, PDE4C, and PDE4D families by extracellular signal-regulated kinase.

[35] Scaffold RACK1 recruits PKC, Src, integrins, and GABA receptors. In particular, PDE4d5 long form that binds to RACK1 can be recruited to RACK1-supported complex in appropriate regions of the cell.

PDE4d

The PDE4D gene encodes 11 splice variants (PDE4d1–PDE4d11) due to alternative splicing. These isoforms can be classified into long (PDE4d$_L$, i.e., PDE4d3 to PDE4d5, PDE4d7 to PDE4d9, and PDE4d11), short (PDE4d$_S$, i.e., PDE4d1), and supershort (PDE4d$_{SS}$, i.e., PDE4d2, PDE4d6, and PDE4d10) categories [1460]. Each isoform is characterized by an isoform-specific N-terminal region. The catalytic unit comprises 3 subdomains that are coordinated by 2 metal ions (Mg^{++} and Zn^{++}). Widespread PDE4d isoforms can reside in the cytosol and associate with membranes.

Phosphorylation of PDE4d$_L$ by ERK1 and ERK2 on catalytic unit (inhibitory) antagonizes that of PKA on UCR1 domain (stimulatory), and conversely the action of PKA neutralizes the effect of ERK enzymes [1460]. These kinases can thus organize feedback loops.

The N-terminal region enables the recruitment of PDE4d isoforms by scaffold proteins, such as β-arrestin, AKAPs, disrupted in schizophrenia DISc1,[36] receptor of activated protein kinase-C RACK1, and nuclear distribution gene-E homolog (NuDe)-like protein NDeL1,[37] thereby determining sites of distinct PDE4 isoforms (local cAMP sinks) to compartmentalize cAMP signaling [1460].

Myomegalin, or phosphodiesterase-4D-interacting protein (PDE4dIP),[38] of the Golgi body and centrosome interacts directly with PDE4d enzymes.

Isoform PDE4d5 that is detected in cardiomyocytes and vascular smooth muscle cells has an additional binding site for β-arrestin in its N-terminus. The βArr–PDE4d5 complex anchors Ub ligase DM2 [1460]. The subsequent transient ubiquitination of PDE4d5 further enhances the βArr–PDE4d5 interaction (autoamplification). This transient ubiquitination can result from the activation of receptors, such as β2-adrenoceptor and vasopressin V$_2$ receptor, that recruit the βArr–PDE4d5–DM2 complex, thereby delivering a cAMP-degrading complex to the plasma membrane, where is synthesized cAMP messenger.

Unlike β2-adrenoceptor that attracts the βArr–DM2–PDE4d5 complex, β1-adrenoceptor directly interacts with PDE4d8 [1460].

In vascular smooth muscle cells, activation of PDE4d increases its affinity for Mg^{++} and decreases cAMP concentration. Isoform PDE4d also controls cAMP level in respiratory conduits. Parasympathetic control of the smooth muscle tone of airways involves: (1) muscarinic M$_1$ and M$_3$ receptors coupled to phospholipase-C and Ca^{++} as well as (2) M$_2$ receptors inhibitory for adenylate cyclase. Released

---

[36] The DISc1 interactome comprises also PDE4b enzyme.

[37] Protein DISc1 dimerizes and oligomerizes. The self-association ability allows the regulation of its affinity for binding partners such as NDeL1 agent. The latter localizes to the centrosome (microtubule organizing center) and acts as a cysteine endopeptidase. Scaffold protein NDeL1 participates in centrosome function and nuclear translocation, as well as in the regulation of the dynein nanomotor during neuronal migration and in cortical neuronal positioning [1460]. Protein NDeL1 can bind directly to members of the 4 PDE4 subfamilies.

[38] A.k.a. cardiomyopathy-associated protein CMyA2.

acetylcholine from parasympathetic nerves induces smooth muscle contraction, with a control by M_2 muscarinic receptors to limit the bronchoconstriction magnitude.

Multiprotein signaling complexes focalize the enzyme activity that transduce actions of many signaling agents. Isoform PDE4d and PKA form a complex coordinated by A-kinase anchoring proteins. In cardiomyocytes, AKAP6 assembles a self-regulatory cAMP signaling module with PKA and PDE4d3 [1461]. Enzyme PDE4 is recruited by β-arrestin close to G-protein-coupled receptors [1462].[39]

In cardiomyocytes, AKAP9 and PDE4d complex with $K_V7.1$ channel to control the basal and β-adrenoceptor-activated channel activity (i_{Ks} current) [1460].

Protein AKAP7 that anchors PKA to the basolateral plasma membrane of epithelial cells and interacts with $Ca_V1.2a$ channels in skeletal muscle cells and cardiomyocytes, thereby facilitating $Ca_V1.2a$ phosphorylation by protein kinase-A, interacts directly with PDE4d3 enzyme. The AKAP7–PDE4d complex may regulate the activity of AKAP-tethered PKA on aquaporin-2+ vesicles, thereby regulating aquaporin-2 transfer and water permeability [1460].

Cardiomyocytes and arterial smooth muscle cells use distinct phosphodiesterase PDE4d splice variants to regulate PKA activity [1463]. In rat cardiomyocytes (hypertrophied or not), PDE4d3 participates in the PKA–AKAP complex, whereas in rat aortic smooth muscle cells, PDE4d3 is not detected in PKA–AKAP complex, but PDE4d8 associates with PKA–AKAP complexes. Both PDE4d8 and PKA are recruited to the leading edge of migrating vascular smooth muscle cells.

In cardiomyocytes, PDE4d3 interacts with the ryanodine RyR2 receptor [1460]. This interaction may yield a negative feedback to limit β-adrenoceptor-primed PKA phosphorylation of RyR2 receptor.

Prostaglandin-E2 is a major pro-inflammatory agent in the cardiovascular apparatus and a potent inducer of cAMP production (Gs–ACase–cAMP–PKA axis), which remains confined along the plasma membrane, via stimulatory G protein coupled to type-E prostaglandin EP4 receptor, the most abundant EP subtype in the myocardium. In addition, PGE2 attenuates the cardiac contractility (negative inotropic effect) caused by adrenergic stimuli [1464]. Prostaglandin-E2 activates PKA that phosphorylates (activates) PDE4d isozyme that degrades cAMP, thereby impeding the transfer of cAMP from the plasma membrane to the sarcoplasmic reticulum. On the one hand, catecholamines stimulate Gs-coupled β-adrenergic receptors, hence cAMP signaling that initiates phosphorylation of PKA substrates on the plasma membrane, sarcoplasmic reticulum, and myofibrils, to support cardiomyocyte contractility. β-Adrenoceptor associates with different PDE4d subtypes, but, once it is liganded, PDE4d isoforms dissociate from activated β-adrenoceptors to allow cAMP diffusion from the plasma membrane to intracellular organelles [1464]. On the other, cAMP produced in response of PGE2

---

[39] β2-Adrenergic receptor stimulation promotes cAMP production. β-Arrestins desensitize β2-adrenergic receptors by binding to receptors and recruiting phosphodiesterases for cAMP degradation, hence restricting cAMP signaling magnitude.

11.1 cAMP Signaling

cannot access the sarcoplasmic reticulum to foster calcium signaling and cardiomyocyte contraction. In particular, PGE2 precludes phosphorylation by PKA of phospholamban primed by β-adrenoceptors. Therefore, a Gs-coupled prostaglandin receptor is able to prevent the intracellular signaling initiated by a Gs-coupled catecholamine receptor via the control of cAMP transport that results from intracellular cAMP degradation. Moreover, PGE2 prevents PDE4 dissociation from β-adrenoceptors [1464]. These 2 types of Gs-coupled receptors — EPs and βARs — colocalize in many cell types, such as neurons and astrocytes, in addition to cardiomyocytes [1464]. In microglial cells, EP promotes and βAR represses inflammation.

Phosphodiesterase-4D complexes with SH3 and multiple ankyrin repeat domain-containing protein SHAnk2 and cystic fibrosis transmembrane conductance regulator (CFTR) that controls phosphorylation by PKA of CFTR [1460].

Subtype PDE4d3 can connect to RapGEF3 and -4, which activate Rap1 and Rap2 GTPases [1460]. Scaffold AKAP6 can tether to PDE4d3, PKA, ERK5, and RapGEF3 and -4 agents. Isozyme PDE4d undergoes inhibitory and stimulatory phosphorylations by ERK and PKA, respectively.

Isoform PDE4d4 interacts directly with certain SH3 domain-containing proteins, such as SRC family kinases (Abl, Fyn, Lyn, and Src) as well as PI3K and fodrin [1460]. It also links to spectrin and, then, may control the phosphorylation by PKA of the microtubule-stabilizing Tau protein.

Inhibitors that are PDE4-selective and inhibit all PDE4 isoforms are used to treat chronic obstructive pulmonary disease and fibrosis, among other diseases [1460].

11.1.5.5 Phosphodiesterase-5 Family (PDE5)

Phosphodiesterase-5 that possesses high affinity-binding sites for cGMP is characterized by a relative specificity for cGMP messenger at low levels. A single Pde5 gene generates 3 variants (PDE5a1-PDE5a3) due to differentially regulated promoters.

Phosphodiesterase-5 lodges in the brain, heart, lung, kidney, liver, skeletal muscle, pancreas, gastrointestinal tract, and placenta [1450].[40]

In the cardiovasular system, PDE5a is synthesized in vascular endothelial and smooth muscle cells, cardiomyocytes, as well as platelets.

Phosphodiesterase-5A participates in the regulation of vascular tone. Endothelial PDE5a resides in or near caveolin-rich membrane rafts. The vasodilatory NO–sGC–cGMP–PKG pathway is associated with caveolae. Agents NOS, sGC, PKG1, and PKA, actually localize to or near caveolae. Following cGMP catabolism by PDE5a, PKG1 activity decays, hence NOS3 phosphorylation (activation) by PKG1 (Ser1177) [1465]. In addition, NOS3 can be strongly phosphorylated by PKG2 kinase.

---

[40] The highest levels of Pde5A mRNA are found in the cerebellum, kidney, and pancreas [1450].

Phosphodiesterase-5 is phosphorylated not only by PKG, but also by PKA. The latter stabilizes its catalytic activity by enhancing the affinity of cGMP binding [1450]. This modification prolongs PDE5 activation (feedback loop initiated by cGMP synthesis via guanylate cyclases). Messenger cGMP operates as a PDE5 feedforward activator.

11.1.5.6 Phosphodiesterase-6 Family (PDE6)

Members of the PDE6 family are photoreceptor phosphodiesterases, as they contribute to the conversion of a light signal into a photoresponse. They are encoded by 3 genes (Pde6A–Pde6C). In addition, PDE6γ and PDE6δ subunits modulate PDE6 activity and localization.

11.1.5.7 Phosphodiesterase-7 Family (PDE7)

Members of the PDE7 family, like PDE4 and PDE8, is highly selective for cAMP, especially at low cAMP concentrations. In humans, 3 splice variants exist (PDE7a1–PDE7a3).

They are synthesized in the brain, heart, lung, kidney, muscle, spleen, and thymus, as well as various types of immunocytes [1450]. However, PDE7b2 and PDE7b3 are restricted to testis and heart, respectively.

11.1.5.8 Phosphodiesterase-8 Family (PDE8)

Members of the PDE8 family are encoded by 2 genes (Pde8A–Pde8B). Both isoforms (PDE8a–PDE8b) have a very high affinity for cAMP (the highest cAMP affinity among PDEs; K_M 40–150 nmol for cAMP, K_M >100 μmol for cGMP [1466]) that confers a specificity for this messenger.

Both PDE8a and PDE8b share C-terminal catalytic domain common to all PDEs, but they differ from other PDEs by their N-terminus. The N-terminus contains a receiver domain (Rec) and a close Period, ARNT, and Sim motif (PAS), a regulatory sequence of several regulators of circadian rhythm.

PDE8a

Subtype PDE8a possesses several variants produced by alternative splicing and alternative start sites [1450]. Isoform PDE8a2 is a spliced variant of the most abundant variant PDE8a1 without the PAS domain; PDE8a3 is a truncated protein without the PAS and REC sequences; PDE8a4 and PDE8a5 are identical truncated proteins that are longer than PDE8a3 and also lacks the PAS and REC regions. Various PDE8b variants also arise from alternative splicing.

Expression of Pde8A mRNA is widespread with its highest levels in the heart, kidney, small intestine, colon, spleen, ovary, and testis [1450]. Protein PDE8a1 has been detected in T lymphocytes. Production of PDE8b is more restricted, mainly in the brain and thyroid. However, alternatively spliced isoform PDE8b1 is detected only in the brain, although PDE8b3 is produced in the brain and thyroid.

Activity of PDE8a depends on divalent cations, either Mg^{++} or Mn^{++} ion. It can be phosphorylated by protein kinase-A and -G. Enzyme PDE8a modulates excitation–contraction coupling in ventriculomyocytes, as its decreases calcium transients and calcium spark frequency [1466].

PDE8b

In humans, full length PDE8b (PDE8b1) contains 885 amino acids. It has a high affinity for cAMP (K_M 101 nmol) [1467]. Several splice variants of PDE8b exist in humans, with specific tissue expression. In particular, variants lacking the PAS domain have been observed.

11.1.5.9 Phosphodiesterase-9 Family (PDE9)

Ubiquitous members of the PDE9 family have the highest affinity for cGMP messenger ($K_{M_{cGMP}}$ 70–170 nmol). The single family member PDE9a coexist with numerous variants after mRNA processing, but only nuclear PDE9a1 and cytosolic PDE9a5 have been characterized [1450].

11.1.5.10 Phosphodiesterase-10 Family (PDE10)

Phosphodiesterase-10 is encoded by a single PDE10A gene. Four variants PDE10a1–PDE10a4) have been identified. Phosphodiesterase-10 hydrolyzes both cAMP and cGMP in vitro, with a higher affinity for cAMP than for cGMP substrate [1450]. In vivo, PDE10 may function as a cAMP-inhibited cGMP phosphodiesterase.

Phosphodiesterase-10 is mainly synthesized in the central nervous system. It also produced in the thyroid and pituitary gland as well as skeletal and cardiac muscles [1450].

Variant PDE10a2 can be phosphorylated by cAMP-dependent protein kinase-A, thereby moving from the Golgi body to the cytosol [1450].

11.1.5.11 Phosphodiesterase-11 Family (PDE11)

A single gene encodes members of the PDE11A family. Four variants (PDE11a1–PDE11a4) have been detected, among which 3 isoforms are truncations of varying lengths of the longest PDE11a4 variant. They hydrolyze both cAMP and cGMP nucleotides ($K_{M_{cAMP}}$ 1 –6 µmol; $K_{M_{cGMP}}$ 0.5–4 µmol) [1450].

In humans, Pde11A1 mRNA is prominent in skeletal muscle and prostate; Pde11A3 mRNA is found specifically in testis; and Pde11A4 mRNA is highly expressed in prostate, but also in the hypophysis, heart, and liver [1450]. In fact, PDE11a variants localize to epithelial, endothelial, and smooth muscle cells, but at the highest levels in the kidney, adrenal gland, colon, skin, and prostate.

11.2 cGMP Signaling

Cyclic guanosine monophosphate (cGMP) is a second messenger implicated in cell growth, smooth muscle cell relaxation, homeostasis, and inflammation, among other tasks [1468].

Two well-known activators of cGMP signaling exist (Fig. 11.1): atrial natriuretic peptide[41] and nitric oxide. Natriuretic peptides bind to a transmembrane receptor, the particulate guanylate cyclase (pGC; Table 11.13).[42] Nitric oxide acts via soluble guanylate cyclase (sGC), especially in smooth muscle cells.[43]

At least 3 types of cGMP-binding effectors transduce cGMP signals: (1) cGMP-modulated cation channels; (2) cGMP-dependent protein kinases (PKG or cGK);[44] and (3) cGMP-regulated phosphodiesterases (PDE).[45] Protein kinase-G1 stimulates myosin phosphatase. It can also interact with inositol (1,4,5)-trisphosphate receptor-associated PKG1β (IRAG), inhibiting intracellular Ca^{++} release. Agent IRAG may be involved in the anti-platelet effects of exogene nitric oxide. Vasodilator-stimulated phosphoprotein (VASP) activated by PKG1 inhibits platelet adhesion to endothelial cells.

In vascular smooth and cardiac striated myocytes, cGMP level is determined by activity of 3 types of guanylate cyclases: (1) receptor for nitric oxide sGC; (2) receptor for A- and B-type natriuretic peptides GCa; and (3) receptor for C-type natriuretic

---

[41] Atrial natriuretic peptide increases endothelium permeability and lowers arterial blood pressure in hypertension. B-type natriuretic peptide, another cGMP signaling generator, is used as a diagnostic index for heart failure.

[42] Particulate guanylate cyclases are plasmalemmal receptors. Particulate guanylate cyclases have intracellular cyclase domain. Isozyme GCa (GuCy2a), or NPR1, binds atrial and B-type natriuretic peptides; GCb (GuCy2b), or NPR2, links C-type natriuretic peptide.

[43] Soluble guanylate cyclases belong to the heme NO/O_2–binding protein class H-NOx. The H-NOx proteins bind both nitric oxide and oxygen. However, sGC is selective for NO, with a NO-sensitive ferrous (Fe^{++}) state and NO-insensitive oxidized ferric (Fe^{+++}) state. Two sGC isoforms, β1α1 and β1α2, have similar enzymatic activities. Soluble guanylate cyclases translocate from the cytosol to the plasma membrane after activation due to heat-shock protein-70 and possibly other ions (Ca^{++} and Mg^{++}) and substances (ATP and GTP).

[44] Three known cGMP-dependent protein kinases exist: cytosolic PKG1, with 2 isoforms PKG1α and PKG1β, and membrane-bound PKG2 type. Kinase PKG1 regulates the vasomotor tone and hence blood pressure. Kinase PKG2 acts in intestinal electrolyte transport and bone formation.

[45] PDEs hydrolyze cyclic adenosine monophosphate and/or cyclic guanosine monophosphate.

11.2 cGMP Signaling

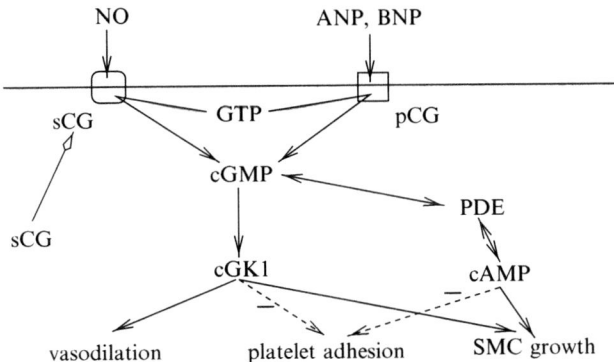

Fig. 11.1 cGMP signaling pathways are involved in cellular functions (vasodilation, clotting, angiogenesis, etc.). Signaling is induced by natriuretic peptides and nitric oxide, via particulate guanylate cyclase (pGC) and soluble guanylate cyclase (sGC). Effectors are cGMP-dependent protein kinases-1 (PKG1) and phosphodiesterases (PDE; adapted from [1468]).

Table 11.13. Types of guanylate cyclases (GC or GuCy; ANPR: atrial natriuretic peptide receptor; GCA: guanylate cyclase activator; GCAP: guanylate cyclase-activating protein; HStaR: heat stable enterotoxin receptor; NPR: natriuretic peptide receptor; RetGC: retinal guanylate cyclase; RoSGC: rod outer segment membrane guanylate cyclase).

Type	Gene	Other aliases and name
Particulate guanylate cyclases		
GC2a	GUCY2A	GuCy2a, GCa
		NPR1, NPRa, ANPRa, ANP$_A$
GC2b	GUCY2B	GuCy2b, GCb
		NPR2, NPRb, ANPRb, ANP$_B$
GC2c	GUCY2C	GuCy2c, GCc, HStaR
GC2d	GUCY2D	GuCy2d, GCd, RetGC1, RoSGC1
GC2e	GUCY2E	GuCy2e, GCe
GC2f	GUCY2F	GuCy2f, GCf, RetGC2, RoSGC2
GC2g	GUCY2G	GuCy2g, GCg
Soluble guanylate cyclase subunits		
GC1α1	GUCY1A3	GC^{S}α1 (α3)
GC1β1	GUCY1B3	GC^{S}β1 (β3)
GC1α2	GUCY1A2	GC^{S}α2
GC1β2	GUCY1B2	GC^{S}β2
Activators		
GCA1a	GUCA1A	GCAP1
GCA1b	GUCA1B	GCAP2
GCA1c	GUCA1C	GCAP3
GCA2a	GUCA2A	GCAP-I, guanylin
GCA2b	GUCA2B	GCAP-II, uroguanylin

peptide GCb. In vascular cells, cGMP is hydrolyzed by phosphodiesterases, mainly PDE5, as well as dual-substrate cGMP–cAMP-hydrolyzing phosphodiesterases PDE1 and PDE2.

In vascular smooth muscle cells, cGMP and its effector PKG1 diminish contraction state, as PKG1 phosphorylates regulators of calcium handling and cytoskeleton. Moreover, PKG1 stimulates growth of vascular smooth muscle cells, as it stimulates sumoylation of the ELk1 transcription factor.

In cardiomyocytes and cardiac fibroblasts, cGMP–PKG1 signaling can counteract trophic action of hormones and growth factors and impedes remodeling in response to pressure overload by repressing some cascades involved in maladaptive cardiac hypertrophy, such as Gq-coupled receptors via PKG1-mediated activation of RGS2 regulator of G-protein signaling. Pulmonary arterial hypertension is in fact associated with an increased expression of PDE5 phosphodiesterase.

Nitric oxide is an intercellular signaling molecule in most cell types, having diverse activities (blood flow regulation, neurotransmission, immune response; Sect. 11.10). Guanylate cyclase-coupled receptor activation by NO binding reversibly triggers a conformational change that transduces the signal with a ligand-concentration dependent intensity. Downstream effectors include kinases, phosphodiesterases, and ion channels. Receptor deactivation follows NO unbinding.

Cyclic guanosine monophosphate is also linked to Rac GTPase for the regulation of actin cytoskeleton during cell migration. Constitutively active Rac increases the activity of transmembrane guanylate cyclases in a cell subjected to a chemotactic signal [1469]. Consequently, concentration in second messenger cGMP rises up to tenfold. Rac effector Ser/Thr P21-activated kinases (PAK1 and PAK2) bind and stimulate the activity of guanylate cyclases. The Rac–PAK–GC–cGMP pathway is involved in fibroblast migration induced by platelet-derived growth factor and lamellipodium formation.

11.3 Signaling via Cell Junctions

11.3.1 Elastin–Laminin Receptor

Certain transmembrane receptors couple the cells to the extracellular matrix and can transduce applied loadings. Elastin strongly increases calcium influx and inhibits calcium efflux in aortic smooth muscle cells [1470]. It also increases sodium influx in monocytes. Elastin-κ induces a vasorelaxation mediated by the elastin–laminin receptor (ELR) and endothelial NO production [1471].

The elastin–laminin receptor is located on both endothelial and smooth muscle cells. Both laminin and elastin bind to cells via the elastin–laminin receptor. The elastin–laminin receptor is a heterotrimer that recognizes several hydrophobic domains on collagen-4, elastin, and laminin. It forms a transmembrane complex with other proteins.

Cyclic stretch of the matrix of cell culture (30-mn duration), which mimics pressure pulse effects on the vessel wall, inhibits the expression of Fos transcription factor, as well as the proliferation of coronary vascular smooth muscle cells grown on elastin matrices [1472]. These effects depend on ELR signaling. On the other hand, cells do not exhibit changes in gene expression or proliferation when the matrix is simply stretched.

11.3.1.1 Collagen Signaling

Heterodimeric integrins operate as receptors for cell adhesion to various constituents of the extracellular matrix (collagens, fibronectin, fibrinogen, laminins, osteopontin, tenascins, thrombospondins, and vitronectin) or cellular receptors such as cell adhesion molecules (e.g., VCAM1 and ICAM). Among integrins, β_1-integrins coupled with α_1, α_2, α_{10}, and α_{11} subunits that link to collagens via their I domain owing to divalent metal Mg^{++} cation are the so-called *collagen receptors*. In addition, $\alpha_X\beta_2$-integrin may also act as a collagen receptor [1473]. Collagen binding is not necessarily specific, as α_1 and α_2 subunits tether laminins, but with a lower affinity.

Integrin-α_1 and -$\alpha_2\beta_1$ are ubiquitous, in particular on the surface of smooth muscle and endothelial cells, whereas α_{10}- and $\alpha_{11}\beta_1$-integrins have a more restricted expression [1473]. Integrin-$\alpha_2\beta_1$ is the single collagen-binding integrin in platelets (GP1a–GP2a). Integrin-$\alpha_{10}\beta_1$ is located with collagen-2 on cartilage cells as well as striated myocytes (heart and skeletal muscle). Integrin-$\alpha_{11}\beta_1$ is detected on mesenchymal cells during embryogenesis and in muscles in adults.

Integrin-$\alpha_1\beta_1$ preferentially attaches to collagen-4 of basement membrane and collagen-13, $\alpha_2\beta_1$- and $\alpha_{11}\beta_1$-integrin to fibrillar collagens, and $\alpha_{10}\beta_1$-integrin to collagen-2 [1473].

After activation by collagen of integrins, integrins recruit cytoplasmic Tyr kinases, in particular focal adhesion kinase. Cytoplasmic tail of β_1-integrin interacts with paxillin and talin that complex with focal adhesion kinase. Autophosphorylation of FAK enables it to phosphorylate its effectors that are mainly cytoskeletal proteins, such as paxillin, talin, α-actinin, tensin, zyxin, VASP, and vinculin. These mediators then recruit adaptors (GRB2, SHC1, and BCAR1 [CAS]), PI3K, and cytosolic Tyr kinases like Src enzyme.

Signaling by FAK also activates small GTPases. Activated Ras then initiates a cascade of Ser/Thr kinases, such as ERK and JNK, to remodel the cytoskeleton. Vinculin and paxillin are scaffold proteins that yield connections to microtubules.

Depending on cell type, different pathways are activated by collagen-binding integrins. Vascular smooth muscle cells do not proliferate in response to collagen. Only collagenase-degraded collagen simulates FAK activation and causes cleavage of paxillin and talin, as well as FAK by calpain.

In platelets, collagen receptor glycoprotein-6, a major thrombocyte aggregation agent, mediates FAK phosphorylation by protein kinase-C [1473].[46] In many cell types, $\alpha_1\beta_1$-integrin stimulation provokes collagen-1 synthesis.

Another type of collagen receptors, glycoprotein-6 (GP6), is a member of the immunoglobulin superfamily that is constitutively associated with Fcγ. It is synthesized only in platelets and their precursors.

11.3.1.2 Collagens, Adhesion Molecules, and Matrix Metallopeptidases

The synthesis and activation of the 26 known matrix metallopeptidases are tightly controlled. Matrix metallopeptidases are released by cells as proenzymes that are afterward activated in the extracellular space mainly by serine peptidases. Yet, MMPs are often sequestered in a deactivating complex with tissue inhibitors of metallopeptidases.

In several cell lines, MMP expression is primed by collagen. Production of MMP1, MMP13, and mt1MMP as well as activation of proMMP2 is triggered by $\alpha_2\beta_1$-integrins [1473].

In human fibroblasts, MMP1 expression depends on PKCζ and P38MAPK kinases. In keratinocytes, proMMP1 complexes with $\alpha_2\beta_1$-integrin, hence competing with collagen. In vascular smooth muscle cells, newly formed adhesive contact between collagen and integrins induces production of proMMP1 that is then secreted and binds to integrin to become active. Activated MMP1 then cleaves collagen-1, thereby lowering its integrin-binding affinity. Focal adhesion complex then disassembles and the cell moves and forms novel cell–matrix adhesions. However, MMP can connect to plasmalemmal receptors that do not bind collagen.

Certain MMPs that are located on plasmalemmal receptors do not bind collagens such as MMP2 that binds to $\alpha_V\beta_3$-integrin, MMP9 to CD44,[47] and MMP7 to heparan sulfate proteoglycans.

Membrane-associated metallopeptidases cause ectodomain shedding of various transmembrane proteins, such as cytokines, growth factors, receptors, and adhesion molecules. They then contribute to the regulation of cell proliferation, migration, and, hence, inflammation and cancer progression.

---

[46] Collagens of the subendothelium, in particular fibrous collagen-1 and -3, are strong platelet agonists that cause thrombocyte shape change, content release, and aggregation, as a consequence of injury to the vessel wall. Glycoprotein-6 recognizes the platelet-activating quaternary structure of collagen, but not glycoproteins GP2b–GP3a, scavenger receptor ScaRb3 (a.k.a. CD36, GP3b, and GP4), and von Willebrand factor [1474]. The latter bridges collagens in the vessel wall and specific receptors on the platelet surface, such as glycoproteic GP1b–GP5–GP9 and GP2b–GP3a ($\alpha_{2B}\beta_3$-integrin) complexes. The GP1bα–GP1bβ–GP5–GP9 receptor not only binds to subendothelial von Willebrand factor, but also thrombin and certain coagulation factors, as well as platelet to endothelial cells via P-selectins to support hemostasis. In addition, this complex links to $\alpha_M\beta_2$-integrin (Mac1) of leukocytes.

[47] Plasmalemmal glycoprotein CD44 contributes to intercellular interactions and cell adhesion and migration. It is receptor for hyaluronic acid that can interact with osteopontin, collagens, and matrix metallopeptidases.

Structurally related, membrane-associated adamlysins also provoke ectodomain cleavage of various substrates. In particuler, ADAM17 causes shedding of transforming growth factor-α, a mitogen that interacts with the epidermal growth factor receptor, as well as other HER family agonists (members of the EGF superfamily), such as amphiregulin and heparin-binding EGF-like growth factor (HB-EGF), thereby controlling the access of generated soluble ligands to receptors of the HER family. Adamlysin ADAM17 also cleaves cytokines and cytokine receptors. Therefore, ectodomain shedding by ADAM17 participates in the regulation of cell proliferation, inflammation, and cancer progression. Ectodomain cleavage is primed once ADAM17 is phosphorylated (activated) by P38MAPK and ERK kinases.

Tissue inhibitors of metallopeptidases not only target matrix metallopeptidases, but also inhibit adamlysins. In basal conditions, ADAM17 resides in the plasma membrane as dimers associated with tissue inhibitor of metallopeptidase TIMP3, which inhibits ADAM17 enzyme. Upon activation of the ERK or P38MAPK pathway, the density of ADAM17 monomers at the cell surface rises and its association with TIMP3 lowers [1475].

11.3.1.3 Collagen Receptors in Wound Healing

In cells at the wound margin, collagen tethers to $\alpha_2\beta_1$-integrin and activates focal adhesion kinase and elicits laminin deposition that triggers FAK signaling via $\alpha_3\beta_1$- and $\alpha_6\beta_4$-integrin in cells more distant from the wound edge [1473].

External regulators, particularly transforming growth factor-β, also modulate keratinocyte migration by promoting collagen and integrin synthesis. Following wound closure, proliferation of dermal fibroblasts terminates with apoptosis that depends on collagen-1 as well as α_1- and $\alpha_2\beta_1$-integrin.

11.3.2 Adhesion Molecules

Integrins, cadherins, and other adhesion molecules interact with growth factor receptors. Cell adhesion is necessary for activation of growth factor receptors, and growth factors are required to stimulate cell adhesion or motility. However, adhesion molecules can trigger ligand-independent activation of growth factor Tyr kinase receptors, translating environmental cues into intracellular signals [1476]. Conversely, growth factors can act on adhesion molecules for adhesion-independent signaling. The receptors for PDGF and VEGF, among others, are activated by integrins [1477,1478]. The EGF receptor associated with E-cadherin can be tyrosine phosphorylated and then activates MAPK and Rac without EGF stimulation [1479].

11.3.2.1 Integrin Receptors

Integrins mediate cell adhesion to the extracellular matrix and transmit signals that stimulate cell spreading, retraction, migration, and proliferation, etc., i.e., processes that range from cell survival to death on the cell level (Vols. 1 – Chap. 7. Plasma Membrane and 2 – Chap. 6. Cell Motility). On the tissue scale, integrins contribute to tissue development, immunity, wound healing, hemostasis and thrombosis, as well as carcinogenesis.

Integrins are heterodimeric, adhesion, and bidirectional signaling receptors that sense chemical and mechanical agents from inside and outside the cell, form integrin clusters, and transmit cues from both side of the plasma membrane, as they connect components of intra- and extracellular medium (Vol. 1 – Chaps. 7. Plasma Membrane and 8. Cell Environment).

On the one hand, activation of bidirectional signaling integrins is regulated by binding of cytoskeletal protein talin and focal adhesion protein kindlin-2 to distinct sites of cytoplasmic tails of β subunit (in--out signaling) [1480, 1481]. Kindlin-2 acts synergistically with talin to activate $\alpha_{2B}\beta_3$-integrins. Moreover, kindlin-2 interacts with integrin-linked kinase complex and mediates its recruitment to focal adhesions to regulate cell spreading and actin cytoskeleton organization.

On the other hand, upon ligand binding, integrins transduce signals into cells by recruiting proteins to their cytoplasmic tails (out--in signaling). Integrin out–in signaling relies on $G\alpha_{13}$ subunit of heterotrimeric guanine nucleotide-binding protein that directly binds to β_3-integrin cytoplasmic domain. Interaction between G13 and $\alpha_{2B}\beta_3$-integrin upon ligand binding to integrin as well as GTP loading on G13 activate Src kinase and RhoA GTPase [1482]. This signaling cascade causes cell spreading and prevents cell retraction.

Integrin adhesome integrates the entire set of interactions involved in integrin-mediated adhesion and signaling, especially responses to mechanical stresses. Integrin adhesome encompasses 6 main types of constituents: (1) actin–integrin set with actin regulators, adaptors, and adhesion molecule-associated proteins; (2) protein Ser/Thr kinases and phosphatases; (3) protein Tyr kinases and phosphatases; (4) monomeric GTPases of the RHO superfamily; (5) phosphoinositides regulated by phosphoinositide kinases and phosphatases; and (6) peptidase calpain and ubiquitin ligase CBL that degrade integrin adhesome proteins.

Integrin adhesome can be considered as a highly dynamic, robust signaling component, as its response depends on stimulus type and it retains its function after experiencing internal and external perturbations.

Small Rap1 GTPase induces the formation of an integrin-activation complex that binds to integrins. It can then mediate protein kinase-C activity associated with the integrin activation, using adaptor talin, a PKC substrate. Small Rap1 GTPase provokes the migration of talin toward the plasma membrane. GTPase Rap1 and protein kinase-Cα then induce, via the Rap1 effector Rap1-interacting adaptor molecule (RIAM) and with recruited talin, the formation of an integrin-activation complex that binds to and activates integrins [1483].

Integrin-mediated cell adhesion regulates numerous cellular responses. Integrins mediate either activation or inhibition of anchorage-dependent receptors. Integrins contribute to signaling from receptor Tyr kinases. On the other hand, integrins can also hinder receptor Tyr kinases [1484].

Signaling cooperation exists between integrins and growth factor receptors. In particular, $\alpha_V\beta_3$-integrin binds to the platelet-derived growth factor receptor and vascular endothelial growth factor receptor-2. The collagen-activated $\alpha_1\beta_1$-integrin attenuates epidermal growth factor receptor signaling via the activation of PTPn2 phosphatase. Phosphatase PTPn2 regulates cell proliferation. Cell adhesion to collagen induces PTPn2 translocation to the cell cortex, where PTPn2 colocalizes with and is activated by α_1-integrin.

Intracellular protein Tyr kinase SYK, a signaling effector of immune receptors, is also involved in integrin-mediated responses. Crosstalk between integrins and immune receptor signaling enables cooperation from the early stage. Both immune pathways and integrin-mediated signaling also work with the same effectors, phospholipase PLCγ2, and lymphocyte cytosolic protein LCP2 adaptor. Integrin-triggered signaling during interactions between lymphocytes and antigen-presenting cells implicates not only kinase SYK, but also transmembrane adaptors such as TYRO protein Tyr kinase-binding protein (TYROBP)[48] Integrin-β_3 clusters bound to fibrinogens on platelet plasma membrane (whereas β_1-integrins link to collagen) phosphorylate (activate) SYK kinase.

11.3.2.2 Catenins

Overexpressed catenin-δ1 disrupts stress fibers and focal adhesions, decreases RhoA functioning, and increases the activity of CDC42 and Rac1, thereby promoting cell migration. Catenin-δ1 binds Vav2 agent.

Catenin-δ1 interacts with transcriptional factor Kaiso [1485].[49] Both catenin-δ1 and Kaiso are increased at the wound border with respect to endothelial cells away from the wound border [1486]. C-terminal Src kinase (CSK) is involved in VE-cadherin signaling [1487]. The association of VE-cadherin and CSK in endothelial cells is increased with elevated cell density, inhibiting cell growth.

---

[48]This membrane receptor component in natural killer and myeloid cells is also called DNAX-activation protein DAP12, killer-activating receptor (KAR)-associated protein KARAP, and polycystic lipomembranous osteodysplasia with sclerosing leukoencephalopathy protein (PLOSL).
[49]Ctnδ1-binding partner Kaiso is a member of the POZ-zinc finger family of transcription factors implicated in development and cancer.

11.3.3 Discoidin Domain Receptors

Discoidin domain receptors (DDR) are receptor Tyr kinases that possess a single transmembrane region and a discoidin motif in their extracellular domain. They form homodimers upon ligand engagement.

Activation of DDR1 or DDR2 by collagen primes a sustained phosphorylation that allows binding of adaptors. Paralog DDR1 autophosphorylates once it is bound to various types of collagens, i.e., collagen-1 to -6 and -8, whereas DDR2 is only activated by fibrillar collagens [1473]. In addition, DDRs have other ligands that can activate these receptors together with or without collagen.

Isoform DDR1 undergoes proteolysis into a membrane-anchored β subunit and a soluble, extracellular α subunit. Five isoforms of DDR1 arise from alternative splicing. Long DDR1c isoform corresponds to the full-length protein, whereas DDR1a and DDR1b isoforms lack 37 or 6 amino acids in the kinase domain, respectively. Both DDR1d and DDR1e are truncated variants that lack the entire kinase region or part of it and the ATP-binding site [1473]. Splice variant DDR1b is the predominant isoform during embryogenesis.

11.3.4 Signaling via Focal Adhesions

Focal adhesions are also important sites of signal transduction. Their components propagate signals that arise from activated integrins following their association with matrix proteins, such as fibronectin, collagen, and laminin. The interaction of integrins with matrix ligands can either generate or modulate signals for motility, cell division, differentiation, and apoptosis [1488].

Integrins and paxillin are implicated in signal transduction. Paxillin binds to β-integrin cytoplasmic tail, vinculin, or other cytoskeletal and signaling proteins [1489]. Paxillin recognizes integrin sequences distinct from α-actinin-binding sites. Paxillin binding is independent of its association with focal adhesion kinase, although both bind to the same region of β_1-integrin. Paxillin yields multiple docking sites for activated FAK and Src kinases [1490]. Various regulatory proteins, such as calpain-2, protein kinase-C, FAK, and Src, control the assembly of focal adhesion [1491]. Focal adhesion disassembly involves microtubules and dynamin that interact with focal adhesion kinase [1492].

11.3.5 Signaling via Gap Junctions

Gap junctions allow communication between adjoining cells. Messenger ATP is released by multiple types of stimuli (mechanical stress, osmotic pressure changes, rise in intracellular concentration of inositol trisphosphate, decay in extracellular calcium ion level, etc.).

Manifold mechanisms of ATP release include vesicular exocytosis, active transport via ABC transporters, diffusion via stretch-activated channels, voltage-dependent anion channels, pores opened by P2X$_7$ receptors, and connexin hemichannels.

Connexin hemichannels, normally closed, are paths for ATP, NAD$^+$, glutamate, prostaglandins, etc. Hemichannels open following membrane depolarization and mechanical stimulation. Inositol trisphosphate activates hemichannel composed of connexin 43. Decreases in extracellular Ca^{++} and Mg^{++} levels potentiate or trigger the opening of hemichannels, releasing particularly ATP and glutamate.

Connexin-32 and -43 have 2 and 1 calmodulin interaction sites, respectively. A Ca^{++}-binding site exists for hemichannels made of connexin-32. Connexin hemichannels open when the cytosolic calcium concentration rises [1493]. However, [Ca^{++}]$_i$ elevation following release from cell storage compartments can close gap junctions.

11.4 Interactions between Ion Channels and Small GTPases

Heterotrimeric GTPases that are activated by heptahelical plasmalemmal receptors (GPCRs or 7TMRs) can regulate ion channels. Small (monomeric) GTPases (Sect. 9.3) that are stimulated by cytoplasmic guanine nucleotide-exchange factors can interact directly with ion channels to regulate their activity (Tables 11.14 to 11.16).

Protein RhoA associates with K$_V$1.2 channel to suppress its activity [1494].

Protein Rab11a binds to epithelial, Ca^{++}-selective members of the transient receptor potential superfamily of cation channels TRPV5 and TRPV6 [1495].

Small Rem GTPase complexes with auxiliary β subunit of voltage-gated Ca$_V$ channel (Sect. 9.3.25), thereby inhibiting Ca$_V$ channels. Small Rho GTPase also regulates Ca$_V$1.2a channels in ventriculomyocytes [370]. Specific GDP-dissociation inhibitor RhoGDIα decreases basal activity of Ca$_V$1.2a channels, but not expression level, without influencing inward rectifier and transient outward K$^+$ channels (K$_V$4). Inhibition of RhoA, but not Rac1 or CDC42, impedes Ca$_V$ activity, i.e., RhoA^{GDP}, but not RhoA^{GTP}, precludes Ca^{++} import via Ca$_V$ channels.

Others small GTPases interact indirectly with ion channels. Small GTPases RhoA and kRas heighten activity of epithelial Na$^+$ channel (ENaC) via PI(4)P5K and PI3K kinases, respectively [1497].

Protein Rac1 mediates rapid vesicular insertion from vesicles held in reserve just under the plasma membrane of transient receptor potential channel TRPC5 into the plasma membrane via PI3K and PI(4)P5-kinase-α following epidermal growth factor stimulation for Ca^{++} influx [1498].

Small Rab27a GTPase regulates ENaC activity, using its effectors synaptotagmin-like protein SLP5 and transport cofactor of the MUNC family of syntaxin-binding protein Munc13-4 [1499].

Different types of small GTPases can have opposite effects. The hypothalamic neuropeptide thyrotropin-releasing hormone (TRH) inhibits the activity of K$_V$11.1

Table 11.14. Regulation of ion channels by small guanosine triphosphatases of the RAB superfamily (Source: [925]; CFTR: cystic fibrosis transmembrane conductance regulator; ENaC: epithelial Na$^+$ channel; SLP: synaptotagmin-like protein; TRPV: transient receptor potential vanilloid channel). Ubiquitous Rab proteins constitute the largest branch of the small GTPase class. With monomeric GTPases of the ARF superfamily, they contribute to intracellular vesicular transport.

Small GTPase	Channel	Effect
Rab3	ENaC	Inhibition via effectors
Rab4	CFTR	Inhibition (basal and cAMP-stimulated activity)
	ENaC	Inhibition via effectors
Rab5	CFTR	Enhanced insertion
Rab7	CFTR	Lysosomal degradation
Rab9	CFTR	Exocytosis
Rab11	CFTR	Endosomal recycling
Rab11a	TRPV5/6	Enhanced translocation
Rab27a	CFTR	Inhibition via effectors
	ENaC	Munc13-4 and SLP5

Table 11.15. Regulation of ion channels by small guanosine triphosphatases of the RAS hyperfamily (Source: [925]; hRas: Harvey rat sarcoma viral oncogene homolog; kRas: Kirsten rat sarcoma viral oncogene homolog). Four major Ras isoforms constitute the P21RAS subfamily (hRas, kRasA, kRasB, and nRas). Effectors of Ras GTPases include Raf kinase (via adaptor GRB2 and RasGEF SOS recruited by RTKs, Src kinases attracted by other receptor types, or transactivation), RalGDS, RIN1, and phosphatidylinositol 3-kinase (PI3K). Members of the RAS hyperfamily can have isoform-specific effects on ion channel activity as well as different effects on distinct channels. Monomeric GTPases Rap1a and Ras have opposite effects on atrial K$^+$ channels coupled to muscarinic M$_2$ receptors and voltage-gated Na$^+$ channels (as well as ^{N}methyl ^{D}aspartic acid [NMDA] and α-amino 3-hydroxy 5-methylisoxazole 4-propionic acid [AMPA]-type glutamate receptors). Subtype hRas inhibits inward rectifier K$_{IR}$1 channel via mitogen-activated protein kinase (MAPK) signaling by supporting removal of channel from the plasma membrane, but excites Ca$_V$3 channel. Aldosterone increases density and activity of kRasA, an activator of epithelial Na$^+$ channel (ENaC) via PI3K, as its product PI(3,4,5)P$_3$ interacts with ENaC channel. Ras-related GTPases of the RGK (Rad–GEM/KIR) family bind to β subunit of Ca$_V$1 channels, thereby inhibiting these channels, with or without attenuation of channel density.

Small GTPase	Channel	Effect
Ras	Atrial K$_{ACh}$	Inhibition
	Na$_V$	Activation
hRas	Ca$_V$3	Stimulation
	K$_{IR}$2	Reduced plasmalemmal insertion
	Na$_V$	Stimulation
kRas	ENaC	Activation via PI3K
Rap1	Na$_V$	Inhibition
	Atrial K$_{ACh}$	Activation
RGK	Ca$_V$1	Inhibition
Rem2	Ca$_V$2.2	Inhibition

11.4 Interactions between Ion Channels and Small GTPases

Table 11.16. Regulation of ion channels by small guanosine triphosphatases of the RHO superfamily (Source: [925]; Ca$_V$: voltage-gated Ca^{++} channel; ENaC: epithelial Na$^+$ channel; NSC: non-selective cation channel; PI(4)P: phosphatidylinositol 4-phosphate; PP1: phosphoprotein phosphatase-1; RoCK: Rho-associated, coiled-coil-containing protein kinase; TRPC: transient receptor potential canonical channel; VRAC: volume-regulated anion channel). The superfamily of Rho GTPases includes Rho, Rac, and CDC42 that are involved in F-actin polymerization, thus channel translocation. They can act via phospholipid messengers and adaptors such as phosphatidylinositol (4,5)-bisphosphate (PI(4,5)P$_2$) that promotes channel insertion into the plasma membrane. Different members of the RHO superfamily, such as Rac1 and RhoA, have opposite effects on K$_V$11.1 channels. Inhibitory Rho and activatory Rac operate via phosphorylation and dephosphorylation, respectively.

Small GTPase	Channel	Effect
CDC42	Ca$_V$	Inhibition
Rac1	ENaC	Enhanced activity via RoCK and PI(4)P5K
	K$_V$11.1	Activation via PP1 and/or PP5
	Ca$_V$	Inhibition
Rho	VRAC	Sensitization
	NSC	Inhibition
RhoA	ENaC	Enhanced activity via RoCK and PI(4)P5K
	K$_V$11.1	Inhibition via kinases
	K$_V$1.2	Inhibition
	Ca$_V$1.x	Stimulation
	TRPC5	Increased density via PI(4)P5-kinase
	K$_{IR}$2.1–K$_{IR}$2.3	Inhibition
RhoQ	CFTR	Augmentation of density

channels in the pituitary gland (hypophysis), whereas thyroid hormone triiodothyronine (T$_3$) antagonizes TRH action, hence stimulating K$_V$11.1 channel activity. Small RhoA GTPase stimulated by G13-protein-coupled receptors upon TRH binding is able to rapidly inhibit K$_V$11.1 channels. On the other hand, Rac activated by RacGEF stimulated by PI3K in T$_3$-primed nuclear hormone receptor signaling stimulates K$_V$11.1 channel [1500].

Neurotransmitters target voltage-gated calcium channels that trigger Ca^{++} influx in neurons. Upon bradykinin excitation, heterotrimeric G-protein subunit G13 inhibits voltage-gated calcium channels. Protein G13 can couple to monomeric RhoA, Rac1, and CDC42 GTPases. Small Rac1 GTPase and/or CDC42 mediate inhibition of Ca$_V$ by neurotransmitters [1501]. Lysophosphatidylcholine (3–50 μmol) generates a non-selective cation current (I$_{NSC}$) in a dose-dependent manner with a lag in guinea pig ventricular myocytes via Gi/o-protein-coupled receptor and small GTPase Rho [1502].

Small GTPases can operate by activating action or permissive effect. Hypoosmotic stimulation of cells rapidly activates compensatory Cl$^-$ and K$^+$ fluxes to limit excessive cell swelling and, ultimately, restore original cell volume. Osmotic stress-induced cell swelling is associated with a rapid, transient remodeling of the actin cytoskeleton. In calf pulmonary artery endothelial as well as human intestinal cells, the Rho–RoCK–myosin light chain phosphorylation axis favors the swelling-triggered outwardly rectifying anion current through volume-regulated anion channels [372, 1504]. However, the Rho pathway sensitizes VRACs to cell swelling via actin filaments. It indeed exerts a permissive effect on Cl^- channel VRAC, i.e., swelling-induced VRAC opening requires functional, but not directly interacting Rho pathway [1506]. The activity of endothelial swelling-activated Cl^- channels also depends on tyrosine phosphorylation, as protein Tyr kinases promote Cl^- flux, whereas protein Tyr phosphatases preclude it [1507].

Small GTPases can enhance activity of plasmalemmal ion channel by heightening their density, as they are involved in cellular exocytosis and membrane insertion of ion channels. On the other hand, small GTPase RhoQ, but not CDC42 or RhoA, interacts with cystic fibrosis transmembrane conductance regulator (CFTR)-associated ligand CAL[50] that reduces plasmalemmal density in CFTRs by targeting CFTR for degradation in lysosome [966]. Active form of RhoJ ($RhoJ^{GTP}$) protects CFTR against CAL-mediated lysosomal degradation.

11.5 Calcium Signaling

Among the second messengers, calcium regulates cellular functions in all cell compartments on time scales ranging from milliseconds to days [1508]. The cell response to stimulation of certain receptor Tyr kinases and G-protein-coupled receptors depends on the amplitude and duration of calcium influx. Calcium flux into the cell can be modulated to ensure the suitable response.

A small calcium proportion binds to effectors, such as annexins, calmodulin, membrane-trafficking synaptotagmin, S100 proteins, and troponin-C (Vol. 5 – Chap. 5. Cardiomyocytes). The calcium-signaling effectors are involved in cellular transport (operating in time of the order of $O[10\,\mu s]$), metabolism (operating in time of the order of $O[s]$), gene transcription (operating in time of the order of $O[mn]$), cell fate and motility (operating in time of the order of $O[h]$), and myocyte contraction (operating in time of the order of $O[10\,ms]$), according to the location, timing, and calcium-bound molecules (Table 11.17). Calcium ions are mostly linked to buffers. Calcium buffers affect both the amplitude and the recovery time of Ca^{++} fluxes. Calcium buffers have different expression patterns, motility, and binding kinetics. Calcium ions mediate or stimulate multiple cellular processes, such as the myocyte contraction.

---

[50] A.k.a. Golgi-associated PDZ and coiled-coil motif-containing protein GoPC1 and PDZ protein interacting specifically with TC10 (RhoJ; alias PIST).

11.5 Calcium Signaling

Table 11.17. Calcium signaling (Source: [10]). The Ca^{++} concentration in a cell at rest equals approximately 100 nmol It augments to at least 500 nmol in a stimulated cell. When the stimulus is removed, the Ca^{++} concentration return to its resting level. Calcium ion is a universal second messenger that triggers many cellular processes over various time scales.

Target process	Time scale
Exocytosis	$O[1\ \mu s]$
Cell contraction	$O[1\ ms]$
Cell metabolism	$O[1\ s]$
Gene transcription	$O[1\ mn]$
Cell proliferation	$O[1\ h]$

Table 11.18. Calcium-induced Ca^{++} release (CICR) and Ca^{++} signal types [10]). Elementary Ca^{++} events, such as Ca^{++} sparks and puffs are produced by ryanodine (RyR) and inositol (1,4,5)-trisphosphate (IP3R) receptors, respectively. Fall in Ca^{++} in the endoplasmic reticulum opens plasmalemmal, store-operated CRAC channels. Resulting Ca^{++} entry through CRAC channels refills intracellular stores and, thus, enables prolonged IP_3-triggered Ca^{++} oscillations. The additional role of CICR is to set up cytoplasmic Ca^{++} waves from initiation sites that propagate through the cytosol to form the global Ca^{++} signal.

Trigger	Target	Effect
Ca^{++}	RyR, IP_3R	Ca^{++} sparks
		Intracellular Ca^{++} wave
IP_3	IP_3R	Ca^{++} puffs
		Intracellular Ca^{++} wave

11.5.1 Calcium-Induced Calcium Release

Calcium-induced Ca^{++} release (CICR) enables the mobilization of Ca^{++} ions from its intracellular stores, primarily the endoplasmic reticulum. Calcium channels in the membrane include ryanodine and IP_3 receptors.

The CICR process transmits information from the plasma membrane to the endoplasmic reticulum via Ca^{++}-releasing RyR and IP_3R receptors. The first mechanism relies on voltage-gated channels that open in response to membrane depolarization to carry a small amount of Ca^{++} ions toward the cytosol (Table 11.18). Calcium ions then diffuses into the cytosol to reach the endoplasmic reticulum and activate RyR and/or IP_3R channels. The second mechanism is based on the IP_3 axis, after synthesis of IP_3 initiated by activated plasmalemmal receptors.

In addition, Ca^{++} released from endoplasmic reticulum RyR and IP_3R channels diffuses along the membrane of the endoplasmic reticulum and can then stimulates neighboring channels to release further Ca^{++} ions, thereby creating *regenerative waves* [10]. Oscillatory Ca^{++} signals with given oscillatory amplitude and frequency for discriminating signaling allow to raise cytoplasmic Ca^{++} transiently, thereby avoiding cell damage caused by sustained elevated Ca^{++} elevation.

Repetitive cytoplasmic Ca^{++} oscillations triggered by IP_3-mediated Ca^{++} release from intracellular stores arise from activated plasmalemmal receptors coupled to the phospholipase-C axis that enables regenerative Ca^{++} release and are supported by store-operated Ca^{++} entry through Ca^{++} release-activated Ca^{++} channels. Upon stimulation of type-1 cysteinyl leukotriene receptors (CysLT1) by leukotriene-C4, large-amplitude Ca^{++} oscillations (rapid, large, all-or-nothing baseline Ca^{++} spikes) and CRAC channel activity result from action of transmembrane endoplasmic reticulum stromal interaction molecule StIM1 sensor and CRAC channel activator [1509]. On the other hand, stimulated FCεR1 receptor by antigens or immunoglobulin-E causes activation by protein Tyr kinase of phospholipase-Cγ and slow Ca^{++} oscillations developed after a longer delay on an elevated background Ca^{++} rise using both StIM2 and StIM1 agents [1509]. Therefore, different stimuli can recruit and activate different combinations of StIM proteins to sustain cytoplasmic Ca^{++} signals.

The kinetics of IP_3 production (fast, transient elevation vs. slow, prolonged rise in IP_3 concentration) and its baseline concentration differ according to the stimulus type. Sensor StIM2 is less efficient, for a given CRAC channel-forming Orai1 density, in Ca^{++} entry. Protein StIM2 has an approximately 2-fold lower affinity for Ca^{++} than StIM1 [1509]. Molecule StIM2 requires a smaller Ca^{++} drop in the endoplasmic reticulum lumen for its activation. Whereas StIM1 activates CRAC channels during strong stimulation, StIM2 is involved in moderate store depletion. Large and quick Ca^{++} release may potentiate calmodulin activity that prevents StIM2 from activating Orai1 channel.

11.5.2 Calcium Signaling and Myocyte Contraction

Myocyte contraction illustrates the role of calcium ions in cell functioning. Several protein sets are involved in such a process: (1) plasmalemmal calcium channnels for influx of calcium ions from and efflux to the extracellular space; (2) calcium channnels in the membrane of intracellular calcium stores, mainly the sarcoplasmic reticulum, to ensure a sufficient amount of calcium into the cytosol for suitable activity; (3) second messenger to release calcium from its intracellular stores; (4) the mitochondrial machinery to synthesize the energy source ATP; (5) buffers; (6) sarcomere effectors; and (7) regulators such as kinases and phosphatases (Fig. 11.2). Other molecules participate in myocyte contraction, especially to coordinate the deformation of the cell compartments. Dystrophin associates with cytoplasmic syntrophin and forms the dystrophin–glycoprotein complex (DGC), which links plasmalemmal β-dystroglycan. The latter binds α2-laminin–merosin of the basal lamina. Dystrophin thus stabilizes the sarcolemma, especially during contraction.

11.5 Calcium Signaling

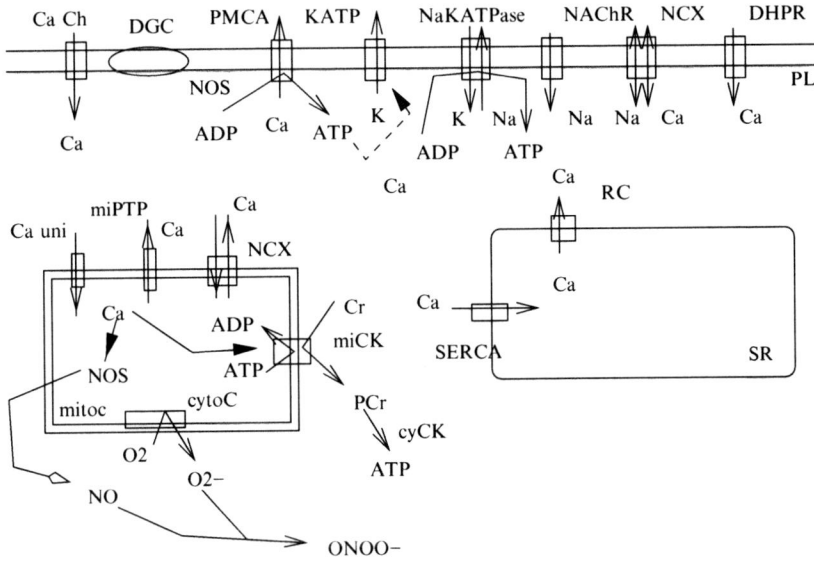

Fig. 11.2 Calcium messengers for myocyte contraction (Source: [1510]). Calcium influxes are due to Ca^{++} channel (Ca Ch), indirectly by nicotinic acetylcholine receptor (NAChR), which drives Na$^+$ influx, thereby stimulating Ca^{++} influx by the sodium–calcium exchanger (NCX), to dihydropyridine receptor (DHPR) of the plasmalemma (PL) and ryanodine receptor of the sarcoplasmic reticulum (SR). A fraction of imported calcium enters in mitochondria (mitoc) by Ca^{++} uniporter (Ca uni). It stimulates nitric oxide synthase (NOS) and mitochondrial creatine kinase (miCK), generating NO and ATP. ATP is indirectly exported from phosphocreatine via cytosolic creatine kinase (cyCK) for contraction and ion ATPase activity. Mitochondrial Ca^{++} is exported by NCX and mitochondrial permeability transition pore (miPTP). Cytosolic Ca^{++} efflux is done by Ca^{++} ATPases of the plasma membrane (PMCA) and the sarcoplasmic reticulum (SERCA), and by NCX. Na$^+$ is exported by sodium–potassium ATPase. Nitric oxide can form with superoxide (O2−), produced by the respiratory chain, peroxynitrite (ONOO−).

11.5.3 Instantaneous and Long-Lasting Responses to Calcium

Calcium signaling is characterized by instantaneous and long-lasting responses. Intracellular calcium controls various events according to its spatial and temporal distributions and magnitude. Cytosolic and plasmalemmal calcium sensors, with high affinity for Ca^{++} (binding Ca^{++} even at small increments in concentration above resting levels) and distinct targets, transduce calcium signals into functional changes. Ubiquitous calmodulin (Cam) is a specific cytosolic Ca^{++} sensor. The calcium–calmodulin complex activates calmodulin-dependent protein kinases. Calmodulin regulates other targets, such as PP3 phosphatase and phosphodiesterases.

Once released in the cytosol, Ca^{++} binds to calmodulin. The Ca^{++}–Cam complex then interacts with other proteins for *instantaneous response*. Instantaneous reaction begins once the cytosolic concentration [Ca^{++}]$_i$ rises

$$([Ca^{++}]_i \sim 10^{-4} [Ca^{++}]_e),$$

and ends as soon as $[Ca^{++}]_i$ returns to its basic level. The release of cAMP is initiated by ligand-loaded G-protein-coupled receptors. Messenger cAMP is produced from adenosine triphosphate after the activation of adenylate cyclase by the receptor-activated Gs protein [1511] (Fig. 11.3).[51]

Messenger cAMP travels in the cytosol and accumulates at specific sites [1513]. The AKAP proteins then recruit PKA to locations of cAMP production, where phosphorylation is confined to a subset of potential substrates. Agent cAMP controls the Ca^{++} influx. In turn, Ca^{++} activates cAMP synthesis.

Sustained contractions of smooth muscle cells triggered by Ca^{++} influx during the instantaneous response result from a *long-lasting response*. When $[Ca^{++}]_i$ increases, the Ca^{++}–Cam complex responsible for the transient reaction interacts with the membrane Ca^{++} pump to augment its functioning. Moreover, Ca^{++}-activated protein kinase-C (PKC) enhances the Ca^{++} pump efficiency. The Ca^{++} efflux thus compensates Ca^{++} influx [1514]. Calcium recycling is improved with an increased submembrane Ca^{++} concentration, which activates membrane-bound PKC (Fig. 11.3). Furthermore, activation of phospholipase-C (PLC) generates inositol trisphosphate (IP$_3$) and diacylglycerol (DAG) from phosphatidylinositol diphosphate (PIP$_2$). Messenger IP$_3$ induces Ca^{++} release from the endoplasmic reticulum. Diacylglycerol remains in the membrane and, as long as its membrane concentration is sufficient,[52] PKC is fixed by DAG to the membrane and is activated. Protein kinase-C thus acts as a transducer during the long-lasting reaction.

11.5.4 Calcium Signaling Components

Several Ca^{++}-regulated proteins maintain low intracellular concentration of Ca^{++} ($[Ca^{++}]_i$) and couple changes in $[Ca^{++}]_i$ to physiological responses. They include Ca^{++} membrane carriers and Ca^{++} effectors. According to the Ca^{++}-binding site, Ca^{++} effectors are grouped into different families.

Effectors characterized by an *EF-hand motif*[53] can be categorized into 2 primary subclasses of proteins: (1) subclass of EF-hand Ca^{++} *sensors* (e.g., calmodulin) that

---

[51] The ligand-bound GPCR catalyzes the exchange of guanosine diphosphate (GDP) for guanosine triphosphate (GTP) on the subunit of the associated G protein, which is activated to initiate or to inhibit signaling cascades. Gs activates adenylate cyclase, which converts ATP to cAMP [1512]. cAMP binds to PKA, the main intracellular effector of cAMP signaling, which is compartmentalized by A-kinase anchoring proteins. cAMP can also activate cyclic nucleotide-gated ion channels, phosphodiesterases and guanine nucleotide-exchange factors.

[52] Long duration fixation of PKC on the plasma membrane provides a memory effect for subsequent amplified response. However, the delay between successive stimuli must not be too long.

[53] The EF hand sequence is a Ca^{++}-binding helix–loop–helix domain. This domain is constituted by 2 α helices linked by a short loop. Orientation of 2 helices E and F can be represented by the spread thumb and forefinger of the human hand. The EF hand region is found in calcium-binding proteins that participate in transduction and modulation of Ca^{++} signals.

11.5 Calcium Signaling

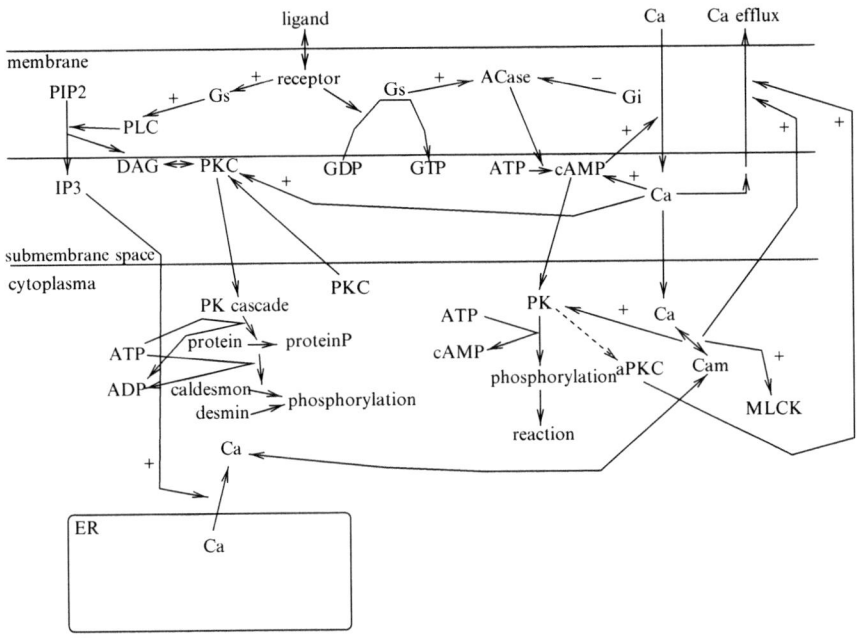

Fig. 11.3 Second messengers (Source: [1514]). The activated receptor activates phospholipase-C (PLC), which cleaves membrane-bound phosphatidylinositol bisphosphate (PIP$_2$) into inositol trisphosphate (IP$_3$) and diacylglycerol (DAG). Messenger IP$_3$ travels to its receptor, a calcium (Ca) channel, on the surface of the endoplasmic reticulum (ER). Calcium is thereby released from its ER store. Cytosolic calcium is then available for binding to calmodulin (Cam) or various cytoskeletal proteins. Protein phosphorylations are catalyzed by Ca–Cam-dependent protein kinases. Mediator ATP is converted into cyclic adenosine monophosphate (cAMP) by adenylate cyclase (AC), stimulated by stimulatory G protein (Gs) and inhibited by inhibitory G protein (Gi). Messenger cAMP activates cAMP-dependent protein kinase-A (PKA). Effectors DAG and Ca^{++} activate protein kinase-C (PKC) at the plasma membrane. Protein phosphorylation by PKC produces cell responses.

transduce Ca^{++} signals and (2) subclass of Ca^{++} signal *modulators* (e.g., calbindin, a 29-kDa calretinin) that modulate the shape and/or duration of Ca^{++} signals.

The superclass of EF-hand proteins that include more than 60 families can be subdivided into 2 classes: (1) class of *canonical EF-hand proteins* (12-residue canonical EF-hand loop), such as calmodulin, calpain, PP3, recoverin, spectrin, troponin-C, etc.; and (2) class of *pseudo-EF-hand proteins*, such as S100 and S100-like proteins. Once bound to Ca^{++}, EF-hand proteins regulate their cellular targets.

11.5.4.1 Channels, Receptors, and Effectors of Calcium Flux

Calcium effects and fluxes result from the combined action of buffers, ion channels, pumps, and exchangers, signaling effectors, and transcription factors

Table 11.19. Calcium signaling components. (**Part 1**) Calcium channels (Source: [1508]). Calcium channels are associated with channel regulators, such as triadin, junctin, sorcin, FKBP12, phospholamban, IP$_3$R-associated PKG substrate IRAG, and adenosylhomocysteinase-like-1. In particular, the ryanodine receptor-2 complex includes stabilizers, such as FKBP12.6, calmodulin, and PP1 and PP2 phosphatases via scaffold proteins spinophilin (PP1$_{r9b}$) and PP2$_{r3a}$, respectively, kinases attached by A-kinase anchoring proteins, and calsequestrin anchored by membrane-bound junctin and triadin. Calcium pumps are characterized by their affinities (functioning thresholds), transport rates, and opening duration. Secretory pathway Ca^{++} ATPases could be responsible for Ca^{++} sequestration into Golgi compartments.

Voltage-gated channels	Ca$_V$1.1–Ca$_V$1.4
	Ca$_V$2.1–Ca$_V$2.3
	Ca$_V$3.1–Ca$_V$3.3
Receptor-operated channels	NMDA-type Glu receptors
Ligand-gated ion channels	(GluNR1, GluNR2A–GluNR2D)
	AMPA-type Glu receptors
	ATP receptor (P2X$_7$)
	nACh receptor
	5HT$_3$
Second messenger-operated channels	Cyclic nucleotide–gated channels
	(CNGA1-CNGA4, CNGB 1, CNGB 3)
	Arachidonate-regulated Ca^{++} channel
	(i_{ARC})
Transient receptor potential ion channels	TRPC1–TRPC7
	TRPV1–TRPV16
	TRPM1–TRPM8
	TRPML, TRPNI

(Tables 11.19 to 11.25). Activated receptors can prime entry into the cell of Ca^{++} ions from the extracellular space and formation of second messengers that release Ca^{++} ions from its intracellular stores. Different cues activate various types of Ca^{++} channels, such as receptor-operated channels, second messenger-operated channels, store-operated channels, thermosensors, and stretch-activated channels. Several types of Ca^{++} channels belong to the superfamily of transient receptor potential protein. They regulates slow cellular processes (smooth muscle cell contraction and cell proliferation).

Signal transmission can tolerate large variations in the expression level of signaling components that are observed in different cell types. Cells can either determine the correct density of signaling components via adequate synthesis and stabilization of produced mediators, or monitor the output of a pathway and use a feedback to adapt concentrations of signaling mediators. Cells monitor Ca^{++} concentrations in the cytosol and endoplasmic reticulum and adjust concentrations of stromal interaction molecule (StIM) as well as of plasma membrane (PMCA) and endoplasmic reticulum (SERCA) Ca^{++} ATPases, among other signaling components [1515]. Cells can sense the state of signaling pathways and use multiple parallel adaptive feedbacks.

11.5 Calcium Signaling

Table 11.20. Calcium signaling components. (**Part 2**) Calcium channels (Cont.; Source: [1508]).

Inositol trisphophate receptors	IP_3R1–IP_3R3
Ryanodine receptors	RyR1-RyR3
Polycystins	PC1–PC2
Calcium pumps	Plasma membrane Ca^{++} ATPases (PMCA1–PMCA4) Sarco(endo)plasmic reticulum Ca^{++} ATPases (SERCA1–SERCA3) Secretory-pathway Ca^{++} ATPases, or Golgi pumps (SPCA1–SPCA2)
Plasmalemmal Na^+–Ca^{++} exchangers	NCX1–NCX3 NCKX1–NCKX4
Mitochondrial Ca^{++} channels	Na^+–Ca^{++} exchangers Ca^{++} uniporter H^+–Ca^{++} exchanger Permeability transition pore

11.5.4.2 Cell-Specific Calcium Signalosomes

Calcium *signalosomes* characterized by their spatiotemporal regulation are adapted to the main cell function (Table 11.26). A Ca^{++} signalosome is composed of various types of effectors that can constitute Ca^{++} signaling *modules*.[54] Inside the cell, Ca^{++} ions can be stored in the endoplasmic reticulum (primarily) and mitochondria. During Ca^{++} influx (in the cytoplasm) through voltage-gated Ca^{++} channel (VGCC or Ca$_V$) as well as receptor (ROC), store (SOC), and second messenger (SMOC)-operated channels, Ca^{++} ions interacts with buffers. Calcium sensors and transducers launch Ca^{++} signaling using a set of effectors. The diversity of a given component heightens the number of signaling pathways.

In the striated myocyte, the Ca^{++} signalosome deliver rapid Ca^{++} pulses for contraction. In T-lymphocytes, Ca^{++} signalosome contains different components to generate much slower repetitive Ca^{++} pulses for cell proliferation required to efficiently struggle against invading pathogens.

---

[54] Examples of calcium signaling modules are: (1) the plasmalemmal Ca^{++} influxers with their triggers (transmembrane potential change, extracellular agonist, or intracellular second messenger); (2) the activated plasmerosome with the StIM–Orai couple (StIM: stromal interaction molecule); (3) the calcium channel-induced Ca^{++} release module with plasmalemmal voltage-gated Ca^{++} channel ($Ca_V^{(PM)}$) coupled to endoplasmic reticulum (ER) ryanodine receptors (RyR$_{(ER)}$); and (4) the PLC–IP_3–$IP_3R^{(ER)}$ axis stimulated by a given plasmalemmal receptors; and (5) the SERCA-based effluxer on the endoplasmic reticulum membrane.

Table 11.21. Calcium signaling components. (**Part 3**) Receptors (Source: [1508]). G-Protein-coupled receptors are associated with G-protein component Gα subtype Gq, G11, G14, and G16, as well as Gβγ dimer. They are controlled by regulators of G-protein signaling RGS1, RGS2, RGS4, and RGS16. They stimulate phospholipase-Cβ. Receptor Tyr kinases can activate phospholipase-Cγ.

Components	Types
G-protein-coupled receptors	Muscarinic receptors (M_1–M_3)
	α1-Adrenoceptors (A–C)
	Endothelin receptors (ET_A–ET_B)
	Angiotensin receptor (AT_1)
	Bradykinin receptors (B_1–B_2)
	Histamine receptor (H_1)
	Serotonin receptors ($5HT_{2A}$–$5HT_{2C}$)
	Leukotrine receptors (BLT, $CysLT_1$–$CysLT_2$)
	Ca^{++}-sensing receptor (CaR)
	Prostanoid receptor (PGF2α)
	Thrombin receptor (PAR_1)
	Bombesin receptors (BRS_1–BRS_2)
	Cholecystokinin receptors (CCK_1–CCK_2)
	Metabotropic glutamate receptors (mGlu1, mGlu5)
	Luteinizing receptor (LSH)
	Neurotensin receptor (Nts_1)
	Oxytocin receptor (OT)
	Substance-P receptor (NK_1)
	Substance-K receptor (NK_2)
	Substance-B receptor (NK_3)
	Thyrotropin-releasing hormone receptor (TRHR)
	Vasopressin receptors (V_{1A}–V_{1B})
Receptor Tyr kinases	Epidermal growth factor receptors (HER1–HER4)
	Platelet-derived growth factor receptors (PDGFRα–PDGFRβ)
	Vascular endothelial growth factor receptors (VEGFR1–VEGFR3

Muscle contraction is triggered by a global elevation in cytosolic Ca^{++} concentration. On the other hand, the release of neurotransmitters is launched by a tiny, localized Ca^{++} pulse delivered directly to storage vesicle by a Ca^{++} sensor linked to exocytotic machinery. Between these 2 extreme cases, many spatiotemporal modalities of Ca^{++} signaling exist.

The major cytosolic buffers in cells are parvalbumin (PAlb) and calbindin-1 (CalB1)[55] Parvalbumin localizes to fast-contracting muscles (at its highest levels)

---

[55] A.k.a. 28-kDa vitamin-D-dependent calbindin-D28k.

Table 11.22. Calcium signaling components. (**Part 4**) Calcium effectors (Source: [1508]).

Phospholipase-C and inositol trisphosphate	PLCβ(1–4), PLCγ(1–2), PLCδ(1–4), PLCε, PLCζ
Ca^{++}-binding proteins	Calmodulin Troponin-C S100A1–S100A14, S100B, S100C, S100P Annexin-1–annexin-10 Neuronal Ca^{++} sensor Visinin-like proteins (ViLiP1–ViLiP3) Hippocalcin, recoverin K_V channel–interacting proteins (KChIP1–KChIP4) Guanylate-cyclase-activating proteins (GCAP1–GCAP3 or GCA1a–GCA1c; phototransduction) Calcium-binding proteins (caldendrins $CaBP1_L$ and $CaBP1_S$ and CaBP2–CaBP5)
Ca^{++}-sensitive ion channels	Ca^{++}-activated K^+ channels (small SK, intermediate IK, and large-conductance BK channels) Cl^+ channel (HClCA1)

Table 11.23. Calcium signaling components. (**Part 5**) Calcium-regulated enzymes (Source: [1508]; Cam: calmodulin, a portmanteau word for calcium-modulated protein). Proline-rich Tyr kinase PYK2 is a Ca^{++}-sensitive protein that acts in osteoclast podosomes.

Ca^{++}-regulated enzymes	Ca^{++}–Cam-dependent protein kinases (CamK1–CamK4) Myosin light-chain kinase (MLCK) Protein kinase-C (PKCα, PKCβ1, PKCβ2, PKCγ) PYK2 Phosphorylase kinase (PhK) Diacylglycerol kinase (DAGK) PIK3C3C (VPS34) IP_3 3-kinase Protein phosphatase-3 (calcineurin) cAMP phosphodiesterase (PDE1a–PDE1c) Adenylate cyclases (AC1, AC3, AC5, AC6, AC8) Nitric oxide synthase (NOS1, NOS3) Miro (mitochondrial motility) Dual oxidases (DuOx1–DuOx2) Ca^{++}-activated peptidases (calpain-1, calpain-2)

Table 11.24. Calcium signaling components. (**Part 6**) Calcium catalytic effectors and ion channel regulators (Source: [1508]; ^{ADP}ribose: adenosine diphosphate–ribose; cADPR: cyclic adenosine diphosphate–ribose; FKBP12 and FKBP12.6: 12- and 12.6-kDa immunosuppressant FK506-binding protein; Miro: mitochondrial Rho GTPase). The cADPR hydrolase, or ^{ADP}ribosyl cyclase-1 (also CD38), is a glycoprotein on the surface of many leukocyte types and at internal sites that functions in cell adhesion and calcium signaling. Aspartyl–asparaginyl β-hydroxylase (AspH), or junctin, is involved in calcium storage in and release from the endoplasmic reticulum (ER) as well as hydroxylation of aspartic acid and asparagine in epidermal growth factor-like domains of proteins. Peptidyl-prolyl cis-trans isomerase FKBP1a (FKBP12) and FKBP1b (FKBP12.6) are members of the immunophilin family that interact with multiple intracellular calcium release channels such as ryanodine receptors. Annexins are Ca^{++}-dependent phospholipid-binding proteins involved in material exo- and endocytosis and organization of vesicles, as well as formation of Ca^{++} channels.

Molecules	Effect
Enzymes	
^{ADP}ribosyl cyclase	Synthesis of Ca^{++}-mobilizing messengers cADPR and NAADP
cADPR hydrolase	Cell adhesion and Ca^{++} signaling
	Generation of NAADP
Regulators of monomeric and heterotrimeric GTPases	
Guanine nucleotide-exchange factors RasGRF1	RAS signaling in the brain (learning and memory)
Regulators of G protein signaling (GTPase-accelerating proteins) (RGS1, RGS2, RGS4, RGS16)	Signal termination
Channel regulators	
Triadin	Calcium-induced calcium release
Junctin	Calcium storage in and release from ER
Sorcin	Binding to annexin-7
FKBP12, FKBP12.6	Interactors of Ca^{++} release channels
Phospholamban	Inhibition of SERCA pump
Molecular tranfer regulators	
Synaptotagmin-1–synaptotagmin-3 Synaptotagmin-5–synaptotagmin-7 Synaptotagmin-9, synaptotagmin-10	Membrane trafficking, calcium sensors and regulators of neurotransmitter release and hormone secretion
Annexins (annexin-A1–annexin-A13)	Intracellular transport Formation of Ca^{++} channels

and in the brain (gabaergic interneurons, Purkinje cells, among others) and some endocrine glands. Parvalbumin is a slow-onset buffer [10]. It has relatively low activation and inactivation rates, thereby being unable to respond to a rapid influx of Ca^{++} ions. However, it can buffer Ca^{++} ions after an initial wave (delayed buffer).

11.5 Calcium Signaling

Table 11.25. Calcium signaling components. (**Part** 7) Transcription factors and buffers (Source: [1508]; GRP78 and GRP94: 78- and 94-kDa glucose-regulatory protein [GRP78 is a.k.a. endoplasmic reticulum luminal Ca^{++}-binding protein (BiP) and glucose-regulated 70-kDa heat shock protein HSPa5; GRP94 as 90-kDa heat shock protein-β1 (HSP90β1) and endoplasmin]; StIM: stromal interaction molecule).

Transcription factors	Nuclear factor of activated T-cells (NFAT1–NFAT4) cAMP response element-binding protein (CREB) Downstream regulatory element modulator (DREAM) CREB-binding protein (CBP)
Cytosolic buffers	Calbindin, calretinin, parvalbumin
Endoplasmic reticulum buffers	Calnexin, calreticulin, calsequestrin, GRP78/94
Endoplasmic reticulum sensor	StIM

Table 11.26. Cell-specific Ca^{++} signalosomes (Source: [10]). Examples of cell-specific Ca^{++} signalosomes with their own spatial and temporal regulation (IP$_3$R: inositol trisphosphate receptor; NCX: Na^+–Ca^{++} exchanger; ^{NMDA}Glu: NMDA-type glutamate receptor; PLC: phospholipase-C; PMCA: plasma membrane Ca^{++} ATPase; RYR: ryanodine receptor; SERCA: sarco(endo)plasmic reticulum Ca^{++} ATPase). Conventionally, whether the Ca^{++} source is the extracellular space or intracellular stores, Ca^{++} influx is referred to as Ca^{++} entry and Ca^{++} release, respectively.

Mediator	Atriomyocyte	Neuron	T lymphocyte
Receptors	ET$_1$, α1AR, ATR	mGluR1, M$_1$	TCR
PLC	PLCβ	PLCβ	PLCγ1
Entry channels	Ca$_V$1.2	Ca$_V$1.2/2.1/2.2 ^{NMDA}Glu	Orai1
Release channels	RyR2, IP$_3$R2	RyR2, IP$_3$R2	IP$_3$R1
PMCA	PMCA1c/1d/2a	PMCA1a/2a/3a	PMCA4b
SERCA	SERCA2a	SERCA2b/3	SERCA2b/3
NCX	NCX1	NCX1/3	
Buffers		Parvalbumin, calbindin-1	
Sensors	Troponin-C, calmodulin	Calmodulin	Calmodulin

Calbindins[56] are classified in subfamilies according to the number of Ca^{++}-binding EF-hand sites. Calbindin-1, encoded by the vitamin-D-responsive CALB1 gene, resides in the kidney as well as neuroendocrine cells. It restricts the magnitude of the elementary Ca^{++} cues generated near Ca^{++} channels. Vitamin-D-dependent calbindin-D9k, or S100 calcium-binding protein-G (S100G), acts in enterocytes. In the kidney and intestine, calbindin facilitates Ca^{++} reabsorption [10].

The major buffers in the lumen of the endoplasmic reticulum are calsequestrin (Csq) in the sarcoplasmic reticulum of myocytes and calreticulin (Crt), or calbindin-2 (CalB2) in the endoplasmic reticulum of non-myocytes. However, calreticulin operates as both a cytosolic and ER luminal buffer. It can indeed be detected in the nucleus and cytoplasm. Calreticulin is a low-affinity Ca^{++}-binding protein. In addition to its buffer role, it acts as a chaperone in conjunction with the chaperone calnexin to ensure a correct folding and subunit assembly of glycoproteins, hence proper protein transfer and secretion [10]. When the Ca^{++} concentration is too high, calreticulin and calnexin inhibit the sarco(endo)plasmic reticulum Ca^{++} ATPase (SERCA pump); when the luminal Ca^{++} concentration diminishes, this inhibition is relieved.

Calsequestrin allows the sarcoplasmic reticulum to store a hugh amount of calcium ions (each molecule of calsequestrin can bind 18 to 50 Ca^{++} ions). It maintains an elevated Ca^{++} concentration in the sarcoplasmic reticulum, which is much higher than in the cytosol. Calsequestrin possesses 2 isoforms: calsequestrin-2 is present in the myocardium and slow skeletal muscle and calsequestrin-1 in the fast skeletal muscle.

Major sensors comprise troponin-C (TnC), calmodulin (Cam), neuronal Ca^{++} sensors (NCS), and S100 proteins. These sensors relay information to numerous effectors, such as Ca^{++}-sensitive K_{Ca} channels, Ca^{++}-sensitive Cl^- channels, Ca^{++}–calmodulin-dependent protein kinases, protein phosphatase-3 (or calcineurin), phosphorylase kinase (PhK),[57] myosin light-chain kinase (MLCK), and RasA4 activator.[58]

11.5.4.3 Calcium Flux from and to Cell Organelles

A set of substances (inositol trisphosphate, cyclic ^{ADP}ribose, nicotinic acid adenine dinucleotide phosphate, and sphingosine 1-phosphate) modulate calcium release from the endoplasmic reticulum and other organelles (Fig. 11.4). Inositol (1,4,5)-

---

[56]Originally named vitamin D-dependent calcium-binding proteins.

[57]This protein Ser/Thr kinase phosphorylates (activates) glycogen phosphorylase to release glucose 1-phosphate from glycogen. It is a homotetramer of tetramers. Each tetramer is composed of 3 regulatory α, β, and δ subunits and 1 catalytic γ subunit. The δ subunit is calmodulin. Inhibitory α and β subunits prevents PhK activity. Phosphorylation of α and β subunits by protein kinase-A counters their repression.

[58]A.k.a. Ca^{++}-promoted Ras inactivator (CAPRI).

11.5 Calcium Signaling

trisphosphate is produced from phosphatidylinositol (4,5)-bisphosphate (PIP$_2$) by different phospholipase-C isoforms. Isozyme PLCβ is activated by G-protein-coupled receptors, PLCγ by receptor Tyr kinases, PLCδ by calcium influx, and PLCε by Ras GTPase. Messenger IP$_3$ stimulates its receptor IP$_3$R at the membrane of the endoplasmic reticulum. Receptor IP$_3$R is modulated by phosphorylation by calcium–calmodulin-dependent kinase-2 and protein kinases PKA, PKC, and PKG, after possible recruitment of corresponding scaffold proteins.

Low concentration in inositol (1,4,5)-trisphosphate causes a minor Ca^{++} flux from its store through a single IP$_3$R. This Ca^{++} release then stimulates additional IP$_3$Rs that generate a Ca^{++} burst. This short, fast flux provokes subsequent Ca^{++} waves, as cytosolic Ca^{++} concentration rises. Recruitment of Ca^{++} egress can result from IP$_3$ concentration elevation. Local interaction between IP$_3$Rs then causes a rapid stimulation and slow inhibition by cytosolic Ca^{++}. The IP$_3$R channels are initially randomly distributed (between-channel distance $\sim$ 1 μm). Rapid, reversible channel oligomerization triggered by low IP$_3$ concentration allows cooperativity [1516]. Oligomeric IP$_3$ receptors (generally tetramers) release Ca^{++} from the endoplasmic reticulum in response to IP$_3$ and Ca^{++} agents. Sensitivity of IP$_3$Rs is modulated by IP$_3$-mediated receptor clustering. At resting cytosolic Ca^{++} concentration, clustered IP$_3$Rs can open independently with lower open probability, shorter open time, and less IP$_3$ sensitivity than isolated IP$_3$Rs. Increasing cytosolic Ca^{++} concentration reverses the functional inhibition induced by channel clustering. Channel clustering reduces the distance between IP$_3$Rs to about 20 nm, so that clustered IP$_3$Rs are immediately exposed to Ca^{++} released by their neighbors. Moreover, in a channel cluster, Ca^{++} can counterbalance attenuated IP$_3$ sensitivity. At rest, IP$_3$ initiates IP$_3$R clustering. At the beginning, Ca^{++} release remains restricted, but IP$_3$Rs are ready for Ca^{++} excitation from adjacent channels for signal amplification. Furthermore, coupled gating enables simultaneous, prolonged open channel states.

Several mechanisms activate plasmalemmal Ca^{++} channels, especially depleted endoplasmic reticulum (capacitative or store-operated Ca^{++} entry. Signaling from the endoplasmic reticulum to the plasma membrane is initiated by membrane Ca^{++} sensors, the stromal interaction molecules StIM1 and StIM2, with Ca^{++}-binding EF-hand motifs directed to endoplasmic reticulum lumen. Dissociation of Ca^{++} from StIMs causes StIM aggregation and accumulation beneath the plasma membrane, where they interact with store-operated plasmalemmal Ca^{++} channels of the Orai family (Orai1–Orai3). Transmembrane proteins StIM1 and StIM2 can strongly, rapidly, and directly interact with Orai1 that has an intrinsic ability to cluster [1517]. Whereas high, fast activation of Orai1 requires StIM1 or StIM2, Orai3 can be activated independently of both StIM1 and StIM2 proteins.

Calcium release-activated Ca^{++} channels generate sustained Ca^{++} signals as: (1) Ca^{++} depletion from the endoplasmic reticulum triggers oligomerization of stromal interaction molecule-1 and its redistribution to junctions between endoplasmic reticulum and plasma membrane (ERPMJ; gap of 10–25 nm) and (2) CRAC channel subunit Orai1 accumulates in the plasma membrane. Non-stimulated Ca^{++}-bound StIM1 moves freely throughout the endoplasmic reticulum membrane. Because

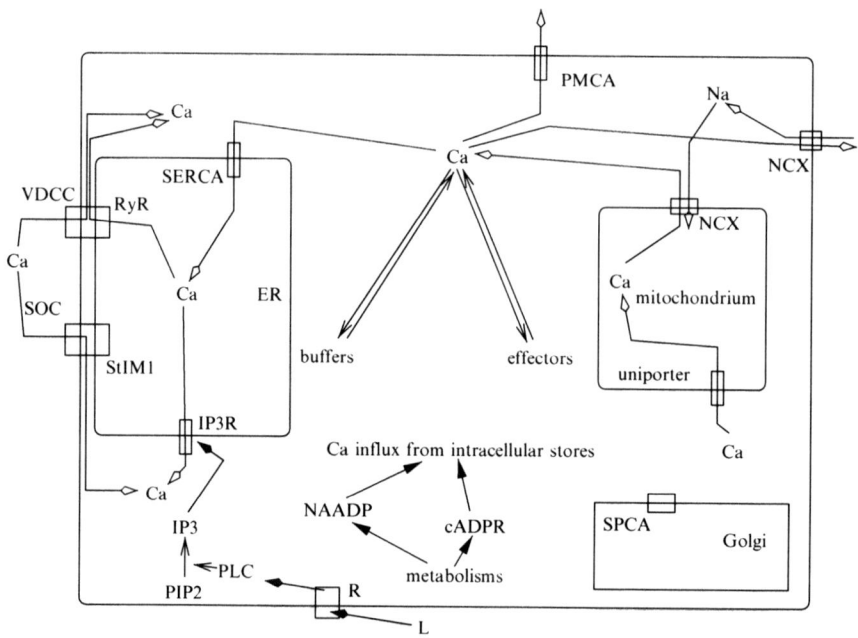

Fig. 11.4 Calcium activities in cells. Ligand (L)-bound receptor (R) causes the entry of calcium from the extracellular space and the formation of second messengers, such as inositol trisphosphate (IP$_3$), cyclic ADP ribose (cADPR), nicotinic acid adenine dinucleotide phosphate (NAADP), and sphingosine-1-phosphate (S1P). The second messenger releases calcium from its intracellular stores, the endoplasmic reticulum (ER; sarcoplasmic reticulum in the myocyte). Agent IP$_3$ is formed from phosphatidylinositol bisphosphate (PIP$_2$) by different isoforms of phospholipase C. It targets its receptor IP$_3$R. Metabolism-linked messengers cyclic ADP ribose (cADPR) and nicotinic acid adenine dinucleotide phosphate (NAADP), generated by ADP ribosyl cyclase from NAD and NADP, respectively, favors Ca^{++} release from its stores. Ca^{++} can also be transported into the cytosol via either ryanodine receptor (RR) associated with plasmalemmal voltage-dependent calcium channels, or by plasmalemmal, non-excitable store-operated channels (SOC; Orai1 channel) coupled to Ca^{++} sensor stromal interaction molecule-1 (StIM1) in regions of the endoplasmic reticulum close to the plasma membrane. Most of the intracellular calcium binds to buffers; the remaining part targets effectors. Afterward, calcium ion leaves its effectors and buffers and is removed from the cell by various exchangers and pumps. Both Na$^+$–Ca^{++} exchanger (NCX) and plasma-membrane Ca^{++}-ATPase (PMCA) expell Ca^{++} in the extracellular medium, and sarco(endo)plasmic reticulum Ca^{++}-ATPase (SERCA) pumps fill the stores. Mitochondria participate in the recovery. They quickly sequester Ca^{++} via uniporters; Ca^{++} is then slowly released into the cytosol to be extruded by SERCAs and PMCAs. (Source: [1508]).

StIM1 binds only a single calcium ion, its oligomerization is required for cooperative, efficient activity after Ca^{++} store depletion. Furthermore, its oligomerization provokes StIM1 accumulation at ERPMJ [1518]. Oligomerization of StIM1 then drives store-operated Ca^{++} influx after having triggered organization and activation of StIM1–Orai1 clusters at ERPMJ junctions.

11.5 Calcium Signaling

Proteins Orai1 and StIM1 that regulate store-operated Ca^{++} entry facilitate cell migration and tumor metastasis (Vol. 2 – Chap. 6. Cell Motility) by regulating focal adhesion turnover that involves protein phosphorylation and proteolysis [1519]. Elevated intracellular Ca^{++} concentration can increase the activity of focal adhesion kinase and calcium-dependent peptidase calpain in focal adhesions. The former activates small Rac GTPase and the latter cleaves talin at adhesion sites. In addition, Ca^{++}-sensitive myosin light-chain kinase and PP3 phosphatase intervene in focal adhesion turnover.

Sensor StIM1 has many binding partners. In fact, it regulate other types of store-operated channels such as members of the TRPC family. Proteins TRPC and Orai can form Ca^{++}-selective, store- (SOC) and receptor-operated (ROC) calcium channels [1520]. Diacylglycerol-responsive channel TRPC can actually be activated by the Gq– and Gi–PLCβ axes. The TRPC–Orai complex participates in Ca^{++} influx with or without activation of store depletion.

Phospholipase-Cγ has dual roles in regulating cellular calcium concentrations. It generates inositol trisphosphate, which releases calcium from intracellular stores. It binds to the transient receptor potential channel TRPC3 and promotes its insertion into the plasma membrane for calcium influx. The general transcription factor GTF2i outside the nucleus[59] inhibits calcium entry into cells by binding phospholipase-Cγ, antagonizing interaction of phospholipase-Cγ with the calcium channel TRPC3. This competition in favor of GTF2i for binding to PLCγ thereby suppresses surface accumulation of TRPC3 channels and hinders calcium influx across the plasma membrane [1521]. Dephosphorylated GTF2i could free PLγ and elevate the density of plasmalemmal TRPC3 receptor.

Ubiquitous *Extended synaptotagmin-like proteins* (ESyt) bind to Ca^{++} in a phospholipid complex of intracellular membrane (ESyt1) and plasmalemmal (ESyt2 and ESyt3) components [1522]. They then serve as calcium sensors with multiple C2 domains. The C2 domain is a protein module used as calcium- and phospholipid-binding sites and/or as between-protein interaction domains.[60]

Calcium depletion from intracellular stores not only activates plasmalemmal store-operated channels, but also regulates formation of cAMP by adenylate cyclase [1523]. Ligand binding to receptors that are coupled to Gq or Gs subunits primes the PLC–IP$_3$ axis and Ca^{++} release from endoplasmic reticulum store through IP$_3$R or excites ACase to trigger the cAMP–PKA pathway, respectively. Lowering concentration of free Ca^{++} in the endoplasmic reticulum, whatever the cytosolic Ca^{++} concentration, leads to recruitment of adenylate cyclases. Translocated Ca^{++} sensor StIM1 that aggregates in a region near the plasma membrane can indeed activates ACase, either alone or in synergy with other ACase activators. Among 9 transmembrane ACases, AC1 and AC8 are major

---

[59] The cytosolic GTF2i fraction remains substantial after growth factor stimulation, although the amount of nuclear GTF2i increases.

[60] Synaptotagmins, ferlins, multiple C2 domain and transmembrane region-containing proteins (MCTP) are components of 3 other families of proteins that contain several C2 domains.

Ca^{++}-activated isoforms, whereas AC5 and AC6 are inhibited by augmented cytosolic Ca^{++} level, particularly after store-operated Ca^{++} entry. Sensor StIM1 may then act as an attractor that facilitates recruitment and activation of ACases that can dimerize (e.g., AC2–AC5 heterodimers), without necessarily binding them. In addition, Ca^{++}-dependent regulation of phosphodiesterases influences cAMP level.

11.5.5 Types of Calcium Signalings

Signaling pathways are characterized by their spatial distribution (cellular compartmentation), associated with involved molecular complexes, and temporal dynamics to keep the specificity of calcium signaling. Quick, brief, localized calcium transients are associated with fast responses; transient, repetitive, distributed calcium oscillations, which can generate calcium waves, trigger slow responses. Calcium flux oscillations of given amplitude and frequency are determined by the stimulus intensity. Furthermore, calcium is able to regulate its own signaling pathways, as it influences the functioning of Ca^{++} channels.

Ca^{++} influx for long-term effect can be triggered by inositol trisphosphate on plasmalemmal IP_3Rs and Orai1 channels (store-operated Ca^{++} release–activated Ca^{++} influx, (Vol. 3 – Chap. 3. Main Classes of Ion Channels and Pumps).

Some activated G-protein-coupled receptors trigger calcium influx (Table 11.20). In the nervous system, both metabotropic glutamate receptor mGluR1 coupled to $G\alpha_q$ and mGluR5 coupled to $G\alpha_{11}$ activate PLCβ, but trigger different types of calcium signaling via IP_3R, generating a single Ca^{++} transient and an oscillatory pattern, respectively [1508]. In pancreatic acini, muscarinic receptors, which are more sensitive to the inhibition of RGS, provoke small, localized Ca^{++} transients, and cholecystokinin receptors cause large, distributed Ca^{++} transients. Among other GPCRs, bradykinin and neurokinin-A receptors give a large, rapid calcium influx. Lysophosphatidic acid, thrombin, and histamine receptors trigger small, slow, persistent calcium fluxes.

The inhibitory subunit of protein phosphatase-1 catalytic subunit $PP1_{r9b}$[61] binds to actin, regulators of G-protein signaling, and G-protein-coupled receptors. Once bound to GPCRs, it reduces the intensity of calcium signaling by GPCRs, such as α1b-adrenergic receptors [1524]. On the other hand, the inhibitory subunit of protein phosphatase-1 $PP1_{r9a}$[62] that does not bind to α1b-adrenoceptors, increases the intensity of calcium signaling by α1b-adrenoceptors. Subunit $PP1_{r9b}$ prevents binding of RGS2 to cytosolic $PP1_{r9a}$ subunit. The latter binds to RGS2 that is thus removed from G-protein-coupled receptors. Agent RGS2 inhibits calcium signaling by GPCRs, especially α1b-adrenergic receptors. Conversely, $PP1_{r9a}$ hindersbinding

---

[61] A.k.a. spinophilin and neurabin-2.
[62] A.k.a. neurabin-1.

11.5 Calcium Signaling

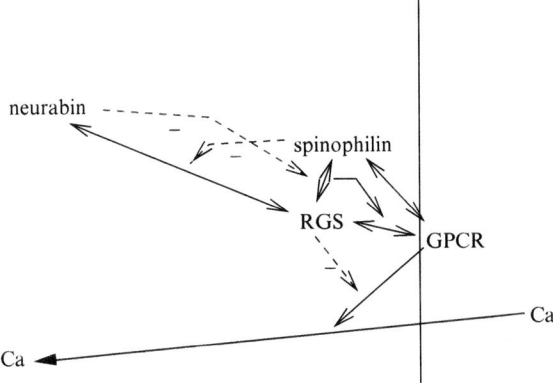

Fig. 11.5 A functional pair of antagonist regulators of GPCR-triggered calcium influx. Both spinophilin (i.e., inhibitory subunit $PP1_{r9b}$ of protein phosphatase-1) and neurabin ($PP1_{r9a}$ inhibitor) bind regulator of G-protein signaling RGS2, which hampers calcium influx by G-protein-coupled receptors. Subunit $PP1_{r9b}$ bound to a GPCR prevents RGS2 binding to cytosolic $PP1_{r9a}$, thus precluding calcium influx. Conversely, $PP1_{r9a}$ hinders RGS2 binding to $PP1_{r9b}$, removing RGS2 away from GPCR, and thereby favoring GPCR-mediated calcium influx.

of RGS2 to $PP1_{r9b}$ associated with G-protein-coupled receptors. Therefore, $PP1_{r9a}$ and $PP1_{r9b}$ form a pair of antagonist regulators that tune the intensity of calcium influx by GPCRs (Fig. 11.5).

11.5.6 Calcium-Mobilizing Mediators cADPR and NAADP

Calcium-mobilizing second messengers include cyclic ^{ADP}ribose (cADPR) and nicotinic acid adenine dinucleotide phosphate (NAADP). The former is one of the Ca^{++} signaling messengers associated with the NAD^+ pathway. Its production is coupled to the cell metabolism, hence available chemical energy. The latter is one of the most potent Ca^{++} signaling messengers that cause Ca^{++} release from intracellular stores.

Two paralogous catalytic transmembrane proteins, ^{ADP}ribosyl cyclase-1 and -2, or cADPR hydrolase-1 and -2,[63] produce NAADP and cADPR, using their ectodomains. This type of enzyme uses NAD^+ to manufacture cADPR and NADP to synthesize NAADP agent. Moreover, synthesis and degradation of cADPR is carried out by the same enzyme that is both a synthase and hydrolase. Hydrolysis of cADPR is inhibited by ATP or NADH [10]. Unlike cADPR, NAADP is degraded to NAAD by phosphatases such as alkaline phosphatase.

---

[63] A.k.a. CD38 and GPI-anchored CD157.

11.5.6.1 Cyclic ^{ADP}Ribose Signaling

The cADPR-mediated control of Ca^{++} release may occur indirectly via the activation of the sarco(endo)plasmic reticulum Ca^{++} ATPase. Increased storage may sensitize release channels such as ryanodine-sensitive channels. Therefore, cADPR messenger operates as a modulator rather than a mediator of Ca^{++} signaling [10].

Modulator cADPR may also enhance the sensitivity of ryanodine receptors. Catabolism of cADPR produces ADPR, a regulator of plasmalemmal melastatin-related TRPM2 transient receptor potential channel.

11.5.6.2 Nicotinic Acid–Adenine Dinucleotide Phosphate Signaling

Nicotinic acid–adenine dinucleotide phosphate (NAADP) uses a NAADP-sensitive store distinct from the endoplasmic reticulum, which may be a lysosome-related organelle. It does not interact with inositol trisphosphate and ryanodine receptors, but with a family of *NAADP-regulated Ca^{++} cation-selective 2-pore channels* (TCP) encoded by the 2 TPCN genes (TCPN1–TCPN2 at least in mice).[64] Subtype TCP1 localizes to the endosomal membrane; TCP2 to the lysosomal membrane [10].

11.5.7 Calcium and Nervous Control of Blood Circulation

11.5.7.1 Neuronal Calcium Sensors

Members of the neuronal calcium sensor (NCS) family are involved in manifold neuronal signaling pathways. The mammalian set of neuronal calcium sensors contains NCS1,[65] 3 visinin-like proteins (ViLiP or VsnL1),[66] neurocalcin, hippocalcin

---

[64] Neither tandem pore domain-containing potassium channels (TAlK1–TALK2, TASK1–TASK5, THIK1–THIK2, TRAAK1, TREK1–TREK2, and TWIK1–TWIK4) encoded by the KCNK gene set (KCNK1–KCNK10, KCNK12–KCNK13, and KCNK15–KCNK18), nor chaperone T-complex protein TCP1 of the chaperonin-containing TCP1 complex (CCT), or TCP1 ring complex (TRiC) involved in folding of newly synthesized polypeptides and protein refolding, as well as assembly of the SMRT-HDAC3 repression complex (SMRT: silencing mediator of retinoic acid and thyroid hormone receptor, or nuclear receptor corepressor NCoR2; HDAC: histone deacetylase).

[65] Agent NCS1 regulates phosphatidylinositol 4-kinase-3β, as well as many other proteins.

[66] Visinin-like protein-1, or ViLiP1, is also referred to as hippocalcin-like protein-3 (HpcaL3) and visinin-like protein-3 as hippocalcin-like protein-1 (HpcaL1),

11.5 Calcium Signaling

Table 11.27. Time scales of calcium-regulated processes in neurons (Source: [1526]).

Neurotransmission	< 1 ms
Channel activity	> 1 ms
Short-term plasticity	> 100 ms
Long-term potentiation	< 10 s
Long-term depression	> 10 s
Gene expression	> 10 s

(HpCa),[67] recoverin,[68] 3 guanylate cyclase-activating proteins (GCAP),[69] and 4 voltage-gated potassium channel-interacting proteins (KChIP1–KchIP4), with several spliced KChIP isoforms.[70]

The activity of the cardiovascular system is controlled by the nervous system. Neuron activities depend on the temporal feature of the calcium signal (Table 11.27). Local increases in intracellular calcium trigger neurotransmitter release less than 100 μs after Ca^{++} influx. Ca^{++}-binding synaptotagmin acts as a Ca^{++} sensor for fast neurotransmission [1526].

---

[67] Ca^{++}-binding hippocalcin is expressed at high levels in hippocampal pyramidal neurons and moderately in certain other neuron types. It undergoes Ca^{++}- and myristoyl-dependent translocation from the cytosol to membranes of the trans-Golgi network and plasma membrane when Ca^{++} concentration increases. This member of the neuronal calcium sensor family participates in olfaction, as it regulates the activities of ciliary adenylate cyclases and particulate guanylate cyclases. This multifunctional modulator contributes to neuronal excitability and long-lasting remodeling associated with memory, as it is involved in the activation of mitogen-activated protein kinases and, subsequently, cAMP-response element-binding protein [1525]. It protects against Ca^{++}-induced cell death, as it interacts with neuronal apoptosis inhibitory protein (NAIP). It is also implicated in hippocampal NMDA-type glutamate receptor-dependent long-term depression, as it binds to β2-adaptin subunit of AP2 adaptor complex that mediates endocytosis of α-amino 3-hydroxy 5-methyl 4-isoxazole propionic acid receptor. It serves as an intermediary between Ca^{++} influx and K^+ channel stimulation.

[68] Recoverin is expressed only in the retina.

[69] Members of the GCAP set exist only in the retina. They are involved in light adaptation during phototransduction.

[70] Members of the KCHIP subfamily are characterized by: (1) differential expression according to the neuron type and regions for a given neuron type (hippocampus, cortex striatum, and cerebellum); and (2) differences in target and location. They augment $K_V 4$ expression, shift the activation voltage threshold, slow the channel inactivation, and accelerate the recovery rate from inactivation. Protein KChIP3 (or calsenilin) interacts with presenilin, like KChIP4 and amyloid precursor protein. It alters Ca^{++} release from the endoplasmic reticulum. Protein KChIP3 is a transcriptional repressor, known as a downstream regulatory element antagonistic modulator (DREAM) at low Ca^{++} concentration. Calcium binding to KChIP3 hampers KChIP3 binding to the downstream regulator element (DRE) of the gene promoter.

11.5.7.2 Calcium Signaling in the Cardiovascular Apparatus

Calcium Signaling in Cardiomyocytes and the Nervous Control

Cardiomyocytes permanently bear calcium influxes and effluxes for contraction and relaxation. Adaptative responses associated with modifications in gene transcription then require changes in Ca^{++} signaling. Calcium ions act indirectly on transcription factors. Phosphatase PP3 dephosphorylates the transcription factor nuclear factor of activated T cells NFAT3, which can enter the nucleus to induce gene transcription, in opposition to the PI3K pathway that inhibits glycogen synthase kinase-3, thus inactivating NFAT3 (Vol. 5 – Chap. 6. Cardiomyocytes).

Elevation of cytosolic calcium levels can be caused by angiotensin-2 and endothelin-1 via GPCR (Gq), and phospholipase-Cβ that generates inositol trisphosphate. Diacylglycerol can target via protein kinase-D histone deacetylases and may recruit the mitogen-activated protein kinase module to activate cAMP response element-binding protein.

The adrenergic pathway enhances calcium signaling via adenylate cyclases that produces cAMP messenger. The latter leads to phosphorylation by PKA of $Ca_V 1.2$ channels and ryanodine receptors RyR2, as well as phospholamban that inhibits SERCA pumps. Calcium ions are involved in cardiac hypertrophy and congestive heart failure. Decay in SERCA activity, at least partially due to enhanced inhibition by phospholamban, is associated with a decline in β-adrenergic signaling.

Calcium Signaling in Vascular Smooth Muscle Cells

Resistance arteries regulate locally blood flow by constricting or relaxing in response to changes in hemodynamic stresses. In particular, increased intraluminal pressure causes gradual depolarization of arterial smooth muscle cells, in which the Ca^{++} concentration subsequently rises and activates myosin light-chain kinase that causes contraction of stress fibers to maintain constant the flow rate (*autoregulation*).

In arterial smooth muscle cells, opening of single or clustered $Ca_V 1.2b$ produces local elevations in intracellular Ca^{++} concentration, the so-called Ca^{++} *sparklets* at hyperpolarized membrane potentials, for which the open probability of $Ca_V 1.2b$ channels is very low.

Another Ca_V type — $Ca_V 1.3$ —, with a voltage dependence of activation more negative than that of $Ca_V 1.2b$ channels, can also cause Ca^{++} sparklets [1527]. Like $Ca_V 1.1$ and $Ca_V 1.2$ channels, $Ca_V 1.3$ may operate with 2 gating modes: short (~ 1.6 ms) and long (~ 9.5 ms) open times. On the other hand, $Ca_V 1.4$ channels may not function with a bimodal gating.

Randomly activating solitary $Ca_V 1.2b$ channels acting in a low-activity mode open rarely and create limited Ca^{++} entry (low-activity Ca^{++} sparklets) [1527]. On the other hand, single or clusters of $Ca_V 1.2b$ channels operating in a high-activity mode generate persistent Ca^{++} sparklets that depend on protein kinase-C (high-activity or persistent Ca^{++} sparklets). Protein kinase-Cα coerces discrete clusters of $Ca_V 1.2b$ channels to work in a high open probability mode, thereby engendering subcellular domains of nearly continuous Ca^{++} entry. Calcium sparklets also depends on activities of protein phosphatases PP2 and PP3.

Like $Ca_V 1.2$ channels, $Ca_V 1.3$ can produce persistent Ca^{++} sparklets. However, $Ca_V 1.2$ channels, not $Ca_V 1.3$ channels, give rise to Ca^{++} sparklets in mouse arterial smooth muscle cells [1527]. The voltage dependences of activation and inactivation of Ca^{++} flux in arterial myocytes ressemble to those of $Ca_V 1.2$, but not $Ca_V 1.3$ currents. Moreover, transcripts for $Ca_V 1.2$, but not $Ca_V 1.3$ protein, are observed in human arterial smooth muscle cells.

Membrane depolarization increases Ca^{++} entry via low- and high-activity Ca^{++} sparklets. Low- and high-activity Ca^{++} sparklets modulate local and global Ca^{++} concentration and effect in arterial smooth muscle cells [1528].

Signaling responsible for endothelium-dependent regulation of vascular smooth muscle tone relies on calcium sparklets in the vascular endothelium of resistance arteries generated by TRPV4 cation channels [1529]. Gating of a single TRPV4 channels within a 4-channel cluster causes vasodilation due to amplification resulting from cooperation of adjoining TRPV4 channels as well as via activation of endothelial intermediate (IK) and small (SK) conductance, Ca^{++}-sensitive K^+ channels. Intermediate conductance (IK) channels colocalize with TRPV4 close to myoendothelial gap junctions. The ionic flux and associated current through activated IK and SK channels can then spread to surrounding smooth muscle cells using myoendothelial gap junctions. Hyperpolarization of smooth muscle cell membrane subsequently induces a vasodilation.

11.5.8 Calcium Signaling and Immunity

Resting lymphocytes have a low concentration of intracellular calcium ions. Upon commitment of antigen receptors, calcium enters into lymphocytes from the extracellular space mainly via store-operated calcium channels. The latter comprises 2 major components: pore-forming calcium release-activated calcium modulator CRACM1 and endoplasmic reticulum-resident sensor of stored calcium, the stromal interaction StIM1 molecule. Upon antigen recognition by lymphocytes, phosphorylated (activated) phospholipase-C generates diacylglycerol and inositol trisphosphate from phosphatidylinositol (4,5)-bisphosphate. Inositol trisphosphate binds to its receptor on the surface of the endoplasmic reticulum to release Ca^{++} from its stores. Calcium influx with protein kinase-C activated by diacylglycerol initiates quick remodeling of actin cytoskeleton to promote T-cell motility, adhesion, and formation of the immunological synapse. Store depletion stimulates SOC channels, such as low-conductance, Ca^{++}-selective calcium release-activated calcium channels for a sustained response that is necessary for the maintenance of immunological synapses [1530].

Other channels can be involved such as canonical transient receptor potential channels. Channel TRPC1 is able to form diverse channels via homo- or heteromeric interactions with TRPC3, TRPC4, and TRPC7 that are relatively selective or non-selective to Ca^{++} [1530]. Some TRPC channels are activated by diacylglycerol.

Diacylglycerol also excites the Ras–MAPK cascade that activates transcription AP1 factor. The latter cooperates with transcription factors NFAT and NFκB to control gene expression. Calcium entry by CRAC channels can activate the Ras–MAPK pathway to enhance AP1 activation [1530]. In activated lymphocytes, Ca^{++} signaling also targets JNK and CamK kinases. A prolonged increase in cytosolic Ca^{++} concentration via CRAC channels is required for activation of phosphatase PP3 that dephosphorylates NFAT factor. Dephosphorylated NFAT enters the nucleus. A persistent elevation in Ca^{++} concentration is needed to prevent PP3 ejection from the nucleus [1531]. Nuclear import of NFAT is impeded by GSK3 kinase. Therefore, NFAT transcriptional activity depends on receptor stimulation and NFAT intranuclear concentration.

Several other cytosolic and nuclear molecules modulate intracellular NFAT location, such as PP3 inhibitors (AKAP, calcineurin (PP3)-binding protein [CaBin], calcineurin homologous protein [CHP], or calcium-binding protein P22, Down syndrome critical region gene product DSCR1, cytoplasmic scaffold proteins Homer-2 and Homer-3, and NFAT kinases GSK3, CK1, and DYRK) [1531, 1532].

Calcium signaling is necessary for T-cell proliferation and cytokine secretion. Composition of CRAC channels in T lymphocyte can vary with its differentiation status. Homologs CRACM1, CRACM2, and CRACM3 are involved at various degrees (CRACM2 to a lesser extent than other types),[71] as well as StIM1, in Ca^{++} influx in developing, mature, and activated T lymphocytes [1533]. T lymphocytes also express Ca_V channels that can participate in response to stimulated T-cell receptors. Phosphatase PP3 that is activated by Ca^{++}–calmodulin complex intervenes in the development of CD4+, CD8+, double-positive thymocytes into CD4+, CD8− and CD4−, CD8+, single-positive thymocytes. Non-store-operated cation channel TRPM7 for Ca^{++} and Mg^{++} ions with an intrinsic kinase activity operates in thymocyte maturation.

Calcium signaling is important in mastocytes that work at the interface of innate and adaptive immune responses. Crosslinking of immunoglobulin-E receptor FcεR1 that is the main activation mechanism of mastocytes initiates intervention of SRC family kinases Lyn and Fyn that trigger the Lyn–SYK–LAT–PLC and Fyn–GAB2–PI3K–PKC cascades to prime Ca^{++} flux and control degranulation. Agents StIM1 and CRACM1 are involved in mastocyte degranulation. Melastatin-related transient receptor potential channel TRPM4, a Ca^{++}-activated non-selective cation channel, diminishes the driving force of Ca^{++} influx through CRAC channels by modulating the membrane potential [1534]. Both StIM1 and CRACM1 are also needed for leukotriene and cytokine secretion as CRAC channels and increase in cytosolic Ca^{++} concentration are involved in the secretion of pro-inflammatory

---

[71] Homolog CRACM2 abounds in mouse thymocytes, but not in humans.

lipid mediators and activation of NFAT and NFκB that regulate the synthesis of cytokines [1535].

11.5.9 Calcium Signaling and Intracellular Transport

Exocytosis comprises multiple Ca^{++}-dependent steps that also involve Ca^{++}-dependent nanomotors, cytoskeleton, and SNARE proteins down to the formation of a *fusion pore*, an aqueous channel that connects the vesicle lumen to the extracellular space. Their controlled opening and closure regulate secretion of the vesicle content. Calcium transients are major players of this process; they are associated with fixed and mobile Ca^{++} buffers, in addition to various types of carriers (channels, pumps, and exchangers).

11.5.9.1 Fast Secretors

In fast secretors such as neurons, pore formation and transmitter release follow (delay $O[1\,\mu s]$–$O[1\,ms]$) elevation of cytoplasmic Ca^{++} concentration. Calcium transients originate from Ca^{++} influx through voltage-gated Ca^{++} channels. Plasmalemmal voltage-gated Ca^{++} channels close to the site of fusion are activated during the prefusion stage in neurons and neuroendocrine cells. Transient calcium signals then propagate throughout the cell, particularly around subplasmalemmal vesicles (speed $\sim 10\,\mu m/s$) [1536].

11.5.9.2 Slow and Non-Excitable Secretors

Slow secretors such as alveolar type-2 pneumocytes that slowly release pulmonary surfactant (choline-based phospholipids and surfactant proteins) stored in secretory vesicles, the so-called ellipsoidal *lamellar granules* or *lamellar bodies* (length 300–400 nm; width 100–150 nm, ≤ 10/cell), which have a scattered intracellular distribution rather than organized in cortical clusters, using Ca^{++} and the actin cytoskeleton.

Lamellar granules fuse with the plasma membrane in a sequential process (duration ~ 20 mn). Each fusion steps is followed by a transient rise of localized cytoplasmic Ca^{++} concentration (decay half-life 3.2 s) that originate at the site of lamellar body fusion [1536]. The major calcium source is the extracellular space, hence the name *fusion-activated Ca^{++} entry*, although Ca^{++} ions are stored in lamellar granules. Besides, type-2 pneumocytes lack voltage-gated Ca^{++} channels.

In isolated rat alveolar type-2 cells, fusion-activated Ca^{++} entry follows initial fusion pore opening with a delay of 200 to 500 ms [1537].

Calcium-induced Ca^{++} release may amplify Ca^{++} influx. A moderate overall increase in cytoplasmic Ca^{++} concentration (slightly above resting values;

~320 nmol/l) is required to induce fusion of lamellar granules with the plasma membrane.

Calcium influx may results from currents through channels of the plasma membrane or channels of the lamellar granule membrane. Whatever its location, channel activation may results from membrane mechanical stress or action of membrane chemical merger. In fact, fusion-activated Ca^{++} entry results from activity of vesicle-associated Ca^{++} channels [1537]. Ionotropic $P2X_4$ receptors indeed reside on lamellar granules.

An actin coat forms around a lamellar granule after fusion; its contraction enables full extraction of the vesicular content. In fact, after fusion of a lamellar granule with the plasma membrane, surfactant, a water-insoluble complex of lipids and proteins remains within the fused vesicle.

During the postfusion phase, when the lumen of the lamellar granule becomes a part of the extracellular space, an elevated cytoplasmic Ca^{++} concentration leads to fusion pore dilation. A fusion-activated Ca^{++} entry supports the postfusion phase of surfactant secretion [1537].

11.5.10 Calcium Signaling and Cell Fate

Promyelocytic leukemia protein (PML),[72] controls calcium signaling at the endoplasmic reticulum, near mitochondria, i.e., at signaling regions involved in endoplasmic reticulum-to-mitochondrion Ca^{++} transport and in induction of apoptosis [1538].[73]

Multiple isoforms of human PML transcripts arise from alternative splicing. Post-translational modifications add further diversity in PML structure and function. These isoforms have both cytoplasmic and nuclear locations. Promyelocytic leukemia protein localizes to nuclear bodies (a.k.a. nuclear dots, PML bodies, and Kremer bodies; number 10–30 per nucleus; size 0.2–1 μm). It resides also in the nucleoplasm as well as in the nucleolus during cell senescence and stress. It also

---

[72] Promyelocytic tumor suppressor was identified as a component of the PML-RARA oncoprotein in acute promyelocytic leukemia (APL). Protein PML enables the assembly of subnuclear structures, PML-nuclear bodies. Protein PML undergoes ubiquitin-mediated degradation after its phosphorylation by CK2 kinase. Sumoylation of PML is carried out by PIAS1 Sumo ligase [1539]. Sumoylation of PML promotes interaction with CK2, ubiquitination, and proteasomal degradation. Kelch-like-20 homolog, an adaptor induced by HIF1 that complexes with cullin-3 to from a Cul3-KlhL20 ubiquitin ligase, in coordination with CDK1, CDK2, and Pin1, causes hypoxia-induced PML proteasomal degradation [1540]. Sumoylation of PML also facilitates its ubiquitination by RNF4 ubiquitin ligase, which also leads to PML degradation. In addition, ubiquitin ligases SIAH and E6-associated protein (E6AP) also target PML to trigger its elimination.

[73] Promyelocytic leukemia protein has multiple functions. It binds to eukaryotic translation initiation factor eIF4e, thereby impeding mRNA nuclear export. It promotes apoptosis, represses cell cycle progression, senescence, and some types of oncogenic transformation, and operates in stem cell renewal and in defense against many viruses.

lodges at sites of contact between the endoplasmic reticulum and mitochondria, where it connects to the IP_3R calcium channel, protein kinase-B, and protein phosphatase-2 [1538].

Protein PML can then modulate phosphorylation of IP_3R and hence calcium release from the endoplasmic reticulum to regulate calcium mobilization into the mitochondrion. It can then trigger early, transcription-independent apoptosis program.

11.6 Oxygen Delivery and Hypoxia Transduction

The cardiovascular and ventilatory apparatus cooperate under the control of the nervous system to ensure the delivery of oxygen and nutrients needed throughout the organism. Oxygen supply is indeed necessary for efficient energy production. Cells use the tricarboxylic acid cycle and oxidative phosphorylation in the mitochondria as well as glycolysis for energy production under aerobic conditions (Vol. 1 – Chap. 4. Cell Structure and Function). Cell survival depends on O_2 delivery.

Oxygen is an electron acceptor during mitochondrial respiration, but the respiratory chain functions within a narrow range of O_2 concentrations. Oxygen is required to generate energy, but energy generation produces potentially toxic oxidants such as reactive oxygen species (Sect. 10.6.1).[74]

Acute hypoxia and hyperoxia disturb cell life. However, the organism adapts to conditions of low or high oxygen by triggering a cascade of events to reprogram cellular oxygen requirements. In particular, multiple mechanisms allow the cell to respond to decreased oxygen levels. The hypoxia response provokes activation of multiple signaling pathways involved in regulation of cell respiration, metabolism, and survival. Under hypoxia, cells shift from an aerobic mode of energy production to an anaerobic mode. They raise glucose uptake and inhibit enzymes leading to the tricarboxylic acid cycle, hence shifting ATP generation from mitochondrial respiration that consumes oxygen to anaerobic glycolysis outside mitochondria (*Pasteur effect*). They also decrease the metabolic rate.

Hypoxia at high altitudes induces 3 main responses: (1) neurotransmitter release by the carotid body to increase breathing; (2) pulmonary vascular constriction in poorest oxygenated regions of the lung;and (3) erythropoietin production (up to

---

[74]Reactive oxygen species include superoxide anion (O_2^-), hydrogen peroxide (H_2O_2), and hydroxyl radical (•OH). Electrons transferred in the mitochondrial respiratory chain react with O_2 to form H_2O using cytochrome-C oxidase. Some electrons escape the respiratory chain, combine with O_2 prematurely and generate superoxide anion, which is converted into hydrogen peroxide by superoxide dismutase.

several 100-fold increase) by renal interstitial fibroblasts to augment erythropoiesis and hemoglobin concentrations in blood. The hypoxia response comprises 2 axes: HIF-dependent and -independent pathways.

11.6.1 Hypoxia-Inducible Factor Axis

The main transcription factor that regulates transcriptional responses to hypoxia is hypoxia-inducible factor (Sect. 10.9.2) that was first found to interact with erythropoietin gene. Hypoxia-inducible factor is a DNA-binding heterodimer.[75]

The HIF1α–HIF1β dimer, i.e., HIF1 transcription factor, binds to *hypoxia response elements* in the regulatory regions of target genes, thereby controlling the expression of a huge number of genes. These target genes vary according to the cell type (*canonical hypoxic signaling*). Both HIF2α and HIF3α are also produced in response to hypoxia. Oxygen-regulated HIF2α also dimerizes with HIF1β. HIF2α is produced not only in endothelial cells, but also in parenchymal and interstitial cells of many organs, such as cardiomyocytes, renal interstitial cells, hepatocytes, duodenal epithelial cells, and astrocytes.[76] Subunit HIF3α inhibits HIF1αcomponent. Under O_2-independent conditions, HIF1α production can be enhanced by the PI3K–PKB–TOR pathway.

Hypoxia-inducible factor regulates the expression of several genes involved in iron homeostasis, such as transferrin, thereby coordinating erthyropoiesis with iron availability. It also favors cell survival [1544] (Table 11.28). It indeed adapts the

---

[75]It is made up of 2 basic helix–loop–helix PAS domain-containing subunits: oxygen-regulated α subunit and constitutively expressed non-O_2-responsive nuclear protein HIF1β or aryl hydrocarbon receptor nuclear translocator (ARNT). Three isoforms of HIFα subunit exist (HIF1α–HIF3α). Subunit HIF2α is also called endothelial PAS domain-containing protein-1. Both HIF1α and HIF2α interact with hypoxia-responsive targets, but these targets can have higher specificity either for HIF2α or HIF1α [1541]. In cultured cardiomyocytes, interleukin-1β, but not tumor-necrosis factor-α, increases HIF2α production. Interleukin-1β also upregulates adrenomedullin expression via HIF2α [1542]. These subunits are encoded by distinct genes; an additional diversity arises from alternative splicing.

[76]Transcript HIF2α mRNA has particularly high levels in tissues that are important for O_2 delivery, such as lungs (pneumocytes-2 and pulmonary endothelial cells), heart, and endothelia [1543].

Table 11.28. Hypoxia-inducible factor intervenes in cell survival via a set of pathways. HIF activation favors cell protection from oxygen deprivation. Glucose transporters and glycolytic enzymes favor anaerobic ATP production. Under hypoxia, HIF1 upregulates the gene expression for [1545]: (1) glucose transporters GluT1 and GluT3, which increase intracellular glucose uptake; (2) glycolytic enzymes that convert glucose into pyruvate; (3) lactate dehydrogenase-A that transforms pyruvate into lactate; (4) pyruvate dehydrogenase kinase-1 that phosphorylates (inactivates) pyruvate dehydrogenase (that transforms pyruvate into acetylCoA, which enters into the mitochondrial tricarboxylic acid cycle); and (5) cytochrome-C oxidase subunit CyOx4-2 that replaces CyOx4-1 and then increases the efficiency of mitochondrial respiration.

↓ cell apoptosis modulation of activity of hypoxia-regulated microRNAs ↓ ROS production
↓ inflammation ↑ antioxidant enzymes (heme oxygenase-1) ↑ inducible nitric oxide synthase ↑ cyclooxygnease-2
↑ glucose uptake (GluT) ↑ glycolysis ↓ ATP depletion
↑ oxygen delivery ↑ erythropoiesis ↑ angiogenesis
↑ extracellular matrix regulation

cellular metabolism to hypoxia by priming glycolysis to compensate energy loss due to reduced oxidative phosphorylation and enhances glucose uptake (Fig. 11.6).[77] HIF also intervenes in mitochondrial respiration,[78] and regulates intracellular pH.[79]

---

[77] Many enzymes that shift the metabolism toward ATP generation by anaerobic glycolysis are regulated by ubiquitous HIF1α, such as phosphoglycerate kinase-1 and lactate dehydrogenase-A. Glucose transporters (GluT1 and GluT3) and other enzymes of gluconeogenesis are also upregulated by hypoxia. In aerobic conditions, glucose undergoes an oxidative metabolism that yields energy at high efficiency (about 30 mol ATP per mol glucose). In anaerobic conditions, glucose enters in glycolytic catabolism that reduces the energy production (2 mol ATP per mol glucose).

[78] Mitochondrial oxygen-consumption is reduced and ROS generation is attenuated during hypoxia. Subunit HIF1α stimulates pyruvate dehydrogenase kinase PDK1, which impedes the conversion of pyruvate into acetylCoA by pyruvate dehydrogenase, hence preventing pyruvate entry into the tricarboxylic acid cycle. Agent HIF1 also modulates mitochondrial respiration by switching the composition of the cytochrome oxidase complex (CyOx4-2 rather than CyOx4-1) in hypoxia. In addition, as mitochondrial respiration decays, cells rely almost exclusively on glucose as fuel via anaerobic glycolysis.

[79] Subunit HIF1α regulates monocarboxylate transporter, a member of the H^+–lactate cotransporter family, which excretes lactic acid from the cytoplasm. Moreover, HIF1α targets Na^+–H^+ exchanger NHE1, carbonic anhydrase-9 and -12, which convert CO_2 into carbonic acid.

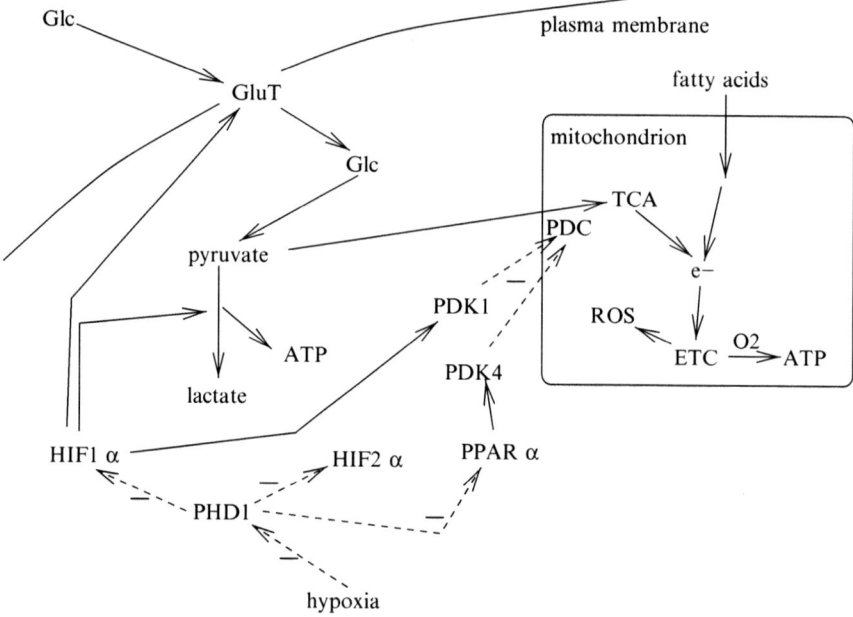

Fig. 11.6 Hypoxia tolerance by reprogramming basal metabolism from oxidative to anaerobic ATP production via inhibition of oxygen-sensitive prolyl hydroxylase domain-containing protein PHD1 that lowers oxygen consumption (Source: [1546]). Subunit HIF2α augments the expression of pyruvate dehydrogenase kinases (PDK) that inhibit the pyruvate dehydrogenase complex (PDC), which controls entry of glucose-derived pyruvate into the mitochondrion. Inside the mitochondrion, pyruvate enters the tricarboxylic acid cycle (TCA) that together with fatty acid oxidation indirectly generates ATP using enzymes of the electron transport chain (ETC) in the presence of oxygen, as well as reactive oxygen species (ROS) produced by the electron transport chain. (When myocytes continue to consume oxygen despite the limited oxygen supply, they generate excessive ROS amounts.) Subunit HIF1α upregulates pyruvate dehydrogenase kinase PDK1 isoform. Subunit HIF2α increases the expression of the transcription factor peroxisome proliferator-activated receptor PPARα, which elevates PDK4 to reduce mitochondrial respiration in striated myocyte.

Subunit HIF1α improves oxygen delivery not only by incresing erythropoiesis, but also stimulating synthesis of various compounds, such as vasoactive molecules (nitric oxide synthase), growth factor (VEGF), and extracellular matrix regulators (urokinase-type plasminogen activator receptor, collagen prolyl 4-hydroxylase, matrix metallopeptidase MMP2, and tissue inhibitor of matrix metallopeptidase TIMP1), hence enhancing angiogenesis (Table 11.29).

The transcriptional response to hypoxia is primarily mediated by HIF1α and HIF2α subunits. Both HIF1α and HIF2α have distinct functions with a functional overlap on targeted genes, which varies according to the cell type. Genes of vascular endothelial growth factor, adrenomedullin, and glucose transporter GluT1 are stimulated by both HIF1α and HIF2α.

11.6 Oxygen Delivery and Hypoxia Transduction

Table 11.29. Regulation by HIF1α subunit of genes of factors involved in successive steps of angiogenesis (Source: [1544]). Growth factor VEGF is upregulated by HIF1α as well as its VEGFR1 receptor, thereby further increasing VEGF activity.

Vasodilation	NOS
Vascular permeability	VEGF
Extracellular matrix remodeling	MMP, uPAR
Endothelial sprouting	Ang2
Cell proliferation	VEGF, FGF, PDGF, SDF1, CXCR4
Inhibitors	Ang1, Tsp1

Subunit HIF1α uniquely upregulates the expression of glycolytic enzymes. Subunit HIF2α targets genes involved in erythropoiesis, angiogenesis, and cell proliferation, such as those that encode embryonic transcription factor Oct4, cyclin-D1, transforming growth factor-α, and erythropoietin. Subunit HIF2α cooperates with ETS1 transcription factor to synergistically activate VEGFR2 expression and with ELk1 transcription factor for erythropoietin, insulin-like growth factor-binding protein IGFBP3, and plasminogen activator inhibitor PAI1. Subunit HIF2α specifically interacts with NFκB essential modulator NEMo (IKKγ). It also regulates the activity of transcription factors, such as Notch and Myc, thus enhancing expression of cyclin-D2.

Activity of HIF1 depends on the availability of HIF1α subunit. In normoxia, continuously synthesized HIFα subunits have a very short half-life. Subunit HIF1α is only detectable during hypoxia. The concentration and activity of HIF1α is controlled by oxygen-dependent prolyl-hydroxylating domain-containing proteins (PHD) and asparaginyl factor-inhibiting HIF1α hydroxylases (FIH1). Under normoxia, prolyl hydroxylase domain-containing protein[80] leads to interaction of newly synthesized, oxygen-sensitive HIF1α subunit with von Hippel-Lindau protein for polyubiquitination and proteasomal degradation.[81] In addition, asparaginyl hydroxylase[82] FIH prevents HIF interactions with CBP and P300 coactivators.

Hypoxia activates HIF1α by suppressing hydroxylation of proline and asparagine residues, thereby preventing HIF1α degradation and favoring its accumulation and dimerization with HIF1β in the nucleus. The dimer then recruits coactivators such as P300 and binds to hypoxia-response elements of target genes.

Many environmental and intracellular factors modulate the activity of prolyl hydroxylase. Nitric oxide at relatively high concentration and ROS inhibit PHD

---

[80] Prolyl hydroxylase domain protein is also called HIF-prolyl hydroxylase. It belongs to the family of dioxygenases that use oxygen and 2-oxoglutarate as cosubstrates, and iron and ascorbate as cofactors [1547].

[81] von Hippel–Lindau protein is the recognition component of the ubiquitin ligase complex, which also contains elongin-B, elongin-C, Cul2, and Rbx components.

[82] Oxygen-dependent asparaginyl hydroxylase FIH is also an iron-dependent dioxygenase.

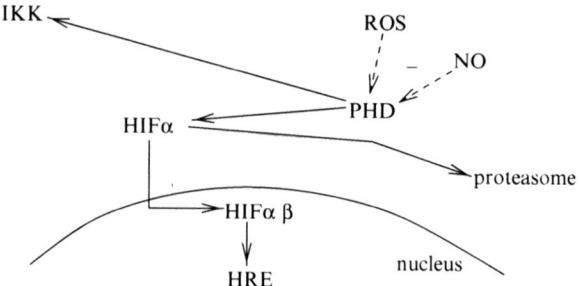

Fig. 11.7 Schematic illustration of HIF activity and regulation by prolyl hydroxylase domain-containing protein (PHD). These enzymes are controlled by NO and ROS agents. Isozyme IKKβ hydroxylated by PHD1 or PHD2 fails to phosphorylate inhibitory factor IκBα, thus being unable to dissociate IκBα from NFκB and to activate NFκB. Hypoxia response element (HRE) are upregulated by HIF factor.

activity [1547].[83] Different PHD isoforms differentially contribute to physiological and pathophysiological processes, such as growth, differentiation and survival at the cell level, and angiogenesis, erythropoiesis, and tumorigenesis at the tissue level. These enzymes can function as tumor suppressors. Among the 4 PHD isoforms (PHD1–PHD3 and transmembrane (endoplasmic reticulum) prolyl 4-hydroxylase [$_{TM}$P4Hor PH4tm]), HIF1α is more efficiently hydroxylated by PHD2 and HIF2α by PHD1 and PHD3 isozymes. These 3 PHDs are widely distributed among different organs, but at different levels (PHD3 is the highest subtype in the heart). They mostly localize to the cytoplasm. They undergo proteasomal degradation. Both PHD2 and PHD3 bear feedback upregulation by HIFs to suppresses further accumulation of HIFα and ensure rapid removal of HIF after reoxygenation.[84] Factor TGFβ1 that is also stimulated by hypoxia inhibits PHD2 transcription via SMAD signaling, thereby counteracting the effect of HIF-induced upregulation of PHD2 enzyme. Prolyl hydroxylases also hydroxylate non-HIF substrates and use hydroxylase-independent mechanisms to modify HIF activity [1547]. They can inhibit HIF1α transcriptional activity without decreasing HIF1α amount. Both PHD1 and PHD2 regulate the transcriptional activity of NFκB, hydroxylating IKKβ (Fig. 11.7).

Factor HIF1α is able to complex with [1545]: (1) HIF1β and P300 coactivator that work in transcription for metabolic adaption to hypoxia on the one hand, and angiogenesis and erythropoiesis to increase O_2 delivery on the other hand; (2) FIH1 to inhibit the interaction with P300; (3) heat shock protein HSP90 to stabilize HIF1α; (4) endoplasmic reticulum lectin ERLec2[85] and prolyl hydroxylase domain-containing protein PHD2 to promote hydroxylation and subsequent degradation;

---

[83]Hypoxia elevates mitochondrial ROS production that subsequently further activates HIF1α.

[84]PHD2 and PHD3 genes contain hypoxia response elements.

[85]A.k.a. Osteosarcoma-amplified protein Os9.

and either (5) spermidine/spermidine acetyltransferase SAT2,[86] von Hippel-Lindau protein and elongin-C ubiquitin-ligase, or RACK1 and elongin-C for ubiquitination and degradation by 26S proteasome.

11.6.1.1 HIF1 and ROS

Hypoxia-induced production of reactive oxygen species in mitochondria by complex-3 of the electron transport chain under hypoxia is necessary and sufficient for HIF1α accumulation [1548]. Mitochondria acting as oxygen sensors thus regulate HIF1 activity.

11.6.1.2 HIF1 and FoxO3a

Loss in phosphatase and tensin homolog deleted on chromosome-10 (PTen) and subsequent inactivation of Forkhead box factor FoxO3a[87] relieve HIF1 inhibition [1549]. Overexpression of P300 reverses FoxO3a-mediated repression of HIF1 activity. Factor FoxO3a interferes with transcriptional coactivator P300, thereby impeding HIF1 transcriptional activity.

11.6.1.3 HIF–Notch Coupling

Hypoxia-inducible factor-1α also interacts with the intracellular domain of Notch (Notch^{ICD}), in addition to the canonical hypoxia-primed HIF pathway. Released Notch^{ICD} interacts with HIF1α that is recruited to Notch-responsive promoters during hypoxia. Consequently, hypoxia promotes the undifferentiated cell state in various stem and precursor cell populations via Notch signaling [1550].

Factor HIF can also enhance expression of Notch target genes, such as HES1 and related HRT2 transcriptional regulators. Moreover, Notch^{ICD} potentiates the recruitment of HIF1α to HIF-responsive promoters [1551].

Factor-inhibiting HIF1 hydroxylase (FIH1) hydroxylates not only HIF, but also Notch intracellular domain. Enzyme FIH1 has a higher affinity for Notch^{ICD} than for HIF1αfactor. Intracellular domain of Notch causes FIH1 to deviate away from HIF1α in the absence of excessive FIH1 amount during hypoxia.

---

[86] A.k.a. polyamine N-acetyltransferase-2,.

[87] Factors FoxO1a (or FKHR), FoxO3a (or FKHRL1), and FoxO4 (or AFX) are effectors of the PKB pathway. Phosphorylation of FoxO factors by PKB promotes their export to the cytoplasm, thereby preventing their transcriptional activity.

11.6.1.4 HIF and RTK-Based Signaling

Heterodimer HIF lowers clathrin-mediated endocytosis at the early endosome sorting stage, thereby prolonging signaling from receptor protein Tyr kinase at the plasma membrane to promote cell survival under hypoxia. In addition, HIF1 and HIF2, during hypoxia, promotes the formation of caveolae, flask-shaped invaginations involved in endocytosis and signaling [1552]. They indeed foster the production of caveolin-1 that binds to EGFR receptors. Caveolin-1 provokes EGFR dimerization within caveolae and subsequent phosphorylation (activation) in the absence of ligand. Caveolin-1 also causes the activation of other RTKs, such as PDGFR and IGF1R receptors.

11.6.2 Hypoxia-Inducible Factor-Independent Axis

11.6.2.1 Target of Rapamycin

Under hypoxia, target of rapamycin activity is impeded to save energy. Stress-response, hypoxia- and DNA damage-inducible transcript-4 (DDIT4)[88] regulates TOR activity via the TSC1–TSC2 inhibitory complex. In addition, promyelocytic leukemia protein (PML) interacts with TOR and sequesters it into nuclear bodies [1362].

11.6.2.2 Hypoxia-Regulated MicroRNAs

Hypoxia during tumorigenesis upregulates a specific group of microRNAs, the hypoxia-regulated microRNAs (HRM).[89] They could affect cell apoptosis and proliferation, hence angiogenesis in cancers. Certain microRNAs can be controlled by transcription factors in response to endogenous and exogenous stimuli. Transcription factors cMyc and E2F activate the oncogenic miR17–miR92 cluster. Transcription factors P53 and NFκB also affect the expression of microRNAs. Hypoxia-inducible factor activation triggers production of several HRMs [1553].

---

[88] A.k.a. protein regulated in development and DNA-damage response-1 (REDD1).

[89] The HRM group includes miR21, -23a, -23b, -24, -26a, -26b, -27a, -30b, -93, -103, -103, -106a, -107, -125b, -181a, -181b, -181c, -192, -195, -210, and -213 [1553]. Conversely, a set of microRNAs are downregulated in hypoxic cells, such as miR15b, -16, -19a, -20a, -20b, -29b, -30b, -30e-5p, -101, -122a, -126, -128, -138, -141, -186, -197, -320, -323, and -326, in one or more cell lines. Members of let7 family either have increased or decreased levels during hypoxia; let7e, let7f, let7g, and let7i production changes vary according to the cell line.

11.6.2.3 Reactive Oxygen Species

Acute hypoxia and hyperoxia yield excessive amounts of reactive oxygen species. These molecules act not only as toxics, but also as signaling molecules (Sect. 10.6.1). Signaling primed by ROS must be protected, whereas ROS toxicity must be prevented. Intracellular ROS concentration must thus be adjusted. Main sources of ROS are plasmalemmal NADPH oxidases and mitochondria. Nuclear factor erythroid-derived-2-like transcription factor NRF2 is involved in oxidative and environmental stress tolerance and oxidant elimination.

Global long-lasting differentiation pathways regulated by P53, peroxisome proliferator-activated receptor-γ coactivator-1α (PGC1α), MyC transcription factor, and forkhead box-O family proteins yield either long-lasting oxidant-protective responses or damaged cell death [1554]. Reactive oxygen species modulate various redox-sensitive signaling pathways, such as cascades triggered by growth factors, according to the metabolic state of the cell. They target protein Tyr and MAPK phosphatases. Receptors of ROS include peroxidases.

Peroxiredoxin-1 is required in P38MAPK activation by H_2O_2 agent. Peroxiredoxin-2 controls H_2O_2 concentration produced during growth factor signaling. For example, peroxiredoxin-2 interacts with activated platelet-derived growth factor receptor. Peroxidase-3 regulates apoptosis by H_2O_2.

11.6.2.4 Chromatin Remodeling

Endothelial cells respond to hypoxia and reduce abundance of endothelial nitric oxide synthase to cause hypoxia-induced vasoconstriction. Vascular endothelium is characterized by synthesis of labeling proteins, such as NOS3, von Willebrand factor, vascular-endothelial (VE)-cadherin (Cdh5), intercellular adhesion molecule ICAM2, angiopoietin TIE1 and TIE2 receptors, vascular endothelial growth factor receptors VEGFR1 and VEGFR2; Notch-4, and EPHb4 receptor. Hypoxia reduces expression of endothelial nitric oxide synthase.

Short-term hypoxia triggers an acute, transient decrease in acetylation and methylation of histones at Nos3 promoter, hence rapid histone eviction from the Nos3 proximal promoter, especially endothelium-specific histone H2A variant, H2Az, which is not detected in vascular smooth muscle cells [1555]. Therefore, hypoxia represses NOS3 transcription.

Nucleosome accessibility depends on hypoxia duration. The shorter the hypoxia, the higher the nucleosome accessibility, the smaller the chromatin remodeling extent owing to chromatin (and nucleosome) remodeler and transcription activator SMARCa4 (switch–sucrose non-fermentable [Swi–SNF]-related, matrix-associated, actin-dependent regulator of chromatin).

After longer durations of hypoxia, histones are reincorporated at the Nos3 promoter, but they lack substantial histone acetylation [1555]. Chromatin remodeling is reversible upon return to normoxia, as histone acetylation, abundance of RNA polymerase-2 at NOS3 promoter, and NOS3 transcription are restored.

11.7 VEGF Signaling

Intracrine VEGF signaling[90] in endothelial cells is compulsory for homeostasis of blood vessels in adults, but dispensable for developmental and pathological angiogenesis [1556]. Moreover, paracrine VEGF signaling specific to endothelial cells does not compensate for the absence of endothelial autocrine VEGF signaling (Fig. 11.8).

Cytoplasmic Tyr kinases neuropilins Nrp1 and Nrp2 are coreceptors for semaphorin-3 and vascular endothelial growth factor isoforms (VEGFa–VEGFd, and PlGF). Due to the existence of different binding sites, semaphorins and VEGF can simultaneously bind Nrps [1558]. Heparan sulfate proteoglycans can intervene in interactions between Nrp and VEGF facors. Neuropilin-1 and -2 homo- or heteromultimerize even in the absence of ligand.

In most cell types, LDL uptake leads to lysosomal degradation of LDL components and release of cholesterol into the cytoplasm. In endothelial cells, lysosomal degradation of LDL cannot be excessive whatever the LDL plasmatic concentration and cellular uptake, because of LDL transcytosis and downregulation of LDL receptors by high intracellular cholesterol content. Vascular endothelial growth factor receptor VEGFR1 (but not VEGFR2) can be activated by low-density lipoprotein via its cognate receptor LDLR in the absence of members of the VEGF ligand family [1559]. LDL binding to LDLR induces VEGFR1 autophosphorylation and LDLR–VEGFR1 co-endocytosis.

Peroxisome proliferator-activated receptor-γ coactivator PGC1α stimulated by a lack of nutrients and oxygen activates the secretion of vascular endothelial growth factor, at least in skeletal muscle in vivo, to elicit vascularization [1560].[91] Release of VEGF primed by PGC1α does not require hypoxia-inducible factors HIF1α and HIF2α and HIF-targeted hypoxic response element. Transcriptional coactivator PGC1α indeed activates estrogen-related receptor-α to trigger VEGF gene activity, hence to stimulate angiogenesis during hypoxia.

11.8 Insulin Pathway

Insulin and insulin-like growth factor IGF1 (Vol. 2 – Chap. 3. Growth Factors) activate a tyrosine phosphorylation cascade from their receptors (Vol. 3 – Chap. 8.

---

[90]Phosphorylation of VEGFR2 can occur either after ligand binding (paracrine signaling) by VEGF produced by pericytes and other cell types associated with the endothelium or intracellularly by VEGF produced by endothelial cells (intracrine signaling). Autocrine VEGF does not contribute to angiogenesis, as vascular density and patterning are identical in the presence or in the absence of autocrine VEGF signaling.

[91]Factor PGC1α is mainly known by its function in cellular energy metabolism and oxidative metabolism, as it regulates oxidative phosphorylation, mitochondrial biogenesis, and cellular respiration. Factor PGC1α participates in the activation of several transcription factors, such as members of the myocyte enhancer factor-2, Forkhead box-O, and nuclear receptor families.

11.8 Insulin Pathway

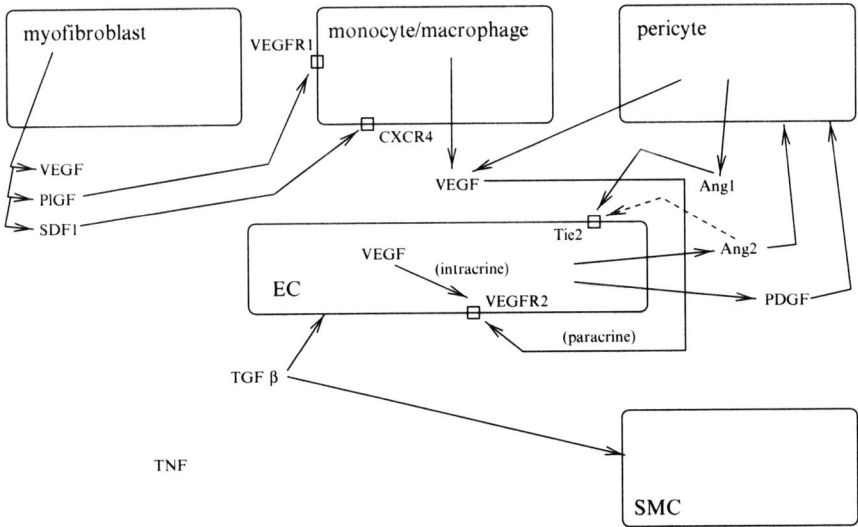

Fig. 11.8 Angiogenesis factors. Vascular growth depends on paracrine signaling of factors such as vascular endothelial and platelet-derived growth factor, transforming growth factor-β, and angiopoietins. VEGF at normal dose is mandatory for the differentiation of endothelial cells and formation and growth of the vasculature. Factor VEGF binds to 2 receptor Tyr kinases VEGFR1 and VEGFR2, as well as to cytoplasmic Tyr kinase neuropilin-1. Both autocrine and paracrine VEGFs synthesized by vascular cells (endothelial and non-endothelial cells) are vital for the maintenance of blood vessels, in particular differentiated endothelia such as fenestrated endothelia of pancreatic islets and renal glomeruli. Paracrine VEGF signaling from adjoining cell types other than endothelial cells is essential for survival, differentiation, and proliferation of endothelial cells and for angiogenesis. On the other hand, autocrine VEGF signaling in endothelial cells is vital for cell survival. By using specific complexes at appropriate sites and regulators, internalized receptors in endosomes may initiate signaling different from that triggered by plasmalemmal receptors. Endothelial cells secrete PDGFb that acts on pericytes, which in turn secrete Ang1 for stabilization of blood vessels. Autocrine production of Ang2 is involved in pericyte detachment. Agents VEGF, placental growth factor (PlGF), and stroma-derived factor-1 (SDF1) produced by perivascular myofibroblasts recruit mononuclear cells into the perivascular space by binding to VEGFR1 and CXCR4, respectively. These cells and related macrophages produce additional angiogenic factors such as VEGFc and VEGFd, which can participate in pathological angiogenesis (Sources: [1556, 1557]).

Receptor Kinases). Insulin[92] is a hormone produced in the pancreas as proinsulin precursor processed by prohormone convertases PC1 and PC2 as well as the exopeptidase carboxypeptidase-E. Insulin is involved in the regulation of carbohydrate and lipid metabolism. In the human brain, insulin supports learning and memory.

---

[92]Latin insula, as insulin is synthesized in β cells of Langerhans islets (islands) in the pancreas in response to increased blood glucose concentration (phase-1 Ca^{++}-mediated secretion) and independently of carbohydrates (phase-2 sustained and slow release).

Phosphorylated effectors include docking proteins such as insulin receptor substrates (IRS1–IRS4). Insulin receptor substrates that are Ser/Thr phosphorylated preclude insulin signaling.

Upon insulin binding to its receptor, IRS1 is phosphorylated and then links to phosphoinositol 3-kinase. The latter transforms PIP_2 into PIP_3 that activates protein kinase-B. This kinase phosphorylates (inactivates) glycogen synthase kinase, thereby impeding glycogen synthase phosphorylation by GSK and increasing glycogen synthesis.

11.8.1 Glucose and Lipid Metabolism

In mammals, insulin provokes glucose uptake from blood particularly in adipocytes, hepatocytes, and skeletal myocytes; in hepatocytes and skeletal myocytes, it can be stored as glycogen; in adipocytes, it promotes lipid (i.e., triglycerides) synthesis. Insulin suppresses hepatic glucose export.

Insulin regulates the transcriptional activity of numerous genes involved in glucose and lipid metabolism in the liver. It represses genes encoding gluconeogenic enzymes.

In the absence of insulin binding, FoxO is a transcriptional activator of multiple insulin-responsive genes (Sect. 10.9.3.3). In response to insulin, phosphatidylinositol 3-kinase activates protein kinase-B. Subsequent phosphorylation by PKB of FoxO proteins (Thr24, Ser256, and Ser319 in FoxO1) reduces, then impedes DNA-binding affinity, and provokes nuclear export.

The FoxO proteins increase insulin sensitivity (in normal conditions, but not diabetes) by upregulating insulin receptor and IRS2 insulin receptor substrate. They also regulate fatty acid oxidation. Fatty acid breakdown and oxidative phosphorylation allow ATP generation. The FoxO factors upregulate fatty acyl-CoA carriers and sterol carrier protein SCP2 for suitable fatty acid processing. However, FoxOs oppose other insulin functions. In addition, the insulin–PI3K signaling stimulates cell cycle progression, preventing FoxO-induced cell cycle arrest.

11.8.2 Other Effects and Partners

Downstream effectors comprise MAPK with transiently (peak at ~ 5 mn) activated ERK1 and ERK2 in adipocytes of brown fat tissue of newborn mice, and PDZ domain-containing protein-11 (PDZD11), an interactor of plasma membrane Ca^{++} ATPase [1561]. Agent PDZD11 also interacts with sarcoplasmatic Ca^{++} ATPase SERCA2 and S100 Ca^{++}-binding proteins.

Using Ca^{++}, insulin acts on the cytoskeleton via phosphorylation of Ca^{++}-calmodulin-dependent myosin light-chain kinase, which subsequently regulates the translocation of GluT4 transporter.

11.8 Insulin Pathway

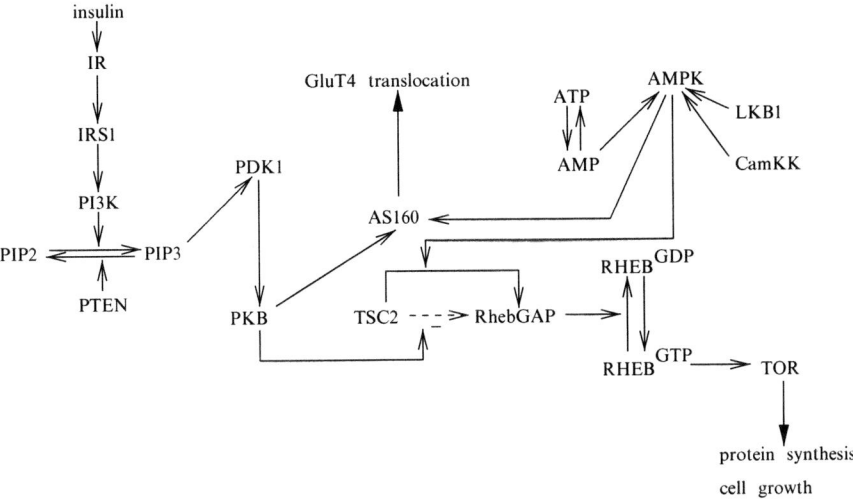

Fig. 11.9 Effects of insulin and AMP-activated protein kinase (AMPK) on the TOR pathway (Source: [1562]). Insulin binds to insulin receptor that phosphorylates insulin receptor substrate-1 (IRS1). The latter then activates phosphatidylinositol 3-kinase (PI3K). This enzyme catalyzes the conversion of phosphatidylinositol (4,5)-bisphosphate (PIP_2) to phosphatidylinositol (3,4,5)-trisphosphate (PIP_3). This reaction is reversed by phosphatase and tensin homolog deleted on chromosome 10 (PTen). Agent PIP_3 primes phosphorylation of protein kinase-B (PKB) by phosphoinositide-dependent kinase-1 (PDK1). Activated PKB phosphorylates tuberous sclerosis-2 (TSC2) that favors small GTPase Ras homolog enriched in brain (RHEB) activity, as TSC2 inhibits RHEB-GTPase activating protein (RhebGAP). Active $RHEB^{GTP}$ activates target of rapamycin (TOR). The latter promotes protein synthesis and cell growth. Conversely, AMPK activated by metabolic stresses is phosphorylated by protein kinase LKB1. Kinase AMPK phosphorylates TSC2 at different sites from those targeted by PKB, thus stimulates RhebGAP that inhibits TOR activation. In addition, in resting muscle, glucose transporter GluT4 mainly localizes to intracellular storage vesicles. Kinase PKB activated by the insulin pathway and activation of the AMPK pathway causes phosphorylation of Akt(PKB) substrate of 160 kDa (AS160). The latter is then dissociated from GluT4 storage vesicle, and cannot convert Rab^{GTP} into Rab^{GDP}. Active Rab elicits docking and/or fusion of GluT4 storage vesicle with the plasma membrane.

Insulin provokes membrane raft formation using caveolin-1, -2, and -60, as well as cavin-2[93] that interacts with PKCδ-binding protein to recruit PKCδ in caveolae. Low-density lipoprotein receptor-related proteins LRP1 and LRP6 are also involved in formation of membrane rafts. Agent LRP6 may yield a potential link between the insulin and canonical Wnt pathway.

Insulin and AMPK kinase have opposite effects on target of rapamycin, i.e., on protein synthesis and cell growth (Fig. 11.9). However, both insulin and AMPK lead to translocation of glucose transporter GluT4 at the plasma membrane.

---

[93] A.k.a. phosphatidylserine-binding protein (PS-P68) and serum deprivation response factor (SDR).

In addition to classical transcriptional regulators such as endoplasmic reticulum-bound sterol regulatory element-binding proteins (SREBP)[94] MicroRNAs miR33a and miR33b that are generated from intronic sequences of the Srebp genes regulate cholesterol homeostasis, fatty acid metabolism, as well as insulin signaling. [1563]. They target: (1) carriers involved in HDL genesis and cellular cholesterol efflux, such as membrane adenosine triphosphate-binding cassette transporters ABCa1 and ABCg1 and intracellular transport protein (in late endosomes and lysosomes) Niemann-Pick protein-C1 (NCP1); (2) enzymes involved in the regulation of fatty acid oxidation, such as carnitine ^{O}octaniltransferase (CrOT), carnitine palmitoyltransferase-1A (CPT1a), hydroxyacylCoA dehydrogenase β subunit (HADHB), sirtuin-6, and AMPK α subunit; (3) components of the insulin pathway such as insulin receptor substrate-2 (IRS2).

The survival function of heat shock factor HSF1 can be controlled by the insulin and insulin-like pathways.[95]

11.9 Inflammasome

Sensing and elimination of invading pathogens is carried out by the immune system specialized in the body's defense against invading microbes. The immune system is composed of: (1) the innate immune system that recognizes and destroys microorganisms and relies on specific clones of lymphocytes to perform the initial immune response and (2) the adaptive immune system that is rapidly activated upon pathogen detection.

Organism components are discriminated from encountered germs via *pathogen-associated molecular patterns* (PAMP), such as lipopolysaccharides (LPS), peptidoglycan, flagellin, and microbial nucleic acids, that are recognized by innate immune receptors, such as Toll-like (TLR) and NOD-like (NLR; or nucleotide [ATP]-binding [for ATPase-induced] oligomerization domain [NOD], ligand-sensing leucine-rich repeat [LRR] protein) receptors [1565].

Whereas TLRs sense PAMPs on cell surface or in endosomes, NLRs detect microbial molecules in the host cytosol. The NLR family is composed of 23 members in humans. Most NLRs have a variable N-terminus with binding sites to attract

---

[94]The SREBP family of basic helix–loop–helix leucine zipper (bHLHLZ) transcription factors consists of SREBP1a, SREBP1c, and SREBP2 that are encoded by 2 genes (Srebp1–Srebp2). They have distinct expression pattern, gene selectivity, and potencies of transactivation. Factor SREBP1c regulates the transcription of genes involved in fatty acid metabolism, such as fatty acid synthase. Protein SREBP2 controls the transcription of cholesterol-related genes, such as 3-hydroxy 3-methylglutarylCoA reductase (HMGCR) and low-density lipoprotein receptor.

[95]Heat shock factors (HSF1–HSF4, and HSFX and HSFY associated with chromosomes X and Y) are transcriptional regulators of genes that encode chaperones and other stress-related proteins involved in cell survival upon exposures to acute stress, such as heat, oxidative stress, toxics, and infections, as well as cancer [1564]. These stress integrators also contribute to development and survival in normal conditions.

adaptors and effectors, especially either a caspase activation and recruitment domain (CARD) or a pyrin domain (PyD), a central nucleotide-binding oligomerization domain (NOD), and a C-terminal region that detects PAMP substances. The pyrin domain recruits procaspase-1 via the adaptor Apoptosis-associated speck-like protein containing a CARD (ASC).

NOD-like receptors NOD1 and NOD2 sense bacterial molecules produced during the synthesis, processing, and degradation of peptidoglycan of bacterium walls and activate nuclear factor-κB and mitogen-activated protein kinases. On the other hand, several NLRs, such as NLRC4, NLRP1, and NLRP3,[96] and neuronal apoptosis inhibitor proteins are involved in the assembly of the sensory *inflammasome*. Different types of inflammasomes exist according to their constituents, activators, and effectors.

Inflammasome is a large proteic complex that assembles in response to infection, alerts the immune system, recruits the cysteine peptidase zymogen procaspase-1, activates caspase-1 by causing its autoproteolysis, thereby causing cleavage of cytokine substrates of caspase-1, and triggers secretion of interleukins IL1β and IL18 as well as pyroptosis (Vol. 2 – Chap. 4. Cell Survival and Death).

Pro-inflammatory caspases, such as caspase-1, -4, and -5, are involved in processing and secretion of pro-inflammatory cytokines. Activated caspase-1 from inactive zymogen cleaves pro-interleukin-1β and -18 into mature forms. Mature IL1β operates in many immune reactions, in particular the recruitment of inflammatory cells, whereas IL18 participates in the production of interferon-γ and enhancement of the cytolytic activity of natural killer cells.

In most cases, an inflammasome contains a NLR protein. The latter has a caspase activation and recruitment domain (CARD) that serves to activate the transcription factors nuclear factor-κB and interferon regulatory factor IRF3, as well as pro-apoptotic and pro-inflammatory caspases. Three inflammasomes have been determined according to NLR components (NLRC4, NLRP1, and NLRP3 inflammasomes). Receptor NLRP2 associates with procaspase-1 and promotes IL1β production.

Inflammasome composed of members of the NLR family and ASC adaptor are able to link NLRs to caspase-1. In fact, some NLR proteins recruit caspase-1 directly, but others such as NLRP3 require an adaptor to interact with caspase-1. Recruitment of caspase-1 leads to its auto-activation by autoproteolysis. This peptidase can then cleave interleukin-1β and -18 into mature, active cytokines.

Guanylate-binding proteins (GBP) are cytoplasmic, interferon-γ-inducible GTPases. In monocytes and macrophages stimulated by bacterial lipopolysaccharide, bacterial nigericin, or intracellular pattern-recognition molecules that recognize bacterial peptidoglycan and bind to NOD-like receptors, tetramers of guanylate-binding protein GBP5 promotes the assembly of the NLRP3-containing inflammasome in response to adequate activation signals, i.e., some, but not all NLRP3agonists, such

---

[96] Proteins NLRC4 and NLRP3 are also called IPAF and cryopyrin or NALP3, respectively.

as adenosine triphosphate [1566]. Indeed, GBP5 fosters the NLRP3–ASC assembly in response to bacteria and their wall components, but neither crystalline nor double-stranded DNA.

Dysregulated activation of caspase-1 and secretion of IL1β leads to several diseases. Mutations in genes encoding components of the inflammasome or inflammasome-associated pathways can generate auto-inflammatory disorders, periodic fever syndromes, vitiligo, and Crohn's disease.

11.10 Mechanotransduction

Membrane tension influences all processes that involve membrane deformations. Membrane tension developed from the cell or the extracellular matrix operates as a mechanical regulator in endo- and exocytosis as well as during cell migration and mitosis. All cells are exposed to mechanical and physical signals. Mechanical stresses can lead to changes in cell membrane tension, thereby affecting the function of membrane lipids and embedded proteins.

Mechanical forces are directly sensed by cells and transduced into chemical cues via mediators, including some types of enzymes (e.g., lysyl oxidase), that crosslink cells and matrix. Mechanical forces cause changes in molecule conformation and structure organization. Target substances encompass mechanosensitive ion channels gated by membrane tension, integrins that transmit forces from the environment to the cytoskeleton and conversely at focal adhesions, cadherin complexes at adherens junctions (Vol. 1 – Chap. 7. Plasma Membrane), as well as G-protein-coupled receptors and receptor protein Tyr kinases (e.g., EGFR and EPH). The organization of receptors depends on mechanoresponsive actin structuring at the cell cortex.

Once they are activated, they trigger signaling. In response to elevated stretch in focal contacts, integrin clustering increases and focal adhesion kinase can enter into action to initiate a cascade of events, such as activation of GTPases of the RHO superfamily that provoke actin remodeling and protein phosphorylations (e.g., via MAPK modules) down to activation of transcription factors to regulate gene expression.

Mechanical forces participate in embryo- and fetogenesis (Vol. 1 – Chap. 1. Cells and Tissues). During the body's development, the physical state and rheology of the extracellular matrix can govern cell migration and differentiation and hence compartmentation. Mechanotransduction occurs on a relatively fast timescale ($O[1\,\text{ms}]$–$O[1\,\text{h}]$).

During migration, cells sense the matrix composition as well as environment stiffness. They anchor and pull on the matrix. Focal adhesions are major mechano-sensing and -transduction sites. Yet, a given surface area of focal adhesions create different forces according to the matrix rheology [1567]. In addition, Ca^{++} channels can participate in the generation of cell tension in response to mechanical signals, as Ca^{++} influx elicits the activity of the contractile actomyosin filament network. Consequently, sensing of and reaction to the matrix rigidity relies not

11.10 Mechanotransduction

only on focal adhesions and their adaptation to the matrix features, but also on the cytoskeleton [1567]. The greater the matrix stiffness, the stronger the traction exerted by cells that reorganize their actin stress fibers. Furthermore, durotaxis is optimal within a range of values of the matrix stiffness.

Tumor cells exploit interstitial flow by generating an autocrine chemokine circuit (using chemokines CCL19 and CCL21 as well as chemokine receptor CCR7) to promote their migration toward draining lymphatic vessels and subsequent metastasis in lymph nodes [1568].

Bone formation and remodeling rely on mechanical loading on the mechanosensory couple constituted by the osteocyte and antagonist osteoclast in the mineralized bone matrix. Osteocytes interact with the extracellular matrix via the mechanosensor and transducer integrins and bridge elements between osteocytes and canalicular wall. Furthermore, mechanical forces cause fluid flow through the canalicular network. Hydrodynamic foces are transmitted to cells. In addition, signaling is transmitted between cells through gap junctions at tips of connecting dendritic processes and connexin-43-based hemichannels at osteocyte bodies. The latter release auto- and paracrine regulators ATP and prostaglandins. Fibronectin receptor synthesized in bone and cartilage, integrin-$\alpha_5\beta_1$, interacts with connexin-43, thereby opening the hemichannel under mechanical stress [1569].[97]

In epithelia, such as renal and respiratory epithelia, as well as vascular endothelia, long-range transport of molecules, including liquid and gas, is facilitated by the development of epithelial tubules (length range $O[100\,\mu m]-O[1\,cm]$; caliber 10–20 μm). These tubules relies on cell interactions with collagen-1. Initiation, once cells are aligned, and maintenance of tubules mainly depends on traction forces exerted by cells rather than on secreted substances such as diffusive morphogens [1570]. Cells use traction force to condense and align collagen fibers that influence cell motion and positioning. Inside cells, PI3K, Rac1, and RoCK enable initiation and maintenance of long, linear tubules.

Mechanical forces can provoke expression of endothelial genes (in particular those encoding von Willebrand factor, VE-cadherin, and PECAM1) and endothelial cell differentiation from embryonic stem cells [1571]. Production of tight junction protein zonula occludens ZO1 and vasodilatory enzymes NOS3 and COx2 also rises, whereas that of vasoconstrictive agent endothelin-1 decays. The glycocalyx component heparan sulfate proteoglycan mediates mechanotransduction in mature endothelial cells as well as embryonic stem cell-derived endothelial cells.

Mechanotransduction especially occurs at the endothelium wetted plasma membrane, as well as within the blood vessel and cardiac wall (Fig. 11.10; Vol. 5 – Chaps. 8. Smooth Muscle Cells and 9. Endothelium). Wall displacements determine the local, time-dependent size of the blood vessel lumen.

Immediately downstream from the ventriculoarterial valves, a small blood bolus (60–100 ml) is expelled in arteries during only a part of the cardiac cycle. Blood

---

[97] Signaling mediated by mechanosensitive ephrins controls the spatial organization of $\alpha_5\beta_1$-integrin clusters at the cell surface and fibronectin matrix in the extracellular matrix [1568].

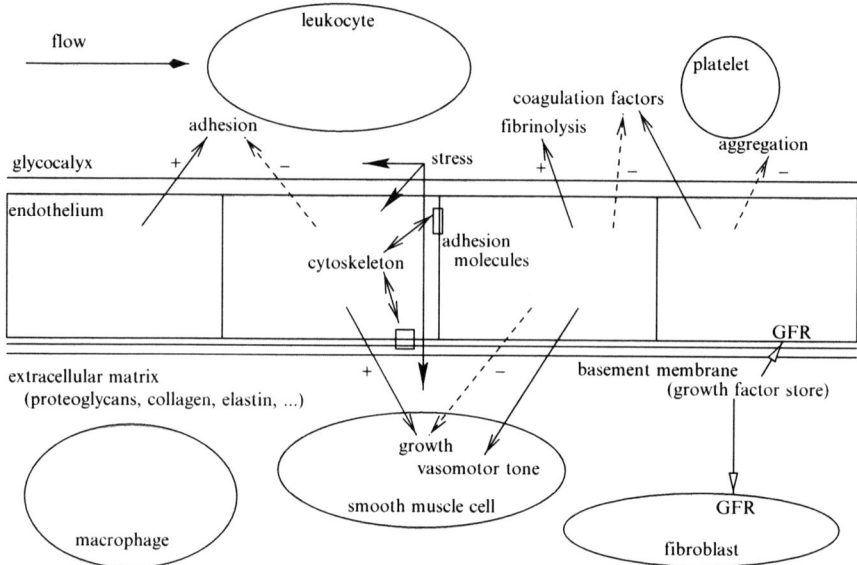

Fig. 11.10 Mechanotransduction and interactions between vascular cells. Time-dependent three-dimensional shearing (torque), stretching (tension), and bending result from loading applied by the flowing blood on the vasculature walls. Shear stress at the wall has a low magnitude but strongly varies both in magnitude and direction during the cardiac cycle, as blood is conveyed in a three-dimensional, unsteady flow. Pressure applied on the vessel wall by the blood has high magnitude and undergoes large amplitude variations during the cardiac cycle. The mechanical loading sensed by the endothelium is transduced into chemical signals that allow the endothelial cell to adapt. Furthermore, cues are sent to adjoining smooth muscle cells to regulate the vasomotor tone. Interactions between flowing molecules (clotting factors) and cells (leukocytes and thrombocytes) and the endothelium to regulate blood coagulation and cell extravasation on the one hand, and between the extracellular matrix and mural cells required for cell growth and tissue maintainance on the other hand are also summarized. Endothelial cells detect hemodynamic stresses via mechanosensors (adhesion molecules, mainly integrins, ion channels, and plasmalemmal receptors, such as GPCRs and RTKs). Signaling pathways augment the activity of transcription factors. Time-dependent hemodynamic stresses on the endothelium as well as within the wall are implicated in: (1) the vascular tone (vessel bore) regulation; (2) wall adaptation (short-term) and remodeling (long-term); (3) tissue evolution (angiogenesis); and (4) vasculature diseases and tumors.

vessel wall deformability allows the transient storage of blood in the elastic arteries (close to the heart) during the ventricular ejection. Consequently, the blood permanently flows in the arterial tree downstream from the heart during the cardiac cycle. The highly distensible venous compartment plays the role of blood reservoir, adapting the operative blood volume to the body's needs.

Artery collapsibility is used to measure the arterial blood pressure in the systemic circulation (Vol. 6 – Chap. 5. Images, Signals, and Measurements). The artery distensibility allows the propagation in finite times of the pressure wave, which runs from the heart toward the microvessels. Vein collapsibility is targeted by treatment of dysfunctional superficial veins of the lower limbs.

11.10 Mechanotransduction

Numerical simulations deal with fluid–structure interaction because of the strong coupling between the blood dynamics and the compliant wall mechanics. Although the blood flow simulations in any explored segment of the vasculature are carried out in a deformable fluid domain, the numerical results remain questionable because: (1) the vasculature wall is represented as a shell, and not as a multilayered structure made of different types of composite materials, within which a region of stress concentration can appear; (2) the material constants remain most often unknown in vivo; and (3) the vessel wall is assumed to be a more or less passive material.

The size of the computational domain depends on the controlled motions of the blood vessel wall. In addition to the flow governing equations (Vols. 1 – Chap. 1. Cells and Tissues and 7 – Chap. 1. Hemodynamics) coupled to the equations of the wall mechanics, the set of equations to be solved can incorporate the equations of a phenomenological model that describes the vascular cell response, i.e., biochemical reaction cascades triggered by stresses imposed by the flowing blood on the flexible wall, between-cell interaction, and relaxation or contraction state of smooth muscle cells.

The blood vessel wall is actually a living tissue that is able to quickly react to the load applied on it by the flowing blood. In a given region of a blood vessel segment, endothelial and smooth muscle cells sense the small-magnitude wall shear stress with its large-amplitude space and time variations (three-dimensional, unsteady flow) and large-magnitude wall stretch generated by blood pressure.

These cells respond with a short-time scale to adapt the vessel caliber according to the loading, especially when changes exceed the limits of the usual stress range. The mechanotransduction pathways determine the local vasomotor tone, i.e., the lumen bore of the reacting blood vessel, and hence affect the wall deformation, taking into account the short-term adaptation to stresses applied at its wetted surface.

A vascular mechanotransduction model relies on the decomposition of the process into successive phases from observed hemodynamic stress field at a given time (input). These phases are defined in the case of sustained hypotension. Phase 1 is related to the effect of the resulting mural stress field (C) on vascular smooth muscle cells, i.e., the density of actin–myosin bridges ($[AM]_{SMC}$). Phase 2 corresponds to the effect of the wall shear stress on endothelial cells, i.e., the amount of released endothelin ($[ET]_{EC}$) from its available pool. The time scale of these 2 steps is $O[1\ mn]$, when vasoactive regulators are supposed to be stored in adequate quentities, i.e., in the absence of synthesis. Phase 3 describes the effect of released endothelin on adjacent smooth muscle cells (paracrine regulation) with a delay (τ_1). Phase 4 refers to the feedback exerted by released endothelin on endothelial cells (autocrine regulation) with a delay (τ_2), i.e., on nitric oxide production by its specific synthase. Phase 5 represents the inhibition exerted by nitric oxide on endothelin.

The local caliber of the vessel lumen is assumed to be proportional to the density of actomyosin bridges. In the simplest model, intermediary stages of signaling cascades between mechanically stressed plasma membrane and formation of actomyosin bridges in vascular smooth muscle cells or the release of vasoactive regulators by vascular endothelial cells canbe neglected as well as those between the binding of vasoactive regulators on their cognate receptors on the plasma membrane or in the cortex of vascular smooth muscle cells and the formation of actomyosin bridges.

In summary, in the case of sustained hypotension:

$$R_h(t)|_x = f_R\left(<\mathbf{C}>, c_{AM_{SMC}}(t)\right),$$
$$c_{AM_{SMC}} = f_{AM}\left(<\mathbf{C}>, c_{ET_{EC}}\right),$$
$$c_{ET_{EC}} = f_{ET}\left(<\mathbf{C}_w>\right),$$
$$c_{NO_{EC}} = f_{ET}\left(c_{ET_{EC(ECM)}}\right), \quad (11.1)$$

11.10.1 Basic Components of Vascular Mechanotransduction

Mechanotransduction at the endothelial interface and in the blood vessel wall involves a microscopic couple made of an endothelial cell and a smooth muscle cell (Fig. 11.11) and relies on a nanoscopic couple made by a locally synthesized vasodilator and its antagonist vasoconstrictor among the entire set of vasoactive substances (Table 11.30; Fig. 11.12).

The dynamics of vascular lumen deformation induced by stress-subjected vascular cells directly or indirectly via vasomotor regulator release can be characterized by: (1) deformation magnitude that regulates the actin–myosin bridge state in the smooth muscle cell; (2) time response of the vasomotor tone control to an increase or decrease in shear stress and/or pressure above or below the ranges of normally experienced space and time variations; (3) sensitivity to rate of change in stress, absolute stress magnitude, or a combination of both.

11.10.2 Mechanotransduction Signaling

Various cell types react to loading with a cell production that depends on the stress direction. Several investigative teams have developed bioreactors (Vol. 5 – Chap. 11. Tissue Growth, Repair, and Remodeling) that comprise microgrooved silicone cell-growing surfaces, which are subjected to stretching. Cultures of confluent, elongated, aligned cells are then obtained. The expression of atrial natriuretic peptide, connexin-43, and N-cadherin in cardiomyocytes increases when the cell is subjected to strain perpendicular to the main cell axis imposed by microgrooves [1572]. However, the expression of these proteins does not significantly vary with respect

11.10 Mechanotransduction

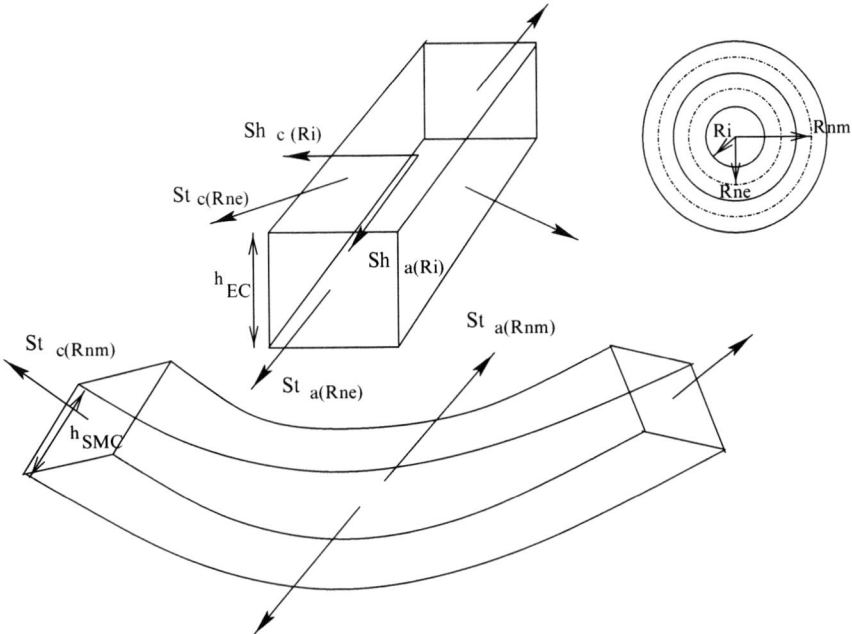

Fig. 11.11 Microscopic couple in vascular mechanotransduction composed of vascular endothelial and smooth muscle cells that are oriented perpendicularly to each other, as the main axes of the endothelial and smooth muscle cell point in the local streamwise and azimuthal direction, respectively. Endothelial and smooth muscle cells can be assumed to be mainly sensitive to wall shear stress and blood pressure-resulting intramural stretch averaged on a local neutral line of their corresponding cell layer within the vessel wall, respectively.

Table 11.30. Vasorelaxants and vasostimulants involved in vascular mechanotransduction that regulate the state of the actin–myosin cytoskeleton of smooth muscle cells. A representative nanoscopic couple is constituted by vasodilator nitric oxide (NO) and vasoconstrictor endothelin-1 (ET1).

Vasoconstrictor	Vasodilator
ET1	NO
5HT	H_2S
ATP	ANP
NPY	Acetylcholine
TXA2	PGI2
$PGF2\alpha$	Adrenomedullin
Apelin	Apelin
Up4a	EDHF

to unstretched cells when the loading direction is parallel to the main cell axis. When the stretching direction is similar to the microgroove axes, tendon fibroblasts growing on microgrooves take on an elongated shape according to the microgroove

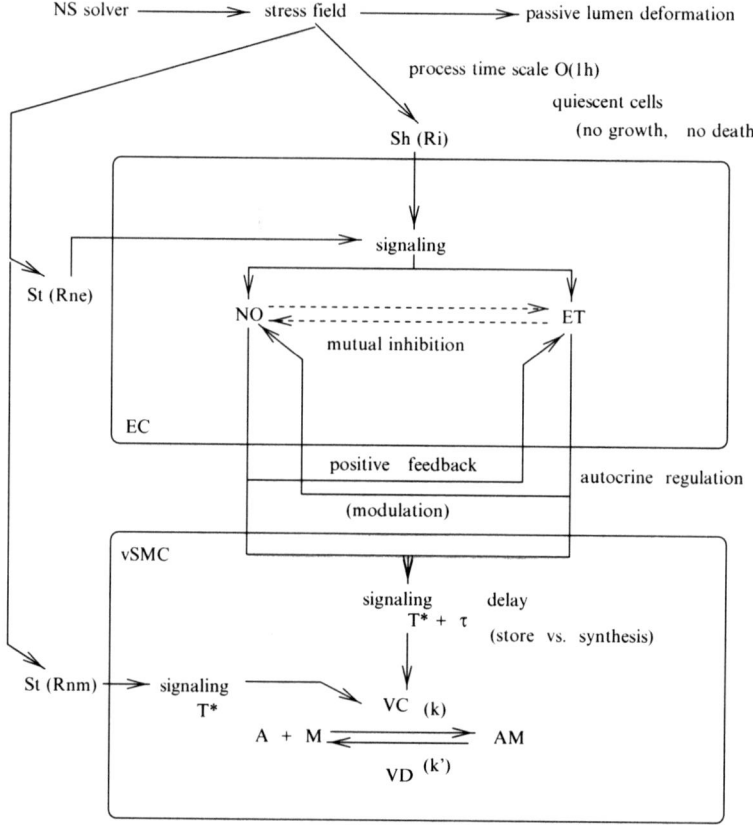

Fig. 11.12 Simple compartment-like model of vascular mechanotransduction based on mutual interactions between nitric oxide (NO) and endothelin (ET). Both molecules are subjected to mutual inhibition within the endothelial cell (EC) and positive feedback via autocrine loop in the gap between it and the smooth muscle cell (vSMC) that modulates the response to shear applied to the endothelial wetted surface (Sh) and circumferential and axial stretches at laterobasal edges of endothelial cell and the entire SMC surface. The endothelial-based response of the smooth muscle cell is delayed with respect to the SMC cytoskeleton regulation by stretch directly applied on SMC surface. This delay depends on whether vasomotor tone regulators are stored or synthesized by endothelial cells. Vasoconstriction (VC) and dilation (VD) have corresponding kinetic coefficients (k and k'). Shear and pressure-induced stretch could target different amount of given mechanosensitive plasmalemmal proteins (e.g., receptors, such as G-protein-coupled receptors and receptor Tyr kinases, and ion channels) as well as distinct receptors (e.g., some types of adhesion molecules).

direction. When the stretching direction is oriented at 45 and 90 degrees with respect to microgrooves, tendon fibroblasts do not change shape. Whatever the stretching direction, α-actin expression by tendon fibroblasts is upregulated in response to 8% cyclic uniaxial stretching [1573]. However, the closer the stretching direction to the microgroove axes, the higher the α-actin production. The longer

Table 11.31. Elements of endothelial cell responses to blood flow.

Receptors	G proteins (Gs), RTK
Ion channels	Ca^{++}, K^+, Na^+, and Cl^- channels
Adhesion molecules	VE-Cadherin, PECAM1, integrin
Transport substances	Caveolin
Second messengers	IP_3, cGMP
Vasoactive compounds	PGI2, NO, ET1, adrenomedullin
Growth factors	PDGF
MAPK production	ERK1/2
Gene expression	Fos, Jun

the stretching duration, the greater the α-actin synthesis. Mesenchymal stem cells differentiate and adapt their protein synthesis according to the mechanical loading in bioreactors [1574].

Mechanical stretching of cultured cardiomyocytes, which leads to cardiomyocyte hypertrophy, activate angiotensin-2 G-protein-coupled receptor AT1R (or AT_1) [1575],[98] The release of angiotensin-2 by stimulated cardiomyocytes can mediate a ligand-dependent stretch-induced response (autocrine control) [1576]. Angiotensin-2 acts as an initial mediator of the stretch-induced hypertrophic response.

Mechanotransduction at the endothelial plasma membrane drives many cell responses (Tables 11.31 and 11.32). Heterotrimeric G protein subunits of the $G\alpha_{q/11}$ and $G\alpha_{i/o}$ subclasses are activated within 1 s of flow onset, yielding the earliest mechanochemical signal transduction events in suddenly stressed endothelium (not the actual endothelium state that has already experienced a long-time stress history at observation time) [1577]. Certain amino acids of transmembrane helices interact to lock the receptor in an inactive state. The receptor activation results from motion of its transmembrane helices embedded in the lipid bilayer solvent.

Calcium ion enables contraction of actomyosin filaments. Upon flow stimulation (an experimental, but not physiological condition, as mural cells are subjected to flow since in utero life), Ca^{++} from the extracellular medium enter the cell through different calcium channels, such as voltage-gated and, predominantly, stretch-activated channels, and afterward store-operated channels. These channels include, in particular, ligand-gated ion channel with high calcium permeability $P2X_4$ as well as vanilloid-related TRPV4 and melastatin-related TRPM7 transient receptor potential channels [1578]. Further Ca^{++} influx results from release through endoplasmic reticulum IP_3 and ryanodine receptors. Many agents influence Ca^{++} influx, such as protein kinase-C and cGMP-dependent protein kinase-G [1578]. When bovine aortic endothelial cells are subjected to flow, the intracellular Ca^{++} concentration rises immediately due to Ca^{++} flux through plasmalemmal channels [1578]. It is then maintained at high levels. On the other hand, calcium

---

[98]Mechanical stretch causes association of the AT_1 receptor with Janus kinase-2 and translocation of G proteins into the cytosol.

Table 11.32. Endothelial cell signaling pathways triggered by blood stresses (RTK: receptor Tyr kinase; ET: endothelin; NO: nitric oxide; MAPK: mitogen-activated protein kinase; ERK: extracellular signal-regulated protein kinase; PI3K: phosphatidylinositol 3-kinase; FAK: focal adhesion kinase; AKAP: A-kinase anchoring protein; NOx: NAD(P)H oxidase; HSPG: heparan sulfate proteoglycan; LKLF: lung Krüppel-like factor; TxnIP: thioredoxin-interacting protein; ROS: reactive oxygen species).

G proteins	Ras, PKC–ET, ERK
RTK	PI3K, MAPK
VEGFR2	Cbl–PLC–PI3K–ERK, PECAM1, VE-cadherin
Ca^{++} channels	NO
MSK^+ channels	NOx, TGFβ
VGK^+ channels	PKA/C–AKAP–SAP–Pyk
Na^+ channels	ERK1/2
VE-cadherin	PI3K–PKB (with VEGFR2, β-catenin, and PTPn11)
PECAM1	ERK; NKκB; PI3K (VEGFR2)
Integrin	Rho–CSK, VEGFR2–SHC, IKK–NKκB, CDC42, Rac
Occludin	Cell junction
FAK	GRB2–Src–Rho–MAPK
Caveolin	NOS, COx, PGI2; ERK1
HSPG	NO
NOx	ERK
Elastin-laminin receptor	Fos
LKLF	NO, ET, adrenomedullin
TNFα	ASK1
ROS	ERK1/2
TxnIP (ASK1)	MAPK, VCAM1
NO	cGMP, Ca^{++} (MLCK), ERK1/2
ET	collagen, Ca^{++}

concentration in the endoplasmic reticulum lowers only after 300 s. The second phase of flow-induced Ca^{++} influx through IP_3 receptors is controlled by Src kinase and phospholipase-C [1578]. Coupled microtubules and actin filaments support mechanical transmission not only to adapt cell shape, but also to open stretch-activated channels and couple these channels to IP_3 receptors of the endoplasmic reticulum. In addition, integrin engagement can cause the association of $\alpha_V\beta_3$-integrin with receptor protein Tyr phosphatase PTPRa, an activator of SRC family kinases.

Mechanical stress across the plasma membrane of endothelial cells can directly activate G-protein-coupled receptors. Hemodynamic shear stress applied to the wetted (apical) plasma membrane of endothelial cells changes the conformation of bradykinin B−2 receptor, independently of ligand binding [1579]. The response time to stimulation by shear stress is about 80 s. Hypotonic stress and membrane fluidizing agents also lead to a significant increase in receptor activity.[99]

In addition to G-protein-coupled receptors, mechanosensitive ion channels, regulator lipids that are involved in opening and closing of mechanosensitive ion channels [1580], and, more generally, any change in protein structure that results from mechanical tension are factors of mechanotransduction.

Traction forces can be associated with phosphorylation of BCAR1 adaptor that can serve as a primary force sensor, transducing mechanical extension and priming phosphorylation of downstream signaling effectors [1581]. Adaptor BCAR1 in a cytoskeletal complex is involved in force-dependent activation of small Rap1 GTPase.

Mechanotransduction time scale can be short. Stress applied via activated integrins causes quick Src activition (<300 ms) at remote cytoplasmic sites, which depends on the cytoskeleton prestress [1582].[100] Activation of Src in the cytoplasm depends on the stress magnitude. Sites of stress-activated Src corresponds to loci of large microtubule deformation. A threshold (applied stress 1.8 Pa) of microtubule deformation exists.[101] Microtubules transmit stresses to activate cytoplasmic proteins. Rapid and long-range (15–60 μm) activation of Src by stress cannot indeed be explained by diffusion- or translocation-based mechanisms. On the other hand, stresses can propagate to remote cytoplasmic sites (>30 μm) in less than 5 ms. Similar kinetics are observed in flow shear-induced Ras activation in endothelial cells. Therefore, mechanotransduction not only operates via stress-induced local conformational changes of plasmalemmal proteins that trigger signaling cascades, but also to cytosolic effector activation related to focal adhesions (integrin stimulation) by microfilaments (actin and myosin-2) and microtubules.

Information on mechanical loads includes stress or strain magnitude, loading rate, and loading time integral. Mechanotransduction that is associated with Jun N-terminal kinase JNK2 relies on load magnitude and rate, but is independent of loading time integral [1583]. On the other hand, mechanotransduction via P70 ribosomal S6 kinase (S6K) is not sensitive to loading rate, but depends on load magnitude and loading time integral.

---

[99] Bradykinin B−2 G-protein-coupled receptor can be stimulated by ligand binding, shear stresses generated by fluid flow, plasmalemmal stretches caused by osmolarity changes, or plasmalemmal fluidity variations, and switches from inactive to active conformation.

[100] Kinase Src is not activated within 12 s of soluble EGF stimulation. Stress-induced signal transduction is at least 40 times faster than growth factor-generated signal transduction.

[101] A 1.8-Pa oscillatory stress applied with magnetic bead via a focal adhesion (~ 3–$5 \mu m^2$) activates Src similarly to 0.4 ng/ml EGF.

Table 11.33. Mechanotransduction types and compartments of the arterial network.

Arterial compartment	Function	Goal
Large arteries	Body' territory irrigation	Regulation of cardiac postload
Small resistive arteries	Organ perfusion	Maintenance of a constant flow rate
Arterioles	Tissue perfusion	Vasodilation caused by increased local metabolism

11.10.3 Mechanotransduction in Arterial Compartments

In the arterial bed, mechanotransduction ensures: (1) an appropriate arterial pressure to limit heart load and (2) an adequate organ perfusion. Three types of mechanotransduction mechanisms can be observed along the arterial bed (Table 11.33). The arterial network can be indeed split into 3 main compartments: (1) large, strongly compliant arteries; (2) small, resistive arteries that belong to the macrocirculation; and (3) arterioles in the microcirculation.

Mechanotransduction in large arteries is supposed to maintain an adequate *postload* for blood ejection from cardiac ventricules. For example, mechanotransduction is aimed at generating a vasodilation to decrease the local resistance to blood flow in the case of elevated arterial pressure. The value of interest in pulsatile flow is the peak value, i.e., the sum of the amplitude of the unsteady component and mean value (steady component). Vascular endothelial and smooth muscle cells can be assumed to sense wall shear stress and luminal pressure-induced azimuthal and longitudinal tension, respectively. To improve the signal-to-noise ratio and to avoid to trigger a premature, inadequate response, these cells integrate the mechanical stress signals over a set of cardiac cycles. Any plasmalemmal mechanosensitive proteins (GPCR, RTK, ion channel, adhesion molecule, etc.) bears the loading that provokes a conformational change. When the integrated value of peak stresses exceed a given threshold, the conformational change liberates sites of signaling initiation, either directly (e.g., catalytic sites of kinases, gates of ion channels, etc.) or indirectly by allosteric effect.

In small arteries that perfuse body's organs, *autoregulation* (Vol. 6 – Chap. 3. Cardiovascular Physiology) is associated with the myogenic tone (Vol. 5 – Chap. 8. Smooth Muscle Cells). Autoregulation results from a paradoxical vasoconstriction (local increase in flow resistance [R]) when the local arterial pressure heightens to maintain a constant flow rate ($q = \Delta p/R$), or vasodilation (local decrease in flow resistance) when the local arterial pressure decays.

11.10 Mechanotransduction

An increased demand that is confined to a restricted region, i.e., a change in metabolic activity that concerns a limited numbers of adjacent groups of working cells, triggers the *hypoxic vasodilation* of arterioles of the microvascular bed. Hypoxic vasodilation that is restricted to a local microvasculature due to local changes in metabolic conditions differs from a response to a global hypoxemia as well as hypoxic potentiation of vasoactivity mediated by some agents that can depend on the partial pressure of oxygen, but not changes in O_2 saturation of hemoglobin. Hypoxic vasodilation is the response to decreased oxygen saturation of hemoglobin. It can promptly raise the local blood perfusion, hence supplying more oxygen, without changes in the upstream arterial condition, such as vascular resistance at a distance from the site of interest, as well as without modifications in the control by the central nervous system and activity of circulating hormones. This process is based on ^{S}nitrosothiols (Sect. 10.3.1).

Chapter 12
Conclusion

Input data for integrative investigation of complex, dynamical biological phenomena and physiological apparatus include knowledge accumulated at various length scales, from molecular biology to physiology on the one hand, and histology to anatomy on the other hand. Tier architecture of living systems is characterized by its communication means and regulation procedures, enabling to integrate environmental changes to adapt. Multiple molecules interact to create the adaptable activity of the cells, tissues, organs, and body.

The huge quantity of these molecules forms a complex reaction set with feedback loops and a hierarchical organization. Studies from molecular cascades primed by mechanical stresses to cell, then to tissues and organs need to be combined to study living systems with complex dynamics; but future investigations are still needed to mimic more accurately system functioning and interaction with the environment, using multiscale modeling. An integrative model also incorporates behavior at various time scales, including response characteristic times, cardiac and breathing cycles (s), and diurnal periodicities (h; Table 12.1), to efficiently describe the structure–function relationships of the explored physiological system.

Mathematical description of large, complex biochemical reaction networks, in which molecules are nodes, and modeling of the dynamics of interactions (scaffolding, reaction, transcription, etc.) relies on computational simulations. A biochemical reaction network is defined by: (1) a set of variables — the state variables — that define the state of the system and (2) rules of temporal changes and possible transport of involved variables. The network behavior can be analyzed in a single cells and extended to multicellular systems, especially in tumor models.

Numerous process can be modeled to quickly assess effects of parameters, all other agents remaining constant, once the mathematical model has been validated. The advantage of mathematical models is they yield complete quantity fields, whereas measurements are made only at some points or correspond to averages of exploration windows of the field of the investigated variable. Moreover, mathematical models and numerical simulations support proper design of experiments.

Table 12.1. Scales from molecules to tissue dynamics.

Element	Space scale	Time scale
Atom	0.1 nm	μs
Molecule	nm	ms
Polymer	50 nm	s
Cell	10–50 μm	mn
Tissue	1–10 cm	hour–day

Cells react to various types of external stimuli, in particular, mechanical stresses. In physiological apparatus aimed at conveying flows, the magnitude and direction of mechanical stresses applied by the flowing fluid on the wetted surface of conduit wall (i.e., vascular endothelium with its glycocalyx or respiratory epithelium with the mucus layer and periciliary fluid) as well as within the vessel wall varies during the cardiac and breathing cycles.

The heart, pump embedded in the blood circuit, generates an unsteady flow with a given frequency spectrum in a network of blood vessels characterized by a complicated architecture and variable geometry both in space and time, in addition to between-subject changes. Vessel geometry varies over short distances. The vascular network of curved blood vessels is composed of successive geometrical singularities, mainly branchings.

The thoracic muscular cage, which covers the distal part of the respiratory tract, cyclically (but not symmetrically) inflates and deflates, thereby lowering and heightening the intrathoracic pressure, and hence alternatively dilating and collapsing lung alveoli and intrathoracic airways and enabling inhalation and exhalation of air. The respiratory tract is characterized by a large wetted surface inside a small volume, especially in its proximal and distal regions, i.e., the nose and thorax. In the nose, turbinates allow heat and water exchange, but render air currents less simple. In addition, the laryngeal constriction, the aperture of which varies during the breathing cycle, provokes air jet. The bronchial tree is a network of successive branchings at inspiration, or junctions at expiration, between short, more or less curved pipes of corrugated walls in large bronchi due to the presence of partial or complete cartilaginous rings in large bronchi.

Therefore, blood and air streams correspond to time-dependent, three-dimensional, developing, as these fluids flow in conduit entrance length, where the boundary layer develop (Vol. 7). Moreover, blood vessels and airways are deformable. Changes in transmural pressure (the pressure difference between the pressure at the wetted surface of the lumen applied by the moving fluid on the deformable conduit wall and the pressure at the external wall side that depends on the activity on the neighbor organs) can also influence the shape of vessel cross-section, especially when it becomes negative (collapsing regime).

In the arterial compartment, the change in cross-section shape can result from taper, when the trunk gives rise to lateral branches before bifurcating. More generally, possible prints of adjacent organs with more or less progressive constriction and enlargment, and adaptation to branching (transition zone) are also responsible for three-dimensional flows. These flows are commonly displayed by virtual transverse currents, even in approximately straight segments up- and downstream from geometrical singularities (end effects), especially when the explored vessel section differ from the local cross-section.

Local changes in the direction of stress components can also be caused by flow separation and flow reversal during the cardiac and respiratory cycles. Flow separation is set by an adverse pressure gradient when inertial forces and fluid vorticity are high enough, especially in branching segments. Due to its time-dependent feature, flow separation regions spread over a variable length during the flow cycle and can move. The location and variable size of the flow separation region depends on the flow distribution between branches that can vary during the flow cycle.

Complete flow reversal happens during alternations from inspiratory decelerating flow phase and expiratory accelerating flow phase and conversely. Partial or complete flow reversal occurs during the diastole of the left ventricle in elastic arteries, such as the aorta, and most of the muscular arteries, such as brachial and femoral arteries (but not in carotid arteries with their particular rheology). In arteries, flow reversal can actually be observed either in a region near the wall, more or less wide with respect to the position of the local center of vessel curvature, or in the entire lumen.

Consequently, the stress field experienced by mural tissues is strongly variable both in time and space. Cellular sensors then process mechanical signals by ensemble averaging not only to raise the signal-to-noise ratio, but also to adequately adapt the local size of the conduit lumen, i.e., the local flow resistance to maintain either flow rate or pressure, only in the case of sustained, abnormal stress field.

Sensing and adaptation to an applied mechanical stress field is not restrained to cells at the interface between a bioconduit and a biofluid or within the wall itself, but to any cell type in any site of the organism that can bear a stress field. Bone remodeling is a typical example of adaptation of a tissue to the gravity and lifestyle. Cells adjust their behavior to their environment rheology, especially when they migrate.

> *" We are threatened with suffering from three directions: from our own body, which is doomed to decay and dissolution and which cannot even do without pain and anxiety as warning signals; from the external world, which may rage against us with overwhelming and merciless forces of destruction; and finally from our relations to other men. The suffering which comes from this last source is perhaps more painful than any other."* (S. Freud, Das Unbehagen in der Kultur, 1930 [Civilization and Its Discontents])

Traditional Chinese medicine defines acupuncture points, or *acupoints*, for therapeutic objectives, more than 2500 years ago. An acupuncture needle is inserted into selected acupoints on the body's surface, on which mechanical (needling with lifting–thrusting cycle or twisting), electrical (electro-needling), or other types (e.g., heat [moxibustion]) of physical stimulations are exerted, in particular to cause analgesia. Afferent fibers of peripheral nerves are stimulated to elicit the **De–Qi** sensation and signal to adequate zones in the central nervous system. Acupoints are loci through which **Qi** is transferred to the body's surface.

Meridians (Chinese: **Jing**) and collaterals (Chinese: **Luo**) are communication paths for **Qi**, the vital energy master of body fluids, which can be transported to acupoints. These spatially restricted sites that do not correspond to a specialized biological tissue, but to localized structural and fonctional units, are loci from which energy pours and pervade into the body's tissues.

Acupuncture is aimed at relieving a pathological state by liberating the sequestered energy and rearranging the balance of **Yin** and **Yang** to ensure homeostasis. A disease is indeed supposed to result from an imbalance between Yin and Yang.

> " *The negative and positive spiritual forces (Kuei-Shen) are the spontaneous activity of the two material forces (Yin and Yang).*" (Chang Tsai [1020-1078])

Yang and Yin are 2 fundamental opposing, complementary, and interdependent forces found in all things in the universe, with traces of one in the other, that support each other and can transform into one another. Nothing in the universe is completely Yin or Yang; everything is a mixture of the two. In particular, Yang may be considered as mental activity in its strength aspect, Yin mental activity in its imaginative aspect; in other words, Yang constructs, Yin instructs, or conversely. Yin is related to static and hypoactive phenomena, Yang to dynamic and hyperactive processes, or conversely.

> " *Greater activity is called major Yang, whereas greater tranquillity is named major Yin. Lesser activity is termed minor Yang, whereas lesser tranquillity is designated as minor Yin... Yang cannot exist by itself; it can exist only when it is supported by Yin. Hence, Yin is the foundation of Yang. Similarly, Yin cannot alone manifest itself; it can manifest itself only when accompanied by Yang. Hence, Yang is the expression of Yin.*" (Shao Yong [1012-1077])

Insertion into the skin of thin needles is the most common technique. Manual manipulation or electrical stimulation is then achieved. Any acupoint localized in the vicinity of bones, aponeuroses, muscles, and tendons that contain neural units with somatosensory receptors. It is characterized by a large density of mastocytes that can secrete high concentrations of vaso- and neuroactive molecules, the latter targeting the central nervous system via both nervous transmission and blood convection. This pool of mastocytes reside close to neurovascular bundles, in a region where capillaries, lymphatic vessels, and nervous structures abound. Free nerve endings and cutaneous receptors (Merkel, Meissner, Ruffini, and Pacinian corpuscles), sarcous sensory receptors (muscle spindles and tendon organs), and their afferent fibers, as well as somatic efferent fibers innervating muscles, small nerve bundles, and plexi are observed in acupoints [1584]. The mechanical stress field can activate Aα, Aβ, and Aδ fibers, as well as C-fibers of nervous structures at acupoints.[1] Other features include a large skin electrical conductance and high ionic concentrations (K$^+$, Ca^{++}, Fe^{++}, Mn^{++}, Zn^{++}, and PO$_4^{3-}$) [1585].[2]

Acupuncture can be assumed to be based on the chemical response especially of mastocytes at acupoints to the sensed mechanical stresses caused by needle motions. Other cells, such as neurons, macrophages, fibroblasts, and lymphocytes can contribute to the emission of local and endocrine signals (Table 12.2).

Mastocytes free numerous types of molecules, such as calcitonin gene-related peptide (CGRP), heparin, histamine, leukotrienes (LTb4, LTc4, LTd4, and LTe4), platelet-activating factor, prostaglandin-E2, serotonin, substance-P, and thromboxane-A2 (Table 12.3; Vol. 3 – Chap. 7. G-Protein-Coupled Receptors). Mastocytes also secrete peptidases (e.g., tryptase), growth factors (e.g., FGF, gmCSF, and NGF), and cytokines (e.g., interleukins and tumor-necrosis factor). Nerve endings are stimulated and release substance-P that further activates mastocytes and triggers the production of nitric oxide.

A self-sustained process is created via the recruitment of circulating mastocytes and excitation of regional pools of mastocytes. This traditional procedure of Chinese medicine relies on intra-, auto-, juxta-, para-, and endocrine signaling aimed at triggering mastocyte chemotaxis and sending messages via: (1) nerves for an immediate (O[1 s–1 mn]), fast, and transient response of the central nervous system responsible of hyperemia in a given local region of the brain, in which neurons then secrete endocannabinoids, enkephalins, endomorphins, dynorphins, and other analgesic subtances, in particular, as well as a permanent response ensured by the continuous flux of activators; and (2) blood and lymph vessels for a delayed, slower, and sustained reaction based on transmission of substances that are conveyed throughout the brain, but preferentially to the highly perfused zone. Target nervous centers then reply by regulating the behavior of proper peripheral organs.

Any mastocyte in the local vasculature moves along the chemoattractant gradient, hence undergoes a transmigration (across blood vessel wall to exit blood, remaining granulated outside a region of triggering mechanical stress ($\mathbf{x} \gg \Delta \mathbf{x}$) and liberating

---

[1] A nerve fiber corresponds to an axon possibly endowed with a myelin sheath, whether the nerve fiver is myelinated or not. Three types of peripheral nerve fibers exist according to their caliber (A–C). The **A type** consists of 4 subtypes of relatively large (1–22 μm), myelinated fibers characterized by a high conduction velocity (5–120 m/s): Aα (afferent or efferent; motor and proprioceptive; bore 13–22 μm; conduction velocity 70–120 m/s); Aβ (afferent or efferent; motor and proprioceptive; caliber 8–13 μm; conduction velocity 30–90 m/s); Aγ (efferent; only motor fibers; caliber 3–7 μm; conduction velocity 15–40 m/s); and Aδ (afferent; only sensory fibers [rapid pain sensation]; caliber 1–5 μm; conduction velocity 5–15 m/s) fibers. Type-A fibers transmit impulses related to muscle, tendon, and joint movement and situation. **B-type** nerve fibers are also myelinated, albeit thinner ($\leq$ 3 μm). These preganglionic fibers of the autonomic nervous system transmit involuntary impulses with a slower conduction velocity (3–15 m/s). **C-type**, thin (0.3–1.3 μm), unmyelinated fibers have a low conduction velocity (0.6–2.3 m/s). They include postganglionic fibers in the autonomic nervous system as well as nerve fibers of the dorsal roots that carry sensory information (heat, pressure, and slowly pain).

[2] Japanese, ryodoraku: good electroconductive linee, proper electrical pathway.

Table 12.2. Signaling mediators released at acupoints (Source:[1586]; CGRP: calcitonin gene-related peptide; MOR: μ-opioid receptor). Mechanical stresses can activate $A\alpha$, $A\beta$, and $A\delta$ fibers, as well as C-fibers of nervous structures at acupoints and augment locally the vascular permeability to accelerate the transfer of mediators to the flowing blood.

Mediators	Releasing cells	Receptors Effects
Acetylcholine	Neuron, keratinocyte	M_2
Adenosine		A_1
ATP	Epidermal cells	P2X, P2Y
Bradykinin	Local cells	B_1, B_2
CGRP	Epidermal cells, T cell, macrophage	
Cytokines (interleukins, tumor-necrosis factor-α		
IL1β/6/8, TNFα	Local cells	Enhanced excitability of afferent fibers
IL4/10	Local cells	Inhibition of inflammatory signals in afferent terminals
β-endorphin	Fibroblasts, leukocytes, keratinocyte, melanocyte	MOR
Histamine	Mastocyte	H_1, H_3
GABA	Macrophage, lymphocyte	$GABA_A$
Glutamate	Macrophage, epidermal cells	
Nitric oxide	Many cell types	Inhibition of substance-P release from nerve terminals Stimulation of acetylcholine and β-endorphin secretion
Noradrenaline	Sympathetic nerve	α2AR
Serotonin	Mastocyte, platelet,	$5HT_1$, $5HT_3$ (afferent nerve)
Somatostatin	Merkel cell, keratinocyte	SstR
Substance-P	Mastocyte, fibroblast, platelet, macrophage, keratinocyte	
Prostaglandins	Local cells	EP

its granule content, once it reaches a region close to the acupoint ($0—\Delta x$), where it can sense a significant magnitude of the mechanical stress. Two mastocyte states indeed exist according to its localization w.r.t. the acupoint (non-degranulated and degranulated).

Chemical mediators are supposed to be quasi-instantaneously released by the mechanotransduction process that mainly relies on an sudden, rapid, and copious calcium entry in the mastocyte cytosol. This calcium wave that gush and pervade the mastocyte enables it to discharge chemoattractants, nerve messengers, cardiovascular stimulants, and endocrine messengers. On the other hand, the regeneration of granules content inside the mastocyte is delayed and slow.

Table 12.3. Released molecules by the mastocyte and their effects. Acupuncture can be modeled by an immediate and a late response. Nerves and mastocytes exchange chemical messengers such as substance-P. The latter stimulates histamine and nitric oxide (NO) release. Calcitonin gene-related peptide (CGRP) causes a vasodilation; nitric oxide cooperates with CGRP to increase its positive inotropic effect that raises the local blood flow in dilated vessels. Histamine is quickly catabolized, thereby acting near the site of release. Resulting vasodilation and increased vessel wall permeability support the transfer of chemical mediators into the blood circulation. The NO concentration rises and enhances the vasodilation. Serotonin has a biphasic effect, as it triggers a vasoconstriction and promotes NO release, hence a subsequent vasodilation. Nerve growth factor (NGF), tumor-necrosis factor (TNF) and interleukins (IL) are potent mastocyte chemoattractants. Mastocyte chemotaxis is supported by matrix degradation by secreted peptidases.

Agent	Effects
CGRP	Vasodilation, positive chronotropy, inotropy, and lusitropy, mastocyte degranulation
Heparin	Blood clot prevention
Histamine	Vasodilation (directly and via NO), nerve stimulation
Leukotrienes	Vasodilation, vascular permeability elevation
IL, NGF, TNF	Chemotaxis
Prostaglandin-D2	Nerve stimulation
Prostaglandin-E2	Vasodilation, inhibition of mediator release
Serotonin	Vasoconstriction followed by NO-mediated vasodilation
Thromboxane-A2	Vasoconstriction, platelet aggregation
Tryptase, chymase	Matrix degradation for enhanced cell migration

Following chemotaxis from regional pools and blood, newly arrived mastocytes at acupoints experience a degranulation triggered by the stress field. The resulting self-sustained process enables the local elevation of vascular permeability for improved cardiac output and enhanced endocrine signaling, vasodilation associated with a resulting increase in blood flow (remote cardiac effect), endocrine signaling to central nervous system that supports a delayed and permanent response from neurons situated in a brain region characterized by hyperemia, which can receive set of action potentials during a long period.

The mathematical model of acupuncture can be represented by the set of equations that incorporates 2 equation related to the populations of non-degranulated and degranulated mastocytes, $n_{nd}(t,x)$ and $n_d(t,x)$ being the density of non-degranulated and degranulated mastocytes; 3 equations that describe the temporal evolution of concentrations of chemoattractant ($c(t,x)$), liberated nerve stimulant ($s_n(t,x)$), and endocrine activator ($s_e(t,x)$) of some sites of the central nervous system, when convection (i.e., Stokes flow of extracellular water triggered by needle motions) in the extracellular matrix remains negligible:

$$\partial_t n_{nd}(t,\mathbf{x}) - \mathcal{D}_m \nabla^2 n_{nd}(t,\mathbf{x})$$
$$+ \nabla \cdot (S_{ca} \nabla c(t,\mathbf{x}) \cdot n_{nd}(t,\mathbf{x})) = -L\Phi(x) n_{nd}(t,\mathbf{x}) + R n_d(t,\mathbf{x}); \quad (12.1)$$

$$\partial_t n_d(t,\mathbf{x}) = L\Phi(x) n_{nd}(t,\mathbf{x}) - R n_d(t,\mathbf{x}); \quad (12.2)$$

$$\partial_t c(t,\mathbf{x}) - \mathcal{D}_c \nabla^2 c(t,\mathbf{x}) = L\Phi(x) n_{nd} - D_c c(t,\mathbf{x}); \quad (12.3)$$

$$\partial_t s_n - \mathcal{D}_n \nabla^2 s_n = L\Phi(x) n_{nd} - D_n s_n; \quad (12.4)$$

$$\partial_t s_e - \mathcal{D}_e \nabla^2 s_e = L\Phi(x) n_{nd} - D_e s_e, \quad (12.5)$$

where $\Phi(x)$ is the magnitude of applied mechanical stress

$$(0 \leq \Phi(x) \leq 1, 0 \leq \mathbf{x} \leq \Delta x);$$

$\mathcal{D}_{m/c/n/e}$ the diffusion coefficient of mastocyte, chemoattractant, nervous messenger, and endocrine mediator; D the degradation rate; L the release rate

$$(L_c \equiv L_n \equiv L_e \equiv L);$$

R the regeneration rate of degranulated masotocytes; S_{ca} the mastocyte sensitivity to chemoattractant, a measure of the chemoattractant power.

References

Chap. 1. Introduction

1. Ezkurdia I, Del Pozo A, Frankish A, Rodriguez JM, Harrow J, Ashman K, Valencia A, Tress ML (2012) Comparative proteomics reveals a significant bias toward alternative protein isoforms with conserved structure and function. Molecular Biology and Evolution (mbe.oxfordjournals.org/content/early/2012/03/22/molbev.mss100.abstract)
2. Chavent G (2010) Nonlinear Least Squares for Inverse Problems, Theoretical Foundations and Step-by-Step Guide for Applications. Springer, New York
3. Bensoussan A (1971) Filtrage optimal des systèmes linéaires [Optimal filtering of linear systems] Dunod, Paris
4. Bertoglio C, Moireau P, Gerbeau JF (2012) Sequential parameter estimation for fluid-structure problems. Application to hemodynamics. International Journal for Numerical Methods in Biomedical Engineering 28:434–455
5. Lombardi D, Iollo A, Colin T, Saut O (2009) Inverse problems in tumor growth modelling (communication at CEMRACS summer school)
6. Lagaert JB (2011) Modélisation de la croissance tumorale [Tumor Growth Modeling], PhD thesis, Bordeaux University

Chap. 2. Signaling Lipids

7. van Meer G, Voelker DR, Feigenson GW (2008) Membrane lipids: where they are and how they behave. Nature Reviews – Molecular Cell Biology 9:112–124
8. D'Arrigo P, Servi S (2010) Synthesis of lysophospholipids. Molecules 15:1354–1377
9. Voet D, Voet JG (2011) Signal Transduction (Chap 19). Biochemistry (4th edition), Wiley, Hoboken, New Jersey
10. Berridge MJ (2009) Module 2: Cell Signalling Pathways. Cell Signalling Biology. Biochemical Journal's Signal Knowledge Environment Portland Press Ltd., London, UK (www.biochemj.org/csb/002/csb002.pdf)
11. Delon C, Manifava M, Wood E, Thompson D, Krugmann S, Pyne S, Ktistakis NT (2004) Sphingosine kinase 1 is an intracellular effector of phosphatidic acid. Journal of Biological Chemistry 279:44763–44774

12. Carricaburu V, Lamia KA, Lo E, Favereaux L, Payrastre B, Cantley LC, Rameh LE (2003) The phosphatidylinositol (PI)-5-phosphate 4-kinase type II enzyme controls insulin signaling by regulating PI-3,4,5-trisphosphate degradation. Proceedings of the National Academy of Sciences of the United States of America 100:9867–9872
13. Maag D, Maxwell MJ, Hardesty DA, Boucher KL, Choudhari N, Hanno AG, Ma JF, Snowman AS, Pietropaoli JW, Xu R, Storm PB, Saiardi A, Snyder SH, Resnick AC (2011) Inositol polyphosphate multikinase is a physiologic PI3-kinase that activates Akt/PKB. Proceedings of the National Academy of Sciences of the United States of America 108: 1391–1396
14. Weirich CS, Erzberger JP, Flick JS, Berger JM, Thorner J, Weis K (2006) Activation of the DExD/H-box protein Dbp5 by the nuclear-pore protein Gle1 and its coactivator InsP6 is required for mRNA export. Nature – Cell Biology 8:668–676
15. Alcázar-Román AR, Tran EJ, Guo S, Wente SR (2006) Inositol hexakisphosphate and Gle1 activate the DEAD-box protein Dbp5 for nuclear mRNA export. Nature – Cell Biology 8: 711–716
16. Macbeth MR, Schubert HL, VanDemark AP, Lingam AT, Hill CP, Bass BL (2005) Inositol hexakisphosphate is bound in the ADAR2 core and required for RNA editing. Science 309:1534–1539
17. Huang YH, Grasis JA, Miller AT, Xu R, Soonthornvacharin S, Andreotti AH, Tsoukas CD, Cooke MP, Sauer K (2007) Positive regulation of Itk PH domain function by soluble IP4. Science 316:886–889
18. Shen X, Xiao H, Ranallo R, Wu WH, Wudagger C (2003) Modulation of ATP-dependent chromatin-remodeling complexes by inositol polyphosphates. Science 299:112–114
19. Mulugu S, Bai W, Fridy PC, Bastidas RJ, Otto JC, Dollins DE, Haystead TA, Ribeiro AA, York JD (2007) A conserved family of enzymes that phosphorylate inositol hexakisphosphate. Science 316:106–109
20. Lee YS, Mulugu S, York JD, O'Shea EK (2007) Regulation of a cyclin-CDK-CDK inhibitor complex by inositol pyrophosphates. Science 316:109–112
21. Saiardi A, Bhandari R, Resnick AC, Snowman AM, Snyder SH (2004) Phosphorylation of proteins by inositol pyrophosphates. Science 306:2101–2105
22. Chakraborty A, Koldobskiy MA, Sixt KM, Juluri KR, Mustafa AK, Snowman AM, van Rossum DB, Patterson RL, Snyder SH (2008) HSP90 regulates cell survival via inositol hexakisphosphate kinase-2. Proceedings of the National Academy of Sciences of the United States of America 105:1134–1139
23. Szado T, Vanderheyden V, Parys JB, De Smedt H, Rietdorf K, Kotelevets L, Chastre E, Khan F, Landegren U, Söderberg O, Bootman MD, Roderick HL (2008) Phosphorylation of inositol 1,4,5-trisphosphate receptors by protein kinase B/Akt inhibits Ca2+ release and apoptosis. Proceedings of the National Academy of Sciences of the United States of America 105: 2427–2432
24. Otto JC, Kelly P, Chiou ST, York JD (2007) Alterations in an inositol phosphate code through synergistic activation of a G protein and inositol phosphate kinases. Proceedings of the National Academy of Sciences of the United States of America 104:15653–15658
25. Zhang C, Majerus PW, Wilson MP (2012) Regulation of inositol 1,3,4-trisphosphate 5/6-kinase (ITPK1) by reversible lysine acetylation. Proceedings of the National Academy of Sciences of the United States of America 109:2290–2295
26. Ohnishi T, Ohba H, Seo KC, Im J, Sato Y, Iwayama Y, Furuichi T, Chung SK, Yoshikawa T (2007) Spatial expression patterns and biochemical properties distinguish a second myo-inositol monophosphatase IMPA2 from IMPA1. Journal of Biological Chemistry 282: 637–646
27. Berggard T, Szczepankiewicz O, Thulin E, Linse S (2002) Myo-inositol monophosphatase is an activated target of calbindin D28k. Journal of Biological Chemistry 277:41954–41959
28. Irvine RF (2002) Nuclear lipid signaling. Science Signaling 150:re13

29. Schouten A, Agianian B, Westerman J, Kroon J, Wirtz KWA, Gros P (2002) Structure of apo-phosphatidylinositol transfer protein α provides insight into membrane association. EMBO Journal 21:2117–2121
30. Woodcock EA, Kistler PM, Ju YK (2009) Phosphoinositide signalling and cardiac arrhythmias. Cardiovascular Research 82:286–295
31. Jao CY, Roth M, Welti R, Salic A (2009) Metabolic labeling and direct imaging of choline phospholipids in vivo. Proceedings of the National Academy of Sciences of the United States of America 106:15332–15337
32. Breslow DK, Collins SR, Bodenmiller B, Aebersold R, Simons K, Shevchenko A, Ejsing CS, Weissman JS (2010) Orm family proteins mediate sphingolipid homeostasis. Nature 463:1048–1053
33. Pavoine C, Pecker F (2009) Sphingomyelinases: their regulation and roles in cardiovascular pathophysiology. Cardiovascular Research 82:175–183
34. Cogolludo A, Moreno L, Frazziano G, Moral-Sanz J, Menendez C, Castañeda J, González C, Villamor E, Perez-Vizcaino F (2009) Activation of neutral sphingomyelinase is involved in acute hypoxic pulmonary vasoconstriction. Cardiovascular Research 82:296–302
35. Karliner JS (2009) Sphingosine kinase regulation and cardioprotection. Cardiovascular Research 82:184–192
36. Means CK, Brown JH (2009) Sphingosine-1-phosphate receptor signalling in the heart. Cardiovascular Research 82:193–200
37. Gellings Lowe N, Swaney JS, Moreno KM, Sabbadini RA (2009) Sphingosine-1-phosphate and sphingosine kinase are critical for transforming growth factor-beta-stimulated collagen production by cardiac fibroblasts. Cardiovascular Research 82:303–312
38. Sanchez T, Skoura A, Wu MT, Casserly B, Harrington EO, Hla T (2007) Induction of vascular permeability by the sphingosine 1-phosphate receptor-2 (S1P2R) and its downstream effectors ROCK and PTEN. Arteriosclerosis, Thrombosis, and Vascular Biology 27:1312–1318
39. Sattler K, Levkau B (2009) Sphingosine-1-phosphate as a mediator of high-density lipoprotein effects in cardiovascular protection. Cardiovascular Research 82:201–211
40. Igarashi J, Michel T (2009) Sphingosine-1-phosphate and modulation of vascular tone. Cardiovascular Research 82:212-220
41. Frias MA, James RW, Gerber-Wicht C, Lang U (2009) Native and reconstituted HDL activate Stat3 in ventricular cardiomyocytes via ERK1/2: Role of sphingosine-1-phosphate. Cardiovascular Research 82:313–323
42. Ke Y, Lei M, Solaro RJ (2008) Regulation of cardiac excitation and contraction by p21 activated kinase-1. Progress in Biophysics and Molecular Biology 98:238–250
43. Jenkins CM, Cedars A, Gross RW (2009) Eicosanoid signalling pathways in the heart. Cardiovascular Research 82:240–249
44. Zhao J, O'Donnell VB, Balzar S, St Croix CM, Trudeau JB, Wenzel SE (2011) 15-Lipoxygenase 1 interacts with phosphatidylethanolamine-binding protein to regulate MAPK signaling in human airway epithelial cells. Proceedings of the National Academy of Sciences of the United States of America 108:14246–14251
45. Jacobs ER, Zeldin DC (2001) The lung HETEs (and EETs) up. American Journal of Physiology – Heart and Circulatory Physiology 280:H1–H10
46. Watanabe K (2011) Recent reports about enzymes related to the synthesis of prostaglandin (PG) F2 (PGF2α and 9α, 11β-PGF2). Journal of Biochemistry 150:593–596
47. Yeaman SJ (2004) Hormone-sensitive lipase – new roles for an old enzyme. Biochemical Journal 379:11–22
48. Okazaki H, Igarashi M, Nishi M, Sekiya M, Tajima M, Takase S, Takanashi M, Ohta K, Tamura Y, Okazaki S, Yahagi N, Ohashi K, Amemiya-Kudo M, Nakagawa Y, Nagai R, Kadowaki T, Osuga J, Ishibashi S (2008) Identification of neutral cholesterol ester hydrolase, a key enzyme removing cholesterol from macrophages. Journal of Biological Chemistry 283:33357–33364

49. Jaye M, Lynch KJ, Krawiec J, Marchadier D, Maugeais C, Doan K, South V, Amin D, Perrone M, Rader DJ (1999) A novel endothelial-derived lipase that modulates HDL metabolism. Nature – Genetics 21:424–428
50. Strauss JG, Zimmermann R, Hrzenjak A, Zhou Y, Kratky D, Levak-Frank S, Kostner GM, Zechner R, Frank S (2002) Endothelial cell-derived lipase mediates uptake and binding of high-density lipoprotein (HDL) particles and the selective uptake of HDL-associated cholesterol esters independent of its enzymic activity. Biochemical Journal 368:69–79
51. Kojima Y, Ishida T, Sun L, Yasuda T, Toh R, Rikitake Y, Fukuda A, Kume N, Koshiyama H, Taniguchi A, Hirata KI (2010) Pitavastatin decreases the expression of endothelial lipase both in vitro and in vivo. Cardiovascular Research 87:385–393
52. Yano M, Matsumura T, Senokuchi T, Ishii N, Murata Y, Taketa K, Motoshima H, Taguchi T, Sonoda K, Kukidome D, Takuwa Y, Kawada T, Brownlee M, Nishikawa T, Araki E (2007) Statins activate peroxisome proliferator-activated receptor γ through extracellular signal-regulated kinase 1/2 and p38 mitogen-activated protein kinase-dependent cyclooxygenase-2 expression in macrophages. Circulation Research 100:1442–1451
53. Favari E, Zanotti I, Zimetti F, Ronda N, Bernini F, Rothblat GH (2004) Probucol inhibits ABCA1-mediated cellular lipid efflux. Arteriosclerosis, Thrombosis, and Vascular Biology 24:2345–2350
54. Aoki J, Inoue A, Makide K, Saiki N, Arai H (2007) Structure and function of extracellular phospholipase A1 belonging to the pancreatic lipase gene family. Biochimie 89:197–204
55. Aoki J, Nagai Y, Hosono H, Inoue K, Arai H (2002) Structure and function of phosphatidylserine-specific phospholipase A1. Biochimica et Biophysica Acta 1582:26–32
56. Wassum KM, Ostlund SB, Maidment NT, Balleine BW (2009) Distinct opioid circuits determine the palatability and the desirability of rewarding events. Proceedings of the National Academy of Sciences of the United States of America 106:12512–12517
57. D'Agostino D, Lowe ME (2004) Pancreatic lipase-related protein 2 is the major colipase-dependent pancreatic lipase in suckling mice. Journal of Nutrition 134:132–134
58. Alexander SPH, Mathie A, Peters JA (2009) Guide to Receptors and Channels (GRAC), 4th edn., British Journal of Pharmacology 158:S1–S254 (www3.interscience.wiley.com/journal/122684220/issue)
59. Burke JE, Dennis EA (2009) Phospholipase A2 structure/function, mechanism, and signaling. Journal of Lipid Research 50:S237–S242
60. Schaloske RH, Dennis EA (2006) The phospholipase A2 superfamily and its group numbering system. Biochimica et Biophysica Acta (BBA) – Molecular and Cell Biology of Lipids 1761:1246–1259
61. Rosa AO, Rapoport SI (2009) Intracellular- and extracellular-derived Ca^{2+} influence phospholipase A_2-mediated fatty acid release from brain phospholipids. Biochimica et Biophysica Acta 1791:697–705
62. Murakami M, Taketomi Y, Sato H, Yamamoto K (2011) Secreted phospholipase-A2 revisited. Journal of Biochemistry 150:233–255
63. Lee JC, Simonyi A, Sun AY, Sun GY (2011) Phospholipases A2 and neural membrane dynamics: implications for Alzheimer's disease. Journal of Neurochemistry 116:813–819
64. Perrin-Cocon L, Agaugué S, Coutant F, Masurel A, Bezzine S, Lambeau G, André P, Lotteau V (2004) Secretory phospholipase A2 induces dendritic cell maturation. European Journal of Immunology 34:2293–2302
65. Ancian P, Lambeau G, Mattéi MG, Lazdunski M (1995) The human 180-kDa receptor for secretory phospholipases A2. Molecular cloning, identification of a secreted soluble form, expression, and chromosomal localization. Journal of Biological Chemistry 270:8963–8970
66. Triggiani M, Granata F, Oriente A, De Marino V, Gentile M, Calabrese C, Palumbo C, Marone G (2000) Secretory phospholipase A2 induce β-glucuronidase release and IL-6 production from human lung macrophages. Journal of Immunology 164:4908–4915
67. Tada K, Murakami M, Kambe T, Kudo I (1998) Induction of cyclooxygenase-2 by secretory phospholipases A2 in nerve growth factor-stimulated rat serosal mast cells is facilitated by interaction with fibroblasts and mediated by a mechanism independent of their enzymatic functions. Journal of Immunology 161:5008–5015

68. Ibeas E, Fuentes L, Martin R, Hernandez M, Nieto ML (2009) Secreted phospholipase A2 type IIA as a mediator connecting innate and adaptive immunity: new role in atherosclerosis. Cardiovascular Research 81(1):54-63
69. Murakami M, Kudo I (2003) New phospholipase A2 isozymes with a potential role in atherosclerosis. Current Opinion in Lipidology 14:431-436
70. Kuksis A, Pruzanski W (2008) Phase composition of lipoprotein SM/cholesterol/PtdCho affects FA specificity of sPLA2s. Journal of Lipid Research 49:2161-2168
71. The HUGO Gene Nomenclature Committee (HGNC; www.genenames.org/genefamilies/PNPLA) Phospholipases
72. Wilson PA, Gardner SD, Lambie NM, Commans SA, Crowther DJ (2006) Characterization of the human patatin-like phospholipase family. Journal of Lipid Research 47:1940-1949
73. Kienesberger PC, Oberer M, Lass A, Zechner R (2009) Mammalian patatin domain containing proteins: a family with diverse lipolytic activities involved in multiple biological functions. Journal of Lipid Research 50:S63-S68
74. Sun GY, Shelat PB, Jensen MB, He Y, Sun AY, Simonyi A (2010) Phospholipases A2 and inflammatory responses in the central nervous system. Neuromolecular Medicine 12:133-148
75. Ohto T, Uozumi N, Hirabayashi T, Shimizu T (2005) Identification of novel cytosolic phospholipase A_2s, murine cPLA$_2\delta$, ϵ, and ζ, which form a gene cluster with cPLA$_2\beta$. Journal of Biological Chemistry 280:24576-24583
76. Hoffmann R, Valencia A (2004) A gene network for navigating the literature. Nature – Genetics 36:664 (Information Hyperlinked over Proteins www.ihop-net.org/)
77. Alberghina M (2010) Phospholipase-A_2: new lessons from endothelial cells. Microvascular Research 80:280-285
78. Suh PG, Park JI, Manzoli L, Cocco L, Peak JC, Katan M, Fukami K, Kataoka T, Yun S, Ryu SH (2008) Multiple roles of phosphoinositide-specific phospholipase C isozymes. BMB Reports 41:415-434
79. Waldo GL, Ricks TK, Hicks SN, Cheever ML, Kawano T, Tsuboi K, Wang X, Montell C, Kozasa T, Sondek J, Harden TK (2010) Kinetic scaffolding mediated by a phospholipase C-β and Gq signaling complex. Science 330:974-980
80. Gutman O, Walliser C, Piechulek T, Gierschik P, Henis YI (2010) Differential regulation of phospholipase C-β2 activity and membrane interaction by Gαq, Gβ1γ2, and Rac2. Journal of Biological Chemistry 285:3905-3915
81. Hunter I, Mascall KS, Ramos JW, Nixon GF (2011) A phospholipase Cγ1-activated pathway regulates transcription in human vascular smooth muscle cells. Cardiovascular Research 90:557-564
82. von Essen MR, Kongsbak M, Schjerling P, Olgaard K, Odum N, Geisler C (2010) Vitamin D controls T cell antigen receptor signaling and activation of human T cells. Nature – Immunology 11:344-349
83. Kobayashi M, Lomasney JW (2010) Phospholipase C δ4. UCSD-Nature Molecule Pages, UCSD-Nature Signaling Gateway (www.signaling-gateway.org)
84. Wing MR, Bourdon DM, Harden TK (2003) PLC-ϵ: a shared effector protein in Ras-, Rho-, and G$\alpha\beta\gamma$-mediated signaling. Molecular Interventions 3:273-280
85. Hu L, Edamatsu H, Takenaka N, Ikuta S, Kataoka T (2010) Crucial role of phospholipase Cepsilon in induction of local skin inflammatory reactions in the elicitation stage of allergic contact hypersensitivity. Journal of Immunology 184:993-1002
86. Stewart AJ, Morgan K, Farquharson C, Millar RP (2007) Phospholipase C-η enzymes as putative protein kinase C and Ca^{2+} signalling components in neuronal and neuroendocrine tissues. Neuroendocrinology 86:243-248
87. Nomikos M, Blayney LM, Larman MG, Campbell K, Rossbach A, Saunders CM, Swann K, Lai FA (2005) Role of phospholipase C-ζ domains in Ca^{2+}-dependent phosphatidylinositol 4,5-bisphosphate hydrolysis and cytoplasmic Ca^{2+} oscillations. Journal of Biological Chemistry 280:31011-31018
88. Martinson EA, Scheible S, Greinacher A, Presek P (1995) Platelet phospholipase D is activated by protein kinase C via an integrin $\alpha_{2b}\beta_3$-independent mechanism. Biochemical Journal 310:623-628

89. Gironcel D, Racaud-Sultan C, Payrastre B, Haricot M, Borchert G, Kieffer N, Breton M, Chap H (1996) $\alpha_{2b}\beta_3$-integrin mediated adhesion of human platelets to a fibrinogen matrix triggers phospholipase C activation and phosphatidylinositol 3',4'-biphosphate accumulation. FEBS Letters 389:253–256
90. Elvers M, Stegner D, Hagedorn I, Kleinschnitz C, Braun A, Kuijpers ME, Boesl M, Chen Q, Heemskerk JW, Stoll G, Frohman MA, Nieswandt B (2010) Impaired $\alpha_{2b}\beta_3$ integrin activation and shear-dependent thrombus formation in mice lacking phospholipase D1. Science Signaling 3:ra1
91. Mahankali M, Peng HJ, Henkels KM, Dinauer MC, Gomez-Cambronero J (2011) Phospholipase D2 (PLD2) is a guanine nucleotide exchange factor (GEF) for the GTPase Rac2. Proceedings of the National Academy of Sciences of the United States of America 108:19617–19622
92. Medina-Tato DA, Ward SG, Watson ML (2007) Phosphoinositide 3-kinase signalling in lung disease: leucocytes and beyond. Immunology 121:448–461
93. Foster FM, Traer CJ, Abraham SM, Fry MJ (2003) The phosphoinositide (PI) 3-kinase family. Journal of Cell Science 116:3037–3040
94. Vanhaesebroeck B, Guillermet-Guibert J, Graupera M, Bilange B (2010) The emerging mechanisms of isoform-specific PI3K signalling. Nature Reviews – Molecular Cell Biology 11:329–341
95. Yuan TL, Choi HS, Matsui A, Benes C, Lifshits E, Luo J, Frangioni JV, Cantley LC (2008) Class 1A PI3K regulates vessel integrity during development and tumorigenesis. Proceedings of the National Academy of Sciences of the United States of America 105:9739–9744
96. Foukas LC, Berenjeno IM, Gray A, Khwaja A, Vanhaesebroeck B (2010) Activity of any class IA PI3K isoform can sustain cell proliferation and survival. Proceedings of the National Academy of Sciences of the United States of America 107:11381–11386
97. Kurig B, Shymanets A, Bohnacker T, Prajwal, Brock C, Ahmadian MR, Schaefer M, Gohla A, Harteneck C, Wymann MP, Jeanclos E, Nürnberg B (2009) Ras is an indispensable coregulator of the class IB phosphoinositide 3-kinase p87/p110γ. Proceedings of the National Academy of Sciences of the United States of America 106:20312–20317
98. Shinohara M, Terada Y, Iwamatsu A, Shinohara A, Mochizuki N, Higuchi M, Gotoh Y, Ihara S, Nagata S, Itoh H, Fukui Y, Jessberger R (2002) SWAP-70 is a guanine-nucleotide-exchange factor that mediates signalling of membrane ruffling. Nature 416:759–763
99. Ihara S, Oka T, Fukui Y (2006) Direct binding of SWAP-70 to non-muscle actin is required for membrane ruffling. Journal of Cell Science 119:500–507
100. Donald S, Humby T, Fyfe I, Segonds-Pichon A, Walker SA, Andrews SR, Coadwell WJ, Emson P, Wilkinson LS, Welch HC (2008) P-Rex2 regulates Purkinje cell dendrite morphology and motor coordination. Proceedings of the National Academy of Sciences of the United States of America 105:4483–4488
101. Oudit GY, Penninger JM (2009) Cardiac regulation by phosphoinositide 3-kinases and PTEN. Cardiovascular Research 82:250–260
102. Foukas LC, Claret M, Pearce W, Okkenhaug K, Meek S, Peskett E, Sancho S, Smith AJ, Withers DJ, Vanhaesebroeck B (2006) Critical role for the p110α phosphoinositide-3-OH kinase in growth and metabolic regulation. Nature 441:366–370
103. Knight ZA, Gonzalez B, Feldman ME, Zunder ER, Goldenberg DD, Williams O, Loewith R, Stokoe D, Balla A, Toth B, Balla T, Weiss WA, Williams RL, Shokat KM (2006) A pharmacological map of the PI3-K family defines a role for p110α in insulin signaling. Cell 125:647–649
104. Jia S, Liu Z, Zhang S, Liu P, Zhang L, Lee SH, Zhang J, Signoretti S, Loda M, Roberts TM, Zhao JJ (2008) Essential roles of PI(3)K-p110β in cell growth, metabolism and tumorigenesis. Nature 454:776–779
105. Kumar A, Fernandez-Capetillo O, Carrera AC (2010) Nuclear phosphoinositide 3-kinase β controls double-strand break DNA repair. Proceedings of the National Academy of Sciences of the United States of America 107:7491–7496

106. Hawkins PT, Stephens LR (2007) PI3Kγ is a key regulator of inflammatory responses and cardiovascular homeostasis. Science 318:64–66
107. Park SJ, Lee KS, Kim SR, Min KH, Moon H, Lee MH, Chung CR, Han HJ, Puri KD, Lee YC (2010) Phosphoinositide 3-kinase δ inhibitor suppresses IL-17 expression in a murine asthma model. European Respiratory Journal (doi:10.1183/09031936.00106609)
108. Marwick JA, Caramori G, Casolari P, Mazzoni F, Kirkham PA, Adcock IM, Chung KF, Papi A (2010) A role for phosphoinositol 3-kinase δ in the impairment of glucocorticoid responsiveness in patients with chronic obstructive pulmonary disease. Journal of Allergy and Clinical Immunology 125:1146–1153
109. Beltran L, Chaussade C, Vanhaesebroeck B, Cutillas PR (2011) Calpain interacts with class IA phosphoinositide 3-kinases regulating their stability and signaling activity. Proceedings of the National Academy of Sciences of the United States of America 108:16217–16222
110. Miller S, Tavshanjian B, Oleksy A, Perisic O, Houseman BT, Shokat KM, Williams RL (2010) Shaping development of autophagy inhibitors with the structure of the lipid kinase Vps34. Science 327:1638–1642
111. Matsunaga K, Saitoh T, Tabata K, Omori H, Satoh T, Kurotori N, Maejima I, Shirahama-Noda K, Ichimura T, Isobe T, Akira S, Noda T, Yoshimori T (2009) Two Beclin 1-binding proteins, Atg14L and Rubicon, reciprocally regulate autophagy at different stages. Nature – Cell Biology 11:385–396
112. Yan Y, Flinn RJ, Wu H, Schnur RS, Backer JM (2009) hVps15, but not Ca^{2+}/CaM, is required for the activity and regulation of hVps34 in mammalian cells. Biochemical Journal 417: 747–755
113. Okada M, Jang SW, Ye K (2008) Akt phosphorylation and nuclear phosphoinositide association mediate mRNA export and cell proliferation activities by ALY. Proceedings of the National Academy of Sciences of the United States of America 105:8649–8654
114. Platanias LC (2005) Mechanisms of type-I and type-II-interferon-mediated signalling. Nature Reviews – Immunology 5:375–386
115. Higuchi M, Onishi K, Kikuchi C, Gotoh Y (2008) Scaffolding function of PAK in the PDK1-Akt pathway. Nature – Cell Biology 10:1356–1364
116. Park SW, Zhou Y, Lee J, Lu A, Sun C, Chung J, Ueki K, Ozcan U (2010) The regulatory subunits of PI3K, p85α and p85β, interact with XBP-1 and increase its nuclear translocation. Nature – Medicine 16:429–437
117. Morello F, Perino A, Hirsch E (2009) Phosphoinositide 3-kinase signalling in the vascular system. Cardiovascular Research 82:261–271
118. Blero D, Payrastre B, Schurmans S, Erneux C (2007) Phosphoinositide phosphatases in a network of signalling reactions. Pflugers Archiv – European Journal of Physiology 455: 31–44
119. Jin N, Chow CY, Liu L, Zolov SN, Bronson R, Davisson M, Petersen JL, Zhang Y, Park S, Duex JE, Goldowitz D, Meisler MH, Weisman LS (2008) VAC14 nucleates a protein complex essential for the acute interconversion of PI3P and PI(3,5)P(2) in yeast and mouse. EMBO Journal 27:3221–3234
120. Balla A, Balla T (2006) Phosphatidylinositol 4-kinases: old enzymes with emerging functions. Trends in Cell Biology 16:351–361
121. Hsuan J, Waugh MG, Minogue S (2008) Phosphatidylinositol 4-kinase type 2α. UCSD-Nature Molecule Pages, UCSD-Nature Signaling Gateway (www.signaling-gateway.org)
122. Berditchevski F, Tolias KF, Wong K, Carpenter CL, Hemler ME (1997) A novel link between integrins, transmembrane-4 superfamily proteins (CD63 and CD81), and phosphatidylinositol 4-kinase. Journal of Biological Chemistry 272:2595–2598
123. Nishikawa K, Toker A, Wong K, Marignani PA, Johannes FJ, Cantley LC (1998) Association of protein kinase Cμ with type II phosphatidylinositol 4-kinase and type I phosphatidylinositol-4-phosphate 5-kinase. Journal of Biological Chemistry 273: 23126–23133
124. Kauffmann-Zeh A, Thomas GM, Ball A, Prosser S, Cunningham E, Cockcroft S, Hsuan JJ (1995) Requirement for phosphatidylinositol transfer protein in epidermal growth factor signaling. Science 268:1188–1190

125. Minogue S, Hsuan J (2008) Phosphatidylinositol 4-kinase type IIβ. UCSD-Nature Molecule Pages, UCSD-Nature Signaling Gateway (www.signaling-gateway.org)
126. Kakuk A, Friedländer E, Vereb G Jr, Kása A, Balla A, Balla T, Heilmeyer LM Jr, Gergely P, Vereb G (2006) Nucleolar localization of phosphatidylinositol 4-kinase PI4K230 in various mammalian cells. Cytometry A 69:1174–1183
127. Balla A, Tuymetova G, Tsiomenko A, Várnai P, Balla T (2005) A plasma membrane pool of phosphatidylinositol 4-phosphate is generated by phosphatidylinositol 4-kinase type-IIIα: studies with the PH domains of the oxysterol binding protein and FAPP1. Molecular Biology of the Cell 16:1282–1295
128. Balla A, Kim YJ, Varnai P, Szentpetery Z, Knight Z, Shokat KM, Balla T (2008) Maintenance of hormone-sensitive phosphoinositide pools in the plasma membrane requires phosphatidylinositol 4-kinase IIIα. Molecular Biology of the Cell 19:711–721
129. Trotard M, Lepère-Douard C, Régeard M, Piquet-Pellorce C, Lavillette D, Cosset FL, Gripon P, Le Seyec J (2009) Kinases required in hepatitis C virus entry and replication highlighted by small interference RNA screening. FASEB Journal
130. Guerriero CJ, Weixel KM, Bruns JR, Weisz OA (2006) Phosphatidylinositol 5-kinase stimulates apical biosynthetic delivery via an Arp2/3-dependent mechanism. Journal of Biological Chemistry 281:15376–15384
131. Clarke JH, Emson PC, Irvine RF (2008) Localization of phosphatidylinositol phosphate kinase IIγ in kidney to a membrane trafficking compartment within specialized cells of the nephron. American Journal of Physiology – Renal Physiology 295:F1422–F1430
132. Loijens JC, Anderson RA (2006) Type I phosphatidylinositol-4-phosphate 5-kinases are distinct members of this novel lipid kinase family. Journal of Biological Chemistry 271: 32937–32943
133. Chao WT, Daquinag AC, Ashcroft F, Kunz J (2010) Type I PIPK-α regulates directed cell migration by modulating Rac1 plasma membrane targeting and activation. Journal of Cell Biology 190:247–262
134. Nakano-Kobayashi A, Yamazaki M, Unoki T, Hongu T, Murata C, Taguchi R, Katada T, Frohman MA, Yokozeki T, Kanaho Y (2007) Role of activation of PIP5Kγ661 by AP-2 complex in synaptic vesicle endocytosis. EMBO Journal 26:1105–1116
135. Rahdar M, Inoue T, Meyer T, Zhang J, Vazquez F, Devreotes PN (2009) A phosphorylation-dependent intramolecular interaction regulates the membrane association and activity of the tumor suppressor PTEN. Proceedings of the National Academy of Sciences of the United States of America 106:480–485
136. Yu J, Zhang SS, Saito K, Williams S, Arimura Y, Ma Y, Ke Y, Baron V, Mercola D, Feng GS, Adamson E, Mustelin T (2009) PTEN regulation by Akt–EGR1–ARF–PTEN axis. EMBO Journal 28:21–33
137. Song MS, Salmena L, Carracedo A, Egia A, Lo-Coco F, Teruya-Feldstein J, Pandolfi PP (2008) The deubiquitinylation and localization of PTEN are regulated by a HAUSP–PML network. Nature 455:813–817
138. Sigal YJ, McDermott MI, Morris AJ (2005) Integral membrane lipid phosphatases/ phosphotransferases: common structure and diverse functions.
139. Begley MJ, Dixon JE (2005) The structure and regulation of myotubularin phosphatases. Current Opinion in Structural Biology 15:614–620
140. Sasaki J, Kofuji S, Itoh R, Momiyama T, Takayama K, Murakami H, Chida S, Tsuya Y, Takasuga S, Eguchi S, Asanuma K, Horie Y, Miura K, Davies EM, Mitchell C, Yamazaki M, Hirai H, Takenawa T, Suzuki A, Sasaki T (2010) The PtdIns(3,4)P$_2$ phosphatase INPP4A is a suppressor of excitotoxic neuronal death. Natur 465:497–501
141. Munday AD, Norris FA, Caldwell KK, Brown S, Majerus PW, Mitchell CA (1999) The inositol polyphosphate 4-phosphatase forms a complex with phosphatidylinositol 3-kinase in human platelet cytosol. Proceedings of the National Academy of Sciences of the United States of America 96:3640–3645
142. Minagawa T, Ijuin T, Mochizuki Y, Takenawa T (2001) Identification and characterization of a sac domain-containing phosphoinositide 5-phosphatase. Journal of Biological Chemistry 276:22011–22015

143. Symons MH, Chuang Y (2011) Synaptojanin 2. UCSD-Nature Molecule Pages, UCSD-Nature Signaling Gateway (www.signaling-gateway.org)
144. Nemoto Y, De Camilli P (1999) Recruitment of an alternatively spliced form of synaptojanin 2 to mitochondria by the interaction with the PDZ domain of a mitochondrial outer membrane protein. EMBO Journal 18:2991–3006
145. Liu Y, Boukhelifa M, Tribble E, Morin-Kensicki E, Uetrecht A, Bear JE, Bankaitis VA (2008) The Sac1 phosphoinositide phosphatase regulates Golgi membrane morphology and mitotic spindle organization in mammals. Molecular Biology of the Cell 19:3080–3096
146. Jin N, Chow CY, Liu L, Zolov SN, Bronson R, Davisson M, Petersen JL, Zhang Y, Park S, Duex JE, Goldowitz D, Meisler MH, Weisman LS (2008) VAC14 nucleates a protein complex essential for the acute interconversion of PI3P and PI(3,5)P(2) in yeast and mouse. EMBO Journal 27:3221–3234
147. Rudge SA, Anderson DM, Emr SD (2004) Vacuole size control: regulation of PtdIns(3,5)P2 levels by the vacuole-associated Vac14-Fig4 complex, a PtdIns(3,5)P2-specific phosphatase. Molecular Biology of the Cell 15:24–36
148. Harris SJ, Parry RV, Westwick J, Ward SG (2008) Phosphoinositide lipid phosphatases: natural regulators of phosphoinositide 3-kinase signaling in T lymphocytes. Journal of Biological Chemistry 283:2465–2469
149. Halaszovich CR, Schreiber DN, Oliver D (2009) Ci-VSP is a depolarization-activated phosphatidylinositol 4,5-bisphosphate and phosphatidylinositol 3,4,5-trisphosphate 5'-phosphatase. Journal of Biological Chemistry 284:2106–2113
150. Brindley DN (2004) Lipid phosphate phosphatases and related proteins: Signaling functions in development, cell division, and cancer. Journal of Cellular Biochemistry 92:900–912
151. Nakanaga K, Hama K, Aoki J (2010) Autotaxin: an LPA producing enzyme with diverse functions. Journal of Biochemistry 148:13–24
152. Goding JW, Grobben B, Slegers H (2003) Physiological and pathophysiological functions of the ecto-nucleotide pyrophosphatase/phosphodiesterase family. Biochimica et Biophysica Acta 1638:1–19

Chap. 3. Preamble to Protein Kinases

153. Ubersax JA, Ferrell JE (2007) Mechanisms of specificity in protein phosphorylation. Nature Reviews – Molecular Cell Biology 8:530–541
154. Manning G, Whyte DB, Martinez R, Hunter T, Sudarsanam S (2002) The protein kinase complement of the human genome. Science 298:1912–1934 (www.kinase.com/human/kinome)
155. Manning BD, Cantley LC (2002) Hitting the target: emerging technologies in the search for kinase substrates. Science STKE 2002:pe49
156. Phylogenetic tree of the human kinome in Signal Transduction Knowledge Environment (www.stke.org and stke.sciencemag.org)
157. Katoh Y, Takemori H, Horike N, Doi J, Muraoka M, Min L, Okamoto M (2004) Salt-inducible kinase (SIK) isoforms: their involvement in steroidogenesis and adipogenesis. Molecular and Cellular Endocrinology 217:109–112
158. LaRonde-LeBlanc N, Wlodawer A (2005) The RIO kinases: An atypical protein kinase family required for ribosome biogenesis and cell cycle progression. Biochimica et Biophysica Acta – Proteins and Proteomics 1754:14–24
159. Mollet J, Delahodde A, Serre V, Chretien D, Schlemmer D, Lombes A, Boddaert N, Desguerre I, de Lonlay P, de Baulny HO, Munnich A, Rötig A (2008) CABC1 gene mutations cause ubiquinone deficiency with cerebellar ataxia and seizures. American Journal of Human Genetics 82:623–630

160. Macinga DR, Cook GM, Poole RK, Rather PN (1998) Identification and characterization of aarF, a locus required for production of ubiquinone in Providencia stuartii and Escherichia coli and for expression of 2′-N-acetyltransferase in P. stuartii. Journal of Bacteriology 180: 128–135
161. Sargent CA, Anderson MJ, Hsieh SL, Kendall E, Gomez-Escobar N, Campbell RD (1994) Characterisation of the novel gene G11 lying adjacent to the complement C4A gene in the human major histocompatibility complex. Human Molecular Genetics 3:481–488
162. Fraser RA, Heard DJ, Adam S, Lavigne AC, Le Douarin B, Tora L, Losson R, Rochette-Egly C, Chambon P (1998) The putative cofactor TIF1α is a protein kinase that is hyperphosphorylated upon interaction with liganded nuclear receptors. Journal of Biological Chemistry 273:16199–16204
163. Allton K, Jain AK, Herz HM, Tsai WW, Jung SY, Qin J, Bergmann A, Johnson RL, Barton MC (2009) Trim24 targets endogenous p53 for degradation. Proceedings of the National Academy of Sciences of the United States of America 106:11612–11616
164. Dupont S, Zacchigna L, Cordenonsi M, Soligo S, Adorno M, Rugge M, Piccolo S (2005) Germ-layer specification and control of cell growth by Ectodermin, a Smad4 ubiquitin ligase. Cell 121:87–99
165. Dupont S, Mamidi A, Cordenonsi M, Montagner M, Zacchigna L, Adorno M, Martello G, Stinchfield MJ, Soligo S, Morsut L, Inui M, Moro S, Modena N, Argenton F, Newfeld SJ, Piccolo S (2009) FAM/USP9x, a deubiquitinating enzyme essential for TGFβ signaling, controls Smad4 monoubiquitination. Cell 136:123–135
166. He W, Dorn DC, Erdjument-Bromage H, Tempst P, Moore MA, Massagué J (2006) Hematopoiesis controlled by distinct TIF1γ and Smad4 branches of the TGFβ pathway. Cell 125:929–941
167. Forrest AR, Taylor DF, Crowe ML, Chalk AM, Waddell NJ, Kolle G, Faulkner GJ, Kodzius R, Katayama S, Wells C, Kai C, Kawai J, Carninci P, Hayashizaki Y, Grimmond SM (2006) Genome-wide review of transcriptional complexity in mouse protein kinases and phosphatases. Genome Biology 7:R5
168. Boudeau J, Miranda-Saavedra D, Barton GJ, Alessi DR (2006) Emerging roles of pseudokinases. Trends in Cell Biology 16:443–452
169. Lange A, Wickström SA, Jakobson M, Zent R, Sainio K, Fässler R (2009) Integrin-linked kinase is an adaptor with essential functions during mouse development. Nature 461:1002–1006
170. Dougherty MK, Ritt DA, Zhou M, Specht SI, Monson DM, Veenstra TD, Morrison DK (2009) KSR2 is a calcineurin substrate that promotes ERK cascade activation in response to calcium signals. Molecular Cell 34:652–662
171. Rajakulendran T, Sicheri F (2010) Allosteric protein kinase regulation by pseudokinases: insights from STRAD. Science Signaling 3:pe8
172. Baas AF, Boudeau J, Sapkota GP, Smit L, Medema R, Morrice NA, Alessi DR, Clevers HC (2003) Activation of the tumour suppressor kinase LKB1 by the STE20-like pseudokinase STRAD. EMBO Journal 22:3062–3072
173. Hawley SA, Boudeau J, Reid JL, Mustard KJ, Udd L, Mäkelä TP, Alessi DR, Hardie DG (2003) Complexes between the LKB1 tumor suppressor, STRAD alpha/beta and MO25 alpha/beta are upstream kinases in the AMP-activated protein kinase cascade. Journal of Biology 2:28
174. Lizcano JM, Göransson O, Toth R, Deak M, Morrice NA, Boudeau J, Hawley SA, Udd L, Mäkelä TP, Hardie DG, Alessi DR (2004) LKB1 is a master kinase that activates 13 kinases of the AMPK subfamily, including MARK/PAR-1. EMBO Journal 23:833–843
175. Kiss-Toth E, Bagstaff SM, Sung HY, Jozsa V, Dempsey C, Caunt JC, Oxley KM, Wyllie DH, Polgar T, Harte M, O'Neill LA, Qwarnstrom EE, Dower SK (2004) Human tribbles, a protein family controlling mitogen-activated protein kinase cascades. Journal of Biological Chemistry 279:42703–42708
176. Kiss-Toth E (2007) Trb1. UCSD-Nature Molecule Pages, UCSD-Nature Signaling Gateway (www.signaling-gateway.org)

177. Hegedus Z, Czibula A, Kiss-Toth E (2006) Tribbles: novel regulators of cell function; evolutionary aspects. Cellular and Molecular Life Sciences 63:1632–1641
178. Sung HY, Guan H, Czibula A, King AR, Eder K, Heath E, Suvarna SK, Dower SK, Wilson AG, Francis SE, Crossman DC, Kiss-Toth E (2007) Human tribbles-1 controls proliferation and chemotaxis of smooth muscle cells via MAPK signaling pathways. Journal of Biological Chemistry 282:18379–18387
179. Kiss-Toth E, Docherty LM (2007) Trb2. UCSD-Nature Molecule Pages, UCSD-Nature Signaling Gateway (www.signaling-gateway.org)

Chap. 4. Cytoplasmic Protein Tyrosine Kinases

180. Filippakopoulos P, Kofler M, Hantschel O, Gish GD, Grebien F, Salah E, Neudecker P, Kay LE, Turk BE, Superti-Furga G, Pawson T, Knapp S (2008) Structural coupling of SH2-kinase domains links Fes and Abl substrate recognition and kinase activation. Cell 134:793–803
181. Pendergast AM, Zipfel P (2006) Abl; Arg. UCSD-Nature Molecule Pages, UCSD-Nature Signaling Gateway (www.signaling-gateway.org)
182. Eiring AM, Harb JG, Neviani P, Garton C, Oaks JJ, Spizzo R, Liu S, Schwind S, Santhanam R, Hickey CJ, Becker H, Chandler JC, Andino R, Cortes J, Hokland P, Huettner CS, Bhatia R, Roy DC, Liebhaber SA, Caligiuri MA, Marcucci G, Garzon R, Croce CM, Calin GA, Perrotti D (2010) miR-328 functions as an RNA decoy to modulate hnRNP E2 regulation of mRNA translation in leukemic blasts. Cell 140:652–665
183. Wang B, Golemis EA, Kruh GD (1997) ArgBP2, a multiple Src homology 3 domain-containing, Arg/Abl-interacting protein, is phosphorylated in v-Abl-transformed cells and localized in stress fibers and cardiocyte Z-disks. Journal of Biological Chemistry 272: 17542–17550
184. Galisteo ML, Yang Y, Ureña J, Schlessinger J (2006) Activation of the nonreceptor protein tyrosine kinase Ack by multiple extracellular stimuli. Proceedings of the National Academy of Sciences of the United States of America 103:9796–9801
185. Satoh T (2005) Ack1. UCSD-Nature Molecule Pages, UCSD-Nature Signaling Gateway (www.signaling-gateway.org)
186. Crompton MR (2005) Brk. UCSD-Nature Molecule Pages, UCSD-Nature Signaling Gateway (www.signaling-gateway.org)
187. Qiu H, Zappacosta F, Su W, Annan RS, Miller WT (2005) Interaction between Brk kinase and insulin receptor substrate-4. Oncogene 24:5656–5664
188. Vang T, Methi T, Veillette A, Mustelin T, Tasken K (2006) Csk. UCSD-Nature Molecule Pages, UCSD-Nature Signaling Gateway (www.signaling-gateway.org)
189. Zhao M, Janas JA, Niki M, Pandolfi PP, Van Aelst L (2006) Dok-1 independently attenuates Ras/mitogen-activated protein kinase and Src/c-myc pathways to inhibit platelet-derived growth factor-induced mitogenesis. Molecular and Cellular Biology 26:2479–2489
190. Lemay S, Davidson D, Latour S, Veillette A (2000) Dok-3, a novel adapter molecule involved in the negative regulation of immunoreceptor signaling. Molecular and Cellular Biology 20:2743–2754
191. Frame MC, Patel H, Serrels B, Lietha D, Eck MJ (2010) The FERM domain: organizing the structure and function of FAK. Nature Reviews – Molecular Cell Biology 11:802–814
192. Cooper J, Li W, You L, Schiavon G, Pepe-Caprio A, Zhou L, Ishii R, Giovannini M, Hanemann CO, Long SB, Erdjument-Bromage H, Zhou P, Tempst P, Giancotti FG (2011) Merlin/NF2 functions upstream of the nuclear E3 ubiquitin ligase CRL4^{DCAF1} to suppress oncogenic gene expression. Science Signaling 4:pt6
193. Koshman YE, Engman SJ, Kim T, Iyengar R, Henderson KK, Samarel AM (2010) Role of FRNK tyrosine phosphorylation in vascular smooth muscle spreading and migration. Cardiovascular Research 85:571–581

194. Hauck CR, Hsia DA, Schlaepfer DD (2002) The focal adhesion kinase – a regulator of cell migration and invasion. IUBMB Life 53:115–119
195. Mitra SK, Hanson DA, Schlaepfer DD (2005) Focal adhesion kinase: in command and control of cell motility. Nature Reviews – Molecular Cell Biology 6:56–68
196. Dalla Costa AP, Clemente CF, Carvalho HF, Carvalheira JB, Nadruz W Jr, Franchini KG (2010) FAK mediates the activation of cardiac fibroblasts induced by mechanical stress through regulation of the mTOR complex. Cardiovascular Research 86:421–431
197. Seko Y, Takahashi N, Sabe H, Tobe K, Kadowaki T, Nagai R (1999) Hypoxia induces activation and subcellular translocation of focal adhesion kinase (p125(FAK)) in cultured rat cardiac myocytes. Biochemical and Biophysical Research Communications 262:290–296
198. Pawson T, Letwin K, Lee T, Hao QL, Heisterkamp N, Groffen J (1989) The FER gene is evolutionarily conserved and encodes a widely expressed member of the FPS/FES protein-tyrosine kinase family. Molecular and Cellular Biology 9:5722–5725
199. Greer P (2002) Closing in on the biological functions of Fps/Fes and Fer. Nature Reviews – Molecular Cell Biology 3:278–289
200. Itoh T, Hasegawa J, Tsujita K, Kanaho Y, Takenawa T (2009) The tyrosine kinase Fer is a downstream target of the PLD-PA pathway that regulates cell migration. Science Signaling 2:ra52
201. Kapus A, Di Ciano C, Sun J, Zhan X, Kim L, Wong TW, Rotstein OD (2000) Cell volume-dependent phosphorylation of proteins of the cortical cytoskeleton and cell-cell contact sites. The role of Fyn and FER kinases. Journal of Biological Chemistry 275:32289–32298
202. Schwartz Y, Ben-Dor I, Navon A, Motro B, Nir U (1998) Tyrosine phosphorylation of the TATA element modulatory factor by the FER nuclear tyrosine kinases. FEBS Letters 434: 339–345
203. Craven RJ, Cance WG, Liu ET (1995) The nuclear tyrosine kinase Rak associates with the retinoblastoma protein pRb. Cancer Research 55:3969–3972
204. Annerén C, Lindholm CK, Kriz V, Welsh M (2003) The FRK/RAK-SHB signaling cascade: a versatile signal-transduction pathway that regulates cell survival, differentiation and proliferation. Current Molecular Medicine 3:313–324
205. Serfas MS, Tyner AL (2003) Brk, Srm, Frk, and Src42A form a distinct family of intracellular Src-like tyrosine kinases. Oncology Research 13:409–419
206. Yim EK, Peng G, Dai H, Hu R, Li K, Lu Y, Mills GB, Meric-Bernstam F, Hennessy BT, Craven RJ, Lin SY (2009) Rak functions as a tumor suppressor by regulating PTEN protein stability and function. Cancer Cell 15:304–314
207. Leonard WJ (2001) Cytokines and immunodeficiency diseases. Nature Reviews – Immunology 1:200–208
208. Kurdi M, Booz GW (2009) JAK redux: a second look at the regulation and role of JAKs in the heart. American Journal of Physiology – Heart and Circulatory Physiology 297: H1545–H1556
209. Yang J, Stark GR (2008) Roles of unphosphorylated STATs in signaling. Cell Research 18:443–451
210. Kile BT, Nicola NA, Alexander WS (2001) Negative regulators of cytokine signaling. International Journal of Hematology 73:292–298
211. Nicholson SE, Willson TA, Farley A, Starr R, Zhang JG, Baca M, Alexander WS, Metcalf D, Hilton DJ, Nicola NA (1999) Mutational analyses of the SOCS proteins suggest a dual domain requirement but distinct mechanisms for inhibition of LIF and IL-6 signal transduction. EMBO Journal 18:375–385
212. Kinjyo I, Hanada T, Inagaki-Ohara K, Mori H, Aki D, Ohishi M, Yoshida H, Kubo M, Yoshimura A (2002) SOCS1/JAB is a negative regulator of LPS-induced macrophage activation. Immunity 17:583–591
213. Gingras S, Parganas E, de Pauw A, Ihle JN, Murray PJ (2004) Re-examination of the role of suppressor of cytokine signaling 1 (SOCS1) in the regulation of toll-like receptor signaling. Journal of Biological Chemistry 79:54702–54707
214. Cohney SJ, Sanden D, Cacalano NA, Yoshimura A, Mui A, Migone TS, Johnston JA (1999) SOCS-3 is tyrosine phosphorylated in response to interleukin-2 and suppresses

STAT5 phosphorylation and lymphocyte proliferation. Molecular and Cellular Biology 19: 4980–4988
215. Croker BA, Krebs DL, Zhang JG, Wormald S, Willson TA, Stanley EG, Robb L, Greenhalgh CJ, Förster I, Clausen BE, Nicola NA, Metcalf D, Hilton DJ, Roberts AW, Alexander WS (2003) SOCS3 negatively regulates IL-6 signaling in vivo. Nature – Immunology 4:540–545
216. Ueki K, Kondo T, Kahn CR (2004) Suppressor of cytokine signaling 1 (SOCS-1) and SOCS-3 cause insulin resistance through inhibition of tyrosine phosphorylation of insulin receptor substrate proteins by discrete mechanisms. Molecular and Cellular Biology 24:5434–5446
217. Mooney RA, Senn J, Cameron S, Inamdar N, Boivin LM, Shang Y, Furlanetto RW (2001) Suppressors of cytokine signaling-1 and -6 associate with and inhibit the insulin receptor. A potential mechanism for cytokine-mediated insulin resistance. Journal of Biological Chemistry 276:25889–25893
218. Cheng HC, Chong YP, Ia KK, Tan O, Mulhern TD (2006) Csk homologous kinase. UCSD-Nature Molecule Pages, UCSD-Nature Signaling Gateway (www.signaling-gateway.org)
219. Jhun BH, Rivnay B, Price D, Avraham H (1995) The MATK tyrosine kinase interacts in a specific and SH2-dependent manner with c-Kit. Journal of Biological Chemistry 270: 9661–9666
220. Yamashita H, Avraham S, Jiang S, Dikic I, Avraham H (1999) The Csk homologous kinase associates with TrkA receptors and is involved in neurite outgrowth of PC12 cells. Journal of Biological Chemistry 274:15059–15065
221. Kohmura N, Yagi T, Tomooka Y, Oyanagi M, Kominami R, Takeda N, Chiba J, Ikawa Y, Aizawa S (1994) A novel nonreceptor tyrosine kinase, Srm: cloning and targeted disruption. Molecular and Cellular Biology 14:6915–6925
222. Mano H, Yamashita Y, Miyazato A, Miura Y, Ozawa K (1996) Tec protein-tyrosine kinase is an effector molecule of Lyn protein-tyrosine kinase. FASEB Journal 10:637-642.
223. Sandilands E, Cans C, Fincham VJ, Brunton VG, Mellor H, Prendergast GC, Norman JC, Superti-Furga G, Frame MC (2004) RhoB and actin polymerization coordinate Src activation with endosome-mediated delivery to the membrane. Developmental Cell 7:855–869
224. Knock GA, Snetkov VA, Shaifta Y, Drndarski S, Ward JPT, Aaronson PI (2008) Role of src-family kinases in hypoxic vasoconstriction of rat pulmonary artery. Cardiovascular Research 80:453–462
225. Seko Y, Tobe K, Takahashi N, Kaburagi Y, Kadowaki T, Yazaki Y (1996) Hypoxia and hypoxia/reoxygenation activate Src family tyrosine kinases and p21ras in cultured rat cardiac myocytes. Biochemical and Biophysical Research Communications 226:530–535
226. Yin H, Chao L, Chao J (2005) Kallikrein/kinin protects against myocardial apoptosis after ischemia/reperfusion via Akt-glycogen synthase kinase-3 and Akt-Bad.14-3-3 signaling pathways. Journal of Biological Chemistry 280:8022–8030
227. Sato H, Sato M, Kanai H, Uchiyama T, Iso T, Ohyama Y, Sakamoto H, Tamura J, Nagai R, Kurabayashi M (2005) Mitochondrial reactive oxygen species and c-Src play a critical role in hypoxic response in vascular smooth muscle cells. Cardiovascular Research 67:714–722
228. Pleiman CM, Clark MR, Gauen LK, Winitz S, Coggeshall KM, Johnson GL, Shaw AS, Cambier JC (1993) Mapping of sites on the Src family protein tyrosine kinases p55blk, p59fyn, and p56lyn which interact with the effector molecules phospholipase C-γ2, microtubule-associated protein kinase, GTPase-activating protein, and phosphatidylinositol 3-kinase. Molecular and Cellular Biology 13:5877–5887
229. Hegde R, Srinivasula SM, Ahmad M, Fernandes-Alnemri T, Alnemri ES (1998) Blk, a BH3-containing mouse protein that interacts with Bcl-2 and Bcl-xL, is a potent death agonist. Journal of Biological Chemistry 273:7783–7786
230. Dymecki SM, Niederhuber JE, Desiderio SV (1990) Specific expression of a tyrosine kinase gene, blk, in B lymphoid cells. Science 247:332–336
231. Lin YH, Shin EJ, Campbell MJ, Niederhuber JE (1995) Transcription of the blk gene in human B lymphocytes is controlled by two promoters. Journal of Biological Chemistry 270:25968–25975

232. Bagheri-Yarmand R, Mandal M, Taludker AH, Wang RA, Vadlamudi RK, Kung HJ, Kumar R (2001) Etk/Bmx tyrosine kinase activates Pak1 and regulates tumorigenicity of breast cancer cells. Journal of Biological Chemistry 276:29403–29409
233. Mitchell-Jordan SA, Holopainen T, Ren S, Wang S, Warburton S, Zhang MJ, Alitalo K, Wang Y, Vondriska TM (2008) Loss of Bmx nonreceptor tyrosine kinase prevents pressure overload-induced cardiac hypertrophy. Circulation Research 103:1359–1362
234. Pan S, An P, Zhang R, He X, Yin G, Min W (2002) Etk/Bmx as a tumor necrosis factor receptor type 2-specific kinase: role in endothelial cell migration and angiogenesis. Molecular and Cellular Biology 22:7512–7523
235. Waldeck-Weiermair M, Zoratti C, Osibow K, Balenga N, Goessnitzer E, Waldhoer M, Malli R, Graier WF (2008) Integrin clustering enables anandamide-induced Ca^{2+} signaling in endothelial cells via GPR55 by protection against CB1-receptor-triggered repression. Journal of Cell Science 121:1704–1717
236. Semaan N, Alsaleh G, Gottenberg JE, Wachsmann D, Sibilia J (2008) Etk/BMX, a Btk family tyrosine kinase, and Mal contribute to the cross-talk between MyD88 and FAK pathways. Journal of Immunology 180:3485–3491
237. Palmer CD, Mutch BE, Workman S, McDaid JP, Horwood NJ, Foxwell BM (2008) Bmx tyrosine kinase regulates TLR4-induced IL-6 production in human macrophages independently of p38 MAPK and NFκB activity. Blood 111:1781–1788
238. Carpenter CL (2004) Btk-dependent regulation of phosphoinositide synthesis. Biochemical Society Transactions 32:326–329
239. Uckun FM (2008) Bruton's tyrosine kinase (BTK) as a dual-function regulator of apoptosis. Biochemical Pharmacology 56:683–691
240. Lowell CA, Berton G (1999) Integrin signal transduction in myeloid leukocytes. Journal of Leukocyte Biology 65:313–320
241. Resh M (2006) Fyn. UCSD-Nature Molecule Pages, UCSD-Nature Signaling Gateway (www.signaling-gateway.org)
242. Banin S, Truong O, Katz DR, Waterfield MD, Brickell PM, Gout I (1996) Wiskott-Aldrich syndrome protein (WASp) is a binding partner for c-Src family protein-tyrosine kinases. Current Biology 6:981–988
243. Stanglmaier M, Warmuth M, Kleinlein I, Reis S, Hallek M (2003) The interaction of the Bcr-Abl tyrosine kinase with the Src kinase Hck is mediated by multiple binding domains. Leukemia 17:283–289
244. Scott MP, Zappacosta F, Kim EY, Annan RS, Miller WT (2002) Identification of novel SH3 domain ligands for the Src family kinase Hck. Wiskott-Aldrich syndrome protein (WASP), WASP-interacting protein (WIP), and ELMO1. Journal of Biological Chemistry 277: 28238–28246
245. Briggs SD, Bryant SS, Jove R, Sanderson SD, Smithgall TE (1995) The Ras GTPase-activating protein (GAP) is an SH3 domain-binding protein and substrate for the Src-related tyrosine kinase, Hck. Journal of Biological Chemistry 270:14718–14724
246. Shivakrupa R, Radha V, Sudhakar Ch, Swarup G (2003) Physical and functional interaction between Hck tyrosine kinase and guanine nucleotide exchange factor C3G results in apoptosis, which is independent of C3G catalytic domain. Journal of Biological Chemistry 278:52188–52194
247. Ward AC, Monkhouse JL, Csar XF, Touw IP, Bello PA (1998) The Src-like tyrosine kinase Hck is activated by granulocyte colony-stimulating factor (G-CSF) and docks to the activated G-CSF receptor. Biochemical and Biophysical Research Communications 251:117–123
248. Poghosyan Z, Robbins SM, Houslay MD, Webster A, Murphy G, Edwards DR (2002) Phosphorylation-dependent interactions between ADAM15 cytoplasmic domain and Src family protein-tyrosine kinases. Journal of Biological Chemistry 277:4999-5007
249. Hao S, August A (2002) The proline rich region of the Tec homology domain of ITK regulates its activity. FEBS Letters 525:53–58
250. Bunnell SC, Henry PA, Kolluri R, Kirchhausen T, Rickles RJ, Berg LJ (1996) Identification of Itk/Tsk Src homology 3 domain ligands. Journal of Biological Chemistry 271:25646–25656

251. BioGRID: General Repository for Interaction Datasets; database of physical and genetic interactions for model organisms (www.thebiogrid.org)
252. Bléry M, Kubagawa H, Chen CC, Vély F, Cooper MD, Vivier E (1998) The paired Ig-like receptor PIR-B is an inhibitory receptor that recruits the protein-tyrosine phosphatase SHP-1. Proceedings of the National Academy of Sciences of the United States of America 95: 2446–2451
253. Belsches AP, Haskell MD, Parsons SJ (1997) Role of c-Src tyrosine kinase in EGF-induced mitogenesis. Frontiers in Bioscience 2:d501–d518
254. Amanchy R, Zhong J, Hong R, Kim JH, Gucek M, Cole RN, Molina H, Pandey A (2009) Identification of c-Src tyrosine kinase substrates in platelet-derived growth factor receptor signaling. Molecular Oncology 3:439–450
255. Goldman TL, Du Y, Buttrick PM, Walker LA (2009) Src kinase expression, phosphorylation and activation in human and bovine left ventricles. FASEB Journal 23:524.7
256. Yang WC, Ghiotto M, Castellano R, Collette Y, Auphan N, Nunès JA, Olive D (2000) Role of Tec kinase in nuclear factor of activated T cells signaling. International Immunology 12: 1547–1552
257. Kane LP, Watkins SC (2005) Dynamic regulation of Tec kinase localization in membrane-proximal vesicles of a T cell clone revealed by total internal reflection fluorescence and confocal microscopy. Journal of Biological Chemistry 280:21949–21954
258. Felices M, Falk M, Kosaka Y, Berg LJ (2007) Tec kinases in T cell and mast cell signaling. Advances in Immunology 93:145–184
259. Smith CI, Islam TC, Mattsson PT, Mohamed AJ, Nore BF, Vihinen M (2001) The Tec family of cytoplasmic tyrosine kinases: mammalian Btk, Bmx, Itk, Tec, Txk and homologs in other species. Bioessays 23:436–446
260. Yang WC, Collette Y, Nunès JA, Olive D (2000) Tec kinases: a family with multiple roles in immunity. Immunity 12:373–382
261. Rajagopal K, Sommers CL, Decker DC, Mitchell EO, Korthauer U, Sperling AI, Kozak CA, Love PE, Bluestone JA (1999) RIBP, a novel Rlk/Txk- and itk-binding adaptor protein that regulates T cell activation. Journal of Experimental Medicine 190:1657–1668
262. Maruyama T, Nara K, Yoshikawa H, Suzuki N (2006) Txk, a member of the non-receptor tyrosine kinase of the Tec family, forms a complex with poly(ADP-ribose) polymerase 1 and elongation factor 1α and regulates interferon-γ gene transcription in Th1 cells. Clinical and Experimental Immunology 147:164–175
263. Chen YH, Lu Q, Goodenough DA, Jeansonne B (2002) Nonreceptor tyrosine kinase c-Yes interacts with occludin during tight junction formation in canine kidney epithelial cells. Molecular Biology of the Cell 13:1227–1237
264. Mócsai A, Ruland J, Tybulewicz VLJ (2010) The SYK tyrosine kinase: a crucial player in diverse biological functions. Nature Reviews – Immunology 10:387–402
265. Geahlen RL (2007) Syk. UCSD-Nature Molecule Pages, UCSD-Nature Signaling Gateway (www.signaling-gateway.org)
266. Tohyama Y, Yamamura H (2009) Protein tyrosine kinase, Syk: a key player in phagocytic cells. Journal of Biochemistry 145:267–273

Chap. 5. Cytosolic Protein Serine/Threonine Kinases

267. Kyriakis JM, Avruch J (2001) Mammalian mitogen-activated protein kinase signal transduction pathways activated by stress and inflammation. Physiological Reviews 81:807–869
268. Chaar Z, O'Reilly P, Gelman I, Sabourin LA (2006) v-Src-dependent down-regulation of the Ste20-like kinase SLK by casein kinase II. Journal of Biological Chemistry 281:28193–28199
269. Sabourin LA, Tamai K, Seale P, Wagner J, Rudnicki MA (2000) Caspase 3 cleavage of the Ste20-related kinase SLK releases and activates an apoptosis-inducing kinase domain and an actin-disassembling region. Molecular and Cellular Biology 20:684–696

270. Oehrl W, Kardinal C, Ruf S, Adermann K, Groffen J, Feng GS, Blenis J, Tan TH, Feller SM (1998) The germinal center kinase (GCK)-related protein kinases HPK1 and KHS are candidates for highly selective signal transducers of Crk family adapter proteins. Oncogene 17:1893–1901
271. Kyriakis JM (1999) Signaling by the germinal center kinase family of protein kinases. Journal of Biological Chemistry 274:5259–5262
272. Chen YR, Tan TH (1999) Mammalian c-Jun N-terminal kinase pathway and STE20-related kinases. Gene Therapy and Molecular Biology 4:83–98
273. Kiefer F, Arnold R (2010) Hpk1. UCSD-Nature Molecule Pages, UCSD-Nature Signaling Gateway (www.signaling-gateway.org)
274. Ensenat D, Yao Z, Wang XS, Kori R, Zhou G, Lee SC, Tan TH (1999) A novel src homology 3 domain-containing adaptor protein, HIP-55, that interacts with hematopoietic progenitor kinase 1. Journal of Biological Chemistry 274:33945–33950
275. Han J, Kori R, Shui JW, Chen YR, Yao Z, Tan TH (2003) The SH3 domain-containing adaptor HIP-55 mediates c-Jun N-terminal kinase activation in T cell receptor signaling. Journal of Biological Chemistry 278:52195–52202
276. Liu SK, Fang N, Koretzky GA, McGlade CJ (1999) The hematopoietic-specific adaptor protein gads functions in T-cell signaling via interactions with the SLP-76 and LAT adaptors. Current Biology 9:67–75
277. Zhou G, Boomer JS, Tan TH (2004) Protein phosphatase 4 is a positive regulator of hematopoietic progenitor kinase 1. Journal of Biological Chemistry 279:49551–49561
278. Findlay GM, Yan L, Procter J, Mieulet V, Lamb RF (2007) A MAP4 kinase related to Ste20 is a nutrient-sensitive regulator of mTOR signalling. Biochemical Journal 403:13–20
279. Chernoff J (2008) Mst1; Mst2. UCSD-Nature Molecule Pages, UCSD-Nature Signaling Gateway (www.signaling-gateway.org)
280. Braun H, Suske G (1998) Combinatorial action of HNF3 and Sp family transcription factors in the activation of the rabbit uteroglobin/CC10 promoter. Journal of Biological Chemistry 273:9821–9828
281. Glantschnig H, Rodan GA, Reszka AA (2002) Mapping of MST1 kinase sites of phosphorylation. Activation and autophosphorylation. Journal of Biological Chemistry 277:42987–42996
282. Nakano K, Kanai-Azuma M, Kanai Y, Moriyama K, Yazaki K, Hayashi Y, Kitamura N (2003) Cofilin phosphorylation and actin polymerization by NRK/NESK, a member of the germinal center kinase family. Experimental Cell Research 287:219–227
283. Fu CA, Shen M, Huang BC, Lasaga J, Payan DG, Luo Y (1999) TNIK, a novel member of the germinal center kinase family that activates the c-Jun N-terminal kinase pathway and regulates the cytoskeleton. Journal of Biological Chemistry 274:30729–30737
284. Kuramochi S, Moriguchi T, Kuida K, Endo J, Semba K, Nishida E, Karasuyama H (1997) LOK is a novel mouse STE20-like protein kinase that is expressed predominantly in lymphocytes. Journal of Biological Chemistry 272:22679–22684
285. Walter SA, Cutler RE Jr, Martinez R, Gishizky M, Hill RJ (2003) Stk10, a new member of the polo-like kinase kinase family highly expressed in hematopoietic tissue. Journal of Biological Chemistry 278:18221–18228
286. Choe KP, Strange K (2010) OXSR1. UCSD-Nature Molecule Pages, UCSD-Nature Signaling Gateway (www.signaling-gateway.org)
287. Lee SJ, Cobb MH, Goldsmith EJ (2009) Crystal structure of domain-swapped STE20 OSR1 kinase domain. Protein Science 18:304–313
288. Mudumana SP, Hentschel D, Liu Y, Vasilyev A, Drummond IA (2008) Odd skipped related 1 reveals a novel role for endoderm in regulating kidney versus vascular cell fate. Development 135:3355–3367
289. Strange K, Denton J, Nehrke K (2006) Ste20-type kinases: evolutionarily conserved regulators of ion transport and cell volume. Physiology 21:61–68
290. Russell JM (2000) Sodium–potassium–chloride cotransport. Physiological Reviews 80:211–276

291. Anselmo AN, Earnest S, Chen W, Juang YC, Kim SC, Zhao Y, Cobb MH (2006) WNK1 and OSR1 regulate the Na^+, K^+, $2Cl^-$ cotransporter in HeLa cells. Proceedings of the National Academy of Sciences of the United States of America 103:10883–10888
292. Ko B, Hoover RS (2009) Molecular physiology of the thiazide-sensitive sodium-chloride cotransporter. Current Opinion in Nephrology and Hypertension 18:421–427
293. Komaba S, Inoue A, Maruta S, Hosoya H, Ikebe M (2003) Determination of human myosin III as a motor protein having a protein kinase activity. Journal of Biological Chemistry 278:21352–21360
294. Dosé AC, Burnside B (2002) A class III myosin expressed in the retina is a potential candidate for Bardet-Biedl syndrome. Genomics 79:621–624
295. Hutchison M, Berman KS, Cobb MH (1998) Isolation of TAO1, a protein kinase that activates MEKs in stress-activated protein kinase cascades. Journal of Biological Chemistry 273:28625–28632
296. Zihni C, Mitsopoulos C, Tavares IA, Ridley AJ, Morris JD (2006) Prostate-derived sterile 20-like kinase 2 (PSK2) regulates apoptotic morphology via C-Jun N-terminal kinase and Rho kinase-1. Journal of Biological Chemistry 281:7317–7323
297. Chen Z, Cobb MH (2001) Regulation of stress-responsive mitogen-activated protein (MAP) kinase pathways by TAO2. Journal of Biological Chemistry 276:16070–16075
298. Tassi E, Biesova Z, Di Fiore PP, Gutkind JS, Wong WT (1999) Human JIK, a novel member of the STE20 kinase family that inhibits JNK and is negatively regulated by epidermal growth factor. Journal of Biological Chemistry 274:33287–33295
299. Hergovich A, Stegert MR, Schmitz D, Hemmings BA (2006) NDR kinases regulate essential cell processes from yeast to humans. Nature Reviews – Molecular Cell Biology 7:253–264
300. Pearce LR, Komander D, Alessi DR (2010) The nuts and bolts of AGC protein kinases. Nature Reviews – Molecular Cell Biology 11:9–22
301. Okuzumi T, Fiedler D, Zhang C, Gray DC, Aizenstein B, Hoffman R, Shokat KM (2009) Inhibitor hijacking of Akt activation. Nature – Chemical Biology 5:484–493
302. Cameron AJM, Escribano C, Saurin AT, Kostelecky B, Parker PJ (2009) PKC maturation is promoted by nucleotide pocket occupation independently of intrinsic kinase activity. Nature – Structural and Molecular Biology 16:624–630
303. Kohler RS, Schmitz D, Cornils H, Hemmings BA, Hergovich A (2010) Differential NDR/LATS interactions with the human MOB family reveal a negative role for human MOB2 in the regulation of human NDR kinases. Molecular and Cellular Biology 30:4507–4520
304. Stokoe D (2007) Pdk1. UCSD-Nature Molecule Pages, UCSD-Nature Signaling Gateway (www.signaling-gateway.org)
305. Masters TA, Calleja V, Armoogum DA, Marsh RJ, Applebee CJ, Laguerre M, Bain AJ, Larijani B (2010) Regulation of 3-phosphoinositide-dependent protein kinase 1 activity by homodimerization in live cells. Science Signaling 3:ra78
306. Ito K, Akazawa H, Tamagawa M, Furukawa K, Ogawa W, Yasuda N, Kudo Y, Liao CH, Yamamoto R, Sato T, Molkentin JD, Kasuga M, Noda T, Nakaya H, Komuro I (2009) PDK1 coordinates survival pathways and β-adrenergic response in the heart. Proceedings of the National Academy of Sciences of the United States of America 106:8689–8694
307. Lignitto L, Carlucci A, Sepe M, Stefan E, Cuomo O, Nisticò R, Scorziello A, Savoia C, Garbi C, Annunziato L, Feliciello A (2011) Control of PKA stability and signalling by the RING ligase praja2. Nature – Cell Biology 13:412–422
308. Masterson LR, Mascioni A, Traaseth NJ, Taylor SS, Veglia G (2008) Allosteric cooperativity in protein kinase A. Proceedings of the National Academy of Sciences of the United States of America 105:506–511
309. Wu J, Brown SHJ, von Daake S, Taylor SS (2007) PKA type IIα holoenzyme reveals a combinatorial strategy for isoform diversity. Science 318:274–279
310. Goel M, Zuo CD, Schilling WP (2010) Role of cAMP/PKA signaling cascade in vasopressin-induced trafficking of TRPC3 channels in principal cells of the collecting duct. American Journal of Physiology – Renal Physiology 298:F988–F996

311. Butterworth MB, Frizzell RA, Johnson JP, Peters KW, Edinger RS (2005) PKA-dependent ENaC trafficking requires the SNARE-binding protein complexin. American Journal of Physiology – Renal Physiology 289:F969–F977
312. Nejsum LN, Zelenina M, Aperia A, Frøkiaer J, Nielsen S (2005) Bidirectional regulation of AQP2 trafficking and recycling: involvement of AQP2-S256 phosphorylation. American Journal of Physiology – Renal Physiology 288:F930–F938
313. Faul C, Dhume A, Schecter AD, Mundel P (2007) Protein kinase A, Ca^{2+}/calmodulin-dependent kinase II, and calcineurin regulate the intracellular trafficking of myopodin between the Z-disc and the nucleus of cardiac myocytes. Molecular and Cellular Biology 27:8215–8227
314. Hallaq H, Yang Z, Viswanathan PC, Fukuda K, Shen W, Wang DW, Wells KS, Zhou J, Yi J, Murray KT (2006) Quantitation of protein kinase A-mediated trafficking of cardiac sodium channels in living cells. Cardiovascular Research 72:250–261
315. Lin L, Sun W, Wikenheiser AM, Kung F, Hoffman DA (2010) KChIP4a regulates Kv4.2 channel trafficking through PKA phosphorylation. Molecular and Cellular Neurosciences 43:315–325
316. Esteban JA, Shi SH, Wilson C, Nuriya M, Huganir RL, Malinow R (2003) PKA phosphorylation of AMPA receptor subunits controls synaptic trafficking underlying plasticity. Nature – Neuroscience 6:136–143
317. Parvathenani LK, Buescher ES, Chacon-Cruz E, Beebe SJ (1998) Type I cAMP-dependent protein kinase delays apoptosis in human neutrophils at a site upstream of caspase-3. Journal of Biological Chemistry 273:6736–6743
318. Prinz A, Herberg FW (2009) Prkx. UCSD-Nature Molecule Pages, UCSD-Nature Signaling Gateway (www.signaling-gateway.org)
319. Semizarov D, Glesne D, Laouar A, Schiebel K, Huberman E (1998) A lineage-specific protein kinase crucial for myeloid maturation. Proceedings of the National Academy of Sciences of the United States of America 95:15412-15417
320. Lasserre R, Guo XJ, Conchonaud F, Hamon Y, Hawchar O, Bernard AM, Soudja SM, Lenne PF, Rigneault H, Olive D, Bismuth G, Nunès JA, Payrastre B, Marguet D, He HT (2008) Raft nanodomains contribute to Akt/PKB plasma membrane recruitment and activation. Nature – Chemical Biology 4:538–547
321. Gonzalez E, McGraw TE (2009) Insulin-modulated Akt subcellular localization determines Akt isoform-specific signaling. Proceedings of the National Academy of Sciences of the United States of America 106:7004–7009
322. Mîinea CP, Sano H, Kane S, Sano E, Fukuda M, Peränen J, Lane WS, Lienhard GE (2005) AS160, the Akt substrate regulating GLUT4 translocation, has a functional Rab GTPase-activating protein domain. Biochemical Journal 391:87–93
323. Xie X, Zhang D, Zhao B, Lu MK, You M, Condorelli G, Wang CY, Guan KL (2011) IκB kinase ε and TANK-binding kinase 1 activate AKT by direct phosphorylation. Proceedings of the National Academy of Sciences of the United States of America 108:6474–6479
324. Miyamoto S, Rubio M, Sussman mA (2009) Nuclear and mitochondrial signalling Akts in cardiomyocytes. Cardiovascular Research 82:272–285
325. Kim HE, Du F, Fang M, Wang X (2005) Formation of apoptosome is initiated by cytochrome c-induced dATP hydrolysis and subsequent nucleotide exchange on Apaf-1. Proceedings of the National Academy of Sciences of the United States of America 102:17545–17550
326. Yoeli-Lerner M, Yiu GK, Rabinovitz I, Erhardt P, Jauliac S, Toker A (2005) Akt blocks breast cancer cell motility and invasion through the transcription factor NFAT. Molecular Cell 20:539–550
327. Irie HY, Pearline RV, Grueneberg D, Hsia M, Ravichandran P, Kothari N, Natesan S, Brugge JS (2005) Distinct roles of Akt1 and Akt2 in regulating cell migration and epithelial-mesenchymal transition. Journal of Cell Biology 171:1023–1034
328. Gottlob K, Majewski N, Kennedy S, Kandel E, Robey RB, Hay N (2001) Inhibition of early apoptotic events by Akt/PKB is dependent on the first committed step of glycolysis and mitochondrial hexokinase. Genes and Development 15:1406–1418

329. Edinger AL, Thompson CB (2002) Akt maintains cell size and survival by increasing mTOR-dependent nutrient uptake. Molecular Biology of the Cell 13:2276–2288
330. Di Lorenzo A, Fernández-Hernando C, Cirino G, Sessa WC (2009) Akt1 is critical for acute inflammation and histamine-mediated vascular leakage. Proceedings of the National Academy of Sciences of the United States of America 106:14552–14557
331. Ezell SA, Polytarchou C, Hatziapostolou M, Guo A, Sanidas I, Bihani T, Comb MJ, Sourvinos G, Tsichlis PN (2012) The protein kinase Akt1 regulates the interferon response through phosphorylation of the transcriptional repressor EMSY. Proceedings of the National Academy of Sciences of the United States of America 109:E613–E621
332. Rosse C, Linch M, Kermorgant S, Cameron AJM, Boeckeler K, Parker PJ (2010) PKC and the control of localized signal dynamics. Nature Reviews – Molecular Cell Biology 11:103–112
333. Palaniyandi SS, Sun L, Ferreira JCB, Mochly-Rosen D (2009) Protein kinase C in heart failure: a therapeutic target? Cardiovascular Research 82:229–239
334. Suzuki T, Elias BC, Seth A, Shen L, Turner JR, Giorgianni F, Desiderio D, Guntaka R, Rao R (2009) PKCη regulates occludin phosphorylation and epithelial tight junction integrity. Proceedings of the National Academy of Sciences of the United States of America 106: 61–66
335. Chapline C, Cottom J, Tobin H, Hulmes J, Crabb J, Jaken S (1998) A major, transformation-sensitive PKC-binding protein is also a PKC substrate involved in cytoskeletal remodeling. Journal of Biological Chemistry 273:19482–19489
336. Mochly-Rosen D, Gordon AS (1998) Anchoring proteins for protein kinase C: a means for isozyme selectivity. FASEB Journal 12:35–42
337. Staudinger J, Lu J, Olson EN (1997) Specific interaction of the PDZ domain protein PICK1 with the COOH terminus of protein kinase C-α. Journal of Biological Chemistry 272: 32019–32024
338. Izumi Y, Hirose T, Tamai Y, Hirai S, Nagashima Y, Fujimoto T, Tabuse Y, Kemphues KJ, Ohno S (1998) An atypical PKC directly associates and colocalizes at the epithelial tight junction with ASIP, a mammalian homologue of Caenorhabditis elegans polarity protein PAR-3. Journal of Cell Biology 143:95–106
339. Joberty G, Petersen C, Gao L, Macara IG (2000) The cell-polarity protein Par6 links Par3 and atypical protein kinase C to Cdc42. Nature – Cell Biology 2:531–539
340. Wooten MW, Geetha T, Seibenhener ML, Babu JR, Diaz-Meco MT, Moscat J (2005) The p62 scaffold regulates nerve growth factor-induced NF-kappaB activation by influencing TRAF6 polyubiquitination. Journal of Biological Chemistry 280:35625–35629
341. Konopatskaya O, Gilio K, Harper MT, Zhao Y, Cosemans JM, Karim ZA, Whiteheart SW, Molkentin JD, Verkade P, Watson SP, Heemskerk JW, Poole AW (2009) PKCα regulates platelet granule secretion and thrombus formation in mice. Journal of Clinical Investigation 119:399–407
342. Rajagopal S, Fang H, Oronce CI, Jhaveri S, Taneja S, Dehlin EM, Snyder SL, Sando JJ, Kamatchi GL (2009) Site-specific regulation of Ca(V)2.2 channels by protein kinase C isozymes βII and ε. Neuroscience 159:618–628
343. Rajagopal S, Fang H, Patanavanich S, Sando JJ, Kamatchi GL (2008) Protein kinase C isozyme-specific potentiation of expressed Ca(V)2.3 currents by acetyl-beta-methylcholine and phorbol-12-myristate, 13-acetate. Brain Research 1210:1–10
344. Steinberg SF (2005) Protein kinase Cδ. UCSD-Nature Molecule Pages, UCSD-Nature Signaling Gateway (www.signaling-gateway.org)
345. Robles-Flores M, Rendon-Huerta E, Gonzalez-Aguilar H, Mendoza-Hernandez G, Islas S, Mendoza V, Ponce-Castaneda MV, Gonzalez-Mariscal L, Lopez-Casillas F (2002) p32 (gC1qBP) is a general protein kinase C (PKC)-binding protein; interaction and cellular localization of P32-PKC complexes in ray hepatocytes. Journal of Biological Chemistry 277:5247–5255
346. Roybal KT, Wülfing C (2010) Inhibiting the inhibitor of the inhibitor: blocking PKC-θ to enhance regulatory T cell function. Science Signaling 3:pe24
347. Moscat J, Diaz-Meco MT (2005) Protein kinase Cζ. UCSD-Nature Molecule Pages, UCSD-Nature Signaling Gateway (www.signaling-gateway.org)

348. Hodgkinson CP, Mander A, Sale GJ (2005) Protein kinase-ζ interacts with munc18c: role in GLUT4 trafficking. Diabetologia 48:1627–1636
349. Westmark PR, Westmark CJ, Wang S, Levenson J, O'Riordan KJ, Burger C, Malter JS (2010) Pin1 and PKMζ sequentially control dendritic protein synthesis. Science Signaling 3:ra18
350. Chang S, Kim JH, Shin J (2002) p62 forms a ternary complex with PKCζ and PAR-4 and antagonizes PAR-4-induced PKCζ inhibition. FEBS Letters 510:57–61
351. Seibenhener ML, Babu JR, Geetha T, Wong HC, Krishna NR, Wooten MW (2004) Sequestosome 1/p62 is a polyubiquitin chain binding protein involved in ubiquitin proteasome degradation. Molecular and Cellular Biology 24: 8055–8068
352. Samuels IS, Seibenhener ML, Neidigh KB, Wooten MW (2001) Nerve growth factor stimulates the interaction of ZIP/p62 with atypical protein kinase C and targets endosomal localization: evidence for regulation of nerve growth factor-induced differentiation. Journal of Cellular Biochemistry 82:452–466
353. Kwan HY, Huang Y, Yao X (2000) Store-operated calcium entry in vascular endothelial cells is inhibited by cGMP via a protein kinase G-dependent mechanism. Journal of Biological Chemistry 275:6758–6763
354. Schwappacher R, Weiske J, Heining E, Ezerski V, Marom B, Henis YI, Huber O, Knaus P (2009) Novel crosstalk to BMP signalling: cGMP-dependent kinase I modulates BMP receptor and SMAD activity. EMBO Journal 28:1537–1550
355. Zheng H, Worrall C, Shen H, Issad T, Seregard S, Girnita A, Girnita L (2012) Selective recruitment of G protein-coupled receptor kinases (GRKs) controls signaling of the insulin-like growth factor 1 receptor. Proceedings of the National Academy of Sciences of the United States of America 109:7055–7060
356. Burgess A, Vigneron S, Brioudes E, Labbé JC, Lorca T, Castro A (2010) Loss of human Greatwall results in G2 arrest and multiple mitotic defects due to deregulation of the cyclin B-Cdc2/PP2A balance. Proceedings of the National Academy of Sciences of the United States of America 107:12564-12569
357. BokochGM (2008) Pak1 UCSD-Nature Molecule Pages, UCSD-Nature Signaling Gateway (www.signaling-gateway.org)
358. Koh CG, Tan EJ, Manser E, Lim L (2002) The p21-activated kinase PAK is negatively regulated by POPX1 and POPX2, a pair of serine/threonine phosphatases of the PP2C family. Current Biology 12:317–321
359. Ching YP, Leong VY, Wong CM, Kung HF (2003) Identification of an autoinhibitory domain of p21-activated protein kinase 5. Journal of Biological Chemistry 278:33621–33624
360. Nekrasova T, Jobes ML, Ting JH, Wagner GC, Minden A (2008) Targeted disruption of the Pak5 and Pak6 genes in mice leads to deficits in learning and locomotion. Developmental Biology 322:95–108
361. Baird D, Feng Q, Cerione RA (2006) Biochemical characterization of the Cool (Cloned-out-of-Library)/Pix (Pak-interactive exchange factor) proteins. Methods in Enzymology 406: 58–69
362. Baldwin A, Grueneberg DA, Hellner K, Sawyer J, Grace M, Li W, Harlow E, Munger K (2010) Kinase requirements in human cells. V. Synthetic lethal interactions between p53 and the protein kinases SGK2 and PAK3. Proceedings of the National Academy of Sciences of the United States of America 107:12463–12468
363. Chernoff J (2007) Pak5. UCSD-Nature Molecule Pages, UCSD-Nature Signaling Gateway (www.signaling-gateway.org)
364. Riento K, Ridley AJ (2003) ROCKs: multifunctional kinases in cell behaviour. Nature Reviews – Molecular Cell Biology 4:446-456
365. Iftinca M, Hamid J, Chen L, Varela D, Tadayonnejad R, Altier C, Turner RW, Zampon GW (2007) Regulation of T-type calcium channels by Rho-associated kinase. Nature – Neuroscience 10:854–860
366. Luykenaar KD, El-Rahman RA, Walsh MP, Welsh DG (2009) Rho-kinase-mediated suppression of KDR current in cerebral arteries requires an intact actin cytoskeleton. American Journal of Physiology – Heart and Circulatory Physiology 296:H917–H926

367. Storey NM, O'Bryan JP, Armstrong DL (2002) Rac and Rho mediate opposing hormonal regulation of the ether-a-go-go-related potassium channel. Current Biology 12:27–33
368. Jones SVP (2003) Role of the small GTPase Rho in modulation of the inwardly rectifying potassium channel Kir2.1. Molecular Pharmacology 64:987–993
369. Luykenaar KD, Welsh DG (2007) Activators of the PKA and PKG pathways attenuate RhoA-mediated suppression of the KDR current in cerebral arteries. American Journal of Physiology – Heart and Circulatory Physiology 292:H2654–H2663
370. Yatani A, Irie K, Otani T, Abdellatif M, Wei L (2005) RhoA GTPase regulates L-type Ca^{2+} currents in cardiac myocytes. American Journal of Physiology – Heart and Circulatory Physiology 288:H650–H659
371. Staruschenko A, Nichols A, Medina JL, Camacho P, Zheleznova NN, Stockand JD (2004) Rho small GTPases activate the epithelial Na^+ channel. Journal of Biological Chemistry 279:49989–49994
372. Nilius B, Voets T, Prenen J, Barth H, Aktories K, Kaibuchi K, Droogmans G, Eggermont J (1999) Role of Rho and Rho kinase in the activation of volume-regulated anion channels in bovine endothelial cells. Journal of Physiology 516:67–74
373. van Nieuw Amerongen GP, van Hinsbergh VWM (2009) Role of ROCK I/II in vascular branching. American Journal of Physiology – Heart and Circulatory Physiology 296: H903–H905
374. Kroll J, Epting D, Kern K, Dietz CT, Feng Y, Hammes HP, Wieland T, Augustin HG (2009) Inhibition of Rho-dependent kinases ROCK I/II activates VEGF-driven retinal neovascularization and sprouting angiogenesis. Journal of Physiology – Heart and Circulatory Physiology 296:H893–H899
375. Fischer RS, Gardel M, Ma X, Adelstein RS, Waterman CM (2009) Local cortical tension by myosin II guides 3D endothelial cell branching. Current Biology 19:260–265
376. Olson MF (2006) Rock1. UCSD-Nature Molecule Pages, UCSD-Nature Signaling Gateway (www.signaling-gateway.org)
377. Olson MF (2007) Rock2. UCSD-Nature Molecule Pages, UCSD-Nature Signaling Gateway (www.signaling-gateway.org)
378. The human kinome, Science's signal transduction knowledge environment (STKE) www.cellsignal.com
379. Hauge C, Frödin M (2006) RSK and MSK in MAP kinase signalling. Journal of Cell Science 119:3021–3023
380. Cargnello M, Roux PP (2011) Activation and function of the MAPKs and their substrates, the MAPK-activated protein kinases. Microbiology and Molecular Biology Reviews 75:50-83
381. Hayashi S, Okada T, Igarashi N, Fujita T, Jahangeer S, Nakamura S (2002) Identification and characterization of RPK118, a novel sphingosine kinase-1-binding protein. Journal of Biological Chemistry 277:33319–33324
382. Liu L, Yang C, Yuan J, Chen X, Xu J, Wei Y, Yang J, Lin G, Yu L (2005) RPK118, a PX domain-containing protein, interacts with peroxiredoxin-3 through pseudo-kinase domains. Molecules and Cells 19:39–45
383. Anjum R, Blenis J (2008) The RSK family of kinases: emerging roles in cellular signalling. Nature Reviews – Molecular Cell Biology 9:747–758
384. Roux PP (2007) Rsk1; Rsk4. UCSD-Nature Molecule Pages, UCSD-Nature Signaling Gateway (www.signaling-gateway.org)
385. Julien LA, Roux PP (2007) Rsk3. UCSD-Nature Molecule Pages, UCSD-Nature Signaling Gateway (www.signaling-gateway.org)
386. Dennis PB, Thomas G (2008) S6K1; S6K2. UCSD-Nature Molecule Pages, UCSD-Nature Signaling Gateway (www.signaling-gateway.org)
387. Shin S, Wolgamott L, Yu Y, Blenis J, Yoon SO (2011) Glycogen synthase kinase (GSK)-3 promotes p70 ribosomal protein S6 kinase (p70S6K) activity and cell proliferation. Proceedings of the National Academy of Sciences of the United States of America 108:E1204–E1213
388. Marklund U, Lightfoot K, Cantrell D (2003) Intracellular location and cell context-dependent function of protein kinase D. Immunity 19:491–501

389. Canagarajah B, Leskow FC, Ho JY, Mischak H, Saidi LF, Kazanietz MG, Hurley JH (2004) Structural mechanism for lipid activation of the Rac-specific GAP, 2-chimaerin. Cell 119:407–418
390. Raval AP, Dave KR, Prado R, Katz LM, Busto R, Sick TJ, Ginsberg MD, Mochly-Rosen D, Perez-Pinzon MA (2005) Protein kinase C delta cleavage initiates an aberrant signal transduction pathway after cardiac arrest and oxygen glucose deprivation. Journal of Cerebral Blood Flow and Metabolism 25:730–741
391. Kermorgant S, Zicha D, Parker PJ (2004) PKC controls HGF-dependent c-Met traffic, signalling and cell migration. EMBO Journal 23:3721–3734
392. Bredt DS, Ferris CD, Snyder SH (1992) Nitric oxide synthase regulatory sites. Phosphorylation by cyclic AMP-dependent protein kinase, protein kinase C, and calcium/calmodulin protein kinase; identification of flavin and calmodulin binding sites. Journal of Biological Chemistry 267:10976–10981
393. Hsu LS, Chen GD, Lee LS, Chi CW, Cheng JF, Chen JY (2001) Human Ca^{2+}/calmodulin-dependent protein kinase kinase β gene encodes multiple isoforms that display distinct kinase activity. Journal of Biological Chemistry 276:31113–31123
394. Condon JC, Pezzi V, Drummond BM, Yin S, Rainey WE (2002) Calmodulin-dependent kinase I regulates adrenal cell expression of aldosterone synthase. Endocrinology 143:3651–3657
395. Matsushita M, Nairn AC (1999) Inhibition of the Ca^{2+}/calmodulin-dependent protein kinase I cascade by cAMP-dependent protein kinase. Journal of Biological Chemistry 274:10086–10093
396. Backs J, Backs T, Neef S, Kreusser MM, Lehmann LH, Patrick DM, Grueter CE, Qi X, Richardson JA, Hill JA, Katus HA, Bassel-Duby R, Maier LS, Olson EN (2009) The δ isoform of CaM kinase II is required for pathological cardiac hypertrophy and remodeling after pressure overload. Proceedings of the National Academy of Sciences of the United States of America 106:2342–2347
397. Komukai K, O-Uchi J, Morimoto S, Kawai M, Hongo K, Yoshimura M, Kurihara S (2010) Role of Ca^{2+}/calmodulin-dependent protein kinase II in the regulation of the cardiac L-type Ca^{2+} current during endothelin-1 stimulation. American Journal of Physiology – Heart and Circulatory Physiology 298:H1902–H1907
398. Anderson KA, Noeldner PK, Reece K, Wadzinski BE, Means AR (2004) Regulation and function of the calcium/calmodulin-dependent protein kinase IV/protein serine/threonine phosphatase 2A signaling complex. Journal of Biological Chemistry 279:31708–31716
399. Shen Q, Rigor RR, Pivetti CD, Wu MH, Yuan SY (2010) Myosin light chain kinase in microvascular endothelial barrier function. Cardiovascular Research 87:272–280
400. Poperechnaya A, Varlamova O, Lin PJ, Stull JT, Bresnick AR (2000) Localization and activity of myosin light chain kinase isoforms during the cell cycle. Journal of Cell Biology 151:697–708'
401. Birukov KG, Csortos C, Marzilli L, Dudek S, Ma SF, Bresnick AR, Verin AD, Cotter RJ, Garcia JG (2001) Differential regulation of alternatively spliced endothelial cell myosin light chain kinase isoforms by $p60^{Src}$. Journal of Biological Chemistry 276:8567–8573
402. Zhang WC, Peng YJ, Zhang GS, He WQ, Qiao YN, Dong YY, Gao YQ, Chen C, Zhang CH, Li W, Shen HH, Ning W, Kamm KE, Stull JT, Gao X, Zhu MS (2010) Myosin light chain kinase is necessary for tonic airway smooth muscle contraction. Journal of Biological Chemistry 285:5522–5531
403. van Riper DA, McDaniel NL, Rembold CM (1997) Myosin light chain kinase phosphorylation in nitrovasodilator induced swine carotid artery relaxation. Biochimica et Biophysica Acta 1355:323–330
404. Sauzeau V, Le Jeune H, Cario-Toumaniantz C, Smolenski A, Lohmann SM, Bertoglio J, Chardin P, Pacaud P, Loirand G (2000) Cyclic GMP-dependent protein kinase signaling pathway inhibits RhoA-induced Ca^{2+} sensitization of contraction in vascular smooth muscle. Journal of Biological Chemistry 275:21722–21729

405. Oakhill JS, Steel R, Chen ZP, Scott JW, Ling N, Tam S, Kemp BE (2011) AMPK is a direct adenylate charge-regulated protein kinase. Science 332:1433–1435
406. Hawley SA, Pan DA, Mustard KJ, Ross L, Bain J, Edelman AM, Frenguelli BG, Hardie DG (2005) Calmodulin-dependent protein kinase kinase-β is an alternative upstream kinase for AMP-activated protein kinase. Cell Metabolism 2:9–19
407. Guarente L (2006) Sirtuins as potential targets for metabolic syndrome. Nature 444:868–874
408. Zheng B, Cantley LC (2007) Regulation of epithelial tight junction assembly and disassembly by AMP-activated protein kinase. Proceedings of the National Academy of Sciences of the United States of America 104:819–822
409. Lee JH, Koh H, Kim M, Kim Y, Lee SY, Karess RE, Lee SH, Shong M, Kim JM, Kim J, Chung J (2007) Energy-dependent regulation of cell structure by AMP-activated protein kinase. Nature 447:1017–1020
410. Ikematsu N, Dallas ML, Ross FA, Lewis RW, Rafferty JN, David JA, Suman R, Peers C, Hardie DG, Evans AM (2011) Phosphorylation of the voltage-gated potassium channel Kv2.1 by AMP-activated protein kinase regulates membrane excitability. Proceedings of the National Academy of Sciences of the United States of America 108:18132–18137
411. Bright NJ, Thornton C, Carling D (2009) The regulation and function of mammalian AMPK-related kinases. Acta Physiologica 196:15–26
412. Vancauwenbergh S, Bollen M (2006) Melk. UCSD-Nature Molecule Pages, UCSD-Nature Signaling Gateway (www.signaling-gateway.org)
413. Drewes G (2006) Mark1; Mark4. UCSD-Nature Molecule Pages, UCSD-Nature Signaling Gateway (www.signaling-gateway.org)
414. Zagrska A, Deak M, Campbell DG, Banerjee S, Hirano M, Aizawa S, Prescott AR, Alessi DR (2010) New roles for the LKB1-NUAK pathway in controlling myosin phosphatase complexes and cell adhesion. Science Signaling 3:ra25
415. Stolz A, Ertych N, Kienitz A, Vogel C, Schneider V, Fritz B, Jacob R, Dittmar G, Weichert W, Petersen I, Bastians H (2010) The CHK–BRCA1 tumour suppressor pathway ensures chromosomal stability in human somatic cells. Nature – Cell Biology 12:492–499
416. Lukas TJ (2006) Dap kinase. UCSD-Nature Molecule Pages, UCSD-Nature Signaling Gateway (www.signaling-gateway.org)
417. de Diego I, Kuper J, Bakalova N, Kursula P, Wilmanns M (2010) Molecular basis of the death-associated protein kinase-calcium/calmodulin regulator complex. Science Signaling 3:ra6
418. Sanjo H, Kawai T, Akira S (1998) DRAKs, novel serine/threonine kinases related to death-associated protein kinase that trigger apoptosis. Journal of Biological Chemistry 273:29066–29071
419. Walsh CM (2007) Drak2. UCSD-Nature Molecule Pages, UCSD-Nature Signaling Gateway (www.signaling-gateway.org)
420. Rozengurt E, Rey O, Waldron RT (2005) Protein kinase D signaling. Journal of Biological Chemistry 280:13205–13208
421. Fielitz J, Kim MS, Shelton JM, Qi X, Hill JA, Richardson JA, Bassel-Duby R, Olson EN (2008) Proceedings of the National Academy of Sciences of the United States of America 105:3059–3063
422. Eiseler T, Döppler H, Yan IK, Kitatani K, Mizuno K, Storz P (2009) Protein kinase D1 regulates cofilin mediated F-actin reorganization and cell motility via slingshot. Nature – Cell Biology 11:545–556
423. Storz P (2006) Protein kinase D2. UCSD-Nature Molecule Pages, UCSD-Nature Signaling Gateway (www.signaling-gateway.org)
424. Storz P (2006) Protein kinase D3. UCSD-Nature Molecule Pages, UCSD-Nature Signaling Gateway (www.signaling-gateway.org)
425. Lee JH, Budanov AV, Park EJ, Birse R, Kim TE, Perkins GA, Ocorr K, Ellisman MH, Bodmer R, Bier E, Karin M (2010) Sestrin as a feedback inhibitor of TOR that prevents age-related pathologies. Science 327:1223–1228
426. Alexander A, Cai SL, Kim J, Nanez A, Sahin M, Maclean KH, Inoki K, Guan KL, Shen J, Person MD, Kusewitt D, Mills GB, Kastan MB, Walker CL (2010) ATM signals to TSC2

in the cytoplasm to regulate mTORC1 in response to ROS. Proceedings of the National Academy of Sciences of the United States of America 107:4153–4158
427. Guo Z, Kozlov S, Lavin MF, Person MD, Paull TT (2010) ATM Activation by oxidative stress. Science 330:517–521
428. Meek K, Dang V, Lees-Miller SP (2008) DNA-PK: the means to justify the ends? Advances in Immunology 99:33–58
429. Goudelock DM, Jiang K, Pereira E, Russell B, Sanchez Y (2003) Regulatory interactions between the checkpoint kinase Chk1 and the proteins of the DNA-dependent protein kinase complex. Journal of Biological Chemistry 278:29940–29947
430. Hsu PP, Kang SA, Rameseder J, Zhang Y, Ottina KA, Lim D, Peterson TR, Choi Y, Gray NS, Yaffe MB, Marto JA, Sabatini DM (2011) The mTOR-regulated phosphoproteome reveals a mechanism of mTORC1-mediated inhibition of growth factor signaling. Science 332:1317–1322
431. Parnell SC, Magenheimer BS, Maser RL, Zien CA, Frischauf AM, Calvet JP (2002) Polycystin-1 activation of c-Jun N-terminal kinase and AP-1 is mediated by heterotrimeric G proteins. Journal of Biological Chemistry 277:19566–19572
432. Boehlke C, Kotsis F, Patel V, Braeg S, Voelker H, Bredt S, Beyer T, Janusch H, Hamann C, Gödel M, Müller K, Herbst M, Hornung M, Doerken M, Köttgen M, Nitschke R, Igarashi P, Walz G, Kuehn EW (2010) Primary cilia regulate mTORC1 activity and cell size through Lkb1. Nature – Cell Biology 12:1115–1122
433. Ramanathan A, Schreiber SL (2009) Direct control of mitochondrial function by mTOR. Proceedings of the National Academy of Sciences of the United States of America 106:22229–22232
434. Yu Y, Yoon SO, Poulogiannis G, Yang Q, Ma XM, Villén J, Kubica N, Hoffman GR, Cantley LC, Gygi SP, Blenis J (2011) Phosphoproteomic analysis identifies Grb10 as an mTORC1 substrate that negatively regulates insulin signaling. Science 332:1322–1326
435. Saci A, Cantley LC, Carpenter CL (2011) Rac1 regulates the activity of mTORC1 and mTORC2 and controls cellular size. Molecular Cell 42:50–61
436. Robitaille AM, Hall MN (2008) mTOR. UCSD-Nature Molecule Pages, UCSD-Nature Signaling Gateway (www.signaling-gateway.org)
437. Dowling RJ, Topisirovic I, Alain T, Bidinosti M, Fonseca BD, Petroulakis E, Wang X, Larsson O, Selvaraj A, Liu Y, Kozma SC, Thomas G, Sonenberg N (2010) mTORC1-Mediated cell proliferation, but not cell growth, controlled by the 4E-BPs. Science 328:1172–1176
438. Li S, Brown MS, Goldstein JL (2010) Bifurcation of insulin signaling pathway in rat liver: mTORC1 required for stimulation of lipogenesis, but not inhibition of gluconeogenesis. Proceedings of the National Academy of Sciences of the United States of America 107:3441–3446
439. Frias MA, Thoreen CC, Jaffe JD, Schroder W, Sculley T, Carr SA, Sabatini DM (2006) mSin1 is necessary for Akt/PKB phosphorylation, and its isoforms define three distinct mTORC2s. Current Biology 16:1865–1870
440. Liu L, Das S, Losert W, Parent CA (2010) mTORC2 regulates neutrophil chemotaxis in a cAMP- and RhoA-dependent fashion. Developmental Cell 19:845–857
441. Xu J, Dang Y, Ren YR, Liu JO (2010) Cholesterol trafficking is required for mTOR activation in endothelial cells. Proceedings of the National Academy of Sciences of the United States of America 107:4764–4769
442. Graves PR, Roach PJ (1995) Role of COOH-terminal phosphorylation in the regulation of casein kinase Iδ. Journal of Biological Chemistry 270:21689–21694
443. Nichols RJ, Traktman P (2004) Characterization of three paralogous members of the Mammalian vaccinia related kinase family. Journal of Biological Chemistry 279:7934–7946
444. Lazo PA, Vega FM, Sevilla A (2005) Vrk1. UCSD-Nature Molecule Pages, UCSD-Nature Signaling Gateway (www.signaling-gateway.org)
445. Blanco S, Lazo PA (2009) Vrk2. UCSD-Nature Molecule Pages, UCSD-Nature Signaling Gateway (www.signaling-gateway.org)

446. Scheeff ED, Eswaran J, Bunkoczi G, Knapp S, Manning G (2009) Structure of the pseudokinase VRK3 reveals a degraded catalytic site, a highly conserved kinase fold, and a putative regulatory binding site. Structure 17:128–138
447. Lazo PA (2009) Vrk3. UCSD-Nature Molecule Pages, UCSD-Nature Signaling Gateway (www.signaling-gateway.org)
448. Vartiainen MK, Sarkkinen EM, Matilainen T, Salminen M, Lappalainen P (2003) Mammals have two twinfilin isoforms whose subcellular localizations and tissue distributions are differentially regulated. Journal of Biological Chemistry 278:34347–34355
449. Kitano-Takahashi M, Morita H, Kondo S, Tomizawa K, Kato R, Tanio M, Shirota Y, Takahashi H, Sugio S, Kohno T (2007) Expression, purification and crystallization of a human tau-tubulin kinase 2 that phosphorylates tau protein. Acta crystallographica. Section F, Structural Biology and Crystallization Communications 63:602–604
450. Sato S, Xu J, Okuyama S, Martinez LB, Walsh SM, Jacobsen MT, Swan RJ, Schlautman JD, Ciborowski P, Ikezu T (2008) Spatial learning impairment, enhanced CDK5/p35 activity, and downregulation of NMDA receptor expression in transgenic mice expressing tau-tubulin kinase 1. Journal of Neuroscience 28:14511–14521
451. Zhang S, Edelmann L, Liu J, Crandall JE, Morabito MA (2008) Cdk5 regulates the phosphorylation of tyrosine 1472 NR2B and the surface expression of NMDA receptors. Journal of Neuroscience 28:415–424
452. Houlden H, Johnson J, Gardner-Thorpe C, Lashley T, Hernandez D, Worth P, Singleton AB, Hilton DA, Holton J, Revesz T, Davis MB, Giunti P, Wood NW (2007) Mutations in TTBK2, encoding a kinase implicated in tau phosphorylation, segregate with spinocerebellar ataxia type 11. Nature – Genetics 39:1434–1436
453. Bouskila M, Esoof N, Gay L, Fang EH, Deak M, Begley MJ, Cantley LC, Prescott A, Storey KG, Alessi DR (2011) TTBK2 kinase substrate specificity and the impact of spinocerebellar-ataxia-causing mutations on expression, activity, localization and development. Biochemical Journal 437:157–167
454. Olsten MEK, Litchfield DW (2006) Casein kinase II α1. UCSD-Nature Molecule Pages, UCSD-Nature Signaling Gateway (www.signaling-gateway.org)
455. Litchfield DW (2005) Casein kinase II α2. UCSD-Nature Molecule Pages, UCSD-Nature Signaling Gateway (www.signaling-gateway.org)
456. Bibby AC, Litchfield DW (2005) Casein kinase II β. UCSD-Nature Molecule Pages, UCSD-Nature Signaling Gateway (www.signaling-gateway.org)
457. Tian B, Yang Q, Mao Z (2009) Phosphorylation of ATM by Cdk5 mediates DNA damage signalling and regulates neuronal death. Nature – Cell Biology 11:211–218
458. Yu DS, Zhao R, Hsu EL, Cayer J, Ye F, Guo Y, Shyr Y, Cortez D (2010) Cyclin-dependent kinase 9–cyclin K functions in the replication stress response. EMBO Reports 11:876–882
459. Li X, Zhang R, Luo D, Park SJ, Wang Q, Kim Y, Min W (2005) Tumor necrosis factor α-induced desumoylation and cytoplasmic translocation of homeodomain-interacting protein kinase 1 are critical for apoptosis signal-regulating kinase 1-JNK/p38 activation. Journal of Biological Chemistry 280:15061–15070
460. Moehlenbrink J, Hofmann TG (2009) Hipk2 UCSD-Nature Molecule Pages, UCSD-Nature Signaling Gateway (www.signaling-gateway.org)
461. Calzado MA, de la Vega L, Möller A, Bowtell DDL, Schmitz ML (2009) An inducible autoregulatory loop between HIPK2 and Siah2 at the apex of the hypoxic response. Nature – Cell Biology 11:85–91
462. Lan HC, Li HJ, Lin G, Lai PY, Chung BC (2007) Cyclic AMP stimulates SF-1-dependent CYP11A1 expression through homeodomain-interacting protein kinase 3-mediated Jun N-terminal kinase and c-Jun phosphorylation. Molecular and Cellular Biology 27:2027–2036
463. Arai S, Matsushita A, Du K, Yagi K, Okazaki Y, Kurokawa R (2007) Novel homeodomain-interacting protein kinase family member, HIPK4, phosphorylates human p53 at serine 9. FEBS Letters 581:5649–5657
464. Boudeau J, Miranda-Saavedra D, Barton GJ, Alessi DR (2006) Emerging roles of pseudokinases. Trends in Cell Biology 16:443–452

465. Wickström SA, Lange A, Montanez E, Fässler R (2010) The ILK/PINCH/parvin complex: the kinase is dead, long live the pseudokinase! EMBO Journal 29:281–291
466. Radeva G, Petrocelli T, Behrend E, Leung-Hagesteijn C, Filmus J, Slingerland J, Dedhar S (1997) Overexpression of the integrin-linked kinase promotes anchorage-independent cell cycle progression. Journal of Biological Chemistry 272:13937–13944
467. Xu Z, Fukuda T, Li Y, Zha X, Qin J, Wu C (2005) Molecular dissection of PINCH-1 reveals a mechanism of coupling and uncoupling of cell shape modulation and survival. Journal of Biological Chemistry 280:27631–27637
468. Tu Y, Li F, Goicoechea S, Wu C (1999) The LIM-only protein PINCH directly interacts with integrin-linked kinase and is recruited to integrin-rich sites in spreading cells. Molecular and Cellular Biology 19:2425–2434
469. Montanez E, Wickström SA, Altstätter J, Chu H, Fässler R (2009) α-Parvin controls vascular mural cell recruitment to vessel wall by regulating RhoA/ROCK signalling. EMBO Journal 28:3132–3144
470. Zhang Y, Chen K, Tu Y, Velyvis A, Yang Y, Qin J, Wu C (2002) Assembly of the PINCH-ILK-CH-ILKBP complex precedes and is essential for localization of each component to cell-matrix adhesion sites. Journal of Cell Science 115:4777–4786
471. Flannery S, Bowie AG (2010) The interleukin-1 receptor-associated kinases: critical regulators of innate immune signalling. Biochemical Pharmacology 80:1981–1991
472. Suzuki N, Suzuki S, Saito T (2005) IRAKs: key regulatory kinases of innate immunity. Current Medicinal Chemistry. Anti-Inflammatory and Anti-Allergy Agents 4:13–20
473. Brissoni B, Agostini L, Kropf M, Martinon F, Swoboda V, Lippens S, Everett H, Aebi N, Janssens S, Meylan E, Felberbaum-Corti M, Hirling H, Gruenberg J, Tschopp J, Burns K (2006) Intracellular trafficking of interleukin-1 receptor I requires Tollip. Current Biology 16:2265–2270
474. Akira S, Takeda K (2004) Toll-like receptor signalling. Nature Reviews – Immunology 4:499–511
475. Smith H, Liu XY, Dai L, Goh ET, Chan AT, Xi J, Seh CC, Qureshi IA, Lescar J, Ruedl C, Gourlay R, Morton S, Hough J, McIver EG, Cohen P, Cheung PC (2011) The role of TBK1 and IKKε in the expression and activation of Pellino-1. Biochemical Journal 434:537–548
476. Moynagh PN (2009) The Pellino family: IRAK E3 ligases with emerging roles in innate immune signalling. Trends in Immunology 30:33–42
477. Wesche H, Gao X, Li X, Kirschning CJ, Stark GR, Cao Z (1999) IRAK-M is a novel member of the Pelle/interleukin-1 receptor-associated kinase (IRAK) family. Journal of Biological Chemistry 274:19403–19410
478. Chen BC, Wu WT, Ho FM, Lin WW (2002) Inhibition of interleukin-1β-induced NF-κB activation by calcium/calmodulin-dependent protein kinase kinase occurs through Akt activation associated with interleukin-1 receptor-associated kinase phosphorylation and uncoupling of MyD88. Journal of Biological Chemistry 277:24169–24179
479. Li S, Strelow A, Fontana EJ, Wesche H (2002) IRAK-4: a novel member of the IRAK family with the properties of an IRAK-kinase. Proceedings of the National Academy of Sciences of the United States of America 99:5567–5572
480. Conze DB, Wu CJ, Thomas JA, Landstrom A, Ashwell JD (2008) Lys63-linked polyubiquitination of IRAK-1 is required for interleukin-1 receptor- and toll-like receptor-mediated NF-kappaB activation. Molecular and Cellular Biology 28:3538–3547
481. Motshwene PG, Moncrieffe MC, Grossmann JG, Kao CC, Ayaluru M, Sandercock AM, Robinson CV, Latz E, Gay NJ (2009) An oligomeric signalling platform formed by the toll-like receptor signal transducers MyD88 and IRAK4. Journal of Biological Chemistry 284:25404–25411
482. Bernard O (2008) LIMK1; LIMK2. UCSD-Nature Molecule Pages, UCSD-Nature Signaling Gateway (www.signaling-gateway.org)
483. Huang TY, DerMardirossian C, Bokoch GM (2006) Cofilin phosphatases and regulation of actin dynamics. Current Opinion in Cell Biology 18:26–31

484. Festjens N, VandenBerghe T, Cornelis S, Vandenabeele P (2007) RIP1, a kinase on the crossroads of a cell's decision to live or die. Cell Death and Differentiation 14:400–410
485. Vandenabeele P, Declercq W, Van Herreweghe F, Vanden Berghe T (2010) The role of the kinases RIP1 and RIP3 in TNF-induced necrosis. Science Signaling 3:re4
486. Plotnikova OV, Pugacheva EN, Dunbrack RL, Golemis EA (2010) Rapid calcium-dependent activation of Aurora-A kinase. Nature – Communications 1:64
487. Dephoure N, Zhou C, Villén J, Beausoleil SA, Bakalarski CE, Elledge SJ, Gygi SP (2008) A quantitative atlas of mitotic phosphorylation. Proceedings of the National Academy of Sciences of the United States of America 105:10762–10767
488. Plotnikova OV, Pugacheva EN, Golemis EA (2011) Aurora A kinase activity influences calcium signaling in kidney cells. Journal of Cell Biology 193:1021–1032
489. Rannou Y, Prigent C (2006) Aurora B. UCSD-Nature Molecule Pages, UCSD-Nature Signaling Gateway (www.signaling-gateway.org)
490. Nigg EA (1998) Polo-like kinases: positive regulators of cell division from start to finish. Current Opinion in Cell Biology 10:776–783
491. Tsvetkov L, Xu X, Li J, Stern DF (2003) Polo-like kinase 1 and Chk2 interact and co-localize to centrosomes and the midbody. Journal of Biological Chemistry 278:8468–8475
492. Johmura Y, Soung NK, Park JE, Yu LR, Zhou M, Bang JK, Kim BY, Veenstra TD, Erikson RL, Lee KS (2011) Regulation of microtubule-based microtubule nucleation by mammalian polo-like kinase 1. Proceedings of the National Academy of Sciences of the United States of America 108:11446–11451
493. Kauselmann G, Weiler M, Wulff P, Jessberger S, Konietzko U, Scafidi J, Staubli U, Bereiter-Hahn J, Strebhardt K, Kuhl D (1999) The polo-like protein kinases Fnk and Snk associate with a Ca^{2+}- and integrin-binding protein and are regulated dynamically with synaptic plasticity. EMBO Journal 18:5528–5539
494. Matsumoto T, Wang PY, Ma W, Sung HJ, Matoba S, Hwang PM (2009) Polo-like kinases mediate cell survival in mitochondrial dysfunction. Proceedings of the National Academy of Sciences of the United States of America 106:14542–14546
495. Xie S, Wu H, Wang Q, Cogswell JP, Husain I, Conn C, Stambrook P, Jhanwar-Uniyal M, Dai W (2001) Plk3 functionally links DNA damage to cell cycle arrest and apoptosis at least in part via the p53 pathway. Journal of Biological Chemistry 276:43305–43312
496. Ouyang B, Li W, Pan H, Meadows J, Hoffmann I, Dai W (1999) The physical association and phosphorylation of Cdc25C protein phosphatase by Prk. Oncogene 18:6029–6036
497. Holtrich U, Wolf G, Yuan J, Bereiter-Hahn J, Karn T, Weiler M, Kauselmann G, Rehli M, Andreesen R, Kaufmann M, Kuhl D, Strebhardt K (2000) Adhesion induced expression of the serine/threonine kinase Fnk in human macrophages. Oncogene 19:4832–4839
498. Kleylein-Sohn J, Westendorf J, Le Clech M, Habedanck R, Stierhof YD, Nigg EA (2007) Plk4-induced centriole biogenesis in human cells. Developmental Cell 13:190–202
499. Lee BH, Chen W, Stippec S, Cobb MH (2007) Biological cross-talk between WNK1 and the transforming growth factor-β–Smad signaling pathway. Journal of Biological Chemistry 282:17985–17996
500. Lee BH, Min X, Heise CJ, Xu BE, Chen S, Shu H, Luby-Phelps K, Goldsmith EJ, Cobb MH (2004) WNK1 phosphorylates synaptotagmin 2 and modulates its membrane binding. Molecular Cell 15:741–751
501. Wang WH, Giebisch G (2009) Regulation of potassium handling in the renal collecting duct. Pflügers Archiv (European Journal of Physiology) 458:157–168
502. Kahle KT, Rinehart J, Giebisch G, Gamba G, Hebert SC, Lifton RP (2008) A novel protein kinase signaling pathway essential for blood pressure regulation in humans. Trends in Endocrinology and Metabolism 19:91–95
503. Choate KA, Kahle KT, Wilson FH, Nelson-Williams C, Lifton RP (2003) WNK1, a kinase mutated in inherited hypertension with hyperkalemia, localizes to diverse Cl^--transporting epithelia. Proceedings of the National Academy of Sciences of the United States of America 100:663–668

504. Delaloy C, Lu J, Houot AM, Disse-Nicodeme S, Gasc JM, Corvol P, Jeunemaitre X (2003) Multiple promoters in the WNK1 gene: one controls expression of a kidney-specific kinase-defective isoform. Molecular and Cellular Biology 23:9208–9221
505. Xu BE, Stippec S, Lenertz L, Lee BH, Zhang W, Lee YK, Cobb MH (2004) WNK1 activates ERK5 by an MEKK2/3-dependent mechanism. Journal of Biological Chemistry 279:7826–7831
506. Xu BE, Stippec S, Chu PY, Lazrak A, Li XJ, Lee BH, English JM, Ortega B, Huang CL, Cobb MH (2005) WNK1 activates SGK1 to regulate the epithelial sodium channel. Proceedings of the National Academy of Sciences of the United States of America 102:10315–10320
507. Xu BE, Stippec S, Lazrak A, Huang CL, Cobb MH (2005) WNK1 activates SGK1 by a phosphatidylinositol 3-kinase-dependent and non-catalytic mechanism. Journal of Biological Chemistry 280:34218–34223
508. Vitari AC, Deak M, Collins BJ, Morrice N, Prescott AR, Phelan A, Humphreys S, Alessi DR (2004) WNK1, the kinase mutated in an inherited high-blood-pressure syndrome, is a novel PKB (protein kinase B)/Akt substrate. Biochemical Journal 378:257–268
509. Moniz S, Matos P, Jordan P (2008) WNK2 modulates MEK1 activity through the Rho GTPase pathway. Cellular Signalling 20:1762–1768
510. Kahle KT, Rinehart J, de Los Heros P, Louvi A, Meade P, Vazquez N, Hebert SC, Gamba G, Gimenez I, Lifton RP (2005) WNK3 modulates transport of Cl$^-$ in and out of cells: implications for control of cell volume and neuronal excitability. Proceedings of the National Academy of Sciences of the United States of America 102:16783–16788
511. San-Cristobal P, de los Heros P, Ponce-Coria J, Moreno E, Gamba G (2008) WNK kinases, renal ion transport and hypertension. American Journal of Nephrology 28:860–870
512. Hoorn EJ, van der Lubbe N, Zietse R (2009) The renal WNK kinase pathway: a new link to hypertension. Nephrology Dialysis Transplantation 24:1074–1077
513. Liapis H, Nag M, Kaji DM (1998) K-Cl cotransporter expression in the human kidney. American Journal of Physiology – Cell Physiology 275:C1432–C1437
514. Yamauchi K, Rai T, Kobayashi K, Sohara E, Suzuki T, Itoh T, Suda S, Hayama A, Sasaki S, Uchida S (2004) Disease-causing mutant WNK4 increases paracellular chloride permeability and phosphorylates claudins. Proceedings of the National Academy of Sciences of the United States of America 101:4690–4694
515. Kahle KT, Gimenez I, Hassan H, Wilson FH, Wong RD, Forbush B, Aronson PS, Lifton RP (2004) WNK4 regulates apical and basolateral Cl$^-$ flux in extrarenal epithelia. Proceedings of the National Academy of Sciences of the United States of America 101:2064–2069
516. Chabwine JN, Talavera K, Verbert L, Eggermont J, Vanderwinden JM, De Smedt H, Van Den Bosch L, Robberecht W, Callewaert G (2009) Differential contribution of the Na$^+$–K$^+$–2Cl$^-$ cotransporter NKCC1 to chloride handling in rat embryonic dorsal root ganglion neurons and motor neurons. FASEB Journal 23:1168–1176
517. Fuse T, Ohmae S, Takemoto-Kimura S, Bito H (2007) DCLK1. UCSD-Nature Molecule Pages, UCSD-Nature Signaling Gateway (www.signaling-gateway.org)
518. Ohmae S, Takemoto-Kimura S, Okamura M, Adachi-Morishima A, Nonaka M, Fuse T, Kida S, Tanji M, Furuyashiki T, Arakawa Y, Narumiya S, Okuno H, Bito H (2006) Molecular identification and characterization of a family of kinases with homology to Ca^{2+}/calmodulin-dependent protein kinases I/IV. Journal of Biological Chemistry 281:20427–20439
519. Higgins JM (2008) Haspin. UCSD-Nature Molecule Pages, UCSD-Nature Signaling Gateway (www.signaling-gateway.org)
520. Partanen JI, Tervonen TA, Myllynen M, Lind E, Imai M, Katajisto P, Dijkgraaf GJ, Kovanen PE, Mäkelä TP, Werb Z, Klefström J (2012) Tumor suppressor function of Liver kinase B1 (Lkb1) is linked to regulation of epithelial integrity. Proceedings of the National Academy of Sciences of the United States of America 109:E388-E397
521. Letwin K, Mizzen L, Motro B, Ben-David Y, Bernstein A, Pawson T (1992) A mammalian dual specificity protein kinase, Nek1, is related to the NIMA cell cycle regulator and highly expressed in meiotic germ cells. EMBO Journal 11:3521–3531

522. Fry AM (2005) Nek2. UCSD-Nature Molecule Pages, UCSD-Nature Signaling Gateway (www.signaling-gateway.org)
523. Roig J (2010) Nek6; Nek7; Nek9. UCSD-Nature Molecule Pages, UCSD-Nature Signaling Gateway (www.signaling-gateway.org)
524. Ishitani T, Hirao T, Suzuki M, Isoda M, Ishitani S, Harigaya K, Kitagawa M, Matsumoto K, Itoh M (2010) Nemo-like kinase suppresses Notch signalling by interfering with formation of the Notch active transcriptional complex. Nature – Cell Biology 12:278–285
525. Looyenga BD, DeHaan AM, MacKeigan JP (2008) Pink1. UCSD-Nature Molecule Pages, UCSD-Nature Signaling Gateway (www.signaling-gateway.org)
526. Billia F, Hauck L, Konecny F, Rao V, Shen J, Mak TW (2011) PTEN-inducible kinase 1 (PINK1)/Park6 is indispensable for normal heart function. Proceedings of the National Academy of Sciences of the United States of America 108:9572–9577
527. Williams BR, Sadler AJ (2006) Pkr. UCSD-Nature Molecule Pages, UCSD-Nature Signaling Gateway (www.signaling-gateway.org)
528. Daher A, Laraki G, Singh M, Melendez-Peña CE, Bannwarth S, Peters AH, Meurs EF, Braun RE, Patel RC, Gatignol A (2009) Molecular and Cellular Biology 29:254–265
529. Elde NC, Child SJ, Geballe AP, Malik HS (2009) Protein kinase R reveals an evolutionary model for defeating viral mimicry. Nature 457:485–489
530. Goh KC, deVeer MJ, Williams BRG (2000) The protein kinase PKR is required for p38 MAPK activation and the innate immune response to bacterial endotoxin. EMBO Journal 19:4292–4297
531. Silva AM, Whitmore M, Xu Z, Jiang Z, Li X, Williams BRG (2004) Protein kinase R (PKR) interacts with and activates mitogen-activated protein kinase kinase 6 (MKK6) in response to double-stranded RNA stimulation. Journal of Biological Chemistry 279:37670–37676
532. Daub H (2005) Srpk1; Srpk2. UCSD-Nature Molecule Pages, UCSD-Nature Signaling Gateway (www.signaling-gateway.org)
533. Delhase M, Kim SY, Lee H, Naiki-Ito A, Chen Y, Ahn ER, Murata K, Kim SJ, Lautsch N, Kobayashi KS, Shirai T, Karin M, Nakanishi M (2012) TANK-binding kinase 1 (TBK1) controls cell survival through PAI-2/serpinB2 and transglutaminase 2. Proceedings of the National Academy of Sciences of the United States of America 109:E177–E186

Chap. 6. Mitogen-Activated Protein Kinase Module

534. Mody A, Weiner J, Ramanathan S (2009) Modularity of MAP kinases allows deformation of their signalling pathways. Nature – Cell Biology 11:484–491
535. Gehart H, Kumpf S, Ittner A, Ricci R (2010) MAPK signalling in cellular metabolism: stress or wellness? EMBO Reports 11:834–840
536. Dhanasekaran N, Reddy EP (1998) Signaling by dual specificity kinases. Oncogene 17: 1447–1455
537. Kyriakis JM, Avruch J (2001) Mammalian mitogen-activated protein kinase signal transduction pathways activated by stress and inflammation. Physiological Reviews 81:807–869
538. Takahashi K, Tanase-Nicola S, Ten Wolde PR (2010) Spatio-temporal correlations can drastically change the response of a MAPK pathway. Proceedings of the National Academy of Sciences of the United States of America 107:2473–2478
539. Lehoux S, Tedgui A (2003) Cellular mechanics and gene expression in blood vessels. Journal of Biomechanics 36:631–643
540. Jauch R, Cho MK, Jake S, Netter C, Schreiter K, Aicher B, Zweckstetter M, Jackle Wahl MC (2006) Mitogen-activated protein kinases interacting kinases are autoinhibited by a reprogrammed activation segment. EMBO Journal 25:4020–4032

541. Liu L, Channavajhala PL, Rao VR, Moutsatsos I, Wu L, Zhang Y, Lin LL, Qiu Y (2009) Proteomic characterization of the dynamic KSR-2 interactome, a signaling scaffold complex in MAPK pathway. Biochimica et Biophysica Acta 1794:1485–1495
542. Nelson ML, Kang HS, Lee GM, Blaszczak AG, Lau DKW, McIntosh LP, Graves BJ (2010) Ras signaling requires dynamic properties of Ets1 for phosphorylation-enhanced binding to coactivator CBP. Proceedings of the National Academy of Sciences of the United States of America 107:10026–10031
543. Waskiewicz AJ, Flynn A, Proud CG, Cooper JA (1997) Mitogen-activated protein kinases activate the serine/threonine kinases Mnk1 and Mnk2. EMBO Journal 16:1909–1920
544. Sabio G, Kennedy NJ, Cavanagh-Kyros J, Jung DY, Ko HJ, Ong H, Barrett T, Kim JK, Davis RJ (2010) Role of muscle c-Jun NH2-terminal kinase 1 in obesity-induced insulin resistance. Molecular and Cellular Biology 30:106–115
545. Seger R, Krebs EG (1995) The MAPK signaling cascade. FASEB (Federation of American Societies for Experimental Biology) Journal 9:726–735
546. Qi M, Elion EA (2005) MAP kinase pathways Journal of Cell Science 118:3569–3572
547. Rauch J, Kolch W (2010) A-Raf. UCSD-Nature Molecule Pages, UCSD-Nature Signaling Gateway (www.signaling-gateway.org)
548. Feng L, Xie X, Ding Q, Luo X, He J, Fan F, Liu W, Wang Z, Chen Y (2007) Spatial regulation of Raf kinase signaling by RKTG. Proceedings of the National Academy of Sciences of the United States of America 104:14348–14353
549. Rajakulendran T, Sahmi M, Lefrancois M, Sicheri F, Therrien M (2009) A dimerization-dependent mechanism drives RAF catalytic activation. Nature 461:542–545
550. Ren JG, Li Z, Sacks DB (2007) IQGAP1 modulates activation of B-Raf. Proceedings of the National Academy of Sciences of the United States of America 104:10465–10469
551. Zimmermann S, Moelling K (1999) Phosphorylation and regulation of Raf by Akt (protein kinase B). Science 286:1741–1744
552. Hatzivassiliou G, Song K, Yen I, Brandhuber BJ, Anderson DJ, Alvarado R, Ludlam MJ, Stokoe D, Gloor SL, Vigers G, Morales T, Aliagas I, Liu B, Sideris S, Hoeflich KP, Jaiswal BS, Seshagiri S, Koeppen H, Belvin M, Friedman LS, Malek S (2010) RAF inhibitors prime wild-type RAF to activate the MAPK pathway and enhance growth. Nature 464:431–435
553. Poulikakos PI, Zhang C, Bollag G, Shokat KM, Rosen N (2010) RAF inhibitors transactivate RAF dimers and ERK signalling in cells with wild-type BRAF. Nature 464:427–430
554. Solit DB, Garraway LA, Pratilas CA, Sawai A, Getz G, Basso A, Ye Q, Lobo JM, She Y, Osman I, Golub TR, Sebolt-Leopold J, Sellers WR, Rosen N (2006) BRAF mutation predicts sensitivity to MEK inhibition. Nature 439:358–362
555. von Kriegsheim A, Pitt A, Grindlay GJ, Kolch W, Dhillon AS (2006) Regulation of the Raf-MEK-ERK pathway by protein phosphatase 5. Nature – Cell Biology 8:1011–1016
556. Craig EA, Stevens MV, Vaillancourt RR, Camenisch TD (2008) MAP3Ks as central regulators of cell fate during development. Developmental Dynamics 23:3102–3114
557. Dorow DS, Devereux L, Dietzsch E, De Kretser T (1993) Identification of a new family of human epithelial protein kinases containing two leucine/isoleucine-zipper domains. European Journal of Biochemistry 213:701–710
558. Gallo KA, Mark MR, Scadden DT, Wang Z, Gu Q, Godowski PJ (1994) Identification and characterization of SPRK, a novel src-homology 3 domain-containing proline-rich kinase with serine/threonine kinase activity. Journal of Biological Chemistry 269:15092–15100
559. Bisson N, Moss T (2009) Mlk1. UCSD-Nature Molecule Pages, UCSD-Nature Signaling Gateway (www.signaling-gateway.org)
560. Bisson N, Moss T (2009) Mlk2. UCSD-Nature Molecule Pages, UCSD-Nature Signaling Gateway (www.signaling-gateway.org)
561. Marcora E, Gowan K, Lee JE (2003) Stimulation of NeuroD activity by huntingtin and huntingtin-associated proteins HAP1 and MLK2. Proceedings of the National Academy of Sciences of the United States of America 100:9578–9583
562. Schachter K, Liou GY, Du Y, Gallo KA (2006) Mlk3. UCSD-Nature Molecule Pages, UCSD-Nature Signaling Gateway (www.signaling-gateway.org)

563. Vito P, Pellegrini L, Guiet C, D'Adamio L (1999) Cloning of AIP1, a novel protein that associates with the apoptosis-linked gene ALG-2 in a Ca^{2+}-dependent reaction. Journal of Biological Chemistry 274:1533–1540
564. Figueroa C, Tarras S, Taylor J, Vojtek AB (2003) Akt2 negatively regulates assembly of the POSH-MLK-JNK signaling complex. Journal of Biological Chemistry 278:47922–47927
565. Couture JP, Blouin R (2009) DLK. UCSD-Nature Molecule Pages, UCSD-Nature Signaling Gateway (www.signaling-gateway.org)
566. Masaki M, Ikeda A, Shiraki E, Oka S, Kawasaki T (2003) Mixed lineage kinase LZK and antioxidant protein-1 activate NF-κB synergistically. European Journal of Biochemistry 270:76–83
567. Ruggieri R (2006) Mltk. UCSD-Nature Molecule Pages, UCSD-Nature Signaling Gateway (www.signaling-gateway.org)
568. Geh EN, Jin C, Xia Y (2010) Map3k1. UCSD-Nature Molecule Pages, UCSD-Nature Signaling Gateway (www.signaling-gateway.org)
569. Ritterhoff S, Farah CM, Grabitzki J, Lochnit G, Skurat AV, Schmitz ML (2010) The WD40-repeat protein Han11 functions as a scaffold protein to control HIPK2 and MEKK1 kinase functions. EMBO Journal 29:3750–3761
570. Miyata Y, Akashi M, Nishida E (1999) Molecular cloning and characterization of a novel member of the MAP kinase superfamily. Genes to Cells 4:299–309
571. Zhou X, Izumi Y, Burg MB, Ferraris JD (2011) Rac1/osmosensing scaffold for MEKK3 contributes via phospholipase C-γ1 to activation of the osmoprotective transcription factor NFAT5. Proceedings of the National Academy of Sciences of the United States of America 108:12155–12160
572. Matsuzawa A, Takeda K, Ichijo H (2010) ASK1. UCSD-Nature Molecule Pages, UCSD-Nature Signaling Gateway (www.signaling-gateway.org)
573. Yoon KW, Cho JH, Lee JK, Kang YH, Chae JS, Kim YM, Kim J, Kim EK, Kim SE, Baik JH, Naik UP, Cho SG, Choi EJ (2009) CIB1 functions as a Ca^{2+}-sensitive modulator of stress-induced signaling by targeting ASK1. Proceedings of the National Academy of Sciences of the United States of America 106:17389–17394
574. Li X, Zhang R, Luo D, Park SJ, Wang Q, Kim Y, Min W (2005) Tumor necrosis factor α-induced desumoylation and cytoplasmic translocation of homeodomain-interacting protein kinase 1 are critical for apoptosis signal-regulating kinase 1-JNK/p38 activation. Journal of Biological Chemistry 280:15061–15070
575. Gan B, Peng X, Nagy T, Alcaraz A, Gu H, Guan JL (2006) Role of FIP200 in cardiac and liver development and its regulation of TNFα and TSC-mTOR signaling pathways. Journal of Cell Biology 175:121–133
576. Takizawa T, Tatematsu C, Nakanishi Y (2002) Double-stranded RNA-activated protein kinase interacts with apoptosis signal-regulating kinase 1. Implications for apoptosis signaling pathways. European Journal of Biochemistry 269:6126–6132
577. Xie D, Gore C, Zhou J, Pong RC, Zhang H, Yu L, Vessella RL, Min W, Hsieh JT (2009) DAB2IP coordinates both PI3K-Akt and ASK1 pathways for cell survival and apoptosis. Proceedings of the National Academy of Sciences of the United States of America 106:19878–19883
578. Cockrell LM, Fu H (2011) Map3k6. UCSD-Nature Molecule Pages, UCSD-Nature Signaling Gateway (www.signaling-gateway.org)
579. Eto N, Miyagishi M, Inagi R, Fujita T, Nangaku M (2009) Mitogen-activated protein 3 kinase 6 mediates angiogenic and tumorigenic effects via vascular endothelial growth factor expression. American Journal of Pathology 174:1553–1563
580. Ninomiya-Tsuji J, Matsumoto K (2006) Tak1. UCSD-Nature Molecule Pages, UCSD-Nature Signaling Gateway (www.signaling-gateway.org)
581. Omori E, Inagaki M, Mishina Y, Matsumoto K, Ninomiya-Tsuji J (2012) Epithelial transforming growth factor β-activated kinase 1 (TAK1) is activated through two independent mechanisms and regulates reactive oxygen species. Proceedings of the National Academy of Sciences of the United States of America 109:3365–3370

582. Li S, Wang L, Dorf ME (2009) PKC phosphorylation of TRAF2 mediates IKKα/β recruitment and K63-linked polyubiquitination. Molecular Cell 33:30–42
583. Fürthauer M, Lin W, Ang SL, Thisse B, Thisse C (2002) Sef is a feedback-induced antagonist of Ras/MAPK-mediated FGF signalling. Nature – Cell Biology 4:170–174
584. Yang X, Kovalenko D, Nadeau RJ, Harkins LK, Mitchell J, Zubanova O, Chen PY, Friesel R (2004) Sef interacts with TAK1 and mediates JNK activation and apoptosis. Journal of Biological Chemistry 279:38099–38102
585. Gantke T, Sriskantharajah S, Ley SC (2011) Regulation and function of TPL-2, an IκB kinase-regulated MAP kinase kinase kinase. Cell Research 21:131–145
586. Régnier CH, Song HY, Gao X, Goeddel DV, Cao Z, Rothe M (1997) Identification and characterization of an IkappaB kinase. Cell 90:373–383
587. Yasuda S, Sugiura H, Yamagata K (2009) Mek3. UCSD-Nature Molecule Pages, UCSD-Nature Signaling Gateway (www.signaling-gateway.org)
588. Asaoka Y, Nishina H (2010) Diverse physiological functions of MKK4 and MKK7 during early embryogenesis. Journal of Biochemistry 148:393–401
589. Abe JI, Yang J (2010) MEK5 UCSD-Nature Molecule Pages, UCSD-Nature Signaling Gateway (www.signaling-gateway.org)
590. Forcales S, Puri PL (2010) MKK6. UCSD-Nature Molecule Pages, UCSD-Nature Signaling Gateway (www.signaling-gateway.org)
591. Kholodenko BN, Hancock JF, Kolch W (2010) Signalling ballet in space and time. Nature Reviews – Molecular Cell Biology 11:414–426
592. McKay MM, Ritt DA, Morrison DK (2009) Signaling dynamics of the KSR1 scaffold complex. Proceedings of the National Academy of Sciences of the United States of America 106:11022–11027
593. Nojima H, Adachi M, Matsui T, Okawa K, Tsukita S, Tsukita S (2008) IQGAP3 regulates cell proliferation through the Ras/ERK signalling cascade. Nature – Cell Biology 10:971–978
594. Vaidyanathan H, Opoku-Ansah J, Pastorino S, Renganathan H, Matter M, Ramo JW (2008) ERK MAP kinase is targeted to RSK2 by the phosphoprotein PEA-15. Proceedings of the National Academy of Sciences of the United States of America 104:19837–19842
595. Nada S, Hondo A, Kasai A, Koike M, Saito K, Uchiyama Y, Okada M (2009) The novel lipid raft adaptor p18 controls endosome dynamics by anchoring the MEK–ERK pathway to late endosomes. EMBO Journal 28:477–489
596. Magee J, Cygler M (2011) Interactions between kinase scaffold MP1/p14 and its endosomal anchoring protein p18. Biochemistry 50:3696–3705
597. Catalanotti F, Reyes G, Jesenberger V, Galabova-Kovacs G, de Matos Simoes R, Carugo O, Baccarini M (2009) A Mek1–Mek2 heterodimer determines the strength and duration of the Erk signal. Nature – Structural and Molecular Biology 16:294–303
598. Rushworth LK, Hindley AD, O'Neill E, Kolch W (2006) Regulation and role of Raf-1/B-Raf heterodimerization. Molecular and Cellular Biology 26:2262–2272
599. Lawrence MC, Jivan A, Shao C, Duan L, Goad D, Zaganjor E, Osborne J, McGlynn K, Stippec S, Earnest S, Chen W, Cobb MH (2008) The roles of MAPKs in disease. Cell Research 18:436–442
600. Traub O, Monia BP, Dean NM, Berk BC (1997) PKC-epsilon is required for mechano-sensitive activation of ERK1/2 in endothelial cells. Journal of Biological Chemistry 272:31251–31257
601. Cai H, Smola U, Wixler V, Eisenmann TI, Diaz MMT, Moscat J, Rapp U, Cooper GM (1997) Role of diacylglycerol-regulated protein kinase C isotypes in growth factor activation of the Raf-1 protein kinase. Molecular and Cell Biology 17:732–741
602. Nishimoto S, Nishida E (2006) MAPK signalling: ERK5 versus ERK1/2. EMBO Reports 7:782–786
603. Emrick MA, Lee T, Starkey PJ, Mumby MC, Resing KA, Ahn NG (2006) The gatekeeper residue controls autoactivation of ERK2 via a pathway of intramolecular connectivity. Proceedings of the National Academy of Sciences of the United States of America 103:18101–18106

604. Chanalaris A, Lawrence KM, Stephanou A, Knight RD, Hsu SY, Hsueh AJ, Latchman DS (2003) Protective effects of the urocortin homologues stresscopin (SCP) and stresscopin-related peptide (SRP) against hypoxia/reoxygenation injury in rat neonatal cardiomyocytes. Journal of Molecular and Cellular Cardiology 35:1295–1305
605. Lee SJ, Pfluger PT, Kim JY, Nogueiras R, Duran A, Pagès G, Pouysségur J, Tschöp MH, Diaz-Meco MT, Moscat J (2010) A functional role for the p62–ERK1 axis in the control of energy homeostasis and adipogenesis. EMBO Reports 11:226–232
606. Meloche S (2006) Erk4. UCSD-Nature Molecule Pages, UCSD-Nature Signaling Gateway (www.signaling-gateway.org)
607. Nishimoto S, Nishida E (2006) MAPK signalling: ERK5 versus ERK1/2. EMBO Reports 7:782–786
608. Chang L, Kamata H, Solinas G, Luo JL, Maeda S, Venuprasad K, Liu YC, Karin M (2006) The E3 ubiquitin ligase itch couples JNK activation to TNFα-induced cell death by inducing c-FLIP$_L$ turnover. Cell 124:601–613
609. Liu J, Zhao Y, Eilers M, Lin A (2009) Miz1 is a signal- and pathway-specific modulator or regulator (SMOR) that suppresses TNF-α-induced JNK1 activation. Proceedings of the National Academy of Sciences of the United States of America 106:18279–18284
610. Haeusgen W, Herdegen T, Waetzig V (2010) Jnk2. UCSD-Nature Molecule Pages, UCSD-Nature Signaling Gateway (www.signaling-gateway.org)
611. Li C, Zhang GY (2011) Jnk3. UCSD-Nature Molecule Pages, UCSD-Nature Signaling Gateway (www.signaling-gateway.org)
612. Chang L, Karin M (2001) Mammalian MAP kinase signalling cascades. Nature 410:37–40
613. Salvador JM, Mittelstadt PR, Guszczynski T, Copeland TD, Yamaguchi H, Appella E, Fornace AJ, Ashwell JD (2005) Alternative p38 activation pathway mediated by T cell receptor-proximal tyrosine kinases. Nature – Immunology 6:390–395
614. Rousseau S (2011) p38 α MAP kinase. UCSD-Nature Molecule Pages, UCSD-Nature Signaling Gateway (www.signaling-gateway.org)
615. Mahlknecht U, Will J, Varin A, Hoelzer D, Herbein G (2004) Histone deacetylase 3, a class I histone deacetylase, suppresses MAPK11-mediated activating transcription factor-2 activation and represses TNF gene expression. Journal of Immunology 173:3979–3990
616. Hou SW, Lepp A, Chen G (2010) p38γ MAP kinase. UCSD-Nature Molecule Pages, UCSD-Nature Signaling Gateway (www.signaling-gateway.org)
617. Chen Z, Chen J, Weng T, Jin N, Liu L (2006) Identification of rat lung–prominent genes by a parallel DNA microarray hybridization. BMC Genomics 7:47
618. Liao P, Wang SQ, Wang S, Zheng M, Zheng M, Zhang SJ, Cheng H, Wang Y, Xiao RP (2002) p38 Mitogen-activated protein kinase mediates a negative inotropic effect in cardiac myocytes. Circulation Research 90:190–196
619. Liao P, Georgakopoulos D, Kovacs A, Zheng M, Lerner D, Pu H, Saffitz J, Chien K, Xiao RP, Kass DA, Wang Y (2002) The in vivo role of p38 MAP kinases in cardiac remodeling and restrictive cardiomyopathy. Proceedings of the National Academy of Sciences of the United States of America 98:12283–12288
620. Martineau LC, McVeigh LI, Jasmin BJ, Kennedy CR (2004) p38 MAP kinase mediates mechanically induced COX-2 and PG EP4 receptor expression in podocytes: implications for the actin cytoskeleton. American Journal of Physiology – Renal Physiology 286:F693–F701
621. Gaestel M (2006) MAPKAP kinases "MKs" two's company, three's a crowd. Nature Reviews – Molecular Cell Biology 7:120–130
622. Wiggin GR, Soloaga A, Foster JM, Murray-Tait V, Cohen P, Arthur JS (2002) MSK1 and MSK2 are required for the mitogen- and stress-induced phosphorylation of CREB and ATF1 in fibroblasts. Molecular and Cellular Biology 22:2871–2881

Chap. 7. Dual-Specificity Protein Kinases

623. Woodgett JR (1990) Molecular cloning and expression of glycogen synthase kinase-3/factor A. EMBO Journal 9:2431–2438
624. Cole A, Frame S, Cohen P (2004) Further evidence that the tyrosine phosphorylation of glycogen synthase kinase-3 (GSK3) in mammalian cells is an autophosphorylation event. Biochemical Journal 377:249–255
625. Vilimek D, Duronio V (2006) Cytokine-stimulated phosphorylation of GSK-3 is primarily dependent upon PKCs, not PKB. Biochemistry and Cell Biology 84:20–29
626. Kaladchibachi SA, Doble B, Anthopoulos N, Woodgett JR, Manoukian AS (2007) Glycogen synthase kinase 3, circadian rhythms, and bipolar disorder: a molecular link in the therapeutic action of lithium. Journal of Circadian Rhythms 5:3
627. Woods YL, Cohen P, Becker W, Jakes R, Goedert M, Wang X, Proud CG (2001) The kinase DYRK phosphorylates protein-synthesis initiation factor eIF2Bε at Ser539 and the microtubule-associated protein tau at Thr212: potential role for DYRK as a glycogen synthase kinase 3-priming kinase. Biochemical Journal 355:609–615
628. Himpel S, Panzer P, Eirmbter K, Czajkowska H, Sayed M, Packman LC, Blundell T, Kentrup H, Grötzinger J, Joost HG, Becker W (2001) Identification of the autophosphorylation sites and characterization of their effects in the protein kinase DYRK1A. Biochemical Journal 359:497–505
629. Becker W (2008) Dyrk1a. UCSD-Nature Molecule Pages, UCSD-Nature Signaling Gateway (www.signaling-gateway.org)
630. Becker W, Friedman EA (2008) Dyrk1b. UCSD-Nature Molecule Pages, UCSD-Nature Signaling Gateway (www.signaling-gateway.org)
631. Taira N, Nihira K, Yamaguchi T, Miki Y, Yoshida K (2007) DYRK2 is targeted to the nucleus and controls p53 via Ser46 phosphorylation in the apoptotic response to DNA damage. Molecular Cell 25:725–738
632. Bogacheva O, Bogachev O, Menon M, Dev A, Houde E, Valoret EI, Prosser HM, Creasy CL, Pickering SJ, Grau E, Rance K, Livi GP, Karur V, Erickson-Miller CL, Wojchowski DM (2008) DYRK3 dual-specificity kinase attenuates erythropoiesis during anemia. Journal of Biological Chemistry 283:36665–36675
633. Sacher F, Möller C, Bone W, Gottwald U, Fritsch M (2007) The expression of the testis-specific Dyrk4 kinase is highly restricted to step 8 spermatids but is not required for male fertility in mice. Molecular and Cellular Endocrinology 267:80–88
634. Rabinow LJ, Uguen P (2005) Clk1; Clk2; Clk3; Clk4. UCSD-Nature Molecule Pages, UCSD-Nature Signaling Gateway (www.signaling-gateway.org)
635. Hanes J, von der Kammer H, Klaudiny J, Scheit KH (1994) Characterization by cDNA cloning of two new human protein kinases. Evidence by sequence comparison of a new family of mammalian protein kinases. Journal of Molecular Biology 244:665–672
636. Rudolph J, Kristjansdottir KS (2004) Myt1. UCSD-Nature Molecule Pages, UCSD-Nature Signaling Gateway (www.signaling-gateway.org)
637. Granovsky AE, Rosner MR (2008) Raf kinase inhibitory protein: a signal transduction modulator and metastasis suppressor. Cell Research 18:452–457
638. Klysik J, Theroux SJ, Sedivy JM, Moffit JS, Boekelheide K (2008) Signaling crossroads: the function of Raf kinase inhibitory protein in cancer, the central nervous system and reproduction. Cell Signalling 20:1–9
639. Goumon Y, Angelone T, Schoentgen F, Chasserot-Golaz S, Almas B, Fukami MM, Langley K, Welters ID, Tota B, Aunis D, Metz-Boutigue MH (2004) The hippocampal cholinergic neurostimulating peptide, the N-terminal fragment of the secreted phosphatidylethanolamine-binding protein, possesses a new biological activity on cardiac physiology. Journal of Biological Chemistry 279:13054–13064
640. Zhu ST, Mc Henry KT, Lane WS, Fenteany G (2005) A chemical inhibitor reveals the role of Raf kinase inhibitor protein in cell migration. Chemistry Biology 12:981–991

Chap. 8. Cytosolic Protein Phosphatases

641. Moorhead GB, Trinkle-Mulcahy L, Ulke-Lemée A (2007) Emerging roles of nuclear protein phosphatases. Nature Reviews – Molecular Cell Biology 8:234–244
642. Conner SH, Kular G, Peggie M, Shepherd S, Schüttelkopf AW, Cohen P, Van Aalten DM (2006) TAK1-binding protein 1 is a pseudophosphatase. Biochemical Journal 399:427–434
643. Bhalla US, Ram PT, Iyengar R (2002) MAP kinase phosphatase as a locus of flexibility in a mitogen-activated protein kinase signaling network. Science 297:1018–1023
644. Wurzenberger C, Gerlich DW (2011) Phosphatases: providing safe passage through mitotic exit. Nature Reviews – Molecular Cell Biology 12:469–482
645. Ceulemans H, Bollen M (2004) Functional diversity of protein phosphatase-1, a cellular economizer and reset button. Physiological Reviews 84:1–39
646. Ragusa MJ, Dancheck B, Critton DA, Nairn AC, Page R, Peti W (2010) Spinophilin directs protein phosphatase 1 specificity by blocking substrate binding sites. Nature – Structural and Molecular Biology 17:459–464
647. Johnson DF, Moorhead G, Caudwell FB, Cohen P, Chen YH, Chen MX, Cohen PT (1996) Identification of protein-phosphatase-1-binding domains on the glycogen and myofibrillar targetting subunits. European Journal of Biochemistry 239:317–325
648. Shimada M, Haruta M, Niida H, Sawamoto K, Nakanishi M (2010) Protein phosphatase 1γ is responsible for dephosphorylation of histone H3 at Thr 11 after DNA damage. EMBO Reports 11:883–889
649. Wang BJ, Tang W, Zhang P, Wei Q (2012) Regulation of the catalytic domain of protein phosphatase 1 by the terminal region of protein phosphatase 2B. Journal of Biochemistry 151:283–290
650. de Souza RP, Rosa DV, Souza BR, Romano-Silva MA (2006) Darpp32. UCSD-Nature Molecule Pages, UCSD-Nature Signaling Gateway (www.signaling-gateway.org)
651. Vancauwenbergh S, Beullens M, Bollen M (2007) Nipp1. UCSD-Nature Molecule Pages, UCSD-Nature Signaling Gateway (www.signaling-gateway.org)
652. Shichi D, Arimura T, Ishikawa T, Kimura A (2010) Heart-specific small subunit of myosin light chain phosphatase activates rho-associated kinase and regulates phosphorylation of myosin phosphatase target subunit 1. Journal of Biological Chemistry 285:33680–33690
653. Janssens V, Goris J (2001) Protein phosphatase 2A: a highly regulated family of serine/threonine phosphatases implicated in cell growth and signalling. Biochemical Journal 353:417–439
654. Chen J, Martin BL, Brautigan DL (1992) Regulation of protein serine-threonine phosphatase type-2A by tyrosine phosphorylation. Science 257:1261–1264
655. Xing Y, Li Z, Chen Y, Stock JB, Jeffrey PD, Shi Y (2008) Structural mechanism of demethylation and inactivation of protein phosphatase 2A. Cell 133:154–163
656. Hastie CJ, Carnegie GK, Morrice N, Cohen PT (2000) A novel 50 kDa protein forms complexes with protein phosphatase 4 and is located at centrosomal microtubule organizing centres. Biochemical Journal 347:845–855
657. Liu E, Knutzen CA, Krauss S, Schweiger S, Chiang GG (2011) Control of mTORC1 signaling by the Opitz syndrome protein MID1. Proceedings of the National Academy of Sciences of the United States of America 108:8680–8685
658. Strack S, Cribbs JT, Gomez L (2004) Critical role for protein phosphatase 2A heterotrimers in mammalian cell survival. Journal of Biological Chemistry 279:47732–47739
659. Tar K, Csortos C, Czikora I, Olah G, Ma SF, Wadgaonkar R, Gergely P, Garcia JG, Verin AD (2006) Role of protein phosphatase 2A in the regulation of endothelial cell cytoskeleton structure. Journal of Cellular Biochemistry 98:931–953
660. Liu Q, Caldwell-Busby J, Molkentin JD (2009) Interaction between TAK1–TAB1–TAB2 and RCAN1–calcineurin defines a signalling nodal control point. Nature – Cell Biology 11:154–161

661. Duan L, Cobb MH (2010) Calcineurin increases glucose activation of ERK1/2 by reversing negative feedback. Proceedings of the National Academy of Sciences of the United States of America 107:22314–22319
662. Brewis ND, Street AJ, Prescott AR, Cohen PT (1993) PPX, a novel protein serine/threonine phosphatase localized to centrosomes. EMBO Journal 12:987–996
663. Lee DH, Pan Y, Kanner S, Sung P, Borowiec JA, Chowdhury D (2010) A PP4 phosphatase complex dephosphorylates RPA2 to facilitate DNA repair via homologous recombination. Nature – Structural and Molecular Biology 17:365–372
664. Yoon YS, Lee MW, Ryu D, Kim JH, Ma H, Seo WY, Kim YN, Kim SS, Lee CH, Hunter T, Choi CS, Montminy MR, Koo SH (2010) Suppressor of MEK null (SMEK)/protein phosphatase 4 catalytic subunit (PP4C) is a key regulator of hepatic gluconeogenesis. Proceedings of the National Academy of Sciences of the United States of America 107:17704–17709
665. Chinkers M (2001) Protein phosphatase 5 in signal transduction. Trends in Endocrinology and Metabolism 12:28–32
666. Becker W, Kentrup H, Klumpp S, Schultz JE, Joost HG (1994) Molecular cloning of a protein serine/threonine phosphatase containing a putative regulatory tetratricopeptide repeat domain. Journal of Biological Chemistry 269:22586–22592
667. Bastians H, Ponstingl H (1996) The novel human protein serine/threonine phosphatase 6 is a functional homologue of budding yeast Sit4p and fission yeast ppe1, which are involved in cell cycle regulation. Journal of Cell Science 109:2865–2874
668. Huang X, Honkanen RE (1998) Molecular cloning, expression, and characterization of a novel human serine/threonine protein phosphatase, PP7, that is homologous to Drosophila retinal degeneration C gene product (rdgC). Journal of Biological Chemistry 273:1462–1468
669. Wang Y, Dow EC, Liang YY, Ramakrishnan R, Liu H, Sung TL, Lin X, Rice AP (2008) Phosphatase PPM1A regulates phosphorylation of Thr-186 in the Cdk9 T-loop. Journal of Biological Chemistry 283:33578–33584
670. Lu G, Wang Y (2008) Functional diversity of mammalian type 2C protein phosphatase isoforms: new tales from an old family. Clinical and Experimental Pharmacology and Physiology 35:107–112
671. Henmi T, Amano K, Nagaura Y, Matsumoto K, Echigo S, Tamura S, Kobayashi T (2009) A mechanism for the suppression of interleukin-1-induced nuclear factor κB activation by protein phosphatase 2Cη-2. Biochemical Journal 423:71–78
672. Voss M, Paterson J, Kelsall IR, Martn-Granados C, Hastie CJ, Peggie MW, Cohen PT (2011) Ppm1E is an in cellulo AMP-activated protein kinase phosphatase. Cellular Signalling 23:114–124
673. Ishida A, Tada Y, Nimura T, Sueyoshi N, Katoh T, Takeuchi M, Fujisawa H, Taniguchi T, Kameshita I (2005) Identification of major Ca^{2+}/calmodulin-dependent protein kinase phosphatase-binding proteins in brain: biochemical analysis of the interaction. Archives of Biochemistry and Biophysics 435:134–146
674. Lee-Hoeflich ST, Pham TQ, Dowbenko D, Munroe X, Lee J, Li L, Zhou W, Haverty PM, Pujara K, Stinson J, Chan SM, Eastham-Anderson J, Pandita A, Seshagiri S, Hoeflich KP, Turashvili G, Gelmon KA, Aparicio SA, DP Davis, Sliwkowski MX, Stern HM (2011) PPM1H is a p27 phosphatase implicated in trastuzumab resistance. Cancer Discovery 1:326–337
675. Sugiura T, Noguchi Y (2009) Substrate-dependent metal preference of PPM1H, a cancer-associated protein phosphatase 2C: comparison with other family members. Biometals 22:469–477
676. Lu G, Sun H, Korge P, Koehler CM, Weiss JN, Wang Y (2009) Functional Characterization of a Mitochondrial Ser/Thr Protein Phosphatase in Cell Death Regulation (Chap. 14, p.255-273). In Allison WS, Murphy AN (Eds) Methods in Enzymology, Vol. 457 "Mitochondrial Function, Part B: Mitochondrial Protein Kinases, Protein Phosphatases and Mitochondrial Diseases", Elsevier, Amsterdam

677. Shimizu K, Okada M, Nagai K, Fukada Y (2003) Suprachiasmatic nucleus circadian oscillatory protein, a novel binding partner of K-Ras in the membrane rafts, negatively regulates MAPK pathway. Journal of Biological Chemistry 278:14920–14925
678. Kato J, Kato M (2010) Crystallization and preliminary crystallographic studies of the catalytic subunits of human pyruvate dehydrogenase phosphatase isoforms 1 and 2. Acta Crystallographica, Section F, Structural Biology and Crystallization Communications 66:342–345
679. Caruso M, Maitan MA, Bifulco G, Miele C, Vigliotta G, Oriente F, Formisano P, Beguinot F (2001) Activation and mitochondrial translocation of protein kinase Cdelta are necessary for insulin stimulation of pyruvate dehydrogenase complex activity in muscle and liver cells. Journal of Biological Chemistry 276:45088–45097
680. Tonks NK (2006) Protein tyrosine phosphatases: from genes, to function, to disease. Nature Reviews – Molecular Cell Biology 7:833–846
681. Reue K, Brindley DN (2008) Thematic Review Series: glycerolipids. Multiple roles for lipins/phosphatidate phosphatase enzymes in lipid metabolism. Journal of Lipid Research 49:2493–2503
682. Östman A, Frijhoff J, Sandin A, Böhmer FD (2011) Regulation of protein tyrosine phosphatases by reversible oxidation. Journal of Biochemistry 150:345–356
683. Gandhi TK, Chandran S, Peri S, Saravana R, Amanchy R, Prasad TS, Pandey A (2005) A bioinformatics analysis of protein tyrosine phosphatases in humans. DNA Research 12: 79–89
684. Chernoff J (2008) Ptp1b. UCSD-Nature Molecule Pages, UCSD-Nature Signaling Gateway (www.signaling-gateway.org)
685. Lund IK, Hansen JA, Andersen HS, Møller NP, Billestrup N (2005) Mechanism of protein tyrosine phosphatase 1B-mediated inhibition of leptin signalling. Journal of Molecular Endocrinology 34:339–351
686. Aoki N, Matsuda T (2000) A cytosolic protein-tyrosine phosphatase PTP1B specifically dephosphorylates and deactivates prolactin-activated STAT5a and STAT5b. Journal of Biological Chemistry 275:39718–39726
687. Mertins P, Eberl HC, Renkawitz J, Olsen JV, Tremblay ML, Mann M, Ullrich A, Daub H (2008) Investigation of protein-tyrosine phosphatase 1B function by quantitative proteomics. Molecular and Cellular Proteomics 7:1763–1777
688. Ravichandran LV, Chen H, Li Y, Quon MJ (2001) Phosphorylation of PTP1B at Ser(50) by Akt impairs its ability to dephosphorylate the insulin receptor. Molecular Endocrinology 15:1768–1780
689. Tiganis T, Bennett AM, Ravichandran KS, Tonks NK (1998) Epidermal growth factor receptor and the adaptor protein p52Shc are specific substrates of T-cell protein tyrosine phosphatase. Molecular and Cellular Biology 18:1622–1634
690. Tiganis T, Kemp BE, Tonks NK (1999) The protein-tyrosine phosphatase TCPTP regulates epidermal growth factor receptor-mediated and phosphatidylinositol 3-kinase-dependent signaling. Journal of Biological Chemistry 274:27768–27775
691. Yamamoto T, Sekine Y, Kashima K, Kubota A, Sato N, Aoki N, Matsuda T (2002) The nuclear isoform of protein-tyrosine phosphatase TC-PTP regulates interleukin-6-mediated signaling pathway through STAT3 dephosphorylation. Biochemical and Biophysical Research Communications 297:811–817
692. Galic S, Klingler-Hoffmann M, Fodero-Tavoletti MT, Puryer MA, Meng TC, Tonks NK, Tiganis T (2003) Regulation of insulin receptor signaling by the protein tyrosine phosphatase TCPTP. Molecular and Cellular Biology 23:2096–2108
693. Arpin M, Algrain M, Louvard D (1994) Membrane-actin microfilament connections: an increasing diversity of players related to band 4.1. Current Opinion in Cell Biology 6:136–141
694. Zhang SH, Kobayashi R, Graves PR, Piwnica-Worms H, Tonks NK (1997) Serine phosphorylation-dependent association of the band 4.1-related protein-tyrosine phosphatase PTPH1 with 14-3-3β protein. Journal of Biological Chemistry 272:27281–27287
695. Zheng Y, Schlondorff J, Blobel CP (2002) Evidence for regulation of the tumor necrosis factor alpha-convertase (TACE) by protein-tyrosine phosphatase PTPH1. Journal of Biological Chemistry 277:42463–42470

696. Jespersen T, Gavillet B, van Bemmelen MX, Cordonier S, Thomas MA, Staub O, Abriel H (2006) Cardiac sodium channel Na$_V$1.5 interacts with and is regulated by the protein tyrosine phosphatase PTPH1. Biochemical and Biophysical Research Communications 348: 1455–1462
697. Sozio MS, Mathis MA, Young JA, Wälchli S, Pitcher LA, Wrage PC, Bartk B, Campbell A, Watts JD, Aebersold R, Hooft van Huijsduijnen R, van Oers NS (2004) PTPH1 is a predominant protein-tyrosine phosphatase capable of interacting with and dephosphorylating the T cell receptor ζ subunit. Journal of Biological Chemistry 279:7760–7769
698. Hironaka K, Umemori H, Tezuka T, Mishina M, Yamamoto T (2000) The protein-tyrosine phosphatase PTPMEG interacts with glutamate receptor δ2 and ε subunits. Journal of Biological Chemistry 275:16167–16173
699. Lombroso PJ, Murdoch G, Lerner M (1991) Molecular characterization of a protein-tyrosine-phosphatase enriched in striatum. Proceedings of the National Academy of Sciences of the United States of America 88:7242–7246
700. Fitzpatrick CJ, Goebel-Goody SM, Liberzon I, Lombroso PJ (2010) STEP. UCSD-Nature Molecule Pages, UCSD-Nature Signaling Gateway (www.signaling-gateway.org)
701. Nguyen TH, Liu J, Lombroso PJ (2002) Striatal enriched phosphatase 61 dephosphorylates Fyn at phosphotyrosine 420. Journal of Biological Chemistry 277:24274–24279
702. Pulido R, Zñiga A, Ullrich A (1998) PTP-SL and STEP protein tyrosine phosphatases regulate the activation of the extracellular signal-regulated kinases ERK1 and ERK2 by association through a kinase interaction motif. EMBO Journal 17:7337–7350
703. Keilhack H, Müller M, Böhmer SA, Frank C, Weidner KM, Birchmeier W, Ligensa T, Berndt A, Kosmehl H, Günther B, Müller T, Birchmeier C, Böhmer FD (2001) Negative regulation of Ros receptor tyrosine kinase signaling. An epithelial function of the SH2 domain protein tyrosine phosphatase SHP-1. Journal of Cell Biology 152:325–334
704. Tenev T, Keilhack H, Tomic S, Stoyanov B, Stein-Gerlach M, Lammers R, Krivtsov AV, Ullrich A, Böhmer FD (1997) Both SH2 domains are involved in interaction of SHP-1 with the epidermal growth factor receptor but cannot confer receptor-directed activity to SHP-1/SHP-2 chimera. Journal of Biological Chemistry 272:5966–5973
705. Kozlowski M, Larose L, Lee F, Le DM, Rottapel R, Siminovitch KA (1998) SHP-1 binds and negatively modulates the c-Kit receptor by interaction with tyrosine 569 in the c-Kit juxtamembrane domain. Molecular and Cellular Biology 18:2089–2099
706. Klingmüller U, Lorenz U, Cantley LC, Neel BG, Lodish HF (1995) Specific recruitment of SH-PTP1 to the erythropoietin receptor causes inactivation of JAK2 and termination of proliferative signals. Cell 80:729–738
707. Yoshida K, Kufe D (2001) Negative regulation of the SHPTP1 protein tyrosine phosphatase by protein kinase Cδ in response to DNA damage. Molecular Pharmacology 60:1431–1438
708. Meyaard L, Adema GJ, Chang C, Woollatt E, Sutherland GR, Lanier LL, Phillips JH (1997) LAIR-1, a novel inhibitory receptor expressed on human mononuclear leukocytes. Immunity 7:283–290
709. Wang PY, Liu P, Weng J, Sontag E, Anderson RG (2003) A cholesterol-regulated PP2A/HePTP complex with dual specificity ERK1/2 phosphatase activity. EMBO Journal 22:2658–2667
710. Pettiford SM, Herbst R (2003) The protein tyrosine phosphatase HePTP regulates nuclear translocation of ERK2 and can modulate megakaryocytic differentiation of K562 cells. Leukemia 17:366–378
711. Kruger JM, Fukushima T, Cherepanov V, Borregaard N, Loeve C, Shek C, Sharma K, Tanswell AK, Chow CW, Downey GP (2002) Protein-tyrosine phosphatase MEG2 is expressed by human neutrophils. Localization to the phagosome and activation by polyphosphoinositides. Journal of Biological Chemistry 277:2620–2628
712. Cho CY, Koo SH, Wang Y, Callaway S, Hedrick S, Mak PA, Orth AP, Peters EC, Saez E, Montminy M, Schultz PG, Chanda SK (2006) Identification of the tyrosine phosphatase PTP-MEG2 as an antagonist of hepatic insulin signaling. Cell Metabolism 3:367–378

713. Moutoussamy S, Renaudie F, Lago F, Kelly PA, Finidori J (1998) Grb10 identified as a potential regulator of growth hormone (GH) signaling by cloning of GH receptor target proteins. Journal of Biological Chemistry 273:15906–15912
714. Lehmann U, Schmitz J, Weissenbach M, Sobota RM, Hortner M, Friederichs K, Behrmann I, Tsiaris W, Sasaki A, Schneider-Mergener J, Yoshimura A, Neel BG, Heinrich PC, Schaper F (2003) SHP2 and SOCS3 contribute to Tyr-759-dependent attenuation of interleukin-6 signaling through gp130. Journal of Biological Chemistry 78:661–671
715. Yin T, Shen R, Feng GS, Yang YC (1997) Molecular characterization of specific interactions between SHP-2 phosphatase and JAK tyrosine kinases. Journal of Biological Chemistry 272:1032–1037
716. Yu CL, Jin YJ, Burakoff SJ (2000) Cytosolic tyrosine dephosphorylation of STAT5. Potential role of SHP-2 in STAT5 regulation. Journal of Biological Chemistry 275:599–604
717. Boudot C, Kadri Z, Petitfrère E, Lambert E, Chrétien S, Mayeux P, Haye B, Billat C (2002) Phosphatidylinositol 3-kinase regulates glycosylphosphatidylinositol hydrolysis through PLC-γ2 activation in erythropoietin-stimulated cells. Cell Signalling 14:869–878
718. Nakamura T, Gulick J, Colbert MC, Robbins J (2009) Protein tyrosine phosphatase activity in the neural crest is essential for normal heart and skull development. Proceedings of the National Academy of Sciences of the United States of America 106:11270–11275
719. Shen Y, Schneider G, Cloutier JF, Veillette A, Schaller MD (1998) Direct association of protein-tyrosine phosphatase PTP-PEST with paxillin. Journal of Biological Chemistry 273:6474–6481
720. Veillette A, Rhee I, Souza CM, Davidson D (2009) PEST family phosphatases in immunity, autoimmunity, and autoinflammatory disorders. Immunological Reviews 228:312–324
721. Gross C, Heumann R, Erdmann KS (2001) The protein kinase C-related kinase PRK2 interacts with the protein tyrosine phosphatase PTP-BL via a novel PDZ domain binding motif. FEBS Letters 496:101–104
722. Wadham C, Gamble JR, Vadas MA, Khew-Goodall Y (2003) The protein tyrosine phosphatase Pez is a major phosphatase of adherens junctions and dephosphorylates beta-catenin. Molecular Biology of the Cell 14:2520–2529
723. Spencer S, Dowbenko D, Cheng J, Li W, Brush J, Utzig S, Simanis V, Lasky LA (2009) PSTPIP: a tyrosine phosphorylated cleavage furrow-associated protein that is a substrate for a PEST tyrosine phosphatase. Journal of Cell Biology 138:845–860
724. Ohsugi M, Kuramochi S, Matsuda S, Yamamoto T (1997) Molecular cloning and characterization of a novel cytoplasmic protein-tyrosine phosphatase that is specifically expressed in spermatocytes. Journal of Biological Chemistry 272:33092–33099
725. Carlucci A, Gedressi C, Lignitto L, Nezi L, Villa-Moruzzi E, Avvedimento EV, Gottesman M, Garbi C, Feliciello A (2008) Protein-tyrosine phosphatase PTPD1 regulates focal adhesion kinase autophosphorylation and cell migration. Journal of Biological Chemistry 283:10919–10929
726. Castiglioni S, Maier JA, Mariotti M (2007) The tyrosine phosphatase HD-PTP: A novel player in endothelial migration. Biochemical and Biophysical Research Communications 364:534–539
727. Mariotti M, Castiglioni S, Garcia-Manteiga JM, Beguinot L, Maier JA (2009) HD-PTP inhibits endothelial migration through its interaction with Src. International Journal of Biochemistry and Cell Biology 41:687–693
728. Takahashi Y, Morales FC, Kreimann EL, Georgescu MM (2006) PTEN tumor suppressor associates with NHERF proteins to attenuate PDGF receptor signaling. EMBO Journal 25:910–920
729. Shen WH, Balajee AS, Wang J, Wu H, Eng C, Pandolfi PP, Yin Y (2007) Essential role for nuclear PTEN in maintaining chromosomal integrity. Cell 128:157–170
730. Miller SJ, Lou DY, Seldin DC, Lane WS, Neel BG (2002) Direct identification of PTEN phosphorylation sites. FEBS Letters 528:145–153

731. Chen H, Duncan IC, Bozorgchami H, Lo SH (2002) Tensin1 and a previously undocumented family member, tensin2, positively regulate cell migration. Proceedings of the National Academy of Sciences of the United States of America 99:733–738
732. Eto M, Kirkbride J, Elliott E, Lo SH, Brautigan DL (2007) Association of the tensin N-terminal protein-tyrosine phosphatase domain with the α isoform of protein phosphatase-1 in focal adhesions. Journal of Biological Chemistry 282:17806–17815
733. Hafizi S, Ibraimi F, Dahlbäck B (2005) C1-TEN is a negative regulator of the Akt/PKB signal transduction pathway and inhibits cell survival, proliferation, and migration. FASEB Journal 19:971–973
734. Tapparel C, Reymond A, Girardet C, Guillou L, Lyle R, Lamon C, Hutter P, Antonarakis SE (2003) The TPTE gene family: cellular expression, subcellular localization and alternative splicing. Gene 323:189–199
735. Liu Y, Shepherd EG, Nelin LD (2007) MAPK phosphatases regulating the immune response. Nature Reviews – Immunology 7:202–212
736. Patterson KI, Brummer T, O'Brien PM, Daly RJ (2009) Dual-specificity phosphatases: critical regulators with diverse cellular targets. Biochemical Journal 418:475–489
737. Bayón Y, Alonso A (2010) Atypical DUSPs: 19 phosphatases in search of a role (Chap. 9). In Lazo PA (Ed.) Emerging Signaling Pathways in Tumor Biology. Transworld Research Network, Kerala, India
738. Dickinson RJ, Keyse SM (2006) Diverse physiological functions for dual-specificity MAP kinase phosphatases. Journal of Cell Science 119:4607–4615
739. Keyse SM (2008) Dual-specificity MAP kinase phosphatases (MKPs) and cancer. Cancer Metastasis Reviews 27:253–261
740. Jeffrey KL, Camps M, Rommel C, Mackay CR (2007) Targeting dual-specificity phosphatases: manipulating MAP kinase signalling and immune responses. Nature Reviews – Drug Discovery 6:391–403
741. Lang R, Hammer M, Mages J (2006) DUSP meet immunology: dual specificity MAPK phosphatases in control of the inflammatory response. Journal of Immunology 177: 7497–7504
742. Chi H, Barry SP, Roth RJ, Wu JJ, Jones EA, Bennett AM, Flavell RA (2006) Dynamic regulation of pro-and anti-inflammatory cytokines by MAPK phosphatase 1 (MKP-1) in innate immune responses. Proceedings of the National Academy of Sciences of the United States of America 103:2274–2279
743. Chun CZ, Kaur S, Samant GV, Wang L, Pramanik K, Garnaas MK, Li K, Field L, Mukhopadhyay D, Ramchandran R (2009) Snrk-1 is involved in multiple steps of angioblast development and acts via notch signaling pathway in artery-vein specification in vertebrates. Blood 113:983–984
744. Pramanik K, Chun CZ, Garnaas MK, Samant GV, Li K, Horswill MA, North PE, Ramchandran R (2009) Dusp-5 and Snrk-1 coordinately function during vascular development and disease. Blood 113:1184–1191
745. Molina G, Vogt A, Bakan A, Dai W, Queiroz de Oliveira P, Znosko W, Smithgall TE, Bahar I, Lazo JS, Day BW, Tsang M (2009) Zebrafish chemical screening reveals an inhibitor of Dusp6 that expands cardiac cell lineages. Nature – Chemical Biology 5:680–687
746. Levy-Nissenbaum O, Sagi-Assif O, Witz IP (2004) Characterization of the dual-specificity phosphatase PYST2 and its transcripts. Genes, Chromosomes and Cancer 39:37–47
747. Pulido R, Muda M (2010) MKP-X. UCSD-Nature Molecule Pages, UCSD-Nature Signaling Gateway (www.signaling-gateway.org)
748. Martell KJ, Seasholtz AF, Kwak SP, Clemens KK, Dixon JE (1995) hVH-5: a protein tyrosine phosphatase abundant in brain that inactivates mitogen-activated protein kinase. Journal of Neurochemistry 65:1823–1833
749. Bernabeu R, Di Scala G, Zwiller J (2000) Odor regulates the expression of the mitogen-activated protein kinase phosphatase gene hVH-5 in bilateral entorhinal cortex-lesioned rats. Brain Research – Molecular Brain Research 5:113–120
750. Pulido R, Muda M (2010) MKP-4. UCSD-Nature Molecule Pages, UCSD-Nature Signaling Gateway (www.signaling-gateway.org)

751. Zhang Y, Blattman JN, Kennedy NJ, Duong J, Nguyen T, Wang Y, Davis RJ, Greenberg PD, Flavell RA, Dong C (2004) Regulation of innate and adaptive immune responses by MAP kinase phosphatase 5. Nature 430:793–797
752. Wolters NM, MacKeigan JP (2007) Mk-styx. UCSD-Nature Molecule Pages, UCSD-Nature Signaling Gateway (www.signaling-gateway.org)
753. Kligys K, Claiborne JN, DeBiase PJ, Hopkinson SB, Wu Y, Mizuno K, Jones JC (2007) The slingshot family of phosphatases mediates Rac1 regulation of cofilin phosphorylation, laminin-332 organization, and motility behavior of keratinocytes. Journal of Biological Chemistry 282:32520–32528
754. Clague MJ, Lorenzo O (2005) The myotubularin family of lipid phosphatases. Traffic 6: 1063–1069
755. Cao C, Backer JM, Laporte J, Bedrick EJ, Wandinger-Ness A (2008) Sequential actions of myotubularin lipid phosphatases regulate endosomal PI(3)P and growth factor receptor trafficking. Molecular Biology of the Cell 19:3334–3346
756. Wishart MJ (2007) Styx. UCSD-Nature Molecule Pages, UCSD-Nature Signaling Gateway (www.signaling-gateway.org)
757. Xiao J, Engel JL, Zhang J, Chen MJ, Manning G, Dixon JE (2011) Structural and functional analysis of PTPMT1, a phosphatase required for cardiolipin synthesis. Proceedings of the National Academy of Sciences of the United States of America 108:11860–11865
758. Gross AW, Dawson JP, Muda M (2011) Yvh1. UCSD-Nature Molecule Pages, UCSD-Nature Signaling Gateway (www.signaling-gateway.org)
759. Ramponi G, Stefani M (1997) Structure and function of the low Mr phosphotyrosine protein phosphatases. Biochimica et Biophysica Acta 1341:137–156
760. Chiarugi P, Cirri P, Raugei G, Manao G, Taddei L, Ramponi G (1996) Low M(r) phosphotyrosine protein phosphatase interacts with the PDGF receptor directly via its catalytic site. Biochemical and Biophysical Research Communications 219:21–25
761. Shimizu H, Toyama O, Shiota M, Kim-Mitsuyama S, Miyazaki H (2005) Protein tyrosine phosphatase LMW-PTP exhibits distinct roles between vascular endothelial and smooth muscle cells. Journal of Receptors and Signal Transduction 25:19–33
762. Nilsson I, Hoffmann I (2000) Cell cycle regulation by the Cdc25 phosphatase family. Progress in Cell Cycle Research 4:107–114
763. Aressy B, Ducommun B (2008) Cell cycle control by the CDC25 phosphatases. Anti-Cancer Agents in Medicinal Chemistry 8:818–824
764. Potapova TA, Daum JR, Byrd KS, Gorbsky GJ (2009) Fine tuning the cell cycle: activation of the Cdk1 inhibitory phosphorylation pathway during mitotic exit. Molecular Biology of the Cell 20:1737–1748
765. Kiyokawa H, Ray D (2008) In vivo roles of CDC25 phosphatases: biological insight into the anti-cancer therapeutic targets. Anti-Cancer Agents in Medicinal Chemistry 8:832–836
766. Fernandez-Vidal A, Mazars A, Manenti S (2008) CDC25A: a rebel within the CDC25 phosphatases family? Anti-Cancer Agents in Medicinal Chemistry 8:825–831
767. Timofeev O, Cizmecioglu O, Hu E, Orlik T, Hoffmann I (2009) Human Cdc25A phosphatase has a non-redundant function in G2 phase by activating Cyclin A-dependent kinases. FEBS Letters 583:841–847
768. Zhang X, Neganova I, Przyborski S, Yang C, Cooke M, Atkinson SP, Anyfantis G, Fenyk S, Keith WN, Hoare SF, Hughes O, Strachan T, Stojkovic M, Hinds PW, Armstrong L, Lako M (2009) A role for NANOG in G1 to S transition in human embryonic stem cells through direct binding of CDK6 and CDC25A. Journal of Cell Biology 184:67–82
769. Lobjois V, Jullien D, Bouché JP, Ducommun B (2009) The polo-like kinase 1 regulates CDC25B-dependent mitosis entry. Biochimica et Biophysica Acta 793:462–468
770. Telles E, Hosing AS, Kundu ST, Venkatraman P, Dalal SN (2009) A novel pocket in 14-3-3epsilon is required to mediate specific complex formation with cdc25C and to inhibit cell cycle progression upon activation of checkpoint pathways. Experimental Cell Research 315:1448–1457

771. Xu PX, Zheng W, Laclef C, Maire P, Maas RL, Peters H, Xu X (2002) Eya1 is required for the morphogenesis of mammalian thymus, parathyroid and thyroid. Development 129: 3033–3044
772. Almo SC, Bonanno JB, Saunder JM, Emtage S, Dilorenzo TP, Malashkevich V, Wasserman SR, Swaminathan S, Eswaramoorthy S, Agarwal R, Kumaran D, Madegowda M, Ragumani S, Patskovsky Y, Alvarado J, Ramagopal UA, Faber-Barata J, Chance MR, Sali A, Fiser A, Zhang ZY, Lawrence DS, Burley SK (2007) Structural genomics of protein phosphatases. Journal of Structural and Functional Genomics 8:121–140

Chap. 9. Guanosine Triphosphatases and Their Regulators

773. Tuma PL, Collins CA (1995) Dynamin forms polymeric complexes in the presence of lipid vesicles. Characterization of chemically cross-linked dynamin molecules. Journal of Biological Chemistry 270:26707–26714
774. Sirajuddin M, Farkasovsky M, Zent E, Wittinghofer A (2009) GTP-induced conformational changes in septins and implications for function. Proceedings of the National Academy of Sciences of the United States of America 106:16592–16597
775. Gasper R, Meyer S, Gotthardt K, Sirajuddin M, Wittinghofer A (2009) It takes two to tango: regulation of G proteins by dimerization. Nature Reviews – Molecular Cell Biology 10: 423–429
776. Noguchi S, Toyoshima K, Yamamoto S, Miyazaki T, Otaka M, Watanabe S, Imai K, Senoo H, Kobayashi R, Jikei M, Kawata Y, Kubota H, Itoh H (2011) Cytosolic chaperonin CCT possesses GTPase activity. American Journal of Molecular Biology 1:123–130
777. http://www.sigmaaldrich.com/
778. Hildebrandt JD (1997) Role of subunit diversity in signaling by heterotrimeric G proteins. Biochemical Pharmacology 54:325–339
779. Hendriks-Balk MC, Peters SLM, Michel MC, Alewijnse AE (2008) Regulation of G protein-coupled receptor signalling: Focus on the cardiovascular system and regulator of G protein signalling proteins. European Journal of Pharmacology 585:278–291
780. Digby GJ, Lober RM, Sethi PR, Lambert NA (2006) Some G protein heterotrimers physically dissociate in living cells. Proceedings of the National Academy of Sciences of the United States of America 103:17789–17794
781. Berlot C (2004) G protein α s. UCSD-Nature Molecule Pages, UCSD-Nature Signaling Gateway (www.signaling-gateway.org)
782. Jiang P, Enomoto A, Jijiwa M, Kato T, Hasegawa T, Ishida M, Sato T, Asai N, Murakumo Y, Takahashi M (2008) An actin-binding protein Girdin regulates the motility of breast cancer cells. Cancer Research 68:1310–1318
783. Garcia-Marcos M, Kietrsunthorn PS, Pavlova Y, Adia MA, Ghosh P, Farquhar MG (2012) Functional characterization of the guanine nucleotide exchange factor (GEF) motif of GIV protein reveals a threshold effect in signaling. Proceedings of the National Academy of Sciences of the United States of America 109:1961–1966
784. Ghosh P, Beas AO, Bornheimer SJ, Garcia-Marcos M, Forry EP, Johannson C, Ear J, Jung BH, Cabrera B, Carethers JM, Farquhar MG (2010) A Gαi-GIV molecular complex binds epidermal growth factor receptor and determines whether cells migrate or proliferate. Molecular Biology of the Cell 21:2338–2354
785. Bajpayee NS, Jiang M (2010) G protein α i1. UCSD-Nature Molecule Pages, UCSD-Nature Signaling Gateway (www.signaling-gateway.org)
786. Kasahara K, Ui M (2011) G protein α o UCSD-Nature Molecule Pages, UCSD-Nature Signaling Gateway (www.signaling-gateway.org)

787. Tang G, Wang Y, Park S, Bajpayee NS, Vi D, Nagaoka Y, Birnbaumer L, Jiang M (2012) Go2 G protein mediates galanin inhibitory effects on insulin release from pancreatic β cells. Proceedings of the National Academy of Sciences of the United States of America 109:2636–2641
788. Kimple M, Manning D (2009) G protein α z. UCSD-Nature Molecule Pages, UCSD-Nature Signaling Gateway (www.signaling-gateway.org)
789. Wettschureck N (2009) G protein α q. UCSD-Nature Molecule Pages, UCSD-Nature Signaling Gateway (www.signaling-gateway.org)
790. Mizuno N, Itoh H (2009) Functions and regulatory mechanisms of Gq-signaling pathways. Neurosignals 17:42–54
791. Kurrasch DM, Huang J, Wilkie TM (2004) G protein-α11. UCSD-Nature Molecule Pages, UCSD-Nature Signaling Gateway (www.signaling-gateway.org)
792. Anger T, Zhang W, Mende U (2004) Differential contribution of GTPase activation and effector antagonism to the inhibitory effect of RGS proteins on Gq-mediated signaling in vivo. Journal of Biological Chemistry 279:3906–3915
793. Booden MA, Siderovski DP, Der CJ (2002) Leukemia-associated Rho guanine nucleotide exchange factor promotes Gαq-coupled activation of RhoA. Molecular and Cellular Biology 22:4053–4061
794. Oligny-Longpré G, Corbani M, Zhou J, Hogue M, Guillon G, Bouvier M (2012) Engagement of β-arrestin by transactivated insulin-like growth factor receptor is needed for V_2 vasopressin receptor-stimulated ERK1/2 activation. Proceedings of the National Academy of Sciences of the United States of America 109:E1028–E1037
795. Zimmerman B, Beautrait A, Aguila B, Charles R, Escher E, Claing A, Bouvier M, Laporte SA (2012) Differential β-arrestin-dependent conformational signaling and cellular responses revealed by angiotensin analogs. Science Signaling 5:ra33
796. Zhang W, Anger T, Su J, Hao J, Xu X, Zhu M, Gach A, Cui L, Liao R, Mende U (2006) Selective loss of fine tuning of Gq/11 signaling by RGS2 protein exacerbates cardiomyocyte hypertrophy. Journal of Biological Chemistry 281:5811–5820
797. Yan Y, Chi PP, Bourne HR (1997) RGS4 inhibits Gq-mediated activation of mitogen-activated protein kinase and phosphoinositide synthesis. Journal of Biological Chemistry 272:11924–11927
798. Zeng H, Zhao D, Yang S, Datta K, Mukhopadhyay D (2003) Heterotrimeric Gαq/Gα11 proteins function upstream of vascular endothelial growth factor (VEGF) receptor-2 (KDR) phosphorylation in vascular permeability factor/VEGF signaling. Journal of Biological Chemistry 278:20738–20745
799. Huang J, Wilkie TM (2006) G protein α 15. UCSD-Nature Molecule Pages, UCSD-Nature Signaling Gateway (www.signaling-gateway.org)
800. Hajicek N, Kozasa T (2008) G protein-α 13. UCSD-Nature Molecule Pages, UCSD-Nature Signaling Gateway (www.signaling-gateway.org)
801. Kurose H (2003) Gα$_{12}$ and Gα$_{13}$ as key regulatory mediator in signal transduction. Life Sciences 74:155–161
802. Siehler S (2009) Regulation of RhoGEF proteins by $G_{12/13}$-coupled receptors. British Journal of Pharmacology 158:41–49
803. Lee WH, Lee CH, Moon A, Dhanasekaran DN, Kim SK (2010) G protein α 12. UCSD-Nature Molecule Pages, UCSD-Nature Signaling Gateway (www.signaling-gateway.org)
804. Kleuss C (2007) G protein-γ 8. UCSD-Nature Molecule Pages, UCSD-Nature Signaling Gateway (www.signaling-gateway.org)
805. Pumiglia KM, LeVine H, Haske T, Habib T, Jove R, Decker SJ (1995) A direct interaction between G-proteinβγsubunits and the Raf-1 protein kinase. Journal of Biological Chemistry 270:14251–14254
806. Wedegaertner PB, Wilson PT, Bourne HR (1995) Lipid modifications of trimeric G proteins. Journal of Biological Chemistry 270:503–506

807. Yan K, Kalyanaraman V, Gautam N (1996) Differential ability to form the G protein βγ complex among members of the β and γ subunit families. Journal of Biological Chemistry 271:7141–7146
808. Mederos y Schnitzler M, Storch U, Meibers S, Nurwakagari P, Breit A, Essin K, Gollasch M, Gudermann T (2008) Gq-coupled receptors as mechanosensors mediating myogenic vasoconstriction. EMBO Journal 27:3092–3103
809. Hippe HJ, Luedde M, Lutz S, Koehler H, Eschenhagen T, Frey N, Katus HA, Wieland T, Niroomand F (2007) Regulation of cardiac cAMP synthesis and contractility by nucleoside diphosphate kinase B/G protein βγ dimer complexes. Circulation Research 100:1191–1199
810. Hippe HJ, Wolf NM, Abu-Taha I, Mehringer R, Just S, Lutz S, Niroomand F, Postel EH, Katus HA, Rottbauer W, Wieland T (2009) The interaction of nucleoside diphosphate kinase B with Gβγ dimers controls heterotrimeric G protein function. Proceedings of the National Academy of Sciences of the United States of America 106:16269–16274
811. Siderovski DP, Willard FS (2005) The GAPs, GEFs, and GDIs of heterotrimeric G-protein alpha subunits. International Journal of Biological Sciences 1:51–66
812. Popov SG, Krishna UM, Falck JR, Wilkie TM (2000) Ca2+/Calmodulin reverses phosphatidylinositol-3,4,5-trisphosphate-dependent inhibition of regulators of G protein-signaling GTPase-activating protein activity. Journal of Biological Chemistry 275:18962–18968
813. Cho H, Kehrl JH (2009) Rgs1; Rgs3. UCSD-Nature Molecule Pages, UCSD-Nature Signaling Gateway (www.signaling-gateway.org)
814. Ocal O, Wilkie TM (2010) Rgs8. UCSD-Nature Molecule Pages, UCSD-Nature Signaling Gateway (www.signaling-gateway.org)
815. Shi CS, Lee SB, Sinnarajah S, Dessauer CW, Rhee SG, Kehrl JH (2001) Regulator of G-protein signaling 3 (RGS3) inhibits Gbeta1gamma2-induced inositol phosphate production, mitogen-activated protein kinase activation, and Akt activation. Journal of Biological Chemistry 276:24293–24300
816. Mittmann C, Schuler C, Chung CH, Hoppner G, Nose M, Kehrl JH, Wieland T (2001) Evidence for a short form of RGS3 preferentially expressed in the human heart. Naunyn Schmiedebergs Archives of Pharmacology 63:456–463
817. Doupnik CA, Xu T, Shinaman JM (2001) Profile of RGS expression in single rat atrial myocytes. Biochimica et Biophysica Acta 1522:97–107
818. Sabri A, Pak E, Alcott SA, Wilson BA, Steinberg SF (2000) Coupling function of endogenous α1- and β-adrenergic receptors in mouse cardiomyocytes. Circulation Research 86:1047–1053
819. Yasutake M, Haworth RS, King A, Avkiran M (1996) Thrombin activates the sarcolemmal Na+-H+ exchanger. Evidence for a receptor-mediated mechanism involving protein kinase C. Circulation Research 79:705–715
820. Tall GG, Krumins AM, Gilman AG (2003) Mammalian Ric-8A (synembryn) is a heterotrimeric Gα protein guanine nucleotide exchange factor. Journal of Biological Chemistry 278:8356–8362
821. Wang SC, Lai HL, Chiu YT, Ou R, Huang CL, Chern Y (2007) Regulation of type V adenylate cyclase by Ric8a, a guanine nucleotide exchange factor. Biochemical Journal 406:383–388
822. Nishimura A, Okamoto M, Sugawara Y, Mizuno N, Yamauchi J, Itoh H (2006) Ric-8A potentiates Gq-mediated signal transduction by acting downstream of G protein-coupled receptor in intact cells. Genes to Cells 11:487–498
823. Cismowski MJ (2006) Non-receptor activators of heterotrimeric G-protein signaling (AGS proteins). Seminars in Cell and Developmental Biology 17:334–344
824. Takesono A, Nowak MW, Cismowski M, Duzic E, Lanier SM (2002) Activator of G-protein signaling 1 blocks GIRK channel activation by a G-protein-coupled receptor: apparent disruption of receptor signaling complexes. Journal of Biological Chemistry 277:13827–13830

References

825. Jaffrey SR, Snowman AM, Eliasson MJ, Cohen NA, Snyder SH (1998) CAPON: a protein associated with neuronal nitric oxide synthase that regulates its interactions with PSD95, Neuron 20:115–124.
826. Takahashi H, Umeda N, Tsutsumi Y, Fukumura R, Ohkaze H, Sujino M, van der Horstd G, Yasuie A, Inouyeb SIT, Fujimoria A, Ohhata T, Arakia R, Abe M (2003) Mouse dexamethasone-induced RAS protein 1 gene is expressed in a circadian rhythmic manner in the suprachiasmatic nucleus. Molecular Brain Research 110:1–6
827. Sato M, Cismowski MJ, Toyota E, Smrcka Av, Lucchesi PA, Chilian WM, Lanier SM (2006) Identification of a receptor-independent activator of G-protein signaling (AGS8) in ischemic heart and its interaction with G$\beta\gamma$. Proceedings of the National Academy of Sciences of the United States of America 103:797–802
828. Groves B, Gong Q, Xu Z, HuntsmanC, Nguyen C, Li D, Ma D (2007) A specific role of AGS3 in the surface expression of plasma membrane proteins. Proceedings of the National Academy of Sciences of the United States of America 104:18103–18108
829. De Vries L, Fischer T, Tronchère H, Brothers GM, Strockbine B, Siderovski DP, Farquhar MG (2000) Activator of G protein signaling 3 is a guanine dissociation inhibitor for Galpha i subunits. Proceedings of the National Academy of Sciences of the United States of America 97:14364–14369
830. Boughton AP, Yang P, Tesmer VM, Ding B, Tesmer JJ, Chen Z (2011) Heterotrimeric G protein $\beta 1\gamma 2$ subunits change orientation upon complex formation with G protein-coupled receptor kinase 2 (GRK2) on a model membrane. Proceedings of the National Academy of Sciences of the United States of America 108:E667–E673
831. Heasman SJ, Ridley AJ (2008) Mammalian Rho GTPases: new insights into their functions from in vivo studies. Nature Reviews – Molecular Cell Biology 9:690–701.
832. Olofsson B (1999) Rho guanine dissociation inhibitors: pivotal molecules in cellular signalling. Cellular Signalling 11:545–554
833. Iden S, Collard JG (2008) Crosstalk between small GTPases and polarity proteins in cell polarization. Nature Reviews – Molecular Cell Biology 9:846–859
834. van Nieuw Amerongen GP, van Hinsbergh VWM (2001) Cytoskeletal effects of Rho-like small guanine nucleotide-binding proteins in the vascular system. Arteriosclerosis, Thrombosis, and Vascular Biology 21:300–311
835. Ménager C, Vassy J, Doliger C, Legrand Y, Karniguian A (1999) Subcellular localization of RhoA and ezrin at membrane ruffles of human endothelial cells: differential role of collagen and fibronectin. Experimental Cell Research 249:221–230
836. Matsumoto Y, Uwatoku T, Oi K, Abe K, Hattori T, Morishige K, Eto Y, Fukumoto Y, Nakamura K, Shibata Y, Matsuda T, Takeshita A, Shimokawa H (2004) Long-term inhibition of Rho-kinase suppresses neointimal formation after stent implantation in porcine coronary arteries: involvement of multiple mechanisms. Arteriosclerosis, Thrombosis, and Vascular Biology 24:181–186
837. Seasholtz TM, Majumdar M, Brown JH (1999) Rho as a mediator of G protein-coupled receptor signaling. Molecular Pharmacology 55:949–956
838. Donaldson JG, Jackson CL (2011) ARF family G proteins and their regulators: roles in membrane transport, development and disease. Nature Reviews – Molecular Cell Biology 12:362–375
839. Kahn RA (2005) Arf1; Arf2. UCSD-Nature Molecule Pages, UCSD-Nature Signaling Gateway (www.signaling-gateway.org)
840. D'Souza-Schorey C, Chavrier P (2006) ARF proteins: roles in membrane traffic and beyond. Nature Reviews – Molecular Cell Biology 7:347–358
841. Turner CE, Brown MC (2001) Cell motility: ARNO and ARF6 at the cutting edge. Current Biology 11:R875–R877
842. Liu Y, Kahn RA, Prestegard JH (2010) Dynamic structure of membrane-anchored Arf·GTP. Nature – Structural and Molecular Biology 17:876–881

843. Leonoudakis D, Conti LR, Radeke CM, McGuire LM, Vandenberg CA (2004) A multiprotein trafficking complex composed of SAP97, CASK, Veli, and Mint1 is associated with inward rectifier Kir2 potassium channels. Journal of Biological Chemistry 279:19051–19063
844. Kahn RA, Cunningham LA (2005) Arf4. UCSD-Nature Molecule Pages, UCSD-Nature Signaling Gateway (www.signaling-gateway.org)
845. Kontani K, Hori Y, Katada T (2009) Arf-like protein 13B. UCSD-Nature Molecule Pages, UCSD-Nature Signaling Gateway (www.signaling-gateway.org)
846. Hörer J, Blum R, Feick P, Nastainczyk W, Schulz I (1999) A comparative study of rat and human Tmp21 (p23) reveals the pseudogene-like features of human Tmp21-II. DNA Sequence 10:121–126
847. Allison AB, Casanova JE (2011) Arf6. UCSD-Nature Molecule Pages, UCSD-Nature Signaling Gateway (www.signaling-gateway.org)
848. Kahn RA, Shrivastava-Ranjan P (2006) Arf-like protein 1. UCSD-Nature Molecule Pages. UCSD-Nature Signaling Gateway (www.signaling-gateway.org)
849. Kahn RA, Bowzard JB (2006) Arf-like protein 2. UCSD-Nature Molecule Pages. UCSD-Nature Signaling Gateway (www.signaling-gateway.org)
850. Li CC, Lee FJS (2010) Arf-like protein 4D. UCSD-Nature Molecule Pages, UCSD-Nature Signaling Gateway (www.signaling-gateway.org)
851. Thompson A, Kanamarlapudi V (2011) Arf-like protein 8A; Arf-like protein 8B. UCSD-Nature Molecule Pages, UCSD-Nature Signaling Gateway (www.signaling-gateway.org)
852. Shin HW, Kobayashi H, Kitamura M, Waguri S, Suganuma T, Uchiyama Y, Nakayama K (2005) Roles of ARFRP1 (ADP-ribosylation factor-related protein 1) in post-Golgi membrane trafficking. Journal of Cell Science 118:4039–4048
853. Schürmann A, Schmidt M, Asmus M, Bayer S, Fliegert F, Koling S, Massmann S, Schilf C, Subauste MC, Voss M, Jakobs KH, Joost HG (1999) The ADP-ribosylation factor (ARF)-related GTPase ARF-related protein binds to the ARF-specific guanine nucleotide exchange factor cytohesin and inhibits the ARF-dependent activation of phospholipase-D. Journal of Biological Chemistry 274:9744–9751
854. Kanamarlapudi V, Wilson LM (2011) ADP-ribosylation factor domain protein 1. UCSD-Nature Molecule Pages, UCSD-Nature Signaling Gateway (www.signaling-gateway.org)
855. Huang M, Weissman JT, Beraud-Dufour S, Luan P, Wang C, Chen W, Aridor M, Wilson IA, Balch WE (2001) Crystal structure of Sar1-GDP at 1.7 A resolution and the role of the NH2 terminus in ER export. Journal of Cell Biology 155:937–948
856. Stenmark H (2009) Rab GTPases as coordinators of vesicle traffic. Nature Reviews – Molecular Cell Biology 10:513–525
857. Draper RK (2007) Rab6a. UCSD-Nature Molecule Pages. UCSD-Nature Signaling Gateway (www.signaling-gateway.org)
858. Schwartz SL, Cao C, Pylypenko O, Rak A, Wandinger-Ness A (2007) Rab GTPases at a glance. Journal of Cell Science 120:3905–3910
859. Pereira-Leal JB, Seabra MC (2001) Evolution of the Rab family of small GTP-binding proteins. Journal of Molecular Biology 4:889–901
860. Buvelot Frei S, Rahl PB, Nussbaum M, Briggs BJ, Calero M, Janeczko S, Regan AD, Chen CZ, Barral Y, Whittaker GR, Collins RN (2006) Bioinformatic and comparative localization of Rab proteins reveals functional insights into the uncharacterized GTPases Ypt10p and Ypt11p. Molecular and Cellular Biology 26:7299–7317
861. Nottingham RM, Pfeffer SR (2009) Defining the boundaries: Rab GEFs and GAPs. Proceedings of the National Academy of Sciences of the United States of America 106:14185–14186
862. Rivera-Molina FE, Novick PJ (2009) A Rab GAP cascade defines the boundary between two Rab GTPases on the secretory pathway. Proceedings of the National Academy of Sciences of the United States of America 106:14408–14413
863. Sivalingam DA, Amirshahi S, Thyagarajan K, Tofig BN, Stein MP (2011) Rab1a. UCSD-Nature Molecule Pages, UCSD-Nature Signaling Gateway (www.signaling-gateway.org)
864. Thyagarajan K, Stein MP (2008) Rab35. UCSD-Nature Molecule Pages. UCSD-Nature Signaling Gateway (www.signaling-gateway.org)

865. Junutula JR, Prekeris R (2009) Rab11a and Rab11b. UCSD-Nature Molecule Pages. UCSD-Nature Signaling Gateway (www.signaling-gateway.org)
866. Prekeris R (2009) Rab11-FIP3. UCSD-Nature Molecule Pages. UCSD-Nature Signaling Gateway (www.signaling-gateway.org)
867. Coppola T, Magnin-Luthi S, Perret-Menoud V, Gattesco S, Schiavo G, Regazzi R (2001) Direct interaction of the Rab3 effector RIM with Ca^{2+} channels, SNAP-25, and synaptotagmin. Journal of Biological Chemistry 276:32756–32762
868. Ward HH, Peterson BR, Wandinger-Ness A (2008) Rab26. UCSD-Nature Molecule Pages, UCSD-Nature Signaling Gateway (www.signaling-gateway.org)
869. Fukuda M, Kuroda TS (2002) Slac2-c (synaptotagmin-like protein homologue lacking C2 domains-c), a novel linker protein that interacts with Rab27, myosin Va/VIIa, and actin. Journal of Biological Chemistry 277:43096–43103
870. Fukuda M (2003) Distinct Rab binding specificity of Rim1, Rim2, rabphilin, and Noc2. Identification of a critical determinant of Rab3A/Rab27A recognition by Rim2. Journal of Biological Chemistry 278:15373–15380
871. Shirakawa R, Higashi T, Tabuchi A, Yoshioka A, Nishioka H, Fukuda M, Kita T, Horiuchi H (2004) Munc13-4 is a GTP-Rab27-binding protein regulating dense core granule secretion in platelets. Journal of Biological Chemistry 279:10730–10737
872. Yi Z, Yokota H, Torii S, Aoki T, Hosaka M, Zhao S, Takata K, Takeuchi T, Izumi T (2002) The Rab27a/granuphilin complex regulates the exocytosis of insulin-containing dense-core granules. Molecular and Cellular Biology 22:1858–1867
873. Sinka R, Gillingham AK, Kondylis V, Munro SJ (2008) Golgi coiled-coil proteins contain multiple binding sites for Rab family G proteins. Journal of Cell Biology 183:607–615
874. Lütcke A, Olkkonen VM, Dupree P, Lütcke H, Simons K, Zerial M (1995) Isolation of a murine cDNA clone encoding Rab19, a novel tissue-specific small GTPase. Gene 155:257–260
875. Thomas C, Rousset R, Noselli S (2009) JNK signalling influences intracellular trafficking during Drosophila morphogenesis through regulation of the novel target gene Rab30. Developmental Biology 331:250–260
876. Dejgaard SY, Murshid A, Erman A, Kizilay O, Verbich D, Lodge R, Dejgaard K, Ly-Hartig TB, Pepperkok R, Simpson JC, Presley JF (2008) Rab18 and Rab43 have key roles in ER-Golgi trafficking. Journal of Cell Science 121:2768–2781
877. Kajiho H, Saito K, Tsujita K, Kontani K, Araki Y, Kurosu H, Katada T (2003) RIN3: a novel Rab5 GEF interacting with amphiphysin II involved in the early endocytic pathway. Journal of Cell Science 116:4159–4168
878. Jones AT (2006) Rab21. UCSD-Nature Molecule Pages. UCSD-Nature Signaling Gateway (www.signaling-gateway.org)
879. Lin MG, Zhong Q (2011) Interaction between small GTPase Rab7 and PI3KC3 links autophagy and endocytosis: A new Rab7 effector protein sheds light on membrane trafficking pathways. Small Gtpases 2(2):85–88
880. Sun Q, Westphal W, Wong KN, Tan I, Zhong Q (2010) Rubicon controls endosome maturation as a Rab7 effector. Proceedings of the National Academy of Sciences of the United States of America 107:19338–19343
881. Pfeffer S (2011) Rab9. UCSD-Nature Molecule Pages, UCSD-Nature Signaling Gateway (www.signaling-gateway.org)
882. Nazarian R, Falcón-Pérez JM, Dell'Angelica EC (2003) Biogenesis of lysosome-related organelles complex 3 (BLOC-3): a complex containing the Hermansky-Pudlak syndrome (HPS) proteins HPS1 and HPS4. Proceedings of the National Academy of Sciences of the United States of America 100:8770–8775
883. Kloer DP, Rojas R, Ivan V, Moriyama K, van Vlijmen T, Murthy N, Ghirlando R, van der Sluijs P, Hurley JH, Bonifacino JS (2010) Assembly of the biogenesis of lysosome-related organelles complex-3 (BLOC-3) and its interaction with Rab9. Journal of Biological Chemistry 285:7794–7804

884. Sun Y, Bilan PJ, Liu Z, Klip A (2010) Rab8A and Rab13 are activated by insulin and regulate GLUT4 translocation in muscle cells. Proceedings of the National Academy of Sciences of the United States of America 107:19909–19914
885. Wandinger-Ness A, Deretic D (2008) Rab8a. UCSD-Nature Molecule Pages. UCSD-Nature Signaling Gateway (www.signaling-gateway.org)
886. Babbey CM, Ahktar N, Wang E, Chen CC, Grant BD, Dunn KW (2006) Rab10 regulates membrane transport through early endosomes of polarized Madin-Darby canine kidney cells. Molecular Biology of the Cell 17:3156–3175
887. Evans TN, Wicking CA (2006) Rab23. UCSD-Nature Molecule Pages. UCSD-Nature Signaling Gateway (www.signaling-gateway.org)
888. Maltese WA (2006) Rab24. UCSD-Nature Molecule Pages. UCSD-Nature Signaling Gateway (www.signaling-gateway.org)
889. Offermanns S, Rosenthal W (Eds.) (2008) Encyclopedia of Molecular Pharmacology (2nd ed.; 1505 p.) Springer, Berlin, Heidelberg, New York
890. Verma SK, Lal H, Golden HB, Gerilechaogetu F, Smith M, Guleria RS, Foster DM, Lu G, Dostal DE (2011) Rac1 and RhoA differentially regulate angiotensinogen gene expression in stretched cardiac fibroblasts. Cardiovascular Research 90:88–96
891. Castillo-Lluva S, Tatham MH, Jones RC, Jaffray EG, Edmondson RD, Hay RT, Malliri A (2010) Sumoylation of the GTPase Rac1 is required for optimal cell migration. Nature – Cell Biology 12:1078–185
892. Diebold BA, Bokoch GM (2008) Rac2. UCSD-Nature Molecule Pages. UCSD-Nature Signaling Gateway (www.signaling-gateway.org)
893. Knaus UG (2006) Rac3. UCSD-Nature Molecule Pages. UCSD-Nature Signaling Gateway (www.signaling-gateway.org)
894. Zhang J, Chang L, Chen C, Zhang M, Luo Y, Hamblin M, Villacorta L, Xiong JW, Chen YE, Zhang J, Zhu X (2011) Rad GTPase inhibits cardiac fibrosis through connective tissue growth factor. Cardiovascular Research 91:90–98
895. Fu M, Zhang J, Tseng YH, Cui T, Zhu X, Xiao Y, Mou Y, De Leon H, Chang MM, Hamamori Y, Kahn CR, Chen YE (2005) Rad GTPase attenuates vascular lesion formation by inhibition of vascular smooth muscle cell migration. Circulation 111:1071–1077
896. Sancak Y, Peterson TR, Shaul YD, Lindquist RA, Thoreen CC, Bar-Peled L, Sabatini DM (2008) The Rag GTPases bind Raptor and mediate amino acid signaling to mTORC1. Science 320:1496–1501
897. Sidhu RS, Clough RR, Bhullar RP (2005) Regulation of phospholipase C-δ1 through direct interactions with the small GTPase Ral and calmodulin. Journal of Biological Chemistry 280:21933–21941
898. Ohta Y, Suzuki N, Nakamura S, Hartwig JH, Stossel TP (1999) The small GTPase RalA targets filamin to induce filopodia. Proceedings of the National Academy of Sciences of the United States of America 96:2122–2128
899. Luo JQ, Liu X, Hammond SM, Colley WC, Feig LA, Frohman MA, Morris AJ, Foster DA (1997) RalA interacts directly with the Arf-responsive, PIP2-dependent phospholipase D1. Biochemical and Biophysical Research Communications 235:854–859
900. Kim JH, Lee SD, Han JM, Lee TG, Kim Y, Park JB, Lambeth JD, Suh PG, Ryu SH (1998) Activation of phospholipase D1 by direct interaction with ADP-ribosylation factor 1 and RalA. FEBS Letters 430:231–235
901. Chien Y, Kim S, Bumeister R, Loo YM, Kwon SW, Johnson CL, Balakireva MG, Romeo Y, Kopelovich L, Gale M Jr, Yeaman C, Camonis JH, Zhao Y, White MA (2006) RalB GTPase-mediated activation of the IκB family kinase TBK1 couples innate immune signaling to tumor cell survival. Cell 127:157–170
902. Vartak N, Bastiaens P (2010) Spatial cycles in G-protein crowd control. EMBO Journal 29:2689–2699
903. Pannekoek WJ, Kooistra MRH, Zwartkruis FJT, Bos JL (2009) Cell–cell junction formation: The role of Rap1 and Rap1 guanine nucleotide exchange factors. Biochimica et Biophysica Acta 1788:790–796

904. Spilker C, Acuña Sanhueza GA, Böckers TM, Kreutz MR, Gundelfinger ED (2007) SPAR2, a novel SPAR-related protein with GAP activity for Rap1 and Rap2. Journal of Neurochemistry 104:187–201
905. Nancy V, Wolthuis RM, de Tand MF, Janoueix-Lerosey I, Bos JL, de Gunzburg J (1999) Identification and characterization of potential effector molecules of the Ras-related GTPase Rap2. Journal of Biological Chemistry 274:8737–8745
906. Janoueix-Lerosey I, Pasheva E, de Tand MF, Tavitian A, de Gunzburg J (1998) Identification of a specific effector of the small GTP-binding protein Rap2. European Journal of Biochemistry 252:290–298
907. Lafuente E, van Puijenbroek A, Krause M, Carman C, Freeman G, Berezovskaya A, Springer T, Gertler F, Boussiotis V (2004) RIAM, an Ena/VASP and profilin ligand, interacts with Rap1-GTP and mediates Rap1-induced adhesion. Developmental Cell 7:585–595
908. Tamada M, Sheetz M, Sawada Y (2004) Activation of a signaling cascade by cytoskeleton stretch. Developmental Cell 7:709–718
909. Crabbe L, Karlseder J (2010) Mammalian Rap1 widens its impact. Nature – Cell Biology 12:733–735
910. Sfeir A, Kabir S, van Overbeek M, Celli GB, de Lange T (2010) Loss of Rap1 induces telomere recombination in the absence of NHEJ or a DNA damage signal. Science 327:1657–1661
911. Martinez P, Thanasoula M, Carlos AR, Gómez-López G, Tejera AM, Schoeftner S, Dominguez O, Pisano DG, Tarsounas M, Blasco MA (2010) Mammalian Rap1 controls telomere function and gene expression through binding to telomeric and extratelomeric sites. Nature – Cell Biology 12:768–780
912. Teo H, Ghosh S, Luesch H, Ghosh A, Wong ET, Malik N, Orth A, de Jesus P, Perry AS, Oliver JD, Tran NL, Speiser LJ, Wong M, Saez E, Schultz P, Chanda SK, Verma IM, Tergaonkar V (2010) Telomere-independent Rap1 is an IKK adaptor and regulates NF-kappaB-dependent gene expression. Nature – Cell Biology 12:758–767
913. Edreira MM, Li S, Hochbaum D, Wong S, Gorfe AA, Ribeiro-Neto F, Woods VL Jr, Altschuler DL (2009) Phosphorylation-induced conformational changes in Rap1b: allosteric effects on switch domains and effector loop. Journal of Biological Chemistry 284:27480–27486
914. Machida N, Umikawa M, Takei K, Sakima N, Myagmar BE, Taira K, Uezato H, Ogawa Y, Kariya K (2004) Mitogen-activated protein kinase kinase kinase kinase 4 as a putative effector of Rap2 to activate the c-Jun N-terminal kinase. Journal of Biological Chemistry 279:15711–15714
915. Greco F, Sinigaglia F, Balduini C, Torti M (2004) Activation of the small GTPase Rap2B in agonist-stimulated human platelets Journal of Thrombosis and Haemostasis 2:2223–2230
916. Guo, Yuan J, Tang W, Chen X, Gu X, Luo K, Wang Y, Wan B, Yu L (2007) Cloning and characterization of the human gene RAP2C, a novel member of Ras family, which activates transcriptional activities of SRE. Molecular Biology Reports 34:137–144
917. Paganini S, Guidetti GF, Catrical S, Trionfini P, Panelli S, Balduini C, Torti M (2006) Identification and biochemical characterization of Rap2C, a new member of the Rap family of small GTP-binding proteins. Biochimie 88:285–295
918. Gureasko J, Galush WJ, Boykevisch S, Sondermann H, Bar-Sagi D, Groves JT, Kuriyan J (2008) Membrane-dependent signal integration by the Ras activator Son of sevenless. Nature – Structural and Molecular Biology 15:452–461
919. Bos JL, de Rooij J, Reedquist KA (2001) Rap1 signalling: adhering to new models. Nature Reviews – Molecular Cell Biology 2:369–377
920. Ahearn IM, Haigis K, Bar-Sagi D, Philips MR (2011) Regulating the regulator: post-translational modification of RAS. Nature Reviews – Molecular Cell Biology 13:39–51
921. Rocks O, Peyker A, Kahms M, Verveer PJ, Koerner C, Lumbierres M, Kuhlmann J, Waldmann H, Wittinghofer A, Bastiaens PI (2005) An acylation cycle regulates localization and activity of palmitoylated Ras isoforms. Science 307:1746–1752.

922. Bos JL (1998) All in the family? New insights and questions regarding interconnectivity of Ras, Rap1 and Ral. EMBO Journal 17:6776–6782
923. Karnoub AE, Weinberg RA (2008) Ras oncogenes: split personalities. Nature Reviews – Molecular Cell Biology 9:517–531
924. Drosten M, Dhawahir A, Sum EYM, Urosevic J, Lechuga CG, Esteban LM, Castellano E, Guerra C, Santos E, Barbacid M (2010) Genetic analysis of Ras signalling pathways in cell proliferation, migration and survival. EMBO Journal 29:1091–1104
925. Pochynyuk O, Stockand JD, Staruschenko A (2007) Ion channel regulation by Ras, Rho, and Rab small GTPases. Experimental Biology and Medicine 232:1258–1265
926. Kennedy MB, Beale HC, Carlisle HJ, Washburn LR (2005) Integration of biochemical signalling in spines. Nature Reviews – Neuroscience. 6:423–434
927. Komatsu M, Ruoslaht E (2005) R-Ras is a global regulator of vascular regeneration that suppresses intimal hyperplasia and tumor angiogenesis. Nature – Medicine 11:1346–1350
928. Abankwa D, Gorfe AA, Inder K, Hancock JF (2010) Ras membrane orientation and nanodomain localization generate isoform diversity. Proceedings of the National Academy of Sciences of the United States of America 107:1130–1135
929. Berthiaume LG (2002) Insider information: how palmitoylation of Ras makes it a signaling double agent. Science STKE 2002:pe41.
930. Tian T, Harding A, Inder K, Plowman S, Parton RG, Hancock JF (2007) Plasma membrane nanoswitches generate high-fidelity Ras signal transduction. Nature – Cell Biology 9: 905–914
931. White MA (2004) H-Ras. UCSD-Nature Molecule Pages, UCSD-Nature Signaling Gateway (www.signaling-gateway.org)
932. Matheny SA, Chen C, Kortum RL, Razidlo GL, Lewis RE, White MA (2004) Ras regulates assembly of mitogenic signalling complexes through the effector protein IMP. Nature 427:256–260
933. Ebnet K, Schulz CU, Meyer Zu Brickwedde MK, Pendl GG, Vestweber D (2000) Junctional adhesion molecule interacts with the PDZ domain-containing proteins AF-6 and ZO-1. Journal of Biological Chemistry 275:27979–27988
934. Mitin NY, Ramocki MB, Zullo AJ, Der CJ, Konieczny SF, Taparowsky EJ (2004) Identification and characterization of rain, a novel Ras-interacting protein with a unique subcellular localization. Journal of Biological Chemistry 279:22353–22361
935. Young NP, Jacks T (2010) Tissue-specific $p19^{Arf}$ regulation dictates the response to oncogenic K-ras. Proceedings of the National Academy of Sciences of the United States of America 107:10184–10189
936. van der Weyden L, Adams DJ (2007) The Ras-association domain family (RASSF) members and their role in human tumourigenesis. Biochimica et Biophysica Acta 1776:58–85
937. Lim KH, Ancrile BB, Kashatus DF, Counter CM (2008) Tumour maintenance is mediated by eNOS. Nature 452:646–649
938. Cantrell DA (2003) GTPases and T cell activation. Immunological Reviews 192:122–130
939. Inoki K, Li Y, Xu T, Guan KL (2003) Rheb GTPase is a direct target of TSC2 GAP activity and regulates mTOR signaling. Genes and Development 17:1829–1834
940. Saito K, Araki Y, Kontani K, Nishina H, Katada T (2005) Novel role of the small GTPase Rheb: its implication in endocytic pathway independent of the activation of mammalian target of rapamycin. Journal of Biochemistry 137:423–430
941. Alberts AS, Bouquin N, Johnston LH, Treisman R (1998) Analysis of RhoA-binding proteins reveals an interaction domain conserved in heterotrimeric G protein βsubunits and the yeast response regulator protein Skn7. Journal of Biological Chemistry 273:8616–8622
942. Schwartz M (2004) Rho signalling at a glance. Journal of Cell Science 117:5457–5458
943. Zhao ZS, Manser E (2005) PAK and other Rho-associated kinases – effectors with surprisingly diverse mechanisms of regulation. Biochemical Journal 386:201–214
944. Mukai H (2003) The structure and function of PKN, a protein kinase having a catalytic domain homologous to that of PKC. Journal of Biochemistry 133:17–27

945. Miyoshi J, Takai Y (2009) RhoA. UCSD-Nature Molecule Pages, UCSD-Nature Signaling Gateway (www.signaling-gateway.org)
946. El Sayegh TY, Arora PD, Ling K, Laschinger C, Janmey PA, Anderson RA, McCulloch CA (2007) Phosphatidylinositol-4,5 bisphosphate produced by PIP5KIγ regulates gelsolin, actin assembly, and adhesion strength of N-cadherin junctions. Molecular Biology of the Cell 18:3026–3038
947. Laudanna C, Bolomini-Vittori M (2008) RhoC. UCSD-Nature Molecule Pages, UCSD-Nature Signaling Gateway (www.signaling-gateway.org)
948. Ramos S, Khademi F, Somesh BP, Rivero F (2002) Genomic organization and expression profile of the small GTPases of the RhoBTB family in human and mouse. Gene 298:147–157
949. Berthold J, Schenkova K, Rivero F (2008) Rho GTPases of the RhoBTB subfamily and tumorigenesis. Acta Pharmacologica Sinica 29:285–295
950. Wilkins A, Carpenter CL (2008) Regulation of RhoBTB2 by the Cul3 ubiquitin ligase complex. Methods in Enzymology 439:103–109
951. Gasman S, Kalaidzidis Y, Zerial M (2003) RhoD regulates endosome dynamics through Diaphanous-related Formin and Src tyrosine kinase. Nature – Cell Biology 5:195–204
952. Tsubakimoto K, Matsumoto K, Abe H, Ishii J, Amano M, Kaibuchi K, Endo T (1999) Small GTPase RhoD suppresses cell migration and cytokinesis. Oncogene 18:2431–2440
953. Gouw LG, Reading NS, Jenson SD, Lim MS, Elenitoba-Johnson KSJ (2005) Expression of the Rho-family GTPase gene RHOF in lymphocyte subsets and malignant lymphomas. British Journal of Haematology 129:531–533
954. Pellegrin S, Mellor H (2005) The Rho family GTPase Rif induces filopodia through mDia2. Current Biology 15:129–133
955. Gauthier-Rouvière C, Vignal E, Mériane M, Roux P, Montcourier P, Fort P (1998) RhoG GTPase controls a pathway that independently activates Rac1 and Cdc42Hs. Molecular Biology of the Cell 9:1379–1394
956. Le Gallic L, Fort P (1997) Structure of the human ARHG locus encoding the Rho/Rac-like RhoG GTPase. Genomics 42:157–160
957. Fueller F, Kubatzky KF (2008) The small GTPase RhoH is an atypical regulator of haematopoietic cells. Cell Communication and Signaling 6:6
958. de Toledo M, Senic-Matuglia F, Salamero J, Uze G, Comunale F, Fort P, Blangy A (2003) The GTP/GDP cycling of rho GTPase TCL is an essential regulator of the early endocytic pathway. Molecular Biology of the Cell 14:4846–4856
959. Nishizuka M, Arimoto E, Tsuchiya T, Nishihara T, Imagawa M (2003) Crucial role of TCL/TC10βL, a subfamily of Rho GTPase, in adipocyte differentiation. Journal of Biological Chemistry 278:15279–15284
960. Murphy GA, Solski PA, Jillian SA, Pérez de la Ossa P, D'Eustachio P, Der CJ, Rush MG (1999) Cellular functions of TC10, a Rho family GTPase: regulation of morphology, signal transduction and cell growth. Oncogene 18:3831–3845
961. Watson RT, Kanzaki M, Pessin J (2004) Regulated membrane trafficking of the insulin-responsive glucose transporter 4 in adipocytes. Endocrine Reviews 25:177–204
962. Chang L, Chiang SH, Saltiel AR (2007) TC10α is required for insulin-stimulated glucose uptake in adipocytes. Endocrinology 148:27–33
963. Tian L, Nelson DL, Stewart DM (2000) Cdc42-interacting protein 4 mediates binding of the Wiskott-Aldrich syndrome protein to microtubules. Journal of Biological Chemistry 275:7854–7861
964. Chang L, Adams RD, Saltiel AR (2002) The TC10-interacting protein CIP4/2 is required for insulin-stimulated Glut4 translocation in 3T3L1 adipocytes. Proceedings of the National Academy of Sciences of the United States of America 99:12835–12840
965. Maffucci T, Brancaccio A, Piccolo E, Stein RC, Falasca M (2003) Insulin induces phosphatidylinositol-3-phosphate formation through TC10 activation. EMBO Journal 22:4178–4189

966. Cheng J, Wang H, Guggino WB (2005) Regulation of cystic fibrosis transmembrane regulator trafficking and protein expression by a Rho family small GTPase TC10. Journal of Biological Chemistry 280:3731–3739
967. Krekelberg JL, Spiker CL, DeRouen AJ, Repasky GA (2009) Tc10. UCSD-Nature Molecule Pages, UCSD-Nature Signaling Gateway (www.signaling-gateway.org)
968. Joberty G, Perlungher RR, Macara IG (1999) The Borgs, a new family of Cdc42 and TC10 GTPase-interacting proteins. Molecular and Cellular Biology 19:6585–6597
969. Abe T, Kato M, Miki H, Takenawa T, Endo T (2003) Small GTPase Tc10 and its homologue RhoT induce N-WASP-mediated long process formation and neurite outgrowth. Journal of Cell Science 116:155–168
970. Fransson S, Ruusala A, Aspenström P (2006) The atypical Rho GTPases Miro-1 and Miro-2 have essential roles in mitochondrial trafficking. Biochemical and Biophysical Research Communications 344:500–510
971. Brazier H, Pawlak G, Vives V, Blangy A (2009) The Rho GTPase Wrch1 regulates osteoclast precursor adhesion and migration. International Journal of Biochemistry and Cell Biology 41:1391–1401
972. Schiavone D, Dewilde S, Vallania F, Turkson J, Di Cunto F, Poli V (2009) The RhoU/Wrch1 Rho GTPase gene is a common transcriptional target of both the gp130/STAT3 and Wnt-1 pathways. Biochemical Journal 421:283–292
973. Sorokina EM, Chernoff J (2005) Rho-GTPases: new members, new pathways. Journal of Cellular Biochemistry 94:225–231
974. Finlin BS, Andres DA (1999) Phosphorylation-dependent association of the Ras-related GTP-binding protein Rem with 14-3-3 proteins. Archives of Biochemistry and Biophysics 368):401–412
975. Reynet C, Kahn CR (1993) Rad: a member of the Ras family overexpressed in muscle of type II diabetic humans. Science 262:1441–1444
976. Chang L, Zhang J, Tseng YH, Xie CQ, Ilany J, Brüning JC, Sun Z, Zhu X, Cui T, Youker KA, Yang Q, Day SM, Kahn CR, Chen YE (2007) Rad GTPase deficiency leads to cardiac hypertrophy. Circulation 116:2976–2983
977. Bannister RA, Colecraft HM, Beam KG (2008) Rem inhibits skeletal muscle EC coupling by reducing the number of functional L-type Ca^{2+} channels. Biophysical Journal 94: 2631–2638
978. Seu L, Pitt GS (2006) Dose-dependent and isoform-specific modulation of Ca^{2+} channels by RGK GTPases. Journal of General Physiology 128:605–613
979. Leyris JP, Gondeau C, Charnet A, Delattre C, Rousset M, Cens T, Charnet P (2009) RGK GTPase-dependent $Ca_V 2.1$ Ca^{2+} channel inhibition is independent of $Ca_V \beta$-subunit-induced current potentiation. FASEB Journal 23:2627–2638
980. Chen H, Puhl HL, Niu SL, Mitchell DC, Ikeda SR (2005) Expression of Rem2, an RGK family small GTPase, reduces N-type calcium current. Journal of Neuroscience 25: 9762–9772
981. Finlin BS, Mosley AL, Crump SM, Correll RN, Ozcan S, Satin J, Andres DA (2005) Regulation of L-type Ca^{2+} channel activity and insulin secretion by the Rem2 GTPase. Journal of Biological Chemistry 280:41864–41871
982. Splingard A, Ménétrey J, Perderiset M, Cicolari J, Regazzoni K, Hamoudi F, Cabanié L, El Marjou A, Wells A, Houdusse A, de Gunzburg J (2007) Biochemical and structural characterization of the gem GTPase. Journal of Biological Chemistry 282:1905–1915
983. Chardin P (2006) Function and regulation of Rnd proteins. Nature Reviews – Molecular Cell Biology 7:54–62
984. Uesugi K, Oinuma I, Katoh H, Negishi M (2009) Different requirement for Rnd GTPases of R-Ras GAP activity of Plexin-C1 and Plexin-D1. Journal of Biological Chemistry 284: 6743–6751
985. Chardin P (2005) Rnd1; Rnd2; Rnd3. UCSD-Nature Molecule Pages, UCSD-Nature Signaling Gateway (www.signaling-gateway.org)

986. Ongusaha PP, Kim HG, Boswell SA, Ridley AJ, Der CJ, Dotto GP, Kim YB, Aaronson SA, Lee SW (2006) RhoE is a pro-survival p53 target gene that inhibits ROCK I-mediated apoptosis in response to genotoxic stress. Current Biology 16:2466–2472
987. Marie-Claire C, Salzmann J, David A, Courtin C, Canestrelli C, Noble F (2007) Rnd family genes are differentially regulated by 3,4-methylenedioxymethamphetamine and cocaine acute treatment in mice brain. Brain Research 1134:12–17
988. Lee CH, Della NG, Chew CE, Zack DJ (1996) Rin, a neuron-specific and calmodulin-binding small G-protein, and Rit define a novel subfamily of ras proteins. Journal of Neuroscience 16:6784–6794
989. Milstein M, Mooser CK, Hu H, Fejzo M, Slamon D, Goodglick L, Dry S, Colicelli J (2007) RIN1 is a breast tumor suppressor gene. Cancer Research 67:11510–11516
990. Shi GX, Jin L, Andres DA (2008) Pituitary adenylate cyclase-activating polypeptide 38-mediated Rin activation requires Src and contributes to the regulation of HSP27 signaling during neuronal differentiation. Molecular and Cellular Biology 28:4940–4951
991. Andres DA, Rudolph J (2008) Rin. UCSD-Nature Molecule Pages, UCSD-Nature Signaling Gateway (www.signaling-gateway.org)
992. Hoshino M, Yoshimori T, Nakamura S (2005) Small GTPase proteins Rin and Rit bind to PAR6 GTP-dependently and regulate cell transformation. Journal of Biological Chemistry 280:22868–22874
993. Ye K, Snyder SH (2004) PIKE GTPase: a novel mediator of phosphoinositide signaling. Journal of Cell Science 117:155–161
994. Yeh BJ, Rutigliano RJ, Deb A, Bar-Sagi D, Lim WA (2007) Rewiring cellular morphology pathways with synthetic guanine nucleotide exchange factors. Nature 447:596–600
995. Wolthuis RM, Bos JL (1999) Ras caught in another affair: the exchange factors for Ral. Current Opinion in Genetics and Development 9:112–117
996. Tall GG, Barbieri MA, Stahl PD, Horazdovsky BF (2001) Ras-activated endocytosis is mediated by the Rab5 guanine nucleotide exchange activity of RIN1. Developmental Cell 1:73–82
997. Lambert JM, Lambert QT, Reuther GW, Malliri A, Siderovski DP, Sondek J, Collard JG, Der CJ (2002) Tiam1 mediates Ras activation of Rac by a PI(3)K-independent mechanism. Nature – Cell Biology 4:621–625
998. Jackson CL (2003) The Sec7 family of ARF guanine-nucleotide exchange factor (Chap. 4) In Kahn RA (Ed.) ARF Family GTPases, Series: Proteins and Cell Regulation (Vol. 1) Kluwer Academic Publishers, Dordrecht, The Netherlands
999. Casanova JE (2007) Regulation of Arf activation: the Sec7 family of guanine nucleotide exchange factors. Traffic 8:1476–1485
1000. Geiger C, Nagel W, Boehm T, van Kooyk Y, Figdor CG, Kremmer E, Hogg N, Zeitlmann L, Dierks H, Weber KS, Kolanus W (2000) Cytohesin-1 regulates beta-2 integrin-mediated adhesion through both ARF-GEF function and interaction with LFA-1. EMBO Journal 19:2525–2536
1001. Melançon P, Zhao X, Lasell TKR (2004) Large Arf GEFs of the Golgi complex (p. 101–119) In Kahn RA (ed.) ARF Family GTPases. Series: Proteins and Cell Regulation (Vol. 1) Kluwer Academic Publishers, Dordrecht, The Netherlands
1002. Claude A, Zhao BP, Kuziemsky CE, Dahan S, Berger SJ, Yan JP, Arnold AD, Sullivan EM, Melanon P (1999) GBF1: A novel Golgi-associated BFA-resistant guanine nucleotide exchange factor that displays specificity for ADP-ribosylation factor 5. Journal of Cell Biology 146:71–84
1003. Niu TK, Pfeifer AC, Lippincott-Schwartz J, Jackson CL (2005) Dynamics of GBF1, a Brefeldin A-sensitive Arf1 exchange factor at the Golgi. Molecular Biology of the Cell 16:1213–1222
1004. Luton F, Klein S, Chauvin JP, Le Bivic A, Bourgoin S, Franco M, Chardin P (2004) EFA6, exchange factor for ARF6, regulates the actin cytoskeleton and associated tight junction in response to E-cadherin engagement. Molecular Biology of the Cell 15:1134–1145

1005. Fukaya M, Kamata A, Hara Y, Tamaki H, Katsumata O, Ito N, Takeda S, Hata Y, Suzuki T, Watanabe M, Harvey RJ, Sakagami H (2011) SynArfGEF is a guanine nucleotide exchange factor for Arf6 and localizes preferentially at post-synaptic specializations of inhibitory synapses. Journal of Neurochemistry 116:1122–1137
1006. Dunphy JL, Ye K, Casanova JE (2007) Nuclear functions of the Arf guanine nucleotide exchange factor BRAG2. Traffic 8:661–672
1007. Otomo A, Hadano S, Okada T, Mizumura H, Kunita R, Nishijima H, Showguchi-Miyata J, Yanagisawa Y, Kohiki E, Suga E, Yasuda M, Osuga H, Nishimoto T, Narumiya S, Ikeda JE (2003) ALS2, a novel guanine nucleotide exchange factor for the small GTPase Rab5, is implicated in endosomal dynamics. Human Molecular Genetics 12:1671–1687
1008. Lai C, Xie C, Shim H, Chandran J, Howell BW, Cai H (2009) Regulation of endosomal motility and degradation by amyotrophic lateral sclerosis 2/alsin. Molecular Brain 2:23
1009. Kunita R, Otomo A, Mizumura H, Suzuki-Utsunomiya K, Hadano S, Ikeda JE (2007) The Rab5 activator ALS2/alsin acts as a novel Rac1 effector through Rac1-activated endocytosis. Journal of Biological Chemistry 282:16599–16611
1010. Penzes P, Cahill ME, Jones KA, Srivastava DP (2008) Convergent CaMK and RacGEF signals control dendritic structure and function. Trends in Cell Biology 18:405–413
1011. Knezevic II, Predescu SA, Neamu RF, Gorovoy MS, Knezevic NM, Easington C, Malik AB, Predescu DN (2009) Tiam1 and Rac1 are required for platelet-activating factor-induced endothelial junctional disassembly and increase in vascular permeability. Journal of Biological Chemistry 284:5381–5394
1012. Sands WA, Woolson HD, Milne GR, Rutherford C, Palmer TM (2006) Exchange protein activated by cyclic AMP (Epac)-mediated induction of suppressor of cytokine signaling 3 (SOCS-3) in vascular endothelial cells. Molecular and Cellular Biology 26:6333–6346
1013. Rooney C, White G, Nazgiewicz A, Woodcock SA, Anderson KI, Ballestrem C, Malliri A (2010) The Rac activator STEF (Tiam2) regulates cell migration by microtubule-mediated focal adhesion disassembly. EMBO Reports 11:292–298
1014. Hernández-Negrete I, Carretero-Ortega J, Rosenfeldt H, Hernández-Garca R, Caldern-Salinas JV, Reyes-Cruz G, Gutkind JS, Vázquez-Prado J (2007) P-Rex1 links mammalian target of rapamycin signaling to Rac activation and cell migration. Journal of Biological Chemistry 282:23708–23715
1015. Pannekoek WJ, Kooistra MRH, Zwartkruis FJT, Bos JL (2009) Cell–cell junction formation: The role of Rap1 and Rap1 guanine nucleotide exchange factors. Biochimica et Biophysica Acta 1788:790–796
1016. Kuiperij HB, de Rooij J, Rehmann H, van Triest M, Wittinghofer A, Bos JL, Zwartkruis FJ (2003) Characterisation of PDZ-GEFs, a family of guanine nucleotide exchange factors specific for Rap1 and Rap2. Biochimica et Biophysica Acta 1593:141–149
1017. Gao X, Satoh T, Liao Y, Song C, Hu CD, Kariya Ki K, Kataoka T (2001) Identification and characterization of RA-GEF-2, a Rap guanine nucleotide exchange factor that serves as a downstream target of M-Ras. Journal of Biological Chemistry 276:42219–42225
1018. Consonni SV, Gloerich M, Spanjaard E, Bos JL (2012) cAMP regulates DEP domain-mediated binding of the guanine nucleotide exchange factor Epac1 to phosphatidic acid at the plasma membrane. Proceedings of the National Academy of Sciences of the United States of America 109:3814–3819
1019. Ichiba T, Hoshi Y, Eto Y, Tajima N, Kuraishi Y (1999) Characterization of GFR, a novel guanine nucleotide exchange factor for Rap1. FEBS Letters 457:85–89
1020. Rebhun JF, Castro AF, Quilliam LA (2000) Identification of guanine nucleotide exchange factors (GEFs) for the Rap1 GTPase. Regulation of MR-GEF by M-Ras-GTP interaction. Journal of Biological Chemistry 275:34901–34908
1021. Chuang TH, Xu X, Quilliam LA, Bokoch GM (1994) SmgGDS stabilizes nucleotide-bound and -free forms of the Rac1 GTP-binding protein and stimulates GTP/GDP exchange through a substituted enzyme mechanism. Biochemical Journal 303:761–767
1022. Caloca MJ, Zugaza JL, Bustelo XR (2003) Exchange factors of the RasGRP family mediate Ras activation in the Golgi. Journal of Biological Chemistry 278:33465–33473

1023. Ebinu JO, Bottorff DA, Chan EY, Stang SL, Dunn RJ, Stone JC (1998) RasGRP, a Ras guanyl nucleotide-releasing protein with calcium- and diacylglycerol-binding motifs. Science 280:1082–1086
1024. Norum JH, Hart K, Levy FO (2003) Ras-dependent ERK activation by the human Gs-coupled serotonin receptors 5-HT4(b) and 5-HT7(a). Journal of Biological Chemistry 278:3098–3104
1025. Ruiz S, Santos E, Bustelo XR (2007) RasGRF2, a guanosine nucleotide exchange factor for Ras GTPases, participates in T-cell signaling responses. Molecular and Cellular Biology 27:8127–8142
1026. Chiu VK, Bivona T, Hach A, Sajous JB, Silletti J, Wiener H, Johnson RL 2nd, Cox AD, Philips MR (2002) Ras signalling on the endoplasmic reticulum and the Golgi. Nature – Cell Biology 4:343–350
1027. Rubio I, Grund S, Song SP, Biskup C, Bandemer S, Fricke M, Förster M, Graziani A, Wittig U, Kliche S (2010) TCR-induced activation of Ras proceeds at the plasma membrane and requires palmitoylation of N-Ras. Journal of Immunology 185:3536–3543
1028. Quilliam LA, Rebhun JF, Castro AF (2002) A growing family of guanine nucleotide exchange factors is responsible for activation of Ras-family GTPases. Progress in Nucleic Acid Research and Molecular Biology 71:391–444
1029. Santos E, Fernández-Medarde A (2009) Rasgrf1. UCSD-Nature Molecule Pages, UCSD-Nature Signaling Gateway (www.signaling-gateway.org)
1030. Anborgh PH, Qian X, Papageorge AG, Vass WC, DeClue JE, Lowy DR (1999) Ras-specific exchange factor GRF: oligomerization through its Dbl homology domain and calcium-dependent activation of Raf. Molecular and Cellular Biology 19:4611–4622
1031. Fan WT, Koch CA, de Hoog CL, Fam NP, Moran MF (1998) The exchange factor Ras-GRF2 activates Ras-dependent and Rac-dependent mitogen-activated protein kinase pathways. Current Biology 8:935–938
1032. Nishihara H, Kobayashi S, Hashimoto Y, Ohba F, Mochizuki N, Kurata T, Nagashima K, Matsuda M (1999) Non-adherent cell-specific expression of DOCK2, a member of the human CDM-family proteins. Biochimica et Biophysica Acta 1452:179–187
1033. Lu M, Ravichandran KS (2006) Dock180–ELMO cooperation in Rac activation. Methods in Enzymology 406:388–402
1034. Makino Y, Tsuda M, Ichihara S, Watanabe T, Sakai M, Sawa H, Nagashima K, Hatakeyama S, Tanaka S (2006) Elmo1 inhibits ubiquitylation of Dock180. Journal of Cell Science 119:923–932
1035. Hajdo-Milasinovic A, Ellenbroek SI, van Es S, van der Vaart B, Collard JG (2007) Rac1 and Rac3 have opposing functions in cell adhesion and differentiation of neuronal cells. Journal of Cell Science 120:555–566
1036. Upadhyay G, Goessling W, North TE, Xavier R, Zon LI, Yajnik V (2008) Molecular association between β-catenin degradation complex and Rac guanine exchange factor DOCK4 is essential for Wnt/β-catenin signaling. Oncogene 27:5845–5855
1037. Garron ML, Arsenieva D, Zhong J, Bloom AB, Lerner A, O'Neill GM, Arold ST (2009) Structural insights into the association between BCAR3 and Cas family members, an atypical complex implicated in anti-oestrogen resistance. Journal of Molecular Biology 386:190–203
1038. Bouton AH, Riggins RB, Bruce-Staskal PJ (2001) Functions of the adapter protein Cas: signal convergence and the determination of cellular responses. Oncogene 20:6448–6458
1039. Dodelet VC, Pazzagli C, Zisch AH, Hauser CA, Pasquale EB (1999) A novel signaling intermediate, SHEP1, directly couples Eph receptors to R-Ras and Rap1A. Journal of Biological Chemistry 274:31941–31946
1040. Schwab ME (2010) Functions of Nogo proteins and their receptors in the nervous system. Nature Reviews – Neuroscience 11:799–811

1041. Near RI, Zhang Y, Makkinje A, Vanden Borre P, Lerner A (2007) AND-34/BCAR3 differs from other NSP homologs in induction of anti-estrogen resistance, cyclin D1 promoter activation and altered breast cancer cell morphology. Journal of Cellular Physiology 212:655–665

1042. Lu Y, Brush J, Stewart TA (1999) NSP1 defines a novel family of adaptor proteins linking integrin and tyrosine kinase receptors to the c-Jun N-terminal kinase/stress-activated protein kinase signaling pathway. Journal of Biological Chemistry 274:10047–10052

1043. GeneCards human gene database. Crown Human Genome Center, Department of Molecular Genetics, the Weizmann Institute of Science (www.genecards.org)

1044. Riggins RB, Quilliam LA, Bouton AH (2003) Synergistic promotion of c-Src activation and cell migration by Cas and AND-34/BCAR3. Journal of Biological Chemistry 278: 28264–28273

1045. Felekkis KN, Narsimhan RP, Near R, Castro AF, Zheng Y, Quilliam LA, Lerner A (2005) AND-34 activates phosphatidylinositol 3-kinase and induces anti-estrogen resistance in a SH2 and GDP exchange factor-like domain-dependent manner. Molecular Cancer Research 3:32–41

1046. Cai D, Felekkis KN, Near RI, O'Neill GM, van Seventer JM, Golemis EA, Lerner A (2003) The GDP exchange factor AND-34 is expressed in B cells, associates with HEF1, and activates Cdc42. Journal of Immunology 170:969–978

1047. Loirand G, Scalbert E, Bril A, Pacaud P (2008) Rho exchange factors in the cardiovascular system. Current Opinion in Pharmacology 8:174–180

1048. Guilluy C, Brégeon J, Toumaniantz G, Rolli-Derkinderen M, Retailleau K, Loufrani L, Henrion D, Scalbert E, Bril A, Torres RM, Offermanns S, Pacaud P, Loirand G (2010) The Rho exchange factor Arhgef1 mediates the effects of angiotensin II on vascular tone and blood pressure. Nature – Medicine 16:183–190

1049. Chikumi H, Barac A, Behbahani B, Gao Y, Teramoto H, Zheng Y, Gutkind JS (2004) Homo- and hetero-oligomerization of PDZ-RhoGEF, LARG and p115RhoGEF by their C-terminal region regulates their in vivo Rho GEF activity and transforming potential. Oncogene 23:233–240

1050. Chow CR, Yau DM, Kozasa T (2011) p115RhoGEF. UCSD-Nature Molecule Pages, UCSD-Nature Signaling Gateway (www.signaling-gateway.org)

1051. Chen Z, Medina F, Liu MY, Thomas C, Sprang SR, Sternweis PC (2010) Activated RhoA binds to the pleckstrin homology (PH) domain of PDZ-RhoGEF, a potential site for autoregulation. Journal of Biological Chemistry 285(27):21070–21081

1052. Feng Q, Baird D, Peng X, Wang J, Ly T, Guan JL, Cerione RA (2006) Cool-1 functions as an essential regulatory node for EGFreceptor- and Src-mediated cell growth. Nature – Cell Biology 8:945–956

1053. Ueda S, Kataoka T, Satoh T (2004) Role of the Sec14-like domain of Dbl family exchange factors in the regulation of Rho family GTPases in different subcellular sites. Cell Signalling 16:899–906

1054. Ieguchi K, Ueda S, Kataoka T, Satoh T (2007) Role of the guanine nucleotide exchange factor Ost in negative regulation of receptor endocytosis by the small GTPase Rac1. Journal of Biological Chemistry 282:23296–23305

1055. Bellanger JM, Astier C, Sardet C, Ohta Y, Stossel TP, Debant A (2000) The Rac1- and RhoG-specific GEF domain of Trio targets filamin to remodel cytoskeletal actin. Nature – Cell Biology 2:888–892

1056. Ma XM, Kiraly DD, Gaier ED, Wang Y, Kim EJ, Levine ES, Eipper BA, Mains RE (2008) Kalirin-7 is required for synaptic structure and function. Journal of Neuroscience 28: 12368–12382

1057. Salazar MA, Kwiatkowski AV, Pellegrini L, Cestra G, Butler MH, Rossman KL, Serna DM, Sondek J, Gertler FB, De Camilli P (2003) Tuba, a novel protein containing bin/amphiphysin/Rvs and Dbl homology domains, links dynamin to regulation of the actin cytoskeleton. Journal of Biological Chemistry 278:49031–49043.

1058. van Horck FP, Ahmadian MR, Haeusler LC, Moolenaar WH, Kranenburg O (2001) Characterization of p190RhoGEF, a RhoA-specific guanine nucleotide exchange factor that interacts with microtubules. Journal of Biological Chemistry 276:4948–4956
1059. Horowitz A (2010) Plekhg5. UCSD-Nature Molecule Pages, UCSD-Nature Signaling Gateway (www.signaling-gateway.org)
1060. Troyanovsky B, Levchenko T, Mnsson G, Matvijenko O, Holmgren L (2001) Angiomotin: an angiostatin binding protein that regulates endothelial cell migration and tube formation. Journal of Cell Biology 152:1247–1254
1061. Bratt A, Birot O, Sinha I, Veitonmäki N, Aase K, Ernkvist M, Holmgren L (2005) Angiomotin regulates endothelial cell-cell junctions and cell motility. Journal of Biological Chemistry 280:34859–34869
1062. Wells CD, Fawcett JP, Traweger A, Yamanaka Y, Goudreault M, Elder K, Kulkarni S, Gish G, Virag C, Lim C, Colwill K, Starostine A, Metalnikov P, Pawson T (2006) A Rich1/Amot complex regulates the Cdc42 GTPase and apical-polarity proteins in epithelial cells. Cell 125:535–548
1063. De Toledo M, Coulon V, Schmidt S, Fort P, Blangy A (2001) The gene for a new brain specific RhoA exchange factor maps to the highly unstable chromosomal region 1p36.2-1p36.3. Oncogene 20:7307–7317
1064. Sosa MS, Lopez-Haber C, Yang C, Wang H, Lemmon MA, Busillo JM, Luo J, Benovic JL, Klein-Szanto A, Yagi H, Gutkind JS, Parsons RE, Kazanietz MG (2010) Identification of the Rac-GEF P-Rex1 as an essential mediator of ErbB signaling in breast cancer. Molecular Cell 40:877–892
1065. Curtis C, Hemmeryckx B, Haataja L, Senadheera D, Groffen J, Heisterkamp N (2004) Scambio, a novel guanine nucleotide exchange factor for Rho. Molecular Cancer 2004; 3:10
1066. Cario-Toumaniantz C, Ferland-McCollough D, Chadeuf G, Toumaniantz G, Rodriguez M, Galizzi JP, Lockart B, Bril A, Scalbert E, Loirand G, Pacaud P (2012) RhoA guanine exchange factor expression profile in arteries: Evidence for a Rho kinase-dependent negative feedback in angiotensin II-dependent hypertension. American Journal of Physiology – Cell Physiology
1067. Rojas JM, Santos E (2006) Ras GEFs and Ras GAPs (Chap. 2). In Channing D (Ed.) RAS Family GTPases, Series: Proteins and Cell Regulation (Vol. 4) Springer, Dordrecht, The Netherlands
1068. Toumaniantz G, Ferland-McCollough D, Cario-Toumaniantz C, Pacaud P, Loirand G (2010) The Rho protein exchange factor Vav3 regulates vascular smooth muscle cell proliferation and migration. Cardiovascular Research 86:131–140
1069. Hunter SG, Zhuang G, Brantley-Sieders D, Swat W, Cowan CW, Chen J (2006) Essential role of Vav family guanine nucleotide exchange factors in EphA receptor-mediated angiogenesis. Molecular and Cellular Biology 26:4830–4842
1070. Sauzeau V, Sevilla MA, Rivas-Elena JV, de álava E, Montero MJ, Lpez-Novoa JM, Bustelo XR (2006) Vav3 proto-oncogene deficiency leads to sympathetic hyperactivity and cardiovascular dysfunction. Nature – Medicine 12:841–845
1071. Bustelo XR (2010) Vav1. UCSD-Nature Molecule Pages, UCSD-Nature Signaling Gateway (www.signaling-gateway.org)
1072. Cantor SB, Urano T, Feig LA (1995) Identification and characterization of Ral-binding protein 1, a potential downstream target of Ral GTPases. Molecular and Cellular Biology 15:4578–4584
1073. Bryant SS, Briggs S, Smithgall TE, Martin GA, McCormick F, Chang JH, Parsons SJ, Jove R (1995) Two SH2 domains of p120 Ras GTPase-activating protein bind synergistically to tyrosine phosphorylated p190 Rho GTPase-activating protein. Journal of Biological Chemistry 270:17947–17952
1074. Kahn RA, Bruford E, Inoue H, Logsdon JM, Nie Z, Premont RT, Randazzo PA, Satake M, Theibert AB, Zapp ML, Cassel d (2008) Consensus nomenclature for the human ArfGAP domain-containing proteins. Journal of Cell Biology 182:1039–1044
1075. Inoue H, Randazzo PA (2007) Arf GAPs and their interacting proteins. Traffic 8:1465–1475

1076. Gosney BJ, Kanamarlapudi V (2011) ADAP2. UCSD-Nature Molecule Pages, UCSD-Nature Signaling Gateway (www.signaling-gateway.org)
1077. Brown MC, Cary LA, Jamieson JS, Cooper JA, Turner CE (2005) Src and FAK kinases cooperate to phosphorylate paxillin kinase linker, stimulate its focal adhesion localization, and regulate cell spreading and protrusiveness. Molecular Biology of the Cell 16:4316–4328
1078. Jamieson JS, Tumbarello DA, Hallé M, Brown MC, Tremblay ML, Turner CE (2005) Paxillin is essential for PTP-PEST-dependent regulation of cell spreading and motility: a role for paxillin kinase linker. Journal of Cell Science 118:5835–5847
1079. Lamorte L, Rodrigues S, Sangwan V, Turner CE, Park M (2003) Crk associates with a multimolecular Paxillin/GIT2/β-PIX complex and promotes Rac-dependent relocalization of Paxillin to focal contacts. Molecular Biology of the Cell 14:2818–2831
1080. Owens SE, Kanamarlapudi V (2010) Arap3. UCSD-Nature Molecule Pages, UCSD-Nature Signaling Gateway (www.signaling-gateway.org)
1081. Davies JC, Kanamarlapudi V (2011) Smap2. UCSD-Nature Molecule Pages, UCSD-Nature Signaling Gateway (www.signaling-gateway.org)
1082. Liu W, Duden R, Phair RD, Lippincott-Schwartz J (2005) ArfGAP1 dynamics and its role in COPI coat assembly on Golgi membranes of living cells. Journal of Cell Biology 168:1053–1063
1083. Natsume W, Tanabe K, Kon S, Yoshida N, Watanabe T, Torii T, Satake M (2006) SMAP2, a novel ARF GTPase-activating protein, interacts with clathrin and clathrin assembly protein and functions on the AP-1-positive early endosome/trans-Golgi network. Molecular Biology of the Cell 17:2592–2603
1084. Nie Z, Fei J, Premont RT, Randazzo PA (2005) The Arf GAPs AGAP1 and AGAP2 distinguish between the adaptor protein complexes AP-1 and AP-3. Journal of Cell Science 118:3555–3566
1085. Fukui K, Sasaki T, Imazumi K, Matsuura Y, Nakanishi H, Takai Y (1997) Isolation and characterization of a GTPase activating protein specific for the Rab3 subfamily of small G proteins. Journal of Biological Chemistry 272:4655–4658
1086. Meng J, Casey PJ (2002) Activation of Gz attenuates Rap1-mediated differentiation of PC12 cells. Journal of Biological Chemistry 277:43417–43424
1087. Hattori M, Tsukamoto N, Nur-e-Kamal MS, Rubinfeld B, Iwai K, Kubota H, Maruta H, Minato N (1995) Molecular cloning of a novel mitogen-inducible nuclear protein with a Ran GTPase-activating domain that affects cell cycle progression. Molecular and Cellular Biology 15:552–560
1088. Sot B, Kötting C, Deaconescu D, Suveyzdis Y, Gerwert K, Wittinghofer A (2010) Unravelling the mechanism of dual-specificity GAPs. EMBO Journal 29:1205–1214
1089. Trov-Marqui AB, Tajara EH (2006) Neurofibromin: a general outlook. Clinical Genetics 70:1–13
1090. Morrison H, Sperka T, Manent J, Giovannini M, Ponta H, Herrlich P (2007) Merlin/neurofibromatosis type 2 suppresses growth by inhibiting the activation of Ras and Rac. Cancer Research 67:520–527
1091. Ozawa T, Araki N, Yunoue S, Tokuo H, Feng L, Patrakitkomjorn S, Hara T, Ichikawa Y, Matsumoto K, Fujii K, Saya H (2005) The neurofibromatosis type 1 gene product neurofibromin enhances cell motility by regulating actin filament dynamics via the Rho-ROCK-LIMK2-cofilin pathway. Journal of Biological Chemistry 280:39524–39533
1092. Sánchez-Margalet V, Najib S (2001) Sam68 is a docking protein linking GAP and PI3K in insulin receptor signaling. Molecular and Cellular Endocrinology 183:113–121
1093. Chow A, Davis AJ, Gawler DJ (2000) Identification of a novel protein complex containing annexin VI, Fyn, Pyk2, and the p120(GAP) C2 domain. FEBS Letters 469:88–92
1094. Koehler JA, Moran MF (2001) RACK1, a protein kinase C scaffolding protein, interacts with the PH domain of p120GAP. Biochemical and Biophysical Research Communications 283:888–895
1095. Tomoda T, Kim JH, Zhan C, Hatten ME (2004) Role of Unc51.1 and its binding partners in CNS axon outgrowth. Genes and Developement 18:541–558

1096. Liu Q, Walker SA, Gao D, Taylor JA, Dai YF, Arkell RS, Bootman MD, Roderick HL, Cullen PJ, Lockyer PJ (2005) CAPRI and RASAL impose different modes of information processing on Ras due to contrasting temporal filtering of Ca^{2+}. Journal of Cell Biology 170:183–190
1097. Oinuma I, Katoh H, Negishi M (2006) Semaphorin 4D/Plexin-B1-mediated R-Ras GAP activity inhibits cell migration by regulating $\beta(1)$ integrin activity. Journal of Cell Biology 173:601–613
1098. Ito Y, Oinuma I, Katoh H, Kaibuchi K, Negishi M (2006) Sema4D/plexin-B1 activates GSK-3β through R-Ras GAP activity, inducing growth cone collapse. EMBO Reports 7:704–709
1099. Uesugi K, Oinuma I, Katoh H, Negishi M (2009) Different requirement for Rnd GTPases of R-Ras GAP activity of Plexin-C1 and Plexin-D1. Journal of Biological Chemistry 284:6743–6751
1100. Jacobs T, Hall C (2005) Rho GAPs – Regulators of Rho GTPases and more (Chap. 5; p. 93–112). In Manser E (Ed.) RHo Family GTPases, Series: Proteins and Cell Regulation (Vol. 3) Springer, Dordrecht, The Netherlands
1101. Qi RZ, Ching YP, Kung HF, Wang JH (2004) α-Chimaerin exists in a functional complex with the Cdk5 kinase in brain. FEBS Letters 561:177–180
1102. Yang C, Liu Y, Lemmon MA, Kazanietz MG (2006) Essential role for Rac in heregulin beta1 mitogenic signaling: a mechanism that involves epidermal growth factor receptor and is independent of ErbB4. Molecular and Cellular Biology 26:831–842
1103. Christerson LB, Gallagher E, Vanderbilt CA, Whitehurst AW, Wells C, Kazempour R, Sternweis PC, Cobb MH (2002) p115 Rho GTPase activating protein interacts with MEKK1. Journal of Cellular Physiology 192:200–208
1104. Ludwig K, Sanchez Manchinelly SA, Su L, Mikawa M, Parsons SJ (2009) p190RhoGAP-A. UCSD-Nature Molecule Pages, UCSD-Nature Signaling Gateway (www.signaling-gateway.org)
1105. Sanchez Manchinelly SA, Ludwig K, Mikawa M, Parsons SJ (2009) p190RhoGAP-B. UCSD-Nature Molecule Pages, UCSD-Nature Signaling Gateway (www.signaling-gateway.org)
1106. Holinstat M, Knezevic N, Broman M, Samarel AM, Malik AB, Mehta D (2006) Suppression of RhoA activity by focal adhesion kinase-induced activation of p190RhoGAP: role in regulation of endothelial permeability. Journal of Biological Chemistry 281:2296–2305
1107. Barberis D, Casazza A, Sordella R, Corso S, Artigiani S, Settleman J, Comoglio PM, Tamagnone L (2005) p190 Rho-GTPase activating protein associates with plexins and it is required for semaphorin signalling. Journal of Cell Science 118:4689–4700
1108. Wennerberg K, Forget MA, Ellerbroek SM, Arthur WT, Burridge K, Settleman J, Der CJ, Hansen SH (2003) Rnd proteins function as RhoA antagonists by activating p190 RhoGAP. Current Biology 13:1106–1115
1109. Kawai K, Iwamae Y, Yamaga M, Kiyota M, Ishii H, Hirata H, Homma Y, Yagisawa H (2009) Focal adhesion-localization of START-GAP1/DLC1 is essential for cell motility and morphology. Genes to Cells 14:227–241
1110. Eberth A, Lundmark R, Gremer L, Dvorsky R, Koessmeier KT, McMahon HT, Ahmadian MR (2009) A BAR domain-mediated autoinhibitory mechanism for RhoGAPs of the GRAF family. Biochemical Journal 417:371–377
1111. Wells CD, Fawcett JP, Traweger A, Yamanaka Y, Goudreault M, Elder K, Kulkarni S, Gish G, Virag C, Lim C, Colwill K, Starostine A, Metalnikov P, Pawson T (2006) A Rich1/Amot complex regulates the Cdc42 GTPase and apical-polarity proteins in epithelial cells. Cell 125:535–548
1112. Richnau N, Aspenström P (2001) Rich, a rho GTPase-activating protein domain-containing protein involved in signaling by Cdc42 and Rac1. Journal of Biological Chemistry 276:35060–35070
1113. Rollason R, Korolchuk V, Hamilton C, Jepson M, Banting G (2009) A CD317/tetherin–RICH2 complex plays a critical role in the organization of the subapical actin cytoskeleton in polarized epithelial cells. Journal of Cell Biology 184:721–736.

1114. Tcherkezian J, Danek EI, Jenna S, Triki I, Lamarche-Vane N (2005) Extracellular signal-regulated kinase 1 interacts with and phosphorylates CdGAP at an important regulatory site. Molecular and Cellular Biology 25:6314–6329

1115. Chiang SH, Hwang J, Legendre M, Zhang M, Kimura A, Saltiel AR (2003) TCGAP, a multidomain Rho GTPase-activating protein involved in insulin-stimulated glucose transport. EMBO Journal 22:2679–2691

1116. Zalcman G, Dorseuil O, Garcia-Ranea JA, Gacon G, Camonis J (1999) RhoGAPS and RhoGDIs: (His)stories of Two families. In Jeanteur P (Ed.) Cytoskeleton and small G proteins Springer, Berlin, Heidelberg

1117. Zhang B, Zheng Y (1998) Regulation of RhoA GTP hydrolysis by the GTPase-activating proteins p190, p50RhoGAP, Bcr, and 3BP-1. Biochemistry 37:5249–5257

1118. Cho YJ, Cunnick JM, Yi SJ, Kaartinen V, Groffen J, Heisterkamp N (2007) Abr and Bcr, two homologous Rac GTPase-activating proteins, control multiple cellular functions of murine macrophages. Molecular and Cellular Biology 27:899–911

1119. Burkard ME, Maciejowski J, Rodriguez-Bravo V, Repka M, Lowery DM, Clauser KR, Zhang C, Shokat KM, Carr SA, Yaffe MB, Jallepalli PV (2009) Plk1 self-organization and priming phosphorylation of HsCYK-4 at the spindle midzone regulate the onset of division in human cells. PLoS Biology 7:e1000111

1120. Jullien-Flores V, Dorseuil O, Romero F, Letourneur F, Saragosti S, Berger R, Tavitian A, Gacon G, Camonis JH (1995) Bridging Ral GTPase to Rho pathways. RLIP76, a Ral effector with CDC42/Rac GTPase-activating protein activity. Journal of Biological Chemistry 270:22473–22477

1121. Saeki N, Tokuo H, Ikebe M (2005) BIG1 is a binding partner of myosin IXb and regulates its Rho-GTPase activating protein activity. Journal of Biological Chemistry 280:10128-10134

1122. Faucherre A, Desbois P, Nagano F, Satre V, Lunardi J, Gacon G, Dorseuil O (2005) Lowe syndrome protein Ocrl1 is translocated to membrane ruffles upon Rac GTPase activation: a new perspective on Lowe syndrome pathophysiology. Human Molecular Genetics 14: 1441–1448

1123. Dirac-Svejstrup AB, Sumizawa T, Pfeffer SR (1997) Identification of a GDI displacement factor that releases endosomal Rab GTPases from Rab-GDI. EMBO Journal 16:465–472

1124. Wu YW, Tan KT, Waldmann H, Goody RS, Alexandrov K (2007) Interaction analysis of prenylated Rab GTPase with Rab escort protein and GDP dissociation inhibitor explains the need for both regulators. Proceedings of the National Academy of Sciences of the United States of America 104:12294–12299

1125. Oesterlin LK, Goody RS, Itzen A (2012) Posttranslational modifications of Rab proteins cause effective displacement of GDP dissociation inhibitor. Proceedings of the National Academy of Sciences of the United States of America 109:5621–5626

1126. Chen Y, Deng Y, Zhang J, Yang L, Xie X, Xu T (2009) GDI-1 preferably interacts with Rab10 in insulin-stimulated GLUT4 translocation. Biochemical Journal 422:229–235

1127. Elfenbein A, Rhodes JM, Meller J, Schwartz MA, Matsuda M, Simons M (2009) Suppression of RhoG activity is mediated by a syndecan 4-synectin-RhoGDI1 complex and is reversed by PKCα in a Rac1 activation pathway. Journal of Cell Biology 186:75–83

1128. Garcia-Mata R, Boulter E, Burridge K (2011) The 'invisible hand': regulation of RHO GTPases by RHOGDIs. Nature Reviews – Molecular Cell Biology 12:493–504

1129. Boulter E, Garcia-Mata R, Guilluy C, Dubash A, Rossi G, Brennwald PJ, Burridge K (2010) Regulation of RhoGTPase crosstalk, degradation and activity by RhoGDI1. Nature – Cell Biology 12:477–483

1130. Shin EY, Shim ES, Lee CS, Kim HK, Kim EG (2009) Phosphorylation of RhoGDI1 by p21-activated kinase 2 mediates basic fibroblast growth factor-stimulated neurite outgrowth in PC12 cells. Biochemical and Biophysical Research Communications 379:384–389

Chap. 10. Other Major Types of Signaling Mediators

1131. Brivanlou AH, Darnell JE (2002) Signal transduction and the control of gene expression. Science 295:813–818
1132. Berridge MJ (2009) Module 4: Sensors and Effectors. Cell Signalling Biology. Biochemical Journal's Signal Knowledge Environment Portland Press Ltd., London, UK (www.biochemj.org/csb/004/csb004.pdf)
1133. Snyder SH, Ferris CD (2000) Novel neurotransmitters and their neuropsychiatric relevance. American Journal of Psychiatry 157:1738–1751
1134. Herring N, Paterson DJ (2001) Nitric oxide-cGMP pathway facilitates acetylcholine release and bradycardia during vagal nerve stimulation in the guinea-pig in vitro. Journal of Physiology 535:507–518
1135. Xue L, Farrugia G, Miller SM, Ferris CD, Snyder SH, Szurszewski JH (2000) Carbon monoxide and nitric oxide as coneurotransmitters in the enteric nervous system: evidence from genomic deletion of biosynthetic enzymes. Proceedings of the National Academy of Sciences of the United States of America 97:1851–1855
1136. Abraham NG, Kappas A (2008) Pharmacological and clinical aspects of heme oxygenase. Pharmacological Reviews 60:79–127
1137. Leffler CW, Parfenova H, Jaggar JH (2011) Carbon monoxide as an endogenous vascular modulator. American Journal of Physiology – Heart and Circulatory Physiology 1:H1–H11
1138. Abraham NG, Kappas A (2005) Heme oxygenase and the cardiovascular-renal system. Free Radical Biology and Medicine 39:1–25
1139. Otterbein LE, Hedblom A, Harris C, Csizmadia E, Gallo D, Wegiel B (2011) Heme oxygenase-1 and carbon monoxide modulate DNA repair through ataxia-telangiectasia mutated (ATM) protein. Proceedings of the National Academy of Sciences of the United States of America 108:14491–14496
1140. Rhodes MA, Carraway MS, Piantadosi CA, Reynolds CM, Cherry AD, Wester TE, Natoli MJ, Massey EW, Moon RE, Suliman HB (2011) Carbon monoxide, skeletal muscle oxidative stress, and mitochondrial biogenesis in humans. American Journal of Physiology – Heart and Circulatory Physiology 297:H392–H399
1141. Gilchrist M, Shore AC, Benjamin N (2011) Inorganic nitrate and nitrite and control of blood pressure. Cardiovascular Research 89:492–498
1142. Förstermann U, Sessa WC (2012) Nitric oxide synthases: regulation and function. European Heart Journal 33:829–837
1143. Brenman JE, Chao DS, Gee SH, McGee AW, Craven SE, Santillano DR, Wu Z, Huang F, Xia H, Peters MF, Froehner SC, Bredt DS (1996) Interaction of nitric oxide synthase with the postsynaptic density protein PSD-95 and alpha1-syntrophin mediated by PDZ domains. Cell 84:757–767
1144. Shaul PW (2002) Regulation of endothelial nitric oxide synthase: location, location, location. Annual Review of Physiology 64:749–774
1145. Stuehr DJ (1999) Mammalian nitric oxide synthases. Biochimica et Biophysica Acta – Bioenergetics 1411:217–230
1146. Chen CA, Wang TY, Varadharaj S, Reyes LA, Hemann C, Talukder MAH, Chen YR, Druhan LJ, Zweier JL (2010) S-glutathionylation uncouples eNOS and regulates its cellular and vascular function. Nature 468:1115–1118
1147. Elrod JW, Calvert JW, Gundewar S, Bryan NS, Lefer DJ (2008) Nitric oxide promotes distant organ protection: evidence for an endocrine role of nitric oxide. Proceedings of the National Academy of Sciences of the United States of America 105:11430–11435
1148. Mitchell BJ, Chen Z, Tiganis T, Stapleton D, Katsis F, Power DA, Sim AT, Kemp BE (2001) Co-ordinated control of endothelial nitric oxide synthase phosphorylation by protein kinase C and the cAMP-dependent protein kinase. Journal of Cell Chemistry 276:17625–17628

1149. Mount PF, Kemp BE, Power DA (2007) Regulation of endothelial and myocardial NO synthesis by multi-site eNOS phosphorylation. Journal of Molecular and Cellular Cardiology 42:271–279
1150. Kolodziejska KE, Burns AR, Moore RH, Stenoien DL, Eissa NT (2005) Regulation of inducible nitric oxide synthase by aggresome formation. Proceedings of the National Academy of Sciences of the United States of America 102:4854–4859
1151. Choy JC, Wang Y, Tellides G, Pober JS (2007) Induction of inducible NO synthase in bystander human T cells increases allogeneic responses in the vasculature. Proceedings of the National Academy of Sciences of the United States of America 104:1313–1318
1152. Malik M, Jividen K, Padmakumar VC, Cataisson C, Li L, Lee J, Howard OM, Yuspa SH (2012) Inducible NOS-induced chloride intracellular channel 4 (CLIC4) nuclear translocation regulates macrophage deactivation. Proceedings of the National Academy of Sciences of the United States of America 109:6130–6135
1153. Leiper J, Nandi M, Torondel B, Murray-Rust J, Malaki M, O'Hara B, Rossiter S, Anthony S, Madhani M, Selwood D, Smith C, Wojciak-Stothard B, Rudiger A, Stidwill R, McDonald NQ, Vallance P (2007) Disruption of methylarginine metabolism impairs vascular homeostasis. Nature – Medicine 13:198–203
1154. Patel RP, Hogg N, Kim-Shapiro DB (2011) The potential role of the red blood cell in nitrite-dependent regulation of blood flow. Cardiovascular Research 89:507–515
1155. Salgado MT, Nagababu E, Rifkind JM (2009) Quantification of intermediates formed during the reduction of nitrite by deoxyhemoglobin. Journal of Biological Chemistry 284:12710–12718
1156. Hendgen-Cotta UB, Merx MW, Shiva S, Schmitz J, Becher S, Klare JP, Steinhoff HJ, Goedecke A, Schrader J, Gladwin MT, Kelm M, Rassaf T (2008) Nitrite reductase activity of myoglobin regulates respiration and cellular viability in myocardial ischemia-reperfusion injury. Proceedings of the National Academy of Sciences of the United States of America 105:10256–10261
1157. Alzawahra WF, Talukder MAH, Liu X, Samouilov A, Zweier JL (2008) Heme proteins mediate the conversion of nitrite to nitric oxide in the vascular wall. American Journal of Physiology – Heart and Circulatory Physiology 295:H499–H508
1158. Allen BW, Stamler JS, Piantadosi CA (2009) Hemoglobin, nitric oxide and molecular mechanisms of hypoxic vasodilation. Trends in Molecular Medicine 15:452–460
1159. Zuckerbraun BS, George P, Gladwin MT (2011) Nitrite in pulmonary arterial hypertension: therapeutic avenues in the setting of dysregulated arginine/nitric oxide synthase signalling. Cardiovascular Research 89:542–552
1160. Carlström M, Larsen FJ, Nyström T, Hezel M, Borniquel S, Weitzberg E, Lundberg JO (2010) Dietary inorganic nitrate reverses features of metabolic syndrome in endothelial nitric oxide synthase-deficient mice. Proceedings of the National Academy of Sciences of the United States of America 107:17716–17720
1161. Segal SS (2005) Regulation of blood flow in the microcirculation. Microcirculation 12:33–45
1162. Diesen DL, Hess DT, Stamler JS (2008) Hypoxic vasodilation by red blood cells: evidence for an s-nitrosothiol-based signal. Circulation Research 103:545–553
1163. Murphy ME, Sies H (1991) Reversible conversion of nitroxyl anion to nitric oxide by superoxide dismutase. Proceedings of the National Academy of Sciences of the United States of America 88:10860–10864
1164. Isbell TS, Sun CW, Wu LC, Teng X, Vitturi DA, Branch BG, Kevil CG, Peng N, Wyss JM, Ambalavanan N, Schwiebert L, Ren J, Pawlik KM, Renfrow MB, Patel RP, Townes TM (2008) SNO-hemoglobin is not essential for red blood cell-dependent hypoxic vasodilation. Nature – Medicine 14:773–777
1165. Favaloro JL, Kemp-Harper BK (2009) Redox variants of nitric oxide (NO* and HNO) elicit vasorelaxation of resistance arteries via distinct mechanisms. American Journal of Physiology – Heart and Circulatory Physiology 296:H1274–H1280

1166. Andrews KL, Irvine JC, Tare M, Apostolopoulos J, Favaloro JL, Triggle CR, Kemp-Harper BK (2009) A role for nitroxyl (HNO) as an endothelium-derived relaxing and hyperpolarizing factor in resistance arteries. British Journal of Pharmacology 157:540–550
1167. Paolocci N, Saavedra WF, Miranda KM, Martignani C, Isoda T, Hare JM, Espey MG, Fukuto JM, Feelisch M, Wink DA, Kass DA (2001) Nitroxyl anion exerts redox-sensitive positive cardiac inotropy in vivo by calcitonin gene-related peptide signaling. Proceedings of the National Academy of Sciences of the United States of America 98:10463–10468
1168. Pacher P, Beckman JS, Liaudet L (2007) Nitric oxide and peroxynitrite in health and disease. Physiological Reviews 87:315–424
1169. Feil R, Kemp-Harper B (2006) cGMP signalling: from bench to bedside. EMBO Reports 7:149–153
1170. Rho EH, Perkins WJ, Lorenz RR, Warner DO, Jones KA (2002) Differential effects of soluble and particulate guanylyl cyclase on Ca^{2+} sensitivity in airway smooth muscle. Journal of Applied Physiology 92:257–263
1171. Lewko B, Wendt U, Szczepanska-Konkel M, Stepinski J, Drewnowska K, Angielski S (1997) Inhibition of endogenous nitric oxide synthesis activates particulate guanylyl cyclase in the rat renal glomeruli. Kidney International 52:654–659
1172. Crecelius AR, Kirby BS, Voyles WF, Dinenno FA (2010) Nitric oxide, but not vasodilating prostaglandins, contributes to the improvement of exercise hyperemia via ascorbic acid in healthy older adults. American Journal of Physiology – Heart and Circulatory Physiology 299:H1633–H1641
1173. Batchelor AM, Bartus K, Reynell C, Constantinou S, Halvey EJ, Held KF, Dostmann WR, Vernon J, Garthwaite J (2010) Exquisite sensitivity to subsecond, picomolar nitric oxide transients conferred on cells by guanylyl cyclase-coupled receptors. Proceedings of the National Academy of Sciences of the United States of America 107:22060–22065
1174. Illi B, Colussi C, Rosati J, Spallotta F, Nanni S, Farsetti A, Capogrossi MC, Gaetano C (2011) NO points to epigenetics in vascular development. Cardiovascular Research 90: 447–456
1175. Hu RG, Sheng J, Qi X, Xu Z, Takahashi TT, Varshavsky A (2005) The N-end rule pathway as a nitric oxide sensor controlling the levels of multiple regulators. Nature 437:981–986
1176. Hamed S, Brenner B, Roguin A (2011) Nitric oxide: a key factor behind the dysfunctionality of endothelial progenitor cells in diabetes mellitus type-2. Cardiovascular Research 91:9–15
1177. LJ Ignarro, JB Adams, PM Horwitz, KS Wood (1986) Activation of soluble guanylate cyclase by NO-hemoproteins involves NO-heme exchange. Comparison of heme-containing and heme-deficient enzyme forms. Journal of Biological Chemistry 261:4997–5002
1178. LJ Ignarro, RE Byrns, GM Buga, KS Wood (1987) Endothelium-derived relaxing factor from pulmonary artery and vein possesses pharmacologic and chemical properties identical to those of nitric oxide radical. Circulation Research 61:866–879
1179. Balligand JL, Cannon PJ (1997) Nitric oxide synthases and cardiac muscle. Autocrine and paracrine influences. Arteriosclerosis, Thrombosis, and Vascular Biology 17:1846–1858
1180. Ormerod JO, Ashrafian H, Maher AR, Arif S, Steeples V, Born GV, Egginton S, Feelisch M, Watkins H, Frenneaux MP (2011) The role of vascular myoglobin in nitrite-mediated blood vessel relaxation. Cardiovascular Research 89:560–565
1181. Maclauchlan S, Yu J, Parrish M, Asoulin TA, Schleicher M, Krady MM, Zeng J, Huang PL, Sessa WC, Kyriakides TR (2011) Endothelial nitric oxide synthase controls the expression of the angiogenesis inhibitor thrombospondin 2. Proceedings of the National Academy of Sciences of the United States of America 108:E1137–E1145
1182. Bauer EM, Qin Y, Miller TW, Bandle RW, Csanyi G, Pagano PJ, Bauer PM, Schnermann J, Roberts DD, Isenberg JS (2010) Thrombospondin-1 supports blood pressure by limiting eNOS activation and endothelial-dependent vasorelaxation. Cardiovascular Research 88:471–481
1183. Schmid-Schönbein GW (2012) Nitric oxide (NO) side of lymphatic flow and immune surveillance. Proceedings of the National Academy of Sciences of the United States of America 109:3–4

1184. Liao S, Cheng G, Conner DA, Huang Y, Kucherlapati RS, Munn LL, Ruddle NH, Jain RK, Fukumura D, Padera TP (2011) Impaired lymphatic contraction associated with immunosuppression. Proceedings of the National Academy of Sciences of the United States of America 108:18784–18789

1185. Su KH, Tsai JY, Kou YR, Chiang AN, Hsiao SH, Wu YL, Hou HH, Pan CC, Shyue SK, Lee TS (2009) Valsartan regulates the interaction of angiotensin II type 1 receptor and endothelial nitric oxide synthase via Src/PI3K/Akt signalling. Cardiovascular Research 82:468–475

1186. Nausch LWM, Ledoux J, Bonev AD, Nelson MT, Dostmann WR (2008) Differential patterning of cGMP in vascular smooth muscle cells revealed by single GFP-linked biosensors. Proceedings of the National Academy of Sciences of the United States of America 105:365–370

1187. Arejian M, Li Y, Anand-Srivastava MB (2009) Nitric oxide attenuates the expression of natriuretic peptide receptor C and associated adenylyl cyclase signaling in aortic vascular smooth muscle cells: role of MAPK. American Journal of Physiology – Heart and Circulatory Physiology 296:H1859–H1867

1188. Popowich DA, Vavra AK, Walsh CP, Bhikhapurwala HA, Rossi NB, Jiang Q, Aalami OO, Kibbe MR (2010) Regulation of reactive oxygen species by p53: implications for nitric oxide-mediated apoptosis. American Journal of Physiology – Heart and Circulatory Physiology 298:H2192-H2200.

1189. Ridnour LA, Windhausen AN, Isenberg JS, Yeung N, Thomas DD, Vitek MP, Roberts DD, Wink DA (2007) Nitric oxide regulates matrix metalloproteinase-9 activity by guanylyl-cyclase-dependent and -independent pathways. Proceedings of the National Academy of Sciences of the United States of America 104:16898–16903

1190. Mattagajasingh I, Kim CK, Naqvi A, Yamamori T, Hoffman TA, Jung SB, DeRicco J, Kasuno K, Irani K (2007) SIRT1 promotes endothelium-dependent vascular relaxation by activating endothelial nitric oxide synthase. Proceedings of the National Academy of Sciences of the United States of America 104:14855–14860

1191. Kashiwagi S, Tsukada K, Xu L, Miyazaki J, Kozin SV, Tyrrell JA, Sessa WC, Gerweck LE, Jain RK, Fukumura D (2008) Perivascular nitric oxide gradients normalize tumor vasculature. Nature – Medicine 14:255–257

1192. Massion PB, Pelat M, Belge C, Balligand JL (2005) Regulation of the mammalian heart function by nitric oxide. Comparative Biochemistry and Physiology; Part A: Molecular and Integrative Physiology 142:144–150

1193. Tamargo J, Caballero R, Gómez R, Delpón E (2010) Cardiac electrophysiological effects of nitric oxide. Cardiovascular Research 87:593–600

1194. Barouch LA, Harrison RW, Skaf MW, Rosas GO, Cappola TP, Kobeissi ZA, Hobai IA, Lemmon CA, Burnett AL, O'Rourke B, Rodriguez ER, Huang PL, Lima JA, Berkowitz DE, Hare JM (2002) Nitric oxide regulates the heart by spatial confinement of nitric oxide synthase isoforms. Nature 416:337–339

1195. Vila-Petroff MG, Younes A, Egan J, Lakatta EG, Sollott SJ (1999) Activation of distinct cAMP-dependent and cGMP-dependent pathways by nitric oxide in cardiac myocytes. Circulation Research 84:1020–103

1196. Zhang Q, Lazar M, Molino B, Rodriguez R, Davidov T, Su J, Tse J, Weiss HR, Scholz PM (2005) Reduction in interaction between cGMP and cAMP in dog ventricular myocytes with hypertrophic failure. American Journal of Physiology – Heart and Circulatory Physiology 289:H1251–H1257

1197. Chang KC, Barth AS, Sasano T, Kizana E, Kashiwakura Y, Zhang Y, Foster B, Marbán E (2008) CAPON modulates cardiac repolarization via neuronal nitric oxide synthase signaling in the heart. Proceedings of the National Academy of Sciences of the United States of America 105:4477–4482

1198. Schroder F, Klein G, Fiedler B, Bastein M, Schnasse N, Hillmer A, Ames S, Gambaryan S, Drexler H, Walter U, Lohmann SM, Wollert KC (2003) Single L-type Ca2+ channel regulation by cGMP-dependent protein kinase type I (PKG I) in adult cardiomyocytes from PKG I transgenic mice. Cardiovascular Research 60:268–277

1199. Kalra D, Baumgarten G, Dibbs Z, Seta Y, Sivasubramanian N, Mann DL (2000) Nitric oxide provokes tumor necrosis factor-alpha expression in adult feline myocardium through a cGMP-dependent pathway. Circulation 102:1302–1307
1200. Wilson Tang WH, Tong W, Shrestha K, Wang Z, Levison BS, Delfraino B, Hu B, Troughton RW, Klein AL, Hazen SL (2008) Differential effects of arginine methylation on diastolic dysfunction and disease progression in patients with chronic systolic heart failure. European Heart Journal 29:2506
1201. Mujoo K, Sharin VG, Bryan NS, Krumenacker JS, Sloan C, Parveen S, Nikonoff LE, Kots AY, Murad F (2008) Role of nitric oxide signaling components in differentiation of embryonic stem cells into myocardial cells. Proceedings of the National Academy of Sciences of the United States of America 105:18924–18929
1202. Yu Q, Gao F, Ma XL (2011) Insulin says NO to cardiovascular disease. Cardiovascular Research 89:516–524
1203. Ortiz PA, Garvin JL (2002) Role of nitric oxide in the regulation of nephron transport. American Journal of Physiology – Renal Physiology 282:F777–F784
1204. Kelm M, Schrader J (1990) Control of coronary vascular tone by nitric oxide. Circulation Research 66:1561–1575
1205. Palmer RM, Ferrige AG, Moncada S (1987) Nitric oxide release accounts for the biological activity of endothelium-derived relaxing factor. Nature 327:524–526
1206. Gladwin MT, Raat NJ, Shiva S, Dezfulian C, Hogg N, Kim-Shapiro DB, Patel RP (2006) Nitrite as a vascular endocrine nitric oxide reservoir that contributes to hypoxic signaling, cytoprotection, and vasodilation. American Journal of Physiology – Heart and Circulatory Physiology 291:H2026–H2035
1207. Stamler JS, Jaraki O, Osborne J, Simon DI, Keaney J, Vita J, Singel D, Valeri CR, Loscalzo J (1992) Nitric oxide circulates in mammalian plasma primarily as an S-nitroso adduct of serum albumin. Proceedings of the National Academy of Sciences of the United States of America 89:7674–7677
1208. Yong QC, Hu LF, Wang S, Huang D, Bian JS (2010) Hydrogen sulfide interacts with nitric oxide in the heart: possible involvement of nitroxyl. Cardiovascular Research 88:482–491
1209. Tanizawa K (2011) Production of H_2S by 3-mercaptopyruvate sulphurtransferase. Journal of Biochemistry 149:357–359
1210. Bearden SE, Beard RS Jr, Pfau JC (2010) Extracellular transsulfuration generates hydrogen sulfide from homocysteine and protects endothelium from redox stress. American Journal of Physiology – Heart and Circulatory Physiology 299:H1568–H1576
1211. Benavides GA, Squadrito GL, Mills RW, Patel HD, Isbell TS, Patel RP, Darley-Usmar VM, Doeller JE, Kraus DW (2007) Hydrogen sulfide mediates the vasoactivity of garlic. Proceedings of the National Academy of Sciences of the United States of America 104:17977–17982
1212. Yang G, Wu L, Jiang B, Yang W, Qi J, Cao K, Meng Q, Mustafa AK, Mu W, Zhang S, Snyder SH, Wang R (2008) H_2S as a physiologic vasorelaxant: hypertension in mice with deletion of cystathionine γ-lyase. Science 322:587–590
1213. Mishra PK, Tyagi N, Sen U, Givvimani S M D, Tyagi SC (2010) H_2S ameliorates oxidative and proteolytic stresses and protects the heart against adverse remodeling in chronic heart failure. American Journal of Physiology – Heart and Circulatory Physiology 298: H451–H456
1214. Meng QH, Yang G, Yang W, Jiang B, Wu L, Wang R (2007) Protective effect of hydrogen sulfide on balloon injury-induced neointima hyperplasia in rat carotid arteries. American Journal of Pathology 170:1406–1414
1215. Cai WJ, Wang MJ, Moore PK, Jin HM, Yao T, Zhu YC (2007) The novel proangiogenic effect of hydrogen sulfide is dependent on Akt phosphorylation. Cardiovascular Research 76:29–40
1216. Fu M, Zhang W, Wu L, Yang G, Li H, Wang R (2012) Hydrogen sulfide (H_2S) metabolism in mitochondria and its regulatory role in energy production. Proceedings of the National Academy of Sciences of the United States of America 109:2943–2948

1217. Lassegue B, Clempus RE (2003) Vascular NAD(P)H oxidases: specific features, expression, and regulation. American Journal of Physiology – Regulatory, Integrative and Comparative Physiology 285:R277–R297
1218. Bedard K, Krause KH (2007) The NOX family of ROS-generating NADPH oxidases: physiology and pathophysiology. Physiological Reviews 87:245–313
1219. Massaro M, Habib A, Lubrano L, Del Turco S, Lazzerini G, Bourcier T, Weksler BB, De Caterina R (2006) The omega-3 fatty acid docosahexaenoate attenuates endothelial cyclooxygenase-2 induction through both NADP(H) oxidase and PKCε inhibition. Proceedings of the National Academy of Sciences of the United States of America 103:15184–15189
1220. Morgan MJ, Kim YS, Liu ZG (2008) TNFα and reactive oxygen species in necrotic cell death. Cell Research 18:343–349
1221. Gao L, Mann GE (2009) Vascular NAD(P)H oxidase activation in diabetes: a double-edged sword in redox signalling. Cardiovascular Research 82:9–20
1222. Bánfi B, Tirone F, Durussel I, Knisz J, Moskwa P, Molnár GZ, Krause KH, Cox JA (2004) Mechanism of Ca^{2+} activation of the NADPH oxidase 5 (NOX5). Journal of Biological Chemistry 279:18583–18591
1223. El Jamali A, Valente AJ, Lechleiter JD, Gamez MJ, Pearson DW, Nauseef WM, Clark RA (2008) Novel redox-dependent regulation of NOX5 by the tyrosine kinase c-Abl. Free Radical Biology and Medicine 44:868–881
1224. Bánfi B, Maturana A, Jaconi S, Arnaudeau S, Laforge T, Sinha B, Ligeti E, Demaurex N, Krause KH (2000) A mammalian H^+ channel generated through alternative splicing of the NADPH oxidase homolog NOH-1. Science 287:138–142
1225. Montezano AC, Burger D, Paravicini TM, Chignalia AZ, Yusuf H, Almasri M, He Y, Callera GE, He G, Krause KH, Lambeth D, Quinn MT, Touyz RM (2010) Nicotinamide adenine dinucleotide phosphate reduced oxidase 5 (Nox5) regulation by angiotensin II and endothelin-1 is mediated via calcium/calmodulin-dependent, rac-1-independent pathways in human endothelial cells. Circulation Research 106:1363–1373
1226. Hilenski LL, Clempus RE, Quinn MT, Lambeth JD, Griendling KK (2004) Distinct subcellular localizations of Nox1 and Nox4 in vascular smooth muscle cells. Arteriosclerosis, Thrombosis, and Vascular Biology 24:677–683
1227. Manea A, Tanase LI, Raicu M, Simionescu M (2010) Jak/STAT signaling pathway regulates Nox1 and Nox4-based NADPH oxidase in human aortic smooth muscle cells. Arteriosclerosis, Thrombosis, and Vascular Biology 30:105–112
1228. Brandes RP (2003) Role of NADPH oxidases in the control of vascular gene expression. Antioxidants and Redox Signaling 5:803–811
1229. Anilkumar N, Weber R, Zhang M, Brewer A, Shah AM (2008) Nox4 and Nox2 NADPH oxidases mediate distinct cellular redox signaling responses to agonist stimulation. Arteriosclerosis, Thrombosis, and Vascular Biology 28:1347–1354
1230. Godbole AS, Lu X, Guo X, Kassab GS (2009) NADPH oxidase has a directional response to shear stress. American Journal of Physiology – Heart and Circulatory Physiology 296: H152–H158
1231. Redmond EM, Cahill PA (2012) The NOX–ROS connection: targeting Nox1 control of N-cadherin shedding in vascular smooth muscle cells. Circulation Research 93:386–387
1232. Niu XL, Madamanchi NR, Vendrov AE, Tchivilev I, Rojas M, Madamanchi C, Brandes RP, Krause KH, Humphries J, Smith A, Burnand KG, Runge MS (2010) Nox activator 1: a potential target for modulation of vascular reactive oxygen species in atherosclerotic arteries. Circulation 121:549–559
1233. Jagadeesha DK, Takapoo M, Banfi B, Bhalla RC, Miller FJ (2012) Nox1 transactivation of epidermal growth factor receptor promotes N-cadherin shedding and smooth muscle cell migration. Circulation Research 93:406–413
1234. Tanner JJ, Parsons ZD, Cummings AH, Zhou H, Gates KS (2011) Redox regulation of protein tyrosine phosphatases: structural and chemical aspects. Antioxidants and Redox Signaling 15:77–97

1235. Sundar IK, Caito S, Yao H, Rahman I (2010) Oxidative stress, thiol redox signaling methods in epigenetics. Methods in Enzymology 474:213–244
1236. Woo HA, Chae HZ, Hwang SC, Yang KS, Kang SW, Kim K, Rhee SG (2003) Reversing the inactivation of peroxiredoxins caused by cysteine sulfinic acid formation. Science 300: 653–656
1237. Kinnula VL, Lehtonen S, Kaarteenaho-Wiik R, Lakari E, Pääkkö P, Kang SW, Rhee SG, Soini Y (2002) Cell specific expression of peroxiredoxins in human lung and pulmonary sarcoidosis. Thorax 57:157–164
1238. Sundar IK, Chung S, Hwang JW, Arunachalam G, Cook S, Yao H, Mazur W, Kinnula VL, Fisher AB, Rahman I (2010) Peroxiredoxin 6 differentially regulates acute and chronic cigarette smoke-mediated lung inflammatory response and injury. Experimental Lung Research 36:451–462
1239. Song JS, Cho HH, Lee BJ, Bae YC, Jung JS (2011) Role of thioredoxin 1 and thioredoxin 2 on proliferation of human adipose tissue-derived mesenchymal stem cells. Stem Cells and Development 20:1529–1537
1240. Brandes RP (2011) Vascular peroxidase 1/peroxidasin: a complex protein with a simple function? Cardiovascular Research 91:1–2
1241. Shi R, Hu C, Yuan Q, Yang T, Peng J, Li Y, Bai Y, Cao Z, Cheng G, Zhang G (2011) Involvement of vascular peroxidase 1 in angiotensin II-induced vascular smooth muscle cell proliferation. Cardiovascular Research 91:27–36
1242. Miki H, Funato Y. (2012) Regulation of intracellular signalling through cysteine oxidation by reactive oxygen species. Journal of Biochemistry 151:255–261
1243. Guo Z, Kozlov S, Lavin MF, Person MD, Paull TT (2010) ATM activation by oxidative stress. Science 330:517–521
1244. Harris I, McCracken S, Mak TW (2012) PKM2: A gatekeeper between growth and survival. Cell Research 22:447–449
1245. Anastasiou D, Poulogiannis G, Asara JM, Boxer MB, Jiang JK, Shen M, Bellinger G, Sasaki AT, Locasale JW, Auld DS, Thomas CJ, Van der Heiden MG, Cantley LC (2011) Inhibition of pyruvate kinase M2 by reactive oxygen species contributes to cellular antioxidant responses. Science 334:1278–1283
1246. Chai Y, Zhang DM, Lin YF (2011) Activation of cGMP-dependent protein kinase stimulates cardiac ATP-sensitive potassium channels via a ROS/calmodulin/CaMKII signaling cascade. PLoS One 6:e18191
1247. Kubo M, Nakaya Y, Matsuoka S, Saito K, Kuroda Y (1994) Atrial natriuretic factor and isosorbide dinitrate modulate the gating of ATP-sensitive K^+ channels in cultured vascular smooth muscle cells. Circulation Research 74:471–476
1248. Denicola A, Souza JM, Radi R (1998) Diffusion of peroxynitrite across erythrocyte membranes. Proceedings of the National Academy of Sciences of the United States of America 95:3566–3571
1249. Lancaster JR (2006) Nitroxidative, nitrosative, and nitrative stress: kinetic predictions of reactive nitrogen species chemistry under biological conditions. Chemical Research in Toxicology 19:1160–1174
1250. Lu L, Li Y, Kim SM, Bossuyt W, Liu P, Qiu Q, Wang Y, Halder G, Finegold MJ, Lee JS, Johnson RL (2010) Hippo signaling is a potent in vivo growth and tumor suppressor pathway in the mammalian liver. Proceedings of the National Academy of Sciences of the United States of America 107:1437–1442.
1251. Saucedo LJ, Edgar BA (2007) Filling out the Hippo pathway. Nature Reviews – Molecular Cell Biology 8:613–621
1252. Ikeda M, Kawata A, Nishikawa M, Tateishi Y, Yamaguchi M, Nakagawa K, Hirabayashi S, Bao Y, Hidaka S, Hirata Y, Hata Y (2009) Hippo pathway-dependent and -independent roles of RASSF6. Science Signaling 2:ra59
1253. Inui M, Martello G, Piccolo S (2010) MicroRNA control of signal transduction. Nature Reviews – Molecular Cell Biology 11:252–263

1254. Varelas X, Miller BW, Sopko R, Song S, Gregorieff A, Fellouse FA, Sakuma R, Pawson T, Hunziker W, McNeill H, Wrana JL, Attisano L (2010) The Hippo pathway regulates Wnt/beta-catenin signaling. Developmental Cell 18:579–591
1255. Zhao L, Jiang S, Hantash BM (2009) TGF-β1 Induces osteogenic differentiation of murine bone marrow stromal cells. Tissue Engineering Part A 16:725–733
1256. Heallen T, Zhang M, Wang J, Bonilla-Claudio M, Klysik E, Johnson RL, Martin JF (2011) Hippo pathway inhibits Wnt signaling to restrain cardiomyocyte proliferation and heart size. Science 332:458–461
1257. Tian Y, Kolb R, Hong JH, Carroll J, Li D, You J, Bronson R, Yaffe MB, Zhou J, Benjamin T (2007) TAZ promotes PC2 degradation through a SCF$^{\beta TRCP}$ E3 ligase complex. Molecular and Cellular Biology 27:6383–6395
1258. Bao Y, Hata Y, Ikeda M, Withanage K (2011) Mammalian Hippo pathway: from development to cancer and beyond. Journal of Biochemistry 149:361–379
1259. Zhang J, Ji JY, Yu M, Overholtzer M, Smolen GA, Wang R, Brugge JS, Dyson NJ, Haber DA (2009) YAP-dependent induction of amphiregulin identifies a non-cell-autonomous component of the Hippo pathway. Nature – Cell Biology 11:1444–1450
1260. Neto-Silva RM, de Beco S, Johnston LA (2010) Evidence for a growth-stabilizing regulatory feedback mechanism between Myc and Yorkie, the Drosophila homolog of Yap. Developmental Cell 19:507–520
1261. Tanoue T, Takeichi M (2005) New insights into Fat cadherins. Journal of Cell Science 118:2347–2353
1262. Lawrence PA, Struhl G, Casal J (2008) Do the protocadherins Fat and Dachsous link up to determine both planar cell polarity and the dimensions of organs? Nature – Cell Biology 10:1379–1382
1263. Ishikawa HO, Takeuchi H, Haltiwanger RS, Irvine KD (2008) Four-jointed is a Golgi kinase that phosphorylates a subset of cadherin domains. Science 321:401–404
1264. De Souza PM, Kankaanranta H, Michael A, Barnes PJ, Giembycz MA, Lindsay MA (2002) Caspase-catalyzed cleavage and activation of Mst1 correlates with eosinophil but not neutrophil apoptosis. Blood 99:3432–3438
1265. Lai ZC, Wei X, Shimizu T, Ramos E, Rohrbaugh M, Nikolaidis N, Ho LL, Li Y (2005) Control of cell proliferation and apoptosis by mob as tumor suppressor, mats. Cell 120:675–685
1266. Chow A, Hao Y, Yang X (2010) Molecular characterization of human homologs of yeast MOB1. International Journal of Cancer 126:2079–2089
1267. Gee ST, Milgram SL, Kramer KL, Conlon FL, Moody SA (2011) Yes-associated protein 65 (YAP) expands neural progenitors and regulates Pax3 expression in the neural plate border zone. PLoS One 6:e20309
1268. Harvey K, Tapon N (2007) The Salvador-Warts-Hippo pathway - an emerging tumour-suppressor network. Nature Reviews – Cancer 7:182–191
1269. Blandino G, Shaul Y, Strano S, Sudol M, Yaffe M (2009) The Hippo tumor suppressor pathway: a brainstorming workshop. Science Signaling 2:mr6
1270. Dupont S, Morsut L, Aragona M, Enzo E, Giulitti S, Cordenonsi M, Zanconato F, Le Digabel J, Forcato M, Bicciato S, Elvassore N, Piccolo S (2011) Role of YAP/TAZ in mechanotransduction. Nature 474:179–183
1271. Maeda T, Chapman DL, Stewart AF (2002) Mammalian vestigial-like 2, a cofactor of TEF-1 and MEF2 transcription factors that promotes skeletal muscle differentiation. Journal of Biological Chemistry 277:48889–48898
1272. Mielcarek M, Günther S, Krüger M, Braun T (2002) VITO-1, a novel vestigial related protein is predominantly expressed in the skeletal muscle lineage. Mechanisms of Development 119:S269–S274
1273. Goulev Y, Fauny JD, Gonzalez-Marti B, Flagiello D, Silber J, Zider A (2008) SCALLOPED interacts with YORKIE, the nuclear effector of the hippo tumor-suppressor pathway in Drosophila. Current Biology 18:435–441

1274. Yu J, Zheng Y, Dong J, Klusza S, Deng WM, Pan D (2010) Kibra functions as a tumor suppressor protein that regulates hippo signaling in conjunction with Merlin and expanded. Developmental Cell 18:288–299
1275. Genevet A, Wehr MC, Brain R, Thompson BJ, Tapon N (2010) Kibra is a regulator of the Salvador/Warts/Hippo signaling network. Developmental Cell 18:300–308
1276. Baumgartner R, Poernbacher I, Buser N, Hafen E, Stocker H (2010) The WW domain protein Kibra acts upstream of hippo in Drosophila. Developmental Cell 18:309–316
1277. Csukai M, Chen CH, De Matteis MA, Mochly-Rosen D (1997) The coatomer protein β'-COP, a selective binding protein (RACK) for protein kinase Cϵ. Journal of Biological Chemistry 272:29200–29206
1278. Jaburek M, Costa AD, Burton JR, Costa CL, Garlid KD (2006) Mitochondrial PKCϵ and mitochondrial ATP-sensitive K^+ channel copurify and coreconstitute to form a functioning signaling module in proteoliposomes. Circulation Research 99:878–883
1279. Schechtman D, Mochly-Rosen D (2001) Adaptor proteins in protein kinase C-mediated signal transduction. Oncogene 20:6339–6347
1280. Yuasa K, Omori K, Yanaka N (2000) Binding and phosphorylation of a novel male germ cell-specific cGMP-dependent protein kinase-anchoring protein by cGMP-dependent protein kinase Iα. Journal of Biological Chemistry 275:4897–4905
1281. Naisbitt S, Valtschanoff J, Allison DW, Sala C, Kim E, Craig AM, Weinberg RJ, Sheng M (2000) Interaction of the postsynaptic density-95/guanylate kinase domain-associated protein complex with a light chain of myosin-V and dynein. Journal of Neuroscience 20:4524–4534
1282. Takeuchi M, Hata Y, Hirao K, Toyoda A, Irie M, Takai Y (1997) SAPAPs. A family of PSD-95/SAP90-associated proteins localized at postsynaptic density. Journal of Biological Chemistry 272:11943–11951
1283. Boeckers TM, Winter C, Smalla KH, Kreutz MR, Bockmann J, Seidenbecher C, Garner CC, Gundelfinger ED (1999) Proline-rich synapse-associated proteins ProSAP1 and ProSAP2 interact with synaptic proteins of the SAPAP/GKAP family. Biochemical and Biophysical Research Communications 264:247–252
1284. Lim S, Naisbitt S, Yoon J, Hwang JI, Suh PG, Sheng M, Kim E (1999) Characterization of the Shank family of synaptic proteins. Multiple genes, alternative splicing, and differential expression in brain and development. Journal of Biological Chemistry 274:29510–29518
1285. Sheng M, Kim E (2000) The Shank family of scaffold proteins. Journal of Cell Science 113:1851–1856
1286. Kwon SK, Woo J, Kim SY, Kim H, Kim E (2010) Trans-synaptic adhesions between netrin-G ligand-3 (NGL-3) and receptor tyrosine phosphatases LAR, protein-tyrosine phosphatase δ (PTPδ), and PTPσ via specific domains regulate excitatory synapse formation.
1287. Carnegie GK, Scott JD (2003) A-kinase anchoring proteins and neuronal signaling mechanisms. Genes and Development 17:1557–1568
1288. Chen L, Kass RS (2005) A-kinase anchoring proteins: different partners, different dance. Nature – Cell Biology 7:1050–1051
1289. Lim CJ, Han J, Yousefi N, Ma Y, Amieux PS, McKnight GS, Taylor SS, Ginsberg MH (2007) α4 Integrins are type I cAMP-dependent protein kinase-anchoring proteins. Nature – Cell Biology 9:415–421
1290. Wong W, Scott JD (2004) AKAP signalling complexes: focal points in space and time, Nature Reviews – Molecular Cell Biology 5:959–970
1291. Diviani D, Dodge-Kafka KL, Li J, Kapiloff MS (2011) A-kinase anchoring proteins: scaffolding proteins in the heart. American Journal of Physiology – Heart and Circulatory Physiology 301:H1742–H1753
1292. Dong F, Feldmesser M, Casadevall A, Rubin CS (1998) Molecular characterization of a cDNA that encodes six isoforms of a novel murine A kinase anchor protein. Journal of Biological Chemistry 273:6533–6541

1293. Brown PR, Miki K, Harper DB, Eddy EM (2003) A-kinase anchoring protein 4 binding proteins in the fibrous sheath of the sperm flagellum. Biology of Reproduction 68: 2241–2248
1294. Hoshi N, Langeberg L, Scott JD (2005) Distinct enzyme combinations in AKAP signalling complexes permit functional diversity. Nature – Cell Biology 7:1066–1073
1295. Dodge-Kafka KL, Soughayer J, Pare GC, Carlisle Michel JJ, Langeberg LK, Kapiloff MS, Scott JD (2005) The protein kinase A anchoring protein mAKAP coordinates two integrated cAMP effector pathways. Nature 437:574–578
1296. Akileswaran L, Taraska JW, Sayer JA, Gettemy JM, Coghlan VM (2001) A-kinase-anchoring protein AKAP95 is targeted to the nuclear matrix and associates with p68 RNA helicase. Journal of Biological Chemistry 276:17448–17454
1297. Arsenijevic T, Degraef C, Dumont JE, Roger PP, Pirson I (2004) A novel partner for D-type cyclins: protein kinase A-anchoring protein AKAP95. Biochemical Journal 378:673–679
1298. Gisler SM, Pribanic S, Bacic D, Forrer P, Gantenbein A, Sabourin LA, Tsuji A, Zhao ZS, Manser E, Biber J, Murer H (2003) PDZK1: I. a major scaffolder in brush borders of proximal tubular cells. Kidney International 64:1733–1745
1299. Tanji C, Yamamoto H, Yorioka N, Kohno N, Kikuchi K, Kikuchi A (2002) A-kinase anchoring protein AKAP220 binds to glycogen synthase kinase-3β (GSK-3β) and mediates protein kinase A-dependent inhibition of GSK-3β. Journal of Biological Chemistry 277: 36955–36961
1300. Gelman IH (2011) Akap12. UCSD-Nature Molecule Pages, UCSD-Nature Signaling Gateway (www.signaling-gateway.org)
1301. Kultgen PL, Byrd SK, Ostrowski LE, Milgram SL (2002) Characterization of an A-kinase anchoring protein in human ciliary axonemes. Molecular Biology of the Cell 13:4156–4166
1302. Gerke V, Moss SE (2002) Annexins: from structure to function. Physiological Reviews 82:331–371
1303. Barwise JL, Walker JH (1996) Annexins II, IV, V and VI relocate in response to rises in intracellular calcium in human foreskin fibroblasts. Journal of Cell Science 109:247–255
1304. Gerke V, Creutz CE, Moss SE (2005) Annexins: linking Ca^{2+} signalling to membrane dynamics. Nature Reviews – Molecular Cell Biology 6:449–461
1305. Le Cabec V, Maridonneau-Parini I (1994) Annexin 3 is associated with cytoplasmic granules in neutrophils and monocytes and translocates to the plasma membrane in activated cells. Biochemical Journal 303:481–487
1306. White IJ, Bailey LM, Aghakhani MR, Moss SE, Futter CE (2006) EGF stimulates annexin 1-dependent inward vesiculation in a multivesicular endosome subpopulation. EMBO Journal 25, 1–12
1307. Herbert SP, Odell AF, Ponnambalam S, Walker JH (2007) The confluence-dependent interaction of cytosolic phospholipase A2-α with annexin A1 regulates endothelial cell prostaglandin E2 generation. Journal of Biological Chemistry 82:34468–34478
1308. Ross TS, Tait JF, Majerus PW (1990) Identity of inositol 1,2-cyclic phosphate 2-phosphohydrolase with lipocortin III. Science 248:605–607
1309. Kaetzel MA, Chan HC, Dubinsky WP, Dedman JR, Nelson DJ (1994) A role for annexin IV in epithelial cell function. Inhibition of calcium-activated chloride conductance. Journal of Biological Chemistry 269:5297–5302
1310. Chan HC, Kaetzel MA, Gotter AL, Dedman JR, Nelson DJ (1994) Annexin IV inhibits calmodulin-dependent protein kinase II-activated chloride conductance. A novel mechanism for ion channel regulation. Journal of Biological Chemistry 269:32464–32468
1311. Trouvé P, Le Drévo MA, Kerbiriou M, Friocourt G, Fichou Y, Gillet D, Férec C (2007) Annexin V is directly involved in cystic fibrosis transmembrane conductance regulator's chloride channel function. Biochimica et Biophysica Acta 1772:1121–1133
1312. Riquelme G, Llanos P, Tischner E, Neil J, Campos B (2004) Annexin-6 modulates the maxi-chloride channel of the apical membrane of syncytiotrophoblast isolated from human placenta. Journal of Biological Chemistry 279:50601–50608

1313. d'Anglemont de Tassigny A, Souktani R, Henry P, Ghaleh B, Berdeaux A (2004) Volume-sensitive chloride channels ($I_{Cl,vol}$) mediate doxorubicin-induced apoptosis through apoptotic volume decrease in cardiomyocytes. Fundamental and Clinical Pharmacology 18:531–538
1314. d'Anglemont de Tassigny A, Berdeaux A, Souktani R, Henry P, Ghaleh B (2008) The volume-sensitive chloride channel inhibitors prevent both contractile dysfunction and apoptosis induced by doxorubicin through PI3kinase, Akt and Erk 1/2. European Journal of Heart Failure 10:39–46
1315. Camors E, Monceau V, Charlemagne D (2005) Annexins and Ca^{2+} handling in the heart. Circulation Research 65:793–802
1316. Vignali M, Hassan AH, Neely KE, Workman JL (2000) ATP-dependent chromatin-remodeling complexes. Molecular and Cellular Biology 20:1899–1910
1317. Ghosh S, Hayden MS (2008) New regulators of NF-κB in inflammation. Nature Reviews – Immunology 8:837–848
1318. Peters RT, Liao SM, Maniatis T (2000) IKKε is part of a novel PMA-inducible IkappaB kinase complex. Molecular Cell 5:513–522
1319. Kamata H, Tsuchiya Y, Asano T (2010) IκBβ is a positive and negative regulator of NF-κB activity during inflammation. Cell Research 20:1178–1180
1320. Xu S, Bayat H, Hou X, Jiang B (2006) Ribosomal S6 kinase-1 modulates interleukin-1 β-induced persistent activation of NF-κB through phosphorylation of IκBβ. American Journal of Physiology – Cell Physiology 291:C1336–C1345
1321. Fiorini E, Schmitz I, Marissen WE, Osborn SL, Touma M, Sasada T, Reche PA, Tibaldi EV, Hussey RE, Kruisbeek AM, Reinherz EL, Clayton LK (2002) Peptide-induced negative selection of thymocytes activates transcription of an NF-κB inhibitor. Molecular Cell 9:637–648
1322. Vertegaal ACO, Kuiperij HB, Yamaoka S, Courtois G, van der Eb AJ, Zantema A (2000) Protein kinase C-α is an upstream activator of the IκBβ complex in the TPA signal transduction pathway to NF-κB in U2OS cells. Cellular Signalling 12:759–768
1323. Gould GW (2011) IKKε: a kinase at the intersection of signaling and membrane traffic. Science Signaling 4:pe30
1324. Sun SC (2011) Non-canonical NFκB signaling pathway. Cell Research 21:71–85
1325. Shih VFS, Tsui R, Caldwell A, Hoffmann A (2011) A single NFκB system for both canonical and non-canonical signaling. Cell Research 21:86–102
1326. Liu S, Chen ZJ (2011) Expanding role of ubiquitination in NFκB signaling. Cell Research 21:6–21
1327. Razani B, Zarnegar B, Ytterberg AJ, Shiba T, Dempsey PW, Ware CF, Loo JA, Cheng G (2010) Negative feedback in noncanonical NF-κB signaling modulates NIK stability through IKKα-mediated phosphorylation. Science Signaling 3:ra41
1328. Miyamoto S (2011) Nuclear initiated NFκB signaling: NEMO and ATM take center stage. Cell Research 21:116–130
1329. Wu ZH, Wong ET, Shi Y, Niu J, Chen Z, Miyamoto S, Tergaonkar V (2010) ATM- and NEMO-dependent ELKS ubiquitination coordinates TAK1-mediated IKK activation in response to genotoxic stress. Molecular Cell 40:75–86
1330. Hinz M, Stilmann M, Arslan SC, Khanna KK, Dittmar G, Scheidereit C (2010) A cytoplasmic ATM-TRAF6-cIAP1 module links nuclear DNA damage signaling to ubiquitin-mediated NF-κB activation. Molecular Cell 40:63–74
1331. Blonska M, Lin X (2011) NFκB signaling pathways regulated by CARMA family of scaffold proteins. Cell Research 21:55–70
1332. Staal J, Bekaert T, Beyaert R (2011) Regulation of NFκB signaling by caspases and MALT1 paracaspase. Cell Research 21:40–54
1333. Yamanaka K, Ishikawa H, Megumi Y, Tokunaga F, Kanie M, Rouault TA, Morishima I, Minato N, Ishimori K, Iwai K (2003) Identification of the ubiquitin-protein ligase that recognizes oxidized IRP2. Nature – Cell Biology 5:336–340

1334. Harhaj EW, Dixit VM (2011) Deubiquitinases in the regulation of NFκB signaling. Cell Research 21:22–39
1335. Tokunaga F, Sakata SI, Saeki Y, Satomi Y, Kirisako T, Kamei K, Nakagawa T, Kato M, Murata S, Yamaoka S, Yamamoto M, Akira S, Takao T, Tanaka K, Iwai K (2009) Involvement of linear polyubiquitylation of NEMO in NF-κB activation. Nature – Cell Biology 11:123–132
1336. Xia ZP, Sun L, Chen X, Pineda G, Jiang X, Adhikari A, Zeng W, Chen ZJ (2009) Direct activation of protein kinases by unanchored polyubiquitin chains. Nature 461:114–119
1337. Gyrd-Hansen M, DardingM, Miasari M, Santoro MM, Zender L, Xue W, Tenev T, da Fonseca PCA, Zvelebil M, Bujnicki JM, Lowe S, Silke J, Meier P (2008) IAPs contain an evolutionarily conserved ubiquitin-binding domain that regulates NF-κB as well as cell survival and oncogenesis. Nature – Cell Biology 10:1309–1317
1338. Stilmann M, Hinz M, Arslan SC, Zimmer A, Schreiber V, Scheidereit C (2009) A nuclear poly(ADP-ribose)-dependent signalosome confers DNA damage-induced IkappaB kinase activation. Molecular Cell 36:365–378
1339. Yazdanpanah B, Wiegmann K, Tchikov V, Krut O, Pongratz C, Schramm M, Kleinridders A, Wunderlich T, Kashkar H, Utermöhlen O, Brüning JC, Schütze S, Krönke M (2009) Riboflavin kinase couples TNF receptor 1 to NADPH oxidase. Nature 460:1159–1163
1340. Morgan MJ, Liu ZG (2011) Crosstalk of reactive oxygen species and NFκB signaling. Cell Research 21:103–115
1341. Perkins ND (2007) Integrating cell-signalling pathways with NF-kappaB and IKK function. Nature Reviews – Molecular Cell Biology 8:49–62
1342. Wang D, You Y, Lin PC, Xue L, Morris SW, Zen H, Wen R, Lin X (2007) Bcl10 plays a critical role in NF-kappaB activation induced by G protein-coupled receptors. Proceedings of the National Academy of Sciences of the United States of America 104:145–150
1343. Shrum CK, Defrancisco D, Meffert MK (2009) Stimulated nuclear translocation of NF-κB and shuttling differentially depend on dynein and the dynactin complex. Proceedings of the National Academy of Sciences of the United States of America 106:2647–2652
1344. Bosisio D, Marazzi I, Agresti A, Shimizu N, Bianchi ME, Natoli G (2006) A hyper-dynamic equilibrium between promoter-bound and nucleoplasmic dimers controls NF-kappaB-dependent gene activity. EMBO Journal 25:798–810
1345. Chew J, Biswas S, Shreeram S, Humaidi M, Wong ET, Dhillion MK, Teo H, Hazra A, Fang CC, Lpez-Collazo E, Bulavin DV, Tergaonkar V (2009) WIP1 phosphatase is a negative regulator of NF-κB signalling. Nature – Cell Biology 11:659–666
1346. Nakajima A, Komazawa-Sakon S, Takekawa M, Sasazuki T, Yeh WC, Yagita H, Okumura K, Nakano H (2006) An antiapoptotic protein, c-FLIPL, directly binds to MKK7 and inhibits the JNK pathway. EMBO Journal 25:5549–5559
1347. Yang Q, Kim YS, Lin Y, Lewis J, Neckers L, Liu ZG (2006) Tumour necrosis factor receptor 1 mediates endoplasmic reticulum stress-induced activation of the MAP kinase JNK. EMBO Reports 7:622–627
1348. Ea CK, Deng L, Xia ZP, Pineda G, Chen ZJ (2006) Activation of IKK by TNF requires site-specific ubiquitination of RIP1 and polyubiquitin binding by NEMO. Molecular Cell 22:245–257
1349. Vallabhapurapu S, Matsuzawa A, Zhang WZ, Tseng PH, Keats JJ, Wang H, Vignali DAA, Bergsagel PL, Karin M (2008) Nonredundant and complementary functions of TRAF2 and TRAF3 in a ubiquitination cascade that activates NIK-dependent alternative NF-κB signaling. Nature – Immunology 9:1364–1370
1350. Zarnegar BJ, Wang Y, Mahoney DJ, Dempsey PW, Cheung HH, He J, Shiba T, Yang X, Yeh WC, Mak TW, Korneluk RG, Cheng G (2008) Noncanonical NF-κB activation requires coordinated assembly of a regulatory complex of the adaptors cIAP1, cIAP2, TRAF2 and TRAF3 and the kinase NIK. Nature – Immunology 9:1371–1378
1351. Chen Y, Rabson AB, Gorski DH (2010) MEOX2 regulates nuclear factor-κB activity in vascular endothelial cells through interactions with p65 and IκBβ. Cardiovascular Research 87:723–731

1352. Wu CJ, Conze DB, Li T, Srinivasula SM, Ashwell JD (2006) Sensing of Lys 63-linked polyubiquitination by NEMO is a key event in NF-kappaB activation. Nature – Cell Biology 8:398–406
1353. Basak S, Kim H, Kearns JD, Tergaonkar V, O'Dea E, Werner SL, Benedict CA, Ware CF, Ghosh G, Verma IM, Hoffmann A (2007) A fourth IκB protein within the NF-κB signaling module. Cell 128:369–381
1354. Gringhuis SI, den Dunnen J, Litjens M, van der Vlist M, Wevers B, Bruijns SCM, Geijtenbeek TBH (2009) Dectin-1 directs T helper cell differentiation by controlling noncanonical NF-κB activation through Raf-1 and Syk. Nature – Immunology 10:203–213
1355. Goto A, Matsushita K, Gesellchen V, El Chamy L, Kuttenkeuler D, Takeuchi O, Hoffmann JA, Akira S, Boutros M, Reichhart JM (2008) Akirins are highly conserved nuclear proteins required for NF-κB-dependent gene expression in drosophila and mice. Nature – Immunology 9:97–104
1356. Shembade N, Ma A, Harhaj EW (2010) Inhibition of NF-κB signaling by A20 through disruption of ubiquitin enzyme complexes. Science 327:1135–1139
1357. Neumann M, Klar S, Wilisch-Neumann A, Hollenbach E, Kavuri S, Leverkus M, Kandolf R, Brunner-Weinzierl MC, Klingel K (2011) Glycogen synthase kinase-3ᾰ3β2 is a crucial mediator of signal-induced RelB degradation. Oncogene 30:2485–2492
1358. Hailfinger S, Nogai H, Pelzer C, Jaworski M, Cabalzar K, Charton JE, Guzzardi M, Décaillet C, Grau M, Dörken B, Lenz P, Lenz G, Thome M (2011) Malt1-dependent RelB cleavage promotes canonical NF-κB activation in lymphocytes and lymphoma cell lines. Proceedings of the National Academy of Sciences of the United States of America 108:14596–14601
1359. Oliver KM, Lenihan CR, Bruning U, Cheong A, Laffey JG, McLoughlin P, Taylor CT, Cummins EP (2012) Hypercapnia induces cleavage and nuclear localization of RelB protein, giving insight into CO_2 sensing and signaling. Journal of Biological Chemistry 287:14004–14011
1360. Cummins EP, Berra E, Comerford KM, Ginouves A, Fitzgerald KT, Seeballuck F, Godson C, Nielsen JE, Moynagh P, Pouyssegur J, Taylor CT (2006) Prolyl hydroxylase-1 negatively regulates IkappaB kinase-beta, giving insight into hypoxia-induced NFkappaB activity. Proceedings of the National Academy of Sciences of the United States of America 103:18154–18159
1361. Rius J, Guma M, Schachtrup C, Akassoglou K, Zinkernagel AS, Nizet V, Johnson RS, Haddad GG, Karin M (2008) NF-κB links innate immunity to the hypoxic response through transcriptional regulation of HIF-1α. Nature 453:807–811
1362. Nakayama K (2009) Cellular signal transduction of the hypoxia response. Journal of Biochemistry 146:757–765
1363. Cockman ME, Lancaster DE, Stolze IP, Hewitson KS, McDonough MA, Coleman ML, Coles CH, Yu X, Hay RT, Ley SC, Pugh CW, Oldham NJ, Masson N, Schofield CJ, Ratcliffe PJ (2006) Posttranslational hydroxylation of ankyrin repeats in IkappaB proteins by the hypoxia-inducible factor (HIF) asparaginyl hydroxylase, factor inhibiting HIF (FIH). Proceedings of the National Academy of Sciences of the United States of America 103:14767–14772
1364. Resnik ER, Herron JM, Lyu SC, Cornfield DN (2007) Developmental regulation of hypoxia-inducible factor 1 and prolyl-hydroxylases in pulmonary vascular smooth muscle cells. Proceedings of the National Academy of Sciences of the United States of America 104:18789–18794
1365. Loor G, Schumacker PT (2008) Role of hypoxia-inducible factor in cell survival during myocardial ischemia–reperfusion. Cell Death and Differentiation 15:686–690
1366. Formenti F, Constantin-Teodosiu D, Emmanuel Y, Cheeseman J, Dorrington KL, Edwards LM, Humphreys SM, Lappin TR, McMullin MF, McNamara CJ, Mills W, Murphy JA, O'Connor DF, Percy MJ, Ratcliffe PJ, Smith TG, Treacy M, Frayn KN, Greenhaff PL, Karpe F, Clarke K, Robbins PA (2010) Regulation of human metabolism by hypoxia-inducible factor. Proceedings of the National Academy of Sciences of the United States of America 107:12722–12727

1367. Wei H, Bedja D, Koitabashi N, Xing D, Chen J, Fox-Talbot K, Rouf R, Chen S, Steenbergen C, Harmon JW, Dietz HC, Gabrielson KL, Kass DA, Semenza GL (2012) Endothelial expression of hypoxia-inducible factor 1 protects the murine heart and aorta from pressure overload by suppression of TGF-β signaling. Proceedings of the National Academy of Sciences of the United States of America 109:E841–E850
1368. Hannenhalli S, Kaestner KH (2009) The evolution of Fox genes and their role in development and disease. Nature Reviews – Genetics 10:233–240
1369. Tuteja G, Kaestner KH (2007) SnapShot: forkhead transcription factors I. Cell 130:1160
1370. Tuteja G, Kaestner KH (2007) Forkhead transcription factors II. Cell 131:192
1371. van der Horst A, Burgering BMT (2007) Stressing the role of FoxO proteins in lifespan and disease. Nature Reviews – Molecular Cell Biology 8:440–450
1372. Ni YG, Wang N, Cao DJ, Sachan N, Morris DJ, Gerard RD, Kuro-o M, Rothermel BA, Hill JA (2007) FoxO transcription factors activate Akt and attenuate insulin signaling in heart by inhibiting protein phosphatases. Proceedings of the National Academy of Sciences of the United States of America 104:20517–20522
1373. Patel PH, Tamanoi F (2006) Increased Rheb-TOR signaling enhances sensitivity of the whole organism to oxidative stress. Journal of Cell Science 119:4285–4292
1374. Yuan Z, Becker EBE, Merlo P, Yamada T, DiBacco S, Konishi Y, Schaefer EM, Bonni A (2008) Activation of FOXO1 by Cdk1 in cycling cells and postmitotic neurons. Science 319:1665–1668
1375. Van der Heide LP, Hoekman MFM, Smidt MP (2004) The ins and outs of FoxO shuttling: mechanisms of FoxO translocation and transcriptional regulation. Biochemical Journal 380:297–309
1376. Ni YG, Berenji K, Wang N, Oh M, Sachan N, Dey A, Cheng J, Lu G, Morris DJ, Castrillon DH, Gerard RD, Rothermel BA, Hill JA (2006) Foxo transcription factors blunt cardiac hypertrophy by inhibiting calcineurin signaling. Circulation 114:1159–1168
1377. Brunet A, Bonni A, Zigmond MJ, Lin MZ, Juo P, Hu LS, Anderson MJ, Arden KC, Blenis J, Greenberg ME (1999) Akt promotes cell survival by phosphorylating and inhibiting a Forkhead transcription factor. Cell 96:857–868
1378. Dobson ME, Tzivion G (2011) Foxo3. UCSD-Nature Molecule Pages, UCSD-Nature Signaling Gateway (www.signaling-gateway.org)
1379. Randell JCW, Bowers JL, Rodrguez HK, Bell SP (2006) Sequential ATP hydrolysis by Cdc6 and ORC directs loading of the Mcm2-7 helicase. Molecular Cell 21:29–39
1380. Zhang Y, Xing Y, Zhang L, Mei Y, Yamamoto K, Mak TW, You H (2012) Regulation of cell cycle progression by forkhead transcription factor FOXO3 through its binding partner DNA replication factor Cdt1. Proceedings of the National Academy of Sciences of the United States of America 109:5717–5722
1381. Skurk C, Izumiya Y, Maatz H, Razeghi P, Shiojima I, Sandri M, Sato K, Zeng L, Schiekofer S, Pimentel D, Lecker S, Taegtmeyer H, Goldberg AL, Walsh K (2005) The FOXO3a transcription factor regulates cardiac myocyte size downstream of AKT signaling. Journal of Biological Chemistry 280:20814–20823
1382. An BS, Tavera-Mendoza LE, Dimitrov V, Wang X, Calderon MR, Wang HJ, White JH (2010) Stimulation of Sirt1-regulated FoxO protein function by the ligand-bound vitamin D receptor. Molecular and Cellular Biology 30:4890–4900
1383. Jacobs FM, van der Heide LP, Wijchers PJ, Burbach JP, Hoekman MF, Smidt MP (2003) FoxO6, a novel member of the FoxO class of transcription factors with distinct shuttling dynamics. Journal of Biological Chemistry 278:35959–35967
1384. van der Heide LP, Jacobs FM, Burbach JP, Hoekman MF, Smidt MP (2005) FoxO6 transcriptional activity is regulated by Thr26 and Ser184, independent of nucleo-cytoplasmic shuttling. Biochemical Journal 391:623–629
1385. Kim DH, Perdomo G, Zhang T, Slusher S, Lee S, Phillips BE, Fan Y, Giannoukakis N, Gramignoli R, Strom S, Ringquist S, Dong HH (2011) FoxO6 integrates insulin signaling with gluconeogenesis in the liver. Diabetes 60:2763–2774

1386. Joerger AC, Rajagopalan S, Natan E, Veprintsev DB, Robinson CV, Fersht AR (2009) Structural evolution of p53, p63, and p73: implication for heterotetramer formation. Proceedings of the National Academy of Sciences of the United States of America 106:17705–17710
1387. Holmberg CI, Tran SE, Eriksson JE, Sistonen L (2002) Multisite phosphorylation provides sophisticated regulation of transcription factors. Trends in Biochemical Sciences 27: 619–627
1388. Huang J, Berger SL (2008) The emerging field of dynamic lysine methylation of non-histone proteins. Current Opinion in Genetics and Development 18:152–158
1389. Jansson M, Durant ST, Cho EC, Sheahan S, Edelmann M, Kessler B, La Thangue NB (2008) Arginine methylation regulates the p53 response. Nature – Cell Biology 10:1431–1439
1390. Riley T, Sontag E, Chen P, Levine A (2008) Transcriptional control of human p53-regulated genes. Nature Reviews – Molecular Cell Biology 9:402–412
1391. Brooks CL, Gu W (2010) New insights into p53 activation. Cell Research 20:614–621
1392. Meyer KD, Lin SC, Bernecky C, Gao Y, Taatjes DJ (2010) p53 Activates transcription by directing structural shifts in Mediator. Nature – Structural and Molecular Biology 17: 753–760
1393. Green DR, Kroemer G (2009) Cytoplasmic functions of the tumour suppressor p53. Nature 458:1127–1130
1394. Geva-Zatorsky N, Dekel E, Batchelor E, Lahav G, Alon U (2010) Fourier analysis and systems identification of the p53 feedback loop. Proceedings of the National Academy of Sciences of the United States of America 107:13550–13555
1395. Shi D, Pop MS, Kulikov R, Love IM, Kung A, Grossman SR (2009) CBP and p300 are cytoplasmic E4 polyubiquitin ligases for p53. Proceedings of the National Academy of Sciences of the United States of America 106:16275–16280
1396. Kulikov R, Letienne J, Kaur M, Grossman SR, Arts J, Blattner C (2010) Mdm2 facilitates the association of p53 with the proteasome. Proceedings of the National Academy of Sciences of the United States of America 107:10038–10043
1397. Ferreon JC, Lee CW, Arai M, Martinez-Yamout HJ, Wright PE (2009) Cooperative regulation of p53 by modulation of ternary complex formation with CBP/p300 and HDM2. Proceedings of the National Academy of Sciences of the United States of America 106:6591–6596
1398. Tai E, Benchimol S (2009) TRIMming p53 for ubiquitination. Proceedings of the National Academy of Sciences of the United States of America 106:11431–11432
1399. Kumamoto K, Spillare EA, Fujita K, Horikawa I, Yamashita T, Appella E, Nagashima M, Takenoshita S, Yokota J, Harris CC (2008) Nutlin-3a activates p53 to both down-regulate inhibitor of growth 2 and up-regulate mir-34a, mir-34b, and mir-34c expression, and induce senescence. Cancer Research 68:3193–3203
1400. Allton K, Jain AK, Herz HM, Tsai WW, Jung SY, Qin J, Bergmann A, Johnson RL, Barton MC (2009) Trim24 targets endogenous p53 for degradation. Proceedings of the National Academy of Sciences of the United States of America 106:11612–11616
1401. Le Cam L, Linares LK, Paul C, Julien E, Lacroix M, Hatchi E, Triboulet R, Bossis G, Shmueli A, Rodriguez MS, Coux O, Sardet C (2006) E4F1 is an atypical ubiquitin ligase that modulates p53 effector functions independently of degradation. Cell 127:775–788
1402. Coutts AS, Weston L, La Thangue NB (2009) A transcription co-factor integrates cell adhesion and motility with the p53 response. Proceedings of the National Academy of Sciences of the United States of America 106:19872–19877
1403. Cho YY, He Z, Zhang Y, Choi HS, Zhu F, Choi BY, Kang BS, Ma WY, Bode AM, Dong Z (2005) The p53 protein is a novel substrate of ribosomal S6 kinase 2 and a critical intermediary for ribosomal S6 kinase 2 and histone H3 interaction. Cancer Research 65:3596–3603
1404. Llanos S, Cuadrado A, Serrano M (2009) MSK2 inhibits p53 activity in the absence of stress. Science Signaling 2:ra57

1405. Lee CW, Ferreon JC, Ferreon AC, Arai M, Wright PE (2010) Graded enhancement of p53 binding to CREB-binding protein (CBP) by multisite phosphorylation. Proceedings of the National Academy of Sciences of the United States of America 107:19290–19295
1406. Drost J, Mantovani F, Tocco F, Elkon R, Comel A, Holstege H, Kerkhoven R, Jonkers J, Voorhoeve PM, Agami R, Del Sal G (2010) BRD7 is a candidate tumour suppressor gene required for p53 function. Nature – Cell Biology 12:380–389
1407. Park SY, Lee JH, Ha M, Nam JW, Kim VN (2008) miR-29 miRNAs activate p53 by targeting p85α and CDC42. Nature – Structural and Molecular Biology 16:23–29
1408. He L, He X, Lim LP, de Stanchina E, Xuan Z, Liang Y, Xue W, Zender L, Magnus J, Ridzon D, Jackson AL, Linsley PS, Chen C, Lowe SW, Cleary MA, Hannon GJ (2007) A microRNA component of the p53 tumour suppressor network. Nature 447:1130–1134
1409. Suzuki HI, Yamagata K, Sugimoto K, Iwamoto T, Kato S, Miyazono K (2009) Modulation of microRNA processing by p53. Nature 460:529-533
1410. Yamakuchi M, Lotterman CD, Bao C, Hruban RH, Karim B, Mendell JT, Huso D, Lowenstein CJ (2010) P53-induced microRNA-107 inhibits HIF-1 and tumor angiogenesis. Proceedings of the National Academy of Sciences of the United States of America 107:6334–6339
1411. Sachdeva M, Zhu S, Wu F, Wu H, Walia V, Kumar S, Elble R, Watabe K, Mo YY (2009) p53 represses c-Myc through induction of the tumor suppressor miR-145. Proceedings of the National Academy of Sciences of the United States of America 106:3207–3212
1412. Vousden KH, Lane DP (2007) p53 in health and disease. Nature Reviews – Molecular Cell Biology 8:275–283
1413. Cordenonsi M, Montagner M, Adorno M, Zacchigna L, Martello G, Mamidi A, Soligo S, Dupont S, Piccolo S (2007) Integration of TGF-beta and Ras/MAPK signaling through p53 phosphorylation. Science 315:840–843
1414. Hu W, Zhang C, Wu R, Sun Y, Levine A, Feng Z (2010) Glutaminase 2, a novel p53 target gene regulating energy metabolism and antioxidant function. Proceedings of the National Academy of Sciences of the United States of America 107:7455–7460
1415. Suzuki S, Tanaka T, Poyurovsky MV, Nagano H, Mayama T, Ohkubo S, Lokshin M, Hosokawa H, Nakayama T, Suzuki Y, Sugano S, Sato E, Nagao T, Yokote K, Tatsuno I, Prives C (2010) Phosphate-activated glutaminase (GLS2), a p53-inducible regulator of glutamine metabolism and reactive oxygen species. Proceedings of the National Academy of Sciences of the United States of America 107:7461–7466
1416. Zhang XP, Liu F, Cheng Z, Wang W (2009) Cell fate decision mediated by p53 pulses. Proceedings of the National Academy of Sciences of the United States of America 106:12245–12250
1417. Cho YJ, Liang P (2008) Killin is a p53-regulated nuclear inhibitor of DNA synthesis. Proceedings of the National Academy of Sciences of the United States of America 105:5396–5401
1418. Sendoel A, Kohler I, Fellmann C, Lowe SW, Hengartner MO (2010) HIF-1 antagonizes p53-mediated apoptosis through a secreted neuronal tyrosinase. Nature 465:577–583
1419. Kim HJ, Lee HJ, Jun JI, Oh Y, Choi SG, Kim H, Chung CW, Kim IK, Park IS, Chae HJ, Kim HR, Jung YK (2009) Intracellular cleavage of osteopontin by caspase-8 modulates hypoxia/reoxygenation cell death through p53. Proceedings of the National Academy of Sciences of the United States of America 106:15326-15331
1420. Koldobskiy MA, Chakraborty A, Werner JK, Snowman AM, Juluri KR, Vandiver MS, Kim S, Heletz S, Snyder SH (2010) p53-mediated apoptosis requires inositol hexakisphosphate kinase-2. Proceedings of the National Academy of Sciences of the United States of America 107:20947–20951
1421. Li L, Deng B, Xing G, Teng Y, Tian C, Cheng X, Yin X, Yang J, Gao X, Zhu Y, Sun Q, Zhang L, Yang X, He F (2007) PACT is a negative regulator of p53 and essential for cell growth and embryonic development. Proceedings of the National Academy of Sciences of the United States of America 104:7951–7956

1422. Jänicke RL, Sohn D, Schulze-Osthoff K (2008) The dark side of a tumor suppressor: anti-apoptotic p53. Cell Death and Differentiation 15:959–976
1423. Tasdemir E, Maiuri MC, Galluzzi L, Vitale I, Djavaheri-Mergny M, D'Amelio M, Criollo A, Morselli E, Zhu C, Harper F, Nannmark U, Samara C, Pinton P, Vicencio JM, Carnuccio R, Moll UM, Madeo F, Paterlini-Brechot P, Rizzuto R, Szabadkai G, Pierron G, Blomgren K, Tavernarakis N, Codogno P, Cecconi F, Kroemer G (2008) Regulation of autophagy by cytoplasmic p53. Nature – Cell Biology 10:676–687
1424. Lee KH, Li M, Michalowski AM, Zhang X, Liao H, Chen L, Xu Y, Wu X, Huang J (2010) A genomewide study identifies the Wnt signaling pathway as a major target of p53 in murine embryonic stem cells. Proceedings of the National Academy of Sciences of the United States of America 107:69–74
1425. Xia M, Land H (2007) Tumor suppressor p53 restricts Ras stimulation of RhoA and cancer cell motility. Nature – Structural and Molecular Biology 14:215–223
1426. Zhang J, Jun Cho S, Chen X (2010) RNPC1, an RNA-binding protein and a target of the p53 family, regulates p63 expression through mRNA stability. Proceedings of the National Academy of Sciences of the United States of America 107:9614–9619
1427. Vikhanskaya F, Toh WH, Dulloo I, Wu Q, Boominathan L, Ng HH, Vousden KH, Sabapathy K (2007) p73 supports cellular growth through c-Jun-dependent AP-1 transactivation. Nature – Cell Biology 9:698–706
1428. Hauck L, Harms C, An J, Rohne J, Gertz K, Dietz K, Endres M, von Harsdorf R (2008) Protein kinase CK2 links extracellular growth factor signaling with the control of p27Kip1 stability in the heart. Nature – Medicine 14:315–324

Chap. 11. Signaling Pathways

1429. Tour O, Adams SR, Kerr RA, Meijer RM, Sejnowski TJ, Tsien RW, Tsien RY (2007) Calcium Green FlAsH as a genetically targeted small-molecule calcium indicator. Nature – Chemical Biology 3:423–431
1430. McNeill H, Woodgett JR (2010) When pathways collide: collaboration and connivance among signalling proteins in development. Nature Reviews – Molecular Cell Biology 11:404–413
1431. Wilson LS, Elbatarny HS, Crawley SW, Bennett BM, Maurice DH (2008) Compartmentation and compartment-specific regulation of PDE5 by protein kinase G allows selective cGMP-mediated regulation of platelet functions. Proceedings of the National Academy of Sciences of the United States of America 105:13650–13655
1432. Willoughby D, Wong W, Schaack J, Scott JD, Cooper DMF (2006) An anchored PKA and PDE4 complex regulates subplasmalemmal cAMP dynamics. EMBO Journal 25:2051–2061
1433. Steinberg SF, Brunton LL (2001) Compartmentation of G protein-coupled signaling pathways in cardiac myocytes. Annual Review of Pharmacology and Toxicology 41:751–773
1434. Devic E, Xiang Y, Gould D, Kobilka B (2001) beta-adrenergic receptor subtype-specific signaling in cardiac myocytes from beta1 and beta2 adrenoceptor knockout mice. Molecular Pharmacology 60:577–583
1435. Jurevicius J, Fischmeister R (1996) cAMP compartmentation is responsible for a local activation of cardiac Ca2+ channels by beta-adrenergic agonists. Proceedings of the National Academy of Sciences of the United States of America 93:295–299
1436. Marx SO, Reiken S, Hisamatsu Y, Jayaraman T, Burkhoff D, Rosemblit N, Marks AR (2000) PKA phosphorylation dissociates FKBP12.6 from the calcium release channel (ryanodine receptor): defective regulation in failing hearts. Cell 101:365–376
1437. Willoughby D, Everett KL, Halls ML, Pacheco J, Skroblin P, Vaca L, Klussmann E, Cooper DMF (2012) Direct binding between Orai1 and AC8 mediates dynamic interplay between Ca^{2+} and cAMP signaling. Science Signaling 5:ra29

1438. Houslay MD, Baillie GS (2003) The role of ERK2 docking and phosphorylation of PDE4 cAMP phosphodiesterase isoforms in mediating cross-talk between the cAMP and ERK signalling pathways. Biochemical Society Transactions 31:1186–1190
1439. Amelio AL, Miraglia LJ, Conkright JJ, Mercer BA, Batalov S, Cavett V, Orth AP, Busby J, Hogenesch JB, Conkright MD (2007) A coactivator trap identifies NONO (p54nrb) as a component of the cAMP-signaling pathway. Proceedings of the National Academy of Sciences of the United States of America 104:20314–20319
1440. Koo SH, Flechner L, Qi L, Zhang X, Screaton RA, Jeffries S, Hedrick S, Xu W, Boussouar F, Brindle P, Takemori H, Montminy M (2005) The CREB coactivator TORC2 is a key regulator of fasting glucose metabolism. Nature 437:1109–1111
1441. Screaton RA, Conkright MD, Katoh Y, Best JL, Canettieri G, Jeffries S, Guzman E, Niessen S, Yates JR 3rd, Takemori H, Okamoto M, Montminy M (2004) The CREB coactivator TORC2 functions as a calcium- and cAMP-sensitive coincidence detector. Cell 119:61–74
1442. Litvin TN, Kamenetsky M, Zarifyan A, Buck J, Levin LR (2003) Kinetic properties of "soluble" adenylyl cyclase. Synergism between calcium and bicarbonate. Journal of Biological Chemistry 278:15922–15926
1443. Geng W, Wang Z, Zhang J, Reed BY, Pak CY, Moe OW (2005) Cloning and characterization of the human soluble adenylyl cyclase. American Journal of Physiology – Cell Physiology 288:C1305–C1316
1444. Buck J, Sinclair ML, Schapal L, Cann MJ, Levin LR (1999) Cytosolic adenylyl cyclase defines a unique signaling molecule in mammals. Proceedings of the National Academy of Sciences of the United States of America 96:79–84
1445. Farrell J, Ramos L, Tresguerres M, Kamenetsky M, Levin LR, Buck J (2008) Somatic "soluble" adenylyl cyclase isoforms are unaffected in Sacy^{tm1Lex}/Sacy^{tm1Lex} "knockout" mice. PLoS ONE 3:e3251
1446. Cooper DMF (2003) Regulation and organization of adenylyl cyclases and cAMP. Biochemical Journal 375:517–529
1447. Lavine N, Ethier N, Oak JN, Pei L, Liu F, Trieu P, Rebois RV, Bouvier M, Hebert TE, Van Tol HH (2002) G protein-coupled receptors form stable complexes with inwardly rectifying potassium channels and adenylyl cyclase. Journal of Biological Chemistry 277:46010–46019
1448. Tovey SC, Dedos SG, Taylor EJ, Church JE, Taylor CW (2008) Selective coupling of type 6 adenylyl cyclase with type 2 IP3 receptors mediates direct sensitization of IP3 receptors by cAMP. Journal of Cell Biology 183:297–311
1449. Conti M, Richter W, Mehats C, Livera G, Park JY, Jin C (2003) Cyclic AMP-specific PDE4 phosphodiesterases as critical components of cyclic AMP signaling. Journal of Biological Chemistry 278:5493–5496
1450. Bender AT, Beavo JA (2006) Cyclic nucleotide phosphodiesterases: molecular regulation to clinical use. Pharmacological Reviews 58:488–520
1451. Furman B, Ong WK, Pyne NJ (2010) Cyclic AMP signaling in pancreatic islets. Advances in Experimental Medicine and Biology 654:281–304
1452. Maclean MR, Johnston ED, McCulloch KM, Pooley L, Houslay MD, Sweeney G (1997) Phosphodiesterase isoforms in the pulmonary arterial circulation of the rat: changes in pulmonary hypertension. Journal of Pharmacology and Experimental Therapeutics 283:619–624
1453. Kessler T, Lugnier C (1995) Rolipram increases cyclic GMP content in L-arginine-treated cultured bovine aortic endothelial cells. European Journal of Pharmacology 290:163–167
1454. Conti M, Jin SL (1999) The molecular biology of cyclic nucleotide phosphodiesterases. Progress in Nucleic Acid Research and Molecular Biology 63:1–38
1455. Soderling SH, Beavo JA (2000) Regulation of cAMP and cGMP signaling: new phosphodiesterases and new functions. Current Opinion in Cell Biology 12:174–179

1456. Shrivastav A, Selvakumar P, Wang JH, Sharma RK (2009) Phosphodiesterase 1A, calmodulin dependent; Phosphodiesterase 1B, calmodulin dependent; Phosphodiesterase 1C, calmodulin dependent. UCSD-Nature Molecule Pages, UCSD-Nature Signaling Gateway (www.signaling-gateway.org)
1457. Goraya TA, Cooper DM (2005) Ca^{2+}-calmodulin-dependent phosphodiesterase (PDE1): current perspectives. Cellular Signalling 17:789–797
1458. Mantzoros CS, Magkos F, Brinkoetter M, Sienkiewicz E, Dardeno TA, Kim SY, Hamnvik OP, Koniaris A (2011) Leptin in human physiology and pathophysiology. American Journal of Physiology – Endocrinology and Metabolism 301:E567–E584
1459. Houslay MD (2001) PDE4 cAMP-specific phosphodiesterases. Progress in Nucleic Acid Research and Molecular Biology 69:249–315
1460. Houslay MD, Conti M (2010) Phosphodiesterase 4D, cAMP specific. UCSD-Nature Molecule Pages, UCSD-Nature Signaling Gateway (www.signaling-gateway.org)
1461. Dodge KL, Khouangsathiene S, Kapiloff MS, Mouton R, Hill EV, Houslay MD, Langeberg LK, Scott JD (2001) mAKAP assembles a protein kinase A/PDE4 phosphodiesterase cAMP signaling module. EMBO Journal 20:1921–1930
1462. Perry SJ, Baillie G, Kohout TA, McPhee I, Magiera MM, Ang KL, Miller WE, McLean AJ, Conti M, Houslay MD, Lefkowitz RJ (2002) Targeting of cyclic AMP degradation to beta 2-adrenergic receptors by beta-arrestins. Science 298:834–836
1463. Raymond DR, Carter RL, Ward CA, Maurice DH (2009) Distinct phosphodiesterase-4D variants integrate into protein kinase A-based signaling complexes in cardiac and vascular myocytes. American Journal of Physiology – Heart and Circulatory Physiology 296:H263–H271
1464. Liu S, Li Y, Kim S, Fu Q, Parikh D, Sridhar B, Shi Q, Zhang X, Guan Y, Chen X, Xiang YK (2012) Phosphodiesterases coordinate cAMP propagation induced by two stimulatory G protein-coupled receptors in hearts. Proceedings of the National Academy of Sciences of the United States of America 109:6578–6583
1465. Gebska MA, Stevenson BK, Hemnes AR, Bivalacqua TJ, Haile A, Hesketh GG, Murray CI, Zaiman AL, Halushka MK, Krongkaew N, Strong TD, Cooke CA, El-Haddad H, Tuder RM, Berkowitz DE, Champion HC (2011) Phosphodiesterase-5A (PDE5A) is localized to the endothelial caveolae and modulates NOS3 activity. Cardiovascular Research 90:353–363
1466. Patrucco E, Kraynik S, Beavo JA (2012) Phosphodiesterase 8A, cAMP-specific. UCSD-Nature Molecule Pages, UCSD-Nature Signaling Gateway (www.signaling-gateway.org)
1467. Omori K (2006) Phosphodiesterase 8B. UCSD-Nature Molecule Pages, UCSD-Nature Signaling Gateway (www.signaling-gateway.org)
1468. Feil R, Kemp-Harper B (2006) cGMP signalling: from bench to bedside. EMBO Reports 7:149–153
1469. Guo D, Tan YC, Wang D, Madhusoodanan KS, Zheng Y, Maack T, Zhang JJ, Huang XY (2007) A Rac-cGMP signaling pathway. Cell 128:341–355
1470. Jacob MP, Fulop T, Foris G, Robert L (1987) Effect of elastin peptides on ion fluxes in mononuclear cells, fibroblasts, and smooth muscle cells. Proceedings of the National Academy of Sciences of the United States of America 84:995–999
1471. Faury, G, Usson Y, Robert-Nicoud M, Robert L, Verdetti J (1998) Nuclear and cytoplasmic free calcium level changes induced by elastin peptides in human endothelial cells. Proceedings of the National Academy of Sciences of the United States of America 95:2967–2972
1472. Spofford CM, Chilian WM (2001) The elastin-laminin receptor functions as a mechanotransducer in vascular smooth muscle. American Journal of Physiology – Heart and Circulation Physiology 280:H1354–H1360
1473. Vogel WF (2001) Collagen-receptor signaling in health and disease. European Journal of Dermatology 11:506–514
1474. Kehrel B, Wierwille S, Clemetson KJ, Anders O, Steiner M, Knight CG, Farndale RW, Okuma M, Barnes MJ (1998) Glycoprotein VI is a major collagen receptor for platelet activation: it recognizes the platelet-activating quaternary structure of collagen, whereas CD36, glycoprotein IIb/IIIa, and von Willebrand factor do not. Blood 91:491–499

1475. Xu P, Liu J, Sakaki-Yumoto M, Derynck R (2012) TACE activation by MAPK-mediated regulation of cell surface dimerization and TIMP3 association. Science Signaling 5:ra34
1476. Comoglio PM, Boccaccio C, Trusolino L (2003) Interactions between growth factor receptors and adhesion molecules: breaking the rules. Current Opinion in Cell Biology 15:565–571
1477. Sundberg C, Rubin K (1996) Stimulation of β_1 integrins on fibroblasts induces PDGF-independent tyrosine phosphorylation of PDGF β receptors. Journal of Cell Biology 132:741–752
1478. Wang JF, Zhang XF, Groopman JE (2001) Stimulation of β_1 integrin induces tyrosine phosphorylation of vascular endothelial growth factor receptor-3 and modulates cell migration. Journal of Biological Chemistry 276:41950–41957
1479. Betson M, Lozano E, Zhang J, Braga VM (2002) Rac activation upon cell-cell contact formation is dependent on signaling from the epidermal growth factor receptor. Journal of Biological Chemistry 277:36962–36969
1480. Ma YQ (2008) Kindlin-2 (Mig-2): a coactivator of $\beta 3$ integrins. Journal of Cell Biology 181:439–446
1481. Montanez E (2008) Kindlin-2 controls bidirectional signaling of integrins. Genes and Development 22:1325–1330
1482. Gong H, Shen B, Flevaris P, Chow C, Lam SCT, Voyno-Yasenetskaya TA, Kozasa T, Du X (2010) G Protein subunit Gα13 binds to integrin $\alpha_{IIb}\beta_3$ and mediates integrin "outside-in" signaling. Science 327:340–343
1483. Han J, Lim CJ, Watanabe N, Soriani A, Ratnikov B, Calderwood DA, Puzon-McLaughlin W, Lafuente EM, Boussiotis VA, Shattil SJ, Ginsberg MH (2006) Reconstructing and deconstructing agonist-induced activation of integrin alphaIIbbeta3. Current Biology 16:1796–1806
1484. Mattila E, Pellinen T, Nevo J, Vuoriluoto K, Arjonen A, Ivaska J (2005) Negative regulation of EGFR signalling through integrin-$\alpha_1\beta_1$-mediated activation of protein tyrosine phosphatase TCPTP. Nature – Cell Biology 7:78–85
1485. Daniel JM, Spring CM, Crawford HC, Reynolds AB, Baig A (2002) The p120ctn-binding partner Kaiso is a bi-modal DNA-binding protein that recognizes both a sequence-specific consensus and methylated CpG dinucleotides. Nucleic Acids Research 30:2911–2919
1486. Kondapalli J, Flozak AS, Albuquerque MLC (2004) Laminar shear stress differentially modulates gene expression of p120 catenin, Kaiso transcription factor, and vascular endothelial cadherin in human coronary artery endothelial cells. Journal of Biological Chemistry 279:11417–11424
1487. Baumeister U, Funke R, Ebnet K, Vorschmitt H, Koch S, Vestweber D (2005) Association of Csk to VE-cadherin and inhibition of cell proliferation. EMBO Journal 24:1686–1695
1488. Howe A, Aplin AE, Alahari SK, Juliano RL (1998) Integrin signaling and cell growth control, Current Opinion in Cell Biology 10:220–231
1489. Schaller MD, Otey CA, Hildebrand JD, Parsons JT (1995) Focal adhesion kinase and paxillin bind to peptides mimicking beta integrin cytoplasmic domains. The Journal of Cell Biology 130:1181–1187
1490. Giancotti FG, Ruoslahti E (1999) Integrin signaling. Science 285:1028–1032
1491. Zamir E, Geiger B (2001) Molecular complexity and dynamics of cell-matrix adhesions. Journal of Cell Science 114:3583–3590
1492. Ezratty EJ, Partridge MA, Gundersen GG (2005) Microtubule-induced focal adhesion disassembly is mediated by dynamin and focal adhesion kinase. Nature – Cell Biology 7:581–590
1493. De Vuyst E, Decrock E, Cabooter L, Dubyak GR, Naus CC, Evans WH, Leybaert L (2006) Intracellular calcium changes trigger connexin 32 hemichannel opening. EMBO Journal 25:34–44
1494. Cachero TG, Morielli AD, Peralta EG (1998) The small GTP-binding protein RhoA regulates a delayed rectifier potassium channel. Cell 93:1077–1085

1495. van de Graaf SF, Chang Q, Mensenkamp AR, Hoenderop JG, Bindels RJ (2006) Direct interaction with Rab11a targets the epithelial Ca^{2+} channels TRPV5 and TRPV6 to the plasma membrane. Molecular and Cellular Biology 26:303–312
1496. Yatani A, Irie K, Otani T, Abdellatif M, Wei L (2005) RhoA GTPase regulates L-type Ca^{2+} currents in cardiac myocytes. American Journal of Physiology – Heart and Circulatory Physiology 288:H650–659
1497. Pochynyuk O, Medina J, Gamper N, Genth H, Stockand JD, Staruschenko A (2006) Rapid translocation and insertion of the epithelial Na^+ channel in response to RhoA signaling. Journal of Biological Chemistry 281:26520–26527
1498. Bezzerides VJ, Ramsey IS, Kotecha S, Greka A, Clapham DE (2004) Rapid vesicular translocation and insertion of TRP channels. Nature – Cell Biology 6:709–720
1499. Saxena SK, Horiuchi H, Fukuda M (2006) Rab27a regulates epithelial sodium channel (ENaC) activity through synaptotagmin-like protein (SLP-5) and Munc13-4 effector mechanism. Biochemical and Biophysical Research Communications 344:651–657
1500. Storey NM, O'Bryan JP, Armstrong DL (2002) Rac and Rho mediate opposing hormonal regulation of the ether-a-go-go-related potassium channel. Current Biology 12:27–33
1501. Wilk-Blaszczak MA, Singer WD, Quill T, Miller B, Frost JA, Sternweis PC, Belardetti F (1997) The monomeric G-proteins Rac1 and/or Cdc42 are required for the inhibition of voltage-dependent calcium current by bradykinin. Journal of Neuroscience 17:4094–4100
1502. Li L, Matsuoka I, Suzuki Y, Watanabe Y, Ishibashi T, Yokoyama K, Maruyama Y, Kimura J (2002) Inhibitory effect of fluvastatin on lysophosphatidylcholine-induced nonselective cation current in Guinea pig ventricular myocytes. Molecular Pharmacology 62:602–607
1503. Cheng J, Wang H, Guggino WB (2005) Regulation of cystic fibrosis transmembrane regulator trafficking and protein expression by a Rho family small GTPase TC10. Journal of Biological Chemistry 280:3731–3739
1504. Tilly BC, Edixhoven MJ, Tertoolen LG, Morii N, Saitoh Y, Narumiya S, de Jonge HR (1996) Activation of the osmo-sensitive chloride conductance involves P21rho and is accompanied by a transient reorganization of the F-actin cytoskeleton. Molecular Biology of the Cell 7:1419–1427
1505. Nilius B, Voets T, Prenen J, Barth H, Aktories K, Kaibuchi K, Droogmans G, Eggermont J (1999) Role of Rho and Rho kinase in the activation of volume-regulated anion channels in bovine endothelial cells. Journal of Physiology 516:67–74
1506. Carton I, Trouet D, Hermans D, Barth H, Aktories K, Droogmans G, Jorgensen NK, Hoffmann EK, Nilius B, Eggermont J (2002) RhoA exerts a permissive effect on volume-regulated anion channels in vascular endothelial cells. American Journal of Physiology – Cell Physiology 283:C115–C125
1507. Voets T, Manolopoulos V, Eggermont J, Ellory C, Droogmans G, Nilius B (1998) Regulation of a swelling-activated chloride current in bovine endothelium by protein tyrosine phosphorylation and G proteins. Journal of Physiology 506:341–352
1508. Berridge MJ, Bootman MD, Roderick HL (2003) Calcium signalling: dynamics, homeostasis and remodelling. Nature Reviews – Molecular Cell Biology 4:517–529
1509. Kar P, Bakowski D, Di Capite J, Nelson C, Parekh AB (2012) Different agonists recruit different stromal interaction molecule proteins to support cytoplasmic Ca^{2+} oscillations and gene expression. Proceedings of the National Academy of Sciences of the United States of America 109:6969–6974
1510. Ruegg UT, Nicolas-Metral V, Challet C, Bernard-Helary K, Dorchies OM, Wagner S, Buetler TM (2002) Pharmacological control of cellular calcium handling in dystrophic skeletal muscle. Neuromuscular Disorders 12:S155-S161
1511. Bockaert J (1986) Les récepteurs membranaires[The membrane receptors]. La Recherche 17:892–900
1512. Levitzki A (1988) From epinephrine to cyclic AMP. Science 241:800–806
1513. Zaccolo M, Pozzan T (2002) Discrete microdomains with high concentration of cAMP in stimulated rat neonatal cardiac myocytes. Science 295:1711–1715

1514. Rasmussen H (1989) The cycling of calcium as an intracellular messenger. Scientific American 261:66–73
1515. Abell E, Ahrends R, Bandara S, Park BO, Teruel MN (2011) Parallel adaptive feedback enhances reliability of the Ca^{2+} signaling system. Proceedings of the National Academy of Sciences of the United States of America 108:14485–14490
1516. Taufiq-Ur-Rahman, Skupin A, Falcke M, Taylor CW (2009) Clustering of InsP3 receptors by InsP3 retunes their regulation by InsP3 and Ca^{2+}. Nature 458:655–659
1517. Wang Y, Deng X, Zhou Y, Hendron E, Mancarella S, Ritchie MF, Tang XD, Baba Y, Kurosaki T, Mori Y, Soboloff J, Gill DL (2009) STIM protein coupling in the activation of Orai channels. Proceedings of the National Academy of Sciences of the United States of America 106:7391–7396
1518. Luik RM, Wang B, Prakriya M, Wu MM, Lewis RS (2008) Oligomerization of STIM1 couples ER calcium depletion to CRAC channel activation. Nature 454:538–542
1519. Yang S, Zhang JJ, Huang XY (2009) Orai1 and STIM1 are critical for breast cancer cell migration and metastasis. Cancer Cell 15:124–134
1520. Liao Y, Plummer NW, George MD, Abramowitz J, Zhu MX, Birnbaumer L (2009) A role for Orai in TRPC-mediated Ca^{2+} entry suggests that a TRPC:Orai complex may mediate store and receptor operated Ca^{2+} entry. Proceedings of the National Academy of Sciences of the United States of America 106:3202–3206
1521. Caraveo G, van Rossum DB, Patterson RL, Snyder SH, Desiderio S (2006) Action of TFII-I outside the nucleus as an inhibitor of agonist-induced calcium entry. Science 314:122–125
1522. Min SW, Chang WP, Sudhof TC (2007) E-Syts, a family of membranous Ca2+-sensor proteins with multiple C2 domains. Proceedings of the National Academy of Sciences of the United States of America 104:3823–3828
1523. Lefkimmiatis K, Srikanthan M, Maiellaro I, Moyer MP, Curci S, Hofer AM (2009) Store-operated cyclic AMP signalling mediated by STIM1. Nature – Cell Biology 11:433–442
1524. Wang X, Zeng W, Kim MS, Allen PB, Greengard P, Muallem S (2007) Spinophilin/neurabin reciprocally regulate signaling intensity by G protein-coupled receptors. EMBO Journal 26:2768–2776
1525. Kobayashi M, Takamatsu K (2009) Hippocalcin UCSD-Nature Molecule Pages, UCSD-Nature Signaling Gateway (www.signaling-gateway.org)
1526. Burgoyne RD (2007) Neuronal calcium sensor proteins: generating diversity in neuronal Ca^{2+} signalling. Nature Reviews – Neuroscience 8:182–193
1527. Navedo MF, Amberg GC, Westenbroek RE, Sinnegger-Brauns MJ, Catterall WA, Striessnig J, Santana LF (2007) $Ca_V1.3$ channels produce persistent calcium sparklets, but $Ca_V1.2$ channels are responsible for sparklets in mouse arterial smooth muscle. American Journal of Physiology – Heart and Circulatory Physiology 293:H1359-H1370
1528. Amberg GC, Navedo MF, Nieves-Cintrón M, Molkentin JD, Santana LF (2007) Calcium sparklets regulate local and global calcium in murine arterial smooth muscle. Journal of Physiology 579:187–201
1529. Sonkusare SK, Bonev AD, Ledoux J, Liedtke W, Kotlikoff MI, Heppner TJ, Hill-Eubanks DC, Nelson MT (2012) Elementary Ca^{2+} signals through endothelial TRPV4 channels regulate vascular function. Science 336:597–601
1530. Vig M, Kinet JP (2009) Calcium signaling in immune cells. Nature – Immunology 10:21–27
1531. Crabtree GR, Olson EN (2002) NFAT signaling: choreographing the social lives of cells. Cell 109:S67–S79
1532. Huang GN, Huso DL, Bouyain S, Tu J, McCorkell KA, May MJ, Zhu Y, Lutz M, Collins S, Dehoff M, Kang S, Whartenby K, Powell J, Leahy D, Worley PF (2008) NFAT binding and regulation of T cell activation by the cytoplasmic scaffolding Homer proteins. Science 319:476–481
1533. Lis A, Peinelt C, Beck A, Parvez S, Monteilh-Zoller M, Fleig A, Penner R (2007) CRACM1, CRACM2, and CRACM3 are store-operated Ca^{2+} channels with distinct functional properties. Current Biology 17:794–800

1534. Launay P, Fleig A, Perraud AL, Scharenberg AM, Penner R, Kinet JP (2002) TRPM4 is a Ca^{2+}-activated nonselective cation channel mediating cell membrane depolarization. Cell 109:397–407
1535. Vig M, DeHaven WI, Bird GS, Billingsley JM, Wang H, Rao PE, Hutchings AB, Jouvin MH, Putney JW, Kinet JP (2008) Defective mast cell effector functions in mice lacking the CRACM1 pore subunit of store-operated calcium release-activated calcium channels. Nature – Immunology 9:89–96
1536. Miklavc P, Frick M, Wittekindt OH, Haller T, Dietl P (2010) Fusion-activated ca^{2+} entry: an "active zone" of elevated Ca^{2+} during the postfusion stage of lamellar body exocytosis in rat type II pneumocytes. PLoS One 5:e10982
1537. Miklavc P, Mair N, Wittekindt OH, Haller T, Dietl P, Felder E, Timmler M, Frick M (2011) Fusion-activated Ca^{2+} entry via vesicular $P2X_4$ receptors promotes fusion pore opening and exocytotic content release in pneumocytes. Proceedings of the National Academy of Sciences of the United States of America 108:14503–14508
1538. Giorgi C, Ito K, Lin HK, Santangelo C, Wieckowski MR, Lebiedzinska M, Bononi A, Bonora M, Duszynski J, Bernardi R, Rizzuto R, Tacchetti C, Pinton P, Pandolfi PP (2010) PML regulates apoptosis at endoplasmic reticulum by modulating calcium release. Science 330:1247–1251
1539. Rabellino A, Carter B, Konstantinidou G, Wu SY, Rimessi A, Byers LA, Heymach JV, Girard L, Chiang CM, Teruya-Feldstein J, Scaglioni PP (2012) The SUMO E3-ligase PIAS1 regulates the tumor suppressor PML and its oncogenic counterpart PML-RARA. Cancer Research 72:2275–2284
1540. Yuan WC, Lee YR, Huang SF, Lin YM, Chen TY, Chung HC, Tsai CH, Chen HY, Chiang CT, Lai CK, Lu LT, Chen CH, Gu DL, Pu YS, Jou YS, Lu KP, Hsiao PW, Shih HM, Chen RH (2011) A Cullin3-KLHL20 ubiquitin ligase-dependent pathway targets PML to potentiate HIF-1 signaling and prostate cancer progression. Cancer Cell 20:214–228
1541. Wiesener MS, Turley H, Allen WE, Willam C, Eckardt KU, Talks KL, Wood SM, Gatter KC, Harris AL, Pugh CW, Ratcliffe PJ, Maxwell PH Induction of endothelial PAS domain protein-1 by hypoxia: characterization and comparison with hypoxia-inducible factor-1α. Blood 92:2260–2268
1542. Tanaka T, Akiyama H, Kanai H, Sato M, Takeda S, Sekiguchi K, Yokoyama T, Kurabayashi M (2002) Endothelial PAS domain protein 1 (EPAS1) induces adrenomedullin gene expression in cardiac myocytes: role of EPAS1 in an inflammatory response in cardiac myocytes. Journal of Molecular and Cellular Cardiology 34:703–707
1543. Patel SA, Simon MC (2008) Biology of hypoxia-inducible factor-2α in development and disease. Cell Death and Differentiation 15:628–634
1544. Weidemann A, Johnson RS (2008) Biology of HIF-1α. Cell Death and Differentiation 15:621–627
1545. Semenza GL (2007) Life with oxygen. Science 318:62–64
1546. Aragonés J, Schneider M, Van Geyte K, Fraisl P, Dresselaers T, Mazzone M, Dirkx R4, Zacchigna S, Lemieux H, Jeoung NH, Lambrechts D, Bishop T, Lafuste P, Diez-Juan A, Harten SK, Van Noten P, De Bock K, Willam C, Tjwa M, Grosfeld A, Navet R, Moons L, Vandendriessche T, Deroose C, Wijeyekoon B, Nuyts J, Jordan B, Silasi-Mansat R, Lupu F, Dewerchin M, Pugh C, Salmon P, Mortelmans L, Gallez B, Gorus F, Buyse J, Sluse F, Harris RA, Gnaiger E, Hespel P, Van Hecke P, Schuit F, Van Veldhoven P, Ratcliffe P, Baes M, Maxwell P, Carmeliet P (2008) Deficiency or inhibition of oxygen sensor Phd1 induces hypoxia tolerance by reprogramming basal metabolism. Nature – Genetics 40:170–180
1547. Fong GH, Takeda K (2008) Role and regulation of prolyl hydroxylase domain proteins. Cell Death and Differentiation 15:635–641
1548. Lin X, David CA, Donnelly JB, Michaelides M, Chandel NS, Huang X, Warrior U, Weinberg F, Tormos KV, Fesik SW, Shen Y (2008) A chemical genomics screen highlights the essential role of mitochondria in HIF-1 regulation. Proceedings of the National Academy of Sciences of the United States of America 105:174–179

1549. Emerling BM, Weinberg F, Liu JL, Mak TW, Chandel NS (2008) PTEN regulates p300-dependent hypoxia-inducible factor 1 transcriptional activity through Forkhead transcription factor 3a (FOXO3a). Proceedings of the National Academy of Sciences of the United States of America 105:2622–2627

1550. Gustafsson MV, Zheng X, Pereira T, Gradin K, Jin S, Lundkvist J, Ruas JL, Poellinger L, Lendahl U, Bondesson M (2005) Hypoxia requires notch signaling to maintain the undifferentiated cell state. Developmental Cell 9:617–628.

1551. Zheng X, Linke S, Dias JM, Zheng X, Gradin K, Wallis TP, Hamilton BR, Gustafsson M, Ruas JL, Wilkins S, Bilton RL, Brismar K, Whitelaw ML, Pereira T, Gorman JJ, Ericson J, Peet DJ, Lendahl U, Poellinger L (2008) Interaction with factor inhibiting HIF-1 defines an additional mode of cross-coupling between the Notch and hypoxia signaling pathways. Proceedings of the National Academy of Sciences of the United States of America 105:3368–3373

1552. Wang Y, Roche O, Xu C, Moriyama EH, Heir P, Chung J, Roos FC, Chen Y, Finak G, Milosevic M, Wilson BC, Teh BT, Park M, Irwin MS, Ohh M (2012) Hypoxia promotes ligand-independent EGF receptor signaling via hypoxia-inducible factor-mediated upregulation of caveolin-1. Proceedings of the National Academy of Sciences of the United States of America 109:4892–4897

1553. Kulshreshtha R, Davuluri RV, Calin GA, Ivan M (2008) A microRNA component of the hypoxic response. Cell Death and Differentiation 15:667–671

1554. D'Autréaux B, Toledano MB (2007) ROS as signalling molecules: mechanisms that generate specificity in ROS homeostasis. Nature Reviews – Molecular Cell Biology 8:813–824

1555. Fish JE, Yan MS, Matouk CC, St Bernard R, Ho JJ, Gavryushova A, Srivastava D, Marsden PA (2010) Hypoxic repression of endothelial nitric-oxide synthase transcription is coupled with eviction of promoter histones. Journal of Biological Chemistry 285:810–826

1556. Lee S, Chen TT, Barber CL, Jordan MC, Murdock J, Desai S, Ferrara N, Nagy A, Roos KP, Iruela-Arispe ML (2007) Autocrine VEGF signaling is required for vascular homeostasis. Cell 130:691–703

1557. Heloterä H, Alitalo K (2007) The VEGF family, the inside story. Cell 130:591–592

1558. Appleton BA, Wu P, Maloney J, Yin JP, Liang WC, Stawicki S, Mortara K, Bowman KK, Elliott JM, Desmarais W, Bazan JF, Bagri A, Tessier-Lavigne M, Koch AW, Wu Y, Watts RJ, Wiesmann C (2007) Structural studies of neuropilin/antibody complexes provide insights into semaphorin and VEGF binding. EMBO Journal 26:4902–4912

1559. Usui R, Shibuya M, Ishibashi S, Maru Y (2007) Ligand-independent activation of vascular endothelial growth factor receptor 1 by low-density lipoprotein. EMBO Reports 8: 1155–1161

1560. Arany Z, Foo SY, Ma Y, Ruas JL, Bommi-Reddy A, Girnun G, Cooper M, Laznik D, Chinsomboon J, Rangwala SM, Baek KH, Rosenzweig A, Spiegelman BM (2008) HIF-independent regulation of VEGF and angiogenesis by the transcriptional coactivator PGC-1α. Nature 451:1008–1012

1561. Krüger M, Kratchmarova I, Blagoev B, Tseng YH, Kahn CR, Mann M (2008) Dissection of the insulin signaling pathway via quantitative phosphoproteomics. Proceedings of the National Academy of Sciences of the United States of America 105:2451–2456

1562. Grahame Hardie D (2007) AMP-activated/SNF1 protein kinases: conserved guardians of cellular energy. Nature Reviews – Molecular Cell Biology 8:774–785

1563. Dávalos A, Goedeke L, Smibert P, Ramírez CM, Warrier NP, Andreo U, Cirera-Salinas D, Rayner K, Suresh U, Pastor-Pareja JC, Esplugues E, Fisher EA, Penalva LO, Moore KJ, Suárez Y, Lai EC, Fernández-Hernando C (2011) miR-33a/b contribute to the regulation of fatty acid metabolism and insulin signaling. Proceedings of the National Academy of Sciences of the United States of America 108:9232–9237

1564. Akerfelt M, Morimoto RI, Sistonen L (2010) Heat shock factors: integrators of cell stress, development and lifespan. Nature Reviews – Molecular Cell Biology 11:545–555

1565. Franchi L, Eigenbrod T, Muñoz-Planillo R, Nuñez G (2009) The inflammasome: a caspase-1-activation platform that regulates immune responses and disease pathogenesis. Nature – Immunology 10:241–247
1566. Shenoy AR, Wellington DA, Kumar P, Kassa H, Booth CJ, Cresswell P, MacMicking JD (2012) GBP5 promotes NLRP3 inflammasome assembly and immunity in mammals. Science 336:481–485
1567. Trichet L, Le Digabel J, Hawkins RJ, Vedula SR, Gupta M, Ribrault C, Hersen P, Voituriez R, Ladoux B (2012) Evidence of a large-scale mechanosensing mechanism for cellular adaptation to substrate stiffness. Proceedings of the National Academy of Sciences of the United States of America 109:6933–6938
1568. DuFort CC, Paszek MJ, Weaver VM (2011) Balancing forces: architectural control of mechanotransduction. Nature Reviews – Molecular Cell Biology 12:308–319
1569. Batra N, Burra S, Siller-Jackson AJ, Gu S, Xia X, Weber GF, Desimone D, Bonewald LF, Lafer EM, Sprague E, Schwartz MA, Jiang JX (2012) Mechanical stress-activated integrin $\alpha_5\beta_1$ induces opening of connexin 43 hemichannels. Proceedings of the National Academy of Sciences of the United States of America 109:3359–3364
1570. Guo CL, Ouyang M, Yu JY, Maslov J, Price A, Shen CY (2012) Long-range mechanical force enables self-assembly of epithelial tubular patterns. Proceedings of the National Academy of Sciences of the United States of America 109:5576–5582
1571. Nikmanesh M, Shi ZD, Tarbell JM (2011) Heparan sulfate proteoglycan mediates shear stress-induced endothelial gene expression in mouse embryonic stem cell-derived endothelial cells. Biotechnology and Bioengineering 109:583–594
1572. Gopalan SM, Flaim C, Bhatia SN, Hoshijima M, Knoell R, Chien KR, Omens JH, McCulloch AD (2003) Anisotropic stretch-induced hypertrophy in neonatal ventricular myocytes micropatterned on deformable elastomers. Biotechnology and Bioengineering 81:578–587
1573. Wang JH, Yang G, Li Z (2005) Controlling cell responses to cyclic mechanical stretching. Annals of Biomedical Engineering 33:337–342
1574. Kurpinski K, Chu J, Hashi C, Li S (2006) Anisotropic mechanosensing by mesenchymal stem cells. Proceedings of the National Academy of Sciences of the United States of America 103:16095–16100
1575. Zou Y, Akazawa H, Qin Y, Sano M, Takano H, Minamino T, Makita N, Iwanaga K, Zhu W, Kudoh S, Toko H, Tamura K, Kihara M, Nagai T, Fukamizu A, Umemura S, Iiri T, Fujita T, Komuro I (2004) Mechanical stress activates angiotensin II type 1 receptor without the involvement of angiotensin II. Nature – Cell Biology 6:499–506
1576. Sadoshima J, Xu Y, Slayter HS, Izumo S (1993) Autocrine release of angiotensin II mediates stretch-induced hypertrophy of cardiac myocytes in vitro. Cell 75:977–984
1577. Gudi SRP, Clark CB, Frangos JA (1996) Fluid flow rapidly activates G proteins in human endothelial cells. Involvement of G proteins in mechanochemical signal transduction. Circulation Research 79:834–839
1578. Liu B, Lu S, Zheng S, Jiang Z, Wang Y (2011) Two distinct phases of calcium signalling under flow. Cardiovascular Research 91:124-1-133
1579. Chachisvilis M, Zhang YL, Frangos JA (2006) G protein-coupled receptors sense fluid shear stress in endothelial cells Proceedings of the National Academy of Sciences of the United States of America 103:15463–15468
1580. Kung C (2005) A possible unifying principle for mechanosensation. Nature 436:647–654.
1581. Sawada Y, Tamada M, Dubin-Thaler BJ, Cherniavskaya O, Sakai R, Tanaka S, Sheetz MP (2006) Force sensing by mechanical extension of the Src family kinase substrate p130Cas. Cell 127:1015–1026
1582. Na S, Collin O, Chowdhury F, Tay B, Ouyang M, Wang Y, Wang N (2008) Rapid signal transduction in living cells is a unique feature of mechanotransduction. Proceedings of the National Academy of Sciences of the United States of America 105:6626–6631
1583. Frey JW, Farley EE, O'Neil TK, Burkholder TJ, Hornberger TA (2009) Evidence that mechanosensors with distinct biomechanical properties allow for specificity in mechanotransduction. Biophysical Journal 97:347–356

Chap. 12. Conclusion

1584. Zhou F, Huang D, Xia Y (2010) Neuroanatomical basis of acupuncture points (Chap. 2; 32-80), In Xia Y, Cao X, Wu G, Cheng J (Eds.) Acupuncture Therapy for Neurological Diseases: A Neurobiological View, Tsinghua University Press, Beijing, China and Springer, Berlin Heidelberg
1585. Fei L, Cheng HS, Cai DH, Yang SX, Xu JR, Chen EY, Dang RS, Dang GH, Shen XY, Tang Y, and Yao W (1998) Experimental exploration and research prospect of physical bases and functional characteristics of meridians. Chinese Science Bulletin 43:1233–1252
1586. Zhang ZJ, Wang XM, McAlonan GM (2012) Neural acupuncture unit: a new concept for interpreting effects and mechanisms of acupuncture. Evidence-Based Complementary and Alternative Medicine

Notation Rules: Aliases and Symbols

Aliases that designate different types of molecules as well as those that do not carry an obvious meaning should be eliminated; they are thus not used in the present text. For example, P35 is an alias for annexin-A1, brain syntaxin-1A, ficolin-2, interleukin-12A, the cyclin-H assembly factor Ménage à trois homolog-1, regulatory subunit-1 of cyclin-dependent kinase CDK5, and uroplakin-3B, among others. It is substituted by AnxA1, Stx1a, Fcn2, IL12a, MAT1, $CDK5_{r1}$, and UPk3b aliases, respectively. Protein P39 corresponds to the subunit D1 of the lysosomal V-type H^+ ATPase (ATP6v0d1), Jun transcription factor, a component of the Activator protein AP1, and regulatory subunit-2 of cyclin-dependent kinase CDK5 ($CDK5_{r2}$). Extracellular signal-regulated protein kinases ERK1 and ERK2, members of the mitogen-activated protein kinase (MAPK) module (last tier), are also abbreviated P44 and P42 (also P40 and P41). However, both P42 and P44 correspond to the 26S protease regulatory AAA ATPase subunit (PSMC6). Alias P42 is also utilized for cyclin-dependent kinase CDK20, cyclin-dependent kinase-like protein CDKL1, and 43-kDa NuP43 nucleoporin. Alias P44 can also refer to interferon-induced protein IFI44 (or microtubule-associated protein MTAP44) and androgen receptor cofactor P44 (a.k.a. methylosome protein MeP50 and WD repeat-containing protein WDR77). In the present text, P38 members (P38α–P38δ) of the mitogen-activated protein kinase modules (i.e., MAPK11–MAPK14)[1] are designated as P38MAPKs to avoid confusion with other molecules, the alias of which is also P38.[2]

---

[1] Protein P38α is also known as MAPK14, cytokine suppressive anti-inflammatory drug (CSAID)-binding protein CSBP, CSBP1, or CSBP2, and stress-activated protein kinase SAPK2a; P38β as MAPK11 and SAPK2b; P38γ as MAPK12, ERK6, and SAPK3; P38δ as MAPK13 and SAPK4.

[2] Alias P38 is used for: (1) mitogen-activated protein kinase MAPK1, extracellular signal-regulated kinase ERK2, as well as P40, P41, and P42; (2) adaptor CRK (chicken tumor virus regulator of kinase, or v-crk sarcoma virus CT10 oncogene homolog); (3) growth factor receptor-binding protein GRB2-related adaptor protein GRAP2 (a.k.a. GRID, GADS, GRB2L, GRF40, GRPL, and Mona); (4) ubiquitin ligase RING finger protein RNF19a, or dorfin; (5) 38-kDa DNA polymerase-δ-interacting protein PolδIP2 (a.k.a. polymerase [DNA-directed] PDIP38 and PolD4); (6) activator of 90-kDa heat shock protein ATPase homolog AHSA1; and (7) aminoacyl

Aliases for Molecules

Aliases include all written variants, i.e., any abbreviation[3] such as acronyms. An *acronym* corresponds to a word made from the initial letters or syllables of nouns that are pronounceable as a word. Acronyms are generally written with all letters in upper case. Yet, some acronyms are treated as words and written in lower case (e.g., laser [originally LASER] is an acronym for light amplification by stimulated emission of radiation, sonar [originally SONAR] for sound navigation and ranging). A substance name can derive from its chemical name (e.g., amphetamine: α-methylphenethylamine).

Acronyms can give rise to molecule names by adding a scientific suffix such as "-in", a common ending of molecule nouns (e.g., sirtuin, a portmanteau, that comes from the alias SIRT, which stands for silent information regulator-2 [two]). Other scientific prefixes and suffixes can be frequently detected throughout the present text. Their meaning is given in appendix Notations – Prefixes and Suffixes, particularly for readers from Asia. Many prefixes are used to specify position, configuration and behavior, quantity, direction and motion, structure, timing, frequency, and speed.

A *portmanteau* is a word that combines initials and some inner letters of at least 2 words (e.g., calmodulin stands for calcium modulated protein; caspase for cysteine-dependent aspartate-specific protease; chanzyme for ion channel and enzyme; chemokine for chemoattractant cytokine;[4] emilin for elastin microfibril interfacer; endorphins and endomorphins for endogenous morphines; ephrin for erythropoietin-producing hepatocyte (EPH) receptor kinase interactor; moesin for membrane-organizing extension spike protein; porin for pore-forming protein; restin for Reed-Steinberg cell-expressed intermediate filament-associated protein, an alias for cytoplasmic linker protein CLiP1 (or CLiP170); serpin for serine protease inhibitor; siglec for sialic acid-binding Ig-like lectin; transceptor for transporter-related receptor; and Prompt for promoter upstream transcript).[5]

Initialisms are abbreviations that are formed from initial letters of a single long noun or several nouns and, instead of being pronounced like an ordinary word, are read letter-by-letter (e.g., DNA that stands deoxyribonucleic acid).

---

tRNA synthase complex-interacting multifunctional protein AIMP2, or tRNA synthase complex component JTV1 [76].

[3]In general, abbreviations exclude the initials of short function words, such as "and", "or", "of", or "to". However, they are sometimes included in acronyms to make them pronounceable (e.g., radar [originally RADAR] for radio detection and ranging). These letters are often written in lower case. In addition, both cardinal (size, molecular weight, etc.) and ordinal (isoform discovery order) numbers in names are represented by digits.

[4]Cytokines are peptidic, proteic, or glycoproteic regulators that are secreted by cells of the immune system. These immunomodulating agents serve as auto- or paracrine signals.

[5]The upper case initial P in Prompt is used to avoid confusion with command-line interpreter prompt or prompt book to direct precise timing of actions on theater stage.

Some abbreviations can give rise to alphabetisms that are written as new words (e.g., Rho-associated, coiled-coil-containing protein kinase [RoCK] that is also called Rho kinase). In biochemistry, multiple-letter abbreviations can also be formed from a single word that can be long (e.g., Cam stands for calmodulin, which is itself a portmanteau word, Trx for thioredoxin, etc.) as well as short (e.g., Ttn for titin, etc.). In addition, single-letter symbols of amino acids are often used to define a molecule alias (e.g., tyrosine can be abbreviated as Tyr or Y, hence SYK stands for spleen tyrosine kinase).

Aliases use, in general, capital letters and can include hyphens and dots. Yet, as a given protein can represent a proto-oncogene[6] encoded by a gene that can give rise to an oncogene (tumor promoter) after gain- or loss-of-function mutations,[7] the same acronym represents 3 different entities.[8]

Besides, a given abbreviation can designate distinct molecules without necessarily erroneous consequence in a given context (e.g., PAR: poly^{ADP}ribose or protease-activated receptor and GCK: germinal center kinases or glucokinase; in the latter case, the glucokinase abbreviation should be written as GcK or, better, GK).

In addition, a large number of aliases that designate a single molecule results from the fact that molecules have been discovered independently several times with possibly updated functions. Some biochemists uppercase the name of a given molecule, whereas others lowercase (e.g., cell division cycle guanosine triphosphatase of the Rho family CDC42 or Cdc42, adaptor growth factor receptor-bound protein GRB2 or Grb2, chicken tumor virus regulator of kinase CRK or Crk, guanine nucleotide-exchange factor Son of sevenless SOS or Sos, etc.). Acronyms

---

[6]In 1911, P. Rous isolated a virus that was capable of generating tumors of connective tissue (sarcomas) in chicken. Proteins were afterward identified, the activity of which, when uncontrolled, can provoke cancer, hence the name oncogene given to genes that encode these proteins. Most of these proteins are enzymes, more precisely kinases. The first oncogene was isolated from the avian Rous virus by D. Stéhelin and called Src (from sarcoma). This investigator demonstrated that the abnormal functioning of the Src protein resulted from mutation of a normal gene, or proto-oncogene, which is involved in cell division.

[7]Loss-of-function mutations cause complete or partial loss of function of gene products that operate as tumor suppressors, whereas gain-of-function mutations generate gene products with new or abnormal function that can then act as oncogenes. Typical tumor-inducing agents are enzymes, mostly regulatory kinases and small guanosine triphosphatases, that favor proliferation of cells, which do normally need to be activated to exert their activities. Once their genes are mutated, these enzymes become constitutively active. Other oncogenes include growth factors (a.k.a. mitogens) and transcription factors. Mutations can also disturb signaling axis regulation, thereby raising protein expression. Last, but not least, chromosomal translocation can also provoke the expression of a constitutively active hybrid protein.

[8]Like Latin-derived shortened expressions – as well as foreign words – that are currently written in italics, genes can be italicized. However, this usage is not required in scientific textbooks published by Springer. Italic characters are then used to highlight words within a text to easily target them. Proteins are currently romanized (ordinary print), but with a capital initial. Nevertheless, names (not aliases) of chemical species are entirely lowercased in most – if not all – scientific articles, except to avoid confusion with a usual word (e.g., hedgehog animal vs. Hedgehog protein and raptor [bird of prey] vs. Raptor molecule).

are then not always capitalized. Printing style of aliases should not only avoid confusion, but also help one in remembering alias meaning.

In the present textbook, choice of lower and upper case letters in molecule aliases is dictated by the following criteria.

(1) An upper case letter is used for initials of words that constitute molecule nouns (e.g., receptor tyrosine kinase RTK). An alias of any compound takes into account added atoms or molecules (e.g., PI: phosphoinositide and PIP: phosphoinositide phosphate) as well as their number (e.g., PIP_2: phosphatidylinositol bisphosphate, DAG: diacylglycerol, and PDE: [cyclic nucleotide] phosphodiesterases).

(2) A lower case letter is used when a single letter denotes a subfamily or an isoform when it is preceded by a capital letter (e.g., PTPRe: protein tyrosine phosphatase receptor-like type-E). Nevertheless, an upper case letter is used in an alias after a single or several lower case letters to distinguish the isoform type (e.g., RhoA isoform and DNA-repair protein RecA for recombination protein-A), but OSM stands for oncostatin-M, not osmole Osm^9 to optimize molecule identification.

These criteria enable to use differently written aliases with the same sequence of letters for distinct molecules (e.g., CLIP for corticotropin-like intermediate peptide, CLiP: cytoplasmic CAP-Gly domain-containing linker protein, and iCliP: intramembrane-cleaving protease).

As the exception proves the rule, current aliases, such as PKA and PLA that designate protein kinase-A and phospholipase-A, respectively, have been kept. Preceded by only 2 upper case letters, a lower case letter that should be used to specify an isoform can bring confusion with acronyms of other protein types (e.g., phospholamban alias PLb).

Nouns (e.g., hormone-like fibroblast growth factor [hFGF] and urokinase-type plasminogen activator [uPA]) or adjectives (e.g., intracellular FGF isoform [iFGF]) that categorize a subtype of a given molecule correspond to a lower case letter to emphasize the molecule species. Hence, an upper case letter with a commonly used hyphen (e.g., I[R]-SMAD that stands for inhibitory [receptor-regulated] SMAD; V-ATPase for vacuolar adenosine triphosphatase; MT1-MMP for membrane type-1 matrix metalloproteinase; and T[V]-SNARE for target [vesicle-associated] soluble ^{N}ethylmaleimide-sensitive factor-attachment protein receptor) is then replaced by a lower case letter (e.g., i[r]SMAD, vATPase, mt1MMP, and t[v]SNARE), as is usual for RNA subtypes (mRNA, rRNA, snRNA, and tRNA for messenger, ribosomal, small nuclear, and transfer RNA, respectively). Similarly, membrane-bound and secreted forms of receptors and coreceptors that can derive from alternative mRNA splicing are defined by a lower case letter (e.g., sFGFR for secreted extracellular FGFR form and sFRP for soluble Frizzled-related protein), as well as eukaryotic translation elongation (eEF) and initiation (eIF) factors.

---

[9]Osmole: the amount of osmotically active particles that exerts an osmotic pressure of 1 atm when dissolved in 22.4 l of solvent at 0 °C.

(3) Although l, r, and t can stand for molecule-like, -related, and -type, respectively, when a chemical is related to another one, in general, upper case letters are used for the sake of homogenity and to clearly distinguish between the letter L and numeral 1 (e.g., KLF: Krüppel-like factor, CTK: C-terminal Src kinase (CSK)-type kinase, and SLA: Src-like adaptor).

(4) An upper case letter is most often used for initials of adjectives contained in the molecule name (e.g., AIP: actin-interacting protein; BAX: BCL2-associated X protein; HIF: hypoxia-inducible factor; KHC: kinesin heavy chain; LAB: linker of activated B lymphocytes; MAPK: mitogen-activated protein kinase; and SNAP: soluble N-ethylmaleimide-sensitive factor-attachment protein);

(5) Lower case letters are used when alias letters do not correspond to initials (e.g., Fox – not fox –: forkhead box), except for portmanteau words that are entirely written in minuscules (e.g., gadkin: γ1-adaptin and kinesin interactor).

This rule applies, whether alias letters do correspond to successive noun letters (e.g., Par: partitioning defective protein and Pax: paxillin, as well as BrK: breast tumor kinase and ChK: checkpoint kinase, whereas CHK denotes C-terminal Src kinase [CSK]-homologous kinase) or not (e.g., Fz: Frizzled and HhIP: Hedgehog-interacting protein),[10] except for composite chemical species (e.g., DAG: diacylglycerol). However, some current usages have been kept for short aliases of chemical species name (e.g., Rho for Ras homolog rather than RHo).

In any case, molecule (super)family (class) aliases as well as those of their members are written in capital letters, such as the IGSF (IGSFi: member i; immunoglobulin), KIF (KIFi; kinesin), SLC (SLCi; solute carrier), TNFSF (TNFSFi; tumor-necrosis factor), and TNFRSF (TNFRSFi; tumor-necrosis factor receptor) superfamily.

Gene names are also written with majuscules when the corresponding protein name contains at least one minuscule, otherwise only the gene name initial is written with an upper case letter that is then followed by lower case letters.

To highlight its function, substrate aliases (e.g., ARF GTPases) contained in a molecule alias are partly written with lower case letters (e.g., ArfRP, ArfGEF, ArfGAP stand for ARF-related protein, ARF guanine-nucleotide exchange factor, and ARF GTPase-activating protein, respectively).

Last, but not least, heavy and pedantic designation of protein isoforms based on roman numerals has been avoided and replaced by usual arabic numerals (e.g., angiotensin-2 rather than angiotensin-II), except for coagulation (or clotting) factors. Moreover, character I can mean either letter I or number 1 without obvious discrimination at first glance (e.g., GAPI that stands for Ras GTPase-activating protein GAP1, but can be used to designate a growth-associated protein inhibitor).

---

[10]The Hedgehog gene was originally identified in the fruit fly Drosophila melanogaster. It encodes a protein involved in the determination of segmental polarity and intercellular signaling during morphogenesis. Homologous gene and protein exist in various vertebrate species. The name of the mammal hedgehog comes from hecg and hegge (dense row of shrubs or low trees), as it resides in hedgerows, and hogg and hogge, due to its pig-like, long projecting nose (snout). The word Hedgehog hence is considered as a seamless whole.

Unnecessary hyphenation in aliases of substances (between an upper case letter, which can define the molecule function, and the chemical alias, or between it and assigned isotype number) has been avoided (). In any case, the Notation section serves not only to define aliases, but also, in some instances, as disambiguation pages.

A space rather than hyphen is used in: (1) structural components at the picoscale (e.g., P loop), nanoscale (e.g., G protein [G standing for guanine nucleotide-binding]), microscale (e.g., H zone, M line, A band, I band, and Z disc of the sarcomere and T tubule of the cardiomyocyte); (2) process stages (e.g., M phase of the cell division cycle); and (3) cell types (e.g., B and T lymphocytes). When these terms are used as adjectives, an hypen is then employed (e.g., P-loop Cys–X_5–Arg (CX_5R) motif, G-protein-coupled receptor, Z-disc ligand, M-phase enzyme, and T-cell activation).

In terms incorporating a Greek letter, similarly, a space is used in: (1) structural components (e.g., α and β chains and subunits); (2) cellular organelles (e.g., α granule); and (3) cell types (e.g., pancreatic β cell). On the other hand, terms are hyphenated when they refer to (1) structural shape (e.g., α-helix and $\alpha(\beta)$-sheet) and (2) molecule subtype (e.g., α-actinin, β-glycan, and γ-secretase).

Symbols for Physical Variables

Unlike substances aliases, symbols for physical quantities are most often represented by a single letter of the Latin or Greek alphabet (i: current; J: flux; L: length; m: mass; p: pressure; P: power; T: temperature; t: time; u: displacement; v: velocity; x: space; λ: wavelength; μ: dynamic viscosity; ρ: mass density; etc.). These symbols are specified using sub- and superscripts (c_p and c_v: heat capacity at constant pressure and volume, respectively; $\mathcal{D}_T$: thermal diffusivity; G_h: hydraulic conductivity; G_T: thermal conductivity; α_k: kinetic energy coefficient; α_m: momentum coefficient; etc.).

A physical quantity associated with a given point in space at a given time can be: (1) a scalar uniquely defined by its magnitude; (2) a vector characterized by a magnitude, a support, and a direction represented by an oriented line segment defined by a unit vector; and (3) a tensor specified by a magnitude and a few directions. To ensure a straightforward meaning of symbols used for scalar, vectorial, and tensorial quantities, bold face upper (**T**) and lower (**v**) case letters are used to denote a tensor and a vector, respectively, whereas both roman (plain, upright)-style upper and lower case letters designate a scalar.

List of Currently Used Prefixes and Suffixes

Prefixes (localization)

"ab-" (Latin) and "apo-" (Greek: απο): away from or off (abluminal: endothelial edge opposite to wetted surface; apolipoproteins: lipid carriers that cause egress [also ingress] from cells; aponeurosis (απονευρωσις; νευρον: sinew, tendon) muscle sheath that limits radial motion and enhances axial contraction; and apoptosis: separation ["-ptosis": fall (πτωσις): as leaves fall away from a tree], a type of programmed cell death)

"acr-" (variant "acro-" [ακρος]): top or apex

"ad-" (adfecto: to reach; adfio: to blow toward; adfluo: to flow toward): toward (ad- becomes "ac-" before c, k, or q; "af-" before f [afferent]; "ag-" before g [agglutination]; "al-" before l; "ap-" before p [approximation]; "as-" before s; and "at-" before t)

"cis-", "juxta-", and "para-" (παρα): near, beside, or alongside

"contra-": opposite side; "ipsi-" (ipse): same side; "latero-": side;

"ecto-" (εκτος), "exo-" (εξο), and "extra-": outside, outer, external, or beyond (exogenous chemicals produced by an external source, or xenobiotics ["xeno-": foreigner])

"endo-" (ενδον) and "intra-": inside (endogenous substances synthesized by the body's cells; endomembranes at organelle surfaces within the cell)

"ep-" (variant "eph-", or "epi-" [επι]): upon (epigenetics refers to the inheritance ("-genetic": ability to procreate [γεννητικος]) of variations in gene expression beyond ("epi-": on, upon, above, close to, beside, near, toward, against, among, beyond, and also) change in the DNA sequence.

"front-" and "pre-": anterior or in front of

"post-": behind

"infra-" and "sub-": under or below

"super-" and "supra-": above

"inter-": between or among

"peri-" (περι): around

"tele-" (τελε): remote

"trans-": across

Prefixes (composition)

"an-" and "aniso-" (ανισος): unequal, uneven, heterogeneous

"iso-" (ισος): equal, alike (isomer [μερος: part, portion]

"mono-" (μονος) and "uni-" (unicus): single

"oligo-" (ολιγος): few, little, small

"multi-" (multus), "pluri-" (plus, plures), and "poly-" (πολυς): many, much

"ultra-": in excess.

Prefixes (quantity)

"demi-" (dimidius) and "hemi-" (ημι): half

"sesqui-": one and a half (half more)

"di-" or "dis-" (δυο; δις) as well as "bi-" or "bis-": 2, twice
"tri" (τρεις, τρι-; tres, tria): 3
"tetra-" (τετρα), "quadri-" (variant: "quadr-" and "quadru-"): 4
"penta-" (πεντας; pentas), "quinqu-", and "quint-": 5
"hexa-" (εξ) and "sexa-": 6
"hepta-" (επτα): 7
"octa-" (οκτα): 8
"nona-" (εννεα): 9 (ninth part)
"deca-" (δεκα): 10
"quadra-" (quadragenarius): 40 (elements)
"quinqua-" (quinquagenarius): 50
"sexa-" (sexagenarius [sex: 6]): 60
"septua-" (septuagenarius [septem: 7]): 70
"nona-" (nonagenarius): 90

Prefixes (motion and direction)

"af-": toward the center (single master object); e.g., nerve and vascular afferents (ferre: to carry) to brain and heart, respectively, rather than toward any slave, supplied tissue from the set of the body's organs; also affector, i.e., chemical messenger that brings a signal to the cell considered as the object of interest, this exploration focus being virtually excised from the organism with its central command system, except received signals
"ef-" (effero: to take away): from the center (efferent; effector, i.e., chemical transmitter recruited by the previous mediator of a signaling cascade at a given locus to possibly translocate to another subcellular compartment)
"antero-" (anterior): before, in front of, facing, or forward
"retro-": behind or backward
"tropo-" (τροπος): duct direction; (tropa: rotation; celestial revolution); e.g., tropomyosin (μυς, musculus: muscle; μυο-: refers to muscle [μυοτρωτος: injured at a muscle])

Prefixes (structure and size)

"macro-" (μακρος): large, long, or big
"mega-" (μεγας): great, large
"meso-" (μεσος): middle
"micro-" (μικρος): small
"nano-" (νανος): dwarf, tiny
"homo-" (ομο-): same (ομολογος: agreeing, corroborating; variant: "homeo-" [homeostasis])
"hetero-" (ετερο-): other

Prefixes (timing)

"ana-" (ανα): culminating (anaphase of the cell division cycle), up, above (ανοδος: a way up, anode [positive electrode; οδος; way, path, road, track])
"ante-": before
"circa-": approximately, around (circadian: approximately one day)
"infra-": below, shorter (infradian: rhythm with lower frequency than that of circadian rhythm, not smaller period)
"inter-": among, between, during
"meta-" (μετα): after, beyond, behind, later; in the middle of (metaphase of the cell division cycle); as well as connected to, but with a change of state (metabolism) and about (metadata)
"post-": after
"pre-": earlier
"pro-" (προ): preceding, first, before (prophase of the cell division cycle)
"telo-" (τελος): end, completion
"ultra-": beyond, longer (ultradian: period smaller than that of 24–28-hour cycle, i.e., frequency greater than that of the circadian rhythm)

Prefixes (functioning modality)

"auto-" (αυτος): same, self
"brady-" (βραδυς): slow (decelerate)
"tachy-" (ταχος): rapid (accelerate)
"amphi-" (αμφι): both (amphiphilic substances are both hydrophilic and lipophilic; amphisomes are generated by both autophagosomes and endosomes)
"ana-" : upward (anabolism) or against (anaphylaxis)
"cata-" (κατα): downward (catabolism, cathode [negative electrode; οδος; way, path, road, track])

List of Currently Used Prefixes and Suffixes

"anti-" (αντι): against
"pro-": favoring
"co-" (coaccedo: add itself to): together
"contra-": adverse, against, beside, next to, opposite
"de-": remove, reduce, separation after association (Latin de; e.g., deoxy-)
"dys-" (δυς): abnormal (δυσαης): ill-blowing)
"equi-" (æque): equal or alike
"hem-" or "hemat-" (αιμα: blood): related to blood
"hypo-" (υπο): under, beneath, and low
"hyper-" (υπερ): above, beyond, and large
"per-": through (e.g., percutaneous) and during (e.g., peroperative)
"pseudo-" (ψευδο): pretended, false
"re-"; again

Scientific suffixes

"-ase": enzyme (synthase, lipase, etc.)
"-ate": salt of a base
"-cyte" (κυτος): cell (erythro- [ερυθρος: red], leuko- [λευκος: light, bright, clear, white], thrombo- [θρομβος: lump, clot], adipo- [adeps: fat; adipalis, adipatus, adipeus, adipinus: fatty], fibro- [fibra: fiber, filament], myo- [μυς: muscle, mouse, mussel], myocardiocyte [κραδια: heart; cardiacus: related to heart, stomach; to have heart trouble, stomach trouble], etc.);
"-crine" (κρινω): to decide, to separate, and to secrete (e.g., endocrine regulator) (ευκρινεω: keep in order)
"-elle": small (organelle in a cell [like an organ in a body])
"-ium", "-ion", "-isk", and "-iscus": little ("-ium": tissue interface and envelope, such as endothelium and pericardium)
"-phil" (φιλια): attracted (αφιλια: want of friends)
"-phob" (φοβια): repulsed (υδροφοβια, hydrophobia [Latin]: horror of water)
"-phore" (φερω): carrier (αμφερω: to bring up)
"-yl" denotes a radical (molecules with unpaired electrons)
"-ploid" (πλοω): double, fold (diploid, twofold; διπλοω: to double; διαπλοω: unfold)
"-emia": in relation to flow (ανεμια: flatulence; ευηνεμια: fair wind), particularly blood condition
"-genesis" (γενεσις): cause, generation, life source, origin, productive force
"-iasis": for diseased condition
"-itis": inflammation
"-lemma" (λεμμα: skin): sheath
"-ole" and "-ule": small (arteriole and venule; variant "-ula" [blastula] and "-ulum")
"-plasma" (πλασμα): anything molded (plasma: creature generated from silt of earth)
"-plasia" (πλασια): formation, molding
"-podium" (ποδος: foot; podium [Latin]: small knoll, small protuberance): protrusion
"-poiesis" (ποιεω): production
"-soma" (σωμα): body
"-sclerosis" (σκλημα): hardness, induration
"-stasis" (στασις): stabilization (αποκαταστασις: restoration; ανυποστασις: migration)
"-stomosis" (στομα: mouth): equipped with an outlet
"-taxy/tactic" (ταχυ: rapid; τακτικος: to maneuver): related to motion (also prefix, i.e., ταχυκινησις: quick motion; ταχυνω: to accelerate; and ταχυπνοια: short breath; not [δια]ταξις: disposition, arrangement)
"-trophy/trophic" (τροφις: well fed): related to growth
"-oma": tumor of
"-pathy" (παθος, παθεια): disease of
"-tomy" (τομια) and "-ectomy": surgical removal (απλοτομια: simple incision; φαρhουγγοτομια: laryngotomy)

List of Aliases

A

$\mathcal{A}$: Avogadro number
$\mathcal{A}(p)$: area–pressure relation
A: Almansi strain tensor
A: cross-sectional area
A: actin-binding site
a: acceleration
a: major semi-axis
AA: arachidonic acid
AAA: ATPase associated with diverse cellular activities
AAA: abdominal aortic aneurysm
AAAP: aneurysm-associated antigenic protein
AAK: adaptin-associated kinase
AATK: apoptosis-associated tyrosine kinase
ABC: ATP-binding cassette transporter (transfer ATPase)
AbI: Abelson kinase interactor
Abl: Abelson leukemia viral proto-oncogene product (NRTK)
ABLIM: actin-binding LIM domain-containing protein
ABP: actin-binding protein
ABR: active breakpoint cluster region (BCR)-related gene product (GEF and GAP)
AC: atrial contraction
ACAP: ArfGAP with coiled-coil, ankyrin repeat, PH domains
ACase: adenylate cyclase
ACi: adenylate cyclase isoform i
ACAT: acylCoA–cholesterol acyltransferase
ACC: acetyl coenzyme-A carboxylase
ACD: adrenocortical dysplasia homolog
ACE: angiotensin-converting enzyme
ACh: acetylcholine
ACK: activated CDC42-associated kinase
ACP1: acid phosphatase-1, soluble (lmwPTP)
ACTH: adrenocorticotropic hormone
^{F}actin: filamentous actin
(Cav–actin: caveolin-associated ^{F}actin)
^{G}actin: monomeric globular actin
AcvR: activin receptor (TGFβ receptor superfamily)
Ad: adrenaline
ADAM: a disintegrin and metallopeptidase (adamalysin)
ADAMTS: a disintegrin and metallopeptidase with thrombospondin
ADAP: adhesion and degranulation-promoting adaptor protein
ADAP: ArfGAP with dual PH domains
ADF: actin-depolymerizing factor (cofilin-related destrin)
ADH: antidiuretic hormone (vasopressin)
ADMA: asymmetric dimethylarginine
ADP: adenosine diphosphate
aDuSP: atypical dual specificity phosphatase
AE: anion exchanger
AEA: N-arachidonoyl ethanolamine (anandamide)
AF: atrial fibrillation
AFAP: ArfGAP with phosphoinositide-binding and PH domains
aFGF: acidic fibroblast growth factor (FGF1)
AGAP: ArfGAP with GTPAse, ankyrin repeat, and PH domains
AGF: autocrine growth factor
AGFG: ArfGAP with FG repeats
Ago: Argonaute protein
AGS: activator of G-protein signaling
AHR: aryl hydrocarbon receptor

AIF: apoptosis-inducing factor
AIP: actin-interacting protein
AIRe: autoimmune regulator
AKAP: A-kinase (PKA)-anchoring protein
ALE: arbitrary Eulerian Lagrangian
ALIX: apoptosis-linked gene-2-interacting protein-X
ALK: anaplastic lymphoma kinase
ALKi: type-i activin receptor-like kinase (TGFβ receptor superfamily)
ALOx5: arachidonate 5-lipoxygenase
ALOx5AP: arachidonate 5-lipoxygenase activation protein
ALP: actinin-associated LIM protein (PDLIM3)
alsin: amyotrophic lateral sclerosis protein (portmanteau)
ALX: adaptor in lymphocytes of unknown function X
AMAP: A multidomain ArfGAP protein
AMBRA: activating molecule in beclin-1-regulated autophagy protein
AMHR: anti-Müllerian hormone receptor (TGFβ receptor superfamily)
AMIS: apical membrane initiation site (lumenogenesis)
AMP: adenosine monophosphate
AMPAR: α-amino 3-hydroxy 5-methyl 4-isoxazole propionic acid receptor
AMPK: AMP-activated protein kinase
AMSH: associated molecule with SH3 domain (deubiquitinase)
AmyR: amylin receptor
Ang: angiopoietin
AngL: angiopoietin-like molecule
Ank: ankyrin
ANP: atrial natriuretic peptide
ANPR (NP$_1$): atrial natriuretic peptide receptor (guanylate cyclase)
ANS: autonomic nervous system
ANT: adenine nucleotide transporter
Anx: annexin
AOC: amine oxidase copper-containing protein
AoV: aortic valve
AP: (clathrin-associated) adaptor proteic complex
AP: Activator protein (transcription factor)
AP: activating enhancer-binding protein
AP4A: diadenosine tetraphosphate
APAF: apoptotic peptidase-activating factor
APAP: ArfGAP with PIX- and paxillin-binding domains
APC: antigen-presenting cell

APC: adenomatous polyposis coli protein (Ub ligase)
APC/C: anaphase-promoting complex (or cyclosome; Ub ligase)
APH: anterior pharynx defective phenotype homolog
aPKC: atypical protein kinase C
APl: action potential
Apn: adiponectin
Apo: apolipoprotein
ApoER: apolipoprotein-E receptor
APPL: adaptor containing phosphoTyr interaction, PH domain, and Leu zipper
APS: adaptor with a PH and SH2 domain
Aqp: aquaporin
AR: adrenergic receptor (adrenoceptor)
AR: androgen receptor (nuclear receptor NR3c4; transcription factor)
AR: area ratio
ARAP: ArfGAP with RhoGAP, ankyrin repeat, PH domains
ARE: activin-response element
ARE: androgen response element
ARE: anti-oxidant response element
Areg: amphiregulin (EGF superfamily member)
ARF: ADP-ribosylation factor
ArfRP: ARF-related protein
ARFTS: CKI2A-locus alternate reading frame tumor suppressor (ARF or p14^{ARF})
ARH: autosomal recessive hypercholesterolemia adaptor (low-density lipoprotein receptor adaptor)
ARH: aplysia Ras-related homolog
ArhGEF: RhoGEF
ARL: ADP-ribosylation factor-like protein
ARNO: ARF nucleotide site opener
ARNT: aryl hydrocarbon nuclear receptor translocator
ARP: absolute refractory period
ARP: actin-related protein
ARPP: cAMP-regulated phosphoprotein
Arr: arrestin
ART: arrestin-related transport adaptor (α-arrestin)
ART: adpribosyltransferase
Artn: artemin
ARVCF: armadillo repeat gene deleted in velocardiofacial syndrome
ARVD: arrythmogenic right ventricular dystrophy
AS: Akt (PKB) substrate

ASAP: artery-specific antigenic protein
ASAP: ArfGAP with SH3, ankyrin repeat, PH domains
ASIC: acid-sensing ion channel
ASK: apoptosis signal-regulating kinase
aSMC: airway smooth muscle cell
ASP: actin-severing protein
AT: antithrombin
ATAA: ascending thoracic aortic aneurysm
ATF: activating transcription factor
AtG: autophagy-related gene product
ATMK: ataxia telangiectasia mutated kinase
ATn: angiotensin
ATng: angiotensinogen
AtOx: anti-oxidant protein (metallochaperone)
ATP: adenosine triphosphate
ATPase: adenosine triphosphatase
ATR ($AT_{1/2}$): angiotensin receptor
ATRK: ataxia telangiectasia and Rad3-related kinase
AVN: atrioventricular node
AVV: atrioventricular valves
AW: analysis window

B

B: Biot-Finger strain tensor
B: bulk modulus
$\mathcal{B}$: bilinear form
b: minor semi-axis
b: body force
$\hat{\mathbf{b}}$: unit binormal
B lymphocyte (B cell): bone marrow lymphocyte
BACE: β-amyloid precursor protein-converting enzyme
BAD: BCL2 antagonist of cell death
BAF: barrier-to-autointegration factor
BAG: BCL2-associated athanogene (chaperone regulator)
BAI: brain-specific angiogenesis inhibitor (adhesion-GPCR)
BAIAP: brain-specific angiogenesis inhibitor-1-associated protein (insulin receptor substrate)
BAK: BCL2-antagonist–killer
(i)BALT: (inducible) bronchus-associated lymphoid tissue
BAMBI: BMP and activin membrane-bound inhibitor homolog
BAnk: B-cell scaffold with ankyrin repeats

Barkor: beclin-1-associated autophagy-related key regulator
BAT: brown adipose tissue
BATF: basic leucine zipper ATF-like transcription factor (B-cell-activating transcription factor)
BAX: BCL2-associated X protein
BBB: blood–brain barrier
BBS: Bardet-Biedl syndrome protein
BBSome: BBS coat complex (transport of membrane proteins into cilium)
BC: boundary condition
bCAM: basal cell adhesion molecule (Lutheran blood group glycoprotein)
BCAP: B-cell adaptor for phosphatidylinositol 3-kinase
BCAR: Breast cancer anti-estrogen resistance docking protein
BCL: B-cell lymphoma (leukemia) protein
BCLxL: B-cell lymphoma extra-large protein
BCR: B-cell receptor
BCR: breakpoint cluster region protein (GAP and GEF)
Bdk: bradykinin
BDNF: brain-derived neurotrophic factor
Be: Bejan number
Becn, beclin: BCL2-interacting protein
BEM: boundary element method
Best: bestrophin
bFGF: basic fibroblast growth factor (FGF2)
BFUe: burst-forming unit erythroid
BFUmeg: burst-forming unit megakaryocyte
BGT: betaine–GABA transporter
BH_4: tetrahydrobiopterin (enzyme cofactor)
BID: BH3-interacting domain death agonist
BIG: brefeldin-A-inhibited GEFs for ARFs
BIK: BCL2-interacting killer
BIM: BH3-containing protein BCL2-like 11 (BCL2L11)
BK: high-conductance, Ca^{++}-activated, voltage-gated K^+ channel
BLK: B-lymphoid tyrosine kinase
Blm: Bloom syndrome, RecQ DNA helicase-like protein
BLnk: B-cell linker protein
BLOC: biogenesis of lysosome-related organelles
BM: basement membrane
BMAL: brain and muscle ARNT-like protein (gene Bmal)
BMAT: bone-marrow adipose tissue
BMF: BCL2 modifying factor
BMP: bone morphogenetic protein (TGFβ superfamily)

BMPR: bone morphogenetic protein receptor
BNIP: BCL2/adenovirus E1B 19-kDa protein-interacting protein
BNP: B-type natriuretic peptide
BMX: bone marrow Tyr kinase gene in chromosome-X product
Bo: Boltzmann constant
BOC: brother of CDO
BOK: BCL2-related ovarian killer
BORG: binder of Rho GTPase
Br: Brinkman number
BRAG: brefeldin-resistant ArfGEF
BrCa: breast cancer-associated (susceptibility) protein (tumor suppressor; DNA-damage repair; a.k.a. FancD1)
BrD: bromodomain-containing protein
BrK: breast tumor kinase
BrSK: brain-selective kinase
BSEP: bile salt export pump
BTF: basic transcription factor
BTK: Bruton Tyr kinase
BUB: budding uninhibited by benzimidazoles

C

C: stress tensor
C: compliance
C: heat capacity
C: chronotropy
Cx: type-x chemokine C (γ)
C_D: drag coefficient
C_f: friction coefficient
C_L: lift coefficient
C_p: pressure coefficient
c: stress vector
c_τ: shear
c_w: wall shear stress
c: concentration
$c(p)$: wave speed
c_p: isobar heat capacity
c_v: isochor heat capacity
C1P: ceramide 1-phosphate
C-terminus: carboxy (carboxyl group COOH)-terminus
C/EBP: CCAAT/enhancer-binding protein
CA: computed angiography
CAi: carbonic anhydrase isoform i
Ca: calcium
Ca$_V$: voltage-gated Ca^{++} channel
Ca$_V$1.x: L-type high-voltage-gated Ca^{++} channel
Ca$_V$2.x: P/Q/R-type Ca^{++} channel
Ca$_V$3.x: T-type low-voltage-gated Ca^{++} channel
CAAT: cationic amino acid transporter
CABG: coronary artery bypass grafting
Cables: CDK5 and Abl enzyme substrate
CAK: CDK-activating kinase (pseudokinase)
Cam: calmodulin (calcium-modulated protein)
CamK: calmodulin-dependent kinase
cAMP: cyclic adenosine monophosphate
CAP: adenylate cyclase-associated protein
CAP: carboxyalkylpyrrole protein adduct
CAP: chromosome-associated protein (BrD4)
CAPN: calpain gene
CaPON: carboxy-terminal PDZ ligand of NOS1 (NOS1AP)
CAR: constitutive androstane receptor (NR1i3)
CaR: calcium-sensing receptor
CARD: caspase activation and recruitment domain-containing protein
CARMA: CARD and membrane-associated guanylate kinase-like (GuK) domain-containing protein
CARP: cell division cycle and apoptosis regulatory protein
CAS: cellular apoptosis susceptibility protein
CAS: CRK-associated substrate (or P130CAS and BCAR1)
CAs: cadherin-associated protein
CASK: calcium–calmodulin-dependent serine kinase (pseudokinase)
CASL: CRK-associated substrate-related protein (CAS2)
CASP: cytohesin-associated scaffold protein
caspase: cysteine-dependent aspartate-specific peptidase
Cav: caveolin
CBF: coronary blood flow
CBF: core-binding factor
CBL: Casitas B-lineage lymphoma adaptor and Ub ligase
CBLb: CBL-related adaptor
CBP: cap-binding protein
CBP: CREB-binding protein
CBP: C-terminal Src kinase-binding protein
CBS: cystathionine β-synthase (H_2S production)
CCD: cortical collecting duct
CCDC: coiled-coil domain-containing protein
CCICR: calcium channel-induced Ca^{++} release
CCK4: colon carcinoma kinase 4 (PTK7)
CCL: chemokine CC-motif ligand
CCN: CyR61, CTGF, and NOv (CCN1–CCN3) family

List of Aliases

Ccn: cyclin
Ccnx–CDK*i*: type-x cyclin–type-*i* cyclin-dependent kinase dimer
CCPg: cell cycle progression protein
CCS: copper chaperone for superoxide dismutase
CCT: chaperonin containing T-complex protein
CCx: type-x chemokine CC (β)
CCR: chemokine CC motif receptor
CD: cluster determinant protein (cluster of differentiation)
CDase: ceramidase
CDC: cell division cycle protein
cDC: classical dendritic cell
CDH: CDC20 homolog
Cdh: cadherin
CDK: cyclin-dependent kinase
Cdm: caldesmon
CDO: cell adhesion molecule-related/downregulated by oncogenes
CE (CsE): cholesteryl esters
CEC: circulating endothelial cell
CELSR: cadherin, EGF-like, LAG-like, and seven-pass receptor
CenP: centromere protein
CEP: carboxyethylpyrrole
CeP: centrosomal protein
CEPC: circulating endothelial progenitor cell
Cer: ceramide
CerK: ceramide kinase
CerT: ceramide transfer protein
CETP: cholesterol ester transfer protein
CFD: computational fluid dynamics
CFLAR: caspase-8 and FADD-like apoptosis regulator
CFTR: cystic fibrosis transmembrane conductance regulator
CFU: colony-forming unit
CFUb: CFU basophil (basophil-committed stem cells)
CFUc: CFU in culture (granulocyte precursors, i.e., CFUgm)
CFUe: CFU erythroid
CFUeo: CFU eosinophil
CFUg: CFU granulocyte
CFUgm: CFU granulocyte–macrophage
CFUgemm: CFU granulocyte–erythroid–macrophage–megakaryocyte
CFUm: CFU macrophage
CFUmeg: CFU megakaryocyte
CFUs: colony-forming unit spleen (pluripotent stem cells)
CG: chromogranin

cGK: cGMP-dependent protein kinase (protein kinase G)
cGMP: cyclic guanosine monophosphate
CGN: cis-Golgi network
CGRP: calcitonin gene-related peptide
chanzyme: ion channel and enzyme
chemokine: chemoattractant cytokine
CHIP: C-terminus heat shock cognate-70-interacting protein
ChK: checkpoint kinase
CHK: CSK homologous kinase
Chn: chimerin (GAP)
CHOP: CCAAT/enhancer-binding protein homologous protein
CHREBP: carbohydrate-responsive element-binding protein
ChT: choline transporter
CI: cardiac index
CICR: calcium-induced calcium release
Cin: chronophin
CIP: CDC42-interacting protein
CIP2a: cancerous inhibitor of protein phosphatase-2A
CIPC: CLOCK-interacting protein, circadian
CIS: cytokine-inducible SH2-containing protein
CITED: CBP/P300-interacting transactivator with glutamic (E) and aspartic acid (D)-rich C-terminus-containing protein
CK: creatine kinase
CK: casein kinase
CKI: cyclin-dependent kinase inhibitor
CLAsP: CLiP-associated protein (microtubule binder)
ClASP: clathrin-associated sorting protein
CLC: cardiotrophin-like cytokine
ClC: voltage-gated chloride channel
ClCa: calcium-activated chloride channel
ClIC: chloride intracellular channel
CLINT: clathrin-interacting protein located in the trans-Golgi network
CLIP: corticotropin-like intermediate peptide
CLiP: cytoplasmic CAP-Gly domain-containing linker protein
iCliP: intramembrane-cleaving peptidase (that clips)
CLK: CDC-like kinase
CINS: Cl$^-$ channel nucleotide-sensitive
CLOCK: circadian locomotor output cycles kaput
CLP: common lymphoid progenitor
CLS: ciliary localization signal
Cmi: chylomicron
CMLP: common myeloid–lymphoid progenitor

CMP: common myeloid progenitor
CMC: cardiomyocyte
Col: collagen
CoLec: collectin
ColF: collagen fiber
CORM: carbon monoxide (CO)-releasing molecule
CNG: cyclic nucleotide-gated channel
CnK: connector enhancer of kinase suppressor of Ras
CNS: central nervous system
CNT: connecting tubule
CNTi: concentrative nucleoside transporter (SLC28ai)
CNTF: ciliary neurotrophic factor
CntnAP: contactin-associated protein
CO: cardiac output
CoA: coenzyme-A
CoBl: Cordon-bleu homolog (actin nucleator)
COLD: chronic obstructive lung disease
COOL: Cloned out of library (RhoGEF6/7)
coSMAD: common (mediator) SMAD (SMAD4)
COx: cyclooxygenase (prostaglandin endoperoxide synthase)
COx17: cytochrome-C oxidase copper chaperone
CoP: coat protein
CoP: constitutive photomorphogenic protein (Ub ligase)
COPD: chronic obstructive pulmonary disease
COUPTF: chicken ovalbumin upstream promoter transcription factor (NR2f1/2)
CP4H: collagen prolyl 4-hydroxylase
CPC: chromosomal passenger complex
CpG: cytidine^p–guanosine oligodeoxynucleotide (motif)
cPKC: conventional protein kinase C
Cpx: complexin
CR: complement component receptor
Cr: creatine
cRABP: cellular retinoic acid-binding protein
cRBP: cellular retinol-binding protein
CRAC: Ca^{++} release-activated Ca^{++} channel
CRACR: CRAC regulator
Crb: Crumbs homolog polarity complex
CRE: cAMP-responsive element
CREB: cAMP-responsive element-binding protein
CRF: corticotropin-releasing factor (family)
CRH: corticotropin-releasing hormone
CRIB: CDC42/Rac interactive-binding protein
CRIK: citron Rho-interacting, Ser/Thr kinase (STK21)

CRK: chicken tumor virus CT10 regulator of kinase
CRKL: CRK avian sarcoma virus CT10 homolog-like
CRL4: cullin-4A RING ubiquitin ligase
CRLR: calcitonin receptor-like receptor
CRP: C-reactive protein
Crt: calreticulin
CRTC: CREB-regulated transcription coactivator
Cry: cryptochrome
Cs: cholesterol
CSBP: cytokine-suppressive anti-inflammatory drug-binding protein
CSE: cystathionine γ-lyase (H_2S production)
CSF: cerebrospinal fluid
CSF: colony-stimulating factor
CSF1: macrophage colony-stimulating factor (mCSF)
CSF2: granulocyte–macrophage colony-stimulating factors (gmCSF and sargramostim)
CSF3: granulocyte colony-stimulating factors (gCSF and filgrastim)
CSK: C-terminal Src kinase
Csk: cytoskeleton
Csq: calsequestrin
CSS: candidate sphingomyelin synthase
CT: cardiotrophin
CT: computed tomography
CTBP: C-terminal-binding protein
CTen: C-terminal tensin-like protein
CTF: C-terminal fragment
CTGF: connective tissue growth factor
CTL: cytotoxic T lymphocyte
CTLA: cytotoxic T-lymphocyte-associated protein
Ctn: catenin
CTr: copper transporter
CtR: calcitonin receptor
CTRC: CREB-regulated transcription coactivator
Cul: cullin
CUT: cryptic unstable transcript
CVI: chronic venous insufficiency
CVLM: caudal ventrolateral medulla
CVP: central venous pressure
CVS: cardiovascular system
Cx: connexin
CXCLi: type-i CXC (C-X-C motif; α) chemokine ligand
CXCRi: type-i CXC (C-X-C motif; α) chemokine receptor

List of Aliases

CX3CLi: type-i CX3C (δ) chemokine ligand
CX3CRi: type-i CX3C (δ) chemokine receptor
cyCK: cytosolic creatine kinase
Cyld: cylindromatosis tumor suppressor protein (deubiquitinase USPL2)
CyP: member of the cytochrome-P450 superfamily
C3G: Crk SH3-binding GEF

D

D: dromotropy
D: vessel distensibility
$\mathcal{D}$: diffusion coefficient
$\mathcal{D}_T$: thermal diffusivity
D: deformation rate tensor
d: displacement vector
D: flexural rigidity
D: demobilization function (from proliferation to quiescence)
d: death, decay, degradation rate
d: duration
Dab: Disabled homolog
DAD: delayed afterdepolarization
DAG: diacylglycerol
DAPC: dystrophin-associated protein complex
DAPK: death-associated protein kinase
DARC: Duffy antigen receptor for chemokine
DAT: dopamine active transporter
DAX: dosage-sensitive sex reversal, adrenal hypoplasia critical region on chromosome X (NR0b1)
DBC: deleted in breast cancer protein
DBF: dumbbell formation kinase (in Saccharomyces cerevisiae; e.g., DBF2)
DBP: albumin D-element binding protein (PAR/b–ZIP family)
DC: dendritic cell
DCA: directional coronary atherectomy
DCAF: DDB1- and Cul4-associated factor
DCC: deleted in colorectal carcinoma (netrin receptor)
DCT: distal convoluted tubule
Dctn: dynactin
DDAH: dimethylarginine dimethylaminohydrolase
DDB: damage-specific DNA-binding protein
DDEF: development and differentiation-enhancing factor (ArfGAP)
DDR: discoidin domain receptor
De: Dean number
Deb: Deborah number

DEC: differentially expressed in chondrocytes (DEC1 and DEC2 are a.k.a bHLHe40 and bHLHe41, bHLHb2 and bHLHb3, or HRT2 and HRT1)
DEC: deleted in esophageal cancer
DEG: delayed-early gene
deoxyHb: deoxyhemoglobin (deoxygenated hemoglobin)
DETC: dendritic epidermal γδ T cell
DH: Dbl homology
DHA: docosahexaenoic acid
DHET: dihydroxyeicosatrienoic acid
DHh: desert Hedgehog
Dia: Diaphanous
DICOM: digital imaging and communication for medicine
DICR: depolarization-induced Ca^{++} release
DISC: death-inducing signaling complex
Dkk: Dickkopf
DLg: Disc large homolog
DLL: Delta-like (Notch) ligand
DLx: distal-less homeobox protein
DM: double minute
DMM: DNA methylation modulator
DMPK: myotonic dystrophy-associated protein kinase
DMT: divalent metal transporter
DN1: double-negative-1 cell
DN2: double-negative-2 cell
DN3: double-negative-3 cell
DNA: deoxyribonucleic acid
DNAPK: DNA-dependent protein kinase
DoC2: double C2-like domain-containing protein
DOCK: dedicator of cytokinesis (GEF)
DOK: downstream of Tyr kinase docking protein
DOR: δ-opioid receptor
DPG: diphosphoglyceric acid
DRAM: damage-regulated modulator of autophagy
DRF: Diaphanous-related formin (for GTPase-triggered actin rearrangement)
DRG: dorsal root ganglion
Drl: Derailed
Dsc: desmocollin
Dsg: desmoglein
Dsh: Disheveled (Wnt-signaling mediator)
DSK: dual-specificity kinase
dsRNA: double-stranded RNA
Dst: dystonin
DUb: deubiquitinase
DuOx: dual oxidase
DUS: Doppler ultrasound

DuSP: dual-specificity phosphatase
DV: dead space volume
Dvl: Disheveled (cytoplasmic phosphoprotein; other alias Dsh)
DVT: deep-vein thrombosis
dynactin: dynein activator
DYRK: dual-specificity Tyr (Y) phosphorylation-regulated kinase

E

E: strain tensor
E: electric field
E: elastic modulus
E: elastance
$\mathcal{E}$: energy
$\{\hat{\mathbf{e}}_i\}_{i=1}^{3}$: basis
e: strain vector
e: specific free energy
E-box: enhancer box sequence of DNA
E2: ubiquitin conjugase
E3: ubiquitin ligase
EAAT: excitatory amino acid (glutamatee–aspartate) transporter
EAD: early afterdepolarization
EAR: V-erbA-related nuclear receptor (NR2f6)
EB: end-binding protein
EBCT: electron beam CT
EBF: early B-cell factor
EC: endothelial cell
Ec: Eckert number
ECA: external carotid artery
ECF: extracellular fluid
ECG: electrocardiogram
ECM: extracellular matrix
ED1L: EGF-like repeat- and discoidin-1-like domain-containing protein
EDGR: endothelial differentiation gene receptor
EDHF: endothelial-derived hyperpolarizing factor
EDIL: EGF-like repeats and discoidin-1 (I)-like domain-containing protein
EDV: end-diastolic volume
EEA: early endosomal antigen
eEF: eukaryotic translation elongation factor
EEL: external elastic lamina
EET: epoxyeicosatrienoic acid
EFA6: exchange factor for ARF6 (ArfGEF)
EF-Tu: elongation factor Tu
EGF: epidermal growth factor
EGFL: EGF-like domain-containing protein

EGFR: epidermal growth factor receptor
EGR: early growth response transcription factor
EHD: C-terminal EGFR substrate-15 homology domain-containing protein
eIF: eukaryotic translation initiation factor
EL: endothelial lipase
ELAM: endothelial–leukocyte adhesion molecules
ELCA: excimer laser coronary angioplasty
ELk: ETS-like transcription factor (ternary complex factor [TCF] subfamily)
ElMo: engulfment and cell motility adaptor
Eln: elastin
ElnF: elastin fiber
ELP: early lymphoid progenitor
EMI: early mitotic inhibitor
EMR: EGF-like module-containing, mucin-like, hormone receptor-like protein
EMT: epithelial–mesenchymal transition
ENA–VASP: Enabled homolog and vasoactive (vasodilator)-stimulated phosphoprotein family
ENaC: epithelial Na^+ channel
EnaH: Enabled homolog
endo-siRNA: endogenous small interfering RNA
ENPP: ectonucleotide pyrophosphatase–phosphodiesterase
Ens: endosulfine
ENT: equilibrative nucleoside transporter
ENTPD: ectonucleoside triphosphate diphosphohydrolase
EPAC: exchange protein activated by cAMP
EPAS: endothelial PAS domain protein
EPC: endothelial progenitor cell
EPCR: endothelial protein-C receptor
EPDC: epicardial-derived cell
Epgn: epigen (EGF superfamily member)
EPH: erythropoietin-producing hepatocyte receptor kinase or pseudokinase (EPHa10 and EPHb6)
ephrin: EPH receptor interactor
Epo: erythropoietin
EPS: epidermal growth factor receptor pathway substrate
ER: endoplasmic reticulum
ERx: type-x estrogen receptor (NR3a1/2)
eRas: embryonic stem cell-expressed Ras (or hRas2)
ErbB: erythroblastoma viral gene product B (HER)

List of Aliases

ERE: estrogen response element (DNA sequence)
Ereg: epiregulin (EGF superfamily member)
eRF: eukaryotic release factor
ERGIC: endoplasmic reticulum–Golgi intermediate compartment
ERK: extracellular signal-regulated protein kinase
ERK1/2: usually refers to ERK1 and ERK2
ERM: ezrin–radixin–moesin
ERMES: endoplasmic reticulum–mitochondrion encounter structure
ERP: effective refractory period
ERR: estrogen-related receptor (NR3b1–NR3b3)
ESCRT: endosomal sorting complex required for transport
ESL: E-selectin ligand
ESRP: epithelial splicing regulatory protein
ESV: end-systolic volume
ET: endothelin
ETP: early thymocyte progenitor
ETR ($ET_{A/B}$): endothelin receptor
ETS: E-twenty six (transcription factor; erythroblastosis virus E26 proto-oncogene product homolog)
ETV: ETS-related translocation variant
EVAR: endovascular aneurysm repair
Exo: exocyst subunit
Ext: exostosin (glycosyltransferase)

F

F: transformation gradient tensor
F: function fraction of proliferating cells
F: erythrocytic rouleau fragmentation rate
f: surface force
$\hat{f}$: fiber direction unit vector
f: binding frequency
f_C: cardiac frequency
f_R: breathing frequency
f: friction shape factor
f_v: head loss per unit length
f_i: molar fraction of gas component i
FA: fatty acid
FABP: fatty acid-binding protein
FABP: filamentous actin-binding protein
FACAP: F-actin complex-associated protein
FAD: flavine adenine dinucleotide
FADD: Fas receptor-associated death domain
FAK: focal adhesion kinase
Fanc: Fanconi anemia protein

FAN: Fanconi anemia-associated nuclease
FAPP: phosphatidylinositol four-phosphate adaptor protein
Fas: death receptor (TNFRSF6a)
FasL: death ligand (TNFSF6)
FAST: Forkhead activin signal transducer
FB: fibroblast
Fbln (Fibl): fibulin
Fbn: fibrillin
FBS: F-box, Sec7 protein (ArfGEF)
FBx: F-box only protein (ArfGEF)
FC: fibrocyte
FCHO: FCH domain only protein
FcαR: Fc receptor for IgA
FcγR: Fc receptor for IgG
FcϵR: Fc receptor for IgE
FCP: TF2F-associating C-terminal domain phosphatase
FDM: finite difference method
FEM: finite element method
FERM: four point-1, ezrin–radixin–moesin domain
FeR: FeS-related Tyr kinase
FeS: feline sarcoma kinase
FFA: free fatty acid
FGF: fibroblast growth factor
FGFR: fibroblast growth factor receptor
FGR: viral feline Gardner-Rasheed sarcoma oncogene homolog kinase
FHL: four-and-a-half LIM-only protein
FHoD: formin homology domain-containing protein (FmnL)
FIH: factor inhibiting HIF1α (asparaginyl hydroxylase)
FIP: family of Rab11-interacting protein
FIP: focal adhesion kinase family-interacting protein
FIT: Fat-inducing transcript
FKBP: FK506-binding protein
FlIP: flice-inhibitory protein
FLK: fetal liver kinase
fMLP: N-formyl methionyl-leucyl-phenylalanine
FN: fibronectin
Fn: fibrin
Fng: fibrinogen
Fos: Finkel Biskis Jinkins murine osteosarcoma virus sarcoma proto-oncogene product
Fox: forkhead box transcription factor
Fpn: ferroportin
FR: flow ratio
FRK: Fyn-related kinase
FrmD: FERM domain-containing adaptor
FRNK: FAK-related non-kinase

FRS: fibroblast growth factor receptor substrate
FSH: follicle-stimulating hormone
FSI: fluid–structure interaction
FVM: finite volume method
FXR: farnesoid X receptor (NR1h4)
Fz: Frizzled (Wnt GPCR)

G

G: Green-Lagrange strain tensor
G: shear modulus
G': storage modulus
G'': loss modulus
$\mathcal{G}$: Gibbs function
$\mathcal{G}$: conductance
$\mathcal{G}_p$: pressure gradient
$\mathcal{G}_h$: perfusion conductivity
$\mathcal{G}_e$: electrical conductivity
$\mathcal{G}_h$: hydraulic conductivity
$\mathcal{G}_T$: thermal conductivity
g: gravity acceleration
g: physical quantity
g: gravity
g: detachment frequency
g: free enthalpy
G protein: guanine nucleotide-binding protein ($G\alpha\beta\gamma$ trimer)
$G\alpha$: α subunit (signaling mediator) of G protein
$G\alpha_i$ (Gi): inhibitory $G\alpha$ subunit
$G\alpha_s$ (Gs): stimulatory $G\alpha$ subunit
$G\alpha_t$ (Gt): transducin, $G\alpha$ subunit of rhodopsin
Gs$_{XL}$: extra-large Gs protein
$G\alpha_{i/o}$ (Gi/o): $G\alpha$ subunit class
$G\alpha_{q/11}$ (Gq/11): $G\alpha$ subunit class
$G\alpha_{12/13}$ (G12/13): $G\alpha$ subunit class
$G\beta\gamma$: dimeric subunit (signaling effector) of G protein
G_{gust}: gustducin, G protein α subunit (Gi/o) of taste receptor
G_{olf}: G protein α subunit (Gs) of olfactory receptor
GAB: GRB2-associated binder
GABA: γ-aminobutyric acid
GABA$_A$: GABA ionotropic receptor (Cl^- channel)
GABA$_B$: GABA metabotropic receptor (GPCR)
GABARAP: GABA$_A$ receptor-associated protein
GaBP: globular actin-binding protein

GADD: growth arrest and DNA-damage-induced protein
gadkin: γ1-adaptin and kinesin interactor
GAG: glycosaminoglycan
GAK: cyclin G-associated kinase
Gal: galanin
GAP: GTPase-activating protein
GAPDH: glyceraldehyde 3-phosphate dehydrogenase
GARP: Golgi-associated retrograde protein complex
GAS: growth arrest-specific non-coding, single-stranded RNA
GAT: γ-aminobutyric acid transporter
GATA: DNA sequence GATA-binding protein (TF)
GBF: Golgi-associated brefeldin-A-resistant guanine nucleotide-exchange factor
GBP: guanylate-binding protein
GCAP: guanylate cyclase-activating protein
GCC: Golgi coiled-coil domain-containing protein
GCK: germinal center kinase
GCKR: GCK-related kinase
GCNF: germ cell nuclear factor (NR6a1)
GCN2: general control non-derepressible 2 (pseudokinase)
gCSF: granulocyte colony-stimulating factor (CSF3)
GD: disialoganglioside
GDP: guanosine diphosphate
GDF: growth differentiation factor
GDF: (Rab)GDI displacement (dissociation) factor
GDI: guanine nucleotide-dissociation inhibitor
GDNF: glial cell line-derived neurotrophic factor
GEF: guanine nucleotide (GDP-to-GTP)-exchange factor
GF: growth factor
GFAP: glial fibrillary acidic protein (intermediate filament)
GFL: GDNF family of ligands
GFP: geodesic front propagation
GFR: growth factor receptor
GFRαi: type-i GDNF family receptor-α
GGA: Golgi-localized γ-adaptin ear-containing Arf-binding protein
Ggust: (G protein) $G\alpha$ subunit gustducin
GH: growth hormone
GHR: growth hormone receptor
GHRH: growth hormone-releasing hormone
GIP: GPCR-interacting protein

GIRK: Gβγ-regulated inwardly rectifying K$^+$ channel
GIT: GPCR kinase-interacting protein
GKAP: G-kinase-anchoring protein
GKAP: glucokinase-associated phosphatase (DuSP12)
GKAP: guanylate kinase-associated protein
GLK: GCK-like kinase
GluK: ionotropic glutamate receptor (kainate type)
GluN: ionotropic glutamate receptor (NMDA type)
GluR: ionotropic glutamate receptor (AMPA type)
GluT: glucose transporter
GlyCAM: glycosylation-dependent cell adhesion molecule
GlyR: glycine receptor (channel)
GlyT: glycine transporter
GM: monosialoganglioside
gmCSF: granulocyte–monocyte colony-stimulating factor (CSF2)
GMP: granulocyte–monocyte progenitor
GMP: guanosine monophosphate
GnRH: gonadotropin-releasing hormone
GP: glycoprotein
Gpc: glypican
GPI: glycosyl-phosphatidylinositol anchor
gpiAP: GPI-anchored protein
GPCR: G-protein-coupled receptor
GPx: glutathione peroxidase
GQ: quadrisialoganglioside
GR: glucocorticoid receptor (NR3c1)
Gr: Graetz number
GRAP: GRB2-related adaptor protein (or GAds)
GRB: growth factor receptor-bound protein
GRC: growth factor-regulated, Ca^{++}-permeable, cation channel (TRPV2)
GRE: glucocorticoid response element (DNA sequence)
GRK: G-protein-coupled receptor kinase
GRP: G-protein-coupled receptor phosphatase
GSH: reduced form of glutathione
GSK: glycogen synthase kinase
GSSG: oxidized form of glutathione (glutathione disulfide)
GT: trisialoganglioside
GTF: general transcription factor
GTP: guanosine triphosphate
GTPase: guanosine triphosphatase
GuCy: guanylate cyclase (CyG)
GWAS: genome-wide association study

H

H: height
$\mathcal{H}$: history function
H: dissipation
h: head loss
h: thickness
h: specific enthalpy
h_m: mass transfer coefficient
h_T: heat transfer coefficient
HA: hyaluronic acid
HAD: haloacid dehalogenase
HAP: huntingtin-associated protein
HAT: histone acetyltransferase
HAAT: heterodimeric amino acid transporter
HAND: heart and neural crest derivatives expressed protein
Hb: hemoglobin
Hb^{SNO}: ^{S}nitrosohemoglobin
HBEGF: heparin-binding EGF-like growth factor
HCK: hematopoietic cell kinase
HCLS: hematopoietic lineage cell-specific Lyn substrate protein
HCN: hyperpolarization-activated, cyclic nucleotide-gated K$^+$ channel
HCNP: hippocampal cholinergic neurostimulatory peptide
HCT: helical CT
HDAC: histone deacetylase complex
HDL: high-density lipoprotein
HDL–C: HDL–cholesterol
HDL–CE: HDL–cholesteryl ester
HDM: human double minute (Ub ligase)
HEET: hydroxyepoxyeicosatrienoic
hemin: heme oxygenase-1 inducer
HERG: human ether-a-go-go related gene
HER: human epidermal growth factor receptor (HER3: pseudokinase)
HES: Hairy enhancer of split
HETE: hydroxyeicosatrienoic acid
HEV: high endothelial venule
HGF: hepatocyte growth factor
HGFA: hepatocyte growth factor activator (serine peptidase)
HGFR: hepatocyte growth factor receptor
HGS: HGF-regulated Tyr kinase substrate (HRS)
HhIP: Hedgehog-interacting protein
HIF: hypoxia-inducible factor
HIP: huntingtin-interacting protein

HIP1R: HIP1-related protein
His: histamine
Hjv: hemojuvelin
HK: hexokinase
HL: hepatic lipase
HMG: high mobility group protein
HMGB: high mobility group box protein
HMGCoAR: hydroxy methyl glutaryl coenzyme-A reductase
HMT: histone methyl transferase
HMWK: high-molecular-weight kininogen
HNF: hepatocyte nuclear factor (NR2a1/2)
HNP: human neutrophil peptide
hnRNP: heterogeneous nuclear ribonucleoprotein
HODE: hydroxy octadecadienoic acid
HOP: HSP70–HSP90 complex-organizing protein
HoPS: homotypic fusion and vacuole protein sorting complex
HotAIR: HOX antisense intergenic RNA (large intergenic non-coding RNA)
HOx: heme oxygenase
Hox: homeobox DNA sequence (encodes homeodomain-containing morphogens)
HpCa: hippocalcin
HPK: hematopoietic progenitor kinase (MAP4K)
hpRNA: long hairpin RNA
hRas: Harvey Ras
HRE: hormone response element (DNA sequence)
HRM: hypoxia-regulated microRNA
hRNP: heterogeneous ribonucleoprotein
HRS: hepatocyte growth factor-regulated Tyr kinase substrate
HRT: Hairy and enhancer of Split-related transcription factor
HS: heparan sulfate
HSC: hematopoietic stem cell
HSC: heat shock cognate
HSER: heat stable enterotoxin receptor (guanylate cyclase 2C)
HSP: heat shock protein (chaperone)
HSPG: heparan sulfate proteoglycan
Ht: hematocrit
HTR: high temperature requirement endopeptidase
HUNK: hormonally up-regulated Neu-associated kinase

I

I: identity tensor
i: current
I: inotropy
IAP: inhibitor of apoptosis protein
IBABP: intestinal bile acid-binding protein
IC: isovolumetric contraction
ICA: internal carotid artery
ICAM: intercellular adhesion molecule (IgCAM member)
IgCAM: immunoglobulin-like cell adhesion molecule
ICF: intracellular fluid
ICliP: intramembrane-cleaving peptidase
ID: inhibitor of DNA binding
IDL: intermediate-density lipoprotein
IDmiR: immediately downregulated microRNA
IDOL: inducible degrader of LDL receptor (Ub ligase)
IEG: immediate-early gene
IEL: internal elastic lamina
IEL: intra-epithelial lymphocyte
IfIH: interferon-induced with helicase-C domain-containing protein
Ifn: interferon
IfnAR: interferon-$\alpha/\beta/\omega$ receptor
IFT: intraflagellar transport complex
Ig: immunoglobulin
IGF: insulin-like growth factor
IGFBP: IGF-binding protein
IgHC: immunoglobulin heavy chain
IgLC: immunoglobulin light chain
iGluR: ionotropic glutamate receptor
IH: intimal hyperplasia
IHh: indian Hedgehog
IK: intermediate-conductance Ca^{++}-activated K^+ channel
IκB: inhibitor of NFκB
IKK: IκB kinase
IL: interleukin
ILC: innate lymphoid cell
iLBP: intracellular lipid-binding protein
ILK: integrin-linked (pseudo)kinase
ILKAP: integrin-linked kinase-associated Ser/Thr phosphatase-2C
IMM: inner mitochondrial membrane
IMP: Impedes mitogenic signal propagation

List of Aliases

INAD1: inactivation no after-potential D protein
InCenP: inner centromere protein
InF: inverted formin
InsIG: insulin-induced gene product (ER anchor)
InsL: insulin-like peptide
InsR (IR): insulin receptor
InsRR: insulin receptor-related receptor
IP: inositol phosphate
IP_3: inositol (1,4,5)-trisphosphate
IP_3R: IP_3 receptor (IP_3-sensitive Ca^{++}-release channel)
IP_4: inositol (1,3,4,5)-tetrakisphosphate
IP_5: inositol pentakisphosphate
IP_6: inositol hexakisphosphate
IPCEF: interaction protein for cytohesin exchange factor
IPOD: (perivacuolar) insoluble protein deposit
IPP: inositol polyphosphate phosphatase
IPP: ILK–PINCH–parvin complex
iPSC: induced pluripotent stem cell
IQGAP: IQ motif-containing GTPase-activating protein (IQ: first 2 amino acids of the motif: isoleucine [I; commonly] and glutamine [Q; invariably]).
IR: isovolumetric relaxation
IRAK: IL1 receptor-associated kinase (IRAK2: pseudokinase)
IRE: irreversible electroporation
IRES: internal ribosome entry site
IRF: interferon-regulatory protein (transcription factor)
IRFF: interferon-regulatory factor family
IRP: iron regulatory protein
IRS: insulin receptor substrate
ISA: intracranial saccular aneurysm
ISG: interferon-stimulated gene product
iSMAD: inhibitory SMAD (SMAD6 or SMAD7)
ITAM: immunoreceptor tyrosine-based activation motif
Itch: Itchy homolog (Ub ligase)
Itg: integrin
ITIM: immunoreceptor tyrosine-based inhibitory motif
ITK: interleukin-2-inducible T-cell kinase
ITPK: inositol trisphosphate kinase
IVC: inferior vena cava
IVP: initial value problem
IVUS: intravascular ultrasound

J

J: flux
J_m: cell surface current density
JAM: junctional adhesion molecule
JaK: Janus (pseudo)kinase
JIP: JNK-interacting protein (MAPK8IP1 and -2)
JMy: junction-mediating and regulatory protein
JNK: Jun N-terminal kinase (MAPK8–MAPK10)
JNKBP: JNK-binding protein;
JNKK: JNK kinase
JSAP: JNK/SAPK-associated protein
Jun: avian sarcoma virus-17 proto-oncogene product (Japanese juunana: seventeen [17]; TF)
JUNQ: juxtanuclear quality-control compartment

K

K: conductivity tensor
K: bending stiffness
κ: reflection coefficient
K_d: dissociation constant (index of ligand–target affinity: ([L][T])/[C]; [L], [T], [C]: molar concentrations of the ligand, target, and created complex, respectively)
K_M: Michaelis constant (chemical reaction kinetics)
K_m: material compressibility
k: cross-section ellipticity
k_{ATP}: myosin ATPase rate
k_B: Boltzmann constant (1.38×10^{-23} J/K)
k_c: spring stiffness
k_m: mass-transfer coefficient
k_P: Planck constant
K_R: resistance coefficient
KaP: karyopherin
K_{ATP}: ATP-sensitive K^+ channel
$K_{Ca}1.x$: BK channel
$K_{Ca}2/3/4.x$: SK channel
$K_{Ca}5.x$: IK channel
K_{IR}: inwardly rectifying K^+ channel

K_V: voltage-gated K^+ channel
KAP: kinesin (KIF)-associated protein
Kap: karyopherin
KAT: lysine (K) acetyltransferase
KCC: K^+–Cl^- cotransporter
KChAP: K^+ channel-associated protein
KChIP: K_V channel-interacting protein
KDELR: KDEL (Lys–Asp–Glu–Leu) endoplasmic reticulum retention receptor
KDR: kinase insert domain receptor
KHC: kinesin heavy chain
KIF: kinesin family
KIR: killer cell immunoglobulin-like receptor
KIT: cellular kinase in tyrosine (SCFR)
Kk: kallikrein
KLC: kinesin light chain
KLF: Krüppel-like factor
KLR: killer cell lectin-like receptor
Kn: Knudsen number
KOR: κ-opioid receptor
kRas: Kirsten Ras
Krt: keratin
KSR: kinase suppressor of Ras (adaptor; pseudokinase)

L

L: velocity gradient tensor
L: inertance
L: length
L_e: entry length
LA: left atrium
LAB: linker of activated B lymphocyte
LAd: LCK-associated adaptor
LAMTOR: late endosomal and lysosomal adaptor, MAPK and TOR activator
LANP: long-acting natriuretic peptide
LAP: leucine-rich repeat and PDZ domain-containing protein (4-member family)
LAP: latency-associated peptide (4 isoforms LAP1–LAP4)
LAP: nuclear lamina-associated polypeptide
LAR: leukocyte common-antigen-related receptor (PTPRF)
LAT: linker of activated T lymphocytes
LaTS: large tumor suppressor
LAX: linker of activated X cells (both B and T cells)
LBR: lamin-B receptor
LCA: left coronary artery
LCAT: lysolecithin cholesterol acyltransferase
LCC: left coronary cusp

LCK: leukocyte-specific cytosolic (non-receptor) Tyr kinase
LCP: lymphocyte cytosolic protein (adaptor SLP76)
LDL: low-density lipoprotein
LDLR: low-density lipoprotein receptor
LDV: laser Doppler velocimetry
Le: entry length
LEF: lymphoid enhancer-binding transcription factor
LGalS: lectin, galactoside-binding, soluble cell adhesion molecule
LGIC: ligand-gated ion channel
LGL: lethal giant larva protein
LH: luteinizing hormone
LIF: leukemia-inhibitory factor
LIFR: leukemia-inhibitory factor receptor
LIMA: LIM domain and actin-binding protein
LIME: LCK-interacting molecule
LIMK: Lin1, Isl1, and Mec3 motif-containing kinase
LIMS: LIM and senescent cell antigen-like-containing domain protein
LiNC: linker of nucleoskeleton and cytoskeleton
lincRNA: large intergenic non-coding RNA
LipC: hepatic lipase
LipD: lipoprotein lipase
LipE: hormone-sensitive lipase
LipG: endothelial lipase
LipH: lipase-H
liprin: LAR PTP-interacting protein
LIR: leukocyte immunoglobulin-like receptor
LIS: lissencephaly protein
LKB: liver kinase-B
LKLF: lung Krüppel-like factor
LLTC: large latent TGFβ complex
LMan: lectin, mannose-binding
LMO: LIM domain-only-7 protein
Lmod: leiomodin (actin nucleator)
LMPP: lymphoid-primed multipotent progenitor
LMR: laser myocardial revascularization
Ln: laminin
LOx: lipoxygenase
LP: lipoprotein
LPA: lysophosphatidic acid
LPase: lipoprotein lipase
lpDC: lamina propria dendritic cell
Lphn: latrophilin (adhesion-GPCR)
LPL: lysophospholipid
LPLase: lysophospholipase
LPP: lipid phosphate phosphatase
LPR: lipid phosphatase-related protein

List of Aliases

LPS: lipopolysaccharide
LQTS: long-QT syndrome
LRAT: lecithin–retinol acyltransferase
LRH: liver receptor homolog (NR5a2)
LRO: lysosome-related organelle
LRP: LDL receptor-related protein
LRRTM: leucine-rich repeat-containing transmembrane protein
LSK: Lin−, SCA1+, KIT+ cell
LST: lethal with Sec-thirteen
LSV: long saphenous vein
LT (Lkt): leukotriene
LTBP: latent TGFβ-binding protein
LTCC: L-type Ca^{++} channel (Ca_V1)
LTK: leukocyte tyrosine kinase
LUbAC: linear ubiquitin chain assembly complex
LV: left ventricle
LVAD: left ventricular assist device
LX: lipoxin
LXR: liver X receptor (NR1h2/3)
LyVE: lymphatic vessel endothelial hyaluronan receptor

M

M: molar mass
M: metabolic rate
$\mathcal{M}$: moment
m: mass
Ma: Mach number
MACF: microtubule-actin crosslinking factor
mAChR: acetylcholine muscarinic receptor (metabotropic; GPCR)
MAD: mothers against decapentaplegic homolog
MAD: mitotic arrest-deficient protein
MAdCAM: mucosal vascular addressin cell adhesion molecule
MAF: musculoaponeurotic fibrosarcoma oncogene homolog (TF)
MAGI: membrane-associated guanylate kinase-related protein with inverted domain organization
MAGP: microfibril-associated glycoprotein
MAGuK: membrane-associated guanylate kinase
MAIT: mucosal-associated invariant T lymphocyte
MALT: mucosa-associated lymphoid tissue
MALT1: mucosa-associated lymphoid tissue lymphoma translocation peptidase

MAO: monoamine oxidase
MAP: microtubule-associated protein
MAP1LC3: microtubule-associated protein-1 light chain-3 (LC3)
mAP: mean arterial pressure
MAPK: mitogen-activated protein kinase
MAP2K: MAPK kinase
MAP3K: MA2KP kinase
MAP3K7IP: MAP3K7-interacting protein
MAPKAPK: MAPK-activated protein kinase
MARCKS: myristoylated alanine-rich C kinase substrate
MaRCo: macrophage receptor with collagenous structure (ScaRa2)
MARK: microtubule affinity-regulating kinase
MASTL: microtubule-associated Ser/Thr kinase-like protein
MAT: ménage à trois
MATK: megakaryocyte-associated Tyr kinase
MAVS: mitochondrial antiviral signaling protein
MAX: MyC-associated factor-X (bHLHd4–bHLHd8)
MBP: myosin-binding protein
MBP: myeloid–B-cell progenitor
MBTPSi: membrane-bound transcription factor peptidase site i
MCAK: mitotic centromere-associated kinesin
MCAM: melanoma cell adhesion molecule
MCD: medullary collecting duct
MCL1: BCL2-related myeloid cell leukemia sequence protein-1
MCLC: stretch-gated Mid1-related chloride channel
MCM: minichromosome maintenance protein
MCP: monocyte chemoattractant protein
mCSF: macrophage colony-stimulating factor (CSF1)
MCT: monocarboxylate–proton cotransporter
mDC: myeloid dendritic cell
MDM: mitochondrial distribution and morphology protein
MDR: multiple drug resistance (ABC transporter)
MEF: myocyte enhancer factor
megCSF: megakaryocyte colony-stimulating factor
MEJ: myoendothelial junction
MELK: maternal embryonic leucine zipper kinase
MEP: megakaryocyte erythroid progenitor
MEP: myeloid–erythroid progenitor
MET: mesenchymal–epithelial transition factor (proto-oncogene; HGFR)

METC: mitochondrial electron transport chain
metHb: methemoglobin
MGIC: mechanogated ion channel
mGluR: metabotropic glutamate receptor
MGP: matrix Gla protein
MHC: major histocompatibility complex
MHC: myosin heavy chain
MyHC or MYH: myosin heavy chain gene
miCK: mitochondrial creatine kinase
MiCU: mitochondrial calcium uptake protein
Mid: midline
MinK: misshapen-like kinase
miR: microRNA
miRNP: microribonucleoprotein
MiRP: MinK-related peptide
MIRR: multichain immune-recognition receptor
MIS: Müllerian inhibiting substance
MIS: mini-invasive surgery
MIS: mitochondrial intermembrane space
MIST: mastocyte immunoreceptor signal transducer
MIT: mini-invasive therapy
MiV: mitral valve
MIZ: Myc-interacting zinc finger protein
MJD: Machado-Joseph disease protein domain-containing peptidase (DUb)
MKL: megakaryoblastic leukemia-1 fusion coactivator
MKnK: MAPK-interacting protein Ser/Thr kinase (MnK)
MKP: mitogen-activated protein kinase phosphatase
MLC: myosin light chain
MLCK: myosin light-chain kinase
MLCP: myosin light-chain phosphatase
MLK: mixed lineage kinase
MLKL: mixed lineage kinase-like pseudokinase
MLL: mixed lineage [myeloid–lymphoid] leukemia factor
MLLT: mixed lineage leukemia translocated protein
MLP: muscle LIM protein
mmCK: myofibrillar creatine kinase
MME: membrane metalloendopeptidase
MMM: maintenance of mitochondrial morphology protein
MMP: matrix metallopeptidase
MO: mouse protein
Mo: monocyte
MOMP: mitochondrial outer membrane permeabilization

MOR: μ-opioid receptor
MP: MAPK partner
MPF: mitosis (maturation)-promoting factor (CcnB–CDK1 complex)
MPG: N-methylpurine (N-methyladenine)-DNA glycosylase
MPO: median preoptic nucleus
Mpo: myeloperoxidase
MP_P: membrane protein, palmitoylated
MPP: multipotent progenitor
MR: mineralocorticoid receptor (NR3c2)
mRas: muscle Ras (or rRas3)
MRCK: myotonic dystrophy kinase-related CDC42-binding kinase
MRI: (nuclear) magnetic resonance imaging
mRNA: messenger RNA
mRNP: messenger ribonucleoprotein
MRTF: myocardin-related transcription factor
MSC: mesenchymal stem cell
MSH: melanocyte-stimulating hormone
MSIC: mechanosensitive ion channel
MSSCT: multi-slice spiral CT
MST: mammalian sterile-twenty-like kinase
MSt1R: macrophage-stimulating-1 factor receptor (RON)
MT: metallothionein
MTM: myotubularin (myotubular myopathy-associated gene product)
mtMMP: membrane-type MMP (mt$i$MMP: type-i mtMMP)
MTMR: myotubularin-related phosphatase
MTOC: microtubule organizing center
MTP: myeloid–T-cell progenitor
MTP: microsomal triglyceride transfer protein
MuRF: muscle-specific RING finger (Ub ligase)
MuSK: muscle-specific kinase
MVB: multivesicular body
MVE: multivesicular endosome (MVB)
MVO2: myocardial oxygen consumption
MWSS: maximal wall shear stress
MyB: myeloblastosis viral oncogene homolog (TF)
MyC: myelocytomatosis viral oncogene homolog (TF)
MyD88: myeloid differentiation primary response gene product-88
MyHC: myosin heavy chain
MyLC or MYL: myosin light-chain gene
MyPT: myosin phosphatase targeting subunit
MyT: myelin transcription factor

N

N: sarcomere number
$\hat{n}$: unit normal vector
n: mole number
n: PAM density with elongation x
n: myosin head density
$\mathcal{N}_A$: Avogadro number
N-terminus: amino (amine group NH_2)-terminus
NAADP: nicotinic acid adenine dinucleotide phosphate
nAChR: acetylcholine nicotinic receptor (ionotropic; LGIC)
NAD: nicotine adenine dinucleotide
NADPH: reduced form of nicotinamide adenine dinucleotide phosphate
NAd: noradrenaline
NAF: nutrient-deprivation autophagy factor
NALT: nasal-associated lymphoid tissu
NAmPT: nicotinamide phosphoribosyltransferase
Nanog: ever young (Gaelic)
NAP: NCK-associated protein (NCKAP)
NAT: nucleobase–ascorbate transporter
NAT1: noradrenaline transporter
Na$_V$ voltage-gated Na$^+$ channel
NBC: Na$^+$–HCO$_3^-$ cotransporters
NCC: non-coronary cusp
NCC: Na$^+$–Cl$^-$ cotransporter
Ncdn: neurochondrin
NCK: non-catalytic region of Tyr kinase adaptor
NCoA: nuclear receptor coactivator
NCoR: nuclear receptor corepressor
NCR: natural cytotoxicity-triggering receptor
ncRNA: non-coding RNA
NCS: neuronal calcium sensor
NCKX: Na$^+$–Ca^{++}–K$^+$ exchanger
NCLX: Na$^+$–Ca^{++}–Li$^+$ exchanger
NCX: Na$^+$–Ca^{++} exchanger
NDCBE: Na$^+$-dependent Cl$^-$–HCO$_3^-$ exchanger
NecL: nectin-like molecule
NEDD: neural precursor cell expressed, developmentally downregulated
NDFIP: NEDD4 family-interacting protein
NeK: never in mitosis gene-A (NIMA)-related kinase
NES: nuclear export signal
NESK: NIK-like embryo-specific kinase
nesprin: nuclear envelope spectrin repeat protein

NeuroD: neurogenic differentiation protein
NF: neurofilament protein (intermediate filament)
NF: neurofibromin (RasGAP)
NFAT: nuclear factor of activated T cells
NFe2: erythroid-derived nuclear factor-2
NFH: neurofilament, heavy polypeptide
NFκB: nuclear factor κ light chain enhancer of activated B cells
NFL: neurofilament, light polypeptide
NFM: neurofilament, medium polypeptide
NGAL: neutrophil gelatinase-associated lipocalin
NGF: nerve growth factor
Ngn: neogenin (netrin receptor)
NHA: Na$^+$–H$^+$ antiporter
NHE: Na$^+$–H$^+$ exchanger
NHERF: NHE regulatory factor
NHR: nuclear hormone receptor
NIc: nucleoporin-interacting protein
NIK: NFκB-inducing kinase
NIK: NCK-interacting kinase
NIP: neointimal proliferation
NK: natural killer cell
NKCC: Na$^+$–Ka$^+$–2Cl$^-$ cotransporter
NKG: NK receptor group
NKT: natural killer T cell
NKx2: NK2 transcription factor-related homeobox protein
NLR: NOD-like receptor (nucleotide-binding oligomerization domain, Leu-rich repeat-containing)
NLS: nuclear localization signal
NMDAR: ^{N}methyl ^{D}aspartate receptor
NmU: neuromedin-U
NO: nitric oxide (nitrogen monoxide)
NOD: nucleotide-binding oligomerization domain
NonO: non-POU domain-containing octamer-binding protein
NOR: neuron-derived orphan receptor (NR4a3)
NOS: nitric oxide synthase
NOS1: neuronal NOS
NOS1AP: NOS1 adaptor protein
NOS2: inducible NOS
NOS3: endothelial NOS
NOx: NAD(P)H oxidase
Noxa: damage (Latin)
NPAS: neuronal PAS domain-containing transcription factor
NPC: nuclear-pore complex
NPC: Niemann-Pick disease type-C protein
NPC1L: Niemann-Pick protein-C1-like
nPKC: novel protein kinase C

NPY: neuropeptide Y
NR: nuclear receptor
NRAP: nebulin-related actinin-binding protein
nRas: neuroblastoma Ras
NRBP: nuclear receptor-binding protein
NRF: nuclear factor erythroid-derived-2 (NF-E2)-related factor
NRF1: nuclear respiratory factor-1
Nrg: neuregulin (EGF superfamily member)
Nrgn: neuroligin
Nrp: neuropilin (VEGF-binding molecule; VEGFR coreceptor)
NRPTP: non-receptor protein Tyr phosphatase
NRSTK: non-receptor Ser/Thr kinase
NRTK: non-receptor Tyr kinase
NRx: nucleoredoxin
Nrxn: neurexin
NSCLC: non-small-cell lung cancer
NSF: N-ethylmaleimide-sensitive factor
NSLTP: non-specific lipid-transfer protein
NST: nucleus of the solitary tract
NT: neurotrophin
NT5E: ecto-5′-nucleotidase
NTCP: sodium–taurocholate cotransporter polypeptide
NTF: N-terminal fragment
NTP: nucleoside triphosphate
NTPase: nucleoside triphosphate hydrolase superfamily member
NTRK: neurotrophic tyrosine receptor kinase (TRK)
NTRKR: neurotrophic Tyr receptor kinase-related protein ($ROR_{(RTK)}$)
NTS: nucleus tractus solitarius
Nu: Nusselt number
NuAK: nuclear AMPK-related kinase
NuP: nucleoporin (nuclear-pore complex protein)
NuRD: nucleosome remodeling and histone deacetylase
NuRR: nuclear receptor-related factor (NR4a2)
nWASP: neuronal WASP

O

$^{O}Glc^{N}Ac$: β^{N}acetyl ^{D}glucosamine
OCRL: oculocerebrorenal syndrome of Lowe phosphatase
Oct: octamer-binding transcription factor
ODE: ordinary differential equation
OGA: $^{O}Glc^{N}$Acase (β^{N}acetylglucosaminidase)
OMCD: outer medullary collecting duct

OMM: outer mitochondrial membrane
ORC: origin recognition complex
ORF: open reading frame
ORP: OSBP-related protein
OSA: obstructive sleep apnea
OSBP: oxysterol-binding protein
OSI: oscillatory shear index
OSM: oncostatin M
OSMR: oncostatin M receptor
OSR (OxSR): oxidative stress-responsive kinase
OTK: off-track (pseudo)kinase
OTU: ovarian tumor superfamily peptidase (deubiquitinase)
OTUB: otubain (Ub thioesterase of the OTU superfamily)
OVLT: organum vasculosum lamina terminalis
oxyHb: oxyhemoglobin (oxygenated hemoglobin)

P

$\mathcal{P}$: permeability
P: power
P: cell division rate
p: production rate
p: pressure
p_i: partial pressure of gas component i
PA: phosphatidic acid
PAAT: proton–amino acid transporter
PACS: phosphofurin acidic cluster sorting protein
PAF: platelet-activating factor
PAFAH: platelet-activating factor acetylhydrolase
PAG: phosphoprotein associated with glycosphingolipid-enriched microdomains
PAH: polycyclic aromatic hydrocarbon
PAH: pulmonary arterial hypertension
PAI: plasminogen activator inhibitor
PAK: P21-activated kinase
PALR: promoter-associated long RNA
PALS: protein associated with Lin-7
PAMP: pathogen-associated molecular pattern
PAMP: proadrenomedullin peptide
PAR: poly^{ADP}ribose
PAR: promoter-associated, non-coding RNA
PARi: type-i peptidase-activated receptor
Par: partitioning defective protein
PARG: polyADPribosyl glycosidase
PARP: polyADPribose polymerase

List of Aliases

PASR: promoter-associated short RNA
PATJ: protein (PALS1) associated to tight junctions
Pax: paxillin
Pax$i$: paired box protein-i (transcription regulator)
PBC: pre-Bötzinger complex (ventilation frequency)
PBIP: Polo box-interacting protein
PC: phosphatidylcholine
PC: polycystin
PC: protein C
PCMRV: phase-contrast MR velocimetry
PCr: phosphocreatine
PCT: proximal convoluted tubule
PCTP: phosphatidylcholine-transfer protein
PD: pharmacodynamics
pDC: plasmacytoid dendritic cell
PdCD: programmed cell death protein
PdCD6IP: PdCD 6-interacting protein
PdCD1Lg: programmed cell death-1 ligand
PDE: phosphodiesterase
PDE: partial differential equation
PDGF: platelet-derived growth factor
PDGFR: platelet-derived growth factor receptor
PDI: protein disulfide isomerase
PDK: phosphoinositide-dependent kinase
PDP: pyruvate dehydrogenase phosphatase
Pe: Péclet number
PE: phosphatidylethanolamine
PE: pulmonary embolism
PEBP: phosphatidylethanolamine-binding protein
PECAM: platelet–endothelial cell adhesion molecule
PEDF: pigment epithelium-derived factor (serpin F1)
PEn2: presenilin enhancer-2
PEO: proepicardial organ
Per: Period homolog
PERK: protein kinase-like endoplasmic reticulum kinase
PERP: P53 apoptosis effector related to peripheral myelin protein PMP22
PET: positron emission tomography
Pex: peroxin
PF: platelet factor
PFK: phosphofructokinase
pFRG: parafacial respiratory group
PG: prostaglandin
PGC: PPARγ coactivator
pGC: particulate guanylate cyclase
PGEA: prostaglandin ethanolamide

PGF: paracrine growth factor
PGG: prostaglandin glycerol ester
PGi2: prostacyclin
PGP: permeability glycoprotein
PGx: type-x (D, E, F, H, I) prostaglandin
PGxS: type-x prostaglandin synthase
PH: pleckstrin homology domain
PHD: prolyl hydroxylase
PhK: phosphorylase kinase
PHLPP: PH domain and Leu-rich repeat protein phosphatase
PI: phosphoinositide (phosphorylated phosphatidylinositol)
PI(4)P: phosphatidylinositol 4-phosphate
PI(i)PiK: phosphatidylinositol i-phosphate i-kinase
PI(i,j)P$_2$: phosphatidylinositol (i,j)-bisphosphate (PIP$_2$)
PI(3,4,5)P$_3$: phosphatidylinositol (3,4,5)-trisphosphate (PIP$_3$)
PI3K: phosphatidylinositol 3-kinase
PI3KAP: PI3K adaptor protein
PIiK: phosphatidylinositol i-kinase
PIAS: protein inhibitor of activated STAT (SUMo ligase)
PIC: pre-initiation complex
PICK: protein that interacts with C-kinase
PIDD: P53-induced protein with a death domain
PIKE: phosphoinositide 3-kinase enhancer (GTPase; ArfGAP)
PIKK: phosphatidylinositol 3-kinase-related kinase (pseudokinase)
PIM: provirus insertion of Molony murine leukemia virus gene product
PIN: peptidyl prolyl isomerase interacting with NIMA
PINCH: particularly interesting new Cys–His protein (or LIMS1)
PInK: PTen-induced kinase
PIP: phosphoinositide monophosphate
PIPiK: phosphatidylinositol phosphate i-kinase
PIP$_2$: phosphatidylinositol bisphosphate
PIP$_3$: phosphatidylinositol triphosphate
PIPP: proline-rich inositol polyphosphate 5-phosphatase
PIR: paired immunoglobulin-like receptor
piRNA: P-element-induced wimpy testis-interacting (PIWI) RNA
PIRT: phosphoinositide-interacting regulator of TRP channels
PITP: phosphatidylinositol-transfer protein
Pitx: pituitary (or paired-like) homeobox transcription factor

PIV: particle image velocimetry
PIX: P21-activated kinase (PAK)-interacting exchange factor (RhoGEF6/7)
PK: pharmacokinetics
PK: protein kinase
PKA: protein kinase A
PKB: protein kinase B
PKC: protein kinase C
PKD: protein kinase D
PKG: protein kinase G
PKL: paxillin kinase linker
PKM: pyruvate kinase muscle isozyme
PKMYT (MYT): membrane-associated Tyr/Thr protein kinase
PKN: protein kinase novel
Pkp: plakophilin
PL: phospholipase
PLA2: phospholipase A2
PLC: phospholipase C
PLD: phospholipase D
PLb: phospholamban
PLd: phospholipid
PlGF: placental growth factor
PLK: Polo-like kinase
PLTP: phospholipid transfer protein
Plxn: plexin
PMCA: plasma membrane Ca^{++} ATPase
PML: promyelocytic leukemia protein
PMR: percutaneous (laser) myocardial revascularization
PMRT: protein arginine methyltransferase
Pn: plasmin
Png: plasminogen
PNS: peripheral nervous system
PoG: proteoglycan
PoM: pore membrane protein
Pon: paraoxonase
POPx: partner of PIX
POSH: scaffold plenty of SH3 domains
POT: Protection of telomeres (single-stranded telomeric DNA-binding protein)
PP: protein phosphatase
PP3: protein phosphatase 3 (PP2b or calcineurin)
PPAR: peroxisome proliferator-activated receptor (NR1c1–3)
PPG: photoplethysmography
PPId: peptidyl prolyl isomerase-D
PPIP: monopyrophosphorylated inositol phosphate
$(PP)_2IP$: bisphosphorylated inositol phosphate
PPK: PIP kinase

PPM: protein phosphatase (magnesium-dependent)
PPR: pathogen-recognition receptor
PPRE: PPAR response element (DNA sequence)
PPTC: protein phosphatase T-cell activation (TAPP2c)
PR: progesterone receptor (NR3c3)
Pr: Prandtl number
PRC: protein regulator of cytokinesis
PRC: Polycomb repressive complex
Prdx: peroxiredoxin
pre-cDC: pre-classical dendritic cell
pre-miR: precursor microRNA
preBotC: pre-Bötzinger complex
preKk: prekallikrein
PREx: PIP_3-dependent Rac exchanger (RacGEF)
PRG: plasticity-related gene product
PRH: prolactin-releasing hormone
pri-miR: primary microRNA
PRL: phosphatase of regenerating liver
Prl: prolactin
PrlR: prolactin receptor
PRMT: protein arginine (R) N-methyltransferase
Prompt: promoter upstream transcript
Protor: protein observed with Rictor
PROX: prospero homeobox gene
Prox: PROX gene product (transcription factor)
PrP: processing protein
PRPK: P53-related protein kinase
PRR: pattern recognition receptor
PRR: prorenin and renin receptor
PS: presenilin
PS: protein S
PSC: pluripotent stem cell
PSD: postsynaptic density adaptor
PsD: postsynaptic density
PSEF: pseudo-strain energy function
PSer: phosphatidylserine
PSGL: P-selectin glycoprotein ligand
PSKh: protein serine kinase H
Psm: proteasome subunit
PSTPIP: Pro–Ser–Thr phosphatase-interacting protein
PTA: plasma thromboplastin antecedent
Ptc: Patched receptor (Hedgehog signaling)
PTCA: percutaneous transluminal coronary angioplasty
PtcH: Patched Hedgehog receptor
PTCRA: PTC rotational burr atherectomy
PtdCho (PC): phosphatidylcholine
PtdEtn (PE): phosphatidylethanolamine

List of Aliases

PtdSer (PS): phosphatidylserine
PtdIns (PI): phosphatidylinositol
PTen: phosphatase and tensin homolog deleted on chromosome ten (phosphatidylinositol 3-phosphatase)
PTFE: polytetrafluoroethylene
PTH: parathyroid hormone
PTHRP: parathyroid hormone-related protein
PTK: protein Tyr kinase
PTK7: pseudokinase (RTK)
PTP: protein Tyr phosphatase
PTP$n i$: protein Tyr phosphatase non-receptor type i
PTPR: protein Tyr phosphatase receptor
PTRF: RNA polymerase-1 and transcript release factor
PUFA: polyunsaturated fatty acid
PUMA: P53-upregulated modulator of apoptosis
PuV: pulmonary valve
PVF: PDGF- and VEGF-related factor
PVNH: paraventricular nucleus of hypothalamus
PVR: pulmonary vascular resistance
PWS: pulse wave speed
Px: pannexin
PXR: pregnane X receptor (NR1i2)
PYK: proline-rich tyrosine kinase
P2X: purinergic ligand-gated channel
P53AIP: P53-regulated apoptosis-inducing protein
p75NtR: pan-neurotrophin receptor

Q

Q: material quantity
Q: thermal energy
Q_c: electric current density
Q_T: thermal energy (heat)
q_T: transfer rate of thermal energy (power)
q_{met}: metabolic heat source
q: flow rate

R

R: resistance
$\mathcal{R}$: local reaction term
R_h: hydraulic radius
R_g: gas constant
R_R: respiratory quotient

R: recruitment function (from quiescence to proliferation)
r: cell renewal rate
r: radial coordinate
RA: right atrium
RAAS: renin–angiotensin–aldosterone system
Rab: Ras from brain
Rab11FIP: Rab11 family-interacting protein
Rac: Ras-related C3-botulinum toxin substrate
RACC: receptor-activated cation channel
RACK: receptor for activated C-kinase
RAD: recombination protein-A (RecA)-homolog DNA-repair protein
Rad: radiation sensitivity protein
Rag: Ras-related GTP-binding protein
Ral: Ras-related protein
RAlBP: retinaldehyde-binding protein
RalGDS: Ral guanine nucleotide-dissociation stimulator
RAMP: (calcitonin receptor-like) receptor-activity-modifying protein
Ran: Ras-related nuclear protein
RANTES: regulated upon activation, normal T-cell expressed, and secreted product (CCL5)
RAP: receptor-associated protein
Rap: Ras-proximate (Ras-related) protein
Raptor: regulatory associated protein of TOR
RAR: retinoic acid receptor (NR1b2/3)
Ras: rat sarcoma viral oncogene homolog (small GTPase)
RasA: Ras p21 protein activator
rasiRNA: repeat-associated small interfering RNA (PIWI)
RASSF: Ras interaction/interference protein RIN1, afadin, and Ras association domain-containing protein family member
RB: retinoblastoma protein
RBC: red blood cell (erythrocyte)
RBP: retinoid-binding protein
RC: ryanodine calcium channel (RyR)
RCA: right coronary artery
RCan: regulator of calcineurin
RCC: right coronary cusp
RCC: regulator of chromosome condensation
Re: Reynolds number
REDD: regulated in development and DNA-damage response gene product
Rel: reticuloendotheliosis proto-oncogene product (TF; member of NFκB)
REP: Rab escort protein
ReR: renin receptor (PRR)

restin: Reed-Steinberg cell-expressed intermediate filament-associated protein (CLiP1)
ReT: rearranged during transfection (receptor Tyr kinase)
RevRE: reverse (Rev)-ErbA (NR1d1/2) response element (DNA sequence)
RFA: radiofrequency ablation
RGL: Ral guanine nucleotide dissociation stimulator-like protein (GEF)
RGS: regulator of G-protein signaling
RHEB: Ras homolog enriched in brain
RHS: equation right-hand side
Rho: Ras homologous
RIAM: Rap1-GTP-interacting adaptor
RIBP: RLK- and ITK-binding protein
RICH: RhoGAP interacting with CIP4 homolog
RICK: receptor for inactive C-kinase
Rictor: rapamycin-insensitive companion of TOR
RIF: Rho in filopodium
RIN: Ras-like protein expressed in neurons (GTPase)
RIn: Ras and Rab interactor (RabGEF)
RIP: regulated intramembrane proteolysis
RIPK: receptor-interacting protein kinase
RISC: RNA-induced silencing complex
RIT: Ras-like protein expressed in many tissues
RKIP: Raf kinase inhibitor protein
RlBP: retinaldehyde-binding protein
RLC: RISC-loading complex
RLK: resting lymphocyte kinase (TXK)
RNA: ribonucleic acid
RNABP: RNA-binding protein
RNase: ribonuclease
RnBP: renin-binding protein
RNF2: RING finger protein-2 (Ub ligase)
RNP: ribonucleoprotein
Robo: roundabout
ROC: receptor-operated channel
RoCK: Rho-associated, coiled-coil-containing protein kinase
ROI: region of interest
ROMK: renal outer medullary potassium channel
ROR: RAR-related orphan receptor (NR1f1–NR1f3)
ROR$_{(RTK)}$: receptor Tyr kinase-like orphan receptor
ROS: reactive oxygen species
Ros: ros UR2 sarcoma virus proto-oncogene product (RTK)

RPIP: Rap2-interacting protein
RPS6: ribosomal protein S6
RPTP: receptor protein Tyr phosphatase
rRas: related Ras
rRNA: ribosomal RNA
RSA: respiratory sinus arrhythmia
RSE: rapid systolic ejection
RSK: P90 ribosomal S6 kinase (^{P90}RSK)
RSKL: ribosomal protein S6 kinase-like (pseudokinase)
rSMAD: receptor-regulated SMAD (SMAD1–SMAD3, SMAD5, and SMAD9)
RSMCS: robot-supported medical and surgical system
RSpo: R-spondin
RSTK: receptor Ser/Thr kinase
RTK: receptor Tyr kinase
RTN: retrotrapezoid nucleus
Rubicon: RUN domain and Cys-rich domain-containing, beclin-1-interacting protein
Runx: Runt-related transcription factor
RV: right ventricle
RVF: rapid ventricular filling
RVLM: rostral ventrolateral medulla
RVMM: rostral ventromedial medulla
RXR: retinoid X receptor (NR2b1–NR2b3)
RYK: receptor-like Tyr (Y) kinase (pseudokinase)
RyR: ryanodine receptor (ryanodine-sensitive Ca^{++}-release channel)

S

S: Cauchy-Green deformation tensor
s: entropy
s: sarcomere length
s: evolution speed
SAA: serum amyloid A
$SAC_{Cl(K)}$: stretch-activated Cl^- (K^+)-selective channel
SAc: suppressor of actin domain-containing 5-phosphatase
sAC: soluble adenylate cyclase
$SACC_{NS}$: stretch-activated cation non-selective channel
SACM1L: suppressor of actin mutation-1-like
SAH: subarachnoid hemorrhage
SAIC: stretch-activated ion channel
SAN: sinoatrial node
SAP: SLAM-associated protein
SAP: stress-activated protein

List of Aliases

SAP*i*: synapse-associated protein *i*
SAPK: stress-activated protein kinase (MAPK)
SAR: secretion-associated and Ras-related protein
SBE: SMAD-binding element
SBF: SET-binding factor
Sc: Schmidt number
SCA: stem cell antigen
SCAMP: secretory carrier membrane protein
SCAP: SREBP cleavage-activating protein (SREBP escort)
SCAR: suppressor of cAMP receptor (WAVe)
ScaR: scavenger receptor
SCF: SKP1–Cul1–F-box Ub-ligase complex
SCF: stem cell factor
SCFR: stem cell factor receptor (KIT)
Scgb: secretoglobin
SCLC: small-cell lung cancer
scLC: squamous-cell lung cancer (NSCLC subtype)
SCN: suprachiasmatic nucleus
SCO: synthesis of cytochrome-C oxidase
SCP (CTDSP): small C-terminal domain (CTD)-containing phosphatase
Scp: stresscopin (urocortin 3)
Scrib: Scribble polarity protein
Sdc: syndecan
SDF: stromal cell-derived factor
SDPR: serum deprivation protein response
SE: systolic ejection
SEF: strain-energy function
SEF: similar expression to FGF genes (inhibitor of RTK signaling)
SEK: SAPK/ERK kinase
Sema: semaphorin (Sema, Ig, transmembrane, and short cytoplasmic domain)
SERCA: sarco(endo)plasmic reticulum calcium ATPase
serpin: serine peptidase inhibitor
SerT: serotonin transporter
SF: steroidogenic factor (NR5a1)
SFK: SRC-family kinase
SFO: subfornical organ
SFPQ: splicing factor proline and glutamine-rich
sFRP: secreted Frizzled-related protein
SftP (SP): surfactant protein
sGC: soluble guanylate cyclase
SGK: serum- and glucocorticoid-regulated kinase
SGlT: Na$^+$–glucose cotransporter (SLC5a)
Sgo: shugoshin (Japanese: guardian spirit)
SH: Src homology domain
Sh: Sherwood number

SH3P: Src homology-3 domain-containing adaptor protein
SHAnk: SH3 and multiple ankyrin repeat domain-containing protein
SHAX: SNF7 (VSP32) homolog associated with ALIX
SHB: Src homology-2 domain-containing adaptor
SHC: Src-homologous and collagen-like substrate
SHC: Src homology-2 domain-containing transforming protein
SHh: sonic Hedgehog
SHIP: SH-containing inositol phosphatase
SHP: SH-containing protein Tyr phosphatase (PTPn6/11)
SHP: small heterodimer partner (NR0b2)
shRNA: small (short) hairpin RNA
SIAH: Seven in absentia homolog (Ub ligase)
siglec: sialic acid-binding Ig-like lectin
SIK: salt-inducible kinase
SIn: stress-activated protein kinase-interacting protein
SIP: steroid receptor coactivator-interacting protein
siRNA: small interfering RNA
SiRP: signal-regulatory protein
SIRT: sirtuin (silent information regulator-2 [two]; histone deacetylase)
SIT: SHP2-interacting transmembrane adaptor
SK: small conductance Ca^{++}-activated K$^+$ channel
SK*i*: sphingosine kinase-*i*
SKIP: sphingosine kinase-1-interacting protein
SKIP: skeletal muscle and kidney-enriched inositol phosphatase
SKP: S-phase kinase-associated protein
SLA: Src-like adaptor
SLAM: signaling lymphocytic activation molecule
SLAMF: SLAM family member
SLAP: Src-like adaptor protein
SLC: solute carrier class member
SLCO: solute carrier organic anion class transporter
SLK: Ste20-like kinase
Sln: sarcolipin
SLPI: secretory leukocyte peptidase inhibitor
SLTC: small latent TGFβ complex
SM: sphingomyelin
SMA: smooth muscle actin
SMAD: small (son of, similar to) mothers against decapentaplegia homolog

SMAP: Small ArfGAP protein, stromal membrane-associated GTPase-activating protein
SMase: sphingomyelinase
SMC: smooth muscle cell
Smo: Smoothened
SMPD: sphingomyelin phosphodiesterase
SMRT: silencing mediator of retinoic acid and thyroid hormone receptor (NCoR2)
SMS: sphingomyelin synthase
SMURF: SMAD ubiquitination regulatory factor
SNAAT: sodium-coupled neutral amino acid transporter
SNAP: soluble N-ethylmaleimide-sensitive factor-attachment protein
SnAP: synaptosomal-associated protein
SNARE: SNAP receptor
SNF7: sucrose non-fermenting (VPS32)
SNIP: SMAD nuclear-interacting protein
snoRNA: small nucleolar RNA
snoRNP: small nucleolar ribonucleoprotein
SNP: single-nucleotide polymorphism
snRNA: small nuclear RNA
snRNP: small nuclear ribonucleoprotein
SNx: sorting nexin
SOC: store-operated Ca^{++} channel
SOCE: store-operated Ca^{++} entry
SOCS: suppressor of cytokine signaling protein
SOD: superoxide dismutase
SorbS: sorbin and SH3 domain-containing adaptor
SOS: Son of sevenless (GEF)
Sost: sclerostin
SostDC: sclerostin domain-containing protein
SOX: sex-determining region Y (SRY)-box gene
Sox: SOX gene product (transcription factor)
SP1: specificity protein (transcription factor)
SPARC: secreted protein acidic and rich in cysteine
SPC: sphingosylphosphorylcholine
SPCA: secretory pathway Ca^{++} ATPase
SPECT: single photon emission CT
Sph: sphingosine
SphK: sphingosine kinase
SPI: spleen focus-forming virus (SFFV) proviral integration proto-oncogene product (transcription factor)
SPInt: serine peptidase inhibitor
SPN: supernormal period
SPP: sphingosine phosphate phosphatase

SpRED: Sprouty-related protein with an EVH1 domain
SPURT: secretory protein in upper respiratory tract
SQTS: short-QT syndrome
SR: sarcoplasmic reticulum
SR: Arg/Ser domain-containing protein (alternative splicing)
SRA: steroid receptor RNA activator
SRC: steroid receptor coactivator
Src: sarcoma-associated (Schmidt-Ruppin A2 viral oncogene homolog) kinase
SREBP: sterol regulatory element-binding protein
SRF: serum response factor
SRM/SMRS: Src-related kinase lacking regulatory and myristylation sites
SRP: stresscopin-related peptide (urocortin 2)
SRPK: splicing factor RS domain-containing protein kinase
SRY: sex-determining region Y
SSAC: shear stress-activated channel
SSE: slow systolic ejection
Ssh: slingshot homolog phosphatase
SSI: STAT-induced STAT inhibitor
ssRNA: single-stranded RNA
Sst: somatostatin
SSV: short saphenous vein
St: Strouhal number
STAM: signal-transducing adaptor molecule
STAMBP: STAM-binding protein (Ub isopeptidase)
StAR: steroidogenic acute regulatory protein
StART: StAR-related lipid transfer protein
STAT: signal transducer and activator of transduction
STEAP: six transmembrane epithelial antigen of the prostate
STICK: substrate that interacts with C-kinase
StIM: stromal interaction molecule
STK: protein Ser/Thr kinase
STK1: stem cell protein Tyr kinase receptor
STLK: Ser/Thr kinase-like (pseudo)kinase
Sto: Stokes number
StRAd: STe20-related adaptor
STRAP: Ser/Thr kinase receptor-associated protein
StRAP: stress-responsive activator of P300
Stx: syntaxin ($SNARE^Q$)
SUMo: small ubiquitin-related modifier
SUn: Sad1 and Unc84 homology protein
SUR: sulfonylurea receptor
SUT: stable unannotated transcript
SV: stroke volume

List of Aliases 1039

SVC: superior vena cava
SVCT: sodium-dependent vitamin-C transporter
SVF: slow ventricular filling
SVP: synaptic vesicle precursor
SVR: systemic vascular resistance
SW: stroke work
SwAP70: 70-kDa switch-associated protein (RacGEF)
SYK: spleen tyrosine kinase
Synj: synaptojanin
Syp: synaptophysin
Syt: synaptotagmin
S1P: sphingosine 1-phosphate
S6K: P70 ribosomal S6 kinase (^{P70}RSK)

T

T: extrastress tensor
T: transition rate from a cell cycle phase to the next
T: temperature
T lymphocyte (T cell): thymic lymphocyte
T_C: cytotoxic T lymphocyte (CD8+ effector T cell; CTL)
T_{C1}: type-1 cytotoxic T lymphocyte
T_{C2}: type-2 cytotoxic T lymphocyte
T_{CM}: central memory T lymphocyte
T_{Conv}: conventional T lymphocyte
T_{Eff}: effector T lymphocyte
T_{EM}: effector memory T lymphocyte
T_{FH}: follicular helper T lymphocyte
T_H: helper T lymphocyte (CD4+ effector T cell)
T_{Hi}: type-i helper T lymphocyte ($i = 1/2/9/17/22$)
T_{H3}: TGFβ-secreting T_{Reg} lymphocyte
T_L: lung transfer capacity (alveolocapillary membrane)
T_{R1}: type-1, IL10-secreting, regulatory T lymphocyte
T_{Reg}: regulatory T lymphocyte
aT_{Reg}: CD45RA−, FoxP3^{hi}, activated T_{Reg} cell
iT_{Reg}: inducible T_{Reg} lymphocyte
nT_{Reg}: naturally occurring (natural) T_{Reg} lymphocyte
rT_{Reg}: CD45RA+, FoxP3^{low}, resting T_{Reg} cell
$\hat{t}$: unit tangent
t: time
TβRi: type-i TGFβ receptor
TAA: thoracic aortic aneurysm
TAB: TAK1-binding protein

TACE: tumor-necrosis factor-α-converting enzyme (ADAM17)
TACE: transarterial chemoembolization
TAF: TBP-associated factor
TAK: TGFβ-activated kinase (MAP3K7)
TALK: TWIK-related alkaline pH-activated K$^+$ channel
TANK: TRAF family member-associated NFκB activator
TASK: TWIK-related acid-sensitive K$^+$ channel
TASR: terminus-associated short RNA
TAP: transporter associated with antigen processing (ABC transporter)
Taz: taffazin
TBC1D: Tre2 (or USP6), BUB2, CDC16 domain-containing RabGAP
TBCK: tubulin-binding cofactor kinase (pseudokinase)
TBK: TANK-binding kinase
TBP: TATA box-binding protein (subclass-4F transcription factor)
TBx: T-box transcription factor
TC: thrombocyte (platelet)
TCA: tricarboxylic acid cycle
TCF: T-cell factor
TCF: ternary complex factor
TcFi: type-i transcription factor
TCP: T-complex protein
TCR: T-cell receptor
TEA: transluminal extraction atherectomy
TEC: Tyr kinase expressed in hepatocellular carcinoma
TEF: thyrotroph embryonic factor (PAR/b–ZIP family)
TEK: Tyr endothelial kinase
TEM: transendothelial migration
Ten: tenascin
TF: transcription factor
Tf: transferrin
TFPI: tissue factor pathway inhibitor
TfR: transferrin receptor
TG: triglyceride (triacylglycerol)
TGF: transforming growth factor
TGFBR: TGFβ receptor gene
TGFβRAP: TGFβ receptor-associated protein
TGN: trans-Golgi network
THETE: trihydroxyeicosatrienoic acid
THIK: tandem pore-domain halothane-inhibited K$^+$ channel
THR: thyroid hormone receptor (NR1a1/2)
TIAM: T-lymphoma invasion and metastasis-inducing protein (RacGEF)
TICE: transintestinal cholesterol efflux

TIE: Tyr kinase with Ig and EGF homology domains (angiopoietin receptor)
TIEG: TGFβ-inducible early gene product
TIF: transcription intermediary factor (kinase and Ub. ligase)
TIGAR: TP53-inducible glycolysis and apoptosis regulator
TIM: T-cell immunoglobulin and mucin domain-containing protein
Tim: timeless homolog
TIMM: translocase of inner mitochondrial membrane
TIMP: tissue inhibitor of metallopeptidase
TIRAP: Toll–IL1R domain-containing adaptor protein
tiRNA: transcription initiation RNA
TJ: tight junction
TKR: Tyr kinase receptor
TLC: total lung capacity
TLR: Toll-like receptor
TLT: TREM-like transcript
TLX: tailless receptor (NR2e1)
TM: thrombomodulin
TMi: transmembrane segment-i of membrane protein
TMC: twisting magnetocytometry
TMePAI: transmembrane prostate androgen-induced protein
TMy: tropomyosin
Tnn (TN): troponin
Tn: thrombin
TNF: tumor-necrosis factor
TNFαIP: tumor-necrosis factor-α-induced protein
TNFR: tumor-necrosis factor receptor
TNFRSF: tumor-necrosis factor receptor superfamily member
TNFSF: tumor-necrosis factor superfamily member
TNK: Tyr kinase inhitor of NFκB
Tns: tensin
TOR: target of rapamycin
TORC: target of rapamycin complex
TORC: transducer of regulated CREB activity (a.k.a. CRTC)
TP: thromboxane-A2 Gq/11-coupled receptor
TP53I: tumor protein P53-inducible protein
tPA: tissue plasminogen activator
Tpo: thrombopoietin
TPPP: tubulin polymerization-promoting protein
TPST: tyrosylprotein sulftotransferase
TR: testicular receptor (NR2c1/2)

TRAAK: TWIK-related arachidonic acid-stimulated K$^+$ channel
TRADD: tumor-necrosis factor receptor-associated death domain adaptor
TRAF: tumor-necrosis factor receptor-associated factor
TRAM: TRIF-related adaptor molecule
transceptor: transporter-related receptor
TRAP: TNF receptor-associated protein (HSP75)
TraPP: transport protein particle
TRAT: T-cell receptor-associated transmembrane adaptor
Trb: Tribbles homolog (pseudokinase)
TRE: TPA-response element (AP1/CREB-binding site on promoters)
TRE: trapped in endoderm
TREK: TWIK-related K$^+$ channel
TREM: triggering receptor expressed on myeloid cells
TRESK: TWIK-related spinal cord K$^+$ channel
TRF: TBP-related factor
TRF: double-stranded telomeric DNA-binding repeat-binding factor
TRH: thyrotropin-releasing hormone
TRIF: Toll–IL1R domain-containing adaptor inducing Ifnβ
TRIM: T-cell receptor-interacting molecule
TRIP: TGFβ receptor-interacting protein (eIF3S2)
TRK: tropomyosin receptor kinase (NTRK)
tRNA: transfer RNA
TRP: transient receptor potential channel
TRPA: ankyrin-like transient receptor potential channel
TRPC: canonical transient receptor potential channel
TRPM: melastatin-related transient receptor potential channel
TRPML: mucolipin-related transient receptor potential channel
TRPN: no mechanoreceptor potential C
TRPP: polycystin-related transient receptor potential channel
TRPV: vanilloid transient receptor potential channel
TrrAP: transactivation (transformation)/transcription domain-associated protein (pseudokinase)
TrV: tricuspid valve
TRx: thioredoxin
TRxIP: thioredoxin-interacting protein
TSC: tuberous sclerosis complex
TSH: thyroid-stimulating hormone

List of Aliases

TSLP: thymic stromal lymphopoietin
Tsp: thrombospondin
Tspan: tetraspanin
TsPO: translocator protein of the outer mitochondrial membrane
tSNARE: target SNARE
tsRNA: tRNA-derived small RNA
tssaRNA: transcription start site-associated RNA
TTbK: Tau-tubulin kinase
TTK: dual-specificity Thr/Tyr kinase
Ttn: titin (pseudokinase)
TUT: terminal uridine transferase
TWIK: tandem of P domains in a weak inwardly rectifying K^+ channel
TxA2: thromboxane A2 (thromboxane)
TxB2: thromboxane B2 (thromboxane metabolite)
TXK: Tyr kinase mutated in X-linked agammaglobulinemia
TxaS: thromboxane-A synthase
TyK: tyrosine kinase
T_3: tri-iodothyronine
T_4: thyroxine
$^+$TP: plus-end-tracking proteins

U

U: right stretch tensor
u: displacement vector
u: electrochemical command
u: specific internal energy
Ub: ubiquitin
UbC: ubiquitin conjugase
UbE2: E2 ubiquitin conjugase
UbE3: E3 ubiquitin ligase
UbL: ubiquitin-like protein
UCH: ubiquitin C-terminal hydrolase (DUb)
Ucn: urocortin
UCP: uncoupling protein
UDP: uridine diphosphate-glucose
UK: urokinase
ULK: uncoordinated-51-like kinase (pseudokinase)
Unc: uncoordinated receptor
uPA: urokinase-type plasminogen activator (urokinase)
uPAR: uPA receptor
uPARAP: uPAR-associated protein (CLec13e)
UPR: unfolded protein response
UPS: ubiquitin-proteasome system
UP4A: uridine adenosine tetraphosphate

Uro: urodilatin
US: ultrasound
USC: unipotential stem cell
USF: upstream stimulatory factor
USI: ultrasound imaging
USP: ubiquitin-specific peptidase (deubiquitinase)
UTP: uridine triphosphate
UTR: untranslated region
UVRAG: ultraviolet wave resistance-associated gene product

V

V: left stretch tensor
V: volume
V_q: cross-sectional average velocity
V_s: specific volume
v: velocity vector
v: recovery variable
V1(2)R: type-1(2) vomeronasal receptor
$V_{1A/1B/2}$: type-1a/1b/2 arginine vasopressin receptor
VAAC: volume-activated anion channel
$VAC_{Cl(K)}$: volume-activated Cl^- (K^+)-selective channel
VACamKL: vesicle-associated CamK-like (pseudokinase)
$VACC_{NS}$: volume-activated cation non-selective channel
VAChT: vesicular acetylcholine transporter
VAIC: volume-activated ion channel
VAMP: vesicle-associated membrane protein (synaptobrevin)
VanGL: Van Gogh (Strabismus)-like protein
VAP: VAMP-associated protein
VASP: vasoactive stimulatory phosphoprotein
VAT: vesicular amine transporter
vATPase: vesicular-type H^+ ATPase
VAV: ventriculoarterial valve
Vav: GEF named from Hebrew sixth letter
VC: vital capacity
VCAM: vascular cell adhesion molecule
VCt: vasoconstriction
VDAC: voltage-dependent anion channel (porin)
VDACL: plasmalemmal, volume- and voltage-dependent, ATP-conductive, large-conductance, anion channel
VDCC: voltage-dependent calcium channel
VDP: vesicle docking protein
VDt: vasodilation

VEGF: vascular endothelial growth factor
VEGFR: vascular endothelial growth factor receptor
VF: ventricular fibrillation
VF: ventricular filling
VGAT: vesicular GABA transporter
VGC: voltage-gated channel
VgL: Vestigial-like protein
VGluT: vesicular glutamate transporter
VHL: von Hippel-Lindau Ub ligase
VIP: vasoactive intestinal peptide
VLDL: very-low-density lipoprotein
VLDLR: very-low-density lipoprotein receptor
VMAT: vesicular monoamine transporter
VN: vitronectin
VPO: vascular peroxidase
VPS: vacuolar protein sorting-associated kinase
VR: venous return
VRAC: volume-regulated anion channel
VRC: ventral respiratory column
VRK: vaccinia-related kinase
VS: vasostatin
vSMC: vascular smooth muscle cell
vSNARE: vesicular SNAP receptor (SNARE)
VSOR: volume-sensitive outwardly rectifying anion channel
VSP: voltage-sensing phosphatase
VVO: vesiculo-vacuolar organelle
vWF: von Willebrand factor

W

W: vorticity tensor
$\mathcal{W}$: strain energy density
W: work, deformation energy
w: weight
w: grid velocity
WASH: WASP and SCAR homolog
WASP: Wiskott-Aldrich syndrome protein
WAT: white adipose tissue
WAVe: WASP-family verprolin homolog
WBC: white blood cell
WDR: WD repeat-containing protein
Wee: small (Scottish)
WHAMM: WASP homolog associated with actin, membranes, and microtubules
WIP: WASP-interacting protein
WIPF: WASP-interacting protein family protein

WIPI: WD repeat domain-containing phosphoinositide-interacting protein
WNK: with no K (Lys) kinase
Wnt: wingless-type
WPWS: Wolff-Parkinson-White syndrome
WNRRTK: Wnt and neurotrophin receptor-related receptor Tyr kinase (ROR$_{(RTK)}$)
WSB: WD-repeat and SOCS box-containing protein (Ub ligase)
WSS: wall shear stress
WSSTG: WSS transverse gradient
WWTR: WW domain-containing transcription regulator

X

X: trajectory
X: reactance
X: Lagrangian position vector
x: position vector
$\{x,y,z\}$: Cartesian coordinates
XBE: X-factor-binding element
XBP: X-box-binding protein (transcription factor)
XIAP: X-linked inhibitor of apoptosis protein (Ub ligase)

Y

Y: admittance coefficient
YAP: Yes-associated protein
YBP: Y-box-binding protein (transcription factor)
YY: yin yang (transcriptional repressor)

Z

Z: impedance
ZAP70: -associated protein 70
ZBTB: zinc finger and BTB (Broad complex, Tramtrack, and bric-à-brac) domain-containing transcription factor
ZnF: zinc finger protein
ZO: zonula occludens

List of Aliases

Miscellaneous

2-5A: 5'-triphosphorylated, (2',5')-phosphodiester-linked oligoadenylate
2AG: 2-arachidonyl glycerol
3DR: three-dimensional reconstruction
3BP2: Abl Src homology-3 domain-binding adaptor
4eBP1: inhibitory eIF4e-binding protein
5HT: serotonin
7TMR: 7-transmembrane receptor (GPCR)

Complementary Lists of Notations

Greek Letters

α: volumic fraction
α: convergence/divergence angle
α: attenuation coefficient
α_k: kinetic energy coefficient
α_m: momentum coefficient
β: inclination angle
$\{\beta_i\}_1^2$: myocyte parameters
β_T: coefficient of thermal expansion
Γ: domain boundary
Γ_L: local reflection coefficient
Γ_G: global reflection coefficient
γ: heat capacity ratio
γ: activation factor
γ_G: amplitude ratio (modulation rate) of G
γ_s: surface tension
$\dot{\gamma}$: shear rate
δ: boundary layer thickness
ϵ_T: emissivity (thermal energy radiation)
ϵ_e: electric permittivity
ε: strain
ε: small quantity
ζ: singular head loss coefficient
ζ: transmural coordinate
$\{\zeta_j\}_1^3$: local coordinate
η: azimuthal spheroidal coordinate
θ: circumferential polar coordinate
θ: $(\hat{\mathbf{e}}_x, \hat{\mathbf{t}})$ angle
κ: wall curvature
κ_c: curvature ratio
κ_d: drag reflection coefficient
κ_h: hindrance coefficient
κ_o: osmotic reflection coefficient
κ_s: size ratio
$\{\kappa_k\}_{k=1}^9$: tube law coefficients
κ_e: correction factor
Λ: head loss coefficient
λ_L: Lamé coefficient
λ: stretch ratio
λ: wavelength
λ_A: area ratio
λ_a: acceleration ratio
λ_L: length ratio
λ_q: flow rate ratio
λ_t: time ratio
λ_v: velocity ratio
μ: dynamic viscosity
μ_L: Lamé coefficient
ν: kinematic viscosity
ν_P: Poisson ratio
Π: osmotic pressure
ρ: mass density
τ: time constant
Φ: potential
$\phi(t)$: creep function
φ: phase
χ: Lagrangian label
χ_i: molar fraction of species i
χ_i: wetted perimeter
$\psi(t)$: relaxation function
Ψ: porosity
ω: angular frequency
Ω: computational domain

Dual Notations

Bφ: basophil
Eφ: eosinophil

Lφ: lymphocyte
Mφ: macrophage
aaMφ: alternatively activated macrophage
caMφ: classically activated macrophage
Nφ: neutrophil
Σc: sympathetic
pΣc: parasympathetic

Subscripts

$_A$: alveolar, atrial
$_{Ao}$: aortic
$_a$: arterial
$_{app}$: apparent
$_{atm}$: atmospheric
$_b$: blood
$_c$: contractile
$_c$: center
$_c$: point-contact
$_D$: Darcy (filtration)
$_d$: diastolic
$_{dyn}$: dynamic
$_E$: expiration, Eulerian
$_e$: external
$_e$: extremum
$_{eff}$: effective
$_f$: fluid
$_g$: grid
$_I$: inspiration
$_i$: internal
$_{inc}$: incremental
$_L$: Lagrangian
$_l$: limit
$_\ell$: line-contact
$_M$: macroscopic
$_m$: mean
$_{max}$: maximum
$_m$: muscular, mouth
$_{met}$: metabolic
$_\mu$: microscopic
$_P$: pulmonary
$_p$: parallel
$_p$: particle
$_q$: quasi-ovalization
$_r$: radial
$_{rel}$: relative
$_S$: systemic
$_s$: solute
$_s$: serial
$_s$: systolic
$_t$: stream division
$_T$: total

$_t$: turbulence
$_t$: time derivative of order 1
$_{tt}$: time derivative of order 2
$_{tis}$: tissue
$_V$: ventricular
$_v$: venous
$_w$: wall
$_w$: water (solvent)
$_\Gamma$: boundary
$_\theta$: azimuthal
$_+$: positive command
$_-$: negative command
$_*$: at interface
$_0$: reference state ($_{\cdot 0}$: unstressed or low shear rate)
$_\infty$: high shear rate

Superscripts

^{a}: active state
^{e}: elastic
^{f}: fluid
^{h}: hypertensive
^{n}: normotensive
^{p}: passive state
^{p}: power
^{s}: solid
^{T}: transpose
^{v}: viscoelastic
*: scale
*: complex variable
$^\prime$: first component of complex elastic and shear moduli
$^{\prime\prime}$: second component of complex elastic and shear moduli
$^-$: static, stationary, steady variable

Mathematical Notations

T: bold face capital letter means tensor
v: bold face lower case letter means vector
S, s: upper or lower case letter means scalar
$\Delta\bullet$: difference
$\delta\bullet$: increment
$d\bullet/dt$: time gradient
∂_t: first-order time partial derivative
∂_{tt}: second-order time partial derivative
∂_i: first-order space partial derivative with respect to spatial coordinate x_i
∇: gradient operator

Complementary Lists of Notations

$\nabla \mathbf{u}$: displacement gradient tensor
$\nabla \mathbf{v}$: velocity gradient tensor
$\nabla \cdot$: divergence operator
∇^2: Laplace operator
$||_+$: positive part
$||_-$: negative part
$\dot{\bullet}$: time derivative
$\bar{\bullet}$: time mean
$\breve{\bullet}$: space averaged
$\langle \bullet \rangle$: ensemble averaged
$\tilde{\bullet}$: dimensionless
$\bullet^+$: normalized ($\in [0,1]$)
$\hat{\bullet}$: peak value
$\bullet_\sim$: modulation amplitude
$\det(\bullet)$: determinant
$\text{cof}(\bullet)$: cofactor
$\text{tr}(\bullet)$: trace

Cranial Nerves

I: olfactory nerve (sensory)
II: optic nerve (sensory)
III: oculomotor nerve (mainly motor)
IV: trochlear nerve (mainly motor)
V: trigeminal nerve (sensory and motor)
VI: abducens nerve (mainly motor)
VII: facial nerve (sensory and motor)
VIII: vestibulocochlear (auditory-vestibular) nerve (mainly sensory)
IX: glossopharyngeal nerve (sensory and motor)
X: vagus nerve (sensory and motor)
XI: cranial accessory nerve (mainly motor)
XII: hypoglossal nerve (mainly motor)

Chemical Notations

$[X]$: concentration of X species
X (x): upper and lower case letters correspond to gene and corresponding protein or conversely (i.e., Fes, FES, and fes designate protein, a proto-oncogene product that acts as a kinase, and corresponding gene and oncogene product, respectively)
$^\bullet$: radical (unpaired electron[s])
Δ^{NT}: truncated form without the N-terminal domain
Δ^{CT}: truncated form without the C-terminal domain
$^{D(L)}X$: D (L)-stereoisomer of amino acids and carbohydrates (chirality prefixes for dextro- [dexter: right] and levorotation [lævus: left]), i.e., dextro(levo)rotatory enantiomer
^GX: globular form of X molecule
$^{F(G)}$actin: polymeric, filamentous (monomeric, globular) actin
$_CX$: carboxy (carboxyl group COOH [C])-terminal cleaved part of X molecule
$_cX$: cytosolic molecule
$_{L,Ac}X$: lysosomal, acidic X molecule (e.g., sphingomyelinase)
$_mX$: membrane-bound molecule
$_NX$: amino (amine group NH_2 [N])-terminal cleaved part of X molecule
$_sX$: secreted form of X molecule
$_{S,Ac}X$: secreted, acidic molecule X (e.g., sphingomyelinase)
$_tX$: truncated isoform
X_i: type-i isoform of the receptor of ligand X (i: integer)
XRi: receptor isoform i of ligand X (i: integer)
X+: molecule X expressed (X-positive)
X^+: cation; also intermediate product X of oxidation (loss of electron) from a reductant (or reducer) by an oxidant (electron acceptor that removes electrons from a reductant)
X−: molecule X absent (X-negative)
X^-: anion; also intermediate product X of reduction (gain of electron) from an oxidant (or oxidizer) by a reductant (electron donor that transfers electrons to an oxidant)
X^A: activator form of molecule X
X^a: active form of molecule X
X^{ECD}: soluble fragment corresponding to the ectodomain of molecule X after extracellular proteolytic cleavage and shedding (possible extracellular messenger or sequestrator)
$X^{(ER)}$: endoplasmic reticulum type of molecule X
small $GTPase^{GTP(GDP)}$: active (inactive) form of small (monomeric), regulatory guanosine triphosphatase
$X^{GTP(GDP)}$: GTP (GDP)-loaded molecule X
X^{ICD}: soluble fragment corresponding to intracellular domain of molecule X after intracellular proteolytic cleavage

(possible messenger and/or transcription factor; e.g., Notch^{ICD}: intracellular Notch fragment)
X^M: methylated molecule X
X^{MT}: mitochondrial type of molecule X
X^P: phosphorylated molecule X
pAA: phosphorylated amino acid (pSer, pThr, and pTyr)
X^{PM}: plasmalemmal type of molecule X
X^R: repressor form of molecule X
X^S: soluble form
X^{SNO}: ^{S}nitrosylated molecule X
X^U: ubiquitinated protein X
X_{alt}: alternative splice variant
X_{FL}: full-length protein X
$X_{h(l,m)MW}$: high (low, mid)-molecular-weight isotype
$X_{L(S)}$: long (short) isoform (splice variants)
X_c: catalytic subunit
X_P: palmitoylated molecule X
X_i: number of molecule or atom (i: integer, often 2 or 3)
$(X_1-X_2)_i$: oligomer made of i complexes constituted of molecules X_1 and X_2 (e.g., histones)
a, c, nX: atypical, conventional, novel molecule X (e.g., PKC)
acX: acetylated molecule X (e.g., acLDL)
al, ac, nX: alkaline, acidic, neutral molecule X (e.g., sphingomyelinase)
asX: alternatively spliced molecule X (e.g., asTF)
cX: cellular, cytosolic, constitutive (e.g., cNOS), or cyclic (e.g., cAMP and cGMP) X molecule
caX: cardiomyocyte isoform (e.g., caMLCK)
dX: deoxyX
eX: endothelial isoform (e.g., eNOS and eMLCK)
hX: human form (ortholog); heart type (e.g., hFABP); hormone-like isoform (FGF)
iX: inhibitory mediator (e.g., iSMAD) or intracellular (e.g., iFGF) or inducible (e.g., iNOS) isoform
kX: renal type (kidney) X molecule
ksX: kidney-specific isoform of X molecule
lX: lysosomal X molecule
lpX: lipoprotein-associated X molecule (e.g., lpPLA2)
mX: mammalian species or membrane-associated X molecule (e.g., mTGFβ)
mtX: mitochondrial type of X molecule
nX: neutral X; neuronal type (e.g., nWASP)

oxX: oxidized X molecule (e.g., oxLDL)
plX: plasmalemmal type of X molecule
rX: receptor-associated mediator or receptor-like enzyme; also regulatory type of molecular species (e.g., rSMAD)
skX: skeletal myocyte isoform (e.g., skMLCK)
smcX: smooth muscle cell isoform (e.g., smcMLCK)
tX: target type of X (e.g., tSNARE); tissue type (e.g., tPA)
tmX: transmembrane type of X
vX: vesicle-associated (e.g., vSNARE) or vacuolar (e.g., vATPase) type of X
GPX: glycoprotein (X: molecule abbreviation or assigned numeral)
Xx: (x: single letter) splice variants
X1: human form (ortholog)
Xi: isoform type i (paralog or splice variant; i: integer)
Xi/j: (i,j: integers) refers to either both isoforms (i.e., Xi and Xj, such as ERK1/2) or heterodimer (i.e., Xi-Xj, such as ARP2/3)
X1/X2: molecular homologs or commonly used aliases (e.g., contactin-1/F3)
PI(i)P, PI(i,j)P$_2$, PI(i,j,k)P$_3$: i,j,k (integers): position(s) of phosphorylated OH groups of the inositol ring of phosphatidylinositol mono-, bis-, and trisphosphates

Amino Acids

Ala (A): alanine
Arg (R): arginine
Asn (N): asparagine
Asp (D): aspartic acid
Asp$^{COO^-}$: aspartate
CysH (C): cysteine
Cys: cystine
Gln (Q): glutamine
Glu (E): glutamic acid
Glu$^{COO^-}$: glutamate
Gly (G): glycine
His (H): histidine
Iso, Ile (I): isoleucine
Leu (L): leucine
Lys (K): lysine
Met (M): methionine

Phe (F): phenylalanine
Pro (P): proline
Ser (S): serine
Thr (T): threonine
Trp (W): tryptophan
Tyr (Y): tyrosine
Val (V): valine

Ions

Asp^-: aspartate (carboxylate anion of aspartic acid)
ADP^{3-}: ADP anion
ATP^{4-}: ATP anion
Ca^{++}: calcium cation
Cl^-: chloride anion
Co^{++}: cobalt cation
Cu^+: copper monovalent cation
Cu^{++}: copper divalent cation
Fe^{++}: ferrous iron cation
Fe^{3+}: ferric iron cation
Glu^-: glutamate (carboxylate anion of glutamic acid)
H^+: hydrogen cation (proton)
H_3O^+: hydronium (oxonium or hydroxonium) cation
HCO_3^-: bicarbonate anion
HPO_4^{2-}: hydrogen phosphate anion
K^+: potassium cation
Mg^{++}: magnesium cation
$^{Mg}ATP^{2-}$: ATP anion
Mn^{++}: manganese cation
Na^+: sodium cation
Ni^{++}: nickel cation (common oxidation state)
OH^-: hydroxide anion
PO_4^{3-}: phosphate anion
SO_4^{2-}: sulfate anion
Zn^{++}: zinc cation (common oxidation state)

Inhaled and Signaling Gas

CO: carbon monoxide (or carbonic oxide; signaling gas and pollutant)
CO_2: carbon dioxide (cell waste)
H_2S: hydrogen sulfide (signaling gas)
He: helium (inert monatomic gas)
N_2: nitrogen (inert diatomic gas)
NO: nitric oxide (or nitrogen monoxide; signaling gas and pollutant)

NO_2: nitrogen dioxide (air pollutant)
O_2: oxygen (cell energy producer)
SO_2: sulfur dioxide (air pollutant)

Nitric Oxide Derivatives

$NO^\bullet$: free radical form
NO^+: nitrosyl or nitrosonium cation
NO^-: nitroxyl or hyponitrite anion (inodilator)
HNO: protonated nitroxyl anion
HNO_2: nitrous acid
NO_2^-: nitrite anion
NO_3^-: nitrate anion

Reactive Oxygen and Nitrogen Species

H_2O_2: hydrogen peroxide
HOCl: hypochlorous acid
N_2O_3: dinitrogen trioxide
$NO_2^\bullet$: nitrogen dioxide
$^1O_2$: singlet oxygen
O_2^-: superoxide ($O_2^{\bullet-}$)
$O=C(O^\bullet)O^-$: carbonate radical
$OH^\bullet$: hydroxyl radical (hydroxide ion neutral form)
$ONOO^-$: peroxynitrite
$RO^\bullet$: alkoxyl
$RO_2^\bullet$: peroxyl

Moieties (R denotes an organic group)

R: alkyl group (with only carbon and hydrogen atoms linked exclusively by single bonds)
$R-CH_3$: methyl group (with 3 forms: methanide anion [CH_3^-], methylium cation [CH_3^+], and methyl radical [$CH_3^\bullet$])
R–CHO: aldehyde group
R–CN: nitrile group
R–CO: acyl group
R–CO–R: carbonyl group
$R-COO^-$: carboxylate group
R–COOH: carboxyl group

R–NC: isonitrile group
R–NCO: isocyanate group
R–NH$_2$: amine group
R–NO: nitroso group
R–NO$_2$: nitro group
R–O: alkoxy group
R=O: oxo group
R–OCN: cyanate group
R–OH: hydroxyl group
R–ONO: nitrosooxy group
R–ONO$_2$: nitrate group
R–OO–R: peroxy group
R–OOH: hydroperoxy group
R–S–R: sulfide group
R–SH: thiol (or sulfhydryl) moiety
R–SN: sulfenyl-amide moiety
R–SNO: nitrosothiol (or thionitrite) moiety
R–SO: sulfinyl
R–SO–R: sulfoxide group
R–SO$_2$: sulfonyl group
R–SO$_2$H: sulfinic acid (sulfinyl moiety)
R–SO$_2$N: sulfonyl-amide moiety
R–SO$_3$H: sulfonic acid (sulfonyl moiety)
R–SOH: sulfenic acid (sulfenyl moiety)
R–SON: sulfinyl-amide moiety
R–SS–R: disulfide group

Time Units

d: day
h: hour
mn: minute
s: second
wk: week

SI-Based and Non-SI Units of Quantity

mmHg: millimeter of mercury (133.322 Pa [$\sim$ 0.1333 kPa])
mmol, nmol, μmol: milli-, nano-, micromoles (amount of a chemical species, one mole containing about $6.02214078 \times 10^{23}$ molecules)
mosm: milliosmole
(osm: number of moles of a osmotically active chemical compound)
kDa: kiloDalton
(Da: atomic or molecular mass unit)
ppm: parts per million
l: liter

Index

A

α-adrenergic receptor 468, 874
α-kinase 125
AaTyK kinase 191
ABC transporter 47, 670, 822, 896
Abl kinase . 126, 140, 198, 431, 433, 437, 571, 807
ABR GEF/GAP 126, 609, 629, 639
ACAP ArfGAP 613, 616
acetylcholine 35, 385, 659, 678, 682
ACh muscarinic receptor 81, 743, 841
acid phosphatase AcP1 458
ACK kinase 143
actin ... 15, 142, 147, 149, 157, 178, 216, 229, 269, 501, 537, 566, 580, 874, 882
actin-related protein 146, 510, 516
actin comet 90
activating transcription factor (ATF) . 318, 321, 350, 357, 373, 374, 783
Activator protein-1 ... 177, 318, 357, 373, 701
adamlysin 426, 851
adaptin-associated kinase 112
adaptor protein 2
ADCK kinase 125
adducin 231
adenylate cyclase 266, 470, 485, 492, 742, 822, 862, 873
adhesion molecule 849
adhesome 852
adipocyte 321
adipokine 764
adiponectin 250, 652, 653
adipose triglyceride lipase 44, 45
ADP 652
^{ADP}ribosyl cyclase 875
Adpribosyl hydrolase 875

adrenaline 472, 838
adrenomedullin 884
afadin/AF6 GEF 537, 545, 548
AFAP ArfGAP 613
AGAP ArfGAP 613, 617
aggresome 662
aging 430, 797
airway epithelium .. 40, 43, 185, 191, 373, 659, 662, 697, 703, 704, 706, 720, 747, 754, 829
airway smooth muscle cell 43, 673, 841
airway surface fluid 43
AKAP 199, 200, 211, 484, 489, 495, 683, 731, 737, 841, 842
akirin 786
aldosterone 191, 243, 295, 298, 301, 838
alveolar macrophage 43, 704, 707
AMAP ArfGAP 613
AMPA-type glutamate receptor 742
amphiphysin 99, 525
amphiregulin 725
AMPK . 45, 248, 305, 412, 454, 660, 676, 798, 816, 895
anandamide 10
androgen receptor (NR3c4) 274
angiogenesis. 33, 104, 159, 229, 278, 372, 434, 608, 611, 641, 675, 677, 758, 803, 886
angiomotin 608
angiotensin .. 89, 105, 166, 191, 198, 242, 301, 359, 370, 428, 479, 532, 602, 603, 609, 611, 696–698, 710, 819, 905
angiotensin-converting enzyme 532
angiotensinogen 532
angiotensin receptor 428, 532
annexin 731, 747

M. Thiriet, *Intracellular Signaling Mediators in the Circulatory and Ventilatory Systems*,
Biomathematical and Biomechanical Modeling of the Circulatory and Ventilatory Systems 4,
DOI 10.1007/978-1-4614-4370-4, © Springer Science+Business Media New York 2013

APAP ArfGAP.....................613, 618
APC Ub ligase 292, 727, 801
apoptosis....59, 178–180, 184–186, 193, 215,
    256, 572, 678, 682, 754, 803
apoptosome...........................205
aquaporin 201, 842
arachidonic acid............... 10, 38, 50, 60
ARAP ArfGAP................613, 619, 629
ARD1 GTPase (TriM23 Ub ligase)...503, 514
ARF1GEF.........................575, 578
ArfGAP 614
ArfGEF..............................575
ArfGEF1/2/3 .... 575, 578
ARF GTPase 90, 502, 572, 812
ARL GTPase........................512
ARNO/cytohesin ArfGEF..........512–515
arrestin 352, 360, 367, 469, 478, 485, 494, 495,
    578, 841, 842
AS160 RabGAP 204, 236
ASAP ArfGAP 613, 620
asthma 29, 40, 43, 80
astrocyte.............................652
ataxia-telangiectasia mutated kinase 348
ATF factor 378
atherosclerosis..... 56, 80, 135, 662, 700, 717
ATM kinase 127, 260, 407, 412, 655, 710, 768,
    775, 804, 807, 813
ATP 97, 666, 673, 854, 862, 899
atrogin (FBxO32) Ub ligase 801
ATR kinase . 127, 260, 407, 412, 655, 804, 807
augmin...............................293
Aurora.......... 289, 304, 386, 396, 560, 727
autoimmunity 758
autophagosome 81, 531
autophagy...........................80, 816
autoregulation 485, 669, 878, 908
Axl (RTK)...........................438

B

β-adrenergic receptor.......35, 534, 746, 824
β-catenin (transcription factor)...........800
B-cell receptor... 158, 180, 435, 611, 764, 771
basement membrane 305
basophil 107
Bayliss effect 485, 878
BCAR/CAS docker .. 142, 147, 166, 431, 585,
    594, 597, 620, 623, 907
BCL2 protein 240
BCR kinase/GEF/GAP 126, 610, 629, 638, 639
bicarbonate......................685, 827
bilirubin 655
biliverdin 651

BLK kinase 158
BLOC complex........................528
blood–brain barrier 295
BMX kinase 146, 159, 224, 434, 645
bone morphogenetic protein 347
BORG...............................566
bradykinin.................... 659, 678, 907
BRAG ArfGEF 576, 579
brain............491, 608, 637, 646, 652, 843
BrCa1 Ub ligase 256, 261
BrD kinase...........................126
bronchoconstriction 42, 231, 248, 842
bronchodilation 42, 674
BrSK kinase 251, 252
BTK kinase 83, 160, 211, 280, 323

C

C/EBP factor..................142, 238, 534
Ca^{++}-induced Ca^{++} release 859
Ca^{++} ATPase......................670, 711
Ca_V channel ... 35, 41, 85, 214, 227, 228, 244,
    474, 522, 534, 568, 649, 675, 681, 682,
    711, 742, 821, 824, 842, 855, 878
cadherin 147, 170, 533, 537, 561, 851
cADPR 875, 876
calbindin........................... 27, 868
calcitonin gene-related peptide........... 672
calcium .. 13, 48, 159, 248, 591, 597, 628, 652,
    658, 659, 673, 675, 685, 688, 690, 693,
    708, 732, 827, 829, 855, 858, 905
calcium oscillation....................874
calcium sensitization 231, 248
calcium spark 654
calcium sparklet 878
calcium transient 158, 682, 874
caldesmon 224
calmodulin .. 61, 137, 242, 245, 406, 412, 413,
    490, 491, 494, 534, 535, 592, 652, 658,
    662, 688, 808, 829, 830, 836, 855, 861,
    863
calmodulin-dependent kinase 26, 61, 175, 242,
    270, 343, 358, 412, 413, 534, 539, 567,
    581, 627, 711, 753, 808, 836, 861, 870,
    880
calnexin 870
calpain 78, 270, 361, 437, 852
calreticulin...........................870
calsequestrin 870
Cam2K 249, 798
cAMP. 137, 587, 649, 683, 733, 743, 821, 822,
    830, 862
cancer......................412, 758, 803

CaPON adaptor........................493
carbon monoxide......................650
cardiolipin...........................454
cardiomyocyte 17, 33, 34, 39, 59, 61, 143, 148, 153, 154, 158, 159, 199, 201, 205, 206, 216, 219, 223, 244, 307, 333, 339, 343, 400, 405, 406, 472, 480, 486, 492, 534, 635, 672, 679, 682, 720, 723, 742, 743, 745, 746, 754–756, 792, 801, 819, 823, 837, 838, 842, 843, 878, 905
CARD scaffold.........................770
carotid body...........................654
casein kinase....267, 330, 357, 397, 436, 652, 723, 798, 813, 819
CASK kinase...........................132
caspase..........178, 184, 185, 228, 726, 812, 897
catalase......................675, 705, 775
catecholamine.........................842
catenin....................... 150, 170, 561
caveola..................660, 680, 682, 829
caveolin.... 486, 565, 651, 659, 680, 752, 895
cavin..................................895
CBL Ub ligase.. 163, 430, 434, 552, 604, 612, 620, 771, 807, 852
CBP (HAT).. 26, 237, 238, 395, 790, 800, 887
CDC-like kinase.......................382
CDC14 phosphatase 448
CDC20.................................724
CDC24 RasGEF.......................589
CDC25 phosphatase..222, 252, 254, 256, 258, 290, 292, 342, 343, 404, 416, 433, 459, 704
CDC25 RasGEF.......................589
CDC42 GTPase ... 77, 84, 143, 223, 336, 510, 515, 541, 552, 553, 604, 637, 639, 853
CDK inhibitor.........................413
cell cycle................ 291, 362, 386, 459
cell division...........................471
cell migration83, 149, 208, 213, 228, 244, 259, 336, 338, 339, 341, 344, 359, 361, 375, 448, 471, 500, 511, 533, 560, 571, 578, 582, 583, 588, 593, 597, 608, 700
cell polarity.................. 212, 305, 515
centaurin..............................572
central nervous system370, 580, 583, 588, 596, 606, 630, 636, 649, 652, 802, 834
ceramide.......................... 29, 336
cerebrospinal fluid.....................652
CFTR channel .. 216, 242, 565, 754, 829, 843, 858
CG3 GEF..............................178
cGMP. 137, 248, 649, 654, 672, 673, 683, 830, 846

chaperone.............................466
chaperonin............................466
checkpoint kinase.... 256, 262, 412, 459, 460, 807
chemokine............................. 33
chemokine receptor....................522
chemotaxis............... 70, 266, 372, 516
chimerin (RhoGAP2/3).................630
chloride...............................754
cholesterol...................... 266, 429
chromatin licensing and DNA replication factor CDT1.......................800
chronic obstructive pulmonary disease 80, 787, 843
chronophin............... 390, 419, 448, 462
chylomicron............................49
ciliary beat frequency 747
ciliary neurotrophic factor 154
ciliated cell...........................263
ciliogenesis...........................529
CIP...................................565
circadian rhythm................... 380, 493
Clara cell.........................219, 707
clathrin..................... 98, 335, 579
claudin...............................302
ClCa channel 753, 870
ClIC channel 662, 745
CnK adaptor..........................550
coat protein................. 99, 508, 515
cofilin . 187, 230, 285, 286, 376, 448, 463, 516
cohesin...............................304
collagen................849, 850, 853, 899
complement..........................172
connective tissue growth factor..........534
connexin...................... 855, 899
copper................................710
COP Ub ligase 134
cortactin.........................149, 622
CRAC channel....................860, 879
creatine kinase 861
CREB factor..............373, 378, 395, 779
CRIK kinase......................232, 554
CRKL adaptor... 178, 322, 340, 552, 585, 622
CRK adaptor.... 142, 178, 322, 340, 585, 594, 597, 619, 622
CRTC factor 251
CSF2.................................837
CSK kinase...............144, 433, 434, 853
cullin Ub ligase.................. 146, 801
cyclic nucleotide-gated channel .. 87, 673, 678, 823
cyclin 366
cyclin-B 459
cyclin-D 724

cyclin-dependent kinase... 141, 222, 224, 270, 273, 317, 371, 384, 397, 411, 418, 448, 459, 460, 631, 641, 797, 819
cyclooxygenase 11, 301, 376, 652, 777
cyclophilin............................. 383
Cyld deubiquitinase... 347, 762, 766, 771, 782
cytochrome-C oxidase 655, 718
cytochrome C..................... 333, 714
cytochrome P450.................... 675, 777
cytohesin ArfGEF................... 576, 577
cytokine receptor 151

D

DabIP RasGAP.................... 344, 629
Dab adaptor........................... 344
Dachsous cadherin..................... 725
DAPK kinase..................... 256, 401
DCLK kinase 304
DDR kinase........................... 854
dendritic cell 786
desmin 224
desmosome 305
diabetes 155, 676, 699
diacylglycerol 10, 48, 137, 259, 590, 591, 732, 862, 880
diacylglycerol kinase............. 17, 591, 643
diacylglycerol receptor 241
diaphanous 532, 554
DIP1 Ub ligase (Mib1).................. 257
Disheveled 709, 723
DLg adaptor 493, 659, 734
DM2 Ub ligase.. 220, 275, 549, 727, 742, 746, 805, 806, 810, 841
DMPK kinase 232
DNAPK kinase 127, 260, 262, 807
DOCK GEF.......... 178, 537, 584, 588, 593
docosahexaenoic acid 52
DOK adaptor.................. 145, 166, 552
dopamine receptor 446
DRAK kinase 257
dual-specificity phosphatase.... 342, 343, 355, 361, 370, 372, 374
Duo (RhoGEF24) 584, 606
dynamin............................. 465
dynein............................... 224
DYRK kinase 338, 380, 798

E

E-twenty six factor (ETS)........... 675, 887
E2F factor 730
E4F transcription factor/Ub ligase 808
E74-like factor (ELF) 225
early growth response factor (EGR)....... 701
ECS Ub ligase......................... 780
EET............................... 41–43
EF-hand motif......................... 862
EFA6 ArfGEF..................... 576, 580
EGF receptor 458, 527
eicosanoid..................... 10, 11, 38
eIF2αK3 kinase (PERK) 783
elastin 848
elastin–laminin receptor 848
electron transport chain.................. 718
ElMo adaptor 593
ENaC channel... 201, 228, 296, 521, 522, 548, 560, 855
endocannabinoid.................. 159, 443
endocytosis 80, 98, 450
endophilin 99
endoplasmic reticulum 590, 651, 680, 687, 696, 697, 859
endoplasmic reticulum stress 342, 816
endosome 360, 434
endothelial dysfunction 663, 699
endothelial lipase..................... 44, 46
endothelin....... 244, 359, 370, 491, 698, 778
endothelin receptor 480
endothelium ... 30, 32, 37, 43, 47, 50, 64, 146, 158, 159, 207, 219, 224, 229, 247, 248, 285, 373, 434, 444, 459, 472, 561, 562, 582, 583, 587, 597, 608, 651, 652, 655, 657, 659, 668, 676, 697, 710, 720, 746, 753, 775, 785, 792, 832, 838, 843, 848, 853, 891, 905
ENPP/NPP phosphatase 102
ENPP2/autotaxin 106
enteric nervous system 649
eosinophil peroxidase 709
EPH receptor.......................... 611
epidermal growth factor 88, 143, 425
epithelial–mesenchymal transition ... 340, 728
epithelial tubule 899
ERK . 17, 33, 60, 61, 64, 67, 78, 147, 191, 239, 241, 248, 323, 332, 359, 361, 390, 405, 427, 428, 430, 445, 455, 457, 479, 495, 537, 587, 590, 632, 637, 675, 678, 697, 709, 710, 720, 743, 746, 761, 792, 825, 840, 851, 894
ERN1 kinase (IRE1) 783
erythrocyte .. 91, 651, 664, 668, 669, 672, 688, 713, 719
erythropoiesis 886
erythropoietin 154, 372, 788, 887
erythropoietin receptor 428
ETS-like factor (ELk/TCF) . 67, 350, 357, 366, 374, 501, 588, 848, 887

ETS variant factor (ETV) 237
ezrin radixin moesin . . 146, 227, 230, 231, 481,
    501, 554, 561, 587, 643

F

farnesylation 549
fatty acid 10
Fat cadherin 723
FBx ArfGEF 576, 581
Fc receptor 428, 770
feedback . . . 17, 18, 77, 93, 153, 199, 204, 217,
    224, 240, 243, 261, 264, 283, 285, 314,
    343, 344, 346, 347, 352, 359–364, 386,
    390–392, 405, 406, 411, 412, 424, 429,
    441, 444, 446, 459, 460, 476, 517, 526,
    540, 554, 561, 583, 603, 609, 652, 657,
    669, 684, 688, 694, 704, 714, 729, 741,
    743, 764, 765, 767, 779, 780, 797, 798,
    805, 814, 819, 829, 838, 841, 842, 844,
    864, 888, 901
feedforward 528, 603, 844
FeR kinase 148
Fer kinase 433
FeS kinase 148
FGR kinase 160
fibrillar adhesion 437
fibrinogen 849
fibroblast 150, 213, 373, 532, 534, 756
fibroblast growth factor 241, 434, 444, 569, 644
fibronectin 157, 849
fibrosis 376
FIH hydrolase 889
filamin 100, 224, 535, 606
flotillin 565
focal adhesion . . . 226, 278, 437, 501, 622, 696,
    853, 854
focal adhesion kinase . . 93, 145, 150, 198, 323,
    430, 431, 434, 436, 618, 634, 849, 851,
    854
Fos factor 237, 350, 357, 374, 849
FoxA factor 185
FoxO factor 84, 265, 320, 796, 889, 891
frequenin 90
FRK kinase 150
Fyn kinase . 158, 162, 164, 216, 280, 427, 433,
    434, 552

G

G-protein-coupled receptor 385, 467, 501, 824,
    842, 862, 905, 906
G-protein signaling modulator 489
GABA 303

GAB adaptor 430
galanin 474
galanin receptor 474
galectin 547
gap junction 854, 899
gas transporter 672
GCK kinase 176, 178
gelsolin 187, 560
Gem GTPase 567
general transcription factor GTF2 . . . . 129, 873
genotoxic stress 368, 768, 810, 814, 817
girdin 471
GIRK channel 35, 493
GKAP (G-kinase-anchoring protein) . 683, 731,
    733
GKAP (guanylate kinase-associated protein)
    734
glial cell 60
Gli transcription factor 725
glucagon 320
glucagon-like peptide 446
glucocorticoid 652
glucocorticoid receptor 105, 662
glucose 207
glucose transporter 40, 100, 204, 217, 522, 529,
    530, 565, 638, 641, 895
glutamate 652
glutamate receptor 572, 874
glutaredoxin 708
glutathiolation 660
glutathione 702, 706
glutathione peroxidase 706
glutathione reductase 706
glycogen 395
glycogen synthase kinase . . 100, 175, 239, 240,
    251, 270, 320, 350, 357, 379, 594, 635,
    745, 787
Golgi body 590
GPCR kinase . . . 194, 199, 214, 220, 385, 469,
    478–480, 492, 494, 495, 618
granulocyte 202
granulocyte–macrophage colony-
    stimulating factor (CSF2) 154
granulocyte colony-stimulating factor (CSF3)
    154
granzyme 228
GRB adaptor 99, 147, 199, 320, 322, 340, 434,
    552, 556, 585
growth and differentiation factor 347
growth hormone 154
GTP 673
GTPase-activating protein 612
guanine nucleotide-dissociation inhibitor . . 640
guanine nucleotide-exchange factor 573

guanosine triphosphatase 465
guanylate cyclase 112, 654, 672, 673
guanylin 301
G protein 323, 467, 822, 862, 873, 905

H

H^+–K^+ ATPase 300
H^+ ATPase 685
H_V channel 696
Haspin kinase 304
HCK kinase 163, 280, 719
HDL 31, 46, 55
healing 448
heart .. 473, 567, 583, 608, 609, 626, 637, 655, 679, 745, 819, 834, 836, 843
heart failure 689
heat shock factor 896
heat shock protein 198, 659
Hedgehog 444, 531, 795
helper T cell 212
hemeprotein 650
heme oxygenase 650–652, 775
hemoglobin 664, 667, 678, 714
heparan sulfate proteoglycan 47, 850, 892
heparin 104
hepatic lipase 44, 45
hepatocyte 265, 321, 721, 823
hepatocyte growth factor 643
hepsin 305
heregulin 632
HER receptor 428, 455, 604, 608, 699
HES factor 889
HETE 40–43, 652
HGF receptor 213
HIF Asn hydroxylase 790
HIF prolyl hydroxylase (PHD) .. 744, 788, 789
HIPK kinase 274, 338, 343
Hippo 185, 194, 196, 720
hippocalcin 876
hippocampal cholinergic neurostimulatory
    peptide 385
histone 891
histone deacetylase 374, 381
HNF factor 381
Homer 494, 572
HoPS complex 526, 528
hormone-sensitive lipase 44, 58, 321
HRT factor 889
HTRa2 peptidase 307
HUWE1 Ub ligase 808
hydrogen peroxide 704
hydrogen sulfide 686, 687
hydroxide 710

hydroxyl radical 710
hypercapnia 787
hyperglycemia 698
hypertension 191, 602, 611, 663, 846
hypertrophy 244, 376, 532, 534, 741, 743, 746, 878, 905
hypochlorous acid 709, 710
hypoxia 61, 148, 157, 275, 364, 374, 400, 654, 664, 668, 669, 788, 816, 883
hypoxia-inducible factor .. 675, 708, 744, 788, 815, 884
hypoxic vasodilation 668
hypoxic pulmonary vasoconstriction 30, 43

I

IAP (BIRC) Ub ligase 332, 340, 344, 346, 767, 768, 775, 783, 786
IGF1R receptor 343
IKK kinase 175, 182, 204, 216, 335, 336, 338–340, 346, 349, 350, 385, 411, 412, 760, 763, 764, 766
ILKAP 409
immunological synapse 879
IMP Ub ligase 548
infarction 80, 241
inflammasome 897
inflammation 758
inhibitor of NFκB kinase 538
initiation factor 380
inositol hexakisphosphate 815
inositol trisphosphate .. 48, 137, 829, 855, 862
inotropy 244, 672
InPP phosphatase 79
insulation 786
insulin 47, 78, 83, 97, 100, 101, 155, 198, 203, 250, 265, 301, 320, 380, 415, 424, 430, 441, 474, 529, 530, 634, 638, 641, 684, 697, 710, 764, 795, 796, 798, 802, 838
insulin-like growth factor 17, 78, 145, 206, 296, 634, 797, 838
insulin receptor 155, 423, 425, 459
insulin receptor substrate 75, 78, 144, 145, 150, 155, 240, 255, 265, 320
insulin resistance 155, 266, 797
integrin .. 33, 70, 145, 172, 241, 286, 501, 522, 532, 537, 544, 545, 554, 597, 628, 677, 849–852, 854, 906, 907
integrin-linked kinase 132, 852
interferon 151, 154, 207, 308
interferon regulatory factor 662, 764
interleukin 80, 154, 761, 762, 764
interleukin receptor 347, 430
intersectin 296

intimal hyperplasia 534, 662, 700
ionotropic glutamate receptor 654, 742
IP$_3$ receptor 704, 883
IPP phosphatase 12, 15, 97
IQGAP 329, 360, 362, 566
IRAK kinase 352
iron 651, 710
Itch Ub ligase . . . 163, 368, 728, 729, 783, 785, 787, 819
ITK kinase 83, 163

J

Janus kinase 132, 151, 167, 170, 425, 430, 602, 603, 662
JNK . . 177, 270, 322, 323, 333, 336, 338, 341, 367, 410, 414, 455, 457, 483, 495, 501, 532, 539, 566, 567, 582, 597, 697, 720, 782, 907
Jun factor 170, 318, 350, 357, 501

K

K$_{2P}$ channel 300, 682, 753
K$_{ACh}$ current 682
K$_{ATP}$ channel 41, 682, 689, 711, 732
K$_{Ca}$ (BK) channel . . 41, 43, 301, 654, 673, 684
K$_{Ca}$ (IK) channel 879
K$_{Ca}$ (SK) channel 879
K$_{Ca}$ channel 450, 670, 870
K$_{IR}$ channel . 18, 20, 35, 85, 87, 228, 296, 472, 474, 494, 548, 682, 711
K$_V$ channel . 18, 35, 85, 87, 228, 250, 501, 560, 670, 681, 734, 742, 745, 842, 855
Kaiso 853
karyopherin 164
KAT Ub ligase 808
KCC cotransporter 188, 190, 301, 302
kidney 132, 608, 637, 685, 742, 745, 843
kinase suppressor of Ras (adaptor) . . 133, 329, 331, 349, 350, 360, 361, 484, 548, 746
kindlin 852

L

lactoperoxidase 709
laforin phosphatase 453
lamellipodium 532, 598, 619
lamin 18
laminin 848, 849
LaTS kinase 194, 196, 723
LAT adaptor 141, 164, 552
LCK kinase 164, 167, 216, 372, 434
LCP2 adaptor 164, 552, 853

LDL 55, 892
lectin 55, 172, 770, 786, 888
leptin 78, 154, 250, 423, 838
leptin receptor 423
leukemia inhibitory factor 154
leukocyte 149, 160
leukotriene 50
LIME interactor 145
LIMK kinase 227, 230, 231, 285, 286, 391, 448, 462, 463, 516
LIMS/PINCH adaptor 278
lipase H 48
lipid droplet 321, 530
lipin phosphatase 419
lipoprotein lipase 44, 47
lipoxygenase 777
liver 250, 608, 637, 651, 668, 709, 843
liver kinase-B (STK11) . . . 237, 249, 251, 253, 254, 262, 305
lmwPTP 458, 635
LUbAC Ub ligase 772
lung 42, 348, 375, 432, 567, 608, 609, 637, 646, 709, 742, 835, 836, 843
lusitropy 672, 682
lymphangiogenesis 171
lymphatic 171
lymphocyte 68, 84, 92, 94, 100, 159, 160, 163, 164, 171, 552, 577, 590, 593, 611, 662, 762, 780, 834, 879
Lyn kinase 165, 280, 400, 710, 719
lysophosphatidic acid . . . 8, 10, 11, 48, 79, 102, 106, 227
lysosome 513, 515

M

macrophage . . 45, 55, 135, 149, 173, 202, 284, 343, 373, 639, 641, 662, 679, 834
macropinocytosis 581
magnesium . . 97, 387, 409, 515, 578, 673, 685, 828, 830, 841, 855
MAGUK 734
MAK kinase 433
malin Ub ligase 453
MALT1 paracaspase 771, 772, 780, 787
manganese 387, 409, 413, 827, 830
mannose 6-phosphate receptor 528
MAP2K 135, 177, 312, 350, 410, 633, 746, 782
MAP3K . . . 177, 192, 313, 314, 324, 412, 414, 633, 708, 782
MAP3K (ASK) 342
MAP3K (MEKK) 337
MAP3K (MLK) 333
MAP3K (TAK1) 346

MAP3K (TAOK) 348
MAP3K (TPL2) 348
MAP4K 180, 322
MAPK . 79, 175, 194, 218, 311, 355, 427, 458, 480, 590, 593, 679, 700, 720, 755, 764, 782, 812, 880, 897
MAPK-interacting kinase 358, 376
MAPKAPK kinase 350, 358, 376
MAPK phosphatase............350, 438, 710
MARK kinase 250, 251, 253, 270
MASK kinase (MST4) 186
MASTL kinase 222
mastocyte 54, 55, 100, 107, 149, 160, 163, 169, 524, 641, 771, 880
MAST kinase 222
MATK kinase 155
matrix metallopeptidase 850
Maxi anion channel 754
mechanical stress . . 8, 315, 656, 658, 659, 698, 852, 878, 882, 898
mechanosensitive receptor 905
mechanotransduction ... 37, 77, 263, 537, 538, 669, 698, 729, 899
MELK kinase..................... 251, 252
membrane raft 169, 203, 752, 843
membrin........................... 509
Meox factor......................... 784
merlin.......................... 146, 224
metabolic syndrome 321, 667
metabotropic glutamate receptor 654
metastasis........................... 562
microphthalmia transcription factor (MiTF) 237
microRNA...... 226, 723, 780, 808, 811, 890, 896
microtubule 565, 607, 828, 907
microtubule-organizing center............516
Mid1 Ub ligase....................... 403
MinK kinase (MAP4K6)............ 184, 186
mitochondrion ... 59, 454, 456, 457, 655, 680, 689, 718
mitogen- and stress-activated kinase . 356, 358, 376–378, 779, 810
mitosis 391, 396, 463
MMP...............634, 677, 679, 689, 700
monocyte 207, 284, 373, 753, 848
Mos kinase 333, 433
MRCK kinase 219, 232
MSL2 Ub ligase 808
mucin 40
mucociliary clearance 747
MyB transcription factor 701
MyC transcription factor . . 305, 350, 372, 662, 723, 800, 891

MyD88 adaptor . 279, 283–285, 347, 442, 443, 762
myddosome 283, 285
myeloperoxidase 709
myocardium...........................213
myocyte enhancer factor (MEF) 350, 357, 359, 366, 374, 375, 381
MyoD factor 381
myogenic response 485, 669, 908
myoglobin.......................677, 714
myopodin............................201
myosin..........158, 216, 224, 228, 229, 639
myosin-3 kinase 192
myosin light-chain kinase . 224, 228, 361, 554, 878, 894
myosin light-chain phosphatase . 228, 400, 554
myotubularin 16, 18, 78, 82, 94, 388, 449
myristoylation 456, 457

N

Na^+–Ca^{++} exchanger 20, 681, 685, 756
Na^+–H^+ exchanger . . 183, 187, 245, 301, 483, 685
Na^+–K^+ ATPase..........300, 680, 681, 685
Na^+–Pi cotransporter 745
Na_V channel 201, 364, 680
NAADP......................... 875, 876
NADPH oxidase ... 33, 60, 641, 655, 690, 703, 710, 775
Nanog factor 461
natriuretic peptide....582, 587, 651, 678, 711, 757, 838, 846
natriuretic peptide receptor 678
natural killer cell 163, 171, 428, 771, 796
natural killer T lymphocyte 577
NCC cotransporter 188, 189, 191, 302
NCK adaptor 340, 493, 556, 618
NDPK kinase 486
necrosome 287
NEDD Ub ligase 150, 296, 460
NEK kinase 305, 433
nephron..........................300, 685
nerve growth factor 572
nervous system 473, 583, 592, 628
nesprin............................ 743
netrin............................735
neurexin............................735
neurocalcin 876
NeuroD factor.........................334
neurofibromin 625, 626
neuroligin..........................735
neuron 60, 579, 591, 594
neuronal calcium sensor 876

neuropilin.............................892
neuroprotectin..........................52
neurotransmitter........................91
neutrophil...70, 149, 163, 173, 207, 247, 266, 373, 641, 753, 829
NFκB..80, 175, 312, 322, 323, 328, 336, 340, 342, 346–349, 364, 385, 414, 501, 538, 582, 647, 662, 709, 712, 720, 758, 880, 890, 897
NFAT factor 206, 237, 258, 341, 382, 406, 611, 741, 878, 880
NHERF...............................495
NHE exchanger........................230
Niemann-Pick type-C protein............896
NIK kinase (MAP3K14)............765, 783
NIK kinase (MAP4K4).............183, 186
nischarin........................224, 286
nitrate............................660, 666
nitrative stress.........................713
nitric oxide .32, 33, 37, 61, 153, 248, 649, 651, 655, 656, 672, 846, 848, 861, 887
nitric oxide synthase...84, 207, 219, 237, 242, 248, 481, 493, 551, 655, 657–659, 662, 734, 777, 788, 843, 891
nitrite.................. 660, 663, 664
nitrosative stress................. 712, 713
nitrosoprotein..........................663
nitrosothiol.......................660, 663
nitrosylation............... 662, 667, 674
nitroxyl...........................670, 687
NKCC cotransporter...... 188–190, 302, 685
NKx factor............................274
NLK kinase................... 306, 346
NMDA-type glutamate receptor..........734
nodal cell.............................649
NOGo................................596
NOS1AP adaptor......................684
NOS uncoupling...............657, 660, 699
Notch............................306, 889
NRF factor...................481, 699, 891
NRK kinase...........................187
NSP GEF............................596
NuAK kinase..........................251
nuclear body (PML)....................882
nuclear estrogen receptor.......225, 374, 457
nuclear respiratory factor (NRF)..........655
nuclear speckle.........................82
nucleoredoxin.........................709
nutlin................................807

O

obesity...............................155
occludin..............................170

octamer-binding transcription factor......789
omegasome............................81
Orai Ca^{++} channel............825, 860, 871
orosomucoid..........................29
osmotic pressure..................341, 371
osmotic stress...337, 338, 341, 348, 373, 374, 857
osteopontin.......................815, 849
oxidative stress.33, 34, 80, 142, 154, 198, 215, 307, 328, 339, 342, 365, 408, 442, 651, 652, 655, 660, 682, 698, 700, 702, 706, 707, 711, 712, 715, 797, 813, 816
OxSR1 kinase....................188, 189
oxygen...............................672

P

P21-activated kinase...84, 159, 175, 208, 223, 253, 254, 285, 286, 289, 323, 412, 413, 474, 516, 532, 541, 566, 618, 643, 737
P27 transcription factor.................819
P2X channel..................343, 882, 905
P2Y GPCR........................60, 539
P300 (HAT)............... 26, 790, 800, 887
P38MAPK..61, 371, 410, 412, 414, 459, 532, 537, 654, 693, 697, 720, 779, 851, 891
P53 transcription factor... 130, 225, 239, 274, 294, 374, 382, 411, 412, 515, 662, 675, 678, 701, 722, 727, 778, 797, 800, 803, 890
P63 transcription factor.................819
P70 ribosomal S6 kinase (S6K)..81, 172, 175, 194, 239, 265, 907
P73 transcription factor............728, 819
P90 ribosomal S6 kinase (RSK) 175, 194, 234, 319, 320, 350, 357, 358, 376, 384, 761, 810
palmitoylation.........................549
Park2 Ub ligase...................286, 807
partitioning-defective protein... 516, 572, 608
parvalbumin..........................866
parvin................................278
paxillin.........147, 363, 614, 622, 849, 854
PDGF receptor........................459
PDHK kinase.........................127
PDK1 kinase...15, 71, 83, 194, 198, 203, 212, 224, 225, 236, 238, 240, 343, 797
Peli Ub ligase.........................281
peptidase.............................385
perilipin..............................321
peripheral nervous system...............649
peroxidase............................709
peroxiredoxin........ 142, 679, 706, 708, 891
peroxynitrite..........................713

PGC factor 655, 891, 892
phagocytosis 80, 161, 172, 532
PHD phosphatase 409
PHLPP phosphatase 409, 414, 727, 798
phosphatidic acid ... 10, 11, 14, 48, 69, 70, 91, 149, 511, 514, 535, 572, 587, 643, 840
phosphatidylinositol transfer protein 89
phosphodiesterase 77, 146, 485, 673, 737, 743, 830, 861
phosphoinositide 12
phospholamban 843, 878
phospholipase A .. 11, 17, 18, 38, 48, 243, 360, 697, 752, 753
phospholipase B 65
phospholipase C 11, 13, 18, 21, 65, 67, 83, 142, 160, 164, 248, 341, 430, 463, 480, 485, 535, 589, 591, 612, 636, 744, 873, 906
phospholipase D .. 69, 149, 508, 511, 514, 535, 560, 578, 697
phosphorylase kinase 870
PI3K 11, 71, 159, 218, 296, 317, 411, 436, 463, 472, 525, 526, 528, 543, 551, 572, 583, 611, 629, 680, 703, 798, 839
PIAS SUMo ligase ... 423, 533, 768, 775, 779, 780, 882
PIKE GTPase/ArfGAP 18, 204, 572
PIKK kinase 127, 260, 768
PIM kinase 205, 254, 258
PInK kinase 307
PI kinase 160, 508, 560
PKMYT kinase 222, 384, 460
PKN kinase 555
plasmin 753
plasminogen 750
platelet ... 66, 70, 95, 173, 213, 219, 423, 472, 524, 539, 672, 754, 823, 838, 843, 850, 853
platelet-activating factor 10, 11, 55, 56, 59, 352, 582
platelet-derived growth factor .. 166, 390, 643, 695, 710
PlekHg GEF 229, 607
plexin 569, 628, 634
PMCA pump 680
PML factor 728, 729, 882
pneumocyte 707, 881
Polo-like kinase .. 187, 291, 304, 384, 459, 797
polyADPribosyl polymerase 768
polycystin 202, 263, 725
POU factor 571
PP1 35, 158, 238, 290, 292, 320, 393, 437, 491, 582, 661, 728, 737, 744, 745, 798, 802, 823, 874

PP2 . 35, 45, 179, 196, 215, 222, 245, 272, 292, 317, 343, 401, 429, 438, 661, 681, 727, 732, 737, 743, 798, 879, 883
PP3 ... 201, 341, 343, 405, 433, 737, 742, 798, 861, 870, 878–880
PP4 .... 181, 407
PP5 .... 343, 408
PP6 408
PP7 408
PPAR factor 366
PPM1 .. 45, 223, 236, 238, 239, 347, 372, 398, 409, 438, 612, 662, 814
PPM1d phosphatase 779, 780, 807
PPMTC7 409
pre-replication complex 800
prestress 907
PREx GEF 77, 583
primary cilium 263
profilin 566
progenitor cell 676
prolactin 151, 154, 423
prostacyclin 33, 655
prostaglandin . 34, 50, 181, 301, 376, 582, 587, 696, 838, 842, 855, 899
protein kinase A ... 35, 45, 144, 175, 194, 199, 248, 251, 270, 300, 397, 428, 433, 539, 660, 676, 681, 684, 694, 698, 731, 737, 745, 823, 826, 836, 842, 844
protein kinase B . 15, 71, 78, 83, 84, 100, 135, 144, 175, 194, 203, 252, 320, 329, 335, 343, 380, 384, 414, 458, 471, 572, 616, 629, 660, 675, 676, 726, 729, 798, 799, 807, 883
protein kinase C . 11, 13, 17, 37, 48, 64, 77, 83, 153, 175, 194, 208, 241, 363, 380, 385, 414, 433, 474, 516, 582, 591, 603, 612, 641, 642, 652, 660, 676, 693, 697, 737, 746, 763, 771, 850, 852, 862, 878, 895, 905
protein kinase D .. 30, 175, 181, 218, 241, 259, 343, 510, 741, 746
protein kinase G . 175, 194, 218, 248, 654, 672, 673, 684, 711, 731, 733, 844, 905
protein kinase R 308
protein kinase X 202
protein kinase Y 203
PrsS23S peptidase 222
pseudokinase 131, 269
pseudophosphatase 94, 449
PSK kinase (MAP3K) 193
PTen phosphatase 12, 15, 18, 92, 150, 230, 416, 436, 703, 704, 710, 798, 889
PTK6 (BrK kinase) 144
PTK9 kinase 125

PTP4a 448
PTPn1 .... 150, 382, 422, 591, 710
PTPn11/SHP2 .... 150, 430, 610, 612
PTPn12 .... 150, 431, 618, 635
PTPn13 .... 225, 432
PTPn14 432
PTPn18 .... 433, 635
PTPn2 .... 424, 853
PTPn20 433
PTPn21 433
PTPn23 434
PTPn3 .... 374, 426
PTPn4 427
PTPn5 .... 162, 427, 438
PTPn6/SHP1 152, 162, 215, 428, 436, 610, 638
PTPn7 .... 428, 438
PTPn8/22 .... 429, 434
PTPn9 429
PTPRa 150, 162, 906
PTPRc 162, 272
PTPRd 736
PTPRf 150, 162, 736
PTPRj 84
PTPRm 732
PTPRr 438
PTPRs 736
PTPRv 433
PTPRz1 635
pulmonary hypertension 30, 832, 837
pulse 813

R
Rab5GEF GEF (alsin) 581, 584
RabGEF 581
Rab GTPase 77, 78, 90, 581, 764, 855
RacGAP 560, 639
RacGEF 581
RACK 146, 211, 214–216, 241, 627, 731, 732, 841
Rac GTPase ... 77, 91, 99, 147, 223, 336, 341, 531, 541, 552, 553, 581, 583, 639, 692, 697, 703, 848, 853, 855
Rad GTPase 436, 534, 567
Raf kinase 40, 124, 164, 185, 216, 231, 235, 273, 317, 323, 327, 329, 345, 350, 360–362, 384, 385, 405, 484, 495, 543, 548, 590, 593, 727, 746, 786
Rag GTPase 265, 266, 534
RalBP 639
RalGDS 584, 586
RalGEF 535, 545, 548
Ral GTPase 317, 535
RanGEF 584

Ran GTPase 536, 584, 768
RapGAP 489
RapGEF 163, 552, 566, 584, 743, 823, 843
Rap GTPase 536, 537, 543, 552, 581, 627, 743, 823, 852
Ras-effector RASSF .. 185, 543, 545, 548, 550, 722, 727
RasA GAP 163, 483, 552, 625, 627
RasGAP 625
RasGDS (GEF) 545
RasGEF 589
RasGRF (GEF) .. 352, 372, 548, 589, 590, 592
RasGRP (GEF) . 214, 548, 552, 588, 589, 591, 612
Ras GTPase .. 74, 77, 323, 329, 415, 540, 589, 818, 855
reactive nitrogen species .. 345, 419, 674, 678, 702, 712
reactive oxygen species .. 33, 37, 60, 158, 163, 173, 241, 261, 343, 347, 366, 419, 583, 651, 655, 660, 674, 690, 700, 702, 775, 782, 797, 840, 883, 889, 891
receptor-operated channel 864, 873
receptor Tyr kinase 597
recoverin 877
regulator of G-protein signaling 214, 215, 473, 474, 476, 477, 480, 484, 487, 676, 874
Rem GTPase 567, 855
renin 301, 532
replication stress 768
reticulon 596
retinoblastoma protein 141, 150
ReT receptor 198
RGL GEF 548, 584
RHEB GTPase 266, 553
RhoGAP 163, 229, 431, 566, 608, 629
RhoGAP10/26/42 (GRAF) 636
RhoGAP13/14/34 (SRGAP) 636
RhoGAP17/44 (RICH) 637
RhoGAP5/35 (P190) 633
RhoGAP7/37/38 (DLC) 635
RhoGEF ... 143, 167, 412, 413, 489, 584, 598, 643
RhoGEF6/7 .... 224, 604
Rho GTPase . 77, 144, 147, 157, 229, 230, 241, 367, 376, 381, 481, 498, 532, 540, 552, 553, 818, 852, 855
RIn GEF 527, 545, 548, 581
RIN GTPase 570
RiOK kinase 128
RIP kinase . 123, 124, 214, 215, 287, 340, 347, 350, 470, 768, 775, 782, 785
RIT GTPase 570, 610
RKIP 361, 385

Rnd GTPase 568, 628, 634, 639
RNF Ub ligase .. 199, 254, 286, 772, 787, 807, 808
Robo 636
RoCK kinase 158, 175, 193, 194, 226, 266, 285, 286, 326, 401, 501, 534, 541, 570
Ros kinase 428
rRasGAP 628
Rubicon 528
ruffle 77
Runx transcription factor 728
ryanodine receptor . 35, 41, 680, 704, 711, 743

S
S100 protein 694, 863
salivary peroxidase 709
salt-inducible kinase 255, 823
scaffold protein 1, 311, 360
Scambio GEF 609
scavenger receptor 46, 47, 677
SCFR receptor 428, 430, 433
SCF Ub ligase .. 275, 348, 725, 763, 764, 772, 801, 819
Sec23/24 GAP 515
second messenger-operated channel 864
secretoglobin 185
Sec GEF 515
Sef inhibitor 445
selectin 753
semaphorin 628, 892
senescence 803
separase 304
septin 466
sequestosome 211, 212, 217, 364
14-3-3 sequestrator 20, 185, 198, 201, 213–216, 236, 238, 244, 252–255, 330–332, 334, 339–343, 345, 363, 366, 376, 381, 426, 447, 459, 461, 491, 494, 567, 591, 629, 639, 694, 729, 741, 746, 792, 797, 800, 804, 812, 814, 816, 826
SERCA pump 35, 680, 682, 687, 744
serotonin 5HT receptor 590
serpin 532, 763, 804
serum response factor (SRF) ... 237, 318, 350, 357, 501
sestrin 261, 817
SGK kinase 194, 296, 300, 641, 798
SH3RF1 Ub ligase (POSH) 335
SHB adaptor 150
SHC adaptor 93, 166, 320, 484
shelterin 538
SHIP phosphatase 15, 96, 100, 145, 390
shugoshin 304

SIAH Ub ligase 274, 744, 790, 807
siglec 428, 643
SIK kinase 251
sinoatrial node 35
SIPA RapGAP 624
sirtuin 679, 797, 800, 802, 809
SIT adaptor 145
SLC transporter . 188, 216, 302, 375, 664, 669, 729, 745
slingshot 259, 286, 447, 463
Slit 636
SLK kinase 178, 187, 343
SMAD factor 130, 725
small ArfGAP 621
SMG1 kinase 260
smooth muscle cell 35, 134, 148, 219, 228, 244, 278, 341, 459, 631, 644
SMURF Ub ligase 272, 642
Snail homolog (Snai) 225, 723
SNARE protein 505
SNRK kinase 444
SOCS 154, 167, 582, 780
soluble adenylate cyclase 827
soluble guanylate cyclase 655, 672
SOS GEF .. 178, 320, 350, 361, 548, 552, 589, 609
Sox transcription factor 723
SP1 transcription factor 675, 701
SpAR RapGAP 624
sphingomyelinase 29
sphingosine 1-phosphate 29, 31, 102, 583
sphingosine kinase 30
spinophilin 495
spliceosome 382, 409, 412
splicing factor 381, 383
Sprouty 381, 444
Src kinase 33, 79, 147, 156–158, 164, 166, 173, 198, 247, 270, 323, 403, 423, 433, 434, 437, 472, 495, 571, 590, 618, 676, 697, 699, 852, 854, 906, 907
SREBP factor 265, 320, 896
SRMS kinase 156
SRPK kinase 309
stathmin 224
STAT factor . 33, 132, 153, 225, 284, 430, 455, 457, 567, 582, 662, 697
stem cell 684
StIM Ca^{++} sensor 860, 871, 873
STK10 kinase 187
STK16 kinase 232
STK19 (G11) kinase 126
STK24 kinase 186
STK25 kinase 186
STK37 kinase (PASK) 189

STK38 kinase (NDR) 194, 196, 727
STK39 kinase (SPAK) 188, 189
STK3 kinase 184, 550
STK4 kinase 185, 550
store-operated channel 219, 711, 825, 860, 864, 871, 873, 879
StRAd adaptor 133
stress-activated MAPK 375
stress fiber 25, 37, 68, 178, 226, 228, 245, 247, 248, 257, 286, 337, 376, 377, 405, 476, 478, 481, 499, 501, 573, 606, 608, 620, 622, 729, 853, 878, 899
stress history 905
stretch-activated ion channels 864
SUMo 413
sumoylation 533
superoxide 704
superoxide dismutase 671, 704
surfactant 185, 417, 881
surfactant protein 881
SWAP70 GEF 77
SYK kinase 171, 435, 786, 853
sympathetic 669
synaptojanin 390
synaptotagmin 296, 522, 523
syndecan 641, 643
synectin 641, 644
syntrophin 659

T

T-cell receptor 68, 100, 180, 182, 350, 369, 370, 407, 426, 428, 446, 590, 611, 762, 764, 771
talin 849, 852
target of rapamycin 80, 84, 127, 148, 239, 260, 262, 404, 583, 797, 895
Tau (MAPT) stabilizer 252–254, 257, 270, 371, 374, 381, 400
TBK1 kinase 309, 764
TEC kinase 83, 100, 157, 167, 433
telomere 538
telosome 538
tenascin 849
tenovin 809
tensin 437
ternary complex factor (TCF) 317, 357
TesK kinase 286, 433, 448, 463
tetraspanin 88
thioredoxin 344, 679, 708
thioredoxin reductase 708
thrombin 95, 696, 697, 700
thrombospondin 677, 849
thromboxane synthase 652

thymocyte 552, 762
thyroid hormone 857
thyroid peroxidase 709
thyroid transcription factor (TTF) 185
thyrotropin-releasing hormone 855
TIAM GEF 214–216, 352, 372, 543, 545, 582, 583
TIF1 kinase/Ub ligase 129
tight junction 170, 212, 247, 250, 580, 608, 637
tissue inhibitor of metallopeptidase 850
tissue plasminogen activator 750
TNFαIP DUb/Ub ligase 766, 780, 786
TNFRSF 126
TNFSF 338, 763
TNF receptor 785
TNIK kinase 187
Toll-like receptor 350, 764
TRAFAC GTPase 465
TRAF Ub ligase . 199, 283, 328, 339, 342, 343, 346, 350, 352, 364, 767, 770, 774, 775, 786
transferrin 522, 564
transforming growth factor 105, 662, 725, 792, 851
transmembrane adenylate cyclase 822
Trb adaptor 134
TriM (TIF) kinase 129
TriM Ub ligase 515
Trio RhoGEF23/kinase 117, 584, 606
TRP channel ... 14, 21, 87, 125, 201, 202, 290, 522, 533, 753, 855, 864, 873, 876, 879, 880, 905
Trrap adaptor 135
tryptase 915
TTbK kinase 270
tubby 480
tubulin 224
tumor-necrosis factor . 154, 312, 374, 652, 697, 764, 782
tumorigenesis 552
tumor cell 472
twifilin 269
TXK kinase 83, 169
TyK2 kinase 455

U

UbE3a Ub ligase 165
ubiquitination 274, 281, 772, 798
UbR1 Ub ligase 366
Ub conjugase 413, 771
uridine triphosphate 228
urocortin 364
uroguanylin 301

urokinase-type plasminogen activator 351
USP deubiquitinase 797, 807, 809

V

vascular endothelial growth factor 37, 229, 345, 480, 788, 792, 892
vascular peroxidase 709
vascular smooth muscle cell ... 37, 43, 67, 219, 400, 479, 534, 597, 602, 609, 611, 689, 695, 709, 761, 790, 832, 837, 839, 841, 843, 848, 878
vascular tone 37, 43, 656, 659
vasoconstriction 30, 43, 50, 157, 231, 486, 655, 878
vasodilation .. 32, 43, 248, 652, 653, 655, 664, 671, 711, 832, 848, 878
vasopressin 191, 201, 301, 479, 685
Vav GEF ... 216, 431, 552, 582, 584, 610, 639
VDAC channel/porin 216
VHL Ub ligase 744, 789, 807, 887
vimentin 224, 231
vinculin 849
visivin 876
vitamin D 68, 801
vitronectin 849
VLDL 49
VLDLR receptor 677
von Willebrand factor 753
VRAC channel 228, 858
VSP phosphatase 18, 101

W

wall shear stress 363, 901, 906
wall tension 901
WASP 146, 163, 164, 516, 566
WAVe 142, 532
Wee kinase 222, 252, 384, 460
WNK kinase 188, 191, 294
Wnt ... 267, 274, 346, 380, 567, 594, 709, 723, 818
WSB1 Ub ligase 275
WWP Ub ligase 819

X

X-box-binding protein (XBP) 84
XIAP Ub ligase 768

Y

Yes kinase 170

Z

ZAP70 kinase ... 141, 171, 372, 434, 455, 552
zinc 456, 720, 830, 841
zonula adherens 247
zonula occludens protein 899